a LANGE medical book

Ganong's
Review of
Medical Physiology

Twenty-Third Edition

Kim E. Barrett, PhD
Professor
Department of Medicine
Dean of Graduate Studies
University of California, San Diego
La Jolla, California

Scott Boitano, PhD
Associate Professor, Physiology
Arizona Respiratory Center
Bio5 Collaborative Research Institute
University of Arizona
Tucson, Arizona

Susan M. Barman, PhD
Professor
Department of Pharmacology/Toxicology
Michigan State University
East Lansing, Michigan

Heddwen L. Brooks, PhD
Associate Professor
Department of Physiology
College of Medicine
University of Arizona
Tucson, Arizona

McGraw Hill Medical

New York Chicago San Francisco Lisbon London Madrid Mexico City
Milan New Delhi San Juan Seoul Singapore Sydney Toronto

Ganong's Review of Medical Physiology, Twenty-Third Edition

Copyright © 2010 by The McGraw-Hill Companies, Inc. All rights reserved. Printed in the United States of America. Except as permitted under the United States Copyright Act of 1976, no part of this publication may be reproduced or distributed in any form or by any means, or stored in a data base or retrieval system, without the prior written permission of the publisher.

1 2 3 4 5 6 7 8 9 0 BAN/BAN 12 11 10 9

ISBN 978-0-07-160567-0
MHID 0-07-160567-3
ISSN 0892-1253

Notice

Medicine is an ever-changing science. As new research and clinical experience broaden our knowledge, changes in treatment and drug therapy are required. The authors and the publisher of this work have checked with sources believed to be reliable in their efforts to provide information that is complete and generally in accord with the standards accepted at the time of publication. However, in view of the possibility of human error or changes in medical sciences, neither the authors nor the publisher nor any other party who has been involved in the preparation or publication of this work warrants that the information contained herein is in every respect accurate or complete, and they disclaim all responsibility for any errors or omissions or for the results obtained from use of the information contained in this work. Readers are encouraged to confirm the information contained herein with other sources. For example and in particular, readers are advised to check the product information sheet included in the package of each drug they plan to administer to be certain that the information contained in this work is accurate and that changes have not been made in the recommended dose or in the contraindications for administration. This recommendation is of particular importance in connection with new or infrequently used drugs.

This book was set in Minion Pro by Silverchair Science + Communications, Inc.
The editors were Michael Weitz and Regina Y. Brown.
The production supervisor was Catherine H. Saggese.
Project management was provided by Jeff Houck, Progressive Information Technologies.
The designer was Elise Lansdon.
The cover art director was Margaret Webster-Shapiro.
Cover Photo: Anterior angled view of the main body systems. Credit: MedicalRF.com / Getty Images.
RR Donnelley was printer and binder.

This book is printed on acid-free paper.

Dedication to

WILLIAM FRANCIS GANONG

William Francis ("Fran") Ganong was an outstanding scientist, educator, and writer. He was completely dedicated to the field of physiology and medical education in general. Chairman of the Department of Physiology at the University of California, San Francisco, for many years, he received numerous teaching awards and loved working with medical students.

Over the course of 40 years and some 22 editions, he was the sole author of the best selling *Review of Medical Physiology*, and a co-author of 5 editions of *Pathophysiology of Disease: An Introduction to Clinical Medicine*. He was one of the "deans" of the Lange group of authors who produced concise medical text and review books that to this day remain extraordinarily popular in print and now in digital formats. Dr. Ganong made a gigantic impact on the education of countless medical students and clinicians.

A general physiologist par excellence and a neuroendocrine physiologist by subspecialty, Fran developed and maintained a rare understanding of the entire field of physiology. This allowed him to write each new edition (every 2 years!) of the *Review of Medical Physiology* as a sole author, a feat remarked on and admired whenever the book came up for discussion among physiologists. He was an excellent writer and far ahead of his time with his objective of distilling a complex subject into a concise presentation. Like his good friend, Dr. Jack Lange, founder of the Lange series of books, Fran took great pride in the many different translations of the *Review of Medical Physiology* and was always delighted to receive a copy of the new edition in any language.

He was a model author, organized, dedicated, and enthusiastic. His book was his pride and joy and like other best-selling authors, he would work on the next edition seemingly every day, updating references, rewriting as needed, and always ready and on time when the next edition was due to the publisher. He did the same with his other book, *Pathophysiology of Disease: An Introduction to Clinical Medicine*, a book that he worked on meticulously in the years following his formal retirement and appointment as an emeritus professor at UCSF.

Fran Ganong will always have a seat at the head table of the greats of the art of medical science education and communication. He died on December 23, 2007. All of us who knew him and worked with him miss him greatly.

Key Features of the 23rd Edition of
Ganong's Review of Medical Physiology

- **Thoroughly updated** to reflect the latest research and developments in important areas such as the cellular basis of neurophysiology

- **Incorporates examples from clinical medicine** throughout the chapters to illustrate important physiologic concepts

- **Delivers more detailed, clinically-relevant, high-yield information per page** than any similar text or review

- **NEW full-color illustrations**—the authors have worked with an outstanding team of medical illustrators, photographers, educators, and students to provide an unmatched collection of 600 illustrations and tables

- **NEW boxed clinical cases**—featuring examples of diseases that illustrate important physiologic principles

- **NEW high-yield board review** questions at the end of each chapter

- **NEW larger 8½ X 11" trim-size** enhances the rich visual content

- **NEW companion online learning center (LangeTextbooks.com)** offers a wealth of innovative learning tools and illustrations

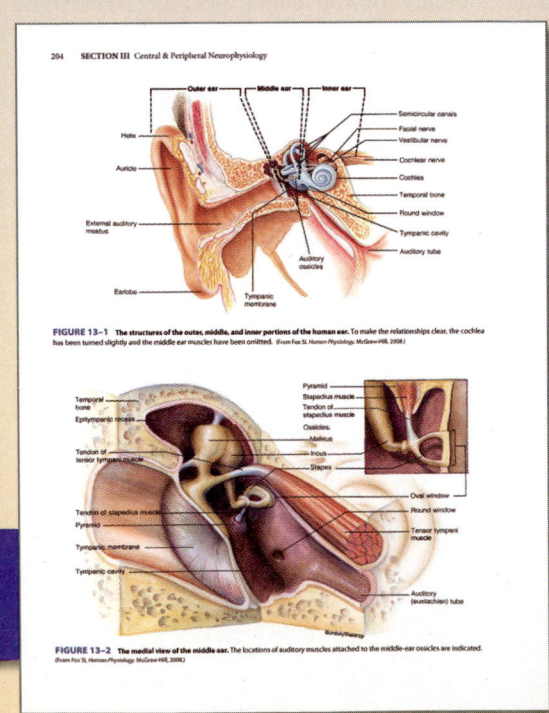

204   **SECTION III** Central & Peripheral Neurophysiology

FIGURE 13–1   The structures of the outer, middle, and inner portions of the human ear. To make the relationships clear, the cochlea has been turned slightly and the middle ear muscles have been omitted. (From Fox SI, Human Physiology. McGraw-Hill, 2008.)

FIGURE 13–2   The medial view of the middle ear. The locations of auditory muscles attached to the middle-ear ossicles are indicated. (From Fox SI, Human Physiology. McGraw-Hill, 2008.)

for Your iPod

NEW iPod-compatible review—Medical PodClass offers audio and text for study on the go

Full-color illustrations enrich the text

Clinical Cases illustrate essential physiologic principles

Summary tables and charts encapsulate important information

Chapters conclude with Chapter Summaries and review questions

About the Authors

KIM E. BARRETT

Kim Barrett received her PhD in biological chemistry from University College London in 1982. Following postdoctoral training at the National Institutes of Health, she joined the faculty at the University of California, San Diego, School of Medicine in 1985, rising to her current rank of Professor of Medicine in 1996. Since 2006, she has also served the University as Dean of Graduate Studies. Her research interests focus on the physiology and pathophysiology of the intestinal epithelium, and how its function is altered by commensal, probiotics, and pathogenic bacteria as well as in specific disease states, such as inflammatory bowel diseases. She has published almost 200 articles, chapters, and reviews, and has received several honors for her research accomplishments including the Bowditch and Davenport Lectureships from the American Physiological Society and the degree of Doctor of Medical Sciences, *honoris causa*, from Queens University, Belfast. She is also a dedicated and award-winning instructor of medical, pharmacy, and graduate students, and has taught various topics in medical and systems physiology to these groups for more than 20 years. Her teaching experiences led her to author a prior volume (*Gastrointestinal Physiology*, McGraw-Hill, 2005) and she is honored to have been invited to take over the helm of Ganong.

SUSAN M. BARMAN

Susan Barman received her PhD in physiology from Loyola University School of Medicine in Maywood, Illinois. Afterward she went to Michigan State University (MSU) where she is currently a Professor in the Department of Pharmacology/ Toxicology and the Neuroscience Program. Dr Barman has had a career-long interest in neural control of cardiorespiratory function with an emphasis on the characterization and origin of the naturally occurring discharges of sympathetic and phrenic nerves. She was a recipient of a prestigious National Institutes of Health MERIT (Method to Extend Research in Time) Award. She is also a recipient of an Outstanding University Woman Faculty Award from the MSU Faculty Professional Women's Association and an MSU College of Human Medicine Distinguished Faculty Award. She has been very active in the American Physiological Society (APS) and recently served on its council. She has also served as Chair of the Central Nervous System Section of APS as well as Chair of both the Women in Physiology and Section Advisory Committees of APS. In her spare time, she enjoys daily walks, aerobic exercising, and mind-challenging activities like puzzles of various sorts.

SCOTT BOITANO

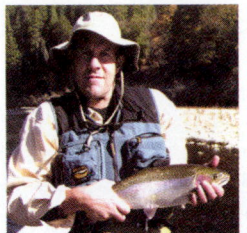

Scott Boitano received his PhD in genetics and cell biology from Washington State University in Pullman, Washington, where he acquired an interest in cellular signaling. He fostered this interest at University of California, Los Angeles, where he focused his research on second messengers and cellular physiology of the lung epithelium. He continued to foster these research interests at the University of Wyoming and at his current positions with the Department of Physiology and the Arizona Respiratory Center, both at the University of Arizona.

HEDDWEN L. BROOKS

Heddwen Brooks received her PhD from Imperial College, University of London and is an Associate Professor in the Department of Physiology at the University of Arizona (UA). Dr Brooks is a renal physiologist and is best known for her development of microarray technology to address in vivo signaling pathways involved in the hormonal regulation of renal function. Dr Brooks' many awards include the American Physiological Society (APS) Lazaro J. Mandel Young Investigator Award, which is for an individual demonstrating outstanding promise in epithelial or renal physiology. She will receive the APS Renal Young Investigator Award at the 2009 annual meeting of the Federation of American Societies for Experimental Biology. Dr Brooks is a member of the APS Renal Steering Section and the APS Committee of Committees. She is on the Editorial Board of the American Journal of Physiology-Renal Physiology (since 2001), and she has also served on study sections of the National Institutes of Health and the American Heart Association.

Contents

Preface

From the Authors

We are very pleased to launch the 23rd edition of *Ganong's Review of Medical Physiology*. The current authors have attempted to maintain the highest standards of excellence, accuracy, and pedagogy developed by Fran Ganong over the 46 years in which he educated countless students worldwide with this textbook.

At the same time, we have been attuned to the evolving needs of both students and professors in medical physiology. Thus, in addition to usual updates on the latest research and developments in areas such as the cellular basis of physiology and neurophysiology, this edition has added both outstanding pedagogy and learning aids for students.

We are truly grateful for the many helpful insights, suggestions, and reviews from around the world that we received from colleagues and students. We hope you enjoy the new features and the 23rd edition!

This edition is a revision of the original works of Dr. Francis Ganong.

New 4 Color Illustrations

- We have worked with a large team of medical illustrators, photographers, educators, and students to build an accurate, up-to-date, and visually appealing new illustration program. Full-color illustrations and tables are provided throughout, which also include detailed figure legends that tell a short story or describes the key point of the illustration.

New 8½ x 11 Format

- Based on student and instructor focus groups, we have increased the trim size, which will provide additional white space and allow our new art program to really show!

New Boxed Clinical Cases

- Highlighted in a shaded background, so students can recognize the boxed clinical cases, examples of diseases illustrating important physiological principles are provided.

New End of Chapter Board Review Questions

- New to this edition, chapters now conclude with board review questions.

New Media

- This new edition has focused on creating new student content that is built upon learning outcomes and assessing student performance. Free with every student copy is an iPod Review Tutorial Product. Questions and art based from each chapter tests students comprehension and is easy to navigate with a simple click of the scroll bar!
- Online Learning Center will provide students and faculty with cases and art and board review questions on a dedicated website.

General Principles & Energy Production in Medical Physiology

OBJECTIVES

After studying this chapter, you should be able to:

- Name the different fluid compartments in the human body.
- Define moles, equivalents, and osmoles.
- Define pH and buffering.
- Understand electrolytes and define diffusion, osmosis, and tonicity.
- Define and explain the resting membrane potential.
- Understand in general terms the basic building blocks of the cell: nucleotides, amino acids, carbohydrates, and fatty acids.
- Understand higher-order structures of the basic building blocks: DNA, RNA, proteins, and lipids.
- Understand the basic contributions of these building blocks to cell structure, function, and energy balance.

INTRODUCTION

In unicellular organisms, all vital processes occur in a single cell. As the evolution of multicellular organisms has progressed, various cell groups organized into tissues and organs have taken over particular functions. In humans and other vertebrate animals, the specialized cell groups include a gastrointestinal system to digest and absorb food; a respiratory system to take up O_2 and eliminate CO_2; a urinary system to remove wastes; a cardiovascular system to distribute nutrients, O_2, and the products of metabolism; a reproductive system to perpetuate the species; and nervous and endocrine systems to coordinate and integrate the functions of the other systems. This book is concerned with the way these systems function and the way each contributes to the functions of the body as a whole.

In this section, general concepts and biophysical and biochemical principles that are basic to the function of all the systems are presented. In the first chapter, the focus is on review of basic biophysical and biochemical principles and the introduction of the molecular building blocks that contribute to cellular physiology. In the second chapter, a review of basic cellular morphology and physiology is presented. In the third chapter, the process of immunity and inflammation, and their link to physiology, are considered.

GENERAL PRINCIPLES

THE BODY AS AN ORGANIZED "SOLUTION"

The cells that make up the bodies of all but the simplest multicellular animals, both aquatic and terrestrial, exist in an "internal sea" of **extracellular fluid (ECF)** enclosed within the integument of the animal. From this fluid, the cells take up O_2 and nutrients; into it, they discharge metabolic waste products. The ECF is more dilute than present-day seawater, but its composition closely resembles that of the primordial oceans in which, presumably, all life originated.

In animals with a closed vascular system, the ECF is divided into two components: the **interstitial fluid** and the circulating **blood plasma.** The plasma and the cellular elements of the blood, principally red blood cells, fill the vascular system, and together they constitute the **total blood volume.** The interstitial fluid is that part of the ECF that is outside the vascular system, bathing the cells. The special fluids considered together as transcellular fluids are discussed in the following text. About a third of the **total body water** is extracellular; the remaining two thirds is intracellular **(intracellular fluid).** In the average young adult male, 18% of the body weight is protein and related substances, 7% is mineral, and 15% is fat. The remaining 60% is water. The distribution of this water is shown in Figure 1–1A.

The intracellular component of the body water accounts for about 40% of body weight and the extracellular component for about 20%. Approximately 25% of the extracellular component is in the vascular system (plasma = 5% of body weight) and 75% outside the blood vessels (interstitial fluid = 15% of body weight). The total blood volume is about 8% of body weight. Flow between these compartments is tightly regulated.

UNITS FOR MEASURING CONCENTRATION OF SOLUTES

In considering the effects of various physiologically important substances and the interactions between them, the number of molecules, electric charges, or particles of a substance per unit volume of a particular body fluid are often more meaningful than simply the weight of the substance per unit volume. For this reason, physiological concentrations are frequently expressed in moles, equivalents, or osmoles.

Moles

A mole is the gram-molecular weight of a substance, ie, the molecular weight of the substance in grams. Each mole (mol) consists of 6×10^{23} molecules. The millimole (mmol) is 1/1000 of a mole, and the micromole (μmol) is 1/1,000,000 of a mole. Thus, 1 mol of NaCl = 23 g + 35.5 g = 58.5 g, and 1 mmol = 58.5 mg. The mole is the standard unit for expressing the amount of substances in the SI unit system.

The molecular weight of a substance is the ratio of the mass of one molecule of the substance to the mass of one twelfth the mass of an atom of carbon-12. Because molecular weight is a ratio, it is dimensionless. The dalton (Da) is a unit of mass equal to one twelfth the mass of an atom of carbon-12. The kilodalton (kDa = 1000 Da) is a useful unit for expressing the molecular mass of proteins. Thus, for example, one can speak of a 64-kDa protein or state that the molecular mass of the protein is 64,000 Da. However, because molecular weight is a dimensionless ratio, it is incorrect to say that the molecular weight of the protein is 64 kDa.

Equivalents

The concept of electrical equivalence is important in physiology because many of the solutes in the body are in the form of charged particles. One equivalent (eq) is 1 mol of an ionized substance divided by its valence. One mole of NaCl dissociates into 1 eq of Na^+ and 1 eq of Cl^-. One equivalent of $Na^+ = 23$ g, but 1 eq of $Ca^{2+} = 40$ g/2 = 20 g. The milliequivalent (meq) is 1/1000 of 1 eq.

Electrical equivalence is not necessarily the same as chemical equivalence. A gram equivalent is the weight of a substance that is chemically equivalent to 8.000 g of oxygen. The normality (N) of a solution is the number of gram equivalents in 1 liter. A 1 N solution of hydrochloric acid contains both H^+ (1 g) and Cl^- (35.5 g) equivalents, = (1 g + 35.5 g)/L = 36.5 g/L.

WATER, ELECTROLYTES, & ACID/BASE

The water molecule (H_2O) is an ideal solvent for physiological reactions. H_2O has a **dipole moment** where oxygen slightly pulls away electrons from the hydrogen atoms and creates a charge separation that makes the molecule **polar.** This allows water to dissolve a variety of charged atoms and molecules. It also allows the H_2O molecule to interact with other H_2O molecules via hydrogen bonding. The resultant hydrogen bond network in water allows for several key properties in physiology: (1) water has a high surface tension, (2) water has a high heat of vaporization and heat capacity, and (3) water has a high dielectric constant. In layman's terms, H_2O is an excellent biological fluid that serves as a solute; it provides optimal heat transfer and conduction of current.

Electrolytes (eg, NaCl) are molecules that dissociate in water to their cation (Na^+) and anion (Cl^-) equivalents. Because of the net charge on water molecules, these electrolytes tend not to reassociate in water. There are many important electrolytes in physiology, notably Na^+, K^+, Ca^{2+}, Mg^{2+}, Cl^-, and HCO_3^-. It is important to note that electrolytes and other charged compounds (eg, proteins) are unevenly distributed in the body fluids (Figure 1–1B). These separations play an important role in physiology.

FIGURE 1–1   **Organization of body fluids and electrolytes into compartments. A)** Body fluids are divided into Intracellular and extracellular fluid compartments (ICF and ECF, respectively). Their contribution to percentage body weight (based on a healthy young adult male; slight variations exist with age and gender) emphasizes the dominance of fluid makeup of the body. Transcellular fluids, which constitute a very small percentage of total body fluids, are not shown. Arrows represent fluid movement between compartments. **B)** Electrolytes and proteins are unequally distributed among the body fluids. This uneven distribution is crucial to physiology. Prot⁻, protein, which tends to have a negative charge at physiologic pH.

pH AND BUFFERING

The maintenance of a stable hydrogen ion concentration ($[H^+]$) in body fluids is essential to life. The **pH** of a solution is defined as the logarithm to the base 10 of the reciprocal of the H^+ concentration ($[H^+]$), ie, the negative logarithm of the $[H^+]$. The pH of water at 25 °C, in which H^+ and OH^- ions are present in equal numbers, is 7.0 (Figure 1–2). For each pH unit less than 7.0, the $[H^+]$ is increased tenfold; for each pH unit above 7.0, it is decreased tenfold. In the plasma of healthy individuals, pH is slightly alkaline, maintained in the narrow range of 7.35 to 7.45. Conversely, gastric fluid pH can be quite acidic (on the order of 2.0) and pancreatic secretions can be quite alkaline (on the order of 8.0). Enzymatic activity and protein structure are frequently sensitive to pH; in any given body or cellular compartment, pH is maintained to allow for maximal enzyme/protein efficiency.

Molecules that act as H^+ donors in solution are considered acids, while those that tend to remove H^+ from solutions are considered bases. Strong acids (eg, HCl) or bases (eg, NaOH) dissociate completely in water and thus can most change the $[H^+]$ in solution. In physiological compounds, most acids or bases are considered "weak," that is, they contribute relatively few H^+ or take away relatively few H^+ from solution. Body pH is stabilized by the **buffering capacity** of the body fluids. A **buffer** is a substance that has the ability to bind or release H^+ in solution, thus keeping the pH of the solution relatively constant despite the addition of considerable quantities of acid or base. Of course there are a number of buffers at work in biological fluids at any given time. All buffer pairs in a homogenous solution are in equilibrium with the same $[H^+]$; this is known as the **isohydric principle.** One outcome of this principle is that by assaying a single buffer system, we can understand a great deal about all of the biological buffers in that system.

When acids are placed into solution, there is a dissociation of some of the component acid (HA) into its proton (H^+) and free acid (A^-). This is frequently written as an equation:

$$HA \leftrightarrows H^+ + A^-.$$

According to the laws of mass action, a relationship for the dissociation can be defined mathematically as:

$$K_a = [H^+] [A^-] / [HA]$$

where K_a is a constant, and the brackets represent concentrations of the individual species. In layman's terms, the product of the proton concentration ($[H^+]$) times the free acid concentration ($[A^-]$) divided by the bound acid concentration ($[HA]$) is a defined constant (K). This can be rearranged to read:

$$[H^+] = K_a [HA]/[A^-]$$

If the logarithm of each side is taken:

$$\log [H^+] = \log K_a + \log[HA]/[A^-]$$

Both sides can be multiplied by –1 to yield:

$$-\log [H^+] = -\log K_a + \log[A^-]/[HA]$$

This can be written in a more conventional form known as the **Henderson Hasselbach equation:**

$$pH = pK_a + \log [A^-]/[HA]$$

This relatively simple equation is quite powerful. One thing that we can discern right away is that the buffering capacity of a particular weak acid is best when the pK_a of that acid is equal to the pH of the solution, or when:

$$[A-] = [HA], pH = pK_a$$

Similar equations can be set up for weak bases. An important buffer in the body is carbonic acid. Carbonic acid is a weak acid, and thus is only partly dissociated into H^+ and bicarbonate:

$$H_2CO_3 \leftrightarrows H^+ + HCO_3^-$$

If H^+ is added to a solution of carbonic acid, the equilibrium shifts to the left and most of the added H^+ is removed from solution. If OH^- is added, H^+ and OH^- combine, taking H^+ out of solution. However, the decrease is countered by more dissociation of H_2CO_3, and the decline in H^+ concentration is minimized. A unique feature of bicarbonate is the linkage between its buffering ability and the ability for the lungs to remove carbon dioxide from the body. Other important biological buffers include phosphates and proteins.

DIFFUSION

Diffusion is the process by which a gas or a substance in a solution expands, because of the motion of its particles, to fill all the available volume. The particles (molecules or atoms) of a substance dissolved in a solvent are in continuous random movement. A given particle is equally likely to move into or

FIGURE 1–2   Proton concentration and pH. Relative proton (H^+) concentrations for solutions on a pH scale are shown. (Redrawn from Alberts B et al: *Molecular Biology of the Cell,* 4th ed. Garland Science, 2002.)

out of an area in which it is present in high concentration. However, because there are more particles in the area of high concentration, the total number of particles moving to areas of lower concentration is greater; that is, there is a **net flux** of solute particles from areas of high to areas of low concentration. The time required for equilibrium by diffusion is proportionate to the square of the diffusion distance. The magnitude of the diffusing tendency from one region to another is directly proportionate to the cross-sectional area across which diffusion is taking place and the **concentration gradient**, or **chemical gradient**, which is the difference in concentration of the diffusing substance divided by the thickness of the boundary **(Fick's law of diffusion)**. Thus,

$$J = -DA \frac{\Delta c}{\Delta x}$$

where J is the net rate of diffusion, D is the diffusion coefficient, A is the area, and $\Delta c/\Delta x$ is the concentration gradient. The minus sign indicates the direction of diffusion. When considering movement of molecules from a higher to a lower concentration, $\Delta c/\Delta x$ is negative, so multiplying by $-DA$ gives a positive value. The permeabilities of the boundaries across which diffusion occurs in the body vary, but diffusion is still a major force affecting the distribution of water and solutes.

OSMOSIS

When a substance is dissolved in water, the concentration of water molecules in the solution is less than that in pure water, because the addition of solute to water results in a solution that occupies a greater volume than does the water alone. If the solution is placed on one side of a membrane that is permeable to water but not to the solute, and an equal volume of water is placed on the other, water molecules diffuse down their concentration (chemical) gradient into the solution (Figure 1–3). This process—the diffusion of **solvent** molecules into a region in which there is a higher concentration of a **solute** to which the membrane is impermeable—is called **osmosis**. It is an important factor in physiologic processes. The tendency for movement of solvent molecules to a region of greater solute concentration can be prevented by applying pressure to the more concentrated solution. The pressure necessary to prevent solvent migration is the **osmotic pressure** of the solution.

Osmotic pressure—like vapor pressure lowering, freezing-point depression, and boiling-point elevation—depends on the number rather than the type of particles in a solution; that is, it is a fundamental colligative property of solutions. In an **ideal solution**, osmotic pressure (P) is related to temperature and volume in the same way as the pressure of a gas:

$$P = \frac{nRT}{V}$$

where n is the number of particles, R is the gas constant, T is the absolute temperature, and V is the volume. If T is held constant, it is clear that the osmotic pressure is proportional to the number of particles in solution per unit volume of solution.

FIGURE 1–3 **Diagrammatic representation of osmosis.** Water molecules are represented by small open circles, solute molecules by large solid circles. In the diagram on the left, water is placed on one side of a membrane permeable to water but not to solute, and an equal volume of a solution of the solute is placed on the other. Water molecules move down their concentration (chemical) gradient into the solution, and, as shown in the diagram on the right, the volume of the solution increases. As indicated by the arrow on the right, the osmotic pressure is the pressure that would have to be applied to prevent the movement of the water molecules.

For this reason, the concentration of osmotically active particles is usually expressed in **osmoles**. One osmole (Osm) equals the gram-molecular weight of a substance divided by the number of freely moving particles that each molecule liberates in solution. For biological solutions, the milliosmole (mOsm; 1/1000 of 1 Osm) is more commonly used.

If a solute is a nonionizing compound such as glucose, the osmotic pressure is a function of the number of glucose molecules present. If the solute ionizes and forms an ideal solution, each ion is an osmotically active particle. For example, NaCl would dissociate into Na^+ and Cl^- ions, so that each mole in solution would supply 2 Osm. One mole of Na_2SO_4 would dissociate into Na^+, Na^+, and SO_4^{2-} supplying 3 Osm. However, the body fluids are not ideal solutions, and although the dissociation of strong electrolytes is complete, the number of particles free to exert an osmotic effect is reduced owing to interactions between the ions. Thus, it is actually the effective concentration (**activity**) in the body fluids rather than the number of equivalents of an electrolyte in solution that determines its osmotic capacity. This is why, for example, 1 mmol of NaCl per liter in the body fluids contributes somewhat less than 2 mOsm of osmotically active particles per liter. The more concentrated the solution, the greater the deviation from an ideal solution.

The osmolal concentration of a substance in a fluid is measured by the degree to which it depresses the freezing point, with 1 mol of an ideal solution depressing the freezing point 1.86 °C. The number of milliosmoles per liter in a solution equals the freezing point depression divided by 0.00186. The **osmolarity** is the number of osmoles per liter of solution (eg, plasma), whereas the **osmolality** is the number of osmoles per kilogram of solvent. Therefore, osmolarity is affected by the volume of the various solutes in the solution and the temperature, while the osmolality is not. Osmotically active substances in the body are dissolved in water, and the density of water is 1, so osmolal concentrations can be expressed as osmoles per liter (Osm/L) of water. In this book, osmolal (rather than osmolar) concentrations are considered, and osmolality is expressed in milliosmoles per liter (of water).

Note that although a homogeneous solution contains osmotically active particles and can be said to have an osmotic pressure, it can exert an osmotic pressure only when it is in contact with another solution across a membrane permeable to the solvent but not to the solute.

OSMOLAL CONCENTRATION OF PLASMA: TONICITY

The freezing point of normal human plasma averages –0.54 °C, which corresponds to an osmolal concentration in plasma of 290 mOsm/L. This is equivalent to an osmotic pressure against pure water of 7.3 atm. The osmolality might be expected to be higher than this, because the sum of all the cation and anion equivalents in plasma is over 300. It is not this high because plasma is not an ideal solution and ionic interactions reduce the number of particles free to exert an osmotic effect. Except when there has been insufficient time after a sudden change in composition for equilibrium to occur, all fluid compartments of the body are in (or nearly in) osmotic equilibrium. The term **tonicity** is used to describe the osmolality of a solution relative to plasma. Solutions that have the same osmolality as plasma are said to be **isotonic;** those with greater osmolality are **hypertonic;** and those with lesser osmolality are **hypotonic.** All solutions that are initially isosmotic with plasma (ie, that have the same actual osmotic pressure or freezing-point depression as plasma) would remain isotonic if it were not for the fact that some solutes diffuse into cells and others are metabolized. Thus, a 0.9% saline solution remains isotonic because there is no net movement of the osmotically active particles in the solution into cells and the particles are not metabolized. On the other hand, a 5% glucose solution is isotonic when initially infused intravenously, but glucose is metabolized, so the net effect is that of infusing a hypotonic solution.

It is important to note the relative contributions of the various plasma components to the total osmolal concentration of plasma. All but about 20 of the 290 mOsm in each liter of normal plasma are contributed by Na^+ and its accompanying anions, principally Cl^- and HCO_3^-. Other cations and anions make a relatively small contribution. Although the concentration of the plasma proteins is large when expressed in grams per liter, they normally contribute less than 2 mOsm/L because of their very high molecular weights. The major nonelectrolytes of plasma are glucose and urea, which in the steady state are in equilibrium with cells. Their contributions to osmolality are normally about 5 mOsm/L each but can become quite large in hyperglycemia or uremia. The total plasma osmolality is important in assessing dehydration, overhydration, and other fluid and electrolyte abnormalities (Clinical Box 1–1).

CLINICAL BOX 1–1

Plasma Osmolality & Disease

Unlike plant cells, which have rigid walls, animal cell membranes are flexible. Therefore, animal cells swell when exposed to extracellular hypotonicity and shrink when exposed to extracellular hypertonicity. Cells contain ion channels and pumps that can be activated to offset moderate changes in osmolality; however, these can be overwhelmed under certain pathologies. Hyperosmolality can cause coma (hyperosmolar coma). Because of the predominant role of the major solutes and the deviation of plasma from an ideal solution, one can ordinarily approximate the plasma osmolality within a few mosm/liter by using the following formula, in which the constants convert the clinical units to millimoles of solute per liter:

$$\text{Osmolality (mOsm/L)} = 2[Na^+] \, (mEq/L) + 0.055[\text{Glucose}] \, (mg/dL) + 0.36[\text{BUN}] \, (mg/dL)$$

BUN is the blood urea nitrogen. The formula is also useful in calling attention to abnormally high concentrations of other solutes. An observed plasma osmolality (measured by freezing-point depression) that greatly exceeds the value predicted by this formula probably indicates the presence of a foreign substance such as ethanol, mannitol (sometimes injected to shrink swollen cells osmotically), or poisons such as ethylene glycol or methanol (components of antifreeze).

NONIONIC DIFFUSION

Some weak acids and bases are quite soluble in cell membranes in the undissociated form, whereas they cannot cross membranes in the charged (ie, dissociated) form. Consequently, if molecules of the undissociated substance diffuse from one side of the membrane to the other and then dissociate, there is appreciable net movement of the undissociated substance from one side of the membrane to the other. This phenomenon is called **nonionic diffusion.**

DONNAN EFFECT

When an ion on one side of a membrane cannot diffuse through the membrane, the distribution of other ions to which the membrane is permeable is affected in a predictable way. For example, the negative charge of a nondiffusible anion hinders diffusion of the diffusible cations and favors diffusion of the diffusible anions. Consider the following situation,

$$\underline{X} \quad \underline{Y}$$

m

K^+	K^+
Cl^-	Cl^-
$Prot^-$	

in which the membrane (m) between compartments X and Y is impermeable to charged proteins (Prot⁻) but freely permeable to K⁺ and Cl⁻. Assume that the concentrations of the anions and of the cations on the two sides are initially equal. Cl⁻ diffuses down its concentration gradient from Y to X, and some K⁺ moves with the negatively charged Cl⁻ because of its opposite charge. Therefore

$$[K^+_x] > [K^+_y]$$

Furthermore,

$$[K^+_x] + [Cl^-_x] + [Prot^-_x] > [K^+_y] + [Cl^-_y]$$

that is, more osmotically active particles are on side X than on side Y.

Donnan and Gibbs showed that in the presence of a nondiffusible ion, the diffusible ions distribute themselves so that at equilibrium their concentration ratios are equal:

$$\frac{[K^+_x]}{[K^+_y]} = \frac{[Cl^-_y]}{[Cl^-_x]}$$

Cross-multiplying,

$$[K^+_x] + [Cl^-_x] = [K^+_y] + [Cl^-_y]$$

This is the **Gibbs–Donnan equation.** It holds for any pair of cations and anions of the same valence.

The Donnan effect on the distribution of ions has three effects in the body introduced here and discussed below. First, because of charged proteins (Prot⁻) in cells, there are more osmotically active particles in cells than in interstitial fluid, and because animal cells have flexible walls, osmosis would make them swell and eventually rupture if it were not for **Na, K ATPase** pumping ions back out of cells. Thus, normal cell volume and pressure depend on Na, K ATPase. Second, because at equilibrium the distribution of permeant ions across the membrane (m in the example used here) is asymmetric, an electrical difference exists across the membrane whose magnitude can be determined by the **Nernst equation.** In the example used here, side X will be negative relative to side Y. The charges line up along the membrane, with the concentration gradient for Cl⁻ exactly balanced by the oppositely directed electrical gradient, and the same holds true for K⁺. Third, because there are more proteins in plasma than in interstitial fluid, there is a Donnan effect on ion movement across the capillary wall.

FORCES ACTING ON IONS

The forces acting across the cell membrane on each ion can be analyzed mathematically. Chloride ions (Cl⁻) are present in higher concentration in the ECF than in the cell interior, and they tend to diffuse along this **concentration gradient** into the cell. The interior of the cell is negative relative to the exterior, and chloride ions are pushed out of the cell along this **electrical gradient.** An equilibrium is reached between Cl⁻ influx and Cl⁻ efflux. The membrane potential at which this equilibrium exists

is the **equilibrium potential.** Its magnitude can be calculated from the Nernst equation, as follows:

$$E_{Cl} = \frac{RT}{FZ_{Cl}} \ln \frac{[Cl_o^-]}{[Cl_i^-]}$$

where

E_{Cl} = equilibrium potential for Cl⁻

R = gas constant

T = absolute temperature

F = the faraday (number of coulombs per mole of charge)

Z_{Cl} = valence of Cl⁻ (–1)

$[Cl_o^-]$ = Cl⁻ concentration outside the cell

$[Cl_i^-]$ = Cl⁻ concentration inside the cell

Converting from the natural log to the base 10 log and replacing some of the constants with numerical values, the equation becomes:

$$E_{Cl} = 61.5 \log \frac{[Cl_i^-]}{[Cl_o^-]} \text{ at 37 °C}$$

Note that in converting to the simplified expression the concentration ratio is reversed because the –1 valence of Cl⁻ has been removed from the expression.

The equilibrium potential for Cl⁻ (E_{Cl}), calculated from the standard values listed in Table 1–1, is –70 mV, a value identical to the measured resting membrane potential of –70 mV. Therefore, no forces other than those represented by the chemical and electrical gradients need be invoked to explain the distribution of Cl⁻ across the membrane.

A similar equilibrium potential can be calculated for K⁺ (E_K):

$$E_K = \frac{RT}{FZ_K} \ln \frac{[K_o^+]}{[K_i^+]} = 61.5 \log \frac{[K_o^+]}{[K_i^+]} \text{ at 37 °C}$$

where

E_K = equilibrium potential for K⁺

Z_K = valence of K⁺ (+1)

$[K_o^+]$ = K⁺ concentration outside the cell

$[K_i^+]$ = K⁺ concentration inside the cell

R, T, and F as above

In this case, the concentration gradient is outward and the electrical gradient inward. In mammalian spinal motor neurons, E_K is –90 mV (Table 1–1). Because the resting membrane potential is –70 mV, there is somewhat more K⁺ in the neurons than can be accounted for by the electrical and chemical gradients.

The situation for Na⁺ is quite different from that for K⁺ and Cl⁻. The direction of the chemical gradient for Na⁺ is inward, to the area where it is in lesser concentration, and the electrical gradient is in the same direction. E_{Na} is +60 mV (Table 1–1). Because neither E_K nor E_{Na} is equal to the membrane potential,

TABLE 1–1 Concentration of some ions inside and outside mammalian spinal motor neurons.

Ion	Concentration (mmol/L of H₂O)		Equilibrium Potential (mV)
	Inside Cell	Outside Cell	
Na⁺	15.0	150.0	+60
K⁺	150.0	5.5	–90
Cl⁻	9.0	125.0	–70

Resting membrane potential = –70 mV

FIGURE 1–4 Energy-rich adenosine derivatives. Adenosine triphosphate is broken down into its backbone purine base and sugar (at right) as well as its high energy phosphate derivatives (across bottom). (Reproduced, with permission, from Murray RK et al: *Harper's Biochemistry*, 26th ed. McGraw-Hill, 2003.)

one would expect the cell to gradually gain Na⁺ and lose K⁺ if only passive electrical and chemical forces were acting across the membrane. However, the intracellular concentration of Na⁺ and K⁺ remain constant because of the action of the Na, K ATPase that actively transports Na⁺ out of the cell and K⁺ into the cell (against their respective electrochemical gradients).

GENESIS OF THE MEMBRANE POTENTIAL

The distribution of ions across the cell membrane and the nature of this membrane provide the explanation for the membrane potential. The concentration gradient for K⁺ facilitates its movement out of the cell via K⁺ channels, but its electrical gradient is in the opposite (inward) direction. Consequently, an equilibrium is reached in which the tendency of K⁺ to move out of the cell is balanced by its tendency to move into the cell, and at that equilibrium there is a slight excess of cations on the outside and anions on the inside. This condition is maintained by Na, K ATPase, which uses the energy of ATP to pump K⁺ back into the cell and keeps the intracellular concentration of Na⁺ low. Because the Na, K ATPase moves three Na⁺ out of the cell for every two K⁺ moved in, it also contributes to the membrane potential, and thus is termed an **electrogenic pump**. It should be emphasized that the number of ions responsible for the membrane potential is a minute fraction of the total number present and that the total concentrations of positive and negative ions are equal everywhere except along the membrane.

ENERGY PRODUCTION

ENERGY TRANSFER

Energy is stored in bonds between phosphoric acid residues and certain organic compounds. Because the energy of bond formation in some of these phosphates is particularly high, relatively large amounts of energy (10–12 kcal/mol) are released when the bond is hydrolyzed. Compounds containing such bonds are called **high-energy phosphate compounds.** Not all organic phosphates are of the high-energy type. Many, like glucose 6-phosphate, are low-energy phosphates that on

hydrolysis liberate 2–3 kcal/mol. Some of the intermediates formed in carbohydrate metabolism are high-energy phosphates, but the most important high-energy phosphate compound is **adenosine triphosphate (ATP).** This ubiquitous molecule (Figure 1–4) is the energy storehouse of the body. On hydrolysis to adenosine diphosphate (ADP), it liberates energy directly to such processes as muscle contraction, active transport, and the synthesis of many chemical compounds. Loss of another phosphate to form adenosine monophosphate (AMP) releases more energy.

Another group of high-energy compounds are the thioesters, the acyl derivatives of mercaptans. **Coenzyme A (CoA)** is a widely distributed mercaptan-containing adenine, ribose, pantothenic acid, and thioethanolamine (Figure 1–5). Reduced CoA (usually abbreviated HS–CoA) reacts with acyl groups (R–CO–) to form R–CO–S–CoA derivatives. A prime example is the reaction of HS-CoA with acetic acid to form acetylcoenzyme A (acetyl-CoA), a compound of pivotal importance in intermediary metabolism. Because acetyl-CoA has a much higher energy content than acetic acid, it combines readily with substances in reactions that would otherwise require outside energy. Acetyl-CoA is therefore often called "active acetate." From the point of view of energetics, formation of 1 mol of any acyl-CoA compound is equivalent to the formation of 1 mol of ATP.

BIOLOGIC OXIDATIONS

Oxidation is the combination of a substance with O₂, or loss of hydrogen, or loss of electrons. The corresponding reverse processes are called **reduction.** Biologic oxidations are catalyzed by specific enzymes. Cofactors (simple ions) or coenzymes (organic, nonprotein substances) are accessory substances that

FIGURE 1–5   **Coenzyme A (CoA) and its derivatives. Left:** Formula of reduced coenzyme A (HS-CoA) with its components highlighted. **Right:** Formula for reaction of CoA with biologically important compounds to form thioesters. R, remainder of molecule.

usually act as carriers for products of the reaction. Unlike the enzymes, the coenzymes may catalyze a variety of reactions.

A number of coenzymes serve as hydrogen acceptors. One common form of biologic oxidation is removal of hydrogen from an R–OH group, forming R=O. In such dehydrogenation reactions, nicotinamide adenine dinucleotide (NAD^+) and dihydronicotinamide adenine dinucleotide phosphate ($NADP^+$) pick up hydrogen, forming dihydronicotinamide adenine dinucleotide (NADH) and dihydronicotinamide adenine dinucleotide phosphate (NADPH) (Figure 1–6). The hydrogen is then transferred to the flavoprotein–cytochrome system, reoxidizing the NAD^+ and $NADP^+$. Flavin adenine dinucleotide (FAD) is formed when riboflavin is phosphorylated, forming flavin mononucleotide (FMN). FMN then combines with AMP, forming the dinucleotide. FAD can accept hydrogens in a similar fashion, forming its hydro (FADH) and dihydro ($FADH_2$) derivatives.

The flavoprotein–cytochrome system is a chain of enzymes that transfers hydrogen to oxygen, forming water. This process occurs in the mitochondria. Each enzyme in the chain is reduced

FIGURE 1–6   **Structures of molecules important in oxidation reduction reactions to produce energy. Top:** Formula of the oxidized form of nicotinamide adenine dinucleotide (NAD^+). Nicotinamide adenine dinucleotide phosphate ($NADP^+$) has an additional phosphate group at the location marked by the asterisk. **Bottom:** Reaction by which NAD^+ and $NADP^+$ become reduced to form NADH and NADPH. R, remainder of molecule; R', hydrogen donor.

FIGURE 1–7 **Simplified diagram of transport of protons across the inner and outer lamellas of the inner mitochondrial membrane.** The electron transport system (flavoprotein-cytochrome system) helps create H^+ movement from the inner to the outer lamella. Return movement of protons down the proton gradient generates ATP.

and then reoxidized as the hydrogen is passed down the line. Each of the enzymes is a protein with an attached nonprotein prosthetic group. The final enzyme in the chain is cytochrome c oxidase, which transfers hydrogens to O_2, forming H_2O. It contains two atoms of Fe and three of Cu and has 13 subunits.

The principal process by which ATP is formed in the body is **oxidative phosphorylation.** This process harnesses the energy from a proton gradient across the mitochondrial membrane to produce the high-energy bond of ATP and is briefly outlined in Figure 1–7. Ninety percent of the O_2 consumption in the basal state is mitochondrial, and 80% of this is coupled to ATP synthesis. About 27% of the ATP is used for protein synthesis, and about 24% is used by Na, K ATPase, 9% by gluconeogenesis, 6% by Ca^{2+} ATPase, 5% by myosin ATPase, and 3% by ureagenesis.

MOLECULAR BUILDING BLOCKS

NUCLEOSIDES, NUCLEOTIDES, & NUCLEIC ACIDS

Nucleosides contain a sugar linked to a nitrogen-containing base. The physiologically important bases, **purines** and **pyrimidines,** have ring structures (Figure 1–8). These structures are bound to ribose or 2-deoxyribose to complete the nucleoside. When inorganic phosphate is added to the nucleoside, a **nucleotide** is formed. Nucleosides and nucleotides form the backbone for RNA and DNA, as well as a variety of coenzymes and regulatory molecules (eg, NAD^+, $NADP^+$, and ATP) of physiological importance (Table 1–2). Nucleic acids in the diet are digested and their constituent purines and pyrimidines absorbed, but most of the purines and pyrimidines are synthesized from amino acids, principally in the liver. The nucleotides and RNA and DNA are then synthesized. RNA is in dynamic equilibrium with the amino acid pool, but DNA, once formed, is metabolically stable throughout life. The purines and pyrimidines released by the breakdown of nucleotides may be reused or catabolized. Minor amounts are excreted unchanged in the urine.

The pyrimidines are catabolized to the β-**amino acids,** β-alanine and β-aminoisobutyrate. These amino acids have their amino group on β-carbon, rather than the α-carbon typical to physiologically active amino acids. Because β-aminoisobutyrate is a product of thymine degradation, it can serve as a measure of DNA turnover. The β-amino acids are further degraded to CO_2 and NH_3.

Uric acid is formed by the breakdown of purines and by direct synthesis from 5-phosphoribosyl pyrophosphate (5-PRPP) and glutamine (Figure 1–9). In humans, uric acid is excreted in the urine, but in other mammals, uric acid is further oxidized to allantoin before excretion. The normal blood uric acid level in humans is approximately 4 mg/dL (0.24 mmol/L). In the kidney, uric acid is filtered, reabsorbed, and secreted. Normally, 98% of the filtered uric acid is reabsorbed and the remaining 2% makes up approximately 20% of the amount excreted. The remaining 80% comes from the tubular secretion. The uric acid excretion on a purine-free diet is about 0.5 g/24 h and on a regular diet about 1 g/24 h. Excess uric acid in the blood or urine is a characteristic of gout (Clinical Box 1–2).

Adenine: 6-Aminopurine

Guanine: 1-Amino-6-oxypurine

Hypoxanthine: 6-Oxypurine

Xanthine: 2,6-Dioxypurine

Purine nucleus

Cytosine: 4-Amino-2-oxypyrimidine

Uracil: 2,4-Dioxypyrimidine

Thymine: 5-Methyl-2,4-dioxypyrimidine

Pyrimidine nucleus

FIGURE 1–8 **Principal physiologically important purines and pyrimidines.** Purine and pyrimidine structures are shown next to representative molecules from each group. Oxypurines and oxypyrimidines may form enol derivatives (hydroxypurines and hydroxypyrimidines) by migration of hydrogen to the oxygen substituents.

TABLE 1–2 Purine- and pyrimidine-containing compounds.

Type of Compound	Components
Nucleoside	Purine or pyrimidine plus ribose or 2-deoxyribose
Nucleotide (mononucleotide)	Nucleoside plus phosphoric acid residue
Nucleic acid	Many nucleotides forming double-helical structures of two polynucleotide chains
Nucleoprotein	Nucleic acid plus one or more simple basic proteins
Contain ribose	Ribonucleic acids (RNA)
Contain 2-deoxyribose	Deoxyribonucleic acids (DNA)

FIGURE 1–9 Synthesis and breakdown of uric acid. Adenosine is converted to hypoxanthine, which is then converted to xanthine, and xanthine is converted to uric acid. The latter two reactions are both catalyzed by xanthine oxidase. Guanosine is converted directly to xanthine, while 5-PRPP and glutamine can be converted to uric acid. An additional oxidation of uric acid to allantoin occurs in some mammals.

DNA

Deoxyribonucleic acid (DNA) is found in bacteria, in the nuclei of eukaryotic cells, and in mitochondria. It is made up of two extremely long nucleotide chains containing the bases adenine (A), guanine (G), thymine (T), and cytosine (C) (Figure 1–10). The chains are bound together by hydrogen bonding between the bases, with adenine bonding to thymine and guanine to cytosine. This stable association forms a double-helical structure (Figure 1–11). The double helical structure of DNA is compacted in the cell by association with **histones,** and further compacted into **chromosomes.** A diploid human cell contains 46 chromosomes.

A fundamental unit of DNA, or a **gene,** can be defined as the sequence of DNA nucleotides that contain the information for the production of an ordered amino acid sequence for a single polypeptide chain. Interestingly, the protein encoded by a single gene may be subsequently divided into several different physiologically active proteins. Information is accumulating at an accelerating rate about the structure of genes and their regulation. The basic structure of a typical eukaryotic gene is shown in diagrammatic form in Figure 1–12. It is made up of a strand of DNA that includes coding and noncoding regions. In eukaryotes, unlike prokaryotes, the portions of the genes that dictate the formation of proteins are usually broken into several segments (**exons**) separated by segments that are not translated (**introns**). Near the transcription start site of the gene is a **promoter,** which is the site at which RNA polymerase and its cofactors bind. It often includes a thymidine–adenine–thymidine–adenine (TATA) sequence (**TATA box**), which ensures that transcription starts at the proper point. Farther out in the 5' region are **regulatory elements,** which include enhancer and silencer sequences. It has been estimated that each gene has an average of five regulatory sites. Regulatory sequences are sometimes found in the 3'-flanking region as well.

Gene mutations occur when the base sequence in the DNA is altered from its original sequence. Such alterations can affect protein structure and be passed on to daughter cells after cell division. **Point mutations** are single base substitutions. A variety of chemical modifications (eg, alkylating or intercalating agents, or ionizing radiation) can lead to changes in DNA sequences and mutations. The collection of genes within the full expression of DNA from an organism is termed its **genome.** An indication of the complexity of DNA in the human haploid genome (the total genetic message) is its size; it is made up of 3×10^9 base pairs that can code for approximately 30,000 genes. This genetic message is the blueprint for

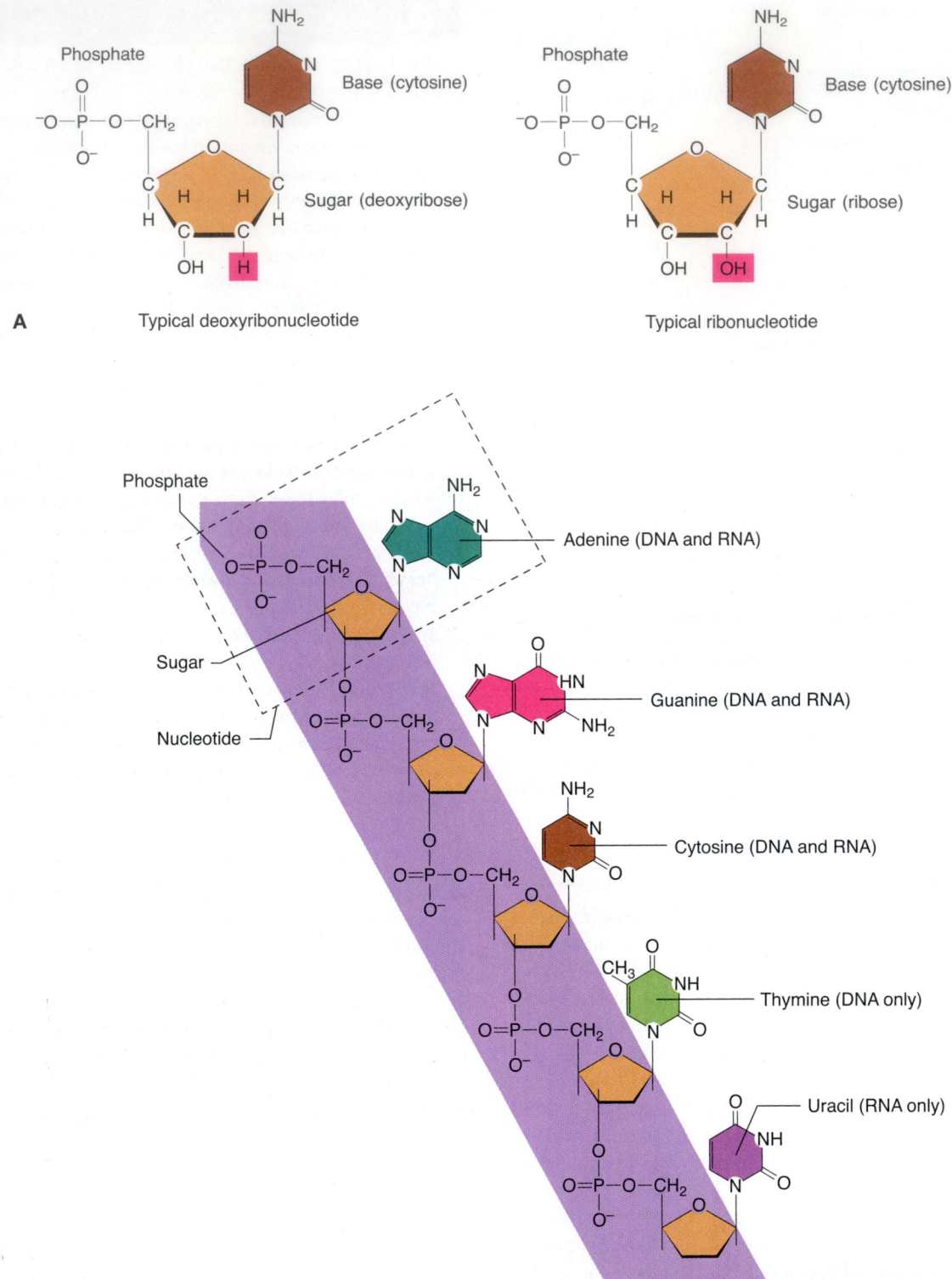

FIGURE 1–10   **Basic structure of nucleotides and nucleic acids. A)** At left, the nucleotide cytosine is shown with deoxyribose and at right with ribose as the principal sugar. **B)** Purine bases adenine and guanine are bound to each other or to pyrimidine bases, cytosine, thymine, or uracil via a phosphodiester backbone between 2′-deoxyribosyl moieties attached to the nucleobases by an N-glycosidic bond. Note that the backbone has a polarity (ie, a 5′ and a 3′ direction). Thymine is only found in DNA, while the uracil is only found in RNA.

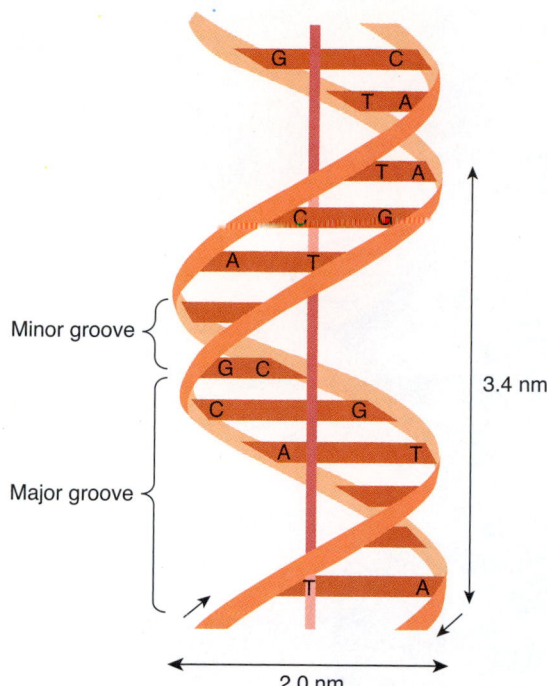

FIGURE 1–11   Double-helical structure of DNA. The compact structure has an approximately 2.0 nm thickness and 3.4 nm between full turns of the helix that contains both major and minor grooves. The structure is maintained in the double helix by hydrogen bonding between purines and pyrimidines across individual strands of DNA. Adenine (A) is bound to thymine (T) and cytosine (C) to guanine (G). (Reproduced with permission from Murray RK et al: *Harper's Biochemistry,* 26th ed. McGraw-Hill, 2003.)

the heritable characteristics of the cell and its descendants. The proteins formed from the DNA blueprint include all the enzymes, and these in turn control the metabolism of the cell.

Each nucleated somatic cell in the body contains the full genetic message, yet there is great differentiation and specialization in the functions of the various types of adult cells. Only small parts of the message are normally transcribed. Thus, the genetic message is normally maintained in a repressed state. However, genes are controlled both spatially and temporally. First, under physiological conditions, the double helix requires highly regulated interaction by proteins to unravel for **replication, transcription,** or both.

REPLICATION: MITOSIS & MEIOSIS

At the time of each somatic cell division (**mitosis**), the two DNA chains separate, each serving as a template for the synthesis of a new complementary chain. DNA polymerase catalyzes this reaction. One of the double helices thus formed goes to one daughter cell and one goes to the other, so the amount of DNA in each daughter cell is the same as that in the parent cell. The life cycle of the cell that begins after mitosis is highly regulated and is termed the **cell cycle** (Figure 1–13). The G_1 (or Gap 1) phase represents a period of cell growth and divides the end of mitosis from the DNA synthesis (or S) phase. Following DNA synthesis, the cell enters another period of cell growth, the G_2 (Gap 2) phase. The ending of this stage is marked by chromosome condensation and the beginning of mitosis (M stage).

In germ cells, reduction division (**meiosis**) takes place during maturation. The net result is that one of each pair of chromosomes ends up in each mature germ cell; consequently, each mature germ cell contains half the amount of chromosomal material found in somatic cells. Therefore, when a sperm unites with an ovum, the resulting zygote has the full complement of DNA, half of which came from the father and half from the mother. The term "ploidy" is sometimes used to refer to the number of chromosomes in cells. Normal resting diploid cells are **euploid** and become **tetraploid** just before division. **Aneuploidy** is the condition in which a cell contains other than the haploid number of chromosomes or an exact multiple of it, and this condition is common in cancerous cells.

RNA

The strands of the DNA double helix not only replicate themselves, but also serve as templates by lining up complementary bases for the formation in the nucleus of **ribonucleic acids** (**RNA**). RNA differs from DNA in that it is single-stranded, has **uracil** in place of thymine, and its sugar moiety is ribose rather than 2'-deoxyribose (Figure 1–13). The production of RNA from DNA is called **transcription.** Transcription can lead to several types of RNA including: **messenger RNA** (**mRNA**), **transfer RNA** (**tRNA**), **ribosomal RNA** (**rRNA**), and other RNAs. Transcription is catalyzed by various forms of **RNA polymerase.**

FIGURE 1–12   Diagram of the components of a typical eukaryotic gene. The region that produces introns and exons is flanked by noncoding regions. The 5'-flanking region contains stretches of DNA that interact with proteins to facilitate or inhibit transcription. The 3'-flanking region contains the poly(A) addition site. (Modified from Murray RK et al: *Harper's Biochemistry,* 26th ed. McGraw-Hill, 2003.)

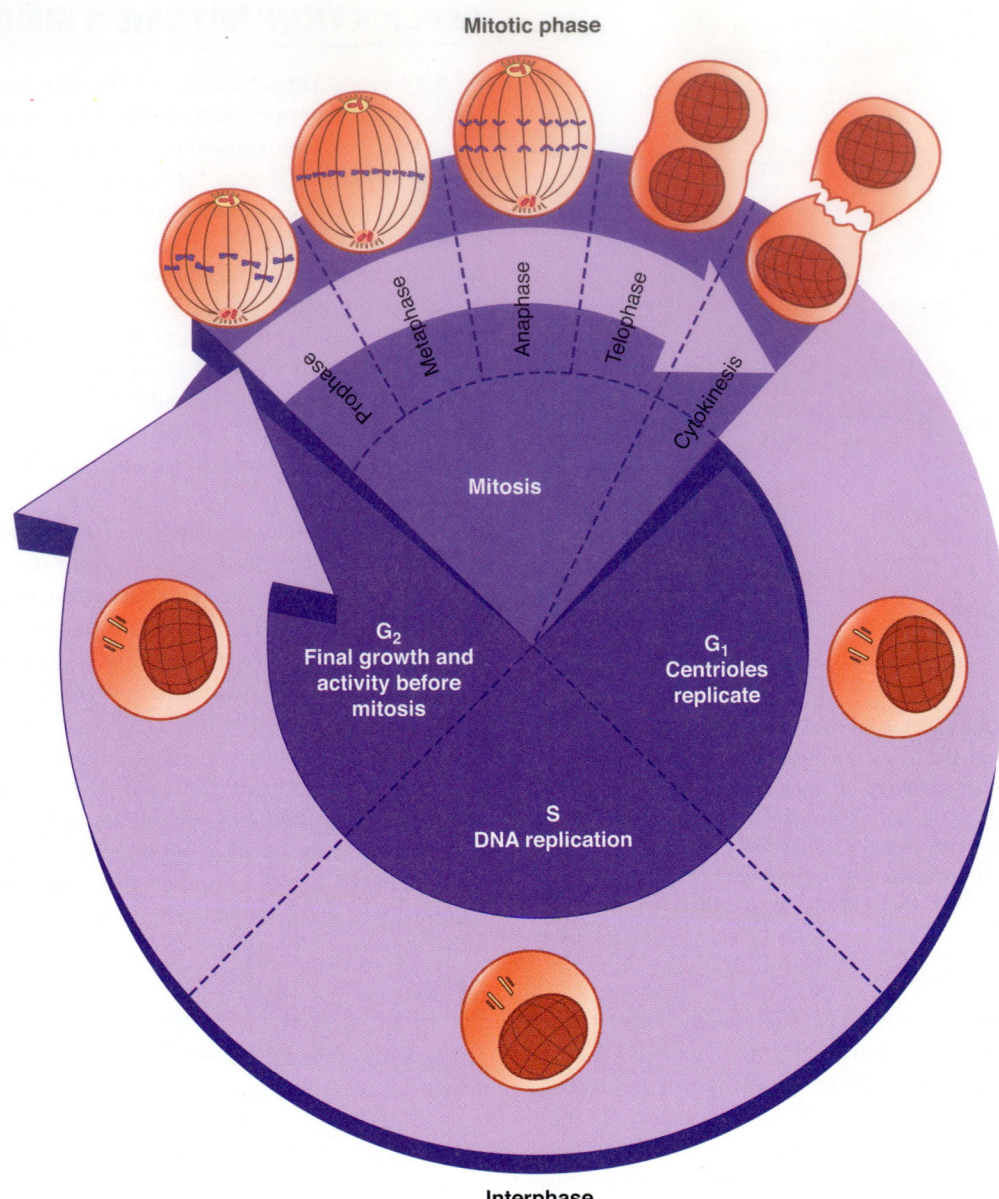

FIGURE 1–13 **Sequence of events during the cell cycle.** Immediately following mitosis (M) the cell enters a gap phase (G1) before a DNA synthesis phase (S) a second gap phase (G2) and back to mitosis. Collectively G1, S, and G2 phases are referred to as interphase (I).

Typical transcription of an mRNA is shown in Figure 1–14. When suitably activated, transcription of the gene into a pre-mRNA starts at the **cap site** and ends about 20 bases beyond the AATAAA sequence. The RNA transcript is capped in the nucleus by addition of 7-methylguanosine triphosphate to the 5' end; this cap is necessary for proper binding to the ribosome. A **poly(A) tail** of about 100 bases is added to the untranslated segment at the 3' end to help maintain the stability of the mRNA. The pre-mRNA formed by capping and addition of the poly(A) tail is then processed by elimination of the introns, and once this posttranscriptional modification is complete, the mature mRNA moves to the cytoplasm. Posttranscriptional modification of the pre-mRNA is a regulated process where differential splicing can occur to form more than one mRNA from a single pre-mRNA. The introns of some genes are eliminated by **spliceosomes,** complex units that are made up of small RNAs and proteins. Other introns are eliminated by **self-splicing** by the RNA they contain. Because of introns and splicing, more than one mRNA can be formed from the same gene.

Most forms of RNA in the cell are involved in **translation,** or protein synthesis. A brief outline of the transition from transcription to translation is shown in Figure 1–15. In the cytoplasm, ribosomes provide a template for tRNA to deliver specific amino acids to a growing polypeptide chain based on specific sequences in mRNA. The mRNA molecules are smaller than the DNA molecules, and each represents a

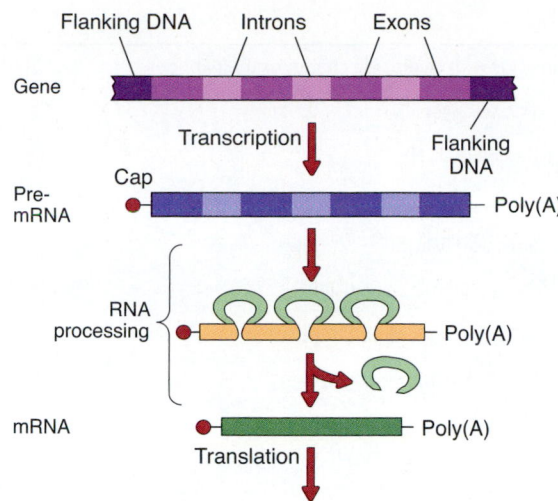

FIGURE 1–14 **Transcription of a typical mRNA.** Steps in transcription from a typical gene to a processed mRNA are shown. Cap, cap site. (Modified from Baxter JD: Principles of endocrinology. In: *Cecil Textbook of Medicine,* 16th ed. Wyngaarden JB, Smith LH Jr (editors). Saunders, 1982.)

transcript of a small segment of the DNA chain. For comparison, the molecules of tRNA contain only 70–80 nitrogenous bases, compared with hundreds in mRNA and 3 billion in DNA.

AMINO ACIDS & PROTEINS

AMINO ACIDS

Amino acids that form the basic building blocks for proteins are identified in Table 1–3. These amino acids are often referred to by their corresponding three-letter, or single-letter abbreviations. Various other important amino acids such as ornithine, 5-hydroxytryptophan, L-dopa, taurine, and thyroxine (T_4) occur in the body but are not found in proteins. In higher animals, the L isomers of the amino acids are the only naturally occurring forms in proteins. The L isomers of hormones such as thyroxine are much more active than the D isomers. The amino acids are acidic, neutral, or basic in reaction, depending on the relative proportions of free acidic (–COOH) or basic (–NH$_2$) groups in the molecule. Some of the amino acids are **nutritionally essential amino acids,** that is, they must be obtained in the diet, because they cannot be made in the body. Arginine and histidine must be provided through diet during times of rapid growth or recovery from illness and are termed **conditionally essential.** All others are **nonessential amino acids** in the sense that they can be synthesized in vivo in amounts sufficient to meet metabolic needs.

FIGURE 1–15 **Diagrammatic outline of transcription to translation.** From the DNA molecule, a messenger RNA is produced and presented to the ribosome. It is at the ribosome where charged tRNA match up with their complementary codons of mRNA to position the amino acid for growth of the polypeptide chain. DNA and RNA are represented as lines with multiple short projections representing the individual bases. Small boxes labeled A represent individual amino acids.

TABLE 1–3 Amino acids found in proteins.*

Amino acids with aliphatic side chains	Amino acids with acidic side chains, or their amides
Alanine (Ala, A)	Aspartic acid (Asp, D)
Valine (Val, V)	Asparagine (Asn, N)
Leucine (Leu, L)	Glutamine (Gln, Q)
Isoleucine (Ile, I)	Glutamic acid (Glu, E)
Hydroxyl-substituted amino acids	γ-Carboxyglutamic acid[b] (Gla)
Serine (Ser, S)	Amino acids with side chains containing basic groups
Threonine (Thr, T)	**Arginine**[c] (Arg, R)
Sulfur-containing amino acids	**Lysine** (Lys, K)
Cysteine (Cys, C)	Hydroxylysine[b] (Hyl)
Methionine (Met, M)	**Histidine**[c] (His, H)
Selenocysteine[a]	Imino acids (contain imino group but no amino group)
Amino acids with aromatic ring side chains	Proline (Pro, P)
Phenylalanine (Phe, F)	4-Hydroxyproline[b] (Hyp)
Tyrosine (Tyr, Y)	3-Hydroxyproline[b]
Tryptophan (Trp, W)	

*Those in bold type are the nutritionally essential amino acids. The generally accepted three-letter and one-letter abbreviations for the amino acids are shown in parentheses.

[a]Selenocysteine is a rare amino acid in which the sulfur of cysteine is replaced by selenium. The codon UGA is usually a stop codon, but in certain situations it codes for selenocysteine.

[b]There are no tRNAs for these four amino acids; they are formed by post-translational modification of the corresponding unmodified amino acid in peptide linkage. There are tRNAs for selenocysteine and the remaining 20 amino acids, and they are incorporated into peptides and proteins under direct genetic control.

[c]Arginine and histidine are sometimes called "conditionally essential"—they are not necessary for maintenance of nitrogen balance, but are needed for normal growth.

THE AMINO ACID POOL

Although small amounts of proteins are absorbed from the gastrointestinal tract and some peptides are also absorbed, most ingested proteins are digested and their constituent amino acids absorbed. The body's own proteins are being continuously hydrolyzed to amino acids and resynthesized. The turnover rate of endogenous proteins averages 80–100 g/d, being highest in the intestinal mucosa and practically nil in the extracellular structural protein, collagen. The amino acids formed by endogenous protein breakdown are identical to those derived from ingested protein. Together they form a common **amino acid pool** that supplies the needs of the body (Figure 1–16).

PROTEINS

Proteins are made up of large numbers of amino acids linked into chains by **peptide bonds** joining the amino group of one amino acid to the carboxyl group of the next (Figure 1–17). In addition, some proteins contain carbohydrates (glycoproteins) and lipids (lipoproteins). Smaller chains of amino acids are called **peptides** or **polypeptides.** The boundaries between peptides, polypeptides, and proteins are not well defined. For

this text, amino acid chains containing 2–10 amino acid residues are called peptides, chains containing more than 10 but fewer than 100 amino acid residues are called polypeptides, and chains containing 100 or more amino acid residues are called proteins.

FIGURE 1–16 Amino acids in the body. There is an extensive network of amino acid turnover in the body. Boxes represent large pools of amino acids and some of the common interchanges are represented by arrows. Note that most amino acids come from the diet and end up in protein, however, a large portion of amino acids are interconverted and can feed into and out of a common metabolic pool through amination reactions.

FIGURE 1–17 **Amino acid structure and formation of peptide bonds.** The dashed line shows where peptide bonds are formed between two amino acids. The highlighted area is released as H_2O. R, remainder of the amino acid. For example, in glycine, R = H; in glutamate, R = —$(CH_2)_2$—COO^-.

The order of the amino acids in the peptide chains is called the **primary structure** of a protein. The chains are twisted and folded in complex ways, and the term **secondary structure** of a protein refers to the spatial arrangement produced by the twisting and folding. A common secondary structure is a regular coil with 3.7 amino acid residues per turn (α-helix). Another common secondary structure is a β-sheet. An antiparallel β-sheet is formed when extended polypeptide chains fold back and forth on one another and hydrogen bonding occurs between the peptide bonds on neighboring chains. Parallel β-sheets between polypeptide chains also occur. The **tertiary structure** of a protein is the arrangement of the twisted chains into layers, crystals, or fibers. Many protein molecules are made of several proteins, or subunits (eg, hemoglobin), and the term **quaternary structure** is used to refer to the arrangement of the subunits into a functional structure.

PROTEIN SYNTHESIS

The process of protein synthesis, **translation,** is the conversion of information encoded in mRNA to a protein (Figure 1–15). As described previously, when a definitive mRNA reaches a ribosome in the cytoplasm, it dictates the formation of a polypeptide chain. Amino acids in the cytoplasm are activated by combination with an enzyme and adenosine monophosphate (adenylate), and each **activated amino acid** then combines with a specific molecule of tRNA. There is at least one tRNA for each of the 20 unmodified amino acids found in large quantities in the body proteins of animals, but some amino acids have more than one tRNA. The tRNA–amino acid–adenylate complex is next attached to the mRNA template, a process that occurs in the ribosomes. The tRNA "recognizes" the proper spot to attach on the mRNA template because it has on its active end a set of three bases that are complementary to a set of three bases in a particular spot on the mRNA chain. The genetic code is made up of such triplets **(codons),** sequences of three purine, pyrimidine, or purine and pyrimidine bases; each codon stands for a particular amino acid.

Translation typically starts in the ribosomes with an AUG (transcribed from ATG in the gene), which codes for methionine. The amino terminal amino acid is then added, and the chain is lengthened one amino acid at a time. The mRNA attaches to the 40S subunit of the ribosome during protein synthesis, the polypeptide chain being formed attaches to the 60S subunit, and the tRNA attaches to both. As the amino acids are added in the order dictated by the codon, the ribosome moves along the mRNA molecule like a bead on a string. Translation stops at one of three stop, or nonsense, codons (UGA, UAA, or UAG), and the polypeptide chain is released. The tRNA molecules are used again. The mRNA molecules are typically reused approximately 10 times before being replaced. It is common to have more than one ribosome on a given mRNA chain at a time. The mRNA chain plus its collection of ribosomes is visible under the electron microscope as an aggregation of ribosomes called a **polyribosome.**

POSTTRANSLATIONAL MODIFICATION

After the polypeptide chain is formed, it "folds" into its biological form and can be further modified to the final protein by one or more of a combination of reactions that include hydroxylation, carboxylation, glycosylation, or phosphorylation of amino acid residues; cleavage of peptide bonds that converts a larger polypeptide to a smaller form; and the further folding, packaging, or folding and packaging of the protein into its ultimate, often complex configuration. Protein folding is a complex process that is dictated primarily by the sequence of the amino acids in the polypeptide chain. In some instances, however, nascent proteins associate with other proteins called **chaperones,** which prevent inappropriate contacts with other proteins and ensure that the final "proper" conformation of the nascent protein is reached.

Proteins also contain information that helps to direct them to individual cell compartments. Many proteins that are going to be secreted or stored in organelles and most transmembrane proteins have at their amino terminal a **signal peptide (leader sequence)** that guides them into the endoplasmic reticulum. The sequence is made up of 15 to 30 predominantly hydrophobic amino acid residues. The signal peptide, once synthesized, binds to a **signal recognition particle (SRP),** a complex molecule made up of six polypeptides and 7S RNA, one of the small RNAs. The SRP stops translation until it binds to a **translocon,** a pore in the endoplasmic reticulum that is a heterotrimeric structure made up of Sec 61 proteins. The ribosome also binds, and the signal peptide leads the growing peptide chain into the cavity of the endoplasmic reticulum (Figure 1–18). The signal

FIGURE 1–18 **Translation of protein into endoplasmic reticulum according to the signal hypothesis.** The ribosomes synthesizing a protein move along the mRNA from the 5' to the 3' end. When the signal peptide of a protein destined for secretion, the cell membrane, or lysosomes emerges from the large unit of the ribosome, it binds to a signal recognition particle (SRP), and this arrests further translation until it binds to the translocon on the endoplasmic reticulum. N, amino end of protein; C, carboxyl end of protein. (Reproduced, with permission, from Perara E, Lingappa VR: Transport of proteins into and across the endoplasmic reticulum membrane. In: *Protein Transfer and Organelle Biogenesis.* Das RC, Robbins PW (editors). Academic Press, 1988.)

peptide is next cleaved from the rest of the peptide by a signal peptidase while the rest of the peptide chain is still being synthesized. SRPs are not the only signals that help to direct proteins to their proper place in or out of the cell; other signal sequences, posttranslational modifications, or both (eg, glycosylation) can serve this function.

PROTEIN DEGRADATION

Like protein synthesis, protein degradation is a carefully regulated, complex process. It has been estimated that overall, up to 30% of newly produced proteins are abnormal, such as can occur during improper folding. Aged normal proteins also need to be removed as they are replaced. Conjugation of proteins to the 74-amino-acid polypeptide **ubiquitin** marks them for degradation. This polypeptide is highly conserved and is present in species ranging from bacteria to humans. The process of binding ubiquitin is called **ubiquitination,** and in some instances, multiple ubiquitin molecules bind **(polyubiquitination).** Ubiquitination of cytoplasmic proteins, including integral proteins of the endoplasmic reticulum, marks the proteins for degradation in multisubunit proteolytic particles, or **proteasomes.** Ubiquitination of membrane proteins, such as the growth hormone receptors, also marks them for degradation, however these can be degraded in lysosomes as well as via the proteasomes.

There is an obvious balance between the rate of production of a protein and its destruction, so ubiquitin conjugation is of major importance in cellular physiology. The rates at which individual proteins are metabolized vary, and the body has mechanisms by which abnormal proteins are recognized and degraded more rapidly than normal body constituents. For example, abnormal hemoglobins are metabolized rapidly in individuals with congenital hemoglobinopathies.

CATABOLISM OF AMINO ACIDS

The short-chain fragments produced by amino acid, carbohydrate, and fat catabolism are very similar (see below). From this **common metabolic pool** of intermediates, carbohydrates, proteins, and fats can be synthesized. These fragments can enter the citric acid cycle, a final common pathway of catabolism, in which they are broken down to hydrogen atoms and CO_2. Interconversion of amino acids involve transfer, removal, or formation of amino groups. **Transamination** reactions, conversion of one amino acid to the corresponding keto acid with simultaneous conversion of another keto acid to an amino acid, occur in many tissues:

$$\text{Alanine} + \alpha\text{-Ketoglutarate} \leftrightarrows \text{Pyruvate} + \text{Glutamate}$$

The **transaminases** involved are also present in the circulation. When damage to many active cells occurs as a result of a pathologic process, serum transaminase levels rise. An example is the rise in **plasma aspartate aminotransferase (AST)** following myocardial infarction.

Oxidative deamination of amino acids occurs in the liver. An imino acid is formed by dehydrogenation, and this compound is hydrolyzed to the corresponding keto acid, with production of NH_4^+:

$$\text{Amino acid} + NAD^+ \rightarrow \text{Imino acid} + NADH + H^+$$

$$\text{Imino acid} + H_2O \rightarrow \text{Keto acid} + NH_4^+$$

Interconversions between the amino acid pool and the common metabolic pool are summarized in Figure 1–19. Leucine, isoleucine, phenylalanine, and tyrosine are said to be **ketogenic** because they are converted to the ketone body acetoacetate (see below). Alanine and many other amino acids are **glucogenic** or **gluconeogenic;** that is, they give rise to compounds that can readily be converted to glucose.

UREA FORMATION

Most of the NH_4^+ formed by deamination of amino acids in the liver is converted to urea, and the urea is excreted in the urine. The NH_4^+ forms carbamoyl phosphate, and in the mitochondria it is transferred to ornithine, forming citrulline. The enzyme involved is ornithine carbamoyltransferase. Citrulline is converted to arginine, after which urea is split off and ornithine is regenerated (urea cycle; Figure 1–20). The overall reaction in the urea cycle consumes 3 ATP (not shown) and thus requires significant energy. Most of the urea is formed in the liver, and in severe liver disease the blood urea nitrogen (BUN) falls and blood NH_3 rises (see Chapter 29). Congenital deficiency of ornithine carbamoyltransferase can also lead to NH_3 intoxication, even in individuals who are heterozygous for this deficiency.

FIGURE 1–19   Involvement of the citric acid cycle in transamination and gluconeogenesis. The bold arrows indicate the main pathway of gluconeogenesis. Note the many entry positions for groups of amino acids into the citric acid cycle. (Reproduced with permission from Murray RK et al: *Harper's Biochemistry,* 26th ed. McGraw-Hill, 2003.)

METABOLIC FUNCTIONS OF AMINO ACIDS

In addition to providing the basic building blocks for proteins, amino acids also have metabolic functions. Thyroid hormones, catecholamines, histamine, serotonin, melatonin, and intermediates in the urea cycle are formed from specific amino acids. Methionine and cysteine provide the sulfur contained in proteins, CoA, taurine, and other biologically important compounds. Methionine is converted into *S*-adenosylmethionine, which is the active methylating agent in the synthesis of compounds such as epinephrine.

CARBOHYDRATES

Carbohydrates are organic molecules made of equal amounts of carbon and H_2O. The simple sugars, or **monosaccharides,** including **pentoses** (5 carbons; eg, ribose) and **hexoses** (6 carbons; eg, glucose) perform both structural (eg, as part of nucleotides discussed previously) and functional roles (eg, inositol 1,4,5 trisphosphate acts as a cellular signaling molecules) in the body. Monosaccharides can be linked together to form disaccharides (eg, sucrose), or polysaccharides (eg, glycogen). The placement of sugar moieties onto proteins (glycoproteins) aids in cellular targeting, and in the case of some

FIGURE 1–20   Urea cycle. The processing of NH_3 to urea for excretion contains several coordinative steps in both the cytoplasm (Cyto) and the mitochondria (Mito). The production of carbamoyl phosphate and its conversion to citrulline occurs in the mitochondria, whereas other processes are in the cytoplasm.

$$
\begin{array}{ccc}
H-C=O & H-C=O & CH_2OH \\
| & | & | \\
H-C-OH & H-C-OH & C=O \\
| & | & | \\
HO-C-H & HO-C-H & HO-C-H \\
| & | & | \\
H-C-OH & HO-C-H & H-C-OH \\
| & | & | \\
H-C-OH & H-C-OH & H-C-OH \\
| & | & | \\
CH_2OH & CH_2OH & CH_2OH \\
\text{D-Glucose} & \text{D-Galactose} & \text{D-Fructose}
\end{array}
$$

FIGURE 1–21   Structures of principal dietary hexoses. Glucose, galactose, and fructose are shown in their naturally occurring D isomers.

receptors, recognition of signaling molecules. In this section we will discuss a major role for carbohydrates in physiology, the production and storage of energy.

Dietary carbohydrates are for the most part polymers of hexoses, of which the most important are glucose, galactose, and fructose (Figure 1–21). Most of the monosaccharides occurring in the body are the D isomers. The principal product of carbohydrate digestion and the principal circulating sugar is glucose. The normal fasting level of plasma glucose in peripheral venous blood is 70 to 110 mg/dL (3.9–6.1 mmol/L). In arterial blood, the plasma glucose level is 15 to 30 mg/dL higher than in venous blood.

Once it enters the cells, glucose is normally phosphorylated to form glucose 6-phosphate. The enzyme that catalyzes this reaction is **hexokinase.** In the liver, there is an additional enzyme called **glucokinase,** which has greater specificity for glucose and which, unlike hexokinase, is increased by insulin and decreased in starvation and diabetes. The glucose 6-phosphate is either polymerized into glycogen or catabolized. The process of glycogen formation is called **glycogenesis,** and glycogen breakdown is called **glycogenolysis.** Glycogen, the storage form of glucose, is present in most body tissues, but the major supplies are in the liver and skeletal muscle. The breakdown of glucose to pyruvate or lactate (or both) is called **glycolysis.** Glucose catabolism proceeds via cleavage through fructose to trioses or via oxidation and decarboxylation to pentoses. The pathway to pyruvate through the trioses is the **Embden–Meyerhof pathway,** and that through 6-phosphogluconate and the pentoses is the **direct oxidative pathway (hexose monophosphate shunt).** Pyruvate is converted to acetyl-CoA. Interconversions between carbohydrate, fat, and protein include conversion of the glycerol from fats to dihydroxyacetone phosphate and conversion of a number of amino acids with carbon skeletons resembling intermediates in the Embden–Meyerhof pathway and citric acid cycle to these intermediates by deamination. In this way, and by conversion of lactate to glucose, nonglucose molecules can be converted to glucose (**gluconeogenesis**). Glucose can be converted to fats through acetyl-CoA, but because the conversion of pyruvate to acetyl-CoA, unlike most reactions in glycolysis, is irreversible, fats are not converted to glucose via this pathway. There is therefore very little net conversion of fats to carbohydrates in

the body because, except for the quantitatively unimportant production from glycerol, there is no pathway for conversion.

CITRIC ACID CYCLE

The **citric acid cycle** (Krebs cycle, tricarboxylic acid cycle) is a sequence of reactions in which acetyl-CoA is metabolized to CO_2 and H atoms. Acetyl-CoA is first condensed with the anion of a four-carbon acid, oxaloacetate, to form citrate and HS-CoA. In a series of seven subsequent reactions, $2CO_2$ molecules are split off, regenerating oxaloacetate (Figure 1–22). Four pairs of H atoms are transferred to the flavoprotein–cytochrome chain, producing 12ATP and $4H_2O$, of which $2H_2O$ is used in the cycle. The citric acid cycle is the common pathway for oxidation to CO_2 and H_2O of carbohydrate, fat, and some amino acids. The major entry into it is through acetyl-CoA, but a number of amino acids can be converted to citric acid cycle intermediates by deamination. The citric acid cycle requires O_2 and does not function under anaerobic conditions.

ENERGY PRODUCTION

The net production of energy-rich phosphate compounds during the metabolism of glucose and glycogen to pyruvate depends on whether metabolism occurs via the Embden–Meyerhof pathway or the hexose monophosphate shunt. By oxidation at the substrate level, the conversion of 1 mol of phosphoglyceraldehyde to phosphoglycerate generates 1 mol of ATP, and the conversion of 1 mol of phosphoenolpyruvate to pyruvate generates another. Because 1 mol of glucose 6-phosphate produces, via the Embden–Meyerhof pathway, 2 mol of phosphoglyceraldehyde, 4 mol of ATP is generated per mole of glucose metabolized to pyruvate. All these reactions occur in the absence of O_2 and consequently represent anaerobic production of energy. However, 1 mol of ATP is used in forming fructose 1,6-diphosphate from fructose 6-phosphate and 1 mol in phosphorylating glucose when it enters the cell. Consequently, when pyruvate is formed anaerobically from glycogen, there is a *net* production of 3 mol of ATP per mole of glucose 6-phosphate; however, when pyruvate is formed from 1 mol of blood glucose, the net gain is only 2 mol of ATP.

A supply of NAD^+ is necessary for the conversion of phosphoglyceraldehyde to phosphoglycerate. Under anaerobic conditions (anaerobic glycolysis), a block of glycolysis at the phosphoglyceraldehyde conversion step might be expected to develop as soon as the available NAD^+ is converted to NADH. However, pyruvate can accept hydrogen from NADH, forming NAD^+ and lactate:

$$\text{Pyruvate} + \text{NADH} \leftrightarrows \text{Lactate} + NAD^+$$

In this way, glucose metabolism and energy production can continue for a while without O_2. The lactate that accumulates is converted back to pyruvate when the O_2 supply is restored, with NADH transferring its hydrogen to the flavoprotein–cytochrome chain.

FIGURE 1–22 Citric acid cycle. The numbers (6C, 5C, etc) indicate the number of carbon atoms in each of the intermediates. The conversion of pyruvate to acetyl-CoA and each turn of the cycle provide four NADH and one $FADH_2$ for oxidation via the flavoprotein-cytochrome chain plus formation of one GTP that is readily converted to ATP.

During aerobic glycolysis, the net production of ATP is 19 times greater than the two ATPs formed under anaerobic conditions. Six ATPs are formed by oxidation via the flavoprotein–cytochrome chain of the two NADHs produced when 2 mol of phosphoglyceraldehyde is converted to phosphoglycerate (Figure 1–22), six ATPs are formed from the two NADHs produced when 2 mol of pyruvate is converted to acetyl-CoA, and 24 ATPs are formed during the subsequent two turns of the citric acid cycle. Of these, 18 are formed by oxidation of six NADHs, 4 by oxidation of two $FADH_2$s, and 2 by oxidation at the substrate level when succinyl-CoA is converted to succinate. This reaction actually produces GTP, but the GTP is converted to ATP. Thus, the net production of ATP per mol of blood glucose metabolized aerobically via the Embden–Meyerhof pathway and citric acid cycle is $2 + [2 \times 3] + [2 \times 3] + [2 \times 12] = 38$.

Glucose oxidation via the hexose monophosphate shunt generates large amounts of NADPH. A supply of this reduced coenzyme is essential for many metabolic processes. The pentoses formed in the process are building blocks for nucleotides (see below). The amount of ATP generated depends on the amount of NADPH converted to NADH and then oxidized.

"DIRECTIONAL-FLOW VALVES"

Metabolism is regulated by a variety of hormones and other factors. To bring about any net change in a particular metabolic process, regulatory factors obviously must drive a chemical reaction in one direction. Most of the reactions in intermediary metabolism are freely reversible, but there are a number of "directional-flow valves," ie, reactions that proceed in one direction under the influence of one enzyme or transport mechanism and in the opposite direction under the influence of another. Five examples in the intermediary metabolism of carbohydrate are shown in Figure 1–23. The different pathways for fatty acid synthesis and catabolism (see below) are another example. Regulatory factors exert their influence on metabolism by acting directly or indirectly at these directional-flow valves.

GLYCOGEN SYNTHESIS & BREAKDOWN

Glycogen is a branched glucose polymer with two types of glycoside linkages: $1:4\alpha$ and $1:6\alpha$ (Figure 1–24). It is synthesized on **glycogenin,** a protein primer, from glucose 1-phosphate via uridine diphosphoglucose (UDPG). The enzyme **glycogen synthase** catalyses the final synthetic step. The availability of

1. Glucose entry into cells and glucose exit from cells

2. Glucose $\xrightleftharpoons[\text{Glucose 6-phosphatase}]{\text{Hexokinase}}$ Glucose 6-phosphate

3. Glucose 1-phosphate $\xrightleftharpoons[\text{Phosphorylase}]{\text{Glycogen synthase}}$ Glycogen

4. Fructose 6-phosphate $\xrightleftharpoons[\substack{\text{Fructose 1,6-}\\\text{biphosphatase}}]{\substack{\text{Phospho-}\\\text{fructokinase}}}$ Fructose 1,6-biphosphate

5. Phosphoenolpyruvate $\xrightarrow{\substack{\text{ADP}\longrightarrow\text{ATP}\\\text{Pyruvate kinase}}}$ Pyruvate

FIGURE 1–23 **Directional flow valves in energy production reactions.** In carbohydrate metabolism there are several reactions that proceed in one direction by one mechanism and in the other direction by a different mechanism, termed "directional-flow valves." Five examples of these reactions are illustrated (numbered at left). The double line in example 5 represents the mitochondrial membrane. Pyruvate is converted to malate in mitochondria, and the malate diffuses out of the mitochondria to the cytosol, where it is converted to phosphoenolpyruvate.

glycogenin is one of the factors determining the amount of glycogen synthesized. The breakdown of glycogen in 1:4α linkage is catalyzed by phosphorylase, whereas another enzyme catalyzes the breakdown of glycogen in 1:6α linkage.

FACTORS DETERMINING THE PLASMA GLUCOSE LEVEL

The plasma glucose level at any given time is determined by the balance between the amount of glucose entering the bloodstream and the amount leaving it. The principal determinants are therefore the dietary intake; the rate of entry into the cells of muscle, adipose tissue, and other organs; and the glucostatic activity of the liver (Figure 1–25). Five percent of ingested glucose is promptly converted into glycogen in the liver, and 30–40% is converted into fat. The remainder is metabolized in muscle and other tissues. During fasting, liver glycogen is broken down and the liver adds glucose to the bloodstream. With more prolonged fasting, glycogen is depleted and there is increased gluconeogenesis from amino acids and glycerol in the liver. Plasma glucose declines modestly to about 60 mg/dL during prolonged starvation in normal individuals, but symptoms of hypoglycemia do not occur because gluconeogenesis prevents any further fall.

FIGURE 1–24 **Glycogen formation and breakdown.** Glycogen is the main storage for glucose in the cell. It is cycled: built up from glucose 6-phosphate when energy is stored and broken down to glucose 6-phosphate when energy is required. Note the intermediate glucose 1-phosphate and enzymatic control by phosphorylase a and glycogen kinase.

FIGURE 1–25   Plasma glucose homeostasis. Notice the glucostatic function of the liver, as well as the loss of glucose in the urine when the renal threshold is exceeded (dashed arrows).

METABOLISM OF HEXOSES OTHER THAN GLUCOSE

Other hexoses that are absorbed from the intestine include galactose, which is liberated by the digestion of lactose and converted to glucose in the body; and fructose, part of which is ingested and part produced by hydrolysis of sucrose. After phosphorylation, galactose reacts with uridine diphosphoglucose (UDPG) to form uridine diphosphogalactose. The uridine diphosphogalactose is converted back to UDPG, and the UDPG functions in glycogen synthesis. This reaction is reversible, and conversion of UDPG to uridine diphosphogalactose provides the galactose necessary for formation of glycolipids and mucoproteins when dietary galactose intake is inadequate. The utilization of galactose, like that of glucose, depends on insulin. In the inborn error of metabolism known as **galactosemia,** there is a congenital deficiency of galactose 1-phosphate uridyl transferase, the enzyme responsible for the reaction between galactose 1-phosphate and UDPG, so that ingested galactose accumulates in the circulation. Serious disturbances of growth and development result. Treatment with galactose-free diets improves this condition without leading to galactose deficiency, because the enzyme necessary for the formation of uridine diphosphogalactose from UDPG is present.

Fructose is converted in part to fructose 6-phosphate and then metabolized via fructose 1,6-diphosphate. The enzyme catalyzing the formation of fructose 6-phosphate is hexokinase, the same enzyme that catalyzes the conversion of glucose to glucose 6-phosphate. However, much more fructose is converted to fructose 1-phosphate in a reaction catalyzed by fructokinase. Most of the fructose 1-phosphate is then split into dihydroxyacetone phosphate and glyceraldehyde. The glyceraldehyde is phosphorylated, and it and the dihydroxyacetone phosphate enter the pathways for glucose metabolism. Because the reactions proceeding through phosphorylation of fructose in the 1 position can occur at a normal rate in the absence of insulin, it has been recommended that fructose be given to diabetics to replenish their carbohydrate stores. However, most of the fructose is metabolized in the intestines and liver, so its value in replenishing carbohydrate elsewhere in the body is limited.

Fructose 6-phosphate can also be phosphorylated in the 2 position, forming fructose 2,6-diphosphate. This compound is an important regulator of hepatic gluconeogenesis. When the fructose 2,6-diphosphate level is high, conversion of fructose 6-phosphate to fructose 1,6-diphosphate is facilitated, and thus breakdown of glucose to pyruvate is increased. A decreased level of fructose 2,6-diphosphate facilitates the reverse reaction and consequently aids gluconeogenesis.

FATTY ACIDS & LIPIDS

The biologically important lipids are the fatty acids and their derivatives, the neutral fats (triglycerides), the phospholipids and related compounds, and the sterols. The triglycerides are made up of three fatty acids bound to glycerol (Table 1–4). Naturally occurring fatty acids contain an even number of carbon atoms. They may be saturated (no double bonds) or unsaturated (dehydrogenated, with various numbers of double bonds). The phospholipids are constituents of cell membranes and provide structural components of the cell membrane, as well as an important source of intra- and intercellular signaling molecules. Fatty acids also are an important source of energy in the body.

FATTY ACID OXIDATION & SYNTHESIS

In the body, fatty acids are broken down to acetyl-CoA, which enters the citric acid cycle. The main breakdown occurs in the mitochondria by β-oxidation. Fatty acid oxidation begins with activation (formation of the CoA derivative) of the fatty acid, a reaction that occurs both inside and outside the mitochondria. Medium- and short-chain fatty acids can enter the mitochondria without difficulty, but long-chain fatty acids must be bound to **carnitine** in ester linkage before they can cross the inner mitochondrial membrane. Carnitine is β-hydroxy-γ-trimethylammonium butyrate, and it is synthesized in the body from lysine and methionine. A translocase moves the fatty acid–carnitine ester into the matrix space. The ester is hydrolyzed, and the carnitine recycles. β-oxidation proceeds by serial removal of two carbon fragments from the fatty acid (Figure 1–26). The energy yield of this process is large. For example, catabolism of 1 mol of a six-carbon fatty acid through the citric acid cycle to CO_2 and H_2O generates 44 mol of ATP, compared with the 38 mol generated by catabolism of 1 mol of the six-carbon carbohydrate glucose.

KETONE BODIES

In many tissues, acetyl-CoA units condense to form acetoacetyl-CoA (Figure 1–27). In the liver, which (unlike other tissues) contains a deacylase, free acetoacetate is formed. This β-keto acid is converted to β-hydroxybutyrate and acetone, and because these compounds are metabolized with difficulty in

TABLE 1–4. Lipids.

Typical fatty acids:

Palmitic acid: $CH_5(CH_2)_{14}-C-OH$ (with O double bond)

Stearic acid: $CH_5(CH_2)_{16}-C-OH$ (with O double bond)

Oleic acid: $CH_5(CH_2)_7CH=CH(CH_2)_7-C-OH$ (with O double bond)
(Unsaturated)

Triglycerides (triacylglycerols): Esters of glycerol and three fatty acids.

$$CH_2-O-C-R$$
$$CH_2-O-C-R + 3H_2O \rightleftharpoons CHOH + 3HO-C-R$$
$$CH_2-O-C-R$$
Triglyceride   Glycerol

R = Aliphatic chain of various lengths and degrees of saturation.

Phospholipids:

  A. Esters of glycerol, two fatty acids, and

    1. Phosphate = phosphatidic acid

    2. Phosphate plus inositol = phosphatidylinositol

    3. Phosphate plus choline = phosphatidylcholine (lecithin)

    4. Phosphate plus ethanolamine = phosphatidyl-ethanolamine (cephalin)

    5. Phosphate plus serine = phosphatidylserine

  B. Other phosphate-containing derivatives of glycerol

  C. Sphingomyelins: Esters of fatty acid, phosphate, choline, and the amino alcohol sphingosine.

Cerebrosides: Compounds containing galactose, fatty acid, and sphingosine.

Sterols: Cholesterol and its derivatives, including steroid hormones, bile acids, and various vitamins.

the liver, they diffuse into the circulation. Acetoacetate is also formed in the liver via the formation of 3-hydroxy-3-methylglutaryl-CoA, and this pathway is quantitatively more important than deacylation. Acetoacetate, β-hydroxybutyrate, and acetone are called **ketone bodies.** Tissues other than liver transfer CoA from succinyl-CoA to acetoacetate and metabolize the "active" acetoacetate to CO_2 and H_2O via the citric acid cycle. Ketone bodies are also metabolized via other pathways. Acetone is discharged in the urine and expired air. An imbalance of ketone bodies can lead to serious health problems (Clinical Box 1–3).

CELLULAR LIPIDS

The lipids in cells are of two main types: **structural lipids,** which are an inherent part of the membranes and other parts of cells; and **neutral fat,** stored in the adipose cells of the fat depots. Neutral fat is mobilized during starvation, but structural lipid is preserved. The fat depots obviously vary in size, but in nonobese individuals they make up about 15% of body weight in men and 21% in women. They are not the inert structures they were once thought to be but, rather, active dynamic tissues undergoing continuous breakdown and resynthesis. In the depots, glucose is metabolized to fatty acids, and neutral fats are synthesized. Neutral fat is also broken down, and free fatty acids are released into the circulation.

A third, special type of lipid is **brown fat,** which makes up a small percentage of total body fat. Brown fat, which is somewhat more abundant in infants but is present in adults as well, is located between the scapulas, at the nape of the neck, along the great vessels in the thorax and abdomen, and in other scattered locations in the body. In brown fat depots, the fat cells as well as the blood vessels have an extensive sympathetic innervation. This is in contrast to white fat depots, in which some fat cells may be innervated but the principal sympathetic innervation is solely on blood vessels. In addition, ordinary lipocytes have only a single large droplet of white fat, whereas brown fat cells contain several small droplets of fat. Brown fat cells also contain many mitochondria. In these mitochondria, an inward proton conductance that generates ATP takes places as usual, but in addition there is a second proton conductance that does not generate ATP. This "short-circuit" conductance depends on a 32-kDa uncoupling protein (UCP1). It causes uncoupling of metabolism and generation of ATP, so that more heat is produced.

PLASMA LIPIDS & LIPID TRANSPORT

The major lipids are relatively insoluble in aqueous solutions and do not circulate in the free form. **Free fatty acids (FFAs)** are bound to albumin, whereas cholesterol, triglycerides, and phospholipids are transported in the form of **lipoprotein** complexes. The complexes greatly increase the solubility of the lipids. The six families of lipoproteins (Table 1–5) are graded in size and lipid content. The density of these lipoproteins is inversely proportionate to their lipid content. In general, the lipoproteins consist of a hydrophobic core of triglycerides and cholesteryl esters surrounded by phospholipids and protein. These lipoproteins can be transported from the intestine to the liver via an **exogenous pathway,** and between other tissues via an **endogenous pathway.**

Dietary lipids are processed by several pancreatic lipases in the intestine to form mixed micelles of predominantly FFA, **2-monoglycerols,** and cholesterol derivatives (see Chapter 27). These micelles additionally can contain important water-insoluble molecules such as **vitamins A, D, E, and K.** These mixed micelles are taken up into cells of the intestinal

Fatty acid

"Active" fatty acid

$$R-CH_2CH_2-\overset{\overset{O}{\|}}{C}-OH + \text{HS-CoA} \xrightarrow[\text{ATP} \rightarrow \text{ADP}]{Mg^{2+}} H_2O + R-CH_2CH_2-\overset{\overset{O}{\|}}{C}-S-CoA$$

Oxidized flavoprotein

↓

Reduced flavoprotein

$$R-\overset{\overset{OH}{|}}{\underset{|}{C}}-CH_2-\overset{\overset{O}{\|}}{C}-S-CoA \longleftarrow H_2O + R-CH=CH-\overset{\overset{O}{\|}}{C}-S-CoA$$

H   β-Hydroxy fatty acid–CoA   α,β-Unsaturated fatty acid–CoA

$NAD^+ \rightarrow NADH + H^+$

$$R-\overset{\overset{O}{\|}}{C}-CH_2-\overset{\overset{O}{\|}}{C}-S-CoA + \text{HS-CoA} \longrightarrow R-\overset{\overset{O}{\|}}{C}-S-CoA + CH_3-\overset{\overset{O}{\|}}{C}-S-CoA$$

β-Keto fatty acid–CoA   "Active" fatty acid + Acetyl–CoA

R = Rest of fatty acid chain.

FIGURE 1–26   Fatty acid oxidation. This process, splitting off two carbon fragments at a time, is repeated to the end of the chain.

$$CH_3-\overset{\overset{O}{\|}}{C}-S-CoA + CH_3-\overset{\overset{O}{\|}}{C}-S-CoA \underset{\beta\text{-Ketothiolase}}{\overset{}{\rightleftharpoons}} CH_3-\overset{\overset{O}{\|}}{C}-CH_2-\overset{\overset{O}{\|}}{C}-S-CoA + \text{HS-CoA}$$

2 Acetyl-CoA   **Acetoacetyl-CoA**

$$CH_3-\overset{\overset{O}{\|}}{C}-CH_2-\overset{\overset{O}{\|}}{C}-S-CoA + H_2O \xrightarrow[\text{(liver only)}]{\text{Deacylase}} CH_3-\overset{\overset{O}{\|}}{C}-CH_2-\overset{\overset{O}{\|}}{C}-O^- + H^+ + \text{HS-CoA}$$

Acetoacetyl-CoA   **Acetoacetate**

- -

$$\text{Acetyl-CoA} + \text{Acetoacetyl-CoA} \rightarrow CH_3-\overset{\overset{OH}{|}}{\underset{\underset{CH_2-COO^-}{|}}{C}}-CH_2-\overset{\overset{O}{\|}}{C}-S-CoA + H^+$$

3-Hydroxy-3-methylglutaryl-CoA
(HMG-CoA)

$$\text{HMG-CoA} \rightarrow \text{Acetoacetate} + H^+ + \text{Acetyl-CoA}$$

- -

Acetoacetate

$$CH_3-\overset{\overset{O}{\|}}{C}-CH_2-\overset{\overset{O}{\|}}{C}-O^- + H^+ \xrightarrow{\text{Tissues except liver}} CO_2 + \text{ATP}$$

$-CO_2$

$$CH_3-\overset{\overset{O}{\|}}{C}-CH_3$$

Acetone

$+2H \quad -2H$

$$CH_3-CHOH-CH_2-\overset{\overset{O}{\|}}{C}-O^- + H^+$$

β-Hydroxybutyrate

FIGURE 1–27   Formation and metabolism of ketone bodies. Note the two pathways for the formation of acetoacetate.

CLINICAL BOX 1–3

Diseases Associated with Imbalance of β-oxidation of Fatty Acids

Ketoacidosis

The normal blood ketone level in humans is low (about 1 mg/dL) and less than 1 mg is excreted per 24 h, because the ketones are normally metabolized as rapidly as they are formed. However, if the entry of acetyl-CoA into the citric acid cycle is depressed because of a decreased supply of the products of glucose metabolism, or if the entry does not increase when the supply of acetyl-CoA increases, acetyl-CoA accumulates, the rate of condensation to acetoacetyl-CoA increases, and more acetoacetate is formed in the liver. The ability of the tissues to oxidize the ketones is soon exceeded, and they accumulate in the bloodstream (ketosis). Two of the three ketone bodies, acetoacetate and β-hydroxybutyrate, are anions of the moderately strong acids acetoacetic acid and β-hydroxybutyric acid. Many of their protons are buffered, reducing the decline in pH that would otherwise occur. However, the buffering capacity can be exceeded, and the metabolic acidosis that develops in conditions such as diabetic ketosis can be severe and even fatal. Three conditions lead to deficient intracellular glucose supplies, and hence to ketoacidosis: starvation; diabetes mellitus; and a high-fat, low-carbohydrate diet. The acetone odor on the breath of children who have been vomiting is due to the ketosis of starvation. Parenteral administration of relatively small amounts of glucose abolishes the ketosis, and it is for this reason that carbohydrate is said to be antiketogenic.

Carnitine Deficiency

Deficient β-oxidation of fatty acids can be produced by carnitine deficiency or genetic defects in the translocase or other enzymes involved in the transfer of long-chain fatty acids into the mitochondria. This causes cardiomyopathy. In addition, it causes **hypoketonemic hypoglycemia** with coma, a serious and often fatal condition triggered by fasting, in which glucose stores are used up because of the lack of fatty acid oxidation to provide energy. Ketone bodies are not formed in normal amounts because of the lack of adequate CoA in the liver.

mucosa where large lipoprotein complexes, **chylomicrons,** are formed. The chylomicrons and their remnants constitute a transport system for ingested exogenous lipids (exogenous pathway). Chylomicrons can enter the circulation via the lymphatic ducts. The chylomicrons are cleared from the circulation by the action of **lipoprotein lipase,** which is located on the surface of the endothelium of the capillaries. The enzyme catalyzes the breakdown of the triglyceride in the chylomicrons to FFA and glycerol, which then enter adipose cells and are reesterified. Alternatively, the FFA can remain in the circulation bound to albumin. Lipoprotein lipase, which requires heparin as a cofactor, also removes triglycerides from circulating **very low density lipoproteins (VLDL).** Chylomicrons depleted of their triglyceride remain in the circulation as cholesterol-rich lipoproteins called **chylomicron remnants,** which are 30 to 80 nm in diameter. The remnants are carried to the liver, where they are internalized and degraded.

TABLE 1–5 The principal lipoproteins.*

Lipoprotein	Size (nm)	Composition (%)					Origin
		Protein	Free Cholesteryl	Cholesterol Esters	Triglyceride	Phospholipid	
Chylomicrons	75–1000	2	2	3	90	3	Intestine
Chylomicron remnants	30–80	…	…	…	…	…	Capillaries
Very low density lipoproteins (VLDL)	30–80	8	4	16	55	17	Liver and intestine
Intermediate-density lipoproteins (IDL)	25–40	10	5	25	40	20	VLDL
Low-density lipoproteins (LDL)	20	20	7	46	6	21	IDL
High-density lipoproteins (HDL)	7.5–10	50	4	16	5	25	Liver and intestine

*The plasma lipids include these components plus free fatty acids from adipose tissue, which circulate bound to albumin.

The endogenous system, made up of VLDL, **intermediate-density lipoproteins (IDL), low-density lipoproteins (LDL),** and **high-density lipoproteins (HDL),** also transports triglycerides and cholesterol throughout the body. VLDL are formed in the liver and transport triglycerides formed from fatty acids and carbohydrates in the liver to extrahepatic tissues. After their triglyceride is largely removed by the action of lipoprotein lipase, they become IDL. The IDL give up phospholipids and, through the action of the plasma enzyme **lecithin-cholesterol acyltransferase (LCAT),** pick up cholesteryl esters formed from cholesterol in the HDL. Some IDL are taken up by the liver. The remaining IDL then lose more triglyceride and protein, probably in the sinusoids of the liver, and become LDL. LDL provide cholesterol to the tissues. The cholesterol is an essential constituent in cell membranes and is used by gland cells to make steroid hormones.

FREE FATTY ACID METABOLISM

In addition to the exogenous and endogenous pathways described above, FFA are also synthesized in the fat depots in which they are stored. They can circulate as lipoproteins bound to albumin and are a major source of energy for many organs. They are used extensively in the heart, but probably all tissues can oxidize FFA to CO_2 and H_2O.

The supply of FFA to the tissues is regulated by two lipases. As noted above, lipoprotein lipase on the surface of the endothelium of the capillaries hydrolyzes the triglycerides in chylomicrons and VLDL, providing FFA and glycerol, which are reassembled into new triglycerides in the fat cells. The intracellular **hormone-sensitive lipase** of adipose tissue catalyzes the breakdown of stored triglycerides into glycerol and fatty acids, with the latter entering the circulation. Hormone-sensitive lipase is increased by fasting and stress and decreased by feeding and insulin. Conversely, feeding increases and fasting and stress decrease the activity of lipoprotein lipase.

CHOLESTEROL METABOLISM

Cholesterol is the precursor of the steroid hormones and bile acids and is an essential constituent of cell membranes. It is found only in animals. Related sterols occur in plants, but plant sterols are not normally absorbed from the gastrointestinal tract. Most of the dietary cholesterol is contained in egg yolks and animal fat.

Cholesterol is absorbed from the intestine and incorporated into the chylomicrons formed in the intestinal mucosa. After the chylomicrons discharge their triglyceride in adipose tissue, the chylomicron remnants bring cholesterol to the liver. The liver and other tissues also synthesize cholesterol. Some of the cholesterol in the liver is excreted in the bile, both in the free form and as bile acids. Some of the biliary cholesterol is reabsorbed from the intestine. Most of the cholesterol in the liver is incorporated into VLDL and circulates in lipoprotein complexes.

The biosynthesis of cholesterol from acetate is summarized in Figure 1–28. Cholesterol feeds back to inhibit its own synthesis by inhibiting **HMG-CoA reductase,** the enzyme that converts 3-hydroxy-3-methylglutaryl-coenzyme A (HMG-CoA) to mevalonic acid. Thus, when dietary cholesterol intake is high, hepatic cholesterol synthesis is decreased, and vice versa. However, the feedback compensation is incomplete, because a diet that is low in cholesterol and saturated fat leads to only a modest decline in circulating plasma cholesterol. The most effective and most commonly used cholesterol-lowering drugs are lovastatin and other **statins,** which reduce cholesterol synthesis by inhibiting HMG-CoA. The relationship between cholesterol and vascular disease is discussed in Clinical Box 1–4.

ESSENTIAL FATTY ACIDS

Animals fed a fat-free diet fail to grow, develop skin and kidney lesions, and become infertile. Adding linolenic, linoleic, and arachidonic acids to the diet cures all the deficiency symptoms. These three acids are polyunsaturated fatty acids and because of their action are called **essential fatty acids.** Similar deficiency symptoms have not been unequivocally demonstrated in humans, but there is reason to believe that some unsaturated fats are essential dietary constituents, especially for children.

FIGURE 1–28 **Biosynthesis of cholesterol.** Six mevalonic acid molecules condense to form squalene, which is then hydroxylated to cholesterol. The dashed arrow indicates feedback inhibition by cholesterol of HMG-CoA reductase, the enzyme that catalyzes mevalonic acid formation.

CLINICAL BOX 1–4

Cholesterol & Atherosclerosis

The interest in cholesterol-lowering drugs stems from the role of cholesterol in the etiology and course of **atherosclerosis.** This extremely widespread disease predisposes to myocardial infarction, cerebral thrombosis, ischemic gangrene of the extremities, and other serious illnesses. It is characterized by infiltration of cholesterol and oxidized cholesterol into macrophages, converting them into foam cells in lesions of the arterial walls. This is followed by a complex sequence of changes involving platelets, macrophages, smooth muscle cells, growth factors, and inflammatory mediators that produces proliferative lesions which eventually ulcerate and may calcify. The lesions distort the vessels and make them rigid. In individuals with elevated plasma cholesterol levels, the incidence of atherosclerosis and its complications is increased. The normal range for plasma cholesterol is said to be 120 to 200 mg/dL, but in men, there is a clear, tight, positive correlation between the death rate from ischemic heart disease and plasma cholesterol levels above 180 mg/dL. Furthermore, it is now clear that lowering plasma cholesterol by diet and drugs slows and may even reverse the progression of atherosclerotic lesions and the complications they cause.

In evaluating plasma cholesterol levels in relation to atherosclerosis, it is important to analyze the LDL and HDL levels as well. LDL delivers cholesterol to peripheral tissues, including atheromatous lesions, and the LDL plasma concentration correlates positively with myocardial infarctions and ischemic strokes. On the other hand, HDL picks up cholesterol from peripheral tissues and transports it to the liver, thus lowering plasma cholesterol. It is interesting that women, who have a lower incidence of myocardial infarction than men, have higher HDL levels. In addition, HDL levels are increased in individuals who exercise and those who drink one or two alcoholic drinks per day, whereas they are decreased in individuals who smoke, are obese, or live sedentary lives. Moderate drinking decreases the incidence of myocardial infarction, and obesity and smoking are risk factors that increase it. Plasma cholesterol and the incidence of cardiovascular diseases are increased in **familial hypercholesterolemia,** due to various loss-of-function mutations in the genes for LDL receptors.

CLINICAL BOX 1–5

Pharmacology of Prostaglandins

Because prostaglandins play a prominent role in the genesis of pain, inflammation, and fever, pharmacologists have long sought drugs to inhibit their synthesis. Glucocorticoids inhibit phospholipase A_2 and thus inhibit the formation of all eicosanoids. A variety of nonsteroidal anti-inflammatory drugs (NSAIDs) inhibit both cyclooxygenases, inhibiting the production of PGH_2 and its derivatives. Aspirin is the best-known of these, but ibuprofen, indomethacin, and others are also used. However, there is evidence that prostaglandins synthesized by COX2 are more involved in the production of pain and inflammation, and prostaglandins synthesized by COX1 are more involved in protecting the gastrointestinal mucosa from ulceration. Drugs such as celecoxib and rofecoxib that selectively inhibit COX2 have been developed, and in clinical use they relieve pain and inflammation, possibly with a significantly lower incidence of gastrointestinal ulceration and its complications than is seen with nonspecific NSAIDs. However, rofecoxib has been withdrawn from the market in the United States because of a reported increase of strokes and heart attacks in individuals using it. More research is underway to better understand all the effects of the COX enzymes, their products, and their inhibitors.

Dehydrogenation of fats is known to occur in the body, but there does not appear to be any synthesis of carbon chains with the arrangement of double bonds found in the essential fatty acids.

EICOSANOIDS

One of the reasons that essential fatty acids are necessary for health is that they are the precursors of prostaglandins, prostacyclin, thromboxanes, lipoxins, leukotrienes, and related compounds. These substances are called **eicosanoids,** reflecting their origin from the 20-carbon (eicosa-) polyunsaturated fatty acid **arachidonic acid (arachidonate)** and the 20-carbon derivatives of linoleic and linolenic acids.

The **prostaglandins** are a series of 20-carbon unsaturated fatty acids containing a cyclopentane ring. They were first isolated from semen but are now known to be synthesized in most and possibly in all organs in the body. Prostaglandin H_2 (PGH_2) is the precursor for various other prostaglandins, thromboxanes, and prostacyclin. Arachidonic acid is formed from tissue phospholipids by **phospholipase A_2.** It is converted to prostaglandin H_2 (PGH_2) by **prostaglandin G/H synthases** 1 and 2. These are bifunctional enzymes that have both cyclooxygenase and peroxidase activity, but they are more commonly known by the names cyclooxygenase 1 (**COX1**) and cyclooxygenase 2 (**COX2**). Their structures are very similar, but COX1 is constitutive whereas COX2 is induced by growth factors, cytokines, and tumor promoters. PGH_2 is converted to prostacyclin, thromboxanes, and prostaglandins by various tissue isomerases. The effects of prostaglandins are multitudinous and varied. They are particularly important in the female reproductive cycle, in parturition, in the cardiovascular system, in inflammatory responses, and in the causation of pain. Drugs that target production of prostaglandins are among the most common over the counter drugs available (Clinical Box 1–5).

Arachidonic acid also serves as a substrate for the production of several physiologically important **leukotrienes** and **lipoxins.** The leukotrienes, thromboxanes, lipoxins, and prostaglandins have been called local hormones. They have short half-lives and are inactivated in many different tissues. They undoubtedly act mainly in the tissues at sites in which they are produced. The leukotrienes are mediators of allergic responses and inflammation. Their release is provoked when specific allergens combine with IgE antibodies on the surfaces of mast cells (see Chapter 3). They produce bronchoconstriction, constrict arterioles, increase vascular permeability, and attract neutrophils and eosinophils to inflammatory sites. Diseases in which they may be involved include asthma, psoriasis, adult respiratory distress syndrome, allergic rhinitis, rheumatoid arthritis, Crohn's disease, and ulcerative colitis.

CHAPTER SUMMARY

- Cells contain approximately one third of the body fluids, while the remaining extracellular fluid is found between cells (interstitial fluid) or in the circulating blood plasma.
- The number of molecules, electrical charges, and particles of substances in solution are important in physiology.
- The high surface tension, high heat capacity, and high electrical capacity allow H_2O to function as an ideal solvent in physiology.
- Biological buffers including bicarbonate, proteins, and phosphates can bind or release protons in solution to help maintain pH. Biological buffering capacity of a weak acid or base is greatest when $pK_a = pH$.
- Fluid and electrolyte balance in the body is related to plasma osmolality. Isotonic solutions have the same osmolality as blood plasma, hypertonic have higher osmolality, while hypotonic have lower osmolality.
- Although the osmolality of solutions can be similar across a plasma membrane, the distribution of individual molecules and distribution of charge across the plasma membrane can be quite different. These are affected by the Gibbs-Donnan equilibrium and can be calculated using the Nernst potential equation.
- There is a distinct difference in concentration of ions in the extracellular and intracellular fluids (concentration gradient). The separation of concentrations of charged species sets up an electrical gradient at the plasma membrane (inside negative). The electrochemical gradient is in large part maintained by the Na, K ATPase.
- Cellular energy can be stored in high-energy phosphate compounds, including adenosine triphosphate (ATP). Coordinated oxidation-reduction reactions allow for production of a proton gradient at the inner mitochondrial membrane that ultimately yields to the production of ATP in the cell.
- Nucleotides made from purine or pyrimidine bases linked to ribose or 2-deoxyribose sugars with inorganic phosphates are the basic building blocks for nucleic acids, DNA, and RNA.
- DNA is a double-stranded structure that contains the fundamental information for an organism. During cell division, DNA is faithfully replicated and a full copy of DNA is in every cell. The fundamental unit of DNA is the gene, which encodes information to make proteins in the cell. Genes are transcribed into messenger RNA, and with the help of ribosomal RNA and transfer RNAs, translated into proteins.
- Amino acids are the basic building blocks for proteins in the cell and can also serve as sources for several biologically active molecules. They exist in an "amino acid pool" that is derived from the diet, protein degradation, and de novo and resynthesis.
- Translation is the process of protein synthesis. After synthesis, proteins can undergo a variety of posttranslational modifications prior to obtaining their fully functional cell state.
- Carbohydrates are organic molecules that contain equal amounts of C and H_2O. Carbohydrates can be attached to proteins (glycoproteins) or fatty acids (glycolipids) and are critically important for the production and storage of cellular and body energy, with major supplies in the form of glycogen in the liver and skeletal muscle. The breakdown of glucose to generate energy, or glycolysis, can occur in the presence or absence of O_2 (aerobic or anaerobically). The net production of ATP during aerobic glycolysis is 19 times higher than anaerobic glycolysis.
- Fatty acids are carboxylic acids with extended hydrocarbon chains. They are an important energy source for cells and their derivatives, including triglycerides, phospholipids and sterols, and have additional important cellular applications. Free fatty acids can be bound to albumin and transported throughout the body. Triglycerides, phospholipids, and cholesterol are transported as lipoprotein complexes.

MULTIPLE-CHOICE QUESTIONS

For all questions, select the single best answer unless otherwise directed.

1. The membrane potential of a particular cell is at the K^+ equilibrium. The intracellular concentration for K^+ is at 150 mmol/L and the extracellular concentration for K^+ is at 5.5 mmol/L. What is the resting potential?
   - A) –70 mv
   - B) –90 mv
   - C) +70 mv
   - D) +90 mv

2. The difference in concentration of H^+ in a solution of pH 2.0 compared with one of pH 7.0 is
   - A) 5-fold.
   - B) 1/5 as much.
   - C) 10^5 fold.
   - D) 10^{-5} as much.

3. Transcription refers to
   - A) the process where an mRNA is used as a template for protein production.
   - B) the process where a DNA sequence is copied into RNA for the purpose of gene expression.
   - C) the process where DNA wraps around histones to form a nucleosome.
   - D) the process of replication of DNA prior to cell division.

4. The primary structure of a protein refers to
   - A) the twist, folds, or twist and folds of the amino acid sequence into stabilized structures within the protein (ie, α-helices and β-sheets).
   - B) the arrangement of subunits to form a functional structure.
   - C) the amino acid sequence of the protein.
   - D) the arrangement of twisted chains and folds within a protein into a stable structure.

5. Fill in the blanks: Glycogen is a storage form of glucose. _____ refers to the process of making glycogen and _____ refers to the process of breakdown of glycogen.

   A) Glycogenolysis, glycogenesis
   B) Glycolysis, glycogenolysis
   C) Glycogenesis, glycogenolysis
   D) Glycogenolysis, glycolysis

6. The major lipoprotein source of the cholesterol used in cells is

   A) chylomicrons.
   B) intermediate-density lipoproteins (IDLs).
   C) albumin-bound free fatty acids.
   D) LDL.
   E) HDL.

7. Which of the following produces the most high-energy phosphate compounds?

   A) aerobic metabolism of 1 mol of glucose
   B) anaerobic metabolism of 1 mol of glucose
   C) metabolism of 1 mol of galactose
   D) metabolism of 1 mol of amino acid
   E) metabolism of 1 mol of long-chain fatty acid

8. When LDL enters cells by receptor-mediated endocytosis, which of the following does *not* occur?

   A) Decrease in the formation of cholesterol from mevalonic acid.
   B) Increase in the intracellular concentration of cholesteryl esters.
   C) Increase in the transfer of cholesterol from the cell to HDL.
   D) Decrease in the rate of synthesis of LDL receptors.
   E) Decrease in the cholesterol in endosomes.

CHAPTER RESOURCES

Alberts B, et al: *Molecular Biology of the Cell,* 5th ed. Garland Science, 2007.

Hille B: *Ionic Channels of Excitable Membranes,* 3rd ed. Sinauer Associates, 2001.

Kandel ER, Schwartz JH, Jessell TM: *Principles of Neural Science,* 4th ed. McGraw-Hill, 2000.

Macdonald RG, Chaney WG: *USMLE Road Map, Biochemistry.* McGraw- Hill, 2007.

Murray RK, et al: *Harper's Biochemistry,* 26th ed. McGraw-Hill, 2003.

Pollard TD, Earnshaw WC: *Cell Biology,* 2nd ed. Saunders, Elsevier, 2008.

Sack GH, Jr. *USMLE Road Map, Genetics.* McGraw Hill, 2008.

Scriver CR, et al (editors): *The Metabolic and Molecular Bases of Inherited Disease,* 8th ed. McGraw-Hill, 2001.

Sperelakis N (editor): *Cell Physiology Sourcebook,* 3rd ed. Academic Press, 2001.

Overview of Cellular Physiology in Medical Physiology

OBJECTIVES

After studying this chapter, you should be able to:

■ Name the prominent cellular organelles and state their functions in cells.

■ Name the building blocks of the cellular cytoskeleton and state their contributions to cell structure and function.

■ Name the intercellular and cellular to extracellular connections.

■ Define the processes of exocytosis and endocytosis, and describe the contribution of each to normal cell function.

■ Define proteins that contribute to membrane permeability and transport.

■ Describe specialized transport and filtration across the capillary wall.

■ Recognize various forms of intercellular communication and describe ways in which chemical messengers (including second messengers) affect cellular physiology.

■ Define cellular homeostasis.

INTRODUCTION

The cell is the fundamental working unit of all organisms. In humans, cells can be highly specialized in both structure and function; alternatively, cells from different organs can share features and function. In the previous chapter, we examined some basic principles of biophysics and the catabolism and metabolism of building blocks found in the cell. In some of those discussions, we examined how the building blocks could contribute to basic cellular physiology (eg, DNA replication, transcription, and translation). In this chapter, we will briefly review more of the fundamental aspects of cellular and molecular physiology. Additional aspects that concern specialization of cellular and molecular physiology are considered in the next chapter concerning immune function and in the relevant chapters on the various organs.

FUNCTIONAL MORPHOLOGY OF THE CELL

A basic knowledge of cell biology is essential to an understanding of the organ systems in the body and the way they function. A key tool for examining cellular constituents is the microscope. A light microscope can resolve structures as close as 0.2 μm, while an electron microscope can resolve structures as close as 0.002 μm. Although cell dimensions are quite variable, this resolution can give us a good look at the inner workings of the cell. The advent of common access to fluorescent, confocal, and other microscopy along with specialized probes for both static and dynamic cellular structures further expanded the examination of cell structure and function. Equally revolutionary advances in the modern biophysical, biochemical, and molecular biology techniques have also greatly contributed to our knowledge of the cell.

The specialization of the cells in the various organs is considerable, and no cell can be called "typical" of all cells in the body. However, a number of structures (**organelles**) are common to most cells. These structures are shown in Figure 2–1. Many of them can be isolated by ultracentrifugation combined with

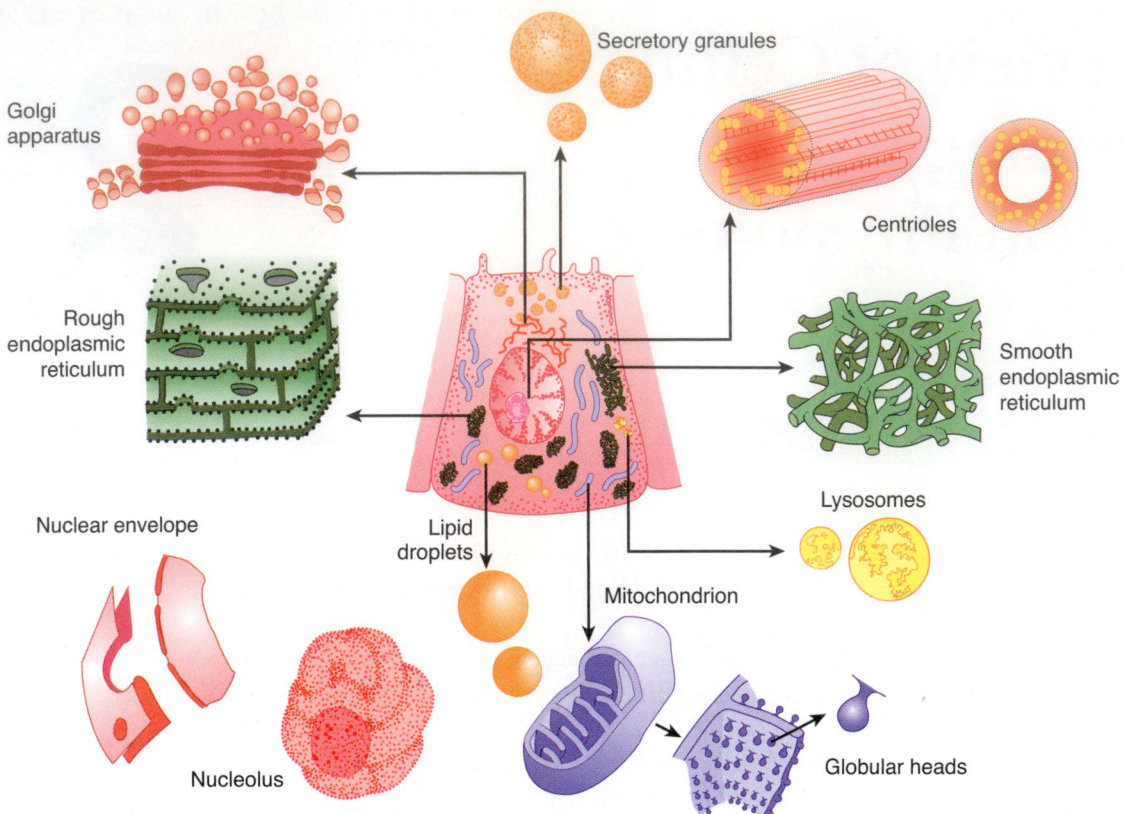

FIGURE 2–1   **Diagram showing a hypothetical cell in the center as seen with the light microscope.** Individual organelles are expanded for closer examination. (Adapted from Bloom and Fawcett. Reproduced with permission from Junqueira LC, Carneiro J, Kelley RO: *Basic Histology,* 9th ed. McGraw-Hill, 1998.)

other techniques. When cells are homogenized and the resulting suspension is centrifuged, the nuclei sediment first, followed by the mitochondria. High-speed centrifugation that generates forces of 100,000 times gravity or more causes a fraction made up of granules called the **microsomes** to sediment. This fraction includes organelles such as the **ribosomes** and **peroxisomes.**

CELL MEMBRANES

The membrane that surrounds the cell is a remarkable structure. It is made up of lipids and proteins and is semipermeable, allowing some substances to pass through it and excluding others. However, its permeability can also be varied because it contains numerous regulated ion channels and other transport proteins that can change the amounts of substances moving across it. It is generally referred to as the **plasma membrane.** The nucleus and other organelles in the cell are bound by similar membranous structures.

Although the chemical structures of membranes and their properties vary considerably from one location to another, they have certain common features. They are generally about 7.5 nm (75 Å) thick. The major lipids are phospholipids such as phosphatidylcholine and phosphatidylethanolamine. The shape of the phospholipid molecule reflects its solubility

properties: the head end of the molecule contains the phosphate portion and is relatively soluble in water (polar, **hydrophilic**) and the tails are relatively insoluble (nonpolar, **hydrophobic**). The possession of both hydrophilic and hydrophobic properties make the lipid an **amphipathic** molecule. In the membrane, the hydrophilic ends of the molecules are exposed to the aqueous environment that bathes the exterior of the cells and the aqueous cytoplasm; the hydrophobic ends meet in the water-poor interior of the membrane (Figure 2–2). In **prokaryotes** (ie, bacteria in which there is no nucleus), the membranes are relatively simple, but in **eukaryotes** (cells containing nuclei), cell membranes contain various glycosphingolipids, sphingomyelin, and cholesterol in addition to phospholipids and phosphatidylcholine.

Many different proteins are embedded in the membrane. They exist as separate globular units and many pass through the membrane **(integral proteins),** whereas others **(peripheral proteins)** stud the inside and outside of the membrane (Figure 2–2). The amount of protein varies significantly with the function of the membrane but makes up on average 50% of the mass of the membrane; that is, there is about one protein molecule per 50 of the much smaller phospholipid molecules. The proteins in the membranes carry out many functions. Some are **cell adhesion molecules** that anchor cells to their neighbors or to basal laminas. Some proteins function as **pumps,** actively

FIGURE 2–2 **Organization of the phospholipid bilayer and associated proteins in a biological membrane.** The phospholipid molecules each have two fatty acid chains (wavy lines) attached to a phosphate head (open circle). Proteins are shown as irregular colored globules. Many are integral proteins, which extend into the membrane, but peripheral proteins are attached to the inside or outside (not shown) of the membrane. Specific protein attachments and cholesterol commonly found in the bilayer are omitted for clarity. (Reproduced with permission from Widmaier EP, Raff H, Strang K: *Vander's Human Physiology: The Mechanisms of Body Function*, 11th ed. McGraw-Hill, 2008.)

transporting ions across the membrane. Other proteins function as **carriers,** transporting substances down electrochemical gradients by facilitated diffusion. Still others are **ion channels,** which, when activated, permit the passage of ions into or out of the cell. The role of the pumps, carriers, and ion channels in transport across the cell membrane is discussed below. Proteins in another group function as **receptors** that bind **ligands** or messenger molecules, initiating physiologic changes inside the cell. Proteins also function as **enzymes,** catalyzing reactions at the surfaces of the membrane. Examples from each of these groups are discussed later in this chapter.

The uncharged, hydrophobic portions of the proteins are usually located in the interior of the membrane, whereas the charged, hydrophilic portions are located on the surfaces. Peripheral proteins are attached to the surfaces of the membrane in various ways. One common way is attachment to glycosylated forms of phosphatidylinositol. Proteins held by these **glycosylphosphatidylinositol (GPI) anchors** (Figure 2–3) include enzymes such as alkaline phosphatase, various antigens, a number of cell adhesion molecules, and three proteins that combat cell lysis by complement. Over 45 GPI-linked cell surface proteins have now been described in humans. Other proteins are **lipidated,** that is, they have specific lipids attached to them (Figure 2–3). Proteins may be **myristolated, palmitoylated,** or **prenylated** (ie, attached to geranylgeranyl or farnesyl groups).

The protein structure—and particularly the enzyme content—of biologic membranes varies not only from cell to cell, but also within the same cell. For example, some of the enzymes embedded in cell membranes are different from those in mitochondrial membranes. In epithelial cells, the enzymes in the cell membrane on the mucosal surface differ from those in the

FIGURE 2–3 **Protein linkages to membrane lipids.** Some are linked by their amino terminals, others by their carboxyl terminals. Many are attached via glycosylated forms of phosphatidylinositol (GPI anchors). (Reproduced with permission from Fuller GM, Shields D: *Molecular Basis of Medical Cell Biology.* McGraw-Hill, 1998.)

FIGURE 2–4 **Components involved in oxidative phosphorylation in mitochondria and their origins.** As enzyme complexes I through IV convert 2-carbon metabolic fragments to CO_2 and H_2O, protons (H^+) are pumped into the intermembrane space. The proteins diffuse back to the matrix space via complex V, ATP synthase (AS), in which ADP is converted to ATP. The enzyme complexes are made up of subunits coded by mitochondrial DNA (mDNA) and nuclear DNA (nDNA), and the figures document the contribution of each DNA to the complexes.

cell membrane on the basal and lateral margins of the cells; that is, the cells are **polarized.** Such polarization makes transport across epithelia possible. The membranes are dynamic structures, and their constituents are being constantly renewed at different rates. Some proteins are anchored to the cytoskeleton, but others move laterally in the membrane.

Underlying most cells is a thin, "fuzzy" layer plus some fibrils that collectively make up the **basement membrane** or, more properly, the **basal lamina.** The basal lamina and, more generally, the extracellular matrix are made up of many proteins that hold cells together, regulate their development, and determine their growth. These include collagens, laminins, fibronectin, tenascin, and various proteoglycans.

MITOCHONDRIA

Over a billion years ago, aerobic bacteria were engulfed by eukaryotic cells and evolved into **mitochondria,** providing the eukaryotic cells with the ability to form the energy-rich compound ATP by **oxidative phosphorylation.** Mitochondria perform other functions, including a role in the regulation of **apoptosis** (programmed cell death), but oxidative phosphorylation is the most crucial. Each eukaryotic cell can have hundreds to thousands of mitochondria. In mammals, they are generally depicted as sausage-shaped organelles (Figure 2–1), but their shape can be quite dynamic. Each has an outer membrane, an intermembrane space, an inner membrane, which is folded to form shelves **(cristae),** and a central matrix space. The enzyme complexes responsible for oxidative phosphorylation are lined up on the cristae (Figure 2–4).

Consistent with their origin from aerobic bacteria, the mitochondria have their own genome. There is much less DNA in the mitochondrial genome than in the nuclear genome, and 99% of the proteins in the mitochondria are the products of nuclear genes, but mitochondrial DNA is responsible for certain key components of the pathway for oxidative phosphorylation. Specifically, human mitochondrial DNA is a double-stranded circular molecule containing approximately

16,500 base pairs (compared with over a billion in nuclear DNA). It codes for 13 protein subunits that are associated with proteins encoded by nuclear genes to form four enzyme complexes plus two ribosomal and 22 transfer RNAs that are needed for protein production by the intramitochondrial ribosomes.

The enzyme complexes responsible for oxidative phosphorylation illustrate the interactions between the products of the mitochondrial genome and the nuclear genome. For example, complex I, reduced nicotinamide adenine dinucleotide dehydrogenase (NADH), is made up of 7 protein subunits coded by mitochondrial DNA and 39 subunits coded by nuclear DNA. The origin of the subunits in the other complexes is shown in Figure 2–4. Complex II, succinate dehydrogenase-ubiquinone oxidoreductase; complex III, ubiquinone-cytochrome c oxidoreductase; and complex IV, cytochrome c oxidase, act with complex I, coenzyme Q, and cytochrome c to convert metabolites to CO_2 and water. Complexes I, III, and IV pump protons (H^+) into the intermembrane space during this electron transfer. The protons then flow down their electrochemical gradient through complex V, ATP synthase, which harnesses this energy to generate ATP.

As zygote mitochondria are derived from the ovum, their inheritance is maternal. This maternal inheritance has been used as a tool to track evolutionary descent. Mitochondria have an ineffective DNA repair system, and the mutation rate for mitochondrial DNA is over 10 times the rate for nuclear DNA. A large number of relatively rare diseases have now been traced to mutations in mitochondrial DNA. These include for the most part disorders of tissues with high metabolic rates in which energy production is defective as a result of abnormalities in the production of ATP.

LYSOSOMES

In the cytoplasm of the cell there are large, somewhat irregular structures surrounded by membrane. The interior of these structures, which are called **lysosomes,** is more acidic than the

TABLE 2–1 Some of the enzymes found in lysosomes and the cell components that are their substrates.

Enzyme	Substrate
Ribonuclease	RNA
Deoxyribonuclease	DNA
Phosphatase	Phosphate esters
Glycosidases	Complex carbohydrates; glycosides and polysaccharides
Arylsulfatases	Sulfate esters
Collagenase	Collagens
Cathepsins	Proteins

CLINICAL BOX 2–1

Lysosomal Diseases

When a lysosomal enzyme is congenitally absent, the lysosomes become engorged with the material the enzyme normally degrades. This eventually leads to one of the **lysosomal storage diseases.** For example, α-galactosidase A deficiency causes Fabry disease, and β-galactocerebrosidase deficiency causes Gaucher disease. These diseases are rare, but they are serious and can be fatal. Another example is the lysosomal storage disease called Tay–Sachs disease, which causes mental retardation and blindness. Tay–Sachs is caused by the loss of hexosaminidase A, a lysosomal enzyme that catalyzes the biodegradation of gangliosides (fatty acid derivatives).

rest of the cytoplasm, and external material such as endocytosed bacteria, as well as worn-out cell components, are digested in them. The interior is kept acidic by the action of a **proton pump,** or **H$^+$, ATPase.** This integral membrane protein uses the energy of ATP to move protons from the cytosol up their electrochemical gradient and keep the lysosome relatively acidic, near pH 5.0. Lysosomes can contain over 40 types of hydrolytic enzymes, some of which are listed in Table 2–1. Not surprisingly, these enzymes are all acid hydrolases, in that they function best at the acidic pH of the lysosomal compartment. This can be a safety feature for the cell; if the lysosome were to break open and release its contents, the enzymes would not be efficient at the near neutral cytosolic pH (7.2), and thus would be unable to digest cytosolic enzymes they may encounter. Diseases associated with lysosomal dysfunction are discussed in Clinical Box 2–1.

PEROXISOMES

Peroxisomes are 0.5 μm in diameter, are surrounded by a membrane, and contain enzymes that can either produce H$_2$O$_2$ (**oxidases**) or break it down (**catalases**). Proteins are directed to the peroxisome by a unique signal sequence with the help of protein chaperones, **peroxins.** The peroxisome membrane contains a number of peroxisome-specific proteins that are concerned with transport of substances into and out of the matrix of the peroxisome. The matrix contains more than 40 enzymes, which operate in concert with enzymes outside the peroxisome to catalyze a variety of anabolic and catabolic reactions (eg, breakdown of lipids). Peroxisomes can form by budding of endoplasmic reticulum, or by division. A number of synthetic compounds were found to cause proliferation of peroxisomes by acting on receptors in the nuclei of cells. These **peroxisome proliferation activated receptors (PPARs)** are members of the nuclear receptor superfamily. When activated, they bind to DNA, producing changes in the production of mRNAs. The known effects for PPARs are extensive and can affect most tissues and organs.

CYTOSKELETON

All cells have a **cytoskeleton,** a system of fibers that not only maintains the structure of the cell but also permits it to change shape and move. The cytoskeleton is made up primarily of **microtubules, intermediate filaments,** and **microfilaments** (Figure 2–5), along with proteins that anchor them and tie them together. In addition, proteins and organelles move along microtubules and microfilaments from one part of the cell to another, propelled by molecular motors.

Microtubules (Figures 2–5 and 2–6) are long, hollow structures with 5-nm walls surrounding a cavity 15 nm in diameter. They are made up of two globular protein subunits: α- and β-tubulin. A third subunit, γ-tubulin, is associated with the production of microtubules by the centrosomes. The α and β subunits form heterodimers, which aggregate to form long tubes made up of stacked rings, with each ring usually containing 13 subunits. The tubules interact with GTP to facilitate their formation. Although microtubule subunits can be added to either end, microtubules are polar with assembly predominating at the "+" end and disassembly predominating at the "−" end. Both processes occur simultaneously in vitro. The growth of microtubules is temperature sensitive (disassembly is favored under cold conditions) as well as under the control of a variety of cellular factors that can directly interact with microtubules in the cell.

Because of their constant assembly and disassembly, microtubules are a dynamic portion of the cell skeleton. They provide the tracks along which several different molecular motors move transport vesicles, organelles such as secretory granules, and mitochondria, from one part of the cell to another. They also form the spindle, which moves the chromosomes in mitosis. Cargo can be transported in either direction on microtubules.

There are several drugs available that disrupt cellular function through interaction with microtubules. Microtubule assembly is prevented by colchicine and vinblastine. The anticancer drug **paclitaxel (Taxol)** binds to microtubules and

Cytoskeletal filaments	Diameter (nm)	Protein subunit
Microfilament	7	Actin
Intermediate filament	10	Several proteins
Microtubule	25	Tubulin

FIGURE 2–5   **Cytoskeletal elements of the cell.** Artistic impressions that depict the major cytoskeletal elements are shown on the left, with basic properties of these elements on the right. (Reproduced with permission from Widmaier EP, Raff H, Strang KT: *Vander's Human Physiology: The Mechanisms of Body Function*, 11th ed. McGraw-Hill, 2008.)

makes them so stable that organelles cannot move. Mitotic spindles cannot form, and the cells die.

Intermediate filaments (Figures 2–5 and 2–6) are 8 to 14 nm in diameter and are made up of various subunits. Some of these filaments connect the nuclear membrane to the cell membrane. They form a flexible scaffolding for the cell and help it resist external pressure. In their absence, cells rupture more easily, and when they are abnormal in humans, blistering of the skin is common. The proteins that make up intermediate filaments are cell-type specific, and are thus frequently used as cellular markers. For example, vimentin is a major intermediate filament in fibroblasts, whereas cytokeratin is expressed in epithelial cells.

Microfilaments (Figures 2–5 and 2–6) are long solid fibers with a 4 to 6 nm diameter that are made up of **actin.** Although actin is most often associated with muscle contraction, it is present in all types of cells. It is the most abundant protein in mammalian cells, sometimes accounting for as much as 15% of the total protein in the cell. Its structure is highly conserved; for example, 88% of the amino acid sequences in yeast and rabbit actin are identical. Actin filaments polymerize and depolymerize in vivo, and it is not uncommon to find polymerization occurring at one end of the filament while depolymerization is occurring at the other end. **Filamentous (F) actin** refers to intact microfilaments and **globular (G) actin** refers to the unpolymerized protein actin subunits. F-actin fibers attach to various parts of the cytoskeleton and can interact directly or indirectly with membrane-bound proteins. They reach to the tips of the microvilli on the epithelial cells of the intestinal mucosa. They are also abundant in the lamellipodia that cells put out when they crawl along surfaces. The actin filaments interact with integrin

FIGURE 2–6   **Microfilaments and microtubules.** Electron micrograph **(Left)** of the cytoplasm of a fibroblast, displaying actin microfilaments (MF) and microtubules (MT). (Reproduced, with permission, from Junqueira LC, Carneiro J: *Basic Histology*, 10th ed. McGraw-Hill, 2003.) Fluorescent micrographs of airway epithelial cells displaying actin microfilaments stained with phalloidin **(Middle)** and microtubules visualized with an antibody to β-tubulin **(Right)**. Both fluorescent micrographs are counterstained with Hoechst dye (blue) to visualize nuclei. Note the distinct differences in cytoskeletal structure.

receptors and form **focal adhesion complexes,** which serve as points of traction with the surface over which the cell pulls itself. In addition, some molecular motors use microfilaments as tracks.

MOLECULAR MOTORS

The molecular motors that move proteins, organelles, and other cell parts (collectively referred to as "cargo") to all parts of the cell are 100 to 500 kDa ATPases. They attach to their cargo at one end of the molecule and to microtubules or actin polymers with the other end, sometimes referred to as the "head." They convert the energy of ATP into movement along the cytoskeleton, taking their cargo with them. There are three super families of molecular motors: **kinesin, dynein,** and **myosin.** Examples of individual proteins from each superfamily are shown in Figure 2–7. It is important to note that there is extensive variation among superfamily members, allowing for specialization of function (eg, choice of cargo, cytoskeletal filament type, and/or direction of movement).

The conventional form of **kinesin** is a doubleheaded molecule that tends to move its cargo toward the "+" ends of microtubules. One head binds to the microtubule and then bends its neck while the other head swings forward and binds, producing almost continuous movement. Some kinesins are associated with mitosis and meiosis. Other kinesins perform different functions, including, in some instances, moving cargo to the "–" end of microtubules. **Dyneins** have two heads, with their neck pieces embedded in a complex of proteins. **Cytoplasmic dyneins** have a function like that of conventional kinesin, except they tend to move particles and membranes to the "–" end of the microtubules. The multiple forms of **myosin** in the body are divided into 18 classes. The heads of myosin molecules bind to actin and produce motion by bending their neck regions (myosin II) or walking along microfilaments, one head after the other (myosin V). In these ways,

they perform functions as diverse as contraction of muscle and cell migration.

CENTROSOMES

Near the nucleus in the cytoplasm of eukaryotic animal cells is a **centrosome.** The centrosome is made up of two **centrioles** and surrounding amorphous **pericentriolar material.** The centrioles are short cylinders arranged so that they are at right angles to each other. Microtubules in groups of three run longitudinally in the walls of each centriole (Figure 2–1). Nine of these triplets are spaced at regular intervals around the circumference.

The centrosomes are **microtubule-organizing centers (MTOCs)** that contain γ-tubulin. The microtubules grow out of this γ-tubulin in the pericentriolar material. When a cell divides, the centrosomes duplicate themselves, and the pairs move apart to the poles of the mitotic spindle, where they monitor the steps in cell division. In multinucleate cells, a centrosome is near each nucleus.

CILIA

Cilia are specialized cellular projections that are used by unicellular organisms to propel themselves through liquid and by multicellular organisms to propel mucus and other substances over the surface of various epithelia. Cilia are functionally indistinct from the eukaryotic flagella of sperm cells. Within the cilium there is an **axoneme** that comprises a unique arrangement of nine outer microtubule doublets and two inner microtubules ("9+2" arrangement). Along this cytoskeleton is **axonemal dynein.** Coordinated dynein-microtubule interactions within the axoneme are the basis of ciliary and sperm movement. At the base of the axoneme and just inside lies the **basal body.** It has nine circumferential triplet microtubules, like a centriole, and there is evidence that basal bodies and centrioles are interconvertible.

FIGURE 2–7   Three examples of molecular motors. Conventional kinesin is shown attached to cargo, in this case a membrane-bound organelle. The way that myosin V "walks" along a microtubule is also shown. Note that the heads of the motors hydrolyze ATP and use the energy to produce motion.

CELL ADHESION MOLECULES

Cells are attached to the basal lamina and to each other by **cell adhesion molecules (CAMs)** that are prominent parts of the intercellular connections described below. These adhesion proteins have attracted great attention in recent years because of their unique structural and signaling functions found to be important in embryonic development and formation of the nervous system and other tissues, in holding tissues together in adults, in inflammation and wound healing, and in the metastasis of tumors. Many CAMs pass through the cell membrane and are anchored to the cytoskeleton inside the cell. Some bind to like molecules on other cells (homophilic binding), whereas others bind to nonself molecules (heterophilic binding). Many bind to **laminins,** a family of large cross-shaped molecules with multiple receptor domains in the extracellular matrix.

Nomenclature in the CAM field is somewhat chaotic, partly because the field is growing so rapidly and partly because of the extensive use of acronyms, as in other areas of modern biology. However, the CAMs can be divided into four broad families: (1) **integrins,** heterodimers that bind to various receptors; (2) adhesion molecules of the **IgG superfamily** of immunoglobulins; (3) **cadherins,** Ca^{2+}-dependent molecules that mediate cell-to-cell adhesion by homophilic reactions; and (4) **selectins,** which have lectin-like domains that bind carbohydrates. Specific functions of some of these molecules are addressed in later chapters.

The CAMs not only fasten cells to their neighbors, but they also transmit signals into and out of the cell. For example, cells that lose their contact with the extracellular matrix via integrins have a higher rate of apoptosis than anchored cells, and interactions between integrins and the cytoskeleton are involved in cell movement.

INTERCELLULAR CONNECTIONS

Intercellular junctions that form between the cells in tissues can be broadly split into two groups: junctions that fasten the cells to one another and to surrounding tissues, and junctions that permit transfer of ions and other molecules from one cell to another. The types of junctions that tie cells together and endow tissues with strength and stability include **tight junctions,** which are also known as the **zonula occludens** (Figure 2–8). The **desmosome** and **zonula adherens** also help to hold cells together, and the **hemidesmosome** and **focal adhesions** attach cells to their basal laminas. The **gap junction** forms a cytoplasmic "tunnel" for diffusion of small molecules (< 1000 Da) between two neighboring cells.

Tight junctions characteristically surround the apical margins of the cells in epithelia such as the intestinal mucosa, the walls of the renal tubules, and the choroid plexus. They are also important to endothelial barrier function. They are made up of ridges—half from one cell and half from the other—which adhere so strongly at cell junctions that they

FIGURE 2–8 **Intercellular junctions in the mucosa of the small intestine.** Tight junctions (zonula occludens), adherens junctions (zonula adherens), desmosomes, gap junctions, and hemidesmosomes are all shown in relative positions in a polarized epithelial cell.

almost obliterate the space between the cells. There are three main families of transmembrane proteins that contribute to tight junctions: **occludin, junctional adhesion molecules (JAMs),** and **claudins;** and several more proteins that interact from the cytosolic side. Tight junctions permit the passage of some ions and solute in between adjacent cells **(paracellular pathway)** and the degree of this "leakiness" varies, depending in part on the protein makeup of the tight junction. Extracellular fluxes of ions and solute across epithelia at these junctions are a significant part of overall ion and solute flux. In addition, tight junctions prevent the movement of proteins in the plane of the membrane, helping to maintain the different distribution of transporters and channels in the apical and basolateral cell membranes that make transport across epithelia possible.

In epithelial cells, each zonula adherens is usually a continuous structure on the basal side of the zonula occludens, and it is a major site of attachment for intracellular microfilaments. It contains cadherins.

Desmosomes are patches characterized by apposed thickenings of the membranes of two adjacent cells. Attached to the thickened area in each cell are intermediate filaments, some running parallel to the membrane and others radiating away from it. Between the two membrane thickenings the intercellular space contains filamentous material that includes cadherins and the extracellular portions of several other transmembrane proteins.

Hemidesmosomes look like half-desmosomes that attach cells to the underlying basal lamina and are connected

FIGURE 2–9 **Gap junction connecting the cytoplasm of two cells. A)** A gap junction plaque, or collection of individual gap junctions, is shown to form multiple pores between cells that allow for the transfer of small molecules. Inset is electron micrograph from rat liver (N. Gilula). **B)** Topographical depiction of individual connexon and corresponding 6 connexin proteins that traverse the membrane. Note that each connexin traverses the membrane four times. (Reproduced with permission from Kandel ER, Schwartz JH, Jessell TM [editors]: *Principles of Neural Science,* 4th ed. McGraw-Hill, 2000.)

intracellularly to intermediate filaments. However, they contain integrins rather than cadherins. Focal adhesions also attach cells to their basal laminas. As noted previously, they are labile structures associated with actin filaments inside the cell, and they play an important role in cell movement.

GAP JUNCTIONS

At gap junctions, the intercellular space narrows from 25 nm to 3 nm, and units called **connexons** in the membrane of each cell

are lined up with one another (Figure 2–9). Each connexon is made up of six protein subunits called **connexins.** They surround a channel that, when lined up with the channel in the corresponding connexon in the adjacent cell, permits substances to pass between the cells without entering the ECF. The diameter of the channel is normally about 2 nm, which permits the passage of ions, sugars, amino acids, and other solutes with molecular weights up to about 1000. Gap junctions thus permit the rapid propagation of electrical activity from cell to cell, as well as the exchange of various chemical messengers. However, the gap junction channels are not simply passive, nonspecific conduits. At least 20 different genes code for connexins in humans, and mutations in these genes can lead to diseases that are highly selective in terms of the tissues involved and the type of communication between cells produced. For instance, X-linked **Charcot–Marie–Tooth disease** is a peripheral neuropathy associated with mutation of one particular connexin gene. Experiments in mice in which particular connexins are deleted by gene manipulation or replaced with different connexins confirm that the particular connexin subunits that make up connexons determine their permeability and selectivity. Recently it has been shown that connexons can be used as channels to release small molecules from the cytosol into the ECF.

NUCLEUS & RELATED STRUCTURES

A nucleus is present in all eukaryotic cells that divide. If a cell is cut in half, the anucleate portion eventually dies without dividing. The nucleus is made up in large part of the **chromosomes,** the structures in the nucleus that carry a complete blueprint for all the heritable species and individual characteristics of the animal. Except in germ cells, the chromosomes occur in pairs, one originally from each parent. Each chromosome is made up of a giant molecule of **DNA.** The DNA strand is about 2 m long, but it can fit in the nucleus because at intervals it is wrapped around a core of histone proteins to form a **nucleosome.** There are about 25 million nucleosomes in each nucleus. Thus, the structure of the chromosomes has been likened to a string of beads. The beads are the nucleosomes, and the linker DNA between them is the string. The whole complex of DNA and proteins is called **chromatin.** During cell division, the coiling around histones is loosened, probably by acetylation of the histones, and pairs of chromosomes become visible, but between cell divisions only clumps of chromatin can be discerned in the nucleus. The ultimate units of heredity are the **genes** on the chromosomes). As discussed in Chapter 1, each gene is a portion of the DNA molecule.

The nucleus of most cells contains a **nucleolus** (Figure 2–1), a patchwork of granules rich in **RNA.** In some cells, the nucleus contains several of these structures. Nucleoli are most prominent and numerous in growing cells. They are the site of synthesis of ribosomes, the structures in the cytoplasm in which proteins are synthesized.

The interior of the nucleus has a skeleton of fine filaments that are attached to the **nuclear membrane,** or **envelope** (Figure 2–1), which surrounds the nucleus. This membrane is a double membrane, and spaces between the two folds are called **perinuclear cisterns.** The membrane is permeable only to small molecules. However, it contains **nuclear pore complexes.** Each complex has eightfold symmetry and is made up of about 100 proteins organized to form a tunnel through which transport of proteins and mRNA occurs. There are many transport pathways, and proteins called **importins** and **exportins** have been isolated and characterized. Much current research is focused on transport into and out of the nucleus, and a more detailed understanding of these processes should emerge in the near future.

ENDOPLASMIC RETICULUM

The **endoplasmic reticulum** is a complex series of tubules in the cytoplasm of the cell (Figure 2–1). The inner limb of its membrane is continuous with a segment of the nuclear membrane, so in effect this part of the nuclear membrane is a cistern of the endoplasmic reticulum. The tubule walls are made up of membrane. In **rough,** or **granular, endoplasmic reticulum,** ribosomes are attached to the cytoplasmic side of the membrane, whereas in **smooth,** or **agranular, endoplasmic reticulum,** ribosomes are absent. Free ribosomes are also found in the cytoplasm. The granular endoplasmic reticulum is concerned with protein synthesis and the initial folding of polypeptide chains with the formation of disulfide bonds. The agranular endoplasmic reticulum is the site of steroid synthesis in steroid-secreting cells and the site of detoxification processes in other cells. A modified endoplasmic reticulum, the sarcoplasmic reticulum, plays an important role in skeletal and cardiac muscle. In particular, the endoplasmic or sarcoplasmic reticulum can sequester Ca^{2+} ions and allow for their release as signaling molecules in the cytosol.

RIBOSOMES

The ribosomes in eukaryotes measure approximately 22×32 nm. Each is made up of a large and a small subunit called, on the basis of their rates of sedimentation in the ultracentrifuge, the 60S and 40S subunits. The ribosomes are complex structures, containing many different proteins and at least three ribosomal RNAs. They are the sites of protein synthesis. The ribosomes that become attached to the endoplasmic reticulum synthesize all transmembrane proteins, most secreted proteins, and most proteins that are stored in the Golgi apparatus, lysosomes, and endosomes. These proteins typically have a hydrophobic **signal peptide** at one end (Figure 2–10). The polypeptide chains that form these proteins are extruded into the endoplasmic reticulum. The free ribosomes synthesize cytoplasmic proteins such as hemoglobin and the proteins found in peroxisomes and mitochondria.

GOLGI APPARATUS & VESICULAR TRAFFIC

The Golgi apparatus is a collection of membrane-enclosed sacs (cisterns) that are stacked like dinner plates (Figure 2–1). There are usually about six sacs in each apparatus, but there may be more. One or more Golgi apparati are present in all eukaryotic cells, usually near the nucleus. Much of the organization of the Golgi is directed at proper glycosylation of proteins and lipids. There are more than 200 enzymes that function to add, remove, or modify sugars from proteins and lipids in the Golgi apparatus.

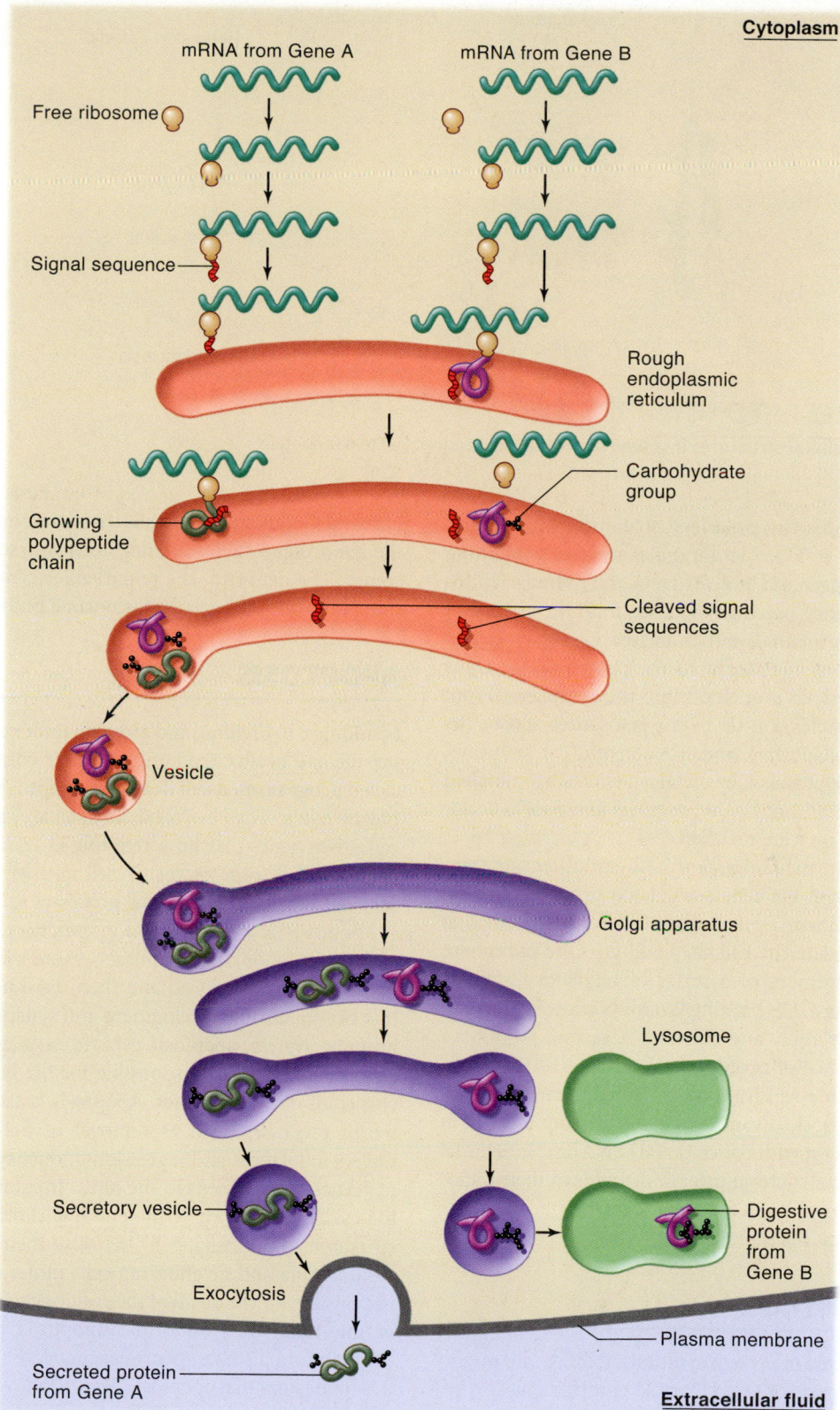

FIGURE 2–10 **Rough endoplasmic reticulum and protein translation.** Messenger RNA and ribosomes meet up in the cytosol for translation. Proteins that have appropriate signal peptides begin translation, then associate with the endoplasmic reticulum (ER) to complete translation. The association of ribosomes is what gives the ER its "rough" appearance. (Reproduced with permission from Widmaier EP, Raff H, Strang KT: *Vander's Human Physiology: The Mechanisms of Body Function,* 11th ed. McGraw-Hill, 2008.)

FIGURE 2–11   **Cellular structures involved in protein processing.** See text for details.

The Golgi apparatus is a polarized structure, with cis and trans sides (Figure 2–11). Membranous vesicles containing newly synthesized proteins bud off from the granular endoplasmic reticulum and fuse with the cistern on the cis side of the apparatus. The proteins are then passed via other vesicles to the middle cisterns and finally to the cistern on the trans side, from which vesicles branch off into the cytoplasm. From the trans Golgi, vesicles shuttle to the lysosomes and to the cell exterior via constitutive and nonconstitutive pathways, both involving **exocytosis.** Conversely, vesicles are pinched off from the cell membrane by **endocytosis** and pass to endosomes. From there, they are recycled.

Vesicular traffic in the Golgi, and between other membranous compartments in the cell, is regulated by a combination of common mechanisms along with special mechanisms that determine where inside the cell they will go. One prominent feature is the involvement of a series of regulatory proteins controlled by GTP or GDP binding (**small G proteins**) associated with vesicle assembly and delivery. A second prominent feature is the presence of proteins called SNAREs (for soluble N-ethylmaleimide-sensitive factor attachment receptor). The v- (for vesicle) SNAREs on vesicle membranes interact in a lock-and-key fashion with t- (for target) SNAREs. Individual vesicles also contain structural protein or lipids in their membrane that help to target them for specific membrane compartments (eg, Golgi sacs, cell membranes).

QUALITY CONTROL

The processes involved in protein synthesis, folding, and migration to the various parts of the cell are so complex that it is remarkable that more errors and abnormalities do not occur. The fact that these processes work as well as they do is because of mechanisms at each level that are responsible for "quality control." Damaged DNA is detected and repaired or bypassed. The various RNAs are also checked during the translation process. Finally, when the protein chains are in the endoplasmic reticu-

lum and Golgi apparatus, defective structures are detected and the abnormal proteins are degraded in lysosomes and proteasomes. The net result is a remarkable accuracy in the production of the proteins needed for normal body function.

APOPTOSIS

In addition to dividing and growing under genetic control, cells can die and be absorbed under genetic control. This process is called **programmed cell death,** or **apoptosis** (Gr. apo "away" + ptosis "fall"). It can be called "cell suicide" in the sense that the cell's own genes play an active role in its demise. It should be distinguished from necrosis ("cell murder"), in which healthy cells are destroyed by external processes such as inflammation.

Apoptosis is a very common process during development and in adulthood. In the central nervous system, large numbers of neurons are produced and then die during the remodeling that occurs during development and synapse formation. In the immune system, apoptosis gets rid of inappropriate clones of immunocytes and is responsible for the lytic effects of glucocorticoids on lymphocytes. Apoptosis is also an important factor in processes such as removal of the webs between the fingers in fetal life and regression of duct systems in the course of sexual development in the fetus. In adults, it participates in the cyclic breakdown of the endometrium that leads to menstruation. In epithelia, cells that lose their connections to the basal lamina and neighboring cells undergo apoptosis. This is responsible for the death of the enterocytes sloughed off the tips of intestinal villi. Abnormal apoptosis probably occurs in autoimmune diseases, neurodegenerative diseases, and cancer. It is interesting that apoptosis occurs in invertebrates, including nematodes and insects. However, its molecular mechanism is much more complex than that in vertebrates.

One final common pathway bringing about apoptosis is activation of **caspases,** a group of cysteine proteases. Many of these have been characterized to date in mammals; 11 have been found in humans. They exist in cells as inactive proenzymes until activated by the cellular machinery. The net result is DNA fragmentation, cytoplasmic and chromatin condensation, and eventually membrane bleb formation, with cell breakup and removal of the debris by phagocytes (see Clinical Box 2–2).

TRANSPORT ACROSS CELL MEMBRANES

There are several mechanisms of transport across cellular membranes. Primary pathways include exocytosis, endocytosis, movement through ion channels, and primary and secondary active transport. Each of these are discussed below.

EXOCYTOSIS

Vesicles containing material for export are targeted to the cell membrane (Figure 2–11), where they bond in a similar manner to that discussed in vesicular traffic between Golgi stacks, via the v-SNARE/t-SNARE arrangement. The area of fusion then breaks down, leaving the contents of the vesicle outside and the cell membrane intact. This is the Ca^{2+}-dependent process of **exocytosis** (Figure 2–12).

Note that secretion from the cell occurs via two pathways (Figure 2–11). In the **nonconstitutive pathway,** proteins from the Golgi apparatus initially enter secretory granules, where processing of prohormones to the mature hormones occurs before exocytosis. The other pathway, the **constitutive pathway,** involves the prompt transport of proteins to the cell membrane in vesicles, with little or no processing or storage. The nonconstitutive pathway is sometimes called the **regulated pathway,** but this term is misleading because the output of proteins by the constitutive pathway is also regulated.

ENDOCYTOSIS

Endocytosis is the reverse of exocytosis. There are various types of endocytosis named for the size of particles being ingested as well as the regulatory requirements for the particular process. These include **phagocytosis, pinocytosis, clathrin-mediated endocytosis, caveolae-dependent uptake,** and **nonclathrin/noncaveolae endocytosis.**

Phagocytosis ("cell eating") is the process by which bacteria, dead tissue, or other bits of microscopic material are engulfed by cells such as the polymorphonuclear leukocytes of the blood. The material makes contact with the cell membrane, which then invaginates. The invagination is pinched off, leaving the engulfed material in the membrane-enclosed vacuole and the cell membrane intact. **Pinocytosis** ("cell drinking") is a similar process with the vesicles much smaller in size and the substances ingested are in solution. The small size membrane that is ingested should not be misconstrued; cells undergoing active pinocytosis (eg, macrophages) can ingest the equivalent of their entire cell membrane in just 1 hour.

CLINICAL BOX 2–2

Molecular Medicine

Fundamental research on molecular aspects of genetics, regulation of gene expression, and protein synthesis has been paying off in clinical medicine at a rapidly accelerating rate.

One early dividend was an understanding of the mechanisms by which antibiotics exert their effects. Almost all act by inhibiting protein synthesis at one or another of the steps described previously. Antiviral drugs act in a similar way; for example, acyclovir and ganciclovir act by inhibiting DNA polymerase. Some of these drugs have this effect primarily in bacteria, but others inhibit protein synthesis in the cells of other animals, including mammals. This fact makes antibiotics of great value for research as well as for treatment of infections.

Single genetic abnormalities that cause over 600 human diseases have now been identified. Many of the diseases are rare, but others are more common and some cause conditions that are severe and eventually fatal. Examples include the defectively regulated Cl^- channel in cystic fibrosis and the unstable **trinucleotide repeats** in various parts of the genome that cause Huntington's disease, the fragile X syndrome, and several other neurologic diseases. Abnormalities in mitochondrial DNA can also cause human diseases such as Leber's hereditary optic neuropathy and some forms of cardiomyopathy. Not surprisingly, genetic aspects of cancer are probably receiving the greatest current attention. Some cancers are caused by **oncogenes,** genes that are carried in the genomes of cancer cells and are responsible for producing their malignant properties. These genes are derived by somatic mutation from closely related **proto-oncogenes,** which are normal genes that control growth. Over 100 oncogenes have been described. Another group of genes produce proteins that suppress tumors, and more than 10 of these **tumor suppressor genes** have been described. The most studied of these is the p53 gene on human chromosome 17. The p53 protein produced by this gene triggers apoptosis. It is also a nuclear transcription factor that appears to increase production of a 21-kDa protein that blocks two cell cycle enzymes, slowing the cycle and permitting repair of mutations and other defects in DNA. The p53 gene is mutated in up to 50% of human cancers, with the production of p53 proteins that fail to slow the cell cycle and permit other mutations in DNA to persist. The accumulated mutations eventually cause cancer.

Clathrin-mediated endocytosis occurs at membrane indentations where the protein **clathrin** accumulates. Clathrin molecules have the shape of triskelions, with three "legs" radiating from a central hub (Figure 2–13). As endocytosis progresses,

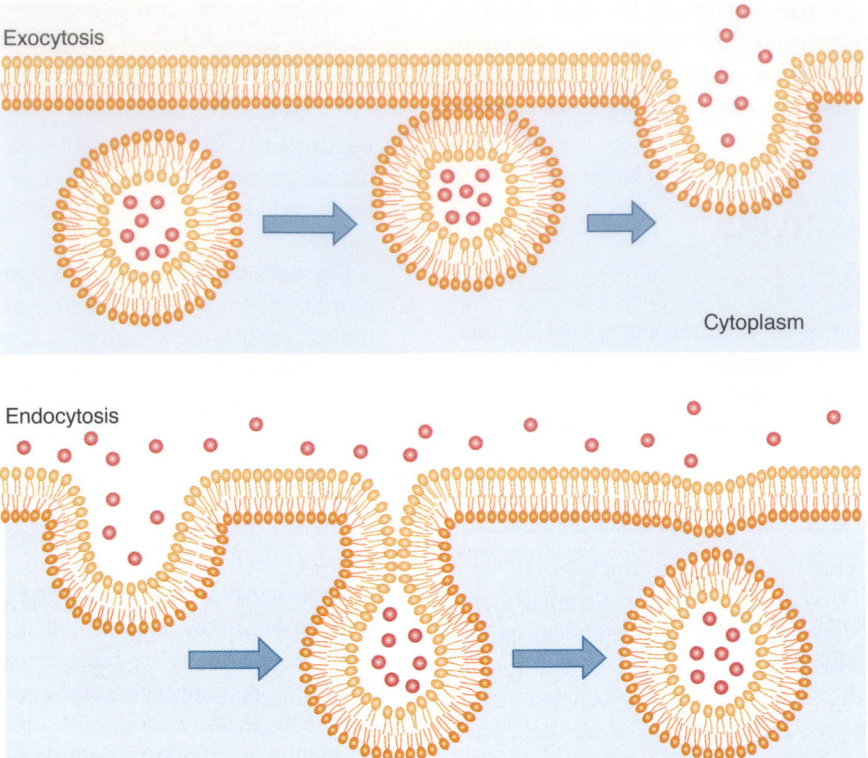

FIGURE 2–12 **Exocytosis and endocytosis.** Note that in exocytosis the cytoplasmic sides of two membranes fuse, whereas in endocytosis two noncytoplasmic sides fuse. (Reproduced with permission from Alberts B et al: *Molecular Biology of the Cell,* 4th ed. Garland Science, 2002.)

the clathrin molecules form a geometric array that surrounds the endocytotic vesicle. At the neck of the vesicle, the GTP binding protein **dynamin** is involved, either directly or indirectly, in pinching off the vesicle. Once the complete vesicle is formed, the clathrin falls off and the three-legged proteins recycle to form another vesicle. The vesicle fuses with and dumps its contents into an **early endosome** (Figure 2–11). From the early endosome, a new vesicle can bud off and return to the cell membrane. Alternatively, the early endosome can become a **late endosome** and fuse with a lysosome (Figure 2–11) in which the contents are digested by the lysosomal proteases. Clathrin-mediated endocytosis is responsible for the internal-

ization of many receptors and the ligands bound to them—including, for example, nerve growth factor and low-density lipoproteins. It also plays a major role in synaptic function.

It is apparent that exocytosis adds to the total amount of membrane surrounding the cell, and if membrane were not removed elsewhere at an equivalent rate, the cell would enlarge. However, removal of cell membrane occurs by endocytosis, and such exocytosis–endocytosis coupling maintains the surface area of the cell at its normal size.

RAFTS & CAVEOLAE

Some areas of the cell membrane are especially rich in cholesterol and sphingolipids and have been called **rafts.** These rafts are probably the precursors of flask-shaped membrane depressions called **caveolae** (little caves) when their walls become infiltrated with a protein called **caveolin** that resembles clathrin. There is considerable debate about the functions of rafts and caveolae, with evidence that they are involved in cholesterol regulation and transcytosis. It is clear, however, that cholesterol can interact directly with caveolin, effectively limiting the protein's ability to move around in the membrane. Internalization via caveolae involves binding of cargo to caveolin and regulation by dynamin. Caveolae are prominent in endothelial cells, where they help in the uptake of nutrients from the blood.

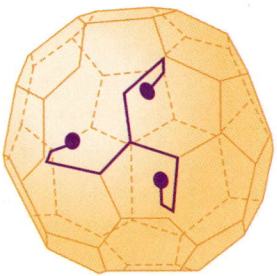

FIGURE 2–13 **Clathrin molecule on the surface of an endocytotic vesicle.** Note the characteristic triskelion shape and the fact that with other clathrin molecules it forms a net supporting the vesicle.

COATS & VESICLE TRANSPORT

It now appears that all vesicles involved in transport have protein coats. In humans, 53 coat complex subunits have been identified. Vesicles that transport proteins from the trans Golgi to lysosomes have **assembly protein 1 (AP-1)** clathrin coats, and endocytotic vesicles that transport to endosomes have AP-2 clathrin coats. Vesicles that transport between the endoplasmic reticulum and the Golgi have coat proteins I and II (COPI and COPII). Certain amino acid sequences or attached groups on the transported proteins target the proteins for particular locations. For example, the amino acid sequence Asn–Pro–any amino acid–Tyr targets transport from the cell surface to the endosomes, and mannose-6-phosphate groups target transfer from the Golgi to mannose-6-phosphate receptors (MPR) on the lysosomes.

Various small G proteins of the Rab family are especially important in vesicular traffic. They appear to guide and facilitate orderly attachments of these vesicles. To illustrate the complexity of directing vesicular traffic, humans have 60 Rab proteins and 35 SNARE proteins.

MEMBRANE PERMEABILITY & MEMBRANE TRANSPORT PROTEINS

An important technique that has permitted major advances in our knowledge about transport proteins is **patch clamping.** A micropipette is placed on the membrane of a cell and forms a tight seal to the membrane. The patch of membrane under the pipette tip usually contains only a few transport proteins, allowing for their detailed biophysical study (Figure 2–14). The cell can be left intact **(cell-attached patch clamp).** Alternatively, the patch can be pulled loose from the cell, forming an **inside-out patch.** A third alternative is to suck out the patch with the micropipette still attached to the rest of the cell membrane, providing direct access to the interior of the cell **(whole cell recording).**

Small, nonpolar molecules (including O_2 and N_2) and small uncharged polar molecules such as CO_2 diffuse across the lipid membranes of cells. However, the membranes have very limited permeability to other substances. Instead, they cross the membranes by endocytosis and exocytosis and by passage through highly specific **transport proteins,** transmembrane proteins that form channels for ions or transport substances such as glucose, urea, and amino acids. The limited permeability applies even to water, with simple diffusion being supplemented throughout the body with various water channels **(aquaporins).** For reference, the sizes of ions and other biologically important substances are summarized in Table 2–2.

Some transport proteins are simple aqueous **ion channels,** though many of these have special features that make them selective for a given substance such as Ca^{2+} or, in the case of aquaporins, for water. These membrane-spanning proteins (or collections of proteins) have tightly regulated pores that can be **gated** opened or closed in response to local changes

FIGURE 2–14 **Patch clamp to investigate transport.** In a patch clamp experiment, a small pipette is carefully maneuvered to seal off a portion of a cell membrane. The pipette has an electrode bathed in an appropriate solution that allows for recording of electrical changes through any pore in the membrane (shown below). The illustrated setup is termed an "inside-out patch" because of the orientation of the membrane with reference to the electrode. Other configurations include cell attached, whole cell, and outside-out patches. (Modified from Ackerman MJ, Clapham DE: Ion channels: Basic science and clinical disease. N Engl J Med 1997;336:1575.)

(Figure 2–15). Some are gated by alterations in membrane potential **(voltage-gated),** whereas others are opened or closed in response to a ligand **(ligand-gated).** The ligand is

TABLE 2–2 **Size of hydrated ions and other substances of biologic interest.**

Substance	Atomic or Molecular Weight	Radius (nm)
Cl^-	35	0.12
K^+	39	0.12
H_2O	18	0.12
Ca^{2+}	40	0.15
Na^+	23	0.18
Urea	60	0.23
Li^+	7	0.24
Glucose	180	0.38
Sucrose	342	0.48
Inulin	5000	0.75
Albumin	69,000	7.50

Data from Moore EW: *Physiology of Intestinal Water and Electrolyte Absorption.* American Gastroenterological Association, 1976.

Closed | Open

A Ligand-gated

Bind ligand

B Phosphorylation-gated

Phosphorylate

Dephosphorylate

C Voltage-gated

Change membrane potential

+ +   + +   − −   − −

− −   − −   + +   + +

D Stretch or pressure-gated

Stretch

Cytoskeleton

FIGURE 2–15   Regulation of gating in ion channels. Several types of gating are shown for ion channels. **A)** Ligand-gated channels open in response to ligand binding. **B)** Protein phosphorylation or dephosphorylation regulate opening and closing of some ion channels. **C)** Changes in membrane potential alter channel openings. **D)** Mechanical stretch of the membrane results in channel opening. (Reproduced with permission from Kandel ER, Schwartz JH, Jessell TM [editors]: *Principles of Neural Science*, 4th ed. McGraw-Hill, 2000.)

the bound molecule from one side of the cell membrane to the other. Molecules move from areas of high concentration to areas of low concentration (down their **chemical gradient**), and cations move to negatively charged areas whereas anions move to positively charged areas (down their **electrical gradient**). When carrier proteins move substances in the direction of their chemical or electrical gradients, no energy input is required and the process is called **facilitated diffusion.** A typical example is glucose transport by the glucose transporter, which moves glucose down its concentration gradient from the ECF to the cytoplasm of the cell. Other carriers transport substances against their electrical and chemical gradients. This form of transport requires energy and is called **active transport.** In animal cells, the energy is provided almost exclusively by hydrolysis of ATP. Not surprisingly, therefore, many carrier molecules are ATPases, enzymes that catalyze the hydrolysis of ATP. One of these ATPases is **sodium–potassium adenosine triphosphatase (Na, K ATPase),** which is also known as the **Na, K pump.** There are also H, K ATPases in the gastric mucosa and the renal tubules. Ca^{2+}ATPase pumps Ca^{2+} out of cells. Proton ATPases acidify many intracellular organelles, including parts of the Golgi complex and lysosomes.

Some of the transport proteins are called **uniports** because they transport only one substance. Others are called **symports** because transport requires the binding of more than one substance to the transport protein and the substances are transported across the membrane together. An example is the symport in the intestinal mucosa that is responsible for the cotransport by facilitated diffusion of Na^+ and glucose from the intestinal lumen into mucosal cells. Other transporters are called **antiports** because they exchange one substance for another.

ION CHANNELS

There are ion channels specific for K^+, Na^+, Ca^{2+}, and Cl^-, as well as channels that are nonselective for cations or anions. Each type of channel exists in multiple forms with diverse properties. Most are made up of identical or very similar subunits. Figure 2–16 shows the multiunit structure of various channels in diagrammatic cross-section.

Most K^+ channels are tetramers, with each of the four subunits forming part of the pore through which K^+ ions pass. Structural analysis of a bacterial voltage-gated K^+ channel indicates that each of the four subunits have a paddle-like extension containing four charges. When the channel is closed, these extensions are near the negatively charged interior of the cell. When the membrane potential is reduced, the paddles containing the charges bend through the membrane to its exterior surface, causing the channel to open. The bacterial K^+ channel is very similar to the voltage-gated K^+ channels in a wide variety of species, including mammals. In the acetylcholine ion channel and other ligand-gated cation or anion channels, five subunits make up the pore. Members of the ClC family of Cl^- channels are dimers, but they have two pores, one in each subunit. Finally, aquaporins are tetramers

often external (eg, a neurotransmitter or a hormone). However, it can also be internal; intracellular Ca^{2+}, cAMP, lipids, or one of the G proteins produced in cells can bind directly to channels and activate them. Some channels are also opened by mechanical stretch, and these mechanosensitive channels play an important role in cell movement.

Other transport proteins are **carriers** that bind ions and other molecules and then change their configuration, moving

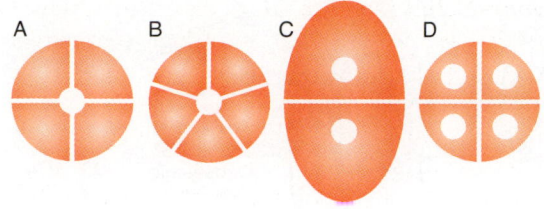

FIGURE 2–16 **Different ways in which ion channels form pores.** Many K⁺ channels are tetramers **(A)**, with each protein subunit forming part of the channel. In ligand-gated cation and anion channels **(B)** such as the acetylcholine receptor, five identical or very similar subunits form the channel. Cl⁻ channels from the ClC family are dimers **(C)**, with an intracellular pore in each subunit. Aquaporin water channels **(D)** are tetramers with an intracellular channel in each subunit. (Reproduced with permission from Jentsch TJ: Chloride channels are different. Nature 2002;415:276.)

with a water pore in each of the subunits. Recently, a number of ion channels with intrinsic enzyme activity have been cloned. More than 30 different voltage-gated or cyclic nucleotide-gated Na⁺ and Ca²⁺ channels of this type have been described. Representative Na⁺, Ca²⁺, and K⁺ channels are shown in extended diagrammatic form in Figure 2–17.

Another family of Na⁺channels with a different structure has been found in the apical membranes of epithelial cells in the kidneys, colon, lungs, and brain. The **epithelial sodium channels (ENaCs)** are made up of three subunits encoded by three different genes. Each of the subunits probably spans the membrane twice, and the amino terminal and carboxyl terminal are located inside the cell. The α subunit transports Na⁺, whereas the β and γ subunits do not. However, the addition of the β and γ subunits increases Na⁺ transport through the α subunit. ENaCs are inhibited by the diuretic amiloride, which binds to the α subunit, and they used to be called **amiloride-inhibitable Na⁺channels.** The ENaCs in the kidney play an important role in the regulation of ECF volume by aldosterone. ENaC knockout mice are born alive but promptly die because they cannot move Na⁺, and hence water, out of their lungs.

Humans have several types of Cl⁻ channels. The ClC dimeric channels are found in plants, bacteria, and animals, and there are nine different ClC genes in humans. Other Cl⁻ channels have the same pentameric form as the acetylcholine receptor; examples include the γ-aminobutyric acid A (GABA_A) and glycine receptors in the central nervous system (CNS). The cystic fibrosis transmembrane conductance regulator (CFTR) that is mutated in cystic fibrosis is also a Cl⁻ channel. Ion channel mutations cause a variety of **channelopathies**—diseases that mostly affect muscle and brain tissue and produce episodic paralyses or convulsions.

Na, K ATPase

As noted previously, Na, K ATPase catalyzes the hydrolysis of ATP to adenosine diphosphate (ADP) and uses the energy to

extrude three Na⁺ from the cell and take two K⁺ into the cell for each molecule of ATP hydrolyzed. It is an **electrogenic pump** in that it moves three positive charges out of the cell for each two that it moves in, and it is therefore said to have a **coupling ratio** of 3:2. It is found in all parts of the body. Its activity is inhibited by ouabain and related digitalis glycosides used in the treatment of heart failure. It is a heterodimer made up of an α subunit with a molecular weight of approximately 100,000 and a β subunit with a molecular weight of approximately 55,000. Both extend through the cell membrane (Figure 2–18). Separation of the subunits eliminates activity. The β subunit is a glycoprotein, whereas Na⁺ and K⁺ transport occur through the α subunit. The β subunit has a single membrane-spanning domain and three extracellular glycosylation sites, all of which appear to have attached carbohydrate residues. These residues account for one third of its molecular weight. The α subunit probably spans the cell membrane 10 times, with the amino and carboxyl terminals both located intracellularly. This subunit has intracellular Na⁺- and ATP-binding sites and a phosphorylation site; it also has extracellular binding sites for K⁺ and ouabain. The endogenous ligand of the ouabain-binding site is unsettled. When Na⁺ binds to the α subunit, ATP also binds and is converted to ADP, with a phosphate being transferred to Asp 376, the phosphorylation site. This causes a change in the configuration of the protein, extruding Na⁺ into the ECF. K⁺ then binds extracellularly, dephosphorylating the α subunit, which returns to its previous conformation, releasing K⁺ into the cytoplasm.

The α and β subunits are heterogeneous, with α₁, α₂, and α₃ subunits and β₁, β₂, and β₃ subunits described so far. The α₁ isoform is found in the membranes of most cells, whereas α₂ is present in muscle, heart, adipose tissue, and brain, and α₃ is present in heart and brain. The β₁ subunit is widely distributed but is absent in certain astrocytes, vestibular cells of the inner ear, and glycolytic fast-twitch muscles. The fast-twitch muscles contain only β₂ subunits. The different α and β subunit structures of Na, K ATPase in various tissues probably represent specialization for specific tissue functions.

REGULATION OF Na, K ATPase ACTIVITY

The amount of Na⁺ normally found in cells is not enough to saturate the pump, so if the Na⁺ increases, more is pumped out. Pump activity is affected by second messenger molecules (eg, cAMP and diacylglycerol [DAG]). The magnitude and direction of the altered pump effects vary with the experimental conditions. Thyroid hormones increase pump activity by a genomic action to increase the formation of Na, K ATPase molecules. Aldosterone also increases the number of pumps, although this effect is probably secondary. Dopamine in the kidney inhibits the pump by phosphorylating it, causing a natriuresis. Insulin increases pump activity, probably by a variety of different mechanisms.

FIGURE 2–17 **Diagrammatic representation of the pore-forming subunits of three ion channels.** The α subunit of the Na^+ and Ca^{2+} channels traverse the membrane 24 times in four repeats of six membrane-spanning units. Each repeat has a "P" loop between membrane spans 5 and 6 that does not traverse the membrane. These P loops are thought to form the pore. Note that span 4 of each repeat is colored in red, representing its net "+" charge. The K^+ channel has only a single repeat of the six spanning regions and P loop. Four K^+ subunits are assembled for a functional K^+ channel. (Reproduced with permission from Kandel ER, Schwartz JH, Jessell TM [editors]: *Principles of Neural Science*, 4th ed. McGraw-Hill, 2000.)

SECONDARY ACTIVE TRANSPORT

In many situations, the active transport of Na^+ is coupled to the transport of other substances (**secondary active transport**). For example, the luminal membranes of mucosal cells in the small intestine contain a symport that transports glucose into the cell only if Na^+ binds to the protein and is transported into the cell at the same time. From the cells, the glucose enters the blood. The electrochemical gradient for Na^+ is maintained by the active transport of Na^+ out of the mucosal cell into ECF. Other examples are shown in Figure 2–19. In the heart, Na,K ATPase indirectly affects Ca^{2+} transport. An antiport in the membranes of cardiac muscle cells normally exchanges intracellular Ca^{2+} for extracellular Na^+.

Active transport of Na^+ and K^+ is one of the major energy-using processes in the body. On the average, it accounts for

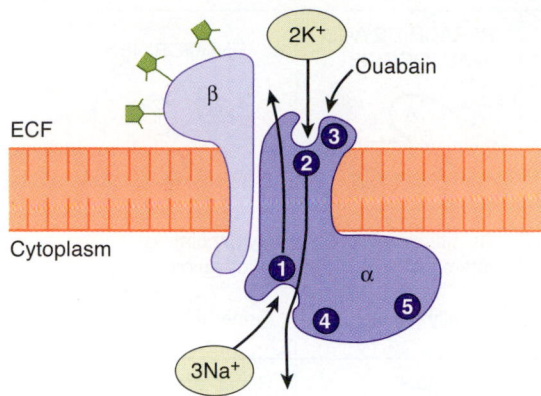

FIGURE 2–18 **Na$^+$–K$^+$ ATPase.** The intracellular portion of the α subunit has a Na$^+$-binding site **(1)**, a phosphorylation site **(4)**, and an ATP-binding site **(5)**. The extracellular portion has a K$^+$-binding site **(2)** and an ouabain-binding site **(3)**. (From Horisberger J-D et al: Structure–function relationship of Na–K-ATPase. Annu Rev Physiol 1991;53:565. Reproduced with permission from the *Annual Review of Physiology*, vol. 53. Copyright © 1991 by Annual Reviews)

FIGURE 2–19 **Composite diagram of main secondary effects of active transport of Na$^+$ and K$^+$.** Na,K ATPase converts the chemical energy of ATP hydrolysis into maintenance of an inward gradient for Na$^+$ and an outward gradient for K$^+$. The energy of the gradients is used for countertransport, cotransport, and maintenance of the membrane potential. Some examples of cotransport and countertransport that use these gradients are shown. (Reproduced with permission from Skou JC: The Na–K pump. News Physiol Sci 1992;7:95.)

about 24% of the energy utilized by cells, and in neurons it accounts for 70%. Thus, it accounts for a large part of the basal metabolism. A major payoff for this energy use is the establishment of the electrochemical gradient in cells.

TRANSPORT ACROSS EPITHELIA

In the gastrointestinal tract, the pulmonary airways, the renal tubules, and other structures, substances enter one side of a cell and exit another, producing movement of the substance from one side of the epithelium to the other. For transepithelial transport to occur, the cells need to be bound by tight junctions and, obviously, have different ion channels and transport proteins in different parts of their membranes. Most of the instances of secondary active transport cited in the preceding paragraph involve transepithelial movement of ions and other molecules.

THE CAPILLARY WALL

FILTRATION

The capillary wall separating plasma from interstitial fluid is different from the cell membranes separating interstitial fluid from intracellular fluid because the pressure difference across it makes **filtration** a significant factor in producing movement of water and solute. By definition, filtration is the process by which fluid is forced through a membrane or other barrier because of a difference in pressure on the two sides.

ONCOTIC PRESSURE

The structure of the capillary wall varies from one vascular bed to another. However, in skeletal muscle and many other organs,

water and relatively small solutes are the only substances that cross the wall with ease. The apertures in the junctions between the endothelial cells are too small to permit plasma proteins and other colloids to pass through in significant quantities. The colloids have a high molecular weight but are present in large amounts. Small amounts cross the capillary wall by vesicular transport, but their effect is slight. Therefore, the capillary wall behaves like a membrane impermeable to colloids, and these exert an osmotic pressure of about 25 mm Hg. The colloid osmotic pressure due to the plasma colloids is called the **oncotic pressure**. Filtration across the capillary membrane as a result of the hydrostatic pressure head in the vascular system is opposed by the oncotic pressure. The way the balance between the hydrostatic and oncotic pressures controls exchanges across the capillary wall is considered in detail in Chapter 32.

TRANSCYTOSIS

Vesicles are present in the cytoplasm of endothelial cells, and tagged protein molecules injected into the bloodstream have been found in the vesicles and in the interstitium. This indicates that small amounts of protein are transported out of capillaries across endothelial cells by endocytosis on the capillary side followed by exocytosis on the interstitial side of the cells. The transport mechanism makes use of coated vesicles that appear to be coated with caveolin and is called **transcytosis, vesicular transport,** or **cytopempsis.**

	GAP JUNCTIONS	SYNAPTIC	PARACRINE AND AUTOCRINE	ENDOCRINE
Message transmission	Directly from cell to cell	Across synaptic cleft	By diffusion in interstitial fluid	By circulating body fluids
Local or general	Local	Local	Locally diffuse	General
Specificity depends on	Anatomic location	Anatomic location and receptors	Receptors	Receptors

FIGURE 2–20 **Intercellular communication by chemical mediators.** A, autocrine; P, paracrine.

INTERCELLULAR COMMUNICATION

Cells communicate with one another via chemical messengers. Within a given tissue, some messengers move from cell to cell via gap junctions without entering the ECF. In addition, cells are affected by chemical messengers secreted into the ECF, or by direct cell–cell contacts. Chemical messengers typically bind to protein receptors on the surface of the cell or, in some instances, in the cytoplasm or the nucleus, triggering sequences of intracellular changes that produce their physiologic effects. Three general types of intercellular communication are mediated by messengers in the ECF: (1) **neural communication,** in which neurotransmitters are released at synaptic junctions from nerve cells and act across a narrow synaptic cleft on a postsynaptic cell; (2) **endocrine communication,** in which hormones and growth factors reach cells via the circulating blood or the lymph; and (3) **paracrine communication,** in which the products of cells diffuse in the ECF to affect neighboring cells that may be some distance away (Figure 2–20). In addition, cells secrete chemical messengers that in some situations bind to receptors on the same cell, that is, the cell that secreted the messenger (**autocrine communication**). The chemical messengers include amines, amino acids, steroids, polypeptides, and in some instances, lipids, purine nucleotides, and pyrimidine nucleotides. It is worth noting that in various parts of the body, the same chemical messenger can function as a neurotransmitter, a paracrine mediator, a hormone secreted by neurons into the blood (neural hormone), and a hormone secreted by gland cells into the blood.

An additional form of intercellular communication is called **juxtacrine communication.** Some cells express multiple repeats of growth factors such as **transforming growth factor alpha (TGFα)** extracellularly on transmembrane proteins that provide an anchor to the cell. Other cells have TGFα receptors. Consequently, TGFα anchored to a cell can bind to a TGFα receptor on another cell, linking the two. This could be important in producing local foci of growth in tissues.

RECEPTORS FOR CHEMICAL MESSENGERS

The recognition of chemical messengers by cells typically begins by interaction with a receptor at that cell. There have been over 20 families of receptors for chemical messengers characterized. These proteins are not static components of the cell, but their numbers increase and decrease in response to various stimuli, and their properties change with changes in physiological conditions. When a hormone or neurotransmitter is present in excess, the number of active receptors generally decreases (**down-regulation**), whereas in the presence of a deficiency of the chemical messenger, there is an increase in the number of active receptors (**up-regulation**). In its actions on the adrenal cortex, angiotensin II is an exception; it increases rather than decreases the number of its receptors in the adrenal. In the case of receptors in the membrane, receptor-mediated endocytosis is responsible for down-regulation in some instances; ligands bind to their receptors, and the ligand–receptor complexes move laterally in the membrane to coated pits, where they are taken into the cell by endocytosis (**internalization**). This decreases the number of receptors in the membrane. Some receptors are recycled after internalization, whereas others are replaced by de novo synthesis in the cell. Another type of down-regulation is **desensitization,** in which receptors are chemically modified in ways that make them less responsive.

MECHANISMS BY WHICH CHEMICAL MESSENGERS ACT

Receptor–ligand interaction is usually just the beginning of the cell response. This event is transduced into secondary responses within the cell that can be divided into four broad categories: (1) ion channel activation, (2) **G-protein** activation, (3) activation of enzyme activity within the cell, or (4) direct activation of transcription. Within each of these groups, responses can be quite varied. Some of the common mechanisms by which

TABLE 2–3 Common mechanisms by which chemical messengers in the ECF bring about changes in cell function.

Mechanism	Examples
Open or close ion channels in cell membrane	Acetylcholine on nicotinic cholinergic receptor, norepinephrine on K+ channel in the heart
Act via cytoplasmic or nuclear receptors to increase transcription of selected mRNAs	Thyroid hormones, retinoic acid, steroid hormones
Activate phospholipase C with intracellular production of DAG, IP$_3$, and other inositol phosphates	Angiotensin II, norepinephrine via α_1-adrenergic receptor, vasopressin via V$_1$ receptor
Activate or inhibit adenylyl cyclase, causing increased or decreased intracellular production of cAMP	Norepinephrine via β_1-adrenergic receptor (increased cAMP); norepinephrine via α_2-adrenergic receptor (decreased cAMP)
Increase cGMP in cell	Atrial natriuretic peptide; nitric oxide
Increase tyrosine kinase activity of cytoplasmic portions of transmembrane receptors	Insulin, epidermal growth factor (EGF), platelet-derived growth factor (PDGF), monocyte colony-stimulating factor (M-CSF)
Increase serine or threonine kinase activity	TGFβ, activin, inhibin

chemical messengers exert their intracellular effects are summarized in Table 2–3. Ligands such as acetylcholine bind directly to ion channels in the cell membrane, changing their conductance. Thyroid and steroid hormones, 1,25-dihydroxycholecalciferol, and retinoids enter cells and act on one or another member of a family of structurally related cytoplasmic or nuclear receptors. The activated receptor binds to DNA and increases transcription of selected mRNAs. Many other ligands in the ECF bind to receptors on the surface of cells and trigger the release of intracellular mediators such as cAMP, IP$_3$, and DAG that initiate changes in cell function. Consequently, the extracellular ligands are called **"first messengers"** and the intracellular mediators are called **"second messengers."** Second messengers bring about many short-term changes in cell function by altering enzyme function, triggering exocytosis, and so on, but they also can lead to the alteration of transcription of various genes. A variety of enzymatic changes, protein–protein interactions or second messenger changes can be activated within a cell in an orderly fashion following receptor recognition of the primary messenger. The resulting **cell signaling pathway** provides amplification of the primary signal and distribution of the signal to appropriate targets within the cell. Extensive cell signaling pathways also provide opportunities for feedback and regulation that can fine tune the signal for the correct physiological response by the cell.

The most predominant posttranslation modification of proteins, phosphorylation, is a common theme in cell signaling pathways. Cellular phosphorylation is under the control of two groups of proteins: **kinases,** enzymes that catalyze the phosphorylation of tyrosine or serine and threonine residues in proteins (or in some cases, in lipids); and **phosphatases,** proteins that remove phosphates from proteins (or lipids). Some of the larger receptor families are themselves kinases. Tyrosine kinase receptors initiate phosphorylation on tyrosine residues on complementary receptors following ligand binding. Serine/threonine kinase receptors initiate phosphorylation on serines or threonines in complementary receptors following ligand binding. Cytokine receptors are directly associated with a group of protein kinases that are activated following cytokine binding. Alternatively, second messengers changes can lead to phosphorylation further downstream in the signaling pathway. More than 300 protein kinases have been described. Some of the principal ones that are important in mammalian cell signaling are summarized in Table 2–4. In general, addition of phosphate groups changes the conformation of the proteins, altering their functions and consequently the functions of the cell. The close relationship between phosphorylation and dephosphorylation of cellular proteins allows for a temporal control of activation of cell signaling pathways. This is sometimes referred to as a **"phosphate timer."**

STIMULATION OF TRANSCRIPTION

The activation of transcription, and subsequent translation, is a common outcome of cellular signaling. There are three

TABLE 2–4 Sample protein kinases.

Phosphorylate serine or threonine residues, or both
Calmodulin-dependent
Myosin light-chain kinase
Phosphorylase kinase
Ca^{2+}/calmodulin kinase I
Ca^{2+}/calmodulin kinase II
Ca^{2+}/calmodulin kinase III
Calcium-phospholipid-dependent
Protein kinase C (seven subspecies)
Cyclic nucleotide-dependent
cAMP-dependent kinase (protein kinase A; two subspecies)
cGMP-dependent kinase
Phosphorylate tyrosine residues
Insulin receptor, EGF receptor, PDGF receptor, and
M-CSF receptor

distinct pathways for primary messengers to alter transcription of cells. First, as is the case with steroid or thyroid hormones, the primary messenger is able to cross the cell membrane and bind to a nuclear receptor, which then can directly interact with DNA to alter gene expression. A second pathway to gene transcription is the activation of cytoplasmic protein kinases that can move to the nucleus to phosphorylate a latent transcription factor for activation. This pathway is a common endpoint of signals that go through the **mitogen activated protein (MAP) kinase** cascade. MAP kinases can be activated following a variety of receptor ligand interactions through second messenger signaling. They comprise a series of three kinases that coordinate a stepwise phosphorylation to activate each protein in series in the cytosol. Phosphorylation of the last MAP kinase in series allows it to migrate to the nucleus where it phosphorylates a latent transcription factor. A third common pathway is the activation of a latent transcription factor in the cytosol, which then migrates to the nucleus and alters transcription. This pathway is shared by a diverse set of transcription factors that include **nuclear factor kappa B (NFκB;** activated following tumor necrosis family receptor binding and others), and **signal transducers of activated transcription (STATs;** activated following cytokine receptor binding). In all cases the binding of the activated transcription factor to DNA increases (or in some cases, decreases) the transcription of mRNAs encoded by the gene to which it binds. The mRNAs are translated in the ribosomes, with the production of increased quantities of proteins that alter cell function.

INTRACELLULAR Ca²⁺ AS A SECOND MESSENGER

Ca^{2+} regulates a very large number of physiological processes that are as diverse as proliferation, neural signaling, learning, contraction, secretion, and fertilization, so regulation of intracellular Ca^{2+} is of great importance. The free Ca^{2+} concentration in the cytoplasm at rest is maintained at about 100 nmol/L. The Ca^{2+} concentration in the interstitial fluid is about 12,000 times the cytoplasmic concentration (ie, 1,200,000 nmol/L), so there is a marked inwardly directed concentration gradient as well as an inwardly directed electrical gradient. Much of the intracellular Ca^{2+} is stored at relatively high concentrations in the endoplasmic reticulum and other organelles (Figure 2–21), and these organelles provide a store from which Ca^{2+} can be mobilized via ligand-gated channels to increase the concentration of free Ca^{2+} in the cytoplasm. Increased cytoplasmic Ca^{2+} binds to and activates calcium-binding proteins. These proteins can have direct effects in cellular physiology, or can activate other proteins, commonly protein kinases, to further cell signaling pathways.

Ca^{2+} can enter the cell from the extracellular fluid, down its electrochemical gradient, through many different Ca^{2+} channels. Some of these are ligand-gated and others are voltage-gated. Stretch-activated channels exist in some cells as well.

FIGURE 2–21   Ca²⁺ handling in mammalian cells. Ca^{2+} is stored in the endoplasmic reticulum and, to a lesser extent, mitochondria and can be released from them to replenish cytoplasmic Ca^{2+}. Calcium-binding proteins (CaBP) bind cytoplasmic Ca^{2+} and, when activated in this fashion, bring about a variety of physiologic effects. Ca^{2+} enters the cells via voltage-gated (volt) and ligand-gated (lig) Ca^{2+} channels and store-operated calcium channels (SOCCs). It is transported out of the cell by Ca, Mg ATPases (not shown), Ca, H ATPase and an Na, Ca antiport. It is also transported into the ER by Ca ATPases.

Many second messengers act by increasing the cytoplasmic Ca^{2+} concentration. The increase is produced by releasing Ca^{2+} from intracellular stores—primarily the endoplasmic reticulum—or by increasing the entry of Ca^{2+} into cells, or by both mechanisms. IP_3 is the major second messenger that causes Ca^{2+} release from the endoplasmic reticulum through the direct activation of a ligand-gated channel, the IP_3 receptor. In effect, the generation of one second messenger (IP_3) can lead to the release of another second messenger (Ca^{2+}). In many tissues, transient release of Ca^{2+} from internal stores into the cytoplasm triggers opening of a population of Ca^{2+} channels in the cell membrane (**store-operated Ca^{2+} channels; SOCCs**). The resulting Ca^{2+} influx replenishes the total intracellular Ca^{2+} supply and refills the endoplasmic reticulum. The exact identity of the SOCCs is still unknown, and there is debate about the signal from the endoplasmic reticulum that opens them.

As with other second messenger molecules, the increase in Ca^{2+} within the cytosol is rapid, and is followed by a rapid decrease. Because the movement of Ca^{2+} outside of the cytosol (ie, across the plasma membrane or the membrane of the internal store) requires that it move up its electrochemical gradient, it requires energy. Ca^{2+} movement out of the cell is facilitated by the plasma membrane Ca^{2+} ATPase. Alternatively, it can be transported by an antiport that exchanges three Na^+ for each Ca^{2+} driven by the energy stored in the Na^+ electrochemical gradient. Ca^{2+} movement into the internal stores is through the action of the **sarcoplasmic or endoplasmic reticulum Ca^{2+} ATPase**, also known as the **SERCA pump**.

CALCIUM-BINDING PROTEINS

Many different Ca^{2+}-binding proteins have been described, including **troponin, calmodulin,** and **calbindin.** Troponin is the

FIGURE 2–22 Structure of calmodulin from bovine brain. Single-letter abbreviations are used for the amino acid residues. Note the four calcium domains (purple residues) flanked on either side by stretches of α helix. (Reproduced with permission from Cheung WY: Calmodulin: An overview. Fed Proc 1982;41:2253.)

Ca^{2+}-binding protein involved in contraction of skeletal muscle (Chapter 5). Calmodulin contains 148 amino acid residues (Figure 2–22) and has four Ca^{2+}-binding domains. It is unique in that amino acid residue 115 is trimethylated, and it is extensively conserved, being found in plants as well as animals. When calmodulin binds Ca^{2+}, it is capable of activating five different calmodulin-dependent kinases (CaMKs; Table 2–4), among other proteins. One of the kinases is **myosin light-chain kinase,** which phosphorylates myosin. This brings about contraction in smooth muscle. CaMKI and CaMKII are concerned with synaptic function, and CaMKIII is concerned with protein synthesis. Another calmodulin-activated protein is **calcineurin,** a phosphatase that inactivates Ca^{2+} channels by dephosphorylating them. It also plays a prominent role in activating T cells and is inhibited by some immunosuppressants.

MECHANISMS OF DIVERSITY OF Ca^{2+} ACTIONS

It may seem difficult to understand how intracellular Ca^{2+} can have so many varied effects as a second messenger. Part of the explanation is that Ca^{2+} may have different effects at low and at high concentrations. The ion may be at high concentration at the site of its release from an organelle or a channel (**Ca^{2+} sparks**) and at a subsequent lower concentration after it diffuses throughout the cell. Some of the changes it produces can outlast the rise in intracellular Ca^{2+} concentration because of the way it binds to some of the Ca^{2+}-binding proteins. In addition, once released, intracellular Ca^{2+} concentrations fre-

quently oscillate at regular intervals, and there is evidence that the frequency and, to a lesser extent, the amplitude of those oscillations codes information for effector mechanisms. Finally, increases in intracellular Ca^{2+} concentration can spread from cell to cell in waves, producing coordinated events such as the rhythmic beating of cilia in airway epithelial cells.

G PROTEINS

A common way to translate a signal to a biologic effect inside cells is by way of nucleotide regulatory proteins that are activated after binding GTP (**G proteins**). When an activating signal reaches a G protein, the protein exchanges GDP for GTP. The GTP–protein complex brings about the activating effect of the G protein. The inherent GTPase activity of the protein then converts GTP to GDP, restoring the G protein to an inactive resting state. G proteins can be divided into two principal groups involved in cell signaling: **small G proteins** and **heterotrimeric G proteins.** Other groups that have similar regulation and are also important to cell physiology include elongation factors, dynamin, and translocation GTPases.

There are six different families of small G proteins (or **small GTPases**) that are all highly regulated. **GTPase activating proteins (GAPs)** tend to inactivate small G proteins by encouraging hydrolysis of GTP to GDP in the central binding site. **Guanine exchange factors (GEFs)** tend to activate small G proteins by encouraging exchange of GDP for GTP in the active site. Some of the small G proteins contain lipid modifications that help to anchor them to membranes, while others are free to diffuse throughout the cytosol. Small G proteins are involved in many cellular functions. Members of the Rab family regulate the rate of vesicle traffic between the endoplasmic reticulum, the Golgi apparatus, lysosomes, endosomes, and the cell membrane. Another family of small GTP-binding proteins, the Rho/Rac family, mediates interactions between the cytoskeleton and cell membrane; and a third family, the Ras family, regulates growth by transmitting signals from the cell membrane to the nucleus.

Another family of G proteins, the larger **heterotrimeric G proteins,** couple cell surface receptors to catalytic units that catalyze the intracellular formation of second messengers or couple the receptors directly to ion channels. Despite the knowledge of the small G proteins described above, the heteromeric G proteins are frequently referred to in the shortened "G protein" form because they were the first to be identified. Heterotrimeric G proteins are made up of three subunits designated α, β, and γ (Figure 2–23). Both the α and the γ subunits have lipid modifications that anchor these proteins to plasma membrane. The α subunit is bound to GDP. When a ligand binds to a G protein-coupled receptor (GPCR), this GDP is exchanged for GTP and the α subunit separates from the combined β and γ subunits. The separated α subunit brings about many biologic effects. The β and γ subunits are tightly bound in the cell and together form a signaling molecule that can also activate a variety of effectors. The intrinsic

FIGURE 2–23   Heterotrimeric G proteins. Top Summary of overall reaction that occurs in the Gα subunit. **Bottom:** When the ligand (square) binds to the G protein-coupled receptor in the cell membrane, GTP replaces GDP on the α subunit. GTP-α separates from the βγ subunit and GTP-α and βγ both activate various effectors, producing physiologic effects. The intrinsic GTPase activity of GTP-α then converts GTP to GDP, and the α, β, and γ subunits reassociate.

GTPase activity of the α subunit then converts GTP to GDP, and this leads to reassociation of the α with the βγ subunit and termination of effector activation. The GTPase activity of the α subunit can be accelerated by a family of **regulators of G protein signaling (RGS)**.

Heterotrimeric G proteins relay signals from over 1000 GPCRs, and their effectors in the cells include ion channels and enzymes (Table 2–5). There are 20 α, 6 β, and 12 γ genes, which allow for over 1400 α, β, and γ combinations. Not all combinations occur in the cell, but over 20 different heterotrimeric G proteins have been well documented in cell signaling. They can be divided into five families, each with a relatively characteristic set of effectors.

G PROTEIN-COUPLED RECEPTORS

All the heterotrimeric **G protein-coupled receptors (GPCRs)** that have been characterized to date are proteins that span the cell membrane seven times. Because of this structure they are alternatively referred to as **seven-helix receptors** or **serpentine receptors**. A very large number have been cloned, and their functions are multiple and diverse. The topological structures of two of them are shown in Figure 2–24. These receptors further assemble into a barrel-like structure. Upon ligand binding, a conformational change activates a resting heterotrimeric G protein associated with the cytoplasmic leaf of the plasma membrane. Activation of a single receptor can result in 1, 10, or more active heterotrimeric G proteins, providing amplification as well as transduction of the first messenger. Bound receptors can be inactivated to limit the amount of cellular signaling. This frequently occurs through phosphorylation of the cytoplasmic side of the receptor.

TABLE 2–5 Some of the ligands for receptors coupled to heterotrimeric G proteins.

Class	Ligand
Neurotransmitters	Epinephrine
	Norepinephrine
	Dopamine
	5-Hydroxytryptamine
	Histamine
	Acetylcholine
	Adenosine
	Opioids
Tachykinins	Substance P
	Neurokinin A
	Neuropeptide K
Other peptides	Angiotensin II
	Arginine vasopressin
	Oxytocin
	VIP, GRP, TRH, PTH
Glycoprotein hormones	TSH, FSH, LH, hCG
Arachidonic acid derivatives	Thromboxane A$_2$
Other	Odorants
	Tastants
	Endothelins
	Platelet-activating factor
	Cannabinoids
	Light

INOSITOL TRISPHOSPHATE & DIACYLGLYCEROL AS SECOND MESSENGERS

The link between membrane binding of a ligand that acts via Ca^{2+} and the prompt increase in the cytoplasmic Ca^{2+} concentration is often **inositol trisphosphate (inositol 1,4,5-trisphosphate; IP$_3$)**. When one of these ligands binds to its receptor, activation of the receptor produces activation of phospholipase C (PLC) on the inner surface of the membrane. Ligands bound to G protein-coupled receptor can do this through the G$_q$ heterotrimeric G proteins, while ligands bound to tyrosine kinase receptors can do this through other cell signaling pathways. PLC has at least eight isoforms; PLC$_β$

FIGURE 2–24   Structures of two G protein-coupled receptors. The individual amino acid residues are identified by their single-letter codes, and the orange residues are sites of phosphorylation. The Y-shaped symbols identify glycosylation sites. Note the extracellular amino terminal, the intracellular carboxyl terminal, and the seven membrane-spanning portions of each protein. (Reproduced with permission from Benovic JL et al: Light-dependent phosphorylation of rhodopsin by β-adrenergic receptor kinase. Reprinted by permission from *Nature* 1986;321:869. Copyright © 1986 by Macmillan Magazines)

is activated by heterotrimeric G proteins, while PLCγ forms are activated through tyrosine kinase receptors. PLC isoforms can catalyze the hydrolysis of the membrane lipid phosphatidyl-inositol 4,5-diphosphate (PIP_2) to form IP_3 and **diacylglycerol (DAG)** (Figure 2–25). The IP_3 diffuses to the endoplasmic reticulum, where it triggers the release of Ca^{2+} into the cytoplasm by binding the IP_3 receptor, a ligand-gated Ca^{2+} channel (Figure 2–26). DAG is also a second messenger; it stays in the cell membrane, where it activates one of several isoforms of **protein kinase C.**

FIGURE 2–25   Metabolism of phosphatidylinositol in cell membranes. Phosphatidylinositol is successively phosphorylated to form phosphatidylinositol 4-phosphate (PIP), then phosphatidylinositol 4,5-bisphosphate (PIP_2). Phospholipase C_β and phospholipase $C\gamma$ catalyze the breakdown of PIP_2 to inositol 1,4,5-trisphosphate (IP_3) and diacylglycerol. Other inositol phosphates and phosphatidylinositol derivatives can also be formed. IP_3 is dephosphorylated to inositol, and diacylglycerol is metabolized to cytosine diphosphate (CDP)-diacylglycerol. CDP-diacylglycerol and inositol then combine to form phosphatidylinositol, completing the cycle. (Modified from Berridge MJ: Inositol triphosphate and diacylglycerol as second messengers. Biochem J 1984;220:345.)

FIGURE 2–26 **Diagrammatic representation of release of inositol triphosphate (IP₃) and diacylglycerol (DAG) as second messengers.** Binding of ligand to G protein-coupled receptor activates phospholipase C (PLC)$_\beta$. Alternatively, activation of receptors with intracellular tyrosine kinase domains can activate PLCγ. The resulting hydrolysis of phosphatidylinositol 4,5-diphosphate (PIP₂) produces IP₃, which releases Ca²⁺ from the endoplasmic reticulum (ER), and DAG, which activates protein kinase C (PKC). CaBP, Ca²⁺-binding proteins. ISF, interstitial fluid.

CYCLIC AMP

Another important second messenger is cyclic adenosine 3',5'-monophosphate (**cyclic AMP** or **cAMP**; Figure 2–27). Cyclic AMP is formed from ATP by the action of the enzyme **adenylyl cyclase** and converted to physiologically inactive 5'AMP by the action of the enzyme **phosphodiesterase.** Some of the phosphodiesterase isoforms that break down cAMP are inhibited by methylxanthines such as caffeine and theophylline. Consequently, these compounds can augment hormonal and transmitter effects mediated via cAMP. Cyclic AMP activates one of the cyclic nucleotide-dependent protein kinases (**protein kinase A, PKA**) that, like protein kinase C, catalyzes the phosphorylation of proteins, changing their conformation and altering their activity. In addition, the active catalytic subunit of PKA moves to the nucleus and phosphorylates the **cAMP-responsive element-binding protein (CREB)**. This transcription factor then binds to DNA and alters transcription of a number of genes.

PRODUCTION OF cAMP BY ADENLYL CYCLASE

Adenylyl cyclase is a transmembrane protein, and it crosses the membrane 12 times. Ten isoforms of this enzyme have been described and each can have distinct regulatory properties, permitting the cAMP pathway to be customized to specific tissue needs. Notably, stimulatory heterotrimeric G proteins (G$_s$) activate, while inhibitory heterotrimeric G proteins (G$_i$) inactivate adenylyl cyclase (Figure 2–28). When the appropriate ligand binds to a stimulatory receptor, a G$_s$ α subunit activates one of the adenylyl cyclases. Conversely, when the

FIGURE 2–27 **Formation and metabolism of cAMP.** The second messenger cAMP is a made from ATP by adenylyl cyclase and broken down into cAMP by phosphodiesterase.

appropriate ligand binds to an inhibitory receptor, a G$_i$ α subunit inhibits adenylyl cyclase. The receptors are specific, responding at low threshold to only one or a select group of related ligands. However, heterotrimeric G proteins mediate the stimulatory and inhibitory effects produced by many different ligands. In addition, cross-talk occurs between the phospholipase C system and the adenylyl cyclase system, as several of the isoforms of adenylyl cyclase are stimulated by calmodulin. Finally, the effects of protein kinase A and protein kinase C are very widespread and can also affect directly, or indirectly, the activity at adenylyl cyclase. The close relationship between activation of G proteins and adenylyl cyclases also allows for spatial regulation of cAMP production. All of these events, and others, allow for fine-tuning the cAMP response for a particular physiological outcome in the cell.

Two bacterial toxins have important effects on adenylyl cyclase that are mediated by G proteins. The A subunit of **cholera toxin** catalyzes the transfer of ADP ribose to an arginine residue in the middle of the α subunit of G$_s$. This inhibits

FIGURE 2–28 **The cAMP system.** Activation of adenylyl cyclase catalyzes the conversion of ATP to cAMP. Cyclic AMP activates protein kinase A, which phosphorylates proteins, producing physiologic effects. Stimulatory ligands bind to stimulatory receptors and activate adenylyl cyclase via G_s. Inhibitory ligands inhibit adenylyl cyclase via inhibitory receptors and G_i. ISF, interstitial fluid.

FIGURE 2–29 **Diagrammatic representation of guanylyl cyclases, tyrosine kinases, and tyrosine phosphatases.** ANP, atrial natriuretic peptide; C, cytoplasm; cyc, guanylyl cyclase domain; EGF, epidermal growth factor; ISF, interstitial fluid; M, cell membrane; PDGF, platelet-derived growth factor; PTK, tyrosine kinase domain; PTP, tyrosine phosphatase domain; ST, *E. coli* enterotoxin. (Modified from Koesling D, Böhme E, Schultz G: Guanylyl cyclases, a growing family of signal transducing enzymes. *FASEB J* 1991;5:2785.)

its GTPase activity, producing prolonged stimulation of adenylyl cyclase. **Pertussis toxin** catalyzes ADP-ribosylation of a cysteine residue near the carboxyl terminal of the α subunit of G_i. This inhibits the function of G_i. In addition to the implications of these alterations in disease, both toxins are used for fundamental research on G protein function. The drug forskolin also stimulates adenylyl cyclase activity by a direct action on the enzyme.

GUANYLYL CYCLASE

Another cyclic nucleotide of physiologic importance is **cyclic guanosine monophosphate** (**cyclic GMP** or **cGMP**). Cyclic GMP is important in vision in both rod and cone cells. In addition, there are cGMP-regulated ion channels, and cGMP activates cGMP-dependent kinase, producing a number of physiologic effects.

Guanylyl cyclases are a family of enzymes that catalyze the formation of cGMP. They exist in two forms (Figure 2–29). One form has an extracellular amino terminal domain that is a receptor, a single transmembrane domain, and a cytoplasmic portion with guanylyl cyclase catalytic activity. Three such guanylyl cyclases have been characterized. Two are receptors for atrial natriuretic peptide (ANP; also known as atrial natriuretic factor), and a third binds an *Escherichia coli* enterotoxin and the gastrointestinal polypeptide guanylin. The other form of guanylyl cyclase is soluble, contains heme, and is not bound to the membrane. There appear to be several isoforms of the intracellular enzyme. They are activated by nitric oxide (NO) and NO-containing compounds.

GROWTH FACTORS

Growth factors have become increasingly important in many different aspects of physiology. They are polypeptides and proteins that are conveniently divided into three groups. One group is made up of agents that foster the multiplication or development of various types of cells; nerve growth factor (NGF), insulin-like growth factor I (IGF-I), activins and inhibins, and epidermal growth factor (EGF) are examples. More than 20 have been described. The cytokines are a second group. These factors are produced by macrophages and lymphocytes, as well as other cells, and are important in regulation of the immune system (see Chapter 3). Again, more than 20 have been described. The third group is made up of the colony-stimulating factors that regulate proliferation and maturation of red and white blood cells.

Receptors for EGF, platelet-derived growth factor (PDGF), and many of the other factors that foster cell multiplication and growth have a single membrane-spanning domain with an intracellular tyrosine kinase domain (Figure 2–29). When ligand binds to a tyrosine kinase receptor, it first causes a dimerization of two similar receptors. The dimerization results in partial activation of the intracellular tyrosine kinase domains and a cross-phosphorylation to fully activate each other. One of the pathways activated by phosphorylation leads, through the small G protein Ras, to MAP kinases, and eventually to the production of transcription factors in the nucleus that alter gene expression (Figure 2–30).

Receptors for cytokines and colony-stimulating factors differ from the other growth factors in that most of them do not have tyrosine kinase domains in their cytoplasmic portions and

FIGURE 2–30 **One of the direct pathways by which growth factors alter gene activity.** TK, tyrosine kinase domain; Grb2, Ras activator controller; Sos, Ras activator; Ras, product of the ras gene; MAP K, mitogen-activated protein kinase; MAP KK, MAP kinase kinase; TF, transcription factors. There is cross-talk between this pathway and the cAMP pathway, as well as cross-talk with the IP$_3$–DAG pathway.

some have little or no cytoplasmic tail. However, they initiate tyrosine kinase activity in the cytoplasm. In particular, they activate the so-called Janus tyrosine kinases (**JAKs**) in the cytoplasm (Figure 2–31). These in turn phosphorylate **STAT** proteins. The phosphorylated STATs form homo- and heterodimers and move to the nucleus, where they act as transcription factors. There are four known mammalian JAKs and seven known STATs. Interestingly, the JAK–STAT pathway can also be activated by growth hormone and is another important direct path from the cell surface to the nucleus. However, it should be emphasized that both the Ras and the JAK–STAT pathways are complex and there is cross-talk between them and other signaling pathways discussed previously.

Finally, note that the whole subject of second messengers and intracellular signaling has become immensely complex, with multiple pathways and interactions. It is only possible in a book such as this to list highlights and present general themes that will aid the reader in understanding the rest of physiology (see Clinical Box 2–3).

HOMEOSTASIS

The actual environment of the cells of the body is the interstitial component of the ECF. Because normal cell function

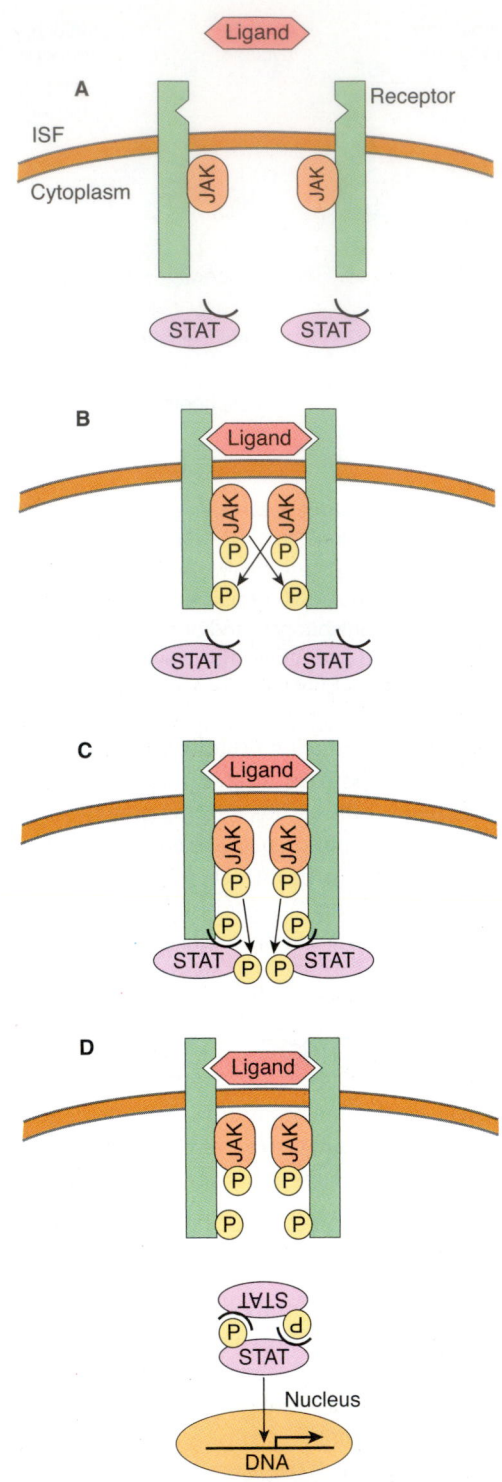

FIGURE 2–31 **Signal transduction via the JAK–STAT pathway. A)** Ligand binding leads to dimerization of receptor. **B)** Activation and tyrosine phosphorylation of JAKs. **C)** JAKs phosphorylate STATs. **D)** STATs dimerize and move to nucleus, where they bind to response elements on DNA. (Modified from Takeda K, Kishimoto T, Akira S: STAT6: Its role in interleukin 4-mediated biological functions. J Mol Med 1997;75:317.)

CLINICAL BOX 2–3

Receptor & G Protein Diseases

Many diseases are being traced to mutations in the genes for receptors. For example, loss-of-function receptor mutations that cause disease have been reported for the 1,25-dihydroxycholecalciferol receptor and the insulin receptor. Certain other diseases are caused by production of antibodies against receptors. Thus, antibodies against thyroid-stimulating hormone (TSH) receptors cause Graves' disease, and antibodies against nicotinic acetylcholine receptors cause myasthenia gravis.

An example of loss of function of a receptor is the type of **nephrogenic diabetes insipidus** that is due to loss of the ability of mutated V_2 vasopressin receptors to mediate concentration of the urine. Mutant receptors can gain as well as lose function. A gain-of-function mutation of the Ca^{2+} receptor causes excess inhibition of parathyroid hormone secretion and **familial hypercalciuric hypocalcemia.** G proteins can also undergo loss-of-function or gain-of-function mutations that cause disease (Table 2–6), In one form of pseudohypoparathyroidism, a mutated $G_s\alpha$ fails to respond to parathyroid hormone, producing the symptoms of hypoparathyroidism without any decline in circulating parathyroid hormone. **Testotoxicosis** is an interesting disease that combines gain and loss of function. In this condition, an activating mutation of $G_s\alpha$ causes excess testosterone secretion and prepubertal sexual maturation. However, this mutation is temperature-sensitive and is active only at the relatively low temperature of the testes (33 °C). At 37 °C, the normal temperature of the rest of the body, it is replaced by loss of function, with the production of hypoparathyroidism and decreased responsiveness to TSH. A different activating mutation in $G_s\alpha$ is associated with the rough-bordered areas of skin pigmentation and hypercortisolism of the McCune–Albright syndrome. This mutation occurs during fetal development, creating a mosaic of normal and abnormal cells. A third mutation in $G_s\alpha$ reduces its intrinsic GTPase activity. As a result, it is much more active than normal, and excess cAMP is produced. This causes hyperplasia and eventually neoplasia in somatotrope cells of the anterior pituitary. Forty percent of somatotrope tumors causing acromegaly have cells containing a somatic mutation of this type.

TABLE 2–6 Examples of abnormalities caused by loss- or gain-of-function mutations of heterotrimeric G protein-coupled receptors and G proteins.

Site	Type of Mutation	Disease
Receptor		
Cone opsins	Loss	Color blindness
Rhodopsin	Loss	Congenital night blindness; two forms of retinitis pigmentosa
V_2 vasopressin	Loss	X-linked nephrogenic diabetes insipidus
ACTH	Loss	Familial glucocorticoid deficiency
LH	Gain	Familial male precocious puberty
TSH	Gain	Familial nonautoimmune hyperthyroidism
TSH	Loss	Familial hypothyroidism
Ca^{2+}	Gain	Familial hypercalciuric hypocalcemia
Thromboxane A_2	Loss	Congenital bleeding
Endothelin B	Loss	Hirschsprung disease
G protein		
$G_s\alpha$	Loss	Pseudohypothyroidism type 1a
$G_s\alpha$	Gain/loss	Testotoxicosis
$G_s\alpha$	Gain (mosaic)	McCune–Albright syndrome
$G_s\alpha$	Gain	Somatotroph adenomas with acromegaly
$G_i\alpha$	Gain	Ovarian and adrenocortical tumors

Modified from Lem J: Diseases of G-protein-coupled signal transduction pathways: The mammalian visual system as a model. *Semin Neurosci* 1998;9:232.

depends on the constancy of this fluid, it is not surprising that in multicellular animals, an immense number of regulatory mechanisms have evolved to maintain it. To describe "the various physiologic arrangements which serve to restore the normal state, once it has been disturbed," W.B. Cannon coined the term **homeostasis.** The buffering properties of the body fluids and the renal and respiratory adjustments to the presence of excess acid or alkali are examples of homeostatic mechanisms. There are countless other examples, and a large part of physiology is concerned with regulatory mechanisms that act to maintain the constancy of the internal environment. Many of these regulatory mechanisms operate on the principle of negative feedback; deviations from a given normal set point are detected by a sensor, and signals from the sensor trigger compensatory changes that continue until the set point is again reached.

CHAPTER SUMMARY

- The cell and the intracellular organelles are surrounded by a semipermeable membrane. Biological membranes have a lipid bilayer with a hydrophobic core and hydrophilic outer regions that provide a barrier between inside and outside compartments as well as a template for biochemical reactions. The membrane is populated by structural and functional proteins that can be integrated into the membrane or be associated with one side of the lipid bilayer. These proteins contribute greatly to the semipermeable properties of biological membrane.

- Mitochondria are organelles that allow for oxidative phosphorylation in eukaryotic cells. They contain their own DNA, however, proteins in the mitochondria are encoded by both mitochondrial and cellular DNA. Mitochondria also are important in specialized cellular signaling.

- Lysosomes and peroxisomes are membrane-bound organelles that contribute to protein and lipid processing. They do this in part by creating acidic (lysosomes) or oxidative (peroxisomes) contents relative to the cell cytosol.

- The cytoskeleton is a network of three types of filaments that provide structural integrity to the cell as well as a means for trafficking of organelles and other structures. Actin is the fundamental building block for thin filaments and represents as much as 15% of cellular protein. Actin filaments are important in cellular contraction, migration, and signaling. Actin filaments also provide the backbone for muscle contraction. Intermediate filaments are primarily structural. Proteins that make up intermediate filaments are cell-type specific. Microtubules are made up of tubulin subunits. Microtubules provide a dynamic structure in cells that allows for movement of cellular components around the cell.

- There are three superfamilies of molecular motor proteins in the cell that use the energy of ATP to generate force, movement, or both. Myosin is the force generator for muscle cell contraction. There are also cellular myosins that interact with the cytoskeleton (primarily thin filaments) to participate in contraction as well as movement of cell contents. Kinesins and cellular dyneins are motor proteins that primarily interact with microtubules to move cargo around the cells.

- Cellular adhesion molecules aid in tethering cells to each other or to the extracellular matrix as well as providing for initiation of cellular signaling. There are four main families of these proteins: integrins, immunoglobulins, cadherins, and selectins.

- Cells contain distinct protein complexes that serve as cellular connections to other cells or the extracellular matrix. Tight junctions provide intercellular connections that link cells into a regulated tissue barrier. Tight junctions also provide a barrier to movement of proteins in the cell membrane and thus, are important to cellular polarization. Gap junctions provide contacts between cells that allow for direct passage of small molecules between two cells. Desmosomes and adherens junctions are specialized structures that hold cells together. Hemidesmosomes and focal adhesions attach cells to their basal lamina.

- The nucleus is an organelle that contains the cellular DNA and is the site of transcription. There are several organelles that emanate from the nucleus, including the endoplasmic reticulum and the Golgi apparatus. These two organelles are important in protein processing and the targeting of proteins to correct compartments within the cell.

- Exocytosis and endocytosis are vesicular fusion events that allow for movement of proteins and lipids between the cell interior, the plasma membrane, and the cell exterior. Exocytosis can be constitutive or nonconstitutive; both are regulated processes that require specialized proteins for vesicular fusion. Endocytosis is the formation of vesicles at the plasma membrane to take material from the extracellular space into the cell interior. Some endocytoses are defined in part by the size of the vesicles formed whereas others are defined by membrane structures that contribute to the endocytosis. All are tightly regulated processes.

- Membranes contain a variety of proteins and protein complexes that allow for transport of small molecules. Aqueous ion channels are membrane-spanning proteins that can be gated open to allow for selective diffusion of ions across membranes and down their electrochemical gradient. Carrier proteins bind to small molecules and undergo conformational changes to deliver small molecules across the membrane. This facilitated transport can be passive or active. Active transport requires energy for transport and is typically provided by ATP hydrolysis.

- Cells can communicate with one another via chemical messengers. Individual messengers (or ligands) typically bind to a plasma membrane receptor to initiate intracellular changes that lead to physiologic changes. Plasma membrane receptor families include ion channels, G protein-coupled receptors, or a variety of enzyme-linked receptors (eg, tyrosine kinase receptors). There are additional cytosolic receptors (eg, steroid receptors) that can bind membrane-permeant compounds. Activation of receptors lead to cellular changes that include changes in membrane potential, activation of heterotrimeric G proteins, increase in second messenger molecules, or initiation of transcription.

- Second messengers are molecules that undergo a rapid concentration changes in the cell following primary messenger recognition. Common second messenger molecules include Ca^{2+}, cyclic adenosine monophosphate (cAMP), cyclic guanine monophosphate (cGMP), inositol trisphosphate (IP_3) and nitric oxide (NO).

MULTIPLE-CHOICE QUESTIONS

For all questions, select the single best answer unless otherwise directed.

1. The electrogenic Na, K ATPase plays a critical role in cellular physiology by
   A) using the energy in ATP to extrude 3 Na^+ out of the cell in exchange for taking two K^+ into the cell.
   B) using the energy in ATP to extrude 3 K^+ out of the cell in exchange for taking two Na^+ into the cell.
   C) using the energy in moving Na^+ into the cell or K^+ outside the cell to make ATP.
   D) using the energy in moving Na^+ outside of the cell or K^+ inside the cell to make ATP.

2. Cell membranes
   A) contain relatively few protein molecules.
   B) contain many carbohydrate molecules.
   C) are freely permeable to electrolytes but not to proteins.
   D) have variable protein and lipid contents depending on their location in the cell.
   E) have a stable composition throughout the life of the cell.

3. Second messengers
   A) are substances that interact with first messengers outside cells.
   B) are substances that bind to first messengers in the cell membrane.
   C) are hormones secreted by cells in response to stimulation by another hormone.
   D) mediate the intracellular responses to many different hormones and neurotransmitters.
   E) are not formed in the brain.

4. The Golgi complex
   A) is an organelle that participates in the breakdown of proteins and lipids.
   B) is an organelle that participates in posttranslational processing of proteins.
   C) is an organelle that participates in energy production.
   D) is an organelle that participates in transcription and translation.
   E) is a subcellular compartment that stores proteins for trafficking to the nucleus.

5. Endocytosis
   A) includes phagocytosis and pinocytosis, but not clathrin-mediated or caveolae-dependent uptake of extracellular contents.
   B) refers to the merging of an intracellular vesicle with the plasma membrane to deliver intracellular contents to the extracellular milieu.
   C) refers to the invagination of the plasma membrane to uptake extracellular contents into the cell.
   D) refers to vesicular trafficking between Golgi stacks.

6. G protein-coupled receptors
   A) are intracellular membrane proteins that help to regulate movement within the cell.
   B) are plasma membrane proteins that couple the extracellular binding of primary signaling molecules to activation of small G proteins.
   C) are plasma membrane proteins that couple the extracellular binding of primary signaling molecules to the activation of heterotrimeric G proteins.
   D) are intracellular proteins that couple the binding of primary messenger molecules with transcription.

7. Gap junctions are intercellular connections that
   A) primarily serve to keep cells separated and allow for transport across a tissue barrier.
   B) serve as a regulated cytoplasmic bridge for sharing of small molecules between cells.
   C) serve as a barrier to prevent protein movement within the cellular membrane.
   D) are cellular components for constitutive exocytosis that occurs between adjacent cells.

CHAPTER RESOURCES

Alberts B et al: *Molecular Biology of the Cell,* 5th ed. Garland Science, 2007.

Cannon WB: *The Wisdom of the Body.* Norton, 1932.

Junqueira LC, Carneiro J, Kelley RO: *Basic Histology,* 9th ed. McGraw-Hill, 1998.

Kandel ER, Schwartz JH, Jessell TM (editors): *Principles of Neural Science,* 4th ed. McGraw-Hill, 2000.

Pollard TD, Earnshaw WC: *Cell Biology,* 2nd ed. Saunders, Elsevier, 2008.

Sperelakis N (editor): *Cell Physiology Sourcebook,* 3rd ed. Academic Press, 2001.

Immunity, Infection, & Inflammation

INTRODUCTION

As an open system, the body is continuously called upon to defend itself from potentially harmful invaders such as bacteria, viruses, and other microbes. This is accomplished by the immune system, which is subdivided into innate and adaptive (or acquired) branches. The immune system is composed of specialized effector cells that sense and respond to foreign antigens and other molecular patterns not found in human tissues. Likewise, the immune system clears the body's own cells that have become senescent or abnormal, such as cancer cells. Finally, occasionally, normal host tissues become the subject of inappropriate immune attack, such as in autoim-mune diseases or in settings where normal cells are harmed as innocent bystanders when the immune system mounts an inflammatory response to an invader. It is beyond the scope of this volume to provide a full treatment of all aspects of modern immunology. Nevertheless, the student of physiology should have a working knowledge of immune functions and their regulation, due to a growing appreciation for the ways in which the immune system can contribute to normal physiological regulation in a variety of tissues, as well as contributions of immune effectors to pathophysiology.

IMMUNE EFFECTOR CELLS

Many immune effector cells circulate in the blood as the white blood cells. In addition, the blood is the conduit for the precursor cells that eventually develop into the immune cells of the tissues. The circulating immunologic cells include **granulocytes (polymorphonuclear leukocytes, PMNs)**, comprising **neutrophils, eosinophils,** and **basophils; lymphocytes;** and **monocytes.** Immune responses in the tissues are further amplified by these cells following their extravascular migration, as well as tissue **macrophages** (derived from monocytes) and **mast cells** (related to basophils). Acting together, these cells provide the body with powerful defenses against tumors and viral, bacterial, and parasitic infections.

GRANULOCYTES

All granulocytes have cytoplasmic granules that contain biologically active substances involved in inflammatory and allergic reactions.

The average half-life of a neutrophil in the circulation is 6 hours. To maintain the normal circulating blood level, it is therefore necessary to produce over 100 billion neutrophils per day. Many neutrophils enter the tissues, particularly if triggered to do so by an infection or by inflammatory cytokines. They are attracted to the endothelial surface by cell adhesion molecules known as selectins, and they roll along it. They then bind firmly to neutrophil adhesion molecules of the integrin family. They next insinuate themselves through the walls of the capillaries between endothelial cells by a process called **diapedesis**. Many of those that leave the circulation enter the gastrointestinal tract and are eventually lost from the body.

Invasion of the body by bacteria triggers the **inflammatory response**. The bone marrow is stimulated to produce and release large numbers of neutrophils. Bacterial products interact with plasma factors and cells to produce agents that attract neutrophils to the infected area **(chemotaxis)**. The chemotactic agents, which are part of a large and expanding family of **chemokines** (see following text), include a component of the complement system (C5a); leukotrienes; and polypeptides from lymphocytes, mast cells, and basophils. Other plasma factors act on the bacteria to make them "tasty" to the phagocytes **(opsonization)**. The principal opsonins that coat the bacteria are immunoglobulins of a particular class (IgG) and complement proteins (see following text). The coated bacteria then bind to receptors on the neutrophil cell membrane. This triggers, via heterotrimeric G protein-mediated responses, increased motor activity of the cell, exocytosis, and the so-called respiratory burst. The increased motor activity leads to prompt ingestion of the bacteria by endocytosis **(phagocytosis)**. By **exocytosis**, neutrophil granules discharge their contents into the phagocytic vacuoles containing the bacteria and also into the interstitial space **(degranulation)**. The granules contain various proteases plus antimicrobial proteins called **defensins**. In addition, the cell membrane-bound enzyme **NADPH oxidase** is activated, with the production of toxic oxygen metabolites. The combination of the toxic oxygen metabolites and the proteolytic enzymes from the granules makes the neutrophil a very effective killing machine.

Activation of NADPH oxidase is associated with a sharp increase in O_2 uptake and metabolism in the neutrophil (the **respiratory burst**) and generation of O_2^- by the following reaction:

$$NADPH + H^+ + 2O_2 + \rightarrow NADP^+ + 2H^+ + 2O_2^-$$

O_2^- is a **free radical** formed by the addition of one electron to O_2. Two O_2^- react with two H^+ to form H_2O_2 in a reaction catalyzed by the cytoplasmic form of superoxide dismutase (SOD-1):

$$O_2^- + O_2^- + H^+ + H^+ \xrightarrow{\text{SOD-1}} \rightarrow H_2O_2 + O_2$$

O_2^- and H_2O_2 are both oxidants that are effective bactericidal agents, but H_2O_2 is converted to H_2O and O_2 by the enzyme **catalase**. The cytoplasmic form of SOD contains both Zn and Cu. It is found in many parts of the body. It is defective as a result of genetic mutation in a familial form of **amyotrophic lateral sclerosis** (ALS; see Chapter 19). Therefore, it may be that O_2^- accumulates in motor neurons and kills them in at least one form of this progressive, fatal disease. Two other forms of SOD encoded by at least one different gene are also found in humans.

Neutrophils also discharge the enzyme **myeloperoxidase,** which catalyzes the conversion of Cl^-, Br^-, I^-, and SCN^- to the corresponding acids (HOCl, HOBr, etc). These acids are also potent oxidants. Because Cl^- is present in greatest abundance in body fluids, the principal product is HOCl.

In addition to myeloperoxidase and defensins, neutrophil granules contain an elastase, two metalloproteinases that attack collagen, and a variety of other proteases that help destroy invading organisms. These enzymes act in a cooperative fashion with the O_2^-, H_2O_2, and HOCl formed by the action of the NADPH oxidase and myeloperoxidase to produce a killing zone around the activated neutrophil. This zone is effective in killing invading organisms, but in certain diseases (eg, rheumatoid arthritis) the neutrophils may also cause local destruction of host tissue.

The movements of the cell in phagocytosis, as well as migration to the site of infection, involve microtubules and microfilaments (see Chapter 1). Proper function of the microfilaments involves the interaction of the actin they contain with myosin-1 on the inside of the cell membrane (see Chapter 1).

Like neutrophils, **eosinophils** have a short half-life in the circulation, are attracted to the surface of endothelial cells by selectins, bind to integrins that attach them to the vessel wall, and enter the tissues by diapedesis. Like neutrophils, they release proteins, cytokines, and chemokines that produce inflammation but are capable of killing invading organisms. However, eosinophils have some selectivity in the way in which they respond and in the killing molecules they secrete. Their maturation and activation in tissues is particularly stimulated by IL-3, IL-5, and GM-CSF (see below). They are especially abundant in the mucosa of the gastrointestinal tract, where they defend against parasites, and in the mucosa of the respiratory and urinary tracts. Circulating eosinophils are increased in allergic diseases such as asthma and in various other respiratory and gastrointestinal diseases.

Basophils also enter tissues and release proteins and cytokines. They resemble but are not identical to mast cells, and like mast cells they contain histamine (see below). They release histamine and other inflammatory mediators when activated by binding of specific antigens to cell-fixed IgE molecules, and are essential for immediate-type hypersensitivity reactions. These range from mild urticaria and rhinitis to severe anaphylactic shock. The antigens that trigger IgE formation and basophil (and mast cell) activation are innocuous to most individuals, and are referred to as allergens.

MAST CELLS

Mast cells are heavily granulated cells of the connective tissue that are abundant in tissues that come into contact with the external environment, such as beneath epithelial surfaces. Their granules contain proteoglycans, histamine, and many proteases. Like basophils, they degranulate when allergens bind to IgE molecules directed against them that previously coat the mast cell surface. They are involved in inflammatory responses initiated by immunoglobulins IgE and IgG (see below). The inflammation combats invading parasites. In addition to this involvement in acquired immunity, they release TNF-α in response to bacterial products by an antibody-independent mechanism, thus participating in the nonspecific **innate immunity** that combats infections prior to the development of an adaptive immune response (see following text). Marked mast cell degranulation produces clinical manifestations of allergy up to and including anaphylaxis.

MONOCYTES

Monocytes enter the blood from the bone marrow and circulate for about 72 hours. They then enter the tissues and become **tissue macrophages** (Figure 3–1). Their life span in the tissues is unknown, but bone marrow transplantation data in humans suggest that they persist for about 3 months. It appears that they do not reenter the circulation. Some of them end up as the multinucleated giant cells seen in chronic inflammatory diseases such as tuberculosis. The tissue macrophages include the Kupffer cells of the liver, pulmonary alveolar macrophages (see Chapter 35), and microglia in the brain, all of which come from the circulation. In the past, they have been called the **reticuloendothelial system,** but the general term **tissue macrophage system** seems more appropriate.

Macrophages are activated by cytokines released from T lymphocytes, among others. Activated macrophages migrate in response to chemotactic stimuli and engulf and kill bacte-

Macrophages

Pseudopods

Bacteria

FIGURE 3–1 **Macrophages contacting bacteria and preparing to engulf them.** Figure is a colorized version of a scanning electron micrograph.

ria by processes generally similar to those occurring in neutrophils. They play a key role in immunity (see below). They also secrete up to 100 different substances, including factors that affect lymphocytes and other cells, prostaglandins of the E series, and clot-promoting factors.

GRANULOCYTE & MACROPHAGE COLONY-STIMULATING FACTORS

The production of white blood cells is regulated with great precision in healthy individuals, and the production of granulocytes is rapidly and dramatically increased in infections. The proliferation and self-renewal of hematopoietic stem cells (HSCs) depends on **stem cell factor (SCF).** Other factors specify particular lineages. The proliferation and maturation of the cells that enter the blood from the marrow are regulated by glycoprotein growth factors or hormones that cause cells in one or more of the committed cell lines to proliferate and mature (Table 3–1). The regulation of erythrocyte production by **erythropoietin** is discussed in Chapter 39. Three additional factors are called **colony-stimulating factors (CSFs),** because they cause appropriate single stem cells to proliferate in soft agar, forming colonies in this culture medium. The factors stimulating the production of committed stem cells include **granulocyte–macrophage CSF (GM-CSF), granulocyte CSF (G-CSF),** and **macrophage CSF (M-CSF).** Interleukins **IL-1** and **IL-6** followed by **IL-3** (Table 3–1) act in sequence to convert pluripotential uncommitted stem cells to committed progenitor cells. IL-3 is also known as **multi-CSF.** Each of the CSFs has a predominant action, but all the CSFs and interleukins also have other overlapping actions. In addition, they activate and sustain mature blood cells. It is interesting in this regard that the genes for many of these factors are located together on the long arm of chromosome 5 and may have originated by duplication of an ancestral gene. It is also interesting that basal hematopoiesis is normal in mice in which the GM-CSF gene is knocked out, indicating that loss of one factor can be compensated for by others. On the other hand, the absence of GM-CSF causes accumulation of surfactant in the lungs (see Chapter 35).

As noted in Chapter 39, erythropoietin is produced in part by kidney cells and is a circulating hormone. The other factors are produced by macrophages, activated T cells, fibroblasts, and endothelial cells. For the most part, the factors act locally in the bone marrow (Clinical Box 3–1).

LYMPHOCYTES

Lymphocytes are key elements in the production of immunity (see below). After birth, some lymphocytes are formed in the bone marrow. However, most are formed in the lymph nodes (Figure 3–2), thymus, and spleen from precursor cells that originally came from the bone marrow and were processed in the thymus or bursal equivalent (see below). Lymphocytes enter the bloodstream for the most part via the lymphatics. At

TABLE 3–1 Hematopoietic growth factors.

Cytokine	Cell Lines Stimulated	Cytokine Source
IL-1	Erythrocyte	Multiple cell types
	Granulocyte	
	Megakaryocyte	
	Monocyte	
IL-3	Erythrocyte	T lymphocytes
	Granulocyte	
	Megakaryocyte	
	Monocyte	
IL-4	Basophil	T lymphocytes
IL-5	Eosinophil	T lymphocytes
IL-6	Erythrocyte	Endothelial cells
	Granulocyte	
	Megakaryocyte	Fibroblasts
	Monocyte	Macrophages
IL-11	Erythrocyte	Fibroblasts
	Granulocyte	Osteoblasts
	Megakaryocyte	
Erythropoietin	Erythrocyte	Kidney
		Kupffer cells of liver
SCF	Erythrocyte	Multiple cell types
	Granulocyte	
	Megakaryocyte	
	Monocyte	
G-CSF	Granulocyte	Endothelial cells
		Fibroblasts
		Monocytes
GM-CSF	Erythrocyte	Endothelial cells
		Fibroblasts
	Granulocyte	Monocytes
	Megakaryocyte	T lymphocytes
M-CSF	Monocyte	Endothelial cells
		Fibroblasts
		Monocytes
Thrombopoietin	Megakaryocyte	Liver, kidney

Key: IL = interleukin; CSF = colony stimulating factor; G = granulocyte; M = macrophage; SCF = stem cell factor.

Reproduced with permission from McPhee SJ, Lingappa VR, Ganong WF (editors): *Pathophysiology of Disease,* 4th ed. McGraw-Hill, 2003.

CLINICAL BOX 3–1

Disorders of Phagocytic Function

More than 15 primary defects in neutrophil function have been described, along with at least 30 other conditions in which there is a secondary depression of the function of neutrophils. Patients with these diseases are prone to infections that are relatively mild when only the neutrophil system is involved, but which can be severe when the monocyte-tissue macrophage system is also involved. In one syndrome (neutrophil hypomotility), actin in the neutrophils does not polymerize normally, and the neutrophils move slowly. In another, there is a congenital deficiency of leukocyte integrins. In a more serious disease (chronic granulomatous disease), there is a failure to generate O_2^- in both neutrophils and monocytes and consequent inability to kill many phagocytosed bacteria. In severe congenital glucose 6-phosphate dehydrogenase deficiency, there are multiple infections because of failure to generate the NADPH necessary for O_2^- production. In congenital myeloperoxidase deficiency, microbial killing power is reduced because hypochlorous acid is not formed.

any given time, only about 2% of the body lymphocytes are in the peripheral blood. Most of the rest are in the lymphoid organs. It has been calculated that in humans, 3.5×10^{10} lymphocytes per day enter the circulation via the thoracic duct alone; however, this count includes cells that reenter the lymphatics and thus traverse the thoracic duct more than once. The effects of adrenocortical hormones on the lymphoid organs, the circulating lymphocytes, and the granulocytes are discussed in Chapter 22.

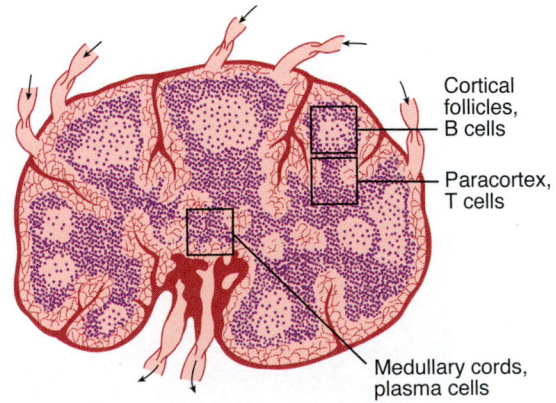

FIGURE 3–2 **Anatomy of a normal lymph node.** (After Chandrasoma. Reproduced with permission from McPhee SJ, Lingappa VR, Ganong WF [editors]: *Pathophysiology of Disease,* 4th ed. McGraw-Hill, 2003.)

IMMUNITY

OVERVIEW

Insects and other invertebrates have only **innate immunity.** This system is triggered by receptors that bind sequences of sugars, fats, or amino acids in common bacteria and activate various defense mechanisms. The receptors are coded in the germ line, and their fundamental structure is not modified by exposure to antigen. The activated defenses include, in various species, release of interferons, phagocytosis, production of antibacterial peptides, activation of the complement system, and several proteolytic cascades. Even plants release antibacterial peptides in response to infection. In vertebrates, innate immunity is also present, but is complemented by **adaptive** or **acquired immunity,** a system in which T and B lymphocytes are activated by very specific antigens. In both innate and acquired immunity, the receptors involved recognize the shape of antigens, not their specific chemical composition. In acquired immunity, activated B lymphocytes form clones that produce more antibodies which attack foreign proteins. After the invasion is repelled, small numbers persist as memory cells so that a second exposure to the same antigen provokes a prompt and magnified immune attack. The genetic event that led to acquired immunity occurred 450 million years ago in the ancestors of jawed vertebrates and was probably insertion of a transposon into the genome in a way that made possible the generation of the immense repertoire of T cell receptors that are present in the body.

In vertebrates, including humans, innate immunity provides the first line of defense against infections, but it also triggers the slower but more specific acquired immune response (Figure 3–3). In vertebrates, natural and acquired immune mechanisms also attack tumors and tissue transplanted from other animals.

Once activated, immune cells communicate by means of cytokines and chemokines. They kill viruses, bacteria, and other foreign cells by secreting other cytokines and activating the complement system.

CYTOKINES

Cytokines are hormonelike molecules that act—generally in a paracrine fashion—to regulate immune responses. They are secreted not only by lymphocytes and macrophages but by endothelial cells, neurons, glial cells, and other types of cells. Most of the cytokines were initially named for their actions, for example, B cell-differentiating factor, B cell-stimulating factor 2. However, the nomenclature has since been rationalized by international agreement to that of the **interleukins.** For example, the name of B cell-differentiating factor was changed to interleukin-4. A number of cytokines selected for their biological and clinical relevance are listed in Table 3–2, but it would be beyond the scope of this text to list all cytokines, which now number more than 100.

Many of the receptors for cytokines and hematopoietic growth factors (see above), as well as the receptors for prolactin

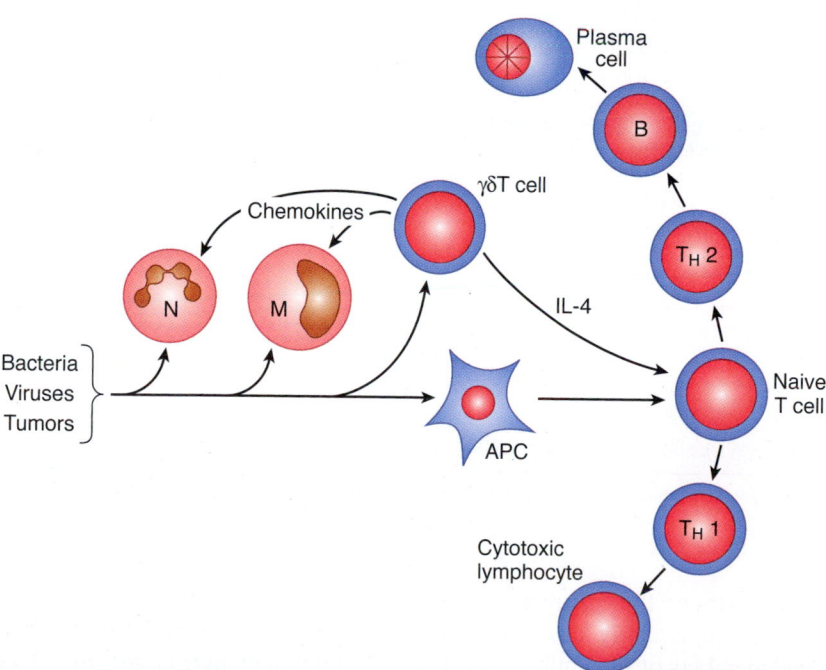

FIGURE 3–3 **How bacteria, viruses, and tumors trigger innate immunity and initiate the acquired immune response.** Arrows indicate mediators/cytokines that act on the target cell shown and/or pathways of differentiation. APC, antigen-presenting cell; M, monocyte; N, neutrophil; TH1 and TH2, helper T cells type 1 and type 2, respectively.

TABLE 3–2   Examples of cytokines and their clinical relevance.

Cytokine	Cellular Sources	Major Activities	Clinical Relevance
Interleukin-1	Macrophages	Activation of T cells and macrophages; promotion of inflammation	Implicated in the pathogenesis of septic shock, rheumatoid arthritis, and atherosclerosis
Interleukin-2	Type 1 (TH1) helper T cells	Activation of lymphocytes, natural killer cells, and macrophages	Used to induce lymphokine-activated killer cells; used in the treatment of metastatic renal-cell carcinoma, melanoma, and various other tumors
Interleukin-4	Type 2 (TH2) helper T cells, mast cells, basophils, and eosinophils	Activation of lymphocytes, monocytes, and IgE class switching	As a result of its ability to stimulate IgE production, plays a part in mast-cell sensitization and thus in allergy and in defense against nematode infections
Interleukin-5	Type 2 (TH2) helper T cells, mast cells, and eosinophils	Differentiation of eosinophils	Monoclonal antibody against interleukin-5 used to inhibit the antigen-induced late-phase eosinophilia in animal models of allergy
Interleukin-6	Type 2 (TH2) helper T cells and macrophages	Activation of lymphocytes; differentiation of B cells; stimulation of the production of acute-phase proteins	Overproduced in Castleman's disease; acts as an autocrine growth factor in myeloma and in mesangial proliferative glomerulonephritis
Interleukin-8	T cells and macrophages	Chemotaxis of neutrophils, basophils, and T cells	Levels are increased in diseases accompanied by neutrophilia, making it a potentially useful marker of disease activity
Interleukin-11	Bone marrow stromal cells	Stimulation of the production of acute-phase proteins	Used to reduce chemotherapy-induced thrombocytopenia in patients with cancer
Interleukin-12	Macrophages and B cells	Stimulation of the production of interferon γ by type 1 (TH1) helper T cells and by natural killer cells; induction of type 1 (TH1) helper T cells	May be useful as an adjuvant for vaccines
Tumor necrosis factor α	Macrophages, natural killer cells, T cells, B cells, and mast cells	Promotion of inflammation	Treatment with antibodies against tumor necrosis factor α beneficial in rheumatoid arthritis
Lymphotoxin (tumor necrosis factor β)	Type 1 (TH1) helper T cells and B cells	Promotion of inflammation	Implicated in the pathogenesis of multiple sclerosis and insulin-dependent diabetes mellitus
Transforming growth factor β	T cells, macrophages, B cells, and mast cells	Immunosuppression	May be useful therapeutic agent in multiple sclerosis and myasthenia gravis
Granulocyte-macrophage colony-stimulating factor	T cells, macrophages, natural killer cells, and B cells	Promotion of the growth of granulocytes and monocytes	Used to reduce neutropenia after chemotherapy for tumors and in ganciclovir-treated patients with AIDS; used to stimulate cell production after bone marrow transplantation
Interferon-α	Virally infected cells	Induction of resistance of cells to viral infection	Used to treat AIDS-related Kaposi sarcoma, melanoma, chronic hepatitis B infection, and chronic hepatitis C infection
Interferon-β	Virally infected cells	Induction of resistance of cells to viral infection	Used to reduce the frequency and severity of relapses in multiple sclerosis
Interferon-γ	Type 1 (TH1) helper T cells and natural killer cells	Activation of macrophages; inhibition of type 2 (TH2) helper T cells	Used to enhance the killing of phagocytosed bacteria in chronic granulomatous disease

Reproduced with permission from Delves PJ, Roitt IM: The immune system. First of two parts. *N Engl J Med* 2000;343:37.

(see Chapter 25), and growth hormone (see Chapter 24) are members of a cytokine-receptor superfamily that has three subfamilies (Figure 3–4). The members of subfamily 1, which includes the receptors for IL-4 and IL-7, are homodimers. The members of subfamily 2, which includes the receptors for IL-3, IL-5, and IL-6, are heterodimers. The receptor for IL-2 and several other cytokines is unique in that it consists of a heterodimer plus an unrelated protein, the so-called Tac antigen. The other members of subfamily 3 have the same γ chain as IL-2R. The extracellular domain of the homodimer and heterodimer subunits all contain four conserved cysteine residues plus a conserved Trp-Ser-X-Trp-Ser domain, and although the

FIGURE 3–4 **Members of one of the cytokine receptor superfamilies, showing shared structural elements.** Note that all the subunits except the α subunit in subfamily 3 have four conserved cysteine residues (open boxes at top) and a Trp-Ser-X-Trp-Ser motif (pink). Many subunits also contain a critical regulatory domain in their cytoplasmic portions (green). CNTF, ciliary neurotrophic factor; LIF, leukemia inhibitory factor; OSM, oncostatin M; PRL, prolactin. (Modified from D'Andrea AD: Cytokine receptors in congenital hematopoietic disease. *N Engl J Med* 1994;330:839.)

intracellular portions do not contain tyrosine kinase catalytic domains, they activate cytoplasmic tyrosine kinases when ligand binds to the receptors.

The effects of the principal cytokines are listed in Table 3–2. Some of them have systemic as well as local paracrine effects. For example, IL-1, IL-6, and tumor necrosis factor α cause fever, and IL-1 increases slow-wave sleep and reduces appetite.

Another superfamily of cytokines is the **chemokine** family. Chemokines are substances that attract neutrophils (see previous text) and other white blood cells to areas of inflammation or immune response. Over 40 have now been identified, and it is clear that they also play a role in the regulation of cell growth and angiogenesis. The chemokine receptors are G protein-coupled receptors that cause, among other things, extension of pseudopodia with migration of the cell toward the source of the chemokine.

THE COMPLEMENT SYSTEM

The cell-killing effects of innate and acquired immunity are mediated in part by a system of more than 30 plasma proteins originally named the **complement system** because they "complemented" the effects of antibodies. Three different pathways or enzyme cascades activate the system: the **classic pathway,** triggered by immune complexes; the **mannose-binding lectin pathway,** triggered when this lectin binds mannose groups in bacteria; and the **alternative** or **properdin pathway,** triggered by contact with various viruses, bacteria, fungi, and tumor cells. The proteins that are produced have three functions: They help kill invading organisms by opsonization, chemotaxis, and eventual lysis of the cells; they serve in part as a bridge from innate to acquired immunity by activating B cells and aiding immune memory; and they help dispose of waste products after apoptosis. Cell lysis, one of the principal ways the complement system kills cells, is brought about by inserting proteins called **perforins** into their cell membranes. These create holes, which permit free flow of ions and thus disruption of membrane polarity.

INNATE IMMUNITY

The cells that mediate innate immunity include neutrophils, macrophages, and **natural killer (NK) cells,** large lymphocytes that are not T cells but are cytotoxic. All these cells respond to lipid and carbohydrate sequences unique to bacterial cell walls and to other substances characteristic of tumor and transplant cells. Many cells that are not professional immunocytes may nevertheless also contribute to innate immune responses, such as endothelial and epithelial cells. The activated cells produce their effects via the release of cytokines, as well as, in some cases, complement and other systems.

An important link in innate immunity in **Drosophila** is a receptor protein named **toll**, which binds fungal antigens and triggers activation of genes coding for antifungal proteins. An expanding list of toll-like receptors (TLRs) have now been identified in humans. One of these, TLR4, binds bacterial lipopolysaccharide and a protein called CD14, and this initiates a cascade of intracellular events that activate transcription of genes for a variety of proteins involved in innate immune responses. This is important because bacterial lipopolysaccharide produced by gram-negative organisms is the cause of septic shock. TLR2 mediates the response to microbial lipoproteins, TLR6 cooperates with TLR2 in recognizing certain peptidoglycans, and TLR9 recognizes the DNA of certain bacteria.

ACQUIRED IMMUNITY

As noted previously, the key to acquired immunity is the ability of lymphocytes to produce antibodies (in the case of B cells) or cell-surface receptors (in the case of T cells) that are specific for one of the many millions of foreign agents that may invade the body. The antigens stimulating production of T cell receptors or antibodies are usually proteins and polypeptides, but antibodies can also be formed against nucleic acids and lipids if these are presented as nucleoproteins and lipoproteins, and antibodies to smaller molecules can be produced experimentally if the molecules are bound to protein. Acquired immunity has two components: humoral immunity and cellular immunity. **Humoral immunity** is mediated by circulating immunoglobulin antibodies in the γ-globulin fraction of the plasma proteins. Immunoglobulins are produced by differentiated forms of B lymphocytes known as **plasma cells,** and they activate the complement system and attack and neutralize antigens. Humoral immunity is a major defense against bacterial infections. **Cellular immunity** is mediated by T lymphocytes. It is responsible for delayed allergic reactions and rejection of transplants of foreign tissue. Cytotoxic T cells attack and destroy cells that have the antigen which activated them. They kill by inserting perforins (see above) and by initiating apoptosis. Cellular immunity constitutes a major defense against infections due to viruses, fungi, and a few bacteria such as the tubercle bacillus. It also helps defend against tumors.

DEVELOPMENT OF THE IMMUNE SYSTEM

During fetal development, and to a much lesser extent during adult life, lymphocyte precursors come from the bone marrow. Those that populate the thymus (Figure 3–5) become transformed by the environment in this organ into T lymphocytes. In birds, the precursors that populate the bursa of Fabricius, a lymphoid structure near the cloaca, become transformed into B lymphocytes. There is no bursa in mammals, and the transformation to B lymphocytes occurs in **bursal equivalents,** that is, the fetal liver and, after birth, the bone marrow. After residence in the thymus or liver, many of the T and B lymphocytes migrate to the lymph nodes.

T and B lymphocytes are morphologically indistinguishable but can be identified by markers on their cell membranes. B cells differentiate into **plasma cells** and **memory B cells.** There are three major types of T cells: **cytotoxic T cells, helper T cells,** and **memory T cells.** There are two subtypes of helper T cells: T helper 1 (TH1) cells secrete IL-2 and γ-interferon and are concerned primarily with cellular immunity; T helper 2 (TH2) cells secrete IL-4 and IL-5 and interact primarily with B cells in relation to humoral immunity. Cytotoxic T cells destroy transplanted and other foreign cells, with their development aided and directed by helper T cells. Markers on the surface of lymphocytes are assigned CD (clusters of differentiation) numbers on the basis of their reactions to a

FIGURE 3–5   Development of the system mediating acquired immunity.

panel of monoclonal antibodies. Most cytotoxic T cells display the glycoprotein CD8, and helper T cells display the glycoprotein CD4. These proteins are closely associated with the T cell receptors and may function as coreceptors. On the basis of differences in their receptors and functions, cytotoxic T cells are divided into αβ and γδ types (see below). Natural killer cells (see above) are also cytotoxic lymphocytes, though they are not T cells. Thus, there are three main types of cytotoxic lymphocytes in the body: αβ T cells, γδ T cells, and NK cells.

MEMORY B CELLS & T CELLS

After exposure to a given antigen, a small number of activated B and T cells persist as memory B and T cells. These cells are readily converted to effector cells by a later encounter with the same antigen. This ability to produce an accelerated response to a second exposure to an antigen is a key characteristic of acquired immunity. The ability persists for long periods of time, and in some instances (eg, immunity to measles) it can be lifelong.

After activation in lymph nodes, lymphocytes disperse widely throughout the body and are especially plentiful in areas where invading organisms enter the body, for example, the mucosa of the respiratory and gastrointestinal tracts. This puts memory cells close to sites of reinfection and may account in part for the rapidity and strength of their response. Chemokines are involved in guiding activated lymphocytes to these locations.

ANTIGEN RECOGNITION

The number of different antigens recognized by lymphocytes in the body is extremely large. The repertoire develops initially without exposure to the antigen. Stem cells differentiate into many million different T and B lymphocytes, each with the ability to respond to a particular antigen. When the antigen first enters the body, it can bind directly to the appropriate receptors on B cells. However, a full antibody response requires that the B cells contact helper T cells. In the case of T cells, the antigen is taken up by an antigen-presenting cell and partially digested. A peptide fragment of it is presented to the appropriate receptors on T cells. In either case, the cells are stimulated to divide, forming **clones** of cells that respond to this antigen (**clonal selection**). Effector cells are also subject to **negative selection,** during which lymphocyte precursors that are reactive with self antigens are normally deleted. This results in immune **tolerance.** It is this latter process that presumably goes awry in autoimmune diseases, where the body reacts to and destroys cells expressing normal proteins, with accompanying inflammation that may lead to tissue destruction.

ANTIGEN PRESENTATION

Antigen-presenting cells (APCs) include specialized cells called **dendritic cells** in the lymph nodes and spleen and the Langerhans dendritic cells in the skin. Macrophages and B cells themselves, and likely many other cell types, can also function as APCs. In APCs, polypeptide products of antigen digestion are coupled to protein products of the **major histocompatibility complex (MHC)** genes and presented on the surface of the cell. The products of the MHC genes are called human leukocyte antigens (HLA).

The genes of the MHC, which are located on the short arm of human chromosome 6, encode glycoproteins and are divided into two classes on the basis of structure and function. Class I antigens are composed of a 45-kDa heavy chain associated noncovalently with β_2-microglobulin encoded by a gene outside the MHC (Figure 3–6). They are found on all nucleated cells. Class II antigens are heterodimers made up of a 29- to 34-kDa α chain associated noncovalently with a 25- to 28-kDa β chain. They are present in antigen-presenting cells, including B cells, and in activated T cells.

The class I MHC proteins (MHC-I proteins) are coupled primarily to peptide fragments generated from proteins synthesized within cells. The peptides to which the host is not tolerant (eg, those from mutant or viral proteins) are recognized by T cells. The digestion of these proteins occurs in

FIGURE 3–6 **Structure of human histocompatibility antigen HLA-A2.** The antigen-binding pocket is at the top and is formed by the α_1 and α_2 parts of the molecule. The α_3 portion and the associated β_2-microglobulin (β_2m) are close to the membrane. The extension of the C terminal from α_3 that provides the transmembrane domain and the small cytoplasmic portion of the molecule have been omitted. (Reproduced with permission from Bjorkman PJ et al: Structure of the human histocompatibility antigen HLA-A2. *Nature* 1987;329:506.)

proteasomes, complexes of proteolytic enzymes that may be produced by genes in the MHC group, and the peptide fragments appear to bind to MHC proteins in the endoplasmic reticulum. The class II MHC proteins (MHC-II proteins) are concerned primarily with peptide products of extracellular antigens, such as bacteria, that enter the cell by endocytosis and are digested in the late endosomes.

T CELL RECEPTORS

The MHC protein–peptide complexes on the surface of the antigen-presenting cells bind to appropriate T cells. Therefore, receptors on the T cells must recognize a very wide variety of complexes. Most of the receptors on circulating T cells are made up of two polypeptide units designated α and β. They form heterodimers that recognize the MHC proteins and the antigen fragments with which they are combined (Figure 3–7). These cells are called αβ T cells. About 10% of the circulating T cells have two different polypeptides designated γ and δ in their receptors, and they are called γδ T cells. These T cells are prominent in the mucosa of the gastrointestinal tract, and there is evidence that they form a link between the innate and acquired immune systems by way of the cytokines they secrete (Figure 3–3).

CD8 occurs on the surface of cytotoxic T cells that bind MHC-I proteins, and CD4 occurs on the surface of helper T cells that bind MHC-II proteins (Figure 3–8). The CD8 and CD4 proteins facilitate the binding of the MHC proteins to the T cell receptors, and they also foster lymphocyte development, but how they produce these effects is unsettled. The

FIGURE 3–8 **Diagrammatic summary of the structure of CD4 and CD8, and their relation to MHC-I and MHC-II proteins.** Note that CD4 is a single protein, whereas CD8 is a heterodimer.

activated CD8 cytotoxic T cells kill their targets directly, whereas the activated CD4 helper T cells secrete cytokines that activate other lymphocytes.

The T cell receptors are surrounded by adhesion molecules and proteins that bind to complementary proteins in the antigen-presenting cell when the two cells transiently join to form the "immunologic synapse" that permits T cell activation to occur. It is now generally accepted that two signals are necessary to produce activation. One is produced by the binding of the digested antigen to the T cell receptor. The other is produced by the joining of the surrounding proteins in the "synapse." If the first signal occurs but the second does not, the T cell is inactivated and becomes unresponsive.

B CELLS

As noted above, B cells can bind antigens directly, but they must contact helper T cells to produce full activation and antibody formation. It is the TH2 subtype that is mainly involved. Helper T cells develop along the TH2 lineage in response to IL-4 (see below). On the other hand, IL-12 promotes the TH1 phenotype. IL-2 acts in an autocrine fashion to cause activated T cells to proliferate. The role of various cytokines in B cell and T cell activation is summarized in Figure 3–9.

The activated B cells proliferate and transform into **memory B cells** (see above) and **plasma cells.** The plasma cells secrete large quantities of antibodies into the general circulation. The antibodies circulate in the globulin fraction of the plasma and, like antibodies elsewhere, are called **immunoglobulins.** The immunoglobulins are actually the secreted form of antigen-binding receptors on the B cell membrane.

FIGURE 3–7 **Interaction between antigen-presenting cell (top) and αβ T lymphocyte (bottom).** The MHC proteins (in this case, MHC-I) and their peptide antigen fragment bind to the α and β units that combine to form the T cell receptor.

FIGURE 3–10 **Typical immunoglobulin G molecule.** Fab, portion of the molecule that is concerned with antigen binding; Fc, effector portion of the molecule. The constant regions are pink and purple, and the variable regions are orange. The constant segment of the heavy chain is subdivided into C_H1, C_H2, and C_H3. SS lines indicate intersegmental disulfide bonds. On the right side, the C labels are omitted to show regions J_H, D, and J_L.

FIGURE 3–9 **Summary of acquired immunity. (1)** An antigen-presenting cell ingests and partially digests an antigen, then presents part of the antigen along with MHC peptides (in this case, MHC II peptides on the cell surface). **(2)** An "immune synapse" forms with a naive CD4 T cell, which is activated to produce IL-2. **(3)** IL-2 acts in an autocrine fashion to cause the cell to multiply, forming a clone. **(4)** The activated CD4 cell may promote B cell activation and production of plasma cells or it may activate a cytotoxic CD8 cell. The CD8 cell can also be activated by forming a synapse with an MCH I antigen-presenting cell. (Reproduced with permission from McPhee SJ, Lingappa VR, Ganong WF [editors]: *Pathophysiology of Disease*, 4th ed. McGraw-Hill, 2003.)

IMMUNOGLOBULINS

Circulating antibodies protect their host by binding to and neutralizing some protein toxins, by blocking the attachment of some viruses and bacteria to cells, by opsonizing bacteria (see above), and by activating complement. Five general types of immunoglobulin antibodies are produced by the lymphocyte–plasma cell system. The basic component of each is a symmetric unit containing four polypeptide chains (Figure 3–10). The two long chains are called **heavy chains,** whereas the two short chains are called **light chains.** There are two types of light chains, k and λ, and eight types of heavy chains. The chains are joined by disulfide bridges that permit mobility, and there are intrachain disulfide bridges as well. In addition, the heavy chains are flexible in a region called the hinge. Each heavy chain has a variable (V) segment in which the amino acid sequence is highly variable, a diversity (D) segment in which the amino acid segment is also highly variable, a joining (J) segment in which the sequence is moderately variable, and a constant (C) segment in which the sequence is constant. Each light chain has a V, a J, and a C segment. The V segments form part of the antigen-binding sites (Fab portion of the molecule [Figure 3–10]). The Fc portion of the molecule is the effector portion, which mediates the reactions initiated by antibodies.

Two of the classes of immunoglobulins contain additional polypeptide components (Table 3–3). In IgMs, five of the basic immunoglobulin units join around a polypeptide called the J chain to form a pentamer. In IgAs, the **secretory immunoglobulins,** the immunoglobulin units form dimers and trimers around a J chain and a polypeptide that comes from epithelial cells, the secretory component (SC).

In the intestine, bacterial and viral antigens are taken up by M cells (see Chapter 27) and passed on to underlying aggregates of lymphoid tissue **(Peyer's patches),** where they activate naive T cells. These lymphocytes then form B cells that infiltrate mucosa of the gastrointestinal, respiratory, genitourinary, and female reproductive tracts and the breast. There they secrete large amounts of IgAs when exposed again to the antigen that initially stimulated them. The epithelial cells produce the SC, which acts as a receptor for and binds the IgA. The resulting secretory immunoglobulin passes through the epithelial cell and is secreted by exocytosis. This system of **secretory immunity** is an important and effective defense mechanism.

GENETIC BASIS OF DIVERSITY IN THE IMMUNE SYSTEM

The genetic mechanism for the production of the immensely large number of different configurations of immunoglobulins produced by human B cells is a fascinating biologic problem.

TABLE 3–3 Human immunoglobulins.[a]

Immunoglobulin	Function	Heavy Chain	Additional Chain	Structure	Plasma Concentration (mg/dL)
IgG	Complement activation	$\gamma_1, \gamma_2, \gamma_3, \gamma_4$		Monomer	1000
IgA	Localized protection in external secretions (tears, intestinal secretions, etc)	α_1, α_2	J, SC	Monomer; dimer with J or SC chain; trimer with J chain	200
IgM	Complement activation	μ	J	Pentamer with J chain	120
IgD	Antigen recognition by B cells	δ		Monomer	3
IgE	Reagin activity; releases histamine from basophils and mast cells	ε		Monomer	0.05

[a]In all instances, the light chains are k or γ.

Diversity is brought about in part by the fact that in immune globulin molecules there are two kinds of light chains and eight kinds of heavy chains. As noted previously, there are areas of great variability (**hypervariable regions**) in each chain. The variable portion of the heavy chains consists of the V, D, and J segments. In the gene family responsible for this region, there are several hundred different coding regions for the V segment, about 20 for the D segment, and 4 for the J segment. During B cell development, one V coding region, one D coding region, and one J coding region are selected at random and recombined to form the gene that produces that particular variable portion. A similar variable recombination takes place in the coding regions responsible for the two variable segments (V and J) in the light chain. In addition, the J segments are variable because the gene segments join in an imprecise and variable fashion (junctional site diversity) and nucleotides are sometimes added (junctional insertion diversity). It has been calculated that these mechanisms permit the production of about 10^{15} different immunoglobulin molecules. Additional variability is added by somatic mutation.

Similar gene rearrangement and joining mechanisms operate to produce the diversity in T cell receptors. In humans, the α subunit has a V region encoded by 1 of about 50 different genes and a J region encoded by 1 of another 50 different genes. The β subunits have a V region encoded by 1 of about 50 genes, a D region encoded by 1 of 2 genes, and a J region encoded by 1 of 13 genes. These variable regions permit the generation of up to an estimated 10^{15} different T cell receptors (Clinical Box 3–2 and Clinical Box 3–3).

A variety of immunodeficiency states can arise from defects in these various stages of B and T lymphocyte maturation. These are summarized in Figure 3–12.

PLATELETS

Platelets are circulating cells that are important mediators of hemostasis. While not immune cells, per se, they often participate in the response to tissue injury in cooperation with in-flammatory cell types (see below). They have a ring of microtubules around their periphery and an extensively invaginated membrane with an intricate canalicular system in contact with the ECF. Their membranes contain receptors for collagen, ADP, vessel wall von Willebrand factor (see below), and fibrinogen. Their cytoplasm contains actin, myosin, glycogen, lysosomes, and two types of granules: (1) dense granules, which contain the nonprotein substances that are secreted in response to platelet activation, including serotonin, ADP, and other adenine nucleotides; and (2) α-granules, which contain secreted proteins other than the hydrolases in lysosomes. These proteins include clotting factors and **platelet-derived growth factor (PDGF)**. PDGF is also produced by macrophages and endothelial cells. It is a dimer made up of A and B subunit polypeptides. Homodimers (AA and BB), as well as the heterodimer (AB), are produced. PDGF stimulates wound healing and is a potent mitogen for vascular smooth muscle. Blood vessel walls as well as platelets contain von Willebrand factor, which, in addition to its role in adhesion, regulates circulating levels of factor VIII (see below).

When a blood vessel wall is injured, platelets adhere to the exposed collagen and **von Willebrand factor** in the wall via receptors on the platelet membrane. Von Willebrand factor is a very large circulating molecule that is produced by endothelial cells. Binding produces platelet activations which release the contents of their granules. The released ADP acts on the ADP receptors in the platelet membranes to produce further accumulation of more platelets (**platelet aggregation**). Humans have at least three different types of platelet ADP receptors: $P2Y_1$, $P2Y_2$, and $P2X_1$. These are obviously attractive targets for drug development, and several new inhibitors have shown promise in the prevention of heart attacks and strokes. Aggregation is also fostered by **platelet-activating factor (PAF)**, a cytokine secreted by neutrophils and monocytes as well as platelets. This compound also has inflammatory activity. It is an ether phospholipid, 1-alkyl-2-acetylglyceryl-3-phosphorylcholine, which is produced from membrane lipids. It acts via a G protein-coupled receptor to

CLINICAL BOX 3–2

Autoimmunity

Sometimes the processes that eliminate antibodies against self antigens fail and a variety of different **autoimmune diseases** are produced. These can be B cell- or T cell-mediated and can be organ-specific or systemic. They include type 1 diabetes mellitus (antibodies against pancreatic islet B cells), myasthenia gravis (antibodies against nicotinic cholinergic receptors), and multiple sclerosis (antibodies against myelin basic protein and several other components of myelin). In some instances, the antibodies are against receptors and are capable of activating those receptors; for example, antibodies against TSH receptors increase thyroid activity and cause Graves' disease (see Chapter 20). Other conditions are due to the production of antibodies against invading organisms that cross-react with normal body constituents **(molecular mimicry).** An example is rheumatic fever following a streptococcal infection; a portion of cardiac myosin resembles a portion of the streptococcal M protein, and antibodies induced by the latter attack the former and damage the heart. Some conditions may be due to **bystander effects,** in which inflammation sensitizes T cells in the neighborhood, causing them to become activated when otherwise they would not respond. However, much is still uncertain about the pathogenesis of autoimmune disease.

CLINICAL BOX 3–3

Tissue Transplantation

The T lymphocyte system is responsible for the rejection of transplanted tissue. When tissues such as skin and kidneys are transplanted from a donor to a recipient of the same species, the transplants "take" and function for a while but then become necrotic and are "rejected" because the recipient develops an immune response to the transplanted tissue. This is generally true even if the donor and recipient are close relatives, and the only transplants that are never rejected are those from an identical twin. A number of treatments have been developed to overcome the rejection of transplanted organs in humans. The goal of treatment is to stop rejection without leaving the patient vulnerable to massive infections. One approach is to kill T lymphocytes by killing all rapidly dividing cells with drugs such as azathioprine, a purine antimetabolite, but this makes patients susceptible to infections and cancer. Another is to administer glucocorticoids, which inhibit cytotoxic T cell proliferation by inhibiting production of IL-2, but these cause osteoporosis, mental changes, and the other facets of Cushing syndrome (see Chapter 22). More recently, immunosuppressive drugs such as **cyclosporine** or **tacrolimus (FK-506)** have found favor. Activation of the T cell receptor normally increases intracellular Ca^{2+}, which acts via calmodulin to activate calcineurin (Figure 3-11). Calcineurin dephosphorylates the transcription factor NF-AT, which moves to the nucleus and increases the activity of genes coding for IL-2 and related stimulatory cytokines. Cyclosporine and tacrolimus prevent the dephosphorylation of NF-AT. However, these drugs inhibit all T cell-mediated immune responses, and cyclosporine causes kidney damage and cancer. A new and promising approach to transplant rejection is the production of T cell unresponsiveness by using drugs that block the costimulation that is required for normal activation (see text). Clinically effective drugs that act in this fashion could be of great value to transplant surgeons.

increase the production of arachidonic acid derivatives, including thromboxane A_2. The role of this compound in the balance between clotting and anticlotting activity at the site of vascular injury is discussed in Chapter 32.

Platelet production is regulated by the colony-stimulating factors that control the production of megakaryocytes, plus **thrombopoietin,** a circulating protein factor. This factor, which facilitates megakaryocyte maturation, is produced constitutively by the liver and kidneys, and there are thrombopoietin receptors on platelets. Consequently, when the number of platelets is low, less is bound and more is available to stimulate production of platelets. Conversely, when the number of platelets is high, more is bound and less is available, producing a form of feedback control of platelet production. The amino terminal portion of the thrombopoietin molecule has the platelet-stimulating activity, whereas the carboxyl terminal portion contains many carbohydrate residues and is concerned with the bioavailability of the molecule.

When the platelet count is low, clot retraction is deficient and there is poor constriction of ruptured vessels. The resulting clinical syndrome **(thrombocytopenic purpura)** is characterized by easy bruisability and multiple subcutaneous hemorrhages. Purpura may also occur when the platelet count is normal, and in some of these cases, the circulating platelets are abnormal **(thrombasthenic purpura).** Individuals with thrombocytosis are predisposed to thrombotic events.

INFLAMMATION & WOUND HEALING

LOCAL INJURY

Inflammation is a complex localized response to foreign substances such as bacteria or in some instances to internally produced substances. It includes a sequence of reactions initially involving cytokines, neutrophils, adhesion molecules, complement, and IgG. PAF, an agent with potent inflammatory effects, also plays a role. Later, monocytes and lymphocytes are involved. Arterioles in the inflamed area dilate, and capillary permeability is increased (see Chapters 33 and 34). When the inflammation occurs in or just under the skin (Figure 3–13), it

FIGURE 3–11   **Action of cyclosporine (CsA) and tacrolimus (TCL) in lymphocytes.** BP, binding protein; CAM, calmodulin.

is characterized by redness, swelling, tenderness, and pain. Elsewhere, it is a key component of asthma, ulcerative colitis, and many other diseases.

Evidence is accumulating that a transcription factor, **nuclear factor-κB,** plays a key role in the inflammatory response. NF-κB is a heterodimer that normally exists in the cytoplasm of cells bound to $I_\kappa B_\alpha$, which renders it inactive. Stimuli such as cytokines, viruses, and oxidants separate NF-κB from $I_\kappa B_\alpha$, which is then degraded. NF-κB moves to the nucleus, where it binds to the DNA of the genes for numerous inflammatory mediators, resulting in their increased production and secretion. Glucocorticoids inhibit the activation of NF-κB by increasing the production of $I_\kappa B_\alpha$, and this is probably the main basis of their anti-inflammatory action (see Chapter 22).

SYSTEMIC RESPONSE TO INJURY

Cytokines produced in response to inflammation and other injuries also produce systemic responses. These include alterations in plasma **acute phase proteins,** defined as proteins whose concentration is increased or decreased by at least 25% following injury. Many of the proteins are of hepatic origin. A number of them are shown in Figure 3–14. The causes of the changes in concentration are incompletely understood, but it can be said that many of the changes make homeostatic sense. Thus, for example, an increase in C-reactive protein activates monocytes and causes further production of cytokines. Other changes that occur in response to injury include somnolence, negative nitrogen balance, and fever.

WOUND HEALING

When tissue is damaged, platelets adhere to exposed matrix via integrins that bind to collagen and laminin (Figure 3–13). Blood coagulation produces thrombin, which promotes platelet aggregation and granule release. The platelet granules generate an inflammatory response. White blood cells are attracted by selectins and bind to integrins on endothelial cells, leading to their extravasation through the blood vessel walls. Cytokines released by the white blood cells and platelets up-regulate integrins on macrophages, which migrate to the area of injury, and on fibroblasts and epithelial cells, which mediate wound healing

FIGURE 3–12   **Sites of congenital blockade of B and T lymphocyte maturation in various immunodeficiency states.** SCID, severe combined immune deficiency. (Modified from Rosen FS, Cooper MD, Wedgwood RJP: The primary immunodeficiencies. *N Engl J Med* 1995;333:431.)

FIGURE 3–13 **Cutaneous wound 3 days after injury, showing the multiple cytokines and growth factors affecting the repair process.** VEGF, vascular endothelial growth factor. For other abbreviations, see Appendix. Note the epidermis growing down under the fibrin clot, restoring skin continuity. (Modified from Singer AJ, Clark RAF: Cutaneous wound healing. *N Engl J Med* 1999;341:738.)

and scar formation. Plasmin aids healing by removing excess fibrin. This aids the migration of keratinocytes into the wound to restore the epithelium under the scab. Collagen proliferates, producing the scar. Wounds gain 20% of their ultimate strength in 3 weeks and later gain more strength, but they never reach more than about 70% of the strength of normal skin.

FIGURE 3–14 **Time course of changes in some major acute phase proteins.** C3, C3 component of complement. (Modified and reproduced with permission from Gitlin JD, Colten HR: Molecular biology of acute phase plasma proteins. In Pick F, et al [editors]: *Lymphokines*, vol 14, pages 123–153. Academic Press, 1987.)

CHAPTER SUMMARY

- Immune and inflammatory responses are mediated by several different cell types—granulocytes, lymphocytes, monocytes, mast cells, tissue macrophages, and antigen presenting cells—that arise predominantly from the bone marrow and may circulate or reside in connective tissues.

- Granulocytes mount phagocytic responses that engulf and destroy bacteria. These are accompanied by the release of reactive oxygen species and other mediators into adjacent tissues that may cause tissue injury.

- Mast cells and basophils underpin allergic reactions to substances that would be treated as innocuous by nonallergic individuals.

- A variety of soluble mediators orchestrate the development of immunologic effector cells and their subsequent immune and inflammatory reactions.

- Innate immunity represents an evolutionarily conserved, primitive response to stereotypical microbial components.

- Acquired immunity is slower to develop than innate immunity, but long-lasting and more effective.

- Genetic rearrangements endow B and T lymphocytes with a vast array of receptors capable of recognizing billions of foreign antigens.

- Self-reactive lymphocytes are normally deleted; a failure of this process leads to autoimmune disease. Disease can also result from abnormal function or development of granulocytes and lymphocytes. In these latter cases, deficient immune responses to microbial threats usually result.

MULTIPLE-CHOICE QUESTIONS

For all questions, select the single best answer unless otherwise directed.

1. In normal human blood
   A) the eosinophil is the most common type of white blood cell.
   B) there are more lymphocytes than neutrophils.
   C) the iron is mostly in hemoglobin.
   D) there are more white cells than red cells.
   E) there are more platelets than red cells.

2. Lymphocytes
   A) all originate from the bone marrow after birth.
   B) are unaffected by hormones.
   C) convert to monocytes in response to antigens.
   D) interact with eosinophils to produce platelets.
   E) are part of the body's defense against cancer.

3. The ability of the blood to phagocytose pathogens and mount a respiratory burst is increased by
   A) interleukin-2 (IL-2).
   B) granulocyte colony-stimulating factor (G-CSF).
   C) erythropoietin.
   D) interleukin-4 (IL-4).
   E) interleukin-5 (IL-5).

4. Cells responsible for innate immunity are activated most commonly by
   A) glucocorticoids.
   B) pollen.
   C) carbohydrate sequences in bacterial cell walls.
   D) eosinophils.
   E) cytoplasmic proteins of bacteria.

CHAPTER RESOURCES

Delibro G: The Robin Hood of antigen presentation. Science 2004;302:485.

Delves PJ, Roitt IM: The immune system. (Two parts.) N Engl J Med 2000;343:37,108.

Dhainaut J-K, Thijs LG, Park G (editors): *Septic Shock*. WB Saunders, 2000.

Ganz T: Defensins and host defense. Science 1999;286:420.

Samstein B, Emond JC: Liver transplant from living related donors. Annu Rev Med 2001;52:147.

Singer AJ, Clark RAF: Cutaneous wound healing. N Engl J Med 1999;341:738

Tedder TF, et al: The selectins: Vascular adhesion molecules. FASEB J 1995;9:866.

Tilney NL: *Transplant: From Myth to Reality*. Yale University Press, 2003.

Walport MJ: Complement. (Two parts) N Engl J Med 2001;344:1058, 1140.

C H A P T E R

4

Excitable Tissue: Nerve

INTRODUCTION

The human central nervous system (CNS) contains about 10^{11} (100 billion) **neurons.** It also contains 10–50 times this number of **glial cells.** The CNS is a complex organ; it has been calculated that 40% of the human genes participate, at least to a degree, in its formation. The neurons, the basic building blocks of the nervous system, have evolved from primitive neuroeffector cells that respond to various stimuli by contracting. In more complex animals, contraction has become the specialized function of muscle cells, whereas integration and transmission of nerve impulses have become the specialized functions of neurons. This chapter describes the cellular components of the CNS and the **excitability** of neurons, which involves the genesis of electrical signals that enable neurons to integrate and transmit impulses (action potentials, receptor potentials, and synaptic potentials).

CELLULAR ELEMENTS IN THE CNS

GLIAL CELLS

For many years following their discovery, glial cells (or glia) were viewed as CNS connective tissue. In fact, the word *glia* is Greek for *glue*. However, today theses cells are recognized for their role in communication within the CNS in partnership with neurons. Unlike neurons, glial cells continue to undergo cell division in adulthood and their ability to proliferate is particularly noticeable after brain injury (eg, stroke).

There are two major types of glial cells in the vertebrate nervous system: microglia and macroglia. **Microglia** are scavenger cells that resemble tissue macrophages and remove debris resulting from injury, infection, and disease (eg, multiple sclerosis, AIDS-related dementia, Parkinson disease, and Alzheimer disease). Microglia arise from macrophages outside of the nervous system and are physiologically and embryologically unrelated to other neural cell types.

There are three types of macroglia: oligodendrocytes, Schwann cells, and astrocytes (Figure 4–1). **Oligodendrocytes** and **Schwann cells** are involved in myelin formation around axons in the CNS and peripheral nervous system, respectively. **Astrocytes,** which are found throughout the brain, are of two subtypes. **Fibrous astrocytes,** which contain many intermediate filaments, are found primarily in white matter. **Protoplasmic astrocytes** are found in gray matter and have a granular cytoplasm. Both types send processes to blood vessels, where they induce capillaries to form the tight junctions making up the **blood–brain barrier.** They also send processes that envelop synapses and the surface of nerve cells. Protoplasmic astrocytes have a membrane potential that varies with the external K^+ concentration but do not generate propagated potentials. They produce substances that are tropic to neurons, and they help maintain the appropriate concentration of ions and neurotransmitters by taking up K^+ and the neurotransmitters glutamate and γ-aminobutyrate (GABA).

NEURONS

Neurons in the mammalian central nervous system come in many different shapes and sizes. Most have the same parts as the typical spinal motor neuron illustrated in Figure 4–2. The cell body **(soma)** contains the nucleus and is the metabolic center of the neuron. Neurons have several processes called **dendrites** that extend outward from the cell body and arborize extensively. Particularly in the cerebral and cerebellar cortex, the dendrites have small knobby projections called **dendritic spines.** A typical neuron also has a long fibrous **axon** that originates from a somewhat thickened area of the cell body, the **axon hillock.** The first portion of the axon is called the **initial segment.** The axon divides into **presynaptic terminals,** each ending in a number of **synaptic knobs** which are also called **terminal buttons** or **boutons.** They contain granules or vesicles in which the synaptic transmitters secreted by the nerves are stored. Based on the number of processes that emanate from their cell body, neurons can be classified as unipolar, bipolar, and multipolar (Figure 4–3).

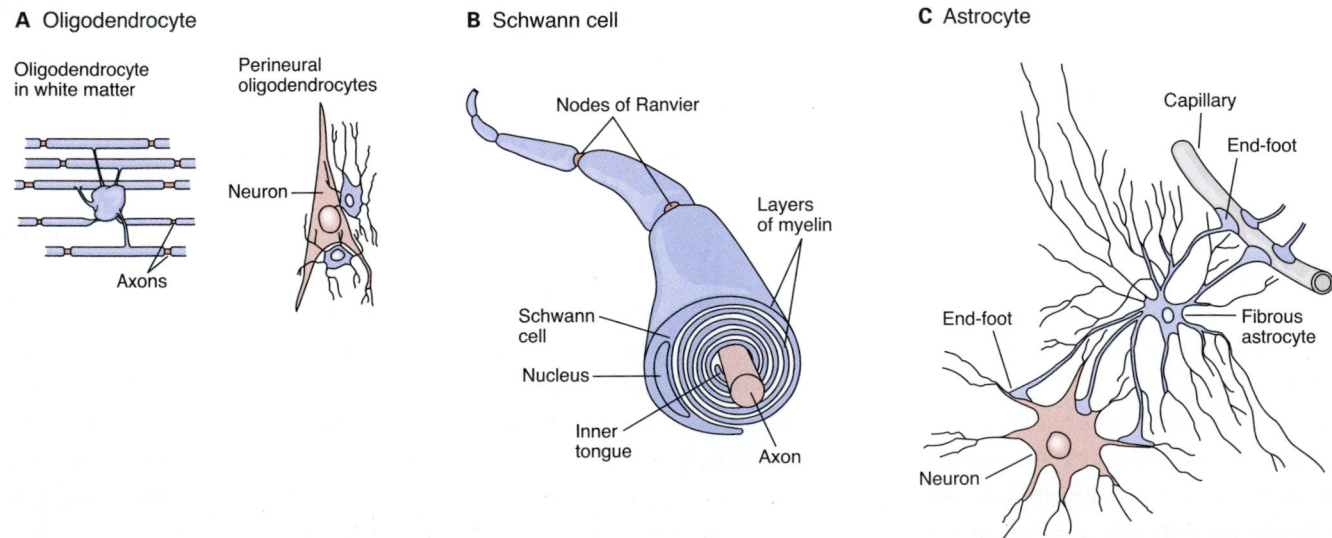

A Oligodendrocyte   **B** Schwann cell   **C** Astrocyte

FIGURE 4–1   **The principal types of glial cells in the nervous system. A)** Oligodendrocytes are small with relatively few processes. Those in the white matter provide myelin, and those in the gray matter support neurons. **B)** Schwann cells provide myelin to the peripheral nervous system. Each cell forms a segment of myelin sheath about 1 mm long; the sheath assumes its form as the inner tongue of the Schwann cell turns around the axon several times, wrapping in concentric layers. Intervals between segments of myelin are the nodes of Ranvier. **C)** Astrocytes are the most common glia in the CNS and are characterized by their starlike shape. They contact both capillaries and neurons and are thought to have a nutritive function. They are also involved in forming the blood–brain barrier. (From Kandel ER, Schwartz JH, Jessell TM [editors]: *Principles of Neural Science,* 4th ed. McGraw-Hill, 2000.)

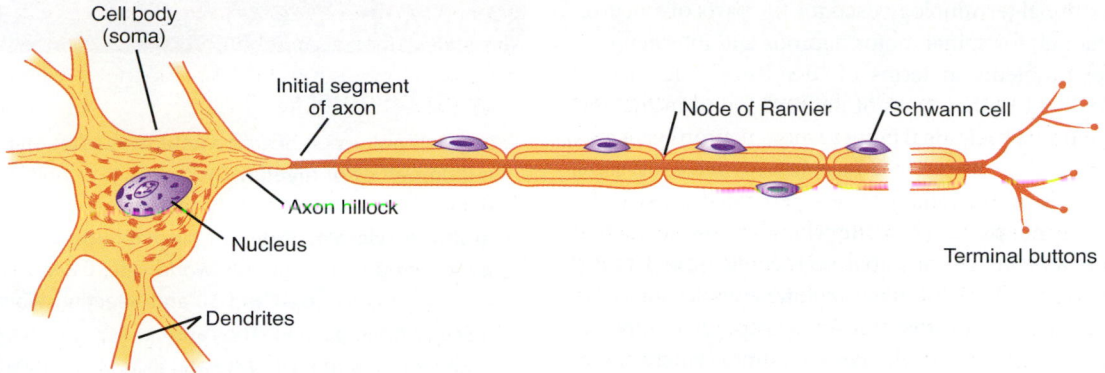

FIGURE 4–2 **Motor neuron with a myelinated axon.** A motor neuron is comprised of a cell body (soma) with a nucleus, several processes called dendrites, and a long fibrous axon that originates from the axon hillock. The first portion of the axon is called the initial segment. A myelin sheath forms from Schwann cells and surrounds the axon except at its ending and at the nodes of Ranvier. Terminal buttons (boutons) are located at the terminal endings.

FIGURE 4–3 **Some of the types of neurons in the mammalian nervous system. A)** Unipolar neurons have one process, with different segments serving as receptive surfaces and releasing terminals. **B)** Bipolar neurons have two specialized processes: a dendrite that carries information to the cell and an axon that transmits information from the cell. **C)** Some sensory neurons are in a subclass of bipolar cells called pseudo-unipolar cells. As the cell develops, a single process splits into two, both of which function as axons—one going to skin or muscle and another to the spinal cord. **D)** Multipolar cells have one axon and many dendrites. Examples include motor neurons, hippocampal pyramidal cells with dendrites in the apex and base, and cerebellar Purkinje cells with an extensive dendritic tree in a single plane. (From Kandel ER, Schwartz JH, Jessell TM [editors]: *Principles of Neural Science*, 4th ed. McGraw-Hill, 2000.)

The conventional terminology used for the parts of a neuron works well enough for spinal motor neurons and interneurons, but there are problems in terms of "dendrites" and "axons" when it is applied to other types of neurons found in the nervous system. From a functional point of view, neurons generally have four important zones: (1) a receptor, or dendritic zone, where multiple local potential changes generated by synaptic connections are integrated; (2) a site where propagated action potentials are generated (the initial segment in spinal motor neurons, the initial node of Ranvier in cutaneous sensory neurons); (3) an axonal process that transmits propagated impulses to the nerve endings; and (4) the nerve endings, where action potentials cause the release of synaptic transmitters. The cell body is often located at the dendritic zone end of the axon, but it can be within the axon (eg, auditory neurons) or attached to the side of the axon (eg, cutaneous neurons). Its location makes no difference as far as the receptor function of the dendritic zone and the transmission function of the axon are concerned.

The axons of many neurons are myelinated, that is, they acquire a sheath of **myelin,** a protein–lipid complex that is wrapped around the axon (Figure 4–2). In the peripheral nervous system, myelin forms when a Schwann cell wraps its membrane around an axon up to 100 times (Figure 4–1). The myelin is then compacted when the extracellular portions of a membrane protein called protein zero (P_0) lock to the extracellular portions of P_0 in the apposing membrane. Various mutations in the gene for P_0 cause peripheral neuropathies; 29 different mutations have been described that cause symptoms ranging from mild to severe. The myelin sheath envelops the axon except at its ending and at the **nodes of Ranvier,** periodic 1-μm constrictions that are about 1 mm apart (Figure 4–2). The insulating function of myelin is discussed later in this chapter. Not all neurons are myelinated; some are **unmyelinated,** that is, simply surrounded by Schwann cells without the wrapping of the Schwann cell membrane that produces myelin around the axon.

In the CNS of mammals, most neurons are myelinated, but the cells that form the myelin are oligodendrocytes rather than Schwann cells (Figure 4–1). Unlike the Schwann cell, which forms the myelin between two nodes of Ranvier on a single neuron, oligodendrocytes emit multiple processes that form myelin on many neighboring axons. In multiple sclerosis, a crippling autoimmune disease, patchy destruction of myelin occurs in the CNS (see Clinical Box 4–1). The loss of myelin is associated with delayed or blocked conduction in the demyelinated axons.

AXONAL TRANSPORT

Neurons are secretory cells, but they differ from other secretory cells in that the secretory zone is generally at the end of the axon, far removed from the cell body. The apparatus for protein synthesis is located for the most part in the cell body, with transport of proteins and polypeptides to the axonal ending by **axoplasmic flow.** Thus, the cell body maintains the functional and an-

atomic integrity of the axon; if the axon is cut, the part distal to the cut degenerates (**wallerian degeneration**). **Orthograde transport** occurs along microtubules that run along the length of the axon and requires two molecular motors, dynein and kinesin (Figure 4–4). Orthograde transport moves from the cell body toward the axon terminals. It has both fast and slow components; **fast axonal transport** occurs at about 400 mm/day, and **slow axonal transport** occurs at 0.5 to 10 mm/day. **Retrograde transport,** which is in the opposite direction (from the

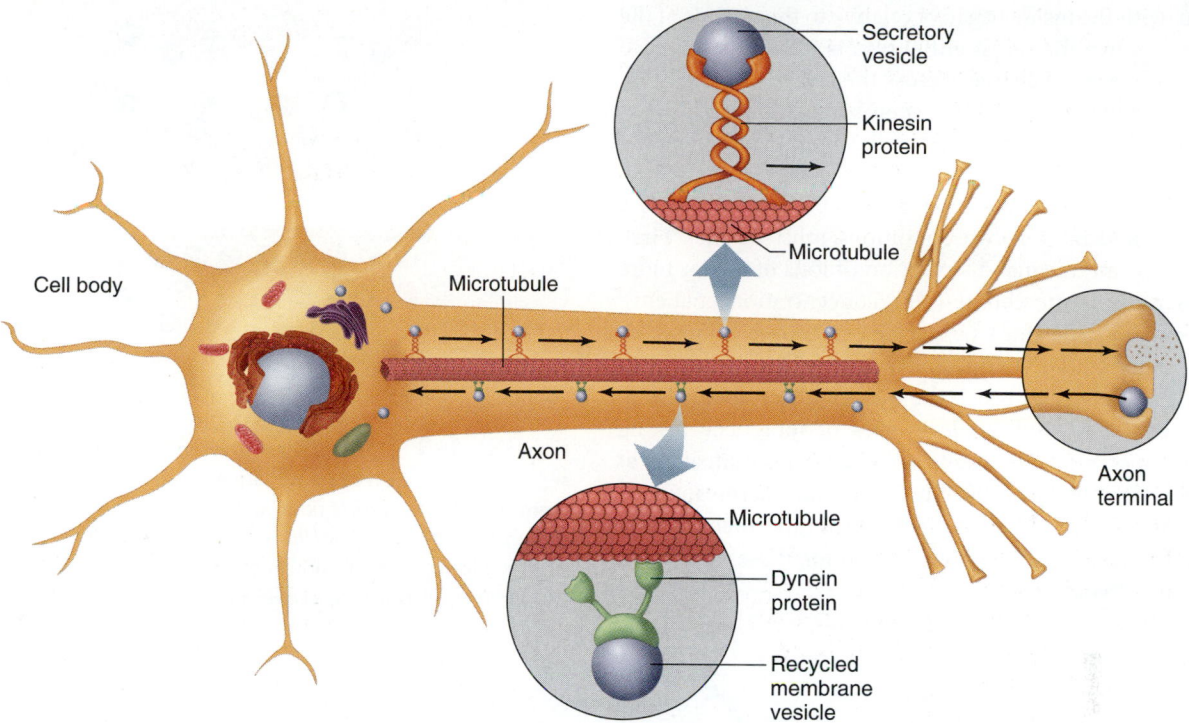

FIGURE 4–4 **Axonal transport along microtubules by dynein and kinesin.** Fast and slow axonal orthograde transport occurs along microtubules that run along the length of the axon from the cell body to the terminal. Retrograde transport occurs from the terminal to the cell body. (From Widmaier EP, Raff H, Strang KT: *Vander's Human Physiology.* McGraw-Hill, 2008.)

nerve ending to the cell body), occurs along microtubules at about 200 mm/day. Synaptic vesicles recycle in the membrane, but some used vesicles are carried back to the cell body and deposited in lysosomes. Some materials taken up at the ending by endocytosis, including **nerve growth factor (NGF)** and various viruses, are also transported back to the cell body. A potentially important exception to these principles seems to occur in some dendrites. In them, single strands of mRNA transported from the cell body make contact with appropriate ribosomes, and protein synthesis appears to create local protein domains.

EXCITATION & CONDUCTION

Nerve cells have a low threshold for excitation. The stimulus may be electrical, chemical, or mechanical. Two types of physicochemical disturbances are produced: local, nonpropagated potentials called, depending on their location, **synaptic, generator,** or **electrotonic potentials;** and propagated potentials, the **action potentials** (or **nerve impulses**). These are the only electrical responses of neurons and other excitable tissues, and they are the main language of the nervous system. They are due to changes in the conduction of ions across the cell membrane that are produced by alterations in ion channels. The electrical events in neurons are rapid, being measured in **milliseconds (ms);** and the potential changes are small, being measured in **millivolts (mV).**

The impulse is normally transmitted **(conducted)** along the axon to its termination. Nerves are not "telephone wires" that transmit impulses passively; conduction of nerve impulses, although rapid, is much slower than that of electricity. Nerve tissue is in fact a relatively poor passive conductor, and it would take a potential of many volts to produce a signal of a fraction of a volt at the other end of a meter-long axon in the absence of active processes in the nerve. Conduction is an active, self-propagating process, and the impulse moves along the nerve at a constant amplitude and velocity. The process is often compared to what happens when a match is applied to one end of a trail of gunpowder; by igniting the powder particles immediately in front of it, the flame moves steadily down the trail to its end as it is extinguished in its progression.

Mammalian neurons are relatively small, but giant unmyelinated nerve cells exist in a number of invertebrate species. Such cells are found, for example, in crabs (*Carcinus*), cuttlefish (*Sepia*), and squid (*Loligo*). The fundamental properties of neurons were first determined in these species and then found to be similar in mammals. The neck region of the muscular mantle of the squid contains single axons up to 1 mm in diameter. The fundamental properties of these long axons are similar to those of mammalian axons.

RESTING MEMBRANE POTENTIAL

When two electrodes are connected through a suitable amplifier and placed on the surface of a single axon, no potential difference is observed. However, if one electrode is inserted into the interior of the cell, a constant **potential difference** is observed, with the inside negative relative to the outside of the cell at rest. A membrane potential results from separation of positive and negative charges across the cell membrane (Figure 4–5). In neurons, the **resting membrane potential** is usually about –70 mV, which is close to the equilibrium potential for K^+ (Figure 4–6).

In order for a potential difference to be present across a membrane lipid bilayer, two conditions must be met. First, there must be an unequal distribution of ions of one or more species across the membrane (ie, a concentration gradient). Two, the membrane must be permeable to one or more of these ion species. The permeability is provided by the existence of channels or pores in the bilayer; these channels are usually permeable to a single species of ions. The resting membrane potential represents an equilibrium situation at which the driving force for the membrane-permeant ions down their concentration gradients across the membrane is equal and opposite to the driving force for these ions down their electrical gradients.

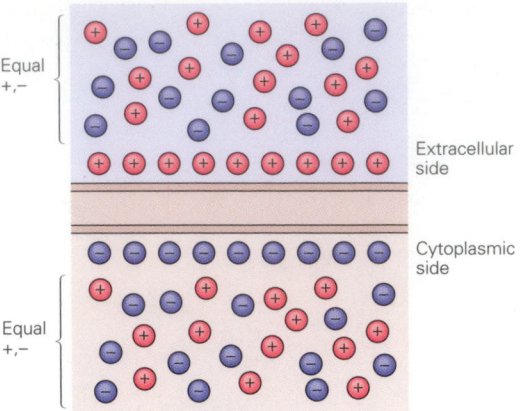

FIGURE 4–5   This membrane potential results from separation of positive and negative charges across the cell membrane. The excess of positive charges (red circles) outside the cell and negative charges (blue circles) inside the cell at rest represents a small fraction of the total number of ions present. (From Kandel ER, Schwartz JH, Jessell TM [editors]: *Principles of Neural Science*, 4th ed. McGraw-Hill, 2000.).

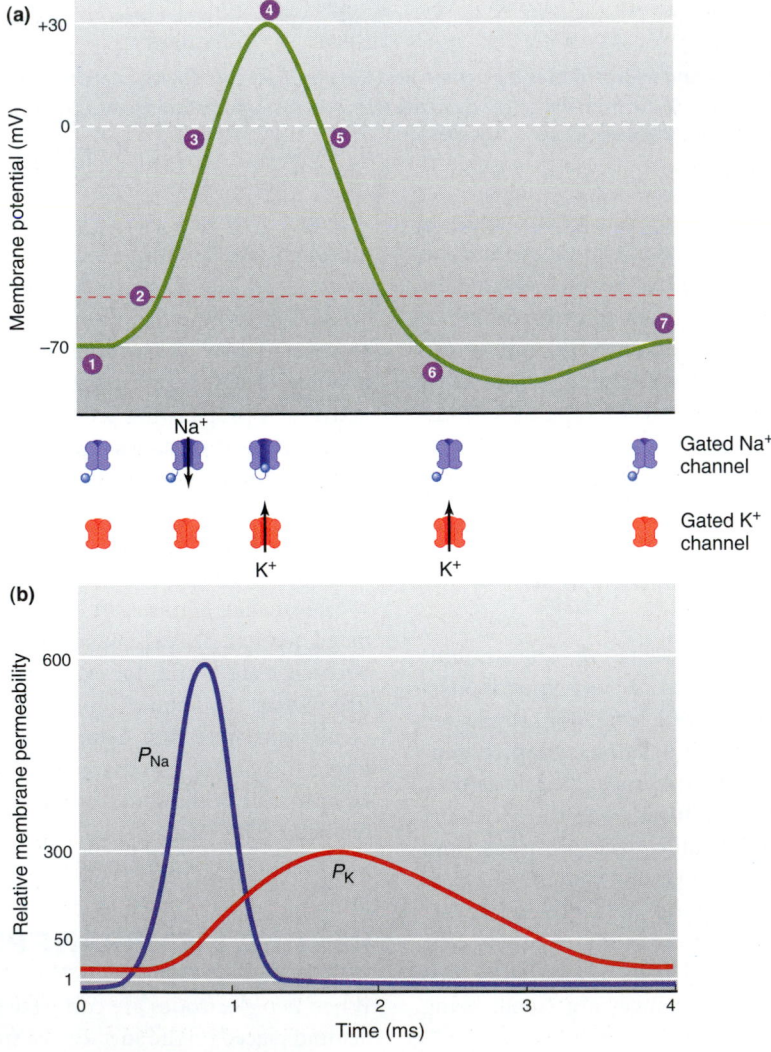

FIGURE 4–6   The changes in **(a)** membrane potential (mV) and **(b)** relative membrane permeability (P) to Na+ and K+ during an action potential. (From Widmaier EP, Raff H, Strang KT: *Vander's Human Physiology*. McGraw-Hill, 2008.)

In neurons, the concentration of K^+ is much higher inside than outside the cell, while the reverse is the case for Na^+. This concentration difference is established by the Na^+-K^+ ATPase. The outward K^+ concentration gradient results in passive movement of K^+ out of the cell when K^+-selective channels are open. Similarly, the inward Na^+ concentration gradient results in passive movement of Na^+ into the cell when Na^+-selective channels are open. Because there are more open K^+ channels than Na^+ channels at rest, the membrane permeability to K^+ is greater. Consequently, the intracellular and extracellular K^+ concentrations are the prime determinants of the resting membrane potential, which is therefore close to the equilibrium potential for K^+. Steady ion leaks cannot continue forever without eventually dissipating the ion gradients. This is prevented by the Na^+-K^+ ATPase, which actively moves Na^+ and K^+ against their electrochemical gradient.

IONIC FLUXES DURING THE ACTION POTENTIAL

The cell membranes of nerves, like those of other cells, contain many different types of ion channels. Some of these are voltage-gated and others are ligand-gated. It is the behavior of these channels, and particularly Na^+ and K^+ channels, which explains the electrical events in nerves.

The changes in membrane conductance of Na^+ and K^+ that occur during the action potentials are shown in Figure 4–6. The conductance of an ion is the reciprocal of its electrical resistance in the membrane and is a measure of the membrane permeability to that ion. In response to a depolarizing stimulus, some of the voltage-gated Na^+ channels become active, and when the **threshold potential** is reached, the voltage-gated Na^+ channels overwhelm the K^+ and other channels and an action potential results (a **positive feedback loop**). The membrane potential moves toward the equilibrium potential for Na^+ (+60 mV) but does not reach it during the action potential, primarily because the increase in Na^+ conductance is short-lived. The Na^+ channels rapidly enter a closed state called the **inactivated state** and remain in this state for a few milliseconds before returning to the resting state, when they again can be activated. In addition, the direction of the electrical gradient for Na^+ is reversed during the **overshoot** because the membrane potential is reversed, and this limits Na^+ influx. A third factor producing **repolarization** is the opening of voltage-gated K^+ channels. This opening is slower and more prolonged than the opening of the Na^+ channels, and consequently, much of the increase in K^+ conductance comes after the increase in Na^+ conductance. The net movement of positive charge out of the cell due to K^+ efflux at this time helps complete the process of repolarization. The slow return of the K^+ channels to the closed state also explains the **after-hyperpolarization,** followed by a return to the resting membrane potential. Thus, voltage-gated K^+ channels bring the action potential to an end and cause closure of their gates through a **negative feedback process.**

Figure 4–7 shows the sequential feedback control in voltage-gated K^+ and Na^+ channels during the action potential.

Decreasing the external Na^+ concentration reduces the size of the action potential but has little effect on the resting membrane potential. The lack of much effect on the resting membrane potential would be predicted, since the permeability of the membrane to Na^+ at rest is relatively low. Conversely, increasing the external K^+ concentration decreases the resting membrane potential.

Although Na^+ enters the nerve cell and K^+ leaves it during the action potential, the number of ions involved is minute relative to the total numbers present. The fact that the nerve gains Na^+ and loses K^+ during activity has been demonstrated experimentally, but significant differences in ion concentrations can be measured only after prolonged, repeated stimulation.

Other ions, notably Ca^{2+}, can affect the membrane potential through both channel movement and membrane interactions. A decrease in extracellular Ca^{2+} concentration increases the excitability of nerve and muscle cells by decreasing the amount of depolarization necessary to initiate the changes in the Na^+ and K^+ conductance that produce the action potential. Conversely, an increase in extracellular Ca^{2+} concentration can stabilize the membrane by decreasing excitability.

DISTRIBUTION OF ION CHANNELS IN MYELINATED NEURONS

The spatial distribution of ion channels along the axon plays a key role in the initiation and regulation of the action potential. Voltage-gated Na^+ channels are highly concentrated in the nodes of Ranvier and the initial segment in myelinated neurons. The initial segment and, in sensory neurons, the first node of Ranvier are the sites where impulses are normally generated, and the other nodes of Ranvier are the sites to which the impulses jump during saltatory conduction. The number of Na^+ channels per square micrometer of membrane in myelinated mammalian neurons has been estimated to be 50–75 in the cell body, 350–500 in the initial segment, less than 25 on the surface of the myelin, 2000–12,000 at the nodes of Ranvier, and 20–75 at the axon terminals. Along the axons of unmyelinated neurons, the number is about 110. In many myelinated neurons, the Na^+ channels are flanked by K^+ channels that are involved in repolarization.

"ALL-OR-NONE" LAW

It is possible to determine the minimal intensity of stimulating current (**threshold intensity**) that, acting for a given duration, will just produce an action potential. The threshold intensity varies with the duration; with weak stimuli it is long, and with strong stimuli it is short. The relation between the strength and the duration of a threshold stimulus is called the **strength–duration curve.** Slowly rising currents fail to fire the nerve because the nerve adapts to the applied stimulus, a process called **adaptation.**

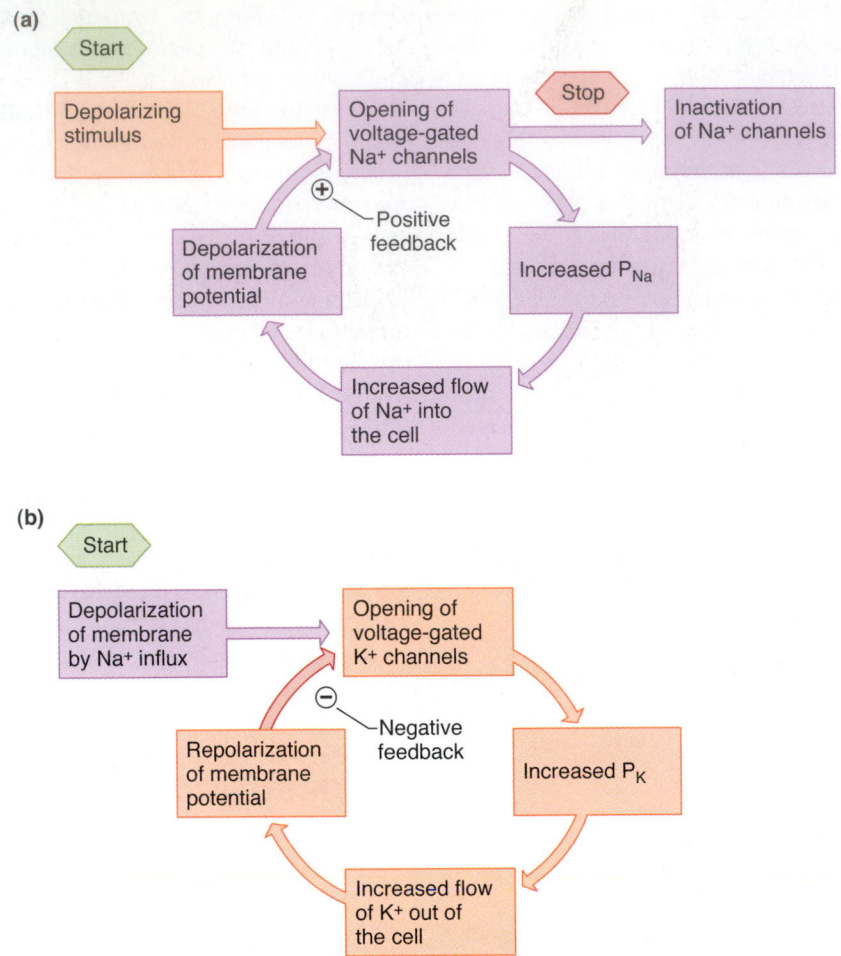

FIGURE 4–7   Feedback control in voltage-gated ion channels in the membrane. (a) Na⁺ channels exert positive feedback. **(b)** K⁺ channels exert negative feedback.   (From Widmaier EP, Raff H, Strang KT: *Vander's Human Physiology*. McGraw-Hill, 2008.)

Once threshold intensity is reached, a full-fledged action potential is produced. Further increases in the intensity of a stimulus produce no increment or other change in the action potential as long as the other experimental conditions remain constant. The action potential fails to occur if the stimulus is subthreshold in magnitude, and it occurs with constant amplitude and form regardless of the strength of the stimulus if the stimulus is at or above threshold intensity. The action potential is therefore "all or none" in character and is said to obey the **all-or-none law.**

ELECTROTONIC POTENTIALS, LOCAL RESPONSE, & FIRING LEVEL

Although subthreshold stimuli do not produce an action potential, they do have an effect on the membrane potential. This can be demonstrated by placing recording electrodes within a few millimeters of a stimulating electrode and applying subthreshold stimuli of fixed duration. Application of such currents leads to a localized depolarizing potential

change that rises sharply and decays exponentially with time. The magnitude of this response drops off rapidly as the distance between the stimulating and recording electrodes is increased. Conversely, an anodal current produces a hyperpolarizing potential change of similar duration. These potential changes are called **electrotonic potentials.** As the strength of the current is increased, the response is greater due to the increasing addition of a **local response** of the membrane (Figure 4–8). Finally, at 7–15 mV of depolarization (potential of –55 mV), the **firing level** is reached and an action potential occurs.

CHANGES IN EXCITABILITY DURING ELECTROTONIC POTENTIALS & THE ACTION POTENTIAL

During the action potential, as well as during electrotonic potentials and the local response, the threshold of the neuron to stimulation changes. Hyperpolarizing responses elevate the threshold, and depolarizing potentials lower it as they move

FIGURE 4–8 **Electrotonic potentials and local response.** The changes in the membrane potential of a neuron following application of stimuli of 0.2, 0.4, 0.6, 0.8, and 1.0 times threshold intensity are shown superimposed on the same time scale. The responses below the horizontal line are those recorded near the anode, and the responses above the line are those recorded near the cathode. The stimulus of threshold intensity was repeated twice. Once it caused a propagated action potential (top line), and once it did not.

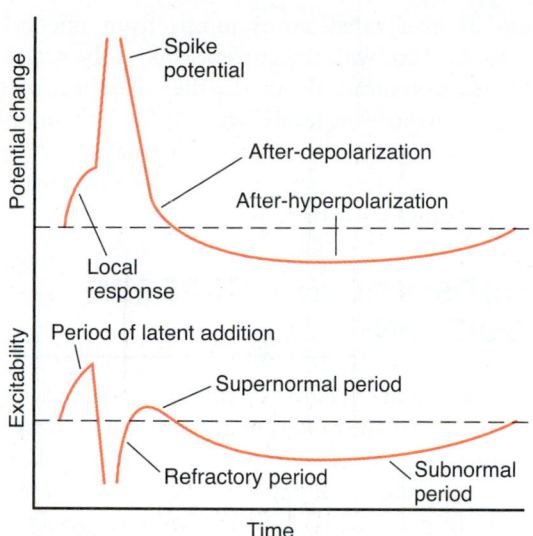

FIGURE 4–9 **Relative changes in excitability of a nerve cell membrane during the passage of an impulse.** Note that excitability is the reciprocal of threshold. (Modified from Morgan CT: *Physiological Psychology.* McGraw-Hill, 1943.)

the membrane potential closer to the firing level. During the local response, the threshold is lowered, but during the rising and much of the falling phases of the spike potential, the neuron is refractory to stimulation. This **refractory period** is divided into an **absolute refractory period,** corresponding to the period from the time the firing level is reached until repolarization is about one-third complete, and a **relative refractory period,** lasting from this point to the start of after-depolarization. During the absolute refractory period, no stimulus, no matter how strong, will excite the nerve, but during the relative refractory period, stronger than normal stimuli can cause excitation. During after-depolarization, the threshold is again decreased, and during after-hyperpolarization, it is increased. These changes in threshold are correlated with the phases of the action potential in Figure 4–9.

ELECTROGENESIS OF THE ACTION POTENTIAL

The nerve cell membrane is polarized at rest, with positive charges lined up along the outside of the membrane and negative charges along the inside. During the action potential, this polarity is abolished and for a brief period is actually reversed (Figure 4–10). Positive charges from the membrane ahead of and behind the action potential flow into the area of negativity represented by the action potential ("current sink"). By drawing off positive charges, this flow decreases the polarity of the membrane ahead of the action potential. Such electrotonic depolarization initiates a local response, and when the firing level is reached, a propagated response occurs that in turn electrotonically depolarizes the membrane in front of it.

SALTATORY CONDUCTION

Conduction in myelinated axons depends on a similar pattern of circular current flow. However, myelin is an effective insulator, and current flow through it is negligible. Instead, depo-

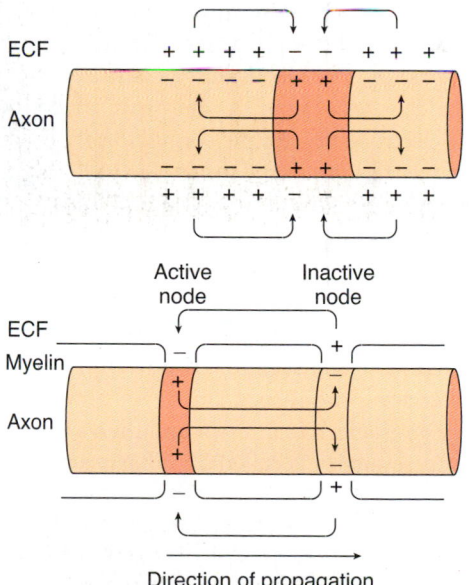

FIGURE 4–10 **Local current flow (movement of positive charges) around an impulse in an axon. Top:** Unmyelinated axon. **Bottom:** Myelinated axon. Positive charges from the membrane ahead of and behind the action potential flow into the area of negativity represented by the action potential ("current sink"). In myelinated axons, depolarization jumps from one node of Ranvier to the next (salutatory conduction).

larization in myelinated axons jumps from one node of Ranvier to the next, with the current sink at the active node serving to electrotonically depolarize the node ahead of the action potential to the firing level (Figure 4–10). This jumping of depolarization from node to node is called **saltatory conduction.** It is a rapid process that allows myelinated axons to conduct up to 50 times faster than the fastest unmyelinated fibers.

ORTHODROMIC & ANTIDROMIC CONDUCTION

An axon can conduct in either direction. When an action potential is initiated in the middle of it, two impulses traveling in opposite directions are set up by electrotonic depolarization on either side of the initial current sink. In the natural situation, impulses pass in one direction only, ie, from synaptic junctions or receptors along axons to their termination. Such conduction is called **orthodromic.** Conduction in the opposite direction is called **antidromic.** Because synapses, unlike axons, permit conduction in one direction only, an antidromic impulse will fail to pass the first synapse they encounter and die out at that point.

BIPHASIC ACTION POTENTIALS

The descriptions of the resting membrane potential and action potential outlined above are based on recording with two electrodes, one in the extracellular space and the other inside it. If both recording electrodes are placed on the surface of the axon, there is no potential difference between them at rest. When the nerve is stimulated and an impulse is conducted past the two electrodes, a characteristic sequence of potential changes results. As the wave of depolarization reaches the electrode nearest the stimulator, this electrode becomes negative relative to the other electrode (Figure 4–11). When the impulse passes to the portion of the nerve between the two electrodes, the potential returns to zero, and then, as it passes the second electrode, the first electrode becomes positive relative to the second. It is conventional to connect the leads in such a way that when the first electrode becomes negative relative to the second, an upward deflection is recorded. Therefore, the record shows an upward deflection followed by an isoelectric interval and then a downward deflection. This sequence is called a **biphasic action potential** (Figure 4–11).

PROPERTIES OF MIXED NERVES

Peripheral nerves in mammals are made up of many axons bound together in a fibrous envelope called the **epineurium.** Potential changes recorded extracellularly from such nerves therefore represent an algebraic summation of the all-or-none action potentials of many axons. The thresholds of the individual axons in the nerve and their distance from the stimulat-

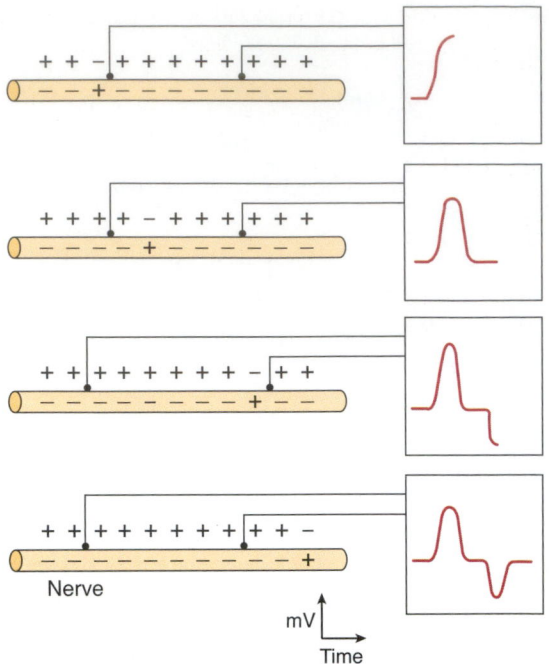

FIGURE 4–11   Biphasic action potential. Both recording electrodes are on the outside of the nerve membrane. It is conventional to connect the leads in such a way that when the first electrode becomes negative relative to the second, an upward deflection is recorded. Therefore, the record shows an upward deflection followed by an isoelectric interval and then a downward deflection.

ing electrodes vary. With subthreshold stimuli, none of the axons are stimulated and no response occurs. When the stimuli are of threshold intensity, axons with low thresholds fire and a small potential change is observed. As the intensity of the stimulating current is increased, the axons with higher thresholds are also discharged. The electrical response increases proportionately until the stimulus is strong enough to excite all of the axons in the nerve. The stimulus that produces excitation of all the axons is the **maximal stimulus,** and application of greater, supramaximal stimuli produces no further increase in the size of the observed potential.

NERVE FIBER TYPES & FUNCTION

After a stimulus is applied to a nerve, there is a **latent period** before the start of the action potential. This interval corresponds to the time it takes the impulse to travel along the axon from the site of stimulation to the recording electrodes. Its duration is proportionate to the distance between the stimulating and recording electrodes and inversely proportionate to the speed of conduction. If the duration of the latent period and the distance between the stimulating and recording electrodes are known, **axonal conduction velocity** can be calculated.

Erlanger and Gasser divided mammalian nerve fibers into A, B, and C groups, further subdividing the A group into α, β, γ, and δ fibers. In Table 4–1, the various fiber types are listed

TABLE 4–1 Nerve fiber types in mammalian nerve.[a]

Fiber Type	Function	Fiber Diameter (μm)	Conduction Velocity (m/s)	Spike Duration (ms)	Absolute Refractory Period (ms)
A					
α	Proprioception; somatic motor	12–20	70–120		
β	Touch, pressure	5–12	30–70	0.4–0.5	0.4–1
γ	Motor to muscle spindles	3–6	15–30		
δ	Pain, cold, touch	2–5	12–30		
B	Preganglionic autonomic	<3	3–15	1.2	1.2
C					
Dorsal root	Pain, temperature, some mechano-reception	0.4–1.2	0.5–2	2	2
Sympathetic	Postganglionic sympathetic	0.3–1.3	0.7–2.3	2	2

[a]A and B fibers are myelinated; C fibers are unmyelinated.

with their diameters, electrical characteristics, and functions. By comparing the neurologic deficits produced by careful dorsal root section and other nerve-cutting experiments with the histologic changes in the nerves, the functions and histologic characteristics of each of the families of axons responsible for the various peaks of the compound action potential have been established. In general, the greater the diameter of a given nerve fiber, the greater its speed of conduction. The large axons are concerned primarily with proprioceptive sensation, somatic motor function, conscious touch, and pressure, while the smaller axons subserve pain and temperature sensations and autonomic function. The dorsal root C fibers conduct some impulses generated by touch and other cutaneous receptors in addition to impulses generated by pain and temperature receptors.

Further research has shown that not all the classically described lettered components are homogeneous, and a numerical system (Ia, Ib, II, III, IV) has been used by some physiologists to classify sensory fibers. Unfortunately, this has led to confusion. A comparison of the number system and the letter system is shown in Table 4–2.

In addition to variations in speed of conduction and fiber diameter, the various classes of fibers in peripheral nerves differ in their sensitivity to hypoxia and anesthetics (Table 4–3). This fact has clinical as well as physiologic significance. Local anesthetics depress transmission in the group C fibers before they affect group A touch fibers. Conversely, pressure on a nerve can cause loss of conduction in large-diameter motor, touch, and pressure fibers while pain sensation remains relatively intact. Patterns of this type are sometimes seen in individuals who sleep with their arms under their heads for long periods, causing compression of the nerves in the arms. Because of the association of deep sleep with alcoholic intoxication, the syndrome is most common on weekends and has acquired the interesting name Saturday night or Sunday morning paralysis.

NEUROTROPHINS
TROPHIC SUPPORT OF NEURONS

A number of proteins necessary for survival and growth of neurons have been isolated and studied. Some of these **neurotrophins** are products of the muscles or other structures that the neurons innervate, but others are produced by astrocytes. These proteins bind to receptors at the endings of a neuron. They are internalized and then transported by retrograde transport to the neuronal cell body, where they foster the production of proteins associated with neuronal development, growth, and survival. Other neurotrophins are produced in neurons and transported in an anterograde fashion to the nerve ending, where they maintain the integrity of the postsynaptic neuron.

TABLE 4–2 Numerical classification sometimes used for sensory neurons.

Number	Origin	Fiber Type
Ia	Muscle spindle, annulo-spiral ending	A α
Ib	Golgi tendon organ	A α
II	Muscle spindle, flower-spray ending; touch, pressure	A β
III	Pain and cold receptors; some touch receptors	A δ
IV	Pain, temperature, and other receptors	Dorsal root C

TABLE 4–3 Relative susceptibility of mammalian A, B, and C nerve fibers to conduction block produced by various agents.

Susceptibility to:	Most Susceptible	Intermediate	Least Susceptible
Hypoxia	B	A	C
Pressure	A	B	C
Local anesthetics	C	B	A

RECEPTORS

Four established neurotrophins and their three high-affinity receptors are listed in Table 4–4. Each of these **trk receptors** dimerizes, and this initiates autophosphorylation in the cytoplasmic tyrosine kinase domains of the receptors. An additional low-affinity NGF receptor that is a 75-kDa protein is called $p75^{NTR}$. This receptor binds all four of the listed neurotrophins with equal affinity. There is some evidence that it can form a heterodimer with trk A monomer and that the dimer has increased affinity and specificity for NGF. However, it now appears that $p75^{NTR}$ receptors can form homodimers that in the absence of trk receptors cause apoptosis, an effect opposite to the usual growth-promoting and nurturing effects of neurotrophins.

ACTIONS

The first neurotrophin to be characterized was NGF, a protein growth factor that is necessary for the growth and maintenance of sympathetic neurons and some sensory neurons. It is present in a broad spectrum of animal species, including humans, and is found in many different tissues. In male mice, there is a particularly high concentration in the submandibular salivary glands, and the level is reduced by castration to that seen in females. The factor is made up of two α, two β, and two γ subunits. The β subunits, each of which has a molecular mass of 13,200 Da, have all the nerve growth-promoting activity, the α subunits have trypsinlike activity, and the γ subunits are serine proteases. The function of the proteases is unknown. The structure of the β unit of NGF resembles that of insulin.

TABLE 4–4 Neurotrophins.

Neurotrophin	Receptor
Nerve growth factor (NGF)	trk A
Brain-derived neurotrophic factor (BDNF)	trk B
Neurotrophin 3 (NT-3)	trk C, less on trk A and trk B
Neurotrophin 4/5 (NT-4/5)	trk B

CLINICAL BOX 4–2

Axonal Regeneration

Peripheral nerve damage is often reversible. Although the axon will degenerate distal to the damage, connective elements of the so-called **distal stump** often survive. **Axonal sprouting** occurs from the proximal stump, growing toward the nerve ending. This results from **growth-promoting factors** secreted by **Schwann cells** that attract axons toward the distal stump. Adhesion molecules of the immunoglobulin superfamily (eg, NgCAM/L1) promote axon growth along cell membranes and extracellular matrices. Inhibitory molecules in the perineurium assure that the regenerating axons grow in a correct trajectory. Denervated distal stumps are able to upregulate production of **neurotrophins** that promote growth. Once the regenerated axon reaches its target, a new functional connection (eg, neuromuscular junction) is formed. Regeneration allows for considerable, although not full, recovery. For example, fine motor control may be permanently impaired because some motor neurons are guided to an inappropriate motor fiber. Nonetheless, recovery of peripheral nerves from damage far surpasses that of central nerve pathways. The proximal stump of a damaged axon in the CNS will form short sprouts, but distant stump recovery is rare, and the damaged axons are unlikely to form new synapses. This is because CNS neurons do not have the growth-promoting chemicals needed for regeneration. In fact, CNS myelin is a potent inhibitor of axonal growth. In addition, following CNS injury several events—**astrocytic proliferation, activation of microglia, scar formation, inflammation,** and **invasion of immune cells**—provide an inappropriate environment for regeneration. Thus, treatment of brain and spinal cord injuries frequently focuses on rehabilitation rather than reversing the nerve damage. New research is aiming to identify ways to initiate and maintain axonal growth, to direct regenerating axons to reconnect with their target neurons, and to reconstitute original neuronal circuitry.

NGF is picked up by neurons and is transported in retrograde fashion from the endings of the neurons to their cell bodies. It is also present in the brain and appears to be responsible for the growth and maintenance of cholinergic neurons in the basal forebrain and striatum. Injection of antiserum against NGF in newborn animals leads to near total destruction of the sympathetic ganglia; it thus produces an **immunosympathectomy.** There is evidence that the maintenance of neurons by NGF is due to a reduction in apoptosis.

Brain-derived neurotrophic factor (BDNF), neurotrophin 3 (NT-3), NT-4/5, and NGF each maintain a different pattern of neurons, although there is some overlap. Disruption of NT-3 by gene knockout causes a marked loss of cutaneous mechanoreceptors, even in heterozygotes. BDNF acts rapidly and can actually depolarize neurons. BDNF-deficient mice lose peripheral sensory neurons and have severe degenerative changes in their vestibular ganglia and blunted long-term potentiation.

OTHER FACTORS AFFECTING NEURONAL GROWTH

The regulation of neuronal growth is a complex process. Schwann cells and astrocytes produce **ciliary neurotrophic factor (CNTF)**. This factor promotes the survival of damaged and embryonic spinal cord neurons and may prove to be of value in treating human diseases in which motor neurons degenerate. **Glial cell line-derived neurotrophic factor (GDNF)** maintains midbrain dopaminergic neurons in vitro. However, GDNF knockouts have dopaminergic neurons that appear normal, but they have no kidneys and fail to develop an enteric nervous system. Another factor that enhances the growth of neurons is **leukemia inhibitory factor (LIF)**. In addition, neurons as well as other cells respond to **insulinlike growth factor I (IGF-I)** and the various forms of **transforming growth factor (TGF), fibroblast growth factor (FGF),** and **platelet-derived growth factor (PDGF)**.

Clinical Box 4–2 compares the ability to regenerate neurons after central and peripheral nerve injury.

CHAPTER SUMMARY

- There are two main types of microglia and macroglia. Microglia are scavenger cells. Macroglia include oligodendrocytes, Schwann cells, and astrocytes. The first two are involved in myelin formation; astrocytes produce substances that are tropic to neurons, and they help maintain the appropriate concentration of ions and neurotransmitters.
- Neurons are composed of a cell body (soma) which is the metabolic center of the neuron, dendrites that extend outward from the cell body and arborize extensively, and a long fibrous axon that originates from a somewhat thickened area of the cell body, the axon hillock.
- The axons of many neurons acquire a sheath of myelin, a protein–lipid complex that is wrapped around the axon. Myelin is an effective insulator, and depolarization in myelinated axons jumps from one node of Ranvier to the next, with the current sink at the active node serving to electrotonically depolarize to the firing level the node ahead of the action potential.
- Orthograde transport occurs along microtubules that run the length of the axon and requires molecular motors, dynein, and kinesin.
- Two types of physicochemical disturbances occur in neurons: local, nonpropagated potentials (synaptic, generator, or electrotonic potentials) and propagated potentials (action potentials).
- In response to a depolarizing stimulus, voltage-gated Na^+ channels become active, and when the threshold potential is reached, an action potential results. The membrane potential moves toward the equilibrium potential for Na^+. The Na^+ channels rapidly enter a closed state (inactivated state) before returning to the resting state. The direction of the electrical gradient for Na^+ is reversed during the overshoot because the membrane potential is reversed, and this limits Na^+ influx. Voltage-gated K^+ channels open and the net movement of positive charge out of the cell helps complete the process of repolarization. The slow return of the K^+ channels to the closed state explains afterhyperpolarization, followed by a return to the resting membrane potential.
- Nerve fibers are divided into different categories based on axonal diameter, conduction velocity, and function.
- Neurotrophins are produced by astrocytes and transported by retrograde transport to the neuronal cell body, where they foster the production of proteins associated with neuronal development, growth, and survival.

MULTIPLE-CHOICE QUESTIONS

For all questions, select the single best answer unless otherwise directed.

1. The distance from between one stimulating electrode to recording electrode is 4.5 cm. When the axon is stimulated, the latent period is 1.5 ms. What is the conduction velocity of the axon?
   A) 15 m/s
   B) 30 m/s
   C) 40 m/s
   D) 67.5 m/s
   E) This cannot be determined from the information given.

2. Which of the following has the slowest conduction velocity?
   A) Aα fibers
   B) Aβ fibers
   C) Aγ fibers
   D) B fibers
   E) C fibers

3. A man falls into a deep sleep with one arm under his head. This arm is paralyzed when he awakens, but it tingles, and pain sensation in it is still intact. The reason for the loss of motor function without loss of pain sensation is that in the nerves to his arm,
   A) A fibers are more susceptible to hypoxia than B fibers.
   B) A fibers are more sensitive to pressure than C fibers.
   C) C fibers are more sensitive to pressure than A fibers.
   D) motor nerves are more affected by sleep than sensory nerves.
   E) sensory nerves are nearer the bone than motor nerves and hence are less affected by pressure.

4. Which part of a neuron has the highest concentration of Na^+ channels per square millimeter of cell membrane?
   A) dendrites
   B) cell body near dendrites
   C) initial segment
   D) axonal membrane under myelin
   E) none of the above

5. Which of the following statements about nerve growth factor is *not* true?
   A) It is made up of three polypeptide subunits.
   B) It facilitates the process of apoptosis.
   C) It is necessary for the growth and development of the sympathetic nervous system.
   D) It is picked up by nerves from the organs they innervate.
   E) It is present in the brain.

CHAPTER RESOURCES

Aidley DJ: *The Physiology of Excitable Cells,* 4th ed. Cambridge University Press, 1998.

Boron WF, Boulpaep EL: *Medical Physiology,* Elsevier, 2005.

Bradbury EJ, McMahon SB: Spinal cord repair strategies: Why do they work? Nat Rev Neurosci 2006;7:644.

Catterall WA: Structure and function of voltage-sensitive ion channels. Science 1988; 242:649.

Hille B: *Ionic Channels of Excitable Membranes,* 3rd ed. Sinauer Associates, 2001.

Kandel ER, Schwartz JH, Jessell TM (editors): *Principles of Neural Science,* 4th ed. McGraw-Hill, 2000.

Nicholls JG, Martin AR, Wallace BG: *From Neuron to Brain: A Cellular and Molecular Approach to the Function of the Nervous System,* 4th ed. Sinauer Associates, 2001.

Thuret S, Moon LDF, Gage FH: Therapeutic interventions after spinal cord injury. Nat Rev Neurosci 2006;7:628.

Volterra A, Meldolesi J: Astrocytes, from brain glue to communication elements: The revolution continues. Nat Rev Neurosci 2005;6:626.

Widmaier EP, Raff H, Strang KT: *Vander's Human Physiology.* McGraw-Hill, 2008.

Excitable Tissue: Muscle

INTRODUCTION

Muscle cells, like neurons, can be excited chemically, electrically, and mechanically to produce an action potential that is transmitted along their cell membranes. Unlike neurons, they respond to stimuli by activating a contractile mechanism. The contractile protein myosin and the cytoskeletal protein actin are abundant in muscle, where they are the primary structural components that bring about contraction.

Muscle is generally divided into three types: **skeletal, cardiac,** and **smooth,** although smooth muscle is not a homogeneous single category. Skeletal muscle makes up the great mass of the somatic musculature. It has well-developed cross-striations, does not normally contract in the absence of nervous stimulation, lacks anatomic and functional connections between individual muscle fibers, and is generally under voluntary control. Cardiac muscle also has cross-striations, but it is functionally syncytial and, although it can be modulated via the autonomic nervous system, it can contract rhythmically in the absence of external innervation owing to the presence in the myocardium of pacemaker cells that discharge spontaneously (see Chapter 30). Smooth muscle lacks cross-striations and can be further subdivided into two broad types: unitary (or visceral) smooth muscle and multiunit smooth muscle. The type found in most hollow viscera is functionally syncytial and contains pacemakers that discharge irregularly. The multiunit type found in the eye and in some other locations is not spontaneously active and resembles skeletal muscle in graded contractile ability.

SKELETAL MUSCLE MORPHOLOGY

ORGANIZATION

Skeletal muscle is made up of individual muscle fibers that are the "building blocks" of the muscular system in the same sense that the neurons are the building blocks of the nervous system. Most skeletal muscles begin and end in tendons, and the muscle fibers are arranged in parallel between the tendinous ends, so that the force of contraction of the units is additive. Each muscle fiber is a single cell that is multinucleated, long, cylindrical, and surrounded by a cell membrane, the **sarcolemma** (Figure 5–1). There are no syncytial bridges between cells. The muscle fibers are made up of myofibrils, which are divisible into individual filaments. These myofilaments contain several proteins that together make up the contractile machinery of the skeletal muscle.

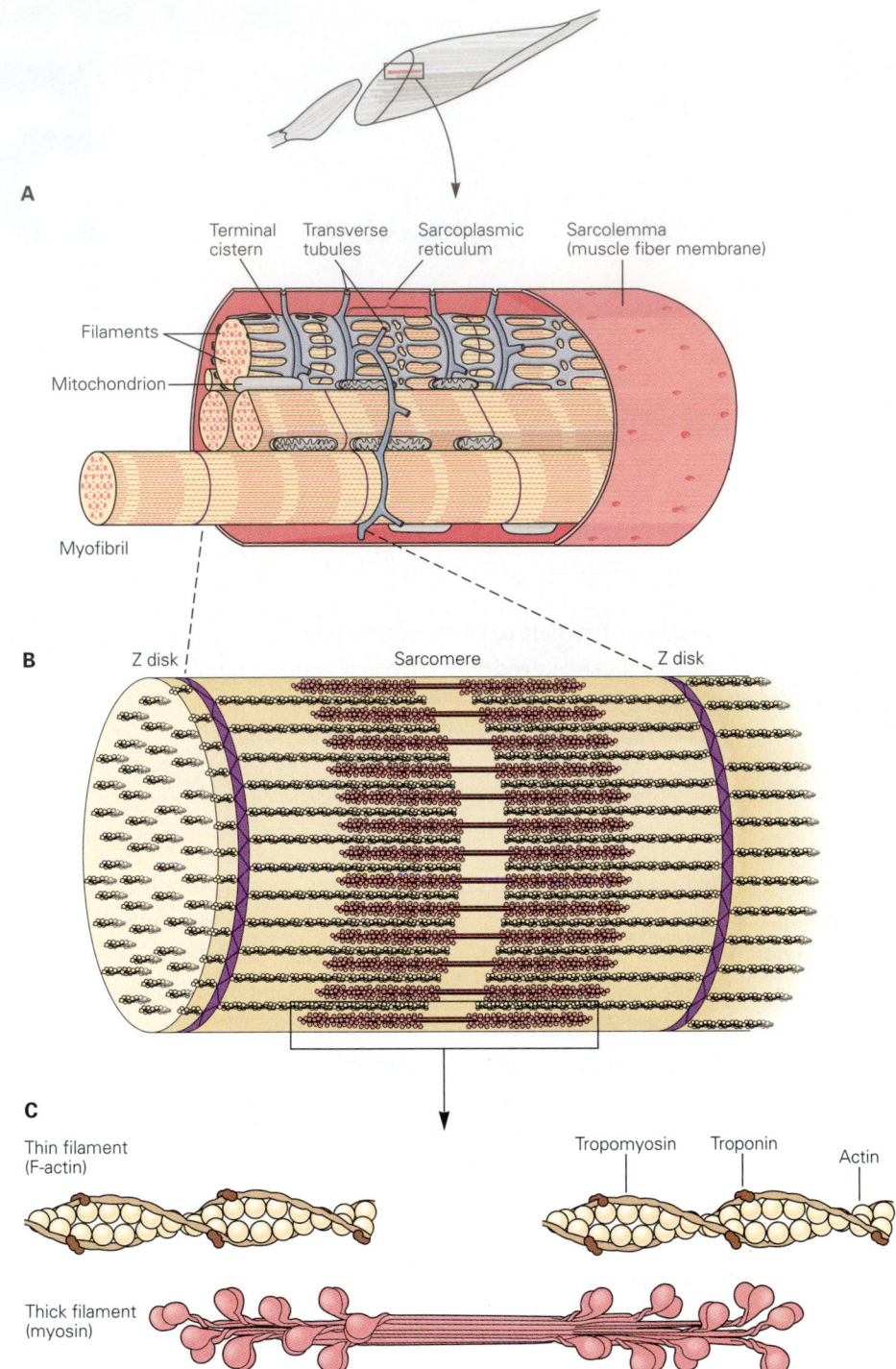

FIGURE 5–1 **Mammalian skeletal muscle.** A single muscle fiber surrounded by its sarcolemma has been cut away to show individual myofibrils. The cut surface of the myofibrils shows the arrays of thick and thin filaments. The sarcoplasmic reticulum with its transverse (T) tubules and terminal cisterns surrounds each myofibril. The T tubules invaginate from the sarcolemma and contact the myofibrils twice in every sarcomere. Mitochondria are found between the myofibrils and a basal lamina surrounds the sarcolemma. (Reproduced with permission from Kandel ER, Schwartz JH, Jessell TM [editors]: *Principles of Neural Science*, 4th ed. McGraw-Hill, 2000.)

The contractile mechanism in skeletal muscle largely depends on the proteins **myosin-II, actin, tropomyosin,** and **troponin.** Troponin is made up of three subunits: **troponin I, troponin T,** and **troponin C.** Other important proteins in muscle are involved in maintaining the proteins that participate in contraction in appropriate structural relation to one another and to the extracellular matrix.

FIGURE 5–2 **Electron micrograph of human gastrocnemius muscle.** The various bands and lines are identified at the top (× 13,500). (Courtesy of Walker SM, Schrodt GR.)

STRIATIONS

Differences in the refractive indexes of the various parts of the muscle fiber are responsible for the characteristic cross-striations seen in skeletal muscle when viewed under the microscope. The parts of the cross-striations are frequently identified by letters (Figure 5–2). The light I band is divided by the dark Z line, and the dark A band has the lighter H band in its center. A transverse M line is seen in the middle of the H band, and this line plus the narrow light areas on either side of it are sometimes called the pseudo-H zone. The area between two adjacent Z lines is called a **sarcomere**. The orderly arrangement of actin, myosin, and related proteins that produces this pattern is shown in Figure 5–3. The thick filaments, which are about twice the diameter of the thin filaments, are made up of myosin; the thin filaments are made up of actin, tropomyosin, and troponin. The thick filaments are lined up to form the A bands, whereas the array of thin filaments extends out of the A band and into the less dense staining I bands. The lighter H bands in the center of the A bands are the regions where, when the muscle is relaxed, the thin filaments do not overlap the thick filaments. The Z lines allow for anchoring of the thin filaments. If a transverse section through the A band is examined under the electron microscope, each thick filament is seen to be surrounded by six thin filaments in a regular hexagonal pattern.

The form of myosin found in muscle is myosin-II, with two globular heads and a long tail. The heads of the myosin

FIGURE 5–3 **A)** Arrangement of thin (actin) and thick (myosin) filaments in skeletal muscle (compare to Figure 5–2). **B)** Sliding of actin on myosin during contraction so that Z lines move closer together. **C)** Detail of relation of myosin to actin in an individual sarcomere, the functional unit of the muscle. **D)** Diagrammatic representation of the arrangement of actin, tropomyosin, and troponin of the thin filaments in relation to a myosin thick filament. The globular heads of myosin interact with the thin filaments to create the contraction. Note that myosin thick filaments reverse polarity at the M line in the middle of the sarcomere, allowing for contraction. (**C** and **D** are modified with permission from Kandel ER, Schwartz JH, Jessell TM [editors]: *Principles of Neural Science*, 4th ed. McGraw-Hill, 2000.)

molecules form cross-bridges with actin. Myosin contains heavy chains and light chains, and its heads are made up of the light chains and the amino terminal portions of the heavy chains. These heads contain an actin-binding site and a catalytic site that hydrolyzes ATP. The myosin molecules are arranged symmetrically on either side of the center of the sarcomere, and it is this arrangement that creates the light areas in the pseudo-H zone. The M line is the site of the reversal of polarity of the myosin molecules in each of the thick filaments. At these points, there are slender cross-connections that hold the thick filaments in proper array. Each thick filament contains several hundred myosin molecules.

The thin filaments are polymers made up of two chains of actin that form a long double helix. Tropomyosin molecules are long filaments located in the groove between the two chains in the actin (Figure 5–3). Each thin filament contains 300 to 400 actin molecules and 40 to 60 tropomyosin molecules. Troponin molecules are small globular units located at intervals along the tropomyosin molecules. Each of the three troponin subunits has a unique function: Troponin T binds the troponin components to tropomyosin; troponin I inhibits the interaction of myosin with actin; and troponin C contains the binding sites for the Ca^{2+} that helps to initiate contraction.

Some additional structural proteins that are important in skeletal muscle function include **actinin, titin,** and **desmin.** Actinin binds actin to the Z lines. Titin, the largest known protein (with a molecular mass near 3,000,000 Da), connects the Z lines to the M lines and provides scaffolding for the sarcomere. It contains two kinds of folded domains that provide muscle with its elasticity. At first when the muscle is stretched there is relatively little resistance as the domains unfold, but with further stretch there is a rapid increase in resistance that protects the structure of the sarcomere. Desmin adds structure to the Z lines in part by binding the Z lines to the plasma membrane. Although these proteins are important in muscle structure/function, by no means do they represent an exhaustive list.

SARCOTUBULAR SYSTEM

The muscle fibrils are surrounded by structures made up of membranes that appear in electron photomicrographs as vesicles and tubules. These structures form the **sarcotubular system,** which is made up of a **T system** and a **sarcoplasmic reticulum.** The T system of transverse tubules, which is continuous with the sarcolemma of the muscle fiber, forms a grid perforated by the individual muscle fibrils (Figure 5–1). The space between the two layers of the T system is an extension of the extracellular space. The sarcoplasmic reticulum, which forms an irregular curtain around each of the fibrils, has enlarged **terminal cisterns** in close contact with the T system at the junctions between the A and I bands. At these points of contact, the arrangement of the central T system with a cistern of the sarcoplasmic reticulum on either side has led to the use of the term **triads** to describe the system. The T system, which is continuous with the sarcolemma, provides a path for the

rapid transmission of the action potential from the cell membrane to all the fibrils in the muscle. The sarcoplasmic reticulum is an important store of Ca^{2+} and also participates in muscle metabolism.

DYSTROPHIN–GLYCOPROTEIN COMPLEX

The large **dystrophin** protein (molecular mass 427,000 Da) forms a rod that connects the thin actin filaments to the transmembrane protein β-**dystroglycan** in the sarcolemma by smaller proteins in the cytoplasm, **syntrophins.** β-dystroglycan is connected to **merosin** (merosin refers to laminins that contain the α2 subunit in their trimeric makeup) in the extracellular matrix by α-**dystroglycan** (Figure 5–4). The dystroglycans are in turn associated with a complex of four transmembrane glycoproteins: α-, β-, γ-, and δ-**sarcoglycan.** This **dystrophin–glycoprotein complex** adds strength to the muscle by providing a scaffolding for the fibrils and connecting them to the extracellular environment. Disruption of the tightly choreographed structure can lead to several different pathologies, or muscular dystrophies (see Clinical Box 5–1).

ELECTRICAL PHENOMENA & IONIC FLUXES

ELECTRICAL CHARACTERISTICS OF SKELETAL MUSCLE

The electrical events in skeletal muscle and the ionic fluxes that underlie them share distinct similarities to those in nerve, with quantitative differences in timing and magnitude. The resting membrane potential of skeletal muscle is about –90 mV. The action potential lasts 2 to 4 ms and is conducted along the muscle fiber at about 5 m/s. The absolute refractory period is 1 to 3 ms long, and the after-polarizations, with their related changes in threshold to electrical stimulation, are relatively prolonged. The initiation of impulses at the myoneural junction is discussed in the next chapter.

ION DISTRIBUTION & FLUXES

The distribution of ions across the muscle fiber membrane is similar to that across the nerve cell membrane. Approximate values for the various ions and their equilibrium potentials are shown in Table 5–1. As in nerves, depolarization is largely a manifestation of Na^+ influx, and repolarization is largely a manifestation of K^+ efflux.

CONTRACTILE RESPONSES

It is important to distinguish between the electrical and mechanical events in skeletal muscle. Although one response

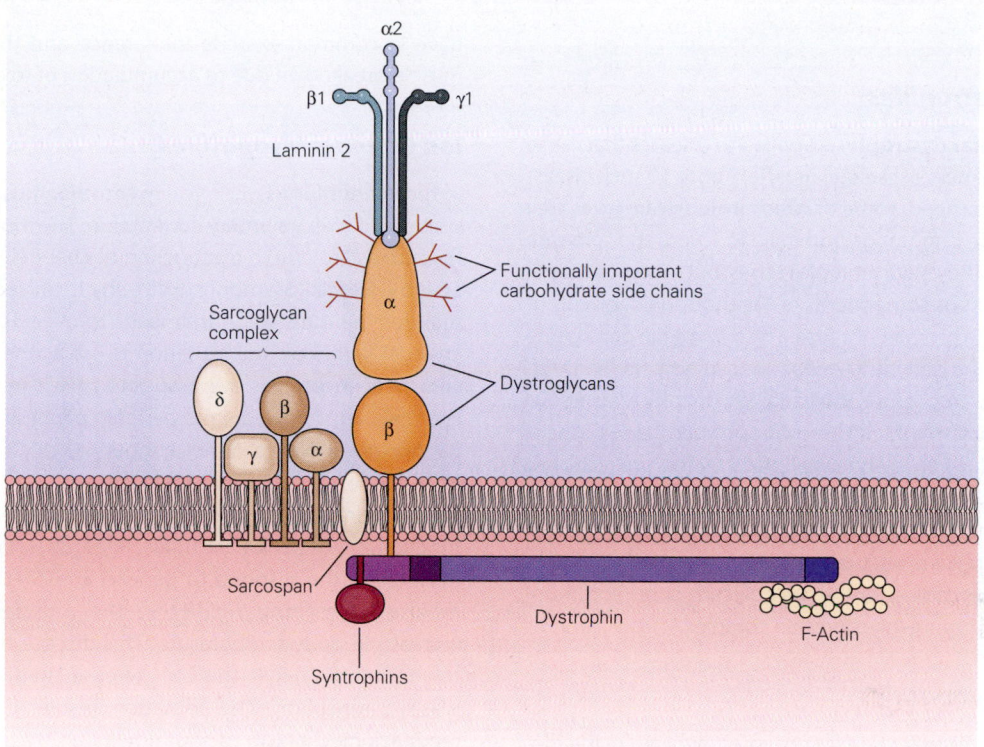

FIGURE 5–4 **The dystrophin–glycoprotein complex.** Dystrophin connects F-actin to the two members of the dystroglycan (DG) complex, α and β-dystroglycan, and these in turn connect to the merosin subunit of laminin 211 in the extracellular matrix. The sarcoglycan complex of four glycoproteins, α-, β-, γ-, and δ-sarcoglycan, sarcospan, and syntropins are all associated with the dystroglycan complex. There are muscle disorders associated with loss, abnormalities, or both of the sarcoglycans and merosin. (Reproduced with permission from Kandel ER, Scwartz JH, Jessell TM [editors]: *Principles of Neural Science*, 4th ed. McGraw-Hill, 2000.)

does not normally occur without the other, their physiologic bases and characteristics are different. Muscle fiber membrane depolarization normally starts at the motor end plate, the specialized structure under the motor nerve ending. The action potential is transmitted along the muscle fiber and initiates the contractile response.

THE MUSCLE TWITCH

A single action potential causes a brief contraction followed by relaxation. This response is called a **muscle twitch.** In Figure 5–5, the action potential and the twitch are plotted on the same time scale. The twitch starts about 2 ms after the start of depolarization of the membrane, before repolarization is complete. The duration of the twitch varies with the type of muscle being tested. "Fast" muscle fibers, primarily those concerned with fine, rapid, precise movement, have twitch durations as short as 7.5 ms. "Slow" muscle fibers, principally those involved in strong, gross, sustained movements, have twitch durations up to 100 ms.

MOLECULAR BASIS OF CONTRACTION

The process by which the contraction of muscle is brought about is a sliding of the thin filaments over the thick filaments. Note that this shortening is not due to changes in the actual lengths of the thick and thin filaments, rather, by their increased overlap within the muscle cell. The width of the A bands is constant, whereas the Z lines move closer together when the muscle contracts and farther apart when it relaxes (Figure 5–3).

The sliding during muscle contraction occurs when the myosin heads bind firmly to actin, bend at the junction of the head with the neck, and then detach. This "power stroke" depends on the simultaneous hydrolysis of ATP. Myosin-II molecules are dimers that have two heads, but only one attaches to actin at any given time. The probable sequence of events of the power stroke is outlined in Figure 5–6. In resting muscle, troponin I is bound to actin and tropomyosin and covers the sites where myosin heads interact with actin. Also at rest, the myosin head contains tightly bound ADP. Following an action potential cytosolic Ca^{2+} is increased and free Ca^{2+} binds to troponin C. This binding results in a weakening of the troponin I interaction with actin and exposes the actin binding site for myosin to

text continues on p. 100

CLINICAL BOX 5–1

Disease of Muscle

Muscular Dystrophies

The term **muscular dystrophy** is applied to diseases that cause progressive weakness of skeletal muscle. About 50 such diseases have been described, some of which include cardiac as well as skeletal muscle. They range from mild to severe and some are eventually fatal. They have multiple causes, but mutations in the genes for the various components of the dystrophin–glycoprotein complex are a prominent cause. The dystrophin gene is one of the largest in the body, and mutations can occur at many different sites in it. **Duchenne muscular dystrophy** is a serious form of dystrophy in which the dystrophin protein is absent from muscle. It is X-linked and usually fatal by the age of 30. In a milder form of the disease, **Becker muscular dystrophy,** dystrophin is present but altered or reduced in amount. Limb-girdle muscular dystrophies of various types are associated with mutations of the genes coding for the sarcoglycans or other components of the dystrophin–glycoprotein complex.

Metabolic Myopathies

Mutations in genes that code for enzymes involved in the metabolism of carbohydrates, fats, and proteins to CO_2 and H_2O in muscle and the production of ATP can cause **metabolic myopathies** (eg, McArdle syndrome). Metabolic myopathies all have in common exercise intolerance and the possibility of muscle breakdown due to accumulation of toxic metabolites.

Ion Channel Myopathies

In the various forms of clinical **myotonia,** muscle relaxation is prolonged after voluntary contraction. The molecular bases of myotonias are due to dysfunction of channels that shape the action potential. Myotonia dystrophy is caused by an autosomal dominant mutation that leads to overexpression of a K^+ channel (although the mutation is *not* at the K^+ channel). A variety of myotonias are associated with mutations in Na^+ channels (eg, hyperkalemic periodic paralysis, paramyotonia congenita, or Na^+ channel congenita) or Cl^- channels (eg, dominant or recessive myotonia congenita).

Malignant hyperthermia is another disease related to dysfunctional muscle ion channels. Patients with malignant hyperthermia can respond to general anesthetics such as halothane by eliciting rigidity in the muscles and a quick increase in body temperature. This disease has been traced to a mutation in RyR, the Ca^{2+} release channel in the sarcoplasmic reticulum. The mutation results in an inefficient feedback mechanism to shut down Ca^{2+} release after stimulation of the RyR, and thus, increased contractility and heat generation.

TABLE 5–1 **Steady-state distribution of ions in the intracellular and extracellular compartments of mammalian skeletal muscle, and the equilibrium potentials for these ions.**

Ion[a]	Concentration (mmol/L)		Equilibrium Potential (mV)
	Intracellular Fluid	Extracellular Fluid	
Na^+	12	145	+65
K^+	155	4	−95
H^+	13×10^{-5}	3.8×10^{-5}	−32
Cl^-	3.8	120	−90
HCO_3^-	8	27	−32
A^-	155	0	...
Membrane potential = −90 mV			

[a]A^- represents organic anions. The value for intracellular Cl^- is calculated from the membrane potential, using the Nernst equation.

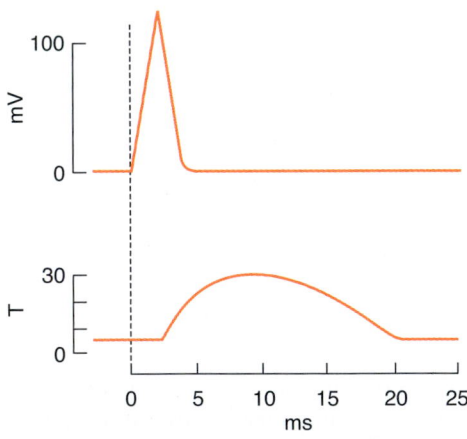

FIGURE 5–5 **The electrical and mechanical responses of a mammalian skeletal muscle fiber to a single maximal stimulus.** The electrical response (mV potential change) and the mechanical response (T, tension in arbitrary units) are plotted on the same abscissa (time). The mechanical response is relatively long-lived compared to the electrical response that initiates contraction.

FIGURE 5–6 **Power stroke of myosin in skeletal muscle. A)** At rest, myosin heads are bound to adenosine diphosphate and are said to be in a "cocked" position in relation to the thin filament, which does not have Ca^{2+} bound to the troponin–tropomyosin complex. **B)** Ca^{2+} bound to the troponin–tropomyosin complex induced a conformational change in the thin filament that allows for myosin heads to cross-bridge with thin filament actin. **C)** Myosin heads rotate, move the attached actin and shorten the muscle fiber, forming the power stroke. **D)** At the end of the power stroke, ATP binds to a now exposed site, and causes a detachment from the actin filament. **E)** ATP is hydrolyzed into ADP and inorganic phosphate (P_i) and this chemical energy is used to "re-cock" the myosin head. (Modified with permission from Kandel ER, Schwartz JH, Jessell TM [editors]: *Principles of Neural Science*, 4th ed. McGraw-Hill, 2000.)

allow for formation of myosin/actin cross-bridges. Upon formation of the cross-bridge, ADP is released, causing a conformational change in the myosin head that moves the thin filament relative to the thick filament, comprising the cross-bridge "power stroke." ATP quickly binds to the free site on the myosin, which leads to a detachment of the myosin head from the thin filament. ATP is hydrolyzed and inorganic phosphate (P_i) released, causing a "re-cocking" of the myosin head and completing the cycle. As long as Ca^{2+} remains elevated and sufficient ATP is available, this cycle repeats. Many myosin heads cycle at or near the same time, and they cycle repeatedly, producing gross muscle contraction. Each power stroke shortens the sarcomere about 10 nm. Each thick filament has about 500 myosin heads, and each head cycles about five times per second during a rapid contraction.

The process by which depolarization of the muscle fiber initiates contraction is called **excitation–contraction coupling.** The action potential is transmitted to all the fibrils in the fiber via the T system (Figure 5–7). It triggers the release of Ca^{2+} from the terminal cisterns, the lateral sacs of the sarcoplasmic reticulum next to the T system. Depolarization of the T tubule membrane activates the sarcoplasmic reticulum via **dihydropyridine receptors (DHPR),** named for the drug dihydropyridine, which blocks them (Figure 5–8). DHPR are voltage-gated Ca^{2+} channels in the T tubule membrane. In cardiac muscle, influx of Ca^{2+} via these channels triggers the release of Ca^{2+} stored in the sarcoplasmic reticulum (calcium-induced calcium release) by activating the **ryanodine receptor (RyR).** The RyR is named after the plant alkaloid ryanodine that was used in its discovery. It is a ligand-gated Ca^{2+} channel with Ca^{2+} as its natural ligand. In skeletal muscle, Ca^{2+} entry from the extracellular fluid (ECF) by this route is not required for Ca^{2+} release. Instead, the DHPR that serves as the voltage sensor unlocks release of Ca^{2+} from the nearby sarcoplasmic reticulum via physical interaction with the RyR. The released Ca^{2+} is quickly amplified through calcium-induced calcium release. Ca^{2+} is reduced in the muscle cell by the sarcoplasmic or endoplasmic reticulum Ca^{2+} ATPase (SERCA) pump. The SERCA pump uses energy from ATP hydrolysis to remove Ca^{2+} from the cytosol back into the terminal cisterns, where it is stored until released by the next action potential. Once the Ca^{2+} concentration outside the reticulum has been lowered sufficiently, chemical interaction between myosin and actin ceases and the muscle relaxes. Note that ATP provides the energy for both contraction (at the myosin head) and relaxation (via SERCA). If transport of Ca^{2+} into the reticulum is inhibited, relaxation does not occur even though there are no more action potentials; the resulting sustained contraction is called a **contracture.**

TYPES OF CONTRACTION

Muscular contraction involves shortening of the contractile elements, but because muscles have elastic and viscous elements in series with the contractile mechanism, it is possible

Steps in contraction[a]

Discharge of motor neuron

Release of transmitter (acetylcholine) at motor end-plate

Binding of acetylcholine to nicotinic acetylcholine receptors

Increased Na^+ and K^+ conductance in end-plate membrane

Generation of end-plate potential

Generation of action potential in muscle fibers

Inward spread of depolarization along T tubules

Release of Ca^{2+} from terminal cisterns of sarcoplasmic reticulum and diffusion to thick and thin filaments

Binding of Ca^{2+} to troponin C, uncovering myosin-binding sites on actin

Formation of cross-linkages between actin and myosin and sliding of thin on thick filaments, producing movement

Steps in relaxation

Ca^{2+} pumped back into sarcoplasmic reticulum

Release of Ca^{2+} from troponin

Cessation of interaction between actin and myosin

[a]Steps 1–6 in contraction are discussed in Chapter 4.

FIGURE 5–7 Flow of information that leads to muscle contraction.

FIGURE 5–8 **Relation of the T tubule (TT) to the sarcoplasmic reticulum in Ca^{2+} transport.** In skeletal muscle, the voltage-gated dihydropyridine receptor in the T tubule triggers Ca^{2+} release from the sarcoplasmic reticulum (SR) via the ryanodine receptor (RyR). Upon sensing a voltage change, there is a physical interaction between the sarcolemmal-bound DHPR and the SR-bound RyR. This interaction gates the RyR and allows for Ca^{2+} release from the SR.

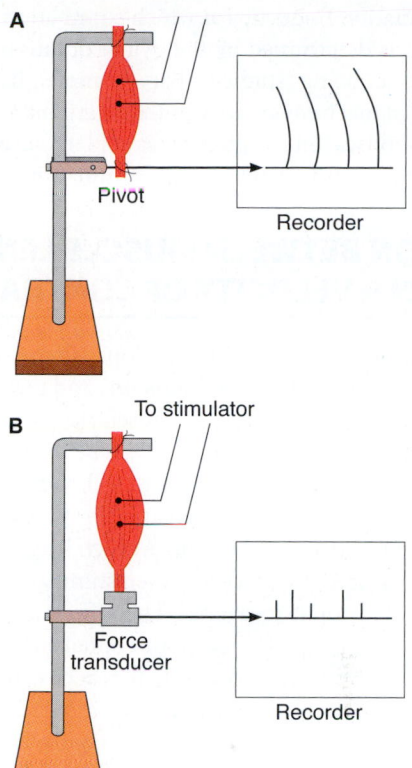

FIGURE 5–9 **A)** Muscle preparation arranged for recording isotonic contractions. **B)** Preparation arranged for recording isometric contractions. In **A,** the muscle is fastened to a writing lever that swings on a pivot. In **B,** it is attached to an electronic transducer that measures the force generated without permitting the muscle to shorten.

for contraction to occur without an appreciable decrease in the length of the whole muscle (Figure 5–9). Such a contraction is called **isometric** ("same measure" or length). Contraction against a constant load with a decrease in muscle length is **isotonic** ("same tension"). Note that because work is the product of force times distance, isotonic contractions do work, whereas isometric contractions do not. In other situations, muscle can do negative work while lengthening against a constant weight.

SUMMATION OF CONTRACTIONS

The electrical response of a muscle fiber to repeated stimulation is like that of nerve. The fiber is electrically refractory only during the rising phase and part of the falling phase of the spike potential. At this time, the contraction initiated by the first stimulus is just beginning. However, because the contractile mechanism does not have a refractory period, repeated stimulation before relaxation has occurred produces additional activation of the contractile elements and a response that is added to the contraction already present. This phenomenon is

known as **summation of contractions.** The tension developed during summation is considerably greater than that during the single muscle twitch. With rapidly repeated stimulation, activation of the contractile mechanism occurs repeatedly before any relaxation has occurred, and the individual responses fuse into one continuous contraction. Such a response is called a **tetanus (tetanic contraction).** It is a **complete tetanus** when no relaxation occurs between stimuli and an **incomplete tetanus** when periods of incomplete relaxation take place between the summated stimuli. During a complete tetanus, the tension developed is about four times that developed by the individual twitch contractions. The development of an incomplete and a complete tetanus in response to stimuli of increasing frequency is shown in Figure 5–10.

FIGURE 5–10 **Tetanus.** Isometric tension of a single muscle fiber during continuously increasing and decreasing stimulation frequency. Dots at the top are at intervals of 0.2 s. Note the development of incomplete and then complete tetanus as stimulation is increased, and the return of incomplete tetanus, then full response, as stimulation frequency is decreased.

The stimulation frequency at which summation of contractions occurs is determined by the twitch duration of the particular muscle being studied. For example, if the twitch duration is 10 ms, frequencies less than 1/10 ms (100/s) cause discrete responses interrupted by complete relaxation, and frequencies greater than 100/s cause summation.

RELATION BETWEEN MUSCLE LENGTH & TENSION & VELOCITY OF CONTRACTION

Both the tension that a muscle develops when stimulated to contract isometrically (the **total tension**) and the **passive tension** exerted by the unstimulated muscle vary with the length of the muscle fiber. This relationship can be studied in a whole skeletal muscle preparation such as that shown in Figure 5–9. The length of the muscle can be varied by changing the distance between its two attachments. At each length, the passive tension is measured, the muscle is then stimulated electrically, and the total tension is measured. The difference between the two values at any length is the amount of tension actually generated by the contractile process, the **active tension.** The records obtained by plotting passive tension and total tension against muscle length are shown in Figure 5–11. Similar curves are obtained when single muscle fibers are studied. The length of the muscle at which the active tension is maximal is usually called its **resting length.** The term comes originally from experiments demonstrating that the length of many of the muscles in the body at rest is the length at which they develop maximal tension.

The observed length–tension relation in skeletal muscle is explained by the sliding filament mechanism of muscle contraction. When the muscle fiber contracts isometrically, the tension developed is proportional to the number of cross-bridges between the actin and the myosin molecules. When muscle is stretched, the overlap between actin and myosin is reduced and the number of cross-linkages is therefore reduced. Conversely, when the muscle is appreciably shorter than resting length, the distance the thin filaments can move is reduced.

The velocity of muscle contraction varies inversely with the load on the muscle. At a given load, the velocity is maximal at the resting length and declines if the muscle is shorter or longer than this length.

FIBER TYPES

Although skeletal muscle fibers resemble one another in a general way, skeletal muscle is a heterogeneous tissue made up of fibers that vary in myosin ATPase activity, contractile speed, and other properties. Muscles are frequently classified into two types, "slow" and "fast." These muscles can contain a mixture of three fiber types: type I (or SO for slow-oxidative); type IIA (FOG for fast-oxidative-glycolytic); or type IIB (FG for fast glycolytic). Some of the properties associated with type I, type IIA, and type IIB fibers are summarized in Table 5–2. Although this classification scheme is valid for muscles across many mammalian species, there are significant variations of fibers within and between muscles. For example, type I fibers in a given muscle can be larger than type IIA fibers from a different muscle in the same animal. Many of the differences in the fibers that make up muscles stem from differences in the proteins within them. Most of these are encoded by multigene families. Ten different **isoforms** of the myosin heavy chains (MHCs) have been characterized. Each of the two types of light chains also have isoforms. It appears that there is only one form of actin, but multiple isoforms of tropomyosin and all three components of troponin.

ENERGY SOURCES & METABOLISM

Muscle contraction requires energy, and muscle has been called "a machine for converting chemical energy into mechanical work." The immediate source of this energy is ATP, and this is formed by the metabolism of carbohydrates and lipids.

PHOSPHORYLCREATINE

ATP is resynthesized from ADP by the addition of a phosphate group. Some of the energy for this endothermic reaction is supplied by the breakdown of glucose to CO_2 and H_2O, but there also exists in muscle another energy-rich phosphate compound that can supply this energy for short periods. This compound is **phosphorylcreatine,** which is hydrolyzed to creatine and phosphate groups with the release of considerable energy (Figure 5–12). At rest, some ATP in the mitochondria transfers its phosphate to creatine, so that a phosphorylcreatine

FIGURE 5–11   Length–tension relationship for the human triceps muscle. The passive tension curve measures the tension exerted by this skeletal muscle at each length when it is not stimulated. The total tension curve represents the tension developed when the muscle contracts isometrically in response to a maximal stimulus. The active tension is the difference between the two.

TABLE 5–2 Classification of fiber types in skeletal muscles.

	Type 1	Type IIA	Type IIB
Other names	Slow, Oxidative (SO)	Fast, Oxidative, Glycolytic (FOG)	Fast, Glycolytic (FG)
Color	Red	Red	White
Myosin ATPase Activity	Slow	Fast	Fast
Ca^{2+}-pumping capacity of sarcoplasmic reticulum	Moderate	High	High
Diameter	Small	Large	Large
Glycolytic capacity	Moderate	High	High
Oxidative capacity	High	Moderate	Low
Associated Motor Unit Type	Slow (S)	Fast Resistant to Fatigue (FR)	Fast Fatigable (FF)
Membrane potential = –90 mV			
Oxidative capacity	High	Moderate	Low

store is built up. During exercise, the phosphorylcreatine is hydrolyzed at the junction between the myosin heads and actin, forming ATP from ADP and thus permitting contraction to continue.

CARBOHYDRATE & LIPID BREAKDOWN

At rest and during light exercise, muscles utilize lipids in the form of free fatty acids as their energy source. As the intensity of exercise increases, lipids alone cannot supply energy fast enough and so use of carbohydrate becomes the predominant component in the muscle fuel mixture. Thus, during exercise, much of the energy for phosphorylcreatine and ATP resynthesis comes from the breakdown of glucose to CO_2 and H_2O. Glucose in the bloodstream enters cells, where it is degraded through a series of chemical reactions to pyruvate. Another

source of intracellular glucose, and consequently of pyruvate, is glycogen, the carbohydrate polymer that is especially abundant in liver and skeletal muscle. When adequate O_2 is present, pyruvate enters the citric acid cycle and is metabolized—through this cycle and the so-called respiratory enzyme pathway—to CO_2 and H_2O. This process is called **aerobic glycolysis.** The metabolism of glucose or glycogen to CO_2 and H_2O forms large quantities of ATP from ADP. If O_2 supplies are insufficient, the pyruvate formed from glucose does not enter the tricarboxylic acid cycle but is reduced to lactate. This process of **anaerobic glycolysis** is associated with the net production of much smaller quantities of energy-rich phosphate bonds, but it does not require the presence of O_2. A brief overview of the various reactions involved in supplying energy to skeletal muscle is shown in Figure 5–13.

FIGURE 5–12 **Creatine, phosphorylcreatine, and creatinine cycling in muscle.** During periods of high activity, cycling of phosphorylcreatine allows for quick release of ATP to sustain muscle activity.

$$ATP + H_2O \rightarrow ADP + H_3PO_4 + 7.3 \text{ kcal}$$

$$\text{Phosphorylcreatine} + ADP \rightleftharpoons \text{Creatine} + ATP$$

Glucose + 2 ATP (or glycogen + 1 ATP)

$$\xrightarrow{\text{Anaerobic}} 2 \text{ Lactic acid} + 4 \text{ ATP}$$

Glucose + 2 ATP (or glycogen + 1 ATP)

$$\xrightarrow{\text{Oxygen}} 6 CO_2 + 6 H_2O + 40 \text{ ATP}$$

$$FFA \xrightarrow{\text{Oxygen}} CO_2 + H_2O + ATP$$

FIGURE 5–13 **ATP turnover in muscle cells.** Energy released by hydrolysis of 1 mol of ATP and reactions responsible for resynthesis of ATP. The amount of ATP formed per mole of free fatty acid (FFA) oxidized is large but varies with the size of the FFA. For example, complete oxidation of 1 mol of palmitic acid generates 140 mol of ATP.

THE OXYGEN DEBT MECHANISM

During exercise, the muscle blood vessels dilate and blood flow is increased so that the available O_2 supply is increased. Up to a point, the increase in O_2 consumption is proportional to the energy expended, and all the energy needs are met by aerobic processes. However, when muscular exertion is very great, aerobic resynthesis of energy stores cannot keep pace with their utilization. Under these conditions, phosphorylcreatine is still used to resynthesize ATP. In addition, some ATP synthesis is accomplished by using the energy released by the anaerobic breakdown of glucose to lactate. Use of the anaerobic pathway is self-limiting because in spite of rapid diffusion of lactate into the bloodstream, enough accumulates in the muscles to eventually exceed the capacity of the tissue buffers and produce an enzyme-inhibiting decline in pH. However, for short periods, the presence of an anaerobic pathway for glucose breakdown permits muscular exertion of a far greater magnitude than would be possible without it. For example, in a 100-m dash that takes 10 s, 85% of the energy consumed is derived anaerobically; in a 2-mi race that takes 10 min, 20% of the energy is derived anaerobically; and in a long-distance race that takes 60 min, only 5% of the energy comes from anaerobic metabolism.

After a period of exertion is over, extra O_2 is consumed to remove the excess lactate, replenish the ATP and phosphorylcreatine stores, and replace the small amounts of O_2 that were released by myoglobin. The amount of extra O_2 consumed is proportional to the extent to which the energy demands during exertion exceeded the capacity for the aerobic synthesis of energy stores, ie, the extent to which an **oxygen debt** was incurred. The O_2 debt is measured experimentally by determining O_2 consumption after exercise until a constant, basal consumption is reached and subtracting the basal consumption from the total. The amount of this debt may be six times the basal O_2 consumption, which indicates that the subject is capable of six times the exertion that would have been possible without it.

RIGOR

When muscle fibers are completely depleted of ATP and phosphorylcreatine, they develop a state of rigidity called **rigor.** When this occurs after death, the condition is called rigor mortis. In rigor, almost all of the myosin heads attach to actin but in an abnormal, fixed, and resistant way.

HEAT PRODUCTION IN MUSCLE

Thermodynamically, the energy supplied to a muscle must equal its energy output. The energy output appears in work done by the muscle, in energy-rich phosphate bonds formed for later use, and in heat. The overall mechanical efficiency of skeletal muscle (work done/total energy expenditure) ranges up to 50% while lifting a weight during isotonic contraction and is essentially 0% during isometric contraction. Energy storage in phosphate bonds is a small factor. Consequently, heat production is considerable. The heat produced in muscle can be measured accurately with suitable thermocouples.

Resting heat, the heat given off at rest, is the external manifestation of basal metabolic processes. The heat produced in excess of resting heat during contraction is called the **initial heat.** This is made up of **activation heat,** the heat that muscle produces whenever it is contracting, and **shortening heat,** which is proportionate in amount to the distance the muscle shortens. Shortening heat is apparently due to some change in the structure of the muscle during shortening.

Following contraction, heat production in excess of resting heat continues for as long as 30 min. This **recovery heat** is the heat liberated by the metabolic processes that restore the muscle to its precontraction state. The recovery heat of muscle is approximately equal to the initial heat; that is, the heat produced during recovery is equal to the heat produced during contraction.

If a muscle that has contracted isotonically is restored to its previous length, extra heat in addition to recovery heat is produced (**relaxation heat**). External work must be done on the muscle to return it to its previous length, and relaxation heat is mainly a manifestation of this work.

PROPERTIES OF SKELETAL MUSCLES IN THE INTACT ORGANISM

EFFECTS OF DENERVATION

In the intact animal or human, healthy skeletal muscle does not contract except in response to stimulation of its motor nerve supply. Destruction of this nerve supply causes muscle atrophy. It also leads to abnormal excitability of the muscle and increases its sensitivity to circulating acetylcholine (denervation hypersensitivity; see Chapter 6). Fine, irregular contractions of individual fibers (**fibrillations**) appear. This is the classic picture of a **lower motor neuron lesion.** If the motor nerve regenerates, the fibrillations disappear. Usually, the contractions are not visible grossly, and they should not be confused with **fasciculations,** which are jerky, visible contractions of groups of muscle fibers that occur as a result of pathologic discharge of spinal motor neurons.

THE MOTOR UNIT

Because the axons of the spinal motor neurons supplying skeletal muscle each branch to innervate several muscle fibers, the smallest possible amount of muscle that can contract in response to the excitation of a single motor neuron is not one muscle fiber but all the fibers supplied by the neuron. Each single motor neuron and the muscle fibers it innervates constitute a **motor unit.** The number of muscle fibers in a motor unit varies. In muscles such as those of the hand and those concerned with motion of the eye (ie, muscles concerned with fine, graded, precise movement), each motor unit innervates very few (on the order of three to six) muscle fibers. On the other hand, values of 600 muscle fibers per motor unit can occur in human leg muscles. The group of muscle fibers that contribute to a motor unit can be intermixed within a muscle. That is, although they contract as a unit, they are not necessarily "neighboring" fibers within the muscle.

Each spinal motor neuron innervates only one kind of muscle fiber, so that all the muscle fibers in a motor unit are of the same type. On the basis of the type of muscle fiber they innervate, and thus on the basis of the duration of their twitch contraction, motor units are divided into S (slow), FR (fast, resistant to fatigue), and FF (fast, fatigable) units. Interestingly, there is also a gradation of innervation of these fibers, with S fibers tending to have a low innervation ratio (ie, small units) and FF fibers tending to have a high innervation ratio (ie, large units). The recruitment of motor units during muscle contraction is not random, rather it follows a general scheme, the **size principle.** In general, a specific muscle action is developed first by the recruitment of S muscle units that contract relatively slowly to produce controlled contraction. Next, FR muscle units are recruited, resulting in more powerful response over a shorter period of time. Lastly, FF muscle units are recruited for the most demanding tasks. For example, in muscles of the leg, the small, slow units are first recruited for standing. As walking motion is initiated, their recruitment of FR units increases. As this motion turns to running or jumping, the FF units are recruited. Of course, there is overlap in recruitment, but, in general, this principle holds true.

The differences between types of muscle units are not inherent but are determined by, among other things, their activity. When the nerve to a slow muscle is cut and the nerve to a fast muscle is spliced to the cut end, the fast nerve grows and innervates the previously slow muscle. However, the muscle becomes fast and corresponding changes take place in its muscle protein isoforms and myosin ATPase activity. This change is due to changes in the pattern of activity of the muscle; in stimulation experiments, changes in the expression of MHC genes and consequently of MHC isoforms can be produced by changes in the pattern of electrical activity used to stimulate the muscle. More commonly, muscle fibers can be altered by a change in activity initiated through exercise (or lack thereof). Increased activity can lead to muscle cell hypertrophy, which allows for increase in contractile strength. Type IIA and IIB fibers are most susceptible to these changes. Alternatively, inactivity can lead to muscle cell atrophy and a loss of contractile strength. Type I fibers—that is, the ones used most often—are most susceptible to these changes.

ELECTROMYOGRAPHY

Activation of motor units can be studied by electromyography, the process of recording the electrical activity of muscle on an

FIGURE 5–14 **Electromyographic tracings from human biceps and triceps muscles during alternate flexion and extension of the elbow.** Note the alternate activation and rest patterns as one muscle is used for flexion and the other for extension. Electrical activity of stimulated muscle can be recorded extracellularly, yielding typical excitation responses after stimulation. (Courtesy of Garoutte BC.)

oscilloscope. This may be done in unanaesthetized humans by using small metal disks on the skin overlying the muscle as the pick-up electrodes or by using hypodermic needle electrodes. The record obtained with such electrodes is the **electromyogram (EMG).** With needle electrodes, it is usually possible to pick up the activity of single muscle fibers. The measured EMG depicts the potential difference between the two electrodes, which is altered by the activation of muscles in between the electrodes. A typical EMG is shown in Figure 5–14.

It has been shown by electromyography that little if any spontaneous activity occurs in the skeletal muscles of normal individuals at rest. With minimal voluntary activity a few motor units discharge, and with increasing voluntary effort, more and more are brought into play to monitor the **recruitment of motor units.** Gradation of muscle response is therefore in part a function of the number of motor units activated. In addition, the frequency of discharge in the individual nerve fibers plays a role, the tension developed during a tetanic contraction being greater than that during individual twitches. The length of the muscle is also a factor. Finally, the motor units fire asynchronously, that is, out of phase with one another. This asynchronous firing causes the individual muscle fiber responses to merge into a smooth contraction of the whole muscle. In summary, EMGs can be used to quickly (and roughly) monitor abnormal electrical activity associated with muscle responses.

THE STRENGTH OF SKELETAL MUSCLES

Human skeletal muscle can exert 3 to 4 kg of tension per square centimeter of cross-sectional area. This figure is about the same as that obtained in a variety of experimental animals and seems to be constant for mammalian species. Because many of the muscles in humans have a relatively large cross-sectional area, the tension they can develop is quite large. The gastrocnemius, for example, not only supports the weight of the whole body during climbing but resists a force several times this great when the foot hits the ground during running or jumping. An even more striking example is the gluteus maximus, which can exert a tension of 1200 kg. The total tension that could be developed if all muscles in the body of an adult man pulled together is approximately 22,000 kg (nearly 25 tons).

BODY MECHANICS

Body movements are generally organized in such a way that they take maximal advantage of the physiologic principles outlined above. For example, the attachments of the muscles in the body are such that many of them are normally at or near their resting length when they start to contract. In muscles that extend over more than one joint, movement at one joint may compensate for movement at another in such a way that relatively little shortening of the muscle occurs during contraction. Nearly isometric contractions of this type permit development of maximal tension per contraction. The hamstring muscles extend from the pelvis over the hip joint and the knee joint to the tibia and fibula. Hamstring contraction produces flexion of the leg on the thigh. If the thigh is flexed on the pelvis at the same time, the lengthening of the hamstrings across the hip joint tends to compensate for the shortening across the knee joint. In the course of various activities, the body moves in a way that takes advantage of this. Such factors as momentum and balance are integrated into body movement in ways that make possible maximal motion with minimal muscular exertion. One net effect is that the stress put on tendons and bones is rarely over 50% of their failure strength, protecting them from damage.

In walking, each limb passes rhythmically through a support or stance phase when the foot is on the ground and a swing phase when the foot is off the ground. The support phases of the two legs overlap, so that two periods of double support occur during each cycle. There is a brief burst of activity in the leg flexors at the start of each step, and then the leg is swung forward with little more active muscular contraction. Therefore, the muscles are active for only a fraction of each step, and walking for long periods causes relatively little fatigue.

A young adult walking at a comfortable pace moves at a velocity of about 80 m/min and generates a power output of 150 to 175 W per step. A group of young adults asked to walk at their most comfortable rate selected a velocity close to 80 m/min, and it was found that they had selected the velocity at which their energy output was minimal. Walking more rapidly or more slowly took more energy.

CARDIAC MUSCLE MORPHOLOGY

The striations in cardiac muscle are similar to those in skeletal muscle, and Z lines are present. Large numbers of elongated mitochondria are in close contact with the muscle fibrils. The muscle fibers branch and interdigitate, but each is a complete unit surrounded by a cell membrane. Where the end of one muscle fiber abuts on another, the membranes of both fibers parallel each other through an extensive series of folds. These areas, which always occur at Z lines, are called **intercalated disks** (Figure 5–15). They provide a strong union between fibers, maintaining cell-to-cell cohesion, so that the pull of one contractile cell can be transmitted along its axis to the next. Along the sides of the muscle fibers next to the disks, the cell membranes of adjacent fibers fuse for considerable distances, forming gap junctions. These junctions provide low-resistance bridges for the spread of excitation from one fiber to another. They permit cardiac muscle to function as if it were a syncytium, even though no protoplasmic bridges are present between cells. The T system in cardiac muscle is located at the Z lines rather than at the A–I junction, where it is located in mammalian skeletal muscle.

ELECTRICAL PROPERTIES

RESTING MEMBRANE & ACTION POTENTIALS

The resting membrane potential of individual mammalian cardiac muscle cells is about –80 mV. Stimulation produces a propagated action potential that is responsible for initiating contraction. Although action potentials vary among the cardiomyocytes in different regions of the heart (discussed in later chapters), the action potential of a typical ventricular cardiomyocyte can be used as an example (Figure 5–16). Depolarization proceeds rapidly and an overshoot of the zero potential is present, as in skeletal muscle and nerve, but this is followed by a plateau before the membrane potential returns to the baseline. In mammalian hearts, depolarization lasts about 2 ms, but the plateau phase and repolarization last 200 ms or more. Repolarization is therefore not complete until the contraction is half over.

As in other excitable tissues, changes in the external K^+ concentration affect the resting membrane potential of cardiac muscle, whereas changes in the external Na^+ concentration affect the magnitude of the action potential. The initial rapid depolarization and the overshoot (phase 0) are due to opening of voltage-gated Na^+ channels similar to that occurring in nerve and skeletal muscle (Figure 5–17). The initial rapid repolarization (phase 1) is due to closure of Na^+ channels and opening of one type of K^+ channel. The subsequent prolonged plateau (phase 2) is due to a slower but prolonged opening of voltage-gated Ca^{2+} channels. Final repolarization (phase 3) to the resting membrane potential (phase 4) is due to closure of the Ca^{2+} channels and a slow, delayed increase of K^+ efflux through various types of K^+ channels. Cardiac myocytes contain at least two types of Ca^{2+} channels (T- and L-types), but the Ca^{2+} current is due mostly to opening of the slower L-type Ca^{2+} channels.

A

Intercalated disk

Nucleus

FIBER

10 μm

Fibrils

Sarcolemma

Sarcoplasmic reticulum

2μm

T system

Capillary

N

N

Terminal cistern

FIBRIL

SARCOMERE

Mitochondria

Intercalated disk

B

FIGURE 5–15 **Cardiac muscle. A)** Electron photomicrograph of cardiac muscle. Note the similarity of the A-I regions seen in the skeletal muscle EM of Figure 3-2. The fuzzy thick lines are intercalated disks and function similarly to the Z-lines but occur at cell membranes (× 12,000). (Reproduced with permission from Bloom W, Fawcett DW: *A Textbook of Histology,* 10th ed. Saunders, 1975.) **B)** Artist interpretation of cardiac muscle as seen under the light microscope **(top)** and the electron microscope **(bottom).** Again, note the similarity to skeletal muscle structure. N, nucleus. (Reproduced with permission from Braunwald E, Ross J, Sonnenblick EH: Mechanisms of contraction of the normal and failing heart. N Engl J Med 1967;277:794. Courtesy of Little, Brown.)

MECHANICAL PROPERTIES

CONTRACTILE RESPONSE

The contractile response of cardiac muscle begins just after the start of depolarization and lasts about 1.5 times as long as the action potential (Figure 5–16). The role of Ca^{2+} in excitation–contraction coupling is similar to its role in skeletal muscle

(see above). However, it is the influx of extracellular Ca^{2+} through the voltage-sensitive DHPR in the T system that triggers calcium-induced calcium release through the RyR at the sarcoplasmic reticulum. Because there is a net influx of Ca^{2+} during activation, there is also a more prominent role for plasma membrane Ca^{2+} ATPases and the Na^+/Ca^{2+} exchanger in recovery of intracellular Ca^{2+} concentrations. Specific effects

FIGURE 5–16 **Comparison of action potentials and contractile response of a mammalian cardiac muscle fiber in a typical ventricular cell.** In the top-most trace, the most commonly viewed surface action potential recording can be seen and it is broken down into four regions: Q, R, S, and T. In the middle trace, the intracellular recording of the action potential shows the quick depolarization and extended recovery. In the bottom trace, the mechanical response is matched to the extracellular and intracellular electrical activities. Note that in the absolute refractory period (ARP), the cardiac myocyte cannot be excited, whereas in the relative refractory period (RRP) minimal excitation can occur.

FIGURE 5–17 **Dissection of the cardiac action potential.** **Top:** The action potential of a cardiac muscle fiber can be broken down into several phases: 0, depolarization; 1, initial rapid repolarization; 2, plateau phase; 3, late rapid repolarization; 4, baseline. **Bottom:** Diagrammatic summary of Na^+, Ca^{2+}, and cumulative K^+ currents during the action potential. As is convention, inward currents are downward, and outward currents are upward.

of drugs that indirectly alter Ca^{2+} concentrations are discussed in Clinical Box 5–2.

During phases 0 to 2 and about half of phase 3 (until the membrane potential reaches approximately –50 mV during repolarization), cardiac muscle cannot be excited again; that is, it is in its **absolute refractory period.** It remains relatively refractory until phase 4. Therefore, tetanus of the type seen in skeletal muscle cannot occur. Of course, tetanization of cardiac muscle for any length of time would have lethal consequences, and in this sense, the fact that cardiac muscle cannot be tetanized is a safety feature.

ISOFORMS

Cardiac muscle is generally slow and has relatively low ATPase activity. Its fibers are dependent on oxidative metabolism and hence on a continuous supply of O_2. The human heart contains both the α and the β isoforms of the myosin heavy chain (α MHC and β MHC). β MHC has lower myosin ATPase activity than α MHC. Both are present in the atria, with the α isoform predominating, whereas the β isoform predominates in the ventricle. The spatial differences in expression contribute to the well-coordinated contraction of the heart.

CORRELATION BETWEEN MUSCLE FIBER LENGTH & TENSION

The relation between initial fiber length and total tension in cardiac muscle is similar to that in skeletal muscle; there is a resting length at which the tension developed on stimulation is maximal. In the body, the initial length of the fibers is determined by the degree of diastolic filling of the heart, and the pressure developed in the ventricle is proportionate to the volume of the ventricle at the end of the filling phase (**Starling's law of the heart**). The developed tension (Figure 5–18) increases as the diastolic volume increases until it reaches a maximum, then tends to decrease. However, unlike skeletal muscle, the decrease in developed tension at high degrees of stretch is not due to a decrease in the number of cross-bridges between actin and myosin, because even severely dilated hearts are not stretched to this degree. The decrease is due instead to beginning disruption of the myocardial fibers.

The force of contraction of cardiac muscle can be also increased by catecholamines, and this increase occurs without a change in muscle length. This positive ionotropic effect of catecholamines is mediated via innervated β_1-adrenergic receptors, cyclic AMP, and their effects on Ca^{2+} homeostasis. The heart also contains noninnervated β_2-adrenergic receptors, which also act via cyclic AMP, but their ionotropic effect is smaller and is maximal in the atria. Cyclic AMP activates protein kinase A, and this leads to phosphorylation of the voltage-dependent Ca^{2+} channels, causing them to spend more time in the open state. Cyclic AMP also increases the active transport of Ca^{2+} to the sarcoplasmic reticulum, thus accelerating relaxation and consequently shortening systole. This is important when the cardiac rate is increased because it permits adequate diastolic filling (see Chapter 31).

FIGURE 5–18   Length–tension relationship for cardiac muscle. Comparison of the systolic intraventricular pressure (top trace) and diastolic intraventricular pressure (bottom trace) display the developed tension in the cardiomyocyte. Values shown are for canine heart.

METABOLISM

Mammalian hearts have an abundant blood supply, numerous mitochondria, and a high content of myoglobin, a muscle pigment that can function as an O_2 storage mechanism. Normally, less than 1% of the total energy liberated is provided by anaerobic metabolism. During hypoxia, this figure may increase to nearly 10%; but under totally anaerobic conditions, the energy liberated is inadequate to sustain ventricular contractions. Under basal conditions, 35% of the caloric needs of the human heart are provided by carbohydrate, 5% by ketones and amino acids, and 60% by fat. However, the proportions of substrates utilized vary greatly with the nutritional state. After ingestion of large amounts of glucose, more lactate and pyruvate are used; during prolonged starvation, more fat is used. Circulating free fatty acids normally account for almost 50% of the lipid utilized. In untreated diabetics, the carbohydrate utilization of cardiac muscle is reduced and that of fat is increased.

SMOOTH MUSCLE MORPHOLOGY

Smooth muscle is distinguished anatomically from skeletal and cardiac muscle because it lacks visible cross-striations. Actin and myosin-II are present, and they slide on each other to produce contraction. However, they are not arranged in regular arrays, as in skeletal and cardiac muscle, and so the striations are absent. Instead of Z lines, there are **dense bodies** in the cytoplasm and attached to the cell membrane, and these are bound by α-actinin to actin filaments. Smooth muscle also contains tropomyosin, but troponin appears to be absent. The isoforms of actin and myosin differ from those in skeletal muscle. A sarcoplasmic reticulum is present, but it is less extensive than those observed in skeletal or cardiac muscle. In general, smooth muscles contain few mitochondria and depend, to a large extent, on glycolysis for their metabolic needs.

TYPES

There is considerable variation in the structure and function of smooth muscle in different parts of the body. In general, smooth muscle can be divided into **unitary** (or **visceral**) **smooth muscle** and **multiunit smooth muscle.** Unitary smooth muscle occurs in large sheets, has many low-resistance gap junctional connections between individual muscle cells, and functions in a syncytial fashion. Unitary smooth muscle is found primarily in the walls of hollow viscera. The musculature of the intestine, the uterus, and the ureters are examples. Multiunit smooth muscle is made up of individual units with few (or no) gap junctional bridges. It is found in structures such as the iris of the eye, in which fine, graded contractions occur. It is not under voluntary control, but it has many functional similarities to skeletal muscle. Each multiunit smooth muscle cell has en passant endings of nerve fibers, but in unitary smooth muscle there are en passant junctions on fewer cells, with excitation spreading to other cells by gap junctions. In addition, these cells respond to hormones and other circulating substances. Blood vessels have both unitary and multiunit smooth muscle in their walls.

ELECTRICAL & MECHANICAL ACTIVITY

Unitary smooth muscle is characterized by the instability of its membrane potential and by the fact that it shows continuous, irregular contractions that are independent of its nerve supply. This maintained state of partial contraction is called **to-nus,** or **tone.** The membrane potential has no true "resting" value, being relatively low when the tissue is active and higher when it is inhibited, but in periods of relative quiescence values for resting potential are on the order of –20 to –65 mV. Smooth muscle cells can display divergent electrical activity (eg, Figure 5–19). There are slow sine wave-like fluctuations a few millivolts in magnitude and spikes that sometimes over-shoot the zero potential line and sometimes do not. In many tissues, the spikes have a duration of about 50 ms, whereas in some tissues the action potentials have a prolonged plateau during repolarization, like the action potentials in cardiac muscle. As in the other muscle types, there are significant con-tributions of K^+, Na^+, and Ca^{2+} channels and Na, K ATPase to this electrical activity. However, discussion of contributions to individual smooth muscle types is beyond the scope of this text.

Because of the continuous activity, it is difficult to study the relation between the electrical and mechanical events in uni-tary smooth muscle, but in some relatively inactive prepara-tions, a single spike can be generated. In such preparations the excitation–contraction coupling in unitary smooth muscle can occur with as much as a 500-ms delay. Thus, it is a very slow process compared with that in skeletal and cardiac mus-cle, in which the time from initial depolarization to initiation of contraction is less than 10 ms. Unlike unitary smooth mus-cle, multiunit smooth muscle is nonsyncytial and contrac-tions do not spread widely through it. Because of this, the contractions of multiunit smooth muscle are more discrete, fine, and localized than those of unitary smooth muscle.

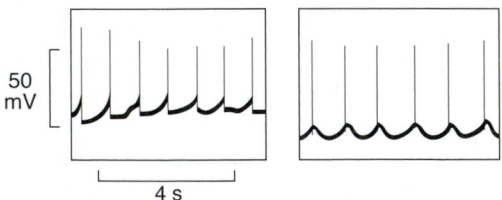

50 mV

4 s

FIGURE 5–19 **Electrical activity of individual smooth muscle cells in the guinea pig taenia coli. Left:** Pacemaker-like activity with spikes firing at each peak. **Right:** Sinusoidal fluctuation of membrane potential with firing on the rising phase of each wave. In other fibers, spikes can occur on the falling phase of sinusoidal fluctuations and there can be mixtures of sinusoidal and pacemaker potentials in the same fiber.

MOLECULAR BASIS OF CONTRACTION

As in skeletal and cardiac muscle, Ca^{2+} plays a prominent role in the initiation of contraction of smooth muscle. How-ever, the source of Ca^{2+} increase can be much different in unitary smooth muscle. Depending on the activating stimu-lus, Ca^{2+} increase can be due to influx through voltage- or ligand-gated plasma membrane channels, efflux from intra-cellular stores through the RyR, efflux from intracellular stores through the **inositol trisphosphate receptor (IP$_3$R)** Ca^{2+} channel, or via a combination of these channels. In ad-dition, the lack of troponin in smooth muscle prevents Ca^{2+} activation via troponin binding. Rather, myosin in smooth muscle must be phosphorylated for activation of the myosin ATPase. Phosphorylation and dephosphorylation of myosin also occur in skeletal muscle, but phosphorylation is not nec-essary for activation of the ATPase. In smooth muscle, Ca^{2+} binds to calmodulin, and the resulting complex activates **cal-modulin-dependent myosin light chain kinase.** This en-zyme catalyzes the phosphorylation of the myosin light chain on serine at position 19. The phosphorylation increases the ATPase activity.

Myosin is dephosphorylated by **myosin light chain phos-phatase** in the cell. However, dephosphorylation of myosin light chain kinase does not necessarily lead to relaxation of the smooth muscle. Various mechanisms are involved. One appears to be a latch bridge mechanism by which myosin cross-bridges remain attached to actin for some time after the cytoplasmic Ca^{2+} concentration falls. This produces sustained contraction with little expenditure of energy, which is espe-cially important in vascular smooth muscle. Relaxation of the muscle presumably occurs when the Ca^{2+}-calmodulin com-plex finally dissociates or when some other mechanism comes into play. The events leading to contraction and relaxation of unitary smooth muscle are summarized in Figure 5–20. The events in multiunit smooth muscle are generally similar.

Unitary smooth muscle is unique in that, unlike other types of muscle, it contracts when stretched in the absence of any extrinsic innervation. Stretch is followed by a decline in mem-brane potential, an increase in the frequency of spikes, and a general increase in tone.

If epinephrine or norepinephrine is added to a preparation of intestinal smooth muscle arranged for recording of intra-cellular potentials in vitro, the membrane potential usually becomes larger, the spikes decrease in frequency, and the muscle relaxes (Figure 5–21). Norepinephrine is the chemical mediator released at noradrenergic nerve endings, and stimu-lation of the noradrenergic nerves to the preparation pro-duces inhibitory potentials. Acetylcholine has an effect opposite to that of norepinephrine on the membrane poten-tial and contractile activity of intestinal smooth muscle. If acetylcholine is added to the fluid bathing a smooth muscle preparation in vitro, the membrane potential decreases and the spikes become more frequent. The muscle becomes more active, with an increase in tonic tension and the number of rhythmic contractions. The effect is mediated by phospholipase

FIGURE 5–20 **Sequence of events in contraction and relaxation of smooth muscle.** Flow chart illustrates many of the molecular changes that occur from the initiation of contraction to its relaxation. Note the distinct differences from skeletal and cardiac muscle excitation.

C, which produces IP$_3$ and allows for Ca^{2+} release through IP$_3$ receptors. In the intact animal, stimulation of cholinergic nerves causes release of acetylcholine, excitatory potentials, and increased intestinal contractions.

FIGURE 5–21 **Effects of various agents on the membrane potential of intestinal smooth muscle.** Drugs and hormones can alter firing of smooth muscle action potentials by raising (top trace) or lowering (bottom trace) resting membrane potential.

CLINICAL BOX 5–3

Common Drugs That Act on Smooth Muscle

Overexcitation of smooth muscle in the airways, such as that observed during an asthma attack, can lead to bronchoconstriction. Inhalers that deliver drugs to the conducting airway are commonly used to offset this smooth muscle bronchoconstriction, as well as other symptoms in the asthmatic airways. The rapid effects of drugs in inhalers are related to smooth muscle relaxation. Rapid response inhaler drugs (eg, ventolin, albuterol, sambuterol) frequently target β-adrenergic receptors in the airway smooth muscle to elicit a relaxation. Although these β-adrenergic receptor agonists targeting the smooth muscle do not treat all symptoms associated with bronchial constriction (eg, inflammation and increased mucus), they are quick and frequently allow for sufficient opening of the conducting airway to restore airflow, and thus allow for other treatments to reduce airway obstruction.

Smooth muscle is also a target for drugs developed to increase blood flow. As discussed in the text, NO is a natural signaling molecule that relaxes smooth muscle by raising cGMP. This signaling pathway is naturally down-regulated by the action of **phosphodiesterase (PDE),** which transforms cGMP into a nonsignaling form, GMP. The drugs sildenafil, tadalafil, and vardenafil are all specific inhibitors of PDE V, an isoform found mainly in the smooth muscle in the corpus cavernosum of the penis (see Chapter 25). Thus, oral administration of these drugs can block the action of PDE V, increasing blood flow in a very limited region in the body and offsetting erectile dysfunction.

Like unitary smooth muscle, multiunit smooth muscle is very sensitive to circulating chemical substances and is normally activated by chemical mediators (acetylcholine and norepinephrine) released at the endings of its motor nerves. Norepinephrine in particular tends to persist in the muscle and to cause repeated firing of the muscle after a single stimulus rather than a single action potential. Therefore, the contractile response produced is usually an irregular tetanus rather than a single twitch. When a single twitch response is obtained, it resembles the twitch contraction of skeletal muscle except that its duration is 10 times as long.

RELAXATION

In addition to cellular mechanisms that increase contraction of smooth muscle, there are cellular mechanisms that lead to its relaxation (Clinical Box 5–3). This is especially important in smooth muscle that surrounds the blood vessels to increase blood flow. It was long known that endothelial cells that line the inside of blood cells could release a substance that relaxed smooth muscle (**endothelial derived relaxation factor,**

EDRF). EDRF was later identified as the gaseous second messenger molecule, **nitric oxide (NO)**. NO produced in endothelial cells is free to diffuse into the smooth muscle for its effects. Once in muscle, NO directly activates a soluble guanylate cyclase to produce another second messenger molecule, **cyclic guanosine monophosphate (cGMP)**. This molecule can activate cGMP-specific protein kinases that can affect ion channels, Ca^{2+} homeostasis, or phosphatases, or all of those mentioned, that lead to smooth muscle relaxation (see Chapters 7 and 33).

FUNCTION OF THE NERVE SUPPLY TO SMOOTH MUSCLE

The effects of acetylcholine and norepinephrine on unitary smooth muscle serve to emphasize two of its important properties: (1) its spontaneous activity in the absence of nervous stimulation, and (2) its sensitivity to chemical agents released from nerves locally or brought to it in the circulation. In mammals, unitary muscle usually has a dual nerve supply from the two divisions of the autonomic nervous system. The function of the nerve supply is not to initiate activity in the muscle but rather to modify it. Stimulation of one division of the autonomic nervous system usually increases smooth muscle activity, whereas stimulation of the other decreases it. However, in some organs, noradrenergic stimulation increases and cholinergic stimulation decreases smooth muscle activity; in others, the reverse is true.

FORCE GENERATION & PLASTICITY OF SMOOTH MUSCLE

Smooth muscle displays a unique economy when compared to skeletal muscle. Despite approximately 20% of the myosin content and a 100-fold difference in ATP use when compared with skeletal muscle, they can generate similar force per cross-sectional area. One of the tradeoffs of obtaining force under these conditions is the noticeably slower contractions when compared to skeletal muscle. There are several known reasons for these noticeable changes, including unique isoforms of myosin and contractile-related proteins expressed in smooth muscle and their distinct regulation (discussed above). The unique architecture of the smooth cell and its coordinated units also likely contribute to these changes.

Another special characteristic of smooth muscle is the variability of the tension it exerts at any given length. If a unitary smooth muscle is stretched, it first exerts increased tension. However, if the muscle is held at the greater length after stretching, the tension gradually decreases. Sometimes the tension falls to or below the level exerted before the muscle was stretched. It is consequently impossible to correlate length and developed tension accurately, and no resting length can be assigned. In some ways, therefore, smooth muscle behaves more like a viscous mass than a rigidly structured tissue, and it is this property that is referred to as the **plasticity** of smooth muscle.

The consequences of plasticity can be demonstrated in humans. For example, the tension exerted by the smooth muscle walls of the bladder can be measured at different degrees of distention as fluid is infused into the bladder via a catheter. Initially, tension increases relatively little as volume is increased because of the plasticity of the bladder wall. However, a point is eventually reached at which the bladder contracts forcefully (see Chapter 38).

CHAPTER SUMMARY

- There are three main types of muscle cells: skeletal, cardiac, and smooth.
- Skeletal muscle is a true syncytium under voluntary control. Skeletal muscles receive electrical stimuli from neurons to elicit contraction: "excitation–contraction coupling." Action potentials in muscle cells are developed largely through coordination of Na^+, K^+, and Ca^{2+} channels. Contraction in skeletal muscle cells is coordinated through Ca^{2+} regulation of the actomyosin system that gives the muscle its classic striated pattern under the microscope.
- There are several different types of skeletal muscle fibers (I, IIA, IIB) that have distinct properties in terms of protein makeup and force generation. Skeletal muscle fibers are arranged into motor units of like fibers within a muscle. Skeletal motor units are recruited in a specific pattern as the need for more force is increased.
- Cardiac muscle is a collection of individual cells (cardiomyocytes) that are linked as a syncytium by gap junctional communication. Cardiac muscle cells also undergo excitation–contraction coupling. Pacemaker cells in the heart can initiate propagated action potentials. Cardiac muscle cells also have a striated, actomyosin system that underlies contraction.
- Smooth muscle cells are largely under control of the autonomic nervous system.
- There are two broad categories of smooth muscle cells: unitary and multiunit. Unitary smooth muscle contraction is synchronized by gap junctional communication to coordinate contraction among many cells. Multiunit smooth muscle contraction is coordinated by motor units, functionally similar to skeletal muscle.
- Smooth muscle cells contract through an actomyosin system, but do not have well-organized striations. Unlike skeletal and cardiac muscle, Ca^{2+} regulation of contraction is primarily through phosphorylation–dephosphorylation reactions.

MULTIPLE-CHOICE QUESTIONS

For all questions, select the single best answer unless otherwise directed.

1. The action potential of skeletal muscle
   A) has a prolonged plateau phase.
   B) spreads inward to all parts of the muscle via the T tubules.
   C) causes the immediate uptake of Ca^{2+} into the lateral sacs of the sarcoplasmic reticulum.
   D) is longer than the action potential of cardiac muscle.
   E) is not essential for contraction.

2. The functions of tropomyosin in skeletal muscle include

    A) sliding on actin to produce shortening.
    B) releasing Ca^{2+} after initiation of contraction.
    C) binding to myosin during contraction.
    D) acting as a "relaxing protein" at rest by covering up the sites where myosin binds to actin.
    E) generating ATP, which it passes to the contractile mechanism.

3. The cross-bridges of the sarcomere in skeletal muscle are made up of

    A) actin.
    B) myosin.
    C) troponin.
    D) tropomyosin.
    E) myelin.

4. The contractile response in skeletal muscle

    A) starts after the action potential is over.
    B) does not last as long as the action potential.
    C) produces more tension when the muscle contracts isometrically than when the muscle contracts isotonically.
    D) produces more work when the muscle contracts isometrically than when the muscle contracts isotonically.
    E) decreases in magnitude with repeated stimulation.

5. Gap junctions

    A) are absent in cardiac muscle.
    B) are present but of little functional importance in cardiac muscle.
    C) are present and provide the pathway for rapid spread of excitation from one cardiac muscle fiber to another.
    D) are absent in smooth muscle.
    E) connect the sarcotubular system to individual skeletal muscle cells.

CHAPTER RESOURCES

Alberts B, et al: *Molecular Biology of the Cell,* 5th ed. Garland Science, 2007.

Fung YC: *Biomechanics,* 2nd ed. Springer, 1993.

Hille B: *Ionic Channels of Excitable Membranes,* 3rd ed. Sinaver Associates, 2001.

Horowitz A: Mechanisms of smooth muscle contraction. Physiol Rev 1996;76:967.

Kandel ER, Schwartz JH, Jessell TM (editors): *Principles of Neural Science,* 4th ed. McGraw-Hill, 2000.

Sperelakis N (editor): *Cell Physiology Sourcebook,* 3rd ed. Academic Press, 2001.

Synaptic & Junctional Transmission

After studying this chapter, you should be able to:

■ Describe the main morphologic features of synapses.

■ Distinguish between chemical and electrical transmission at synapses.

■ Define convergence and divergence in neural networks, and discuss their implications.

■ Describe fast and slow excitatory and inhibitory postsynaptic potentials, outline the ionic fluxes that underlie them, and explain how the potentials interact to generate action potentials.

■ Define and give examples of direct inhibition, indirect inhibition, presynaptic inhibition, and postsynaptic inhibition.

■ Describe the neuromuscular junction, and explain how action potentials in the motor neuron at the junction lead to contraction of the skeletal muscle.

■ Define and explain denervation hypersensitivity.

INTRODUCTION

The all-or-none type of conduction seen in axons and skeletal muscle has been discussed in Chapters 4 and 5. Impulses are transmitted from one nerve cell to another cell at **synapses** (Figure 6–1). These are the junctions where the axon or some other portion of one cell (the **presynaptic cell**) terminates on the dendrites, soma, or axon of another neuron (Figure 6–2) or, in some cases, a muscle or gland cell (the **postsynaptic cell**). Cell-to-cell communication occurs across either a **chemical** or **electrical synapse.** At chemical synapses, a **synaptic cleft** separates the terminal of the presynaptic cell from the postsynaptic cell. An impulse in the presynaptic axon causes secretion of a chemical that diffuses across the synaptic cleft and binds to receptors on the surface of the postsynaptic cell. This triggers events that open or close channels in the membrane of the postsynaptic cell. In electrical synapses, the membranes of the presynaptic and postsynaptic neurons come close together, and gap junctions form between the cells (see Chapter 2). Like the intercellular junctions in other tissues, these junctions form low-resistance bridges through which ions can pass with relative ease. There are also a few conjoint synapses in which transmission is both electrical and chemical.

Regardless of the type of synapse, transmission is not a simple jumping of an action potential from the presynaptic to the postsynaptic cell. The effects of discharge at individual synaptic endings can be excitatory or inhibitory, and when the postsynaptic cell is a neuron, the summation of all the excitatory and inhibitory effects determines whether an action potential is generated. Thus, synaptic transmission is a complex process that permits the grading and adjustment of neural activity necessary for normal function. Because most synaptic transmission is chemical, consideration in this chapter is limited to chemical transmission unless otherwise specified.

Transmission from nerve to muscle resembles chemical synaptic transmission from one neuron to another. The **neuromuscular junction,** the specialized area where a motor nerve terminates on a skeletal muscle fiber, is the site of a stereotyped transmission process. The contacts between autonomic neurons and smooth and cardiac muscle are less specialized, and transmission in these locations is a more diffuse process. These forms of transmission are also considered in this chapter.

FIGURE 6–1 **Synapses on a typical motor neuron.** The neuron has dendrites **(1)**, an axon **(2)**, and a prominent nucleus **(3).** Note that rough endoplasmic reticulum extends into the dendrites but not into the axon. Many different axons converge on the neuron, and their terminal boutons form axodendritic **(4)** and axosomatic **(5)** synapses. **(6)** Myelin sheath. (Reproduced with permission from Krstic RV: *Ultrastructure of the Mammalian Cell.* Springer, 1979.)

FIGURE 6–2 **Electron photomicrograph of synaptic knob (S) ending on the shaft of a dendrite (D) in the central nervous system.** P, postsynaptic density; M, mitochondrion. (×56,000). (Courtesy of DM McDonald.)

SYNAPTIC TRANSMISSION: FUNCTIONAL ANATOMY

TYPES OF SYNAPSES

The anatomic structure of synapses varies considerably in the different parts of the mammalian nervous system. The ends of the presynaptic fibers are generally enlarged to form **terminal boutons (synaptic knobs)** (Figure 6–2). In the cerebral and cerebellar cortex, endings are commonly located on dendrites and frequently on **dendritic spines,** which are small knobs projecting from dendrites (Figure 6–3). In some instances, the terminal branches of the axon of the presynaptic neuron form a basket or net around the soma of the postsynaptic cell (basket cells of the cerebellum and autonomic ganglia). In other locations, they intertwine with the dendrites of the postsynaptic cell (climbing fibers of the cerebellum) or end on the dendrites directly (apical dendrites of cortical pyramidal cells). Some end on axons of postsynaptic neurons (axoaxonal endings). On average, each neuron divides to form over 2000 synaptic endings, and because the human central nervous system (CNS) has 10^{11} neurons, it follows that there are about 2×10^{14} synapses. Ob-

viously, therefore, communication between neurons is extremely complex. It should be noted as well that synapses are dynamic structures, increasing and decreasing in complexity and number with use and experience.

It has been calculated that in the cerebral cortex, 98% of the synapses are on dendrites and only 2% are on cell bodies. In the spinal cord, the proportion of endings on dendrites is less; there are about 8000 endings on the dendrites of a typical spinal neuron and about 2000 on the cell body, making the soma appear encrusted with endings.

PRESYNAPTIC & POSTSYNAPTIC STRUCTURE & FUNCTION

Each presynaptic terminal of a chemical synapse is separated from the postsynaptic structure by a synaptic cleft that is 20 to 40 nm wide. Across the synaptic cleft are many neurotransmitter receptors in the postsynaptic membrane, and usually a postsynaptic thickening called the **postsynaptic density** (Figures 6–2 and 6–3). The postsynaptic density is an ordered complex of specific receptors, binding proteins, and enzymes induced by postsynaptic effects.

Inside the presynaptic terminal are many mitochondria, as well as many membrane-enclosed vesicles, which contain neurotransmitters. There are three kinds of **synaptic vesicles:** small, clear synaptic vesicles that contain acetylcholine, glycine, GABA, or glutamate; small vesicles with a dense core that contain catecholamines; and large vesicles with a dense core that contain

FIGURE 6–3 **Axodendritic, axoaxonal, and axosomatic synapses.** Many presynaptic neurons terminate on dendritic spines, as shown at the top, but some also end directly on the shafts of dendrites. Note the presence of clear and granulated synaptic vesicles in endings and clustering of clear vesicles at active zones.

neuropeptides. The vesicles and the proteins contained in their walls are synthesized in the neuronal cell body and transported along the axon to the endings by fast axoplasmic transport. The neuropeptides in the large dense-core vesicles must also be produced by the protein-synthesizing machinery in the cell body. However, the small clear vesicles and the small dense-core vesicles recycle in the nerve ending. These vesicles fuse with the cell membrane and release transmitters through exocytosis and are then recovered by endocytosis to be refilled locally. In some instances, they enter endosomes and are budded off the endosome and refilled, starting the cycle over again. The steps involved are shown in Figure 6–4. More commonly, however, the synaptic vesicle discharges its contents through a small hole

in the cell membrane, then the opening reseals rapidly and the main vesicle stays inside the cell (kiss-and-run discharge). In this way, the full endocytotic process is short-circuited.

The large dense-core vesicles are located throughout the presynaptic terminals that contain them and release their neuropeptide contents by exocytosis from all parts of the terminal. On the other hand, the small vesicles are located near the synaptic cleft and fuse to the membrane, discharging their contents very rapidly into the cleft at areas of membrane thickening called **active zones** (Figure 6–3). The active zones contain many proteins and rows of calcium channels.

The Ca^{2+} that triggers exocytosis of transmitters enters the presynaptic neurons, and transmitter release starts within 200 μs. Therefore, it is not surprising that the voltage-gated Ca^{2+} channels are very close to the release sites at the active zones. In addition, for the transmitter to be effective on the postsynaptic neuron requires proximity of release to the postsynaptic receptors. This orderly organization of the synapse depends in part on **neurexins,** proteins bound to the membrane of the presynaptic neuron that bind neurexin receptors in the membrane of the postsynaptic neuron. In many vertebrates, neurexins are produced by a single gene that codes for the α isoform. However, in mice and humans they are encoded by three genes, and both α and β isoforms are produced. Each of the genes has two regulatory regions and extensive alternative splicing of their mRNAs. In this way, over 1000 different neurexins are produced. This raises the possibility that the neurexins not only hold synapses together, but also provide a mechanism for the production of synaptic specificity.

As noted in Chapter 2, vesicle budding, fusion, and discharge of contents with subsequent retrieval of vesicle membrane are fundamental processes occurring in most, if not all, cells. Thus, neurotransmitter secretion at synapses and the accompanying membrane retrieval are specialized forms of the general processes of exocytosis and endocytosis. The details of the processes by which synaptic vesicles fuse with the cell membrane are still being worked out. They involve the **v-snare** protein **synaptobrevin** in the vesicle membrane locking with the **t-snare** protein **syntaxin** in the cell membrane; a multiprotein complex regulated by small GTPases such as rab3 is also involved in the process (Figure 6–5). The synapse begins in the presynaptic and not in the postsynaptic cell. The one-way gate at the synapses is necessary for orderly neural function.

Clinical Box 6–1 describes the how neurotoxins can disrupt transmitter release in either the CNS or at the neuromuscular junction.

ELECTRICAL EVENTS IN POSTSYNAPTIC NEURONS

EXCITATORY & INHIBITORY POSTSYNAPTIC POTENTIALS

Penetration of an α-motor neuron is a good example of the techniques used to study postsynaptic electrical activity. It is

FIGURE 6–4 **Small synaptic vesicle cycle in presynaptic nerve terminals.** Vesicles bud off the early endosome and then fill with neurotransmitter (NT; top left). They then move to the plasma membrane, dock, and become primed. Upon arrival of an action potential at the ending,
Ca²⁺ influx triggers fusion and exocytosis of the granule contents to the synaptic cleft. The vesicle wall is then coated with clathrin and taken up
by endocytosis. In the cytoplasm, it fuses with the early endosome, and the cycle is ready to repeat. (Reproduced with permission from Sdhof TC: The
synaptic vesicle cycle: A cascade of proteinprotein interactions. Nature 1995;375:645. Copyright by Macmillan Magazines.)

achieved by advancing a microelectrode through the ventral
portion of the spinal cord. Puncture of a cell membrane is signaled by the appearance of a steady 70-mV potential difference between the microelectrode and an electrode outside the
cell. The cell can be identified as a spinal motor neuron by
stimulating the appropriate ventral root and observing the
electrical activity of the cell. Such stimulation initiates an antidromic impulse (see Chapter 4) that is conducted to the soma

and stops at this point. Therefore, the presence of an action
potential in the cell after antidromic stimulation indicates that
the cell that has been penetrated is an α-motor neuron. Stimulation of a dorsal root afferent (sensory neuron) can be used
to study both excitatory and inhibitory events in α-motor
neurons (Figure 6–6).

When an impulse reaches the presynaptic terminals, an
interval of at least 0.5 ms, the **synaptic delay,** occurs before a
response is obtained in the postsynaptic neuron. It is due to
the time it takes for the synaptic mediator to be released and
to act on the membrane of the postsynaptic cell. Because of it,
conduction along a chain of neurons is slower if many synapses are in the chain than if there are only a few. Because the
minimum time for transmission across one synapse is 0.5 ms,
it is also possible to determine whether a given reflex pathway
is monosynaptic or polysynaptic (contains more than one
synapse) by measuring the delay in transmission from the
dorsal to the ventral root across the spinal cord.

A single stimulus applied to the sensory nerves characteristically does not lead to the formation of a propagated action
potential in the postsynaptic neuron. Instead, the stimulation
produces either a transient partial depolarization or a transient hyperpolarization. The initial depolarizing response
produced by a single stimulus to the proper input begins
about 0.5 ms after the afferent impulse enters the spinal
cord. It reaches its peak 11.5 ms later and then declines
exponentially. During this potential, the excitability of the
neuron to other stimuli is increased, and consequently the
potential is called an **excitatory postsynaptic potential**
(EPSP) (Figure 6–6).

FIGURE 6–5 **Main proteins that interact to produce synaptic**
vesicle docking and fusion in nerve endings. (Reproduced with
permission from Ferro-Novick S, John R: Vesicle fusion from yeast to man. Nature
1994;370:191. Copyright by Macmillan Magazines.)

CLINICAL BOX 6-1

Botulinum and Tetanus Toxins

Several deadly toxins which block neurotransmitter release are zinc endopeptidases that cleave and hence inactivate proteins in the fusion–exocytosis complex. **Tetanus toxin** and **botulinum toxins** B, D, F, and G act on synaptobrevin, and botulinum toxin C acts on syntaxin. Botulinum toxins A and B act on SNAP-25. Clinically, tetanus toxin causes spastic paralysis by blocking presynaptic transmitter release in the CNS, and botulism causes flaccid paralysis by blocking the release of acetylcholine at the neuromuscular junction. On the positive side, however, local injection of small doses of botulinum toxin (botox) has proved effective in the treatment of a wide variety of conditions characterized by muscle hyperactivity. Examples include injection into the lower esophageal sphincter to relieve achalasia and injection into facial muscles to remove wrinkles.

A Stretch reflex circuit for knee jerk

B Experimental setup for recording from cells in the circuit

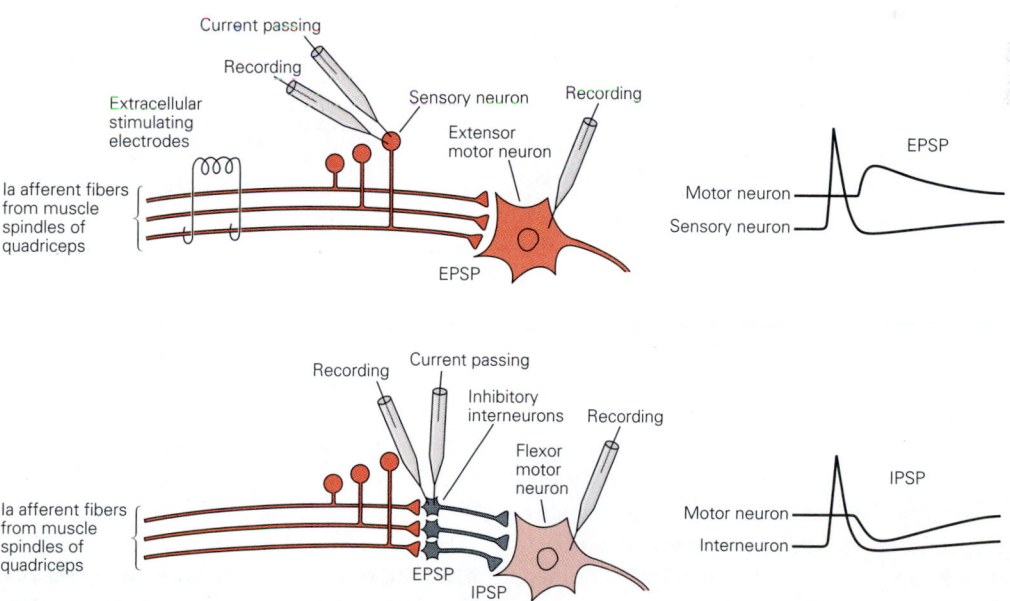

FIGURE 6–6 **Excitatory and inhibitory synaptic connections mediating the stretch reflex provide an example of typical circuits within the CNS. A)** The stretch receptor sensory neuron of the quadriceps muscle makes an excitatory connection with the extensor motor neuron of the same muscle and an inhibitory interneuron projecting to flexor motor neurons supplying the antagonistic hamstring muscle. **B)** Experimental setup to study excitation and inhibition of the extensor motor neuron. Top panel shows two approaches to elicit an excitatory (depolarizing) postsynaptic potential or EPSP in the extensor motor neuron–electrical stimulation of the whole Ia afferent nerve using extracellular electrodes and intracellular current passing through an electrode inserted into the cell body of a sensory neuron. Bottom panel shows that current passing through an inhibitory interneuron elicits an inhibitory (hyperpolarizing) postsynaptic potential or IPSP in the flexor motor neuron. (From Kandel ER, Schwartz JH, Jessell TM [editors]: *Principles of Neural Science*, 4th ed. McGraw-Hill, 2000.)

The EPSP is produced by depolarization of the postsynaptic cell membrane immediately under the presynaptic ending. The excitatory transmitter opens Na^+ or Ca^{2+} ion channels in the postsynaptic membrane, producing an inward current. The area of current flow thus created is so small that it does not drain off enough positive charge to depolarize the whole membrane. Instead, an EPSP is inscribed. The EPSP due to activity in one synaptic knob is small, but the depolarizations produced by each of the active knobs summate.

EPSPs are produced by stimulation of some inputs, but stimulation of other inputs produces hyperpolarizing responses. Like the EPSPs, they peak 11.5 ms after the stimulus and decrease exponentially. During this potential, the excitability of the neuron to other stimuli is decreased; consequently, it is called an **inhibitory postsynaptic potential (IPSP)** (Figure 6–6).

An IPSP can be produced by a localized increase in Cl^- transport. When an inhibitory synaptic knob becomes active, the released transmitter triggers the opening of Cl^- channels in the area of the postsynaptic cell membrane under the knob. Cl^- moves down its concentration gradient. The net effect is the transfer of negative charge into the cell, so that the membrane potential increases.

The decreased excitability of the nerve cell during the IPSP is due to movement of the membrane potential away from the firing level. Consequently, more excitatory (depolarizing) activity is necessary to reach the firing level. The fact that an IPSP is mediated by Cl^- can be demonstrated by repeating the stimulus while varying the resting membrane potential of the postsynaptic cell. When the membrane potential is at E_{Cl}, the potential disappears (Figure 6–7), and at more negative membrane potentials, it becomes positive **(reversal potential)**.

Because IPSPs are net hyperpolarizations, they can be produced by alterations in other ion channels in the neuron. For example, they can be produced by opening of K^+ channels, with movement of K^+ out of the postsynaptic cell, or by closure of Na^+ or Ca^{2+} channels.

TEMPORAL & SPATIAL SUMMATION

Summation may be temporal or spatial (Figure 6–8). **Temporal summation** occurs if repeated afferent stimuli cause new EPSPs before previous EPSPs have decayed. A longer time constant for the EPSP allows for a greater opportunity for summation. When activity is present in more than one synaptic knob at the same time, **spatial summation** occurs and activity in one synaptic knob summates with activity in another to approach the firing level. The EPSP is therefore not an all-or-none response but is proportionate in size to the strength of the afferent stimulus.

Spatial summation of IPSPs also occurs, as shown by the increasing size of the response, as the strength of an inhibitory afferent volley is increased. Temporal summation of IPSPs also occurs.

FIGURE 6–7 IPSP is due to increased Cl influx during stimulation. This can be demonstrated by repeating the stimulus while varying the resting membrane potential (RMP) of the postsynaptic cell. When the membrane potential is at E_{Cl}, the potential disappears, and at more negative membrane potentials, it becomes positive (reversal potential).

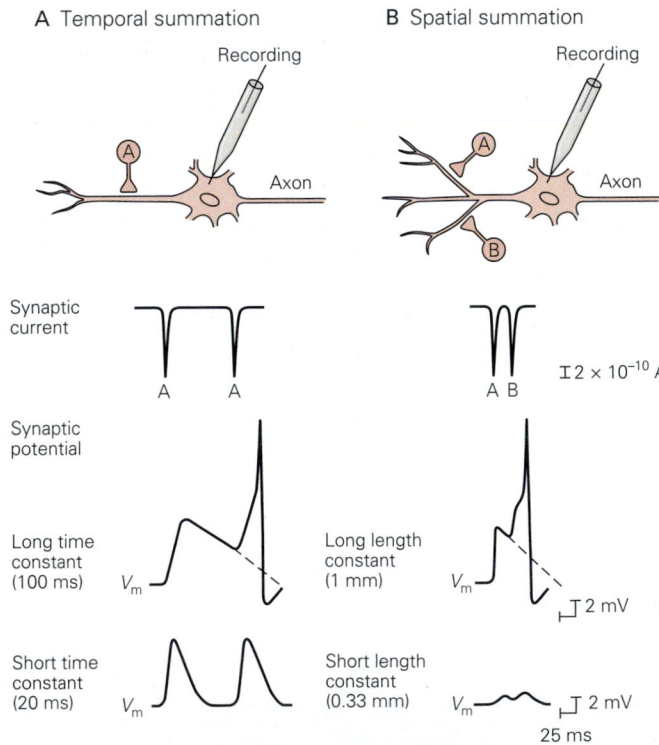

FIGURE 6–8 Central neurons integrate a variety of synaptic inputs through temporal and spatial summation. A) The time constant of the postsynaptic neuron affects the amplitude of the depolarization caused by consecutive EPSPs produced by a single presynaptic neuron. **B)** The length constant of a postsynaptic cell affects the amplitude of two EPSPs produced by two presynaptic neurons, A and B. (From Kandel ER, Schwartz JH, Jessell TM [editors]: *Principles of Neural Science*, 4th ed. McGraw-Hill, 2000.)

SLOW POSTSYNAPTIC POTENTIALS

In addition to the EPSPs and IPSPs described previously, slow EPSPs and IPSPs have been described in autonomic ganglia, cardiac and smooth muscle, and cortical neurons. These postsynaptic potentials have a latency of 100 to 500 ms and last several seconds. The slow EPSPs are generally due to decreases in K^+ conductance, and the slow IPSPs are due to increases in K^+ conductance.

GENERATION OF THE ACTION POTENTIAL IN THE POSTSYNAPTIC NEURON

The constant interplay of excitatory and inhibitory activity on the postsynaptic neuron produces a fluctuating membrane potential that is the algebraic sum of the hyperpolarizing and depolarizing activity. The soma of the neuron thus acts as a sort of integrator. When the 10 to 15 mV of depolarization sufficient to reach the firing level is attained, a propagated spike results. However, the discharge of the neuron is slightly more complicated than this. In motor neurons, the portion of the cell with the lowest threshold for the production of a full-fledged action potential is the **initial segment,** the portion of the axon at and just beyond the axon hillock. This unmyelinated segment is depolarized or hyperpolarized electrotonically by the current sinks and sources under the excitatory and inhibitory synaptic knobs. It is the first part of the neuron to fire, and its discharge is propagated in two directions: down the axon and back into the soma. Retrograde firing of the soma in this fashion probably has value in wiping the slate clean for subsequent renewal of the interplay of excitatory and inhibitory activity on the cell.

FUNCTION OF THE DENDRITES

For many years, the standard view has been that dendrites are simply the sites of current sources or sinks that electrotonically change the membrane potential at the initial segment; that is, they are merely extensions of the soma that expand the area available for integration. When the dendritic tree of a neuron is extensive and has multiple presynaptic knobs ending on it, there is room for a great interplay of inhibitory and excitatory activity.

It is now well established that dendrites contribute to neural function in more complex ways. Action potentials can be recorded in dendrites. In many instances, these are initiated in the initial segment and conducted in a retrograde fashion, but propagated action potentials are initiated in some dendrites. Further research has demonstrated the malleability of dendritic spines. Dendritic spines appear, change, and even disappear over a time scale of minutes and hours, not days and months. Also, although protein synthesis occurs mainly in the soma with its nucleus, strands of mRNA migrate into the dendrites. There, each can become associated with a single ribosome in a dendritic spine and produce proteins, which alters the effects of input from individual synapses on the spine. Changes in dendritic spines have been implicated in motivation, learning, and long-term memory.

ELECTRICAL TRANSMISSION

At synaptic junctions where transmission is electrical, the impulse reaching the presynaptic terminal generates an EPSP in the postsynaptic cell that, because of the low-resistance bridge between the two, has a much shorter latency than the EPSP at a synapse where transmission is chemical. In conjoint synapses, both a short-latency response and a longer-latency, chemically mediated postsynaptic response take place.

INHIBITION & FACILITATION AT SYNAPSES

DIRECT & INDIRECT INHIBITION

Inhibition in the CNS can be postsynaptic or presynaptic. **Postsynaptic inhibition** during the course of an IPSP is called **direct inhibition** because it is not a consequence of previous discharges of the postsynaptic neuron. There are various forms of **indirect inhibition,** which is inhibition due to the effects of previous postsynaptic neuron discharge. For example, the postsynaptic cell can be refractory to excitation because it has just fired and is in its refractory period. During after-hyperpolarization it is also less excitable. In spinal neurons, especially after repeated firing, this after-hyperpolarization may be large and prolonged.

POSTSYNAPTIC INHIBITION IN THE SPINAL CORD

Various pathways in the nervous system are known to mediate postsynaptic inhibition, and one illustrative example is presented here. Afferent fibers from the muscle spindles (stretch receptors) in skeletal muscle project directly to the spinal motor neurons of the motor units supplying the same muscle (Figure 6–6). Impulses in this afferent fiber cause EPSPs and, with summation, propagated responses in the postsynaptic motor neurons. At the same time, IPSPs are produced in motor neurons supplying the antagonistic muscles which have an inhibitory interneuron interposed between the afferent fiber and the motor neuron. Therefore, activity in the afferent fibers from the muscle spindles excites the motor neurons supplying the muscle from which the impulses come, and inhibits those supplying its antagonists (**reciprocal innervation**).

PRESYNAPTIC INHIBITION & FACILITATION

Another type of inhibition occurring in the CNS is **presynaptic inhibition,** a process mediated by neurons whose terminals are on excitatory endings, forming **axoaxonal synapses** (Figure 6–3). The neurons responsible for postsynaptic and presynaptic inhibition are compared in Figure 6–9. Three mechanisms of presynaptic inhibition have been described. First, activation of the presynaptic receptors increases Cl^- conductance, and this has been shown to decrease the size of the action potentials reaching the excitatory ending (Figure 6–10). This in turn reduces Ca^{2+} entry and consequently the amount of excitatory transmitter released. Voltage-gated K^+ channels are also opened, and the resulting K^+ efflux also decreases the Ca^{2+} influx. Finally, there is evidence for direct inhibition of transmitter release independent of Ca^{2+} influx into the excitatory ending.

The first transmitter shown to produce presynaptic inhibition was GABA. Acting via $GABA_A$ receptors, GABA increases Cl^- conductance. $GABA_B$ receptors are also present in the spinal cord and appear to mediate presynaptic inhibition via a G protein that produces an increase in K^+ conductance. Baclofen, a $GABA_B$ agonist, is effective in the treatment of the spasticity of spinal cord injury and multiple sclerosis, particularly when administered intrathecally via an implanted pump. Other transmitters also mediate presynaptic inhibition by G protein-mediated effects on Ca^{2+} channels and K^+ channels.

Conversely, **presynaptic facilitation** is produced when the action potential is prolonged (Figure 6–10) and the Ca^{2+} channels are open for a longer period. The molecular events responsible for the production of presynaptic facilitation mediated by serotonin in the sea snail *Aplysia* have been worked out in detail. Serotonin released at an axoaxonal ending increases intraneuronal cAMP levels, and the resulting phosphorylation

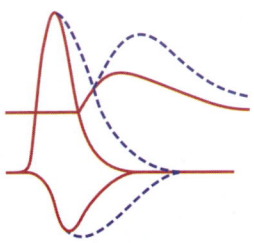

FIGURE 6–10   Effects of presynaptic inhibition and facilitation on the action potential and the Ca^{2+} current in the presynaptic neuron and the EPSP in the postsynaptic neuron. In each case, the solid lines are the controls and the dashed lines the records obtained during inhibition or facilitation. (Modified from Kandel ER, Schwartz JH, Jessell TM [editors]: *Principles of Neural Science,* 4th ed. McGraw-Hill, 2000.)

of one group of K^+ channels closes the channels, slowing repolarization and prolonging the action potential.

ORGANIZATION OF INHIBITORY SYSTEMS

Presynaptic and postsynaptic inhibition are usually produced by stimulation of certain systems converging on a given postsynaptic neuron (afferent inhibition). Neurons may also inhibit themselves in a negative feedback fashion (negative feedback inhibition). For instance, each spinal motor neuron regularly gives off a recurrent collateral that synapses with an inhibitory interneuron, which terminates on the cell body of the spinal neuron and other spinal motor neurons (Figure 6–11). This particular inhibitory neuron is sometimes called a Renshaw cell after its discoverer. Impulses generated in the motor neuron activate the inhibitory interneuron to secrete inhibitory mediators, and this slows or stops the discharge of the motor neuron. Similar inhibition via recurrent collaterals is seen in the cerebral cortex and limbic system. Presynaptic inhibition due to descending pathways that terminate on afferent pathways in the dorsal horn may be involved in the gating of pain transmission.

Another type of inhibition is seen in the cerebellum. In this part of the brain, stimulation of basket cells produces IPSPs in the Purkinje cells. However, the basket cells and the Purkinje

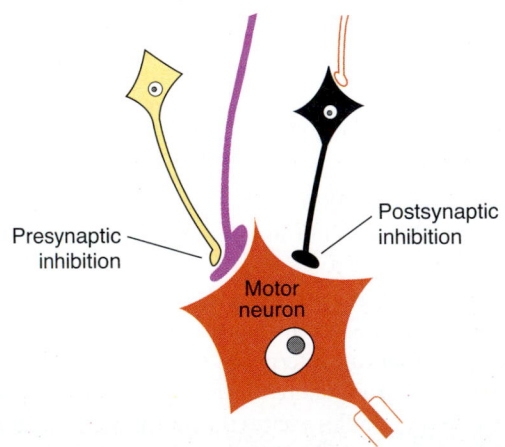

FIGURE 6–9   Arrangement of neurons producing presynaptic and postsynaptic inhibition. The neuron producing presynaptic inhibition is shown ending on an excitatory synaptic knob. Many of these neurons actually end higher up along the axon of the excitatory cell.

FIGURE 6–11   **Negative feedback inhibition of a spinal motor neuron via an inhibitory interneuron (Renshaw cell).**

FIGURE 6–12   **Simple nerve net.** Neurons A, B, and C have excitatory endings on neurons X, Y, and Z.

cells are excited by the same parallel-fiber excitatory input. This arrangement, which has been called feed-forward inhibition, presumably limits the duration of the excitation produced by any given afferent volley.

SUMMATION & OCCLUSION

As noted above, the axons of most neurons discharge onto many other neurons. Conversely, any given neuron receives input from many other neurons (convergence).

In the hypothetical nerve net shown in Figure 6–12, neurons A and B converge on X, and neuron B diverges on X and Y. A stimulus applied to A or to B will set up an EPSP in X. If A and B are stimulated at the same time and action potentials are produced, two areas of depolarization will be produced in X and their actions will sum. The resultant EPSP in X will be twice as large as that produced by stimulation of A or B alone, and the membrane potential may well reach the firing level of X. The effect of the depolarization caused by the impulse in A adds to that due to activity in B, and vice versa; spatial summation has taken place. In this case, Y has not fired, but its excitability has been increased, and it is easier for activity in neuron C to fire Y during the EPSP. Y is therefore said to be in the **subliminal fringe** of X. More generally stated, neurons are in the subliminal fringe if they are not discharged by an afferent volley (not in the **discharge zone**) but do have their excitability increased. The neurons that have few active knobs ending on them are in the subliminal fringe, and those with many are in the discharge zone. Inhibitory impulses show similar temporal and spatial facilitation and subliminal fringe effects.

If action potentials are produced repeatedly in neuron B, X and Y will discharge as a result of temporal summation of the EPSPs that are produced. If C is stimulated repeatedly, Y and Z will discharge. If B and C are fired repeatedly at the same time, X, Y, and Z will discharge. Thus, the response to stimu-

lation of B and C together is not as great as the sum of responses to stimulation of B and C separately, because B and C both end on neuron Y. This decrease in expected response, due to presynaptic fibers sharing postsynaptic neurons, is called **occlusion.**

NEUROMUSCULAR TRANSMISSION: NEUROMUSCULAR JUNCTION

ANATOMY

As the axon supplying a skeletal muscle fiber approaches its termination, it loses its myelin sheath and divides into a number of terminal boutons, or endfeet (Figure 6–13). The endfeet contain many small, clear vesicles that contain acetylcholine, the transmitter at these junctions. The endings fit into **junctional folds,** which are depressions in the **motor end plate,** the thickened portion of the muscle membrane at the junction. The space between the nerve and the thickened muscle membrane is comparable to the synaptic cleft at synapses. The whole structure is known as the **neuromuscular,** or **myoneural, junction.** Only one nerve fiber ends on each end plate, with no convergence of multiple inputs.

SEQUENCE OF EVENTS DURING TRANSMISSION

The events occurring during transmission of impulses from the motor nerve to the muscle are somewhat similar to those occurring at neuron-to-neuron synapses (Figure 6–14). The impulse arriving in the end of the motor neuron increases the permeability of its endings to Ca^{2+}. Ca^{2+} enters the endings and triggers a marked increase in exocytosis of the acetylcholine-containing vesicles. The acetylcholine diffuses to the

(a)

(b)

Motor nerve fiber
Myelin
Axon terminal
Schwann cell
Synaptic vesicles
(containing ACh)
Active zone
Sarcolemma
Synaptic
cleft
Junctional
folds
Nucleus of muscle fiber
Region of
sarcolemma
with ACh receptors

FIGURE 6–13   The neuromuscular junction. (a) Scanning electronmicrograph showing branching of motor axons with terminals embedded in grooves in the muscle fiber's surface. **(b)** Structure of a neuromuscular junction.   (From Widmaier EP, Raff H, Strang KT: *Vanders Human Physiology.* McGraw-Hill, 2008.)

muscle-type nicotinic acetylcholine receptors, which are concentrated at the tops of the junctional folds of the membrane of the motor end plate. Binding of acetylcholine to these receptors increases the Na^+ and K^+ conductance of the membrane, and the resultant influx of Na^+ produces a depolarizing potential, the **end plate potential.** The current sink created by this local potential depolarizes the adjacent muscle membrane to its firing level. Acetylcholine is then removed from the synaptic cleft by acetylcholinesterase, which is present in high concentration at the neuromuscular junction. Action potentials are generated on either side of the end plate and are conducted away from the end plate in both directions along the muscle fiber. The muscle action potential, in turn, initiates muscle contraction, as described in Chapter 5.

END PLATE POTENTIAL

An average human end plate contains about 15 to 40 million acetylcholine receptors. Each nerve impulse releases about 60 acetylcholine vesicles, and each vesicle contains about 10,000 molecules of the neurotransmitter. This amount is enough to activate about 10 times the number of acetylcholine receptors needed to produce a full end plate potential. Therefore, a propagated response in the muscle is regularly produced, and this large response obscures the end plate potential. However, the end plate potential can be seen if the tenfold safety factor is overcome and the potential is reduced to a size that is insuf-

ficient to fire the adjacent muscle membrane. This can be accomplished by administration of small doses of curare, a drug that competes with acetylcholine for binding to muscle-type nicotinic acetylcholine receptors. The response is then recorded only at the end plate region and decreases exponentially away from it. Under these conditions, end plate potentials can be shown to undergo temporal summation.

QUANTAL RELEASE OF TRANSMITTER

Small quanta (packets) of acetylcholine are released randomly from the nerve cell membrane at rest. Each produces a minute depolarizing spike called a **miniature end plate potential,** which is about 0.5 mV in amplitude. The size of the quanta of acetylcholine released in this way varies directly with the Ca^{2+} concentration and inversely with the Mg^{2+} concentration at the end plate. When a nerve impulse reaches the ending, the number of quanta released increases by several orders of magnitude, and the result is the large end plate potential that exceeds the firing level of the muscle fiber.

Quantal release of acetylcholine similar to that seen at the myoneural junction has been observed at other cholinergic synapses, and quantal release of other transmitters probably occurs at noradrenergic, glutaminergic, and other synaptic junctions.

Two diseases of the neuromuscular junction, myasthenia gravis and Lambert-Eaton syndrome, are described in Clinical Box 6–2 and Clinical Box 6–3, respectively.

FIGURE 6–14 **Events at the neuromuscular junction that lead to an action potential in the muscle fiber plasma membrane.** Although potassium exits the muscle cell when Ach receptors are open, sodium entry and depolarization dominate. (From Widmaier EP, Raff H, Strang KT: *Vanders Human Physiology.* McGraw-Hill, 2008.)

NERVE ENDINGS IN SMOOTH & CARDIAC MUSCLE

ANATOMY

The postganglionic neurons in the various smooth muscles that have been studied in detail branch extensively and come in close contact with the muscle cells (Figure 6–15). Some of these nerve fibers contain clear vesicles and are cholinergic, whereas others contain the characteristic dense-core vesicles that contain norepinephrine. There are no recognizable end plates or other postsynaptic specializations. The nerve fibers run along the membranes of the muscle cells and sometimes groove their surfaces. The multiple branches of the noradrenergic and, presumably, the cholinergic neurons are beaded with enlargements (**varicosities**) and contain synaptic vesicles (Figure 6–15). In noradrenergic neurons, the varicosities are about 5 μm apart, with up to 20,000 varicosities per neuron. Transmitter is apparently liberated at each varicosity, that is, at many locations along each axon. This arrangement permits one neuron to innervate many effector cells. The type of contact in which a neuron forms a synapse on the surface of another neuron or a smooth muscle cell and then passes on to make similar contacts with other cells is called a **synapse en passant.**

In the heart, cholinergic and noradrenergic nerve fibers end on the sinoatrial node, the atrioventricular node, and the bundle of His. Noradrenergic fibers also innervate the ventricular muscle. The exact nature of the endings on nodal tissue is not known. In the ventricle, the contacts between the noradrenergic fibers and the cardiac muscle fibers resemble those found in smooth muscle.

JUNCTIONAL POTENTIALS

In smooth muscles in which noradrenergic discharge is excitatory, stimulation of the noradrenergic nerves produces discrete partial depolarizations that look like small end plate potentials and are called **excitatory junction potentials (EJPs).** These potentials summate with repeated stimuli. Similar EJPs are seen in tissues excited by cholinergic discharges. In tissues inhibited by noradrenergic stimuli, hyperpolarizing **inhibitory junction potentials (IJPs)** are produced by stimulation of the noradrenergic nerves. Junctional potentials spread electrotonically.

DENERVATION HYPERSENSITIVITY

When the motor nerve to skeletal muscle is cut and allowed to degenerate, the muscle gradually becomes extremely sensitive to acetylcholine. This **denervation hypersensitivity** or **super-** sensitivity is also seen in smooth muscle. Smooth muscle, unlike skeletal muscle, does not atrophy when denervated, but it becomes hyperresponsive to the chemical mediator that normally activates it. A good example of denervation hypersensitivity is the response of the denervated iris. If the postganglionic sympathetic nerves to one iris are cut in an experimental animal and, after several weeks, norepinephrine (the transmitter released by sympathetic postganglionic neurons) is injected intravenously, the denervated pupil dilates widely. A much smaller, less prolonged response is observed on the intact side.

The reactions triggered by section of an axon are summarized in Figure 6–16. Hypersensitivity of the postsynaptic structure to the transmitter previously secreted by the axon endings is a general phenomenon, largely due to the synthesis or activation of more receptors. There is in addition orthograde degeneration (**wallerian degeneration**) and retrograde degeneration of the axon stump to the nearest collateral (**sustaining collateral**). A series of changes occur in the cell body that include a decrease in Nissl substance (chromatolysis). The nerve then starts to regrow, with multiple small branches projecting along the path the axon previously followed (regenerative sprouting). Axons sometimes grow back to their original targets, especially in locations like the neuromuscular junction. However, nerve regeneration is generally limited because axons often become entangled in the area of tissue damage at the site where they were disrupted. This

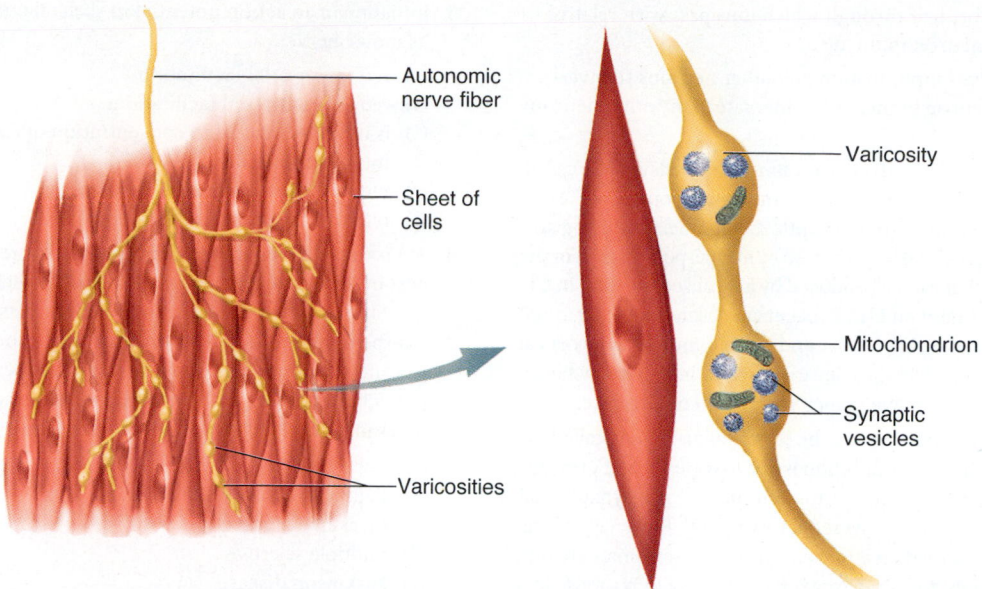

FIGURE 6–15 **Endings of postganglionic autonomic neurons on smooth muscle.** Neurotransmitter, released from varicosities along the branched axon, diffuses to receptors on smooth muscle cell plasma membranes. (From Widmaier EP, Raff H, Strang KT: *Vanders Human Physiology.* McGraw-Hill, 2008.)

difficulty has been reduced by administration of neurotrophins. For example, sensory neurons torn when dorsal nerve roots are avulsed from the spinal cord regrow and form functional connections in the spinal cord if experimental animals are treated with NGF, neurotrophin 3, or GDNF.

Hypersensitivity is limited to the structures immediately innervated by the destroyed neurons and fails to develop in neu-

rons and muscle farther downstream. Suprasegmental spinal cord lesions do not lead to hypersensitivity of the paralyzed skeletal muscles to acetylcholine, and destruction of the preganglionic autonomic nerves to visceral structures does not cause hypersensitivity of the denervated viscera. This fact has practical implications in the treatment of diseases due to spasm of the blood vessels in the extremities. For example, if the upper extremity is sympathectomized by removing the upper part of the ganglionic chain and the stellate ganglion, the hypersensitive smooth muscle in the vessel walls is stimulated by circulating norepinephrine, and episodic vasospasm continues to occur. However, if preganglionic sympathectomy of the arm is performed by cutting the ganglion chain below the third ganglion (to interrupt ascending preganglionic fibers) and the white rami of the first three thoracic nerves, no hypersensitivity results.

Denervation hypersensitivity has multiple causes. As noted in Chapter 2, a deficiency of a given chemical messenger generally produces an upregulation of its receptors. Another factor is lack of reuptake of secreted neurotransmitters.

CHAPTER SUMMARY

- Presynaptic terminals are separated from the postsynaptic structure by a synaptic cleft. The postsynaptic membrane contains many neurotransmitter receptors and usually a postsynaptic thickening called the postsynaptic density.

- At chemical synapses, an impulse in the presynaptic axon causes secretion of a chemical that diffuses across the synaptic cleft and binds to postsynaptic receptors, triggering events that open or close channels in the membrane of the postsynaptic cell. At electrical synapses, the membranes of the presynaptic and postsynaptic neurons come close together, and gap junctions form

FIGURE 6–16 **Summary of changes occurring in a neuron and the structure it innervates when its axon is crushed or cut at the point marked X.** Hypersensitivity of the postsynaptic structure to the transmitter previously secreted by the axon occurs largely due to the synthesis or activation of more receptors. There is both orthograde (wallerian) degeneration from the point of damage to the terminal and retrograde degeneration of the axon stump to the nearest collateral (sustaining collateral). Changes also occur in the cell body, including chromatolysis. The nerve starts to regrow, with multiple small branches projecting along the path the axon previously followed (regenerative sprouting).

low-resistance bridges through which ions pass with relative ease from one neuron to the next.

■ A neuron receives input from many other neurons (convergence), and a neuron branches to innervate many other neurons (divergence).

■ An EPSP is produced by depolarization of the postsynaptic cell after a latency of 0.5 ms; the excitatory transmitter opens Na^+ or Ca^{2+} ion channels in the postsynaptic membrane, producing an inward current. An IPSP is produced by a hyperpolarization of the postsynaptic cell; it can be produced by a localized increase in Cl^- transport. Slow EPSPs and IPSPs occur after a latency of 100 to 500 ms in autonomic ganglia, cardiac and smooth muscle, and cortical neurons. The slow EPSPs are due to decreases in K^+ conductance, and the slow IPSPs are due to increases in K^+ conductance.

■ Postsynaptic inhibition during the course of an IPSP is called direct inhibition. Indirect inhibition is due to the effects of previous postsynaptic neuron discharge; for example, the postsynaptic cell cannot be activated during its refractory period. Presynaptic inhibition is a process mediated by neurons whose terminals are on excitatory endings, forming axoaxonal synapses; in response to activation of the presynaptic terminal. Activation of the presynaptic receptors can increase Cl^- conductance, decreasing the size of the action potentials reaching the excitatory ending, and reducing Ca^{2+} entry and the amount of excitatory transmitter released.

MULTIPLE-CHOICE QUESTIONS

For all questions, select the single best answer unless otherwise directed.

1. Fast inhibitory postsynaptic potentials (IPSPs)

    A) are a consequence of decreased Cl^- conductance.
    B) occur in skeletal muscle.
    C) can be produced by an increase in Na^+ conductance.
    D) can be produced by an increase in Ca^{2+} conductance.
    E) interact with other fast and slow potentials to move the membrane potential of the postsynaptic neuron toward or away from the firing level.

2. Fast excitatory postsynaptic potentials (EPSPs)

    A) are a consequence of decreased Cl^- conductance.
    B) occur in skeletal muscle.
    C) can be produced by an increase in Na^+ conductance.
    D) can be produced by a decrease in Ca^{2+} conductance.
    E) all of the above

3. Initiation of an action potential in skeletal muscle by stimulating its motor nerve

    A) requires spatial facilitation.
    B) requires temporal facilitation.
    C) is inhibited by a high concentration of Ca^{2+} at the neuromuscular junction.
    D) requires the release of norepinephrine.
    E) requires the release of acetylcholine.

4. A 35-year-old woman sees her physician to report muscle weakness in the extraocular eye muscles and muscles of the extremities. She states that she feels fine when she gets up in the morning, but the weakness begins soon after she becomes active. The weakness is improved by rest. Sensation appears normal. The physician treats her with an anticholinesterase inhibitor, and she notes immediate return of muscle strength. Her physician diagnoses her with

    A) Lambert–Eaton syndrome.
    B) myasthenia gravis.
    C) multiple sclerosis.
    D) Parkinson disease.
    E) muscular dystrophy.

CHAPTER RESOURCES

Boron WF, Boulpaep EL: *Medical Physiology,* Elsevier, 2005.

Hille B: *Ionic Channels of Excitable Membranes,* 3rd ed. Sinauer Associates, 2001.

Jessell TM, Kandel ER: Synaptic transmission: A bidirectional and a self-modifiable form of cellcell communication. Cell 1993;72(Suppl):1.

Kandel ER, Schwartz JH, Jessell TM (editors): *Principles of Neural Science,* 4th ed. McGraw-Hill, 2000.

McPhee SJ, Ganong WF: *Pathophysiology of Disease. An Introduction to Clinical Medicine,* 5th ed. McGraw-Hill, 2006.

Squire LR, et al (editors): *Fundamental Neuroscience,* 3rd ed., Academic Press, 2008.

Unwin N: Neurotransmitter action: Opening of ligand-gated ion channels. Cell 1993; 72(Suppl):31.

Van der Kloot W, Molg J: Quantal acetylcholine release at the vertebrate neuromuscular junction. Physiol Rev 1994;74:899.

Neurotransmitters & Neuromodulators

INTRODUCTION

The fact that transmission at most synapses is chemical is of great physiologic and pharmacologic importance. Nerve endings have been called biological transducers that convert electrical energy into chemical energy. In broad terms, this conversion process involves the synthesis of the **neurotransmitters,** their storage in synaptic vesicles, and their release by the nerve impulses into the synaptic cleft. The secreted transmitters then act on appropriate receptors on the membrane of the postsynaptic cell and are rapidly removed from the synaptic cleft by diffusion, metabolism, and, in many instances, reuptake into the presynaptic neuron.

Some chemicals released by neurons have little or no direct effects on their own but can modify the effects of neurotransmitters. These chemicals are called **neuromodulators.** All these processes, plus the postreceptor events in the postsynaptic neuron, are regulated by many physiologic factors and at least in theory can be altered by drugs. Therefore, pharmacologists (in theory) should be able to develop drugs that regulate not only somatic and visceral motor activity but also emotions, behavior, and all the other complex functions of the brain.

CHEMICAL TRANSMISSION OF SYNAPTIC ACTIVITY

CHEMISTRY OF TRANSMITTERS

One suspects that a substance is a neurotransmitter if it is unevenly distributed in the nervous system and its distribution parallels that of its receptors and synthesizing and catabolizing enzymes. Additional evidence includes demonstration that it is released from appropriate brain regions in vitro and that it produces effects on single target neurons when applied to their membranes by means of a micropipette (microiontophoresis).

Many transmitters and enzymes involved in their synthesis and catabolism have been localized in nerve endings by **immuno-histochemistry,** a technique in which antibodies to a given substance are labeled and applied to brain and other tissues. The antibodies bind to the substance, and the location of the substance is then determined by locating the label with the light microscope or electron microscope. **In situ hybridization histochemistry,** which permits localization of the mRNAs for particular synthesizing enzymes or receptors, has also been a valuable tool.

Identified neurotransmitters and neuromodulators can be divided into two major categories: small-molecule transmitters and large-molecule transmitters. Small-molecule transmitters include **monoamines** (eg, acetylcholine, serotonin, histamine), **catecholamines** (dopamine, norepinephrine, and epinephrine), and **amino acids** (eg, glutamate, GABA, glycine). Large-molecule transmitters include a large number of peptides called **neuropeptides** including substance P, enkephalin, vasopressin, and a host of others. In general, neuropeptides are colocalized with one of the small-molecule neurotransmitters (Table 7–1).

There are also other substances thought to be released into the synaptic cleft to act as either a transmitter or modulator of synaptic transmission. These include purine derivatives like adenosine and adenosine triphosphate (ATP) and a gaseous molecule, nitric oxide (NO).

Figure 7–1 shows the biosynthesis of some common small-molecule transmitters released by neurons in the central or peripheral nervous system. Figure 7–2 shows the location of major groups of neurons that contain norepinephrine, epinephrine, dopamine, and acetylcholine. These are some of the major neuromodulatory systems.

RECEPTORS

Cloning and other molecular biology techniques have permitted spectacular advances in knowledge about the structure and function of receptors for neurotransmitters and other chemical messengers. The individual receptors, along with their

ligands, are discussed in the following parts of this chapter. However, five themes have emerged that should be mentioned in this introductory discussion.

First, in every instance studied in detail to date, it has become clear that each ligand has many subtypes of receptors. Thus, for example, norepinephrine acts on α_1 and α_2 receptors, and three of each subtype have been cloned. In addition, there are β_1, β_2, and β_3 receptors. Obviously, this multiplies the possible effects of a given ligand and makes its effects in a given cell more selective.

Second, there are receptors on the presynaptic as well as the postsynaptic elements for many secreted transmitters. These **presynaptic receptors,** or **autoreceptors,** often inhibit further secretion of the ligand, providing feedback control. For example, norepinephrine acts on α_2 presynaptic receptors to inhibit norepinephrine secretion. However, autoreceptors can also facilitate the release of neurotransmitters.

Third, although there are many ligands and many subtypes of receptors for each ligand, the receptors tend to group in large families as far as structure and function are concerned. Many receptors act via trimeric G proteins and protein kinases to produce their effects. Others are ion channels. The receptors for a group of selected, established neurotransmitters and neuromodulators are listed in Table 7–2, along with their principal second messengers and, where established, their net effect on ion channels. It should be noted that this table is an oversimplification. For example, activation of α_2-adrenergic receptors decreases intracellular cAMP concentrations, but there is evidence that the G protein activated by α_2-adrenergic presynaptic receptors also acts directly on Ca^{2+} channels to inhibit norepinephrine release by decreasing Ca^{2+} increases.

Fourth, receptors are concentrated in clusters in postsynaptic structures close to the endings of neurons that secrete the neurotransmitters specific for them. This is generally due to the presence of specific binding proteins for them. In the case of nicotinic acetylcholine receptors at the neuromuscular junction, the protein is **rapsyn,** and in the case of excitatory glutamatergic receptors, a family of **PB2-binding proteins** is involved. GABA$_A$ receptors are associated with the protein **gephyrin,** which also binds glycine receptors, and GABA$_C$ receptors are bound to the cytoskeleton in the retina by the protein **MAP-1B.** At least in the case of GABA$_A$ receptors, the binding protein gephyrin is located in clumps in the postsynaptic membrane. With activity, the free receptors move rapidly to the gephyrin and bind to it, creating membrane clusters. Gephyrin binding slows and restricts their further movement. Presumably, during neural inactivity, the receptors are unbound and move again.

Fifth, prolonged exposure to their ligands causes most receptors to become unresponsive, that is, to undergo **desensitization.** This can be of two types: **homologous desensitization,** with loss of responsiveness only to the particular ligand and maintained responsiveness of the cell to other ligands; and **heterologous desensitization,** in which the cell becomes unresponsive to other ligands as well. Desensitization in β-adrenergic receptors has been studied in considerable detail. One form involves phosphorylation of the carboxyl terminal

TABLE 7–1 Examples of colocalization of small-molecule transmitters with neuropeptides.

Small-Molecule Transmitter	Neuropeptide
Monoamines	
Acetylcholine	Enkephalin, calcitonin-gene-related peptide, galanin, gonadotropin-releasing hormone, neurotensin, somatostatin, substance P, vasoactive intestinal polypeptide
Serotonin	Cholecystokinin, enkephalin, neuropeptide Y, substance P, vasoactive intestinal polypeptide
Catecholamines	
Dopamine	Cholecystokinin, enkephalin, neurotensin
Norepinephrine	Enkephalin, neuropeptide Y, neurotensin, somatostatin, vasopressin
Epinephrine	Enkephalin, neuropeptide Y, neurotensin, substance P
Amino Acids	
Glutamate	Substance P
Glycine	Neurotensin
GABA	Cholecystokinin, enkephalin, somatostatin, substance P, thyrotropin-releasing hormone

A AMINO ACIDS

B MONOAMINES

C CATECHOLAMINES

FIGURE 7–1 **Biosynthesis of some common small molecule transmitters.** (Reproduced with permission from Boron WF, Boulpaep EL: *Medical Physiology.* Elsevier, 2005.)

FIGURE 7–2 **Four diffusely connected systems of central neurons using modulatory transmitters. A)** Norepinephrine-containing neurons. **B)** Serotonin-containing neurons. **C)** Dopamine-containing neurons. **D)** Acetylcholine-containing neurons. (Reproduced with permission from Boron WF, Boulpaep EL: *Medical Physiology.* Elsevier, 2005.)

region of the receptor by a specific β-adrenergic receptor kinase (β-**ARK**) or binding β-**arrestins.** Four β-arrestins have been described in mammals. Two are expressed in rods and cones of the retina and inhibit visual responses. The other two, β-arrestin 1 and β-arrestin 2, are more ubiquitous. They desensitize β-adrenegic receptors, but they also inhibit other heterotrimeric G protein-coupled receptors. In addition, they foster endocytosis of ligands, adding to desensitization.

REUPTAKE

Neurotransmitters are transported from the synaptic cleft back into the cytoplasm of the neurons that secreted them, a process referred to as **reuptake** (Figure 7–3). The high-affinity reuptake systems employ two families of transporter proteins. One family has 12 transmembrane domains and cotransports the transmitter with Na^+ and Cl^-. Members of this family include transporters for norepinephrine, dopamine, serotonin, GABA, and glycine, as well as transporters for proline, taurine, and the acetylcholine precursor choline. In addition, there

TABLE 7–2 Mechanism of action of selected small-molecule transmitters.

Transmitter	Receptor	Second Messenger	Net Channel Effects
Monoamines			
Acetylcholine	Nicotinic		$\uparrow Na^+, K^+$
	M_1, M_3, M_5	$\uparrow IP_3$, DAG	$\uparrow Ca^{2+}$
	M_2, M_4	$\downarrow$ Cyclic AMP	$\uparrow K^+$
Serotonin	$5HT_{1A}$	$\downarrow$ Cyclic AMP	$\uparrow K^+$
	$5HT_{1B}$	$\downarrow$ Cyclic AMP	
	$5HT_{1D}$	$\downarrow$ Cyclic AMP	$\downarrow K^+$
	$5HT_{2A}$	$\uparrow IP_3$, DAG	$\downarrow K^+$
	$5HT_{2C}$	$\uparrow IP_3$, DAG	
	$5HT_3$		$\uparrow Na^+$
	$5HT_4$	$\uparrow$ Cyclic AMP	
Catecholamines			
Dopamine	D_1, D_5	$\uparrow$ Cyclic AMP	
	D_2	$\downarrow$ Cyclic AMP	$\uparrow K^+, \downarrow Ca^{2+}$
	D_3, D_4	$\downarrow$ Cyclic AMP	
Norepinephrine	α_1	$\uparrow IP_3$, DAG	$\downarrow K^+$
	α_2	$\downarrow$ Cyclic AMP	$\uparrow K^+, \downarrow Ca^{2+}$
	β_1	$\uparrow$ Cyclic AMP	
	β_2	$\uparrow$ Cyclic AMP	
	β_3	$\uparrow$ Cyclic AMP	
Amino Acids			
Glutamate	Metabotropic[a]		
	Ionotropic		
	AMPA, Kainate		$\uparrow Na^+, K^+$
	NMDA		$\uparrow Na^+, K^+, Ca^{2+}$
GABA	$GABA_A$		$\uparrow Cl^-$
	$GABA_B$	$\uparrow IP_3$, DAG	$\uparrow K^+, \downarrow Ca^{2+}$
Glycine	Glycine		$\uparrow Cl^-$

[a]Eleven subtypes identified; all decrease cAMP or increase IP3 and DAG, except one, which increases cAMP.

may be an epinephrine transporter. The other family is made up of at least three transporters that mediate glutamate uptake by neurons and two that transport glutamate into astrocytes. These glutamate transporters are coupled to the cotransport of Na^+ and the countertransport of K^+, and they are not dependent on Cl^- transport. There is a debate about their structure, and they may have 6, 8, or 10 transmembrane domains. One of them transports glutamate into glia rather than neurons (see Chapter 4).

There are in addition two vesicular monoamine transporters, VMAT1 and VMAT2, that transport neurotransmitters from the cytoplasm to synaptic vesicles. They are coded by different genes but have extensive homology. Both have a broad specificity, moving dopamine, norepinephrine, epinephrine, serotonin, and histamine from the cytoplasm into secretory granules. Both are inhibited by reserpine,

which accounts for the marked monoamine depletion produced by this drug. Like the neurotransmitter membrane transporter family, they have 12 transmembrane domains, but they have little homology to the other transporters. There is also a vesicular GABA transporter (VGAT) that moves GABA and glycine into vesicles and a vesicular acetylcholine transporter.

Reuptake is a major factor in terminating the action of transmitters, and when it is inhibited, the effects of transmitter release are increased and prolonged. This has clinical consequences. For example, several effective antidepressant drugs are inhibitors of the reuptake of amine transmitters, and cocaine is believed to inhibit dopamine reuptake. Glutamate uptake into neurons and glia is important because glutamate is an excitotoxin that can kill cells by overstimulating them (see Clinical Box 7–1). There is evidence that during ischemia

FIGURE 7–3   **Fate of monoamines secreted at synaptic junctions.** In each monoamine-secreting neuron, the monoamine is synthesized in the cytoplasm and the secretory granules **(1)** and its concentration in secretory granules is maintained **(2)** by the two vesicular monoamine transporters (VMAT). The monoamine is secreted by exocytosis of the granules **(3),** and it acts **(4)** on receptors (Y-shaped structures labeled R). Many of these receptors are postsynaptic, but some are presynaptic and some are located on glia. In addition, there is extensive reuptake into the cytoplasm of the presynaptic terminal **(5)** via the monoamine neurotransmitter transporter (NTT) for the monoamine that is synthesized in the neuron. (Reproduced with permission from Hoffman BJ, et al: Distribution of monoamine neurotransmitter transporters in the rat brain. Front Neuroendocrinol 1998;19:187.)

and anoxia, loss of neurons is increased because glutamate reuptake is inhibited.

SMALL-MOLECULE TRANSMITTERS

Synaptic physiology is a rapidly expanding, complex field that cannot be covered in detail in this book. However, it is appropriate to summarize information about the principal neurotransmitters and their receptors.

MONOAMINES

Acetylcholine

Acetylcholine, which is the acetyl ester of choline, is largely enclosed in small, clear synaptic vesicles in high concentration in the terminal boutons of neurons that release acetyl-choline (**cholinergic** neurons). Synthesis of acetylcholine involves the reaction of choline with acetate (Figure 7–1). Acetylcholine is the transmitter at the neuromuscular junction, in autonomic ganglia, and in postganglionic parasympathetic nerve-target organ junctions and some postganglionic sympathetic nerve-target junctions. It is also found within the brain, including the basal forebrain complex and pontomesencephalic cholinergic complex (Figure 7–2). These systems may be involved in regulation of sleep-wake states, learning, and memory.

Cholinergic neurons actively take up choline via a transporter (Figure 7–4). Choline is also synthesized in neurons. The acetate is activated by the combination of acetate groups with reduced coenzyme A. The reaction between active acetate (acetyl-coenzyme A, acetyl-CoA) and choline is catalyzed by the enzyme **choline acetyltransferase.** This enzyme is found in high concentration in the cytoplasm of cholinergic

CLINICAL BOX 7–1

Excitotoxins

Glutamate is usually cleared from the brain's extracellular fluid by Na⁺-dependent uptake systems in neurons and glia, keeping only micromolar levels of the chemical in the extracellular fluid despite millimolar levels inside neurons. However, excessive levels of glutamate occur in response to ischemia, anoxia, hypoglycemia, or trauma. Glutamate and some of its synthetic congeners are unique in that when they act on neuronal cell bodies, they can produce so much Ca^{2+} influx that neurons die. This is the reason why microinjection of these **excitotoxins** is used in research to produce discrete lesions that destroy neuronal cell bodies without affecting neighboring axons. Evidence is accumulating that excitotoxins play a significant role in the damage done to the brain by a **stroke.** When a cerebral artery is occluded, the cells in the severely ischemic area die. Surrounding partially ischemic cells may survive but lose their ability to maintain the transmembrane Na⁺ gradient. The elevated levels of intracellular Na⁺ prevent the ability of **astrocytes** to remove glutamate from the brain's extracellular fluid. Therefore, glutamate accumulates to the point that excitotoxic damage and cell death occurs in the **penumbra,** the region around the completely infarcted area.

FIGURE 7–4 **Biochemical events at cholinergic endings.** ACh, acetylcholine; ASE, acetylcholinesterase; X, receptor.

nerve endings. Acetylcholine is then taken up into synaptic vesicles by a vesicular transporter, VAChT.

Cholinesterases

Acetylcholine must be rapidly removed from the synapse if repolarization is to occur. The removal occurs by way of hydrolysis of acetylcholine to choline and acetate, a reaction catalyzed by the enzyme **acetylcholinesterase.** This enzyme is also called **true** or **specific cholinesterase.** Its greatest affinity is for acetylcholine, but it also hydrolyzes other choline esters. There are a variety of esterases in the body. One found in plasma is capable of hydrolyzing acetylcholine but has different properties from acetylcholinesterase. It is therefore called **pseudocholinesterase** or **nonspecific cholinesterase.** The plasma moiety is partly under endocrine control and is affected by variations in liver function. On the other hand, the specific cholinesterase molecules are clustered in the postsynaptic membrane of cholinergic synapses. Hydrolysis of acetylcholine by this enzyme is rapid enough to explain the observed changes in Na⁺ conductance and electrical activity during synaptic transmission.

Acetylcholine Receptors

Historically, acetylcholine receptors have been divided into two main types on the basis of their pharmacologic properties. Muscarine, the alkaloid responsible for the toxicity of

toadstools, has little effect on the receptors in autonomic ganglia but mimics the stimulatory action of acetylcholine on smooth muscle and glands. These actions of acetylcholine are therefore called **muscarinic actions,** and the receptors involved are **muscarinic cholinergic receptors.** They are blocked by the drug atropine. In sympathetic ganglia, small amounts of acetylcholine stimulate postganglionic neurons and large amounts block transmission of impulses from preganglionic to postganglionic neurons. These actions are unaffected by atropine but mimicked by nicotine. Consequently, these actions of acetylcholine are **nicotinic actions** and the receptors are **nicotinic cholinergic receptors.** Nicotinic receptors are subdivided into those found in muscle at neuromuscular junctions and those found in autonomic ganglia and the central nervous system. Both muscarinic and nicotinic acetylcholine receptors are found in large numbers in the brain.

The nicotinic acetylcholine receptors are members of a superfamily of ligand-gated ion channels that also includes the $GABA_A$ and glycine receptors and some of the glutamate receptors. They are made up of multiple subunits coded by different genes. Each nicotinic cholinergic receptor is made up of five subunits that form a central channel which, when the receptor is activated, permits the passage of Na⁺ and other cations. The 5 subunits come from a menu of 16 known subunits, α_1–α_9, β_2–β_5, γ, δ, and ε, coded by 16 different genes. Some of the receptors are homomeric—for example, those that contain five α_7 subunits—but most are heteromeric. The muscle type nicotinic receptor found in the fetus is made up of two α_1 subunits, a β_1 subunit, a γ subunit, and a δ subunit (Figure 7–5). In adult mammals, the γ subunit is replaced by a δ subunit, which decreases the channel open time but increases its

FIGURE 7–5 **Three-dimensional model of the nicotinic acetylcholine-gated ion channel.** The receptor–channel complex consists of five subunits, all of which contribute to forming the pore. When two molecules of acetylcholine bind to portions of the α-subunits exposed to the membrane surface, the receptor–channel changes conformation. This opens the pore in the portion of the channel emnbedded in the lipid bilayer, and both K$^+$ and Na$^+$ flow through the open channel down their electrochemical gradient. (From Kandel ER, Schwartz JH, Jessell TM [editors]: *Principles of Neural Science,* 4th ed. McGraw-Hill, 2000.)

conductance. The nicotinic cholinergic receptors in autonomic ganglia are heteromers that usually contain α$_3$ subunits in combination with others, and the nicotinic receptors in the brain are made up of many other subunits. Many of the nicotinic cholinergic receptors in the brain are located presynaptically on glutamate-secreting axon terminals, and they facilitate the release of this transmitter. However, others are postsynaptic. Some are located on structures other than neurons, and some seem to be free in the interstitial fluid, that is, they are perisynaptic in location.

Each α subunit has a binding site for acetylcholine, and when an acetylcholine molecule binds to each of them, they induce a confirmational change in the protein so that the channel opens. This increases the conductance of Na$^+$ and other cations, and the resulting influx of Na$^+$ produces a depolarizing potential. A prominent feature of neuronal nicotinic cholinergic receptors is their high permeability to Ca^{2+}.

Muscarinic cholinergic receptors are very different from nicotinic cholinergic receptors. Five types, encoded by five separate genes, have been cloned. The exact status of M$_5$ is uncertain, but the remaining four receptors are coupled via G proteins to adenylyl cyclase, K$^+$ channels, and/or phospholipase C (Table 7–2). The nomenclature of these receptors has not been standardized, but the receptor designated M$_1$ in Table 7–2 is abundant in the brain. The M$_2$ receptor is found in the heart. The M$_4$ receptor is found in pancreatic acinar and islet tissue, where it mediates increased secretion of pancreatic enzymes and insulin. The M$_3$ and M$_4$ receptors are associated with smooth muscle.

Serotonin

Serotonin (5-hydroxytryptamine; 5-HT) is present in highest concentration in blood platelets and in the gastrointestinal tract, where it is found in the enterochromaffin cells and the myenteric plexus. It is also found within the brain stem in the midline raphé nuclei which project to portions of the hypothalamus, the limbic system, the neocortex, the cerebellum, and the spinal cord (Figure 7–2).

Serotonin is formed in the body by hydroxylation and decarboxylation of the essential amino acid tryptophan (Figures 7–1 and 7–6). After release from serotonergic neurons, much of the released serotonin is recaptured by an active reuptake mechanism and inactivated by monoamine oxidase (MAO) to form 5-hydroxyindoleacetic acid (5-HIAA). This substance is the principal urinary metabolite of serotonin, and urinary output of 5-HIAA is used as an index of the rate of serotonin metabolism in the body.

Tryptophan hydroxylase in the human CNS is slightly different from the tryptophan hydroxylase in peripheral tissues, and is coded by a different gene. This is presumably why knockout of the *TPH1* gene, which codes for tryptophan hydroxylase in peripheral tissues, has much less effect on brain serotonin production than on peripheral serotonin production.

As described in Clinical Box 7–2, there is evidence for a relationship between behavior and brain serotonin content.

Serotonergic Receptors

The number of cloned and characterized serotonin receptors has increased rapidly. There are at least seven types of 5-HT

FIGURE 7–6 **Biochemical events at serotonergic synapses.** 5-HTP, 5-hydroxytryptophan; 5-HT, 5-hydroxytryptamine (serotonin); 5-HIAA, 5-hydroxyindoleacetic acid; X, serotonin receptor. For clarity, the presynaptic receptors have been omitted.

receptors (from 5-HT$_1$ through 5-HT$_7$ receptors). Within the 5-HT$_1$ group are the 5-HT$_{1A}$, 5-HT$_{1B}$, 5-HT$_{1D}$, 5-HT$_{1E}$, and 5-HT$_{1F}$ subtypes. Within the 5-HT$_2$ group there are 5-HT$_{2A}$, 5-HT$_{2B}$, and 5-HT$_{2C}$ subtypes. There are two 5-HT$_5$ subtypes: 5-HT$_{5A}$ and 5-HT$_{5B}$. Most of these are G protein-coupled receptors and affect adenylyl cyclase or phospholipase C (Table 7–2). However, the 5-HT$_3$ receptors, like nicotinic cholinergic receptors, are ligand-gated ion channels. Some of the serotonin receptors are presynaptic, and others are postsynaptic.

5-HT$_{2A}$ receptors mediate platelet aggregation and smooth muscle contraction. Mice in which the gene for 5-HT$_{2C}$ receptors has been knocked out are obese as a result of increased food intake despite normal responses to leptin, and they are prone to fatal seizures. 5-HT$_3$ receptors are present in the gastrointestinal tract and the area postrema and are related to vomiting. 5-HT$_4$ receptors are also present in the gastrointestinal tract, where they facilitate secretion and peristalsis, and in the brain. 5-HT$_6$ and 5-HT$_7$ receptors in the brain are distributed throughout the limbic system, and the 5-HT$_6$ receptors have a high affinity for antidepressant drugs.

Histamine

Histaminergic neurons have their cell bodies in the tuberomammillary nucleus of the posterior hypothalamus, and their axons project to all parts of the brain, including the cerebral cortex and the spinal cord. Histamine is also found in cells in the gastric mucosa and in heparin-containing cells called **mast cells** that are plentiful in the anterior and posterior lobes of the pituitary gland as well as at body surfaces.

Histamine is formed by decarboxylation of the amino acid histidine (Figure 7–1). Histamine is converted to methylhistamine or, alternatively, to imidazoleacetic acid. The latter reaction is quantitatively less important in humans. It requires the enzyme **diamine oxidase (histaminase)** rather than MAO, even though MAO catalyzes the oxidation of methylhistamine to methylimidazoleacetic acid.

The three known types of histamine receptors—H$_1$, H$_2$, and H$_3$—are all found in both peripheral tissues and the brain. Most, if not all, of the H$_3$ receptors are presynaptic, and they mediate inhibition of the release of histamine and other transmitters via a G protein. H$_1$ receptors activate phospholipase C, and H$_2$ receptors increase the intracellular cAMP concentration. The function of this diffuse histaminergic system is unknown, but evidence links brain histamine to arousal, sexual behavior, blood pressure, drinking, pain thresholds, and regulation of the secretion of several anterior pituitary hormones.

CATECHOLAMINES

Norepinephrine & Epinephrine

The chemical transmitter present at most sympathetic postganglionic endings is norepinephrine. It is stored in the synaptic knobs of the neurons that secrete it in characteristic small vesicles that have a dense core (granulated vesicles; see above). Norepinephrine and its methyl derivative, epinephrine, are secreted by the adrenal medulla, but epinephrine is not a mediator at postganglionic sympathetic endings. As discussed in Chapter 6, each sympathetic postganglionic neuron has multiple varicosities along its course, and each of these varicosities appears to be a site at which norepinephrine is secreted.

There are also norepinephrine-secreting and epinephrine-secreting neurons in the brain. Norepinephrine-secreting neurons are properly called **noradrenergic neurons,** although the term **adrenergic neurons** is also applied. However, it seems appropriate to reserve the latter term for epinephrine-secreting neurons. The cell bodies of the norepinephrine-containing neurons are located in the locus ceruleus and other medullary and pontine nuclei (Figure 7–2). From the locus ceruleus, the axons of the noradrenergic neurons form the locus ceruleus system. They descend into the spinal cord, enter the cerebellum, and ascend to innervate the paraventricular, supraoptic, and periventricular nuclei of the hypothalamus, the thalamus, the basal telencephalon, and the entire neocortex.

Biosynthesis & Release of Catecholamines

The principal **catecholamines** found in the body—norepinephrine, epinephrine, and dopamine—are formed by hydroxylation and decarboxylation of the amino acid tyrosine (Figure 7–1). Some of the tyrosine is formed from phenylalanine, but most is of dietary origin. **Phenylalanine hydroxylase** is found primarily in the liver (see Clinical Box 7–3). Tyrosine is transported into catecholamine-secreting neurons and adrenal medullary cells by a concentrating mechanism. It is converted to dopa and then to dopamine in the cytoplasm of the cells by **tyrosine hydroxylase** and **dopa decarboxylase.** The decarboxylase, which is also called aromatic L-amino acid decarboxylase, is very similar but probably not identical to 5-hydroxytryptophan decarboxylase. The dopamine then enters the granulated vesicles, within which it is converted to norepinephrine by **dopamine β-hydroxylase (DBH).** L-Dopa is the isomer involved, but the norepinephrine that is formed is in the D configuration. The rate-limiting step in synthesis is the conversion of tyrosine to dopa. Tyrosine hydroxylase, which catalyzes this step, is subject to feedback inhibition by dopamine and norepinephrine, thus providing internal control of the synthetic process. The cofactor for tyrosine hydroxylase is

---

---

tetrahydrobiopterin, which is converted to dihydrobiopterin when tyrosine is converted to dopa.

Some neurons and adrenal medullary cells also contain the cytoplasmic enzyme **phenylethanolamine-*N*-methyltransferase (PNMT),** which catalyzes the conversion of norepinephrine to epinephrine. In these cells, norepinephrine apparently leaves the vesicles, is converted to epinephrine, and then enters other storage vesicles.

In granulated vesicles, norepinephrine and epinephrine are bound to ATP and associated with a protein called **chromogranin A.** In some but not all noradrenergic neurons, the large granulated vesicles also contain neuropeptide Y. Chromogranin A is a 49-kDa acid protein that is also found in many other neuroendocrine cells and neurons. Six related **chromogranins** have been identified. They have been claimed to have multiple intracellular and extracellular functions. Their level in the plasma is elevated in patients with a variety of tumors and in essential hypertension, in which they probably reflect increased sympathetic activity. However, their specific functions remain unsettled.

The catecholamines are transported into the granulated vesicles by two vesicular transporters, and these transporters are inhibited by the drug reserpine.

Catecholamines are released from autonomic neurons and adrenal medullary cells by exocytosis. Because they are present in the granulated vesicles, ATP, chromogranin A, and the dopamine β hydroxylase that is not membrane-bound are released with norepinephrine and epinephrine. The half-life of circulating dopamine β-hydroxylase is much longer than that of the catecholamines, and circulating levels of this substance are affected by genetic and other factors in addition to the rate of sympathetic activity.

Catabolism of Catecholamines

Norepinephrine, like other amine and amino acid transmitters, is removed from the synaptic cleft by binding to postsynaptic receptors, binding to presynaptic receptors (Figure 7–3), reuptake into the presynaptic neurons, or catabolism. Reuptake is a major mechanism in the case of norepinephrine, and the hypersensitivity of sympathetically denervated structures is explained in part on this basis. After the noradrenergic neurons are cut, their endings degenerate with loss of reuptake in them. Consequently, more norepinephrine from other sources is available to stimulate the receptors on the autonomic effectors.

Epinephrine and norepinephrine are metabolized to biologically inactive products by oxidation and methylation. The former reaction is catalyzed by MAO and the latter by **catechol-*O*-methyltransferase (COMT)**. MAO is located on the outer surface of the mitochondria. It has two isoforms, MAO-A and MAO-B, which differ in substrate specificity and sensitivity to drugs. Both are found in neurons. MAO is widely distributed, being particularly plentiful in the nerve endings at which catecholamines are secreted. COMT is also widely distributed, particularly in the liver, kidneys, and smooth muscles. In the brain, it is present in glial cells, and small amounts are found in postsynaptic neurons, but none is found in presynaptic noradrenergic neurons. Consequently, catecholamine metabolism has two different patterns.

Extracellular epinephrine and norepinephrine are for the most part O-methylated, and measurement of the concentrations of the O-methylated derivatives normetanephrine and metanephrine in the urine is a good index of the rate of secretion of norepinephrine and epinephrine. The O-methylated derivatives that are not excreted are largely oxidized, and 3-methoxy-4-hydroxymandelic acid (vanillylmandelic acid, VMA) is the most plentiful catecholamine metabolite in the urine. Small amounts of the O-methylated derivatives are also conjugated to sulfate and glucuronide.

In the noradrenergic nerve terminals, on the other hand, some of the norepinephrine is constantly being converted by intracellular MAO (Figure 7–7) to the physiologically inactive deaminated derivatives, 3,4-dihydroxymandelic acid (DOMA) and its corresponding glycol (DHPG). These are subsequently converted to their corresponding O-methyl derivatives, VMA and 3-methoxy-4-hydroxyphenylglycol (MHPG).

α & β Receptors

Epinephrine and norepinephrine both act on α and β receptors, with norepinephrine having a greater affinity for α-adrenergic

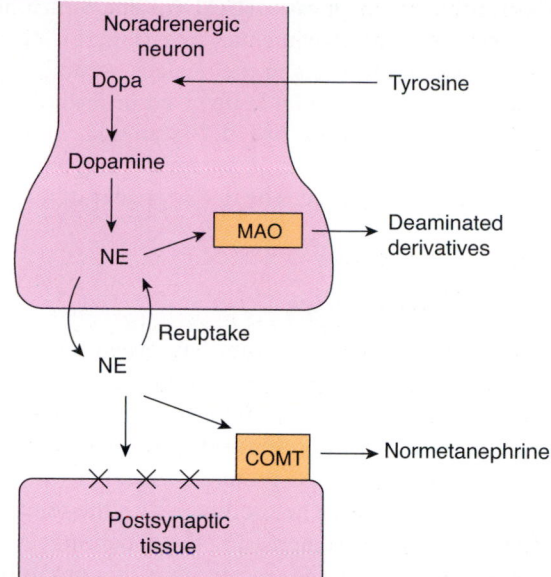

FIGURE 7–7 **Biochemical events at noradrenergic endings.** NE, norepinephrine; COMT, catechol-*O*-methyltransferase; MAO, monoamine oxidase; X, receptor. For clarity, the presynaptic receptors have been omitted. Note that MAO is intracellular, so that norepinephrine is being constantly deaminated in noradrenergic endings. COMT acts primarily on secreted norepinephrine.

receptors and epinephrine for β-adrenergic receptors. As noted previously, the α and β receptors are typical G protein-coupled receptors, and each has multiple forms. They are closely related to the cloned receptors for dopamine and serotonin and to muscarinic acetylcholine receptors.

Clonidine lowers blood pressure when administered centrally. It is an α_2 agonist and was initially thought to act on presynaptic α_2 receptors, reducing central norepinephrine discharge. However, its structure resembles that of **imidazoline,** and it binds to imidazoline receptors with higher affinity than to α_2 adrenergic receptors. A subsequent search led to the discovery that imidazoline receptors occur in the nucleus tractus solitarius and the ventrolateral medulla. Administration of imidazolines lowers blood pressure and has a depressive effect. However, the full significance of these observations remains to be explored.

Dopamine

In certain parts of the brain, catecholamine synthesis stops at dopamine (Figure 7–1) which can then be secreted into the synaptic cleft. Active reuptake of dopamine occurs via a Na^+- and Cl^--dependent dopamine transporter. Dopamine is metabolized to inactive compounds by MAO and COMT in a manner analogous to the inactivation of norepinephrine. 3,4-Dihydroxyphenylacetic acid (DOPAC) and homovanillic acid (HVA) are conjugated, primarily to sulfate.

Dopaminergic neurons are located in several brain regions including the **nigrostriatal system,** which projects from the substantia nigra to the striatum and is involved in motor control, and the **mesocortical system,** which arises primarily in the ventral tegmental area (Figure 7–2). The mesocortical system projects to the nucleus accumbens and limbic subcortical areas, and it is involved in reward behavior and addiction. Studies by PET scanning in normal humans show that a steady loss of dopamine receptors occurs in the basal ganglia with age. The loss is greater in men than in women.

Dopamine Receptors

Five different dopamine receptors have been cloned, and several of these exist in multiple forms. This provides for variety in the type of responses produced by dopamine. Most, but perhaps not all, of the responses to these receptors are mediated by heterotrimeric G proteins. One of the two forms of D_2 receptors can form a heterodimer with the somatostatin SST5 receptor, further increasing the dopamine response menu. Overstimulation of D_2 receptors is thought to be related to schizophrenia (see Clinical Box 7–4). D_3 receptors are highly localized, especially to the nucleus accumbens (Figure 7–2). D_4 receptors have a greater affinity than the other dopamine receptors for the "atypical" antipsychotic drug **clozapine,** which is effective in schizophrenia but produces fewer extrapyramidal side effects than the other major tranquilizers do.

EXCITATORY & INHIBITORY AMINO ACIDS

Glutamate

The amino acid **glutamate** is the main excitatory transmitter in the brain and spinal cord, and it has been calculated that it is the transmitter responsible for 75% of the excitatory transmission in the brain. Glutamate is formed by reductive amination of the Krebs cycle intermediate α-ketoglutarate in the cytoplasm. The reaction is reversible, but in glutaminergic neurons, glutamate is concentrated in synaptic vesicles by the vesicle-bound transporter **BPN1.** The cytoplasmic store of glutamine is enriched by three transporters that import glutamate from the interstitial fluid, and two additional transporters carry glutamate into astrocytes, where it is converted to glutamine and passed on to glutaminergic neurons. The interaction of astrocytes and glutaminergic neurons is shown in Figure 7–8. Released glutamate is taken up by astrocytes and converted to glutamine, which passes back to the neurons and is converted back to glutamate, which is released as the synaptic transmitter. Uptake into neurons and astrocytes is the main mechanism for removal of glutamate from synapses.

Glutamate Receptors

Glutamate receptors are of two types: **metabotropic receptors** and **ionotropic receptors.** The metabotropic receptors are G protein-coupled receptors that increase intracellular IP_3 and

FIGURE 7–8   The glutamate–glutamine cycle through glutaminergic neurons and astrocytes. Glutamate released into the synaptic cleft is taken up by a Na^+-dependent glutamate transporter, and in the astrocyte it is converted to glutamine. The glutamine enters the neuron and is converted to glutamate. Glucose is transported out of capillaries and enters astrocytes and neurons. In astrocytes, it is metabolized to lactate, producing two ATPs. One of these powers the conversion of glutamate to glutamine, and the other is used by Na^+–K^+ ATPase to transport three Na^+ out of the cell in exchange for two K^+. In neurons, the glucose is metabolized further through the citric acid cycle, producing 34 ATPs.

DAG levels or decrease intracellular cAMP levels. Eleven subtypes have been identified (Table 7–2). They are both presynaptic and postsynaptic, and they are widely distributed in the brain. They appear to be involved in the production of synaptic plasticity, particularly in the hippocampus and the cerebellum. Knockout of the gene for one of these receptors, one of the forms of mGluR1, causes severe motor incoordination and deficits in spatial learning.

The ionotropic receptors are ligand-gated ion channels that resemble nicotinic cholinergic receptors and GABA and glycine receptors. There are three general types, each named for the congeners of glutamate to which they respond in maximum fashion. These are the **kainate receptors** (kainate is an acid isolated from seaweed), **AMPA receptors** (for α-amino-3-hydroxy-5-methylisoxazole-4-propionate), and **NMDA receptors** (for *N*-methyl-D-aspartate). Four AMPA, five kainate, and six NMDA subunits have been identified, and each is coded by a different gene. The receptors were initially thought to be pentamers, but some may be tetramers, and their exact stoichiometry is unsettled.

The kainate receptors are simple ion channels that, when open, permit Na^+ influx and K^+ efflux. There are two populations of AMPA receptors: one is a simple Na^+ channel and one also passes Ca^{2+}. The balance between the two in a given synapse can be shifted by activity.

The NMDA receptor is also a cation channel, but it permits passage of relatively large amounts of Ca^{2+}, and it is unique in several ways (Figure 7–9). First, glycine facilitates its function by binding to it, and glycine appears to be essential for its normal response to glutamate. Second, when glutamate binds to it, it opens, but at normal membrane potentials, its channel is blocked by a Mg^{2+} ion. This block is removed only when the neuron containing the receptor is partially depolarized by activation of AMPA or other channels that produce rapid depolarization via other synaptic circuits. Third, phencyclidine and ketamine, which produce amnesia and a feeling of dissociation from the environment, bind to another site inside the channel. Most target neurons for glutamate have both AMPA and NMDA receptors. Kainate receptors are located presynaptically on GABA-secreting nerve endings and postsynaptically at various localized sites in the brain. Kainate and AMPA receptors are found in glia as well as neurons, but it appears that NMDA receptors occur only in neurons.

The concentration of NMDA receptors in the hippocampus is high, and blockade of these receptors prevents **long-term potentiation,** a long-lasting facilitation of transmission in neural pathways following a brief period of high-frequency stimulation. Thus, these receptors may well be involved in memory and learning.

GABA

GABA is the major inhibitory mediator in the brain, including being responsible for presynaptic inhibition. GABA, which exists as β-aminobutyrate in the body fluids, is formed

FIGURE 7–9 Diagrammatic representation of the NMDA receptor. When glycine and glutamate bind to the receptor, the closed ion channel **(left)** opens, but at the resting membrane potential, the channel is blocked by Mg^{2+} **(right).** This block is removed if partial depolarization is produced by other inputs to the neuron containing the receptor, and Ca^{2+} and Na^+ enter the neuron. Blockade can also be produced by the drug dizocilpine maleate (MK-801).

by decarboxylation of glutamate (Figure 7–1). The enzyme that catalyzes this reaction is **glutamate decarboxylase (GAD),** which is present in nerve endings in many parts of the brain. GABA is metabolized primarily by transamination to succinic semialdehyde and thence to succinate in the citric acid cycle. **GABA transaminase (GABA-T)** is the enzyme that catalyzes the transamination. Pyridoxal phosphate, a derivative of the B complex vitamin pyridoxine, is a cofactor for GAD and GABA-T. In addition, there is an active reuptake of GABA via the GABA transporter. A vesicular GABA transporter (VGAT) transports GABA and glycine into secretory vesicles.

GABA Receptors

Three subtypes of GABA receptors have been identified: $GABA_A$, $GABA_B$, and $GABA_C$. The $GABA_A$ and $GABA_B$ receptors are widely distributed in the CNS, whereas in adult vertebrates the $GABA_C$ receptors are found almost exclusively in the retina. The $GABA_A$ and $GABA_C$ receptors are ion channels made up of five subunits surrounding a pore, like the nicotinic acetylcholine receptors and many of the glutamate receptors. In this case, the ion is Cl^- (Figure 7–10). The $GABA_B$ receptors are metabotropic and are coupled to heterotrimeric G proteins that increase conductance in K^+ channels, inhibit adenylyl cyclase, and inhibit Ca^{2+} influx. Increases in Cl^- influx and K^+ efflux and decreases in Ca^{2+} influx all hyperpolarize neurons, producing an IPSP. The G protein mediation of $GABA_B$ receptor effects is unique in that a G protein heterodimer, rather than a single protein, is involved.

The $GABA_C$ receptors are relatively simple in that they are pentamers of three ρ subunits in various combinations. On

FIGURE 7–10 **Diagram of GABA$_A$ and GABA$_B$ receptors, showing their principal actions.** The G protein that mediates the effects of GABA$_B$ receptors is a heterodimer. (Reproduced with permission from Bowery NG, Brown DA: The cloning of GABA$_B$ receptors. Nature 1997;386:223. Copyright © 1997 by Macmillan Magazines.)

the other hand, the GABA$_A$ receptors are pentamers made up of various combinations of six α subunits, four β, four γ, one δ, and one ε. This endows them with considerably different properties from one location to another.

An observation of considerable interest is that there is a chronic low-level stimulation of GABA$_A$ receptors in the CNS that is aided by GABA in the interstitial fluid. This background stimulation cuts down on the "noise" caused by incidental discharge of the billions of neural units and greatly improves the signal-to-noise ratio in the brain. It may be that this GABA discharge declines with advancing age, resulting in a loss of specificity of responses of visual neurons. Support for this hypothesis comes from studies in which microinjection of GABA in older monkeys resulted in restoration of the specificity of visual neurons.

The increase in Cl$^-$ conductance produced by GABA$_A$ receptors is potentiated by benzodiazepines, drugs that have marked anti-anxiety activity and are also effective muscle relaxants, anticonvulsants, and sedatives. Benzodiazepines bind to the α subunits. Diazepam and other benzodiazepines are used throughout the world. At least in part, barbiturates and alcohol also act by facilitating Cl$^-$ conductance through the Cl$^-$ channel. Metabolites of the steroid hormones progesterone and deoxycorticosterone bind to GABA$_A$ receptors and increase Cl$^-$ conductance. It has been known for many years that progesterone and deoxycorticosterone are sleep-inducing and anesthetic in large doses, and these effects are due to their action on GABA$_A$ receptors.

A second class of benzodiazepine receptors is found in steroid-secreting endocrine glands and other peripheral tissues, and hence these receptors are called **peripheral benzodiazepine receptors.** They may be involved in steroid biosynthesis,

possibly performing a function like that of the StAR protein in moving steroids into the mitochondria. Another possibility is a role in the regulation of cell proliferation. Peripheral-type benzodiazepine receptors are also present in astrocytes in the brain, and they are found in brain tumors.

Glycine

Glycine has both excitatory and inhibitory effects in the CNS. When it binds to NMDA receptors, it makes them more sensitive. It appears to spill over from synaptic junctions into the interstitial fluid, and in the spinal cord, for example, this glycine may facilitate pain transmission by NMDA receptors in the dorsal horn. However, glycine is also responsible in part for direct inhibition, primarily in the brain stem and spinal cord. Like GABA, it acts by increasing Cl$^-$ conductance. Its action is antagonized by strychnine. The clinical picture of convulsions and muscular hyperactivity produced by strychnine emphasizes the importance of postsynaptic inhibition in normal neural function. The glycine receptor responsible for inhibition is a Cl$^-$ channel. It is a pentamer made up of two subunits: the ligand-binding α subunit and the structural β subunit. Recently, solid evidence has been presented that three kinds of neurons are responsible for direct inhibition in the spinal cord: neurons that secrete glycine, neurons that secrete GABA, and neurons that secrete both. Presumably, neurons that secrete only glycine have the glycine transporter GLYT2, those that secrete only GABA have GAD, and those that secrete glycine and GABA have both. This third type of neuron is of special interest because the neurons seem to have glycine and GABA in the same vesicles.

Anesthesia

Although general anesthetics have been used for millennia, little has been understood about their mechanisms of action. However, it now appears that alcohols, barbiturates, and many volatile inhaled anesthetics as well act on ion channel receptors and specifically on $GABA_A$ and glycine receptors to increase Cl^- conductance. Regional variation in anesthetic actions in the CNS seems to parallel the variation in subtypes of $GABA_A$ receptors. Other inhaled anesthetics do not act by increasing GABA receptor activity, but appear to act by inhibiting NMDA and AMPA receptors instead.

In contrast to general anesthetics, local anesthetics produce anesthesia by blocking conduction in peripheral nerves via reversibly binding to and inactivating Na^+ channels. Na^+ influx through these channels normally causes depolarization of nerve cell membranes and propagation of impulses toward the nerve terminal. When depolarization and propagation are interrupted, the individual loses sensation in the area supplied by the nerve.

LARGE-MOLECULE TRANSMITTERS: NEUROPEPTIDES

Substance P & Other Tachykinins

Substance P is a polypeptide containing 11 amino acid residues that is found in the intestine, various peripheral nerves, and many parts of the CNS. It is one of a family of six mammalian polypeptides called tachykinins that differ at the amino terminal end but have in common the carboxyl terminal sequence of Phe-X-Gly-LeuMet-NH_2, where X is Val, His, Lys, or Phe. The members of the family are listed in Table 7–3. There are many related tachykinins in other vertebrates and in invertebrates.

The mammalian tachykinins are encoded by two genes. The **neurokinin B gene** encodes only one known polypeptide, neurokinin B. The **substance P/neurokinin A gene** encodes the remaining five polypeptides. Three are formed by alternative processing of the primary RNA and two by post-translational processing.

There are three neurokinin receptors. Two of these, the substance P and the neuropeptide K receptors, are G protein-cou-

pled receptors. Activation of the substance P receptor causes activation of phospholipase C and increased formation of IP_3 and DAG.

Substance P is found in high concentration in the endings of primary afferent neurons in the spinal cord, and it is probably the mediator at the first synapse in the pathways for pain transmission in the dorsal horn. It is also found in high concentrations in the nigrostriatal system, where its concentration is proportional to that of dopamine, and in the hypothalamus, where it may play a role in neuroendocrine regulation. Upon injection into the skin, it causes redness and swelling, and it is probably the mediator released by nerve fibers that is responsible for the axon reflex. In the intestine, it is involved in peristalsis. It has recently been reported that a centrally active NK-1 receptor antagonist has antidepressant activity in humans. This antidepressant effect takes time to develop, like the effect of the antidepressants that affect brain monoamine metabolism, but the NK-1 inhibitor does not alter brain monoamines in experimental animals. The functions of the other tachykinins are unsettled.

Opioid Peptides

The brain and the gastrointestinal tract contain receptors that bind morphine. The search for endogenous ligands for these receptors led to the discovery of two closely related pentapeptides (**enkephalins;** Table 7–4) that bind to these opioid receptors.

TABLE 7–3 Mammalian tachykinins.

Gene	Polypeptide Products	Receptors
SP/NKA	Substance P	Substance P (NK-1)
	Neurokinin A	
	Neuropeptide K	Neuropeptide K (NK-2)
	Neuropeptide α	
	Neurokinin A (3–10)	
NKB	Neurokinin B	Neurokinin B (NK-3)

TABLE 7–4 Opioid peptides and their precursors.

Precursor	Opioid Peptides	Structures
Proenkephalin	Met-enkephalin	Tyr-Gly-Gly-Phe-Met
	Leu-enkephalin	Tyr-Gly-Gly-Phe-Leu
	Octapeptide	Tyr-Gly-Gly-Phe-Met-Arg-Gly-Leu
	Heptapeptide	Tyr-Gly-Gly-Phe-Met-Arg-Phe
Proopiomel-anocortin	β-Endorphin	Tyr-Gly-Glu-Phe-Met-Thr-Ser-Lys-Ser-Gln-Thr-Pro-Leu-Val-Thr-Leu-Phe-Lys-Asn-Ala-Ile-Val-Lys-Asn-Ala-His-Lys-Lys-Gly-Gln
Prodynorphin	Dynorphin 1–8	Tyr-Gly-Gly-Phe-Leu-Arg-Arg-Ile
	Dynorphin 1–17	Tyr-Gly-Gly-Phe-Leu-Arg-Arg-Ile-Arg-Pro-Lys-Leu-Lys-Trp-Asp-Asn-Gln
	α-Neoendorphin	Tyr-Gly-Gly-Phe-Leu-Arg-Lys-Tyr-Pro-Lys
	β-Neoendorphin	Tyr-Gly-Gly-Phe-Leu-Arg-Lys-Tyr-Pro

One contains methionine (**met-enkephalin**), and one contains leucine (**leu-enkephalin**). These and other peptides that bind to opioid receptors are called **opioid peptides.** The enkephalins are found in nerve endings in the gastrointestinal tract and many different parts of the brain, and they appear to function as synaptic transmitters. They are found in the substantia gelatinosa and have analgesic activity when injected into the brain stem. They also decrease intestinal motility.

Like other small peptides, the opioid peptides are synthesized as part of larger precursor molecules. More than 20 active opioid peptides have been identified. Unlike other peptides, however, the opioid peptides have a number of different precursors. Each has a prepro form and a pro form from which the signal peptide has been cleaved. The three precursors that have been characterized, and the opioid peptides they produce, are shown in Table 7–4. **Proenkephalin** was first identified in the adrenal medulla, but it is also the precursor for met-enkephalin and leu-enkephalin in the brain. Each proenkephalin molecule contains four met-enkephalins, one leuenkephalin, one octapeptide, and one heptapeptide. **Proopiomelanocortin,** a large precursor molecule found in the anterior and intermediate lobes of the pituitary gland and the brain, contains β-endorphin, a polypeptide of 31 amino acid residues that has metenkephalin at its amino terminal. There are separate enkephalin-secreting and β endorphin-secreting systems of neurons in the brain. β-Endorphin is also secreted into the bloodstream by the pituitary gland. A third precursor molecule is **prodynorphin,** a protein that contains three leu-enkephalin residues associated with dynorphin and neoendorphin. Dynorphin 1-17 is found in the duodenum and dynorphin 1-8 in the posterior pituitary and hypothalamus. Alpha- and β-neoendorphins are also found in the hypothalamus. The reasons for the existence of multiple opioid peptide precursors and for the presence of the peptides in the circulation as well as in the brain and the gastrointestinal tract are presently unknown.

Enkephalins are metabolized primarily by two peptidases: enkephalinase A, which splits the Gly-Phe bond, and enkephalinase B, which splits the Gly-Gly bond. Aminopeptidase, which splits the Tyr-Gly bond, also contributes to their metabolism.

Opioid receptors have been studied in detail, and three are now established: μ, κ, and δ. They differ in physiologic effects (Table 7–5), distribution in the brain and elsewhere, and affinity for various opioid peptides. All three are G protein-coupled receptors, and all inhibit adenylyl cyclase. In mice in which the μ receptors have been knocked out, morphine fails to produce analgesia, withdrawal symptoms, and self-administration of nicotine. Selective knockout of the other system fails to produce this blockade. Activation of μ receptors increases K^+ conductance, hyperpolarizing central neurons and primary afferents. Activation of κ receptors and δ receptors closes Ca^{2+} channels.

The affinities of individual ligands for the three types of receptors are summarized in Figure 7–11. Endorphins bind only to μ receptors, the main receptors that mediate analgesia. Other opioid peptides bind to multiple opioid receptors.

TABLE 7–5 Physiologic effects produced by stimulation of opiate receptors.

Receptor	Effect
μ	Analgesia
	Site of action of morphine
	Respiratory depression
	Constipation
	Euphoria
	Sedation
	Increased secretion of growth hormone and prolactin
	Meiosis
κ	Analgesia
	Diuresis
	Sedation
	Meiosis
	Dysphoria
δ	Analgesia

Other Polypeptides

Numerous other polypeptides are found in the brain. For example, somatostatin is found in various parts of the brain, where it apparently functions as a neurotransmitter with effects on sensory input, locomotor activity, and cognitive function. In the hypothalamus, this growth hormone-inhibiting hormone is secreted into the portal hypophysial vessels; in the endocrine pancreas, it inhibits insulin secretion and the secretion of other pancreatic hormones; and in the gastrointestinal tract, it is an important inhibitory gastrointestinal regulator. A family of five different somatostatin receptors have been

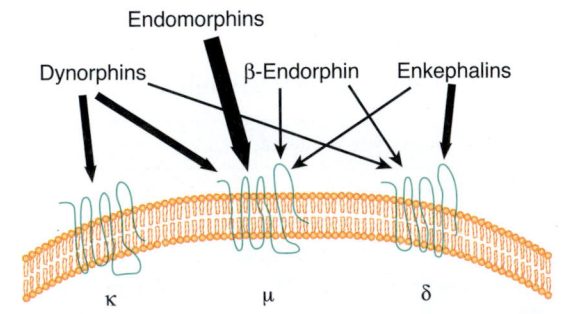

FIGURE 7–11 **Opioid receptors.** The ligands for the κ, μ, and δ receptors are shown with the width of the arrows proportionate to the affinity of the receptor for each ligand. (Reproduced with permission from Julius DJ: Another spark for the masses? Nature 1997;386:442. Copyright © 1997 by Macmillan Magazines.)

identified (SSTR1 through SSTR5). All are G protein-coupled receptors. They inhibit adenylyl cyclase and exert various other effects on intracellular messenger systems. It appears that SSTR2 mediates cognitive effects and inhibition of growth hormone secretion, whereas SSTR5 mediates the inhibition of insulin secretion.

Vasopressin and oxytocin are not only secreted as hormones but also are present in neurons that project to the brain stem and spinal cord. The brain contains bradykinin, angiotensin II, and endothelin. The gastrointestinal hormones VIP, CCK-4, and CCK-8 are also found in the brain. There are two kinds of CCK receptors in the brain, CCK-A and CCK-B. CCK-8 acts at both binding sites, whereas CCK-4 acts at the CCK-B sites. Gastrin, neurotensin, galanin, and gastrin-releasing peptide are also found in the gastrointestinal tract and brain. Neurotensin and VIP receptors have been cloned and shown to be G protein-coupled receptors. The hypothalamus contains both gastrin 17 and gastrin 34. VIP produces vasodilation and is found in vasomotor nerve fibers. The functions of these peptides in the nervous system are unknown.

Calcitonin gene-related peptide (CGRP) is a polypeptide that exists in two forms in rats and humans: CGRPα and CGRPβ. In humans, these two forms differ by only three amino acid residues, yet they are encoded by different genes. In rats, and presumably in humans, CGRPβ is present in the gastrointestinal tract, whereas CGRPβ is found in primary afferent neurons, neurons that project which taste impulses to the thalamus, and neurons in the medial forebrain bundle. It is also present along with substance P in the branches of primary afferent neurons that end near blood vessels. CGRP-like immunoreactivity is present in the circulation, and injection of CGRP causes vasodilation. CGRPα and the calcium-lowering hormone calcitonin are both products of the calcitonin gene. In the thyroid gland, splicing produces the mRNA that codes for calcitonin, whereas in the brain, alternative splicing produces the mRNA that codes for CGRPα. CGRP has little effect on Ca^{2+} metabolism, and calcitonin is only a weak vasodilator.

Neuropeptide Y is a polypeptide containing 36 amino acid residues that acts on at least two of the four known G protein-coupled receptors: Y_1, Y_2, Y_4, and Y_5. Neuropeptide Y is found throughout the brain and the autonomic nervous system. When injected into the hypothalamus, this polypeptide increases food intake, and inhibitors of neuropeptide Y synthesis decrease food intake. Neuropeptide Y-containing neurons have their cell bodies in the arcuate nuclei and project to the paraventricular nuclei.

OTHER CHEMICAL TRANSMITTERS

Purine & Pyrimidine Transmitters

After extended debate, it now seems clear that ATP, uridine, adenosine, and adenosine metabolites are neurotransmitters or neuromodulators. Adenosine is a neuromodulator that acts as a general CNS depressant and has additional widespread effects throughout the body. It acts on four receptors: A_1, A_{2A},

A_{2B}, and A_3. All are G protein-coupled receptors and increase (A_{2A} and A_{2B}) or decrease (A_1 and A_3) cAMP concentrations. The stimulatory effects of coffee and tea are due to blockade of adenosine receptors by caffeine and theophylline. Currently, there is considerable interest in the potential use of A_1 antagonists to decrease excessive glutamate release and thus to minimize the effects of strokes.

ATP is also becoming established as a transmitter, and it has widespread receptor-mediated effects in the body. It appears that soluble nucleotidases are released with ATP, and these accelerate its removal after it has produced its effects. ATP has now been shown to mediate rapid synaptic responses in the autonomic nervous system and a fast response in the habenula. ATP binds to P2X receptors which are ligand-gated ion channel receptors; seven subtypes ($P2X_1$–$P2X_7$) have been identified. P2X receptors have widespread distributions throughout the body; for example, $P2X_1$ and $P2X_2$ receptors are present in the dorsal horn, indicating a role for ATP in sensory transmission. ATP also binds to P2Y receptors which are G protein-coupled receptors. There are eight subtypes of P2Y receptors: $P2Y_1$, $P2Y_2$, $P2Y_4$, $P2Y_6$, $P2Y_{11}$, $P2Y_{12}$, $P2Y_{13}$, and $P2Y_{14}$.

Cannabinoids

Two receptors with a high affinity for Δ^9-tetrahydrocannabinol (THC), the psychoactive ingredient in marijuana, have been cloned. The CB_1 receptor triggers a G protein-mediated decrease in intracellular cAMP levels and is common in central pain pathways as well as in parts of the cerebellum, hippocampus, and cerebral cortex. The endogenous ligand for the receptor is **anandamide,** a derivative of arachidonic acid. This compound mimics the euphoria, calmness, dream states, drowsiness, and analgesia produced by marijuana. There are also CB_1 receptors in peripheral tissues, and blockade of these receptors reduces the vasodilator effect of anandamide. However, it appears that the vasodilator effect is indirect. A CB_2 receptor has also been cloned, and its endogenous ligand may be **palmitoylethanolamide (PEA).** However, the physiologic role of this compound is unsettled.

Gases

Nitric oxide (NO), a compound released by the endothelium of blood vessels as endothelium-derived relaxing factor (EDRF), is also produced in the brain. It is synthesized from arginine, a reaction catalyzed in the brain by one of the three forms of NO synthase. It activates guanylyl cyclase and, unlike other transmitters, it is a gas, which crosses cell membranes with ease and binds directly to guanylyl cyclase. It may be the signal by which postsynaptic neurons communicate with presynaptic endings in long-term potentiation and long-term depression. NO synthase requires NADPH, and it is now known that NADPH-diaphorase (NDP), for which a histochemical stain has been available for many years, is NO synthase.

Other Substances

Prostaglandins are derivatives of arachidonic acid found in the nervous system. They are present in nerve-ending fractions of brain homogenates and are released from neural tissue in vitro. A putative prostaglandin transporter with 12 membrane-spanning domains has been described. However, prostaglandins appear to exert their effects by modulating reactions mediated by cAMP rather than by functioning as synaptic transmitters.

Many steroids are **neuroactive steroids;** that is, they affect brain function, although they are not neurotransmitters in the usual sense. Circulating steroids enter the brain with ease, and neurons have numerous sex steroid and glucocorticoid receptors. In addition to acting in the established fashion by binding to DNA (genomic effects), some steroids seem to act rapidly by a direct effect on cell membranes (nongenomic effects). The effects of steroids on GABA receptors have been discussed previously. Evidence has now accumulated that the brain can produce some hormonally active steroids from simpler steroid precursors, and the term **neurosteroids** has been coined to refer to these products. Progesterone facilitates the formation of myelin, but the exact role of most steroids in the regulation of brain function remains to be determined.

CHAPTER SUMMARY

- Neurotransmitters and neuromodulators are divided into two major categories: small-molecule transmitters (monoamines, catecholamines, and amino acids) and large-molecule transmitters (neuropeptides). Usually neuropeptides are colocalized with one of the small-molecule neurotransmitters.

- Monoamines include acetylcholine, serotonin, and histamine. Catecholamines include norepinephrine, epinephrine, and dopamine. Amino acids include glutamate, GABA, and glycine.

- Acetylcholine is found at the neuromuscular junction, in autonomic ganglia, and in postganglionic parasympathetic nerve-target organ junctions and some postganglionic sympathetic nerve-target junctions. It is also found in the basal forebrain complex and pontomesencephalic cholinergic complex. There are two major types of cholinergic receptors: muscarinic (G protein-coupled receptors) and nicotinic (ligand-gated ion channel receptors).

- Serotonin (5-HT) is found within the brain stem in the midline raphé nuclei which project to portions of the hypothalamus, the limbic system, the neocortex, the cerebellum, and the spinal cord. There are at least seven types of 5-HT receptors, and many of these contain subtypes. Most are G protein-coupled receptors.

- Norepinephrine-containing neurons are in the locus ceruleus and other medullary and pontine nuclei. Some neurons also contain PNMT, which catalyzes the conversion of norepinephrine to epinephrine. Epinephrine and norepinephrine act on α and β receptors, with norepinephrine having a greater affinity for α-adrenergic receptors and epinephrine for β-adrenergic receptors. They are G protein-coupled receptors, and each has multiple forms.

- The amino acid glutamate is the main excitatory transmitter in the CNS. There are two major types of gluatamate receptors: metabotropic (G protein-coupled receptors) and ionotropic (ligand-gated ion channels receptors, including kainite, AMPA, and NMDA).

- GABA is the major inhibitory mediator in the brain. Three subtypes of GABA receptors have been identified: $GABA_A$ and $GABA_C$ (ligand-gated ion channel) and $GABA_B$ (G protein-coupled). The $GABA_A$ and $GABA_B$ receptors are widely distributed in the CNS.

- There are three types of G protein-coupled opioid receptors (μ, κ, and δ) that differ in physiological effects, distribution in the brain and elsewhere, and affinity for various opioid peptides.

MULTIPLE-CHOICE QUESTIONS

For all questions, select the single best answer unless otherwise directed.

1. Which of the following is a ligand-gated ion channel?
   A) VIP receptor
   B) norepinephrine receptor
   C) $GABA_A$ receptor
   D) $GABA_B$ receptor
   E) metabotropic glutamate receptor

2. Which of the following synaptic transmitters is *not* a peptide, polypeptide, or protein?
   A) substance P
   B) met-enkephalin
   C) β-endorphin
   D) serotonin
   E) dynorphin

3. Activation of which of the following receptors would be expected to decrease anxiety?
   A) nicotinic cholinergic receptors
   B) glutamate receptors
   C) $GABA_A$ receptors
   D) glucocorticoid receptors
   E) α_1-adrenergic receptors

4. Which of the following receptors is coupled to a heterotrimeric G protein?
   A) glycine receptor
   B) $GABA_B$ receptor
   C) nicotinic acetylcholine receptor at myoneural junction
   D) $5-HT_3$ receptor
   E) ANP receptor

5. Which of the following would *not* be expected to enhance noradrenergic transmission?
   A) A drug that increases the entry of arginine into neurons.
   B) A drug that enhances tyrosine hydroxylase activity.
   C) A drug that enhances dopamine β-hydroxylase activity.
   D) A drug that inhibits monoamine oxidase.
   E) A drug that inhibits norepinephrine reuptake.

CHAPTER RESOURCES

Boron WF, Boulpaep EL: *Medical Physiology.* Elsevier, 2005.
Cooper JR, Bloom FE, Roth RH: *The Biochemical Basis of Neuropharmacology,* 8th ed. Oxford University Press, 2002.
Fink KB, Göthert M: 5-HT receptor regulation of neurotransmitter release. Pharmacol Rev 2007;59:360.

Kandel ER, Schwartz JH, Jessell TM (editors): *Principles of Neural Science,* 4th ed. McGraw-Hill, 2000.

Monaghan DT, Bridges RJ, Cotman CW: The excitatory amino acid receptors: Their classes, pharmacology, and distinct properties in the function of the central nervous system. Ann Rev Pharmacol Toxicol 1989;29:365.

Nadeau SE, et al: *Medical Neuroscience,* Sauders, 2004.

Olsen RW: The molecular mechanism of action of general anesthetics: Structural aspects of interactions with $GABA_A$ receptors. Toxicol Lett 1998;100:193.

Owens DF, Kriegstein AR: Is there more to GABA than synaptic inhibition? Nat Rev Neurosci 2002;3:715.

Squire LR, et al (editors): *Fundamental Neuroscience,* 3rd ed. Academic Press, 2008.

Properties of Sensory Receptors

INTRODUCTION

Information about the internal and external environment activates the CNS via a variety of **sensory receptors.** These receptors are transducers that convert various forms of energy in the environment into action potentials in neurons. The characteristics of these receptors, the way they generate impulses in afferent neurons, and the general principles or "laws" that apply to sensation are considered in this chapter. Emphasis is placed on receptors mediating the sensation of touch, and later chapters focus on other sensory processes.

We learn in elementary school that there are "five senses," but this dictum takes into account only some of the senses that reach our consciousness. In addition, some sensory receptors relay information that does not reach consciousness. For example, the muscle spindles provide information about muscle length, and other receptors provide information about arterial blood pressure, the temperature of the blood in the head, and the pH of the cerebrospinal fluid. The list of senses in Table 8–1 is somewhat simplified. The rods and cones, for example, respond maximally to light of different wavelengths, and three different types of cones are present, one for each of the three primary colors. There are five different modalities of taste: sweet, salt, sour, bitter, and umami. Sounds of different pitches are heard primarily because different groups of hair cells in the cochlea are activated maximally by sound waves of different frequencies. Whether these various responses to light, taste, and sound should be considered separate senses is a semantic question that in the present context is largely academic.

SENSE RECEPTORS & SENSE ORGANS

It is worth noting that the term *receptor* is used in physiology to refer not only to sensory receptors but also, in a very different sense, to proteins that bind neurotransmitters, hormones, and other substances with great affinity and specificity as a first step in initiating specific physiologic responses.

CLASSIFICATION OF SENSORY RECEPTORS

Numerous attempts have been made to classify sensory receptors, but none has been entirely successful. One classification divides them into (1) teleceptors ("distance receivers"), which are concerned with events at a distance; (2) exteroceptors, which are concerned with the external environment near at hand; (3) interoceptors, which are concerned with the internal

TABLE 8–1 Principle sensory modalities.

Sensory System	Modality	Stimulus Energy	Receptor Class	Receptor Cell Types
Somatosensory	Touch	Tap, flutter 5–40 Hz	Cutaneous mechanoreceptor	Meissner corpuscles
Somatosensory	Touch	Motion	Cutaneous mechanoreceptor	Hair follicle receptors
Somatosensory	Touch	Deep pressure, vibration 60–300 Hz	Cutaneous mechanoreceptor	Pacinian corpuscles
Somatosensory	Touch	Touch, pressure	Cutaneous mechanoreceptor	Merkel cells
Somatosensory	Touch	Sustained pressure	Cutaneous mechanoreceptor	Ruffini corpuscles
Somatosensory	Proprioception	Stretch	Mechanoreceptor	Muscle spindles
Somatosensory	Proprioception	Tension	Mechanoreceptor	Golgi tendon organ
Somatosensory	Temperature	Thermal	Thermoreceptor	Cold and warm receptors
Somatosensory	Pain	Chemical, thermal, and mechanical	Chemoreceptor, thermoreceptor, and mechanoreceptor	Polymodal receptors or chemical, thermal, and mechanical nociceptors
Somatosensory	Itch	Chemical	Chemoreceptor	Chemical nociceptor
Visual	Vision	Light	Photoreceptor	Rods, cones
Auditory	Hearing	Sound	Mechanoreceptor	Hair cells (cochlea)
Vestibular	Balance	Angular acceleration	Mechanoreceptor	Hair cells (semicircular canals)
Vestibular	Balance	Linear acceleration, gravity	Mechanoreceptor	Hair cells (otolith organs)
Olfactory	Smell	Chemical	Chemoreceptor	Olfactory sensory neuron
Gustatory	Taste	Chemical	Chemoreceptor	Taste buds

environment; and (4) proprioceptors, which provide information about the position of the body in space at any given instant. However, the conscious component of proprioception ("body image") is actually synthesized from information coming not only from receptors in and around joints but also from cutaneous touch and pressure receptors.

Other special terms are frequently used to identify sensory receptors. The cutaneous receptors for touch and pressure are **mechanoreceptors.** Potentially harmful stimuli such as pain, extreme heat, and extreme cold are said to be mediated by **nociceptors.** The term **chemoreceptor** is used to refer to receptors stimulated by a change in the chemical composition of the environment in which they are located. These include receptors for taste and smell as well as visceral receptors such as those sensitive to changes in the plasma level of O_2, pH, and osmolality. **Photoreceptors** are those in the rods and cones in the retina that respond to light.

SENSE ORGANS

Sensory receptors can be specialized dendritic endings of afferent nerve fibers, and they are often associated with nonneural cells that surround it, forming a **sense organ.** Touch and pressure are sensed by four types of mechanoreceptors (Figure

8–1). **Meissner corpuscles** are dendrites encapsulated in connective tissue and respond to changes in texture and slow vibrations. **Merkel cells** are expanded dendritic endings, and they respond to sustained pressure and touch. **Ruffini corpuscles** are enlarged dendritic endings with elongated capsules, and they respond to sustained pressure. **Pacinian corpuscles** consist of unmyelinated dendritic endings of a sensory nerve fiber, 2 µm in diameter, encapsulated by concentric lamellae of connective tissue that give the organ the appearance of a cocktail onion. Theses receptors respond to deep pressure and fast vibration.

The Na^+ channel BNC1 is closely associated with touch receptors. This channel is one of the **degenerins,** so called because when they are hyperexpressed, they cause the neurons they are in to degenerate. However, it is not known if BNC1 is part of the receptor complex or the neural fiber at the point of initiation of the spike potential. The receptor may be opened mechanically by pressure on the skin.

Some sensory receptors are not specialized organs but rather are free nerve endings. Pain and temperature sensations arise from unmyelinated dendrites of sensory neurons located around hair follicles throughout the glaborous and hairy skin as well as deep tissue.

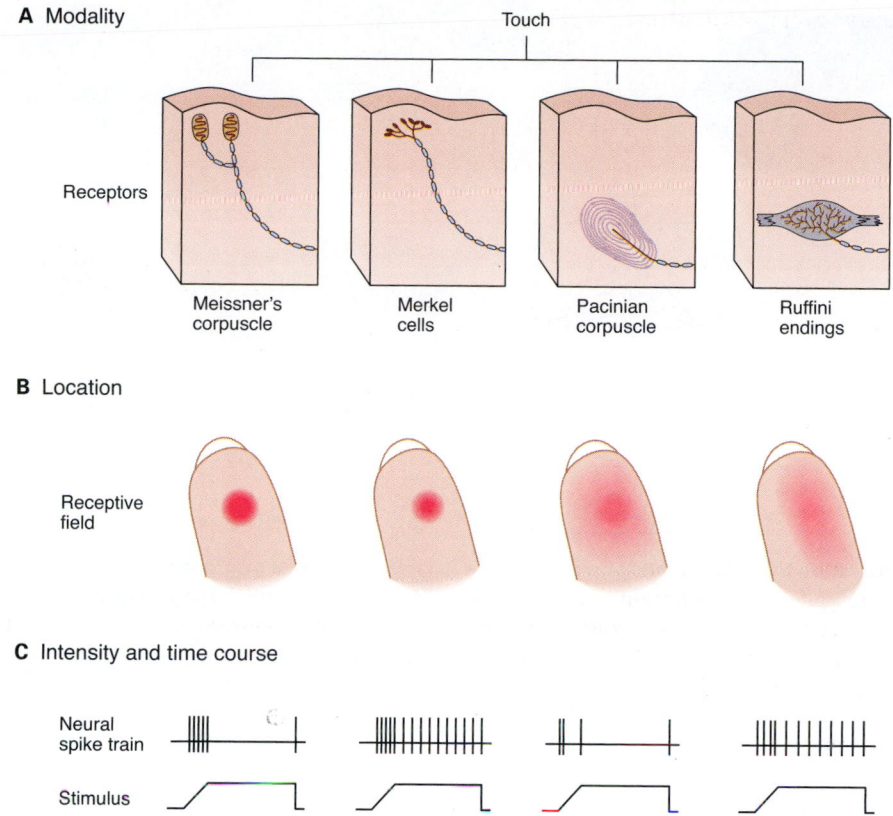

FIGURE 8–1 **Sensory systems encode four elementary attributes of stimuli: modality, location (receptive field), intensity, and duration (timing). A)** The human hand has four types of mechanoreceptors; their combined activation produces the sensation of contact with an object. Selective activation of Merkel cells and Ruffini endings causes sensation of steady pressure; selective activation of Meissner's and Pacinian corpuscles causes tingling and vibratory sensation. **B)** Location of a stimulus is encoded by spatial distribution of the population of receptors activated. A receptor fires only when the skin close to its sensory terminals is touched. These receptive fields of mechanoreceptors (shown as red areas on fingertips) differ in size and response to touch. Merkel cells and Meissner's corpuscles provide the most precise localization as they have the smallest receptive fields and are most sensitive to pressure applied by a small probe. **C)** Stimulus intensity is signaled by firing rates of individual receptors; duration of stimulus is signaled by time course of firing. The spike trains indicate action potentials elicited by pressure from a small probe at the center of each receptive field. Meissner's and Pacinian corpuscles adapt rapidly, the others adapt slowly.   (From Kandel ER, Schwartz JH, Jessell TM [editors]: *Principles of Neural Science*, 4th ed. McGraw-Hill, 2000.)

GENERATION OF IMPULSES IN CUTANEOUS RECEPTORS

PACINIAN CORPUSCLES

The way receptors generate action potentials in the sensory nerves that innervate them varies with the complexity of the sense organ. In the skin, the Pacinian corpuscle has been studied in some detail. As noted above, the Pacinian corpuscles are touch receptors. Because of their relatively large size and accessibility, they can be isolated, studied with microelectrodes, and subjected to microdissection. The myelin sheath of the sensory nerve begins inside the corpuscle (Figure 8–2). The first node of Ranvier is also located inside, whereas the second is usually near the point at which the nerve fiber leaves the corpuscle.

GENERATOR POTENTIALS

Recording electrodes can be placed on the sensory nerve as it leaves a Pacinian corpuscle and graded pressure applied to the corpuscle. When a small amount of pressure is applied, a nonpropagated depolarizing potential resembling an EPSP is recorded. This is called the **generator potential** or **receptor potential** (Figure 8–2). As the pressure is increased, the magnitude of the receptor potential increases. When the magnitude of the generator potential is about 10 mV, an action potential is generated in the sensory nerve. As the pressure is further increased, the generator potential becomes even larger and the sensory nerve fires repetitively.

SOURCE OF THE GENERATOR POTENTIAL

By microdissection techniques, it has been shown that removal of the connective tissue lamellas from the unmyelinated nerve

FIGURE 8–2 **Demonstration that the generator potential in a Pacinian corpuscle originates in the unmyelinated nerve terminal.**
(1) The electrical responses to a pressure of 1× (record a), 2× (b), 3× (c), and 4× (d) were recorded. The strongest stimulus produced an action potential in the sensory nerve (e). **(2)** Similar responses persisted after removal of the connective tissue capsule, except that the responses were more prolonged because of partial loss of adaptation. **(3)** The generator responses persisted but the action potential was absent when the first node of Ranvier was blocked by pressure or with narcotics (arrow). **(4)** All responses disappeared when the sensory nerve was cut and allowed to degenerate before the experiment.

ending in a Pacinian corpuscle does not abolish the generator potential. When the first node of Ranvier is blocked by pressure or narcotics, the generator potential is unaffected but conducted impulses are abolished (Figure 8–2). When the sensory nerve is sectioned and the nonmyelinated terminal is allowed to degenerate, no generator potential is formed. These and other experiments have established that the generator potential is produced in the unmyelinated nerve terminal. The receptor therefore converts mechanical energy into an electrical response, the magnitude of which is proportionate to the intensity of the stimulus. The generator potential in turn depolarizes the sensory nerve at the first node of Ranvier. Once the firing level is reached, an action potential is produced and the membrane repolarizes. If the generator potential is great enough, the neuron fires again as soon as it repolarizes, and it continues to fire as long as the generator potential is large enough to bring the membrane potential of the node to the firing level. Thus, the node converts the graded response of the receptor into action potentials, the frequency of which is proportionate to the magnitude of the applied stimuli.

SENSORY CODING

Converting a receptor stimulus to a recognizable sensation is termed **sensory coding.** All sensory systems code for four elementary attributes of a stimulus: modality, location, intensity, and duration. **Modality** is the type of energy transmitted by the stimulus. **Location** is the site on the body or space where the stimulus originated. **Intensity** is signaled by the response amplitude or frequency of action potential generation. **Duration** refers to the time from start to end of a response in the receptor.

These attributes of sensory coding are shown for the modality of touch in Figure 8–1.

MODALITY

Humans have four basic classes of receptors based on their sensitivity to one predominant form of energy: mechanical, thermal, electromagnetic, or chemical. The particular form of energy to which a receptor is most sensitive is called its **adequate stimulus.** The adequate stimulus for the rods and cones in the eye, for example, is light (an example of electromagnetic energy). Receptors do respond to forms of energy other than their adequate stimuli, but the threshold for these nonspecific responses is much higher. Pressure on the eyeball will stimulate the rods and cones, for example, but the threshold of these receptors to pressure is much higher than the threshold of the pressure receptors in the skin.

LOCATION

The term **sensory unit** is applied to a single sensory axon and all its peripheral branches. These branches vary in number but may be numerous, especially in the cutaneous senses. The **receptive field** of a sensory unit is the spatial distribution from which a stimulus produces a response in that unit (Figure 8–1). Representation of the senses in the skin is punctate. If the skin is carefully mapped, millimeter by millimeter, with a fine hair, a sensation of touch is evoked from spots overlying these touch receptors. None is evoked from the intervening areas. Similarly, temperature sensations and pain are produced by stimulation of

CLINICAL BOX 8–1

Two-Point Discrimination

The size of the receptive fields for light touch can be measured by the **two-point threshold test.** In this procedure, the two points on a pair of calipers are simultaneously positioned on the skin and one determines the minimum distance between the two caliper points that can be perceived as separate points of stimulation. This is called the **two-point discrimination threshold.** If the distance is very small, each caliper point is touching the receptive field of only one sensory neuron. If the distance between stimulation points is less than this threshold, only one point of stimulation can be felt. Thus, the two-point discrimination threshold is a measure of **tactile acuity.** The magnitude of two-point discrimination thresholds varies from place to place on the body and is smallest where touch receptors are most abundant. Stimulus points on the back, for instance, must be separated by at least 65 mm before they can be distinguished as separate, whereas on the fingertips two stimuli are recognized if they are separated by as little as 2 mm. Blind individuals benefit from the tactile acuity of fingertips to facilitate the ability to read Braille; the dots forming Braille symbols are separated by 2.5 mm. Two-point discrimination is used to test the integrity of the **dorsal column (medial lemniscus) system,** the central pathway for touch and proprioception.

the skin only over the spots where the receptors for these modalities are located. In the cornea and adjacent sclera of the eye, the surface area supplied by a single sensory unit is 50–200 mm^2. Generally, the areas supplied by one unit overlap and interdigitate with the areas supplied by others.

One of the most important mechanisms that enable localization of a stimulus site is **lateral inhibition.** Information from sensory neurons whose receptors are at the peripheral edge of the stimulus is inhibited compared to information from the sensory neurons at the center of the stimulus. Thus, lateral inhibition enhances the contrast between the center and periphery of a stimulated area and increases the ability of the brain to localize a sensory input. Lateral inhibition underlies **two-point discrimination** (see Clinical Box 8–1).

INTENSITY

The intensity of sensation is determined by the amplitude of the stimulus applied to the receptor. This is illustrated in Figure 8–3. As a greater pressure is applied to the skin, the receptor potential in the mechanoreceptor increases (not shown), and the frequency of the action potentials in a single axon transmitting information to the CNS is also increased. In addition to increasing the firing rate in a single axon, the greater intensity of stimulation also will recruit more receptors into the receptive field.

It has long been taught that the magnitude of the sensation felt is proportional to the log of the intensity of the stimulus (**Weber–Fechner law**). It now appears, however, that a power function more accurately describes this relation. In other words,

$$R = KS^A$$

where R is the sensation felt, S is the intensity of the stimulus, and, for any specific sensory modality, K and A are constants. The frequency of the action potentials generated in a sensory nerve fiber is also related to the intensity of the initiating stimulus by a power function. An example of this relation is shown is shown in Figure 8–4, in which the calculated exponent is 0.52. However, the relation between direct stimulation of a sensory nerve and the sensation felt is linear. Consequently, it appears that for any given sensory modality, the relation between sensation and stimulus intensity is determined primarily by the properties of the peripheral receptors.

DURATION

When a maintained stimulus of constant strength is applied to a receptor, the frequency of the action potentials in its sensory nerve declines over time. This phenomenon is known as **adaptation** or **desensitization.** The degree to which adaptation occurs varies from one sense to another. Receptors can be classified into **rapidly adapting (phasic) receptors** and **slowly adapting (tonic) receptors.** This is illustrated for different types of touch receptors in Figure 8–1. Meissner and Pacinian corpuscles are examples of rapidly adapting receptors, and Merkel cells and Ruffini endings are examples of slowly adapting receptors. Other examples of slowly adapting receptors are muscle spindles and nociceptors. Different types of sensory adaptation appear to have some value to the individual. Light touch would be distracting if it were persistent; and, conversely, slow adaptation of spindle input is needed to maintain posture. Similarly, input from nociceptors provides a warning that would lose its value if it adapted and disappeared.

SENSORY INFORMATION

The speed of conduction and other characteristics of sensory nerve fibers vary, but action potentials are similar in all nerves. The action potentials in the nerve from a touch receptor, for example, are essentially identical to those in the nerve from a warmth receptor. This raises the question of why stimulation of a touch receptor causes a sensation of touch and not of warmth. It also raises the question of how it is possible to tell whether the touch is light or heavy.

LAW OF SPECIFIC NERVE ENERGIES

The sensation evoked by impulses generated in a receptor depends in part on the specific part of the brain they ultimately activate. The specific sensory pathways are discrete from sense

FIGURE 8–3 **Relationship between stimulus and impulse frequency in an afferent fiber.** Action potentials in an afferent fiber from a mechanoreceptor of a single sensory unit increase in frequency as branches of the afferent neuron are stimulated by pressure of increasing magnitude. (From Widmaier EP, Raff H, Strang KT: *Vander's Human Physiology.* McGraw-Hill, 2008.)

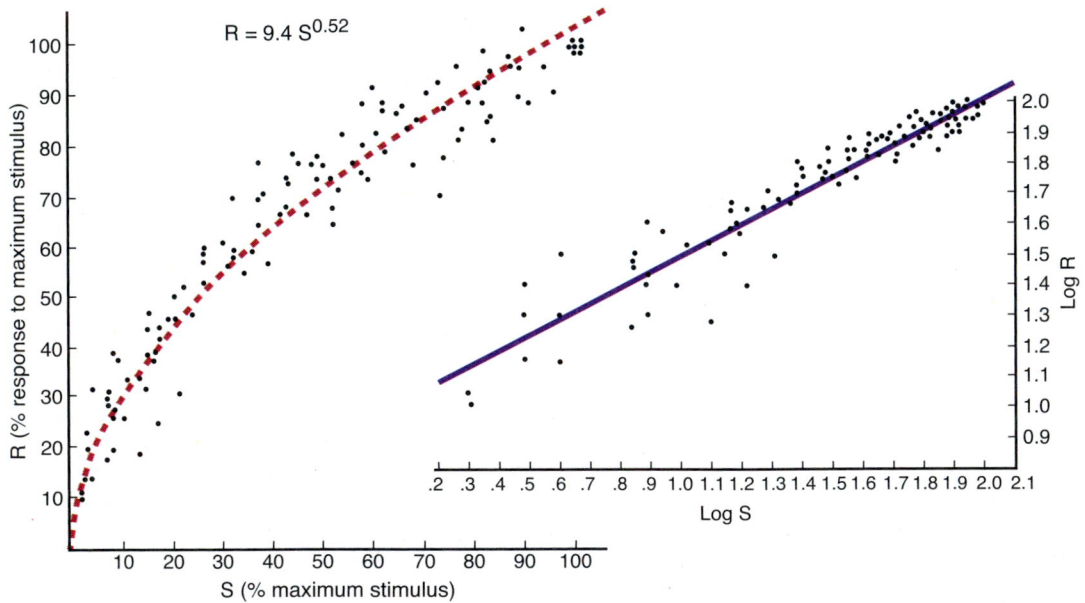

FIGURE 8–4 **Relation between magnitude of touch stimulus (S) and frequency of action potentials in sensory nerve fibers (R).** Dots are individual values from cats and are plotted on linear coordinates **(left)** and log–log coordinates **(right).** The equation shows the calculated power function relationship between R and S. (Reproduced, with permission, from Werner G, Mountcastle VB: Neural activity in mechanoreceptive cutaneous afferents. Stimulus–response relations, Weber functions, and information transmission. J Neurophysiol 1965;28:359.)

CLINICAL BOX 8–2

Vibratory Sensibility

Vibratory sensibility is tested by applying a vibrating (128-Hz) tuning fork to the skin on the fingertip, tip of the toe, or bony prominences of the toes. The normal response is a "buzzing" sensation. The sensation is most marked over bones. The term **pallesthesia** is also used to describe this ability to feel mechanical vibrations. The receptors involved are the receptors for touch, especially **Pacinian corpuscles,** but a time factor is also necessary. A pattern of rhythmic pressure stimuli is interpreted as vibration. The impulses responsible for the vibrating sensation are carried in the **dorsal columns.** Degeneration of this part of the spinal cord occurs in poorly controlled diabetes, pernicious anemia, vitamin B_{12} deficiencies, or early tabes dorsalis. Elevation of the threshold for vibratory stimuli is an early symptom of this degeneration. Vibratory sensation and proprioception are closely related; when one is diminished, so is the other.

CLINICAL BOX 8–3

Stereognosis

Stereognosis is the perception of the form and nature of an object without looking at it. Normal persons can readily identify objects such as keys and coins of various denominations. This ability depends on relatively intact touch and pressure sensation and is compromised when the dorsal columns are damaged. The inability to identify an object by touch is called **tactile agnosia.** It also has a large cortical component; impaired stereognosis is an early sign of damage to the cerebral cortex and sometimes occurs in the absence of any detectable defect in touch and pressure sensation when there is a lesion in the parietal lobe posterior to the postcentral gyrus. Stereoagnosia can also be expressed by the failure to identify an object by sight (**visual agnosia**), the inability to identify sounds or words (**auditory agnosia**) or color (**color agnosia**), or the inability to identify the location or position of an extremity (**position agnosia**).

organ to cortex. Therefore, when the nerve pathways from a particular sense organ are stimulated, the sensation evoked is that for which the receptor is specialized no matter how or where along the pathway the activity is initiated. This principle, first enunciated by Müller in 1835, has been given the rather cumbersome name of the **law of specific nerve energies.** For example, if the sensory nerve from a Pacinian corpuscle in the hand is stimulated by pressure at the elbow or by irritation from a tumor in the brachial plexus, the sensation evoked is touch. Similarly, if a fine enough electrode could be inserted into the appropriate fibers of the dorsal columns of the spinal cord, the thalamus, or the postcentral gyrus of the cerebral cortex, the sensation produced by stimulation would be touch. The general principle of specific nerve energies remains one of the cornerstones of sensory physiology.

LAW OF PROJECTION

No matter where a particular sensory pathway is stimulated along its course to the cortex, the conscious sensation produced is referred to the location of the receptor. This principle is called the **law of projection.** Cortical stimulation experiments during neurosurgical procedures on conscious patients illustrate this phenomenon. For example, when the cortical receiving area for impulses from the left hand is stimulated, the patient reports sensation in the left hand, not in the head.

RECRUITMENT OF SENSORY UNITS

As the strength of a stimulus is increased, it tends to spread over a large area and generally not only activates the sense organs immediately in contact with it but also "recruits" those in the surrounding area. Furthermore, weak stimuli activate the receptors

with the lowest thresholds, and stronger stimuli also activate those with higher thresholds. Some of the receptors activated are part of the same sensory unit, and impulse frequency in the unit therefore increases. Because of overlap and interdigitation of one unit with another, however, receptors of other units are also stimulated, and consequently more units fire. In this way, more afferent pathways are activated, which is interpreted in the brain as an increase in intensity of the sensation.

NEUROLOGICAL EXAM

The sensory component of a neurological exam includes an assessment of various sensory modalities including touch, proprioception, vibratory sense, and pain. Clinical Box 8–2 describes the test for vibratory sensibility. Cortical sensory function can be tested by placing familiar objects in a patient's hands and asking him or her to identify it with the eyes closed (see Clinical Box 8–3).

CHAPTER SUMMARY

- Sensory receptors are commonly classified as mechanoreceptors, nociceptors, chemoreceptors, or photoreceptors.
- Touch and pressure are sensed by four types of mechanoreceptors: Meissner's corpuscles (respond to changes in texture and slow vibrations), Merkel's cells (respond to sustained pressure and touch), Ruffini corpuscles (respond to sustained pressure), and Pacinian corpuscles (respond to deep pressure and fast vibrations).
- Nociceptors and thermoreceptors are free nerve endings on unmyelinated or lightly myelinated fibers in hairy and glaborous skin and deep tissues.
- The generator or receptor potential is the nonpropagated depolarizing potential recorded in a sensory organ after an adequate

stimulus is applied. As the stimulus is increased, the magnitude of the receptor potential increases. When it reaches a critical threshold, an action potential is generated in the sensory nerve.

■ Converting a receptor stimulus to a recognizable sensation is termed sensory coding. All sensory systems code for four elementary attributes of a stimulus: modality, location, intensity, and duration.

MULTIPLE-CHOICE QUESTIONS

For all questions, select the single best answer unless otherwise directed.

1. Pacinian corpuscles are
   A) a type of thermoreceptor.
   B) usually innervated by Aδ nerve fibers.
   C) rapidly adapting touch receptors.
   D) slowly adapting touch receptors.
   E) nociceptors.

2. Adaptation to a sensory stimulus produces
   A) a diminished sensation when other types of sensory stimuli are withdrawn.
   B) a more intense sensation when a given stimulus is applied repeatedly.
   C) a sensation localized to the hand when the nerves of the brachial plexus are stimulated.
   D) a diminished sensation when a given stimulus is applied repeatedly over time.
   E) a decreased firing rate in the sensory nerve from the receptor when one's attention is directed to another matter.

3. Sensory systems code for the following attributes of a stimulus:
   A) modality, location, intensity, and duration
   B) threshold, receptive field, adaptation, and discrimination
   C) touch, taste, hearing, and smell
   D) threshold, laterality, sensation, and duration
   E) sensitization, discrimination, energy, and projection

4. In which of the following is the frequency of stimulation not linearly related to the strength of the sensation felt?
   A) sensory area of the cerebral cortex
   B) specific projection nuclei of the thalamus
   C) lateral spinothalamic tract
   D) dorsal horn
   E) cutaneous receptors

5. Which of the following receptors and sense organs are *incorrectly* paired?
   A) rods and cones : eye
   B) receptors sensitive to sodium : taste buds
   C) hair cells : olfactory epithelium
   D) receptors sensitive to stretch : carotid sinus
   E) glomus cells : carotid body

6. Which best describes the law of specific nerve energies?
   A) No matter where a particular sensory pathway is stimulated along its course to the cortex, the conscious sensation produced is referred to the location of the receptor.
   B) A nerve can only be stimulated by electrical energy.
   C) Receptors can respond to forms of energy other than their adequate stimuli, but the threshold for these nonspecific responses is much higher.
   D) For any given sensory modality, the specific relationship between sensation and stimulus intensity is determined by the properties of the peripheral receptors.
   E) The sensation evoked by impulses generated in a receptor depends in part on the specific part of the brain they ultimately activate.

7. Which of the following does *not* contain cation channels that are activated by mechanical distortion, producing depolarization?
   A) olfactory receptors
   B) Pacinian corpuscles
   C) hair cells in cochlea
   D) hair cells in semicircular canals
   E) hair cells in utricle

CHAPTER RESOURCES

Barlow HB, Mollon JD (editors): *The Senses.* Cambridge University Press, 1982.

Bell J, Bolanowski S, Holmes MH: The structure and function of Pacinian corpuscles: A review. Prog Neurobiol 1994;42:79.

Haines DE (editor): *Fundamental Neuroscience for Basic and Clinical Applications,* 3rd ed. Elsevier, 2006.

Iggo A (editor): *Handbook of Sensory Physiology.* Vol 2, *Somatosensory System.* Springer-Verlag, 1973.

Kandel ER, Schwartz JH, Jessell TM (editors): *Principles of Neural Science,* 4th ed. McGraw-Hill, 2000.

Mountcastle VB: *Perceptual Neuroscience.* Harvard University Press, 1999.

Squire LR, et al (editors): *Fundamental Neuroscience,* 3rd ed. Academic Press, 2008.

Reflexes

INTRODUCTION

The basic unit of integrated reflex activity is the **reflex arc.** This arc consists of a sense organ, an afferent neuron, one or more synapses within a central integrating station, an efferent neuron, and an effector. In mammals, the connection between afferent and efferent somatic neurons is generally in the brain or spinal cord. The afferent neurons enter via the dorsal roots or cranial nerves and have their cell bodies in the dorsal root ganglia or in the homologous ganglia on the cranial nerves. The efferent fibers leave via the ventral roots or corresponding motor cranial nerves. The principle that in the spinal cord the dorsal roots are sensory and the ventral roots are motor is known as the **Bell–Magendie law.**

Activity in the reflex arc starts in a sensory receptor with a receptor potential whose magnitude is proportional to the strength of the stimulus (Figure 9–1). This generates all-or-none action potentials in the afferent nerve, the number of action potentials being proportional to the size of the generator potential. In the central nervous system (CNS), the responses are again graded in terms of excitatory postsynaptic potentials (EPSPs) and inhibitory postsynaptic potentials (IPSPs) at the synaptic junctions. All-or-none responses are generated in the efferent nerve. When these reach the effector, they again set up a graded response. When the effector is smooth muscle, responses summate to produce action potentials in the smooth muscle, but when the effector is skeletal muscle, the graded response is always adequate to produce action potentials that bring about muscle contraction. The connection between the afferent and efferent neurons is usually in the CNS, and activity in the reflex arc is modified by the multiple inputs converging on the efferent neurons or at any synaptic station within the reflex loop.

The simplest reflex arc is one with a single synapse between the afferent and efferent neurons. Such arcs are **monosynaptic,** and reflexes occurring in them are called **monosynaptic reflexes.** Reflex arcs in which one or more interneuron is interposed between the afferent and efferent neurons are called **polysynaptic reflexes.** There can be anywhere from two to hundreds of synapses in a polysynaptic reflex arc.

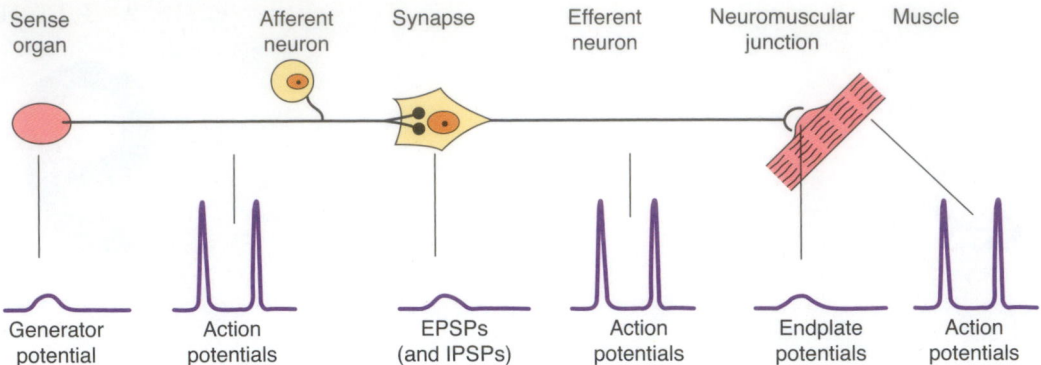

FIGURE 9–1   The reflex arc. Note that at the receptor and in the CNS a nonpropagated graded response occurs that is proportionate to the magnitude of the stimulus. The response at the neuromuscular junction is also graded, though under normal conditions it is always large enough to produce a response in skeletal muscle. On the other hand, in the portions of the arc specialized for transmission (afferent and efferent axons, muscle membrane), the responses are all-or-none action potentials.

MONOSYNAPTIC REFLEXES: THE STRETCH REFLEX

When a skeletal muscle with an intact nerve supply is stretched, it contracts. This response is called the **stretch reflex.** The stimulus that initiates the reflex is stretch of the muscle, and the response is contraction of the muscle being stretched. The sense organ is a small encapsulated spindlelike or fusiform shaped structure called the muscle spindle, located within the fleshy part of the muscle. The impulses originating from the spindle are transmitted to the CNS by fast sensory fibers that pass directly to the motor neurons which supply the same muscle. The neurotransmitter at the central synapse is glutamate. The stretch reflex is the best known and studied monosynaptic reflex and is typified by the **knee jerk reflex** (see Clinical Box 9–1).

STRUCTURE OF MUSCLE SPINDLES

Each muscle spindle has three essential elements: (1) a group of specialized intrafusal muscle fibers with contractile polar ends and a noncontractile center, (2) large diameter myelinated afferent nerves (types Ia and II) originating in the central portion of the intrafusal fibers, and (3) small diameter myelinated efferent nerves supplying the polar contractile regions of the intrafusal fibers (Figure 9–2A). It is important to understand the relationship of these elements to each other and to the muscle itself to appreciate the role of this sense organ in signaling changes in the length of the muscle in which it is located. Changes in muscle length are associated with changes in joint angle; thus muscle spindles provide information on position (ie, **proprioception**).

The **intrafusal fibers** are positioned in parallel to the **extrafusal fibers** (the regular contractile units of the muscle) with the ends of the spindle capsule attached to the tendons at either end of the muscle. Intrafusal fibers do not contribute to the overall contractile force of the muscle, but rather serve a

pure sensory function. There are two types of intrafusal fibers in mammalian muscle spindles. The first type contains many nuclei in a dilated central area and is called a **nuclear bag fiber** (Figure 9–2B). There are two subtypes of nuclear bag fibers, **dynamic** and **static.** Typically, there are two or three nuclear bag fibers per spindle. The second intrafusal fiber type, the **nuclear chain fiber,** is thinner and shorter and lacks a definite bag. Each spindle has about five of these fibers.

There are two kinds of sensory endings in each spindle, a single **primary (group Ia) ending** and up to eight **secondary (group II) endings.** The Ia afferent fiber wraps around the center of the dynamic and static nuclear bag fibers and nuclear chain fibers. Group II sensory fibers are located adjacent to the centers of the static nuclear bag and nuclear chain fibers; these fibers do not innervate the dynamic nuclear bag fibers. Ia afferents are very sensitive to the velocity of the change in muscle length during a stretch (**dynamic response**); thus they provide information about the speed of movements and allow for quick corrective movements. The steady-state (tonic) activity of group Ia and II afferents provide information on steady-state length of the muscle (**static response**). The top trace in Figure 9–2C shows the dynamic and static components of activity in a Ia afferent during muscle stretch. Note that they discharge most rapidly while the muscle is being stretched (shaded area of graphs) and less rapidly during sustained stretch.

The spindles have a motor nerve supply of their own. These nerves are 3–6 μm in diameter, constitute about 30% of the fibers in the ventral roots, and are called γ-**motor neurons.** There are two types of γ-motor neurons: **dynamic,** which supply the dynamic nuclear bag fibers and **static,** which supply the static nuclear bag fibers and the nuclear chain fibers. Activation of dynamic γ-motor neurons increases the dynamic sensitivity of the group Ia endings. Activation of the static γ-motor neurons increases the tonic level of activity in both group Ia and II endings, decreases the dynamic sensitivity of group Ia afferents, and can prevent silencing of Ia afferents during muscle stretch (Figure 9–2C).

CLINICAL BOX 9–1

Knee Jerk Reflex

Tapping the patellar tendon elicits the **knee jerk,** a stretch reflex of the quadriceps femoris muscle, because the tap on the tendon stretches the muscle. A similar contraction is observed if the quadriceps is stretched manually. Stretch reflexes can also be elicited from most of the large muscles of the body. Tapping on the tendon of the triceps brachii, for example, causes an extensor response at the elbow as a result of reflex contraction of the triceps; tapping on the Achilles tendon causes an ankle jerk due to reflex contraction of the gastrocnemius; and tapping on the side of the face causes a stretch reflex in the masseter. The knee jerk reflex is an example of a **deep tendon reflex (DTR)** in a neurological exam and is graded on the following scale: 0 (absent), 1+ (hypoactive), 2+ (brisk, normal), 3+ (hyperactive without clonus), 4+ (hyperactive with mild clonus), and 5+ (hyperactive with sustained clonus). Absence of the knee jerk can signify an abnormality anywhere within the reflex arc, including the muscle spindle, the Ia afferent nerve fibers, or the motor neurons to the quadriceps muscle. The most common cause is a peripheral neuropathy from such things as diabetes, alcoholism, and toxins. A hyperactive reflex can signify an interruption of corticospinal and other descending pathways that influence the reflex arc.

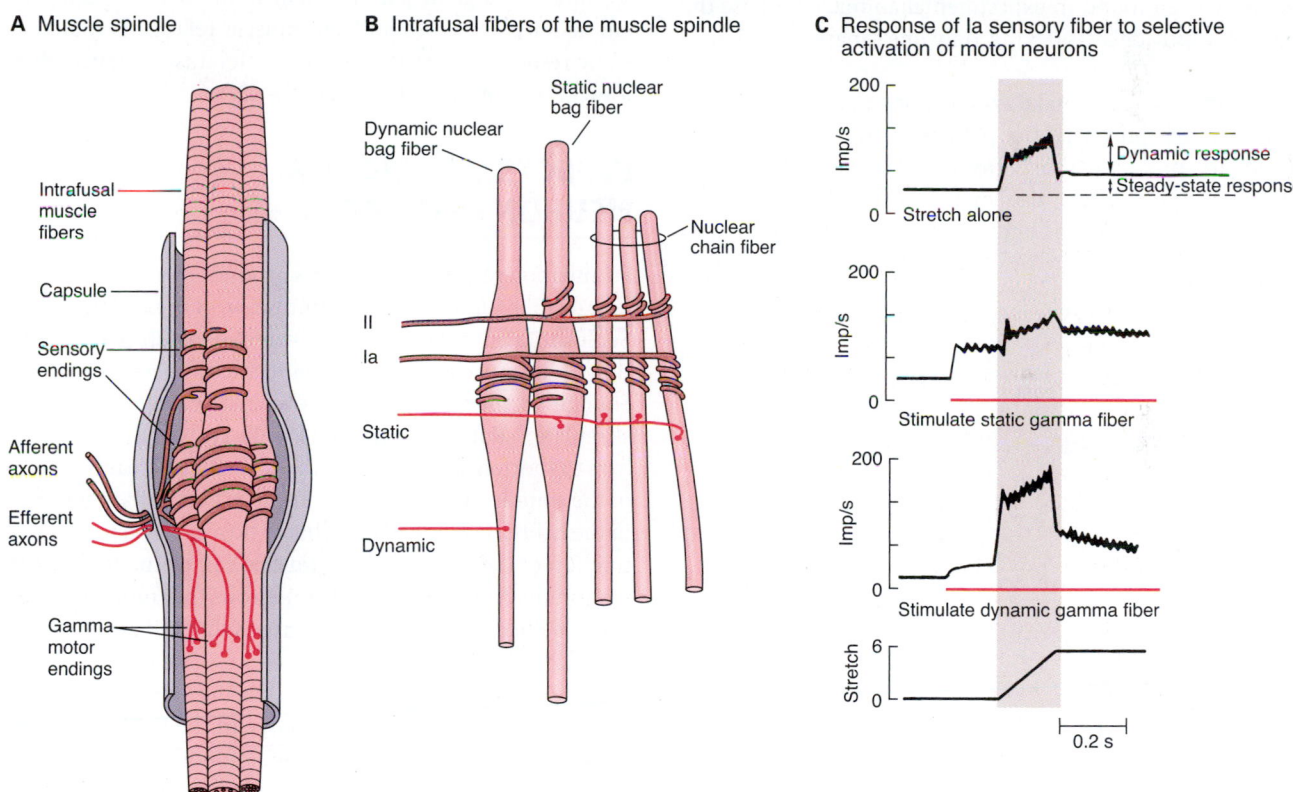

A Muscle spindle

Intrafusal muscle fibers
Capsule
Sensory endings
Afferent axons
Efferent axons
Gamma motor endings

B Intrafusal fibers of the muscle spindle

Static nuclear bag fiber
Dynamic nuclear bag fiber
Nuclear chain fiber
II
Ia
Static
Dynamic

C Response of Ia sensory fiber to selective activation of motor neurons

Imp/s
200
0 Stretch alone
Dynamic response
Steady-state response

Imp/s
200
0 Stimulate static gamma fiber

Imp/s
200
0 Stimulate dynamic gamma fiber

Stretch
6
0

0.2 s

FIGURE 9–2 Mammalian muscle spindle. A) Diagrammatic representation of the main components of mammalian muscle spindle including intrafusal muscle fibers, afferent sensory fiber endings, and efferent motor fibers (γ-motor neurons). **B)** Three types of intrafusal muscle fibers: dynamic nuclear bag, static nuclear bag, and nuclear chain fibers. A single Ia afferent fiber innervates all three types of fibers to form a primary sensory ending. A group II sensory fiber innervates nuclear chain and static bag fibers to form a secondary sensory ending. Dynamic γ-motor neurons innervate dynamic bag fibers; static γ-motor neurons innervate combinations of chain and static bag fibers. **C)** Comparison of discharge pattern of Ia afferent activity during stretch alone and during stimulation of static or dynamic γ-motor neurons. Without γ-stimulation, Ia fibers show a small dynamic response to muscle stretch and a modest increase in steady-state firing. When static γ-motor neurons are activated, the steady-state response increases and the dynamic response decreases. When dynamic γ-motor neurons are activated, the dynamic response is markedly increased but the steady-state response gradually returns to its original level. (From Kandel ER, Schwartz JH, Jessell TM [editors]: *Principles of Neural Science,* 4th ed. McGraw-Hill, 2000.)

CENTRAL CONNECTIONS OF AFFERENT FIBERS

Ia fibers end directly on motor neurons supplying the extrafusal fibers of the same muscle (Figure 9–3). The time between the application of the stimulus and the response is called the **reaction time.** In humans, the reaction time for a stretch reflex such as the knee jerk is 19–24 ms. Weak stimulation of the sensory nerve from the muscle, known to stimulate only Ia fibers, causes a contractile response with a similar latency. Because the conduction velocities of the afferent and efferent fiber types are known and the distance from the muscle to the spinal cord can be measured, it is possible to calculate how much of the reaction time was taken up by conduction to and from the spinal cord. When this value is subtracted from the reaction time, the remainder, called the **central delay,** is the time taken for the reflex activity to traverse the spinal cord. In humans, the central delay for the knee jerk is 0.6–0.9 ms, and figures of similar magnitude have been found in experimental animals. Because the minimal synaptic delay is 0.5 ms, only one synapse could have been traversed.

Muscle spindles also make connections that cause muscle contraction via polysynaptic pathways, and the afferents involved are probably those from the secondary endings. However, group II fibers also make monosynaptic connections to the motor neurons and make a small contribution to the stretch reflex.

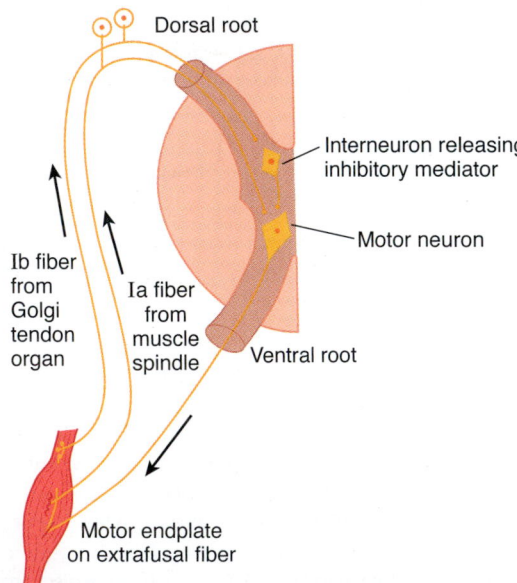

FIGURE 9–3 **Diagram illustrating the pathways responsible for the stretch reflex and the inverse stretch reflex.** Stretch stimulates the muscle spindle, which activates Ia fibers that excite the motor neuron. Stretch also stimulates the Golgi tendon organ, which activates Ib fibers that excite an interneuron that releases the inhibitory mediator glycine. With strong stretch, the resulting hyperpolarization of the motor neuron is so great that it stops discharging.

FUNCTION OF MUSCLE SPINDLES

When the muscle spindle is stretched, its sensory endings are distorted and receptor potentials are generated. These in turn set up action potentials in the sensory fibers at a frequency proportional to the degree of stretching. Because the spindle is in parallel with the extrafusal fibers, when the muscle is passively stretched, the spindles are also stretched, referred to as "loading the spindle." This initiates reflex contraction of the extrafusal fibers in the muscle. On the other hand, the spindle afferents characteristically stop firing when the muscle is made to contract by electrical stimulation of the α-motor neurons to the extrafusal fibers because the muscle shortens while the spindle is unloaded (Figure 9–4).

Thus, the spindle and its reflex connections constitute a feedback device that operates to maintain muscle length; if the muscle is stretched, spindle discharge increases and reflex shortening is produced, whereas if the muscle is shortened without a change in γ-motor neuron discharge, spindle afferent activity decreases and the muscle relaxes. Dynamic and static responses of muscle spindle afferents influence **physiological tremor** (see Clinical Box 9–2).

EFFECTS OF γ-MOTOR NEURON DISCHARGE

Stimulation of γ-motor neurons produces a very different picture from that produced by stimulation of the extrafusal fibers. Such stimulation does not lead directly to detectable contraction of the muscles because the intrafusal fibers are not strong enough or plentiful enough to cause shortening. However, stimulation does cause the contractile ends of the intrafusal fibers to shorten and therefore stretches the nuclear bag portion of the spindles, deforming the endings and initiating impulses in the Ia fibers (Figure 9–4). This in turn can lead to reflex contraction of the muscle. Thus, muscle can be made to contract via stimulation of the α-motor neurons that innervate the extrafusal fibers or the γ-motor neurons that initiate contraction indirectly via the stretch reflex.

If the whole muscle is stretched during stimulation of the γ-motor neurons, the rate of discharge in the Ia fibers is further increased (Figure 9–4). Increased γ-motor neuron activity thus increases **spindle sensitivity** during stretch.

In response to descending excitatory input to spinal motor circuits, both α- and γ-motor neurons are activated. Because of this "α–γ coactivation," intrafusal and extrafusal fibers shorten together, and spindle afferent activity can occur throughout the period of muscle contraction. In this way, the spindle remains capable of responding to stretch and reflexly adjusting α-motor neuron discharge.

Spindle

Tendon — Extrafusal fiber

Sensory nerve

Impulses in sensory nerve

Muscle at rest

Muscle stretched

Muscle contracted

Increased γ efferent discharge

Increased γ efferent discharge—muscle stretched

FIGURE 9–4   Effect of various conditions on muscle spindle discharge. When the whole muscle is stretched, the muscle spindle is also stretched and its sensory endings are activated at a frequency proportional to the degree of stretching ("loading the spindle"). Spindle afferents stop firing when the muscle contracts ("unloading the spindle"). Stimulation of γ-motor neurons cause the contractile ends of the intrafusal fibers to shorten. This stretches the nuclear bag region, initiating impulses in sensory fibers. If the whole muscle is stretched during stimulation of the γ-motor neurons, the rate of discharge in sensory fibers is further increased.

CLINICAL BOX 9–2

Physiological Tremor

The response of the Ia sensory fiber endings to the dynamic (phasic) as well as the static events in the muscle is important because the prompt, marked phasic response helps to dampen oscillations caused by conduction delays in the feedback loop regulating muscle length. Normally a small oscillation occurs in this feedback loop. This **physiologic tremor** has a low amplitude (barely visible to the naked eye) and a frequency of approximately 10 Hz. Physiological tremor is a normal phenomenon which affects everyone while maintaining posture or during movements. However, the tremor would be worse if it were not for the sensitivity of the spindle to velocity of stretch. It can become exaggerated in some situations such as when we are anxious or tired or because of drug toxicity. Numerous factors contribute to the genesis of physiological tremor. It is likely dependent on not only central (**inferior olive**) sources but also from peripheral factors including motor unit firing rates, reflexes, and mechanical resonance.

CONTROL OF γ-MOTOR NEURON DISCHARGE

The γ-motor neurons are regulated to a large degree by descending tracts from a number of areas in the brain. Via these pathways, the sensitivity of the muscle spindles and hence the threshold of the stretch reflexes in various parts of the body can be adjusted and shifted to meet the needs of postural control.

Other factors also influence γ-motor neuron discharge. Anxiety causes an increased discharge, a fact that probably explains the hyperactive tendon reflexes sometimes seen in anxious patients. In addition, unexpected movement is associated with a greater efferent discharge. Stimulation of the skin, especially by noxious agents, increases γ-motor neuron discharge to ipsilateral flexor muscle spindles while decreasing that to extensors and produces the opposite pattern in the opposite limb. It is well known that trying to pull the hands apart when the flexed fingers are hooked together facilitates the knee jerk reflex (Jendrassik's maneuver), and this may also be due to increased γ-motor neuron discharge initiated by afferent impulses from the hands.

RECIPROCAL INNERVATION

When a stretch reflex occurs, the muscles that antagonize the action of the muscle involved (antagonists) relax. This phenomenon is said to be due to **reciprocal innervation.** Impulses in the Ia fibers from the muscle spindles of the protagonist muscle cause postsynaptic inhibition of the motor neurons to the antagonists. The pathway mediating this effect is bisynaptic. A collateral from each Ia fiber passes in the spinal cord to an inhibitory interneuron that synapses on a motor neuron supplying the

FIGURE 9–5   **Golgi tendon organ.** (Reproduced, with permission, from Goss CM [editor]: *Gray's Anatomy of the Human Body,* 29th ed. Lea & Febiger, 1973.)

antagonist muscles. This example of postsynaptic inhibition is discussed in Chapter 6, and the pathway is illustrated in Figure 6–6.

INVERSE STRETCH REFLEX

Up to a point, the harder a muscle is stretched, the stronger is the reflex contraction. However, when the tension becomes great enough, contraction suddenly ceases and the muscle relaxes. This relaxation in response to strong stretch is called the **inverse stretch reflex** or **autogenic inhibition.**

The receptor for the inverse stretch reflex is in the **Golgi tendon organ** (Figure 9–5). This organ consists of a netlike collection of knobby nerve endings among the fascicles of a tendon. There are 3–25 muscle fibers per tendon organ. The fibers from the Golgi tendon organs make up the Ib group of myelinated, rapidly conducting sensory nerve fibers. Stimulation of these Ib fibers leads to the production of IPSPs on the motor neurons that supply the muscle from which the fibers arise. The Ib fibers end in the spinal cord on inhibitory interneurons that in turn

terminate directly on the motor neurons (Figure 9–3). They also make excitatory connections with motor neurons supplying antagonists to the muscle.

Because the Golgi tendon organs, unlike the spindles, are in series with the muscle fibers, they are stimulated by both passive stretch and active contraction of the muscle. The threshold of the Golgi tendon organs is low. The degree of stimulation by passive stretch is not great because the more elastic muscle fibers take up much of the stretch, and this is why it takes a strong stretch to produce relaxation. However, discharge is regularly produced by contraction of the muscle, and the Golgi tendon organ thus functions as a transducer in a feedback circuit that regulates muscle force in a fashion analogous to the spindle feedback circuit that regulates muscle length.

The importance of the primary endings in the spindles and the Golgi tendon organs in regulating the velocity of the muscle contraction, muscle length, and muscle force is illustrated by the fact that that section of the afferent nerves to an arm causes the limb to hang loosely in a semiparalyzed state. The organization of the system is shown in Figure 9–6. The interaction of

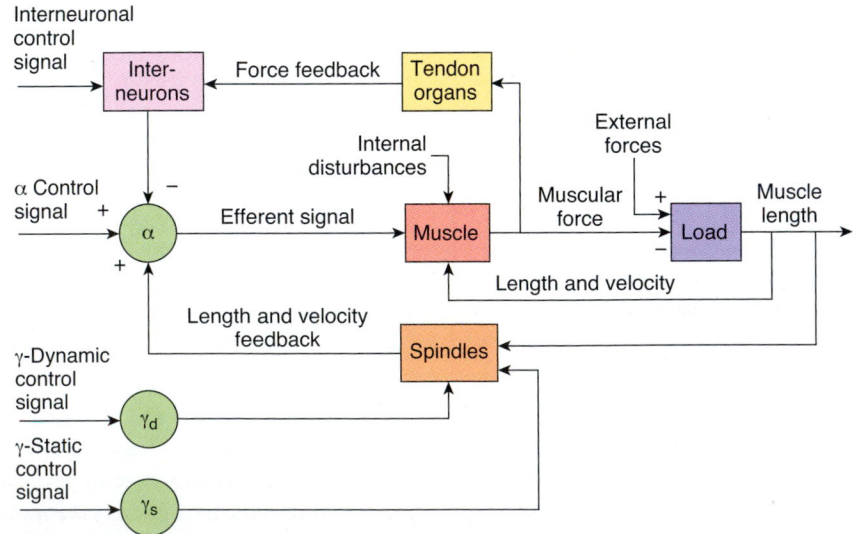

FIGURE 9–6   Block diagram of peripheral motor control system. The dashed line indicates the nonneural feedback from muscle that limits length and velocity via the inherent mechanical properties of muscle. γ_d, dynamic γ-motor neurons; γ_s, static γ-motor neurons. (Reproduced, with permission, from Houk J in: *Medical Physiology,* 13th ed. Mount-Castle VB [editor]. Mosby, 1974.)

Clonus

A characteristic of states in which increased γ-motor neuron discharge is present is **clonus.** This neurologic sign is the occurrence of regular, repetitive, rhythmic contractions of a muscle subjected to sudden, maintained stretch. Only sustained clonus with five or more beats is considered abnormal. Ankle clonus is a typical example. This is initiated by brisk, maintained dorsiflexion of the foot, and the response is rhythmic plantar flexion at the ankle. The **stretch reflex–inverse stretch reflex sequence** may contribute to this response. However, it can occur on the basis of synchronized motor neuron discharge without Golgi tendon organ discharge. The spindles of the tested muscle are hyperactive, and the burst of impulses from them discharges all the motor neurons supplying the muscle at once. The consequent muscle contraction stops spindle discharge. However, the stretch has been maintained, and as soon as the muscle relaxes it is again stretched and the spindles stimulated. Clonus may also occur after disruption of descending cortical input to a spinal glycinergic inhibitory interneuron called the **Renshaw cell.** This cell receives excitatory input from α-motor neurons via an axon collateral (and in turn it inhibits the same). In addition, cortical fibers activating ankle flexors contact Renshaw cells (as well as type Ia inhibitory interneurons) that inhibit the antagonistic ankle extensors. This circuitry prevents reflex stimulation of the extensors when flexors are active. Therefore, when the descending cortical fibers are damaged **(upper motor neuron lesion),** the inhibition of antagonists is absent. The result is repetitive, sequential contraction of ankle flexors and extensors (clonus). Clonus may be seen in patients with amyotrophic lateral sclerosis, stroke, multiple sclerosis, spinal cord damage, and hepatic encephalopathy.

ation is clearly seen. Passive flexion of the elbow, for example, meets immediate resistance as a result of the stretch reflex in the triceps muscle. Further stretch activates the inverse stretch reflex. The resistance to flexion suddenly collapses, and the arm flexes. Continued passive flexion stretches the muscle again, and the sequence may be repeated. This sequence of resistance followed by give when a limb is moved passively is known as the **clasp-knife effect** because of its resemblance to the closing of a pocket knife. It is also known as the **lengthening reaction** because it is the response of a spastic muscle to lengthening.

POLYSYNAPTIC REFLEXES: THE WITHDRAWAL REFLEX

Polysynaptic reflex paths branch in a complex fashion (Figure 9–7). The number of synapses in each of their branches varies. Because of the synaptic delay at each synapse, activity in the branches with fewer synapses reaches the motor neurons first, followed by activity in the longer pathways. This causes prolonged bombardment of the motor neurons from a single stimulus and consequently prolonged responses. Furthermore, some of the branch pathways turn back on themselves, permitting activity to reverberate until it becomes unable to cause a propagated transsynaptic response and dies out. Such **reverberating circuits** are common in the brain and spinal cord.

WITHDRAWAL REFLEX

The withdrawal reflex is a typical polysynaptic reflex that occurs in response to a usually painful stimulation of the skin or subcutaneous tissues and muscle. The response is flexor muscle contraction and inhibition of extensor muscles, so that the body part stimulated is flexed and withdrawn from the stimulus. When a strong stimulus is applied to a limb, the response includes not only flexion and withdrawal of that limb but also

spindle discharge, tendon organ discharge, and reciprocal innervation determines the rate of discharge of α-motor neurons (see Clinical Box 9–3).

MUSCLE TONE

The resistance of a muscle to stretch is often referred to as its **tone** or **tonus.** If the motor nerve to a muscle is cut, the muscle offers very little resistance and is said to be **flaccid.** A **hypertonic (spastic)** muscle is one in which the resistance to stretch is high because of hyperactive stretch reflexes. Somewhere between the states of flaccidity and spasticity is the ill-defined area of normal tone. The muscles are generally **hypotonic** when the rate of γ-motor neuron discharge is low and hypertonic when it is high.

When the muscles are hypertonic, the sequence of moderate stretch → muscle contraction, strong stretch → muscle relax-

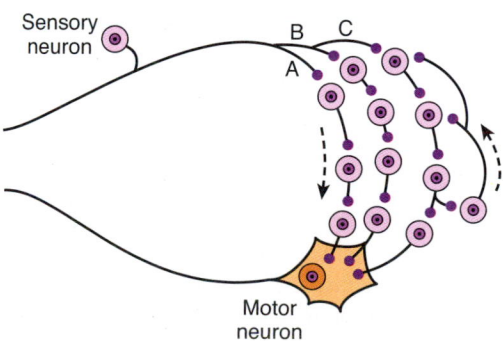

FIGURE 9–7 **Diagram of polysynaptic connections between afferent and efferent neurons in the spinal cord.** The dorsal root fiber activates pathway A with three interneurons, pathway B with four interneurons, and pathway C with four interneurons. Note that one of the interneurons in pathway C connects to a neuron that doubles back to other interneurons, forming reverberating circuits.

extension of the opposite limb. This **crossed extensor response** is properly part of the withdrawal reflex. Strong stimuli in experimental animals generate activity in the interneuron pool that spreads to all four extremities. This is difficult to demonstrate in normal animals but is easily demonstrated in an animal in which the modulating effects of impulses from the brain have been abolished by prior section of the spinal cord (**spinal animal**). For example, when the hind limb of a spinal cat is pinched, the stimulated limb is withdrawn, the opposite hind limb extended, the ipsilateral forelimb extended, and the contralateral forelimb flexed. This spread of excitatory impulses up and down the spinal cord to more and more motor neurons is called **irradiation of the stimulus,** and the increase in the number of active motor units is called **recruitment of motor units.**

IMPORTANCE OF THE WITHDRAWAL REFLEX

Flexor responses can be produced by innocuous stimulation of the skin or by stretch of the muscle, but strong flexor responses with withdrawal are initiated only by stimuli that are noxious or at least potentially harmful to the animal. These stimuli are therefore called **nociceptive stimuli.** Sherrington pointed out the survival value of the withdrawal response. Flexion of the stimulated limb gets it away from the source of irritation, and extension of the other limb supports the body. The pattern assumed by all four extremities puts the animal in position to run away from the offending stimulus. Withdrawal reflexes are **prepotent;** that is, they preempt the spinal pathways from any other reflex activity taking place at the moment.

Many of the characteristics of polysynaptic reflexes can be demonstrated by studying the withdrawal reflex. A weak noxious stimulus to one foot evokes a minimal flexion response; stronger stimuli produce greater and greater flexion as the stimulus irradiates to more and more of the motor neuron pool supplying the muscles of the limb. Stronger stimuli also cause a more prolonged response. A weak stimulus causes one quick flexion movement; a strong stimulus causes prolonged flexion and sometimes a series of flexion movements. This prolonged response is due to prolonged, repeated firing of the motor neurons. The repeated firing is called **after-discharge** and is due to continued bombardment of motor neurons by impulses arriving by complicated and circuitous polysynaptic paths.

As the strength of a noxious stimulus is increased, the reaction time is shortened. Spatial and temporal facilitation occurs at synapses in the polysynaptic pathway. Stronger stimuli produce more action potentials per second in the active branches and cause more branches to become active; summation of the EPSPs to the firing level therefore occurs more rapidly.

FRACTIONATION & OCCLUSION

Another characteristic of the withdrawal response is the fact that supramaximal stimulation of any of the sensory nerves from a limb never produces as strong a contraction of the flex-

or muscles as that elicited by direct electrical stimulation of the muscles themselves. This indicates that the afferent inputs **fractionate** the motor neuron pool; that is, each input goes to only part of the motor neuron pool for the flexors of that particular extremity. On the other hand, if all the sensory inputs are dissected out and stimulated one after the other, the sum of the tension developed by stimulation of each is greater than that produced by direct electrical stimulation of the muscle or stimulation of all inputs at once. This indicates that the various afferent inputs share some of the motor neurons and that **occlusion** occurs when all inputs are stimulated at once.

GENERAL PROPERTIES OF REFLEXES

It is apparent from the preceding description of the properties of monosynaptic and polysynaptic reflexes that reflex activity is stereotyped and specific in terms of both the stimulus and the response; a particular stimulus elicits a particular response. The fact that reflex responses are stereotyped does not exclude the possibility of their being modified by experience. Reflexes are adaptable and can be modified to perform motor tasks and maintain balance. Descending inputs from higher brain regions play an important role in modulating and adapting spinal reflexes.

ADEQUATE STIMULUS

The stimulus that triggers a reflex is generally very precise. This stimulus is called the **adequate stimulus** for the particular reflex. A dramatic example is the scratch reflex in the dog. This spinal reflex is adequately stimulated by multiple linear touch stimuli such as those produced by an insect crawling across the skin. The response is vigorous scratching of the area stimulated. If the multiple touch stimuli are widely separated or not in a line, the adequate stimulus is not produced and no scratching occurs. Fleas crawl, but they also jump from place to place. This jumping separates the touch stimuli so that an adequate stimulus for the scratch reflex is not produced. It is doubtful if the flea population would survive long without the ability to jump.

FINAL COMMON PATH

The motor neurons that supply the extrafusal fibers in skeletal muscles are the efferent side of many reflex arcs. All neural influences affecting muscular contraction ultimately funnel through them to the muscles, and they are therefore called the **final common paths.** Numerous inputs converge on them. Indeed, the surface of the average motor neuron and its dendrites accommodates about 10,000 synaptic knobs. At least five inputs go from the same spinal segment to a typical spinal motor neuron. In addition to these, there are excitatory and inhibitory inputs, generally relayed via interneurons, from other levels of the spinal cord and multiple long-descending tracts from the brain. All of these pathways converge on and determine the activity in the final common paths.

CENTRAL EXCITATORY & INHIBITORY STATES

The spread up and down the spinal cord of subliminal fringe effects from excitatory stimulation has already been mentioned. Direct and presynaptic inhibitory effects can also be widespread. These effects are generally transient. However, the spinal cord also shows prolonged changes in excitability, possibly because of activity in reverberating circuits or prolonged effects of synaptic mediators. The terms **central excitatory state** and **central inhibitory state** have been used to describe prolonged states in which excitatory influences overbalance inhibitory influences and vice versa. When the central excitatory state is marked, excitatory impulses irradiate not only to many somatic areas of the spinal cord but also to autonomic areas. In chronically paraplegic humans, for example, a mild noxious stimulus may cause, in addition to prolonged withdrawal-extension patterns in all four limbs, urination, defecation, sweating, and blood pressure fluctuations (**mass reflex**).

CHAPTER SUMMARY

- A reflex arc consists of a sense organ, an afferent neuron, one or more synapses within a central integrating station, an efferent neuron, and an effector response.

- A muscle spindle is a group of specialized intrafusal muscle fibers with contractile polar ends and a noncontractile center that is located in parallel to the extrafusal muscle fibers and is innervated by types Ia and II afferent fibers and γ-motor neurons. Muscle stretch activates the muscle spindle to initiate reflex contraction of the extrafusal muscle fibers in the same muscle (stretch reflex).

- A Golgi tendon organ is a netlike collection of knobby nerve endings among the fascicles of a tendon that is located in series with extrafusal muscle fibers and innervated by type Ib afferents. They are stimulated by both passive stretch and active contraction of the muscle to relax the muscle (inverse stretch reflex) and function as a transducer to regulate muscle force.

- A collateral from an Ia afferent branches to terminate on an inhibitory interneuron that synapses on an antagonistic muscle (reciprocal innervation) to relax that muscle when the agonist contracts. Clonus is the occurrence of regular, rhythmic contractions of a muscle subjected to sudden, maintained stretch. A sequence of increased resistance followed by reduced resistance when a limb is moved passively is known as the lengthening reaction.

MULTIPLE-CHOICE QUESTIONS

For all questions, select the single best answer unless otherwise directed.

1. The inverse stretch reflex
   A) has a lower threshold than the stretch reflex.
   B) is a monosynaptic reflex.
   C) is a disynaptic reflex with a single interneuron inserted between the afferent and efferent limbs.
   D) is a polysynaptic reflex with many interneurons inserted between the afferent and efferent limbs.
   E) requires the discharge of central neurons that release acetylcholine.

2. When γ-motor neuron discharge increases at the same time as α-motor neuron discharge to muscle,
   A) prompt inhibition of discharge in spindle Ia afferents takes place.
   B) the contraction of the muscle is prolonged.
   C) the muscle will not contract.
   D) the number of impulses in spindle Ia afferents is smaller than when α discharge alone is increased.
   E) the number of impulses in spindle Ia afferents is greater than when α discharge alone is increased.

3. Which of the following is *not* characteristic of a reflex?
   A) Modification by impulses from various parts of the CNS
   B) May involve simultaneous contraction of some muscles and relaxation of others
   C) Chronically suppressed after spinal cord transection
   D) Always involves transmission across at least one synapse
   E) Frequently occurs without conscious perception

4. Withdrawal reflexes are *not*
   A) initiated by nociceptive stimuli.
   B) prepotent.
   C) prolonged if the stimulus is strong.
   D) an example of a flexor reflex.
   E) accompanied by the same response on both sides of the body.

CHAPTER RESOURCES

Haines DE (editor): *Fundamental Neuroscience for Basic and Clinical Applications,* 3rd ed. Elsevier, 2006.

Hulliger M: The mammalian muscle spindle and its central control. Rev Physiol Biochem Pharmacol 1984;101:1.

Hunt CC: Mammalian muscle spindle: Peripheral mechanisms. Physiol Rev 1990;70: 643.

Jankowska E: Interneuronal relay in spinal pathways from proprioceptors. Prog Neurobiol 1992;38:335.

Kandel ER, Schwartz JH, Jessell TM (editors): *Principles of Neural Science,* 4th ed. McGraw-Hill, 2000.

Lundberg A: Multisensory control of spinal reflex pathways. Prog Brain Res 1979;50:11.

Matthews PBC: *Mammalian Muscle Receptors and Their Central Actions,* Williams & Wilkins, 1972.

Pain & Temperature

O B J E C T I V E S

After studying this chapter, you should be able to:

- Name the types of peripheral nerve fibers and receptor types that mediate warmth, cold, and nociception.
- Explain the difference between pain and nociception.
- Explain the differences between fast and slow pain and acute and chronic pain.
- Explain hyperalgesia and allodynia.
- Describe and explain referred pain.

INTRODUCTION

One of the most common reasons an individual seeks the advice of a physician is because he or she is in pain. Pain was called by Sherrington, "the physical adjunct of an imperative protective reflex." Painful stimuli generally initiate potent withdrawal and avoidance responses. Pain differs from other sensations in that it sounds a warning that something is wrong, preempts other signals, and is associated with an unpleasant affect. It turns out to be immensely complex because when pain is prolonged and tissue is damaged, central nociceptor pathways are sensitized and reorganized.

NOCICEPTORS & THERMORECEPTORS

Pain and temperature sensations arise from unmyelinated dendrites of sensory neurons located around hair follicles throughout the glabrous and hairy skin as well as deep tissue. Impulses from **nociceptors** (pain) are transmitted via two fiber types. One system comprises thinly myelinated Aδ fibers (2–5 μm in diameter) which conduct at rates of 12–30 m/s. The other is unmyelinated C fibers (0.4–1.2 μm in diameter) which conduct at low rates of 0.5–2 m/s. **Thermoreceptors** also span these two fiber types. Cold receptors are on dendritic endings of Aδ fibers and C fibers, whereas warmth (heat) receptors are on C fibers.

Mechanical nociceptors respond to strong pressure (eg, from a sharp object). **Thermal nociceptors** are activated by skin temperatures above 45 °C or by severe cold. **Chemically sensitive nociceptors** respond to various agents like bradykinin, histamine, high acidity, and environmental irritants. **Polymodal nociceptors** respond to combinations of these stimuli.

167

Mapping experiments show that the skin has discrete cold-sensitive and heat-sensitive spots. There are 4 to 10 times as many cold-sensitive as heat-sensitive spots. The threshold for activation of **warmth receptors** is 30 °C, and they increase their firing rate up to 46 °C. **Cold receptors** are inactive at temperatures of 40 °C, but then steadily increase their firing rate as skin temperature falls to about 24 °C. As skin temperature further decreases, the firing rate of cold receptors decreases until the temperature reaches 10 °C. Below that temperature, they are inactive and the cold becomes an effective local anesthetic.

Because the sense organs are located subepithelially, it is the temperature of the subcutaneous tissues that determines the responses. Cool metal objects feel colder than wooden objects of the same temperature because the metal conducts heat away from the skin more rapidly, cooling the subcutaneous tissues to a greater degree.

A major advance in this field has been the cloning of three thermoreceptors and nociceptors. The receptor for moderate cold is the **cold- and menthol-sensitive receptor 1 (CMR 1).** Two types of **vanilloid receptors** respond to noxious heat (**VR1** and **VRL-1**). Vanillins are a group of compounds, including capsaicin, that cause pain. The VR1 receptors respond not only to capsaicin but also to protons and to potentially harmful temperatures above 43 °C. **VRL-1,** which responds to temperatures above 50 °C but not to capsaicin, has been isolated from C fibers. There may be many types of receptors on single peripheral C fiber endings, so single fibers can respond to many different noxious stimuli. However, the different properties of the VR1 and the VRL-1 receptors make it likely that there are many different nociceptor C fibers systems as well.

CMR1, VR1, and VRL1 are members of the **transient receptor potential (TRP)** family of excitatory ion channels. VR1 has a PIP_2 binding site, and when the amount of PIP_2 bound is decreased, the sensitivity of the receptors is increased. Aside from the fact that activation of the cool receptor causes an influx of Ca^{2+}, little is known about the ionic basis of the initial depolarization they produce. In the cutaneous receptors in general, depolarization could be due to inhibition of K^+ channels, activation of Na^+ channels, or inhibition of the Na^+–K^+ pump, but the distinction between these possibilities has not been made.

CLASSIFICATION OF PAIN

For scientific and clinical purposes, **pain** is defined by the International Association for the Study of Pain (IASP) as, "an unpleasant sensory and emotional experience associated with actual or potential tissue damage, or described in terms of such damage." This is to be distinguished from the term **nociception** which the IASP defines as the unconscious activity induced by a harmful stimulus applied to sense receptors.

Pain is sometimes classified as fast and slow pain. A painful stimulus causes a "bright," sharp, localized sensation (**fast pain**) followed by a dull, intense, diffuse, and unpleasant feeling (**slow pain**). Evidence suggests that fast pain is due to activity in the

---

CLINICAL BOX 10–1

Itch & Tickle

Itching (**pruritus**) is not much of a problem for normal individuals, but severe itching that is difficult to treat occurs in diseases such as chronic renal failure, some forms of liver disease, atopic dermatitis, and HIV infection. Especially in areas where many naked endings of unmyelinated nerve fibers occur, itch spots can be identified on the skin by careful mapping. In addition, itch-specific fibers have been demonstrated in the ventrolateral spinothalamic tract. This and other evidence implicate the existence of an itch-specific path. Relatively mild stimulation, especially if produced by something that moves across the skin, produces itch and tickle. Scratching relieves itching because it activates large, fast-conducting afferents that gate transmission in the dorsal horn in a manner analogous to the inhibition of pain by stimulation of similar afferents. It is interesting that a tickling sensation is usually regarded as pleasurable, whereas itching is annoying and pain is unpleasant. Itching can be produced not only by repeated local mechanical stimulation of the skin but also by a variety of chemical agents. **Histamine** produces intense itching, and injuries cause its liberation in the skin. However, in most instances of itching, endogenous histamine does not appear to be the responsible agent; doses of histamine that are too small to produce itching still produce redness and swelling on injection into the skin, and severe itching frequently occurs without any visible change in the skin. The **kinins** cause severe itching.

Aδ pain fibers, whereas slow pain is due to activity in the C pain fibers. **Itch** and **tickle** are related to pain sensation (see Clinical Box 10–1).

Pain is frequently classified as **physiologic** or **acute pain** and **pathologic** or **chronic pain,** which includes **inflammatory pain** and **neuropathic pain.** Acute pain typically has a sudden onset and recedes during the healing process. Acute pain can be considered as "good pain" as it serves an important protective mechanism. The withdrawal reflex is an example of this protective role of pain.

Chronic pain can be considered "bad pain" because it persists long after recovery from an injury and is often refractory to common analgesic agents, including nonsteroidal anti-inflammatory drugs (NSAIDs) and opiates. Chronic pain can result from nerve injury (**neuropathic pain**) including diabetic neuropathy, toxin-induced nerve damage, and ischemia. **Causalgia** is a type of neuropathic pain (see Clinical Box 10–2).

Pain is often accompanied by **hyperalgesia** and **allodynia.** Hyperalgesia is an exaggerated response to a noxious stimulus, whereas allodynia is a sensation of pain in response to an innocuous stimulus. An example of the latter is the painful sensation from a warm shower when the skin is damaged by sunburn.

Neuropathic Pain

Neuropathic pain may occur when nerve fibers are injured. Commonly, it is excruciating and a difficult condition to treat. It occurs in various forms in humans. For example, in **causalgia,** spontaneous burning pain occurs long after seemingly trivial injuries. The pain is often accompanied by **hyperalgesia** and **allodynia. Reflex sympathetic dystrophy** is often present as well. In this condition, the skin in the affected area is thin and shiny, and there is increased hair growth. Research in animals indicates that nerve injury leads to sprouting and eventual overgrowth of noradrenergic sympathetic nerve fibers into the dorsal root ganglia of the sensory nerves from the injured area. Sympathetic discharge then brings on pain. Thus, it appears that the periphery has been short-circuited and that the relevant altered fibers are being stimulated by norepinephrine at the dorsal root ganglion level. Alpha-adrenergic blockade produces relief of causalgia-type pain in humans, though for unknown reasons α_1-adrenergic blockers are more effective than α_2-adrenergic blocking agents. Treatment of painful sensory neuropathy is a major challenge and current therapies are often inadequate.

Hyperalgesia and allodynia signify increased sensitivity of nociceptive afferent fibers. Figure 10–1 shows how chemicals released at the site of injury can further activate nociceptors leading to inflammatory pain. Injured cells release chemicals such as K^+ that depolarize nerve terminals, making nociceptors more responsive. Injured cells also release bradykinin and Substance P, which can further sensitize nociceptive terminals. Histamine is released from mast cells, serotonin (5-HT) from platelets, and prostaglandins from cell membranes, all contributing to the inflammatory process and they activate or sensitize the nociceptors. Some released substances act by releasing another one (eg, bradykinin activates both Aδ and C fibers and increases synthesis and release of prostaglandins). Prostaglandin E$_2$ (a cyclooxygenase metabolite of arachidonic acid) is released from damaged cells and produces hyperalgesia. This is why aspirin and other NSAIDs (inhibitors of cyclooxygenase) alleviate pain.

DEEP PAIN

The main difference between superficial and deep sensibility is the different nature of the pain evoked by noxious stimuli. This is probably due to a relative deficiency of Aδ nerve fibers in deep structures, so there is little rapid, bright pain. In addition, deep pain and visceral pain are poorly localized, nauseating, and frequently associated with sweating and changes in blood pressure. Pain can be elicited experimentally from the periosteum and ligaments by injecting hypertonic saline into them. The pain produced in this fashion initiates reflex contraction of nearby skeletal muscles. This reflex contraction is similar to the muscle spasm associated with injuries to bones, tendons, and joints. The steadily contracting muscles become ischemic, and ischemia stimulates the pain receptors in the muscles (see Clinical Box 10–3). The pain in turn initiates more spasm, setting up a vicious cycle.

VISCERAL PAIN

In addition to being poorly localized, unpleasant, and associated with nausea and autonomic symptoms, visceral pain often radiates or is referred to other areas.

FIGURE 10–1 **In response to tissue injury, chemical mediators can sensitize and activate nociceptors.** These factors contribute to hyperalgesia and allodynia. Tissue injury releases bradykinin and prostaglandins that sensitize or activate nociceptors, which in turn releases substance P and calcitonin gene-related peptide (CGRP). Substance P acts on mast cells to cause degranulation and release histamine, which activates nociceptors. Substance P causes plasma extravasation and CGRP dilates blood vessels; the resulting edema causes additional release of bradykinin. Serotonin (5-HT) is released from platelets and activates nociceptors. (From Kandel ER, Schwartz JH, Jessell TM [editors]: *Principles of Neural Science.* McGraw-Hill, 2000.)

Muscle Pain

If a muscle contracts rhythmically in the presence of an adequate blood supply, pain does not usually result. However, if the blood supply to a muscle is occluded, contraction soon causes pain. The pain persists after the contraction until blood flow is reestablished. These observations are difficult to interpret except in terms of the release during contraction of a chemical agent (Lewis's **"P factor"**) that causes pain when its local concentration is high enough. When the blood supply is restored, the material is washed out or metabolized. The identity of the P factor is not settled, but it could be K^+. Clinically, the substernal pain that develops when the myocardium becomes ischemic during exertion (**angina pectoris**) is a classic example of the accumulation of P factor in a muscle. Angina is relieved by rest because this decreases the myocardial O_2 requirement and permits the blood supply to remove the factor. **Intermittent claudication,** the pain produced in the leg muscles of persons with occlusive vascular disease, is another example. It characteristically comes on while the patient is walking and disappears on stopping. Visceral pain, like deep somatic pain, initiates reflex contraction of nearby skeletal muscle. This reflex spasm is usually in the abdominal wall and makes the abdominal wall rigid. It is most marked when visceral inflammatory processes involve the peritoneum. However, it can occur without such involvement. The spasm protects the underlying inflamed structures from inadvertent trauma. Indeed, this reflex spasm is sometimes called **"guarding."**

The autonomic nervous system, like the somatic, has afferent components, central integrating stations, and effector pathways. The receptors for pain and the other sensory modalities present in the viscera are similar to those in skin, but there are marked differences in their distribution. There are no proprioceptors in the viscera, and few temperature and touch receptors. Nociceptors are present, although they are more sparsely distributed than in somatic structures.

Afferent fibers from visceral structures reach the CNS via sympathetic and parasympathetic nerves. Their cell bodies are located in the dorsal roots and the homologous cranial nerve ganglia. Specifically, there are visceral afferents in the facial, glossopharyngeal, and vagus nerves; in the thoracic and upper lumbar dorsal roots; and in the sacral roots (Figure 10–2). There may also be visceral afferent fibers from the eye in the trigeminal nerve.

As almost everyone knows from personal experience, visceral pain can be very severe. The receptors in the walls of the hollow viscera are especially sensitive to distention of these organs. Such distention can be produced experimentally in the gastrointestinal tract by inflation of a swallowed balloon attached to a tube. This produces pain that waxes and wanes (intestinal colic) as the intestine contracts and relaxes on the balloon. Similar colic is produced in intestinal obstruction by the contractions of the dilated intestine above the obstruction. When a visceral organ is inflamed or hyperemic, relatively minor stimuli cause severe pain. This is probably a form of hyperalgesia.

REFERRED PAIN

Irritation of a visceral organ frequently produces pain that is felt not at that site but in some somatic structure that may be a considerable distance away. Such pain is said to be referred to the somatic structure. Obviously, knowledge of **referred pain** and the common sites of pain referral from each of the viscera is of great importance to the physician. Perhaps the best-known example is referral of cardiac pain to the inner aspect of the left arm. Other examples include pain in the tip of the shoulder caused by irritation of the central portion of the diaphragm and pain in the testicle due to distention of the ureter. Additional instances abound in the practices of medicine, surgery, and dentistry. However, sites of reference are not stereotyped, and unusual reference sites occur with considerable frequency. Cardiac pain, for instance, may be referred to the right arm, the abdominal region, or even the back and neck.

When pain is referred, it is usually to a structure that developed from the same embryonic segment or dermatome as the structure in which the pain originates. This principle is called the **dermatomal rule.** For example, the heart and the arm have the same segmental origin, and the testicle has migrated with its nerve supply from the primitive urogenital ridge from which the kidney and ureter have developed.

The basis for referred pain may be convergence of somatic and visceral pain fibers on the same second-order neurons in the dorsal horn that project to the thalamus and then to the somatosensory cortex (Figure 10–3). This is called the **convergence–projection theory.** Somatic and visceral neurons converge in lamina I–VI of the ipsilateral dorsal horn, but neurons in lamina VII receive afferents from both sides of the body—a requirement if convergence is to explain referral to the side opposite that of the source of pain. The somatic nociceptive fibers normally do not activate the second-order neurons, but when the visceral stimulus is prolonged, facilitation of the somatic fiber endings occurs. They now stimulate the second-order neurons, and of course the brain cannot determine whether the stimulus came from the viscera or from the area of referral.

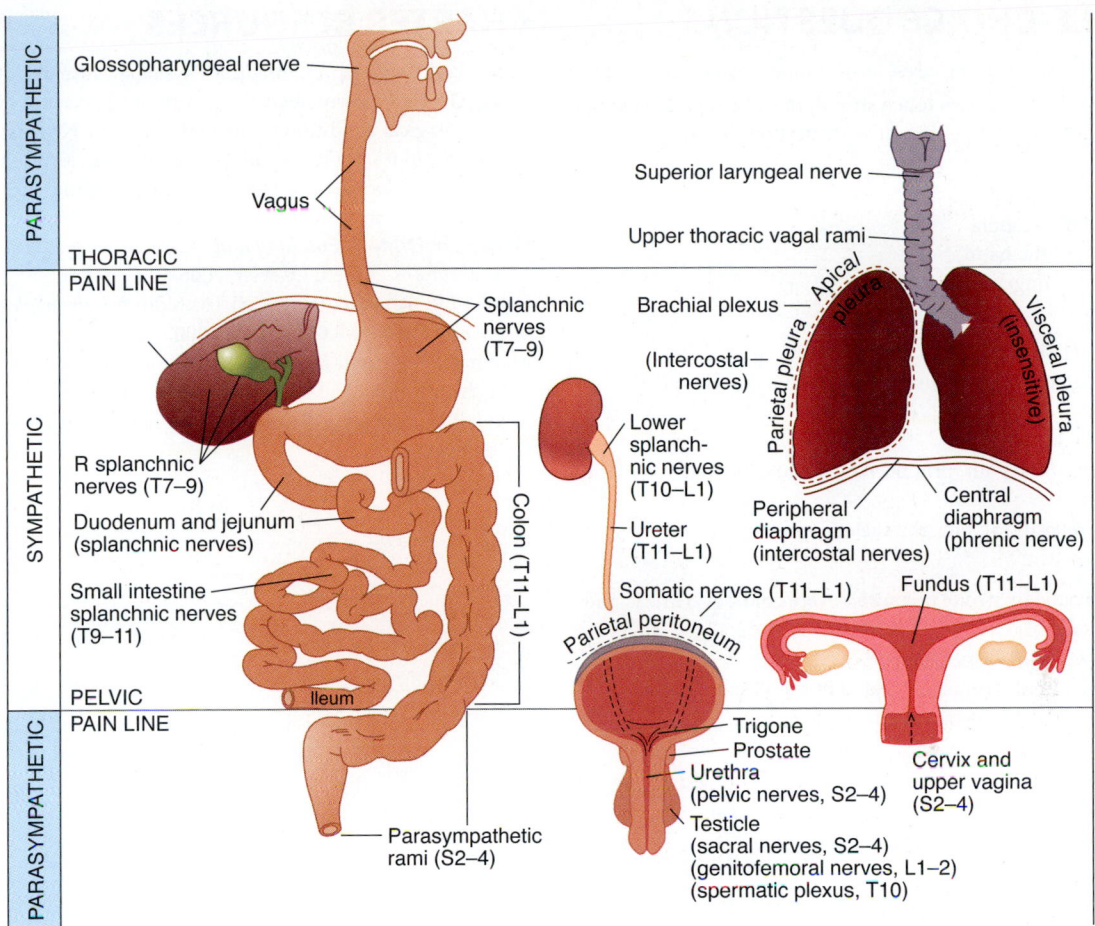

FIGURE 10–2   **Pain innervation of the viscera.** Pain afferents from structures between the pain lines reach the CNS via sympathetic pathways, whereas, they traverse parasympathetic pathways from structures above the thoracic pain line and below the pelvic pain line.   (After White JC. Reproduced with permission from Ruch TC: In *Physiology and Biophysics*, 19th ed. Ruch TC, Patton HD [editors]. Saunders, 1965.)

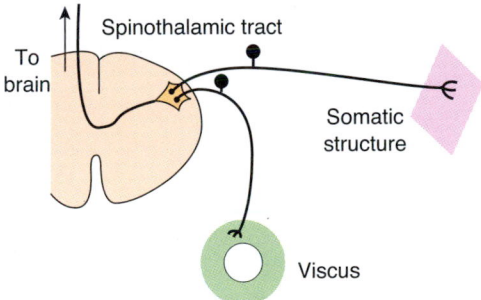

FIGURE 10–3   **Diagram of the way in which convergence of somatic and visceral nociceptive fibers in lamina VII of the dorsal horn may cause referred pain.** When a visceral stimulus is prolonged, somatic fiber facilitation occurs. This leads to activation of spinothalamic tract neurons, and of course the brain cannot determine whether the stimulus came from the viscera or from the somatic area.

CHAPTER SUMMARY

- Pain impulses are transmitted via lightly myelinated Aδ and unmyelinated C fibers. Cold receptors are on dendritic endings of Aδ fibers and C fibers, whereas heat receptors are on C fibers.

- Pain is an unpleasant sensory and emotional experience associated with actual or potential tissue damage, or described in terms of such damage, whereas nociception is the unconscious activity induced by a harmful stimulus applied to sense receptors.

- Fast pain is mediated by Aδ fibers and causes sharp, localized sensation. Slow pain is mediated by C fibers and causes a dull, intense, diffuse, and unpleasant feeling.

- Acute pain has a sudden onset, recedes during the healing process, and serves as an important protective mechanism. Chronic pain is persistent and caused by nerve damage; it is often refractory to NSAIDs and opiates.

- Hyperalgesia is an exaggerated response to a noxious stimulus; allodynia is a sensation of pain in response to an innocuous stimulus.

- Referred pain is pain that originates in a visceral organ but is sensed at a somatic site. It may be due to convergence of somatic and visceral nociceptive afferent fibers on the same second-order neurons in the spinal dorsal horn that project to the thalamus and then to the somatosensory cortex.

MULTIPLE-CHOICE QUESTIONS

For all questions, select the single best answer unless otherwise directed.

1. The distance by which two touch stimuli must be separated to be perceived as two separate stimuli is greatest on the
    A) lips.
    B) palm of the hand.
    C) back of the scapula.
    D) dorsum of the hand.
    E) tips of the fingers.

2. Visceral pain
    A) shows relatively rapid adaptation.
    B) is mediated by B fibers in the dorsal roots of the spinal nerves.
    C) is poorly localized.
    D) resembles "fast pain" produced by noxious stimulation of the skin.
    E) causes relaxation of nearby skeletal muscles.

3. Nociceptors
    A) are activated by strong pressure, severe cold, severe heat, and chemicals.
    B) are absent in visceral organs.
    C) are specialized structures located in the skin and joints.
    D) are innervated by group II afferents.
    E) all of the above

4. Thermoreceptors
    A) are activated only by severe cold or severe heat.
    B) are located on superficial layers of the skin.
    C) are a subtype of nociceptors.
    D) are on dendritic endings of $A\delta$ fibers and C fibers.
    E) all of the above

CHAPTER RESOURCES

Boron WF, Boulpaep EL: *Medical Physiology,* Elsevier, 2005.

Craig AD: How do you feel? Interoception: The sense of the physiological condition of the body. Nat Rev Neurosci 2002;3:655.

Garry EM, Jones E, Fleetwood-Walker SM: Nociception in vertebrates: Key receptors participating in spinal mechanisms of chronic pain in animals. Brain Res Rev 2004;46: 216.

Haines DE (editor): *Fundamental Neuroscience for Basic and Clinical Applications,* 3rd ed. Elsevier, 2006.

Kandel ER, Schwartz JH, Jessell TM (editors): *Principles of Neural Science,* 4th ed. McGraw-Hill, 2000.

Marchand F, Perretti M, McMahon SB: Role of the immune system in chronic pain. Nat Rev Neurosci 2005;6:521.

Mendell JR, Sahenk Z: Painful sensory neuropathy. N Engl J Med 2003;348:1243.

Somatosensory Pathways

OBJECTIVES

After studying this chapter, you should be able to:

- Compare the pathway that mediates sensory input from touch, proprioceptive, and vibratory senses to that mediating information from pain and thermoreceptors.
- Describe the somatotopic organization of ascending sensory pathways.
- Describe descending pathways that modulate transmission in pain pathways.
- List some drugs that have been used for relief of pain, and give the rationale for their use and their clinical effectiveness.

INTRODUCTION

Earlier chapters have described properties of receptors mediating the sensations of touch, vibration, proprioception, temperature, and pain. This chapter will review central ascending pathways that transmit and process the information from peripheral receptors to the cerebral cortex as well as describe some deficits in sensation resulting from lesions at various steps within the ascending systems. Also, various ways to modulate pain transmission will be described.

DORSAL HORN

The dorsal horns are divided on the basis of histologic characteristics into laminae I–VII, with I being the most superficial and VII the deepest. Lamina VII receives afferents from both sides of the body, whereas the other laminae receive only unilateral input. Lamina II and part of lamina III make up the **substantia gelatinosa,** a lightly stained area near the top of each dorsal horn. Three types of primary afferent fibers (with cell bodies in the dorsal root ganglia) mediate cutaneous sensation: (1) large myelinated Aα and Aβ fibers that transmit impulses generated by mechanical stimuli; (2) small myelinated Aδ fibers, some of which transmit impulses from cold receptors and nociceptors that mediate pain and some of which transmit impulses from mechanoreceptors; and (3) small unmyelinated C fibers that are concerned primarily with pain and temperature. However, a few C fibers also transmit impulses from mechanoreceptors. The orderly distribution of these fibers in various layers of the dorsal horn is shown in Figure 11–1.

DORSAL COLUMN PATHWAY

The principal direct pathways to the cerebral cortex for touch, vibratory sense, and proprioception (position sense) are shown in Figure 11–2. Fibers mediating these sensations ascend ipsilaterally in the dorsal columns to the medulla, where they synapse in the **gracilus** and **cuneate nuclei.** The second-order neurons from these nuclei cross the midline and ascend in the **medial lemniscus** to end in the contralateral **ventral posterior lateral (VPL) nucleus** and related specific sensory relay nuclei of the thalamus. This ascending system is called the **dorsal column** or **medial lemniscal system.** The fibers within the dorsal column pathway are joined in the brain stem by fibers mediating sensation from the head. Touch and proprioception are relayed mostly via the main sensory and mesencephalic nuclei of the trigeminal nerve.

SOMATOTOPIC ORGANIZATION

Within the dorsal columns, fibers arising from different levels of the cord are somatotopically organized. Specifically, fibers

FIGURE 11–1   **Schematic representation of the terminations of the three types of primary afferent neurons in the various layers of the dorsal horn of the spinal cord.**

from the sacral cord are positioned most medially and those from the cervical cord are most lateral. This arrangement continues in the medulla with lower body (eg, foot) representation in the gracilus nucleus and upper body (eg, finger) representation in cuneate nucleus. The medial lemniscus is organized dorsal to ventral representing from neck to foot.

Somatotopic organization continues through the thalamus and cortex. VPL thalamic neurons carrying sensory information project in a highly specific way to the two somatic sensory areas of the cortex: **somatic sensory area I (SI)** in the postcentral gyrus and **somatic sensory area II (SII)** in the wall of the sylvian fissure. In addition, SI projects to SII. SI corresponds to **Brodmann's areas 3, 2, and 1.** Brodmann was a histologist who painstakingly divided the cerebral cortex into numbered areas based on their histologic characteristics.

The arrangement of projections to SI is such that the parts of the body are represented in order along the postcentral gyrus, with the legs on top and the head at the foot of the gyrus

FIGURE 11–2   **Ascending tracts carrying sensory information from peripheral receptors to the cerebral cortex. (a)** Dorsal-column pathway mediating touch, vibratory sense, and proprioception. **(b)** Ventrolateral spinothalamic tract mediating pain and temperature. (From Fox SI, *Human Physiology.* McGraw-Hill, 2008.)

FIGURE 11–3 **Brain areas concerned with somatic sensation, and some of the cortical receiving areas for other sensory modalities in the human brain.** The numbers are those of Brodmann's cortical areas. The primary auditory area is actually located in the sylvian fissure on the top of the superior temporal gyrus and is not normally visible in a lateral view of the cortex.

(Figure 11–3). Not only is there detailed localization of the fibers from the various parts of the body in the postcentral gyrus, but also the size of the cortical receiving area for impulses from a particular part of the body is proportionate to the use of the part. The relative sizes of the cortical receiving areas are shown dramatically in Figure 11–4, in which the proportions of the **homunculus** have been distorted to correspond to the size of the cortical receiving areas for each. Note that the cortical areas for sensation from the trunk and back are small, whereas very large areas are concerned with impulses from the hand and the parts of the mouth concerned with speech.

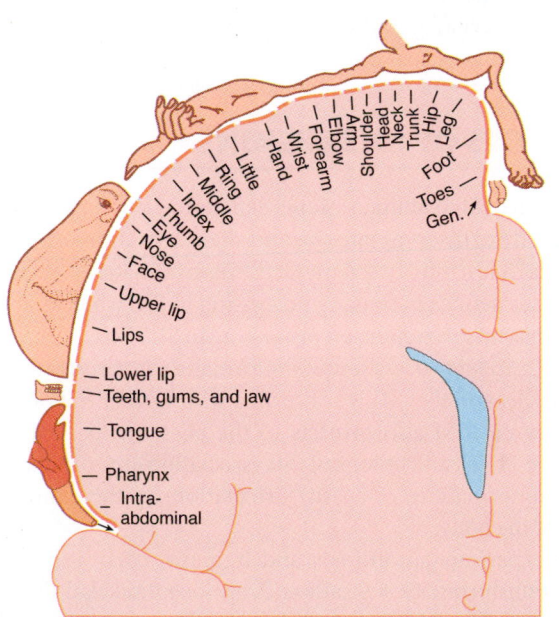

FIGURE 11–4 **Sensory homunculus, drawn overlying a coronal section through the postcentral gyrus.** Gen., genitalia. (Reproduced, with permission, from Penfield W, Rasmussen G: *The Cerebral Cortex of Man.* Macmillan, 1950.)

Studies of the sensory receiving area emphasize the very discrete nature of the point-for-point localization of peripheral areas in the cortex and provide further evidence for the general validity of the law of specific nerve energies (see Chapter 8). Stimulation of the various parts of the postcentral gyrus gives rise to sensations projected to appropriate parts of the body. The sensations produced are usually numbness, tingling, or a sense of movement, but with fine enough electrodes it has been possible to produce relatively pure sensations of touch, warmth, and cold. The cells in the postcentral gyrus are organized in vertical columns, like cells in the visual cortex. The cells in a given column are all activated by afferents from a given part of the body, and all respond to the same sensory modality.

SII is located in the superior wall of the sylvian fissure, the fissure that separates the temporal from the frontal and parietal lobes. The head is represented at the inferior end of the postcentral gyrus, and the feet at the bottom of the sylvian fissure. The representation of the body parts is not as complete or detailed as it is in the postcentral gyrus.

Conscious awareness of the positions of the various parts of the body in space depends in part on impulses from sense organs in and around the joints. The organs involved are slowly adapting spray endings, structures that resemble Golgi tendon organs, and probably Pacinian corpuscles in the synovia and ligaments. Impulses from these organs, touch receptors in the skin and other tissues, and muscle spindles are synthesized in the cortex into a conscious picture of the position of the body in space. Microelectrode studies indicate that many of the neurons in the sensory cortex respond to particular movements, not just to touch or static position.

VENTROLATERAL SPINOTHALAMIC TRACT

Fibers from nociceptors and thermoreceptors synapse on neurons in the dorsal horn (Figure 11–1). Aδ fibers terminate primarily on neurons in laminae I and V, whereas the dorsal root C fibers terminate on neurons in laminae I and II. The synaptic transmitter secreted by afferent fibers subserving fast mild pain is **glutamate,** and the transmitter subserving slow severe pain is **substance P**.

The axons from these neurons cross the midline and ascend in the ventrolateral quadrant of the spinal cord, where they form the **ventrolateral spinothalamic tract** (Figure 11–2). Fibers within this tract synapse in the VPL. Other dorsal horn neurons that receive nociceptive input synapse in the reticular formation of the brain stem (**spinoreticular pathway**) and then project to the centrolateral nucleus of the thalamus.

Positron emission tomographic (PET) and functional magnetic resonance imaging (fMRI) studies in normal humans indicate that pain activates cortical areas SI, SII, and the cingulate gyrus on the side opposite the stimulus. In addition, the mediofrontal cortex, the insular cortex, and the cerebellum are activated. These technologies were important in distinguishing two components of pain pathways. From VPL nuclei in the thalamus, fibers project to SI and SII. This is the pathway responsible for the **discriminative** aspect of pain, and is also called the **neospinothalamic tract.** In contrast, the pathway that includes synapses in the brain stem reticular formation and centrolateral thalamic nucleus projects to the frontal lobe, limbic system, and insula. This pathway mediates the **motivational-affect** component of pain and is called the **paleospinothalamic tract.**

In the central nervous system (CNS), visceral sensation travels along the same pathways as somatic sensation in the spinothalamic tracts and thalamic radiations, and the cortical receiving areas for visceral sensation are intermixed with the somatic receiving areas.

CORTICAL PLASTICITY

It is now clear that the extensive neuronal connections described above are not innate and immutable but can be changed relatively rapidly by experience to reflect the use of the represented area. Clinical Box 11–1 describes remarkable changes in cortical and thalamic organization that occur in response to limb amputation to lead to the phenomenon of **phantom limb pain.**

Numerous animal studies point to dramatic reorganization of cortical structures. If a digit is amputated in a monkey, the cortical representation of the neighboring digits spreads into the cortical area that was formerly occupied by the representation of the amputated digit. Conversely, if the cortical area representing a digit is removed, the somatosensory map of the digit moves to the surrounding cortex. Extensive, long-term deafferentation of limbs leads to even more dramatic shifts in somatosensory representation in the cortex, with, for example, the limb cortical area responding to touching the face. The explanation of these shifts appears to be that cortical connections of sensory units to the cortex have extensive convergence and divergence, with connections that can become weak with disuse and strong with use.

Plasticity of this type occurs not only with input from cutaneous receptors but also with input in other sensory systems. For example, in cats with small lesions of the retina, the cortical area for the blinded spot begins to respond to light striking other areas of the retina. Development of the adult pattern of retinal projections to the visual cortex is another example of this plasticity. At a more extreme level, experimentally routing visual input to the auditory cortex during development creates visual receptive fields in the auditory system.

PET scanning in humans also documents plastic changes, sometimes from one sensory modality to another. Thus, for example, tactile and auditory stimuli increase metabolic activity in the visual cortex in blind individuals. Conversely, deaf individuals respond faster and more accurately than normal individuals to moving stimuli in the visual periphery. Plasticity also occurs in the motor cortex. These findings illustrate the malleability of the brain and its ability to adapt.

CLINICAL BOX 11–1

Phantom Limb Pain

In 1551, a military surgeon, Ambroise Pare, wrote, ". . . the patients, long after the amputation is made, say they still feel pain in the amputated part. Of this they complain strongly, a thing worthy of wonder and almost incredible to people who have not experienced this." This is perhaps the earliest description of **phantom limb pain.** Between 50% and 80% of amputees experience phantom sensations, usually pain, in the region of their amputated limb. Phantom sensations may also occur after the removal of body parts other than the limbs, for example, after amputation of the breast, extraction of a tooth (**phantom tooth pain),** or removal of an eye (**phantom eye syndrome).** Numerous theories have been evoked to explain this phenomenon. The current theory is based on evidence that the brain can reorganize if sensory input is cut off. The **ventral posterior thalamic nucleus** is one example where this change can occur. In patients who have had their leg amputated, single neuron recordings show that the thalamic region that once received input from the leg and foot now respond to stimulation of the stump (thigh). Others have demonstrated remapping of the somatosensory cortex. For example, in some individuals who have had an arm amputated, stroking different parts of the face can lead to the feeling of being touched in the area of the missing limb. Spinal cord stimulation has been shown to be an effective therapy for phantom pain. Electric current is passed through an electrode that is placed next to the spinal cord to stimulate spinal pathways. This interferes with the impulses ascending to the brain and lessens the pain felt in the phantom limb. Instead, amputees feel a tingling sensation in the phantom limb.

EFFECTS OF CNS LESIONS

Ablation of SI in animals causes deficits in position sense and in the ability to discriminate size and shape. Ablation of SII causes deficits in learning based on tactile discrimination. Ablation of SI causes deficits in sensory processing in SII, whereas ablation of SII has no gross effect on processing in SI. Thus, it seems clear that SI and SII process sensory information in series rather than in parallel and that SII is concerned with further elaboration of sensory data. SI also projects to the posterior parietal cortex (Figure 11–3), and lesions of this association area produce complex abnormalities of spatial orientation on the contralateral side of the body.

In experimental animals and humans, cortical lesions do not abolish somatic sensation. Proprioception and fine touch are most affected by cortical lesions. Temperature sensibility is less affected, and pain sensibility is only slightly altered.

Only very extensive lesions completely interrupt touch sensation. When the dorsal columns are destroyed, vibratory sensation

CLINICAL BOX 11-2

Brown–Séquard Syndrome

A functional hemisection of the spinal cord causes a characteristic and easily recognized clinical picture that reflects damage to ascending sensory (dorsal-column pathway, ventrolateral spinothalamic tract) and descending motor (corticospinal tract) pathways, which is called the Brown–Séquard syndrome. The lesion to fasciculus gracilus or fasciculus cuneatus leads to ipsilateral loss of discriminative touch, vibration, and proprioception below the level of lesion. The loss of the spinothalamic tract leads to contralateral loss of pain and temperature sensation beginning one or two segments below the lesion. Damage to the corticospinal tract produces weakness and spasticity in certain muscle groups on the same side of the body. Although a precise spinal hemisection is rare, the syndrome is fairly common because it can be caused by spinal cord tumor, trauma, degenerative disc disease, and ischemia.

and proprioception are reduced, the touch threshold is elevated, and the number of touch-sensitive areas in the skin is decreased. In addition, localization of touch sensation is impaired. An increase in touch threshold and a decrease in the number of touch spots in the skin are also observed after interrupting the spinothalamic tract, but the touch deficit is slight and touch localization remains normal. The information carried in the lemniscal system is concerned with the detailed localization, spatial form, and temporal pattern of tactile stimuli. The information carried in the spinothalamic tracts, on the other hand, is concerned with poorly localized, gross tactile sensations. Clinical Box 11–2 describes the characteristic changes in sensory (and motor) functions that occur in response to spinal hemisection.

Proprioceptive information is transmitted up the spinal cord in the dorsal columns. A good deal of the proprioceptive input goes to the cerebellum, but some passes via the medial lemniscus and thalamic radiations to the cortex. Diseases of the dorsal columns produce ataxia because of the interruption of proprioceptive input to the cerebellum.

MODULATION OF PAIN TRANSMISSION

STRESS-INDUCED ANALGESIA

It is well known that soldiers wounded in the heat of battle often feel no pain until the battle is over (**stress-induced analgesia**). Many people have learned from practical experience that touching or shaking an injured area decreases the pain due to the injury. Stimulation with an electric vibrator at the site of pain also gives some relief. The relief may result from inhibition of pain pathways in the dorsal horn gate by stimulation of large-diameter touch-pressure afferents. Figure 11–1 shows that collaterals from these myelinated afferent fibers synapse in the dorsal horn. These collaterals may modify the input from nociceptive afferent terminals that also synapse in the dorsal horn. This is called the **gate-control hypothesis.**

The same mechanism is probably responsible for the efficacy of counterirritants. Stimulation of the skin over an area of visceral inflammation produces some relief of the pain due to the visceral disease. The old-fashioned mustard plaster works on this principle.

Surgical procedures undertaken to relieve severe pain include cutting the nerve from the site of injury or **ventrolateral cordotomy,** in which the spinothalamic tracts are carefully cut. However, the effects of these procedures are transient at best if the periphery has been short-circuited by sympathetic or other reorganization of the central pathways.

MORPHINE & ENKEPHALINS

Pain can often be handled by administration of analgesic drugs in adequate doses, though this is not always the case. The most effective of these agents is morphine. Morphine is particularly effective when given intrathecally. The receptors that bind morphine and the body's own morphines, the opioid peptides, are found in the midbrain, brain stem, and spinal cord.

There are at least three nonmutually exclusive sites at which opioids can act to produce analgesia: peripherally, at the site of an injury; in the dorsal horn, where nociceptive fibers synapse on dorsal root ganglion cells; and at more rostral sites in the brain stem. Figure 11–5 shows various modes of action of opiates to decrease transmission in pain pathways. Opioid receptors are produced in dorsal root ganglion cells and migrate both peripherally and centrally along their nerve fibers. In the periphery, inflammation causes the production of opioid peptides by immune cells, and these presumably act on the receptors in the afferent nerve fibers to reduce the pain that would otherwise be felt. The opioid receptors in the dorsal horn region could act presynaptically to decrease release of substance P, although presynaptic nerve endings have not been identified. Finally, injections of morphine into the periaqueductal gray matter of the midbrain relieve pain by activating descending pathways that produce inhibition of primary afferent transmission in the dorsal horn. There is evidence that this activation occurs via projections from the periaqueductal gray matter to the nearby raphé magnus nucleus and that descending serotonergic fibers from this nucleus mediate the inhibition.

Chronic use of morphine to relieve pain can cause patients to develop resistance to the drug, requiring progressively higher doses for pain relief. This **acquired tolerance** is different from **addiction,** which refers to a psychological craving. Psychological addiction rarely occurs when morphine is used to treat chronic pain, provided the patient does not have a history of drug abuse. Clinical Box 11–3 describes mechanisms involved in motivation and addiction.

FIGURE 11–5   **Local-circuit interneurons in the superficial dorsal horn of the spinal cord integrate descending and afferent pathways. A)** Possible interactions of nociceptive afferent fibers, interneurons, and descending fibers in the dorsal horn. Nociceptive fibers terminate on second-order spinothalamic projection neurons. Enkephalin (ENK)-containing interneurons exert both presynaptic and postsynaptic inhibitory actions. Serotonergic and noradrenergic neurons in the brain stem activate opioid interneurons and suppress the activity spinothalamic projection neurons. **B₁)** Activation of nociceptors releases glutamate and neuropeptides from sensory terminals, depolarizing and activating projection neurons. **B₂)** Opiates decrease Ca²⁺ influx leading to a decrease in the duration of nociceptor action potentials and a decreased release of transmitter. Also, opiates hyperpolarize the membrane of dorsal horn neurons by activating K⁺ conductance and decrease the amplitude of the EPSP produced by stimulation of nociceptors. (From Kandel ER, Schwartz JH, Jessell TM [editors]: *Principles of Neural Science,* 4th ed. McGraw-Hill, 2000.)

Despite intensive study, relatively little is known about the brain mechanisms that cause tolerance and dependence. However, the two can be separated. Absence of β-**arrestin-2** blocks tolerance but has no effect on dependence. β-Arrestin-2 is a member of a family of proteins that inhibit heterotrimeric G proteins by phosphorylating them.

Acupuncture at a location distant from the site of a pain may act by releasing endorphins. Acupuncture at the site of the pain appears to act primarily in the same way as touching or shaking (gate-control mechanism). A component of stress-induced analgesia appears to be mediated by endogenous opioids, because in experimental animals some forms of stress-induced analgesia are prevented by naloxone, a morphine antagonist. However, other forms are unaffected, and so other components are also involved.

ACETYLCHOLINE

Epibatidine, a cholinergic agonist first isolated from the skin of a frog, is a potent nonopioid analgesic agent, and even more

Motivation & Addiction

Forebrain neurons in the **ventral tegmental area** and **nucleus acumbens** are thought to be involved in motivated behaviors such as reward, laughter, pleasure, addiction, and fear. These areas have been referred to as the brain's **reward center** or **pleasure center. Addiction,** defined as the repeated compulsive use of a substance despite negative health consequences, can be produced by a variety of different drugs. According to the World Health Organization, over 76 million people worldwide suffer from alcohol abuse, and over 15 million suffer from drug abuse. Not surprisingly, alcohol and drug addiction are associated with the reward system. The **mesocortical dopaminergic neurons** that project from the midbrain to the **nucleus accumbens** and the frontal cortex are also involved. The best studied addictive drugs are opiates such as morphine and heroin, cocaine, amphetamine, ethyl alcohol, cannabinoids from marijuana, and nicotine. These drugs affect the brain in different ways, but all have in common the fact that they increase the amount of dopamine available to act on D_3 **receptors** in the nucleus accumbens. Thus, acutely they stimulate the reward system of the brain. On the other hand, long-term addiction involves the development of **tolerance,** that is, the need for increasing amounts of a drug to produce a high. In addition, withdrawal produces psychologic and physical symptoms. Injections of β-noradrenergic antagonists or $α_2$-noradrenergic agonists in the bed nucleus of the stria terminalis reduce the symptoms of opioid **withdrawal,** and so do bilateral lesions of the lateral tegmental noradrenergic fibers. One of the characteristics of addiction is the tendency of addicts to relapse after treatment. For opiate addicts, for example, the relapse rate in the first year is about 80%. Relapse often occurs on exposure to sights, sounds, and situations that were previously associated with drug use. An interesting observation that may be relevant in this regard is that as little as a single dose of an addictive drug facilitates release of excitatory neurotransmitters in brain areas concerned with memory. The medial frontal cortex, the hippocampus, and the amygdala are concerned with memory, and they all project via excitatory glutamatergic pathways to the nucleus accumbens.

potent synthetic congeners of this compound have been developed. Their effects are blocked by cholinergic blocking drugs, and as yet there is no evidence that they are addictive. Conversely, the analgesic effect of nicotine is reduced in mice lacking the $α^4$ and $β^2$ nicotine cholinergic receptor subunits. These observations make it clear that a nicotinic cholinergic mechanism is involved in the regulation of pain, although its exact role remains to be determined.

CANNABINOIDS

The cannabinoids anandamide and palmitoylethanolamide (PEA) are produced endogenously and bind to CB_1 and CB_2 receptors, respectively. Anandamide has been shown to have an analgesic effect, and there are anandamide-containing neurons in the periaqueductal gray and other areas concerned with pain. When PEA is administered, it acts peripherally to augment the analgesic effects of anandamide.

CHAPTER SUMMARY

- Discriminitive touch, proprioception, and vibratory sensations are relayed via the dorsal column (medial lemniscus) pathway to SI. Pain and temperature sensations are mediated via the ventrolateralspinothalmic tract to SI.
- The ascending pathways mediating sensation are organized somatotopically all the way from the spinal cord to SI.
- Descending pathways from the mesencephalic periaqueductal gray inhibit transmission in nociceptive pathways. This descending pathway includes a synapse in the ventromedial medulla (raphé nucleus) and the release of endogenous opiates.
- Morphine is an effective antinociceptive agent that binds to endogenous opiate receptors in the midbrain, brain stem, and spinal cord.
- Anandamide is an endogenous cannabinoid that binds to CB_1 receptors and acts centrally as an analgesic agent.

MULTIPLE-CHOICE QUESTIONS

For all questions, select the single best answer unless otherwise directed.

1. A ventrolateral cordotomy is performed that produces relief of pain in the right leg. It is effective because it interrupts the
   A) left dorsal column.
   B) left ventral spinothalamic tract.
   C) right lateral spinothalamic tract.
   D) left lateral spinothalamic tract.
   E) right corticospinal tract.

2. Which of the following does *not* exert an analgesic effect?
   A) morphine
   B) cholinergic antagonists
   C) adrenergic antagonists
   D) substance P antagonists
   E) anandamide

3. A 40-year-old man loses his right hand in a farm accident. Four years later, he has episodes of severe pain in the missing hand (phantom limb pain). A detailed PET scan study of his cerebral cortex might be expected to show
   A) expansion of the right hand area in his right somatic sensory area I (SI).
   B) expansion of the right-hand area in his left SI.
   C) a metabolically inactive spot where his hand area in his left SI would normally be.
   D) projection of fibers from neighboring sensory areas into the right-hand area of his right SI.
   E) projection of fibers from neighboring sensory areas into the right-hand area of his left SI.

4. A 50-year-old woman undergoes a neurological exam that indicates loss of pain and temperature sensitivity, vibratory sense, and proprioception in both legs. These symptoms could be explained by

   A) a tumor on the medial lemniscal pathway in the sacral spinal cord.
   B) a peripheral neuropathy.
   C) a large tumor in the sacral dorsal horn.
   D) a large tumor affecting the posterior paracentral gyri.
   E) a large tumor in the ventral posterolateral and posteromedial thalamic nuclei.

CHAPTER RESOURCES

Baron R, Maier C: Phantom limb pain: Are cutaneous nociceptors and spinothalamic neurons involved in the signaling and maintenance of spontaneous and touch-evoked pain? A case report. Pain 1995;60:223.

Blumenfeld H: *Neuroanatomy Through Clinical Cases.* Sinauer Associates, 2002.

Haines DE (editor): *Fundamental Neuroscience for Basic and Clinical Applications,* 3rd ed. Elsevier, 2006.

Herman J: Phantom limb: From medical knowledge to folk wisdom and back. Ann Int Med 1998;128:76.

Hopkins K: Show me where it hurts: Tracing the pathways of pain. J NIH Res 1997;9:37.

Kandel ER, Schwartz JH, Jessell TM (editors): *Principles of Neural Science,* 4th ed. McGraw-Hill, 2000.

Melzack R: The tragedy of needless pain. Sci Am 1991;262:27.

Penfield W, Rasmussen T: *The Cerebral Cortex of Man: A Clinical Study of Localization of Function.* Macmillan, 1950.

Willis WD: The somatosensory system, with emphasis on structures important for pain. Brain Res Rev 2007;55:297.

Vision

OBJECTIVES

After studying this chapter, you should be able to:

- Describe the various parts of the eye and list the functions of each.
- Trace the neural pathways that transmit visual information from the rods and cones to the visual cortex.
- Explain how light rays in the environment are brought to a focus on the retina and the role of accommodation in this process.
- Define hyperopia, myopia, astigmatism, presbyopia, and strabismus.
- Describe the electrical responses produced by rods and cones, and explain how these responses are produced.
- Describe the electrical responses and function of bipolar, horizontal, amacrine, and ganglion cells.
- Describe the responses of cells in the visual cortex and the functional organization of the dorsal and ventral pathways to the parietal cortex.
- Define and explain dark adaptation and visual acuity.
- Describe the neural pathways involved in color vision.
- Name the four types of eye movements and the function of each.

INTRODUCTION

The eyes are complex sense organs that have evolved from primitive light-sensitive spots on the surface of invertebrates. Within its protective casing, each eye has a layer of receptors, a lens system that focuses light on these receptors, and a sys-tem of nerves that conducts impulses from the receptors to the brain. The way these components operate to set up con-scious visual images is the subject of this chapter.

ANATOMIC CONSIDERATIONS

The principal structures of the eye are shown in Figure 12–1. The outer protective layer of the eyeball, the **sclera,** is modified anteriorly to form the transparent **cornea,** through which light rays enter the eye. Inside the sclera is the **choroid,** a layer that contains many of the blood vessels that nourish the structures in the eyeball. Lining the posterior two thirds of the choroid is the **retina,** the neural tissue containing the receptor cells.

The **crystalline lens** is a transparent structure held in place by a circular **lens suspensary ligament (zonule).** The zonule is attached to the thickened anterior part of the choroid, the **ciliary body.** The ciliary body contains circular muscle fibers and longitudinal muscle fibers that attach near the corneo-scleral junction. In front of the lens is the pigmented and opaque **iris,** the colored portion of the eye. The iris contains circular muscle fibers that constrict and radial fibers that dilate the **pupil.** Variations in the diameter of the pupil can

FIGURE 12–1

The internal anatomy of the eye. (From Fox SI, *Human Physiology.* McGraw-Hill, 2008.)

produce up to fivefold changes in the amount of light reaching the retina.

The space between the lens and the retina is filled primarily with a clear gelatinous material called the **vitreous (vitreous humor). Aqueous humor,** a clear liquid that nourishes the cornea and lens, is produced in the ciliary body by diffusion and active transport from plasma. It flows through the pupil and fills the anterior chamber of the eye. It is normally reabsorbed through a network of trabeculae into the **canal of Schlemm,** a venous channel at the junction between the iris and the cornea (anterior chamber angle). Obstruction of this outlet leads to increased intraocular pressure (see Clinical Box 12–1).

RETINA

The retina extends anteriorly almost to the ciliary body. It is organized in 10 layers and contains the **rods** and **cones,** which are the visual receptors, plus four types of neurons: **bipolar cells, ganglion cells, horizontal cells,** and **amacrine cells** (Figure 12–2). There are many different synaptic transmitters. The rods and cones, which are next to the choroid, synapse with bipolar cells, and the bipolar cells synapse with ganglion cells. About 12 different types of bipolar cells occur, based on morphology and function. The axons of the ganglion cells converge and leave the eye as the optic nerve. Horizontal cells connect receptor cells to the other receptor cells in the outer plexiform layer. Amacrine cells connect ganglion cells to one another in the inner plexiform layer via processes of varying length and patterns. At least 29 types of amacrine cells have been described on the basis of their connections. Gap junctions also connect retinal neurons to one another, and the permeability of these gap junctions is regulated.

Because the receptor layer of the retina rests on the **pigment epithelium** next to the choroid, light rays must pass through the ganglion cell and bipolar cell layers to reach the

rods and cones. The pigment epithelium absorbs light rays, preventing the reflection of rays back through the retina. Such reflection would produce blurring of the visual images.

The neural elements of the retina are bound together by glial cells called **Müller cells.** The processes of these cells form an internal limiting membrane on the inner surface of the retina and an external limiting membrane in the receptor layer.

The optic nerve leaves the eye and the retinal blood vessels enter it at a point 3 mm medial to and slightly above the posterior pole of the globe. This region is visible through the ophthalmoscope as the **optic disk** (Figure 12–3). There are no visual receptors over the disk, and consequently this spot is blind (the **blind spot**).

Near the posterior pole of the eye is a yellowish pigmented spot, the **macula lutea.** This marks the location of the **fovea**

Pigment epithelium

Rod and cone
　Outer segments

　Inner segments

Outer nuclear layer

Outer plexiform layer

Inner nuclear layer

Inner plexiform layer

Ganglion cell layer

Optic nerve fibers

FIGURE 12–2 **Neural components of the extrafoveal portion of the retina.** C, cone; R, rod; MB, RB, and FB, midget, rod, and flat bipolar cells; DG and MG, diffuse and midget ganglion cells; H, horizontal cells; A, amacrine cells. (Modified from Dowling JE, Boycott BB: Organization of the primate retina: Electron microscopy. Proc R Soc Lond Ser B [Biol] 1966;166:80.)

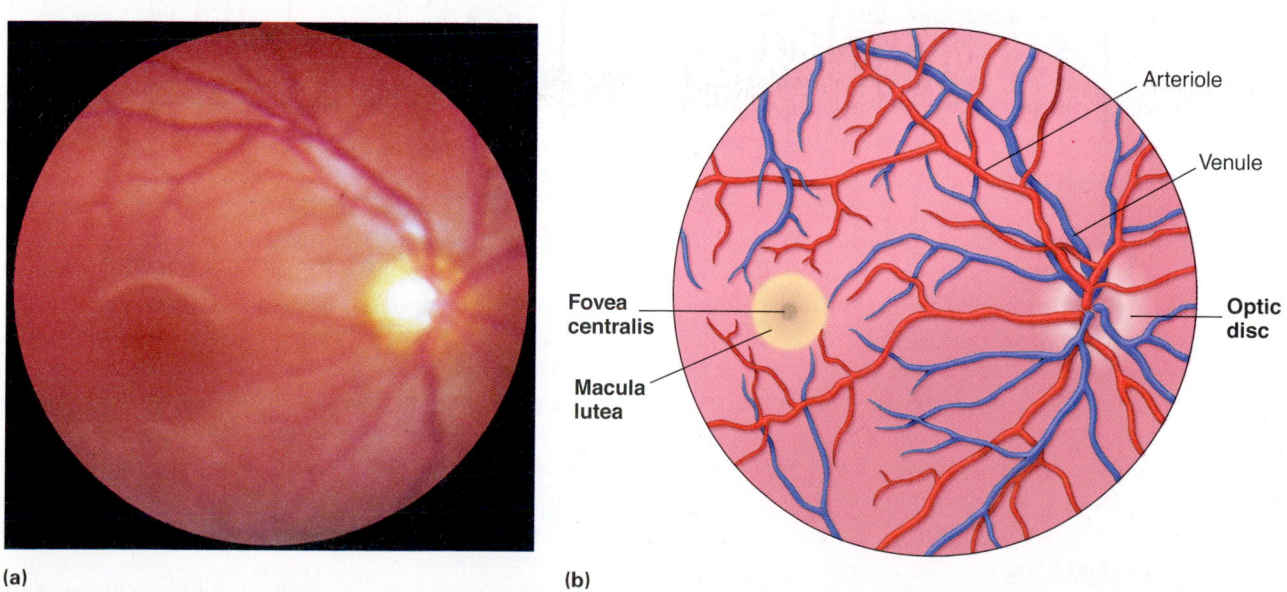

(a)　　　　　(b)

FIGURE 12–3 **Retina seen through the ophthalmoscope in a normal human. (a)** A photograph and **(b)** an illustration of the optic fundus (back of the eye). Optic nerve fibers leave the eyeball at the optic disc to form the optic nerve. The arteries, arterioles, and veins in the superficial layers of the retina near its vitreous surface can be seen through the ophthalmoscope. (From Fox SI, *Human Physiology.* McGraw-Hill, 2008.)

Visual Acuity

Visual acuity is the degree to which the details and contours of objects are perceived, and it is usually defined in terms of the shortest distance by which two lines can be separated and still be perceived as two lines. Clinically, visual acuity is often determined by the use of the familiar **Snellen letter charts** viewed at a distance of 20 ft (6 m). The individual being tested reads aloud the smallest line distinguishable. The results are expressed as a fraction. The numerator of the fraction is 20, the distance at which the subject reads the chart. The denominator is the greatest distance from the chart at which a normal individual can read the smallest line. Normal visual acuity is 20/20; a subject with 20/15 visual acuity has better than normal vision (not farsightedness); and one with 20/100 visual acuity has subnormal vision. The Snellen charts are designed so that the height of the letters in the smallest line a normal individual can read at 20 ft subtends a visual angle of 5 minutes. Each of the lines is separated by 1 minute of arc. Thus, the minimum separable in a normal individual corresponds to a visual angle of about 1 minute. Visual acuity is a complex phenomenon and is influenced by a large variety of factors, including optical factors (eg, the state of the image-forming mechanisms of the eye), retinal factors (eg, the state of the cones), and stimulus factors (eg, illumination, brightness of the stimulus, contrast between the stimulus and the background, length of time the subject is exposed to the stimulus).

centralis, a thinned-out, rod-free portion of the retina that is present in humans and other primates. In it, the cones are densely packed, and each synapses to a single bipolar cell, which, in turn, synapses on a single ganglion cell, providing a direct pathway to the brain. There are very few overlying cells and no blood vessels. Consequently, the fovea is the point where **visual acuity** is greatest (see Clinical Box 12–2). When attention is attracted to or fixed on an object, the eyes are normally moved so that light rays coming from the object fall on the fovea.

The arteries, arterioles, and veins in the superficial layers of the retina near its vitreous surface can be seen through the ophthalmoscope. Because this is the one place in the body where arterioles are readily visible, ophthalmoscopic examination is of great value in the diagnosis and evaluation of diabetes mellitus, hypertension, and other diseases that affect blood vessels. The retinal vessels supply the bipolar and ganglion cells, but the receptors are nourished, for the most part, by the capillary plexus in the choroid. This is why retinal detachment is so damaging to the receptor cells.

NEURAL PATHWAYS

The axons of the ganglion cells pass caudally in the **optic nerve** and **optic tract** to end in the **lateral geniculate body** in the thalamus (Figure 12–4). The fibers from each nasal hemiretina decussate in the **optic chiasm.** In the geniculate body, the fibers from the nasal half of one retina and the temporal half of the other synapse on the cells whose axons form the **geniculocalcarine tract.** This tract passes to the occipital lobe of the cerebral cortex. The effects of lesions in these pathways on visual function are discussed below.

The primary visual receiving area (**primary visual cortex,** Brodmann's area 17; also known as V1), is located principally on the sides of the calcarine fissure (Figure 12–5). The organization of the primary visual cortex is discussed below.

Some ganglion cell axons pass from the lateral geniculate nucleus to the pretectal region of the midbrain and the superior colliculus, where they form connections that mediate pupillary reflexes and eye movements. The frontal cortex is also concerned with eye movement, and especially its refinement. The bilateral **frontal eye fields** in this part of the cortex are concerned with control of saccades, and an area just anterior to these fields is concerned with vergence and the near response. The frontal areas concerned with vision probably project to the nucleus reticularis tegmentalis pontinus, and from there to the other brain stem nuclei mentioned above.

Other axons pass directly from the optic chiasm to the suprachiasmatic nuclei in the hypothalamus, where they form connections that synchronize a variety of endocrine and other circadian rhythms with the light–dark cycle.

The brain areas activated by visual stimuli have been investigated in monkeys and humans by positron emission tomography (PET) and other imaging techniques. Activation occurs not only in the occipital lobe but also in parts of the inferior temporal cortex, the posteroinferior parietal cortex, portions of the frontal lobe, and the amygdala. The subcortical structures activated in addition to the lateral geniculate body include the superior colliculus, pulvinar, caudate nucleus, putamen, and claustrum.

RECEPTORS

Each rod and cone is divided into an outer segment, an inner segment that includes a nuclear region, and a synaptic zone (Figure 12–6). The outer segments are modified cilia and are made up of regular stacks of flattened saccules or disks composed of membrane. These saccules and disks contain the photosensitive compounds that react to light, initiating action potentials in the visual pathways. The inner segments are rich in mitochondria. The rods are named for the thin, rodlike appearance of their outer segments. Cones generally have thick inner segments and conical outer segments, although their morphology varies from place to place in the retina. In cones, the saccules are formed in the outer segments by infoldings of the cell membrane, but in rods the disks are separated from the cell membrane.

Rod outer segments are being constantly renewed by formation of new disks at the inner edge of the segment and phagocytosis of old disks from the outer tip by cells of the

FIGURE 12–4 **Visual pathways.** Transection of the pathways at the locations indicated by the letters causes the visual field defects shown in the diagrams on the right. The fibers from the nasal half of each retina decussate in the optic chiasm, so that the fibers in the optic tracts are those from the temporal half of one retina and the nasal half of the other. A lesion that interrupts one optic nerve causes blindness in that eye (A). A lesion in one optic tract causes blindness in half of the visual field (C) and is called homonymous (same side of both visual fields) hemianopia (half-blindness). Lesions affecting the optic chiasm destroy fibers from both nasal hemiretinas and produce a heteronymous (opposite sides of the visual fields) hemianopia (B). Occipital lesions may spare the fibers from the macula (as in D) because of the separation in the brain of these fibers from the others subserving vision (see Figure 12–5).

FIGURE 12–5 **Medial view of the human right cerebral hemisphere showing projection of the retina on the primary visual cortex (Brodmann's area 17; also known as V1) in the occipital cortex around the calcarine fissure.** The geniculocalcarine fibers from the medial half of the lateral geniculate terminate on the superior lip of the calcarine fissure, and those from the lateral half terminate on the inferior lip. Also, the fibers from the lateral geniculate body that relay macular vision separate from those that relay peripheral vision and end more posteriorly on the lips of the calcarine fissure.

pigment epithelium. Cone renewal is a more diffuse process and appears to occur at multiple sites in the outer segments.

In the extrafoveal portions of the retina, rods predominate (Figure 12–7), and there is a good deal of convergence. Flat bipolar cells (Figure 12–2) make synaptic contact with several cones, and rod bipolar cells make synaptic contact with several rods. Because there are approximately 6 million cones and 120 million rods in each human eye but only 1.2 million nerve fibers in each optic nerve, the overall convergence of receptors through bipolar cells on ganglion cells is about 105:1. However, there is divergence from this point on. There are twice as many fibers in the geniculocalcarine tracts as in the optic nerves, and in the visual cortex the number of neurons concerned with vision is 1000 times the number of fibers in the optic nerves.

PROTECTION

The eye is well protected from injury by the bony walls of the orbit. The cornea is moistened and kept clear by tears that

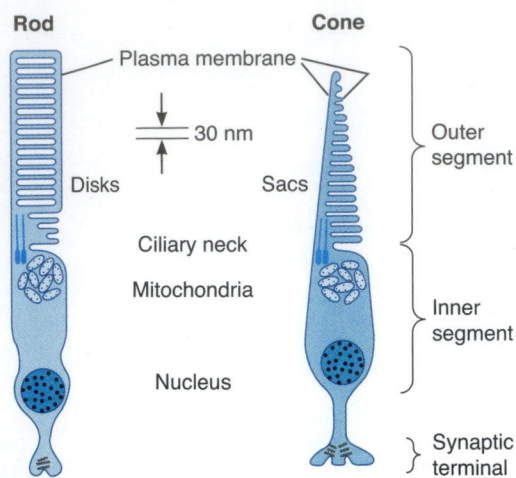

FIGURE 12–6 **Schematic diagram of a rod and a cone.** Each rod and cone is divided into an outer segment, an inner segment with a nuclear region, and a synaptic zone. The saccules and disks in the outer segment contain photosensitive compounds that react to light to initiate action potentials in the visual pathways. (Reproduced with permission from Lamb TD: Electrical responses of photoreceptors. In: *Recent Advances in Physiology.* No.10. Baker PF [editor]. Churchill Livingstone, 1984.)

course from the **lacrimal gland** in the upper portion of each orbit across the surface of the eye to empty via the **lacrimal duct** into the nose. Blinking helps keep the cornea moist.

One of the most important characteristics of the visual system is its ability to function over a wide range of light intensity. When one goes from near darkness to bright sunlight, light intensity increases by 10 log units, that is, by a factor of 10 billion. One factor reducing the fluctuation in intensity is

the diameter of the pupil; when this is reduced from 8 mm to 2 mm, its area decreases by a factor of 16 and light intensity at the retina is reduced by more than 1 log unit.

Another factor in reacting to fluctuations in intensity is the presence of two types of receptors. The rods are extremely sensitive to light and are the receptors for night vision (**scotopic vision**). The scotopic visual apparatus is incapable of resolving the details and boundaries of objects or determining their color. The cones have a much higher threshold, but the cone system has a much greater acuity and is the system responsible for vision in bright light (**photopic vision**) and for color vision. There are thus two kinds of inputs to the central nervous system (CNS) from the eye: input from the rods and input from the cones. The existence of these two kinds of input, each working maximally under different conditions of illumination, is called the **duplicity theory.**

THE IMAGE-FORMING MECHANISM

The eyes convert energy in the visible spectrum into action potentials in the optic nerve. The wavelengths of visible light range from approximately 397–723 nm. The images of objects in the environment are focused on the retina. The light rays striking the retina generate potentials in the rods and cones. Impulses initiated in the retina are conducted to the cerebral cortex, where they produce the sensation of vision.

PRINCIPLES OF OPTICS

Light rays are bent when they pass from a medium of one density into a medium of a different density, except when they strike perpendicular to the interface (Figure 12–8). The bending of light rays is called **refraction** and is the mechanism that

FIGURE 12–7   **Rod and cone density along the horizontal meridian through the human retina.** A plot of the relative acuity of vision in the various parts of the light-adapted eye would parallel the cone density curve; a similar plot of relative acuity of the dark-adapted eye would parallel the rod density curve.

(a)

(b)

FIGURE 12–8 **Focusing point sources of light. (a)** When diverging light rays enter a dense medium at an angle to its convex surface, refraction bends them inward. **(b)** Refraction of light by the lens system. For simplicity, refraction is shown only at the corneal surface (site of greatest refraction) although it also occurs in the lens and elsewhere. Incoming light from *a* (above) and *b* (below) is bent in opposite directions, resulting in *b'* being above *a'* on the retina.   (From Widmaier EP, Raff H, Strang KT: *Vander's Human Physiology,* 11th ed. McGraw-Hill, 2008.)

allows one to focus an accurate image onto the retina. Parallel light rays striking a biconvex lens are refracted to a point (**principal focus**) behind the lens. The principal focus is on a line passing through the centers of curvature of the lens, the **principal axis.** The distance between the lens and the principal focus is the **principal focal distance.** For practical purposes, light rays from an object that strike a lens more than 6 m (20 ft) away are considered to be parallel. The rays from an object closer than 6 m are diverging and are therefore brought to a focus farther back on the principal axis than the principal focus. Biconcave lenses cause light rays to diverge.

Refractive power is greatest when the curvature of a lens is greatest. The refractive power of a lens is conveniently measured in **diopters,** the number of diopters being the reciprocal of the principal focal distance in meters. For example, a lens with a principal focal distance of 0.25 m has a refractive power of 1/0.25, or 4 diopters. The human eye has a refractive power of approximately 60 diopters at rest.

In the eye, light is actually refracted at the anterior surface of the cornea and at the anterior and posterior surfaces of the lens. The process of refraction can be represented diagrammatically, however, without introducing any appreciable error,

by drawing the rays of light as if all refraction occurs at the anterior surface of the cornea (Figure 12–8). It should be noted that the retinal image is inverted. The connections of the retinal receptors are such that from birth any inverted image on the retina is viewed right side up and projected to the visual field on the side opposite to the retinal area stimulated. This perception is present in infants and is innate. If retinal images are turned right side up by means of special lenses, the objects viewed look as if they are upside down.

COMMON DEFECTS OF THE IMAGE-FORMING MECHANISM

In some individuals, the eyeball is shorter than normal and the parallel rays of light are brought to a focus behind the retina. This abnormality is called **hyperopia** or farsightedness (Figure 12–9). Sustained accommodation, even when viewing distant objects, can partially compensate for the defect, but the prolonged muscular effort is tiring and may cause headaches and blurring of vision. The prolonged convergence of the visual axes associated with the accommodation may lead eventually to

FIGURE 12–9 **Common defects of the optical system of the eye.** In hyperopia (farsightedness), the eyeball is too short and light rays come to a focus behind the retina. A biconvex lens corrects this by adding to the refractive power of the lens of the eye. In myopia (nearsightedness), the eyeball is too long and light rays focus in front of the retina. Placing a biconcave lens in front of the eye causes the light rays to diverge slightly before striking the eye, so that they are brought to a focus on the retina. (From Widmaier EP, Raff H, Strang KT: *Vander's Human Physiology*, 11th ed. McGraw-Hill, 2008.)

CLINICAL BOX 12–3

Strabismus & Amblyopia

Strabismus is a misalignment of the eyes and one of the most common eye problems in children, affecting about 4% of children under 6 years of age. It is characterized by one or both eyes turning inward (crossed-eyes), outward (wall eyes), upward, or downward. In some cases, more than one of these conditions is present. Strabismus is also commonly called "wandering eye" or "crossed-eyes." It occurs when visual images do not fall on corresponding retinal points. When visual images chronically fall on noncorresponding points in the two retinas in young children, one is eventually suppressed **(suppression scotoma).** This suppression is a cortical phenomenon, and it usually does not develop in adults. It is important to institute treatment before age 6 in affected children, because if the suppression persists, the loss of visual acuity in the eye generating the suppressed image is permanent.

A similar suppression with subsequent permanent loss of visual acuity can occur in children in whom vision in one eye is blurred or distorted owing to a refractive error. The loss of vision in these cases is called **amblyopia ex anopsia,** a term that refers to uncorrectable loss of visual acuity that is not directly due to organic disease of the eye. Typically, an affected child has one weak eye with poor vision and one strong eye with normal vision. It affects about 3% of the general population. Amblyopia is also referred to as "lazy eye," and it often co-exists with strabismus. Some types of strabismus can be corrected by surgical shortening of some of the eye muscles, by eye muscle training exercises, and by the use of glasses with prisms that bend the light rays sufficiently to compensate for the abnormal position of the eyeball. However, subtle defects in **depth perception** persist. It has been suggested that congenital abnormalities of the visual tracking mechanisms may cause both strabismus and the defective depth perception. In infant monkeys, covering one eye with a patch for 3 months causes a loss of ocular dominance columns; input from the remaining eye spreads to take over all the cortical cells, and the patched eye becomes functionally blind. Comparable changes may occur in children with strabismus.

---

squint (**strabismus;** see Clinical Box 12–3). The defect can be corrected by using glasses with convex lenses, which aid the refractive power of the eye in shortening the focal distance.

In **myopia** (nearsightedness), the anteroposterior diameter of the eyeball is too long (Figure 12–9). Myopia is said to be genetic in origin. However, there is a positive correlation between sleeping in a lighted room before the age of 2 and the subsequent development of myopia. Thus, the shape of the eye appears to be determined in part by the refraction presented to it. In young adult humans the extensive close work involved in activities such as studying accelerates the development of myopia. This defect

can be corrected by glasses with biconcave lenses, which make parallel light rays diverge slightly before they strike the eye.

Astigmatism is a common condition in which the curvature of the cornea is not uniform (Figure 12–9). When the curvature in one meridian is different from that in others, light rays in that meridian are refracted to a different focus, so that part of the retinal image is blurred. A similar defect may be produced if the lens is pushed out of alignment or the curvature of the lens is not uniform, but these conditions are rare. Astigmatism can usually be corrected with cylindric lenses placed in such a way that they equalize the refraction in all meridians.

ACCOMMODATION

When the ciliary muscle is relaxed, parallel light rays striking the optically normal (**emmetropic**) eye are brought to a focus on the retina. As long as this relaxation is maintained, rays from objects closer than 6 m from the observer are brought to a focus behind the retina, and consequently the objects appear blurred. The problem of bringing diverging rays from close objects to a focus on the retina can be solved by increasing the distance between the lens and the retina or by increasing the curvature or refractive power of the lens. In bony fish, the problem is solved by increasing the length of the eyeball, a solution analogous to the manner in which the images of objects closer than 6 m are focused on the film of a camera by moving the lens away from the film. In mammals, the problem is solved by increasing the curvature of the lens.

The process by which the curvature of the lens is increased is called **accommodation.** At rest, the lens is held under tension by the lens ligaments. Because the lens substance is malleable and the lens capsule has considerable elasticity, the lens is pulled into a flattened shape. When the gaze is directed at a near object, the ciliary muscle contracts. This decreases the distance between the edges of the ciliary body and relaxes the lens ligaments, so that the lens springs into a more convex shape (Figure 12–10). The change is greatest in the anterior surface of the lens. In young individuals, the change in shape may add as many as 12 diopters to the refractive power of the eye. The relaxation of the lens ligaments produced by contraction of the ciliary muscle is due partly to the sphincterlike action of the circular muscle fibers in the ciliary body and partly to the contraction of longitudinal muscle fibers that attach anteriorly, near the corneoscleral junction. When these fibers contract, they pull the whole ciliary body forward and inward. This motion brings the edges of the ciliary body closer together. Changes in accommodation with age are described in Clinical Box 12–4.

In addition to accommodation, the visual axes converge and the pupil constricts when an individual looks at a near object. This three-part response—accommodation, convergence of the visual axes, and pupillary constriction—is called the **near response.**

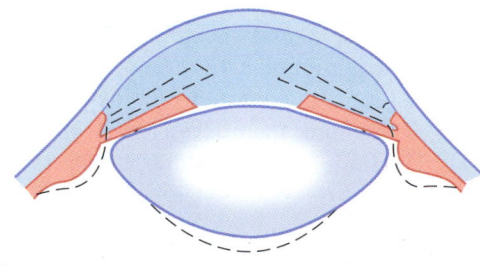

FIGURE 12–10   Accommodation. The solid lines represent the shape of the lens, iris, and ciliary body at rest, and the dashed lines represent the shape during accommodation. When gaze is directed at a near object, ciliary muscles contract. This decreases the distance between the edges of the ciliary body and relaxes the lens ligaments, and the lens becomes more convex.

OTHER PUPILLARY REFLEXES

When light is directed into one eye, the pupil constricts (**pupillary light reflex**). The pupil of the other eye also constricts (**consensual light reflex**). The optic nerve fibers that carry the impulses initiating these pupillary responses leave the optic nerves near the lateral geniculate bodies. On each side, they enter the midbrain via the brachium of the superior colliculus and terminate in the pretectal nucleus. From this nucleus, the second-order neurons project to the ipsilateral and contralateral **Edinger–Westphal nucleus.** The third-order neurons pass from this nucleus to the ciliary ganglion in the **oculomotor nerve,** and the fourth-order neurons pass from this ganglion to the ciliary body. This pathway is dorsal to the pathway for the near response. Consequently, the light response is sometimes lost while the response to accommodation remains intact (**Argyll Robertson pupil**). One cause of this abnormality is CNS syphilis, but the Argyll Robertson pupil is also seen in other diseases producing selective lesions in the midbrain.

THE PHOTORECEPTOR MECHANISM

ELECTRICAL RESPONSES

The potential changes that initiate action potentials in the retina are generated by the action of light on photosensitive compounds in the rods and cones. When light is absorbed by these substances, their structure changes, and this triggers a sequence of events that initiates neural activity.

The eye is unique in that the receptor potentials of the photoreceptors and the electrical responses of most of the other neural elements in the retina are local, graded potentials, and it is only in the ganglion cells that all-or-none action potentials transmitted over appreciable distances are generated. The responses of the rods, cones, and horizontal cells are hyperpolarizing (Figure 12–11), and the responses of the bipolar cells are either hyperpolarizing or depolarizing, whereas amacrine cells produce depolarizing potentials and spikes that may act as generator potentials for the propagated spikes produced in the ganglion cells.

The cone receptor potential has a sharp onset and offset, whereas the rod receptor potential has a sharp onset and slow offset. The curves relating the amplitude of receptor potentials to stimulus intensity have similar shapes in rods and cones, but the rods are much more sensitive. Therefore, rod responses are proportionate to stimulus intensity at levels of illumination that are below the threshold for cones. On the other hand, cone responses are proportionate to stimulus intensity at high levels of illumination when the rod responses are maximal and cannot change. This is why cones generate good responses to changes in light intensity above background but do not represent absolute illumination well, whereas rods detect absolute illumination.

IONIC BASIS OF PHOTORECEPTOR POTENTIALS

Na^+ channels in the outer segments of the rods and cones are open in the dark, so current flows from the inner to the outer segment (Figure 12–12). Current also flows to the synaptic ending of the photoreceptor. The Na^+–K^+ pump in the inner segment maintains ionic equilibrium. Release of synaptic transmitter is steady in the dark. When light strikes the outer segment, the reactions that are initiated close some of the Na^+ channels, and the result is a hyperpolarizing receptor potential. The hyperpolarization reduces the release of synaptic transmitter, and this generates a signal in the bipolar cells that ultimately leads to action potentials in ganglion cells. The action potentials are transmitted to the brain.

PHOTOSENSITIVE COMPOUNDS

The photosensitive compounds in the rods and cones of the eyes of humans and most other mammals are made up of a protein called an **opsin,** and retinene$_1$, the aldehyde of vitamin

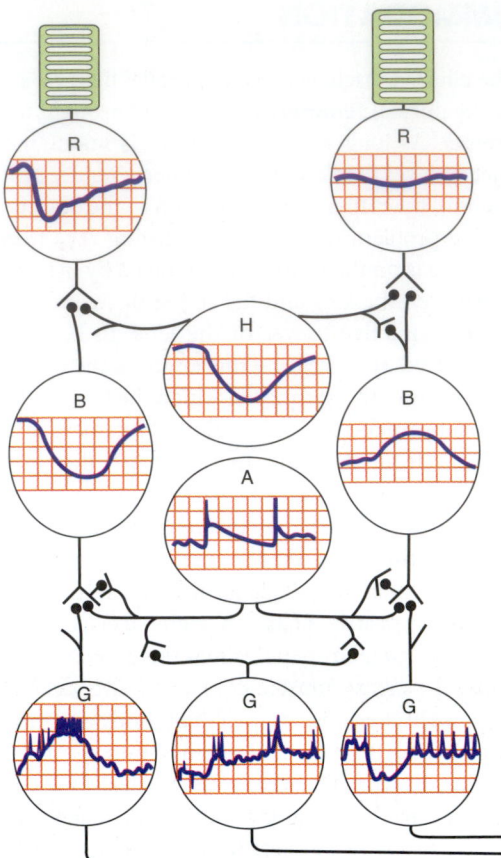

FIGURE 12–11   Intracellularly recorded responses of cells in the retina to light. The synaptic connections of the cells are also indicated. The eye is unique in that the receptor potentials of the photoreceptors and the electrical responses of most of the other neural elements in the retina are local, graded potentials. The rod (R) on the left is receiving a light flash, whereas the rod on the right is receiving steady, low-intensity illumination. The responses of rods and horizontal cells (H) are hyperpolarizing, responses of bipolar cells (B) are either hyperpolarizing or depolarizing, and amacrine (A) cells produce depolarizing potentials and spikes that may act as generator potentials for propagated spikes of ganglion cells (G). (Reproduced with permission from Dowling JE: Organization of vertebrate retinas. Invest Ophthalmol 1970;9:655.)

A$_1$. The term retinene$_1$ is used to distinguish this compound from retinene$_2$, which is found in the eyes of some animal species. Because the retinenes are aldehydes, they are also called **retinals.** The A vitamins themselves are alcohols and are therefore called **retinols** (see Clinical Box 12–5).

RHODOPSIN

The photosensitive pigment in the rods is called **rhodopsin (visual purple).** Its opsin is called **scotopsin.** Rhodopsin has a peak sensitivity to light at a wavelength of 505 nm.

Human rhodopsin has a molecular weight of 41,000. It is found in the membranes of the rod disks and makes up 90% of the total protein in these membranes. It is one of the many receptors coupled to G proteins. Retinene$_1$ is parallel to the surface of

FIGURE 12–12 **Effect of light on current flow in visual receptors.** In the dark, Na$^+$ channels in the outer segment are held open by cGMP. Light leads to increased conversion of cGMP to 5'-GMP, and some of the channels close. This produces hyperpolarization of the synaptic terminal of the photoreceptor.

the membrane (Figure 12–13) and is attached to a lysine residue at position 296 in the seventh transmembrane domain.

In the dark, the retinene$_1$ in rhodopsin is in the 11-*cis* configuration. The only action of light is to change the shape of the retinene, converting it to the all-*trans* isomer. This, in turn, alters the configuration of the opsin, and the opsin change activates the associated heterotrimeric G protein, which in this case is called **transducin** or Gt$_1$. The G protein exchanges GDP for GTP, and the α subunit separates. This subunit remains active until its intrinsic GTPase activity hydrolyzes the GTP. Termination of the activity of transducin is also accelerated by its binding of β-arrestin.

The α subunit activates cGMP phosphodiesterase, which converts cGMP to 5'-GMP (Figure 12–14). cGMP normally acts directly on Na$^+$ channels to maintain them in the open position, so the decline in the cytoplasmic cGMP concentration causes some Na$^+$ channels to close. This produces the hyperpolarizing potential. This cascade of reactions occurs very rapidly and amplifies the light signal. The amplification helps explain the remarkable sensitivity of rod photoreceptors; these receptors are capable of producing a detectable response to as little as one photon of light.

After retinene$_1$ is converted to the all-*trans* configuration, it separates from the opsin (bleaching). Some of the all-*trans* retinene is converted back to the 11-*cis* retinene by retinal isomerase, and then again associates with scotopsin, replenishing the rhodopsin supply. Some 11-*cis* retinene is also synthesized from vitamin A. All of these reactions, except the formation of the all-*trans* isomer of retinene$_1$, are independent of the light intensity, proceeding equally well in light or

CLINICAL BOX 12–5

Vitamin Deficiencies

In view of the importance of **vitamin A** in the synthesis of **retinene$_1$**, it is not surprising that a deficiency in this vitamin produces visual abnormalities. Among these, one of the earliest to appear is night blindness **(nyctalopia).** Vitamin A deficiency also contributes to blindness by causing the eye to become very dry, which damages the cornea **(xerophthalmia)** and retina. Vitamin A first alters rod function, but concomitant cone degeneration occurs as vitamin A deficiency develops. Vitamin A deficiency is due to inadequate intake of foods high in vitamin A (liver, kidney, whole eggs, milk, cream, and cheese) or **beta-carotene,** a precursor of vitamin A, found in dark green leafy vegetables and yellow or orange fruits and vegetables. Vitamin A deficiency is rare in the United States, but it is still a major public health problem in the developing world. Annually, about 80,000 individuals worldwide (mostly children in underdeveloped countries) lose their sight from severe vitamin A deficiency. Prolonged deficiency is associated with anatomic changes in the rods and cones followed by degeneration of the neural layers of the retina. Treatment with vitamin A can restore retinal function if given before the receptors are destroyed. Other vitamins, especially those of the B complex, are also necessary for the normal functioning of the retina and other neural tissues.

FIGURE 12–13 **Diagrammatic representation of the structure of rhodopsin, showing the position of retinene$_1$ (R) in the rod disk membrane.** Retinene$_1$ is parallel to the surface of the membrane and is attached to a lysine residue at position 296 in the seventh transmembrane domain.

FIGURE 12–14 **Initial steps in phototransduction in rods.** Light activates rhodopsin, which activates transducin to bind GTP. This activates phosphodiesterase, which catalyzes the conversion of cGMP to 5'-GMP. The resulting decrease in the cytoplasmic cGMP concentration causes cGMP-gated ion channels to close.

darkness. The amount of rhodopsin in the receptors therefore varies inversely with the incident light level.

CONE PIGMENTS

Primates have three different kinds of cones. These receptors subserve color vision and respond maximally to light at wavelengths of 440, 535, and 565 nm. Each contains retinene$_1$ and an opsin. The opsin resembles rhodopsin and spans the cone membrane seven times but has a characteristic structure in each type of cone. The cell membrane of cones is invaginated to form the saccules, but the cones have no separate intracellular disks like those in rods. The details of the responses of cones to light are probably similar to those in rods. Light activates retinene$_1$, and this activates G_{t2}, a G protein that differs somewhat from rod transducin. G_{t2} in turn activates phosphodiesterase, catalyzing the conversion of cGMP to 5'-GMP. This results in closure of Na$^+$ channels between the extracellular fluid and the cone cytoplasm, a decrease in intracellular Na$^+$ concentration, and hyperpolarization of the cone synaptic terminals.

The sequence of events in photoreceptors by which incident light leads to production of a signal in the next succeeding neural unit in the retina is summarized in Figure 12–15.

RESYNTHESIS OF CYCLIC GMP

Light reduces the concentration of Ca^{2+} as well as that of Na$^+$ in photoreceptors. The resulting decrease in Ca^{2+} concentration activates guanylyl cyclase, which generates more cGMP. It also inhibits the light-activated phosphodiesterase. Both actions speed recovery, restoring the Na$^+$ channels to their open position.

FIGURE 12–15 **Sequence of events involved in phototransduction in rods and cones.**

MELANOPSIN

A small number of photoreceptors contain **melanopsin** rather than rhodopsin or cone pigments. The axons of these neurons project to the suprachiasmatic nuclei and the part of the lateral geniculate nuclei that controls the pupillary responses to light. When the gene for melanopsin is knocked out, circadian photoentrainment is abolished. The papillary light responses are reduced, and they are abolished when the rods and cones are also inactivated. Thus, a part of the pupillary responses and all the circadian entrainment responses to light–dark changes are controlled by a system separate from the rod and cone systems.

PROCESSING OF VISUAL INFORMATION IN THE RETINA

In a sense, the processing of visual information in the retina involves the formation of three images. The first image, formed by the action of light on the photoreceptors, is changed to a second image in the bipolar cells, and this in turn is converted to a third image in the ganglion cells. In the formation of the second image, the signal is altered by the horizontal cells, and in the formation of the third, it is altered by the amacrine cells. There is little change in the impulse pattern in the lateral geniculate bodies, so the third image reaches the occipital cortex.

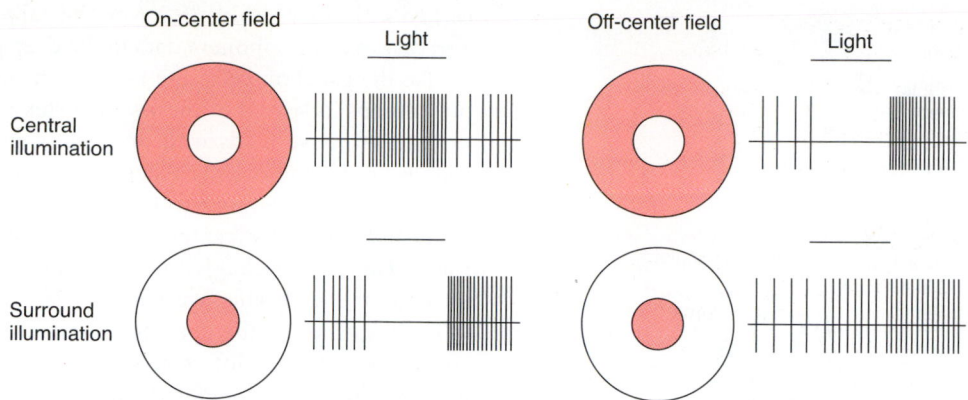

FIGURE 12–16 **Responses of retinal ganglion cells to light on the portions of their receptive fields indicated in white.** Beside each receptive field diagram is a diagram of the ganglion cell response, indicated by extracellularly recorded action potentials. Note that in three of the four situations, there is increased discharge when the light is turned off. (Modified from Kandel E, Schwartz JH, Jessell TM [editors]: *Principles of Neural Science,* 4th ed. McGraw-Hill, 2000.)

A characteristic of the bipolar and ganglion cells (as well as the lateral geniculate cells and the cells in layer 4 of the visual cortex) is that they respond best to a small, circular stimulus and that, within their receptive field, an annulus of light around the center (surround illumination) inhibits the response to the central spot (Figure 12–16). The center can be excitatory with an inhibitory surround (an **"on-center" cell**) or inhibitory with an excitatory surround (an **"off-center" cell**). The inhibition of the center response by the surround is probably due to inhibitory feedback from one photoreceptor to another mediated via horizontal cells. Thus, activation of nearby photoreceptors by addition of the annulus triggers horizontal cell hyperpolarization, which in turn inhibits the response of the centrally activated photoreceptors. The inhibition of the response to central illumination by an increase in surrounding illumination is an example of **lateral inhibition**—that form of inhibition in which activation of a particular neural unit is associated with inhibition of the activity of nearby units. It is a general phenomenon in mammalian sensory systems and helps to sharpen the edges of a stimulus and improve discrimination.

A remarkable degree of processing of visual input occurs in the retina, largely via amacrine cells. For example, movement of an object within the visual field is separated from movement of the background caused by changes in posture and movement of the eyes. This was demonstrated by recording from optic neurons. When an object moved at a different speed or in a different direction than the background, an impulse was generated. However, when the object moved like the background, inhibition occurred and no optic nerve signal was generated.

At least in some vertebrates, dopamine secreted between the inner nuclear and the inner plexiform layers of the retina (Figure 12–2) diffuses throughout the retina and affects the structure of gap junctions. These junctions allow current to pass freely through horizontal cells in the dark, enlarging the receptive fields of the photoreceptors. Light reduces the current flow, decoupling the horizontal cells, and this decoupling appears to be due to increased release of dopamine in daylight.

RESPONSES IN THE VISUAL PATHWAYS & CORTEX

PATHWAYS TO THE CORTEX

The axons of retinal ganglion cells project a detailed spatial representation of the retina on the lateral geniculate body. Each geniculate body contains six well-defined layers (Figure 12–17). Layers 3–6 have small cells and are called parvocellular, whereas layers 1 and 2 have large cells and are called magnocellular. On each side, layers 1, 4, and 6 receive input from the contralateral eye, whereas layers 2, 3, and 5 receive input from the ipsilateral eye. In each layer, there is a precise point-for-point representation of the retina, and all six layers are in register so that along a line perpendicular to the layers, the receptive fields of the cells in each layer are almost identical. It is worth noting that only 10–20% of the input to the lateral geniculate nucleus comes from the retina. Major inputs also occur from the visual cortex and other brain regions. The feedback pathway from the visual cortex has been shown to be involved in visual processing related to the perception of orientation and motion.

Two kinds of ganglion cells can be distinguished in the retina: large ganglion cells (magno, or M cells), which add responses from different kinds of cones and are concerned with movement and stereopsis; and small ganglion cells (parvo, or P cells), which subtract input from one type of cone from input from another and are concerned with color, texture, and shape. The M ganglion cells project to the magnocellular portion of the lateral geniculate, whereas the P ganglion cells project to the parvocellular portion. From the lateral geniculate nucleus, a magnocellular pathway and a parvocellular pathway project to the visual cortex. The magnocellular pathway, from layers 1 and 2, carries signals for detection of movement, depth, and flicker. The parvocellular pathway, from layers 3–6, carries signals for color vision, texture, shape, and fine detail.

Cells in the interlaminar region of the lateral geniculate nucleus also receive input from P ganglion cells, probably via

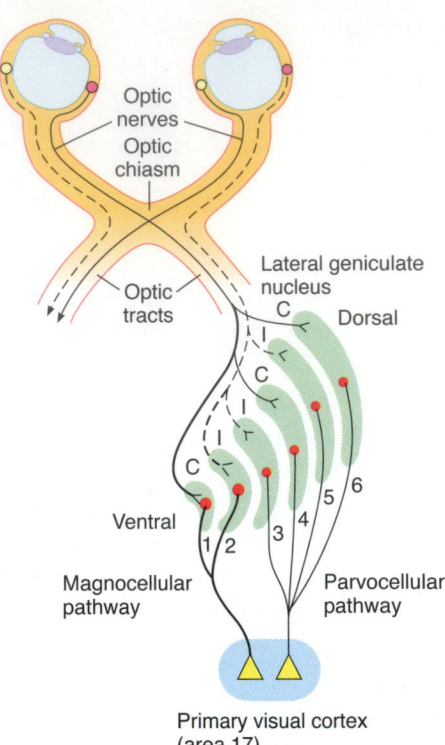

FIGURE 12–17 **Ganglion cell projections from the right hemiretina of each eye to the right lateral geniculate body and from this nucleus to the right primary visual cortex.** Note the six layers of the geniculate. P ganglion cells project to layers 3–6, and M ganglion cells project to layers 1 and 2. The ipsilateral (I) and contralateral (C) eyes project to alternate layers. Not shown are the interlaminar area cells, which project via a separate component of the P pathway to blobs in the visual cortex. (Modified from Kandel ER, Schwartz JH, Jessell TM [editors]: *Principles of Neural Science,* 4th ed. McGraw-Hill, 2000.)

dendrites of interlaminar cells that penetrate the parvocellular layers. They project via a separate component of the P pathway to the blobs in the visual cortex.

PRIMARY VISUAL CORTEX

Just as the ganglion cell axons project a detailed spatial representation of the retina on the lateral geniculate body, the lateral geniculate body projects a similar point-for-point representation on the primary visual cortex (Figure 12–5). In the visual cortex, many nerve cells are associated with each incoming fiber. Like the rest of the neocortex, the visual cortex has six layers. The axons from the lateral geniculate nucleus that form the magnocellular pathway end in layer 4, specifically in its deepest part, layer 4C. Many of the axons that form the parvocellular pathway also end in layer 4C. However, the axons from the interlaminar region end in layers 2 and 3.

Layers 2 and 3 of the cortex contain clusters of cells about 0.2 mm in diameter that, unlike the neighboring cells, contain a high concentration of the mitochondrial enzyme cytochrome oxidase. The clusters have been named **blobs.** They are arranged in a mosaic in the visual cortex and are concerned with color vision. However, the parvocellular pathway also carries color opponent data to the deep part of layer 4.

Like the ganglion cells, the lateral geniculate neurons and the neurons in layer 4 of the visual cortex respond to stimuli in their receptive fields with on centers and inhibitory surrounds or off centers and excitatory surrounds. A bar of light covering the center is an effective stimulus for them because it stimulates the entire center and relatively little of the surround. However, the bar has no preferred orientation and, as a stimulus, is equally effective at any angle.

The responses of the neurons in other layers of the visual cortex are strikingly different. So-called **simple cells** respond to bars of light, lines, or edges, but only when they have a particular orientation. When, for example, a bar of light is rotated as little as 10 degrees from the preferred orientation, the firing rate of the simple cell is usually decreased, and if the stimulus is rotated much more, the response disappears. There are also **complex cells,** which resemble simple cells in requiring a preferred orientation of a linear stimulus but are less dependent upon the location of a stimulus in the visual field than the simple cells and the cells in layer 4. They often respond maximally when a linear stimulus is moved laterally without a change in its orientation. They probably receive input from the simple cells.

The visual cortex, like the somatosensory cortex, is arranged in vertical columns that are concerned with orientation (**orientation columns**). Each is about 1 mm in diameter. However, the orientation preferences of neighboring columns differ in a systematic way; as one moves from column to column across the cortex, sequential changes occur in orientation preference of 5–10 degrees. Thus, it seems likely that for each ganglion cell receptive field in the visual field, there is a collection of columns in a small area of visual cortex representing the possible preferred orientations at small intervals throughout the full 360 degrees. The simple and complex cells have been called **feature detectors** because they respond to and analyze certain features of the stimulus. Feature detectors are also found in the cortical areas for other sensory modalities.

The orientation columns can be mapped with the aid of radioactive 2-deoxyglucose. The uptake of this glucose derivative is proportionate to the activity of neurons. When this technique is employed in animals exposed to uniformly oriented visual stimuli such as vertical lines, the brain shows a remarkable array of intricately curved but evenly spaced orientation columns over a large area of the visual cortex.

Another feature of the visual cortex is the presence of **ocular dominance columns.** The geniculate cells and the cells in layer 4 receive input from only one eye, and the layer 4 cells alternate with cells receiving input from the other eye. If a large amount of a radioactive amino acid is injected into one eye, the amino acid is incorporated into protein and transported by axoplasmic flow to the ganglion cell terminals, across the geniculate synapses, and along the geniculocalcarine fibers to the visual cortex. In layer 4, labeled endings from the injected eye alternate with unlabeled endings from the uninjected eye. The result, when viewed from above, is a vivid pattern of stripes that

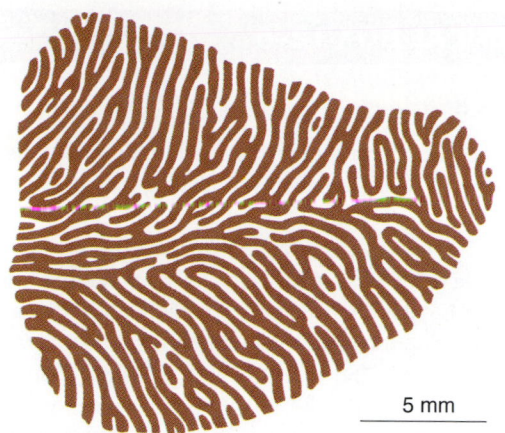

FIGURE 12–18 **Reconstruction of ocular dominance columns in a subdivision of layer 4 of a portion of the right visual cortex of a rhesus monkey.** Dark stripes represent one eye, light stripes the other. (Reproduced with permission from LeVay S, Hubel DH, Wiesel TN: The pattern of ocular dominance columns in macaque visual cortex revealed by a reduced silver stain. J Comp Neurol 1975;159:559.)

covers much of the visual cortex (Figure 12–18) and is separate from and independent of the grid of orientation columns.

About half the simple and complex cells receive an input from both eyes. The inputs are identical or nearly so in terms of the portion of the visual field involved and the preferred orientation. However, they differ in strength, so that between the cells to which the input comes totally from the ipsilateral or the contralateral eye, there is a spectrum of cells influenced to different degrees by both eyes.

Thus, the primary visual cortex segregates information about color from that concerned with form and movement, combines the input from the two eyes, and converts the visual world into short line segments of various orientations.

OTHER CORTICAL AREAS CONCERNED WITH VISION

As mentioned above, the primary visual cortex (V1) projects to many other parts of the occipital lobes and other parts of the brain. These are often identified by number (V2, V3, etc) or by letters (LO, MT, etc). The distribution of some of these in the human brain is shown in Figure 12–19, and their putative functions are listed in Table 12–1. Studies of these areas have been carried out in monkeys trained to do various tasks and then fitted with implanted microelectrodes. In addition, the availability of PET and functional magnetic resonance imaging (fMRI) scanning has made it possible to conduct sophisticated experiments on visual cognition and other cortical visual functions in normal, conscious humans. The visual projections from V1 can be divided roughly into a **dorsal** or **parietal pathway,** concerned primarily with motion, and a **ventral** or **temporal pathway,** concerned with shape and recognition of forms and faces. In addition, connections to the sensory areas are important. For example, in the occipital cortex, visual responses to an object are better if the object is felt at the same time. There are many other relevant connections to other systems.

It is apparent from the preceding paragraphs that parallel processing of visual information occurs along multiple paths. In some as yet unknown way, all the information is eventually pulled together into what we experience as a conscious visual image.

COLOR VISION

CHARACTERISTICS OF COLOR

Colors have three attributes: **hue, intensity,** and **saturation** (degree of freedom from dilution with white). For any color there is

FIGURE 12–19 Some of the main areas to which the primary visual cortex (V1) projects in the human brain. Lateral and medial views. See also Table 8–1. (Modified from Logothetis N: Vision: A window on consciousness. Sci Am [Nov] 1999;281:99.)

TABLE 12–1 Functions of visual projection areas in the human brain.

V1	Primary visual cortex; receives input from lateral geniculate nucleus, begins processing in terms of orientation, edges, etc
V2, V3, VP	Continued processing, larger visual fields
V3A	Motion
V4v	Unknown
MT/V5	Motion; control of movement
LO	Recognition of large objects
V7	Unknown
V8	Color vision

Modified from Logothetis N: Vision: a window on consciousness. Sci Am (Nov) 1999;281:99.

a **complementary color** that, when properly mixed with it, produces a sensation of white. Black is the sensation produced by the absence of light, but it is probably a positive sensation because the blind eye does not "see black;" rather, it "sees nothing."

Another observation of basic importance is the demonstration that the sensation of white, any spectral color, and even the extraspectral color, purple, can be produced by mixing various proportions of red light (wavelength 723–647 nm), green light (575–492 nm), and blue light (492–450 nm). Red, green, and blue are therefore called the **primary colors.** A third important point is that the color perceived depends in part on the color of other objects in the visual field. Thus, for example, a red object is seen as red if the field is illuminated with green or blue light, but as pale pink or white if the field is illuminated with red light. Clinical Box 12–6 describes color blindness.

RETINAL MECHANISMS

The **Young–Helmholtz theory** of color vision in humans postulates the existence of three kinds of cones, each containing a different photopigment and that are maximally sensitive to one of the three primary colors, with the sensation of any given color being determined by the relative frequency of the impulses from each of these cone systems. The correctness of this theory has been demonstrated by the identification and chemical characterization of each of the three pigments (Figure 12–20). One pigment (the blue-sensitive, or short-wave, pigment) absorbs light maximally in the blue-violet portion of the spectrum. Another (the green-sensitive, or middle-wave, pigment) absorbs maximally in the green portion. The third (the red-sensitive, or long-wave, pigment) absorbs maximally in the yellow portion. Blue, green, and red are the primary colors, but the cones with their maximal sensitivity in the yellow portion of the spectrum are sensitive enough in the red portion to respond to red light at a lower threshold than green. This is all the Young–Helmholtz theory requires.

Color Blindness

The most common test for **color blindness** uses the **Ishihara charts,** which are plates containing figures made up of colored spots on a background of similarly shaped colored spots. The figures are intentionally made up of colors that are liable to look the same as the background to an individual who is color blind. Some color-blind individuals are unable to distinguish certain colors, whereas others have only a color weakness. The prefixes "prot-," "deuter-," and "trit-" refer to defects of the red, green, and blue cone systems, respectively. Individuals with normal color vision are called **trichromats. Dichromats** are individuals with only two cone systems; they may have protanopia, deuteranopia, or tritanopia. **Monochromats** have only one cone system. Dichromats can match their color spectrum by mixing only two primary colors, and monochromats match theirs by varying the intensity of only one. Abnormal color vision is present as an inherited abnormality in Caucasian populations in about 8% of the males and 0.4% of the females. Tritanopia is rare and shows no sexual selectivity. However, about 2% of the color-blind males are dichromats who have protanopia or deuteranopia, and about 6% are anomalous trichromats in whom the red-sensitive or the green-sensitive pigment is shifted in its spectral sensitivity. These abnormalities are inherited as recessive and X-linked characteristics. Color blindness is present in males if the X chromosome has the abnormal gene. Females show a defect only when both X chromosomes contain the abnormal gene. However, female children of a man with X-linked color blindness are carriers of the color blindness and pass the defect on to half of their sons. Therefore, X-linked color blindness skips generations and appears in males of every second generation. Color blindness can also occur in individuals with lesions of area V8 of the visual cortex since this region appears to be uniquely concerned with color vision in humans. This deficit is called **achromatopsia.** Transient blue-green color weakness occurs as a side effect in individuals taking sildenafil (Viagra) for the treatment of erectile dysfunction because the drug inhibits the retinal as well as the penile form of phosphodiesterase.

The gene for human rhodopsin is on chromosome 3, and the gene for the blue-sensitive S cone pigment is on chromosome 7. The other two cone pigments are encoded by genes arranged in tandem on the q arm of the X chromosome. The green-sensitive M and red-sensitive L pigments are very similar in structure; their opsins show 96% homology of amino acid sequences, whereas each of these pigments has only about 43% homology with the opsin of blue-sensitive pigment, and all three have about 41% homology with rhodopsin. Many mammals are **dichromats;** that is, they have only two cone pigments, a short-wave and a long-wave pigment.

FIGURE 12–20 **Absorption spectra of the three cone pigments in the human retina.** The S pigment that peaks at 440 nm senses blue, and the M pigment that peaks at 535 nm senses green. The remaining L pigment peaks in the yellow portion of the spectrum, at 565 nm, but its spectrum extends far enough into the long wavelengths to sense red. (Reproduced with permission from Michael CR: Color vision. N Engl J Med 1973;288:724.)

Old World monkeys, apes, and humans are **trichromats,** with separate middle- and long-wave pigments—in all probability because there was duplication of the ancestral long-wave gene followed by divergence.

There are variations in the red, long-wave pigment in humans. It has been known for some time that responses to the **Rayleigh match,** the amounts of red and green light that a subject mixes to match a monochromatic orange, are bimodal. This correlates with new evidence that 62% of otherwise color-normal individuals have serine at site 180 of their long-wave cone opsin, whereas 38% have alanine. The absorption curve of the subjects with serine at position 180 peaks at 556.7 nm, and they are more sensitive to red light, whereas the absorption curve of the subjects with alanine at position 180 peaks at 552.4 nm.

NEURAL MECHANISMS

Color is mediated by ganglion cells that subtract or add input from one type of cone to input from another type. Processing in the ganglion cells and the lateral geniculate nucleus produces impulses that pass along three types of neural pathways that project to V1: a red–green pathway that signals differences between L- and M-cone responses, a blue–yellow pathway that signals differences between S-cone and the sum of L- and M-cone responses, and a luminance pathway that signals the sum of L- and M-cone responses. These pathways project to the blobs and the deep portion of layer 4C of V1. From the blobs and layer 4, color information is projected to V8. However, it is not known how V8 converts color input into the sensation of color.

OTHER ASPECTS OF VISUAL FUNCTION

DARK ADAPTATION

If a person spends a considerable length of time in brightly lighted surroundings and then moves to a dimly lighted environment, the retinas slowly become more sensitive to light as the individual becomes "accustomed to the dark." This decline in visual threshold is known as **dark adaptation.** It is nearly maximal in about 20 minutes, although some further decline occurs over longer periods. On the other hand, when one passes suddenly from a dim to a brightly lighted environment, the light seems intensely and even uncomfortably bright until the eyes adapt to the increased illumination and the visual threshold rises. This adaptation occurs over a period of about 5 minutes and is called **light adaptation,** although, strictly speaking, it is merely the disappearance of dark adaptation.

The dark adaptation response actually has two components. The first drop in visual threshold, rapid but small in magnitude, is known to be due to dark adaptation of the cones because when only the foveal, rod-free portion of the retina is tested, the decline proceeds no further. In the peripheral portions of the retina, a further drop occurs as a result of adaptation of the rods. The total change in threshold between the light-adapted and the fully dark-adapted eye is very great.

Radiologists, aircraft pilots, and others who need maximal visual sensitivity in dim light can avoid having to wait 20 minutes in the dark to become dark-adapted if they wear red goggles when in bright light. Light wavelengths in the red end of the spectrum stimulate the rods to only a slight degree while permitting the cones to function reasonably well. Therefore, a person wearing red glasses can see in bright light during the time it takes for the rods to become dark-adapted.

The time required for dark adaptation is determined in part by the time required to build up the rhodopsin stores. In bright light, much of the pigment is continuously being broken down, and some time is required in dim light for accumulation of the amounts necessary for optimal rod function. However, dark adaptation also occurs in the cones, and additional factors are undoubtedly involved.

CRITICAL FUSION FREQUENCY

The time-resolving ability of the eye is determined by measuring the **critical fusion frequency (CFF),** the rate at which stimuli can be presented and still be perceived as separate stimuli. Stimuli presented at a higher rate than the CFF are perceived as continuous stimuli. Motion pictures move because the frames are presented at a rate above the CFF, and movies begin to flicker when the projector slows down.

VISUAL FIELDS & BINOCULAR VISION

The visual field of each eye is the portion of the external world visible out of that eye. Theoretically, it should be circular, but actually it is cut off medially by the nose and superiorly by the roof of the orbit (Figure 12–21). Mapping the visual fields is important in neurologic diagnosis. The peripheral portions of the visual fields are mapped with an instrument called a **perimeter,** and the process is referred to as **perimetry.** One eye is covered while the other is fixed on a central point. A small

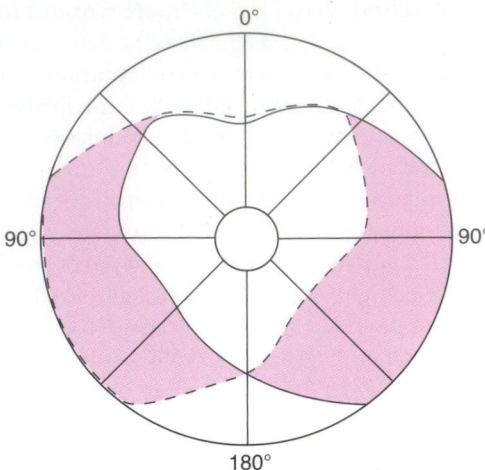

FIGURE 12–21   Monocular and binocular visual fields. The dashed line encloses the visual field of the left eye; the solid line, that of the right eye. The common area (heart-shaped clear zone in the center) is viewed with binocular vision. The colored areas are viewed with monocular vision.

target is moved toward this central point along selected meridians, and, along each, the location where the target first becomes visible is plotted in degrees of arc away from the central point (Figure 12–21). The central visual fields are mapped with a **tangent screen,** a black felt screen across which a white target is moved. By noting the locations where the target disappears and reappears, the blind spot and any **objective scotomas** (blind spots due to disease) can be outlined.

The central parts of the visual fields of the two eyes coincide; therefore, anything in this portion of the field is viewed with **binocular vision.** The impulses set up in the two retinas by light rays from an object are fused at the cortical level into a single image **(fusion).** The points on the retina on which the image of an object must fall if it is to be seen binocularly as a single object are called **corresponding points.** If one eye is gently pushed out of the line while staring fixedly at an object in the center of the visual field, double vision **(diplopia)** results; the image on the retina of the eye that is displaced no longer falls on the corresponding point. When visual images no longer fall on corresponding retinal points, strabismus occurs (see Clinical Box 12–3).

Binocular vision has an important role in the perception of depth. However, depth perception also has numerous monocular components, such as the relative sizes of objects, the degree one looks down at them, their shadows, and, for moving objects, their movement relative to one another (movement parallax).

EFFECT OF LESIONS IN THE OPTIC PATHWAYS

The anatomy of the pathways from the eyes to the brain is shown in Figure 12–4. Lesions along these pathways can be lo-

calized with a high degree of accuracy by the effects they produce in the visual fields.

The fibers from the nasal half of each retina decussate in the optic chiasm, so that the fibers in the optic tracts are those from the temporal half of one retina and the nasal half of the other. In other words, each optic tract subserves half of the field of vision. Therefore, a lesion that interrupts one optic nerve causes blindness in that eye, but a lesion in one optic tract causes blindness in half of the visual field (Figure 12–4). This defect is classified as a **homonymous** (same side of both visual fields) **hemianopia** (half-blindness). Lesions affecting the optic chiasm, such as pituitary tumors expanding out of the sella turcica, cause destruction of the fibers from both nasal hemiretinas and produce a **heteronymous** (opposite sides of the visual fields) **hemianopia.** Because the fibers from the maculas are located posteriorly in the optic chiasm, hemianopic scotomas develop before vision in the two hemiretinas is completely lost. Selective visual field defects are further classified as bitemporal, binasal, and right or left.

The optic nerve fibers from the upper retinal quadrants subserving vision in the lower half of the visual field terminate in the medial half of the lateral geniculate body, whereas the fibers from the lower retinal quadrants terminate in the lateral half. The geniculocalcarine fibers from the medial half of the lateral geniculate terminate on the superior lip of the calcarine fissure, and those from the lateral half terminate on the inferior lip. Furthermore, the fibers from the lateral geniculate body that subserve macular vision separate from those that subserve peripheral vision and end more posteriorly on the lips of the calcarine fissure (Figure 12–5). Because of this anatomic arrangement, occipital lobe lesions may produce discrete quadrantic visual field defects (upper and lower quadrants of each half visual field). **Macular sparing,** that is, loss of peripheral vision with intact macular vision, is also common with occipital lesions (Figure 12–4), because the macular representation is separate from that of the peripheral fields and very large relative to that of the peripheral fields. Therefore, occipital lesions must extend considerable distances to destroy macular as well as peripheral vision. Bilateral destruction of the occipital cortex in humans causes subjective blindness. However, there is appreciable **blindsight,** that is, residual responses to visual stimuli even though they do not reach consciousness. For example, when these individuals are asked to guess where a stimulus is located during perimetry, they respond with much more accuracy than can be explained by chance. They are also capable of considerable discrimination of movement, flicker, orientation, and even color. Similar biasing of responses can be produced by stimuli in the blind areas in patients with hemianopia due to lesions in the visual cortex.

The fibers to the pretectal region that subserve the reflex pupillary constriction produced by shining a light into the eye leave the optic tracts near the geniculate bodies. Therefore, blindness with preservation of the pupillary light reflex is usually due to bilateral lesions behind the optic tract.

EYE MOVEMENTS

The eye is moved within the orbit by six ocular muscles (Figure 12–22). These are innervated by the oculomotor, trochlear, and abducens nerves. Because the oblique muscles pull medially, their actions vary with the position of the eye. When the eye is turned nasally, the inferior oblique elevates it and the superior oblique depresses it. When it is turned laterally, the superior rectus elevates it and the inferior rectus depresses it.

Because much of the visual field is binocular, it is clear that a very high order of coordination of the movements of the two eyes is necessary if visual images are to fall at all times on corresponding points in the two retinas and diplopia is to be avoided.

There are four types of eye movements, each controlled by a different neural system but sharing the same final common path, the motor neurons that supply the external ocular muscles. **Saccades,** sudden jerky movements, occur as the gaze shifts from one object to another. They bring new objects of interest onto the fovea and reduce adaptation in the visual pathway that would occur if gaze were fixed on a single object for long periods. **Smooth pursuit movements** are tracking movements of the eyes as they follow moving objects. **Vestibular movements,** adjustments that occur in response to stimuli initiated in the semicircular canals, maintain visual fixation as the head moves. **Convergence movements** bring the visual axes toward each other as attention is focused on objects near the observer. The similarity to a human-made tracking system on an unstable platform such as a ship is apparent: saccadic movements seek out visual targets, pursuit movements follow them as they move about, and vestibular movements stabilize the tracking device as the platform on which the device is mounted (ie, the head) moves about. In primates, these eye movements depend on an intact visual cortex. Saccades are programmed in the frontal cortex and the superior colliculi and pursuit movements in the cerebellum.

SUPERIOR COLLICULI

The superior colliculi, which regulate saccades, are innervated by M fibers from the retina. They also receive extensive innervation from the cerebral cortex. Each superior colliculus has a map of visual space plus a map of the body surface and a map for sound in space. A motor map projects to the regions of the brain stem that control eye movements. There are also projections via the tectopontine tract to the cerebellum and via the tectospinal tract to areas concerned with reflex movements of the head and neck. The superior colliculi are constantly active positioning the eyes, and they have one of the highest rates of blood flow and metabolism of any region in the brain.

CHAPTER SUMMARY

- The major parts of the eye are the sclera (protective covering), cornea (transfer light rays), choroids (nourishment), retina (receptor cells), lens, and iris.

- The visual pathway is from the rods and cones to bipolar cells to ganglion cells then via the optic tract to the thalamic lateral geniculate body to the occipital lobe of the cerebral cortex. The fibers from each nasal hemiretina decussate in the optic chiasm; the fibers from the nasal half of one retina and the temporal half of the other synapse on the cells whose axons form the geniculolocalcarine tract.

- The bending of light rays (refraction) allows one to focus an accurate image onto the retina. Light is refracted at the anterior surface of the cornea and at the anterior and posterior surfaces of the lens. To bring diverging rays from close objects to a focus on the retina, the curvature of the lens is increased, a process called accommodation.

- In hyperopia (farsightedness), the eyeball is too short and light rays come to a focus behind the retina. In myopia (nearsightedness), the anteroposterior diameter of the eyeball is too long. Astigmatism is a common condition in which the curvature of the cornea is not uniform. Presbyopia is the loss of accommodation for near vision. Strabismus is squinting in an attempt to correct visual acuity.

- Na^+ channels in the outer segments of the rods and cones are open in the dark, so current flows from the inner to the outer segment. When light strikes the outer segment, some of the Na^+ channels are closed and the cells are hyperpolarized.

- In response to light, horizontal cells are hyperpolarized, bipolar cells are either hyperpolarized or depolarized, and amacrine cells are depolarized and develop spikes that may act as generator

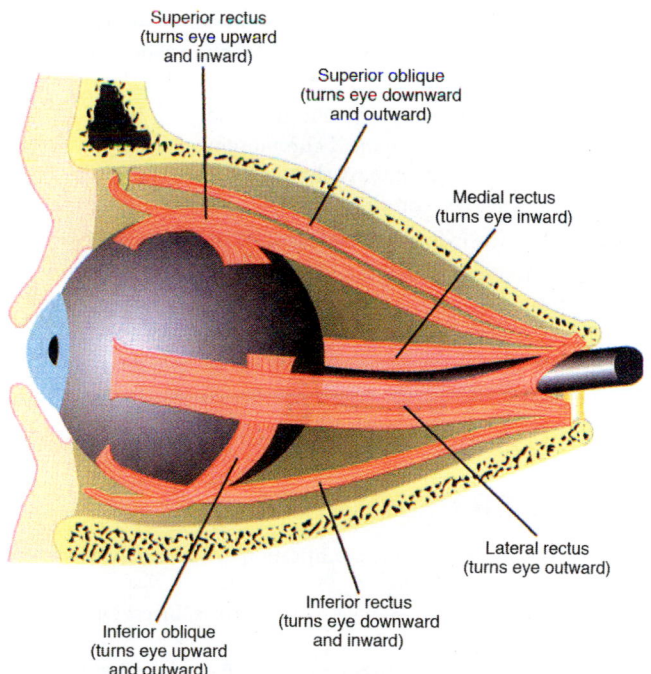

FIGURE 12–22 Extraocular muscles subserving the six cardinal positions of gaze. The eye is adducted by the medial rectus and abducted by the lateral rectus. The adducted eye is elevated by the inferior oblique and depressed by the superior oblique; the abducted eye is elevated by the superior rectus and depressed by the inferior rectus. (From Squire LR, et al [editors]: *Fundamental Neuroscience*, 3rd ed. Academic Press, 2008.)

Superior rectus
(turns eye upward and inward)

Superior oblique
(turns eye downward and outward)

Medial rectus
(turns eye inward)

Lateral rectus
(turns eye outward)

Inferior rectus
(turns eye downward and inward)

Inferior oblique
(turns eye upward and outward)

potentials for the propagated spikes produced in the ganglion cells.

- Neurons in layer 4 of the visual cortex respond to stimuli in their receptive fields with on centers and inhibitory surrounds or off centers and excitatory surrounds. Neurons in other layers are called simple cells if they respond to bars of light, lines, or edges, but only when they have a particular orientation. Complex cells also require a preferred orientation of a linear stimulus but are less dependent on the location of a stimulus in the visual field.

- Projections from V1 can be divided into a dorsal or parietal pathway (concerned primarily with motion) and a ventral or temporal pathway (concerned with shape and recognition of forms and faces).

- The decline in visual threshold after spending long periods of time in a dimly lit room is called dark adaptation.

- The Young–Helmholtz theory of color vision postulates the existence of three kinds of cones, each containing a different photopigment and that are maximally sensitive to one of the three primary colors, with the sensation of any given color being determined by the relative frequency of the impulses from each of these cone systems.

- Saccades (sudden jerky movements) occur as the gaze shifts from one object to another, and they reduce adaptation in the visual pathway that would occur if gaze were fixed on a single object for long periods. Smooth pursuit movements are tracking movements of the eyes as they follow moving objects. Vestibular movements occur in response to stimuli in the semicircular canals to maintain visual fixation as the head moves. Convergence movements bring the visual axes toward each other as attention is focused on objects near the observer.

MULTIPLE-CHOICE QUESTIONS

For all questions, select the single best answer unless otherwise directed.

1. A visual exam in an 80-year-old man shows he has a reduced ability to see objects in the upper and lower quadrants of the left visual fields of both eyes but some vision remains in the central regions of the visual field. The diagnosis is
   A) central scotoma.
   B) heteronymous hemianopia with macular sparing.
   C) lesion of the optic chiasm.
   D) homonymous hemianopia with macular sparing.
   E) retinopathy.

2. Visual accommodation involves
   A) increased tension on the lens ligaments.
   B) a decrease in the curvature of the lens.
   C) relaxation of the sphincter muscle of the iris.
   D) contraction of the ciliary muscle.
   E) increased intraocular pressure.

3. The fovea of the eye
   A) has the lowest light threshold.
   B) is the region of highest visual acuity.
   C) contains only red and green cones.
   D) contains only rods.
   E) is situated over the head of the optic nerve.

4. Which of the following parts of the eye has the greatest concentration of rods?
   A) ciliary body
   B) iris
   C) optic disk
   D) fovea
   E) parafoveal region

5. The correct sequence of events involved in phototransduction in rods and cones in response to light is:
   A) activation of transducin, decreased release of glutamate, structural changes in rhodopsin, closure of Na^+ channels, and decrease in intracellular cGMP.
   B) decreased release of glutamate, activation of transducin, closure of Na^+ channels, decrease in intracellular cGMP, and structural changes in rhodopsin.
   C) structural changes in rhodopsin, decrease in intracellular cGMP, decreased release of glutamate, closure of Na^+ channels, and activation of transducin.
   D) structural changes in rhodopsin, activation of transducin, decrease in intracellular cGMP, closure of Na^+ channels, and decreased release of glutamate.
   E) activation of transducin, structural changes in rhodopsin, closure of Na^+ channels, decrease in intracellular cGMP, and decreased release of glutamate.

6. Vitamin A is a precursor for the synthesis of
   A) somatostatin.
   B) $retinene_1$.
   C) the pigment of the iris.
   D) scotopsin.
   E) aqueous humor.

7. Abnormal color vision is 20 times more common in men than women because most cases are caused by an abnormal
   A) dominant gene on the Y chromosome.
   B) recessive gene on the Y chromosome.
   C) dominant gene on the X chromosome.
   D) recessive gene on the X chromosome.
   E) recessive gene on chromosome 22.

8. Which of the following is *not* involved in color vision?
   A) activation of a pathway that signals differences between S cone responses and the sum of L and M cone responses
   B) geniculate layers 3–6
   C) P pathway
   D) area V3A of visual cortex
   E) area V8 of visual cortex

CHAPTER RESOURCES

Chiu C, Weliky M: Synaptic modification by vision. Science 2003;300:1890.

Dowling JE: Organization of vertebrate retinas. Invest Ophthalmol 1970;9:655

Dowling JE: *The Retina: An Approachable Part of the Brain.* Belknap, 1987.

Gegenfurtner KR, Kiper DC: Color vision. Annu Rev Neurosci 2003;26:181.

Hubel DH: *Eye, Brain, and Vision.* Scientific American Library, 1988.

Kandel ER, Schwartz JH, Jessell TM (editors): *Principles of Neural Science,* 4th ed. McGraw-Hill, 2000.

Lamb TD: Electrical responses of photoreceptors. In: *Recent Advances in Physiology.* No.10. Baker PF (editor). Churchill Livingstone, 1984.

LeVay S, Hubel DH, Wiesel TN: The pattern of ocular dominance columns in macaque visual cortex revealed by a reduced silver stain. J Comp Neurol 1975;159:559.

Logothetis N: Vision: A window on consciousness. Sci Am 1999;281:99.

Oyster CW: *The Human Eye: Structure and Function.* Sinauer, 1999.

Squire LR, et al (editors): *Fundamental Neuroscience,* 3rd ed. Academic Press, 2008.

Widmaier EP, Raff H, Strang KT: *Vander's Human Physiology,* 11th ed. McGraw-Hill, 2008.

13

Hearing & Equilibrium

OBJECTIVES

After studying this chapter, you should be able to:

- Describe the components and functions of the external, middle, and inner ear.
- Describe the way that movements of molecules in the air are converted into impulses generated in hair cells in the cochlea.
- Trace the path of auditory impulses in the neural pathways from the cochlear hair cells to the auditory cortex, and discuss the function of the auditory cortex.
- Explain how pitch, loudness, and timbre are coded in the auditory pathways.
- Describe the various forms of deafness.
- Explain how the receptors in the semicircular canals detect rotational acceleration and how the receptors in the saccule and utricle detect linear acceleration.
- List the major sensory inputs that provide the information which is synthesized in the brain into the sense of position in space.

INTRODUCTION

Receptors for two sensory modalities, hearing and equilibrium, are housed in the ear. The external ear, the middle ear, and the cochlea of the inner ear are concerned with hearing. The semicircular canals, the utricle, and the saccule of the inner ear are concerned with equilibrium. Receptors in the semicircular canals detect rotational acceleration, receptors in the utricle detect linear acceleration in the horizontal direction, and receptors in the saccule detect linear acceleration in the vertical direction. The receptors for hearing and equilibrium are hair cells, six groups of which are present in each inner ear: one in each of the three semicircular canals, one in the utricle, one in the saccule, and one in the cochlea.

ANATOMIC CONSIDERATIONS

EXTERNAL & MIDDLE EAR

The external ear funnels sound waves to the **external auditory meatus** (Figure 13–1)**.** In some animals, the ears can be moved like radar antennas to seek out sound. From the external auditory meatus, sound waves pass inward to the **tympanic membrane** (eardrum).

The middle ear is an air-filled cavity in the temporal bone that opens via the **auditory (eustachian) tube** into the nasopharynx and through the nasopharynx to the exterior. The tube is usually closed, but during swallowing, chewing, and yawning it opens, keeping the air pressure on the two sides of the eardrum equalized. The three **auditory ossicles,** the **malleus, incus,** and **stapes,** are located in the middle ear (Figure 13–2). The **manubrium** (handle of the malleus) is attached to the back of the tympanic membrane. Its head is attached to the wall of the middle ear, and its short process is attached to the incus, which in turn articulates with the head of the stapes. The stapes is named for its resemblance to a stirrup. Its **foot plate** is attached by an annular ligament to the walls of the **oval window.** Two small skeletal muscles, the **tensor tympani** and the **stapedius,** are also located in the middle ear. Contraction of

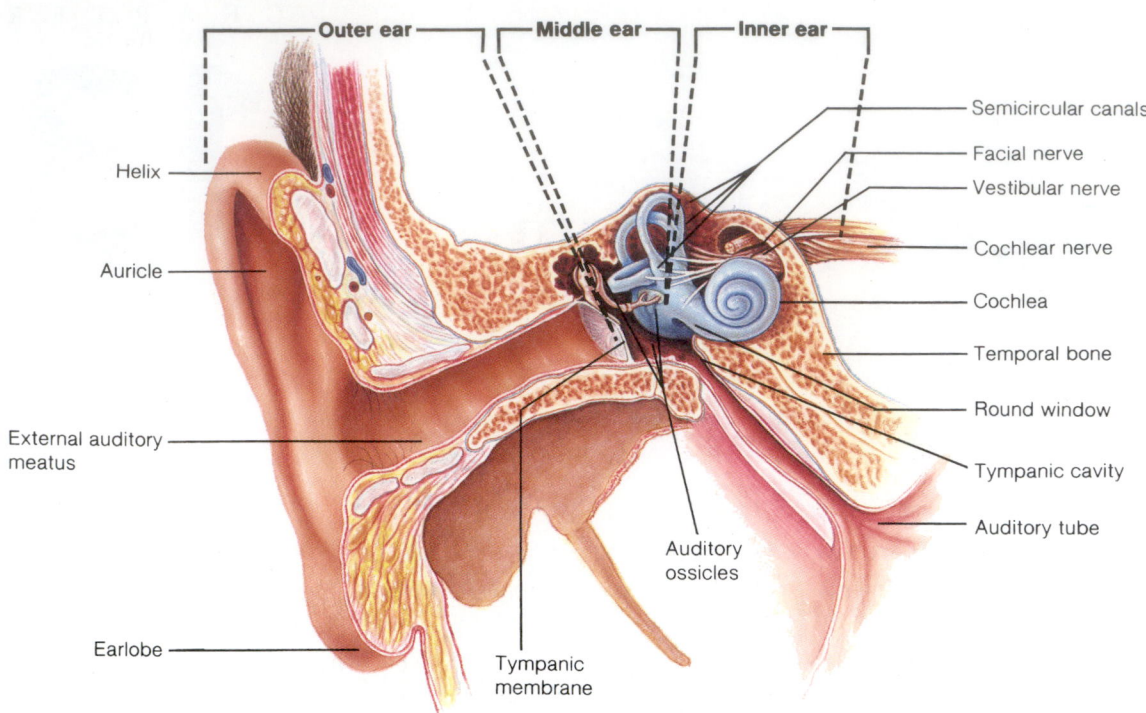

FIGURE 13–1   **The structures of the outer, middle, and inner portions of the human ear.** To make the relationships clear, the cochlea has been turned slightly and the middle ear muscles have been omitted.   (From Fox SI, *Human Physiology.* McGraw-Hill, 2008.)

FIGURE 13–2   **The medial view of the middle ear.** The locations of auditory muscles attached to the middle-ear ossicles are indicated. (From Fox SI, *Human Physiology.* McGraw-Hill, 2008.)

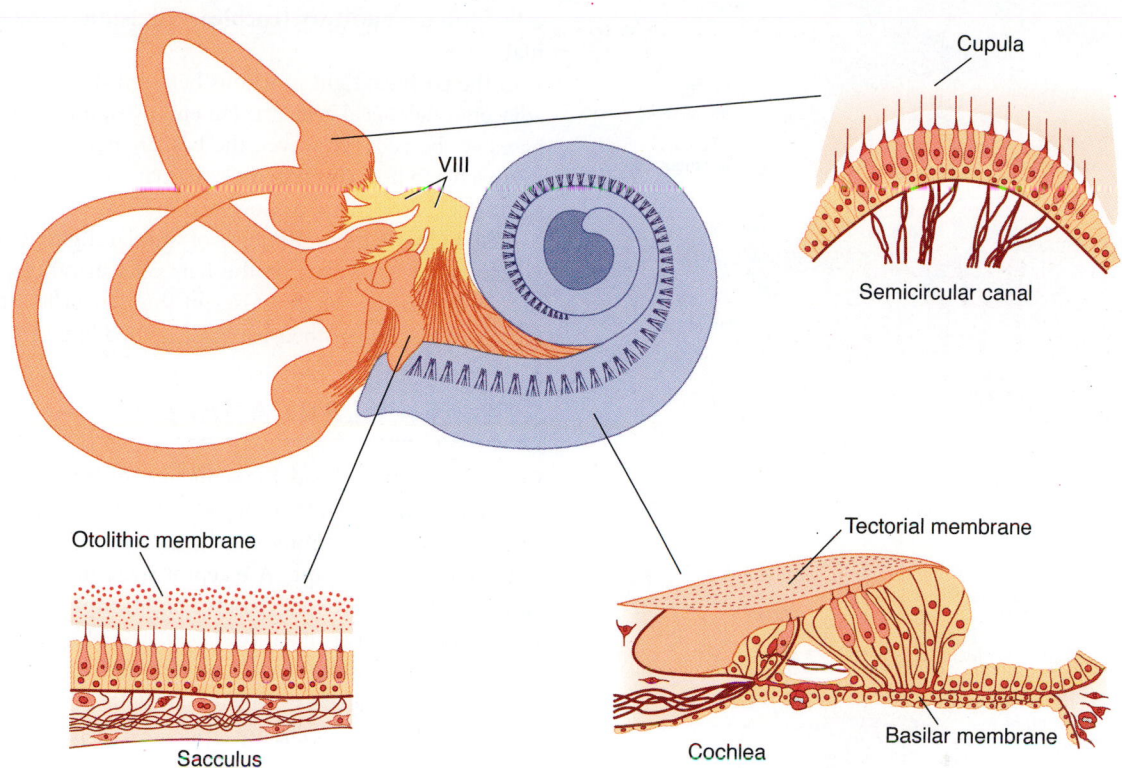

FIGURE 13-3 **Schematic of the human inner ear showing the membranous labyrinth with enlargements of the structures in which hair cells are embedded.** The membranous labyrinth is suspended in perilymph and filled with K$^+$-rich endolymph which bathes the receptors. Hair cells (darkened for emphasis) occur in different arrays characteristic of the receptor organs. The three semicircular canals are sensitive to angular accelerations which deflect the gelatinous cupula and associated hair cells. In the cochlea, hair cells spiral along the basilar membrane in the organ of Corti. Airborne sounds set the eardrum in motion, which is conveyed to the cochlea by bones of the middle ear. This flexes the membrane up and down. Hair cells in the organ of Corti are stimulated by shearing motion. The otolithic organs (saccule and utricle) are sensitive to linear acceleration in vertical and horizontal planes. Hair cells are attached to the otolithic membrane. VIII, eighth cranial nerve, with auditory and vestibular divisions. (Reproduced with permission from Hudspeth AJ: How the ear's works work. Nature 1989;341:397. Copyright © 1989 by Macmillan Magazines.)

the former pulls the manubrium of the malleus medially and decreases the vibrations of the tympanic membrane; contraction of the latter pulls the foot plate of the stapes out of the oval window. The functions of the ossicles and the muscles are considered in more detail below.

INNER EAR

The inner ear (**labyrinth**) is made up of two parts, one within the other. The **bony labyrinth** is a series of channels in the petrous portion of the **temporal bone.** Inside these channels, surrounded by a fluid called **perilymph,** is the **membranous labyrinth** (Figure 13–3). This membranous structure more or less duplicates the shape of the bony channels. It is filled with a K$^+$-rich fluid called **endolymph,** and there is no communication between the spaces filled with endolymph and those filled with perilymph.

COCHLEA

The cochlear portion of the labyrinth is a coiled tube which in humans is 35 mm long and makes a two and three quarter turns.

Throughout its length, the basilar membrane and Reissner's membrane divide it into three chambers or **scalae** (Figure 13–4). The upper **scala vestibuli** and the lower **scala tympani** contain perilymph and communicate with each other at the apex of the cochlea through a small opening called the **helicotrema.** At the base of the cochlea, the scala vestibuli ends at the oval window, which is closed by the footplate of the stapes. The scala tympani ends at the **round window,** a foramen on the medial wall of the middle ear that is closed by the flexible **secondary tympanic membrane.** The **scala media,** the middle cochlear chamber, is continuous with the membranous labyrinth and does not communicate with the other two scalae.

ORGAN OF CORTI

Located on the basilar membrane is the **organ of Corti,** the structure that contains the hair cells, which are the auditory receptors. This organ extends from the apex to the base of the cochlea and consequently has a spiral shape. The processes of the hair cells pierce the tough, membrane-like **reticular lamina** that is supported by the **pillar cells** or **rods of Corti** (Figure

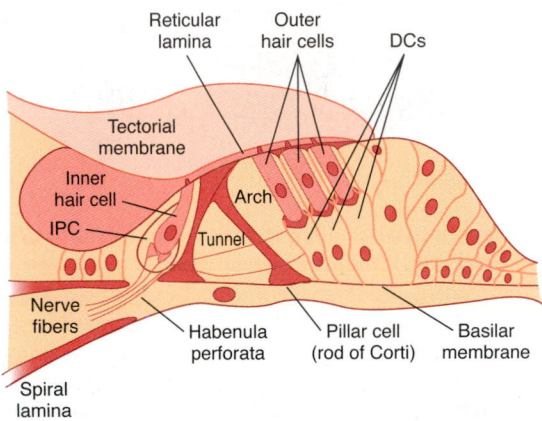

FIGURE 13–4 **Top: Cross-section of the cochlea, showing the organ of Corti and the three scalae of the cochlea. Bottom: Structure of the organ of Corti, as it appears in the basal turn of the cochlea.** DC, outer phalangeal cells (Deiters' cells) supporting outer hair cells; IPC, inner phalangeal cell supporting inner hair cell. (Reproduced with permission from Pickels JO: *An Introduction to the Physiology of Hearing,* 2nd ed. Academic Press, 1988.)

13–4). The hair cells are arranged in four rows: three rows of **outer hair cells** lateral to the tunnel formed by the rods of Corti, and one row of **inner hair cells** medial to the tunnel. There are 20,000 outer hair cells and 3500 inner hair cells in each human cochlea. Covering the rows of hair cells is a thin, viscous, but elastic **tectorial membrane** in which the tips of the hairs of the outer but not the inner hair cells are embedded. The cell bodies of the sensory neurons that arborize around the bases of the hair cells are located in the **spiral ganglion** within the **modiolus,** the bony core around which the cochlea is wound. Ninety to 95% of these sensory neurons innervate the inner hair cells; only 5–10% innervate the more numerous outer hair cells, and each sensory neuron innervates several outer hair cells. By contrast, most of the efferent fibers in the auditory nerve terminate on the outer rather than inner hair cells. The axons of the afferent neurons that innervate the hair cells form the **auditory (cochlear) division** of the eighth cranial nerve.

In the cochlea, tight junctions between the hair cells and the adjacent phalangeal cells prevent endolymph from reaching the bases of the cells. However, the basilar membrane is relatively permeable to perilymph in the scala tympani, and consequently, the tunnel of the organ of Corti and the bases of the hair cells are bathed in perilymph. Because of similar tight junctions, the arrangement is similar for the hair cells in other parts of the inner ear; that is, the processes of the hair cells are bathed in endolymph, whereas their bases are bathed in perilymph.

SEMICIRCULAR CANALS

On each side of the head, the semicircular canals are perpendicular to each other, so that they are oriented in the three planes of space. Inside the bony canals, the membranous canals are suspended in perilymph. A receptor structure, the **crista ampullaris,** is located in the expanded end (**ampulla**) of each of the membranous canals. Each crista consists of hair cells and supporting (sustentacular) cells surmounted by a gelatinous partition (**cupula**) that closes off the ampulla (Figure 13–3). The processes of the hair cells are embedded in the cupula, and the bases of the hair cells are in close contact with the afferent fibers of the **vestibular division** of the eighth cranial nerve.

UTRICLE & SACCULE

Within each membranous labyrinth, on the floor of the utricle, is an **otolithic organ (macula).** Another macula is located on the wall of the saccule in a semivertical position. The maculae contain supporting cells and hair cells, surmounted by an otolithic membrane in which are embedded crystals of calcium carbonate, the **otoliths** (Figure 13–3). The otoliths, which are also called **otoconia** or **ear dust,** range from 3 to 19 μm in length in humans and are more dense than the endolymph. The processes of the hair cells are embedded in the membrane. The nerve fibers from the hair cells join those from the cristae in the vestibular division of the eighth cranial nerve.

HAIR CELLS

STRUCTURE

As noted above, the sensory receptors in the ear consist of six patches of hair cells in the membranous labyrinth. The hair cells in the organ of Corti signal hearing; the hair cells in the utricle signal horizontal acceleration; the hair cells in the saccule signal vertical acceleration; and a patch in each of the three semicircular canals signal rotational acceleration. These hair cells have a common structure (Figure 13–5). Each is embedded in an epithelium made up of supporting cells, with the basal end in close contact with afferent neurons. Projecting from the apical end are 30 to 150 rod-shaped processes, or hairs.

FIGURE 13–5 **Left:** Structure of a hair cell in the saccule. Hair cells in the membranous labyrinth of the ear have a common structure, and each is within an epithelium of supporting cells (SC) surmounted by an otolithic membrane (OM) embedded with crystals of calcium carbonate, the otoliths (OT). Projecting from the apical end are rod-shaped processes, or hair cells (RC), in contact with afferent (A) and efferent (E) nerve fibers. Except in the cochlea, one of these, **kinocilium** (K), is a true but nonmotile cilium with nine pairs of microtubules around its circumference and a central pair of microtubules. The other processes, **stereocilia** (S), are found in all hair cells; they have cores of actin filaments coated with isoforms of myosin. Within the clump of processes on each cell there is an orderly structure. Along an axis toward the kinocilium, the stereocilia increase progressively in height; along the perpendicular axis, all the stereocilia are the same height. (Reproduced with permission from Hillman DE: Morphology of peripheral and central vestibular systems. In: Llinas R, Precht W [editors]: *Frog Neurobiology.* Springer, 1976.) **Right:** Scanning electron photomicrograph of processes on a hair cell in the saccule. The otolithic membrane has been removed. The small projections around the hair cell are microvilli on supporting cells. (Courtesy of AJ Hudspeth.)

Except in the cochlea, one of these, the **kinocilium,** is a true but nonmotile cilium with nine pairs of microtubules around its circumference and a central pair of microtubules. It is one of the largest processes and has a clubbed end. The kinocilium is lost from the hair cells of the cochlea in adult mammals. However, the other processes, which are called **stereocilia,** are present in all hair cells. They have cores composed of parallel filaments of actin. The actin is coated with various isoforms of myosin. Within the clump of processes on each cell there is an orderly structure. Along an axis toward the kinocilium, the stereocilia increase progressively in height; along the perpendicular axis, all the stereocilia are the same height.

ELECTRICAL RESPONSES

The resting membrane potential of the hair cells is about –60 mV. When the stereocilia are pushed toward the kinocilium, the membrane potential is decreased to about –50 mV. When the bundle of processes is pushed in the opposite direction, the cell is hyperpolarized. Displacing the processes in a direction perpendicular to this axis provides no change in membrane potential, and displacing the processes in directions that are intermediate between these two directions produces depolarization or hyperpolarization that is proportionate to the degree to which the direction is toward or away from the kinocilium. Thus, the hair processes provide a mechanism for generating changes in membrane potential proportional to the direction and distance the hair moves.

GENESIS OF ACTION POTENTIALS IN AFFERENT NERVE FIBERS

Very fine processes called **tip links** (Figure 13–6) tie the tip of each stereocilium to the side of its higher neighbor, and at the junction are cation channels in the higher process that appear to be mechanically sensitive. When the shorter stereocilia are

FIGURE 13–6   **Schematic representation of the role of tip links in the responses of hair cells.** When a stereocilium is pushed toward a taller stereocilium, the tip line is stretched and opens an ion channel in its taller neighbor. The channel next is presumably moved down the taller stereocilium by a molecular motor, so the tension on the tip link is released. When the hairs return to the resting position, the motor moves back up the stereocilium. (Modified from Kandel ER, Schwartz JH, Jessel TM [editors]: *Principles of Neuroscience,* 4th ed. McGraw-Hill, 2000.)

pushed toward the higher, the open time of these channels increases. K^+—the most abundant cation in endolymph—and Ca^{2+} enter via the channel and produce depolarization. There is still considerable uncertainty about subsequent events. However, one hypothesis is that a molecular motor in the higher neighbor next moves the channel toward the base, releasing tension in the tip link (Figure 13–6). This causes the channel to close and permits restoration of the resting state. The motor apparently is myosin-based. Depolarization of hair cells causes them to release a neurotransmitter, probably glutamate, which initiates depolarization of neighboring afferent neurons.

The K^+ that enters hair cells via the mechanically sensitive cation channels is recycled (Figure 13–7). It enters supporting cells and then passes on to other supporting cells by way of tight junctions. In the cochlea, it eventually reaches the stria vascularis and is secreted back into the endolymph, completing the cycle.

The processes of the hair cells project into the endolymph whereas the bases are bathed in perilymph. This arrangement is necessary for the normal production of generator potentials. The perilymph is formed mainly from plasma. On the other hand, endolymph is formed in the scala media by the stria vascularis and has a high concentration of K^+ and a low concentration of Na^+ (Figure 13–7). Cells in the stria vascularis have a high concentration of Na^+–K^+ pump. In addition, it appears that a unique electrogenic K^+ pump in the stria vascularis accounts for the fact that the scala media is electrically positive by 85 mV relative to the scala vestibuli and scala tympani.

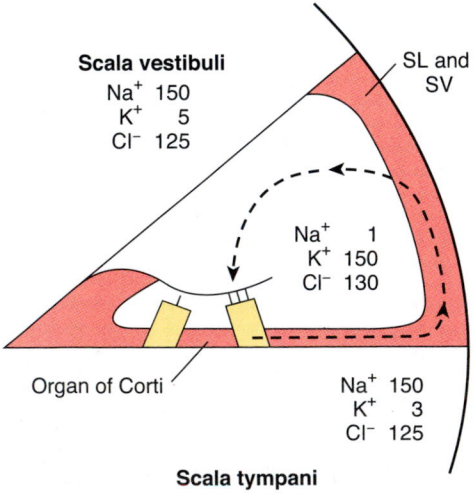

FIGURE 13–7   **Ionic composition of perilymph in the scala vestibuli, endolymph in the scala media, and perilymph in the scala tympani.** SL, spiral ligament. SV, stria vascularis. The dashed arrow indicates the path by which K^+ recycles from the hair cells to the supporting cells to the spiral ligament and is then secreted back into the endolymph by cells in the stria vascularis.

HEARING

SOUND WAVES

Sound is the sensation produced when longitudinal vibrations of the molecules in the external environment—that is, alternate phases of condensation and rarefaction of the molecules—strike the tympanic membrane. A plot of these movements as changes in pressure on the tympanic membrane per unit of time is a series of waves (Figure 13–8); such movements in the environment are generally called sound waves. The waves travel through air at a speed of approximately 344 m/s (770 mph) at 20 °C at sea level. The speed of sound increases with temperature and with altitude. Other media in which humans occasionally find themselves also conduct sound waves but at different speeds. For example, the speed of sound is 1450 m/s at 20 °C in fresh water and is even greater in salt water. It is said that the whistle of the blue whale is as loud as 188 decibels and is audible for 500 miles.

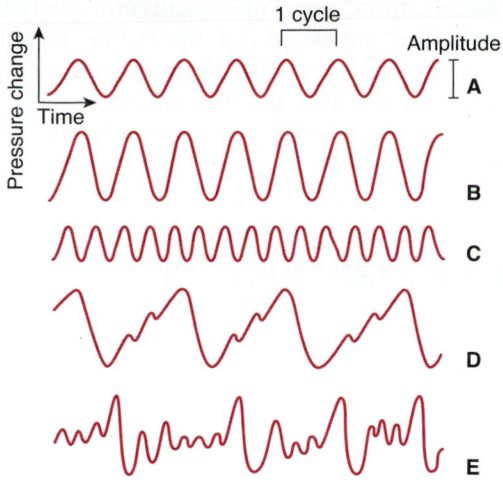

FIGURE 13–8 **Characteristics of sound waves. A** is the record of a pure tone. **B** has a greater amplitude and is louder than **A. C** has the same amplitude as **A** but a greater frequency, and its pitch is higher. **D** is a complex wave form that is regularly repeated. Such patterns are perceived as musical sounds, whereas waves like that shown in **E**, which have no regular pattern, are perceived as noise.

Generally speaking, the **loudness** of a sound is correlated with the **amplitude** of a sound wave and its **pitch** with the **frequency** (number of waves per unit of time). The greater the amplitude, the louder the sound; and the greater the frequency, the higher the pitch. Sound waves that have repeating patterns, even though the individual waves are complex, are perceived as musical sounds; aperiodic nonrepeating vibrations cause a sensation of noise. Most musical sounds are made up of a wave with a primary frequency that determines the pitch of the sound plus a number of harmonic vibrations **(overtones)** that give the sound its characteristic **timbre** (quality). Variations in timbre permit us to identify the sounds of the various musical instruments even though they are playing notes of the same pitch.

The amplitude of a sound wave can be expressed in terms of the maximum pressure change at the eardrum, but a relative scale is more convenient. The **decibel scale** is such a scale. The intensity of a sound in **bels** is the logarithm of the ratio of the intensity of that sound and a standard sound. A decibel (dB) is 0.1 bel. The standard sound reference level adopted by the Acoustical Society of America corresponds to 0 dB at a pressure level of $0.000204 \times$ dyne/cm^2, a value that is just at the auditory threshold for the average human. A value of 0 dB does not mean the absence of sound but a sound level of an intensity equal to that of the standard. The 0- to 140-dB range from threshold pressure to a pressure that is potentially damaging to the organ of Corti actually represents a 10^7 (10 million)-fold variation in sound pressure. Put another way, atmospheric pressure at sea level is 15 lb/in^2 or 1 bar, and the range from the threshold of hearing to potential damage to the cochlea is 0.0002 to 2000 μbar.

A range of 120 to 160 dB (eg, firearms, jackhammer, jet plane on take off) is classified as painful; 90 to 110 dB (eg, subway, bass drum, chain saw, lawn mower) is classified as extremely high; 60 to 80 dB (eg, alarm clock, busy traffic, dishwasher, conversation) is classified as very loud; 40 to 50 dB (eg, moderate rainfall, normal room noise) is moderate; and 30 dB (eg, whisper, library) is faint.

The sound frequencies audible to humans range from about 20 to a maximum of 20,000 cycles per second (cps, Hz). In bats and dogs, much higher frequencies are audible. The threshold of the human ear varies with the pitch of the sound (Figure 13–9), the greatest sensitivity being in the 1000- to 4000-Hz range. The pitch of the average male voice in conversation is about 120 Hz and that of the average female voice about 250 Hz. The number of pitches that can be distinguished by an average individual is about 2000, but trained musicians can improve on this figure considerably. Pitch discrimination is best in the 1000- to 3000-Hz range and is poor at high and low pitches.

The presence of one sound decreases an individual's ability to hear other sounds, a phenomenon known as **masking**. It is believed to be due to the relative or absolute refractoriness of previously stimulated auditory receptors and nerve fibers to other stimuli. The degree to which a given tone masks others is related to its pitch. The masking effect of the background noise in all but the most carefully soundproofed environments raises the auditory threshold by a definite and measurable amount.

SOUND TRANSMISSION

The ear converts sound waves in the external environment into action potentials in the auditory nerves. The waves are transformed by the eardrum and auditory ossicles into movements of the foot plate of the stapes. These movements set up

FIGURE 13–9 **Human audibility curve.** The middle curve is that obtained by audiometry under the usual conditions. The lower curve is that obtained under ideal conditions. At about 140 decibels (top curve), sounds are felt as well as heard.

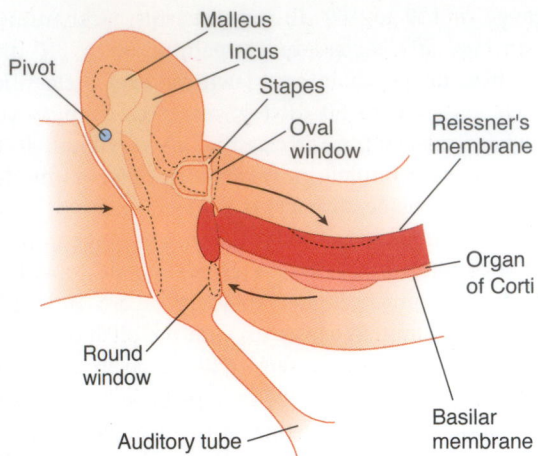

FIGURE 13–10 **Schematic representation of the auditory ossicles and the way their movement translates movements of the tympanic membrane into a wave in the fluid of the inner ear.** The wave is dissipated at the round window. The movements of the ossicles, the membranous labyrinth, and the round window are indicated by dashed lines.

waves in the fluid of the inner ear (Figure 13–10). The action of the waves on the organ of Corti generates action potentials in the nerve fibers.

In response to the pressure changes produced by sound waves on its external surface, the tympanic membrane moves in and out. The membrane therefore functions as a **resonator** that reproduces the vibrations of the sound source. It stops vibrating almost immediately when the sound wave stops. The motions of the tympanic membrane are imparted to the manubrium of the malleus. The malleus rocks on an axis through the junction of its long and short processes, so that the short process transmits the vibrations of the manubrium to the incus. The incus moves in such a way that the vibrations are transmitted to the head of the stapes. Movements of the head of the stapes swing its foot plate to and fro like a door hinged at the posterior edge of the oval window. The auditory ossicles thus function as a lever system that converts the resonant vibrations of the tympanic membrane into movements of the stapes against the perilymph-filled scala vestibuli of the cochlea (Figure 13–10). This system increases the sound pressure that arrives at the oval window, because the lever action of the malleus and incus multiplies the force 1.3 times and the area of the tympanic membrane is much greater than the area of the foot plate of the stapes. Some sound energy is lost as a result of resistance, but it has been calculated that at frequencies below 3000 Hz, 60% of the sound energy incident on the tympanic membrane is transmitted to the fluid in the cochlea.

TYMPANIC REFLEX

When the middle ear muscles (tensor tympani and stapedius) contract, they pull the manubrium of the malleus inward and the footplate of the stapes outward (Figure 13–2). This de-creases sound transmission. Loud sounds initiate a reflex contraction of these muscles called the **tympanic reflex.** Its function is protective, preventing strong sound waves from causing excessive stimulation of the auditory receptors. However, the reaction time for the reflex is 40 to 160 ms, so it does not protect against brief intense stimulation such as that produced by gunshots.

BONE & AIR CONDUCTION

Conduction of sound waves to the fluid of the inner ear via the tympanic membrane and the auditory ossicles, the main pathway for normal hearing, is called **ossicular conduction.** Sound waves also initiate vibrations of the secondary tympanic membrane that closes the round window. This process, unimportant in normal hearing, is **air conduction.** A third type of conduction, **bone conduction,** is the transmission of vibrations of the bones of the skull to the fluid of the inner ear. Considerable bone conduction occurs when tuning forks or other vibrating bodies are applied directly to the skull. This route also plays a role in transmission of extremely loud sounds.

TRAVELING WAVES

The movements of the foot plate of the stapes set up a series of traveling waves in the perilymph of the scala vestibuli. A diagram of such a wave is shown in Figure 13–11. As the wave moves up the cochlea, its height increases to a maximum and then drops off rapidly. The distance from the stapes to this point of maximum height varies with the frequency of the vibrations initiating the wave. High-pitched sounds generate waves that reach maximum height near the base of the cochlea; low-pitched sounds generate waves that peak near the apex. The

FIGURE 13–11 **Traveling waves. Top:** The solid and the short-dashed lines represent the wave at two instants of time. The long-dashed line shows the "envelope" of the wave formed by connecting the wave peaks at successive instants. **Bottom:** Displacement of the basilar membrane by the waves generated by stapes vibration of the frequencies shown at the top of each curve.

bony walls of the scala vestibuli are rigid, but Reissner's membrane is flexible. The basilar membrane is not under tension, and it also is readily depressed into the scala tympani by the peaks of waves in the scala vestibuli. Displacements of the fluid in the scala tympani are dissipated into air at the round window. Therefore, sound produces distortion of the basilar membrane, and the site at which this distortion is maximal is determined by the frequency of the sound wave. The tops of the hair cells in the organ of Corti are held rigid by the reticular lamina, and the hairs of the outer hair cells are embedded in the tectorial membrane (Figure 13–4). When the stapes moves, both membranes move in the same direction, but they are hinged on different axes, so a shearing motion bends the hairs. The hairs of the inner hair cells are not attached to the tectorial membrane, but they are apparently bent by fluid moving between the tectorial membrane and the underlying hair cells.

FUNCTIONS OF THE INNER & OUTER HAIR CELLS

The inner hair cells are the primary sensory cells that generate action potentials in the auditory nerves, and presumably they are stimulated by the fluid movements noted above.

The outer hair cells, on the other hand, have a different function. These respond to sound, like the inner hair cells, but depolarization makes them shorten and hyperpolarization makes them lengthen. They do this over a very flexible part of the basal membrane, and this action somehow increases the amplitude and clarity of sounds. These changes in outer hair cells occur in parallel with changes in **prestin,** a membrane protein, and this protein may well be the motor protein of outer hair cells.

The outer hair cells receive cholinergic innervation via an efferent component of the auditory nerve, and acetylcholine hyperpolarizes the cells. However, the physiologic function of this innervation is unknown.

ACTION POTENTIALS IN AUDITORY NERVE FIBERS

The frequency of the action potentials in single auditory nerve fibers is proportional to the loudness of the sound stimuli. At low sound intensities, each axon discharges to sounds of only one frequency, and this frequency varies from axon to axon depending on the part of the cochlea from which the fiber originates. At higher sound intensities, the individual axons discharge to a wider spectrum of sound frequencies, particularly to frequencies lower than that at which threshold simulation occurs.

The major determinant of the pitch perceived when a sound wave strikes the ear is the place in the organ of Corti that is maximally stimulated. The traveling wave set up by a tone produces peak depression of the basilar membrane, and consequently maximal receptor stimulation, at one point. As noted above, the distance between this point and the stapes is inversely related to the pitch of the sound, with low tones pro-

ducing maximal stimulation at the apex of the cochlea and high tones producing maximal stimulation at the base. The pathways from the various parts of the cochlea to the brain are distinct. An additional factor involved in pitch perception at sound frequencies of less than 2000 Hz may be the pattern of the action potentials in the auditory nerve. When the frequency is low enough, the nerve fibers begin to respond with an impulse to each cycle of a sound wave. The importance of this **volley effect,** however, is limited; the frequency of the action potentials in a given auditory nerve fiber determines principally the loudness, rather than the pitch, of a sound.

Although the pitch of a sound depends primarily on the frequency of the sound wave, loudness also plays a part; low tones (below 500 Hz) seem lower and high tones (above 4000 Hz) seem higher as their loudness increases. Duration also affects pitch to a minor degree. The pitch of a tone cannot be perceived unless it lasts for more than 0.01 s, and with durations between 0.01 and 0.1 s, pitch rises as duration increases. Finally, the pitch of complex sounds that include harmonics of a given frequency is still perceived even when the primary frequency (missing fundamental) is absent.

CENTRAL PATHWAY

The afferent fibers in the auditory division of the eighth cranial nerve end in **dorsal** and **ventral cochlear nuclei** (Figure 13–12). From there, auditory impulses pass by various routes to the **inferior colliculi,** the centers for auditory reflexes, and via the **medial geniculate body** in the thalamus to the **auditory cortex.** Other impulses enter the reticular formation. Information from both ears converges on each superior olive, and beyond this, most of the neurons respond to inputs from both sides. The primary auditory cortex is Brodmann's area 41 (see Figure 13–13). In humans, low tones are represented anterolaterally and high tones posteromedially in the auditory cortex.

In the primary auditory cortex, most neurons respond to inputs from both ears, but strips of cells are stimulated by input from the contralateral ear and inhibited by input from the ipsilateral ear. There are several additional auditory receiving areas, just as there are several receiving areas for cutaneous sensation. The auditory association areas adjacent to the primary auditory receiving areas are widespread.

The **olivocochlear bundle** is a prominent bundle of efferent fibers in each auditory nerve that arises from both ipsilateral and contralateral superior olivary complexes and ends primarily around the bases of the outer hair cells of the organ of Corti.

AUDITORY RESPONSES OF NEURONS IN THE MEDULLA OBLONGATA

The responses of individual second-order neurons in the cochlear nuclei to sound stimuli are like those of the individual auditory nerve fibers. The frequency at which sounds of the

FIGURE 13–12 Simplified diagram of main auditory (left) and vestibular (right) pathways superimposed on a dorsal view of the **brain stem.** Cerebellum and cerebral cortex have been removed.

lowest intensity evoke a response varies from unit to unit; with increased sound intensities, the band of frequencies to which a response occurs becomes wider. The major difference between the responses of the first- and second-order neurons is the presence of a sharper "cutoff" on the low-frequency side in the medullary neurons. This greater specificity of the second-order neurons is probably due to an inhibitory process in the brain stem.

OTHER CORTICAL AREAS CONCERNED WITH AUDITION

The increasing availability of positron emission tomography (PET) scanning and functional magnetic resonance imaging (fMRI) has led to rapid increases in knowledge about auditory association areas in humans. The auditory pathways in the cortex resemble the visual pathways in that increasingly complex processing of auditory information takes place along them. An interesting observation is that although the auditory areas look very much the same on the two sides of the brain, there is marked hemispheric specialization. For example, Brodmann's area 22 is concerned with the processing of auditory signals related to speech. During language processing, it is much more active on the left side than on the right side. Area 22 on the right

side is more concerned with melody, pitch, and sound intensity. The auditory pathways are also very plastic, and, like the visual and somasthetic pathways, they are modified by experience. Examples of auditory plasticity in humans include the observation that in individuals who become deaf before language skills are fully developed, viewing sign language activates auditory association areas. Conversely, individuals who become blind early in life are demonstrably better at localizing sound than individuals with normal eyesight.

Musicians provide additional examples of cortical plasticity. In these individuals, the size of the auditory areas activated by musical tones is increased. In addition, violinists have altered somatosensory representation of the areas to which the fingers they use in playing their instruments project. Musicians also have larger cerebellums than nonmusicians, presumably because of learned precise finger movements.

A portion of the posterior superior temporal gyrus known as the **planum temporale** (Figure 13–13) is regularly larger in the left than in the right cerebral hemisphere, particularly in right-handed individuals. This area appears to be involved in language-related auditory processing. A curious observation, which is presently unexplained, is that the planum temporale is even larger than normal on the left side in musicians and others who have perfect pitch.

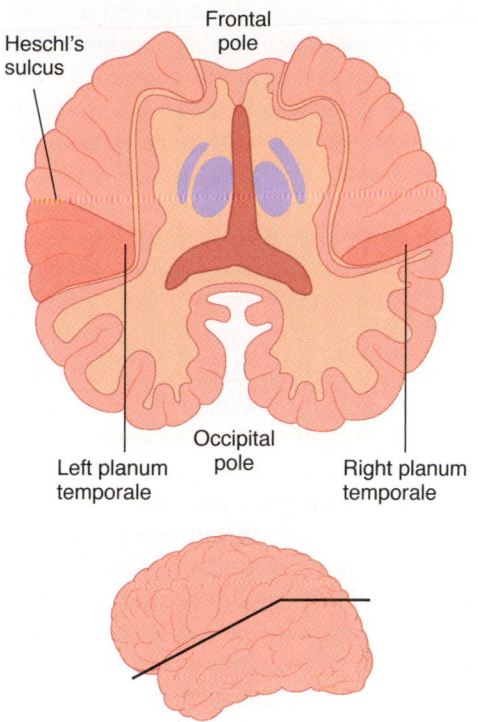

FIGURE 13–13 **Left and right planum temporale in a brain sectioned horizontally along the plane of the sylvian fissure.** Plane of section shown in the insert at the bottom. (Reproduced with permission from Kandel ER, Schwartz JH, Jessel TM [editors]: *Principles of Neural Science,* 3rd ed. McGraw-Hill, 1991.)

SOUND LOCALIZATION

Determination of the direction from which a sound emanates in the horizontal plane depends on detecting the difference in time between the arrival of the stimulus in the two ears and the consequent difference in phase of the sound waves on the two sides; it also depends on the fact that the sound is louder on the side closest to the source. The detectable time difference, which can be as little as 20 μs, is said to be the most important factor at frequencies below 3000 Hz and the loudness difference the most important at frequencies above 3000 Hz. Neurons in the auditory cortex that receive input from both ears respond maximally or minimally when the time of arrival of a stimulus at one ear is delayed by a fixed period relative to the time of arrival at the other ear. This fixed period varies from neuron to neuron.

Sounds coming from directly in front of the individual differ in quality from those coming from behind because each pinna (the visible portion of the exterior ear) is turned slightly forward. In addition, reflections of the sound waves from the pinnal surface change as sounds move up or down, and the change in the sound waves is the primary factor in locating sounds in the vertical plane. Sound localization is markedly disrupted by lesions of the auditory cortex.

AUDIOMETRY

Auditory acuity is commonly measured with an **audiometer.** This device presents the subject with pure tones of various frequencies through earphones. At each frequency, the threshold intensity is determined and plotted on a graph as a percentage of normal hearing. This provides an objective measurement of the degree of deafness and a picture of the tonal range most affected.

DEAFNESS

Hearing loss is the most common sensory defect in humans. According to the World Health Organization, over 270 million people worldwide have moderate to profound hearing loss, with one fourth of these cases beginning in childhood. **Presbycusis,** the gradual hearing loss associated with aging, affects more than one-third of those over 75 and is probably due to gradual cumulative loss of hair cells and neurons. In most cases, hearing loss is a multifactorial disorder caused by both genetic and environmental factors. Genetic factors contributing to deafness are described in Clinical Box 13–1.

Deafness can be divided into two major categories: conductive (or conduction) and sensorineural hearing loss. **Conductive deafness** refers to impaired sound transmission in the external or middle ear and impacts all sound frequencies. Among the causes of conduction deafness are plugging of the external auditory canals with wax (cerumen) or foreign bodies, otitis externa (inflammation of the outer ear, "swimmer's ear") and otitis media (inflammation of the middle ear) causing fluid accumulation, perforation of the eardrum, and osteosclerosis in which bone is resorbed and replaced with sclerotic bone that grows over the oval window.

Sensorineural deafness is most commonly the result of loss of cochlear hair cells but can also be due to problems with the eighth cranial nerve or within central auditory pathways. It often impairs the ability to hear certain pitches while others are unaffected. Aminoglycoside antibiotics such as streptomycin and gentamicin obstruct the mechanosensitive channels in the stereocilia of hair cells and can cause the cells to degenerate, producing sensorineural hearing loss and abnormal vestibular function. Damage to the outer hair cells by prolonged exposure to noise is associated with hearing loss. Other causes include tumors of the eighth cranial nerve and cerebellopontine angle and vascular damage in the medulla.

Conduction and sensorineural deafness can be differentiated by simple tests with a tuning fork. Three of these tests, named for the individuals who developed them, are outlined in Table 13–1. The Weber and Schwabach tests demonstrate the important masking effect of environmental noise on the auditory threshold.

VESTIBULAR SYSTEM

The vestibular system can be divided into the **vestibular apparatus** and central **vestibular nuclei.** The vestibular apparatus

Genetic Mutations Contributing to Deafness

Single-gene mutations have been shown to cause hearing loss. This type of hearing loss is a monogenic disorder with an autosomal dominant, autosomal recessive, X-linked, or mitochondrial mode of inheritance. Monogenic forms of deafness can be defined as **syndromic** (hearing loss associated with other abnormalities) or **nonsyndromic** (only hearing loss). About 0.1% of newborns have genetic mutations leading to deafness. Nonsyndromic deafness due to genetic mutations can first appear in adults rather than in children and may account for many of the 16% of all adults who have significant hearing impairment. It is now estimated that the products of 100 or more genes are essential for normal hearing, and deafness loci have been described in all but 5 of the 24 human chromosomes. The most common mutation leading to congenital hearing loss is that of the protein connexin 26. This defect prevents the normal recycling of K^+ through the sustenacular cells. Mutations in three nonmuscle myosins also cause deafness. These are myosin-VIIa, associated with the actin in the hair cell processes; myosin-Ib, which is probably part of the "adaptation motor" that adjusts tension on the tip links; and myosin-VI, which is essential in some way for the formation of normal cilia. Deafness is also associated with mutant forms of α-tectin, one of the major proteins in the tectorial membrane. An example of syndromic deafness is **Pendred syndrome,** in which a mutant sulfate transport protein causes deafness and goiter. Another example is one form of the **long QT syndrome** in which one of the K^+ channel proteins, **KVLQT1,** is mutated. In the stria vascularis, the normal form of this protein is essential for maintaining the high K^+ concentration in endolymph, and in the heart it helps maintain a normal QT interval. Individuals who are homozygous for mutant KVLQT1 are deaf and predisposed to the ventricular arrhythmias and sudden death that characterize the long QT syndrome. Mutations of the membrane protein **barttin** can cause deafness as well as the renal manifestations of Bartter syndrome.

within the inner ear detects head motion and position and transduces this information to a neural signal (Figure 13–3). The vestibular nuclei are primarily concerned with maintaining the position of the head in space. The tracts that descend from these nuclei mediate head-on-neck and head-on-body adjustments.

CENTRAL PATHWAY

The cell bodies of the 19,000 neurons supplying the cristae and maculae on each side are located in the vestibular ganglion. Each vestibular nerve terminates in the ipsilateral four-part vestibular nucleus and in the flocculonodular lobe of the cerebellum (Figure 13–12). Fibers from the semicircular canals end primarily in the superior and medial divisions of the vestibular nucleus and project mainly to nuclei controlling eye movement. Fibers from the utricle and saccule end predominantly in the lateral division (Deiters nucleus), which projects to the spinal cord. They also end on neurons that project to the cerebellum and the reticular formation. The vestibular nuclei also project to the thalamus and from there to two parts of the primary somatosensory cortex. The ascending connections to cranial nerve nuclei are largely concerned with eye movements.

RESPONSES TO ROTATIONAL ACCELERATION

Rotational acceleration in the plane of a given semicircular canal stimulates its crista. The endolymph, because of its inertia, is displaced in a direction opposite to the direction of rotation. The fluid pushes on the cupula, deforming it. This bends the processes of the hair cells (Figure 13–3). When a constant speed of rotation is reached, the fluid spins at the same rate as the body and the cupula swings back into the upright position. When rotation is stopped, deceleration produces displacement of the endolymph in the direction of the rotation, and the cupula is deformed in a direction opposite to that during acceleration. It returns to mid position in 25 to 30 s. Movement of the cupula in one direction commonly causes an

TABLE 13–1 Common tests with a tuning fork to distinguish between sensorineural and conduction deafness.

	Weber	Rinne	Schwabach
Method	Base of vibrating tuning fork placed on vertex of skull.	Base of vibrating tuning fork placed on mastoid process until subject no longer hears it, then held in air next to ear.	Bone conduction of patient compared with that of normal subject.
Normal	Hears equally on both sides.	Hears vibration in air after bone conduction is over.	
Conduction deafness (one ear)	Sound louder in diseased ear because masking effect of environmental noise is absent on diseased side.	Vibrations in air not heard after bone conduction is over.	Bone conduction better than normal (conduction defect excludes masking noise).
Sensorineural deafness (one ear)	Sound louder in normal ear.	Vibration heard in air after bone conduction is over, as long as nerve deafness is partial.	Bone conduction worse than normal.

FIGURE 13–14 Ampullary responses to rotation. Average time course of impulse discharge from the ampulla of two semicircular canals during rotational acceleration, steady rotation, and deceleration. Movement of the cupula in one direction increases the firing rate of single nerve fibers from the crista, and movement in the opposite direction inhibits neural activity. (Reproduced with permission from Adrian ED: Discharge from vestibular receptors in the cat. J Physiol [Lond] 1943;101:389.)

increase in the firing rate of single nerve fibers from the crista, whereas movement in the opposite direction commonly inhibits neural activity (Figure 13–14).

Rotation causes maximal stimulation of the semicircular canals most nearly in the plane of rotation. Because the canals on one side of the head are a mirror image of those on the other side, the endolymph is displaced toward the ampulla on one side and away from it on the other. The pattern of stimulation reaching the brain therefore varies with the direction as well as the plane of rotation. Linear acceleration probably fails to displace the cupula and therefore does not stimulate the cristae. However, there is considerable evidence that when one part of the labyrinth is destroyed, other parts take over its functions. Clinical Box 13–2 describes the characteristic eye movements that occur during a period of rotation.

RESPONSES TO LINEAR ACCELERATION

In mammals, the utricular and saccular maculae respond to linear acceleration. In general, the utricle responds to horizontal acceleration and the saccule to vertical acceleration. The otoliths are more dense than the endolymph, and acceleration in any direction causes them to be displaced in the opposite direction, distorting the hair cell processes and generating activity in the nerve fibers. The maculae also discharge tonically in the absence of head movement, because of the pull of gravity on the otoliths.

The impulses generated from these receptors are partly responsible for **labyrinth righting reflexes.** These reflexes are a series of responses integrated for the most part in the nuclei of the midbrain. The stimulus for the reflex is tilting of the head, which stimulates the otolithic organs; the response is compensatory contraction of the neck muscles to keep the head level. In cats, dogs, and primates, visual cues can initiate **optical righting reflexes** that right the animal in the absence of labyrinthine or body stimulation. In humans, the operation

of these reflexes maintains the head in a stable position and the eyes fixed on visual targets despite movements of the body and the jerks and jolts of everyday life. The responses are initiated by vestibular stimulation, stretching of neck muscles, and movement of visual images on the retina, and the responses are the **vestibulo-ocular reflex** and other remarkably precise reflex contractions of the neck and extraocular muscles.

CLINICAL BOX 13–2

Nystagmus

The characteristic jerky movement of the eye observed at the start and end of a period of rotation is called nystagmus. It is actually a reflex that maintains visual fixation on stationary points while the body rotates, although it is not initiated by visual impulses and is present in blind individuals. When rotation starts, the eyes move slowly in a direction opposite to the direction of rotation, maintaining visual fixation (vestibulo-ocular reflex, VOR). When the limit of this movement is reached, the eyes quickly snap back to a new fixation point and then again move slowly in the other direction. The slow component is initiated by impulses from the vestibular labyrinths; the quick component is triggered by a center in the brain stem. Nystagmus is frequently horizontal (ie, the eyes move in the horizontal plane), but it can also be vertical (when the head is tipped sidewise during rotation) or rotatory (when the head is tipped forward). By convention, the direction of eye movement in nystagmus is identified by the direction of the quick component. The direction of the quick component during rotation is the same as that of the rotation, but the postrotatory nystagmus that occurs owing to displacement of the cupula when rotation is stopped is in the opposite direction. Clinically, nystagmus is seen at rest in patients with lesions of the brain stem. Nystagmus can persist for hours at rest in patients with acute temporal bone fracture affecting semicircular canals or after damage to the flocculonodular lobe or midline structures such as the fastigial nucleus. Nystagmus can be used as a diagnostic indicator of the integrity of the vestibular system. Caloric stimulation can be used to test the function of the vestibular labyrinth. The semicircular canals are stimulated by instilling warm (40 °C) or cold (30 °C) water into the external auditory meatus. The temperature difference sets up convection currents in the endolymph, with consequent motion of the cupula. In normal subjects, warm water causes nystagmus that bears toward the stimulus, whereas cold water induces nystagmus that bears toward the opposite ear. This test is given the mnemonic COWS (Cold water nystagmus is Opposite sides, Warm water nystagmus is Same side). In the case of a unilateral lesion in the vestibular pathway, nystagmus is reduced or absent on the side of the lesion. To avoid nystagmus, vertigo, and nausea when irrigating the ear canals in the treatment of ear infections, it is important to be sure that the fluid used is at body temperature.

CLINICAL BOX 13–3

Vestibular Disorders

Vestibular balance disorders are the ninth most common reason for visits to a primary care physician. It is one of the most common reasons elderly people seek medical advice. Patients often describe balance problems in terms of vertigo, dizziness, lightheadedness, and motion sickness. Neither lightheadedness nor dizziness is necessarily a symptom of vestibular problems, but **vertigo** is a prominent symptom of a disorder of the inner ear or vestibular system, especially when one labyrinth is inflamed. **Benign paroxysmal positional vertigo** is the most common vestibular disorder characterized by episodes of vertigo that occur with particular changes in body position (eg, turning over in bed, bending over). One possible cause is that **otoconia** from the utricle separate from the otolith membrane and become lodged in the cupula of the posterior semicircular canal. This causes abnormal deflections when the head changes position relative to gravity.

Ménière disease is an abnormality of the inner ear causing vertigo or severe dizziness, **tinnitus,** fluctuating hearing loss, and the sensation of pressure or pain in the affected ear lasting several hours. Symptoms can occur suddenly and recur daily or very rarely. The hearing loss is initially transient but can become permanent. The pathophysiology likely involves an immune reaction. An inflammatory response can increase fluid volume within the membranous labyrinth, causing it to rupture and allowing the endolymph and perilymph to mix together. There is no cure for Ménière disease but the symptoms can be controlled by reducing the fluid retention through dietary changes (low-salt or salt-free diet, no caffeine, no alcohol) or medication.

The nausea, blood pressure changes, sweating, pallor, and vomiting that are the well-known symptoms of **motion sickness** are produced by excessive vestibular stimulation and occurs when conflicting information is fed into the vestibular and other sensory systems. The symptoms are probably due to reflexes mediated via vestibular connections in the brain stem and the flocculonodular lobe of the cerebellum. **Space motion sickness**— the nausea, vomiting, and vertigo experienced by astronauts—develops when they are first exposed to microgravity and often wears off after a few days of space flight. It can then recur with reentry, as the force of gravity increases again. It is believed to be due to mismatches in neural input created by changes in the input from some parts of the vestibular apparatus and other gravity sensors without corresponding changes in the other spatial orientation inputs.

Although most of the responses to stimulation of the maculae are reflex in nature, vestibular impulses also reach the cerebral cortex. These impulses are presumably responsible for conscious perception of motion and supply part of the information necessary for orientation in space. **Vertigo** is the sensation of rotation in the absence of actual rotation and is a prominent symptom when one labyrinth is inflamed.

SPATIAL ORIENTATION

Orientation in space depends in part on input from the vestibular receptors, but visual cues are also important. Pertinent information is also supplied by impulses from proprioceptors in joint capsules, which supply data about the relative position of the various parts of the body, and impulses from cutaneous exteroceptors, especially touch and pressure receptors. These four inputs are synthesized at a cortical level into a continuous picture of the individual's orientation in space. Clinical Box 13–3 describes some common vestibular disorders.

CHAPTER SUMMARY

- The external ear funnels sound waves to the external auditory meatus and tympanic membrane. From there, sound waves pass through three auditory ossicles (malleus, incus, and stapes) in the middle ear. The inner ear, or labyrinth, contains the cochlea and organ of Corti.

- The hair cells in the organ of Corti signal hearing. The stereocilia provide a mechanism for generating changes in membrane potential proportional to the direction and distance the hair moves. Sound is the sensation produced when longitudinal vibrations of air molecules strike the tympanic membrane.

- The activity within the auditory pathway passes from the eighth cranial nerve afferent fibers to the dorsal and ventral cochlear nuclei to the inferior colliculi to the thalamic medial geniculate body and then to the auditory cortex.

- Loudness is correlated with the amplitude of a sound wave, pitch with the frequency, and timbre with harmonic vibrations.

- Conductive deafness is due to impaired sound transmission in the external or middle ear and impacts all sound frequencies. Sensorineural deafness is usually due to loss of cochlear hair cells but can also occur after damage to the eighth cranial nerve or central auditory pathways.

- Rotational acceleration stimulates the crista in the semicircular, displacing the endolymph in a direction opposite to the direction of rotation, deforming the cupula and bending the hair cell. The utricle responds to horizontal acceleration and the saccule to vertical acceleration. Acceleration in any direction displaces the otoliths, distorting the hair cell processes and generating neural.

- Spatial orientation is dependent on input from vestibular receptors, visual cues, proprioceptors in joint capsules, and cutaneous touch and pressure receptors.

MULTIPLE-CHOICE QUESTIONS

For all questions, select the single best answer unless otherwise directed.

1. A 40-year-old male, employed as a road construction worker for nearly 20 years, went to his physician to report that he recently began to notice difficulty hearing during normal conversations. A Weber test showed that sound from a vibrating tuning fork was localized to the right ear. A Schwabach test showed that bone conduction was below normal. A Rinne test showed that both air and bone conduction were abnormal, but air conduction lasted longer than bone conduction. The diagnosis was:

   A) sensorial deafness in both ears
   B) conduction deafness in the right ear
   C) sensorial deafness in the right ear
   D) conduction deafness in the left ear
   E) sensorineural deafness in the left ear

2. What would the diagnosis be if a patient had the following test results? Weber test showed that sound from a vibrating tuning fork was louder than normal; Schwabach test showed that bone conduction was better than normal; and Rinne test showed that air conduction did not outlast bone conduction.

   A sensorial deafness in both ears
   B) conduction deafness in both ears
   C) normal hearing
   D) both sensorial and conduction deafness
   E) a possible tumor on the eighth cranial nerve

3. Postrotatory nystagmus is caused by continued movement of

   A) aqueous humor over the ciliary body in the eye.
   B) cerebrospinal fluid over the parts of the brain stem that contain the vestibular nuclei.
   C) endolymph in the semicircular canals, with consequent bending of the cupula and stimulation of hair cells.
   D) endolymph toward the helicotrema.
   E) perilymph over hair cells that have their processes embedded in the tectorial membrane.

4. Some diseases damage the hair cells in the ear. When the damage to the outer hair cells is greater than the damage to the inner hair cells,

   A) perception of vertical acceleration is disrupted.
   B) K^+ concentration in endolymph is decreased.
   C) K^+ concentration in perilymph is decreased.
   D) there is severe hearing loss.
   E) affected hair cells fail to shorten when exposed to sound.

5. Which of the following are *incorrectly* paired?

   A) tympanic membrane : manubrium of malleus
   B) helicotrema : apex of cochlea
   C) footplate of stapes : oval window
   D) otoliths : semicircular canals
   E) basilar membrane : organ of Corti

6. The direction of nystagmus is vertical when a subject is rotated

   A) after warm water is put in one ear.
   B) with the head tipped backward.
   C) after cold water is put in both ears.
   D) with the head tipped sideways.
   E) after section of one vestibular nerve.

7. In the utricle, tip links in hair cells are involved in

   A) formation of perilymph.
   B) depolarization of the stria vascularis.
   C) movements of the basement membrane.
   D) perception of sound.
   E) regulation of distortion-activated ion channels.

8. A patient enters the hospital for evaluation of deafness. He is found to also have an elevated plasma renin, although his blood pressure is 118/75 mm Hg. Mutation of what single gene may explain these findings?

   A) the gene for barttin
   B) the gene for Na^+ channel
   C) the gene for renin
   D) the gene for cystic fibrosis transmembrane conductance regulator
   E) the gene for tyrosine hydroxylase

CHAPTER RESOURCES

Baloh RW, Halmagyi M: *Disorders of the Vestibular System.* Oxford University Press, 1996.

Fox SI: *Human Physiology.* McGraw-Hill, 2008.

Haines DE (editor): *Fundamental Neuroscience for Basic and Clinical Applications,* 3rd ed. Elsevier, 2006.

Highstein SM, Fay RR, Popper AN (editors): *The Vestibular System.* Springer, 2004.

Hudspeth AJ: The cellular basis of hearing: The biophysics of hair cells. Science 1985;230:745.

Hudspeth AJ: How the ear's works work. Nature 1989;341:397.

Kandel ER, Schwartz JH, Jessell TM (editors): *Principles of Neural Science,* 4th ed. McGraw-Hill, 2000.

Oertel D, Fay RR, Popper AN (editors): *Integrative Functions in the Mammalian Auditory Pathway.* Springer, 2002.

Pickles JO: *An Introduction to the Physiology of Hearing,* 2nd ed. Academic Press, 1988.

Squire LR, et al (editors): *Fundamental Neuroscience,* 3rd ed. Academic Press, 2008.

Willems PJ: Genetic causes of hearing loss. NE J Med 2000;342:1101.

Smell & Taste

After studying this chapter, you should be able to:

- Describe the basic features of the neural elements in the olfactory epithelium and olfactory bulb.
- Describe signal transduction in odorant receptors.
- Outline the pathway by which impulses generated in the olfactory epithelium reach the olfactory cortex.
- Describe the location and cellular composition of taste buds.
- Name the five major taste receptors and signal transduction mechanisms in these receptors.
- Outline the pathways by which impulses generated in taste receptors reach the insular cortex.

INTRODUCTION

Smell and taste are generally classified as visceral senses because of their close association with gastrointestinal function. Physiologically, they are related to each other. The flavors of various foods are in large part a combination of their taste and smell. Consequently, food may taste "different" if one has a cold that depresses the sense of smell. Both smell and taste receptors are **chemoreceptors** that are stimulated by molecules in solution in mucus in the nose and saliva in the mouth. Because stimuli arrive from an external source, they are also classified as **exteroceptors.**

SMELL

OLFACTORY EPITHELIUM

The **olfactory sensory neurons** are located in a specialized portion of the nasal mucosa, the yellowish pigmented **olfactory epithelium.** In dogs and other animals in which the sense of smell is highly developed (macrosmatic animals), the area covered by this membrane is large; in microsmatic animals, such as humans, it is small. In humans, it covers an area of 5 cm^2 in the roof of the nasal cavity near the septum (Figure 14–1). The human olfactory epithelium contains 10 to 20 million bipolar olfactory sensory neurons interspersed with glia-like **supporting (sustentacular) cells** and **basal stem cells.** The olfactory epithelium is said to be the place in the body where the nervous system is closest to the external world. Each neuron has a short, thick dendrite that projects into the nasal cavity where it terminates in a knob containing 10 to 20 **cilia** (Figure 14–2). The cilia are unmyelinated processes about 2 μm long and 0.1 μm in diameter and contain specific receptors for odorants **(odorant receptors).** The axons of the olfactory sensory neurons pass through the cribriform plate of the ethmoid bone and enter the olfactory bulbs (Figure 14–1).

New olfactory sensory neurons are generated by basal stem cells as needed to replace those damaged by exposure to the environment. The olfactory renewal process is carefully regulated, and there is evidence that in this situation a bone morphogenic protein (BMP) exerts an inhibitory effect. BMPs are a large family of growth factors originally described as promoters of bone growth but now known to act on most tissues in the body during development, including many types of nerve cells.

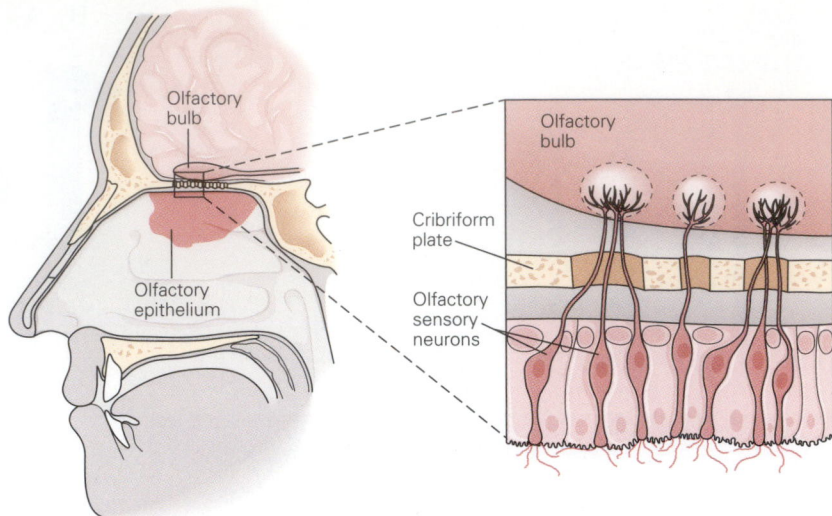

FIGURE 14–1   **Olfactory sensory neurons embedded within the olfactory epithelium in the dorsal posterior recess of the nasal cavity.** These neurons project axons to the olfactory bulb of the brain, a small ovoid structure that rests on the cribriform plate of the ethmoid bone.   (From Kandel ER, Schwartz JH, Jessell TM [editors]: *Principles of Neural Science,* 4th ed. McGraw-Hill, 2000.)

OLFACTORY BULBS

In the olfactory bulbs, the axons of the olfactory sensory neurons (first cranial nerve) contact the primary dendrites of the **mitral cells** and **tufted cells** (Figure 14–3) to form anatomically discrete synaptic units called **olfactory glomeruli.** The tufted cells are smaller than the mitral cells and have thinner axons, but both types send axons into the olfactory cortex, and they appear to be similar from a functional point of view. In addition to mitral and tufted cells, the olfactory bulbs contain

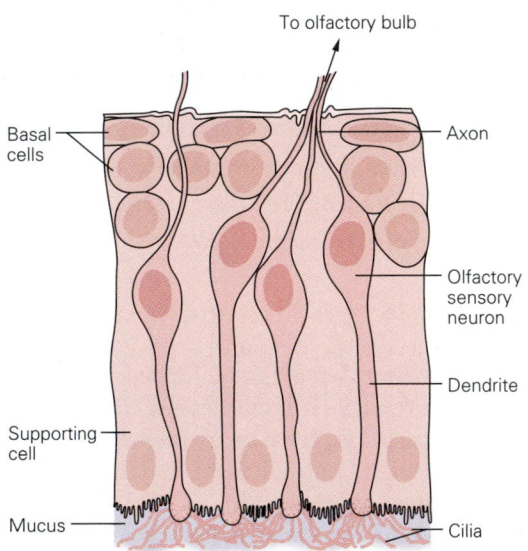

FIGURE 14–2   **Structure of the olfactory epithelium.** There are three cell types: olfactory sensory neurons, supporting cells, and basal stem cells at the base of the epithelium. Each sensory neuron has a dendrite that projects to the epithelial surface. Numerous cilia protrude into the mucosal layer lining the nasal lumen. A single axon projects from each neuron to the olfactory bulb. Odorants bind to specific odorant receptors on the cilia and initiate a cascade of events leading to generation of action potentials in the sensory axon.   (Modified from Kandel ER, Schwartz JH, Jessell TM [editors]: *Principles of Neural Science.* McGraw-Hill, 2000.)

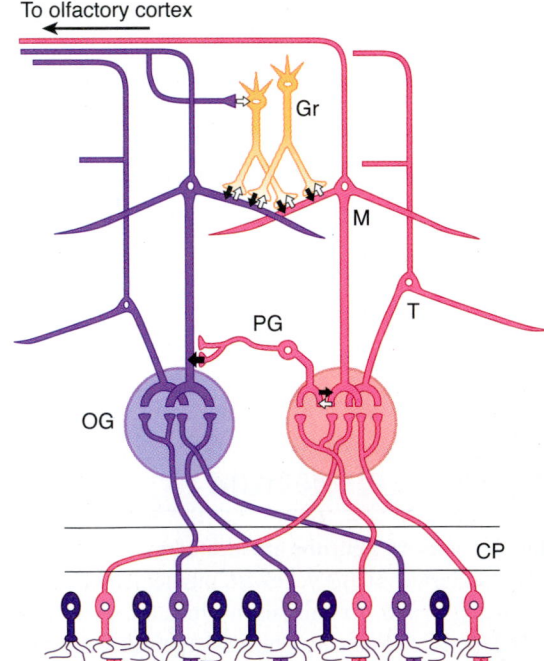

FIGURE 14–3   **Basic neural circuits in the olfactory bulb.** Note that olfactory receptor cells with one type of odorant receptor project to one olfactory glomerulus (OG) and olfactory receptor cells with another type of receptor project to a different olfactory glomerulus. CP, cribriform plate; PG, periglomerular cell; M, mitral cell; T, tufted cell; Gr, granule cell.   (Modified from Mori K, Nagao H, Yoshihara Y: The olfactory bulb: Coding and processing of odor molecular information. Science 1999;286:711.)

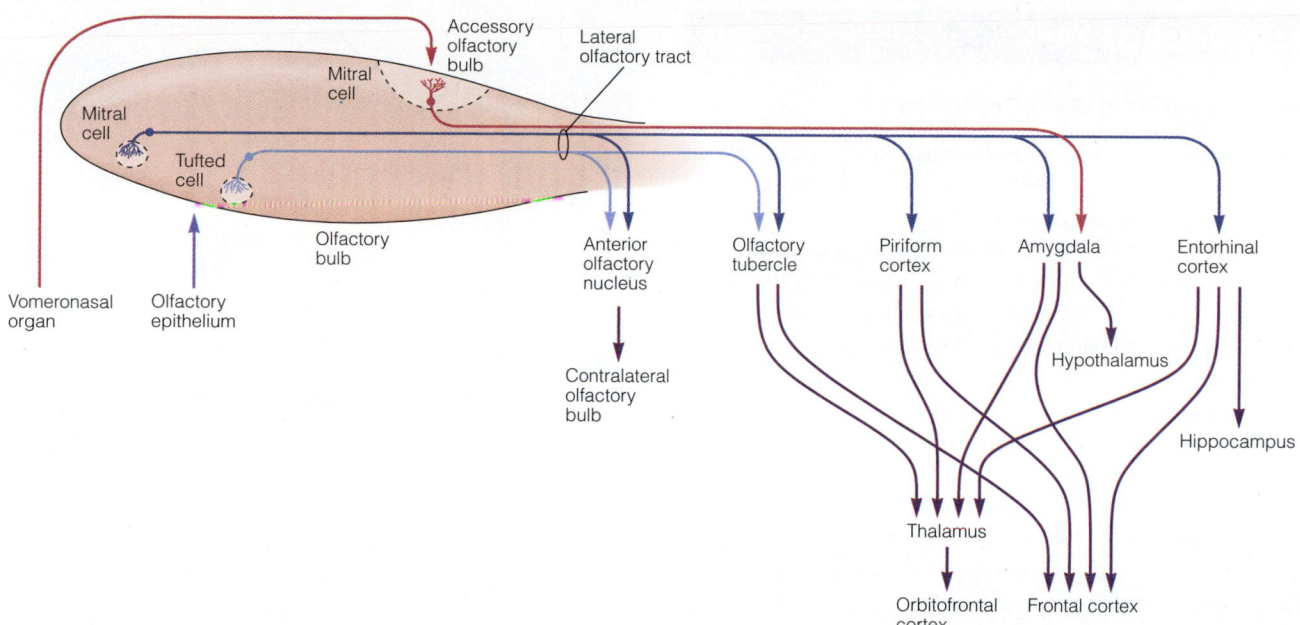

FIGURE 14–4   **Diagram of the olfactory pathway.** Information is transmitted from the olfactory bulb by axons of mitral and tufted relay neurons in the lateral olfactory tract. Mitral cells project to five regions of the olfactory cortex: anterior olfactory nucleus, olfactory tubercle, piriform cortex, and parts of the amygdala and entorhinal cortex. Tufted cells project to anterior olfactory nucleus and olfactory tubercle; mitral cells in the accessory olfactory bulb project only to the amygdala. Conscious discrimination of odor depends on the neocortex (orbitofrontal and frontal cortices). Emotive aspects of olfaction derive from limbic projections (amygdala and hypothalamus).   (From Kandel ER, Schwartz JH, Jessell TM [editors]: *Principles of Neural Science,* 4th ed. McGraw-Hill, 2000.)

periglomerular cells, which are inhibitory neurons connecting one glomerulus to another, and **granule cells,** which have no axons and make reciprocal synapses with the lateral dendrites of the mitral and tufted cells (Figure 14–3). At these synapses, the mitral or tufted cell excites the granule cell by releasing glutamate, and the granule cell in turn inhibits the mitral or tufted cell by releasing GABA.

OLFACTORY CORTEX

The axons of the mitral and tufted cells pass posteriorly through the **lateral olfactory stria** to terminate on apical dendrites of pyramidal cells in five regions of the **olfactory cortex: anterior olfactory nucleus, olfactory tubercle, piriform cortex, amygdala,** and **entorhinal cortex** (Figure 14–4). From these regions, information travels directly to the frontal cortex or via the thalamus to the orbitofrontal cortex. Conscious discrimination of odors is dependent on the pathway to the orbitofrontal cortex. The orbitofrontal activation is generally greater on the right side than the left; thus, cortical representation of olfaction is asymmetric. The pathway to the amygdala is probably involved with the emotional responses to olfactory stimuli, and the pathway to the entorhinal cortex is concerned with olfactory memories.

OLFACTORY THRESHOLDS & DISCRIMINATION

The olfactory epithelium is covered by a thin layer of mucus secreted by the supporting cells and Bowman glands, which lie beneath the epithelium. The mucus bathes the odorant receptors on the cilia and provides the appropriate molecular and ionic environment for odor detection.

The olfactory thresholds for substances shown in Table 14–1 illustrate the remarkable sensitivity of the odorant receptors. For

TABLE 14–1   Some olfactory thresholds.

Substance	mg/L of Air
Ethyl ether	5.83
Chloroform	3.30
Pyridine	0.03
Oil of peppermint	0.02
Lodoform	0.02
Butyric acid	0.009
Propyl mercaptan	0.006
Artificial musk	0.00004
Methyl mercaptan	0.0000004

(a)

(b)

FIGURE 14–5   Signal transduction in an odorant receptor. Olfactory receptors are G protein-coupled receptors that dissociate upon binding to the odorant. The α-subunit of G proteins activates adenylate cyclase to catalyze production of cAMP. cAMP acts as a second messenger to open cation channels. Inward diffusion of Na^+ and Ca^{2+} produces depolarization. (From Fox SI: *Human Physiology.* McGraw-Hill, 2008.)

example, methyl mercaptan, one of the substances in garlic, can be smelled at a concentration of less than 500 pg/L of air. In addition, olfactory discrimination is remarkable; for example, humans can recognize more than 10,000 different odors. On the other hand, determination of differences in the intensity of any given odor is poor. The concentration of an odor-producing substance must be changed by about 30% before a difference can be detected. The comparable visual discrimination threshold is a 1% change in light intensity. The direction from which a smell comes may be indicated by the slight difference in the time of arrival of odoriferous molecules in the two nostrils.

Odor-producing molecules are generally small, containing from 3 to 20 carbon atoms, and molecules with the same number of carbon atoms but different structural configurations have different odors. Relatively high water and lipid solubility are characteristic of substances with strong odors. Some common abnormalities in odor detection are described in Clinical Box 14–1.

SIGNAL TRANSDUCTION

The olfactory system has received considerable attention in recent years because of the intriguing biologic question of how a simple sense organ such as the olfactory epithelium and its brain representation, which apparently lacks a high degree of complexity, can mediate discrimination of more than 10,000 different odors. One part of the answer to this question is that there are many different odorant receptors.

The genes that code for about 1000 different types of odorant receptors make up the largest gene family so far described in mammals—larger than the immunoglobulin and T-cell receptor gene families combined. The amino acid sequences of odorant receptors are very diverse, but all the odorant receptors are coupled to heterotrimeric G proteins. When an

odorant molecule binds to its receptor, the G protein subunits (α, β, γ) dissociate (Figure 14–5). The α-subunit activates adenylate cyclase to catalyze the production of cAMP, which acts as a second messenger to open cation channels, causing an inward-directed Ca^{2+} current. This produces the graded receptor potential, which then leads to an action potential in the olfactory nerve.

A second part of the answer to the question of how 10,000 different odors can be detected lies in the neural organization of the olfactory pathway. Although there are millions of olfactory sensory neurons, each expresses only one of the 1000 different odorant receptors. Each neuron projects to one or two glomeruli (Figure 14–3). This provides a distinct two-dimensional map in the olfactory bulb that is unique to the odorant. The mitral cells with their glomeruli project to different parts of the olfactory cortex.

The olfactory glomeruli demonstrate lateral inhibition mediated by periglomerular cells and granule cells. This sharpens and focuses olfactory signals. In addition, the extracellular field potential in each glomerulus oscillates, and the granule cells appear to regulate the frequency of the oscillation. The exact function of the oscillation is unknown, but it probably also helps to focus the olfactory signals reaching the cortex.

ODORANT-BINDING PROTEINS

In contrast to the low threshold for olfactory stimulation when the olfactory epithelium is intact, single olfactory receptors that have been patch-clamped have a relatively high threshold and a long latency. In addition, lipophilic odor-producing molecules must traverse the hydrophilic mucus in the nose to reach the receptors. These facts led to the suggestion that the olfactory mucus might contain one or more **odorant-binding proteins (OBP)** that concentrate the odorants and transfer them to the receptors. An 18-kDa OBP that is unique to the nasal cavity has been isolated, and other related proteins probably exist. The protein has considerable homology to other proteins in the body that are known to be carriers for small lipophilic molecules. A similar binding protein appears to be associated with taste.

VOMERONASAL ORGAN

In rodents and various other mammals, the nasal cavity contains another patch of olfactory epithelium located along the nasal septum in a well-developed **vomeronasal organ.** This structure is concerned with the perception of odors that act as **pheromones.** Vomeronasal sensory neurons project to the **accessory olfactory bulb** and from there primarily to areas in the amygdala and hypothalamus that are concerned with reproduction and ingestive behavior. Vomeronasal input has major effects on these functions. An example is pregnancy block in mice; the pheromones of a male from a different strain prevent pregnancy as a result of mating with that male, but mating with a mouse of the same strain does not produce blockade. The vomeronasal organ has about 100 G protein-coupled odorant receptors that differ in structure from those in the rest of the olfactory epithelium.

The organ is not well developed in humans, but an anatomically separate and biochemically unique area of olfactory epithelium occurs in a pit in the anterior third of the nasal septum, which appears to be a homologous structure. There is evidence for the existence of pheromones in humans, and there is a close relationship between smell and sexual function. Perfume advertisements bear witness to this. The sense of smell is said to be more acute in women than in men, and in women it is most acute at the time of ovulation. Smell, and to a lesser extent, taste, have a unique ability to trigger long-term memories, a fact noted by novelists and documented by experimental psychologists.

SNIFFING

The portion of the nasal cavity containing the olfactory receptors is poorly ventilated in humans. Most of the air normally moves smoothly over the turbinates with each respiratory cycle, although eddy currents pass some air over the olfactory epithelium. These eddy currents are probably set up by convection as cool air strikes the warm mucosal surfaces. The amount of air reaching this region is greatly increased by sniffing, an action that includes contraction of the lower part of the nares on the septum, deflecting the airstream upward. Sniffing is a semireflex response that usually occurs when a new odor attracts attention.

ROLE OF PAIN FIBERS IN THE NOSE

Naked endings of many trigeminal pain fibers are found in the olfactory epithelium. They are stimulated by irritating substances and leads to the characteristic "odor" of such substances as peppermint, menthol, and chlorine. Activation of these endings by nasal irritants also initiates sneezing, lacrimation, respiratory inhibition, and other reflexes.

ADAPTATION

It is common knowledge that when one is continuously exposed to even the most disagreeable odor, perception of the odor decreases and eventually ceases. This sometimes beneficent phenomenon is due to the fairly rapid adaptation, or desensitization, that occurs in the olfactory system. It is mediated by Ca^{2+} acting via calmodulin on **cyclic nucleotide-gated (CNG)** ion channels. When the CNG A4 subunit is knocked out, adaptation is slowed.

TASTE

TASTE BUDS

The specialized sense organ for taste (gustation) consists of approximately 10,000 **taste buds,** which are ovoid bodies measuring 50–70 μm. There are four morphologically distinct types of cells within each taste bud: basal cells, dark cells, light cells, and intermediate cells (Figure 14–6). The latter three cell types are all referred to as **Type I, II, and III taste cells.** They are the sensory neurons that respond to taste stimuli or **tastants.** The three cell types may represent various stages of differentiation of developing taste cells, with the light cells being the most mature. Alternatively, each cell type might represent different cell lineages. The apical ends of taste cells have microvilli that project into the taste pore, a small opening on the dorsal surface of the tongue where tastes cells are exposed to the oral contents. Each taste bud is innervated by about 50 nerve fibers, and conversely, each nerve fiber receives input from an average of five taste buds. The basal cells arise from the epithelial cells surrounding the taste bud. They differentiate into new taste cells, and the old cells are continuously replaced with a half-time of about 10 days. If the sensory nerve is cut, the taste buds it innervates degenerate and eventually disappear.

In humans, the taste buds are located in the mucosa of the epiglottis, palate, and pharynx and in the walls of **papillae of the tongue** (Figure 14–6). The **fungiform papillae** are rounded structures most numerous near the tip of the tongue; the **circumvallate papillae** are prominent structures arranged in a V on the back of the tongue; the **foliate papillae** are on

FIGURE 14–6   **Taste buds located in papillae of the human tongue. A)** Taste buds on the anterior two-thirds of the tongue are innervated by the chorda tympani branch of the facial nerve; those on the posterior one-third of the tongue are innervated by the lingual branch of the glossopharyngeal nerve. **B)** The three major types of papillae (circumvallate, foliate, and fungiform) are located on specific parts of the tongue. **C)** Taste buds are composed of basal stem cells and three types of taste cells (dark, light, and intermediate). Taste cells extend from the base of the taste bud to the taste pore, where microvilli contact tastants dissolved in saliva and mucus. (Modified from Kandel ER, Schwartz JH, Jessell TM [editors]: *Principles of Neural Science*, 4th ed. McGraw-Hill, 2000.)

the posterior edge of the tongue. Each fungiform papilla has up to five taste buds, mostly located at the top of the papilla, while each vallate and foliate papilla contain up to 100 taste buds, mostly located along the sides of the papillae.

TASTE PATHWAYS

The sensory nerve fibers from the taste buds on the anterior two-thirds of the tongue travel in the chorda tympani branch of the facial nerve, and those from the posterior third of the tongue reach the brain stem via the glossopharyngeal nerve (Figure 14–7). The fibers from areas other than the tongue (eg, pharynx) reach the brain stem via the vagus nerve. On each side, the myelinated but relatively slowly conducting taste fibers in these three nerves unite in the gustatory portion of the **nucleus of the solitary tract (NTS)** in the medulla oblongata (Figure 14–7). From there, axons of second-order neurons ascend in the ipsilateral medial lemniscus and, in primates, pass directly to the ventral posteromedial nucleus of the thalamus. From the thalamus, the axons of the third-order neurons pass to neurons in the anterior insula and the frontal operculum in the ipsilateral cerebral cortex. This region is rostral to the face area of the postcentral gyrus, which is probably the area that mediates conscious perception of taste and taste discrimination.

BASIC TASTE MODALITIES

Humans have five established basic tastes: **sweet, sour, bitter, salt,** and **umami.** It used to be thought that the surface of the tongue had special areas for each of the first four of these sensations, but it is now clear that all tastants are sensed from all parts of the tongue and adjacent structures. Afferent nerves to the NTS contain fibers from all types of taste receptors, without any clear localization of types.

The fifth taste sense, umami, was recently added to the four classic tastes. This taste has actually been known for almost 100 years, and it became established once its receptor was identified. It is triggered by glutamate and particularly by the monosodium glutamate (MSG) used so extensively in Asian cooking. The taste is pleasant and sweet but differs from the standard sweet taste.

TASTE RECEPTORS & TRANSDUCTION

The putative receptors for taste are shown diagrammatically in Figure 14–8. The salty taste is triggered by NaCl. Salt-sensitive taste is mediated by a Na$^+$-selective channel known as **ENaC,** the amiloride-sensitive epithelial sodium channel. The entry of Na$^+$ into the salt receptors depolarizes the membrane, generating the receptor potential. In humans, the amiloride sensitivity of salt taste is less pronounced than in some species,

FIGURE 14–7 **Diagram of taste pathways.** Signals from the taste buds travel via different nerves to gustatory areas of the nucleus of the solitary tract which relays information to the thalamus; the thalamus projects to the gustatory cortex. (Modified from Kandel ER, Schwartz JH, Jessell TM [editors]: *Principles of Neural Science*, 4th ed. McGraw-Hill, 2000.)

FIGURE 14–8 **Signal transduction in taste receptors.** Salt-sensitive taste is mediated by a Na⁺-selective channel (ENaC); sour taste is me-diated by H⁺ ions permeable to ENaCs; umami taste is mediated by glutamate acting on a metabotropic receptor, mGluR4; bitter taste is mediated by the T2R family of G protein-coupled receptors; sweet taste may be dependent on the T1R3 family of G protein-coupled receptors which couple to the G protein gustducin. (Modified from Lindemann B: Receptors and transduction in taste. Nature 2001;413:219.)

suggesting that there are additional mechanisms to activate salt-sensitive receptors.

The sour taste is triggered by protons (H^+ ions). ENaCs permit the entry of protons and may contribute to the sensation of sour taste. The H^+ ions can also bind to and block a K^+-sensitive channel. The fall in K^+ permeability can depolarize the membrane. Also, **HCN**, a hyperpolarization-activated cyclic nucleotide-gated cation channel, and other mechanisms may contribute to sour transduction.

Umami taste is due to activation of a truncated metabotropic glutamate receptor, **mGluR4,** in the taste buds. The way activation of the receptor produces depolarization is unsettled. Glutamate in food may also activate ionotropic glutamate receptors to depolarize umami receptors.

Bitter taste is produced by a variety of unrelated compounds. Many of these are poisons, and bitter taste serves as a warning to avoid them. Some bitter compounds bind to and block K^+-selective channels. Many G protein-linked receptors in the human genome are taste receptors (T2R family) and are stimulated by bitter substances such as strychnine. In some cases, these receptors couple to the heterotrimeric G protein, **gustducin.** Gustducin lowers cAMP and increases the formation of inositol phosphates which could lead to depolarization. Some bitter compounds are membrane permeable and may not involve G proteins; quinine is an example.

Substances that taste sweet also act via the G protein gustducin. The T1R3 family of G protein-coupled receptors is expressed by about 20% of taste cells, some of which also express gustducin. Sugars taste sweet, but so do compounds such as saccharin that have an entirely different structure. It appears at present that natural sugars such as sucrose and synthetic sweeteners act via different receptors on gustducin. Like the bitter-responsive receptors, sweet-responsive receptors act via cyclic nucleotides and inositol phosphate metabolism.

TASTE THRESHOLDS & INTENSITY DISCRIMINATIONS

The ability of humans to discriminate differences in the intensity of tastes, like intensity discrimination in olfaction, is relatively

TABLE 14–2 Some taste thresholds.

Substance	Taste	Threshold Concentration (μmol/L)
Hydrochloric acid	Sour	100
Sodium chloride	Salt	2000
Strychnine hydrochloride	Bitter	1.6
Glucose	Sweet	80,000
Sucrose	Sweet	10,000
Saccharin	Sweet	23

CLINICAL BOX 14–2

Abnormalities in Taste Detection

Ageusia (absence of the sense of taste) and **hypogeusia** (diminished taste sensitivity) can be caused by damage to the lingual or glossopharyngeal nerve. Neurological disorders such as vestibular schwannoma, Bell palsy, familial dysautonomia, multiple sclerosis, and certain infections (eg, primary amoeboid meningoencephalopathy) can also cause problems with taste sensitivity. Ageusia can also be an adverse side effect of various drugs including cisplatin and captopril or vitamin B_3 or zinc deficiencies. Aging and tobacco abuse also contribute to diminished taste. **Dysgeusia** or **parageusia** (unpleasant perception of taste) causes a metallic, salty, foul, or rancid taste. In many cases, dysgeusia is a temporary problem. Factors contributing to ageusia or hypogeusia can also lead to abnormal taste sensitivity.

crude. A 30% change in the concentration of the substance being tasted is necessary before an intensity difference can be detected. The threshold concentrations of substances to which the taste buds respond vary with the particular substance (Table 14–2).

A protein that binds taste-producing molecules has been cloned. It is produced by **Ebner gland** that secretes mucus into the cleft around vallate papillae (Figure 14–6) and probably has a concentrating and transport function similar to that of the OBP described for olfaction. Some common abnormalities in taste detection are described in Clinical Box 14–2.

VARIATION & AFTER EFFECTS

Taste exhibits after reactions and contrast phenomena that are similar in some ways to visual after images and contrasts. Some of these are chemical "tricks," but others may be true central phenomena. A taste modifier protein, **miraculin,** has been discovered in a plant. When applied to the tongue, this protein makes acids taste sweet.

Animals, including humans, form particularly strong aversions to novel foods if eating the food is followed by illness. The survival value of such aversions is apparent in terms of avoiding poisons.

CHAPTER SUMMARY

- Olfactory sensory neurons, supporting (sustentacular) cells, and basal stem cells are located in the olfactory epithelium within the upper portion of the nasal cavity.
- The cilia located on the dendritic knob of the olfactory sensory neuron contain odorant receptors which are coupled to heterotrimeric G proteins.
- Axons of olfactory sensory neurons contact the dendrites of mitral and tufted cells in the olfactory bulbs to form olfactory glomeruli.

- Information from the olfactory bulb travels via the lateral olfactory stria directly to the olfactory cortex, including the anterior olfactory nucleus, olfactory tubercle, piriform cortex, amygdala, and entorhinal cortex.

- Taste buds are the specialized sense organs for taste and are comprised of basal stem cells and three types of taste cells (dark cells, light cells, and intermediate cells). The three types of taste cells may represent various stages of differentiation of developing taste cells, with the light cells being the most mature. Taste buds are located in the mucosa of the epiglottis, palate, and pharynx and in the walls of papillae of the tongue.

- There are taste receptors for sweet, sour, bitter, salt, and umami. Signal transduction mechanisms include passage through ion channels, binding to and blocking ion channels, and second messenger systems.

- The afferents from taste buds in the tongue travel via the seventh, ninth, and tenth cranial nerves to synapse in the nucleus of the tractus solitarius. From there, axons ascend via the ipsilateral medial lemniscus to the ventral posteromedial nucleus of the thalamus, and on to the anterior insula and frontal operculum in the ipsilateral cerebral cortex.

MULTIPLE-CHOICE QUESTIONS

For all questions, select the single best answer unless otherwise directed.

1. Odorant receptors are
   A) located in the olfactory bulb.
   B) located on dendrites of mitral and tufted cells.
   C) located on neurons that project directly to the olfactory cortex.
   D) located on neurons in the olfactory epithelium that project to mitral cells and from there directly to the olfactory cortex.
   E) located on sustentacular cells that project to the olfactory bulb.

2. Taste receptors
   A) for sweet, sour, bitter, salt, and umami are spatially separated on the surface of the tongue.
   B) are synonymous with taste buds.
   C) are a type of chemoreceptor.
   D) are innervated by afferents in the facial, trigeminal, and glossopharyngeal nerves.
   E) all of the above

3. Which of the following does *not* increase the ability to discriminate many different odors?
   A) many different receptors
   B) pattern of olfactory receptors activated by a given odorant
   C) projection of different mitral cell axons to different parts of the brain
   D) high β-arrestin content in olfactory neurons
   E) sniffing

4. As a result of an automobile accident, a 10-year-old boy suffered damage to the brain including the periamygdaloid, piriform, and entorhinal cortices. Which of the following sensory deficits is he most likely to experience?
   A) visual disturbance
   B) hyperosmia
   C) auditory problems
   D) taste and odor abnormalities
   E) no major sensory deficits

5. Which of the following are *incorrectly* paired?
   A) ENaC : sour
   B) α-gustducin : bitter taste
   C) nucleus tractus solitarius : blood pressure
   D) Heschel sulcus : smell
   E) Ebner glands : taste acuity

6. Which of the following is true about olfactory transmission?
   A) An olfactory sensory neuron expresses a wide range of odorant receptors.
   B) Lateral inhibition within the olfactory glomeruli reduces the ability to distinguish between different types of odorant receptors.
   C) Conscious discrimination of odors is dependent on the pathway to the orbitofrontal cortex.
   D) Olfaction is closely related to gustation because odorant and gustatory receptors use the same central pathways.
   E) all of the above

7. Which of the following is not true about gustatory sensation?
   A) The sensory nerve fibers from the taste buds on the anterior two-thirds of the tongue travel in the chorda tympani branch of the facial nerve.
   B) The sensory nerve fibers from the taste buds on the posterior third of the tongue travel in the petrosal branch of the glossopharyngeal nerve.
   C) The pathway from taste buds on the left side of the tongue is transmitted ipsilaterally to the cerebral cortex.
   D) Sustentacular cells in the taste buds serve as stem cells to permit growth of new taste buds.
   E) The pathway from taste receptors includes synapses in the nucleus of the tractus solitarius in the brain stem and ventral posterior medial nucleus in the thalamus.

8. A 20-year-old woman was diagnosed with Bell palsy (damage to facial nerve). Which of the following symptoms is she likely to exhibit?
   A) loss of sense of taste
   B) facial twitching
   C) droopy eyelid
   D) ipsilateral facial paralysis
   E) all of the above

CHAPTER RESOURCES

Adler E, et al: A novel family of mammalian taste receptors. Cell 2000;100:693.

Anholt RRH: Odor recognition and olfactory transduction: The new frontier. Chem Senses 1991;16:421.

Boron WF, Boulpaep EL: *Medical Physiology.* Elsevier, 2005.

Gilbertson TA, Damak S, Margolskee RF: The molecular physiology of taste transduction. Curr Opin Neurobiol 2000;10:519.

Gold GH: Controversial issues in vertebrate olfactory transduction. Annu Rev Physiol 1999;61:857.

Herness HM, Gilbertson TA: Cellular mechanisms of taste transduction. Annu Rev Physiol 1999;61:873.

Kandel ER, Schwartz JH, Jessell TM (editors): *Principles of Neural Science,* 4th ed. McGraw-Hill, 2000.

Lindemann B: Receptors and transduction in taste. Nature 2001;413:219.

Mombaerts P: Genes and ligands for odorant, vomeronasal and taste receptors. Nature Rev Neurosci 2004;5:263.

Ronnett GV, Moon C: G proteins and olfactory signal transduction. Annu Rev Physiol 2002;64:189.

Shepherd GM, Singer MS, Greer CA: Olfactory receptors: A large gene family with broad affinities and multiple functions (Review). Neuroscientist 1996;2:262.

Stern P, Marks J (editors): Making sense of scents. Science 1999;286:703.

Electrical Activity of the Brain, Sleep–Wake States, & Circadian Rhythms

INTRODUCTION

Most of the various sensory pathways described in Chapters 11–14 relay impulses from sense organs via three- and four-neuron chains to particular loci in the cerebral cortex. The impulses are responsible for perception and localization of individual sensations. However, they must be processed in the awake brain to be perceived. At least in mammals, there is a spectrum of behavioral states ranging from deep sleep through light sleep, REM sleep, and the two awake states: relaxed awareness and awareness with concentrated attention. Discrete patterns of brain electrical activity correlate with each of these states. Feedback oscillations within the cerebral cortex and between the thalamus and the cortex serve as producers of this activity and possible determinants of the behavioral state. Arousal can be produced by sensory stimulation and by impulses ascending in the reticular core of the midbrain. Many of these activities have rhythmic fluctuations that are approximately 24 h in length; that is, they are **circadian.**

THALAMUS, CEREBRAL CORTEX, & RETICULAR FORMATION

THALAMIC NUCLEI

The thalamus is a large collection of neuronal groups within the diencephalons; it participates in sensory, motor, and limbic functions. Virtually all information that reaches the cortex is processed by the thalamus, leading to its being called the "gateway" to the cerebral cortex.

The thalamus can be divided into nuclei that project diffusely to wide regions of the neocortex and nuclei that project to specific discrete portions of the neocortex and limbic system. The nuclei that project to wide regions of the neocortex are the **midline and intralaminar nuclei.** The nuclei that project to specific areas include the specific sensory relay nuclei and the nuclei concerned with efferent control mechanisms. The **specific sensory relay nuclei** include the medial and lateral geniculate bodies, which relay auditory and visual impulses to the auditory and visual cortices; and the ventral posterior lateral (VPL) and ventral posteromedial, which relay somatosensory information to the postcentral gyrus. The ventral anterior and ventral lateral nuclei are concerned with motor function. They receive input from the basal ganglia and the cerebellum and project to the motor cortex. The anterior nuclei receive afferents from the mamillary bodies and project to the limbic cortex, which may be involved in memory and emotion. Most of the thalamic nuclei described are excitatory neurons that release glutamate. The thalamus also contains inhibitory neurons in the **thalamic reticular nucleus.** These neurons release GABA, and unlike the other thalamic neurons just described, their axons do not project to the cortex. Rather, they are thalamic interneurons and modulate the responses of other thalamic neurons to input coming from the cortex.

CORTICAL ORGANIZATION

The neocortex is generally arranged in six layers (Figure 15–1). The most common neuronal type is the pyramidal cell with an extensive vertical dendritic tree (Figures 15–1 and 15–2) that may reach to the cortical surface. Their cell bodies can be found in all cortical layers except layer I. The axons of these cells usually give off recurrent collaterals that turn back and synapse on the superficial portions of the dendritic trees. Afferents from the specific nuclei of the thalamus terminate primarily in cortical layer IV, whereas the nonspecific afferents are distributed to layers I–IV. Pyramidal neurons are the only projection neurons of the cortex, and they are excitatory neurons that release glutamate at their terminals. The other cortical cell types are local circuit neurons (interneurons) which have been classified based on their shape, pattern of projection, and neurotransmitter. Inhibitory interneurons (basket cells and chandelier cells) release GABA as their neurotransmitter. Basket cells have long axonal endings that surround the soma of pyramidal neurons;

FIGURE 15–1   **Structure of the cerebral cortex.** The cortical layers are indicated by the numbers. Golgi stain shows neuronal cell bodies and dendrites, Nissl stain shows cell bodies, and Weigert myelin sheath stain shows myelinated nerve fibers.   (Modified from Ranson SW, Clark SL: *The Anatomy of the Nervous System*, 10th ed. Saunders, 1959.)

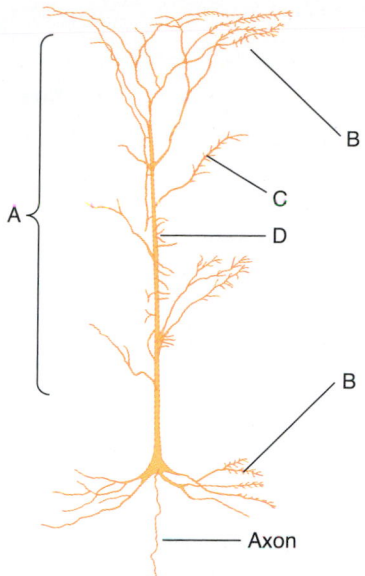

FIGURE 15–2 **Neocortical pyramidal cell, showing the distribution of neurons that terminate on it.** A denotes nonspecific afferents from the reticular formation and the thalamus; **B** denotes recurrent collaterals of pyramidal cell axons; **C** denotes commissural fibers from mirror image sites in the contralateral hemisphere; **D** denotes specific afferents from thalamic sensory relay nuclei. (Modified from Chow KL, Leiman AL: The structural and functional organization of the neocortex. Neurosci Res Program Bull 1970;8:157.)

they account for most inhibitory synapses on the pyramidal soma and dendrites. Chandelier cells are a powerful source of inhibition of pyramidal neurons because they have axonal endings that terminate exclusively on the initial segment of the pyramidal cell axon. Their terminal boutons form short vertical rows that resemble candlesticks, thus accounting for their name. Spiny stellate cells are excitatory interneurons that release glutamate as a neurotransmitter. These cells are located primarily in layer IV and are a major recipient of sensory information arising from the thalamus; they are an example of a multipolar neuron (Chapter 4) with local dendritic and axonal arborizations.

In addition to being organized into layers, the cortex is also organized into columns. Neurons within a column have similar response properties, suggesting they comprise a local processing network (eg, orientation and ocular dominance columns in the visual cortex).

RETICULAR FORMATION & RETICULAR ACTIVATING SYSTEM

The **reticular formation,** the phylogenetically old reticular core of the brain, occupies the midventral portion of the medulla and midbrain. It is primarily an anatomic area made up of various neural clusters and fibers with discrete functions. For example, it contains the cell bodies and fibers of many of the serotonergic, noradrenergic, adrenergic, and cholinergic

FIGURE 15–3 **Diagram showing the ascending reticular system in the human midbrain, its projections to the intralaminar nuclei of the thalamus, and the output from the intralaminar nuclei to many parts of the cerebral cortex.** Activation of these areas is shown by PET scans when subjects shift from a relaxed awake state to an attention-demanding task.

systems. It also contains many of the areas concerned with regulation of heart rate, blood pressure, and respiration. Some of the descending fibers in it inhibit transmission in sensory and motor pathways in the spinal cord; various reticular areas and the pathways from them are concerned with spasticity and adjustment of stretch reflexes. The **reticular activating system (RAS)** and related components of the brain concerned with consciousness and sleep are considered in this chapter.

The RAS is a complex polysynaptic pathway arising from the brain stem reticular formation with projections to the intralaminar and reticular nuclei of the thalamus which, in turn, project diffusely and nonspecifically to wide regions of the cortex (Figure 15–3). Collaterals funnel into it not only from the long ascending sensory tracts but also from the trigeminal, auditory, visual, and olfactory systems. The complexity of the neuron net and the degree of convergence in it abolish modality specificity, and most reticular neurons are activated with equal facility by different sensory stimuli. The system is therefore **nonspecific,** whereas the classic sensory pathways are **specific** in that the fibers in them are activated by only one type of sensory stimulation.

EVOKED CORTICAL POTENTIALS

The electrical events that occur in the cortex after stimulation of a sense organ can be monitored with an exploring electrode connected to another electrode at an indifferent point some distance away. A characteristic response is seen in animals under barbiturate anesthesia, which eliminates much of the background electrical activity. If the exploring electrode is over the primary receiving area for a particular sense, a surface-positive wave appears with a latency of 5 to 12 ms. This is followed by a small negative wave, and then a larger, more prolonged positive deflection frequently occurs with a latency of 20 to 80 ms. The first positive–negative wave sequence is the

primary evoked potential; the second is the **diffuse secondary response.**

The primary evoked potential is highly specific in its location and can be observed only where the pathways from a particular sense organ end. An electrode on the pial surface of the cortex samples activity to a depth of only 0.3–0.6 mm. The primary response is negative rather than positive when it is recorded with a microelectrode inserted in layers II–VI of the underlying cortex, and the negative wave within the cortex is followed by a positive wave. The negative–positive sequence indicates depolarization on the dendrites and somas of the cells in the cortex, followed by hyperpolarization. The positive–negative wave sequence recorded from the surface of the cortex occurs because the superficial cortical layers are positive relative to the initial negativity, then negative relative to the deep hyperpolarization. In unanesthetized animals or humans, the primary evoked potential is largely obscured by the spontaneous activity of the brain, but it can be demonstrated by superimposing multiple traces so that the background activity is averaged out. It is somewhat more diffuse in unanesthetized animals but still well localized compared with the diffuse secondary response.

The surface-positive diffuse secondary response, unlike the primary, is not highly localized. It appears at the same time over most of the cortex and is due to activity in projections from the midline and related thalamic nuclei.

PHYSIOLOGIC BASIS OF THE ELECTROENCEPHALOGRAM

The background electrical activity of the brain in unanesthetized animals was first described in the 19th century. Subsequently, it was analyzed in systematic fashion by the German psychiatrist Hans Berger, who introduced the term **electroencephalogram (EEG)** to denote the record of the variations in brain potential. The EEG can be recorded with scalp electrodes through the unopened skull or with electrodes on or in the brain. The term **electrocorticogram (ECoG)** is used for the record obtained with electrodes on the pial surface of the cortex.

EEG records may be **bipolar** or **unipolar.** Bipolar records show fluctuations in the potential difference between two cortical electrodes; unipolar records show the potential difference between a cortical electrode and a theoretically indifferent electrode on some part of the body distant from the cortex.

CORTICAL DIPOLES

The EEG recorded from the scalp is a measure of the summation of dendritic postsynaptic potentials rather than action potentials (Figure 15–4). The dendrites of the cortical cells are a forest of similarly oriented, densely packed units in the superficial layers of the cerebral cortex (Figure 15–1). Propagated potentials can be generated in dendrites. In addition, recurrent axon collaterals end on dendrites in the superficial layers. As ex-

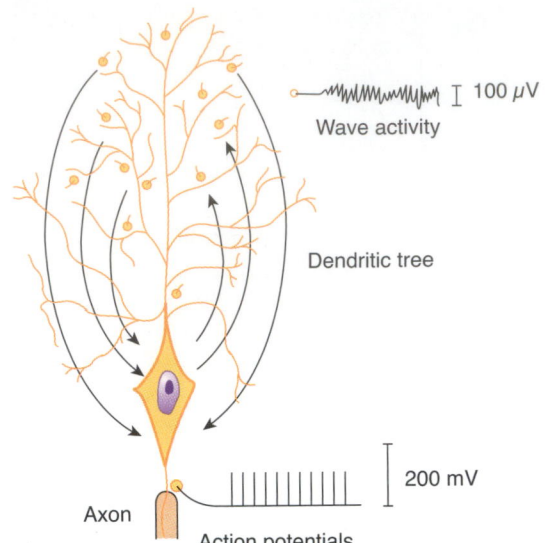

FIGURE 15–4 **Diagrammatic comparison of the electrical responses of the axon and the dendrites of a large cortical neuron.** Current flow to and from active synaptic knobs on the dendrites produces wave activity, while all-or-none action potentials are transmitted along the axon.

citatory and inhibitory endings on the dendrites of each cell become active, current flows into and out of these current sinks and sources from the rest of the dendritic processes and the cell body. The cell–dendrite relationship is therefore that of a constantly shifting dipole. Current flow in this dipole produces wave-like potential fluctuations in a volume conductor (Figure 15–4). When the sum of the dendritic activity is negative relative to the cell, the cell is depolarized and hyperexcitable; when it is positive, the cell is hyperpolarized and less excitable. The cerebellar cortex and the hippocampus are two other parts of the central nervous system (CNS) where many complex, parallel dendritic processes are located subpially over a layer of cells. In both areas, characteristic rhythmic fluctuations occur in surface potential similar to that observed in the cortical EEG.

CLINICAL USES OF THE EEG

The EEG is sometimes of value in localizing pathologic processes. When a collection of fluid overlies a portion of the cortex, activity over this area may be damped. This fact may aid in diagnosing and localizing conditions such as subdural hematomas. Lesions in the cerebral cortex cause local formation of irregular or slow waves that can be picked up in the EEG leads. Epileptogenic foci sometimes generate high-voltage waves that can be localized.

Epilepsy is a syndrome with multiple causes. In some forms, characteristic EEG patterns occur during seizures; between attacks, however, abnormalities are often difficult to demonstrate. Seizures are divided into **partial (focal) seizures** and **generalized seizures.**

Partial seizures originate in a small group of neurons and can result from head injury, brain infection, stroke, or tumor, but often the cause is unknown. Symptoms depend on the seizure focus. They are further subdivided into **simple partial seizures** (without loss of consciousness) and **complex partial seizures** (with altered consciousness). An example of a partial seizure is localized jerking movements in one hand progressing to clonic movements of the entire arm. **Auras** typically precede the onset of a partial seizure and include abnormal sensations. The time after the seizure until normal neurological function returns is called the **postictal period.**

Generalized seizures are associated with widespread electrical activity and involve both hemispheres simultaneously. They are further subdivided into **convulsive** and **nonconvulsive** categories depending on whether tonic or clonic movements occur. **Absence seizures** (formerly called petit mal seizures) are one of the forms of nonconvulsive generalized seizures characterized by a momentary loss of consciousness. They are associated with 3/s doublets, each consisting of a typical spike and rounded wave, and lasting about 10 s (Figure 15–5). They are not accompanied by auras or postictal periods.

The most common convulsive generalized seizure is **tonic–clonic seizure** (formerly called grand mal seizure). This is associated with sudden onset of contraction of limb muscles **(tonic phase)** lasting about 30 s, followed by a clonic phase with symmetric jerking of the limbs as a result of alternating contraction and relaxation **(clonic phase)** lasting 1–2 min. There is fast EEG activity during the tonic phase. Slow waves, each preceded by a spike, occur at the time of each clonic jerk. For a while after the attack, slow waves are present.

Recent research provides insight into a possible role of release of glutamate from astrocytes in the pathophysiology of epilepsy. Also, there is evidence to support the view that reorganization of astrocytes along with dendritic sprouting and new synapse formation form the structural basis for recurrent excitation in the epileptic brain. Clinical Box 15–1 describes information regarding the role of genetic mutations in some forms of epilepsy.

FIGURE 15–5   Absence seizures. Record of four cortical EEG leads from a 6-year-old boy who, during the recording, had one of his "blank spells" in which he was transiently unaware of his surroundings and blinked his eyelids. Time is indicated by the horizontal calibration line. (Reproduced with permission from Waxman SG: *Neuroanatomy with Clinical Correlations,* 25th ed. McGraw-Hill, 2003.)

SLEEP–WAKE CYCLE

ALPHA, BETA, & GAMMA RHYTHMS

In adult humans who are awake but at rest with the mind wandering and the eyes closed, the most prominent component of the EEG is a fairly regular pattern of waves at a frequency of 8–13 Hz and amplitude of 50–100 µV when recorded from the scalp. This pattern is the **alpha rhythm** (Figure 15–6). It is most marked in the parietal and occipital lobes and is associated with decreased levels of attention. A similar rhythm has been observed in a wide variety of mammalian species. There are some minor variations from species to species, but in all mammals the pattern is remarkably similar (see Clinical Box 15–2).

When attention is focused on something, the alpha rhythm is replaced by an irregular 13–30 Hz low-voltage activity, the **beta rhythm** (Figure 15–6). This phenomenon is called **alpha block** and can be produced by any form of sensory stimulation or mental concentration, such as solving arithmetic problems. Another term for this phenomenon is the **arousal** or **alerting response,** because it is correlated with the aroused, alert state. It has also been called **desynchronization,** because it represents breaking up of the obviously synchronized neural

(a)   Alpha rhythm (relaxed with eyes closed)

(b)   Beta rhythm (alert)

Time ⟶

FIGURE 15–6   **EEG records showing the alpha and beta rhythms.** When attention is focused on something, the 8–13 Hz alpha rhythm is replaced by an irregular 13–30 Hz low-voltage activity, the beta rhythm.   (From Widmaier EP, Raff H, Strang KT: *Vander's Human Physiology,* 11th ed. McGraw-Hill, 2008.)

activity necessary to produce regular waves. However, the rapid EEG activity seen in the alert state is also synchronized, but at a higher rate. Therefore, the term *desynchronization* is misleading. **Gamma oscillations** at 30–80 Hz are often seen when an individual is aroused and focuses attention on something. This is often replaced by irregular fast activity as the individual initiates motor activity in response to the stimulus.

SLEEP STAGES

There are two kinds of sleep: **rapid eye movement (REM) sleep** and **non-REM (NREM),** or **slow-wave sleep.** REM sleep is so named because of the characteristic eye movements that occur during this stage of sleep. NREM sleep is divided into four stages (Figure 15–7). A person falling asleep first enters stage 1, the EEG begins to show a low-voltage, mixed frequency pattern. A **theta rhythm** (4–7 Hz) can be seen at this early stage of slow-wave sleep. Throughout NREM sleep, there is some activity of

skeletal muscle but no eye movements occur. Stage 2 is marked by the appearance of sinusoidal waves called **sleep spindles** (12–14 Hz) and occasional high voltage biphasic waves called **K complexes.** In stage 3, a high-amplitude **delta rhythm** (0.5–4 Hz) dominates the EEG waves. Maximum slowing with large waves is seen in stage 4. Thus, the characteristic of deep sleep is a pattern of rhythmic slow waves, indicating marked **synchronization;** it is sometimes referred to as **slow-wave sleep.** Whereas theta and delta rhythms are normal during sleep, their appearance during wakefulness is a sign of brain dysfunction.

REM SLEEP

The high-amplitude slow waves seen in the EEG during sleep are periodically replaced by rapid, low-voltage EEG activity, which resembles that seen in the awake, aroused state and in stage 1 sleep (Figure 15–7). For this reason, REM sleep is also called **paradoxical sleep.** However, sleep is not interrupted; indeed, the threshold for arousal by sensory stimuli and by

	Awake	Sleep stage 1	2	3	4	REM

EOG

EMG

EEG

50 µV
1 s

FIGURE 15–7   **EEG and muscle activity during various stages of the sleep–wake cycle.** NREM sleep has four stages. Stage 1 is characterized by a slight slowing of the EEG. Stage 2 has high-amplitude K complexes and spindles. Stages 3 and 4 have slow, high-amplitude delta waves. REM sleep is characterized by eye movements, loss of muscle tone, and a low-amplitude, high-frequency activity pattern. The higher voltage activity in the EOG tracings during stages 2 and 3 reflect high amplitude EEG activity in the prefrontal areas rather than eye movements. EOG, electrooculogram registering eye movements; EMG, electromyogram registering skeletal muscle activity.   (From Kandel ER, Schwartz JH, Jessell TM [editors]: *Principles of Neural Science,* 4th ed. McGraw-Hill, 2000.)

stimulation of the reticular formation is elevated. Rapid, roving movements of the eyes occur during paradoxical sleep, and it is for this reason that it is also called REM sleep. Another characteristic of REM sleep is the occurrence of large phasic potentials that originate in the cholinergic neurons in the pons and pass rapidly to the lateral geniculate body and from there to the occipital cortex. They are called **pontogeniculo-occipital (PGO) spikes.** The tone of the skeletal muscles in the neck is markedly reduced during REM sleep.

Humans aroused at a time when they show the EEG characteristics of REM sleep generally report that they were dreaming, whereas individuals awakened from slow-wave sleep do not. This observation and other evidence indicate that REM sleep and dreaming are closely associated.

Positron emission tomography (PET) scans of humans in REM sleep show increased activity in the pontine area, amygdala, and anterior cingulate gyrus, but decreased activity in the prefrontal and parietal cortex. Activity in visual association areas is increased, but there is a decrease in the primary visual cortex. This is consistent with increased emotion and operation of a closed neural system cut off from the areas that relate brain activity to the external world.

DISTRIBUTION OF SLEEP STAGES

In a typical night of sleep, a young adult first enters NREM sleep, passes through stages 1 and 2, and spends 70–100 minutes in stages 3 and 4. Sleep then lightens, and a REM period follows. This cycle is repeated at intervals of about 90 minutes throughout the night (Figure 15–8). The cycles are similar, though there is less stage 3 and 4 sleep and more REM sleep toward morning. Thus, four to six REM periods occur per night. REM sleep occupies 80% of total sleep time in premature infants (Figure 15–9) and 50% in full-term neonates. Thereafter, the proportion of REM sleep falls rapidly and plateaus at about 25% until it falls further in old age. Children have more total sleep time and stage 4 sleep than adults.

THALAMOCORTICAL LOOP

A circuit linking the cortex and thalamus is thought to be important in generating patterns of brain activity in sleep–wake states. Figure 15–10 shows properties of activity in such a

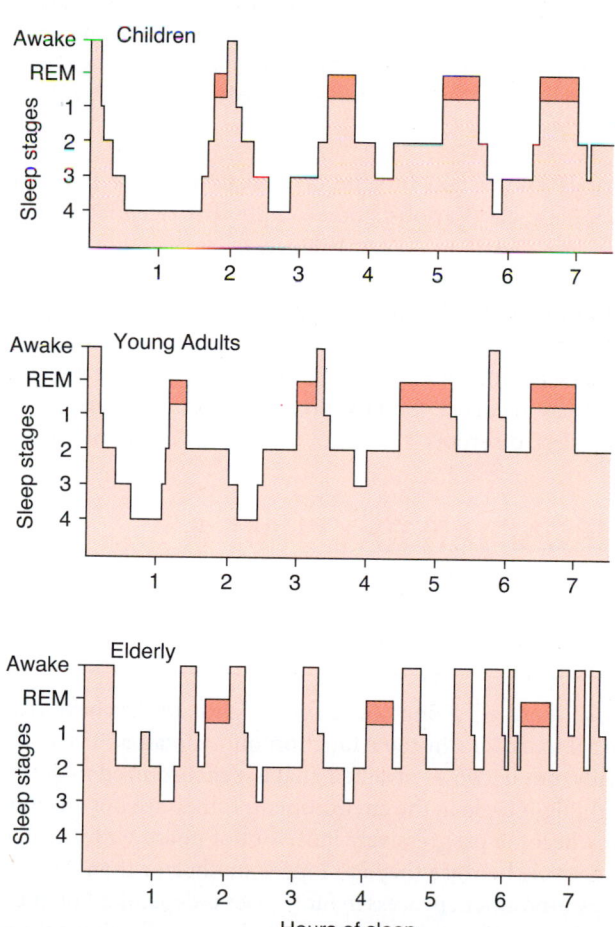

FIGURE 15–8 **Normal sleep cycles at various ages.** REM sleep is indicated by the darker colored areas. (Reproduced with permission from Kales AM, Kales JD: Sleep disorders. N Engl J Med 1974;290:487.)

FIGURE 15–9 **Changes in human sleep pattern with age.** Each plot shows data points for the ages of 6, 10, 21, 30, 69, and 84 years. (Data from Kandel ER, Schwartz JH, Jessel TM [editors]: *Principles of Neural Science,* 4th ed. McGraw-Hill, 2000.)

FIGURE 15–10   **Correlation between behavioral states, EEG, and single-cell responses in the cerebral cortex and thalamus.** The EEG is characterized by high-frequency oscillations in the awake state and low-frequency rhythms during sleep. Thalamic and cortical neurons can also show different patterns of rhythmic activity. Thalamocortical neurons show slow rhythmic oscillations during deep sleep, and fire tonic trains of action potentials in the awake state. Most pyramidal neurons in the cortex generate only tonic trains of action potentials, although others may participate in the generation of high frequency rhythms through activation of rhythmic bursts of spikes. The thalamus and cerebral cortex are connected together in a loop. (Modified from McCormick DA: Are thalamocortical rhythms the Rosetta stone of a subset of neurological disorders? Nat Med 1999;5:1349.)

thalamocortical circuit hypothesized to be involved in generating rhythmic activity. Although not shown, inhibitory thalamic reticular neurons are elements of this network. The EEG shows the characteristic awake, light sleep, and deep sleep patterns of activity described above. Likewise, recordings from individual thalamic and cortical neurons show different patterns of rhythmic activity. In the waking state, corticocortical and thalamocortical networks generate higher-frequency rhythmic activity (30–80 Hz; gamma rhythm). This rhythm may be generated within the cells and networks of the cerebral cortex or within thalamocortical loops. The gamma rhythm has been suggested as a mechanism to "bind" together diverse sensory information into a single percept and action, but this theory is still controversial. In fact, disturbances in the integrity of this thalamocortical loop and its interaction with other brain structures may underlie some neurological disorders, including seizure activity.

IMPORTANCE OF SLEEP

Sleep has persisted throughout evolution of mammals and birds, so it is likely that it is functionally important. Indeed, if humans are awakened every time they show REM sleep, then permitted to sleep without interruption, they show a great deal more than the normal amount of REM sleep for a few nights. Relatively prolonged REM deprivation does not seem to have adverse psychological effects. Rats deprived of sleep for long

periods lose weight in spite of increased caloric intake and eventually die. Various studies imply that sleep is needed to maintain metabolic-caloric balance, thermal equilibrium, and immune competence.

In experimental animals, sleep is necessary for learning and memory consolidation. Learning sessions do not improve performance until a period of slow-wave or slow-wave plus REM sleep has occurred. Clinical Box 15–3 describes several common sleep disorders.

CIRCADIAN RHYTHMS & THE SLEEP–WAKE CYCLE

CIRCADIAN RHYTHMS

Most, if not all, living cells in plants and animals have rhythmic fluctuations in their function on a circadian cycle. Normally they become entrained, that is, synchronized to the day–night light cycle in the environment. If they are not entrained, they become progressively more out of phase with the light–dark cycle because they are longer or shorter than 24 hours. The entrainment process in most cases is dependent on the **suprachiasmatic nuclei (SCN)** located bilaterally above the optic chiasm (Figure 15–11). These nuclei receive information about the light–dark cycle via a special neural pathway, the **retinohypothalamic fibers.** Efferents from the SCN initiate

Sleep Disorders

Narcolepsy is a chronic neurological disorder caused by the brain's inability to regulate sleep–wake cycles normally in which there is a sudden loss of voluntary muscle tone **(cataplexy)**, an eventual irresistible urge to sleep during daytime, and possibly also brief episodes of total paralysis at the beginning or end of sleep. Narcolepsy is characterized by a sudden onset of REM sleep, unlike normal sleep which begins with NREM, slow-wave sleep. The prevalence of narcolepsy ranges from 1 in 600 in Japan to 1 in 500,000 in Israel, with 1 in 1000 Americans being affected. Narcolepsy has a familial incidence strongly associated with a class II antigen of the major histocompatibility complex on chromosome 6 at the HLA-DR2 or HLA-DQW1 locus, implying a genetic susceptibility to narcolepsy. The HLA complexes are interrelated genes that regulate the immune system. Brains from humans with narcolepsy often contain fewer **hypocretin (orexin)**-producing neurons in the hypothalamus. It is thought that the HLA complex may increase susceptibility to an immune attack on these neurons, leading to their degeneration.

Obstructive sleep apnea (OSA) is the most common cause of daytime sleepiness due to fragmented sleep at night and affects about 24% of middle-aged men and 9% of women in the United States. Breathing ceases for more than 10 s during frequent episodes of obstruction of the upper airway (especially the pharynx) due to reduction in muscle tone. The apnea causes brief arousals from sleep in order to reestablish upper airway tone. Snoring is a common patient complaint. There is actually not a reduction in total sleep time, but

individuals with OSA experience a much greater time in stage 1 NREM sleep (from an average of 10% of total sleep to 30–50%) and a marked reduction in slow-wave sleep (stages 3 and 4 NREM sleep). The pathophysiology of OSA includes both a reduction in neuromuscular tone at the onset of sleep and a change in the central respiratory drive.

Periodic limb movement disorder (PLMD) is a stereotypical rhythmic extension of the big toe and dorsiflexion of the ankle and knee during sleep lasting for about 0.5 to 10 s and recurring at intervals of 20 to 90 s. Movements can actually range from shallow, continual movement of the ankle or toes, to wild and strenuous kicking and flailing of the legs and arms. Electromyograph (EMG) recordings show bursts of activity during the first hours of NREM sleep associated with brief EEG signs of arousal. The duration of stage 1 NREM sleep may be increased and that of stages 3 and 4 may be decreased compared to age-matched controls. PLMD is reported to occur in 5% of individuals between the ages of 30 and 50 years and increases to 44% of those over the age of 65. PLMD is similar to **restless leg syndrome** in which individuals have an irresistible urge to move their legs while at rest all day long.

Sleepwalking **(somnambulism)**, bed-wetting **(nocturnal enuresis)**, and **night terrors** are referred to as **parasomnias,** which are sleep disorders associated with arousal from NREM and REM sleep. Episodes of sleepwalking are more common in children than in adults and occur predominantly in males. They may last several minutes. Somnambulists walk with their eyes open and avoid obstacles, but when awakened they cannot recall the episodes.

---

neural and humoral signals that entrain a wide variety of well-known circadian rhythms including the sleep–wake cycle and the secretion of the pineal hormone melatonin.

Evidence suggests that the SCN have two peaks of circadian activity. This may correlate with the observation that exposure to bright light can either advance, delay, or have no effect on the sleep–wake cycle in humans depending on the time of day when it is experienced. During the usual daytime it has no effect, but just after dark it delays the onset of the sleep period, and just before dawn it accelerates the onset of the next sleep period. Injections of melatonin have similar effects. In experimental animals, exposure to light turns on immediate-early genes in the SCN, but only at times during the circadian cycle when light is capable of influencing entrainment. Stimulation during the day is ineffective.

NEUROCHEMICAL MECHANISMS PROMOTING SLEEP & AROUSAL

Transitions between sleep and wakefulness manifest a circadian rhythm consisting of an average of 8 h of sleep and 16 h of wakefulness. Nuclei in both the brain stem and hypothalamus are critical for the transitions between these states of consciousness. A classic study by Moruzzi and Magoun in 1949 showed that high-frequency stimulation of the midbrain reticular formation (the RAS) produces the EEG alerting response and arouses a sleeping animal. Damage to the area produces a comatose state. Electrical stimulation of the posterior hypothalamus also produces arousal similar to that elicited by stimulation of the midbrain, while electrical stimulation of the anterior hypothalamus and adjacent basal forebrain region induces sleep.

As described above, the brainstem RAS is composed of several groups of neurons which release **norepinephrine, serotonin,** or **acetylcholine.** The locations of these neuronal populations are shown in Figure 7–2. In the case of the forebrain neurons

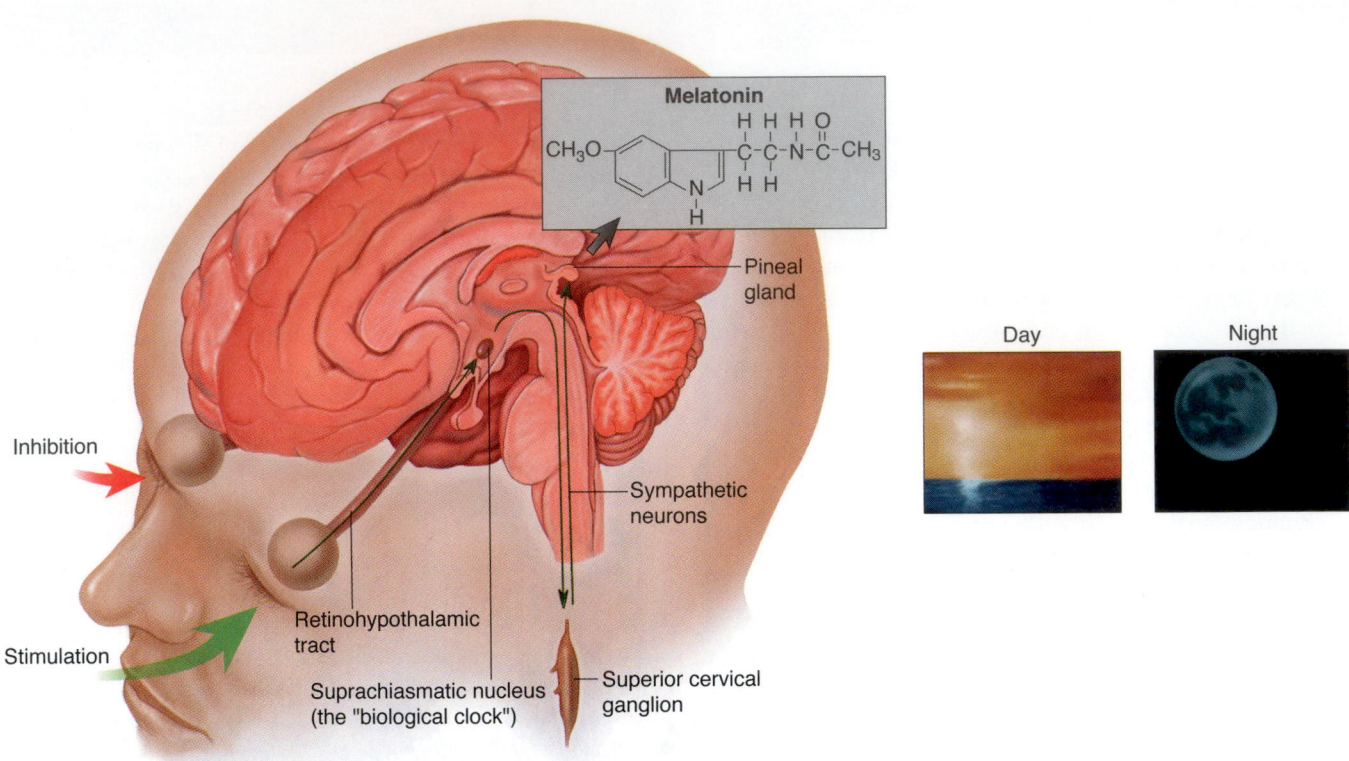

FIGURE 15–11   **Secretion of melatonin.** Retinohypothalamic fibers synapse in the suprachiasmatic nuclei (SCN), and there are connections from the SCN to sympathetic preganglionic neurons in the spinal cord that project to the superior cervical ganglion. Postganglionic neurons project from this ganglion to the pineal gland that secretes melatonin. The cyclic activity of SCN sets up a circadian rhythm for melatonin release. This rhythm is entrained to light/dark cycles by neurons in the retina.   (From Fox SI: *Human Physiology.* McGraw-Hill, 2008.)

involved in control of the sleep–wake cycles, **preoptic neurons** in the hypothalamus release **GABA** and **posterior hypothalamic neurons** release **histamine.**

One theory regarding the basis for transitions from sleep to wakefulness involves alternating reciprocal activity of different groups of RAS neurons. In this model (Figure 15–12), wakefulness and REM sleep are at opposite extremes. When the activity of norepinephrine- and serotonin-containing neurons (locus coeruleus and raphé nuclei) is dominant, there is a reduced level of activity in acetylcholine-containing neurons in the pontine reticular formation. This pattern of activity contributes to the appearance of the awake state. The reverse of this pattern leads to REM sleep. When there is a more even balance in the activity of the aminergic and cholinergic neurons, NREM sleep occurs.

In addition, an increased release of GABA and reduced release of histamine increase the likelihood of NREM sleep via deactivation of the thalamus and cortex. Wakefulness occurs when GABA release is reduced and histamine release is increased.

MELATONIN AND THE SLEEP–WAKE STATE

In addition to the previously described neurochemical mechanisms promoting changes in the sleep–wake state, **melatonin** release from the richly vascularized **pineal gland** plays a role in sleep mechanisms (Figure 15–11). The pineal arises from the roof of the third ventricle in the diencephalon and is encapsulated by the meninges. The pineal stroma contains glial cells and pinealocytes with features suggesting that they have a secretory function. Like other endocrine glands, it has highly permeable fenestrated capillaries. In infants, the pineal is large and the cells tend to be arranged in alveoli. It begins to involute before puberty and small concretions of calcium phosphate and carbonate (**pineal sand**) appear in the tissue. Because the concretions are radiopaque, the pineal is often visible on x-ray films of the skull in adults. Displacement of a calcified pineal from its normal position indicates the presence of a space-occupying lesion such as a tumor in the brain.

Melatonin and the enzymes responsible for its synthesis from serotonin by N-acetylation and O-methylation are present in pineal pinealocytes, and the hormone is secreted by them into the blood and the cerebrospinal fluid (Figure 15–13). Two

FIGURE 15–12   **A model of how alternating activity of brain stem and hypothalamic neurons may influence the different states of consciousness.** (From Widmaier EP, Raff H, Strang KT: *Vander's Human Physiology*, 11th ed. McGraw-Hill, 2008.)

FIGURE 15–13   **Diurnal rhythms of compounds involved in melatonin synthesis in the pineal.** Melatonin and the enzymes responsible for its synthesis from serotonin are found in pineal pinealocytes; melatonin is secreted into the bloodstream. Melatonin synthesis and secretion are increased during the dark period (shaded area) and maintained at a low level during the light period.

form of 6-hydroxy conjugates and 6-sulfatoxymelatonin. The pathway by which the brain metabolizes melatonin is unsettled but may involve cleavage of the indole nucleus.

The discharge of the sympathetic nerves to the pineal is entrained to the light–dark cycle in the environment via the retinohypothalamic nerve fibers to the SCN. From the hypothalamus, descending pathways converge onto preganglionic sympathetic neurons that in turn innervate the superior cervical ganglion, the site of origin of the postganglionic neurons to the pineal gland.

CHAPTER SUMMARY

- The major rhythms in the EEG are alpha (8–13 Hz), beta (13–30 Hz), theta (4–7 Hz), delta (0.5–4 Hz), and gamma (30–80 Hz) oscillations.
- The EEG is of some value in localizing pathologic processes, and it is useful in characterizing different types of epilepsy.
- Throughout NREM sleep, there is some activity of skeletal muscle. A theta rhythm can be seen during stage 1 of sleep. Stage 2 is marked by the appearance of sleep spindles and occasional K complexes. In stage 3, a delta rhythm is dominant. Maximum slowing with slow waves is seen in stage 4.
- REM sleep is characterized by low-voltage, high-frequency EEG activity and rapid, roving movements of the eyes.
- A young adult typically passes through stages 1 and 2, and spends 70–100 min in stages 3 and 4. Sleep then lightens, and a REM period follows. This cycle repeats at 90-min intervals throughout the night. REM sleep occupies 50% of total sleep time in full-term neonates; this proportion declines rapidly and plateaus at about 25% until it falls further in old age.
- Transitions from sleep to wakefulness may involve alternating reciprocal activity of different groups of RAS neurons. When the activity of norepinephrine- and serotonin-containing

melatonin-binding sites have been characterized: a high-affinity ML1 site and a low affinity ML2 site. Two subtypes of the ML1 receptor have been cloned: Mel 1a and Mel 1b. All the receptors are coupled to G proteins, with ML1 receptors inhibiting adenylyl cyclase and ML2 receptors stimulating phosphoinositide hydrolysis.

The diurnal change in melatonin secretion may function as a timing signal to coordinate events with the light–dark cycle in the environment. Melatonin synthesis and secretion are increased during the dark period of the day and maintained at a low level during daylight hours (Figure 15–13). This diurnal variation in secretion is brought about by norepinephrine secreted by the postganglionic sympathetic nerves that innervate the pineal gland (Figure 15–11). Norepinephrine acts via β-adrenergic receptors to increase intracellular cAMP, and the cAMP in turn produces a marked increase in *N*-acetyltransferase activity. This results in increased melatonin synthesis and secretion. Circulating melatonin is rapidly metabolized in the liver by 6-hydroxylation followed by conjugation, and over 90% of the melatonin that appears in the urine is in the neurons is dominant, the activity in acetylcholine-containing neurons is reduced, leading to the appearance of wakefulness. The reverse of this pattern leads to REM sleep. Also, wakefulness occurs when GABA release is reduced and histamine release is increased.

■ The entrainment of biological processes to the light–dark cycle is regulated by the SCN.

■ The diurnal change in melatonin secretion from serotonin in the pineal gland may function as a timing signal to coordinate events with the light–dark cycle, including the sleep–wake cycle.

MULTIPLE-CHOICE QUESTIONS

For all questions, select the single best answer unless otherwise directed.

1. In a healthy, alert adult sitting with the eyes closed, the dominant EEG rhythm observed with electrodes over the occipital lobes is
   A) delta (0.5–4 Hz).
   B) theta (4–7 Hz).
   C) alpha (8–13 Hz).
   D) beta (18–30 Hz).
   E) fast, irregular low-voltage activity.

2. Which of the following pattern of changes in central neurotransmitters/neuromodulators are associated with the transition from NREM to wakefulness?
   A) decrease in norepinephrine, increase in epinephrine, increase in acetylcholine, decrease in histamine, and decrease in GABA
   B) decrease in norepinephrine, increase in epinephrine, increase in acetylcholine, decrease in histamine, and increase in GABA
   C) decrease in norepinephrine, decrease in epinephrine, increase in acetylcholine, increase in histamine, and increase in GABA
   D) increase in norepinephrine, increase in epinephrine, decrease in acetylcholine, increase in histamine, and decrease in GABA
   E) increase in norepinephrine, decrease in epinephrine, decrease in acetylcholine, increase in histamine, and decrease in GABA

3. A gamma rhythm (30–80 Hz)
   A) is characteristic of seizure activity.
   B) is seen in an individual who is awake but not focused.
   C) may be a mechanism to bind together sensory information into a single percept and action.
   D) is independent of thalamocortical loops.
   E) is generated in the hippocampus.

4. Melatonin secretion would probably not be increased by
   A) stimulation of the superior cervical ganglia.
   B) intravenous infusion of tryptophan.
   C) intravenous infusion of epinephrine.
   D) stimulation of the optic nerve.
   E) induction of pineal hydroxyindole-*O*-methyltransferase.

5. Absence seizures
   A) are a form of nonconvulsive generalized seizures accompanied by momentary loss of consciousness.
   B) are a form of complex partial seizures accompanied by momentary loss of consciousness.
   C) are a form of nonconvulsive generalized seizures without a loss of consciousness.
   D) are a form of simple partial seizures without a loss of consciousness.
   E) are a form of convulsive generalized seizures accompanied by momentary loss of consciousness.

6. Narcolepsy is triggered by abnormalities in the
   A) skeletal muscles.
   B) medulla oblongata.
   C) hypothalamus.
   D) olfactory bulb.
   E) neocortex.

CHAPTER RESOURCES

Blackman S: *Consciousness: An Introduction.* Oxford University Press, 2004.

Kandel ER, Schwartz JH, Jessell TM (editors): *Principles of Neural Science,* 4th ed. McGraw-Hill, 2000.

McCormick DA, Contreras D: Of the cellular and network bases of epileptic seizures. Annu Rev Physiol 2001;63:815.

Moruzzi G, Magoun HW: Brain stem reticular formation and activation of the EEG. Electroencephalogr Clin Neurophysiol 1949;1:455.

Oberheim NA, et al: Loss of astrocytic domain organization in the epileptic brain. J Neurosci 2008;28:3264.

Siegel JM: Narcolepsy. Sci Am 2000;282:76.

Squire LR, et al (editors): *Fundamental Neuroscience,* 3rd ed. Academic Press, 2008.

Steinlein O: Genetic mechanisms that underlie epilepsy. Nat Rev Neurosci 2004;5:400.

Steriade M, McCarley RW: *Brain Stem Control of Wakefulness and Sleep.* Plenum, 1990.

Steriade M, Paré D: *Gating in Cerebral Networks.* Cambridge University Press, 2007.

Thorpy M (editor): *Handbook of Sleep Disorders.* Marcel Dekker, 1990.

Waxman SG: *Neuroanatomy with Clinical Correlations,* 25th ed. McGraw-Hill, 2003.

Control of Posture & Movement

OBJECTIVES

After studying this chapter, you should be able to:

- Describe how skilled movements are planned and carried out.
- Name the posture-regulating parts of the central nervous system and discuss the role of each.
- Define spinal shock and describe the initial and long-term changes in spinal reflexes that follow transection of the spinal cord.
- Define decerebrate and decorticate rigidity, and comment on the cause and physiologic significance of each.
- Describe the basal ganglia and list the pathways that interconnect them, along with the neurotransmitters in each pathway.
- Describe and explain the symptoms of Parkinson disease and Huntington disease.
- List the pathways to and from the cerebellum and the connections of each within the cerebellum.
- Discuss the functions of the cerebellum and the neurologic abnormalities produced by diseases of this part of the brain.

INTRODUCTION

Somatic motor activity depends ultimately on the pattern and rate of discharge of the spinal motor neurons and homologous neurons in the motor nuclei of the cranial nerves. These neurons, the final common paths to skeletal muscle, are bombarded by impulses from an immense array of descending pathways, other spinal neurons, and peripheral afferents. Some of these inputs end directly on α-motor neurons, but many exert their effects via interneurons or via γ-motor neurons to the muscle spindles and back through the Ia afferent fibers to the spinal cord. It is the integrated activity of these multiple inputs from spinal, medullary, midbrain, and cortical levels that regulates the posture of the body and makes coordinated movement possible.

The inputs converging on motor neurons subserve three functions: they bring about voluntary activity, they adjust body posture to provide a stable background for movement, and they coordinate the action of the various muscles to make movements smooth and precise. The patterns of voluntary activity are planned within the brain, and the commands are sent to the muscles primarily via the corticospinal and corticobulbar systems. Posture is continually adjusted not only before but also during movement by descending brain stem pathways and peripheral afferents. Movement is smoothed and coordinated by the medial and intermediate portions of the cerebellum (spinocerebellum) and its connections. The basal ganglia and the lateral portions of the cerebellum (cerebrocerebellum) are part of a feedback circuit to the premotor and motor cortex that is concerned with planning and organizing voluntary movement.

FIGURE 16–1   Control of voluntary movement. Commands for voluntary movement originate in cortical association areas. The cortex, basal ganglia, and cerebellum work cooperatively to plan movements. Movement executed by the cortex is relayed via the corticospinal tracts and corticobulbar tracts to motor neurons. The cerebellum provides feedback to adjust and smooth movement.

GENERAL PRINCIPLES

ORGANIZATION

There are two types of motor output: reflexive (involuntary) and voluntary. A subdivision of reflex responses includes some rhythmic movements such as swallowing, chewing, scratching, and walking, which are largely involuntary but subject to voluntary adjustment and control.

To move a limb, the brain must plan a movement, arrange appropriate motion at many different joints at the same time, and adjust the motion by comparing plan with performance. The motor system "learns by doing" and performance improves with repetition. This involves synaptic plasticity.

There is considerable evidence for the general motor control scheme shown in Figure 16–1. Commands for voluntary movement originate in cortical association areas. The movements are planned in the cortex as well as in the basal ganglia and the lateral portions of the cerebellar hemispheres, as indicated by increased electrical activity before the movement. The basal ganglia and cerebellum funnel information to the premotor and motor cortex by way of the thalamus. Motor commands from the motor cortex are relayed in large part via the corticospinal tracts to the spinal cord and the corresponding corticobulbar tracts to motor neurons in the brain stem. However, collaterals from these pathways and a few direct connections from the motor cortex end on brain stem nuclei, which also project to motor neurons in the brain stem and spinal cord. These pathways can also mediate voluntary movement. Movement sets up alterations in sensory input from the special senses and from muscles, tendons, joints, and the skin. This feedback information, which adjusts and smoothes movement, is relayed directly to the motor cortex and to the spinocerebellum. The spinocerebellum projects in turn to the brain stem. The main brain stem pathways that are concerned with posture and coordination are the rubrospinal, reticulospinal, tectospinal, and vestibulospinal tracts.

CONTROL OF AXIAL & DISTAL MUSCLES

Within the brain stem and spinal cord, pathways and neurons that are concerned with the control of skeletal muscles of the trunk and proximal portions of the limbs are located medially or ventrally, whereas pathways and neurons that are concerned with the control of skeletal muscles in the distal portions of the limbs are located laterally. The axial muscles are concerned with postural adjustments and gross movements, whereas the distal limb muscles mediate fine, skilled movements. Thus, for example, neurons in the medial portion of the ventral horn innervate proximal limb muscles, particularly the flexors, whereas lateral ventral horn neurons innervate distal limb muscles. Similarly, the ventral corticospinal tract and medial descending brain stem pathways (tectospinal, reticulospinal, and vestibulospinal tracts) are concerned with adjustments of proximal muscles and posture, whereas the lateral corticospinal and rubrospinal tracts are concerned with distal limb muscles and, particularly in the case of the lateral corticospinal tract, with skilled voluntary movements. Phylogenetically, the lateral pathways are newer. More details about these motor pathways are provided below.

CORTICOSPINAL & CORTICOBULBAR TRACTS

DESCENDING PROJECTIONS

The axons of neurons from the motor cortex that project to spinal motor neurons form the **corticospinal tracts,** a large bundle of about 1 million fibers. About 80% of these fibers cross the midline in the medullary pyramids to form the **lateral corticospinal tract** (Figure 16–2). The remaining 20% make up the **ventral corticospinal tract,** which does not cross the midline until it reaches the level of the spinal cord at which it terminates. Lateral corticospinal tract neurons make monosynaptic connections to motor neurons, especially those concerned with skilled movements. Corticospinal tract neurons also synapse on spinal interneurons antecedent to motor

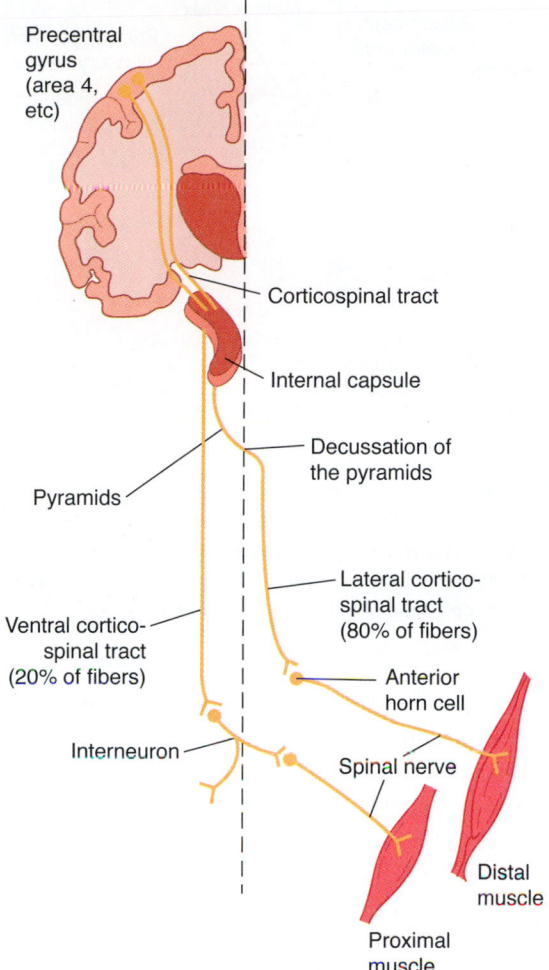

FIGURE 16–2 The corticospinal tracts. This tract originates in the precentral gyrus and passes through the internal capsule. Most fibers decussate in the pyramids and descend in the lateral white matter of the spinal cord to form the lateral division of the tract which can make monosynaptic connections with spinal motor neurons. The ventral division of the tract remains uncrossed until reaching the spinal cord where axons terminate on spinal interneurons antecedent to motor neurons.

neurons; this indirect pathway is important in coordinating groups of muscles.

The trajectory from the cortex to the spinal cord passes through the corona radiata to the posterior limb of the internal capsule. Within the midbrain they traverse the cerebral peduncle and the basilar pons until they reach the medullary pyramids on their way to the spinal cord.

The **corticobulbar tract** is composed of the fibers that pass from the motor cortex to motor neurons in the trigeminal, facial, and hypoglossal nuclei. Corticobulbar neurons end either directly on the cranial nerve nuclei or on their antecedent interneurons within the brain stem. Their axons traverse through the genu of the internal capsule, the cerebral peduncle (medial to corticospinal tract neurons), to descend with corticospinal tract fibers in the pons and medulla.

The motor system can be divided into lower and upper motor neurons. **Lower motor neurons** refer to the spinal and cranial motor neurons that directly innervate skeletal muscles. **Upper motor neurons** are those in the cortex and brain stem that activate the lower motor neurons. The pathophysiological responses to damage to lower and upper motor neurons are very distinctive (see Clinical Box 16–1).

ORIGINS OF CORTICOSPINAL & CORTICOBULBAR TRACTS

Corticospinal and corticobulbar tract neurons are pyramidal shaped and located in layer V of the cerebral cortex (see Chapter 11). The cortical areas from which these tracts originate were identified on the basis of electrical stimulation that produced prompt discrete movement. Figure 16–3 shows the major cortical regions involved in motor control. About 31% of the corticospinal tract neurons are from the **primary motor cortex (M1; Brodmann's area 4).** This region is in the precentral gyrus of the frontal lobe, extending into the central sulcus. The **premotor cortex** and **supplementary motor cortex (Brodmann's area 6)** account for 29% of the corticospinal tract neurons. The premotor area is anterior to the precentral gyrus, on the lateral and medial cortical surface; and the supplementary motor area is on and above the superior bank of the cingulate sulcus on the medial side of the hemisphere. The other 40% of corticospinal tract neurons originate in the **parietal lobe (Brodmann's area 5, 7)** and **primary somatosensory area (Brodmann's area 3, 1, 2)** in the postcentral gyrus.

MOTOR CORTEX & VOLUNTARY MOVEMENT

PRIMARY MOTOR CORTEX

By means of stimulation experiments in patients undergoing craniotomy under local anesthesia, it has been possible to outline most of the motor projections from M1. These have been confirmed in unanesthetized unoperated humans by PET scan and fMRI (Figure 16–4). The various parts of the body are represented in the precentral gyrus, with the feet at the top of the gyrus and the face at the bottom (Figure 16–5). The facial area is represented bilaterally, but the rest of the representation is generally unilateral, with the cortical motor area controlling the musculature on the opposite side of the body. The cortical representation of each body part is proportionate in size to the skill with which the part is used in fine, voluntary movement. The areas involved in speech and hand movements are especially large in the cortex; use of the pharynx, lips, and tongue to form words and of the fingers and apposable thumbs to manipulate the environment are activities in which humans are especially skilled. A somatotopic organization continues throughout the corticospinal and corticobulbar pathways.

Lower versus Upper Motor Neuron Damage

Lower motor neurons are those whose axons terminate on skeletal muscles. Damage to these neurons is associated with **flaccid paralysis, muscular atrophy, fasciculations** (visible muscle twitches that appear as flickers under the skin), **hypotonia** (decreased muscle tone), and **hyporeflexia** or **areflexia**. An example of a disease that leads to lower motor neuron damage is **amyotrophic lateral sclerosis (ALS)**. "Amyotrophic" means "no muscle nourishment" and describes the atrophy that muscles undergo because of disuse. "Sclerosis" refers to the hardness felt when a pathologist examines the spinal cord on autopsy; the hardness is due to proliferation of astrocytes and scarring of the lateral columns of the spinal cord. ALS is a selective, progressive degeneration of α-motor neurons. This fatal disease is also known as **Lou Gehrig disease** because Gehrig, a famous American baseball player, died of it. The worldwide annual incidence of ALS has been estimated to be 0.5–3 cases per 100,000 people. Most cases are sporadic, but 5–10% of the cases are familial. Forty percent of the familial cases have a mutation in the gene for Cu/Zn superoxide dismutase *(SOD-1)* on chromosome 21. SOD is a free radical scavenger that reduces oxidative stress. A defective *SOD-1* gene permits free radicals to accumulate and kill neurons. The disease has no racial, socioeconomic, or ethnic boundaries. The life expectancy of ALS patients is usually 3–5 years after diagnosis. ALS is most commonly diagnosed in middle age and affects men more often than women. The worldwide incidence of ALS is 2 per 100,000 of total population. The causes of ALS are unclear, but possibilities include viruses, neurotoxins, heavy metals, DNA defects (especially in familial ALS), immune system abnormalities, and enzyme abnormalities.

Upper motor neurons typically refer to corticospinal tract neurons that innervate spinal motor neurons, but they can also include brain stem neurons that control spinal motor neurons. Damage to these neurons initially causes muscles to become weak and flaccid but eventually leads to **spasticity, hypertonia** (increased resistance to passive movement), **hyperactive stretch reflexes,** and abnormal plantar extensor reflex **(Babinski sign)**. The Babinski sign is dorsiflexion of the great toe and fanning of the other toes when the lateral aspect of the sole of the foot is scratched. In adults, the normal response to this stimulation is plantar flexion in all the toes. The Babinski sign is believed to be a flexor withdrawal reflex that is normally held in check by the lateral corticospinal system. It is of value in the localization of disease processes, but its physiologic significance is unknown. However, in infants whose corticospinal tracts are not well developed, dorsiflexion of the great toe and fanning of the other toes is the natural response to stimuli applied to the sole of the foot.

FIGURE 16–3   A view of the human cerebral cortex, showing the motor cortex (Brodmann's area 4) and other areas concerned with control of voluntary movement, along with the numbers assigned to the regions by Brodmann. (Reproduced with permission from Kandel ER, Schwartz JH, Jessell TM [editors]: *Principles of Neural Science,* 4th ed. McGraw-Hill, 2000.)

The conditions under which the human stimulation studies were performed precluded stimulation of the banks of the sulci and other inaccessible areas. Meticulous study has shown that in monkeys there is a regular representation of the body, with the axial musculature and the proximal portions of the limbs represented along the anterior edge of the precentral gyrus and the distal part of the limbs along the posterior edge.

The cells in the cortical motor areas are arranged in columns. The ability to elicit discrete movements of a single muscle by electrical stimulation of a column within M1 led to the view that this area was responsible for control of individual muscles. More recent work has shown that neurons in several cortical columns project to the same muscle; in fact, most stimuli activate more than one muscle. Moreover, the cells in each column receive fairly extensive sensory input from the

FIGURE 16–4   **Hand area of motor cortex demonstrated by functional magnetic resonance imaging (fMRI) in a 7-year-old boy.** Changes in activity associated with squeezing a rubber ball with the right hand are shown in white and with the left hand in black. (Reproduced with permission from Waxman SG: *Neuroanatomy with Clinical Correlations,* 25th ed. McGraw-Hill, 2003.)

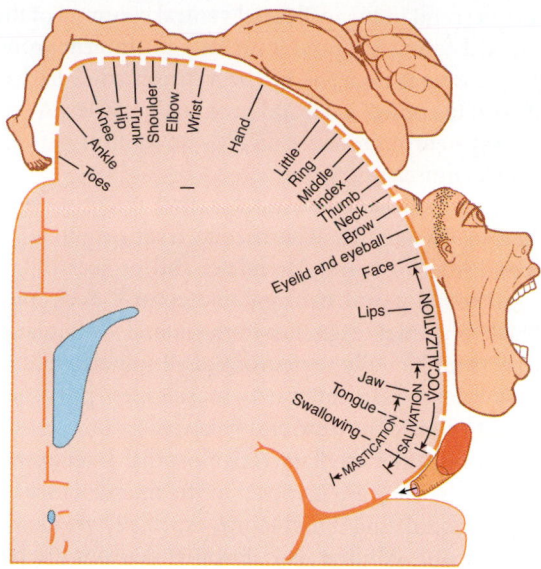

FIGURE 16–5 **Motor homunculus.** The figure represents, on a coronal section of the precentral gyrus, the location of the cortical representation of the various parts. The size of the various parts is proportionate to the cortical area devoted to them. Compare with Figure 11–4. (Reproduced with permission from Penfield W, Rasmussen G: *The Cerebral Cortex of Man.* Macmillan, 1950.)

peripheral area in which they produce movement, providing the basis for feedback control of movement. Some of this input may be direct and some is relayed from other parts of the cortex. The current view is that M1 neurons represent movements of groups of muscles for different tasks.

SUPPLEMENTARY MOTOR AREA

For the most part, the supplementary motor area projects to the motor cortex. This region also contains a map of the body, but it is less precise than in M1. It appears to be involved primarily in organizing or planning motor sequences, while M1 executes the movements. Lesions of this area in monkeys produce awkwardness in performing complex activities and difficulty with bimanual coordination.

When human subjects count to themselves without speaking, the motor cortex is quiescent, but when they speak the numbers aloud as they count, blood flow increases in M1 and the supplementary motor area. Thus, the supplementary motor area as well as M1 is involved in voluntary movement when the movements being performed are complex and involve planning. Blood flow increases whether or not a planned movement is carried out. The increase occurs whether the movement is performed by the contralateral or the ipsilateral hand.

PREMOTOR CORTEX

The premotor cortex, which also contains a somatotopic map, receives input from sensory regions of the parietal cortex and

projects to M1, the spinal cord, and the brain stem reticular formation. Its function is still incompletely understood, but it may be concerned with setting posture at the start of a planned movement and with getting the individual prepared to move. It is most involved in control of proximal limb muscles needed to orient the body for movement.

POSTERIOR PARIETAL CORTEX

In addition to providing fibers that run in the corticospinal and corticobulbar tracts, the somatic sensory area and related portions of the posterior parietal lobe project to the premotor area. Lesions of the somatic sensory area cause defects in motor performance that are characterized by inability to execute learned sequences of movements such as eating with a knife and fork. Some of the neurons in area 5 (Figure 16–3) are concerned with aiming the hands toward an object and manipulating it, whereas some of the neurons in area 7 are concerned with hand–eye coordination.

ROLE IN MOVEMENT

The corticospinal and corticobulbar system is the primary pathway for the initiation of skilled voluntary movement. This does not mean that movement—even skilled movement—is impossible without it. Nonmammalian vertebrates have essentially no corticospinal and corticobulbar system, but they move with great agility. Cats and dogs stand, walk, and run after complete destruction of this system. Only in primates are relatively marked deficits produced.

Careful section of the pyramids producing highly selective destruction of the lateral corticospinal tract in laboratory primates produces prompt and sustained loss of the ability to grasp small objects between two fingers and to make isolated movements of the wrists. However, the animal can still use the hand in a gross fashion and can stand and walk. These deficits are consistent with loss of control of the distal musculature of the limbs, which is concerned with fine-skilled movements. On the other hand, lesions of the ventral corticospinal tract produce axial muscle deficits that cause difficulty with balance, walking, and climbing.

PLASTICITY

A striking discovery made possible by PET and fMRI is that the motor cortex shows the same kind of plasticity as the sensory cortex (Chapter 11). For example, the finger areas of the contralateral motor cortex enlarge as a pattern of rapid finger movement is learned with the fingers of one hand; this change is detectable at 1 week and maximal at 4 weeks. Cortical areas of output to other muscles also increase in size when motor learning involves these muscles. When a small focal ischemic lesion is produced in the hand area of the motor cortex of monkeys, the hand area may reappear, with return of motor function, in an adjacent undamaged part of the cortex. Thus, the maps of the motor cortex are not immutable, and they change with experience.

BRAIN STEM PATHWAYS INVOLVED IN POSTURE AND VOLUNTARY MOVEMENT

As mentioned above, spinal motor neurons are organized such that those innervating the most proximal muscles are located most medially and those innervating the more distal muscles are located more laterally. This organization is also reflected in descending brain stem pathways (Figure 16–6).

MEDIAL BRAIN STEM PATHWAYS

The medial brain stem pathways, which work in concert with the ventral corticospinal tract, are the **pontine and medullary reticulospinal, vestibulospinal,** and **tectospinal tracts.** These pathways descend in the ipsilateral ventral columns of the spinal cord and terminate predominantly on interneurons and long propriospinal neurons in the ventromedial part of the ventral horn to control axial and proximal muscles. A few medial pathway neurons synapse directly on motor neurons controlling axial muscles.

The medial and lateral vestibulospinal tracts were briefly described in Chapter 13. The medial tract originates in the medial and inferior vestibular nuclei and projects bilaterally to cervical spinal motor neurons that control neck musculature. The lateral tract originates in the lateral vestibular nuclei and projects ipsilaterally to neurons at all spinal levels. It activates motor neurons to antigravity muscles (eg, proximal limb extensors) to control posture and balance.

The pontine and medullary reticulospinal tracts project to all spinal levels. They are involved in the maintenance of posture and in modulating muscle tone, especially via an input to γ-motor neurons. Pontine reticulospinal neurons are primarily excitatory and medullary reticulospinal neurons are primarily inhibitory.

A Medial brain stem pathways

B Lateral brain stem pathways

Tectum

Medial reticular formation

Tectospinal tract

Reticulospinal tract

Lateral and medial vestibular nuclei

Vestibulospinal tracts

Red nucleus (magnocellular part)

Rubrospinal tract

FIGURE 16–6   Medial and lateral descending brain stem pathways involved in motor control. A) Medial pathways (reticulospinal, vestibulospinal, and tectospinal) terminate in ventromedial area of spinal gray matter and control axial and proximal muscles. **B)** Lateral pathway (rubrospinal) terminates in dorsolateral area of spinal gray matter and controls distal muscles. (From Kandel ER, Schwartz JH, Jessell TM [editors]: *Principles of Neural Science,* 4th ed. McGraw-Hill, 2000.)

The tectospinal tract originates in the superior colliculus of the midbrain. It projects to the contralateral cervical spinal cord to control head and eye movements.

LATERAL BRAIN STEM PATHWAY

The main control of distal muscles arise from the lateral corticospinal tract, but neurons within the red nucleus of the midbrain cross the midline and project to interneurons in the dorsolateral part of the spinal ventral horn to also influence motor neurons that control distal limb muscles. This **rubrospinal tract** excites flexor motor neurons and inhibits extensor motor neurons. This pathway is not very prominent in humans, but it may play a role in the posture typical of decorticate rigidity (see below).

POSTURE-REGULATING SYSTEMS

INTEGRATION

In the intact animal, individual motor responses are submerged in the total pattern of motor activity. When the neural axis is transected, the activities integrated below the section are cut off, or released, from the control of higher brain centers and often appear to be accentuated. Release of this type, long a cardinal principle in neurology, may be due in some situations to removal of an inhibitory control by higher neural centers. A more important cause of the apparent hyperactivity is loss of differentiation of the reaction so that it no longer fits into the broader pattern of motor activity. An additional factor may be denervation hypersensitivity of the centers below the transection, but the role of this component remains to be determined.

Animal experimentation has led to information on the role of cortical and brain stem mechanisms involved in control of voluntary movement and posture. The deficits in motor control seen after various lesions mimic those seen in humans with damage in the same structures.

DECEREBRATION

A complete transection of the brain stem between the superior and inferior colliculi permits the brain stem pathways to function independent of their input from higher brain structures. This is called a **midcollicular decerebration** and is diagrammed in Figure 16–7 by the dashed line labeled A. This lesion interrupts all input from the cortex (corticospinal and corticobulbar tracts) and red nucleus (rubrospinal tract), primarily to distal muscles of the extremities. The excitatory and inhibitory reticulospinal pathways (primarily to postural extensor muscles) remain intact. The dominance of drive from ascending sensory pathways to the excitatory reticulospinal pathway leads to hyperactivity in extensor muscles in all four extremities which is called **decerebrate rigidity**. This resembles what ensues after **supratentorial lesions** in humans cause **uncal**

herniation. Uncal herniation can occur in patients with large tumors or a hemorrhage in the cerebral hemisphere. Figure 16–8A shows the posture typical of such a patient. Clinical Box 16–2 describes complications related to uncal herniation.

In midcollicular decerebrate cats, section of dorsal roots to a limb (dashed line labeled B in Figure 16–7) immediately eliminates the hyperactivity of extensor muscles. This suggests that decerebrate rigidity is spasticity due to facilitation of the myotatic stretch reflex. That is, the excitatory input from the reticulospinal pathway activates γ-motor neurons which indirectly activate α-motor neurons (via Ia spindle afferent activity). This is called the **gamma loop**.

The exact site of origin within the cerebral cortex of the fibers that inhibit stretch reflexes is unknown. Under certain conditions, stimulation of the anterior edge of the precentral gyrus can cause inhibition of stretch reflexes and cortically evoked movements. This region, which also projects to the basal ganglia, has been named area 4s, or the **suppressor strip.**

There is also evidence that decerebrate rigidity leads to direct activation of α-motor neurons. If the anterior lobe of the cerebellum is removed in a decerebrate animal (dashed line labeled C in Figure 16–7), extensor muscle hyperactivity is exaggerated (**decerebellate rigidity**). This cut eliminates cortical inhibition of the cerebellar fastigial nucleus and secondarily increases excitation to vestibular nuclei. Subsequent dorsal root section does not reverse the rigidity, thus it was due to activation of α-motor neurons independent of the gamma loop.

DECORTICATION

Removal of the cerebral cortex (**decortication;** dashed line labeled D in Figure 16–7) produces **decorticate rigidity** which is characterized by flexion of the upper extremities at the elbow and extensor hyperactivity in the lower extremities (Figure 16–8B). The flexion can be explained by rubrospinal excitation of flexor muscles in the upper extremities; the hyperextension of lower extremities is due to the same changes that occur after midcollicular decerebration.

Decorticate rigidity is seen on the hemiplegic side in humans after hemorrhages or thromboses in the internal capsule. Probably because of their anatomy, the small arteries in the internal capsule are especially prone to rupture or thrombotic obstruction, so this type of decorticate rigidity is fairly common. Sixty percent of intracerebral hemorrhages occur in the internal capsule, as opposed to 10% in the cerebral cortex, 10% in the pons, 10% in the thalamus, and 10% in the cerebellum.

SPINAL INTEGRATION

The responses of animals and humans to **spinal cord injury (SCI)** illustrate the integration of reflexes at the spinal level. The deficits seen after SCI vary, of course, depending on the level of the injury. Clinical Box 16–3 provides information on long-term problems related to SCI and recent advancements in treatment options.

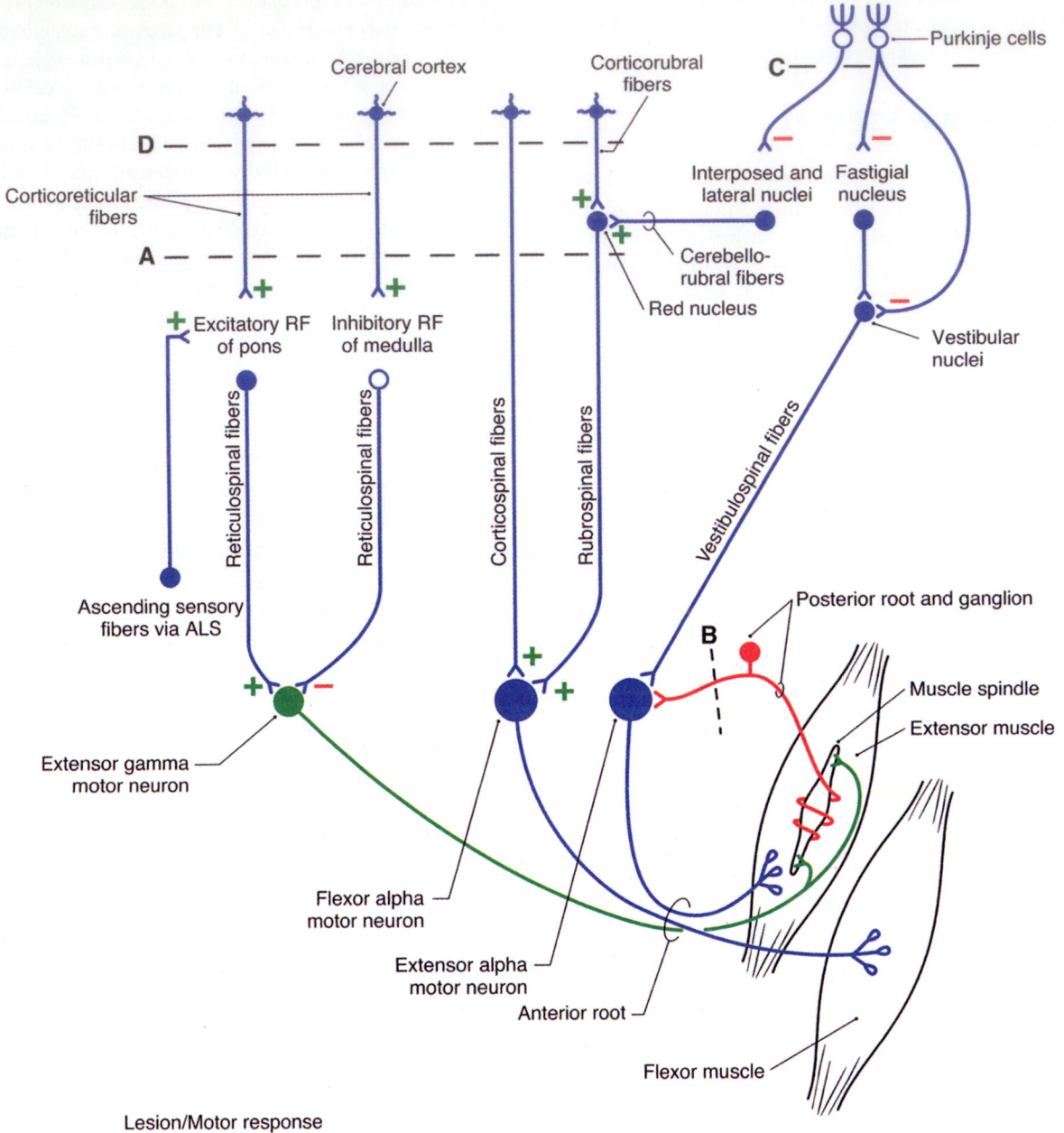

Lesion/Motor response

    A = Extensor rigidity in all limbs, decerebrate rigidity/posturing
  A+B = Relaxation of extensor rigidity in limb with sectioned root
  A+C = Slight enhancement of decerebrate rigidity compared to A
A+C+B = No relaxation of decerebrate rigidity
    D = Flexion of upper limbs, extension of lower limbs, decorticate rigidity/posturing

FIGURE 16–7   **A circuit drawing representing lesions produced in experimental animals to replicate decerebrate and decorticate deficits seen in humans.** Bilateral transections are indicated by dashed lines A, B, C, and D. Decerebration is at a midcollicular level **(A),** decortication is rostral to the superior colliculus, dorsal roots sectioned for one extremity **(B),** and removal of anterior lobe of cerebellum **(C).** The objective was to identify anatomic substrates responsible for decerebrate or decorticate rigidity/posturing seen in humans with lesions that either isolate the forebrain from the brain stem or separate rostral from caudal brain stem and spinal cord.   (Reproduced with permission from Haines DE [editor]: *Fundamental Neuroscience for Basic and Clinical Applications,* 3rd ed. Elsevier, 2006.)

CLINICAL BOX 16–2

Uncal Herniation

Space-occupying lesions from large tumors, hemorrhages, strokes, or abscesses in the cerebral hemisphere can drive the uncus of the temporal lobe over the edge of the cerebellar tentorium, compressing the ipsilateral cranial nerve III (uncal herniation). Before the herniation these patients experience a decreased level of consciousness, lethargy, poorly reactive pupils, deviation of the eye to a "down and out" position, hyperactive reflexes, and a bilateral Babinski sign (due to compression of the ipsilateral corticospinal tract). After the brain herniates, the patients are decerebrate and comatose, have fixed and dilated pupils, and eye movements are absent. Once damage extends to the midbrain, a Cheyne–Stokes respiratory pattern develops, characterized by a pattern of waxing-and-waning depth of respiration with interposed periods of apnea. Eventually, medullary function is lost, breathing ceases, and recovery is unlikely. Hemispheric masses closer to the midline compress the thalamic reticular formation and can cause coma before eye findings develop (central herniation). As the mass enlarges, midbrain function is affected, the pupils dilate, and a decerebrate posture ensues. With progressive herniation, pontine vestibular and then medullary respiratory functions are lost.

SPINAL SHOCK

In all vertebrates, transection of the spinal cord is followed by a period of **spinal shock** during which all spinal reflex responses are profoundly depressed. Subsequently, reflex responses return and become hyperactive. The duration of spinal shock is proportionate to the degree of encephalization of motor function in the various species. In frogs and rats it lasts for minutes; in dogs and cats it lasts for 1 to 2 h; in monkeys it lasts for days; and in humans it usually lasts for a minimum of 2 wk.

The cause of spinal shock is uncertain. Cessation of tonic bombardment of spinal neurons by excitatory impulses in descending pathways undoubtedly plays a role, but the subsequent return of reflexes and their eventual hyperactivity also have to be explained. The recovery of reflex excitability may be due to the development of denervation hypersensitivity to the mediators released by the remaining spinal excitatory endings. Another possibility for which there is some evidence is the sprouting of collaterals from existing neurons, with the formation of additional excitatory endings on interneurons and motor neurons.

The first reflex response to appear as spinal shock wears off in humans is often a slight contraction of the leg flexors and adductors in response to a noxious stimulus. In some patients, the knee jerk reflex recovers first. The interval between cord transection and the return of reflex activity is about 2 weeks in the absence of any complications, but if complications are present it is much longer. It is not known why infection, malnutrition, and other complications of SCI inhibit spinal reflex activity. Once the spinal reflexes begin to reappear after spinal shock, their threshold steadily drops.

LOCOMOTION GENERATOR

Circuits intrinsic to the spinal cord can produce walking movements when stimulated in a suitable fashion even after spinal cord transection in cats and dogs. There are two **locomotor pattern generators** in the spinal cord: one in the cervical region and one in the lumbar region. However, this does not mean that spinal animals or humans can walk without stimulation; the pattern generator has to be turned on by tonic discharge of a discrete area in the midbrain, the mesencephalic

FIGURE 16–8 **Decerebrate and decorticate postures. A)** Damage to lower midbrain and upper pons causes decerebrate posturing in which lower extremities are extended with toes pointed inward and upper extremities extended with fingers flexed and forearms pronate. Neck and head are extended. **B)** Damage to upper midbrain may cause decorticate posturing in which upper limbs are flexed, lower limbs are extended with toes pointed slightly inward, and head is extended. (Modified from Kandel ER, Schwartz JH, Jessell TM [editors]: *Principles of Neural Science*, 4th ed. McGraw-Hill, 2000.)

A Upper pontine damage

B Upper midbrain damage

Spinal Cord Injury

It has been estimated that the worldwide annual incidence of sustaining **spinal cord injury (SCI)** is between 10 and 83 per million of the population. Leading causes are vehicle accidents, violence, and sports injuries. The mean age of patients who sustain an SCI is 33 years old, and men outnumber women with a nearly 4 to 1 ratio. Approximately 52% of SCI cases result in quadriplegia and about 42% lead to paraplegia. In quadriplegic humans, the threshold of the withdrawal reflex is very low; even minor noxious stimuli may cause not only prolonged withdrawal of one extremity but marked flexion–extension patterns in the other three limbs. Stretch reflexes are also hyperactive. Afferent stimuli irradiate from one reflex center to another after SCI. When even a relatively minor noxious stimulus is applied to the skin, it may activate autonomic neurons and produce evacuation of the bladder and rectum, sweating, pallor, and blood pressure swings in addition to the withdrawal response. This distressing **mass reflex** can sometimes be used to give paraplegic patients a degree of bladder and bowel control. They can be trained to initiate urination and defecation by stroking or pinching their thighs, thus producing an intentional mass reflex. If the cord section is incomplete, the flexor spasms initiated by noxious stimuli can be associated with bursts of pain that are particularly bothersome. They can be treated with considerable success with baclofen, a GABA$_B$ receptor agonist that crosses the blood–brain barrier and facilitates inhibition.

Treatment of SCI patients presents complex problems. Administration of large doses of **glucocorticoids** has been shown to foster recovery and minimize loss of function after SCI. They need to be given soon after the injury and then discontinued because of the well-established deleterious effects of long-term steroid treatment. Their immediate value is likely due to reduction of the inflammatory response in the damaged tissue. Due to immobilization, SCI patients develop a negative nitrogen balance and catabolize large amounts of body protein. Their body weight compresses the circulation to the skin over bony prominences, causing **decubitus ulcers** to form. The ulcers heal poorly and are prone to infection because of body protein depletion. The tissues that are broken down include the protein matrix of bone and this, plus the immobilization, cause Ca^{2+} to be released in large amounts, leading to **hypercalcemia, hypercalciuria,** and formation of **calcium stones** in the urinary tract. The combination of stones and bladder paralysis cause urinary stasis, which predisposes to **urinary tract infection,** the most common complication of SCI. The search continues for ways to get axons of neurons in the spinal cord to regenerate. Administration of **neurotrophins** shows some promise in experimental animals, and so does implantation of **embryonic stem cells** at the site of injury. Another possibility being explored is bypassing the site of SCI with **brain–computer interface devices.** However, these novel approaches are a long way from routine clinical use.

locomotor region, and, of course, this is only possible in patients with incomplete spinal cord transection. Interestingly, the generators can also be turned on in experimental animals by administration of the norepinephrine precursor L-dopa (levodopa) after complete section of the spinal cord. Progress is being made in teaching humans with SCI to take a few steps by placing them, with support, on a treadmill.

BASAL GANGLIA

ANATOMIC CONSIDERATIONS

The term **basal ganglia** (or **basal nuclei**) is generally applied to five interactive structures on each side of the brain (Figure 16–9). These are the **caudate nucleus, putamen,** and **globus pallidus** (three large nuclear masses underlying the cortical mantle), the **subthalamic nucleus,** and **substantia nigra.** The globus pallidus is divided into external and internal segments (GPe and GPi). The substantia nigra is divided into a **pars compacta** and a **pars reticulata.** The caudate nucleus and putamen are commonly called the **striatum;** the putamen and globus pallidus are sometimes called the **lenticular nucleus.**

The main inputs to the basal ganglia terminate in the striatum (Figure 16–10). They include the excitatory **corticostriate pathway** from M1 and premotor cortex. There is also a projection from intralaminar nuclei of the thalamus to the striatum (**thalamostriatal pathway).**

The connections between the parts of the basal ganglia include a dopaminergic nigrostriatal projection from the substantia nigra pars compacta to the striatum and a corresponding GABAergic projection from the striatum to substantia nigra pars reticulata. The striatum projects to both GPe and GPi. GPe projects to the subthalamic nucleus, which in turn projects to both GPe and GPi.

The principal output from the basal ganglia is from GPi via the **thalamic fasciculus** to the ventral lateral, ventral anterior, and centromedian nuclei of the thalamus. From the thalamic nuclei, fibers project to the prefrontal and premotor cortex. The substantia nigra also projects to the thalamus. These connections, along with the probable synaptic transmitters involved, are summarized in Figure 16–10.

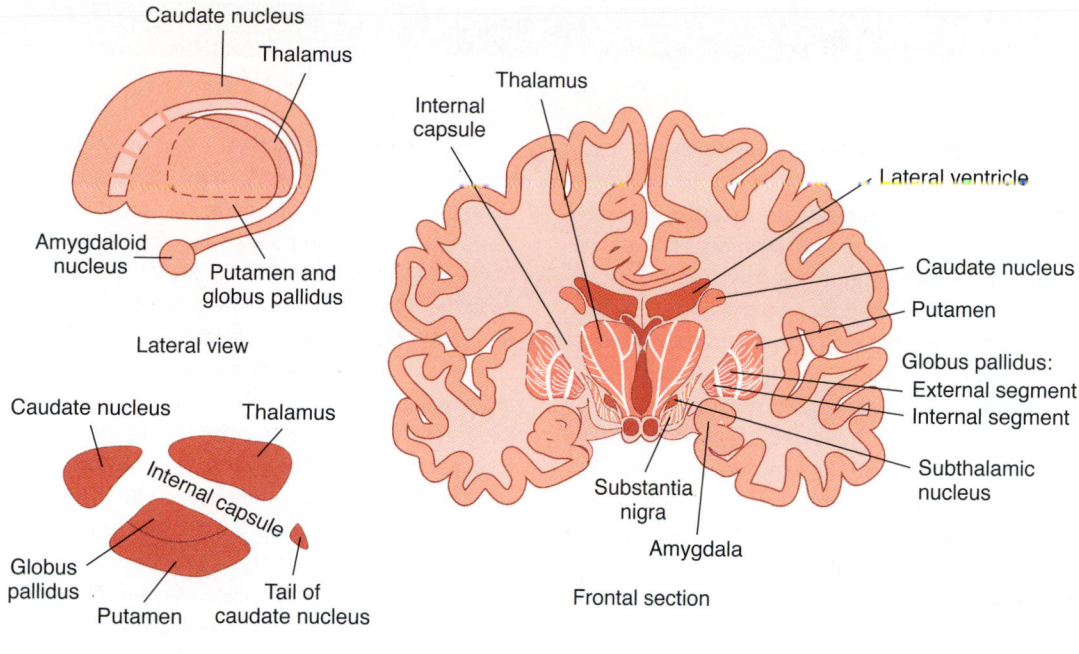

FIGURE 16–9 **The basal ganglia.** The basal ganglia are composed of the caudate nucleus, putamen, and globus pallidus and the functionally related subthalamic nucleus and substantia nigra. The frontal (coronal) section shows the location of the basal ganglia in relation to surrounding structures.

The main feature of the connections of the basal ganglia is that the cerebral cortex projects to the striatum, the striatum to GPi, GPi to the thalamus, and the thalamus back to the cortex, completing a loop. The output from GPi to the thalamus is inhibitory, whereas the output from the thalamus to the cerebral cortex is excitatory.

FIGURE 16–10 **Diagrammatic representation of the principal connections of the basal ganglia.** Solid lines indicate excitatory pathways, dashed lines inhibitory pathways. The transmitters are indicated in the pathways, where they are known. Glu, glutamate; DA, dopamine. Acetylcholine is the transmitter produced by interneurons in the striatum. SNPR, substantia nigra, pars reticulata; SNPC, substantia nigra, pars compacta; ES, external segment; IS, internal segment; PPN, pedunculopontine nuclei. The subthalamic nucleus also projects to the pars compacta of the substantia nigra; this pathway has been omitted for clarity.

The striatum is made up of two parts that differ histologically: a unique mosaic of **patches** or **striosomes** (an area with little acetylcholinesterase) and a **matrix** (an area high in acetylcholinesterase). The neurons of the corticostriate projection that originate in the deep portion of layer V of the cortex terminate in the patches, whereas the neurons that originate in layers II and III and the superficial part of layer V end primarily in the matrix. Neurons with their cell bodies in patches project in large part to dopaminergic neurons in the substantia nigra pars compacta, whereas many of the neurons with their cell bodies in the matrix project to GABAergic neurons in the substantia nigra pars reticulata.

FUNCTION

Neurons in the basal ganglia, like those in the lateral portions of the cerebellar hemispheres, discharge before movements begin. This observation, plus careful analysis of the effects of diseases of the basal ganglion in humans and the effects of drugs that destroy dopaminergic neurons in animals, have led to the idea that the basal ganglia are involved in the planning and programming of movement or, more broadly, in the processes by which an abstract thought is converted into voluntary action (Figure 16–1). They influence the motor cortex via the thalamus, and the corticospinal pathways provide the final common pathway to motor neurons. In addition, GPi projects to nuclei in the brain stem, and from there to motor neurons in the brain stem and spinal cord. The basal ganglia, particularly

Basal Ganglia Diseases

The initial detectable damage in **Huntington disease** is to medium spiny neurons in the striatum. This loss of this GABAergic pathway to the globus pallidus external segment releases inhibition, permitting the hyperkinetic features of the disease to develop. An early sign is a jerky trajectory of the hand when reaching to touch a spot, especially toward the end of the reach. Later, hyperkinetic **choreiform movements** appear and gradually increase until they incapacitate the patient. Speech becomes slurred and then incomprehensible, and a progressive dementia is followed by death, usually within 10–15 years after the onset of symptoms. Huntington disease affects 5 out of 100,000 people worldwide. It is inherited as an autosomal dominant disorder, and its onset is usually between the ages of 30 and 50. The abnormal gene responsible for the disease is located near the end of the short arm of chromosome 4. It normally contains 11–34 cytosine-adenine-guanine (CAG) repeats, each coding for glutamine. In patients with Huntington disease, this number is increased to 42–86 or more copies, and the greater the number of repeats, the earlier the age of onset and the more rapid the progression of the disease. The gene codes for **huntingtin,** a protein of unknown function. Poorly soluble protein aggregates, which are toxic, form in cell nuclei and elsewhere. However, the correlation between aggregates and symptoms is less than perfect. It appears that a loss of the function of huntingtin occurs that is proportionate to the size of the CAG insert. At present, no effective treatment is available, and the disease is uniformly fatal. However, there are a few glimmers of hope. In animal models of the disease, intrastriatal grafting of fetal striatal tissue improves cognitive performance. In addition, tissue caspase-1 activity is increased in the brains of humans and animals with the disease and in mice in which the gene for this apoptosis-regulating enzyme has been knocked out, progression of the disease is slowed.

Another basal ganglia disorder is **Wilson disease** (or **hepatolenticular degeneration**), which is a rare disorder of copper metabolism which has an onset between 6 to 25 years of age, affecting about four times as many females as males. Wilson disease affects about 30,000 people worldwide. It is a genetic autosomal recessive disorder due to a mutation on the long arm of chromosome 13q. It affects the copper-transporting ATPase gene (*ATP7B*) in the liver, leading to an accumulation of copper in the liver and resultant progressive liver damage. About 1% of the population carries a single abnormal copy of this gene but do not develop any symptoms. A child who inherits the gene from both parents may develop the disease. In affected individuals, copper accumulates in the periphery of the cornea in the eye accounting for the characteristic yellow **Kayser–Fleischer rings.** The dominant neuronal pathology is degeneration of the putamen, a part of the **lenticular nucleus.** Motor disturbances include "wing-beating" tremor or **asterixis, dysarthria,** unsteady gait, and rigidity. Treatment is to reduce the copper in the body.

Another disease commonly referred to as a disease of the basal ganglia is **tardive dyskinesia.** This disease indeed involves the basal ganglia, but it is caused by medical treatment of another disorder with **neuroleptic drugs** such as phenothiazides or haloperidol. Therefore, tardive dyskinesia is iatrogenic in origin. Long-term use of these drugs may produce biochemical abnormalities in the striatum. The motor disturbances include either temporary or permanent uncontrolled involuntary movements of the face and tongue and cogwheel rigidity. The neuroleptic drugs act via blockade of dopaminergic transmission. Prolonged drug use leads to hypersensitivity of D_3 dopaminergic receptors and an imbalance in nigrostriatal influences on motor control.

---

the caudate nuclei, also play a role in some cognitive processes. Possibly because of the interconnections of this nucleus with the frontal portions of the neocortex, lesions of the caudate nuclei disrupt performance on tests involving object reversal and delayed alternation. In addition, lesions of the head of the left but not the right caudate nucleus and nearby white matter in humans are associated with a dysarthric form of aphasia that resembles Wernicke aphasia.

DISEASES OF THE BASAL GANGLIA IN HUMANS

Three distinct biochemical pathways in the basal ganglia normally operate in a balanced fashion: (1) the nigrostriatal dopa-

minergic system, (2) the intrastriatal cholinergic system, and (3) the GABAergic system, which projects from the striatum to the globus pallidus and substantia nigra. When one or more of these pathways become dysfunctional, characteristic motor abnormalities occur. Diseases of the basal ganglia lead to two general types of disorders: **hyperkinetic** and **hypokinetic.** The hyperkinetic conditions are those in which movement is excessive and abnormal, including chorea, athetosis, and ballism. Hypokinetic abnormalities include akinesia and bradykinesia.

Chorea is characterized by rapid, involuntary "dancing" movements. **Athetosis** is characterized by continuous, slow writhing movements. Choreiform and athetotic movements have been likened to the start of voluntary movements occurring in an involuntary, disorganized way. In **ballism,** involuntary

TABLE 16–1 Examples of trinucleotide repeat diseases.

Disease	Expanded Trinucleotide Repeat	Affected Protein
Huntington disease	CAG	Huntingtin
Spinocerebellar ataxia, types 1, 2, 3, 7	CAG	Ataxin 1, 2, 3, 7
Spinocerebellar ataxia, type 6	CAG	α_{1A} subunit of Ca^{2+} channel
Dentatorubral-pallidoluysian atrophy	CAG	Atrophin
Spinobulbar muscular atrophy	CAG	Androgen receptor
Fragile X syndrome	CGG	FMR-1
Myotonic dystrophy	CTG	DM protein kinase
Friedreich ataxia	GAA	Frataxin

flailing, intense, and violent movements occur. **Akinesia** is difficulty in initiating movement and decreased spontaneous movement. **Bradykinesia** is slowness of movement.

In addition to Parkinson disease, which is described below, there are several other disorders known to involve a malfunction within the basal ganglia. A few of these are described in Clinical Box 16–4. **Huntington disease** is one of an increasing number of human genetic diseases affecting the nervous system that are characterized by **trinucleotide repeat** expansion. Most of these involve cytosine-adenine-guanine (CAG) repeats (Table 16–1), but one involves CGG repeats and another involves CTG repeats. All of these are in exons; however, a GAA repeat in an intron is associated with Friedreich's ataxia. There is also preliminary evidence that increased numbers of a 12-nucleotide repeat are associated with a rare form of epilepsy.

PARKINSON DISEASE (PARALYSIS AGITANS)

Parkinson disease has both hypokinetic and hyperkinetic features. It was originally described by James Parkinson and is named for him. Parkinson disease is the first disease identified as being due to a deficiency in a specific neurotransmitter. In the 1960s, Parkinson disease was shown to result from the degeneration of dopaminergic neurons in the substantia nigra pars compacta.

The fibers to the putamen are most severely affected. Parkinsonism now occurs in sporadic idiopathic form in many middle-aged and elderly individuals and is one of the most common neurodegenerative diseases. It is estimated to occur in 1–2% of individuals over age 65. Dopaminergic neurons and dopamine receptors are steadily lost with age in the basal ganglia in normal individuals, and an acceleration of these

losses apparently precipitates parkinsonism. Symptoms appear when 60–80% of the nigrostriatal dopaminergic neurons degenerate.

Parkinsonism is also seen as a complication of treatment with the phenothiazine group of tranquilizer drugs and other drugs that block D_2 receptors. It can be produced in rapid and dramatic form by injection of 1-methyl-4-phenyl-1,2,5,6-tetrahydropyridine (MPTP). This effect was discovered by chance when a drug dealer in northern California supplied some of his clients with a homemade preparation of synthetic heroin that contained MPTP. MPTP is a prodrug that is metabolized in astrocytes by the enzyme MOA-B to produce a potent oxidant, 1-methyl-4-phenylpyridinium (MPP^+). In rodents, MPP^+ is rapidly removed from the brain, but in primates it is removed more slowly and is taken up by the dopamine transporter into dopaminergic neurons in the substantia nigra, which it destroys without affecting other dopaminergic neurons to any appreciable degree. Consequently, MPTP can be used to produce parkinsonism in monkeys, and its availability has accelerated research on the function of the basal ganglia.

The hypokinetic features of Parkinson disease are akinesia and bradykinesia, and the hyperkinetic features are **cogwheel rigidity** and **tremor at rest.** The absence of motor activity and the difficulty in initiating voluntary movements are striking. There is a decrease in the normal, unconscious movements such as swinging of the arms during walking, the panorama of facial expressions related to the emotional content of thought and speech, and the multiple "fidgety" actions and gestures that occur in all of us. The rigidity is different from spasticity because motor neuron discharge increases to both the agonist and antagonist muscles. Passive motion of an extremity meets with a plastic, dead-feeling resistance that has been likened to bending a lead pipe and is therefore called **lead pipe rigidity.** Sometimes a series of "catches" takes place during passive motion (cogwheel rigidity), but the sudden loss of resistance seen in a spastic extremity is absent. The tremor, which is present at rest and disappears with activity, is due to regular, alternating 8-Hz contractions of antagonistic muscles.

A current view of the pathogenesis of the movement disorders in Parkinson disease is shown in Figure 16–11. In normal individuals, basal ganglia output is inhibitory via GABAergic nerve fibers. The dopaminergic neurons that project from the substantia nigra to the putamen normally have two effects: they stimulate the D_1 dopamine receptors, which inhibit GPi via direct GABAergic receptors, and they inhibit D_2 receptors, which also inhibit the GPi. In addition, the inhibition reduces the excitatory discharge from the subthalamic nucleus to the GPi. This balance between inhibition and excitation somehow maintains normal motor function. In Parkinson disease, the dopaminergic input to the putamen is lost. This results in decreased inhibition and increased excitation from the STN to the GPi. The overall increase in inhibitory output to the thalamus and brain stem disorganizes movement.

FIGURE 16–11   Probable basal ganglia-thalamocortical circuitry in Parkinson disease. Solid arrows indicate excitatory outputs and dashed arrows inhibitory outputs. The strength of each output is indicated by the width of the arrow. GPe, external segment of the globus pallidus; GPi, internal segment of the globus pallidus; SNC, pars compacta of the substantia nigra; STN, subthalamic nucleus; PPN, pedunculopontine nuclei; Thal, thalamus. See text for details.   (Modified from Grafton SC, DeLong M: Tracing the brain circuitry with functional imaging. Nat Med 1997;3:602.)

Treatment

An important consideration in Parkinson disease is the balance between the excitatory discharge of cholinergic interneurons and the inhibitory dopaminergic input in the striatum. Some improvement is produced by decreasing the cholinergic influence with anticholinergic drugs. More dramatic improvement is produced by administration of L-dopa **(levodopa).** Unlike dopamine, this dopamine precursor crosses the blood–brain barrier and helps repair the dopamine deficiency. However, the degeneration of these neurons continues and in 5 to 7 y the beneficial effects of L-dopa disappear.

Surgical treatment by making lesions in GPi **(pallidotomy)** or in the subthalamic nucleus helps to restore the output balance toward normal (Figure 16–11). Surgical outcomes have been further improved by implanting electrodes attached to subcutaneous stimulators and administering high-frequency current. This produces temporary disruption of circuits at the electrode tip on demand.

Another surgical approach is to implant dopamine-secreting tissue in or near the basal ganglia. Transplants of the patient's own adrenal medullary tissue or carotid body works for a while, apparently by functioning as a sort of dopamine minipump, but long-term results have been disappointing. Results with transplantation of fetal striatal tissue have been better, and there is evidence that the transplanted cells not only survive but make appropriate connections in the host's basal ganglia. However, some patients with transplants develop severe involuntary movements **(dyskinesias).**

In monkeys with experimental parkinsonism, neurotrophic factors benefit the nigrostriatal neurons, and local injection of GDNF attached to a lentivirus vector so that it penetrates cells has produced promising results.

FIGURE 16–12   Diagrammatic representation of the principal parts of the brain. The parts are distorted to show the cerebellar peduncles and the way the cerebellum, pons, and middle peduncle form a "napkin ring" around the brain stem.   (Reproduced with permission, from Goss CM [editor]: *Gray's Anatomy of the Human Body,* 27th ed. Lea & Febiger, 1959.)

Familial cases of Parkinson disease occur, but these are uncommon. The genes for at least five proteins can be mutated. These proteins appear to be involved in ubiquitination. Two of the proteins, α-**synuclein** and **barkin,** interact and are found in **Lewy bodies.** The Lewy bodies are inclusion bodies in neurons that occur in all forms of Parkinson disease. However, the significance of these findings is still unsettled.

CEREBELLUM

ANATOMIC DIVISIONS

The cerebellum sits astride the main sensory and motor systems in the brain stem (Figure 16–12). It is connected to the brain stem on each side by a **superior peduncle** (brachium conjunctivum), **middle peduncle** (brachium pontis), and **inferior peduncle** (restiform body). The medial **vermis** and lateral **cerebellar hemispheres** are more extensively folded and fissured than the cerebral cortex (Figure 16–13). The cerebellum weighs only 10% as much as the cerebral cortex, but its surface area is about 75% of that of the cerebral cortex. Anatomically, the cerebellum is divided into three parts by two transverse fissures. The posterolateral fissure separates the medial nodulus and the lateral flocculus on either side from the rest of the cerebellum, and the primary fissure divides the remainder into an anterior and a posterior lobe. Lesser fissures divide the vermis into smaller sections, so that it contains 10 primary lobules numbered I–X from superior to inferior. These lobules are identified by name and number in Figure 16–13.

ORGANIZATION

The cerebellum has an external **cerebellar cortex** separated by white matter from the **deep cerebellar nuclei.** Its primary

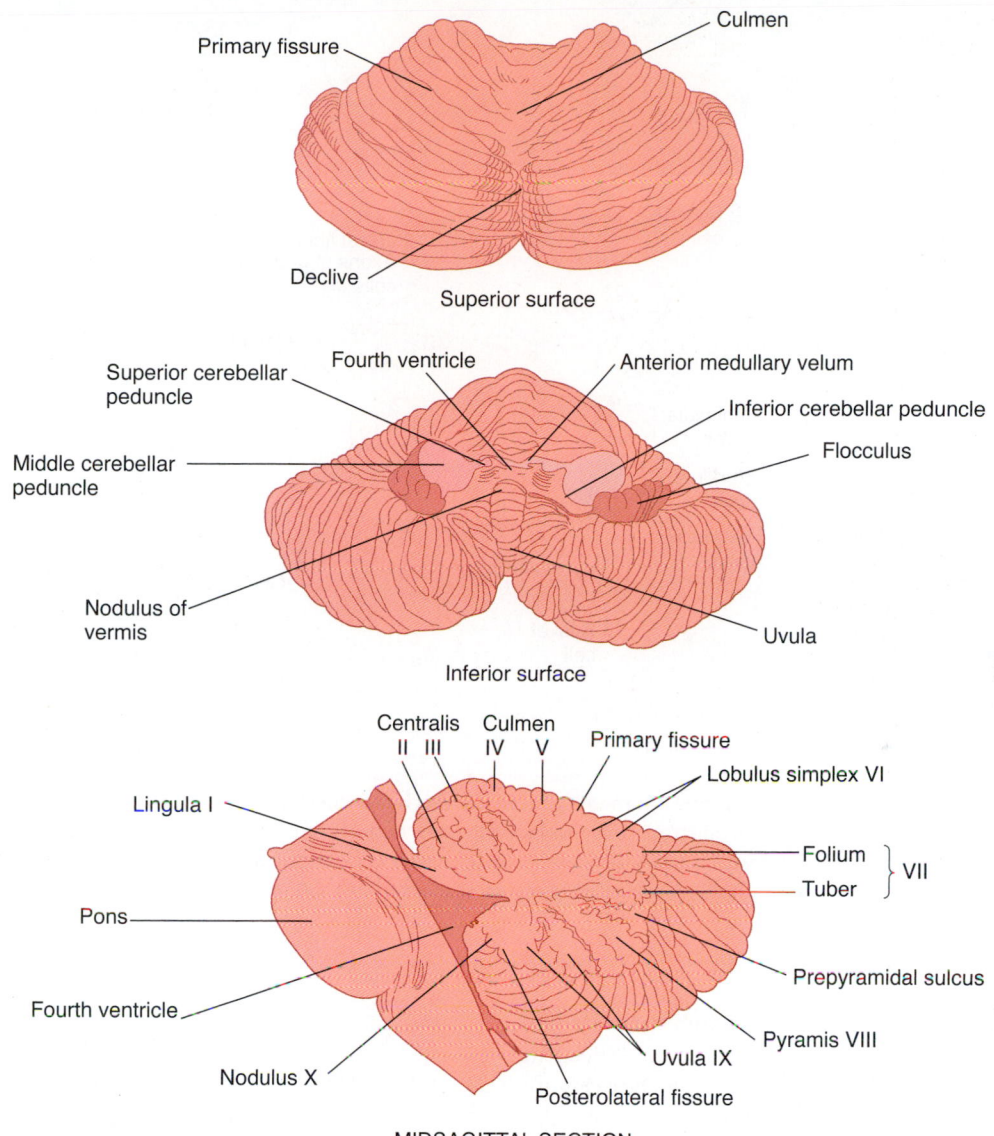

FIGURE 16–13 **Superior and inferior views and sagittal section of the human cerebellum.** The 10 principal lobules are identified by name and by number (I–X).

afferent inputs, the mossy and climbing fibers, send collaterals to the deep nuclei and pass to the cortex. There are four deep nuclei: the **dentate,** the **globose,** the **emboliform,** and the **fastigial** nuclei. The globose and the emboliform nuclei are sometimes lumped together as the **interpositus nucleus.** The cerebellar cortex contains five types of neurons: Purkinje, granule, basket, stellate, and Golgi cells. It has three layers (Figure 16–14): an external molecular layer, a Purkinje cell layer that is only one cell thick, and an internal granular layer. The **Purkinje cells** are among the biggest neurons in the body. They have very extensive dendritic arbors that extend throughout the molecular layer. Their axons, which are the only output from the cerebellar cortex, generally pass to the deep nuclei. The cerebellar cortex also contains **granule cells,** which receive input from the mossy fibers and innervate the

Purkinje cells. The granule cells have their cell bodies in the granular layer. Each sends an axon to the molecular layer, where the axon bifurcates to form a T. The branches of the T are straight and run long distances. Consequently, they are called **parallel fibers.** The dendritic trees of the Purkinje cells are markedly flattened (Figure 16–14) and oriented at right angles to the parallel fibers. The parallel fibers thus make synaptic contact with the dendrites of many Purkinje cells, and the parallel fibers and Purkinje dendritic trees form a grid of remarkably regular proportions.

The other three types of neurons in the cerebellar cortex are in effect inhibitory interneurons. **Basket cells** (Figure 16–14) are located in the molecular layer. They receive input from the parallel fibers and each projects to many Purkinje cells (Figure 16–15). Their axons form a basket around the cell body

FIGURE 16–14   **Location and structure of five neuronal types in the cerebellar cortex.** Drawings are based on Golgi-stained preparations. Purkinje cells (1) have processes aligned in one plane; their axons are the only output from the cerebellum. Axons of granule cells (4) traverse and make connections with Purkinje cell processes in molecular layer. Golgi (2), basket (3), and stellate (5) cells have characteristic positions, shapes, branching patterns, and synaptic connections. (Reproduced with permission from Kuffler SW, Nicholls JG, Martin AR: *From Neuron to Brain,* 2nd ed. Sinauer, 1984.)

and axon hillock of each Purkinje cell they innervate. **Stellate cells** are similar to the basket cells but more superficial in location. **Golgi cells** are located in the granular layer (Figure 16–14). Their dendrites, which project into the molecular layer, receive input from the parallel fibers (Figure 16–15). Their cell bodies receive input via collaterals from the incoming mossy fibers and the Purkinje cells. Their axons project to the dendrites of the granule cells.

The two main inputs to the cerebellar cortex are **climbing fibers** and **mossy fibers.** Both are excitatory (Figure 16–15). The climbing fibers come from a single source, the inferior olivary nuclei. Each projects to the primary dendrites of a Purkinje cell, around which it entwines like a climbing plant. Proprioceptive input to the inferior olivary nuclei comes from all over the body. On the other hand, the mossy fibers provide direct proprioceptive input from all parts of the body plus input from the cerebral cortex via the pontine nuclei to the cerebellar cortex. They end on the dendrites of granule cells in complex synaptic groupings called **glomeruli.** The glomeruli also contain the inhibitory endings of the Golgi cells mentioned above.

The fundamental circuits of the cerebellar cortex are thus relatively simple (Figure 16–15). Climbing fiber inputs exert a strong excitatory effect on single Purkinje cells, whereas mossy

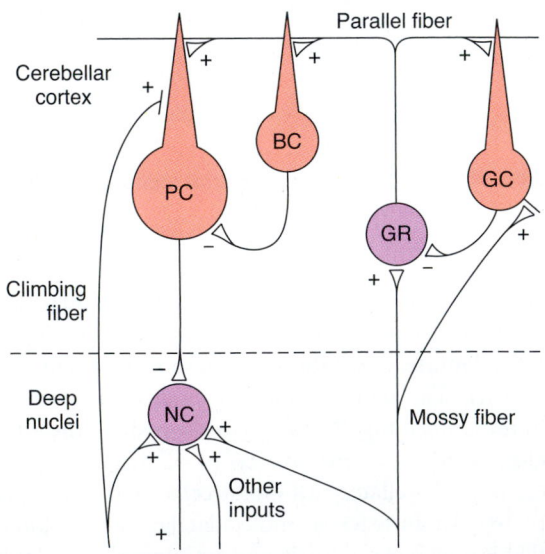

FIGURE 16–15   **Diagram of neural connections in the cerebellum.** Plus (+) and minus (−) signs indicate whether endings are excitatory or inhibitory. BC, basket cell; GC, Golgi cell; GR, granule cell; NC, cell in deep nucleus; PC, Purkinje cell. Note that PCs and BCs are inhibitory. The connections of the stellate cells, which are not shown, are similar to those of the basket cells, except that they end for the most part on Purkinje cell dendrites.

TABLE 16–2 Function of principal afferent systems to the cerebellum.[a]

Afferent Tracts	Transmits
Vestibulocerebellar	Vestibular impulses from labyrinths, direct and via vestibular nuclei
Dorsal spinocerebellar	Proprioceptive and exteroceptive impulses from body
Ventral spinocerebellar	Proprioceptive and exteroceptive impulses from body
Cuneocerebellar	Proprioceptive impulses, especially from head and neck
Tectocerebellar	Auditory and visual impulses via inferior and superior colliculi
Pontocerebellar	Impulses from motor and other parts of cerebral cortex via pontine nuclei
Olivocerebellar	Proprioceptive input from whole body via relay in inferior olive

[a]The olivocerebellar pathway projects to the cerebellar cortex via climbing fibers; the rest of the listed paths project via mossy fibers. Several other pathways transmit impulses from nuclei in the brain stem to the cerebellar cortex and to the deep nuclei, including a serotonergic input from the raphé nuclei to the granular and molecular layers and a noradrenergic input from the locus ceruleus to all three layers.

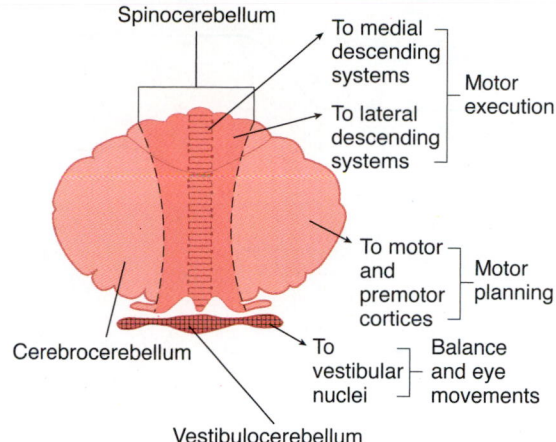

FIGURE 16–16 Functional divisions of the cerebellum. (Modified from Kandel ER, Schwartz JH, Jessell TM [editors]: *Principles of Neural Science,* 4th ed. McGraw-Hill, 2000.)

fiber inputs exert a weak excitatory effect on many Purkinje cells via the granule cells. The basket and stellate cells are also excited by granule cells via the parallel fibers, and their output inhibits Purkinje cell discharge (feed-forward inhibition). Golgi cells are excited by the mossy fiber collaterals, Purkinje cell collaterals, and parallel fibers, and they inhibit transmission from mossy fibers to granule cells. The transmitter secreted by the stellate, basket, Golgi, and Purkinje cells is GABA, whereas the granule cells secrete glutamate. GABA acts via $GABA_A$ receptors, but the combinations of subunits in these receptors vary from one cell type to the next. The granule cell is unique in that it appears to be the only type of neuron in the CNS that has a $GABA_A$ receptor containing the α6 subunit.

The output of the Purkinje cells is in turn inhibitory to the deep cerebellar nuclei. As noted above, these nuclei also receive excitatory inputs via collaterals from the mossy and climbing fibers. It is interesting, in view of their inhibitory Purkinje cell input, that the output of the deep cerebellar nuclei to the brain stem and thalamus is always excitatory. Thus, almost all the cerebellar circuitry seems to be concerned solely with modulating or timing the excitatory output of the deep cerebellar nuclei to the brain stem and thalamus. The primary afferent systems that converge to form the mossy fiber or climbing fiber input to the cerebellum are summarized in Table 16–2.

FUNCTIONAL DIVISIONS

From a functional point of view, the cerebellum is divided into three parts (Figure 16–16). The nodulus in the vermis and the flanking flocculus in the hemisphere on each side form the **vestibulocerebellum** (or **flocculonodular lobe**). This lobe, which is phylogenetically the oldest part of the cerebellum, has vestibular connections and is concerned with equilibrium and eye movements. The rest of the vermis and the adjacent medial portions of the hemispheres form the **spinocerebellum,** the region that receives proprioceptive input from the body as well as a copy of the "motor plan" from the motor cortex. By comparing plan with performance, it smoothes and coordinates movements that are ongoing. The vermis projects to the brain stem area concerned with control of axial and proximal limb muscles (medial brain stem pathways), whereas the hemispheres project the brain stem areas concerned with control of distal limb muscles (lateral brain stem pathways). The lateral portions of the cerebellar hemispheres are called the **cerebrocerebellum.** They are the newest from a phylogenetic point of view, reaching their greatest development in humans. They interact with the motor cortex in planning and programming movements.

Most of the vestibulocerebellar output passes directly to the brain stem, but the rest of the cerebellar cortex projects to the deep nuclei, which in turn project to the brain stem. The deep nuclei provide the only output for the spinocerebellum and the cerebrocerebellum. The medial portion of the spinocerebellum projects to the fastigial nuclei and from there to the brain stem. The adjacent hemispheric portions of the spinocerebellum project to the emboliform and globose nuclei and from there to the brain stem. The cerebrocerebellum projects to the dentate nucleus and from there either directly or indirectly to the ventrolateral nucleus of the thalamus.

MECHANISMS

Although the functions of the flocculonodular lobe, spinocerebellum, and cerebrocerebellum are relatively clear and the cerebellar circuits are simple, the exact ways their different parts carry out their functions are still unknown. The relation of the

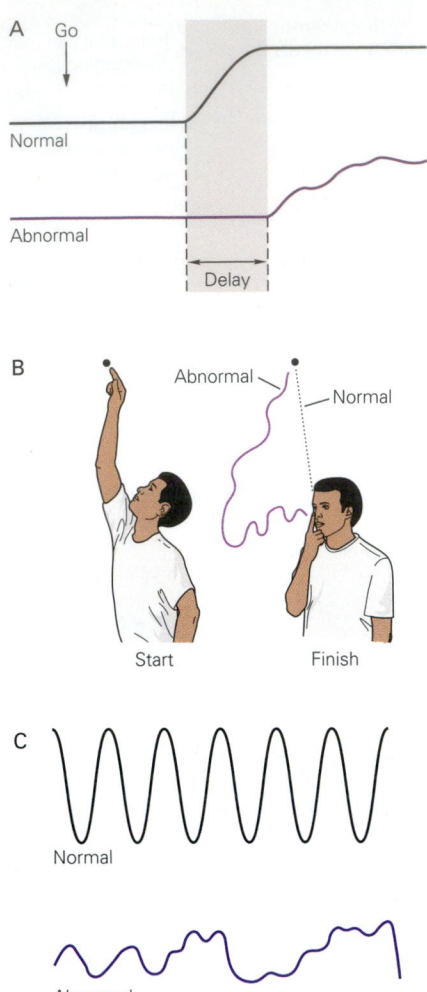

FIGURE 16–17   **Typical defects associated with cerebellar disease. A)** Lesion of the right cerebellar hemisphere delays initiation of movement. The patient is told to clench both hands simultaneously; right hand clenches later than left (shown by recordings from a pressure bulb transducer squeezed by the patient). **B)** Dysmetria and decomposition of movement shown by patient moving his arm from a raised position to his nose. Tremor increases on approaching the nose. **C)** Dysdiadochokinesia occurs in the abnormal position trace of hand and forearm as a cerebellar subject tries alternately to pronate and supinate forearm while flexing and extending elbow as rapidly as possible. (From Kandel ER, Schwartz JH, Jessell TM [editors]: *Principles of Neural Science,* 4th ed. McGraw-Hill, 2000.)

electrical events in the cerebellum to its function in motor control is another interesting problem. The cerebellar cortex has a basic, 150 to 300/s, 200-μV electrical rhythm and, superimposed on this, a 1000 to 2000/s component of smaller amplitude. The frequency of the basic rhythm is thus more than 10 times greater than that of the similarly recorded cerebral cortical alpha rhythm. Incoming stimuli generally alter the amplitude of the cerebellar rhythm like a broadcast signal modulating a carrier frequency in radio transmission. However, the significance of these electrical phenomena in terms of cerebellar function is unknown.

CEREBELLAR DISEASE

Damage to the cerebellum leads to several characteristic abnormalities, including **hypotonia, ataxia,** and **intention tremor.** Figure 16–17 illustrates some of these abnormalities. Most abnormalities are apparent during movement. The marked ataxia is characterized as incoordination due to errors in the rate, range, force, and direction of movement. Voluntary movements are also highly abnormal. For example, attempting to touch an object with a finger results in overshooting to one side or the other. This **dysmetria,** which is also called **past-pointing,** promptly initiates a gross corrective action, but the correction overshoots to the other side. Consequently, the finger oscillates back and forth. This oscillation is the **intention tremor** of cerebellar disease. Another characteristic of cerebellar disease is inability to "put on the brakes," that is, to stop movement promptly. Normally, for example, flexion of the forearm against resistance is quickly checked when the resistance force is suddenly broken off. The patient with cerebellar disease cannot brake the movement of the limb, and the forearm flies backward in a wide arc. This abnormal response is known as the **rebound phenomenon,** and similar impairment is detectable in other motor activities. This is one of the important reasons these patients show **dysdiadochokinesia,** the inability to perform rapidly alternating opposite movements such as repeated pronation and supination of the hands. Finally, patients with cerebellar disease have difficulty performing actions that involve simultaneous motion at more than one joint. They dissect such movements and carry them out one joint at a time, a phenomenon known as **decomposition of movement.**

Other signs of cerebellar deficit in humans provide additional illustrations of the importance of the cerebellum in the control of movement. Ataxia is manifest not only in the wide-based, unsteady, "drunken" gait of patients, but also in defects of the skilled movements involved in the production of speech, so that slurred, **scanning speech** results.

Motor abnormalities associated with cerebellar damage vary depending on the region involved. The major dysfunction seen after damage to the vestibulocerebellum is ataxia, dysequilibrium, and nystagmus. Damage to the vermis and fastigial nucleus (part of the spinocerebellum) leads to disturbances in control of axial and trunk muscles during attempted antigravity postures and scanning speech. Degeneration of this portion of the cerebellum can result from thiamine deficiency in alcoholics or malnourished individuals. The major dysfunction seen after damage to the cerebrocerebellum is delays in initiating movements and decomposition of movement.

THE CEREBELLUM & LEARNING

The cerebellum is concerned with learned adjustments that make coordination easier when a given task is performed over and over. As a motor task is learned, activity in the brain shifts from the prefrontal areas to the parietal and motor cortex and the cerebellum. The basis of the learning in the cerebellum is probably the input via the olivary nuclei. It is worth noting that each Purkinje cell receives inputs from 250,000 to 1 million mossy fibers, but each has only a single climbing fiber from the inferior olive, and this fiber makes 2000–3000 synapses on the Purkinje cell. Climbing fiber activation produces a large, complex spike in the Purkinje cell and this spike in some way produces long-term modification of the pattern of mossy fiber input to that particular Purkinje cell. Climbing fiber activity is increased when a new movement is being learned, and selective lesions of the olivary complex abolish the ability to produce long-term adjustments in certain motor responses.

CHAPTER SUMMARY

- The ventral corticospinal tract and medial descending brain stem pathways (tectospinal, reticulospinal, and vestibulospinal tracts) regulate proximal muscles and posture. The lateral corticospinal and rubrospinal tracts control distal limb muscles and skilled voluntary movements.
- Spinal cord transection is followed by a period of spinal shock during which all spinal reflex responses are profoundly depressed.
- Decerebrate rigidity leads to hyperactivity in extensor muscles in all four extremities; it is actually spasticity due to facilitation of the myotatic stretch reflex. Decorticate posturing or decorticate rigidity is flexion of the upper extremities at the elbow and extensor hyperactivity in the lower extremities.
- The basal ganglia include the caudate nucleus, putamen, globus pallidus, subthalamic nucleus, and substantia nigra. The connections between the parts of the basal ganglia include a dopaminergic nigrostriatal projection from the substantia nigra to the striatum and a GABAergic projection from the striatum to substantia nigra.
- Parkinson disease is due to degeneration of the nigrostriatal dopaminergic neurons and is characterized by akinesia, bradykinesia, cogwheel rigidity, and tremor at rest. Huntington disease is characterized by choreiform movements due to the loss of the GABAergic pathway to the globus pallidus.
- The cerebellar cortex contains five types of neurons: Purkinje, granule, basket, stellate, and Golgi cells. The two main inputs to the cerebellar cortex are climbing fibers and mossy fibers. Purkinje cells are the only output from the cerebellar cortex and they generally project to the deep nuclei.
- Damage to the cerebellum leads to several characteristic abnormalities, including hypotonia, ataxia, and intention tremor.

MULTIPLE-CHOICE QUESTIONS

For all questions, select the single best answer unless otherwise directed.

1. A primary function of the basal ganglia is
   A) sensory integration.
   B) short-term memory.
   C) planning voluntary movement.
   D) neuroendocrine control.
   E) slow-wave sleep.

2. The therapeutic effect of L-dopa in patients with Parkinson disease eventually wears off because
   A) antibodies to dopamine receptors develop.
   B) inhibitory pathways grow into the basal ganglia from the frontal lobe.
   C) there is an increase in circulating α-synuclein.
   D) the normal action of nerve growth factor (NGF) is disrupted.
   E) the dopaminergic neurons in the substantia nigra continue to degenerate.

3. Increased neural activity before a skilled voluntary movement is *first* seen in the
   A) spinal motor neurons.
   B) precentral motor cortex.
   C) midbrain.
   D) cerebellum.
   E) cortical association areas.

4. After falling down a flight of stairs, a young woman is found to have partial loss of voluntary movement on the right side of her body and loss of pain and temperature sensation on the left side below the midthoracic region. It is probable that she has a lesion
   A) transecting the left half of the spinal cord in the lumbar region.
   B) transecting the left half of the spinal cord in the upper thoracic region.
   C) transecting sensory and motor pathways on the right side of the pons.
   D) transecting the right half of the spinal cord in the upper thoracic region.
   E) transecting the dorsal half of the spinal cord in the upper thoracic region.

5. Patients with transected spinal cords frequently have a negative nitrogen balance because
   A) they develop hypercalcemia, and this causes dissolution of the protein in bone.
   B) they are paralyzed below the level of the transection.
   C) they lack the afferent input that normally maintains growth hormone secretion.
   D) they have difficulty voiding, and this causes nitrogen to accumulate in the urine in the bladder.
   E) their corticotropin response to stress is reduced.

6. Which of the following diseases is *not* known to be caused by overexpression of a trinucleotide repeat?
   A) Alzheimer disease
   B) Fragile X syndrome
   C) Spinocerebellar ataxia, type 3
   D) Huntington disease
   E) Friedreich ataxia

CHAPTER RESOURCES

Alexi T, et al: Neuroprotective strategies for basal ganglia degeneration: Parkinson's and Huntington's diseases. Prog Neurobiol 2000;60:409.

De Zeeuw CI, Strata P, Voogd J: *The Cerebellum: From Structure to Control.* Elsevier, 1997.

Ditunno JF Jr, Formal CF: Chronic spinal cord injury. N Engl J Med 1994; 330:550.

Graybiel AM, Delong MR, Kitai ST: *The Basal Ganglia VI*. Springer, 2003.

Haines DE (editor): *Fundamental Neuroscience for Basic and Clinical Applications*, 3rd ed. Elsevier, 2006.

He SQ, Dum RP, Strick PL: Topographic organization of corticospinal projections from the frontal lobe: Motor areas on the lateral surface of the hemisphere. J Neurosci 1993; 13: 952.

Holstege G, Kuypers HGJM: The anatomy of brain stem pathways to the spinal cord in cat. A labeled amino acid tracing study. Prog Brain Res 1982;57:145.

Jueptner M, Weiller C: A review of differences between basal ganglia and cerebellar control of movements as revealed by functional imaging studies. Brain 1998;121:1437.

Kandel ER, Schwartz JH, Jessell TM (editors): *Principles of Neural Science*, 4th ed. McGraw-Hill, 2000.

Lemon RN: Descending pathways in motor control. Annu Rev Neurosci 2008;31:195.

Manto MU, Pandolfo M: *The Cerebellum and its Disorders*. Cambridge University Press, 2001.

McDonald JW, et al: Transplanted embryonic stem cells survive, differentiate and promote recovery in injured rat spinal cord. Nature Med 1999;5:1410.

Nicholls JG, Martin AR, Wallace BG: *From Neuron to Brain: A Cellular and Molecular Approach to the Function of the Nervous System*, 4th ed. Sinauer, 2001.

Nudo RJ: Postinfarct cortical plasticity and behavioral recovery. Stroke 2007;38:840.

Ramer LM, Ramer MS, Steeves JD: Setting the stage for functional repair of spinal cord injuries: a cast of thousands. Spinal Cord 2005;43:134.

The Autonomic Nervous System

After studying this chapter, you should be able to:

■ Describe the location of the cell bodies and axonal trajectories of preganglionic sympathetic and parasympathetic neurons.

■ Describe the location and trajectories of postganglionic sympathetic and parasympathetic neurons.

■ Name the neurotransmitters that are released by preganglionic autonomic neurons, postganglionic sympathetic neurons, postganglionic parasympathetic neurons, and adrenal medullary cells.

■ Outline the functions of the autonomic nervous system.

■ List the ways that drugs act to increase or decrease the activity of the components of the autonomic nervous system.

■ Describe the location of neurons that provide input to sympathetic preganglionic neurons.

■ Describe the composition and functions of the enteric nervous system.

INTRODUCTION

The autonomic nervous system (ANS) is the part of the nervous system that is responsible for homeostasis. Except for skeletal muscle, which gets its innervation from the somatomotor nervous system, innervation to all other organs is supplied by the ANS. Nerve terminals are located in smooth muscle (eg, blood vessels, gut wall, urinary bladder), cardiac muscle, and glands (eg, sweat glands, salivary glands). Although survival is possible without an ANS, the ability to adapt to environmental stressors and other challenges is severely compromised (see Clinical Box 17–1). The ANS has two major divisions: the **sympathetic** and **parasympathetic** nervous systems. As will be described, some target organs are innervated by both divisions and others are controlled by only one. In addition, the ANS includes the **enteric nervous system** within the gastrointestinal tract. The classic definition of the ANS is the preganglionic and postganglionic neurons within the sympathetic and parasympathetic divisions. This would be equivalent to defining the somatomotor nervous system as the cranial and spinal motor neurons. A modern definition of the ANS takes into account the descending pathways from several forebrain and brain stem regions as well as visceral afferent pathways that set the level of activity in sympathetic and parasympathetic nerves. This is analogous to including the many descending and ascending pathways that influence the activity of somatic motor neurons as elements of the somatomotor nervous system.

Multiple System Atrophy & Shy–Drager Syndrome

Multiple system atrophy (MSA) is a neurodegenerative disorder associated with autonomic failure due to loss of preganglionic autonomic neurons in the spinal cord and brain stem. In the absence of an autonomic nervous system, it is difficult to regulate body temperature, fluid and electrolyte balance, and blood pressure. In addition to these autonomic abnormalities, MSA presents with cerebellar, basal ganglia, locus coeruleus, inferior olivary nucleus, and pyramidal tract deficits. MSA is defined as "a sporadic, progressive, adult onset disorder characterized by autonomic dysfunction, parkinsonism, and cerebellar ataxia in any combination." **Shy–Drager syndrome** is a subtype of MSA in which autonomic failure dominates. The pathological hallmark of MSA is cytoplasmic and nuclear inclusions in oligodendrocytes and neurons in central motor and autonomic areas. There is also depletion of monoaminergic, cholinergic, and peptidergic markers in several brain regions and in the cerebrospinal fluid. Basal levels of sympathetic activity and plasma norepinephrine levels are normal in MSA patients, but they fail to increase in response to standing or other stimuli and leads to severe **orthostatic hypotension.** In addition to the fall in blood pressure, orthostatic hypotension leads to dizziness, dimness of vision, and even fainting. MSA is also accompanied by parasympathetic dysfunction, including urinary and sexual dysfunction. MSA is most often diagnosed in individuals between 50 and 70 years of age; it affects more men than women. Erectile dysfunction is often the first symptom of the disease. There are also abnormalities in baroreceptor reflex and respiratory control mechanisms. Although autonomic abnormalities are often the first symptoms, 75% of patients with MSA also experience motor disturbances.

ANATOMIC ORGANIZATION OF AUTONOMIC OUTFLOW

GENERAL FEATURES

Figure 17–1 compares some fundamental characteristics of the innervation to skeletal muscles and innervation to smooth muscle, cardiac muscle, and glands. As discussed in earlier chapters, the final common pathway linking the central nervous system (CNS) to skeletal muscles is the α-motor neuron. Similarly, sympathetic and parasympathetic neurons serve as the final common pathway from the CNS to visceral targets. However, in marked contrast to the somatomotor nervous system, the peripheral motor portions of the ANS are made up of two neurons: **preganglionic** and **postganglionic neurons.**

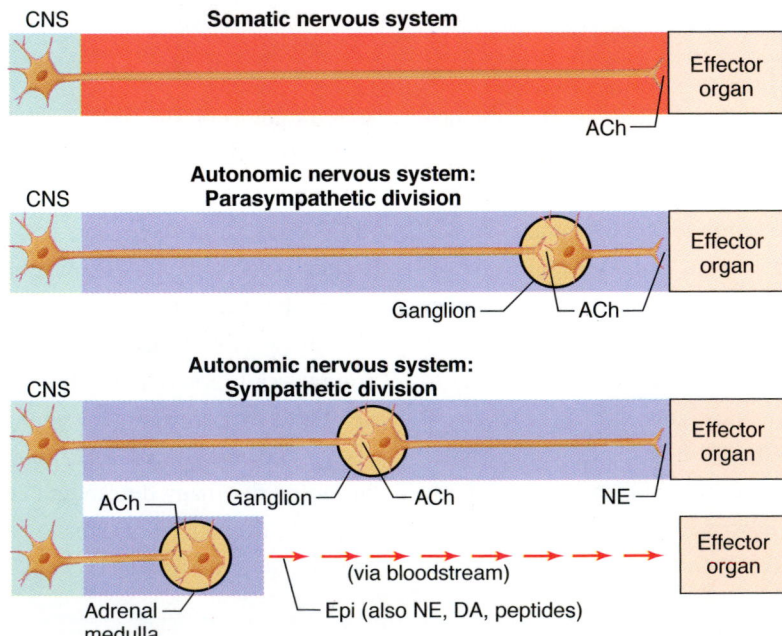

FIGURE 17–1   Comparison of peripheral organization and transmitters released by somatomotor and autonomic nervous systems **(NS).** ACh, acetylcholine; DA, dopamine; NE, norepinephrine; Epi, epinephrine. (From Widmaier EP, Raff H, Strang KT: *Vander's Human Physiology.* McGraw-Hill, 2008.)

The cell bodies of the preganglionic neurons are located in the intermediolateral column (IML) of the spinal cord and in motor nuclei of some cranial nerves. In contrast to the large diameter and rapidly conducting α-motor neurons, preganglionic axons are small-diameter, myelinated, relatively slowly conducting B fibers. A preganglionic axon diverges to an average of eight or nine postganglionic neurons. In this way, autonomic output is diffused. The axons of the postganglionic neurons are mostly unmyelinated C fibers and terminate on the visceral effectors.

SYMPATHETIC DIVISION

In contrast to α-motor neurons, which are located at all spinal segments, sympathetic preganglionic neurons are located in the IML of only the first thoracic to the third or fourth lumbar segments. This is why the sympathetic nervous system is sometimes called the thoracolumbar division of the ANS. The axons of the sympathetic preganglionic neurons leave the spinal cord at the level at which their cell bodies are located and exit via the ventral root along with axons of α- and γ-motor neurons (Figure 17–2). They then separate from the ventral root via the **white rami communicans** and project to the adjacent **sympathetic paravertebral ganglion,** where some of them end on the cell bodies of the postganglionic neurons. Paravertebral ganglia are located adjacent to each thoracic and upper lumbar spinal segments; in addition, there are a few ganglia adjacent to the cervical and sacral spinal segments. These ganglia form the **sympathetic chain** bilaterally. The ganglia are connected to each other via the axons of preganglionic neurons that travel rostrally or caudally to terminate on postganglionic neurons located at some distance. This arrangement is seen in Figures 17–2 and 17–3.

Some preganglionic neurons pass through the paravertebral ganglion chain and end on postganglionic neurons located in **prevertebral** (or **collateral**) **ganglia** close to the viscera, including the celiac, superior mesenteric, and inferior mesenteric ganglia (Figure 17–3). There are also preganglionic neurons whose axons terminate directly on the effector organ, the adrenal gland.

The axons of some of the postganglionic neurons leave the chain ganglia and reenter the spinal nerves via the **gray rami**

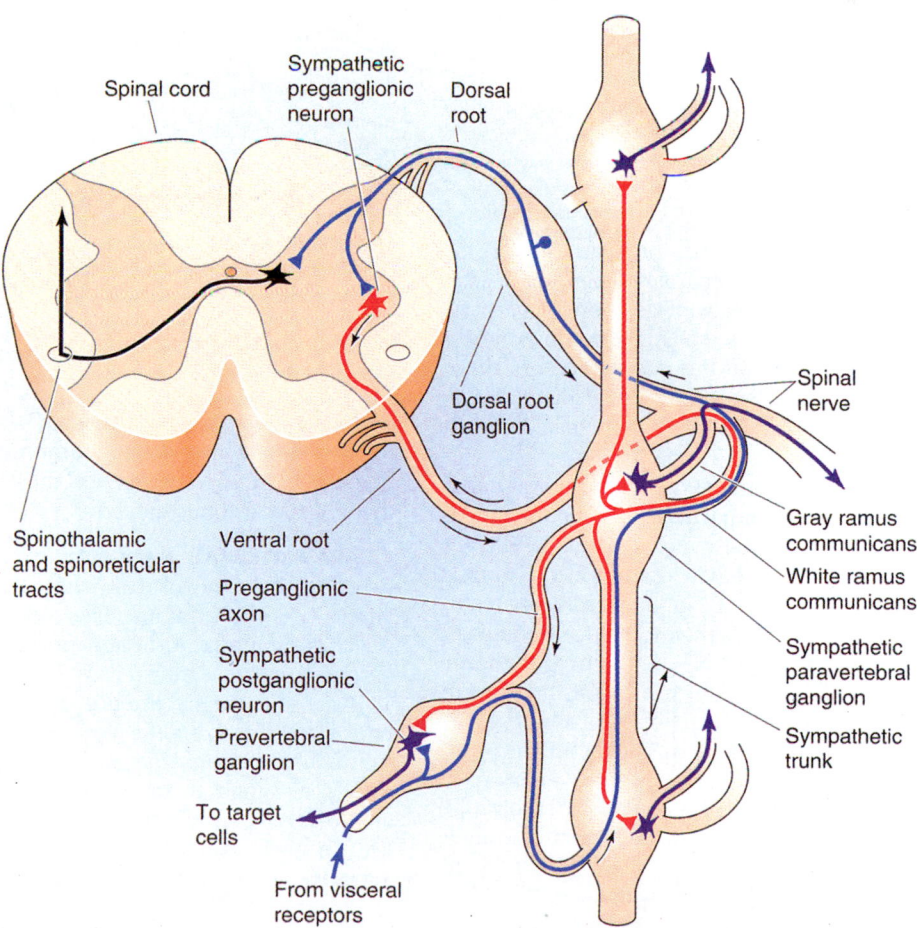

FIGURE 17–2 **Projection of sympathetic preganglionic and postganglionic fibers.** The drawing shows the thoracic spinal cord, paravertebral, and prevertebral ganglia. Preganglionic neurons are shown in red, postganglionic neurons in dark blue, afferent sensory pathways in blue, and interneurons in black. (Reproduced with permission from Boron WF, Boulpaep EL: *Medical Physiology.* Elsevier, 2005.)

FIGURE 17–3 **Organization of sympathetic (left) and parasympathetic (right) nervous systems.** Preganglionic sympathetic and parasympathetic neurons are shown in red and orange, respectively; postganglionic sympathetic and parasympathetic neurons in blue and green, respectively. (Reproduced with permission from Boron WF, Boulpaep EL: *Medical Physiology.* Elsevier, 2005.)

communicans and are distributed to autonomic effectors in the areas supplied by these spinal nerves (Figure 17–2). These postganglionic sympathetic nerves terminate mainly on smooth muscle (eg, blood vessels, hair follicles, airways) and on sweat glands in the limbs. Other postganglionic fibers leave the chain ganglia to enter the thoracic cavity to terminate in visceral organs. Postganglionic fibers from prevertebral ganglia also terminate in visceral targets.

PARASYMPATHETIC DIVISION

The parasympathetic nervous system is sometimes called the **craniosacral division** of the ANS because of the location of its preganglionic neurons (Figure 17–3). The parasympathetic nerves supply the visceral structures in the head via the oculomotor, facial, and glossopharyngeal nerves, and those in the thorax and upper abdomen via the vagus nerves. The sacral outflow supplies the pelvic viscera via branches of the second to fourth sacral spinal nerves. Parasympathetic preganglionic fibers synapse on ganglia cells clustered within the walls of visceral organs; thus these parasympathetic postganglionic fibers are very short.

CHEMICAL TRANSMISSION AT AUTONOMIC JUNCTIONS

ACETYLCHOLINE & NOREPINEPHRINE

The first evidence for chemical neurotransmission was provided by a simple yet dramatic study of heart rate control by the parasympathetic nervous system performed by Otto Loewi in 1920 (Clinical Box 17–2). Transmission at the synaptic junctions between pre- and postganglionic neurons and between the postganglionic neurons and the autonomic effectors is chemi-

cally mediated. The principal transmitter agents involved are **acetylcholine** and **norepinephrine** (Figures 17–1 and 17–4).

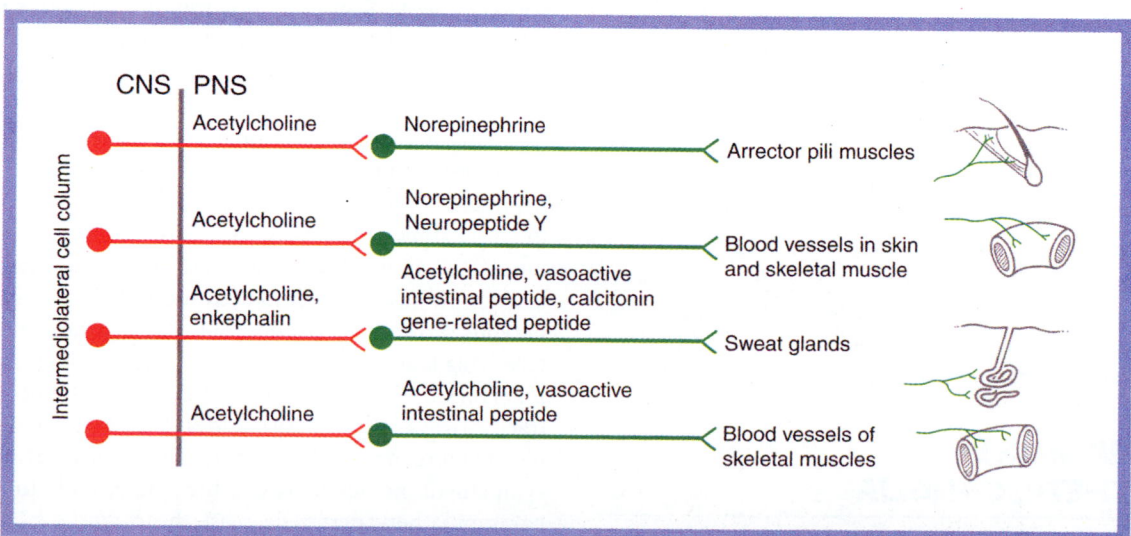

FIGURE 17–4 **Chemical coding of sympathetic preganglionic and postganglionic neurons.** CNS, central nervous system; PNS, peripheral nervous system. (Reproduced with permission from Haines DE [editor]: *Fundamental Neuroscience for Basic and Clinical Applications,* 3rd ed. Elsevier, 2006.)

The neurons that are cholinergic (ie, release acetylcholine) are (1) all preganglionic neurons, (2) all parasympathetic postganglionic neurons, (3) sympathetic postganglionic neurons that innervate sweat glands, and (4) sympathetic postganglionic neurons that end on blood vessels in some skeletal muscles and produce vasodilation when stimulated (sympathetic vasodilator nerves). The remaining sympathetic postganglionic neurons are noradrenergic (ie, release norepinephrine). The adrenal medulla is essentially a sympathetic ganglion in which the postganglionic cells have lost their axons and secrete norepinephrine and epinephrine directly into the bloodstream. The cholinergic preganglionic neurons to these cells have consequently become the secretomotor nerve supply of this gland.

Transmission in autonomic ganglia is mediated primarily by N_2 nicotinic cholinergic receptors that are blocked by hexamethonium. This is in contrast to the N_1 nicotinic cholinergic receptors at the neuromuscular junction, which are blocked by D-tubocurare. The release of acetylcholine from postganglionic fibers acts on muscarinic receptors, which are blocked by atropine. The release of norepinephrine from sympathetic postganglionic fibers acts on α_1, β_1, or β_2 adrenoreceptors, depending on the target organ. Table 17–1 shows the types of receptors at various junctions within the autonomic nervous system.

In addition to these "classical neurotransmitters," some autonomic fibers also release neuropeptides. Figure 17–4 shows some examples for sympathetic postganglionic fibers. The small granulated vesicles in postganglionic noradrenergic neurons contain ATP and norepinephrine, and the large granulated vesicles contain neuropeptide Y. There is evidence that low-frequency stimulation promotes release of ATP, whereas high-frequency stimulation causes release of neuropeptide Y. The viscera contains purinergic receptors, and evidence is accumulating that ATP is a mediator in the autonomic nervous system along with norepinephrine. However, its exact role is unsettled.

Acetylcholine does not usually circulate in the blood, and the effects of localized cholinergic discharge are generally discrete and of short duration because of the high concentration of acetylcholinesterase at cholinergic nerve endings. Norepinephrine spreads farther and has a more prolonged action than acetylcholine. Norepinephrine, epinephrine, and dopamine are all found in plasma. The epinephrine and some of the dopamine come from the adrenal medulla, but most of the norepinephrine diffuses into the bloodstream from noradrenergic nerve endings. Metabolites of norepinephrine and dopamine also enter the circulation, some from the sympathetic nerve endings and some from smooth muscle cells (Figure 17–5). It is worth noting that even when monoamine oxidase (MAO) and catechol-O-methyltransferase (COMT) are both inhibited, the metabolism of norepinephrine is still rapid. However, inhibition of reuptake prolongs its half-life.

TRANSMISSION IN SYMPATHETIC GANGLIA

At least in experimental animals, the responses produced in postganglionic neurons by stimulation of their preganglionic innervation include both a rapid depolarization (**fast excitatory postsynaptic potential [EPSP]**) that generates action potentials and a prolonged excitatory postsynaptic potential (**slow EPSP**). The slow response apparently modulates and regulates transmission through the sympathetic ganglia. As just described, the initial depolarization is produced by acetylcholine via the N_2 nicotinic receptor. The slow EPSP is produced by acetylcholine acting on a muscarinic receptor on the membrane of the postganglionic neuron.

The junctions in the peripheral autonomic motor pathways are a logical site for pharmacologic manipulation of visceral function. The transmitter agents are synthesized, stored in the nerve endings, and released near the neurons, muscle cells, or gland cells on which they act. They bind to receptors on these cells, thus initiating their characteristic actions, and they are then removed from the area by reuptake or metabolism. Each of these steps can be stimulated or inhibited, with predictable consequences.

Some of the drugs and toxins that affect the activity of the autonomic nervous system and the mechanisms by which they produce their effects are listed in Table 17–2. Compounds with muscarinic actions include congeners of acetylcholine and drugs that inhibit acetylcholinesterase. Among the latter are the insecticide parathion and diisopropyl fluorophosphate (DFP), a component of the so-called nerve gases, which kill by producing massive inhibition of acetylcholinesterase.

RESPONSES OF EFFECTOR ORGANS TO AUTONOMIC NERVE IMPULSES

GENERAL PRINCIPLES

The effects of stimulation of the noradrenergic and cholinergic postganglionic nerve fibers are indicated in Figure 17–3 and Table 17–1. These findings point out another difference between the ANS and the somatomotor nervous system. The release of acetylcholine by α-motor neurons only leads to contraction of skeletal muscles. In contrast, release of acetylcholine onto smooth muscle of some organs leads to contraction (eg, walls of the gastrointestinal tract) while release onto other organs leads to relaxation (eg, sphincters in the gastrointestinal tract). The only way to relax a skeletal muscle is to inhibit the discharges of the α-motor neurons; but for some targets innervated by the ANS, one can shift from contraction to relaxation by switching from activation of the parasympathetic nervous system to activation of the sympathetic nervous system. This is the case for the many organs which receive dual innervation with antagonistic effects, including the digestive tract, airways, and urinary bladder. The heart is another example of an organ with dual antagonistic control. Stimulation of sympathetic nerves increases heart rate, and stimulation of parasympathetic nerves decreases heart rate.

In other cases, the effects of sympathetic and parasympathetic activation can be considered complementary. An example

TABLE 17–1 Responses of some effector organs to autonomic nerve activity.

Effector Organs	Parasympathetic Nervous System	Sympathetic Nervous System Receptor Type	Sympathetic Nervous System Response
Eyes			
Radial muscle of iris	—[a]	α_1	Contraction (mydriasis)
Sphincter muscle of iris	Contraction (miosis)		—
Ciliary muscle	Contraction for near vision		—
Heart			
S–A node	Decreases heart rate	β_1	Increases heart rate
Atria & ventricle	Decreases contractility	β_1, β_2	Increases contractility
AV node & Purkinje	Decreases conduction velocity	β_1, β_2	Increases conduction velocity
Arterioles			
Coronary	—	α_1, α_2	Constriction
		β_2	Dilation
Skin	—	α_1, α_2	Constriction
Skeletal muscle	—	α_1	Constriction
		β_2, M	Dilation
Abdominal viscera	—	α_1	Constriction
Salivary glands	Dilation	α_1, α_2	Constriction
Renal	—	α_1	Constriction
Systemic veins	—	α_1, α_2	Constriction
		β_2	Dilation
Lungs			
Bronchial muscle	Contraction	β_2	Relaxation
Stomach			
Motility and tone	Increases	$\alpha_1, \alpha_2, \beta_2$	Decreases
Sphincters	Relaxation	α_1	Contraction
Secretion	Stimulation	?	Inhibition
Intestine			
Motility and tone	Increases	$\alpha_1, \alpha_2, \beta_1, \beta_2$	Decreases
Sphincters	Relaxation	α_1	Contraction (usually)
Secretion	Stimulation	α_2	Inhibition
Gall bladder	Contraction	β_2	Relaxation
Urinary bladder			
Detrusor	Contraction	β_2	Relaxation
Sphincter	Relaxation	α_1	Contraction

(continued)

TABLE 17–1 Responses of some effector organs to autonomic nerve activity. (Continued)

Effector Organs	Parasympathetic Nervous System	Sympathetic Nervous System	
		Receptor Type	Response
Uterus	Variable	α_1	Contraction (pregnant)
		β_2	Relaxation
Male sex organs	Erection	α_1	Ejaculation
Skin			
Pilomotor muscles	—	α_1	Contraction
Sweat glands	—	α_1	Slight, localized secretion[b]
		M	Generalized abundant, dilute secretion
Liver	—	α_1, β_2	Glycogenolysis
Pancreas			
Exocrine glands	Increases secretion	α	Decreases secretion
Endocrine glands	—	α_2	Inhibits secretion
Salivary glands	Profuse, watery secretion	α_1	Thick, viscous secretion
		β	Amylase secretion
Lacrimal glands	Secretion		—
Adipose tissue	—	α_2, β_3	Lipolysis

[a]A dash means these cells are not innervated by this division of the autonomic nervous system.

[b]On palms of hands and in some other locations ("adrenergic sweating").

Modified from Hardman JG, Limbird LE, Gilman AG (editors): *Goodman and Gilman's The Pharmacological Basis of Therapeutics,* 10th ed. McGraw-Hill, 2001.

is the innervation of salivary glands. Parasympathetic activation causes release of watery saliva, while sympathetic activation causes the production of thick, viscous saliva.

The two divisions of the ANS can also act in a synergistic or cooperative manner in the control of some functions. One example is the control of pupil diameter in the eye. Both sympathetic and parasympathetic innervations are excitatory, but the former contracts the radial muscle to cause mydriasis and the latter contracts the sphincter (or constrictor) muscle to cause meiosis. Another example is the synergistic actions of these nerves on sexual function. Activation of parasympathetic nerves to the penis increases blood flow and leads to erection while activation of sympathetic nerves to the penis causes ejaculation.

There are also several organs that are innervated by only one division of the ANS. In addition to the adrenal gland, most blood vessels, the pilomotor muscles in the skin (hair follicles), and sweat glands are innervated exclusively by sympathetic nerves. The lacrimal muscle (tear gland), ciliary muscle (for accommodation for near vision), and the sublingual salivary gland are innervated exclusively by parasympathetic nerves.

PARASYMPATHETIC CHOLINERGIC & SYMPATHETIC NORADRENERGIC DISCHARGE

In a general way, the functions promoted by activity in the cholinergic division of the autonomic nervous system are those concerned with the vegetative aspects of day-to-day living. For example, parasympathetic action favors digestion and absorption of food by increasing the activity of the intestinal musculature, increasing gastric secretion, and relaxing the pyloric sphincter. For this reason, the cholinergic division is sometimes called the anabolic nervous system.

The sympathetic (noradrenergic) division discharges as a unit in emergency situations and can be called the catabolic nervous system. The effects of this discharge prepares the individual to cope with an emergency. Sympathetic activity dilates the pupils (letting more light into the eyes), accelerates the heartbeat and raises the blood pressure (providing better perfusion of the vital organs and muscles), and constricts the blood vessels of the skin (which limits bleeding from wounds).

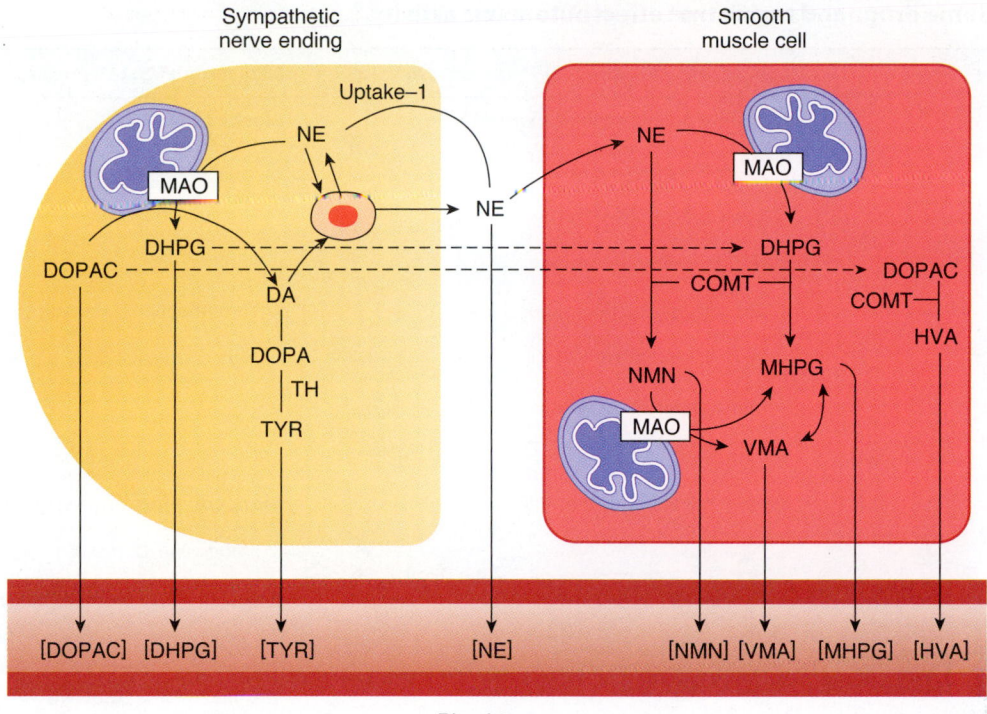

FIGURE 17–5 **Catecholamine metabolism in the sympathetic nervous system.** COMT, catechol-*O*-methyltransferase; DA, dopamine; DHPG, dihydroxyphenylglycol; DOPA, dihydroxyphenylalanine; DOPAC, dihydroxyphenylacetic acid; HVA, homovanillic acid; MHPG, 3-methoxy-4-hydroxyphenylglycol; MOA, monoamine oxidase; NE, norepinephrine; NMN, normetanephrine; TH, tyrosine hydroxylase; TYR, tyrosine; VMA, vanillylmandelic acid. (Courtesy of DS Goldstein.)

Noradrenergic discharge also leads to elevated plasma glucose and free fatty acid levels (supplying more energy). On the basis of effects like these, Walter Cannon called the emergency-induced discharge of the noradrenergic nervous system the "preparation for flight or fight."

The emphasis on mass discharge in stressful situations should not obscure the fact that the sympathetic fibers also subserve other functions. For example, tonic sympathetic discharge to the arterioles maintains arterial pressure, and variations in this tonic discharge are the mechanism by which carotid sinus feedback regulation of blood pressure is effected. In addition, sympathetic discharge is decreased in fasting animals and increased when fasted animals are refed. These changes may explain the decrease in blood pressure and metabolic rate produced by fasting and the opposite changes produced by feeding.

DESCENDING INPUT TO AUTONOMIC PREGANGLIONIC NEURONS

As is the case for α-motor neurons, the activity of autonomic nerves is dependent on both reflexes (eg, baroreceptor and chemoreceptor reflexes) and descending excitatory and inhibitory input from several brain regions. Figure 17–6 shows the source of some forebrain and brain stem descending inputs to autonomic preganglionic neurons. For example, a major source of excitatory drive to sympathetic preganglionic neurons comes from the rostral ventrolateral medulla. Although not shown, medullary raphé neurons project to the spinal cord to inhibit or excite sympathetic activity. In addition to these direct pathways to preganglionic neurons, there are many brain stem nuclei that feed into these pathways. This is analogous to the control of somatomotor function by areas such as the basal ganglia.

ENTERIC NERVOUS SYSTEM

The enteric nervous system, which can be considered as the third division of the ANS, is located within the wall of the digestive tract, all the way from the esophagus to the anus. It is comprised of two well-organized neural plexuses. The **myenteric plexus** is located between longitudinal and circular layers of muscle; it is involved in control of digestive tract motility. The **submucosal plexus** is located between the circular muscle and the luminal mucosa; it senses the environment of the lumen and regulates gastrointestinal blood flow and epithelial cell function.

The enteric nervous system contains as many neurons as the entire spinal cord. It is sometimes referred to as a "mini-brain" as it contains all the elements of a nervous system including sensory neurons, interneurons, and motor neurons.

TABLE 17–2 Some drugs and toxins that affect autonomic activity.[a]

Site of Action	Compounds That Augment Autonomic Activity	Compounds That Depress Autonomic Activity
Autonomic ganglia	**Stimulate postganglionic neurons**	**Block conduction**
	Nicotine	Hexamethonium (C-6)
	Low concentration of acetylcholine	Mecamylamine (Inversine)
	Inhibit acetylcholinesterase	Pentolinium
	DFP (diisopropyl fluorophosphate)	Trimethaphan (Arfonad)
	Physostigmine (Eserine)	High concentration of acetylcholine
	Neostigmine (Prostigmin)	
	Parathion	
Postganglionic sympathetic terminals	**Release norepinephrine**	**Block norepinephrine synthesis**
	Tyramine	Metyrosine (Demser)
	Ephedrine	**Interfere with norepinephrine storage**
	Amphetamine	Reserpine
		Guanethidine[b] (Ismelin)
		Prevent norepinephrine release
		Bretylium (Bretylol)
		Guanethidine[b] (Ismelin)
		Form false transmitters
		Methyldopa (Aldomet)
Muscarinic receptors		Atropine, scopolamine
α adrenergic receptors	**Stimulate α_1 receptors**	**Block α receptors**
	Methoxamine (Vasoxyl)	Phenoxybenzamine (Dibenzyline)
	Phenylephrine (Neosynephrine)	Phentolamine (Regitine)
		Prazosin (Minipress) blocks α_1
		Yohimbine blocks α_2
β adrenergic receptors	**Stimulate β receptors**	**Block β receptors**
	Isoproterenol (Isuprel)	Propranolol (Inderal) blocks β_1 and β_2
		Atenolol (Tenormin) blocks β_1
		Butoxamine blocks β_2

[a]Only the principal actions are listed.

[b]Guanethidine is believed to have two principal actions.

It contains sensory neurons innervating receptors in the mucosa that respond to mechanical, thermal, osmotic, and chemical stimuli. Motor neurons control motility, secretion, and absorption by acting on smooth muscle and secretory cells. Interneurons integrate information from sensory neurons and feedback to the enteric motor neurons.

Parasympathetic and sympathetic nerves connect the central nervous system to the enteric nervous system or directly to the digestive tract. Although the enteric nervous system can function autonomously, normal digestive function often requires communication between the central nervous system and the enteric nervous system.

FIGURE 17–6 **Pathways that control autonomic responses.** Direct projections (solid lines) to autonomic preganglionic neurons include the hypothalamic paraventricular nucleus, parabrachial nucleus, nucleus of the solitary tract, ventrolateral medulla, and medullary raphé (not shown). Indirect projections (dashed lines) include the cerebral cortex, amygdala, and periaqueductal grey matter. (From Kandel ER, Schwartz JH, Jessell TM [editors]: *Principles of Neural Science,* 4th ed. McGraw-Hill, 2000.)

CHAPTER SUMMARY

- Preganglionic sympathetic neurons are located in the IML of the thoracolumbar spinal cord and project to postganglionic neurons in the paravertebral or prevertebral ganglia or the adrenal medulla. Preganglionic parasympathetic neurons are located in motor nuclei of cranial nerves III, VII, IX, and X and the sacral IML.

- Nerve terminals of postganglionic neurons are located in smooth muscle (eg, blood vessels, gut wall, urinary bladder), cardiac muscle, and glands (eg, sweat gland, salivary glands).

- Acetylcholine is released at nerve terminals of all preganglionic neurons, postganglionic parasympathetic neurons, and a few postganglionic sympathetic neurons (sweat glands, sympathetic vasodilator fibers). The remaining sympathetic postganglionic neurons release norepinephrine.

- Sympathetic activity prepares the individual to cope with an emergency by accelerating the heartbeat, raising blood pressure (perfusion of the vital organs), and constricting the blood vessels of the skin (limits bleeding from wounds). Parasympathetic activity is concerned with the vegetative aspects of day-to-day living and favors digestion and absorption of food by increasing the activity of the intestinal musculature, increasing gastric secretion, and relaxing the pyloric sphincter.

- Ganglionic transmission is blocked by N2 nicotinic antagonists. Postganglionic cholinergic transmission is blocked by muscarinic antagonists. Postganglionic adrenergic transmission is blocked by antagonists of α_1, β_1, or β_2 adrenoreceptors, depending on the target organ.

- The enteric nervous system is located within the wall of the digestive tract and is composed of the myenteric plexus (control of digestive tract motility) and the submucosal plexus (regulates gastrointestinal blood flow and epithelial cell function).

MULTIPLE-CHOICE QUESTIONS

For all questions, select the single best answer unless otherwise directed.

1. Which of the following drugs would *not* be expected to increase sympathetic discharge or mimic the effects of increased sympathetic discharge?

   A) Prazosin
   B) Neostigmine
   C) Amphetamine
   D) Isoproterenol
   E) Methoxamine

2. Sympathetic nerve activity
   A) is essential for survival.
   B) causes contraction of some smooth muscles and relaxation of others.
   C) causes relaxation of the radial muscle of the eye to dilate the pupil.
   D) relaxes smooth muscle of the gastrointestinal wall and gastrointestinal sphincter.
   E) all of the above

3. Parasympathetic nerve activity
   A) is essential for survival.
   B) affects only smooth muscles and glands.
   C) causes contraction of the radial muscle of the eye to allow accommodation for near vision.
   D) contracts smooth muscle of the gastrointestinal wall and relaxes the gastrointestinal sphincter.
   E) all of the above

4. Which of the following is correctly paired?
   A) sinoatrial node : nicotinic cholinergic receptors
   B) autonomic ganglia : muscarinic cholinergic receptors
   C) pilomotor smooth muscle : β_2-adrenergic receptors
   D) vasculature of some skeletal muscles : muscarinic cholinergic receptors
   E) sweat glands : α_2-adrenergic receptors

CHAPTER RESOURCES

Benarroch EE: *Central Autonomic Network. Functional Organization and Clinical Correlations.* Futura Publishing, 1997.

Boron WF, Boulpaep EL: *Medical Physiology.* Elsevier, 2005.

Brodal P: *The Central Nervous System. Structure and Function.* Oxford University Press, 1998.

Elvin LG, Lindh B, Hokfelt T: The chemical neuroanatomy of sympathetic ganglia. Annu Rev Neurosci 1993;16:471.

Jänig W: *The Integrative Action of the Autonomic Nervous System. Neurobiology of Homeostasis.* Cambridge University Press, 2006.

Kandel ER, Schwartz JH, Jessell TM (editors): *Principles of Neural Science,* 4th ed. McGraw-Hill, 2000.

Loewy AD, Spyer KM (editors): *Central Regulation of Autonomic Function.* Oxford University Press, 1990.

Pick J: *The Autonomic Nervous System.* Lippincott, 1970.

Squire LR, et al (editors): *Fundamental Neuroscience,* 3rd ed. Academic Press, 2008.

Hypothalamic Regulation of Hormonal Functions

INTRODUCTION

Many of the complex autonomic mechanisms that maintain the chemical constancy and temperature of the internal environment are integrated in the hypothalamus. The hypothalamus also functions with the limbic system as a unit that regulates emotional and instinctual behavior.

HYPOTHALAMUS: ANATOMIC CONSIDERATIONS

The hypothalamus (Figure 18–1) is the portion of the anterior end of the diencephalon that lies below the hypothalamic sulcus and in front of the interpeduncular nuclei. It is divided into a variety of nuclei and nuclear areas.

AFFERENT & EFFERENT CONNECTIONS OF THE HYPOTHALAMUS

The principal afferent and efferent neural pathways to and from the hypothalamus are mostly unmyelinated. Many connect the hypothalamus to the limbic system. Important connections also exist between the hypothalamus and nuclei in the midbrain tegmentum, pons, and hindbrain.

Norepinephrine-secreting neurons with their cell bodies in the hindbrain end in many different parts of the hypothalamus

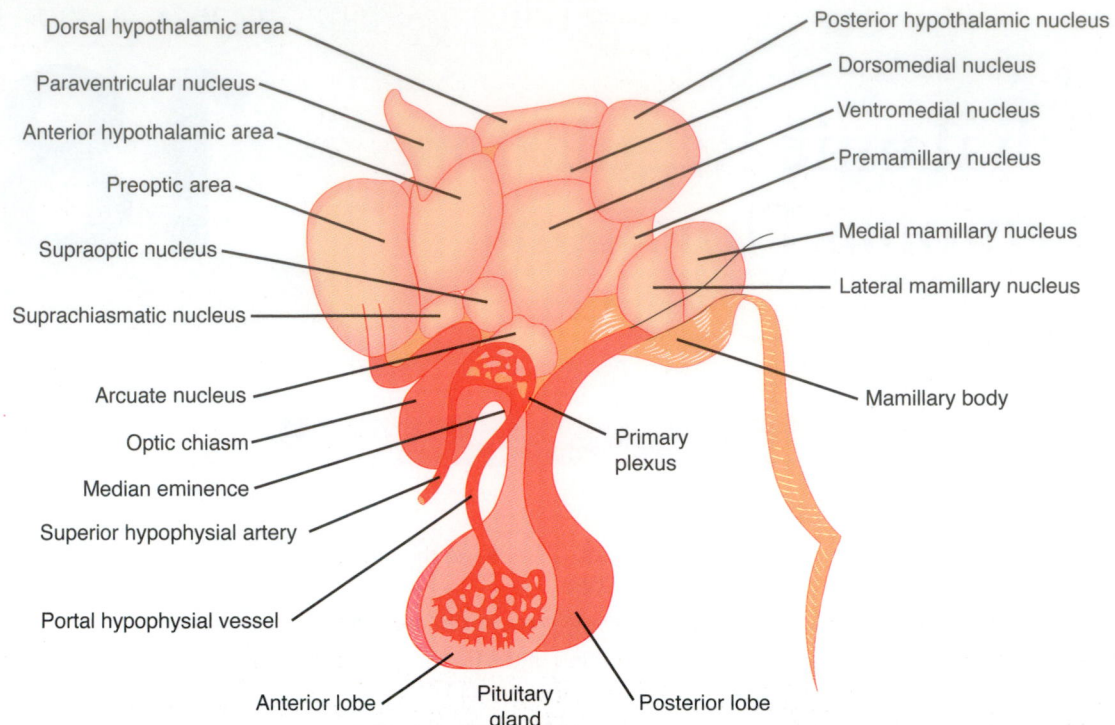

FIGURE 18–1   **Human hypothalamus, with a superimposed diagrammatic representation of the portal hypophysial vessels.**

(see Figure 15–5). Paraventricular neurons that probably secrete oxytocin and vasopressin project in turn to the hindbrain and the spinal cord. Neurons that secrete epinephrine have their cell bodies in the hindbrain and end in the ventral hypothalamus.

An intrahypothalamic system of dopamine-secreting neurons have their cell bodies in the arcuate nucleus and end on or near the capillaries that form the portal vessels in the median eminence. Serotonin-secreting neurons project to the hypothalamus from the raphe nuclei.

RELATION TO THE PITUITARY GLAND

There are neural connections between the hypothalamus and the posterior lobe of the pituitary gland and vascular connections between the hypothalamus and the anterior lobe. Embryologically, the posterior pituitary arises as an evagination of the floor of the third ventricle. It is made up in large part of the endings of axons that arise from cell bodies in the supraoptic and paraventricular nuclei and pass to the posterior pituitary (Figure 18–2) via the **hypothalamohypophysial tract.** Most of the supraoptic fibers end in the posterior lobe itself, whereas some of the paraventricular fibers end in the median eminence. The anterior and intermediate lobes of the pituitary arise in the embryo from the Rathke pouch, an evagination from the roof of the pharynx (see Figure 24–1). Sympathetic nerve fibers reach the anterior lobe from its capsule, and parasympathetic fibers reach it from the petrosal nerves, but few if any nerve fibers pass to it from the hypothalamus. However, the **portal hypophysial**

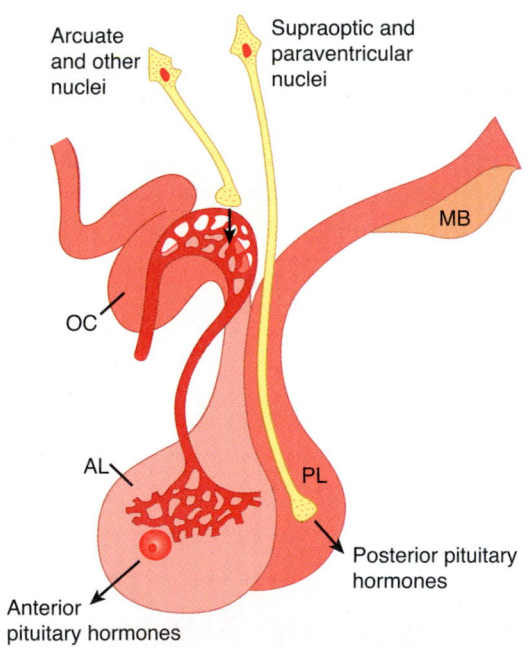

FIGURE 18–2   **Secretion of hypothalamic hormones.** The hormones of the posterior lobe (PL) are released into the general circulation from the endings of supraoptic and paraventricular neurons, whereas hypophysiotropic hormones are secreted into the portal hypophysial circulation from the endings of arcuate and other hypothalamic neurons. AL, anterior lobe; MB, mamillary bodies; OC, optic chiasm.

vessels form a direct vascular link between the hypothalamus and the anterior pituitary. Arterial twigs from the carotid arteries and circle of Willis form a network of fenestrated capillaries called the **primary plexus** on the ventral surface of the hypothalamus (Figure 18–1). Capillary loops also penetrate the median eminence. The capillaries drain into the sinusoidal portal hypophysial vessels that carry blood down the pituitary stalk to the capillaries of the anterior pituitary. This system begins and ends in capillaries without going through the heart and is therefore a true portal system. In birds and some mammals, including humans, there is no other anterior hypophysial arterial supply except capsular vessels and anastomotic connections from the capillaries of the posterior pituitary. The **median eminence** is generally defined as the portion of the ventral hypothalamus from which the portal vessels arise. This region is outside the blood–brain barrier (see Chapter 34).

HYPOTHALAMIC FUNCTION

The major functions of the hypothalamus are summarized in Table 18–1. Some are fairly clear-cut visceral reflexes, and others include complex behavioral and emotional reactions; however, all involve a particular response to a particular stimulus. It is important to keep this in mind in considering hypothalamic function.

RELATION TO AUTONOMIC FUNCTION

Many years ago, Sherrington called the hypothalamus "the head ganglion of the autonomic system." Stimulation of the hypothalamus produces autonomic responses, but the hypothalamus does not seem to be concerned with the regulation of visceral function per se. Rather, the autonomic responses triggered in the

TABLE 18–1 Summary of principal hypothalamic regulatory mechanisms.

Function	Afferents from	Integrating Areas
Temperature regulation	Temperature receptors in the skin, deep tissues, spinal cord, hypothalamus, and other parts of the brain	Anterior hypothalamus, response to heat; posterior hypothalamus, response to cold
Neuroendocrine control of:		
Catecholamines	Limbic areas concerned with emotion	Dorsal and posterior hypothalamus
Vasopressin	Osmoreceptors, "volume receptors," others	Supraoptic and paraventricular nuclei
Oxytocin	Touch receptors in breast, uterus, genitalia	Supraoptic and paraventricular nuclei
Thyroid-stimulating hormone (thyrotropin, TSH) via TRH	Temperature receptors in infants, perhaps others	Paraventricular nuclei and neighboring areas
Adrenocorticotropic hormone (ACTH) and β-lipotropin (β-LPH) via CRH	Limbic system (emotional stimuli); reticular formation ("systemic" stimuli); hypothalamic and anterior pituitary cells sensitive to circulating blood cortisol level; suprachiasmatic nuclei (diurnal rhythm)	Paraventricular nuclei
Follicle-stimulating hormone (FSH) and luteinizing hormone (LH) via GnRH	Hypothalamic cells sensitive to estrogens, eyes, touch receptors in skin and genitalia of reflex ovulating species	Preoptic area; other areas
Prolactin via PIH and PRH	Touch receptors in breasts, other unknown receptors	Arcuate nucleus; other areas (hypothalamus inhibits secretion)
Growth hormone via somatostatin and GRH	Unknown receptors	Periventricular nucleus, arcuate nucleus
"Appetitive" behavior		
Thirst	Osmoreceptors, probably located in the organum vasculosum of the lamina terminalis; angiotensin II uptake in the subfornical organ	Lateral superior hypothalamus
Hunger	Glucostat cells sensitive to rate of glucose utilization; leptin receptors; receptors for other polypeptides	Ventromedial, arcuate, and paraventricular nuclei; lateral hypothalamus
Sexual behavior	Cells sensitive to circulating estrogen and androgen, others	Anterior ventral hypothalamus plus, in the male, piriform cortex
Defensive reactions (fear, rage)	Sense organs and neocortex, paths unknown	Diffuse, in limbic system and hypothalamus
Control of body rhythms	Retina via retinohypothalamic fibers	Suprachiasmatic nuclei

hypothalamus are part of more complex phenomena such as eating, and emotions such as rage. For example, stimulation of various parts of the hypothalamus, especially the lateral areas, produces diffuse sympathetic discharge and increased adrenal medullary secretion, the mass sympathetic discharge seen in animals exposed to stress (the flight or fight reaction; see Chapter 17).

It has been claimed that separate hypothalamic areas control epinephrine and norepinephrine secretion. Differential secretion of one or the other of these adrenal medullary catecholamines does occur in certain situations (see Chapter 22), but the selective increases are small.

Body weight depends on the balance between caloric intake and utilization of calories. Obesity results when the former exceeds the latter. The hypothalamus and related parts of the brain play a key role in the regulation of food intake. Obesity is considered in detail in Chapter 27, and the relation of obesity to diabetes mellitus is discussed in Chapter 21.

Hypothalamic regulation of sleep and circadian rhythms are discussed in Chapter 15.

THIRST

Another appetitive mechanism under hypothalamic control is thirst. Drinking is regulated by plasma osmolality and extracellular fluid (ECF) volume in much the same fashion as vasopressin secretion. Water intake is increased by increased effective osmotic pressure of the plasma (Figure 18–3), by decreases in ECF volume, and by psychologic and other factors. Osmolality acts via **osmoreceptors,** receptors that sense the osmolality of the body fluids. These osmoreceptors are located in the anterior hypothalamus.

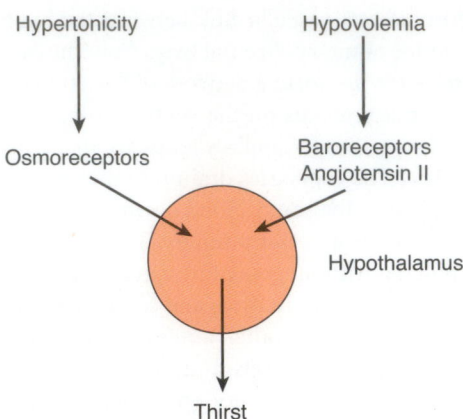

FIGURE 18–4 Diagrammatic representation of the way in which changes in plasma osmolality and changes in ECF volume affect thirst by separate pathways.

Decreases in ECF volume also stimulate thirst by a pathway independent of that mediating thirst in response to increased plasma osmolality (Figure 18–4). Thus, hemorrhage causes increased drinking even if there is no change in the osmolality of the plasma. The effect of ECF volume depletion on thirst is mediated in part via the renin–angiotensin system (see Chapter 39). Renin secretion is increased by hypovolemia and results in an increase in circulating angiotensin II. The angiotensin II acts on the **subfornical organ,** a specialized receptor area in the diencephalon (see Figure 34–7), to stimulate the neural areas concerned with thirst. Some evidence suggests that it acts on the **organum vasculosum of the lamina terminalis (OVLT)** as well. These areas are highly permeable and are two of the circumventricular organs located outside the blood–brain barrier (see Chapter 34). However, drugs that block the action of angiotensin II do not completely block the thirst response to hypovolemia, and it appears that the baroreceptors in the heart and blood vessels are also involved.

The intake of liquids is increased during eating (**prandial drinking**). The increase has been called a learned or habit response, but it has not been investigated in detail. One factor is an increase in plasma osmolality that occurs as food is absorbed. Another may be an action of one or more gastrointestinal hormones on the hypothalamus.

When the sensation of thirst is obtunded, either by direct damage to the diencephalon or by depressed or altered states of consciousness, patients stop drinking adequate amounts of fluid. Dehydration results if appropriate measures are not instituted to maintain water balance. If the protein intake is high, the products of protein metabolism cause an osmotic diuresis (see Chapter 38), and the amounts of water required to maintain hydration are large. Most cases of **hypernatremia** are actually due to simple dehydration in patients with psychoses or hypothalamic disease who do not or cannot increase their water intake when their thirst mechanism is stimulated. Lesions of the anterior communicating artery can also obtund thirst because branches of this artery supply the hypothalamic areas concerned with thirst.

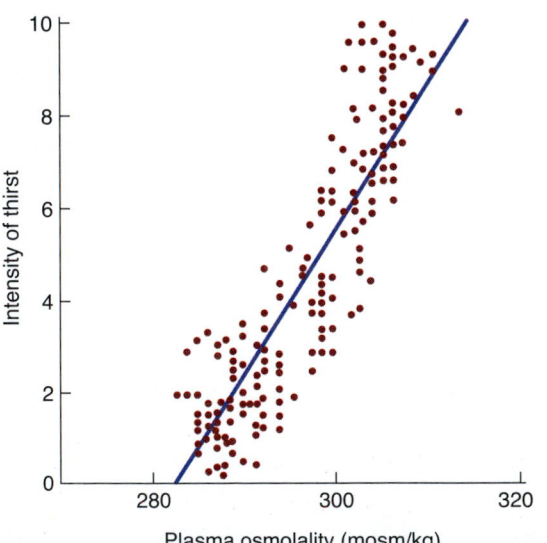

FIGURE 18–3 **Relation of plasma osmolality to thirst in healthy adult humans during infusion of hypertonic saline.** The intensity of thirst is measured on a special analog scale. (Reproduced with permission from Thompson CJ et al: The osmotic thresholds for thirst and vasopressin release are similar in healthy humans. Clin Sci Lond 1986;71:651.)

OTHER FACTORS REGULATING WATER INTAKE

A number of other well-established factors contribute to the regulation of water intake. Psychologic and social factors are important. Dryness of the pharyngeal mucous membrane causes a sensation of thirst. Patients in whom fluid intake must be restricted sometimes get appreciable relief of thirst by sucking ice chips or a wet cloth.

Dehydrated dogs, cats, camels, and some other animals rapidly drink just enough water to make up their water deficit. They stop drinking before the water is absorbed (while their plasma is still hypertonic), so some kind of pharyngeal gastrointestinal "metering" must be involved. Some evidence suggests that humans have a similar metering ability, though it is not well developed.

CONTROL OF POSTERIOR PITUITARY SECRETION

VASOPRESSIN & OXYTOCIN

In most mammals, the hormones secreted by the posterior pituitary gland are **arginine vasopressin (AVP)** and **oxytocin.** In hippopotami and most pigs, arginine in the vasopressin molecule is replaced by lysine to form **lysine vasopressin.** The posterior pituitaries of some species of pigs and marsupials contain a mixture of arginine and lysine vasopressin. The posterior lobe hormones are nonapeptides with a disulfide ring at one end (Figure 18–5).

BIOSYNTHESIS, INTRANEURONAL TRANSPORT, & SECRETION

The hormones of the posterior pituitary gland are synthesized in the cell bodies of the magnocellular neurons in the supraoptic

FIGURE 18–5   Arginine vasopressin and oxytocin.

and paraventricular nuclei and transported down the axons of these neurons to their endings in the posterior lobe, where they are secreted in response to electrical activity in the endings. Some of the neurons make oxytocin and others make vasopressin, and oxytocin-containing and vasopressin-containing cells are found in both nuclei.

Oxytocin and vasopressin are typical **neural hormones,** that is, hormones secreted into the circulation by nerve cells. This type of neural regulation is compared with other types in Figure 18–6. The term **neurosecretion** was originally coined to describe the secretion of hormones by neurons, but the term is somewhat misleading because it appears that all neurons secrete chemical messengers (see Chapter 7).

Like other peptide hormones, the posterior lobe hormones are synthesized as part of larger precursor molecules. Vasopressin and oxytocin each have a characteristic **neurophysin** associated with them in the granules in the neurons that secrete them—neurophysin I in the case of oxytocin and neurophysin II in the case of vasopressin. The neurophysins were originally thought to be binding polypeptides, but it now appears that they are simply parts of the precursor molecules. The precursor for arginine vasopressin, **prepropressophysin,** contains a 19-amino-acid residue leader sequence followed by

FIGURE 18–6   Neural control mechanisms. In the two situations on the left, neurotransmitters act at nerve endings on muscle; in the two in the middle, neurotransmitters regulate the secretion of endocrine glands; and in the two on the right, neurons secrete hormones into the hypophysial portal or general circulation.

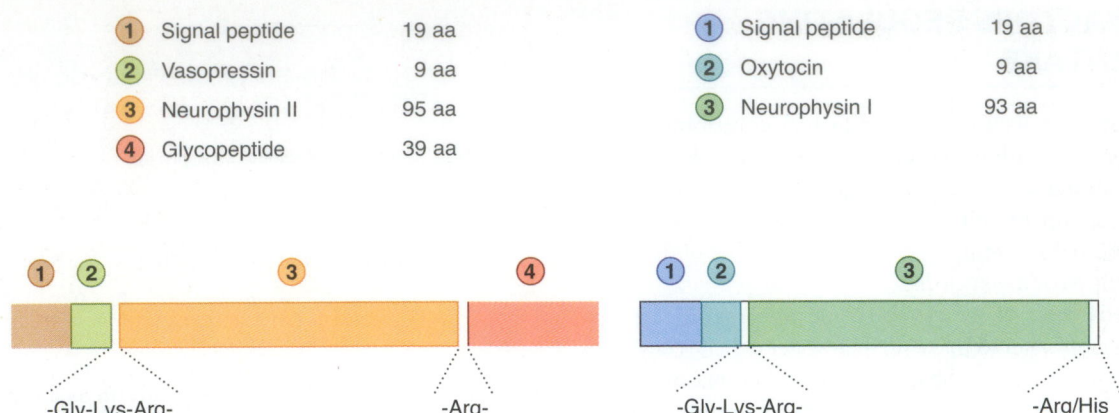

①	Signal peptide	19 aa
②	Vasopressin	9 aa
③	Neurophysin II	95 aa
④	Glycopeptide	39 aa

①	Signal peptide	19 aa
②	Oxytocin	9 aa
③	Neurophysin I	93 aa

FIGURE 18–7   **Structure of bovine prepropressophysin (left) and prepro-oxyphysin (right).** Gly in the 10 position of both peptides is necessary for amidation of the Gly residue in position 9. aa, amino acid residues. (Reproduced with permission from Richter D: Molecular events in expression of vasopressin and oxytocin and their cognate receptors. Am J Physiol 1988;255:F207.)

arginine vasopressin, neurophysin II, and a glycopeptide (Figure 18–7). **Prepro-oxyphysin,** the precursor for oxytocin, is a similar but smaller molecule that lacks the glycopeptide.

The precursor molecules are synthesized in the ribosomes of the cell bodies of the neurons. They have their leader sequences removed in the endoplasmic reticulum, are packaged into secretory granules in the Golgi apparatus, and are transported down the axons by axoplasmic flow to the endings in the posterior pituitary. The secretory granules, called **Herring bodies,** are easy to stain in tissue sections, and they have been extensively studied. Cleavage of the precursor molecules occurs as they are being transported, and the storage granules in the endings contain free vasopressin or oxytocin and the corresponding neurophysin. In the case of vasopressin, the glycopeptide is also present. All these products are secreted, but the functions of the components other than the established posterior pituitary hormones are unknown.

ELECTRICAL ACTIVITY OF MAGNOCELLULAR NEURONS

The oxytocin-secreting and vasopressin-secreting neurons also generate and conduct action potentials, and action potentials reaching their endings trigger release of hormone from them by Ca^{2+}-dependent exocytosis. At least in anesthetized rats, these neurons are silent at rest or discharge at low, irregular rates (0.1–3 spikes/s). However, their response to stimulation varies (Figure 18–8). Stimulation of the nipples causes a synchronous, high-frequency discharge of the oxytocin neurons after an appreciable latency. This discharge causes release of a pulse of oxytocin and consequent milk ejection in postpartum females. On the other hand, stimulation of the vasopressin-secreting neurons by a stimulus such as hemorrhage causes an initial steady increase in firing rate followed by a prolonged pattern of phasic discharge in which periods of high-frequency discharge alternate with periods of electrical quiescence (**phasic bursting**). These phasic bursts are gener-

ally not synchronous in different vasopressin-secreting neurons. They are well suited to maintain a prolonged increase in the output of vasopressin, as opposed to the synchronous, relatively short, high-frequency discharge of oxytocin-secreting neurons in response to stimulation of the nipples.

FIGURE 18–8   **Responses of magnocellular neurons to stimulation.** The tracings show individual extracellularly recorded action potentials, discharge rates, and intramammary duct pressure. **A)** Response of an oxytocin-secreting neuron. HFD, high-frequency discharge; ME, milk ejection. Stimulation of nipples started before the onset of recording. **B)** Responses of a vasopressin-secreting neuron, showing no change in the slow firing rate in response to stimulation of nipples and a prompt increase in the firing rate when 5 mL of blood was drawn, followed by typical phasic discharge. (Modified from Wakerly JB: Hypothalamic neurosecretory function: Insights from electrophysiological studies of the magno-cellular nuclei. IBRO News 1985;4:15.)

VASOPRESSIN & OXYTOCIN IN OTHER LOCATIONS

Vasopressin-secreting neurons are found in the suprachiasmatic nuclei, and vasopressin and oxytocin are also found in the endings of neurons that project from the paraventricular nuclei to the brain stem and spinal cord. These neurons appear to be involved in cardiovascular control. In addition, vasopressin and oxytocin are synthesized in the gonads and the adrenal cortex, and oxytocin is present in the thymus. The functions of the peptides in these organs are unsettled.

Vasopressin Receptors

There are at least three kinds of vasopressin receptors: V_{1A}, V_{1B}, and V_2. All are G protein-coupled. The V_{1A} and V_{1B} receptors act through phosphatidylinositol hydrolysis to increase the intracellular Ca^{2+} concentration. The V_2 receptors act through G_s to increase cAMP levels.

Effects of Vasopressin

Because one of its principal physiologic effects is the retention of water by the kidney, vasopressin is often called the **antidiuretic hormone (ADH).** It increases the permeability of the collecting ducts of the kidney so that water enters the hypertonic interstitium of the renal pyramids (see Chapter 38). The urine becomes concentrated and its volume decreases. The overall effect is therefore retention of water in excess of solute; consequently, the effective osmotic pressure of the body fluids is decreased. In the absence of vasopressin, the urine is hypotonic to plasma, urine volume is increased, and there is a net water loss. Consequently, the osmolality of the body fluid rises.

Effects of Oxytocin

In humans, oxytocin acts primarily on the breasts and uterus, though it appears to be involved in luteolysis as well (see Chapter 25). A G protein-coupled serpentine oxytocin receptor has been identified in human myometrium, and a similar or identical receptor is found in mammary tissue and the ovary. It triggers increases in intracellular Ca^{2+} levels.

The Milk Ejection Reflex

Oxytocin causes contraction of the **myoepithelial cells,** smooth-muscle-like cells that line the ducts of the breast. This squeezes the milk out of the alveoli of the lactating breast into the large ducts (sinuses) and thence out of the nipple (**milk ejection**). Many hormones acting in concert are responsible for breast growth and the secretion of milk into the ducts (see Chapter 25), but milk ejection in most species requires oxytocin.

Milk ejection is normally initiated by a neuroendocrine reflex. The receptors involved are the touch receptors, which are plentiful in the breast—especially around the nipple. Impulses generated in these receptors are relayed from the somatic touch pathways to the supraoptic and paraventricular nuclei. Discharge of the oxytocin-containing neurons causes secretion of oxytocin from the posterior pituitary (Figure 18–8). The infant suckling at the breast stimulates the touch receptors, the nuclei are stimulated, oxytocin is released, and the milk is expressed into the sinuses, ready to flow into the mouth of the waiting infant. In lactating women, genital stimulation and emotional stimuli also produce oxytocin secretion, sometimes causing milk to spurt from the breasts.

Other Actions of Oxytocin

Oxytocin causes contraction of the smooth muscle of the uterus. The sensitivity of the uterine musculature to oxytocin is enhanced by estrogen and inhibited by progesterone. The inhibitory effect of progesterone is due to a direct action of the steroid on uterine oxytocin receptors. In late pregnancy, the uterus becomes very sensitive to oxytocin coincident with a marked increase in the number of oxytocin receptors and oxytocin receptor mRNA (see Chapter 25). Oxytocin secretion is increased during labor. After dilation of the cervix, descent of the fetus down the birth canal initiates impulses in the afferent nerves that are relayed to the supraoptic and paraventricular nuclei, causing secretion of sufficient oxytocin to enhance labor (Figure 25-32). The amount of oxytocin in plasma is normal at the onset of labor. It is possible that the marked increase in oxytocin receptors at this time causes normal oxytocin levels to initiate contractions, setting up a positive feedback. However, the amount of oxytocin in the uterus is also increased, and locally produced oxytocin may also play a role.

Oxytocin may also act on the nonpregnant uterus to facilitate sperm transport. The passage of sperm up the female genital tract to the uterine tubes, where fertilization normally takes place, depends not only on the motile powers of the sperm but also, at least in some species, on uterine contractions. The genital stimulation involved in coitus releases oxytocin, but it has not been proved that it is oxytocin which initiates the rather specialized uterine contractions that transport the sperm. The secretion of oxytocin is increased by stressful stimuli and, like that of vasopressin, is inhibited by alcohol.

Circulating oxytocin increases at the time of ejaculation in males, and it is possible that this increase causes increased contraction of the smooth muscle of the vas deferens, propelling sperm toward the urethra.

CONTROL OF ANTERIOR PITUITARY SECRETION

ANTERIOR PITUITARY HORMONES

The anterior pituitary secretes six hormones: **adrenocorticotropic hormone (corticotropin, ACTH), thyroid-stimulating hormone (thyrotropin, TSH), growth hormone, follicle-stimulating hormone (FSH), luteinizing hormone (LH),**

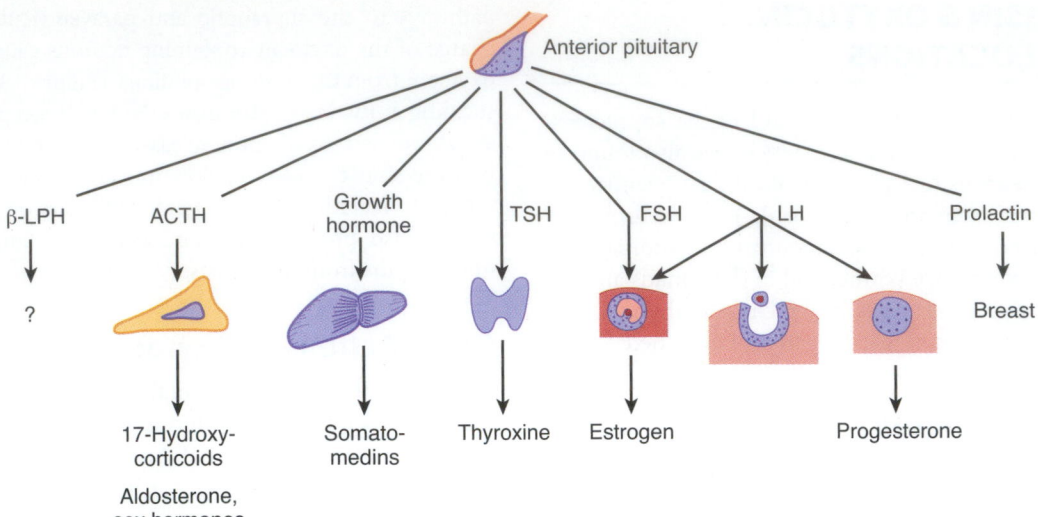

FIGURE 18–9   **Anterior pituitary hormones.** In women, FSH and LH act in sequence on the ovary to produce growth of the ovarian follicle, ovulation, and formation and maintenance of the corpus luteum. Prolactin stimulates lactation. In men, FSH and LH control the functions of the testes.

and **prolactin (PRL).** An additional polypeptide, β-lipotropin (β-LPH), is secreted with ACTH, but its physiologic role is unknown. The actions of the anterior pituitary hormones are summarized in Figure 18–9. The hormones are discussed in detail in the chapters on the endocrine system. The hypothalamus plays an important stimulatory role in regulating the secretion of ACTH, β-LPH, TSH, growth hormone, FSH, and LH. It also regulates prolactin secretion, but its effect is predominantly inhibitory rather than stimulatory.

NATURE OF HYPOTHALAMIC CONTROL

Anterior pituitary secretion is controlled by chemical agents carried in the portal hypophysial vessels from the hypothalamus to the pituitary. These substances used to be called releasing and inhibiting factors, but now they are commonly called **hypophysiotropic hormones.** The latter term seems appropriate since they are secreted into the bloodstream and act at a distance from their site of origin. Small amounts escape into

the general circulation, but they are in high concentration in portal hypophysial blood.

HYPOPHYSIOTROPIC HORMONES

There are six established hypothalamic releasing and inhibiting hormones (Figure 18–10): **corticotropin-releasing hormone (CRH); thyrotropin-releasing hormone (TRH); growth hormone-releasing hormone (GRH); growth hormone-inhibiting hormone (GIH),** now generally called **somatostatin; luteinizing hormone-releasing hormone (LHRH),** now generally known as **gonadotropin-releasing hormone (GnRH);** and **prolactin-inhibiting hormone (PIH).** In addition, hypothalamic extracts contain prolactin-releasing activity, and a **prolactin-releasing hormone (PRH)** has been postulated to exist. TRH, VIP, and several other polypeptides found in the hypothalamus stimulate prolactin secretion, but it is uncertain whether one or more of these peptides is the physiologic PRH. Recently, an orphan receptor was isolated from the anterior pituitary, and

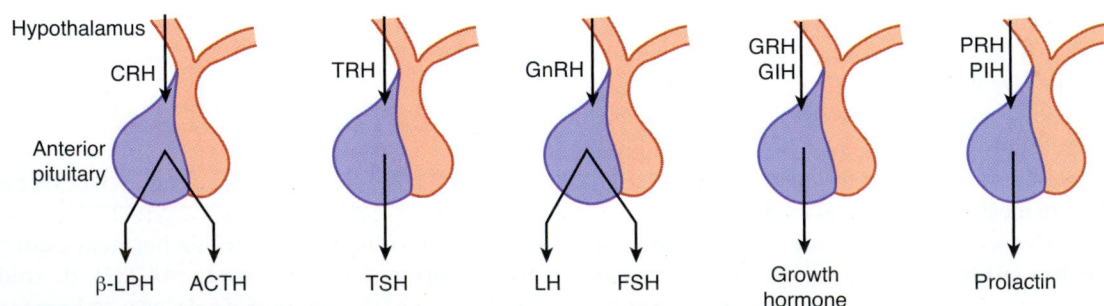

FIGURE 18–10   Effects of hypophysiotropic hormones on the secretion of anterior pituitary hormones.

TRH	(pyro)Glu-His-Pro-NH$_2$
GnRH	(pyro)Glu-His-Trp-Ser-Tyr-Gly-Leu-Arg-Pro-Gly-NH$_2$
Somatostatin	Ala-Gly-Cys-Lys-Asn-Phe-Phe-Trp-Lys-Thr-Phe-Thr-Ser-Cys
CRH	Ser-Glu-Glu-Pro-Pro-Ile-Ser-Leu-Asp-Leu-Thr-Phe-His-Leu-Leu-Arg-Glu-Val-Leu-Glu-Met-Ala-Arg-Ala-Glu-Gln-Leu- Ala-Gln-Gln-Ala-His-Ser-Asn-Arg-Lys-Leu-Met-Glu-Ile-Ile-NH$_2$
GRH	Tyr-Ala-Asp-Ala-Ile-Phe-Thr-Asn-Ser-Tyr-Arg-Lys-Val-Leu-Gly-Gln-Leu-Ser-Ala-Arg-Lys-Leu-Leu-Gln-Asp-Ile-Met- Ser-Arg-Gln-Gln-Gly-Glu-Ser-Asn-Gln-Glu-Arg-Gly-Ala-Arg-Ala-Arg-Leu-NH$_2$
PIH	Dopamine

FIGURE 18–11 **Structure of hypophysiotropic hormones in humans.** Preprosomatostatin is processed to a tetradecapeptide (somatostatin 14, [SS14], shown above) and also to a polypeptide containing 28 amino acid residues (SS28).

the search for its ligand led to the isolation of a 31-amino-acid polypeptide from the human hypothalamus. This polypeptide stimulated prolactin secretion by an action on the anterior pituitary receptor, but additional research is needed to determine if it is the physiologic PRH. GnRH stimulates the secretion of FSH as well as that of LH, and it seems unlikely that a separate follicle-stimulating hormone-releasing hormone exists.

The structures of the six established hypophysiotropic hormones are shown in Figure 18–11. The structures of the genes and preprohormones for TRH, GnRH, somatostatin, CRH, and GRH are known. PreproTRH contains six copies of TRH. Several other preprohormones may contain other hormonally active peptides in addition to the hypophysiotropic hormones.

The area from which the hypothalamic releasing and inhibiting hormones are secreted is the median eminence of the hypothalamus. This region contains few nerve cell bodies, but many nerve endings are in close proximity to the capillary loops from which the portal vessels originate.

The locations of the cell bodies of the neurons that project to the external layer of the median eminence and secrete the hypophysiotropic hormones are shown in Figure 18–12, which also shows the location of the neurons secreting oxytocin and vasopressin. The GnRH-secreting neurons are primarily in the medial preoptic area, the somatostatin-secreting neurons are in the periventricular nuclei, the TRH-secreting and CRH-secreting neurons are in the medial parts of the paraventricular nuclei, and the GRH-secreting and dopamine-secreting neurons are in the arcuate nuclei.

Most, if not all, of the hypophysiotropic hormones affect the secretion of more than one anterior pituitary hormone (Figure 18–10). The FSH-stimulating activity of GnRH has been mentioned previously. TRH stimulates the secretion of prolactin as well as TSH. Somatostatin inhibits the secretion of TSH as well as growth hormone. It does not normally inhibit the secretion of the other anterior pituitary hormones, but it inhibits the abnormally elevated secretion of ACTH in patients with Nelson's syndrome. CRH stimulates the secretion of ACTH and β-LPH.

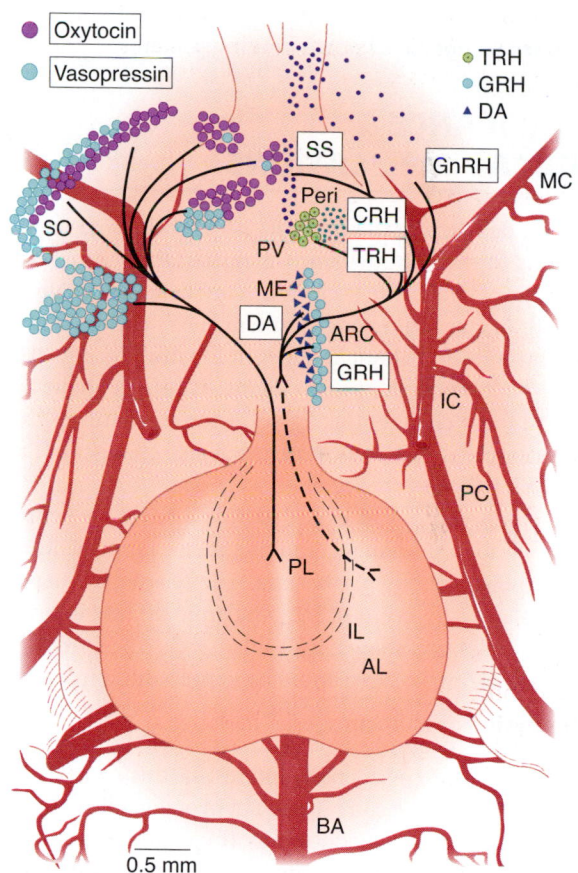

FIGURE 18–12 **Location of cell bodies of hypophysiotropic hormone-secreting neurons projected on a ventral view of the hypothalamus and pituitary of the rat.** AL, anterior lobe; ARC, arcuate nucleus; BA, basilar artery; DA, dopamine; IC, internal carotid artery; IL, intermediate lobe; MC, middle cerebral artery; ME, median eminence; PC, posterior cerebral artery; Peri, periventricular nucleus; PL, posterior lobe; PV, paraventricular nucleus; SO, supraoptic nucleus. The names of the hormones are enclosed in boxes. (Courtesy of LW Swanson and ET Cunningham Jr.)

Hypophysiotropic hormones function as neurotransmitters in other parts of the brain, the retina, and the autonomic nervous system (see Chapter 7). In addition, somatostatin is found in the pancreatic islets (see Chapter 21), GRH is secreted by pancreatic tumors, and somatostatin and TRH are found in the gastrointestinal tract (see Chapter 26).

Receptors for most of the hypophysiotropic hormones are serpentine and coupled to G proteins. There are two human CRH receptors: hCRH-RI, and hCRHRII. The latter differs from the former in having a 29-amino-acid insert in its first cytoplasmic loop. The physiologic role of hCRH-RII is unsettled, though it is found in many parts of the brain. In addition, a **CRH-binding protein** in the peripheral circulation inactivates CRH. It is also found in the cytoplasm of corticotropes in the anterior pituitary, and in this location it might play a role in receptor internalization. However, the exact physiologic role of this protein is unknown. Other hypophysiotropic hormones do not have known binding proteins.

SIGNIFICANCE & CLINICAL IMPLICATIONS

Research delineating the multiple neuroendocrine regulatory functions of the hypothalamus is important because it helps explain how endocrine secretion is made appropriate to the demands of a changing environment. The nervous system receives information about changes in the internal and external environment from the sense organs. It brings about adjustments to these changes through effector mechanisms that include not only somatic movement but also changes in the rate at which hormones are secreted.

The manifestations of hypothalamic disease are neurologic defects, endocrine changes, and metabolic abnormalities such as hyperphagia and hyperthermia. The relative frequencies of the signs and symptoms of hypothalamic disease in one large series of cases are shown in Table 18–2. The possibility of hypothalamic pathology should be kept in mind in evaluating all patients with pituitary dysfunction, especially those with isolated deficiencies of single pituitary tropic hormones.

A condition of considerable interest in this context is **Kallmann syndrome,** the combination of hypogonadism due to low levels of circulating gonadotropins (**hypogonadotropic hypogonadism**) with partial or complete loss of the sense of smell (**hyposmia** or **anosmia**). Embryologically, GnRH neurons develop in the nose and migrate up the olfactory nerves and then through the brain to the hypothalamus. If this migration is prevented by congenital abnormalities in the olfactory pathways, the GnRH neurons do not reach the hypothalamus and pubertal maturation of the gonads fails to occur. The syndrome is most common in men, and the cause in many cases is mutation of the *KALIG1* gene, a gene on the X chromosome that codes for what is apparently an adhesion molecule necessary for normal development of the olfactory nerve on which the GnRH neurons migrate into the brain. However, the condition also occurs in women and can be due to other genetic abnormalities.

TABLE 18–2 Symptoms and signs in 60 autopsied patients with hypothalamic disease.

Symptoms and Signs	Percentage of Cases
Endocrine and metabolic findings	
Precocious puberty	40
Hypogonadism	32
Diabetes insipidus	35
Obesity	25
Abnormalities of temperature regulation	22
Emaciation	18
Bulimia	8
Anorexia	7
Neurologic findings	
Eye signs	78
Pyramidal and sensory deficits	75
Headache	65
Extrapyramidal signs	62
Vomiting	40
Psychic disturbances, rage attacks, etc	35
Somnolence	30
Convulsions	15

Data from Bauer HG: Endocrine and other clinical manifestations of hypothalamic disease. J Clin Endocrinol 1954;14:13. See also Kahana L, et al: Endocrine manifestations of intracranial extrasellar lesions. J Clin Endocrinol 1962;22:304.

TEMPERATURE REGULATION

In the body, heat is produced by muscular exercise, assimilation of food, and all the vital processes that contribute to the basal metabolic rate (see Chapter 27). It is lost from the body by radiation, conduction, and vaporization of water in the respiratory passages and on the skin. Small amounts of heat are also removed in the urine and feces. The balance between heat production and heat loss determines the body temperature. Because the speed of chemical reactions varies with the temperature and because the enzyme systems of the body have narrow temperature ranges in which their function is optimal, normal body function depends on a relatively constant body temperature.

Invertebrates generally cannot adjust their body temperatures and so are at the mercy of the environment. In vertebrates, mechanisms for maintaining body temperature by adjusting heat production and heat loss have evolved. In reptiles, amphibians, and fish, the adjusting mechanisms are relatively rudimentary, and these species are called "cold-blooded" (**poikilothermic**) because their body temperature fluctuates over a considerable range. In birds and mammals, the "warm-blooded" (**homeothermic**) animals, a group of reflex responses that are primarily integrated in the hypothalamus, operate to maintain body temperature within a narrow range in spite of wide fluctuations in environmental temperature. The hibernating mammals are a partial exception. While awake they are homeothermic, but during hibernation their body temperature falls.

NORMAL BODY TEMPERATURE

In homeothermic animals, the actual temperature at which the body is maintained varies from species to species and, to a lesser degree, from individual to individual. In humans, the traditional normal value for the oral temperature is 37 °C (98.6 °F), but in one large series of normal young adults, the morning oral temperature averaged 36.7 °C, with a standard deviation of 0.2 °C. Therefore, 95% of all young adults would be expected to have a morning oral temperature of 36.3–37.1 °C (97.3–98.8 °F; mean ± 1.96 standard deviations; see Appendix). Various parts of the body are at different temperatures, and the magnitude of the temperature difference between the parts varies with the environmental temperature (Figure 18–13). The extremities are generally cooler than the rest of the body. The temperature of the scrotum is carefully regulated at 32 °C. The rectal temperature is representative of the temperature at the core of the body and varies least with changes in environmental temperature. The oral temperature is normally 0.5 °C lower than the rectal temperature, but it is affected by many factors, including ingestion of hot or cold fluids, gum chewing, smoking, and mouth breathing.

The normal human core temperature undergoes a regular circadian fluctuation of 0.5–0.7 °C. In individuals who sleep at

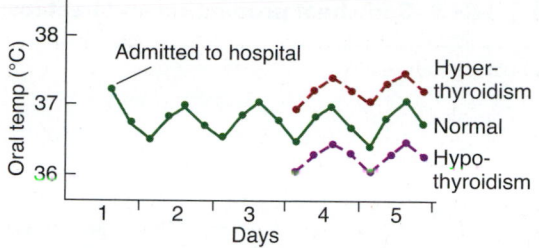

FIGURE 18–14 **Typical temperature chart of a hospitalized patient who does not have a febrile disease.** Note the slight rise in temperature, due to excitement and apprehension, at the time of admission to the hospital, and the regular circadian temperature cycle.

night and are awake during the day (even when hospitalized at bed rest), it is lowest at about 6:00 AM and highest in the evenings (Figure 18–14). It is lowest during sleep, is slightly higher in the awake but relaxed state, and rises with activity. In women, an additional monthly cycle of temperature variation is characterized by a rise in basal temperature at the time of ovulation (Figure 25–38). Temperature regulation is less precise in young children and they may normally have a temperature that is 0.5 ° or so above the established norm for adults.

During exercise, the heat produced by muscular contraction accumulates in the body and the rectal temperature normally rises as high as 40 °C (104 °F). This rise is due in part to the inability of the heat-dissipating mechanisms to handle the greatly increased amount of heat produced, but evidence suggests that in addition there is an elevation of the body temperature at which the heat-dissipating mechanisms are activated during exercise. Body temperature also rises slightly during emotional excitement, probably owing to unconscious tensing of the muscles. It is chronically elevated by as much as 0.5 °C when the metabolic rate is high, as in hyperthyroidism, and lowered when the metabolic rate is low, as in hypothyroidism (Figure 18–14). Some apparently normal adults chronically have a temperature above the normal range (constitutional hyperthermia).

HEAT PRODUCTION

Heat production and energy balance are discussed in Chapter 27. A variety of basic chemical reactions contribute to body heat production at all times. Ingestion of food increases heat production because of the specific dynamic action of the food (see Chapter 27), but the major source of heat is the contraction of skeletal muscle (Table 18–3). Heat production can be varied by endocrine mechanisms in the absence of food intake or muscular exertion. Epinephrine and norepinephrine produce a rapid but short-lived increase in heat production; thyroid hormones produce a slowly developing but prolonged increase. Furthermore, sympathetic discharge decreases during fasting and is increased by feeding.

A source of considerable heat, particularly in infants, is **brown fat**. This fat has a high rate of metabolism and its thermogenic function has been likened to that of an electric blanket.

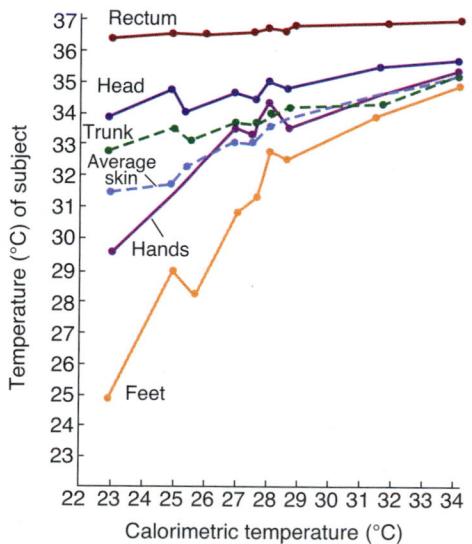

FIGURE 18–13 **Temperatures of various parts of the body of a naked subject at various ambient temperatures in a calorimeter.** (Redrawn and reproduced, with permission, from Hardy JD, DuBois EF: Basal metabolism, radiation, convection and vaporization at temperatures of 22–35 °C. J Nutr 1938;15:477.)

TABLE 18–3 Body heat production and heat loss.

Body heat is produced by:	
Basic metabolic processes	
Food intake (specific dynamic action)	
Muscular activity	
Body heat is lost by:	**Percentage of heat lost at 21 °C**
Radiation and conduction	70
Vaporization of sweat	27
Respiration	2
Urination and defecation	1

HEAT LOSS

The processes by which heat is lost from the body when the environmental temperature is below body temperature are listed in Table 18–3. **Conduction** is heat exchange between objects or substances at different temperatures that are in contact with one another. A basic characteristic of matter is that its molecules are in motion, with the amount of motion proportionate to the temperature. These molecules collide with the molecules in cooler objects, transferring thermal energy to them. The amount of heat transferred is proportionate to the temperature difference between the objects in contact **(thermal gradient)**. Conduction is aided by **convection,** the movement of molecules away from the area of contact. Thus, for example, an object in contact with air at a different temperature changes the specific gravity of the air, and because warm air rises and cool air falls, a new supply of air is brought into contact with the object. Of course, convection is greatly aided if the object moves about in the medium or the medium moves past the object, for example, if a subject swims through water or a fan blows air through a room. **Radiation** is the transfer of heat by infrared electromagnetic radiation from one object to another at a different temperature with which it is not in contact. When an individual is in a cold environment, heat is lost by conduction to the surrounding air and by radiation to cool objects in the vicinity. Conversely, of course, heat is transferred to an individual and the heat load is increased by these processes when the environmental temperature is above body temperature. Note that because of radiation, an individual can feel chilly in a room with cold walls even though the room is relatively warm. On a cold but sunny day, the heat of the sun reflected off bright objects exerts an appreciable warming effect. It is the heat reflected from the snow, for example, that makes it possible to ski in fairly light clothes even though the air temperature is below freezing.

Because conduction occurs from the surface of one object to the surface of another, the temperature of the skin deter-

mines to a large extent the degree to which body heat is lost or gained. The amount of heat reaching the skin from the deep tissues can be varied by changing the blood flow to the skin. When the cutaneous vessels are dilated, warm blood wells into the skin, whereas in the maximally vasoconstricted state, heat is held centrally in the body. The rate at which heat is transferred from the deep tissues to the skin is called the **tissue conductance.** Birds have a layer of feathers next to the skin, and most mammals have a significant layer of hair or fur. Heat is conducted from the skin to the air trapped in this layer and from the trapped air to the exterior. When the thickness of the trapped layer is increased by fluffing the feathers or erection of the hairs **(horripilation),** heat transfer across the layer is reduced and heat losses (or, in a hot environment, heat gains) are decreased. "Goose pimples" are the result of horripilation in humans; they are the visible manifestation of cold-induced contraction of the piloerector muscles attached to the rather meager hair supply. Humans usually supplement this layer of hair with one or more layers of clothes. Heat is conducted from the skin to the layer of air trapped by the clothes, from the inside of the clothes to the outside, and from the outside of the clothes to the exterior. The magnitude of the heat transfer across the clothing, a function of its texture and thickness, is the most important determinant of how warm or cool the clothes feel, but other factors, especially the size of the trapped layer of warm air, are important also. Dark clothes absorb radiated heat and light-colored clothes reflect it back to the exterior.

The other major process transferring heat from the body in humans and other animals that sweat is vaporization of water on the skin and mucous membranes of the mouth and respiratory passages. Vaporization of 1 g of water removes about 0.6 kcal of heat. A certain amount of water is vaporized at all times. This **insensible water loss** amounts to 50 mL/h in humans. When sweat secretion is increased, the degree to which the sweat vaporizes depends on the humidity of the environment. It is common knowledge that one feels hotter on a humid day. This is due in part to the decreased vaporization of sweat, but even under conditions in which vaporization of sweat is complete, an individual in a humid environment feels warmer than an individual in a dry environment. The reason for this difference is unknown, but it seems related to the fact that in the humid environment sweat spreads over a greater area of skin before it evaporates. During muscular exertion in a hot environment, sweat secretion reaches values as high as 1600 mL/h, and in a dry atmosphere, most of this sweat is vaporized. Heat loss by vaporization of water therefore varies from 30 to over 900 kcal/h.

Some mammals lose heat by **panting.** This rapid, shallow breathing greatly increases the amount of water vaporization in the mouth and respiratory passages and therefore the amount of heat lost. Because the breathing is shallow, it produces relatively little change in the composition of alveolar air (see Chapter 35).

The relative contribution of each of the processes that transfer heat away from the body (Table 18–3) varies with the environmental temperature. At 21 °C, vaporization is a minor component in humans at rest. As the environmental temperature approaches body temperature, radiation losses decline and vaporization losses increase.

TEMPERATURE-REGULATING MECHANISMS

The reflex and semireflex thermoregulatory responses in humans are listed in Table 18–4. They include autonomic, somatic, endocrine, and behavioral changes. One group of responses increases heat loss and decreases heat production; the other decreases heat loss and increases heat production. In general, exposure to heat stimulates the former group of responses and inhibits the latter, whereas exposure to cold does the opposite.

Curling up "in a ball" is a common reaction to cold in animals and has a counterpart in the position some people assume on climbing into a cold bed. Curling up decreases the body surface exposed to the environment. Shivering is an involuntary response of the skeletal muscles, but cold also causes a semiconscious general increase in motor activity. Examples include foot stamping and dancing up and down on a cold day. Increased catecholamine secretion is an important endocrine response to cold. Mice unable to make norepinephrine and epinephrine because their dopamine β-hydroxylase

TABLE 18–4 Temperature-regulating mechanisms.

Mechanisms activated by cold
Shivering
Hunger
Increased voluntary activity
Increased secretion of norepinephrine and epinephrine
Decreased heat loss
Cutaneous vasoconstriction
Curling up
Horripilation
Mechanisms activated by heat
Increased heat loss
Cutaneous vasodilation
Sweating
Increased respiration
Decreased heat production
Anorexia
Apathy and inertia

gene is knocked out do not tolerate cold; they have deficient vasoconstriction and are unable to increase thermogenesis in brown adipose tissue through UCP 1. TSH secretion is increased by cold and decreased by heat in laboratory animals, but the change in TSH secretion produced by cold in adult humans is small and of questionable significance. It is common knowledge that activity is decreased in hot weather—the "it's too hot to move" reaction.

Thermoregulatory adjustments involve local responses as well as more general reflex responses. When cutaneous blood vessels are cooled they become more sensitive to catecholamines and the arterioles and venules constrict. This local effect of cold directs blood away from the skin. Another heat-conserving mechanism that is important in animals living in cold water is heat transfer from arterial to venous blood in the limbs. The deep veins (**venae comitantes**) run alongside the arteries supplying the limbs and heat is transferred from the warm arterial blood going to the limbs to the cold venous blood coming from the extremities (**countercurrent exchange;** see Chapter 38). This keeps the tips of the extremities cold but conserves body heat.

The reflex responses activated by cold are controlled from the posterior hypothalamus. Those activated by warmth are controlled primarily from the anterior hypothalamus, although some thermoregulation against heat still occurs after decerebration at the level of the rostral midbrain. Stimulation of the anterior hypothalamus causes cutaneous vasodilation and sweating, and lesions in this region cause hyperthermia, with rectal temperatures sometimes reaching 43 °C (109.4 °F). Posterior hypothalamic stimulation causes shivering, and the body temperature of animals with posterior hypothalamic lesions falls toward that of the environment.

AFFERENTS

The hypothalamus is said to integrate body temperature information from sensory receptors (primarily cold receptors) in the skin, deep tissues, spinal cord, extrahypothalamic portions of the brain, and the hypothalamus itself. Each of these five inputs contributes about 20% of the information that is integrated. There are threshold core temperatures for each of the main temperature-regulating responses and when the threshold is reached the response begins. The threshold is 37 °C for sweating and vasodilation, 36.8 °C for vasoconstriction, 36 °C for nonshivering thermogenesis, and 35.5 °C for shivering.

FEVER

Fever is perhaps the oldest and most universally known hallmark of disease. It occurs not only in mammals but also in birds, reptiles, amphibia, and fish. When it occurs in homeothermic animals, the thermoregulatory mechanisms behave as if they were adjusted to maintain body temperature at a higher than normal level, that is, "as if the thermostat had been reset" to a new point above 37 °C. The temperature receptors then signal that the actual temperature is below the new set point, and the temperature-raising mechanisms are activated. This usually produces chilly sensations due to cutaneous vasoconstriction and occasionally enough shivering to produce a shaking chill. However, the nature of the response depends on the ambient temperature. The temperature rise in experimental animals injected with a pyrogen is due mostly to increased heat production if they are in a cold environment and mostly to decreased heat loss if they are in a warm environment.

The pathogenesis of fever is summarized in Figure 18–15. Toxins from bacteria such as endotoxin act on monocytes, macrophages, and Kupffer cells to produce cytokines that act as **endogenous pyrogens (EPs).** There is good evidence that IL-1β, IL-6, β-IFN, γ-IFN, and TNF-α (see Chapter 3) can act independently to produce fever. These cytokines are polypeptides and it is unlikely that circulating cytokines penetrate the brain. Instead, evidence suggests that they act on the OVLT, one of the circumventricular organs (see Chapter 34). This in turn activates the preoptic area of the hypothalamus. Cytokines are also produced by cells in the central nervous system (CNS) when these are stimulated by infection, and these may act directly on the thermoregulatory centers.

The fever produced by cytokines is probably due to local release of prostaglandins in the hypothalamus. Intrahypothalamic injection of prostaglandins produces fever. In addition, the antipyretic effect of aspirin is exerted directly on the hypothalamus, and aspirin inhibits prostaglandin synthesis. PGE_2 is one of the prostaglandins that causes fever. It acts on four subtypes of prostaglandin receptors—EP_1, EP_2, EP_3, and EP_4—and knockout of the EP_3 receptor impairs the febrile response to PGE_2, IL-1β, and bacterial lipopolysaccharide (LPS).

The benefit of fever to the organism is uncertain. It is presumably beneficial because it has evolved and persisted as a response to infections and other diseases. Many microorganisms grow best within a relatively narrow temperature range and a rise in temperature inhibits their growth. In addition, antibody production is increased when body temperature is elevated. Before the advent of antibiotics, fevers were artificially induced for the treatment of neurosyphilis and proved to be beneficial. Hyperthermia benefits individuals infected with anthrax, pneumococcal pneumonia, leprosy, and various fungal, rickettsial, and viral diseases. Hyperthermia also slows the growth of some tumors. However, very high temperatures are harmful. A rectal temperature over 41 °C (106 °F) for prolonged periods results in some permanent brain damage. When the temperature is over 43 °C, heat stroke develops and death is common.

In **malignant hyperthermia,** various mutations of the gene coding for the ryanodine receptor (see Chapter 5) lead to excess Ca^{2+} release during muscle contraction triggered by stress. This in turn leads to contractures of the muscles, increased muscle metabolism, and a great increase in heat production in muscle. The increased heat production causes a marked rise in body temperature that is fatal if not treated.

Periodic fevers also occur in humans with mutations in the gene for **pyrin,** a protein found in neutrophils; the gene for mevalonate kinase, an enzyme involved in cholesterol synthesis; and the gene for the type 1 TNF receptor, which is involved in inflammatory responses. However, how any of these three mutant gene products cause fever is unknown.

HYPOTHERMIA

In hibernating mammals, body temperature drops to low levels without causing any demonstrable ill effects on subsequent arousal. This observation led to experiments on induced hypothermia. When the skin or the blood is cooled enough to lower the body temperature in nonhibernating animals and in humans, metabolic and physiologic processes slow down. Respiration and heart rate are very slow, blood pressure is low, and consciousness is lost. At rectal temperatures of about 28 °C, the ability to spontaneously return the temperature to normal is lost, but the individual continues to survive and, if rewarmed with external heat, returns to a normal state. If care is taken to prevent the formation of ice crystals in the tissues, the body temperature of experimental animals can be lowered to subfreezing levels without producing any detectable damage after subsequent rewarming.

Humans tolerate body temperatures of 21–24 °C (70–75 °F) without permanent ill effects, and induced hypothermia has been used in surgery. On the other hand, accidental hypothermia due to prolonged exposure to cold air or cold water is a serious condition and requires careful monitoring and prompt rewarming.

CHAPTER SUMMARY

- Neural connections run between the hypothalamus and the posterior lobe of the pituitary gland, and vascular connections between the hypothalamus and the anterior lobe of the pituitary.
- In most mammals, the hormones secreted by the posterior pituitary gland are vasopressin and oxytocin. Vasopressin increases

FIGURE 18–15 **Pathogenesis of fever.**

Endotoxin
Inflammation
Other pyrogenic stimuli
↓
Monocytes
Macrophages
Kupffer cells
↓ Cytokines
Preoptic area
of hypothalamus
↓ Prostaglandins
Raise temperature
set point
↓
Fever

the permeability of the collecting ducts of the kidney to water, thus concentrating the urine. Oxytocin acts on the breasts (lactation) and the uterus (contraction).

- The anterior pituitary secretes six hormones: adrenocortico-tropic hormone (corticotropin, ACTH), thyroid-stimulating hormone (thyrotropin, TSH), growth hormone, follicle-stimulating hormone (FSH), luteinizing hormone (LH), and prolactin (PRL).

- Other complex autonomic mechanisms that maintain the chemical constancy and temperature of the internal environment are integrated in the hypothalamus.

MULTIPLE-CHOICE QUESTIONS

For all questions, select the single best answer unless otherwise directed.

1. Thirst is stimulated by

   A) increases in plasma osmolality and volume.
   B) an increase in plasma osmolality and a decrease in plasma volume.
   C) a decrease in plasma osmolality and an increase in plasma volume.
   D) decreases in plasma osmolality and volume.
   E) injection of vasopressin into the hypothalamus.

2. When an individual is naked in a room in which the air temperature is 21 °C (69.8 °F) and the humidity 80%, the greatest amount of heat is lost from the body by

   A) elevated metabolism.
   B) respiration.
   C) urination.
   D) vaporization of sweat.
   E) radiation and conduction.

In questions 3–8, select A if the item is associated with (a) below, B if the item is associated with (b) below, C if the item is associated with both (a) and (b), and D if the item is associated with neither (a) nor (b).

   (a) V_{1A} vasopressin receptors
   (b) V_2 vasopressin receptors

3. Activation of G_s
4. Vasoconstriction
5. Increase in intracellular inositol triphosphate
6. Movement of aquaporin
7. Proteinuria
8. Milk ejection

CHAPTER RESOURCES

Brunton PJ, Russell JA, Douglas AJ: Adaptive responses of the maternal hypothalamic-pituitary-adrenal axis during pregnancy and lactation. J Neuroendocrinol. 2008;20:764.

Lamberts SWJ, Hofland LJ, Nobels FRE: Neuroendocrine tumor markers. Front Neuroendocrinol 2001;22:309.

Loh JA, Verbalis JG: Disorders of water and salt metabolism associated with pituitary disease. Endocrinol Metab Clin 2008;37:213.

McKinley MS, Johnson AK: The physiologic regulation of thirst and fluid intake. News Physiol Sci 2004;19:1.

Learning, Memory, Language, & Speech

INTRODUCTION

A revolution in our understanding of brain function in humans has been brought about by the development and widespread availability of **positron emission tomographic (PET) scanning, functional magnetic resonance imaging (fMRI),** and related techniques. PET is often used to measure local glucose metabolism, which is proportionate to neural activity, and fMRI is used to measure local amounts of oxygenated blood. These techniques make it possible to determine the activity of the various parts of the brain in completely intact normal humans and in humans with many different diseases. They have been used to study not only simple responses but complex aspects of learning, memory, and perception. An example of the use of PET scans to study the functions of the cerebral cortex in processing words is shown

in Figure 19–1. Different portions of the cortex are activated when hearing, seeing, speaking, or generating words.

Other techniques that have provided information on cortical function include stimulation of the exposed cerebral cortex in conscious humans undergoing neurosurgical procedures and, in a few instances, studies with chronically implanted electrodes. Valuable information has also been obtained from investigations in laboratory primates, but it is worth remembering that in addition to the difficulties in communicating with them, the brain of the rhesus monkey is only one-fourth the size of the brain of the chimpanzee, our nearest primate relative, and the chimpanzee brain is in turn one-fourth the size of the human brain.

LEARNING & MEMORY

A characteristic of animals and particularly of humans is the ability to alter behavior on the basis of experience.

Learning is acquisition of the information that makes this possible and **memory** is the retention and storage of that information. The two are obviously closely related and should be considered together.

FIGURE 19–1 Drawings of PET scans of the left cerebral hemisphere showing areas of greatest neuronal activation when subjects performed various language-based activities. (From Widmaier EP, Raff H, Strang KT: *Vander's Human Physiology.* McGraw-Hill, 2008.)

become unconscious and automatic. **Priming** is facilitation of recognition of words or objects by prior exposure to them. An example is improved recall of a word when presented with the first few letters of it. In **nonassociative learning,** the organism learns about a single stimulus. In **associative learning,** the organism learns about the relation of one stimulus to another.

Explicit memory and many forms of implicit memory involve (1) **short-term memory,** which lasts seconds to hours, during which processing in the hippocampus and elsewhere lays down long-term changes in synaptic strength; and (2) **long-term memory,** which stores memories for years and sometimes for life. During short-term memory, the memory traces are subject to disruption by trauma and various drugs, whereas long-term memory traces are remarkably resistant to disruption. **Working memory** is a form of short-term memory that keeps information available, usually for very short periods, while the individual plans action based on it.

FORMS OF MEMORY

From a physiologic point of view, memory is appropriately divided into explicit and implicit forms (Figure 19–2). **Explicit** or **declarative memory** is associated with consciousness—or at least awareness—and is dependent on the hippocampus and other parts of the medial temporal lobes of the brain for its retention. Clinical Box 19–1 describes how tracking a patient with brain damage has led to an awareness of the role of the temporal lobe in declarative memory. **Implicit** or **nondeclarative memory** does not involve awareness, and its retention does not usually involve processing in the hippocampus.

Explicit memory is divided into **episodic memory** for events and **semantic memory** for facts (eg, words, rules, and language). Explicit memories initially required for activities such as riding a bicycle can become implicit once the task is thoroughly learned.

Implicit memory is subdivided into four types. **Procedural memory** includes skills and habits, which, once acquired,

NEURAL BASIS OF MEMORY

The key to memory is alteration in the strength of selected synaptic connections. In all but the simplest of cases, the alteration involves protein synthesis and activation of genes. This occurs during the change from short-term working memory to long-term memory. In animals, acquisition of long-term learned responses is prevented if, within 5 min after each training session, the animals are anesthetized, given electroshock, subjected to hypothermia, or given drugs, antibodies, or oligonucleotides that block the synthesis of proteins. If these interventions are performed 4 h after the training sessions, there is no effect on acquisition.

The human counterpart of this phenomenon is the loss of memory for the events immediately preceding brain concussion or electroshock therapy (**retrograde amnesia**). This amnesia encompasses longer periods than it does in experimental animals—sometimes many days—but remote memories remain intact.

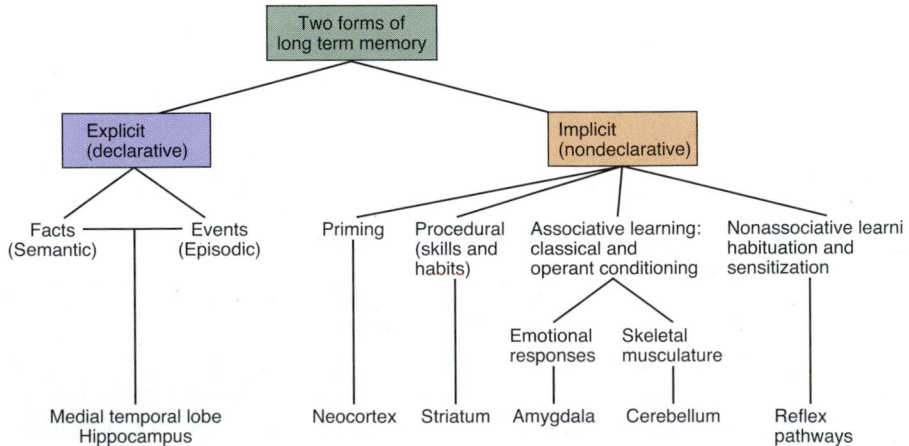

FIGURE 19–2 Forms of long-term memory. (Modified from Kandel ER, Schwartz JH, Jessell TM [editors]: *Principles of Neural Science,* 4th ed. McGraw-Hill, 2000.)

CLINICAL BOX 19–1

The Case of HM: Defining a Link between Brain Function & Memory

HM is an anonymous patient who suffered from bilateral temporal lobe seizures that began following a bicycle accident at age 9. His case has been studied by many scientists and has led to a greater understanding of the link between the **temporal lobe** and **declarative memory.** HM had partial seizures for many years, and then several tonic-clonic seizures by age 16. In 1953, at the age of 27, HM underwent bilateral surgical removal of the amygdala, large portions of the hippocampal formation, and portions of the association area of the temporal cortex. HM's seizures were better controlled after surgery, but removal of the temporal lobes led to devastating memory deficits. He maintained **long-term memory** for events that occurred prior to surgery, but he suffered from **anterograde amnesia.** His **short-term memory** was intact, but he could not commit new events to long-term memory. He had normal procedural memory, and he could learn new puzzles and motor tasks. His case is the first to bring attention to the critical role of temporal lobes in formation of long-term declarative memories and to implicate this region in the conversion of short-term to long-term memories. Later work showed that the **hippocampus** is the primary structure within the temporal lobe involved in this conversion. Because HM retained memories from before surgery, his case also shows that the hippocampus is not involved in the storage of declarative memory. An audio-recording from the 1990s of HM talking to scientists was released in 2007 and is available at http://www.npr.org/templates/story/story.php?storyId=7584970.

SYNAPTIC PLASTICITY & LEARNING

Short- and long-term changes in synaptic function can occur as a result of the history of discharge at a synapse; that is, synaptic conduction can be strengthened or weakened on the basis of past experience. These changes are of great interest because they represent forms of learning and memory. They can be presynaptic or postsynaptic in location.

One form of plastic change is **posttetanic potentiation,** the production of enhanced postsynaptic potentials in response to stimulation. This enhancement lasts up to 60 seconds and occurs after a brief (tetanizing) train of stimuli in the presynaptic neuron. The tetanizing stimulation causes Ca^{2+} to accumulate in the presynaptic neuron to such a degree that the intracellular binding sites that keep cytoplasmic Ca^{2+} low are overwhelmed.

Habituation is a simple form of learning in which a neutral stimulus is repeated many times. The first time it is applied it is novel and evokes a reaction (the orienting reflex or "what is it?" response). However, it evokes less and less electrical response as it is repeated. Eventually, the subject becomes habituated to the stimulus and ignores it. This is associated with decreased release of neurotransmitter from the presynaptic terminal because of decreased intracellular Ca^{2+}. The decrease in intracellular Ca^{2+} is due to a gradual inactivation of Ca^{2+} channels. It can be short term, or it can be prolonged if exposure to the benign stimulus is repeated many times. Habituation is a classic example of nonassociative learning.

Sensitization is in a sense the opposite of habituation. Sensitization is the prolonged occurrence of augmented postsynaptic responses after a stimulus to which one has become habituated is paired once or several times with a noxious stimulus. At least in the sea snail *Aplysia,* the noxious stimulus causes discharge of serotonergic neurons that end on the presynaptic endings of sensory neurons. Thus, sensitization is due to presynaptic facilitation. Sensitization may occur as a transient response, or if it is reinforced by additional pairings of the noxious stimulus and the initial stimulus, it can exhibit features of short-term or long-term memory. The short-term prolongation of sensitization is due to a Ca^{2+}-mediated change in adenylyl cyclase that leads to a greater production of cAMP. The long-term potentiation also involves protein synthesis and growth of the presynaptic and postsynaptic neurons and their connections.

Long-term potentiation (LTP) is a rapidly developing persistent enhancement of the postsynaptic potential response to presynaptic stimulation after a brief period of rapidly repeated stimulation of the presynaptic neuron. It resembles posttetanic potentiation but is much more prolonged and can last for days. Unlike posttetanic potentiation, it is initiated by an increase in intracellular Ca^{2+} in the postsynaptic rather than the presynaptic neuron. It occurs in many parts of the nervous system but has been studied in greatest detail in the hippocampus. There are two forms in the hippocampus: mossy fiber LTP, which is presynaptic and independent of N-methyl-D-aspartate (NMDA) receptors; and Schaffer collateral LTP, which is postsynaptic and NMDA receptor-dependent. The hypothetical basis of the latter form is summarized in Figure 19–3. The basis of mossy fiber LTP is unsettled, although it appears to include cAMP and I_h, a hyperpolarization-activated cation channel. Other parts of the nervous system have not been as well studied, but it is interesting that NMDA-independent LTP can be produced in GABAergic neurons in the amygdala.

Long-term depression (LTD) was first noted in the hippocampus but was subsequently shown to be present throughout the brain in the same fibers as LTP. LTD is the opposite of LTP. It resembles LTP in many ways, but it is characterized by a decrease in synaptic strength. It is produced by slower stimulation of presynaptic neurons and is associated with a smaller rise in intracellular Ca^{2+} than occurs in LTP. In the cerebellum, its occurrence appears to require the phosphorylation of the GluR2 subunit of the α-amino-3-hydroxy-5-methylisoxazole-4 propionic acid (AMPA) receptors. It may be involved in the mechanism by which learning occurs in the cerebellum.

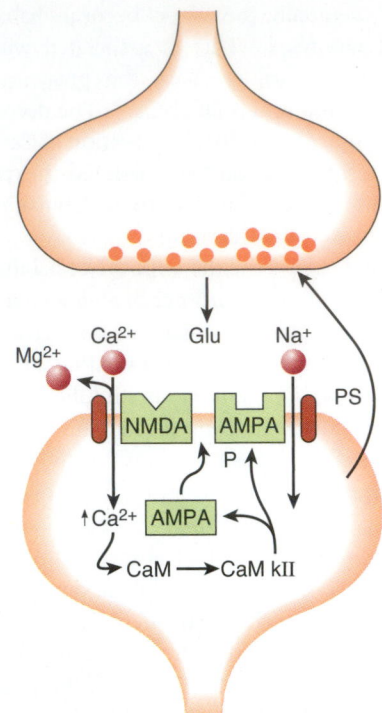

FIGURE 19–3   Production of LTP in Schaffer collaterals in the hippocampus. Glutamate (Glu) released from the presynaptic neuron binds to AMPA and NMDA receptors in the membrane of the postsynaptic neuron. The depolarization triggered by activation of the AMPA receptors relieves the Mg^{2+} block in the NMDA receptor channel, and Ca^{2+} enters the neuron with Na^+. The increase in cytoplasmic Ca^{2+} activates calmodulin (CaM), which in turn activates Ca^{2+}/calmodulin kinase II (CaM kII). The kinase phosphorylates the AMPA receptors (P), increasing their conductance, and moves more AMPA receptors into the synaptic cell membrane from cytoplasmic storage sites. In addition, a chemical signal (PS) may pass to the presynaptic neuron, producing a long-term increase in the quantal release of glutamate. (Courtesy of R Nicoll.)

CONDITIONED REFLEXES

A classic example of associative learning is a **conditioned reflex.** A conditioned reflex is a reflex response to a stimulus that previously elicited little or no response, acquired by repeatedly pairing the stimulus with another stimulus that normally does produce the response. In Pavlov's classic experiments, the salivation normally induced by placing meat in the mouth of a dog was studied. A bell was rung just before the meat was placed in the dog's mouth, and this was repeated a number of times until the animal would salivate when the bell was rung even though no meat was placed in its mouth. In this experiment, the meat placed in the mouth was the **unconditioned stimulus (US),** the stimulus that normally produces a particular innate response. The **conditioned stimulus (CS)** was the bell ringing. After the CS and US had been paired a sufficient number of times, the CS produced the response originally evoked only by the US. The CS had to precede the US. An immense number of somatic, visceral, and other neural changes can be made to occur as conditioned reflex responses.

Conditioning of visceral responses is often called **biofeedback.** The changes that can be produced include alterations in heart rate and blood pressure. Conditioned decreases in blood pressure have been advocated for the treatment of hypertension; however, the depressor response produced in this fashion is small.

INTERCORTICAL TRANSFER OF MEMORY

If a cat or monkey is conditioned to respond to a visual stimulus with one eye covered and then tested with the blindfold transferred to the other eye, it performs the conditioned response. This is true even if the optic chiasm has been cut, making the visual input from each eye go only to the ipsilateral cortex. If, in addition to the optic chiasm, the anterior and posterior commissures and the corpus callosum are sectioned ("split-brain animal"), no memory transfer occurs. Partial callosal section experiments indicate that the memory transfer occurs in the anterior portion of the corpus callosum. Similar results have been obtained in humans in whom the corpus callosum is congenitally absent or in whom it has been sectioned surgically in an effort to control epileptic seizures. This demonstrates that the neural coding necessary for "remembering with one eye what has been learned with the other" has been transferred to the opposite cortex via the commissures. Evidence suggests that similar transfer of information is acquired through other sensory pathways.

WORKING MEMORY

As noted above, working memory keeps incoming information available for a short time while deciding what to do with it. It is that form of memory which permits us, for example, to look up a telephone number, then remember the number while we pick up the telephone and dial the number. It consists of what has been called a **central executive** located in the prefrontal cortex, and two "rehearsal systems:" a **verbal system** for retaining verbal memories and a parallel **visuospatial system** for retaining visual and spatial aspects of objects. The executive steers information into these rehearsal systems.

HIPPOCAMPUS & MEDIAL TEMPORAL LOBE

Working memory areas are connected to the hippocampus and the adjacent parahippocampal portions of the medial temporal cortex (Figure 19–4). In humans, bilateral destruction of the ventral hippocampus, or Alzheimer disease and similar disease processes that destroy its CA1 neurons, cause striking defects in short-term memory, as do bilateral lesions of the same area in monkeys. Humans with such destruction have intact working memory and remote memory. Their

FIGURE 19–4 **Areas concerned with encoding explicit memories.** The prefrontal cortex and the parahippocampal cortex of the brain are active during the encoding of memories. (Modified from Russ MD: Memories are made of this. Science 1998;281:1151.)

implicit memory processes are generally intact. They perform adequately in terms of conscious memory as long as they concentrate on what they are doing. However, if they are distracted for even a very short period, all memory of what they were doing and what they proposed to do is lost. They are thus capable of new learning and retain old prelesion memories, but they cannot form new long-term memories.

The hippocampus is closely associated with the overlying parahippocampal cortex in the medial frontal lobe (Figure 19–4). Memory processes have now been studied not only with fMRI but with measurement of evoked potentials (event-related potentials; ERPs) in epileptic patients with implanted electrodes. When subjects recall words, activity in their left frontal lobe and their left parahippocampal cortex increases, but when they recall pictures or scenes, activity takes place in their right frontal lobe and the parahippocampal cortex on both sides.

The connections of the hippocampus to the diencephalon are also involved in memory. Some people with alcoholism-related brain damage develop impairment of recent memory, and the memory loss correlates well with the presence of pathologic changes in the mamillary bodies, which have extensive efferent connections to the hippocampus via the fornix. The mamillary bodies project to the anterior thalamus via the mamillothalamic tract, and in monkeys, lesions of the thalamus cause loss of recent memory. From the thalamus, the fibers concerned with memory project to the prefrontal cortex and from there to the basal forebrain. From the basal forebrain, a diffuse cholinergic projection goes to all of the neocortex, the amygdala, and the hippocampus from the **nucleus basalis of Meynert.** Severe loss of these fibers occurs in Alzheimer disease.

The amygdala is closely associated with the hippocampus and is concerned with encoding and recalling emotionally charged memories. During retrieval of fearful memories, the theta rhythms of the amygdala and the hippocampus become

synchronized. In normal humans, events associated with strong emotions are remembered better than events without an emotional charge, but in patients with bilateral lesions of the amygdala, this difference is absent.

Confabulation is an interesting though poorly understood condition that sometimes occurs in individuals with lesions of the ventromedial portions of the frontal lobes. These individuals perform poorly on memory tests, but they spontaneously describe events that never occurred. This has been called "honest lying."

NEW BRAIN CELLS?

It is now established that the traditional view that brain cells are not added after birth is wrong; new neurons form from stem cells throughout life in two areas: the olfactory bulb and the hippocampus. This is a process called **neurogenesis.** There is evidence implicating a role of neurogenesis in the hippocampus with learning and memory. A reduction in the number of new neurons formed reduces at least one form of hippocampal memory production. However, a great deal more is still to be done before the relation of new cells to memory processing can be considered established.

LONG-TERM MEMORY

While the encoding process for short-term explicit memory involves the hippocampus, long-term memories are stored in various parts of the neocortex. Apparently, the various parts of the memories—visual, olfactory, auditory, etc—are located in the cortical regions concerned with these functions, and the pieces are tied together by long-term changes in the strength of transmission at relevant synaptic junctions so that all the components are brought to consciousness when the memory is recalled.

Once long-term memories have been established, they can be recalled or accessed by a large number of different associations. For example, the memory of a vivid scene can be evoked not only by a similar scene but also by a sound or smell associated with the scene and by words such as "scene," "vivid," and "view." Thus, each stored memory must have multiple routes or keys. Furthermore, many memories have an emotional component or "color," that is, in simplest terms, memories can be pleasant or unpleasant.

STRANGENESS & FAMILIARITY

It is interesting that stimulation of some parts of the temporal lobes in humans causes a change in interpretation of one's surroundings. For example, when the stimulus is applied, the subject may feel strange in a familiar place or may feel that what is happening now has happened before. The occurrence of a sense of familiarity or a sense of strangeness in appropriate situations probably helps the normal individual adjust to the environment. In strange surroundings, one is alert and on guard, whereas in familiar surroundings, vigilance is relaxed. An inappropriate feeling of familiarity with new events or in new surroundings is known clinically as the **déjà vu phenomenon,** from the French words meaning "already seen." The phenomenon occurs from time to time in normal individuals, but it also may occur as an aura (a sensation immediately preceding a seizure) in patients with temporal lobe epilepsy.

SUMMARY

In summary, much is still to be learned about the encoding of explicit memory. However, according to current views, information from the senses is temporarily stored in various areas of the prefrontal cortex as working memory. It is also passed to the medial temporal lobe, and specifically to the parahippocampal gyrus. From there, it enters the hippocampus and is processed in a way that is not yet fully understood. At this time, the activity is vulnerable, as described above. Output from the hippocampus leaves via the subiculum and the entorhinal cortex and somehow binds together and strengthens circuits in many different neocortical areas, forming over time the stable remote memories that can now be triggered by many different cues.

ALZHEIMER DISEASE & SENILE DEMENTIA

Alzheimer disease is the most common age-related neurodegenerative disorder. Memory decline initially manifests as a loss of episodic memory, which impedes recollection of recent events. Loss of short-term memory is followed by general loss of cognitive and other brain functions, the need for constant care, and, eventually, death.

It was originally characterized in middle-aged people, and similar deterioration in elderly individuals is technically **senile dementia** of the alzheimer type, though it is frequently just called Alzheimer disease. Most cases are sporadic, but some are familial. Senile dementia can be caused by vascular disease and other disorders, but Alzheimer disease is the most common cause, accounting for 50–60% of the cases. It is present in about 17% of the population aged 65–69, but its incidence increases steadily with age, and in those who are 95 and older, the incidence is 40–50%. Thus, Alzheimer disease plus the other forms of senile dementia are a major medical problem.

Figure 19–5 summarizes some of the risk factors, pathogenic processes, and clinical signs linked to cellular abnormalities that occur in Alzheimer disease. The cytopathologic hallmarks of Alzheimer disease are intracellular **neurofibrillary tangles,** made up in part of hyperphosphorylated forms of the tau protein that normally binds to microtubules, and extracellular **senile plaques,** which have a core of β-**amyloid peptides (Aβ)** surrounded by altered nerve fibers and reactive glial cells. Figure 19–6 compares a normal nerve cell to one showing abnormalities associated with Alzheimer disease.

The Aβ peptides are products of a normal protein, **amyloid precursor protein (APP),** a transmembrane protein that projects into the extracellular fluid (ECF) from all nerve cells. This protein is hydrolyzed at three different sites by α-secretase, β-secretase, and γ-secretase, respectively. When APP is hydrolyzed by α-secretase, nontoxic peptide products are produced. However, when it is hydrolyzed by β-secretase and γ-secretase, polypeptides with 40 to 42 amino acids are produced; the actual length varies because of variation in the site at which γ-secretase cuts the protein chain. These polypeptides are toxic, the most toxic being $A\beta\sigma_{1-42}$. The polypeptides form extracellular aggregates, which can stick to AMPA receptors and Ca^{2+} ion channels, increasing Ca^{2+} influx. The polypeptides also initiate an inflammatory response, with production of intracellular tangles. The damaged cells eventually die.

An interesting finding that may well have broad physiologic implications is the observation—now confirmed in a rigorous prospective study—that frequent effortful mental activities, such as doing difficult crossword puzzles and playing board games, slow the onset of cognitive dementia due to Alzheimer disease and vascular disease. The explanation for this "use it or

FIGURE 19–5 Relationships of risk factors, pathogenic processes, and clinical signs to cellular abnormalities in the brain during **Alzheimer disease.** (From Kandel ER, Schwartz JH, Jessell TM [editors]: *Principles of Neural Science,* 4th ed. McGraw-Hill, 2000.)

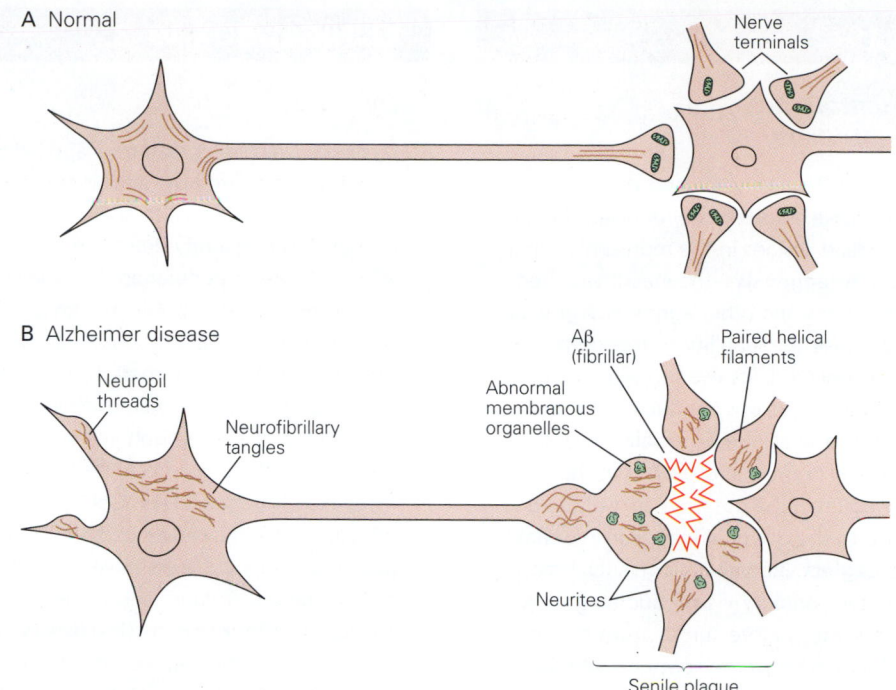

FIGURE 19–6 **Comparison of a normal neuron and one with abnormalities associated with Alzheimer disease.** (From Kandel ER, Schwartz JH, Jessell TM [editors]: *Principles of Neural Science,* 4th ed. McGraw-Hill, 2000.)

lose it" phenomenon is as yet unknown, but it certainly suggests that the hippocampus and its connections have plasticity like other parts of the brain and skeletal and cardiac muscles.

LANGUAGE & SPEECH

Memory and learning are functions of large parts of the brain, but the centers controlling some of the other "higher functions of the nervous system," particularly the mechanisms related to language, are more or less localized to the neocortex. Speech and other intellectual functions are especially well developed in humans—the animal species in which the neocortical mantle is most highly developed.

COMPLEMENTARY SPECIALIZATION OF THE HEMISPHERES VERSUS "CEREBRAL DOMINANCE"

One group of functions more or less localized to the neocortex in humans consists of those related to language, that is, understanding the spoken and printed word and expressing ideas in speech and writing. It is a well-established fact that human language functions depend more on one cerebral hemisphere than on the other. This hemisphere is concerned with categorization and symbolization and has often been called the **dominant hemisphere.** However, it is clear that the other hemisphere is not simply less developed or "nondominant;" instead, it is specialized in the area of spatiotemporal relations. It is this hemi-

sphere that is concerned, for example, with the identification of objects by their form and the recognition of musical themes. It also plays a primary role in the recognition of faces. Consequently, the concept of "cerebral dominance" and a dominant and nondominant hemisphere has been replaced by a concept of complementary specialization of the hemispheres, one for sequential-analytic processes (the **categorical hemisphere**) and one for visuospatial relations (the **representational hemisphere**). The categorical hemisphere is concerned with language functions, but hemispheric specialization is also present in monkeys, so it antedates the evolution of language. Clinical Box 19–2 describes deficits that occur in subjects with representational or categorical hemisphere lesions.

Hemispheric specialization is related to handedness. Handedness appears to be genetically determined. In 96% of right-handed individuals, who constitute 91% of the human population, the left hemisphere is the dominant or categorical hemisphere, and in the remaining 4%, the right hemisphere is dominant. In approximately 15% of left-handed individuals, the right hemisphere is the categorical hemisphere and in 15%, there is no clear lateralization. However, in the remaining 70% of left-handers, the left hemisphere is the categorical hemisphere. It is interesting that learning disabilities such as **dyslexia** (see Clinical Box 19–3), an impaired ability to learn to read, are 12 times as common in left-handers as they are in right-handers, possibly because some fundamental abnormality in the left hemisphere led to a switch in handedness early in development. However, the spatial talents of left-handers may be well above average; a disproportionately large number

Lesions of Representational & Categorical Hemispheres

Lesions in the categorical hemisphere produce language disorders, whereas extensive lesions in the representational hemisphere do not. Instead, lesions in the representational hemisphere produce **astereognosis**—the inability to identify objects by feeling them—and other agnosias. **Agnosia** is the general term used for the inability to recognize objects by a particular sensory modality even though the sensory modality itself is intact. Lesions producing these defects are generally in the parietal lobe. Especially when they are in the representational hemisphere, lesions of the inferior parietal lobule, a region in the posterior part of the parietal lobe that is close to the occipital lobe, cause **unilateral inattention** and **neglect.** Individuals with such lesions do not have any apparent primary visual, auditory, or somatesthetic defects, but they ignore stimuli from the contralateral portion of their bodies or the space around these portions. This leads to failure to care for half their bodies and, in extreme cases, to situations in which individuals shave half their faces, dress half their bodies, or read half of each page. This inability to put together a picture of visual space on one side is due to a shift in visual attention to the side of the brain lesion and can be improved, if not totally corrected, by wearing eyeglasses that contain prisms. Hemispheric specialization extends to other parts of the cortex as well. Patients with lesions in the categorical hemisphere are disturbed about their disability and often depressed, whereas patients with lesions in the representational hemisphere are sometimes unconcerned and even euphoric. Lesions of different parts of the categorical hemisphere produce **fluent, nonfluent,** and **anomic aphasias** (see text for more details). Although aphasias are produced by lesions of the categorical hemisphere, lesions in the representational hemisphere also have effects. For example, they may impair the ability to tell a story or make a joke. They may also impair a subject's ability to get the point of a joke and, more broadly, to comprehend the meaning of differences in inflection and the "color" of speech. This is one more example of the way the hemispheres are specialized rather than simply being dominant and nondominant.

Dyslexia

Dyslexia, which is a broad term applied to impaired ability to read, is characterized by difficulties with learning how to decode at the word level, to spell, and to read accurately and fluently. It is frequently due to an inherited abnormality that affects 5% of the population. Many individuals with dyslexic symptoms also have problems with short-term memory skills and problems processing spoken language. Although its precise cause is unknown, there is evidence that dyslexia is of neurological origin. Acquired dyslexias occur due to brain damage in the left hemisphere's key language areas. Also, in many cases, there is a decreased blood flow in the angular gyrus in the categorical hemisphere. There are numerous theories to explain the causes of dyslexia. The **phonological hypothesis** is that dyslexics have a specific impairment in the representation, storage, and/or retrieval of speech sounds. The **rapid auditory processing theory** proposes that the primary deficit is the perception of short or rapidly varying sounds. The **visual theory** is that a defect in the magnocellular portion of the visual system slows processing and also leads to phonemic deficit. More selective speech defects have also been described. For example, lesions limited to the left temporal pole (area 38) cause inability to retrieve names of places and persons but preserves the ability to retrieve common nouns, that is, the names of nonunique objects. The ability to retrieve verbs and adjectives is also intact.

exclusively a human trait. Imaging studies show that other portions of the upper surface of the left temporal lobe are larger in right-handed individuals, the right frontal lobe is normally thicker than the left, and the left occipital lobe is wider and protrudes across the midline. Chemical differences also exist between the two sides of the brain. For example, the concentration of dopamine is higher in the nigrostriatal pathway on the left side in right-handed humans but higher on the right in left-handers. The physiologic significance of these differences is unknown.

In patients with schizophrenia, MRI studies have demonstrated reduced volumes of gray matter on the left side in the anterior hippocampus, amygdala, parahippocampal gyrus, and posterior superior temporal gyrus. The degree of reduction in the left superior temporal gyrus correlates with the degree of disordered thinking in the disease. There are also apparent abnormalities of dopaminergic systems and cerebral blood flow in this disease.

PHYSIOLOGY OF LANGUAGE

Language is one of the fundamental bases of human intelligence and a key part of human culture. The primary brain areas concerned with language are arrayed along and near the

of artists, musicians, and mathematicians are left-handed. For unknown reasons, left-handers have slightly but significantly shorter life spans than right-handers.

Some anatomic differences between the two hemispheres may correlate with the functional differences. The **planum temporale,** an area of the superior temporal gyrus that is involved in language-related auditory processing, is regularly larger on the left side than the right. It is also larger on the left in the brain of chimpanzees, even though language is almost sylvian fissure (lateral cerebral sulcus) of the categorical hemisphere. A region at the posterior end of the superior temporal gyrus called **Wernicke's area** (Figure 19–7) is concerned with comprehension of auditory and visual information. It projects via the **arcuate fasciculus** to **Broca's area** (area 44) in the frontal lobe immediately in front of the inferior end of the motor cortex. Broca's area processes the information received from Wernicke's area into a detailed and coordinated pattern for vocalization and then projects the pattern via a speech articulation area in the insula to the motor cortex, which initiates the appropriate movements of the lips, tongue, and larynx to produce speech. The probable sequence of events that occurs when a subject names a visual object is shown in Figure 19–8. The angular gyrus behind Wernicke's area appears to process information from words that are read in such a way that they can be converted into the auditory forms of the words in Wernicke's area.

It is interesting that in individuals who learn a second language in adulthood, fMRI reveals that the portion of Broca's area concerned with it is adjacent to but separate from the area concerned with the native language. However, in children who learn two languages early in life, only a single area is involved with both. It is well known, of course, that children acquire fluency in a second language more easily than adults.

LANGUAGE DISORDERS

Aphasias are abnormalities of language functions that are not due to defects of vision or hearing or to motor paralysis. They are caused by lesions in the categorical hemisphere (see Clinical Box 19–2). The most common cause is embolism or thrombosis of a cerebral blood vessel. Many different classifications of the aphasias have been published, but a convenient classification divides them into **fluent, nonfluent,** and **anomic aphasias.** In nonfluent aphasia, the lesion is in Broca's area (Table 19–1). Speech is slow, and words are hard to come by. Patients with severe damage to this area are limited to two or three words with which to express the whole range of meaning and emotion.

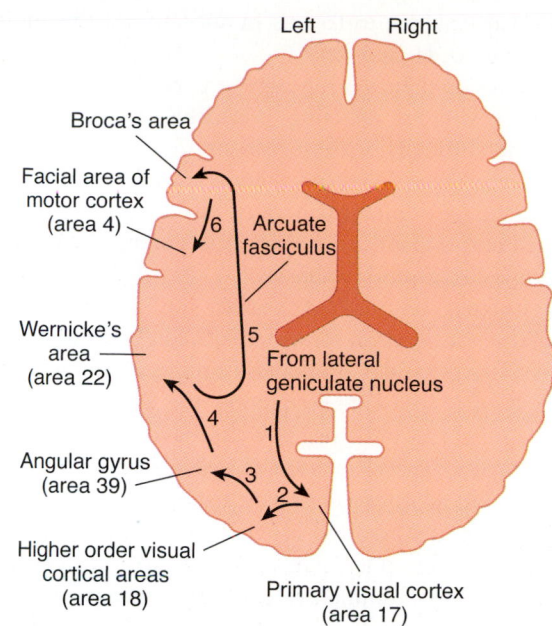

FIGURE 19–8 Path taken by impulses when a subject names a visual object projected on a horizontal section of the human brain.

Sometimes the words retained are those that were being spoken at the time of the injury or vascular accident that caused the aphasia.

In one form of fluent aphasia, the lesion is in Wernicke's area. In this condition, speech itself is normal and sometimes the patients talk excessively. However, what they say is full of jargon and neologisms that make little sense. The patient also fails to comprehend the meaning of spoken or written words, so other aspects of the use of language are compromised.

Another form of fluent aphasia is a condition in which patients can speak relatively well and have good auditory comprehension but cannot put parts of words together or conjure up words. This is called **conduction aphasia** because it was thought to be due to lesions of the arcuate fasciculus connecting Wernicke's and Broca's areas. However, it now

FIGURE 19–7 Location of some of the areas in the categorical hemisphere that are concerned with language functions.

TABLE 19–1 Aphasias. Characteristic responses of patients with lesions in various areas when shown a picture of a chair.

Type of Aphasia and Site of Lesion	Characteristic Naming Errors
Nonfluent (Broca's area)	"Tssair"
Fluent (Wernicke's area)	"Stool" or "choss" (neologism)
Fluent (areas 40, 41, and 42; conduction aphasia)	"Flair . . . no, swair . . . tair."
Anomic (angular gyrus)	"I know what it is . . . I have a lot of them."

Modified from Goodglass H: Disorders of naming following brain injury. Am Sci 1980;68:647.

appears that it is due to lesions in and around the auditory cortex (areas 40, 41, and 42).

When a lesion damages the angular gyrus in the categorical hemisphere without affecting Wernicke's or Broca's areas, there is no difficulty with speech or the understanding of auditory information; instead there is trouble understanding written language or pictures, because visual information is not processed and transmitted to Wernicke's area. The result is a condition called **anomic aphasia.**

The isolated lesions that cause the selective defects described above occur in some patients, but brain destruction is often more general. Consequently, more than one form of aphasia is often present. Frequently, the aphasia is general **(global),** involving both receptive and expressive functions. In this situation, speech is scant as well as nonfluent. Writing is abnormal in all aphasias in which speech is abnormal, but the neural circuits involved are unknown. In addition, deaf subjects who develop a lesion in the categorical hemisphere lose their ability to communicate in sign language.

Stuttering has been found to be associated with right cerebral dominance and widespread overactivity in the cerebral cortex and cerebellum. This includes increased activity of the supplementary motor area. Stimulation of part of this area has been reported to produce **laughter,** with the duration and intensity of the laughter proportionate to the intensity of the stimulus.

RECOGNITION OF FACES

An important part of the visual input goes to the inferior temporal lobe, where representations of objects, particularly faces, are stored (Figure 19–9). Faces are particularly important in distinguishing friends from foes and the emotional state of those seen. In humans, storage and recognition of faces is more strongly represented in the right inferior temporal lobe in right-handed individuals, though the left lobe is also active. Lesions in this area cause **prosopagnosia,** the inability to recognize faces. Patients with this abnormality can recognize forms and reproduce them. They can recognize people by

FIGURE 19–9 **Areas in the right cerebral hemisphere, in right-handed individuals, that are concerned with recognition of faces.** (Modified from Szpir M: Accustomed to your face. Am Sci 1992;80:539.)

Labels: Stores biographical information; Connects facial features to biographical information; Extracts facial features

their voices, and many of them show autonomic responses when they see familiar as opposed to unfamiliar faces. However, they cannot identify the familiar faces they see. The left hemisphere is also involved, but the role of the right hemisphere is primary. The presence of an autonomic response to a familiar face in the absence of recognition has been explained by postulating the existence of a separate dorsal pathway for processing information about faces that leads to recognition at only a subconscious level.

LOCALIZATION OF OTHER FUNCTIONS

Use of fMRI and PET scanning combined with study of patients with strokes and head injuries has provided further insights—or at least glimpses—into the ways serial processing of sensory information produce cognition, reasoning, comprehension, and language. Analysis of the brain regions involved in arithmetic calculations has highlighted two areas. In the inferior portion of the left frontal lobe is an area concerned with number facts and exact calculations. Frontal lobe lesions can cause **acalculia,** a selective impairment of mathematical ability. There are areas around the intraparietal sulci of both parietal lobes that are concerned with visuospatial representations of numbers and, presumably, finger counting.

Two right-sided subcortical structures play a role in accurate navigation in humans. One is the right hippocampus, which is concerned with learning where places are located, and the other is the right caudate nucleus, which facilitates movement to the places. Men have larger brains than women and are said to have superior spatial skills and ability to navigate.

Other defects seen in patients with localized cortical lesions include, for example, the inability to name animals, though the ability to name other living things and objects is intact. One patient with a left parietal lesion had difficulty with the second half but not the first half of words. Some patients with parietooccipital lesions write only with consonants and omit vowels. The pattern that emerges from studies of this type is one of precise sequential processing of information in localized brain areas. Additional research of this type should greatly expand our understanding of the functions of the neocortex.

CHAPTER SUMMARY

- Long-term memory is divided into explicit (declarative) and implicit (nondeclarative). Explicit is further subdivided into semantic and episodic. Implicit is further subdivided into priming, procedural, associative learning, and nonassociative learning.
- Synaptic plasticity is the ability of neural tissue to change as reflected by LTP (an increased effectiveness of synaptic activity) or LTD (a reduced effectiveness of synaptic activity) after continued use.
- Hippocampal and other temporal lobe structures and association cortex are involved in declarative memory.
- Alzheimer disease is characterized by progressive loss of short-term memory followed by general loss of cognitive function.

The cytopathologic hallmarks of Alzheimer disease are intracellular neurofibrillary tangles and extracellular senile plaques.

■ Categorical and representational hemispheres are for sequential-analytic processes and visuospatial relations, respectively. Lesions in the categorical hemisphere produce language disorders, whereas lesions in the representational hemisphere produce astereognosis.

■ Aphasias are abnormalities of language functions and are caused by lesions in the categorical hemisphere. They are classified as fluent (Wernicke's area; areas 40, 41, 42), nonfluent (Broca's area), and anomic (angular gyrus) based on the location of brain lesions.

MULTIPLE-CHOICE QUESTIONS

For all questions, select the single best answer unless otherwise directed.

1. The representational hemisphere
   A) is the right cerebral hemisphere in most right-handed individuals.
   B) is the left cerebral hemisphere in most left-handed individuals.
   C) includes the part of the brain concerned with language functions.
   D) is the site of lesions in most patients with aphasia.
   E) is morphologically identical to the opposite nonrepresentational hemisphere.

2. The optic chiasm and corpus callosum are sectioned in a dog, and with the right eye covered, the animal is trained to bark when it sees a red square. The right eye is then uncovered and the left eye covered. The animal will now
   A) fail to respond to the red square because the square does not produce impulses that reach the right occipital cortex.
   B) fail to respond to the red square because the animal has bitemporal hemianopia.
   C) fail to respond to the red square if the posterior commissure is also sectioned.
   D) respond to the red square only after retraining.
   E) respond promptly to the red square in spite of the lack of input to the left occipital cortex.

3. The effects of bilateral loss of hippocampal function include
   A) disappearance of remote memories.
   B) loss of working memory.
   C) loss of the ability to encode events of the recent past in long-term memory.
   D) loss of the ability to recall faces and forms but not the ability to recall printed or spoken words.
   E) production of inappropriate emotional responses when recalling events of the recent past.

4. Which of the following are *incorrectly* paired?
   A) lesion of the parietal lobe of the representational hemisphere : unilateral inattention and neglect
   B) loss of cholinergic neurons in the nucleus basalis of Meynert and related areas of the forebrain : loss of recent memory
   C) lesions of mamillary bodies : loss of recent memory
   D) lesion of the angular gyrus in the categorical hemisphere : nonfluent aphasia
   E) lesion of Broca's area in the categorical hemisphere : slow speech

5. The representational hemisphere is better than the categorical hemisphere at
   A) language functions.
   B) recognition of objects by their form.
   C) understanding printed words.
   D) understanding spoken words.
   E) mathematical calculations.

6. A lesion of Wernicke's area (the posterior end of the superior temporal gyrus) in the categorical hemisphere causes patients to
   A) lose short-term memory.
   B) speak in a slow, halting voice.
   C) experience déjà vu.
   D) talk rapidly but make little sense.
   E) lose the ability to recognize faces.

7. Which of the following is most likely *not* to be involved in production of LTP?
   A) NO
   B) Ca^{2+}
   C) NMDA receptors
   D) membrane hyperpolarization
   E) membrane depolarization

CHAPTER RESOURCES

Andersen P, Morris R, Amaral D, Bliss T, O'Keefe J: *The Hippocampus Book.* Oxford University Press, New York, 2007.
Bird CM, Burgess N: The hippocampus and memory: Insights from spatial processing. Nature Rev Neurosci 2008;9:182.
Charlton MH: Apasia: A review. Am J Psychiatry 1963;119:872.
Goodglass H: *Understanding Aphasia.* Academic Press, 1993.
Ingram VM: Alzheimer's disease. Am Scientist 2003;91:312.
Kandel ER: The molecular biology of memory: A dialogue between genes and synapses. Science 2001;294:1028.
Kandel ER, Schwartz JH, Jessell TM (editors): *Principles of Neural Science,* 4th ed. McGraw-Hill, 2000.
LaFerla FM, Green KN, Oddo S: Intracellular amyloid-β in Alzheimer's disease. Nature Rev Neurosci 2007;8:499.
Ramus F: Developmental dyslexia: Specific phonological defect or general sensorimotor dysfunction. Curr Opin Neurobiol 2003;13:212.
Russ MD: Memories are made of this. Science 1998;281:1151.
Selkoe DJ: Translating cell biology into therapeutic advances in Alzheimer's disease. Nature 1999;399 (Suppl): A23.
Shaywitz S: Dyslexia. N Engl J Med 1998;338:307.
Squire LR, Stark CE, Clark RE: The medial temporal lobe. Annu Rev Neurosci 2004; 27:279.
Squire LR, Zola SM: Structure and function of declarative and nondeclarative memory systems. Proc Natl Acad Sci 1996;93:13515.

C H A P T E R

20

The Thyroid Gland

OBJECTIVES

After studying this chapter, you should be able to:

- Describe the structure of the thyroid gland and how it relates to its function.
- Define the chemical nature of the thyroid hormones and how they are synthesized.
- Understand the critical role of iodine in the thyroid gland and how its transport is controlled.
- Describe the role of protein binding in the transport of thyroid hormones and peripheral metabolism.
- Identify the role of the hypothalamus and pituitary in regulating thyroid function.
- Define the effects of the thyroid hormones in homeostasis and development.
- Understand the basis of conditions where thyroid function is abnormal and how they can be treated.

INTRODUCTION

The thyroid gland is one of the larger endocrine glands of the body. The gland has two primary functions. The first is to secrete the thyroid hormones, which maintain the level of metabolism in the tissues that is optimal for their normal function. Thyroid hormones stimulate O_2 consumption by most of the cells in the body, help regulate lipid and carbohydrate metabolism, and thereby influence body mass and mentation. Consequences of thyroid gland dysfunction depend on the life stage at which they occur. The thyroid is not essential for life, but its absence or hypofunction during fetal and neonatal life results in severe mental retardation and dwarfism. In adults, hypothyroidism is accompanied by mental and physical slowing and poor resistance to cold. Conversely, excess thyroid secretion leads to body wasting, nervousness, tachycardia, tremor, and excess heat production. Thyroid function is controlled by the thyroid-stimulating hormone (TSH, thyrotropin) of the anterior pituitary. The secretion of this hormone is in turn increased by thyrotropin-releasing hormone (TRH) from the hypothalamus and is also subject to negative feedback control by high circulating levels of thyroid hormones acting on the anterior pituitary and the hypothalamus.

The second function of the thyroid gland is to secrete calcitonin, a hormone that regulates circulating levels of calcium. This function of the thyroid gland is discussed in Chapter 23 in the broader context of whole body calcium homeostasis.

ANATOMIC CONSIDERATIONS

The thyroid is a butterfly-shaped gland that straddles the trachea in the front of the neck. It develops from an evagination of the floor of the pharynx, and a **thyroglossal duct** marking the path of the thyroid from the tongue to the neck sometimes persists in the adult. The two lobes of the human thyroid are connected by a bridge of tissue, the **thyroid isthmus,** and there is sometimes a **pyramidal lobe** arising from the isthmus in front of the larynx (Figure 20–1). The gland is well vascularized, and the thyroid has one of the highest rates of blood flow per gram of tissue of any organ in the body.

The portion of the thyroid concerned with the production of thyroid hormone consists of multiple **acini (follicles).** Each spherical follicle is surrounded by a single layer of polarized epithelial cells and filled with pink-staining proteinaceous material called **colloid.** Colloid consists predominantly of the glycoprotein, thyroglobulin. When the gland is inactive, the colloid is abundant, the follicles are large, and the cells lining them are flat. When the gland is active, the follicles are small, the cells are cuboid or columnar, and areas where the colloid is being actively reabsorbed into the thyrocytes are visible as "reabsorption lacunae" (Figure 20–2).

Microvilli project into the colloid from the apexes of the thyroid cells and canaliculi extend into them. The endoplasmic reticulum is prominent, a feature common to most glandular cells, and secretory granules containing thyroglobulin are seen (Figure 20–3). The individual thyroid cells rest on a basal lamina that separates them from the adjacent capillaries. The

FIGURE 20–2   Thyroid histology. Note the small, punched-out "reabsorption lacunae" in the colloid next to the cells in the active gland.

capillaries are fenestrated, like those of other endocrine glands (see Chapter 32).

FORMATION & SECRETION OF THYROID HORMONES

CHEMISTRY

The primary hormone secreted by the thyroid is **thyroxine (T_4),** along with much lesser amounts of **triiodothyronine**

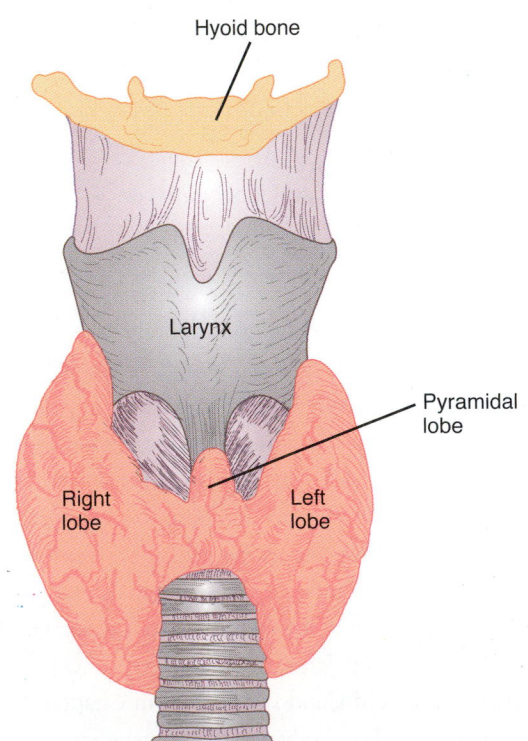

FIGURE 20–1   The human thyroid.

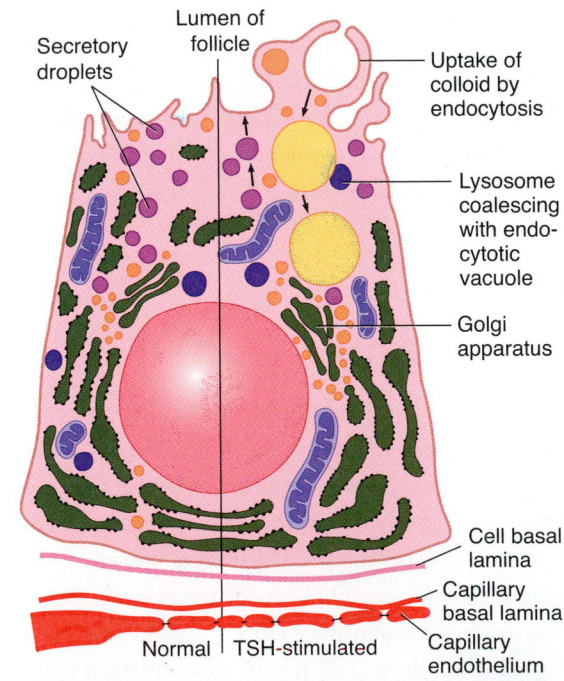

FIGURE 20–3   Thyroid cell. Left: Normal pattern. **Right:** After TSH stimulation. The arrows on the right show the secretion of thyroglobulin into the colloid. On the right, endocytosis of the colloid and merging of a colloid-containing vacuole with a lysosome are also shown. The cell rests on a capillary with gaps (fenestrations) in the endothelial wall.

HO— (ring structure with positions 3', 5') —O— (ring structure with positions 3, 5) —CH$_2$—CH—C—OH
with NH$_2$ and O

3,5,3',5',-Tetraiodothyronine (thyroxine, T$_4$)

HO— (ring structure) —O— (ring structure) —CH$_2$—CH—C—OH
with NH$_2$ and O

3,5,3',-Triiodothyronine (T$_3$)

FIGURE 20–4 **Thyroid hormones.** The numbers in the rings in the T$_4$ formula indicate the number of positions in the molecule. RT$_3$ is 3,3',5'-triiodothyronine.

FIGURE 20–5 **Iodine metabolism.**

(T$_3$). T$_3$ has much greater biological activity than T$_4$ and is specifically generated at its site of action in peripheral tissues by deiodination of T$_4$ (see below). Both hormones are iodine-containing amino acids (Figure 20–4). Small amounts of reverse triiodothyronine (3,3',5'-triiodothyronine, RT$_3$) and other compounds are also found in thyroid venous blood. RT$_3$ is not biologically active.

IODINE HOMEOSTASIS

Iodine is an essential raw material for thyroid hormone synthesis. Dietary iodide is absorbed by the intestine and enters the circulation; its subsequent fate is summarized in Figure 20–5. The minimum daily iodine intake that will maintain normal thyroid function is 150 μg in adults. In most developed countries, supplementation of table salt means that the average dietary intake is approximately 500 μg/d. The principal organs that take up circulating I$^-$ are the thyroid, which uses it to make thyroid hormones, and the kidneys, which excrete it in the urine. About 120 μg/d enter the thyroid at normal rates of thyroid hormone synthesis and secretion. The thyroid secretes 80 μg/d in the form of T$_3$ and T$_4$, while 40 μg/d diffuses back into the extracellular fluid (ECF). Circulating T$_3$ and T$_4$ are metabolized in the liver and other tissues, with the release of a further 60 μg of I$^-$ per day into the ECF. Some thyroid hormone derivatives are excreted in the bile, and some of the iodine in them is reabsorbed (enterohepatic circulation), but there is a net loss of I$^-$ in the stool of approximately 20 μg/d. The total amount of I$^-$ entering the ECF is thus 500 + 40 + 60, or 600 μg/d; 20% of this I$^-$ enters the thyroid, whereas 80% is excreted in the urine.

IODIDE TRANSPORT
ACROSS THYROCYTES

The basolateral membranes of thyrocytes facing the capillaries contain a **symporter** that transports two Na$^+$ ions and one I$^-$

ion into the cell with each cycle, against the electrochemical gradient for I$^-$. This Na$^+$/I$^-$ symporter (**NIS**) is capable of producing intracellular I$^-$ concentrations that are 20 to 40 times as great as the concentration in plasma. The process involved is secondary active transport (see Chapter 2), with the energy provided by active transport of Na$^+$ out of thyroid cells by Na, K ATPase. NIS is regulated both by transcriptional means and by active trafficking into and out of the thyrocyte basolateral membrane; in particular, thyroid stimulating hormone (TSH; see below) induces both NIS expression and the retention of NIS in the basolateral membrane where it can mediate sustained iodide uptake.

Iodide must also exit the thyrocyte across the apical membrane to access the colloid, where the initial steps of thyroid hormone synthesis occur. This transport step is believed to be mediated, at least in part, by a Cl$^-$/I$^-$ exchanger known as **pendrin**. This protein was first identified as the product of the gene responsible for the Pendred syndrome, whose patients suffer from thyroid dysfunction and deafness. Pendrin (SLC26A4) is one member of the larger family of SLC26 anion exchangers.

The relation of thyroid function to iodide is unique. As discussed in more detail below, iodide is essential for normal thyroid function, but iodide deficiency and iodide excess both inhibit thyroid function.

The salivary glands, the gastric mucosa, the placenta, the ciliary body of the eye, the choroid plexus, the mammary glands, and certain cancers derived from these tissues also express NIS and can transport iodide against a concentration gradient, but the transporter in these tissues is not affected by TSH. The physiologic significance of all these extrathyroidal iodide-concentrating mechanisms is obscure, but they may provide pathways for radioablation of NIS-expressing cancer cells using iodide radioisotopes. This approach is also useful for the ablation of thyroid cancers.

THYROID HORMONE SYNTHESIS & SECRETION

At the interface between the thyrocyte and the colloid, iodide undergoes a process referred to as organification. First, it is oxidized to iodine, and then incorporated into the carbon 3 position of tyrosine residues that are part of the thyroglobulin molecule in the colloid (Figure 20–6). **Thyroglobulin** is a glycoprotein made up of two subunits and has a molecular weight of 660 kDa. It contains 10% carbohydrate by weight. It also contains 123 tyrosine residues, but only 4 to 8 of these are normally incorporated into thyroid hormones. Thyroglobulin is synthesized in the thyroid cells and secreted into the colloid by exocytosis of granules. The oxidation and reaction of iodide with the secreted thyroglobulin is mediated by **thyroid peroxidase,** a membrane-bound enzyme found in the thyrocyte apical membrane. The thyroid hormones so produced remain part of the thyroglobulin molecule until needed. As such, colloid represents a reservoir of thyroid hormones, and humans can ingest a diet completely devoid of iodide for up to 2 months before a decline in circulating thyroid hormone levels is seen. When there is a need for thyroid hormone secretion, colloid is internalized by the thyrocytes by endocytosis, and directed toward lysosomal degradation. Thus, the peptide

bonds of thyroglobulin are hydrolyzed, and free T_4 and T_3 are discharged into cytosol and thence to the capillaries (see below). Thyrocytes thus have four functions: They collect and transport iodine, they synthesize thyroglobulin and secrete it into the colloid, they fix iodine to the thyroglobulin to generate thyroid hormones, and they remove the thyroid hormones from thyroglobulin and secrete them into the circulation.

Thyroid hormone synthesis is a multistep process. Thyroid peroxidase generates reactive iodine species that can attack thyroglobulin. The first product is monoiodotyrosine (MIT). MIT is next iodinated on the carbon 5 position to form diiodotyrosine (DIT). Two DIT molecules then undergo an oxidative condensation to form T_4 with the elimination of the alanine side chain from the molecule that forms the outer ring. There are two theories of how this **coupling reaction** occurs. One holds that the coupling occurs with both DIT molecules attached to thyroglobulin (intramolecular coupling). The other holds that the DIT that forms the outer ring is first detached from thyroglobulin (intermolecular coupling). In either case, thyroid peroxidase is involved in coupling as well as iodination. T_3 is formed by condensation of MIT with DIT. A small amount of RT_3 is also formed, probably by condensation of DIT with MIT. In the normal human thyroid, the average distribution of iodinated compounds is

FIGURE 20–6 **Outline of thyroid hormone biosynthesis.** Iodination of tyrosine takes place at the apical border of the thyroid cells while the molecules are bound in peptide linkage in thyroglobulin.

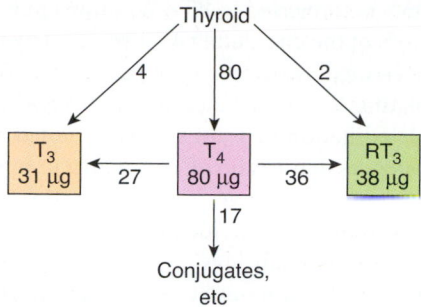

FIGURE 20–7 **Secretion and interconversion of thyroid hormones in normal adult humans.** Figures are in micrograms per day. Note that most of the T_3 and RT_3 are formed from T_4 deiodination in the tissues and only small amounts are secreted by the thyroid.

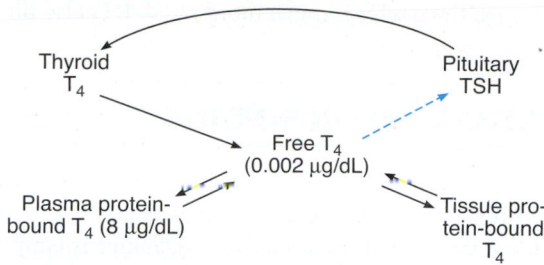

FIGURE 20–8 **Regulation of thyroid hormone synthesis.**

23% MIT, 33% DIT, 35% T_4, and 7% T_3. Only traces of RT_3 and other components are present.

The human thyroid secretes about 80 µg (103 nmol) of T_4, 4 µg (7 nmol) of T_3, and 2 µg (3.5 nmol) of RT_3 per day (Figure 20–7). MIT and DIT are not secreted. These iodinated tyrosines are deiodinated by a microsomal **iodotyrosine deiodinase.** This represents a mechanism to recover iodine and bound tyrosines and recycle them for additional rounds of hormone synthesis. The iodine liberated by deiodination of MIT and DIT is reutilized in the gland and normally provides about twice as much iodide for hormone synthesis as NIS does. In patients with congenital absence of the iodotyrosine deiodinase, MIT and DIT appear in the urine and there are symptoms of iodine deficiency (see below). Iodinated thyronines are resistant to the activity of iodotyrosine deiodinase, thus allowing T_4 and T_3 to pass into the circulation.

TRANSPORT & METABOLISM OF THYROID HORMONES

PROTEIN BINDING

The normal total **plasma T_4** level in adults is approximately 8 µg/dL (103 nmol/L), and the **plasma T_3** level is approximately 0.15 µg/dL (2.3 nmol/L). T_4 and T_3 are relatively lipophilic; thus, their free forms in plasma are in equilibrium with a much larger pool of protein-bound thyroid hormones in plasma and in tissues. Free thyroid hormones are added to the circulating pool by the thyroid. It is the free thyroid hormones in plasma that are physiologically active and that feed back to inhibit pituitary secretion of TSH (Figure 20–8). The function of protein-binding appears to be maintenance of a large pool of hormone that can readily be mobilized as needed. In addition, at least for T_3, hormone binding prevents excess uptake by the first cells encountered and promotes uniform tissue distribution. Both total T_4 and T_3 can be measured by radioimmunoassay. There are also direct assays that specifically measure only the free forms of the hormones. The latter are the more

clinically relevant measures given that these are the active forms, and also due to both acquired and congenital variations in the concentrations of binding proteins between individuals.

The plasma proteins that bind thyroid hormones are **albumin,** a prealbumin called **transthyretin** (formerly called **thyroxine-binding prealbumin**), and a globulin known as **thyroxine-binding globulin (TBG).** Of the three proteins, albumin has the largest **capacity** to bind T_4 (ie, it can bind the most T_4 before becoming saturated) and TBG has the smallest capacity. However, the **affinities** of the proteins for T_4 (ie, the avidity with which they bind T_4 under physiologic conditions) are such that most of the circulating T_4 is bound to TBG (Table 20–1), with over a third of the binding sites on the protein occupied. Smaller amounts of T_4 are bound to transthyretin and albumin. The half-life of transthyretin is 2 d, that of TBG is 5 d, and that of albumin is 13 d.

Normally, 99.98% of the T_4 in plasma is bound; the free T_4 level is only about 2 ng/dL. There is very little T_4 in the urine. Its biologic half-life is long (about 6–7 d), and its volume of distribution is less than that of ECF (10 L, or about 15% of body weight). All of these properties are characteristic of a substance that is strongly bound to protein.

T_3 is not bound to quite as great an extent; of the 0.15 µg/dL normally found in plasma, 0.2% (0.3 ng/dL) is free. The remaining 99.8% is protein-bound, 46% to TBG and most of the remainder to albumin, with very little binding to transthyretin (Table 20–1). The lesser binding of T_3 correlates with the facts that T_3 has a shorter half-life than T_4 and that

TABLE 20–1 **Binding of thyroid hormones to plasma proteins in normal adult humans.**

Protein	Plasma Concentration (mg/dL)	Amount of Circulating Hormone Bound (%)	
		T_4	T_3
Thyroxine-binding globulin (TBG)	2	67	46
Transthyretin (thyroxine-binding prealbumin, TBPA)	15	20	1
Albumin	3500	13	53

its action on the tissues is much more rapid. RT_3 also binds to TBG.

FLUCTUATIONS IN BINDING

When a sudden, sustained increase in the concentration of thyroid-binding proteins in the plasma takes place, the concentration of free thyroid hormones falls. This change is temporary, however, because the decrease in the concentration of free thyroid hormones in the circulation stimulates TSH secretion, which in turn causes an increase in the production of free thyroid hormones. A new equilibrium is eventually reached at which the total quantity of thyroid hormones in the blood is elevated but the concentration of free hormones, the rate of their metabolism, and the rate of TSH secretion are normal. Corresponding changes in the opposite direction occur when the concentration of thyroid-binding protein is reduced. Consequently, patients with elevated or decreased concentrations of binding proteins, particularly TBG, are typically neither hyper- nor hypothyroid; that is, they are **euthyroid.**

TBG levels are elevated in estrogen-treated patients and during pregnancy, as well as after treatment with various drugs (Table 20–2). They are depressed by glucocorticoids, androgens, the weak androgen danazol, and the cancer chemotherapeutic agent L-asparaginase. A number of other drugs, including salicylates, the anti-convulsant phenytoin, and the cancer chemotherapeutic agents mitotane (o, p'-DDD) and 5-fluorouracil inhibit binding of T_4 and T_3 to TBG and consequently produce changes similar to those produced by a decrease in TBG concentration. Changes in total plasma T_4 and T_3 can also be produced by changes in plasma concentrations of albumin and prealbumin.

METABOLISM OF THYROID HORMONES

T_4 and T_3 are deiodinated in the liver, the kidneys, and many other tissues. These deiodination reactions serve not only to catabolize the hormones, but also to provide a local supply specifically of T_3, which is believed to be the primary mediator of the physiological effects of thyroid secretion. One third of the circulating T_4 is normally converted to T_3 in adult humans, and 45% is converted to RT_3. As shown in Figure 20–7, only about 13% of the circulating T_3 is secreted by the thyroid while 87% is formed by deiodination of T_4; similarly, only 5% of the circulating RT_3 is secreted by the thyroid and 95% is formed by deiodination of T_4. It should be noted as well that marked differences in the ratio of T_3 to T_4 occur in various tissues. Two tissues that have very high T_3/T_4 ratios are the pituitary and the cerebral cortex, due to the expression of specific deiodinases, as discussed below.

Three different deiodinases act on thyroid hormones: D_1, D_2, and D_3. All are unique in that they contain the rare amino acid selenocysteine, with selenium in place of sulfur, which is essential for their enzymatic activity. D_1 is present in high concentrations in the liver, kidneys, thyroid, and pituitary. It appears primarily to be responsible for monitoring the formation of T_3 from T_4 in the periphery. D_2 is present in the brain, pituitary, and brown fat. It also contributes to the formation of T_3. In the brain, it is located in astroglia and produces a supply of T_3 to neurons. D_3 is also present in the brain and in reproductive tissues. It acts only on the 5 position of T_4 and T_3 and is probably the main source of RT_3 in the blood and tissues. Overall, the deiodinases appear to be responsible for maintaining differences in T_3/T_4 ratios in the various tissues in the body. In the brain, in particular, high levels of deiodinase activity ensure an ample supply of active T_3.

Some of the T_4 and T_3 is further converted to deiodotyrosines by deiodinases. T_4 and T_3 are also conjugated in the liver to form sulfates and glucuronides. These conjugates enter the bile and pass into the intestine. The thyroid conjugates are hydrolyzed, and some are reabsorbed (enterohepatic circulation), but some are excreted in the stool. In addition, some T_4 and T_3 passes directly from the circulation to the intestinal lumen. The iodide lost by these routes amounts to about 4% of the total daily iodide loss.

FLUCTUATIONS IN DEIODINATION

Much more RT_3 and much less T_3 are formed during fetal life, and the ratio shifts to that of adults about 6 wk after birth. Various drugs inhibit deiodinases, producing a fall in plasma T_3

TABLE 20–2 Effect of variations in the concentrations of thyroid hormone-binding proteins in the plasma on various parameters of thyroid function after equilibrium has been reached.

Condition	Concentrations of Binding Proteins	Total Plasma T_4, T_3, RT_3	Free Plasma T_4, T_3, RT_3	Plasma TSH	Clinical State
Hyperthyroidism	Normal	High	High	Low	Hyperthyroid
Hypothyroidism	Normal	Low	Low	High	Hypothyroid
Estrogens, methadone, heroin, major tranquilizers, clofibrate	High	High	Normal	Normal	Euthyroid
Glucocorticoids, androgens, danazol, asparaginase	Low	Low	Normal	Normal	Euthyroid

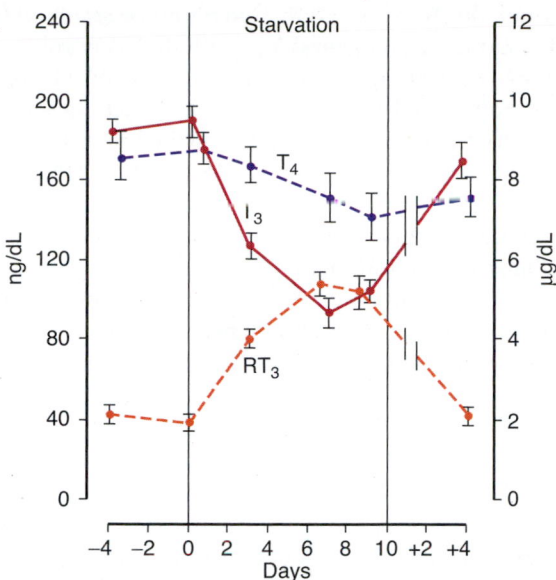

FIGURE 20–9 **Effect of starvation on plasma levels of T_4, T_3, and RT_3 in humans.** Similar changes occur in wasting diseases. The scale for T_3 and RT_3 is on the left and the scale for T_4 is on the right. (Reproduced with permission from Burger AG: New aspects of the peripheral action of thyroid hormones. Triangle, Sandoz J Med Sci 1983;22:175. Copyright © 1983 Sandoz Ltd., Basel, Switzerland.)

levels and a reciprocal rise in RT_3. Selenium deficiency has the same effect. A wide variety of nonthyroidal illnesses also suppress deiodinases. These include burns, trauma, advanced cancer, cirrhosis, renal failure, myocardial infarction, and febrile states. The low-T_3 state produced by these conditions disappears with recovery. It is difficult to decide whether individuals with the low-T_3 state produced by drugs and illness have mild hypothyroidism.

Diet also has a clear-cut effect on conversion of T_4 to T_3. In fasted individuals, plasma T_3 is reduced by 10–20% within 24 h and by about 50% in 3 to 7 d, with a corresponding rise in RT_3 (Figure 20–9). Free and bound T_4 levels remain essentially normal. During more prolonged starvation, RT_3 returns to normal but T_3 remains depressed. At the same time, the basal metabolic rate (BMR) falls and urinary nitrogen excretion, an index of protein breakdown, is decreased. Thus, the decline in T_3 conserves calories and protein. Conversely, overfeeding increases T_3 and reduces RT_3.

REGULATION OF THYROID SECRETION

Thyroid function is regulated primarily by variations in the circulating level of pituitary TSH (Figure 20–8). TSH secretion is increased by the hypothalamic hormone thyrotropin-releasing hormone (TRH; see Chapter 18) and inhibited in a negative feedback fashion by circulating free T_4 and T_3. The effect of T_4 is enhanced by production of T_3 in the cytoplasm of the pituitary cells by the 5′-D_2 they contain. TSH secretion is also

inhibited by stress, and in experimental animals it is increased by cold and decreased by warmth.

CHEMISTRY & METABOLISM OF TSH

Human TSH is a glycoprotein that contains 211 amino acid residues. It is made up of two subunits, designated α and β. The α subunit is encoded by a gene on chromosome 6 and the β subunit by a gene on chromosome 1. The α and β subunits become noncovalently linked in the pituitary thyrotropes. TSH-α is identical to the α subunit of LH, FSH, and hCG-α (see Chapters 24 and 25). The functional specificity of TSH is conferred by the β subunit. The structure of TSH varies from species to species, but other mammalian TSHs are biologically active in humans.

The biologic half-life of human TSH is about 60 min. TSH is degraded for the most part in the kidneys and to a lesser extent in the liver. Secretion is pulsatile, and mean output starts to rise at about 9:00 PM, peaks at midnight, and then declines during the day. The normal secretion rate is about 110 μg/d. The average plasma level is about 2 μg/mL (Figure 20–10).

Because the α subunit in hCG is the same as that in TSH, large amounts of hCG can activate thyroid receptors nonspecifically. In some patients with benign or malignant tumors of placental origin, plasma hCG levels can rise so high that they produce mild hyperthyroidism.

EFFECTS OF TSH ON THE THYROID

When the pituitary is removed, thyroid function is depressed and the gland atrophies; when TSH is administered, thyroid

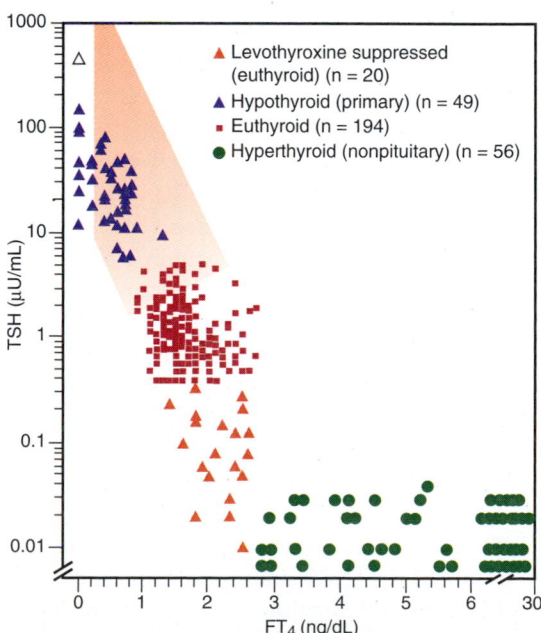

FIGURE 20–10 **Relation between plasma TSH, measured by a highly sensitive radioimmunoassay, and plasma free T_4, measured by dialysis (FT_4).** Note that the TSH scale is a log scale.

function is stimulated. Within a few minutes after the injection of TSH, there are increases in iodide binding; synthesis of T_3, T_4, and iodotyrosines; secretion of thyroglobulin into the colloid; and endocytosis of colloid. Iodide trapping is increased in a few hours; blood flow increases; and, with chronic TSH treatment, the cells hypertrophy and the weight of the gland increases.

Whenever TSH stimulation is prolonged, the thyroid becomes detectably enlarged. Enlargement of the thyroid is called a **goiter.**

TSH RECEPTORS

The TSH receptor is a typical G protein-coupled, seven-transmembrane segment receptor that activates adenylyl cyclase through G_s. It also activates phospholipase C (PLC). Like other glycoprotein hormone receptors, it has an extended, glycosylated extracellular domain.

OTHER FACTORS AFFECTING THYROID GROWTH

In addition to TSH receptors, thyrocytes express receptors for insulin-like growth factor I (IGF-I), EGF, and other growth factors. IGF-I and EGF promote growth, whereas interferon γ and tumor necrosis factor α inhibit growth. The exact physiologic role of these factors in the thyroid has not been established, but the effect of the cytokines implies that thyroid function might be inhibited in the setting of chronic inflammation, which could contribute to cachexia, or weight loss.

CONTROL MECHANISMS

The mechanisms regulating thyroid secretion are summarized in Figure 20–8. The negative feedback effect of thyroid hormones on TSH secretion is exerted in part at the hypothalamic level, but it is also due in large part to an action on the pituitary, since T_4 and T_3 block the increase in TSH secretion produced by TRH. Infusion of either T_4 or T_3 reduces the circulating level of TSH, which declines measurably within 1 hour. In experimental animals, there is an initial rise in pituitary TSH content before the decline, indicating that thyroid hormones inhibit secretion before they inhibit synthesis. The effects on secretion and synthesis of TSH both appear to depend on protein synthesis, even though the former is relatively rapid.

The day-to-day maintenance of thyroid secretion depends on the feedback interplay of thyroid hormones with TSH and TRH (Figure 20–8). The adjustments that appear to be mediated via TRH include the increased secretion of thyroid hormones produced by cold and, presumably, the decrease produced by heat. It is worth noting that although cold produces clear-cut increases in circulating TSH in experimental animals and human infants, the rise produced by cold in adult humans is negligible. Consequently, in adults, increased heat

production due to increased thyroid hormone secretion (**thyroid hormone thermogenesis**) plays little if any role in the response to cold. Stress has an inhibitory effect on TRH secretion. Dopamine and somatostatin act at the pituitary level to inhibit TSH secretion, but it is not known whether they play a physiologic role in the regulation of TSH secretion. Glucocorticoids also inhibit TSH secretion.

The amount of thyroid hormone necessary to maintain normal cellular function in thyroidectomized individuals used to be defined as the amount necessary to normalize the BMR, but it is now defined as the amount necessary to return plasma TSH to normal. Indeed, with the accuracy and sensitivity of modern assays for TSH and the marked inverse correlation between plasma free thyroid hormone levels and plasma TSH, measurement of TSH is now widely regarded as one of the best tests of thyroid function. The amount of T_4 that normalizes plasma TSH in athyreotic individuals averages 112 μg of T_4 by mouth per day in adults. About 80% of this dose is absorbed from the gastrointestinal tract. It produces a slightly greater than normal FT_4I but a normal FT_3I, indicating that in humans, unlike some experimental animals, it is circulating T_3 rather than T_4 that is the principal feedback regulator of TSH secretion (see Clinical Boxes 20–1 and 20–2).

EFFECTS OF THYROID HORMONES

Some of the widespread effects of thyroid hormones in the body are secondary to stimulation of O_2 consumption (**calorigenic action**), although the hormones also affect growth and development in mammals, help regulate lipid metabolism, and increase the absorption of carbohydrates from the intestine (Table 20–5 on page 311). They also increase the dissociation of oxygen from hemoglobin by increasing red cell 2,3-diphosphoglycerate (DPG) (see Chapter 36).

MECHANISM OF ACTION

Thyroid hormones enter cells and T_3 binds to thyroid receptors (TR) in the nuclei. T_4 can also bind, but not as avidly. The hormone-receptor complex then binds to DNA via zinc fingers and increases (or in some cases, decreases) the expression of a variety of different genes that code for proteins that regulate cell function (see Chapter 1). Thus, the nuclear receptors for thyroid hormones are members of the superfamily of hormone-sensitive nuclear transcription factors.

There are two human TR genes: an α receptor gene on chromosome 17 and a β receptor gene on chromosome 3. By alternative splicing, each forms at least two different mRNAs and therefore two different receptor proteins. TRβ2 is found only in the brain, but TRα1, TRα2, and TRβ1 are widely distributed. TRα2 differs from the other three in that it does not bind T3 and its function is not yet fully established. TRs bind to DNA as monomers, homodimers, and heterodimers with other nuclear receptors, particularly the retinoid X receptor

CLINICAL BOX 20–1

Reduced Thyroid Function

The syndrome of adult **hypothyroidism** is generally called **myxedema,** although this term is also used to refer specifically to the skin changes in the syndrome. Hypothyroidism may be the end result of a number of diseases of the thyroid gland, or it may be secondary to pituitary or hypothalamic failure. In the latter two conditions, the thyroid remains able to respond to TSH. Thyroid function may be reduced by a number of conditions (Table 20–3). For example, when the dietary iodine intake falls below 50 μg/d, thyroid hormone synthesis is inadequate and secretion declines. As a result of increased TSH secretion, the thyroid hypertrophies, producing an **iodine deficiency goiter** that may become very large. Such "endemic goiters" have been substantially reduced by the practice of adding iodide to table salt. Drugs may also inhibit thyroid function. Most do so either by interfering with the iodide-trapping mechanism or by blocking the organic binding of iodine. In either case, TSH secretion is stimulated by the decline in circulating thyroid hormones, and a goiter is produced. The **thioureylenes,** a group of compounds related to thiourea, inhibit the iodination of monoiodotyrosine and block the coupling reaction. The two used clinically are propylthiouracil and methimazole (Figure 20–11). Iodination of tyrosine is inhibited because propylthiouracil and methimazole compete with tyrosine residues for iodine and become iodinated. In addition, propylthiouracil but not methimazole inhibits D_2 deiodinase, reducing the conversion of T_4 to T_3 in many extrathyroidal tissues. Paradoxically, another substance that inhibits thyroid function under certain conditions is iodide itself. In normal individuals, large doses of iodide act directly on the thyroid to produce a mild and transient inhibition of organic binding of iodide and hence of hormone synthesis. This inhibition is known as the **Wolff–Chaikoff effect.**

In completely athyreotic adults, the BMR falls to about 40%. The hair is coarse and sparse, the skin is dry and yellowish (carotenemia), and cold is poorly tolerated. Mentation is slow, memory is poor, and in some patients there are severe mental symptoms ("myxedema madness"). Plasma cholesterol is elevated. Children who are hypothyroid from birth or before are called **cretins.** They are dwarfed and mentally retarded. Worldwide, congenital hypothyroidism is one of the most common causes of preventable mental retardation. The main causes are included in Table 20–3. They include not only maternal iodine deficiency and various congenital abnormalities of the fetal hypothalamo–pituitary–thyroid axis, but also maternal antithyroid antibodies that cross the placenta and damage the fetal thyroid. T_4 crosses the placenta, and unless the mother is hypothyroid, growth and development are normal until birth. If treatment is started at birth, the prognosis for normal growth and development is good, and mental retardation can generally be avoided; for this reason, screening tests for congenital hypothyroidism are becoming routine. When the mother is hypothyroid as well, as in the case of iodine deficiency, the mental deficiency is more severe and less responsive to treatment after birth. It has been estimated that 20 million people in the world now have various degrees of brain damage caused by iodine deficiency in utero.

Uptake of tracer doses of radioactive iodine can be used to assess thyroid function (contrast this with the use of large doses to ablate thyroid tissue in cases of hyperthyroidism (Clinical Box 20–2). An analysis of the kinetics of iodine handling also provides insights into the basic physiology of the gland (Figure 20–12).

(RXR). The TR/RXR heterodimer does not bind 9-*cis* retinoic acid, the usual ligand for RXR, but TR binding to DNA is greatly enhanced in response to thyroid hormones when the receptor is in the form of this heterodimer. There are also coactivator and corepressor proteins that affect the actions of

TRs. Presumably, this complexity underlies the ability of thyroid hormones to produce many different effects in the body.

In most of its actions, T_3 acts more rapidly and is three to five times more potent than T_4 (Figure 20–13). This is because T_3 is less tightly bound to plasma proteins than is T_4, but binds more avidly to thyroid hormone receptors. RT_3 is inert (see Clinical Box 20–3).

TABLE 20–3 Causes of congenital hypothyroidism.

Maternal iodine deficiency
Fetal thyroid dysgenesis
Inborn errors of thyroid hormone synthesis
Maternal antithyroid antibodies that cross the placenta
Fetal hypopituitary hypothyroidism

FIGURE 20–11 Structure of commonly used thioureylenes.

CLINICAL BOX 20–2

Hyperthyroidism

The symptoms of an overactive thyroid gland follow logically from the actions of thyroid hormone discussed in this chapter. Thus, hyperthyroidism is characterized by nervousness; weight loss; hyperphagia; heat intolerance; increased pulse pressure; a fine tremor of the outstretched fingers; warm, soft skin; sweating; and a BMR from +10 to as high as +100. It has various causes (Table 20–4); however, the most common cause is **Graves disease (Graves hyperthyroidism),** which accounts for 60–80% of the cases. This is an autoimmune disease, more common in women, in which antibodies to the TSH receptor stimulate the receptor. This produces marked T_4 and T_3 secretion and enlargement of the thyroid gland (goiter). However, due to the feedback effects of T_4 and T_3, plasma TSH is low, not high. Another hallmark of Graves disease is the occurrence of swelling of tissues in the orbits, producing protrusion of the eyeballs **(exophthalmos).** This occurs in 50% of patients and often precedes the development of obvious hyperthyroidism. Other antithyroid antibodies are present in Graves disease, including antibodies to thyroglobulin and thyroid peroxidase. In Hashimoto thyroiditis, autoimmune antibodies ultimately destroy the thyroid, but during the early stage the inflammation of the gland causes excess thyroid hormone secretion and thyrotoxicosis similar to that seen in Graves disease. In general, some of the symptoms of hyperthyroidism can be controlled by the thioureylene drugs discussed above, or by the administration of radioactive iodine that destroys part of the gland.

FIGURE 20–12   Distribution of radioactive iodine in individuals on a relatively low-iodine diet. Percentages are plotted against time after an oral dose of radioactive iodine. In hyperthyroidism, plasma radioactivity falls rapidly and then rises again as a result of release of labeled T_4 and T_3 from the thyroid.

CALORIGENIC ACTION

T_4 and T_3 increase the O_2 consumption of almost all metabolically active tissues. The exceptions are the adult brain, testes, uterus, lymph nodes, spleen, and anterior pituitary. T_4 actually depresses the O_2 consumption of the anterior pituitary, presumably because it inhibits TSH secretion. The increase in metabolic rate produced by a single dose of T_4 becomes measurable after a latent period of several hours and lasts 6 days or more.

Some of the calorigenic effect of thyroid hormones is due to metabolism of the fatty acids they mobilize. In addition, thyroid hormones increase the activity of the membrane-bound Na, K ATPase in many tissues.

EFFECTS SECONDARY TO CALORIGENESIS

When the metabolic rate is increased by T_4 and T_3 in adults, nitrogen excretion is increased; if food intake is not increased, endogenous protein and fat stores are catabolized and weight is lost. In hypothyroid children, small doses of thyroid hormones cause a positive nitrogen balance because they stimulate growth, but large doses cause protein catabolism similar to that produced in the adult. The potassium liberated during protein catabolism appears in the urine, and there is also an increase in urinary hexosamine and uric acid excretion.

When the metabolic rate is increased, the need for all vitamins is increased and vitamin deficiency syndromes may be precipitated. Thyroid hormones are necessary for hepatic conversion of carotene to vitamin A, and the accumulation of carotene in the bloodstream **(carotenemia)** in hypothyroidism is responsible for the yellowish tint of the skin. Carotenemia can

TABLE 20–4 Causes of hyperthyroidism.

Thyroid overactivity
Solitary toxic adenoma
Toxic multinodular goiter
Hashimoto thyroiditis
TSH-secreting pituitary tumor
Mutations causing constitutive activation of TSH receptor
Other rare causes
Extrathyroidal
Administration of T_3 or T_4 (factitious or iatrogenic hyperthyroidism)
Ectopic thyroid tissue

be distinguished from jaundice because in the former condition the scleras are not yellow.

The skin normally contains a variety of proteins combined with polysaccharides, hyaluronic acid, and chondroitin sulfu-

TABLE 20–5 Physiologic effects of thyroid hormones.

Target Tissue	Effect	Mechanism
Heart	Chronotropic	Increased number of β-adrenergic receptors
	Inotropic	Enhanced responses to circulating catecholamines
		Increased proportion of α-myosin heavy chain (with higher ATPase activity)
Adipose tissue	Catabolic	Stimulated lipolysis
Muscle	Catabolic	Increased protein breakdown
Bone	Developmental	Promote normal growth and skeletal development
Nervous system	Developmental	Promote normal brain development
Gut	Metabolic	Increased rate of carbohydrate absorption
Lipoprotein	Metabolic	Formation of LDL receptors
Other	Calorigenic	Stimulated oxygen consumption by metabolically active tissues (exceptions: testes, uterus, lymph nodes, spleen, anterior pituitary)
		Increased metabolic rate

Modified and reproduced with permission from McPhee SJ, Lingarra VR, Ganong WF (editors): *Pathophysiology of Disease*, 4th ed. McGraw-Hill, 2003.

CLINICAL BOX 20–3

Thyroid Hormone Resistance

Some mutations in the gene that codes for TRβ are associated with resistance to the effects of T_3 and T_4. Most commonly, there is resistance to thyroid hormones in the peripheral tissues and the anterior pituitary gland. Patients with this abnormality are usually not clinically hypothyroid, because they maintain plasma levels of T_3 and T_4 that are high enough to overcome the resistance, and hTRα is unaffected. However, plasma TSH is inappropriately high given the high circulating T_3 and T_4 levels and is difficult to suppress with exogenous thyroid hormone. Some patients have thyroid hormone resistance only in the pituitary. They have hypermetabolism and elevated plasma T_3 and T_4 levels with normal, nonsuppressible levels of TSH. A few patients apparently have peripheral resistance with normal pituitary sensitivity. They have hypometabolism despite normal plasma levels of T_3, T_4, and TSH, and they require large doses of thyroid hormones to increase their metabolic rate. An interesting finding is that **attention deficit hyperactivity disorder,** a condition frequently diagnosed in children who are overactive and impulsive, is much more common in individuals with thyroid hormone resistance than in the general population. This suggests that hTRβ may play a special role in brain development.

ric acid. In hypothyroidism, these complexes accumulate, promoting water retention and the characteristic puffiness of the skin (myxedema). When thyroid hormones are administered, the proteins are metabolized, and diuresis continues until the myxedema is cleared.

Milk secretion is decreased in hypothyroidism and stimulated by thyroid hormones, a fact sometimes put to practical use in the dairy industry. Thyroid hormones do not stimulate

FIGURE 20–13 **Calorigenic responses of thyroidectomized rats to subcutaneous injections of T_4 and T_3.** (Redrawn and reproduced with permission from Barker SB: Peripheral actions of thyroid hormones. Fed Proc 1962;21:635.)

the metabolism of the uterus but are essential for normal menstrual cycles and fertility.

EFFECTS ON THE CARDIOVASCULAR SYSTEM

Large doses of thyroid hormones cause enough extra heat production to lead to a slight rise in body temperatures (Chapter 18), which in turn activates heat-dissipating mechanisms. Peripheral resistance decreases because of cutaneous vasodilation, and this increases levels of renal Na^+ and water absorption, expanding blood volume. Cardiac output is increased by the direct action of thyroid hormones, as well as that of catecholamines, on the heart, so that pulse pressure and cardiac rate are increased and circulation time is shortened.

T_3 is not formed from T_4 in myocytes to any degree, but circulatory T_3 enters the myocytes, combines with its receptors, and enters the nucleus, where it promotes the expression of some genes and inhibits the expression of others. Those that are enhanced include the genes for α-myosin heavy chain, sarcoplasmic reticulum Ca^{2+} ATPase, β-adrenergic receptors, G proteins, Na, K ATPase, and certain K^+ channels. Those that are inhibited include the genes for β-myosin heavy chain, phospholamban, two types of adenylyl cyclase, T_3 nuclear receptors, and NCX, the Na^+–Ca^{2+} exchanger. The net result is increased heart rate and force of contraction.

The heart contains two myosin heavy chain (MHC) isoforms, α-MHC and β-MHC. They are encoded by two highly homologous genes located on the short arm of chromosome 17. Each myosin molecule consists of two heavy chains and two pairs of light chains (see Chapter 5). The myosin containing β-MHC has less ATPase activity than the myosin containing α-MHC. α-MHC predominates in the atria in adults, and its level is increased by treatment with thyroid hormone. This increases the speed of cardiac contraction. Conversely, expression of the α-MHC gene is depressed and that of the β-MHC gene is enhanced in hypothyroidism.

EFFECTS ON THE NERVOUS SYSTEM

In hypothyroidism, mentation is slow and the cerebrospinal fluid (CSF) protein level elevated. Thyroid hormones reverse these changes, and large doses cause rapid mentation, irritability, and restlessness. Overall, cerebral blood flow and glucose and O_2 consumption by the brain are normal in adult hypo- and hyperthyroidism. However, thyroid hormones enter the brain in adults and are found in gray matter in numerous different locations. In addition, astrocytes in the brain convert T_4 to T_3, and there is a sharp increase in brain D_2 activity after thyroidectomy that is reversed within 4 h by a single intravenous dose of T_3. Some of the effects of thyroid hormones on the brain are probably secondary to increased responsiveness to catecholamines, with consequent increased activation of the reticular activating system (see Chapter 15). In addition, thyroid hormones have marked effects on brain development. The parts of the central nervous system (CNS) most affected are the cerebral cortex and the basal ganglia. In addition, the cochlea is also affected. Consequently, thyroid hormone deficiency during development causes mental retardation, motor rigidity, and deaf–mutism. Deficiencies in thyroid hormone synthesis secondary to a failure of thyrocytes to transport iodide presumably also contribute to deafness in Pendred syndrome, discussed above.

Thyroid hormones also exert effects on reflexes. The reaction time of stretch reflexes (see Chapter 9) is shortened in hyperthyroidism and prolonged in hypothyroidism. Measurement of the reaction time of the ankle jerk (Achilles reflex) has attracted attention as a clinical test for evaluating thyroid function, but this reaction time is also affected by other diseases and thus is not a specific assessment of thyroid activity.

RELATION TO CATECHOLAMINES

The actions of thyroid hormones and the catecholamines norepinephrine and epinephrine are intimately interrelated. Epinephrine increases the metabolic rate, stimulates the nervous system, and produces cardiovascular effects similar to those of thyroid hormones, although the duration of these actions is brief. Norepinephrine has generally similar actions. The toxicity of the catecholamines is markedly increased in rats treated with T_4. Although plasma catecholamine levels are normal in hyperthyroidism, the cardiovascular effects, tremulousness, and sweating produced by thyroid hormones can be reduced or abolished by sympathectomy. They can also be reduced by drugs such as propranolol that block β-adrenergic receptors. Indeed, propranolol and other β blockers are used extensively in the treatment of thyrotoxicosis and in the treatment of the severe exacerbations of hyperthyroidism called **thyroid storms.** However, even though β blockers are weak inhibitors of extrathyroidal conversion of T_4 to T_3, and consequently may produce a small fall in plasma T_3, they have little effect on the other actions of thyroid hormones. Presumably, the functional synergism observed between catecholamines and thyroid hormones, particularly in pathological settings, arises from their overlapping biological functions as well as the ability of thyroid hormones to increase expression of catecholamine receptors and the signaling effectors to which they are linked.

EFFECTS ON SKELETAL MUSCLE

Muscle weakness occurs in most patients with hyperthyroidism **(thyrotoxic myopathy),** and when the hyperthyroidism is severe and prolonged, the myopathy may be severe. The muscle weakness may be due in part to increased protein catabolism. Thyroid hormones affect the expression of the MHC genes in skeletal as well as cardiac muscle (see Chapter 5). However, the effects produced are complex and their relation to the myopathy is not established. Hypothyroidism is also associated with muscle weakness, cramps, and stiffness.

EFFECTS ON CARBOHYDRATE METABOLISM

Thyroid hormones increase the rate of absorption of carbohydrates from the gastrointestinal tract, an action that is probably independent of their calorigenic action. In hyperthyroidism, therefore, the plasma glucose level rises rapidly after a carbohydrate meal, sometimes exceeding the renal threshold. However, it falls again at a rapid rate.

EFFECTS ON CHOLESTEROL METABOLISM

Thyroid hormones lower circulating cholesterol levels. The plasma cholesterol level drops before the metabolic rate rises, which indicates that this action is independent of the stimulation of O_2 consumption. The decrease in plasma cholesterol concentration is due to increased formation of low-density lipoprotein (LDL) receptors in the liver, resulting in increased hepatic removal of cholesterol from the circulation. Despite considerable effort, however, it has not been possible to produce a clinically useful thyroid hormone analog that lowers plasma cholesterol without increasing metabolism.

EFFECTS ON GROWTH

Thyroid hormones are essential for normal growth and skeletal maturation (see Chapter 23). In hypothyroid children, bone growth is slowed and epiphysial closure delayed. In the absence of thyroid hormones, growth hormone secretion is also depressed. This further impairs growth and development, since thyroid hormones normally potentiate the effect of growth hormone on tissues.

CHAPTER SUMMARY

- The thyroid gland transports and fixes iodide to amino acids present in thyroglobulin to generate the thyroid hormones thyroxine (T_4) and triiodothyronine (T_3).
- Synthesis and secretion of thyroid hormones is stimulated by thyroid-stimulating hormone (TSH) from the pituitary, which in turn is released in response to thyrotropin-releasing hormone (TRH) from the hypothalamus. These releasing factors are controlled by changes in whole body status (eg, exposure to cold or stress).
- Thyroid hormones circulate in the plasma predominantly in protein-bound forms. Only the free hormones are biologically active, and both feed back to reduce secretion of TSH.
- Thyroid hormones exert their effects by entering cells and binding to thyroid receptors. The liganded forms of thyroid receptors are nuclear transcription factors that alter gene expression.
- Thyroid hormones stimulate metabolic rate, calorigenesis, cardiac function, and normal mentation, and interact synergistically with catecholamines. Thyroid hormones also play critical roles in development, particularly of the nervous system, and growth.
- Disease results with both under- and overactivity of the thyroid gland. Hypothyroidism is accompanied by mental and physical slowing in adults, and by mental retardation and dwarfism if it occurs in neonatal life. Overactivity of the thyroid gland, which most commonly is caused by autoantibodies that trigger secretion (Graves disease) results in body wasting, nervousness, and tachycardia.

MULTIPLE-CHOICE QUESTIONS

For all questions, select the single best answer unless otherwise directed.

1. In which of the following conditions is it *most* likely that the TSH response to TRH will be reduced?
   A) hypothyroidism due to tissue resistance to thyroid hormone
   B) hypothyroidism due to disease destroying the thyroid gland
   C) hyperthyroidism due to circulating antithyroid antibodies with TSH activity
   D) hyperthyroidism due to diffuse hyperplasia of thyrotropes of the anterior pituitary
   E) iodine deficiency

2. A young woman has puffy skin and a hoarse voice. Her plasma TSH concentration is low but increases markedly when she is given TRH. She probably has
   A) hyperthyroidism due to a thyroid tumor.
   B) hypothyroidism due to a primary abnormality in the thyroid gland.
   C) hypothyroidism due to a primary abnormality in the pituitary gland.
   D) hypothyroidism due to a primary abnormality in the hypothalamus.
   E) hyperthyroidism due to a primary abnormality in the hypothalamus.

3. The enzyme primarily responsible for the conversion of T_4 to T_3 in the periphery is
   A) D_1 thyroid deiodinase.
   B) D_2 thyroid deiodinase.
   C) D_3 thyroid deiodinase.
   D) thyroid peroxidase.
   E) none of the above

4. The metabolic rate is *least* affected by an increase in the plasma level of
   A) TSH.
   B) TRH.
   C) TBG.
   D) free T_4.
   E) free T_3.

5. Which of the following is *not* essential for normal biosynthesis of thyroid hormones?
   A) iodine
   B) ferritin
   C) thyroglobulin
   D) protein synthesis
   E) TSH

6. Which of the following would be *least* affected by injections of TSH?

    A) thyroidal uptake of iodine
    B) synthesis of thyroglobulin
    C) cyclic adenosine monophosphate (AMP) in thyroid cells
    D) cyclic guanosine monophosphate (GMP) in thyroid cells
    E) size of the thyroid

7. Hypothyroidism due to disease of the thyroid gland is associated with increased plasma levels of

    A) cholesterol.
    B) albumin.
    C) RT_3.
    D) iodide.
    E) TBG.

8. Thyroid hormone receptors bind to DNA in which of the following forms?

    A) a heterodimer with the prolactin receptor
    B) a heterodimer with the growth hormone receptor
    C) a heterodimer with the retinoid X receptor
    D) a heterodimer with the insulin receptor
    E) a heterodimer with the progesterone receptor

9. Increasing intracellular I^- due to the action of NIS is an example of

    A) endocytosis.
    B) passive diffusion.
    C) Na^+ and K^+ cotransport.
    D) primary active transport.
    E) secondary active transport.

CHAPTER RESOURCES

Brent GA: Graves' disease. N Engl J Med 2008;358:2594.

Dohan O, Carrasco N: Advances in Na^+/I^- symporter (NIS) research in the thyroid and beyond. Mol Cell Endocrinol 2003;213:59.

Glaser B: Pendred syndrome. Pediatr Endocrinol Rev 2003;1(Suppl 2):199.

Peeters RP, van der Deure WM, Visser TJ: Genetic variation in thyroid hormone pathway genes: Polymorphisms in the TSH receptor and the iodothyronine deiodinases. Eur J Endocrinol 2006;155:655.

Endocrine Functions of the Pancreas & Regulation of Carbohydrate Metabolism

OBJECTIVES

After reading this chapter, you should be able to:

- List the hormones that affect the plasma glucose concentration and briefly describe the action of each.
- Describe the structure of the pancreatic islets and name the hormones secreted by each of the cell types in the islets.
- Describe the structure of insulin and outline the steps involved in its biosynthesis and release into the bloodstream.
- List the consequences of insulin deficiency and explain how each of these abnormalities is produced.
- Describe insulin receptors, the way they mediate the effects of insulin, and the way they are regulated.
- Describe the types of glucose transporters found in the body and the function of each.
- List the major factors that affect the secretion of insulin.
- Describe the structure of glucagon and other physiologically active peptides produced from its precursor.
- List the physiologically significant effects of glucagon and the factors that regulate glucagon secretion.
- Describe the physiologic effects of somatostatin in the pancreas.
- Outline the mechanisms by which thyroid hormones, adrenal glucocorticoids, catecholamines, and growth hormone affect carbohydrate metabolism.
- Understand the major differences between type 1 and type 2 diabetes.

INTRODUCTION

At least four polypeptides with regulatory activity are secreted by the islets of Langerhans in the pancreas. Two of these are hormones **insulin** and **glucagon,** and have important functions in the regulation of the intermediary metabolism of carbohydrates, proteins, and fats. The third polypeptide, **somatostatin,** plays a role in the regulation of islet cell secretion, and the fourth, **pancreatic polypeptide,** is probably concerned primarily with the regulation of HCO_3^- secretion to the intestine. Glucagon, somatostatin, and possibly pancreatic polypeptide are also secreted by cells in the mucosa of the gastrointestinal tract.

Insulin is anabolic, increasing the storage of glucose, fatty acids, and amino acids. Glucagon is catabolic, mobilizing glucose, fatty acids, and the amino acids from stores into the bloodstream. The two hormones are thus reciprocal in their overall action and are reciprocally secreted in most circumstances. Insulin excess causes hypoglycemia, which leads to convulsions and coma. Insulin deficiency, either absolute or relative, causes **diabetes mellitus** (chronic elevated blood glucose), a complex and debilitating disease that if untreated is eventually fatal. Glucagon deficiency can cause hypoglycemia,

315

and glucagon excess makes diabetes worse. Excess pancreatic production of somatostatin causes hyperglycemia and other manifestations of diabetes.

A variety of other hormones also have important roles in the regulation of carbohydrate metabolism.

ISLET CELL STRUCTURE

The islets of Langerhans (Figure 21–1) are ovoid, 76- × 175-μm collections of cells. The islets are scattered throughout the pancreas, although they are more plentiful in the tail than in the body and head. β-islets make up about 2% of the volume of the gland, whereas the exocrine portion of the pancreas (see Chapter 26) makes up 80%, and ducts and blood vessels make up the remainder. Humans have 1 to 2 million islets. Each has a copious blood supply; blood from the islets, like that from the gastrointestinal tract (but unlike that from any other endocrine organs) drains into the hepatic portal vein.

The cells in the islets can be divided into types on the basis of their staining properties and morphology. Humans have at least four distinct cell types: A, B, D, and F cells. A, B, and D cells are also called α, β, and δ cells. However, this leads to confusion in view of the use of Greek letters to refer to other structures in the body, particularly adrenergic receptors (see Chapter 7). The A cells secrete glucagon, the B cells secrete insulin, the D cells secrete somatostatin, and the F cells secrete pancreatic polypeptide. The B cells, which are the most common and account for 60–75% of the cells in the islets, are generally located in the center of each islet. They tend to be surrounded by the A cells, which make up 20% of the total, and the less common D and F cells. The islets in the tail, the body, and the anterior and superior part of the head of the human pancreas have many A cells and few if any F cells in the outer rim, whereas in rats and probably in humans, the islets in the posterior part of the head of the pancreas have a relatively large number of F cells and few A cells.

The A-cell-rich (glucagon-rich) islets arise embryologically from the dorsal pancreatic bud, and the F-cell-rich (pancreatic polypeptide-rich) islets arise from the ventral pancreatic bud. These buds arise separately from the duodenum.

The B cell granules are packets of insulin in the cell cytoplasm. The shape of the packets varies from species to species; in humans, some are round whereas others are rectangular (Figure 21–2). In the B cells, the insulin molecule forms polymers and also complexes with zinc. The differences in the shape of the packets are probably due to differences in the size of polymers or zinc aggregates of insulin. The A granules, which contain glucagon, are relatively uniform from species to species (Figure 21–3). The D cells also contain large numbers of relatively homogeneous granules.

STRUCTURE, BIOSYNTHESIS, & SECRETION OF INSULIN

STRUCTURE & SPECIES SPECIFICITY

Insulin is a polypeptide containing two chains of amino acids linked by disulfide bridges (Table 21–1). Minor differences occur in the amino acid composition of the molecule from species to species. The differences are generally not sufficient to affect the biologic activity of a particular insulin

FIGURE 21–1   Islet of Langerhans in the rat pancreas. Darkly stained cells are B cells. Surrounding pancreatic acinar tissue is light-colored (× 400). (Courtesy of LL Bennett.)

FIGURE 21–2   Electron micrograph of two adjoining B cells in a human pancreatic islet. The B granules are the crystals in the membrane-lined vesicles. They vary in shape from rhombic to round (× 26,000). (Courtesy of A Like. Reproduced, with permission, from Fawcett DW: *Bloom and Fawcett, A Textbook of Histology,* 11th ed. Saunders, 1986.)

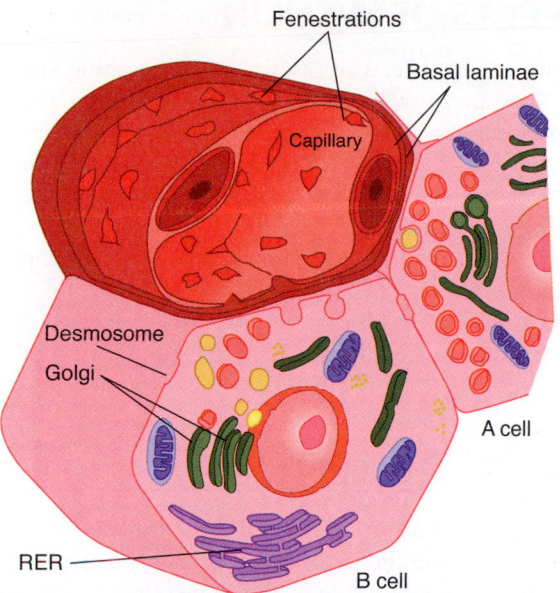

FIGURE 21–3   A and B cells, showing their relation to blood vessels. RER, rough endoplasmic reticulum. Insulin from the B cell and glucagon from the A cell are secreted by exocytosis and cross the basal lamina of the cell and the basal lamina of the capillary before entering the lumen of the fenestrated capillary. (Reproduced with permission from Junqueira IC, Carneiro J: *Basic Histology: Text and Atlas,* 10th ed. McGraw-Hill, 2003.)

in heterologous species but are sufficient to make the insulin antigenic. If insulin of one species is injected for a prolonged period into another species, the anti-insulin antibodies formed inhibit the injected insulin. Almost all humans who have received commercial bovine insulin for more than 2 mo have antibodies against bovine insulin, but the titer is usually low. Porcine insulin differs from human insulin by only one amino acid residue and has low antigenicity. Human insulin produced in bacteria by recombinant DNA technology is now widely used to avoid antibody formation.

BIOSYNTHESIS & SECRETION

Insulin is synthesized in the rough endoplasmic reticulum of the B cells (Figure 21–3). It is then transported to the Golgi apparatus, where it is packaged into membrane-bound granules. These granules move to the plasma membrane by a process involving microtubules, and their contents are expelled by exocytosis (see Chapter 2). The insulin then crosses the basal lamina of the B cell and a neighboring capillary and the fenestrated endothelium of the capillary to reach the bloodstream. The fenestrations are discussed in detail in Chapter 32.

Like other polypeptide hormones and related proteins that enter the endoplasmic reticulum, insulin is synthesized as part of a larger preprohormone (see Chapter 1). The gene for insulin is located on the short arm of chromosome 11 in humans. It has two introns and three exons. **Preproinsulin**

TABLE 21–1   Structure of human insulin (molecular weight 5808) and (below) variations in this structure in other mammalian species.[a]

Species	Variations from Human Amino Acid Sequence	
	A Chain Position 8 9 10	**B Chain Position 30**
Pig, dog, sperm whale	Thr-Ser-Ile	Ala
Rabbit	Thr-Ser-Ile	Ser
Cattle, goat	Ala-Ser-Val	Ala
Sheep	Ala-Gly-Val	Ala
Horse	Thr-Gly-Ile	Ala
Sei whale	Ala-Ser-Thr	Ala

[a] In the rat, the islet cells secrete two slightly different insulins, and in certain fish four different chains are found.

has a 23-amino-acid signal peptide removed as it enters the endoplasmic reticulum. The remainder of the molecule is then folded, and the disulfide bonds are formed to make **proinsulin.** The peptide segment connecting the A and B chains, the **connecting peptide (C peptide),** facilitates the folding and then is detached in the granules before secretion. Two proteases are involved in processing the proinsulin; to date it has no other established physiologic activity. Normally, 90–97% of the product released from the B cells is insulin along with equimolar amounts of C peptide. The rest is mostly proinsulin. C peptide can be measured by radioimmunoassay, and its level in blood provides an index of B cell function in patients receiving exogenous insulin.

FATE OF SECRETED INSULIN

INSULIN & INSULINLIKE ACTIVITY IN BLOOD

Plasma contains a number of substances with insulin-like activity in addition to insulin (Table 21–2). The activity that is not suppressed by anti-insulin antibodies has been called **nonsuppressible insulin-like activity (NSILA).** Most, if not all, of this activity persists after pancreatectomy and is due to the insulinlike growth factors **IGF-I** and **IGF-II** (see Chapter 24). These IGFs are polypeptides. Small amounts are free in the plasma (low-molecular-weight fraction), but large amounts are bound to proteins (high-molecular-weight fraction).

One may well ask why pancreatectomy causes diabetes mellitus when NSILA persists in the plasma. However, the insulinlike activities of IGF-I and IGF-II are weak compared to that of insulin and likely play other specific functions.

METABOLISM

The half-life of insulin in the circulation in humans is about 5 min. Insulin binds to insulin receptors, and some is internalized. It is destroyed by proteases in the endosomes formed by the endocytotic process.

TABLE 21–2 Substances with insulin-like activity in human plasma.

Insulin
Proinsulin
Nonsuppressible insulin-like activity (NSILA)
Low-molecular-weight fraction
IGF-I
IGF-II
High-molecular-weight fraction (mostly IGF bound to protein)

EFFECTS OF INSULIN

The physiologic effects of insulin are far-reaching and complex. They are conveniently divided into rapid, intermediate, and delayed actions, as listed in Table 21–3. The best known is the hypoglycemic effect, but there are additional effects on amino acid and electrolyte transport, many enzymes, and growth. The net effect of the hormone is storage of carbohydrate, protein, and fat. Therefore, insulin is appropriately called the "hormone of abundance."

The actions of insulin on adipose tissue; skeletal, cardiac, and smooth muscle; and the liver are summarized in Table 21–4.

GLUCOSE TRANSPORTERS

Glucose enters cells by **facilitated diffusion** (see Chapter 1) or, in the intestine and kidneys, by secondary active transport with Na^+. In muscle, adipose, and some other tissues, insulin stimulates glucose entry into cells by increasing the number of glucose transporters in the cell membranes.

The glucose transporters (GLUTs) that are responsible for facilitated diffusion of glucose across cell membranes are a family of closely related proteins that span the cell membrane 12 times and have their amino and carboxyl terminals inside the cell. They differ from and have no homology with the sodium-dependent glucose transporters, SGLT 1 and SGLT 2, responsible for the secondary active transport of glucose in the intestine (see Chapter 27) and renal tubules (see Chapter 38), although the SGLTs also have 12 transmembrane domains.

Seven different glucose transporters, named GLUT 1–7 in order of discovery, have been characterized (Table 21–5). They

TABLE 21–3 Principal actions of insulin.

Rapid (seconds)
Increased transport of glucose, amino acids, and K^+ into insulin-sensitive cells
Intermediate (minutes)
Stimulation of protein synthesis
Inhibition of protein degradation
Activation of glycolytic enzymes and glycogen synthase
Inhibition of phosphorylase and gluconeogenic enzymes
Delayed (hours)
Increase in mRNAs for lipogenic and other enzymes

Courtesy of ID Goldfine.

TABLE 21–4 Effects of insulin on various tissues.

Adipose tissue
Increased glucose entry
Increased fatty acid synthesis
Increased glycerol phosphate synthesis
Increased triglyceride deposition
Activation of lipoprotein lipase
Inhibition of hormone-sensitive lipase
Increased K⁺ uptake
Muscle
Increased glucose entry
Increased glycogen synthesis
Increased amino acid uptake
Increased protein synthesis in ribosomes
Decreased protein catabolism
Decreased release of gluconeogenic amino acids
Increased ketone uptake
Increased K⁺ uptake
Liver
Decreased ketogenesis
Increased protein synthesis
Increased lipid synthesis
Decreased glucose output due to decreased gluconeogenesis, increased glycogen synthesis, and increased glycolysis
General
Increased cell growth

contain 492 to 524 amino acid residues and their affinity for glucose varies. Each transporter appears to have evolved for special tasks. GLUT 4 is the transporter in muscle and adipose tissue that is stimulated by insulin. A pool of GLUT 4 molecules is maintained within vesicles in the cytoplasm of insulin-sensitive cells. When the insulin receptors of these cells are activated, the vesicles move rapidly to the cell membrane and fuse with it, inserting the transporters into the cell membrane (Figure 21–4). When insulin action ceases, the transporter-containing patches of membrane are endocytosed and the vesicles are ready for the next exposure to insulin. Activation of the insulin receptor brings about the movement of the vesicles to the cell membrane by activating phosphatidylinositol 3-kinase (Figure 21–4), but how this activation triggers vesicle movement is still unsettled. Most of the other GLUT trans-porters that are not insulin-sensitive appear to be constitutively expressed in the cell membrane.

In the tissues in which insulin increases the number of glucose transporters in the cell membranes, the rate of phosphorylation of the glucose, once it has entered the cells, is regulated by other hormones. Growth hormone and cortisol both inhibit phosphorylation in certain tissues. Transport is normally so rapid that it is not a rate-limiting step in glucose metabolism. However, it is rate-limiting in the B cells.

Insulin also increases the entry of glucose into liver cells, but it does not exert this effect by increasing the number of GLUT 4 transporters in the cell membranes. Instead, it induces glucokinase, and this increases the phosphorylation of glucose, so that the intracellular free glucose concentration stays low, facilitating the entry of glucose into the cell.

Insulin-sensitive tissues also contain a population of GLUT 4 vesicles that move into the cell membrane in response to exercise, a process that occurs independent of the action of insulin. This is why exercise lowers blood sugar. A 5′-AMP-activated kinase may be responsible for the insertion of these vesicles into the cell membrane.

INSULIN PREPARATIONS

The maximal decline in plasma glucose occurs 30 min after intravenous injection of insulin. After subcutaneous administration, the maximal fall occurs in 2 to 3 h. A wide variety of insulin preparations are now available commercially. These include insulins that have been complexed with protamine and other polypeptides to delay absorption and degradation, and synthetic insulins in which there have been changes in amino acid residues. In general, they fall into three categories: rapid, intermediate-acting, and long-acting (24–36 h).

RELATION TO POTASSIUM

Insulin causes K⁺ to enter cells, with a resultant lowering of the extracellular K⁺ concentration. Infusions of insulin and glucose significantly lower the plasma K⁺ level in normal individuals and are very effective for the temporary relief of hyperkalemia in patients with renal failure. **Hypokalemia** often develops when patients with diabetic acidosis are treated with insulin. The reason for the intracellular migration of K⁺ is still uncertain. However, insulin increases the activity of Na⁺–K⁺ ATPase in cell membranes, so that more K⁺ is pumped into cells.

OTHER ACTIONS

The hypoglycemic and other effects of insulin are summarized in temporal terms in Table 21–3, and the net effects on various tissues are summarized in Table 21–4. The action on glycogen synthase fosters glycogen storage, and the actions on glycolytic enzymes favor glucose metabolism to two carbon fragments (see Chapter 1), with resulting promotion of lipogenesis. Stim-

TABLE 21–5 Glucose transporters in mammals.

	Function	K_m (mM)[a]	Major Sites of Expression
Secondary active transport (Na[1]-glucose cotransport)			
SGLT 1	Absorption of glucose	0.1–1.0	Small intestine, renal tubules
SGLT 2	Absorption of glucose	1.6	Renal tubules
Facilitated diffusion			
GLUT 1	Basal glucose uptake	1–2	Placenta, blood-brain barrier, brain, red cells, kidneys, colon, many other organs
GLUT 2	B-cell glucose sensor; transport out of intestinal and renal epithelial cells	12–20	B cells of islets, liver, epithelial cells of small intestine, kidneys
GLUT 3	Basal glucose uptake	<1	Brain, placenta, kidneys, many other organs
GLUT 4	Insulin-stimulated glucose uptake	5	Skeletal and cardiac muscle, adipose tissue, other tissues
GLUT 5	Fructose transport	1–2	Jejunum, sperm
GLUT 6	None	—	Pseudogene
GLUT 7	Glucose 6-phosphate ransporter in endoplasmic reticulum	—	Liver, ? other tissues

[a]The K_m is the glucose concentration at which transport is half-maximal.
Modified from Stephens JM, Pilch PF: The metabolic regulation and vesicular transport of GLUT 4, the major insulin-responsive glucose transporter. Endocr Rev 1995;16:529.

ulation of protein synthesis from amino acids entering the cells and inhibition of protein degradation foster growth.

The anabolic effect of insulin is aided by the protein-sparing action of adequate intracellular glucose supplies. Failure to grow is a symptom of diabetes in children, and insulin stimulates the growth of immature hypophysectomized rats to almost the same degree as growth hormone.

MECHANISM OF ACTION

INSULIN RECEPTORS

Insulin receptors are found on many different cells in the body, including cells in which insulin does not increase glucose uptake.

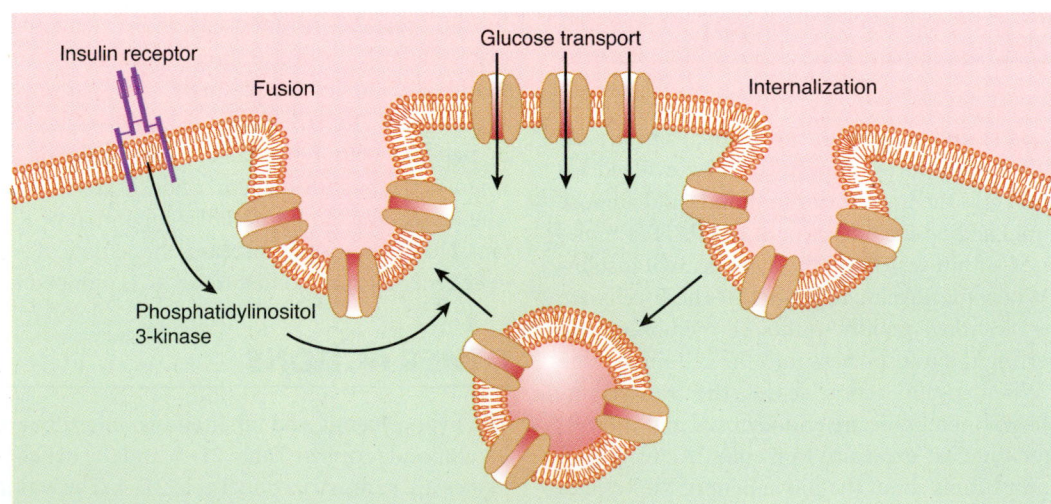

FIGURE 21–4   Cycling of GLUT 4 transporters through endosomes in insulin-sensitive tissues. Activation of the insulin receptor causes activation of phosphatidylinositol 3-kinase, which speeds translocation of the GLUT 4-containing endosomes into the cell membrane. The GLUT 4 transporters then mediate glucose transport into the cell.

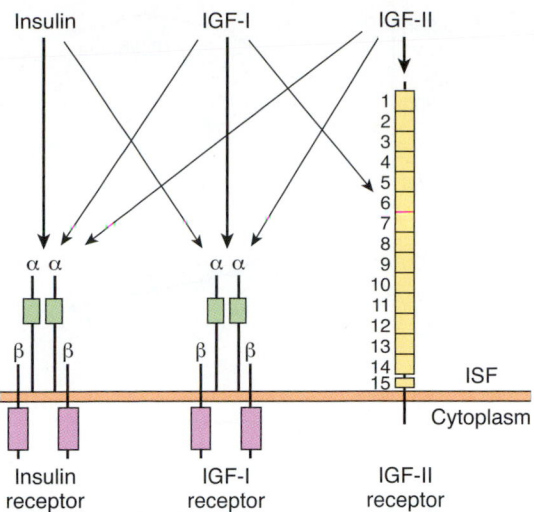

FIGURE 21–5 **Insulin, IGF-I, and IGF-II receptors.** Each hormone binds primarily to its own receptor, but insulin also binds to the IGF-I receptor, and IGF-I and IGF-II bind to all three. The purple boxes are intracellular tyrosine kinase domains. Note the marked similarity between the insulin receptor and the IGF-I receptor; also note the 15 repeat sequences in the extracellular portion of the IGF-II receptor. ISF, interstitial fluid.

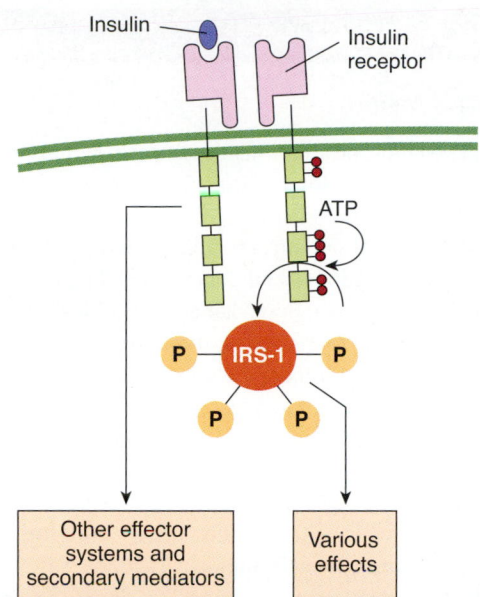

FIGURE 21–6 **Intracellular responses triggered by insulin binding to the insulin receptor.** Red balls and balls labeled P represent phosphate groups. IRS-1, insulin receptor substrate-1.

The insulin receptor, which has a molecular weight of approximately 340,000, is a tetramer made up of two α and two β glycoprotein subunits (Figure 21–5). All these are synthesized on a single mRNA and then proteolytically separated and bound to each other by disulfide bonds. The gene for the insulin receptor has 22 exons and in humans is located on chromosome 19. The α subunits bind insulin and are extracellular, whereas the β subunits span the membrane. The intracellular portions of the β subunits have tyrosine kinase activity. The α and β subunits are both glycosylated, with sugar residues extending into the interstitial fluid.

Binding of insulin triggers the tyrosine kinase activity of the β subunits, producing autophosphorylation of the β subunits on tyrosine residues. The autophosphorylation, which is necessary for insulin to exert its biologic effects, triggers phosphorylation of some cytoplasmic proteins and dephosphorylation of others, mostly on serine and threonine residues. Insulin receptor substrate (IRS-1) mediates some of the effects in humans but there are other effector systems as well (Figure 21–6). For example, mice in which the insulin receptor gene is knocked out show marked growth retardation in utero, have abnormalities of the central nervous system (CNS) and skin, and die at birth of respiratory failure, whereas IRS-1 knockouts show only moderate growth retardation in utero, survive, and are insulin-resistant but otherwise nearly normal.

The growth-promoting protein anabolic effects of insulin are mediated via **phosphatidylinositol 3-kinase (PI3K),** and evidence indicates that in invertebrates, this pathway is involved in the growth of nerve cells and axon guidance in the visual system.

It is interesting to compare the insulin receptor with other related receptors. The insulin receptor is very similar to the receptor for IGF-I but different from the receptor for IGF-II (Figure 21–5). Other receptors for growth factors and receptors for various oncogenes also are tyrosine kinases. However, the amino acid composition of these receptors is quite different.

When insulin binds to its receptors, they aggregate in patches and are taken into the cell by receptor-mediated endocytosis (see Chapter 2). Eventually, the insulin–receptor complexes enter lysosomes, where the receptors are broken down or recycled. The half-life of the insulin receptor is about 7 h.

CONSEQUENCES OF INSULIN DEFICIENCY

The far-reaching physiologic effects of insulin are highlighted by a consideration of the extensive and serious consequences of insulin deficiency (Clinical Box 21–1).

In humans, insulin deficiency is a common pathologic condition. In animals, it can be produced by pancreatectomy; by administration of alloxan, streptozocin, or other toxins that in appropriate doses cause selective destruction of the B cells of the pancreatic islets; by administration of drugs that inhibit insulin secretion; and by administration of anti-insulin antibodies. Strains of mice, rats, hamsters, guinea pigs, miniature swine, and monkeys that have a high incidence of spontaneous diabetes mellitus have also been described.

GLUCOSE TOLERANCE

In diabetes, glucose piles up in the bloodstream, especially after meals. If a glucose load is given to a diabetic, the plasma

Diabetes Mellitus

The constellation of abnormalities caused by insulin deficiency is called **diabetes mellitus.** Greek and Roman physicians used the term "diabetes" to refer to conditions in which the cardinal finding was a large urine volume, and two types were distinguished: "diabetes mellitus," in which the urine tasted sweet; and "diabetes insipidus," in which the urine had little taste. Today, the term "diabetes insipidus" is reserved for conditions in which there is a deficiency of the production or action of vasopressin (see Chapter 39), and the unmodified word "diabetes" is generally used as a synonym for diabetes mellitus.

The cause of clinical diabetes is always a deficiency of the effects of insulin at the tissue level. **Type 1 diabetes,** or **insulin-dependent diabetes mellitus (IDDM),** is due to insulin deficiency caused by autoimmune destruction of the B cells in the pancreatic islets, and it accounts for 3–5% of cases and usually presents in children. **Type 2 diabetes,** or **non-insulin-dependent diabetes mellitus (NIDDM),** is characterized by the dysregulation of insulin release from the B cells, along with insulin resistance in peripheral tissues such as skeletal muscle, brain, and liver. Type 2 diabetes usually presents in overweight or obese adults.

Diabetes is characterized by polyuria (passage of large volumes of urine), polydipsia (excessive drinking), weight loss in spite of polyphagia (increased appetite), hyperglycemia, glycosuria, ketosis, acidosis, and coma. Widespread biochemical abnormalities are present, but the fundamental defects to which most of the abnormalities can be traced are (1) reduced entry of glucose into various "peripheral" tissues and (2) increased liberation of glucose into the circulation from the liver. Therefore there is an extracellular glucose excess and, in many cells, an intracellular glucose deficiency—a situation that has been called "starvation in the midst of plenty." Also, the entry of amino acids into muscle is decreased and lipolysis is increased.

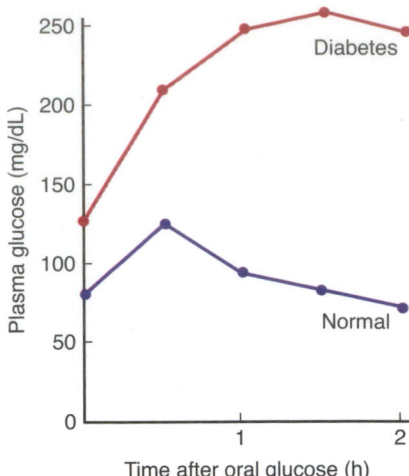

FIGURE 21–7   **Oral glucose tolerance test.** Adults are given 75 g of glucose in 300 mL of water. In normal individuals, the fasting venous plasma glucose is less than 115 mg/dL, the 2-hour value is less than 140 mg/dL, and no value is greater than 200 mg/dL. Diabetes mellitus is present if the 2-hour value and one other value are greater than 200 mg/dL. Impaired glucose tolerance is diagnosed when the values are above the upper limits of normal but below the values diagnostic of diabetes.

The second and the major cause of hyperglycemia in diabetes is derangement of the glucostatic function of the liver (see Chapter 29). The liver takes up glucose from the bloodstream and stores it as glycogen, but because the liver contains glucose 6-phosphatase it also discharges glucose into the bloodstream. Insulin facilitates glycogen synthesis and inhibits hepatic glucose output. When the plasma glucose is high, insulin secretion is normally increased and hepatic glucogenesis is decreased. This response does not occur in type 1 diabetes (as insulin is absent) and in type 2 diabetes (as tissues are insulin resistant). Glucagon can contribute to hyperglycemia as it stimulates gluconeogenesis. Glucose output by the

glucose rises higher and returns to the baseline more slowly than it does in normal individuals. The response to a standard oral test dose of glucose, the **oral glucose tolerance test,** is used in the clinical diagnosis of diabetes (Figure 21–7).

Impaired glucose tolerance in diabetes is due in part to reduced entry of glucose into cells (**decreased peripheral utilization**). In the absence of insulin, the entry of glucose into skeletal, cardiac, and smooth muscle and other tissues is decreased (Figure 21–8). Glucose uptake by the liver is also reduced, but the effect is indirect. Intestinal absorption of glucose is unaffected, as is its reabsorption from the urine by the cells of the proximal tubules of the kidneys. Glucose uptake by most of the brain and the red blood cells is also normal.

FIGURE 21–8   **Disordered plasma glucose homeostasis in insulin deficiency.** The heavy arrows indicate reactions that are accentuated. The rectangles across arrows indicate reactions that are blocked.

liver can be stimulated by catecholamines, cortisol, and growth hormone (ie, during a stress response).

EFFECTS OF HYPERGLYCEMIA

Hyperglycemia by itself can cause symptoms resulting from the hyperosmolality of the blood. In addition, there is glycosuria because the renal capacity for glucose reabsorption is exceeded. Excretion of the osmotically active glucose molecules entails the loss of large amounts of water (osmotic diuresis; see Chapter 38). The resultant dehydration activates the mechanisms regulating water intake, leading to polydipsia. There is an appreciable urinary loss of Na^+ and K^+ as well. For every gram of glucose excreted, 4.1 kcal is lost from the body. Increasing the oral caloric intake to cover this loss simply raises the plasma glucose further and increases the glycosuria, so mobilization of endogenous protein and fat stores and weight loss are not prevented.

When plasma glucose is episodically elevated over time, small amounts of hemoglobin A are nonenzymatically glycated to form **HbA$_{Ic}$** (see Chapter 32). Careful control of the diabetes with insulin reduces the amount formed and consequently HbA$_{Ic}$ concentration is measured clinically as an integrated index of diabetic control for the 4- to 6-wk period before the measurement.

The role of chronic hyperglycemia in production of the long-term complications of diabetes is discussed below.

EFFECTS OF INTRACELLULAR GLUCOSE DEFICIENCY

The plethora of glucose outside the cells in diabetes contrasts with the intracellular deficit. Glucose catabolism is normally a major source of energy for cellular processes, and in diabetes energy requirements can be met only by drawing on protein and fat reserves. Mechanisms are activated that greatly increase the catabolism of protein and fat, and one of the consequences of increased fat catabolism is ketosis.

Deficient glucose utilization and deficient hormone sensing (insulin, leptin, CCK) in the cells of the hypothalamus that regulate satiety are the probable causes of hyperphagia in diabetes. The feeding area of the hypothalamus is not inhibited and thus satiety is not sensed so food intake is increased.

Glycogen depletion is a common consequence of intracellular glucose deficit, and the glycogen content of liver and skeletal muscle in diabetic animals is usually reduced.

CHANGES IN PROTEIN METABOLISM

In diabetes, the rate at which amino acids are catabolized to CO_2 and H_2O is increased. In addition, more amino acids are converted to glucose in the liver. The increased gluconeogenesis has many causes. Glucagon stimulates gluconeogenesis, and hyperglucagonemia is generally present in diabetes. Adre-

nal glucocorticoids also contribute to increased gluconeogenesis when they are elevated in severely ill diabetics. The supply of amino acids is increased for gluconeogenesis because, in the absence of insulin, less protein synthesis occurs in muscle and hence blood amino acid levels rise. Alanine is particularly easily converted to glucose. In addition, the activity of the enzymes that catalyze the conversion of pyruvate and other two-carbon metabolic fragments to glucose is increased. These include phosphoenolpyruvate carboxykinase, which facilitates the conversion of oxaloacetate to phosphoenolpyruvate (see Chapter 1). They also include fructose 1,6-diphosphatase, which catalyzes the conversion of fructose diphosphate to fructose 6-phosphate, and glucose 6-phosphatase, which controls the entry of glucose into the circulation from the liver. Increased acetyl-CoA increases pyruvate carboxylase activity, and insulin deficiency increases the supply of acetyl-CoA because lipogenesis is decreased. Pyruvate carboxylase catalyzes the conversion of pyruvate to oxaloacetate (see Figure 1–22).

In diabetes, the net effect of accelerated protein conversion to CO_2, H_2O, and glucose, plus diminished protein synthesis, is protein depletion and wasting. Protein depletion from any cause is associated with poor "resistance" to infections.

FAT METABOLISM IN DIABETES

The principal abnormalities of fat metabolism in diabetes are acceleration of lipid catabolism, with increased formation of ketone bodies, and decreased synthesis of fatty acids and triglycerides. The manifestations of the disordered lipid metabolism are so prominent that diabetes has been called "more a disease of lipid than of carbohydrate metabolism."

Fifty percent of an ingested glucose load is normally burned to CO_2 and H_2O; 5% is converted to glycogen; and 30–40% is converted to fat in the fat depots. In diabetes, less than 5% of ingested glucose is converted to fat, despite a decrease in the amount burned to CO_2 and H_2O, and no change in the amount converted to glycogen. Therefore, glucose accumulates in the bloodstream and spills over into the urine.

The role of lipoprotein lipase and hormone-sensitive lipase in the regulation of the metabolism of fat depots is discussed in Chapter 1. In diabetes, conversion of glucose to fatty acids in the depots is decreased because of the intracellular glucose deficiency. Insulin inhibits the hormone-sensitive lipase in adipose tissue, and, in the absence of this hormone, the plasma level of **free fatty acids** (NEFA, UFA, FFA) is more than doubled. The increased glucagon also contributes to the mobilization of FFA. Thus, the FFA level parallels the plasma glucose level in diabetes and in some ways is a better indicator of the severity of the diabetic state. In the liver and other tissues, the fatty acids are catabolized to acetyl-CoA. Some of the acetyl-CoA is burned along with amino acid residues to yield CO_2 and H_2O in the citric acid cycle. However, the supply exceeds the capacity of the tissues to catabolize the acetyl-CoA.

In addition to the previously mentioned increase in gluconeogenesis and marked outpouring of glucose into the circu-

Ketosis

When excess acetyl-CoA is present in the body, some of it is converted to acetoacetyl-CoA and then, in the liver, to acetoacetate. Acetoacetate and its derivatives, acetone and β-hydroxybutyrate, enter the circulation in large quantities (see Chapter 1).

These circulating ketone bodies are an important source of energy in fasting. Half of the metabolic rate in fasted normal dogs is said to be due to metabolism of ketones. The rate of ketone utilization in diabetics is also appreciable. It has been calculated that the maximal rate at which fat can be catabolized without significant ketosis is 2.5 g/kg body weight/d in diabetic humans. In untreated diabetes, production is much greater than this, and ketone bodies pile up in the bloodstream.

lation, the conversion of acetyl-CoA to malonyl-CoA and thence to fatty acids is markedly impaired. This is due to a deficiency of acetyl-CoA carboxylase, the enzyme that catalyzes the conversion. The excess acetyl-CoA is converted to ketone bodies.

In uncontrolled diabetes, the plasma concentration of triglycerides and chylomicrons as well as FFA is increased, and the plasma is often lipemic. The rise in these constituents is due mainly to decreased removal of triglycerides into the fat depots. The decreased activity of lipoprotein lipase contributes to this decreased removal (Clinical Box 21–2).

ACIDOSIS

As noted in Chapter 1, acetoacetate and β-hydroxybutyrate are anions of the fairly strong acids acetoacetic acid and β-hydroxybutyric acids. The hydrogen ions from these acids are buffered, but the buffering capacity is soon exceeded if production is increased. The resulting acidosis stimulates respiration, producing the rapid, deep respiration described by Kussmaul as "air hunger" and named (for him) **Kussmaul breathing.** The urine becomes acidic. However, when the ability of the kidneys to replace the plasma cations accompanying the organic anions with H^+ and NH_4^+ is exceeded, Na^+ and K^+ are lost in the urine. The electrolyte and water losses lead to dehydration, hypovolemia, and hypotension. Finally, the acidosis and dehydration depress consciousness to the point of coma. Diabetic acidosis is a medical emergency. Now that the infections that used to complicate the disease can be controlled with antibiotics, acidosis is the most common cause of early death in clinical diabetes.

In severe acidosis, total body Na^+ is markedly depleted, and when Na^+ loss exceeds water loss, plasma Na^+ may also be low. Total body K^+ is also low, but the plasma K^+ is usually normal, partly because extracellular fluid (ECF) volume is reduced and partly because K^+ moves from cells to ECF when the ECF H^+

concentration is high. Another factor tending to maintain the plasma K^+ is the lack of insulin-induced entry of K^+ into cells.

COMA

Coma in diabetes can be due to acidosis and dehydration. However, the plasma glucose can be elevated to such a degree that independent of plasma pH, the hyperosmolarity of the plasma causes unconsciousness (**hyperosmolar coma**). Accumulation of lactate in the blood (**lactic acidosis**) may also complicate diabetic ketoacidosis if the tissues become hypoxic, and lactic acidosis may itself cause coma. Brain edema occurs in about 1% of children with ketoacidosis, and it can cause coma. Its cause is unsettled, but it is a serious complication, with a mortality rate of about 25%.

CHOLESTEROL METABOLISM

In diabetes, the plasma cholesterol level is usually elevated and this plays a role in the accelerated development of the atherosclerotic vascular disease that is a major long-term complication of diabetes in humans. The rise in plasma cholesterol level is due to an increase in the plasma concentration of very low-density lipoprotein (VLDL) and low-density lipoprotein (LDL) (see Chapter 1). These in turn may be due to increased hepatic production of VLDL or decreased removal of VLDL and LDL from the circulation.

SUMMARY

Because of the complexities of the metabolic abnormalities in diabetes, a summary is in order. One of the key features of insulin deficiency (Figure 21–9) is decreased entry of glucose into many tissues (decreased peripheral utilization). Also, the net release of glucose from the liver is increased (increased production), due in part to glucagon excess. The resultant hyperglycemia leads to glycosuria and a dehydrating osmotic diuresis. Dehydration leads to polydipsia. In the face of intracellular glucose deficiency, appetite is stimulated, glucose is formed from protein (gluconeogenesis), and energy supplies are maintained by metabolism of proteins and fats. Weight loss, debilitating protein deficiency, and inanition are the result.

Fat catabolism is increased and the system is flooded with triglycerides and FFA. Fat synthesis is inhibited and the overloaded catabolic pathways cannot handle the excess acetyl-CoA that is formed. In the liver, the acetyl-CoA is converted to ketone bodies. Two of these are organic acids, and metabolic acidosis develops as ketones accumulate. Na^+ and K^+ depletion is added to the acidosis because these plasma cations are excreted with the organic anions not covered by the H^+ and NH_4^+ secreted by the kidneys. Finally, the acidotic, hypovolemic, hypotensive, depleted animal or patient becomes comatose because of the toxic effects of acidosis, dehydra-

FIGURE 21–9 **Effects of insulin deficiency.** (Courtesy of RJ Havel.)

FIGURE 21–10 **Plasma glucose levels at which various effects of hypoglycemia appear.**

tion, and hyperosmolarity on the nervous system and dies if treatment is not instituted.

All of these abnormalities are corrected by administration of insulin. Although emergency treatment of acidosis also includes administration of alkali to combat the acidosis and parenteral water, Na^+, and K^+ to replenish body stores, only insulin repairs the fundamental defects in a way that permits a return to normal.

INSULIN EXCESS

SYMPTOMS

All the known consequences of insulin excess are manifestations, directly or indirectly, of the effects of hypoglycemia on the nervous system. Except in individuals who have been fasting for some time, glucose is the only fuel used in appreciable quantities by the brain. The carbohydrate reserves in neural tissue are very limited and normal function depends on a continuous glucose supply. As the plasma glucose level falls, the first symptoms are palpitations, sweating, and nervousness due to autonomic discharge. These appear at plasma glucose values slightly lower than the value at which autonomic activation first begins, because the threshold for symptoms is slightly above the threshold for initial activation. At lower plasma glucose levels, so-called **neuroglycopenic symptoms** begin to appear. These include hunger as well as confusion and the other cognitive abnormalities. At even lower plasma glucose levels, lethargy, coma, convulsions, and eventually death occur. Obviously, the onset of hypoglycemic symptoms calls for prompt treatment with glucose or glucose-containing drinks such as orange juice. Although a dramatic disappearance of symptoms is the usual response, abnormalities ranging from intellectual dulling to coma may persist if the hypoglycemia was severe or prolonged.

COMPENSATORY MECHANISMS

One important compensation for hypoglycemia is cessation of the secretion of endogenous insulin. Inhibition of insulin secretion is complete at a plasma glucose level of about 80 mg/dL (Figures 21–10 and 21–11). In addition, hypoglycemia triggers increased secretion of at least four counter-regulatory hormones: glucagon, epinephrine, growth hormone, and cortisol. The epinephrine response is reduced during sleep. Glucagon and epinephrine increase the hepatic output of glucose by increasing glycogenolysis. Growth hormone decreases the

FIGURE 21–11 **Mean rates of insulin and glucagon delivery from an artificial pancreas at various plasma glucose levels.** The device was programmed to establish and maintain various plasma glucose levels in insulin-requiring diabetic humans, and the values for hormone output approximate the output of the normal human pancreas. The shape of the insulin curve also resembles the insulin response of incubated B cells to graded concentrations of glucose. (Reproduced with permission from Marliss EB, et al: Normalization of glycemia in diabetics during meals with insulin and glucagon delivery by the artificial pancreas. Diabetes 1977;26:663.)

utilization of glucose in various peripheral tissues, and cortisol has a similar action. The keys to counter-regulation appear to be epinephrine and glucagon: if the plasma concentration of either increases, the decline in the plasma glucose level is reversed; but if both fail to increase, there is little if any compensatory rise in the plasma glucose level. The actions of the other hormones are supplementary.

Note that the autonomic discharge and release of counter-regulatory hormones normally occurs at a higher plasma glucose level than the cognitive deficits and other more serious CNS changes (Figure 21–10). For diabetics treated with insulin, the symptoms caused by the autonomic discharge serve as a warning to seek glucose replacement. However, particularly in long-term diabetics who have been tightly regulated, the autonomic symptoms may not occur, and the resulting **hypoglycemia unawareness** can be a clinical problem of some magnitude.

REGULATION OF INSULIN SECRETION

The normal concentration of insulin measured by radioimmunoassay in the peripheral venous plasma of fasting normal humans is 0–70 µU/mL (0–502 pmol/L). The amount of insulin secreted in the basal state is about 1 U/h, with a fivefold to tenfold increase following ingestion of food. Therefore, the average amount secreted per day in a normal human is about 40 U (287 nmol).

Factors that stimulate and inhibit insulin secretion are summarized in Table 21–6.

EFFECTS OF THE PLASMA GLUCOSE LEVEL

It has been known for many years that glucose acts directly on pancreatic B cells to increase insulin secretion. The response to glucose is biphasic; there is a rapid but short-lived increase in secretion followed by a more slowly developing prolonged increase (Figure 21–12).

Glucose enters the B cells via GLUT 2 transporters and is phosphorylated by glucokinase then metabolized to pyruvate in the cytoplasm (Figure 21–13). The pyruvate enters the mitochondria and is metabolized to CO_2 and H_2O via the citric acid cycle with the formation of ATP by oxidative phosphorylation. The ATP enters the cytoplasm, where it inhibits ATP-sensitive K^+ channels, reducing K^+ efflux. This depolarizes the B cell, and Ca^{2+} enters the cell via voltage-gated Ca^{2+} channels. The Ca^{2+} influx causes exocytosis of a readily releasable pool of insulin-containing secretory granules, producing the initial spike of insulin secretion.

Metabolism of pyruvate via the citric acid cycle also causes an increase in intracellular glutamate. The glutamate appears to act on a second pool of secretory granules, committing them to the releasable form. The action of glutamate may be to decrease the pH in the secretory granules, a necessary step

TABLE 21–6   Factors affecting insulin secretion.

Stimulators	Inhibitors
Glucose	Somatostatin
Mannose	2-Deoxyglucose
Amino acids (leucine, arginine, others)	Mannoheptulose
Intestinal hormones (GIP, GLP-1 [7–36], gastrin, secretin, CCK; others?)	α-Adrenergic stimulators (norepinephrine, epinephrine)
β-Keto acids	β-Adrenergic blockers (propranolol)
Acetylcholine	
Glucagon	Galanin
Cyclic AMP and various cAMP-generating substances	Diazoxide
	Thiazide diuretics
β-Adrenergic stimulators	K^+ depletion
Theophylline	Phenytoin
Sulfonylureas	Alloxan
	Microtubule inhibitors
	Insulin

in their maturation. The release of these granules then produces the prolonged second phase of the insulin response to glucose. Thus, glutamate appears to act as an intracellular second messenger that primes secretory granules for secretion.

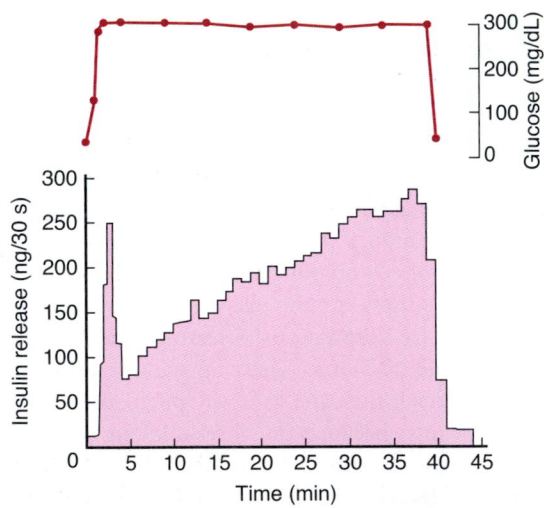

FIGURE 21–12   Insulin secretion from perfused rat pancreas in response to sustained glucose infusion. Values are means of three preparations. The top record shows the glucose concentration in the effluent perfusion mixture. (Reproduced with permission, from Curry DL Bennett LL, Grodsky GM: Dynamics of insulin secretion by the perfused rat pancreas. Endocrinology 1968;83:572.)

FIGURE 21–13 **Insulin secretion.** Glucose enters B cells by GLUT 2 transporters. It is phosphorylated and metabolized to pyruvate (Pyr) in the cytoplasm. The Pyr enters the mitochondria and is metabolized via the citric acid cycle. The ATP formed by oxidative phosphorylation inhibits ATP-sensitive K^+ channels, reducing K^+ efflux. This depolarizes the B cell, and Ca^{2+} influx is increased. The Ca^{2+} stimulates release of insulin by exocytosis. Glutamate (Glu) is also formed, and this primes secretory granules, preparing them for exocytosis.

The feedback control of plasma glucose on insulin secretion normally operates with great precision so that plasma glucose and insulin levels parallel each other with remarkable consistency.

PROTEIN & FAT DERIVATIVES

Insulin stimulates the incorporation of amino acids into proteins and combats the fat catabolism that produces the β-keto acids. Therefore, it is not surprising that arginine, leucine, and certain other amino acids stimulate insulin secretion, as do β-keto acids such as acetoacetate. Like glucose, these compounds generate ATP when metabolized, and this closes ATP-sensitive K^+ channels in the B cells. In addition, L-arginine is the precursor of NO, and NO stimulates insulin secretion.

ORAL HYPOGLYCEMIC AGENTS

Tolbutamide and other sulfonylurea derivatives such as acetohexamide, tolazamide, glipizide, and glyburide are orally active hypoglycemic agents that lower blood glucose by increasing the secretion of insulin. They only work in patients with some remaining B cells and are ineffective after pancreatectomy or in type 1 diabetes. They bind to the ATP-inhibited K^+ channels in the B cell membranes and inhibit channel activity, depolarizing the B cell membrane and increasing Ca^{2+} influx and hence insulin release, independent of increases in plasma glucose.

Persistent hyperinsulinemic hypoglycemia of infancy is a condition in which plasma insulin is elevated despite the hypoglycemia. The condition is caused by mutations in the genes for various enzymes in B cells that decrease K^+ efflux via the ATP-sensible K^+ channels. Treatment consists of administration of diazoxide, a drug that increases the activity of the K^+ channels or, in more severe cases, subtotal pancreatectomy.

The biguanide metformin is an oral hypoglycemic agent that acts in the absence of insulin. Metformin acts primarily by reducing gluconeogenesis and therefore decreasing hepatic glucose output. It is sometimes combined with a sulfonylurea in the treatment of type 2 diabetes. Metformin can cause lactic acidosis, but the incidence is usually low.

Troglitazone (Rezulin) and related **thiazolidinediones** are also used in the treatment of diabetes because they increase insulin-mediated peripheral glucose disposal, thus reducing insulin resistance. They bind to and activate peroxisome proliferator-activated receptor γ (PPARγ) in the nucleus of cells. Activation of this receptor, which is a member of the superfamily of hormone-sensitive nuclear transcription factors, has a unique ability to normalize a variety of metabolic functions.

CYCLIC AMP & INSULIN SECRETION

Stimuli that increase cAMP levels in B cells increase insulin secretion, including β-adrenergic agonists, glucagon, and phosphodiesterase inhibitors such as theophylline.

Catecholamines have a dual effect on insulin secretion; they inhibit insulin secretion via α_2-adrenergic receptors and stimulate insulin secretion via β-adrenergic receptors. The net effect of epinephrine and norepinephrine is usually inhibition. However, if catecholamines are infused after administration of α-adrenergic blocking drugs, the inhibition is converted to stimulation.

EFFECT OF AUTONOMIC NERVES

Branches of the right vagus nerve innervate the pancreatic islets, and stimulation of this parasympathetic pathway causes increased insulin secretion via M_4 receptors (see Table 7–2). Atropine blocks the response and acetylcholine stimulates insulin secretion. The effect of acetylcholine, like that of glucose, is due to increased cytoplasmic Ca^{2+}, but acetylcholine activates phospholipase C, with the released IP_3 releasing the Ca^{2+} from the endoplasmic reticulum.

Stimulation of the sympathetic nerves to the pancreas inhibits insulin secretion. The inhibition is produced by released norepinephrine acting on α_2-adrenergic receptors. However, if α-adrenergic receptors are blocked, stimulation of the sympathetic nerves causes increased insulin secretion mediated by β_2-adrenergic receptors. The polypeptide galanin is found in some of the autonomic nerves innervating the islets, and galanin inhibits insulin secretion by activating the K^+ channels that are inhibited by ATP. Thus, although the denervated pancreas responds to glucose, the autonomic innervation of the pancreas is involved in the overall regulation of insulin secretion.

INTESTINAL HORMONES

Orally administered glucose exerts a greater insulin-stimulating effect than intravenously administered glucose, and orally administered amino acids also produce a greater insulin response than intravenous amino acids. These observations led to exploration of the possibility that a substance secreted by the gastrointestinal mucosa stimulated insulin secretion. Glucagon, glucagon derivatives, secretin, cholecystokinin (CCK), gastrin, and gastric inhibitory peptide (GIP) all have such an action (see Chapter 26), and CCK potentiates the insulin-stimulating effects of amino acids. However, GIP is the only one of these peptides that produces stimulation when administered in doses that reflect blood GIP levels produced by an oral glucose load.

Recently, attention has focused on glucagon-like polypeptide 1 (7–36) (GLP-1 [7–36]) as an additional gut factor that stimulates insulin secretion. This polypeptide is a product of preproglucagon.

B cells have GLP-1 (7–36) receptors as well as GIP receptors, and GLP-1 (7–36) is a more potent insulinotropic hormone than GIP. GIP and GLP-1 (7–36) both appear to act by increasing Ca^{2+} influx through voltage-gated Ca^{2+} channels.

The possible roles of pancreatic somatostatin and glucagon in the regulation of insulin secretion are discussed below (Clinical Box 21–3).

LONG-TERM CHANGES IN B CELL RESPONSES

The magnitude of the insulin response to a given stimulus is determined in part by the secretory history of the B cells. Individuals fed a high-carbohydrate diet for several weeks not only have higher fasting plasma insulin levels but also show a greater secretory response to a glucose load than individuals fed an isocaloric low-carbohydrate diet.

Although the B cells respond to stimulation with hypertrophy like other endocrine cells, they become exhausted and stop secreting (**B cell exhaustion**) when the stimulation is marked or prolonged. The pancreatic reserve is large and it is difficult to produce B cell exhaustion in normal animals, but if the pancreatic reserve is reduced by partial pancreatectomy, exhaustion of the remaining B cells can be initiated by any procedure that chronically raises the plasma glucose level. For example, diabetes can be produced in animals with limited pancreatic reserves by anterior pituitary extracts, growth hormone, thyroid hormones, or the prolonged continuous infusion of glucose alone. The diabetes precipitated by hormones in animals is at first reversible, but with prolonged treatment it becomes permanent. The transient diabetes is usually named for the agent producing it, for example, "hypophysial diabetes" or "thyroid diabetes." Permanent diabetes persisting after treatment has been discontinued is indicated by the prefix meta-,

CLINICAL BOX 21–3

Effects of K^+ Depletion

K^+ depletion decreases insulin secretion, and K^+-depleted patients, for example, patients with primary hyperaldosteronism (see Chapter 22), develop diabetic glucose tolerance curves. These curves are restored to normal by K^+ repletion. The thiazide diuretics, which cause loss of K^+ as well as Na^+ in the urine (see Chapter 38), decrease glucose tolerance and make diabetes worse. They apparently exert this effect primarily because of their K^+-depleting effects, although some of them also cause pancreatic islet cell damage.

for example, **"metahypophysial diabetes"** or **"metathyroid diabetes."** When insulin is administered along with the diabetogenic hormones, the B cells are protected, probably because the plasma glucose is lowered, and diabetes does not develop.

It is interesting in this regard that genetic factors may be involved in the control of B cell reserve. In mice in which the gene for IRS-1 has been knocked out (see above), a robust compensatory B cell response occurs. However, in IRS-2 knockouts, the compensation is reduced and a more severe diabetic phenotype is produced.

GLUCAGON

CHEMISTRY

Human glucagon, a linear polypeptide with a molecular weight of 3485, is produced by the A cells of the pancreatic islets and the upper gastrointestinal tract. It contains 29 amino acid residues. All mammalian glucagons appear to have the same structure. Human preproglucagon (Figure 21–14) is a 179-amino-acid protein that is found in pancreatic A cells, in L cells in the lower gastrointestinal tract, and in the brain. It is the product of a single mRNA, but it is processed differently in different tissues. In A cells, it is processed primarily to glucagon and **major proglucagon fragment (MPGF)**. In L cells, it is processed primarily to **glicentin**, a polypeptide that consists of glucagon extended by additional amino acid residues at either end, plus **glucagon-like polypeptides 1 and 2 (GLP-1** and **GLP-2)**. Some **oxyntomodulin** is also formed, and in both A and L cells, residual **glicentin-related polypeptide (GRPP)** is left. Glicentin has some glucagon activity. GLP-1 and GLP-2 have no definite biologic activity by themselves. However, GLP-1 is processed further by removal of its amino-terminal amino acid residues and the product, **GLP-1 (7–36),** is a potent stimulator of insulin secretion that also increases glucose utilization (see above). GLP-1 and GLP-2 are also produced in the brain. The function of GLP-1 in this location is uncertain, but GLP-2 appears to be the mediator in a pathway from the nucleus tractus solitarius (NTS) to the dorsomedial nuclei of the hypothalamus, and injection of GLP-2 lowers food intake. Oxyntomodulin inhibits gastric acid secretion, though its

FIGURE 21–14 **Posttranslational processing of preproglucagon in A and L cells.** S, signal peptide; GRPP, glicentin-related polypeptide; GLP, glucagon-like polypeptide; Oxy, oxynto-modulin; MPGF, major proglucagon fragment. (Modified from Drucker, DJ: Glucagon and glucagon-like peptides. Pancreas 1990;5:484.)

physiologic role is unsettled, and GRPP does not have any established physiologic effects.

ACTION

Glucagon is glycogenolytic, gluconeogenic, lipolytic, and ketogenic. It acts on G-protein coupled receptors with a molecular weight of about 190,000. In the liver, it acts via G_s to activate adenylyl cyclase and increase intracellular cAMP. This leads via protein kinase A to activation of phosphorylase and therefore to increased breakdown of glycogen and an increase in plasma glucose. However, glucagon acts on different glucagon receptors located on the same hepatic cells to activate phospholipase C, and the resulting increase in cytoplasmic Ca^{2+} also stimulates glycogenolysis. Protein kinase A also decreases the metabolism of glucose 6-phosphate (Figure 21–15) by inhibiting the conversion of phosphoenolpyruvate to pyruvate. It also decreases the concentration of fructose 2,6-diphosphate and this in turn inhibits the conversion of fructose 6-phosphate to fructose 1,6-diphosphate. The resultant buildup of glucose 6-phosphate leads to increased glucose synthesis and release.

Glucagon does not cause glycogenolysis in muscle. It increases gluconeogenesis from available amino acids in the liver and elevates the metabolic rate. It increases ketone body formation by decreasing malonyl-CoA levels in the liver. Its lipolytic activity, which leads in turn to increased ketogenesis, is discussed in Chapter 1. The calorigenic action of glucagon is not due to the hyperglycemia per se but probably to the increased hepatic deamination of amino acids.

Large doses of exogenous glucagon exert a positively inotropic effect on the heart (see Chapter 31) without producing increased myocardial excitability, presumably because they increase myocardial cAMP. Use of this hormone in the treatment of heart disease has been advocated, but there is no evidence for a physiologic role of glucagon in the regulation of cardiac function. Glucagon also stimulates the secretion of growth hormone, insulin, and pancreatic somatostatin.

FIGURE 21–15 **Mechanisms by which glucagon increases glucose output from the liver.** Solid arrows indicate facilitation; dashed arrows indicate inhibition.

METABOLISM

Glucagon has a half-life in the circulation of 5 to 10 min. It is degraded by many tissues but particularly by the liver. Because glucagon is secreted into the portal vein and reaches the liver before it reaches the peripheral circulation, peripheral blood levels are relatively low. The rise in peripheral blood glucagon levels produced by excitatory stimuli is exaggerated in patients with cirrhosis, presumably because of decreased hepatic degradation of the hormone.

REGULATION OF SECRETION

The principal factors known to affect glucagon secretion are summarized in Table 21–7. Secretion is increased by hypoglycemia and decreased by a rise in plasma glucose. Pancreatic B cells contain GABA, and evidence suggests that coincident with the increased insulin secretion produced by hyperglycemia, GABA is released and acts on the A cells to inhibit glucagon secretion by activating $GABA_A$ receptors. The $GABA_A$ receptors are Cl^- channels, and the resulting Cl^- influx hyperpolarizes the A cells.

Secretion is also increased by stimulation of the sympathetic nerves to the pancreas, and this sympathetic effect is mediated via β-adrenergic receptors and cAMP. It appears that the A cells are like the B cells in that stimulation of β-adrenergic receptors increases secretion and stimulation of α-adrenergic receptors inhibits secretion. However, the pancreatic response to sympathetic stimulation in the absence of blocking drugs is increased secretion of glucagon, so the effect of β-receptors predominates in the glucagon-secreting cells. The stimulatory effects of various stresses and possibly of exercise and infection are mediated at least in part via the

TABLE 21–7 Adipokines.

Agent	Effect on Insulin Resistance
Leptin	Decreases
TNFα	Increases
Adiponectin	Decreases
Resistin	Increases

TABLE 21–8 Factors affecting glucagon secretion.

Stimulators	Inhibitors
Amino acids (particularly the glucogenic amino acids: alanine, serine, glycine, cysteine, and threonine)	Glucose
CCK, gastrin	Somatostatin
Cortisol	Secretin
Exercise	FFA
Infections	Ketones
Other stresses	Insulin
β-Adrenergic stimulators	Phenytoin
Theophylline	α-Adrenergic stimulators
Acetylcholine	GABA

sympathetic nervous system. Vagal stimulation also increases glucagon secretion.

A protein meal and infusion of various amino acids increase glucagon secretion. It seems appropriate that the glucogenic amino acids are particularly potent in this regard, since these are the amino acids that are converted to glucose in the liver under the influence of glucagon. The increase in glucagon secretion following a protein meal is also valuable, since the amino acids stimulate insulin secretion and the secreted glucagon prevents the development of hypoglycemia while the insulin promotes storage of the absorbed carbohydrates and lipids. Glucagon secretion increases during starvation. It reaches a peak on the third day of a fast, at the time of maximal gluconeogenesis. Thereafter, the plasma glucagon level declines as fatty acids and ketones become the major sources of energy.

During exercise, there is an increase in glucose utilization that is balanced by an increase in glucose production caused by an increase in circulating glucagon levels.

The glucagon response to oral administration of amino acids is greater than the response to intravenous infusion of amino acids, suggesting that a glucagon-stimulating factor is secreted from the gastrointestinal mucosa. CCK and gastrin increase glucagon secretion, whereas secretin inhibits it. Because CCK and gastrin secretion are both increased by a protein meal, either hormone could be the gastrointestinal mediator of the glucagon response. The inhibition produced by somatostatin is discussed below.

Glucagon secretion is also inhibited by FFA and ketones. However, this inhibition can be overridden, since plasma glucagon levels are high in diabetic ketoacidosis.

INSULIN–GLUCAGON MOLAR RATIOS

As noted previously, insulin is glycogenic, antigluconeogenetic, antilipolytic, and antiketotic in its actions. It thus favors storage of absorbed nutrients and is a "hormone of energy storage." Glucagon, on the other hand, is glycogenolytic, gluconeogenetic, lipolytic, and ketogenic. It mobilizes energy stores and is a "hormone of energy release." Because of their opposite effects, the blood levels of both hormones must be considered in any given situation. It is convenient to think in terms of the molar ratios of these hormones.

The insulin–glucagon molar ratios fluctuate markedly because the secretion of glucagon and insulin are both modi-

fied by the conditions that preceded the application of any given stimulus (Table 21–8). Thus, for example, the insulin–glucagon molar ratio on a balanced diet is approximately 2.3. An infusion of arginine increases the secretion of both hormones and raises the ratio to 3.0. After 3 days of starvation, the ratio falls to 0.4, and an infusion of arginine in this state lowers the ratio to 0.3. Conversely, the ratio is 25 in individuals receiving a constant infusion of glucose and rises to 170 on ingestion of a protein meal during the infusion. The rise occurs because insulin secretion rises sharply, while the usual glucagon response to a protein meal is abolished. Thus, when energy is needed during starvation, the insulin–glucagon molar ratio is low, favoring glycogen breakdown and gluconeogenesis; conversely, when the need for energy mobilization is low, the ratio is high, favoring the deposition of glycogen, protein, and fat.

OTHER ISLET CELL HORMONES

In addition to insulin and glucagon, the pancreatic islets secrete somatostatin and pancreatic polypeptide into the bloodstream. In addition, somatostatin may be involved in regulatory processes within the islets that adjust the pattern of hormones secreted in response to various stimuli.

SOMATOSTATIN

Somatostatin and its receptors are discussed in Chapter 7. Somatostatin 14 (SS 14) and its amino terminal-extended form somatostatin 28 (SS 28) are found in the D cells of pancreatic islets. Both forms inhibit the secretion of insulin, glucagon, and pancreatic polypeptide and act locally within the pancreatic islets in a paracrine fashion. SS 28 is more active than SS 14 in inhibiting insulin secretion, and it apparently acts via the SSTR5 receptor (see Chapter 7). Patients with somatostatin-secreting pancreatic tumors (**somatostatinomas**) develop hyperglycemia and other manifestations of diabetes that disappear when the tumor is removed. They also develop dyspepsia due to slow gastric emptying and decreased gastric acid secretion, and gallstones, which are precipitated by decreased gallbladder contraction due to inhibition of CCK secretion. The secretion of pancreatic somatostatin is increased by several of the same stimuli that increase insulin secretion, that is, glucose and amino acids, particularly arginine and leucine. It is also increased by CCK. Somatostatin is released from the pancreas and the gastrointestinal tract into the peripheral blood.

PANCREATIC POLYPEPTIDE

Human pancreatic polypeptide is a linear polypeptide that contains 36 amino acid residues and is produced by F cells in the islets. It is closely related to two other 36-amino acid polypeptides, **polypeptide YY,** a gastrointestinal peptide (see Chapter 26), and **neuropeptide Y,** which is found in the brain and the autonomic nervous system (see Chapter 7). All end in tyrosine and are amidated at their carboxyl terminal. At least in part, pancreatic polypeptide secretion is under cholinergic control; plasma levels fall after administration of atropine. Its secretion is increased by a meal containing protein and by fasting, exercise, and acute hypoglycemia. Secretion is decreased by somatostatin and intravenous glucose. Infusions of leucine, arginine, and alanine do not affect it, so the stimulatory effect of a protein meal may be mediated indirectly. Pancreatic polypeptide slows the absorption of food in humans, and it may smooth out the peaks and valleys of absorption. However, its exact physiologic function is still uncertain.

ORGANIZATION OF THE PANCREATIC ISLETS

The presence in the pancreatic islets of hormones that affect the secretion of other islet hormones suggests that the islets function as secretory units in the regulation of nutrient homeostasis. Somatostatin inhibits the secretion of insulin, glucagon, and pancreatic polypeptide (Figure 21–16); insulin inhibits the secretion of glucagon; and glucagon stimulates the secretion of insulin and somatostatin. As noted above, A and

D cells and pancreatic polypeptide-secreting cells are generally located around the periphery of the islets, with the B cells in the center. There are clearly two types of islets, glucagon-rich islets and pancreatic polypeptide-rich islets, but the functional significance of this separation is not known. The islet cell hormones released into the ECF probably diffuse to other islet cells and influence their function (paracrine communication; see Chapter 26). It has been demonstrated that gap junctions are present between A, B, and D cells and that these permit the passage of ions and other small molecules from one cell to another, which could coordinate their secretory functions.

EFFECTS OF OTHER HORMONES & EXERCISE ON CARBOHYDRATE METABOLISM

Exercise has direct effects on carbohydrate metabolism. Many hormones in addition to insulin, IGF-I, IGF-II, glucagon, and somatostatin also have important roles in the regulation of carbohydrate metabolism. They include epinephrine, thyroid hormones, glucocorticoids, and growth hormone. The other functions of these hormones are considered elsewhere, but it seems wise to summarize their effects on carbohydrate metabolism in the context of the present chapter.

EXERCISE

The entry of glucose into skeletal muscle is increased during exercise in the absence of insulin by causing an insulin-independent increase in the number of GLUT 4 transporters in muscle cell membranes (see above). This increase in glucose entry persists for several hours after exercise, and regular exercise training can also produce prolonged increases in insulin sensitivity. Exercise can precipitate hypoglycemia in diabetics not only because of the increase in muscle uptake of glucose but also because absorption of injected insulin is more rapid during exercise. Patients with diabetes should take in extra calories or reduce their insulin dosage when they exercise.

CATECHOLAMINES

The activation of phosphorylase in liver by catecholamines is discussed in Chapter 1. Activation occurs via β-adrenergic receptors, which increase intracellular cAMP, and α-adrenergic receptors, which increase intracellular Ca^{2+}. Hepatic glucose output is increased, producing hyperglycemia. In muscle, the phosphorylase is also activated via cAMP and presumably via Ca^{2+}, but the glucose 6-phosphate formed can be catabolized only to pyruvate because of the absence of glucose 6-phosphatase. For reasons that are not entirely clear, large amounts of pyruvate are converted to lactate, which diffuses from the muscle into the circulation (Figure 21–17). The lactate is oxidized in the liver to pyruvate and converted to glycogen.

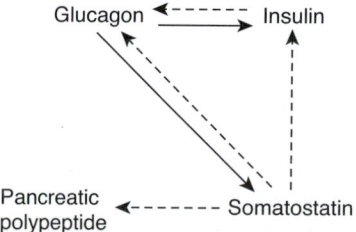

FIGURE 21–16 **Effects of islet cell hormones on the secretion of other islet cell hormones.** Solid arrows indicate stimulation; dashed arrows indicate inhibition.

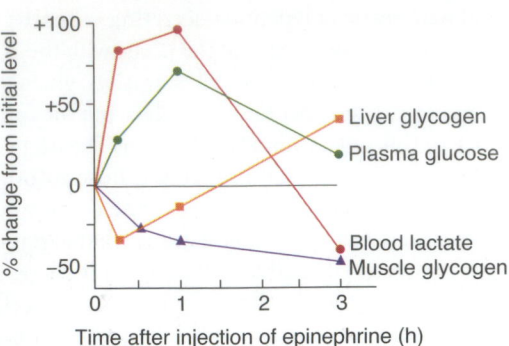

FIGURE 21–17 **Effect of epinephrine on tissue glycogen, plasma glucose, and blood lactate levels in fed rats.** (Reproduced with permission from Ruch TC, Patton HD [editors]: *Physiology and Biophysics,* 20th ed. Vol. 3. Saunders, 1973.)

Therefore, the response to an injection of epinephrine is an initial glycogenolysis followed by a rise in hepatic glycogen content. Lactate oxidation may be responsible for the calorigenic effect of epinephrine (see Chapter 22). Epinephrine and norepinephrine also liberate FFA into the circulation, and epinephrine decreases peripheral utilization of glucose.

THYROID HORMONES

Thyroid hormones make experimental diabetes worse; thyrotoxicosis aggravates clinical diabetes; and metathyroid diabetes can be produced in animals with decreased pancreatic reserve. The principal diabetogenic effect of thyroid hormones is to increase absorption of glucose from the intestine, but the hormones also cause (probably by potentiating the effects of catecholamines) some degree of hepatic glycogen depletion. Glycogen-depleted liver cells are easily damaged. When the liver is damaged, the glucose tolerance curve is diabetic because the liver takes up less of the absorbed glucose. Thyroid hormones may also accelerate the degradation of insulin. All these actions have a hyperglycemic effect and, if the pancreatic reserve is low, may lead to B cell exhaustion.

ADRENAL GLUCOCORTICOIDS

Glucocorticoids from the adrenal cortex (see Chapter 22) elevate blood glucose and produce a diabetic type of glucose tolerance curve. In humans, this effect may occur only in individuals with a genetic predisposition to diabetes. Glucose tolerance is reduced in 80% of patients with Cushing syndrome (see Chapter 22), and 20% of these patients have frank diabetes. The glucocorticoids are necessary for glucagon to exert its gluconeogenic action during fasting. They are gluconeogenic themselves, but their role is mainly permissive. In adrenal insufficiency, the blood glucose is normal as long as food intake is maintained, but fasting precipitates hypoglycemia and collapse. The plasma-glucose-lowering effect of insulin is greatly enhanced in patients with adrenal insufficiency. In animals with experimental diabetes, adrenalectomy markedly ameliorates the diabetes. The major diabetogenic effects are an increase in protein catabolism with increased gluconeogenesis in the liver; increased hepatic glycogenesis and ketogenesis; and a decrease in peripheral glucose utilization relative to the blood insulin level that may be due to inhibition of glucose phosphorylation.

GROWTH HORMONE

Human growth hormone makes clinical diabetes worse, and 25% of patients with growth hormone-secreting tumors of the anterior pituitary have diabetes. Hypophysectomy ameliorates diabetes and decreases insulin resistance even more than adrenalectomy, whereas growth hormone treatment increases insulin resistance.

The effects of growth hormone are partly direct and partly mediated via IGF-I (see Chapter 24). Growth hormone mobilizes FFA from adipose tissue, thus favoring ketogenesis. It decreases glucose uptake into some tissues ("anti-insulin action"), increases hepatic glucose output, and may decrease tissue binding of insulin. Indeed, it has been suggested that the ketosis and decreased glucose tolerance produced by starvation are due to hypersecretion of growth hormone. Growth hormone does not stimulate insulin secretion directly, but the hyperglycemia it produces secondarily stimulates the pancreas and may eventually exhaust the B cells.

HYPOGLYCEMIA & DIABETES MELLITUS IN HUMANS

HYPOGLYCEMIA

"Insulin reactions" are common in type 1 diabetics and occasional hypoglycemic episodes are the price of good diabetic control in most diabetics. Glucose uptake by skeletal muscle and absorption of injected insulin both increase during exercise (see above).

Symptomatic hypoglycemia also occurs in nondiabetics, and a review of some of the more important causes serves to emphasize the variables affecting plasma glucose homeostasis. Chronic mild hypoglycemia can cause incoordination and slurred speech, and the condition can be mistaken for drunkenness. Mental aberrations and convulsions in the absence of frank coma also occur. When the level of insulin secretion is chronically elevated by an **insulinoma,** a rare, insulin-secreting tumor of the pancreas, symptoms are most common in the morning. This is because a night of fasting has depleted hepatic glycogen reserves. However, symptoms can develop at any time, and in such patients, the diagnosis may be missed. Some cases of insulinoma have been erroneously diagnosed as epilepsy or psychosis. Hypoglycemia also occurs in some patients with large malignant tumors that do not involve the pancreatic islets, and the hypoglycemia in these cases is apparently due to excess secretion of IGF-II.

As noted above, the autonomic discharge caused by lowered blood glucose that produces shakiness, sweating, anxiety, and hunger normally occurs at plasma glucose levels that are higher than the glucose levels that cause cognitive dysfunction, thereby serving as a warning to ingest sugar. However, in some individuals, these warning symptoms fail to occur before the cognitive symptoms, due to cerebral dysfunction (desensitization), and this **hypoglycemia unawareness** is potentially dangerous. The condition is prone to develop in patients with insulinomas and in diabetics receiving intensive insulin therapy, so it appears that repeated bouts of hypoglycemia cause the eventual development of hypoglycemia unawareness. If blood sugar rises again for some time, the warning symptoms again appear at a higher plasma glucose level than cognitive abnormalities and coma. The reason why prolonged hypoglycemia causes loss of the warning symptoms is unsettled.

In liver disease, the glucose tolerance curve is diabetic but the fasting plasma glucose level is low (Figure 21–18). In **functional hypoglycemia,** the plasma glucose rise is normal after a test dose of glucose, but the subsequent fall overshoots to hypoglycemic levels, producing symptoms 3 to 4 h after meals. This pattern is sometimes seen in individuals who later develop diabetes. Patients with this syndrome should be distinguished from the more numerous patients with similar symptoms due to psychologic or other problems who do not have hypoglycemia when blood is drawn during the symptomatic episode. It has been postulated that the overshoot of the plasma glucose is due to insulin secretion stimulated by impulses in the right vagus, but cholinergic blocking agents do not routinely correct the abnormality. In some thyrotoxic patients and in patients who have had gastrectomies or other operations that speed the passage of food into the intestine, glucose absorption is abnormally rapid. The plasma glucose rises to a high, early peak, but it then falls rapidly to hypoglycemic levels because the wave of hyperglycemia evokes a

FIGURE 21–18 **Typical glucose tolerance curves after an oral glucose load in liver disease and in conditions causing excessively rapid absorption of glucose from the intestine.** The horizontal line is the approximate plasma glucose level at which hypoglycemic symptoms may appear.

greater than normal rise in insulin secretion. Symptoms characteristically occur about 2 h after meals.

DIABETES MELLITUS

The incidence of diabetes mellitus in the human population has reached epidemic proportions worldwide and it is increasing at a rapid rate. In 2000, there were an estimated 150 million cases in the world; this number is projected to increase to 221 million by 2010. Ninety percent of the present cases are type 2 diabetes, and most of the increase will be in type 2, paralleling the increase in the incidence of obesity.

Diabetes is sometimes complicated by acidosis and coma, and in long-standing diabetes additional complications occur. These include microvascular, macrovascular, and neuropathic disease. The microvascular abnormalities are proliferative scarring of the retina (**diabetic retinopathy**) leading to blindness; and renal disease (**diabetic nephropathy**) leading to renal failure. The macrovascular abnormalities are due to accelerated atherosclerosis, which is secondary to increased plasma LDL. The result is an increased incidence of stroke and myocardial infarction. The neuropathic abnormalities (**diabetic neuropathy**) involve the autonomic nervous system and peripheral nerves. The neuropathy plus the atherosclerotic circulatory insufficiency in the extremities and reduced resistance to infection can lead to chronic ulceration and gangrene, particularly in the feet.

The ultimate cause of the microvascular and neuropathic complications is chronic hyperglycemia, and tight control of the diabetes reduces their incidence. Intracellular hyperglycemia activates the enzyme aldose reductase. This increases the formation of sorbitol in cells, which in turn reduces cellular Na^+–K^+ ATPase. In addition, intracellular glucose can be converted to so-called Amadori products, and these in turn can form **advanced glycosylation end products (AGEs),** which cross-link matrix proteins. This damages blood vessels. The AGEs also interfere with leukocyte responses to infection.

TYPES OF DIABETES

The cause of clinical diabetes is always a deficiency of the effects of insulin at the tissue level, but the deficiency may be relative. One of the common forms, **type 1,** or **insulin-dependent diabetes mellitus (IDDM),** is due to insulin deficiency caused by autoimmune destruction of the B cells in the pancreatic islets; the A, D, and F cells remain intact. The second common form, **type 2,** or **non-insulin-dependent diabetes mellitus (NIDDM),** is characterized by insulin resistance.

In addition, some cases of diabetes are due to other diseases or conditions such as chronic pancreatitis, total pancreatectomy, Cushing syndrome (see Chapter 22), and acromegaly (see Chapter 24). These make up 5% of the total cases and are sometimes classified as **secondary diabetes.**

Type 1 diabetes usually develops before the age of 40 and hence is called **juvenile diabetes.** Patients with this disease are not obese and they have a high incidence of ketosis and acidosis. Various anti-B cell antibodies are present in plasma, but the current thinking is that type 1 diabetes is primarily a T lymphocyte-mediated disease. Definite genetic susceptibility is present as well; if one identical twin develops the disease, the chances are 1 in 3 that the other twin will also do so. In other words, the **concordance rate** is about 33%. The main genetic abnormality is in the major histocompatibility complex on chromosome 6, making individuals with certain types of histocompatibility antigens (see Chapter 3) much more prone to develop the disease. Other genes are also involved.

Immunosuppression with drugs such as cyclosporine ameliorate type 1 diabetes if given early in the disease before all B cells are lost. Attempts have been made to treat type 1 diabetes by transplanting pancreatic tissue or isolated islet cells, but results to date have been poor, largely because B cells are easily damaged and it is difficult to transplant enough of them to normalize glucose responses.

As mentioned above, type 2 is the most common type of diabetes and is usually associated with obesity. It usually develops after age 40 and is not associated with total loss of the ability to secrete insulin. It has an insidious onset, is rarely associated with ketosis, and is usually associated with normal B cell morphology and insulin content if the B cells have not become exhausted. The genetic component in type 2 diabetes is actually stronger than the genetic component in type 1 diabetes; in identical twins, the concordance rate is higher, ranging in some studies to nearly 100%.

In some patients, type 2 diabetes is due to defects in identified genes. Over 60 of these defects have been described. They include defects in glucokinase (about 1% of the cases), the insulin molecule itself (about 0.5% of the cases), the insulin receptor (about 1% of the cases), GLUT 4 (about 1% of the cases), or IRS-1 (about 15% of the cases). In maturity-onset

diabetes occurring in young individuals (MODY), which accounts for about 1% of the cases of type 2 diabetes, loss-of-function mutations have been described in six different genes. Five code for transcription factors affecting the production of enzymes involved in glucose metabolism. The sixth is the gene for glucokinase (Figure 21–13), the enzyme that controls the rate of glucose phosphorylation and hence its metabolism in the B cells. However, the vast majority of cases of type 2 diabetes are almost certainly polygenic in origin, and the actual genes involved are still unknown.

OBESITY, THE METABOLIC SYNDROME, & TYPE 2 DIABETES

Obesity is increasing in incidence, and relates to the regulation of food intake and energy balance and overall nutrition. It deserves additional consideration in this chapter because of its special relation to disordered carbohydrate metabolism and diabetes. As body weight increases, insulin resistance increases, that is, there is a decreased ability of insulin to move glucose into fat and muscle and to shut off glucose release from the liver. Weight reduction decreases insulin resistance. Associated with obesity there is hyperinsulinemia, dyslipidemia (characterized by high circulating triglycerides and low high-density lipoprotein [HDL]), and accelerated development of atherosclerosis. This combination of findings is commonly called the **metabolic syndrome,** or **syndrome X.** Some of the patients with the syndrome are prediabetic, whereas others have type 2 diabetes. It has not been proved but it is logical to assume that the hyperinsulinemia is a compensatory response to the increased insulin resistance and that frank diabetes develops in individuals with reduced B cell reserves.

These observations and other data strongly suggest that fat produces a chemical signal or signals that act on muscles and the liver to increase insulin resistance. Evidence for this includes the recent observation that when glucose transporters are selectively knocked out in adipose tissue, there is an associated decrease in glucose transport in muscle in vivo, but when the muscles of those animals are tested in vitro their transport is normal.

One possible signal is the circulating free fatty acid level, which is elevated in many insulin-resistant states. Other possibilities are peptides and proteins secreted by fat cells. It is now clear that white fat depots are not inert lumps but are actually endocrine tissues that secrete not only leptin but also other hormones that affect fat metabolism. The most intensively studied of these **adipokines** are listed in Table 21–9. Some of the adipokines decrease, rather than increase, insulin resistance. Leptin and adiponectin, for example, decrease insulin resistance, whereas resistin increases insulin resistance. Further complicating the situation, marked insulin resistance is present in the rare metabolic disease **congenital lipodystrophy,** in which fat depots fail to develop. This resistance is reduced by leptin and adiponectin. Finally, a variety of knockouts of intracellular second messengers have been reported to

TABLE 21–9 Insulin-glucagon molar ratios (I/G) in blood in various conditions.

Condition	Hepatic Glucose Storage (S) or Production (P)[a]	I/G
Glucose availability		
Large carbohydrate meal	4+ (S)	70
Intravenous glucose	2+ (S)	25
Small meal	1+ (S)	7
Glucose need		
Overnight fast	1+ (P)	2.3
Low-carbohydrate diet	2+ (P)	1.8
Starvation	4+ (P)	0.4

[a]1+ to 4+ indicate relative magnitude.

Courtesy of RH Unger.

increase insulin resistance. It is unclear how, or indeed if, these findings fit together to provide an explanation of the relation of obesity to insulin tolerance, but the topic is obviously an important one and it is under intensive investigation.

CHAPTER SUMMARY

- Four polypeptides with hormonal activity are secreted by the pancreas: insulin, glucagon, somatostatin, and pancreatic polypeptide.
- Insulin increases the entry of glucose into cells. In skeletal muscle cell it increases the number of GLUT 4 transporters in the cell membranes. In liver it induces glucokinase, which increases the phosphorylation of glucose, facilitating the entry of glucose into the cell.
- Insulin causes K^+ to enter cells, with a resultant lowering of the extracellular K^+ concentration. Insulin increases the activity of Na^+–K^+ ATPase in cell membranes, so that more K^+ is pumped into cells. Hypokalemia often develops when patients with diabetic acidosis are treated with insulin.
- Insulin receptors are found on many different cells in the body and have two subunits, α and β. Binding of insulin to its receptor triggers a signaling pathway that involves autophosphorylation of the β subunits on tyrosine residues. This triggers phosphorylation of some cytoplasmic proteins and dephosphorylation of others, mostly on serine and threonine residues.
- The constellation of abnormalities caused by insulin deficiency is called diabetes mellitus. Type 1 diabetes is due to insulin deficiency caused by autoimmune destruction of the B cells in the pancreatic islets; Type 2 diabetes is characterized by the dysregulation of insulin release from the B cells, along with insulin resistance in peripheral tissues such as skeletal muscle, brain, and liver.

MULTIPLE-CHOICE QUESTIONS

For all questions, select the single best answer unless otherwise directed.

1. Which of the following are *incorrectly* paired?
   A) B cells : insulin
   B) D cells : somatostatin
   C) A cells : glucagon
   D) pancreatic exocrine cells : chymotrypsinogen
   E) F cells : gastrin

2. Which of the following are *incorrectly* paired?
   A) epinephrine : increased glycogenolysis in skeletal muscle
   B) insulin : increased protein synthesis
   C) glucagon : increased gluconeogenesis
   D) progesterone : increased plasma glucose level
   E) growth hormone : increased plasma glucose level

3. Which of the following would be *least* likely to be seen 14 days after a rat is injected with a drug that kills all of its pancreatic B cells?
   A) a rise in the plasma H^+ concentration
   B) a rise in the plasma glucagon concentration
   C) a fall in the plasma HCO_3^+ concentration
   D) a fall in the plasma amino acid concentration
   E) a rise in plasma osmolality

4. When the plasma glucose concentration falls to low levels, a number of different hormones help combat the hypoglycemia. After intravenous administration of a large dose of insulin, the return of a low blood sugar level to normal is delayed in
   A) adrenal medullary insufficiency.
   B) glucagon deficiency.
   C) combined adrenal medullary insufficiency and glucagon deficiency.
   D) thyrotoxicosis.
   E) acromegaly.

5. Insulin increases the entry of glucose into
   A) all tissues.
   B) renal tubular cells.
   C) the mucosa of the small intestine.
   D) most neurons in the cerebral cortex.
   E) skeletal muscle.

6. Glucagon increases glycogenolysis in liver cells but ACTH does not because
   A) cortisol increases the plasma glucose level.
   B) liver cells have an adenylyl cyclase different from that in adrenocortical cells.
   C) ACTH cannot enter the nucleus of liver cells.
   D) the membranes of liver cells contain receptors different from those in adrenocortical cells.
   E) liver cells contain a protein that inhibits the action of ACTH.

7. A meal rich in proteins containing the amino acids that stimulate insulin secretion but low in carbohydrates does not cause hypoglycemia because
   A) the meal causes a compensatory increase in T_4 secretion.
   B) cortisol in the circulation prevents glucose from entering muscle.
   C) glucagon secretion is also stimulated by the meal.
   D) the amino acids in the meal are promptly converted to glucose.
   E) insulin does not bind to insulin receptors if the plasma concentration of amino acids is elevated.

CHAPTER RESOURCES

Bannerjee RK, et al: Regulation of fasted blood glucose by resistin. Science 2004;303:1195.

Gehlert DR: Multiple receptors for the pancreatic polypeptide (PP-fold) family: Physiological implications. Proc Soc Exper Biol Med 1998;218:7.

Harmel AP, Mothur R: *Davidson's Diabetes Mellitus,* 5th ed. Elsvier, 2004.

Kjos SL, Buchanan TA: Gestational diabetes mellitus. N Engl J Med 1999;341:1749.

Kulkarni RN, Kahn CR: HNFs-linking the liver and pancreatic islets in diabetes. Science 2004;303:1311.

Larsen PR, et al (editors): *Williams Textbook of Endocrinology,* 9th ed. Saunders, 2003.

Lechner D, Habner JF: Stem cells for the treatment of diabetes mellitus. Endocrinology Rounds 2003;2:issue 2.

LeRoith D: Insulin-like growth factors. N Engl J Med 1997;336:633.

Meigs JB, Avruch J: The metabolic syndrome. Endocrinology Rounds 2003;2:issue 5.

Sealey RJ (basic research), Rolls BJ (clinical research), Hensrud DD (clinical practice): Three perspectives on obesity. Endocrine News 2004;29:7.

The Adrenal Medulla & Adrenal Cortex

OBJECTIVES

After reading this chapter, you should be able to:

- Name the three catecholamines secreted by the adrenal medulla and summarize their biosynthesis, metabolism, and function.
- List the stimuli that increase adrenal medullary secretion.
- Differentiate between C_{18}, C_{19}, and C_{21} steroids and give examples of each.
- Outline the steps involved in steroid biosynthesis in the adrenal cortex.
- Name the plasma proteins that bind adrenocortical steroids and discuss their physiologic role.
- Name the major site of adrenocortical hormone metabolism and the principal metabolites produced from glucocorticoids, adrenal androgens, and aldosterone.
- Describe the mechanisms by which glucocorticoids and aldosterone produce changes in cellular function.
- List and briefly describe the physiologic and pharmacologic effects of glucocorticoids.
- Contrast the physiologic and pathologic effects of adrenal androgens.
- Describe the mechanisms that regulate secretion of glucocorticoids and adrenal sex hormones.
- List the actions of aldosterone and describe the mechanisms that regulate aldosterone secretion.
- Describe the main features of the diseases caused by excess or deficiency of each of the hormones of the adrenal gland.

INTRODUCTION

There are two endocrine organs in the adrenal gland, one surrounding the other. The main secretions of the inner **adrenal medulla** (Figure 22–1) are the catecholamines **epinephrine, norepinephrine,** and **dopamine;** the outer **adrenal cortex** secretes steroid hormones.

The adrenal medulla is in effect a sympathetic ganglion in which the postganglionic neurons have lost their axons and become secretory cells. The cells secrete when stimulated by the preganglionic nerve fibers that reach the gland via the splanchnic nerves. Adrenal medullary hormones work mostly to prepare the body for emergencies, the "fight-or-flight" responses.

The adrenal cortex secretes **glucocorticoids,** steroids with widespread effects on the metabolism of carbohydrate and protein; a **mineralocorticoid** essential to the maintenance of Na^+ balance and extracellular fluid (ECF) volume; and **sex hormones** that exert effects on reproductive function. Of these, the mineralocorticoids and the glucocorticoids are necessary for survival. Adrenocortical secretion is controlled primarily by adrenocorticotropic hormone (ACTH) from the anterior pituitary, but mineralocorticoid secretion is also subject to independent control by circulating factors, of which the most important is **angiotensin II,** a peptide formed in the bloodstream by the action of **renin.**

FIGURE 22–1   Human adrenal glands. Adrenocortical tissue is yellow; adrenal medullary tissue is orange. Note the location of the adrenals at the superior pole of each kidney. Also shown are extra-adrenal sites (grey) at which cortical and medullary tissue is sometimes found. (Reproduced with permission from Williams RH: *Textbook of Endocrinology*, 4th ed. Williams RH [editor]: Saunders, 1968.)

ADRENAL MORPHOLOGY

The adrenal medulla, which constitutes 28% of the mass of the adrenal gland, is made up of interlacing cords of densely inner-vated granule-containing cells that abut on venous sinuses. Two cell types can be distinguished morphologically: an epineph-rine-secreting type that has larger, less dense granules; and a norepinephrine-secreting type in which smaller, very dense granules fail to fill the vesicles in which they are contained. In humans, 90% of the cells are the epinephrine-secreting type and 10% are the norepinephrine-secreting type. The type of cell that secretes dopamine is unknown. **Paraganglia,** small groups of cells resembling those in the adrenal medulla, are found near the thoracic and abdominal sympathetic ganglia (Figure 22–1).

In adult mammals, the adrenal cortex is divided into three zones (Figure 22–2). The outer **zona glomerulosa** is made up of whorls of cells that are continuous with the columns of cells that form the **zona fasciculata.** These columns are separated by venous sinuses. The inner portion of the zona fasciculata merges into the **zona reticularis,** where the cell columns become inter-laced in a network. The zona glomerulosa makes up 15% of the mass of the adrenal gland; the zona fasciculata, 50%; and the zona reticularis, 7%. The adrenocortical cells contain abundant lipid, especially in the outer portion of the zona fasciculata. All three cortical zones secrete **corticosterone,** but the active enzy-matic mechanism for aldosterone biosynthesis is limited to the zona glomerulosa, whereas the enzymatic mechanisms for form-ing cortisol and sex hormones are found in the two inner zones. Furthermore, subspecialization occurs within the inner two

zones, the zona fasciculata, secreting mostly glucocorticoids and the zona reticularis, secreting mainly sex hormones.

Arterial blood reaches the adrenal from many small branches of the phrenic and renal arteries and the aorta. From a plexus in the capsule, blood flows through the cortex to the sinusoids of the medulla. The medulla is also supplied by a few arterioles that pass directly to it from the capsule. In most species, including humans, blood from the medulla flows into a central adrenal vein. The blood flow through the adrenal is large, as it is in most endocrine glands.

During fetal life, the human adrenal is large and under pitu-itary control, but the three zones of the permanent cortex repre-sent only 20% of the gland. The remaining 80% is the large **fetal adrenal cortex,** which undergoes rapid degeneration at the time of birth. A major function of this fetal adrenal is synthesis and secretion of sulfate conjugates of androgens that are converted in the placenta to estrogens (see Chapter 25). No structure is com-parable to the human fetal adrenal in laboratory animals.

An important function of the zona glomerulosa, in addition to aldosterone synthesis, is the formation of new cortical cells. The adrenal medulla does not regenerate, but when the inner two zones of the cortex are removed, a new zona fasciculata and zona reticularis regenerate from glomerular cells attached to the capsule. Small capsular remnants regrow large pieces of adreno-cortical tissue. Immediately after hypophysectomy, the zona fasciculata and zona reticularis begin to atrophy, whereas the zona glomerulosa is unchanged because of the action of angiotensin II on this zone. The ability to secrete aldosterone and conserve Na^+ is normal for some time after hypophysectomy, but in long-standing hypopituitarism, aldos-terone deficiency may develop, apparently because of the absence of a pituitary factor that maintains the responsiveness of the zona glomerulosa. Injections of ACTH and stimuli that cause endogenous ACTH secretion produce hypertrophy of the zona fasciculata and zona reticularis but actually decrease, rather than increase, the size of the zona glomerulosa.

The cells of the adrenal cortex contain large amounts of smooth endoplasmic reticulum, which is involved in the ster-oid-forming process. Other steps in steroid biosynthesis occur in the mitochondria. The structure of steroid-secreting cells is very similar throughout the body. The typical features of such cells are shown in Figure 22–3.

ADRENAL MEDULLA: STRUCTURE & FUNCTION OF MEDULLARY HORMONES

CATECHOLAMINES

Norepinephrine, epinephrine, and dopamine are secreted by the adrenal medulla. Cats and some other species secrete mainly nor-epinephrine, but in dogs and humans, most of the catecholamine output in the adrenal vein is epinephrine. Norepinephrine also enters the circulation from noradrenergic nerve endings.

Zona
glomerulosa

→ Aldosterone

Zona
fasciculata

→ Cortisol
and
androgens

Zona
reticularis

Cortex

Medulla

→ Epinephrine
and
norepinephrine

— Cortex

— Medulla

FIGURE 22–2 **Section through an adrenal gland showing both the medulla and the zones of the cortex, as well as the hormones they secrete.** (Reproduced with permission from Widmaier EP, Raff H, Strang KT: *Vander's Human Physiology: The Mechanisms of Body Function*, 11th ed. McGraw-Hill, 2008.)

The structures of norepinephrine, epinephrine, and dopamine and the pathways for their biosynthesis and metabolism are discussed in Chapter 7. Norepinephrine is formed by hydroxylation and decarboxylation of tyrosine, and epinephrine by methylation of norepinephrine. Phenylethanolamine-*N*-methyltransferase (PNMT), the enzyme that catalyzes the formation of epinephrine from norepinephrine, is found in appreciable quantities only in the brain and the adrenal medulla. Adrenal medullary PNMT is induced by glucocorticoids. Although relatively large amounts are required, the glucocorticoid concentration is high in the blood draining from the cortex to the medulla. After hypophysectomy, the glucocorticoid concentration of this blood falls and epinephrine synthesis is decreased. In addition, glucocorticoids are apparently necessary for the normal development of the adrenal medulla; in 21β-hydroxylase deficiency, glucocorticoid secretion is reduced during fetal life and the adrenal medulla is dysplastic. In untreated 21β-hydroxylase deficiency, circulating catecholamines are low after birth.

In plasma, about 95% of the dopamine and 70% of the norepinephrine and epinephrine are conjugated to sulfate. Sulfate conjugates are inactive and their function is unsettled. In recumbent humans, the normal plasma level of free norepinephrine is about 300 pg/mL (1.8 nmol/L). On standing, the level increases 50–100% (Figure 22–4). The plasma norepinephrine level is gener-

ally unchanged after adrenalectomy, but the free epinephrine level, which is normally about 30 pg/mL (0.16 nmol/L), falls to essentially zero. The epinephrine found in tissues other than the adrenal medulla and the brain is for the most part absorbed from the bloodstream rather than synthesized in situ. Interestingly, low levels of epinephrine reappear in the blood some time after bilateral adrenalectomy, and these levels are regulated like those secreted by the adrenal medulla. They may come from cells such as the intrinsic cardiac adrenergic (ICA) cells (see Chapter 17), but their exact source is unknown.

The plasma free dopamine level is about 35 pg/mL (0.23 nmol/L), and appreciable quantities of dopamine are present in the urine. Half the plasma dopamine comes from the adrenal medulla, whereas the remaining half presumably comes from the sympathetic ganglia or other components of the autonomic nervous system.

The catecholamines have a half-life of about 2 min in the circulation. For the most part, they are methoxylated and then oxidized to 3-methoxy-4-hydroxymandelic acid (vanillylmandelic acid [VMA]; see Chapter 7). About 50% of the secreted catecholamines appear in the urine as free or conjugated metanephrine and normetanephrine, and 35% as VMA. Only small amounts of free norepinephrine and epinephrine are excreted. In normal humans, about 30 µg of norepinephrine,

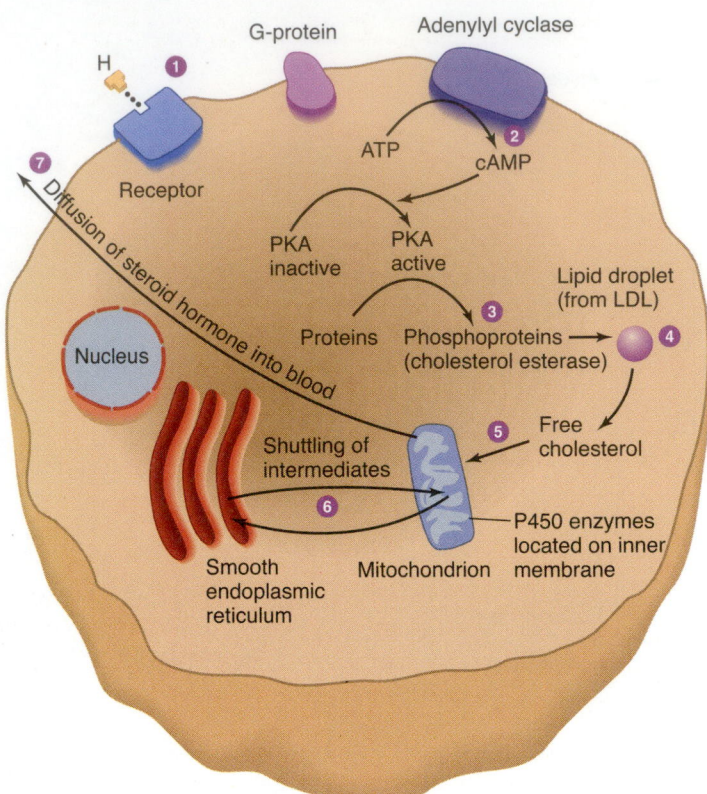

FIGURE 22–3 Schematic overview of the structures of steroid-secreting cells and the intracellular pathway of steroid synthesis. PKA: protein kinase A; LDL: low-density lipoprotein. (Reproduced with permission from Widmaier EP, Raff H, Strang KT: *Vander's Human Physiology: The Mechanisms of Body Function*, 11th ed. McGraw-Hill, 2008.)

6 μg of epinephrine, and 700 μg of VMA are excreted per day.

OTHER SUBSTANCES SECRETED BY THE ADRENAL MEDULLA

In the medulla, norepinephrine and epinephrine are stored in granules with ATP. The granules also contain chromogranin A (see Chapter 7). Secretion is initiated by acetylcholine released from the preganglionic neurons that innervate the secretory cells. Acetylcholine activates cation channels allowing Ca^{2+} to enter the cells from the extracellular fluid (ECF) and trigger the exocytosis of the granules. In this fashion, catecholamines, (adenosine triphosphate) ATP, and proteins from the granules are all released into the blood together.

Epinephrine-containing cells of the medulla also contain and secrete opioid peptides (see Chapter 7). The precursor molecule is preproenkephalin. Most of the circulating metenkephalin comes from the adrenal medulla. The circulating opioid peptides do not cross the blood–brain barrier.

Adrenomedullin, a vasodepressor polypeptide found in the adrenal medulla, is discussed in Chapter 33.

EFFECTS OF EPINEPHRINE & NOREPINEPHRINE

In addition to mimicking the effects of noradrenergic nervous discharge, norepinephrine and epinephrine exert metabolic effects that include glycogenolysis in liver and skeletal muscle,

mobilization of free fatty acids (FFA), increased plasma lactate, and stimulation of the metabolic rate. The effects of norepinephrine and epinephrine are brought about by actions on two classes of receptors: α- and β-adrenergic receptors. Alpha receptors are subdivided into two groups, α_1 and α_2 receptors, and β receptors into β_1, β_2, and β_3 receptors, as outlined in Chapter 4. There are three subtypes of α_1 receptors and three subtypes of α_2 receptors (see Table 7–2).

Norepinephrine and epinephrine both increase the force and rate of contraction of the isolated heart. These responses are mediated by β_1 receptors. The catecholamines also increase myocardial excitability, causing extrasystoles and, occasionally, more serious cardiac arrhythmias. Norepinephrine produces vasoconstriction in most if not all organs via α_1 receptors, but epinephrine dilates the blood vessels in skeletal muscle and the liver via β_2 receptors. This usually overbalances the vasoconstriction produced by epinephrine elsewhere, and the total peripheral resistance drops. When norepinephrine is infused slowly in normal animals or humans, the systolic and diastolic blood pressures rise. The **hypertension** stimulates the carotid and aortic baroreceptors, producing reflex bradycardia that overrides the direct cardioacceleratory effect of norepinephrine. Consequently, cardiac output per minute falls. Epinephrine causes a widening of the pulse pressure, but because baroreceptor stimulation is insufficient to obscure the direct effect of the hormone on the heart, cardiac rate and output increase. These changes are summarized in Figure 22–5.

Catecholamines increase alertness (see Chapter 15). Epinephrine and norepinephrine are equally potent in this

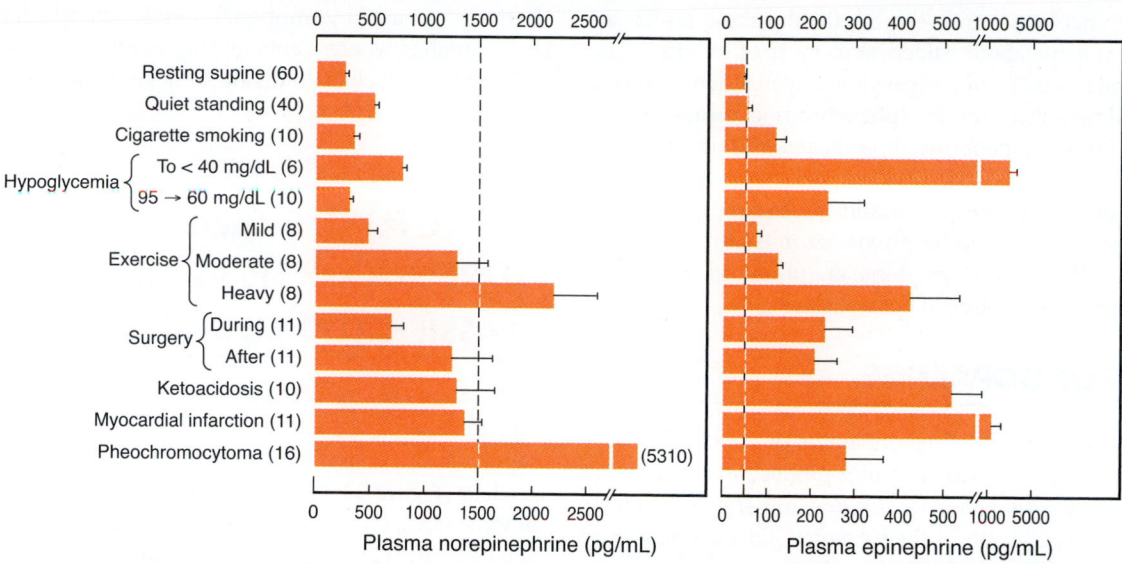

FIGURE 22–4 **Norepinephrine and epinephrine levels in human venous blood in various physiologic and pathologic states.** Note that the horizontal scales are different. The numbers to the left in parentheses are the numbers of subjects tested. In each case, the vertical dashed line identifies the threshold plasma concentration at which detectable physiologic changes are observed. (Modified and reproduced with permission from Cryer PE: Physiology and pathophysiology of the human sympathoadrenal neuroendocrine system. N Engl J Med 1980;303:436.)

regard, although in humans epinephrine usually evokes more anxiety and fear.

The catecholamines have several different actions that affect blood glucose. Epinephrine and norepinephrine both cause glycogenolysis. They produce this effect via β-adrenergic receptors that increase cyclic adenosine monophosphate (cAMP), with activation of phosphorylase, and via α-adrenergic receptors that increase intracellular Ca^{2+} (see Chapter 7). In addition, the catecholamines increase the secretion of insulin and glucagon via β-adrenergic mechanisms and inhibit the secretion of these hormones via α-adrenergic mechanisms.

Norepinephrine and epinephrine also produce a prompt rise in the metabolic rate that is independent of the liver and a smaller, delayed rise that is abolished by hepatectomy and coincides with the rise in blood lactate concentration. The initial rise in metabolic rate may be due to cutaneous vasoconstriction, which decreases heat loss and leads to a rise in body temperature, or to increased muscular activity, or both. The second rise is probably due to oxidation of lactate in the liver. Mice unable to make norepinephrine or epinephrine because their dopamine β-hydroxylase gene is knocked out are intolerant to cold, but surprisingly, their basal metabolic rate is elevated. The cause of this elevation is unknown.

When injected, epinephrine and norepinephrine cause an initial rise in plasma K^+ because of release of K^+ from the liver and then a prolonged fall in plasma K^+ because of an increased entry of K^+ into skeletal muscle that is mediated by $β_2$-adrenergic receptors. Some evidence suggests that activation of α receptors opposes this effect.

The increases in plasma norepinephrine and epinephrine that are needed to produce the various effects listed above have been determined by infusion of catecholamines in resting humans. In general, the threshold for the cardiovascular and

the metabolic effects of norepinephrine is about 1500 pg/mL, that is, about five times the resting value (Figure 22–4). Epinephrine, on the other hand, produces tachycardia when the plasma level is about 50 pg/mL, that is, about twice the resting value. The threshold for increased systolic blood pressure and lipolysis is about 75 pg/mL; the threshold for hyperglycemia, increased plasma lactate, and decreased diastolic blood pressure is about 150 pg/mL; and the threshold for the α-mediated decrease in insulin secretion is about 400 pg/mL. Plasma epinephrine often exceeds these thresholds. On the other hand,

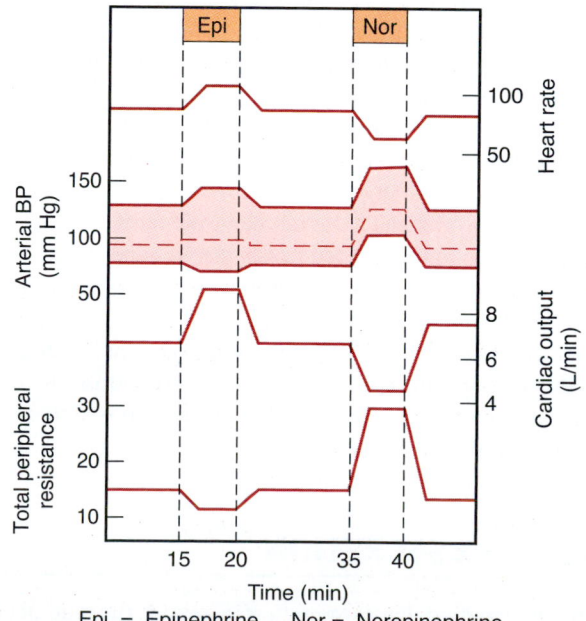

FIGURE 22–5 **Circulatory changes produced in humans by the slow intravenous infusion of epinephrine and norepinephrine.**

plasma norepinephrine rarely exceeds the threshold for its cardiovascular and metabolic effects, and most of its effects are due to its local release from postganglionic sympathetic neurons. Most adrenal medullary tumors (**pheochromocytomas**) secrete norepinephrine, or epinephrine, or both, and produce sustained hypertension. However, 15% of epinephrine-secreting tumors secrete this catecholamine episodically, producing intermittent bouts of palpitations, headache, glycosuria, and extreme systolic hypertension. These same symptoms are produced by intravenous injection of a large dose of epinephrine.

EFFECTS OF DOPAMINE

The physiologic function of the dopamine in the circulation is unknown. However, injected dopamine produces renal vasodilation, probably by acting on a specific dopaminergic receptor. It also produces vasodilation in the mesentery. Elsewhere, it produces vasoconstriction, probably by releasing norepinephrine, and it has a positively inotropic effect on the heart by an action on β_1-adrenergic receptors. The net effect of moderate doses of dopamine is an increase in systolic pressure and no change in diastolic pressure. Because of these actions, dopamine is useful in the treatment of traumatic and cardiogenic shock (see Chapter 33).

Dopamine is made in the renal cortex. It causes natriuresis and may exert this effect by inhibiting renal Na^+–K^+ ATPase.

REGULATION OF ADRENAL MEDULLARY SECRETION

NEURAL CONTROL

Certain drugs act directly on the adrenal medulla, but physiologic stimuli affect medullary secretion through the nervous system. Catecholamine secretion is low in basal states, but the secretion of epinephrine and, to a lesser extent, that of norepinephrine is reduced even further during sleep.

Increased adrenal medullary secretion is part of the diffuse sympathetic discharge provoked in emergency situations, which Cannon called the "emergency function of the sympathoadrenal system." The ways in which this discharge prepares the individual for flight or fight are described in Chapter 17, and the increases in plasma catecholamines under various conditions are shown in Figure 22–4.

The metabolic effects of circulating catecholamines are probably important, especially in certain situations. The calorigenic action of catecholamines in animals exposed to cold is an example, and so is the glycogenolytic effect (see Chapter 21) in combating hypoglycemia.

SELECTIVE SECRETION

When adrenal medullary secretion is increased, the ratio of norepinephrine to epinephrine in the adrenal effluent is generally unchanged. However, norepinephrine secretion tends to be selectively increased by emotional stresses with which the individual is familiar, whereas epinephrine secretion rises selectively in situations in which the individual does not know what to expect.

ADRENAL CORTEX: STRUCTURE & BIOSYNTHESIS OF ADRENOCORTICAL HORMONES

CLASSIFICATION & STRUCTURE

The hormones of the adrenal cortex are derivatives of cholesterol. Like cholesterol, bile acids, vitamin D, and ovarian and testicular steroids, they contain the **cyclopentanoperhydrophenanthrene nucleus** (Figure 22–6). Gonadal and adrenocortical steroids are of three types: C_{21} steroids, which have a two-carbon side chain at position 17; C_{19} steroids, which have a keto or hydroxyl group at position 17; and C_{18} steroids, which, in addition to a 17-keto or hydroxyl group, have no angular methyl group attached to position 10. The adrenal cortex secretes primarily C_{21} and C_{19} steroids. Most of the C_{19} steroids have a keto group at position 17 and are therefore called **17-ketosteroids.** The C_{21} steroids that have a

FIGURE 22–6   Basic structure of adrenocortical and gonadal steroids. The letters in the formula for cholesterol identify the four basic rings, and the numbers identify the positions in the molecule. As shown here, the angular methyl groups (positions 18 and 19) are usually indicated simply by straight lines.

hydroxyl group at the 17 position in addition to the side chain are often called 17-hydroxycorticoids or 17-hydroxycorticosteroids.

The C_{19} steroids have androgenic activity. The C_{21} steroids are classified, using Selye's terminology, as mineralocorticoids or glucocorticoids. All secreted C_{21} steroids have both mineralocorticoid and glucocorticoid activity; **mineralocorticoids** are those in which effects on Na^+ and K^+ excretion predominate and **glucocorticoids** are those in which effects on glucose and protein metabolism predominate.

The details of steroid nomenclature and isomerism can be found elsewhere. However, it is pertinent to mention that the Greek letter Δ indicates a double bond and that the groups that lie above the plane of each of the steroid rings are indicated by the Greek letter α and a solid line (—OH), whereas those that lie below the plane are indicated by α and a dashed line (- - -OH). Thus, the C_{21} steroids secreted by the adrenal have a Δ^4-3-keto configuration in the A ring. In most naturally occurring adrenal steroids, 17-hydroxy groups are in the α configuration, whereas 3-, 11-, and 21-hydroxy groups are in the β configuration. The 18-aldehyde configuration on naturally occurring aldosterone is the D form. L-Aldosterone is physiologically inactive.

SECRETED STEROIDS

Innumerable steroids have been isolated from adrenal tissue, but the only steroids normally secreted in physiologically significant amounts are the mineralocorticoid **aldosterone**, the glucocorticoids **cortisol** and **corticosterone**, and the androgens **dehydroepiandrosterone (DHEA)** and **androstenedi-**

one. The structures of these steroids are shown in Figures 22–7 and 22–8. **Deoxycorticosterone** is a mineralocorticoid that is normally secreted in about the same amount as aldosterone (Table 22–1) but has only 3% of the mineralocorticoid activity of aldosterone. Its effect on mineral metabolism is usually negligible, but in diseases in which its secretion is increased, its effect can be appreciable. Most of the estrogens that are not formed in the ovaries are produced in the circulation from adrenal androstenedione. Almost all the dehydroepiandrosterone is secreted conjugated with sulfate, although most if not all of the other steroids are secreted in the free, unconjugated form.

The secretion rate for individual steroids can be determined by injecting a very small dose of isotopically labeled steroid and determining the degree to which the radioactive steroid excreted in the urine is diluted by unlabeled secreted hormone. This technique is used to measure the output of many different hormones (see Clinical Box 22–1).

SPECIES DIFFERENCES

In all species from amphibia to humans, the major C_{21} steroid hormones secreted by adrenocortical tissue appear to be aldosterone, cortisol, and corticosterone, although the ratio of cortisol to corticosterone varies. Birds, mice, and rats secrete corticosterone almost exclusively; dogs secrete approximately equal amounts of the two glucocorticoids; and cats, sheep, monkeys, and humans secrete predominantly cortisol. In humans, the ratio of secreted cortisol to corticosterone is approximately 7:1.

FIGURE 22–7 **Outline of hormone biosynthesis in the zona fasciculata and zona reticularis of the adrenal cortex.** The major secretory products are underlined. The enzymes for the reactions are shown on the left and at the top of the chart. When a particular enzyme is deficient, hormone production is blocked at the points indicated by the shaded bars.

FIGURE 22–8   Hormone synthesis in the zona glomerulosa.
The zona glomerulosa lacks 17α-hydroxylase activity, and only the zona glomerulosa can convert corticosterone to aldosterone because it is the only zone that normally contains aldosterone synthase. ANG II, angiotensin II.

CLINICAL BOX 22–1

Synthetic Steroids

As with many other naturally occurring substances, the activity of adrenocortical steroids can be increased by altering their structure. A number of synthetic steroids are available that have many times the activity of cortisol. The relative glucocorticoid and mineralocorticoid potencies of the natural steroids are compared with those of the synthetic steroids 9α-fluorocortisol, prednisolone, and dexamethasone in Table 22–2. The potency of dexamethasone is due to its high affinity for glucocorticoid receptors and its long half-life. Prednisolone also has a long half-life.

STEROID BIOSYNTHESIS

The major paths by which the naturally occurring adrenocortical hormones are synthesized in the body are summarized in Figures 22–7 and 22–8. The precursor of all steroids is cholesterol. Some of the cholesterol is synthesized from acetate, but most of it is taken up from LDL in the circulation. LDL receptors are especially abundant in adrenocortical cells. The cholesterol is esterified and stored in lipid droplets. **Cholesterol ester hydrolase** catalyzes the formation of free cholesterol in the lipid droplets (Figure 22–9). The cholesterol is transported to mitochondria by a sterol carrier protein. In the mitochondria, it is converted to pregnenolone in a reaction catalyzed by an enzyme known as **cholesterol desmolase** or **side-chain cleavage enzyme.** This enzyme, like most of the enzymes involved in steroid biosynthesis, is a member of the cytochrome P450 superfamily and is also known as **P450scc** or **CYP11A1.** For convenience, the various

TABLE 22–1   Principal adrenocortical hormones in adult humans.[a]

Name	Synonyms	Average Plasma Concentration (Free and Bound)[a] (μg/dL)	Average Amount Secreted (mg/24 h)
Cortisol	Compound F, hydrocortisone	13.9	10
Corticosterone	Compound B	0.4	3
Aldosterone		0.0006	0.15
Deoxycorticos-terone	DOC	0.0006	0.20
Dehydroepi-androsterone sulfate	DHEAS	175.0	20

[a]All plasma concentration values except DHEAS are fasting morning values after overnight recumbency.

TABLE 22–2   Relative potencies of corticosteroids compared with cortisol.[a]

Steroid	Glucocorticoid Activity	Mineralocorticoid Activity
Cortisol	1.0	1.0
Corticosterone	0.3	15
Aldosterone	0.3	3000
Deoxycorticosterone	0.2	100
Costisone	0.7	0.8
Prednisolone	4	0.8
9α-Fluorocortisol	10	125
Dexamethasone	25	~0

[a]Values are approximations based on liver glycogen deposition or anti-inflammatory assays for glucocorticoid activity, and effect on urinary Na^+/K^+ or maintenance of adrena-lectomized animals for mineralocorticoid activity. The last three steroids listed are synthetic compounds that do not occur naturally.

FIGURE 22–9 **Mechanism of action of ACTH on cortisol-secreting cells in the inner two zones of the adrenal cortex.** When ACTH binds to its receptor (R), adenylyl cyclase (AC) is activated via Gs. The resulting increase in cAMP activates protein kinase A, and the kinase phosphorylates cholesteryl ester hydrolase (CEH), increasing its activity. Consequently, more free cholesterol is formed and converted to pregnenolone. Note that in the subsequent steps in steroid biosynthesis, products are shuttled between the mitochondria and the smooth endoplasmic reticulum (SER). Corticosterone is also synthesized and secreted.

names of the enzymes involved in adrenocortical steroid biosynthesis are summarized in Table 22–3.

Pregnenolone moves to the smooth endoplasmic reticulum, where some of it is dehydrogenated to form progesterone in a reaction catalyzed by **3β-hydroxysteroid dehydrogenase.** This enzyme has a molecular weight of 46,000 and is not a cytochrome P450. It also catalyzes the conversion of 17α-hydroxypregnenolone to 17α-hydroxyprogesterone, and dehydroepiandrosterone to androstenedione (Figure 22–7) in the smooth endoplasmic reticulum. The 17α-hydroxypregnenolone and the 17α-hydroxyprogesterone are formed from pregnenolone and progesterone, respectively (Figure 22–7) by the action of **17α-hydroxylase.** This is another mitochondrial P450, and it is also known as **P450c17** or **CYP17.**

TABLE 22–3 Nomenclature for adrenal steroidogenic enzymes and their location in adrenal cells.

Trivial Name	P450	CYP	Location
Cholesterol desmolase; side-chain cleavage enzyme	P450scc	CYP11A1	Mitochondria
3β-Hydroxysteroid dehydrogenase	...	...	SER
17α-Hydroxylase, 17,20-lyase	P450c17	CYP17	Mitochondria
21β-Hydroxylase	P450c21	CYP21A2	SER
11β-Hydroxylase	P450c11	CYP11B1	Mitochondria
Aldosterone synthase	P450c11AS	CYP11B2	Mitochondria

SER = smooth endoplasmic reticulum.

Located in another part of the same enzyme is **17,20-lyase** activity that breaks the 17,20 bond, converting 17α-pregnenolone and 17α-progesterone to the C$_{19}$ steroids dehydroepiandrosterone and androstenedione.

Hydroxylation of progesterone to 11-deoxycorticosterone and of 17α-hydroxyprogesterone to 11-deoxycortisol occurs in the smooth endoplasmic reticulum. These reactions are catalyzed by 21β-hydroxylase, a cytochrome P450 that is also known as **P450c21** or **CYP21A2.**

11-deoxycorticosterone and the 11-deoxycortisol move back to the mitochondria, where they are 11-hydroxylated to form corticosterone and cortisol. These reactions occur in the zona fasciculata and zona reticularis and are catalyzed by 11β-hydroxylase, a cytochrome P450 also known as **P450c11** or **CYP11B1.**

In the zona glomerulosa there is no 11β-hydroxylase but a closely related enzyme called **aldosterone synthase** is present. This cytochrome P450 is 95% identical to 11β-hydroxylase and is also known as **P450c11AS** or **CYP11B2.** The genes that code CYP11B1 and CYP11B2 are both located on chromosome 8. However, aldosterone synthase is normally found only in the zona glomerulosa. The zona glomerulosa also lacks 17α-hydroxylase. This is why the zona glomerulosa makes aldosterone but fails to make cortisol or sex hormones.

Furthermore, subspecialization occurs within the inner two zones. The zona fasciculata has more 3β-hydroxysteroid dehydrogenase activity than the zona reticularis, and the zona reticularis has more of the cofactors required for the expression of the 17,20-lyase activity of 17α-hydroxylase. Therefore, the zona fasciculata makes more cortisol and corticosterone, and the zona reticularis makes more androgens. Most of the dehydroepiandrosterone that is formed is converted to dehydroepiandrosterone sulfate by **adrenal sulfokinase,** and this enzyme is localized in the zona reticularis as well.

ACTION OF ACTH

ACTH binds to high-affinity receptors on the plasma membrane of adrenocortical cells. This activates adenylyl cyclase via G$_s$. The resulting reactions (Figure 22–9) lead to a prompt increase in the formation of pregnenolone and its derivatives, with secretion of the latter. Over longer periods, ACTH also increases the synthesis of the P450s involved in the synthesis of glucocorticoids.

ACTIONS OF ANGIOTENSIN II

Angiotensin II binds to AT$_1$ receptors (see Chapter 39) in the zona glomerulosa which act via a G protein to activate phospholipase C. The resulting increase in protein kinase C fosters the conversion of cholesterol to pregnenolone (Figure 22–8) and facilitates the action of aldosterone synthase, resulting in increased secretion of aldosterone.

ENZYME DEFICIENCIES

The consequences of inhibiting any of the enzyme systems involved in steroid biosynthesis can be predicted from Figures 22–7 and 22–8. Congenital defects in the enzymes lead to deficient cortisol secretion and the syndrome of **congenital adrenal hyperplasia.** The hyperplasia is due to increased ACTH secretion. Cholesterol desmolase deficiency is fatal in utero because it prevents the placenta from making the progesterone necessary for pregnancy to continue. A cause of severe congenital adrenal hyperplasia in newborns is a loss of function mutation of the gene for the **steroidogenic acute regulatory (StAR) protein.** This protein is essential in the adrenals and gonads but not in the placenta for the normal movement of cholesterol into the mitochondria to reach cholesterol desmolase, which is located on the matrix space side of the internal mitochondrial membrane. In its absence, only small amounts of steroids are formed. The degree of ACTH stimulation is marked, resulting eventually in accumulation of large numbers of lipoid droplets in the adrenal. For this reason, the condition is called **congenital lipoid adrenal hyperplasia.** Because androgens are not formed, female genitalia develop regardless of genetic sex (see Chapter 25). In 3β hydroxysteroid dehydrogenase deficiency, another rare condition, DHEA secretion is increased. This steroid is a weak androgen that can cause some masculinization in females with the disease, but it is not adequate to produce full masculinization of the genitalia in genetic males. Consequently, hypospadias is common. In fully developed 17α-hydroxylase deficiency, a third rare condition due to a mutated gene for **CYP17,** no sex hormones are produced, so female external genitalia are present. However, the pathway leading to corticosterone and aldosterone is intact, and elevated levels of 11-deoxycorticosterone and other mineralocorticoids produce hypertension and hypokalemia. Cortisol is deficient, but this is partially compensated by the glucocorticoid activity of corticosterone.

Unlike the defects discussed in the preceding paragraph, 21β-hydroxylase deficiency is common, accounting for 90% or more of the enzyme deficiency cases. The 21β-hydroxylase gene, which is in the human leukocyte antigen (HLA) complex of genes on the short arm of chromosome 6 (see Chapter 3) is one of the most polymorphic in the human genome. Mutations occur at many different sites in the gene, and the abnormalities that are produced therefore range from mild to severe. Production of cortisol and aldosterone are generally reduced, so ACTH secretion and consequently production of precurose steroids are increased. These steroids are converted to androgens, producing **virilization.** The characteristic pattern that develops in females in the absence of treatment is the **adrenogenital syndrome.** Masculization may not be marked until later in life and mild cases can be detected only by laboratory tests. In 75% of the cases, aldosterone deficiency causes appreciable loss of Na$^+$ (**salt-losing form** of adrenal hyperplasia). The resulting hypovolemia can be severe.

In 11β-hydroxylase deficiency, virilization plus excess secretion of 11-deoxycortisol and 11-deoxycorticosterone take place. Because the former is an active mineralocorticoid, patients with this condition also have salt and water retention and, in two-thirds of the cases, hypertension (**hypertensive form** of congenital adrenal hyperplasia).

Glucocorticoid treatment is indicated in all of the virilizing forms of congenital adrenal hyperplasia because it repairs the glucocorticoid deficit and inhibits ACTH secretion, reducing the abnormal secretion of androgens and other steroids.

Expression of the cytochrome P450 enzymes responsible for steroid hormone biosynthesis depends on **steroid factor-1 (SF-1),** an orphan nuclear receptor. If *Ft2-F1*, the gene for SF-1, is knocked out, gonads as well as adrenals fail to develop and additional abnormalities are present at the pituitary and hypothalamic level.

TRANSPORT, METABOLISM, & EXCRETION OF ADRENOCORTICAL HORMONES

GLUCOCORTICOID BINDING

Cortisol is bound in the circulation to an α globulin called **transcortin** or **corticosteroid-binding globulin (CBG).** A minor degree of binding to albumin also takes place (see Table 25–5). Corticosterone is similarly bound, but to a lesser degree. The half-life of cortisol in the circulation is therefore longer (about 60–90 min) than that of corticosterone (50 min). Bound steroids are physiologically inactive. In addition, relatively little free cortisol and corticosterone are found in the urine because of protein binding.

The equilibrium between cortisol and its binding protein and the implications of binding in terms of tissue supplies and ACTH secretion are summarized in Figure 22–10. The bound cortisol functions as a circulating reservoir of hormone that keeps a supply of free cortisol available to the tissues. The relationship is similar to that of T$_4$ and its binding protein (see Chapter 20). At normal levels of total plasma cortisol (13.5 µg/dL or 375 nmol/L), very little free cortisol is present in the plasma, but the binding sites on CBG become saturated when the total plasma cortisol

FIGURE 22–10   The interrelationships of free and bound cortisol. The dashed arrow indicates that cortisol inhibits ACTH secretion. The value for free cortisol is an approximation; in most studies, it is calculated by subtracting the protein-bound cortisol from the total plasma cortisol.

exceeds 20 μg/dL. At higher plasma levels, binding to albumin increases, but the main increase is in the unbound fraction.

CBG is synthesized in the liver and its production is increased by estrogen. CBG levels are elevated during pregnancy and depressed in cirrhosis, nephrosis, and multiple myeloma. When the CBG level rises, more cortisol is bound, and initially the free cortisol level drops. This stimulates ACTH secretion, and more cortisol is secreted until a new equilibrium is reached at which the bound cortisol is elevated but the free cortisol is normal. Changes in the opposite direction occur when the CBG level falls. This explains why pregnant women have high total plasma cortisol levels without symptoms of glucocorticoid excess and, conversely, why some patients with nephrosis have low total plasma cortisol without symptoms of glucocorticoid deficiency.

METABOLISM & EXCRETION OF GLUCOCORTICOIDS

Cortisol is metabolized in the liver, which is the principal site of glucocorticoid catabolism. Most of the cortisol is reduced to dihydrocortisol and then to tetrahydrocortisol, which is conjugated to glucuronic acid (Figure 22–11). The glucuronyl transferase system responsible for this conversion also catalyzes the formation of the glucuronides of bilirubin (see Chapter 29) and a number of hormones and drugs. Competitive inhibition takes place between these substrates for the enzyme system.

The liver and other tissues contain the enzyme 11β hydroxysteroid dehydrogenase. There are at least two forms of this enzyme. Type 1 catalyzes the conversion of cortisol to cortisone and the reverse reaction, though it functions primarily as a reductase, forming cortisol from corticosterone. Type 2 catalyzes almost exclusively the one-way conversion of cortisol to cortisone. Cortisone is an active glucocorticoid because it is converted to cortisol, and it is well known because of its extensive use in medicine. It is not secreted in appreciable quantities by the adrenal glands. Little, if any, of the cortisone formed in the liver enters the circulation, because it is promptly reduced and conjugated to form tetrahydrocortisone glucuronide. The tetrahydroglucuronide derivatives ("conjugates") of cortisol and corticosterone are freely soluble. They enter the circulation, where they do not become bound to protein. They are rapidly excreted in the urine.

About 10% of the secreted cortisol is converted in the liver to the 17-ketosteroid derivatives of cortisol and cortisone. The ketosteroids are conjugated for the most part to sulfate and then excreted in the urine. Other metabolites, including 20-hydroxy derivatives, are formed. There is an enterohepatic circulation of glucocorticoids and about 15% of the secreted cortisol is excreted in the stool. The metabolism of corticosterone is similar to that of cortisol, except that it does not form a 17-ketosteroid derivative (see Clinical Box 22–2).

ALDOSTERONE

Aldosterone is bound to protein to only a slight extent, and its half-life is short (about 20 min). The amount secreted is small

FIGURE 22–11 Outline of hepatic metabolism of cortisol.

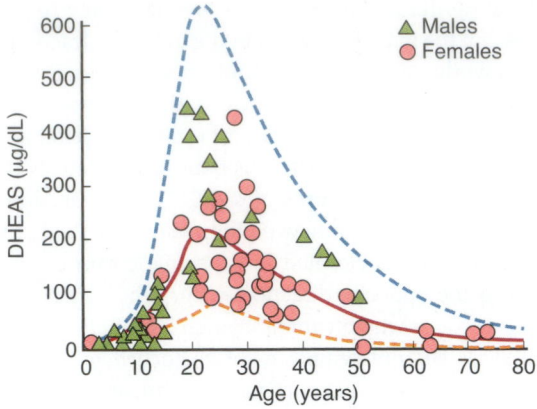

FIGURE 22–12   **Change in serum dehydroepiandrosterone sulfate (DHEAS) with age.** The middle line is the mean, and the dashed lines identify ±1.96 standard deviations.   (Reproduced, with permission, from Smith MR, et al: A radioimmunoassay for the estimation of serum dehydroepiandrosterone sulfate in normal and pathological sera. Clin Chim Acta 1975;65:5.)

(Table 22–1), and the total plasma aldosterone level in humans is normally about 0.006 μg/dL (0.17 nmol/L), compared with a cortisol level (bound and free) of about 13.5 μg/dL (375 nmol/L). Much of the aldosterone is converted in the liver to the tetrahydroglucuronide derivative, but some is changed in the liver and in the kidneys to an 18-glucuronide. This glucuronide, which is unlike the breakdown products of other steroids, is converted to free aldosterone by hydrolysis at pH 1.0, and it is therefore often referred to as the "acid-labile conjugate." Less than 1% of the secreted aldosterone appears in the urine in the free form. Another 5% is in the form of the acid-labile conjugate, and up to 40% is in the form of the tetrahydroglucuronide.

17-KETOSTEROIDS

The major adrenal androgen is the 17-ketosteroid dehydroepiandrosterone, although androstenedione is also secreted. The 11-hydroxy derivative of androstenedione and the 17-ketosteroids formed from cortisol and cortisone by side chain cleavage in the liver are the only 17-ketosteroids that have an =O or an —OH group in the 11 position ("11-oxy-17-keto-steroids"). Testosterone is also converted to 17-ketosteroids. Because the daily 17-ketosteroid excretion in normal adults is 15 mg in men and 10 mg in women, about two thirds of the urinary ketosteroids in men are secreted by the adrenal or formed from cortisol in the liver and about one third are of testicular origin.

Etiocholanolone, one of the metabolites of the adrenal androgens and testosterone, can cause fever when it is unconjugated (see Chapter 18). Certain individuals have episodic bouts of fever due to periodic accumulation in the blood of unconjugated etiocholanolone ("etiocholanolone fever").

EFFECTS OF ADRENAL ANDROGENS & ESTROGENS

ANDROGENS

Androgens are the hormones that exert masculinizing effects and they promote protein anabolism and growth (see Chapter 25). Testosterone from the testes is the most active androgen

and the adrenal androgens have less than 20% of its activity. Secretion of the adrenal androgens is controlled acutely by ACTH and not by gonadotropins. However, the concentration of dehydroepiandrosterone sulfate (DHEAS) increases until it peaks at about 225 mg/dL in the early 20s, then falls to very low values in old age (Figure 22–12). These long-term changes are not due to changes in ACTH secretion and appear to be due instead to a rise and then a gradual fall in the lyase activity of 17α-hydroxylase.

All but about 0.3% of the circulating DHEA is conjugated to sulfate (DHEAS). The secretion of adrenal androgens is nearly as great in castrated males and females as it is in normal males, so it is clear that these hormones exert very little masculinizing effect when secreted in normal amounts. However, they can produce appreciable masculinization when secreted in excessive amounts. In adult males, excess adrenal androgens merely accentuate existing characteristics, but in prepubertal boys they can cause precocious development of the secondary sex characteristics without testicular growth (**precocious pseudopuberty**). In females they cause female pseudo-hermaphroditism and the adrenogenital syndrome. Some health practitioners recommend injections of dehydroepiandrosterone to combat the effects of aging (see Chapter 1), but results to date are controversial at best.

ESTROGENS

The adrenal androgen androstenedione is converted to testosterone and to estrogens (aromatized) in fat and other peripheral tissues. This is an important source of estrogens in men and postmenopausal women (see Chapter 25).

PHYSIOLOGIC EFFECTS OF GLUCOCORTICOIDS

ADRENAL INSUFFICIENCY

In untreated adrenal insufficiency, Na^+ loss and shock occurs due to the lack of mineralocorticoid activity, as well as abnormalities of water, carbohydrate, protein, and fat metabolism due to the lack of glucocorticoids. These metabolic abnormalities are eventually fatal despite mineralocorticoid treatment. Small amounts of glucocorticoids correct the metabolic abnormalities, in part directly and in part by permitting other reactions to occur. It is important to separate these physiologic actions of glucocorticoids from the quite different effects produced by large amounts of the hormones.

MECHANISM OF ACTION

The multiple effects of glucocorticoids are triggered by binding to glucocorticoid receptors, and the steroid–receptor complexes act as transcription factors that promote the transcription of certain segments of DNA (see Chapter 1). This, in turn, leads via the appropriate mRNAs to synthesis of enzymes that alter cell function. In addition, it seems likely that glucocorticoids have nongenomic actions.

EFFECTS ON INTERMEDIARY METABOLISM

The actions of glucocorticoids on the intermediary metabolism of carbohydrate, protein, and fat are discussed in Chapter 21. They include increased protein catabolism and increased hepatic glycogenesis and gluconeogenesis. Glucose 6-phosphatase activity is increased, and the plasma glucose level rises. Glucocorticoids exert an anti-insulin action in peripheral tissues and make diabetes worse. However, the brain and the heart are spared, so the increase in plasma glucose provides extra glucose to these vital organs. In diabetics, glucocorticoids raise plasma lipid levels and increase ketone body formation, but in normal individuals, the increase in insulin secretion provoked by the rise in plasma glucose obscures these actions. In adrenal insufficiency, the plasma glucose level is normal as long as an adequate caloric intake is maintained, but fasting causes hypoglycemia that can be fatal. The adrenal cortex is not essential for the ketogenic response to fasting.

PERMISSIVE ACTION

Small amounts of glucocorticoids must be present for a number of metabolic reactions to occur, although the glucocorticoids do not produce the reactions by themselves. This effect is called their **permissive action.** Permissive effects include the requirement for glucocorticoids to be present for glucagon and catecholamines to exert their calorigenic effects (see above and Chapter 21), for catecholamines to exert their lipolytic effects, and for catecholamines to produce pressor responses and bronchodilation.

EFFECTS ON ACTH SECRETION

Glucocorticoids inhibit ACTH secretion, and ACTH secretion is increased in adrenalectomized animals. The consequences of the feedback action of cortisol on ACTH secretion are discussed below in the section on regulation of glucocorticoid secretion.

VASCULAR REACTIVITY

In adrenally insufficient animals, vascular smooth muscle becomes unresponsive to norepinephrine and epinephrine. The capillaries dilate and, terminally, become permeable to colloidal dyes. Failure to respond to the norepinephrine liberated at noradrenergic nerve endings probably impairs vascular compensation for the hypovolemia of adrenal insufficiency and promotes vascular collapse. Glucocorticoids restore vascular reactivity.

EFFECTS ON THE NERVOUS SYSTEM

Changes in the nervous system in adrenal insufficiency that are reversed only by glucocorticoids include the appearance of electroencephalographic waves slower than the normal α rhythm and personality changes. The latter, which are mild, include irritability, apprehension, and inability to concentrate.

EFFECTS ON WATER METABOLISM

Adrenal insufficiency is characterized by an inability to excrete a water load, causing the possibility of water intoxication. Only glucocorticoids repair this deficit. In patients with adrenal insufficiency who have not received glucocorticoids, glucose infusion may cause high fever ("glucose fever") followed by collapse and death. Presumably, the glucose is metabolized, the water dilutes the plasma, and the resultant osmotic gradient between the plasma and the cells causes the cells of the thermoregulatory centers in the hypothalamus to swell to such an extent that their function is disrupted.

The cause of defective water excretion in adrenal insufficiency is unsettled. Plasma vasopressin levels are elevated in adrenal insufficiency and reduced by glucocorticoid treatment. The glomerular filtration rate is low, and this probably contributes to the reduction in water excretion. The selective effect of glucocorticoids on the abnormal water excretion is consistent with this possibility, because even though the mineralocorticoids improve filtration by restoring plasma volume, the glucocorticoids raise the glomerular filtration rate to a much greater degree.

TABLE 22–4 Typical effects of cortisol on the white and red blood cell counts in humans (cells/μL).

Cell	Normal	Cortisol-Treated
White blood cells		
Total	9000	10,000
PMNs	5760	8330
Lymphocytes	2370	1080
Eosinophils	270	20
Basophils	60	30
Monocytes	450	540
Red blood cells	5 million	5.2 million

EFFECTS ON THE BLOOD CELLS & LYMPHATIC ORGANS

Glucocorticoids decrease the number of circulating eosinophils by increasing their sequestration in the spleen and lungs. Glucocorticoids also lower the number of basophils in the circulation and increase the number of neutrophils, platelets, and red blood cells (Table 22–4).

Glucocorticoids decrease the circulating lymphocyte count and the size of the lymph nodes and thymus by inhibiting lymphocyte mitotic activity. They reduce secretion of cytokines by inhibiting the effect of NF-κB on the nucleus. The reduced secretion of the cytokine IL-2 leads to reduced proliferation of lymphocytes (see Chapter 3), and these cells undergo apoptosis.

RESISTANCE TO STRESS

The term **stress** as used in biology has been defined as any change in the environment that changes or threatens to change an existing optimal steady state. Most, if not all, of these stresses activate processes at the molecular, cellular, or systemic level that tend to restore the previous state, that is, they are homeostatic reactions. Some, but not all, of the stresses stimulate ACTH secretion. The increase in ACTH secretion is essential for survival when the stress is severe. If animals are then hypophysectomized, or adrenalectomized but treated with maintenance doses of glucocorticoids, they die when exposed to the same stress.

The reason an elevated circulating ACTH, and hence glucocorticoid level, is essential for resisting stress remains for the most part unknown. Most of the stressful stimuli that increase ACTH secretion also activate the sympathetic nervous system, and part of the function of circulating glucocorticoids may be maintenance of vascular reactivity to catecholamines. Glucocorticoids are also necessary for the catecholamines to exert their full FFA-mobilizing action, and the FFAs are an important emergency energy supply. However, sympathectomized animals tolerate a variety of stresses with relative impunity. Another theory holds that glucocorticoids prevent other stress-induced changes from becoming excessive. At present, all that can be said is that stress causes increases in plasma glucocorticoids to high "pharmacologic" levels that in the short run are life-saving.

It should also be noted that the increase in ACTH, which is beneficial in the short term, becomes harmful and disruptive in the long term, causing among other things, the abnormalities of Cushing syndrome.

PHARMACOLOGIC & PATHOLOGIC EFFECTS OF GLUCOCORTICOIDS

CUSHING SYNDROME

The clinical picture produced by prolonged increases in plasma glucocorticoids was described by Harvey Cushing and is called **Cushing syndrome** (Figure 22–13). It may be **ACTH-independent** or **ACTH-dependent.** The causes of ACTH-independent Cushing syndrome include glucocorticoid-secreting adrenal tumors, adrenal hyperplasia, and prolonged administration of exogenous glucocorticoids for diseases such as rheumatoid arthritis. Rare but interesting ACTH-independent cases have been reported in which adrenocortical cells abnormally express receptors for gastric inhibitory polypeptide (GIP) (see Chapter 26), vasopressin (see Chapter 39), β-adrenergic agonists, IL-1, or gonadotropin-releasing hormone (GnRH; see Chapter 25), causing these peptides to increase glucocorticoid secretion. The causes of ACTH-dependent Cushing syndrome include ACTH-secreting tumors of the anterior pituitary gland and tumors of other organs, usually the

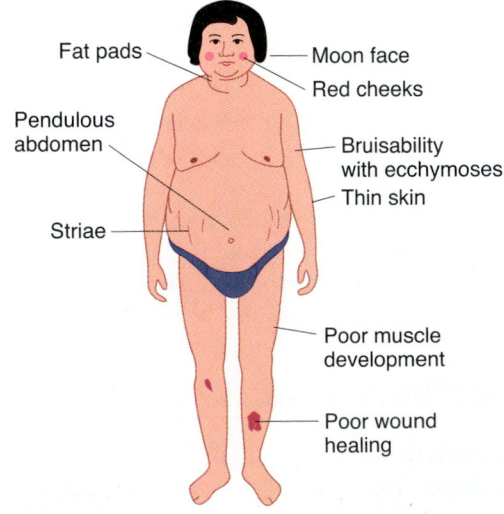

FIGURE 22–13 Typical findings in Cushing syndrome.
(Reproduced with permission from Forsham PH, Di Raimondo VC: *Traumatic Medicine and Surgery for the Attorney.* Butterworth, 1960.)

lungs, that secrete ACTH (ectopic ACTH syndrome) or corticotropin releasing hormone (CRH). Cushing syndrome due to anterior pituitary tumors is often called **Cushing disease** because these tumors were the cause of the cases described by Cushing. However, it is confusing to speak of Cushing disease as a subtype of Cushing syndrome, and the distinction seems to be of little more than historical value.

Patients with Cushing syndrome are protein-depleted as a result of excess protein catabolism. The skin and subcutaneous tissues are therefore thin and the muscles are poorly developed. Wounds heal poorly, and minor injuries cause bruises and ecchymoses. The hair is thin and scraggly. Many patients with the disease have some increase in facial hair and acne, but this is caused by the increased secretion of adrenal androgens and often accompanies the increase in glucocorticoid secretion.

Body fat is redistributed in a characteristic way. The extremities are thin, but fat collects in the abdominal wall, face, and upper back, where it produces a "buffalo hump." As the thin skin of the abdomen is stretched by the increased subcutaneous fat depots, the subdermal tissues rupture to form prominent reddish purple **striae.** These scars are seen normally whenever a rapid stretching of skin occurs, but in normal individuals the striae are usually inconspicuous and lack the intense purplish color.

Many of the amino acids liberated from catabolized proteins are converted into glucose in the liver and the resultant hyperglycemia and decreased peripheral utilization of glucose may be sufficient to precipitate insulin-resistant diabetes mellitus, especially in patients genetically predisposed to diabetes. Hyperlipemia and ketosis are associated with the diabetes, but acidosis is usually not severe.

The glucocorticoids are present in such large amounts in Cushing syndrome that they may exert a significant mineralocorticoid action. Deoxycorticosterone secretion is also elevated in cases due to ACTH hypersecretion. The salt and water retention plus the facial obesity cause the characteristic plethoric, rounded "moon-faced" appearance, and there may be significant K^+ depletion and weakness. About 85% of patients with Cushing syndrome are hypertensive. The hypertension may be due to increased deoxycorticosterone secretion, increased angiotensinogen secretion, or a direct glucocorticoid effect on blood vessels (see Chapter 33).

Glucocorticoid excess leads to bone dissolution by decreasing bone formation and increasing bone resorption. This leads to **osteoporosis,** a loss of bone mass that leads eventually to collapse of vertebral bodies and other fractures. The mechanisms by which glucocorticoids produce their effects on bone are discussed in Chapter 23.

Glucocorticoids in excess accelerate the basic electroencephalographic rhythms and produce mental aberrations ranging from increased appetite, insomnia, and euphoria to frank toxic psychoses. As noted above, glucocorticoid deficiency is also associated with mental symptoms, but the symptoms produced by glucocorticoid excess are more severe.

ANTI-INFLAMMATORY & ANTI-ALLERGIC EFFECTS OF GLUCOCORTICOIDS

Glucocorticoids inhibit the inflammatory response to tissue injury. The glucocorticoids also suppress manifestations of allergic disease that are due to the release of histamine from tissues. Both of these effects require high levels of circulating glucocorticoids and cannot be produced by administering steroids without producing the other manifestations of glucocorticoid excess. Furthermore, large doses of exogenous glucocorticoids inhibit ACTH secretion to the point that severe adrenal insufficiency can be a dangerous problem when therapy is stopped. However, local administration of glucocorticoids, for example, by injection into an inflamed joint or near an irritated nerve, produces a high local concentration of the steroid, often without enough systemic absorption to cause serious side effects.

The actions of glucocorticoids in patients with bacterial infections are dramatic but dangerous. For example, in pneumococcal pneumonia or active tuberculosis, the febrile reaction, the toxicity, and the lung symptoms disappear, but unless antibiotics are given at the same time, the bacteria spread throughout the body. It is important to remember that the symptoms are the warning that disease is present; when these symptoms are masked by treatment with glucocorticoids, there may be serious and even fatal delays in diagnosis and the institution of treatment with antimicrobial drugs.

The role of NF-κB in the anti-inflammatory and anti-allergic effects of glucocorticoids has been mentioned above and is discussed in Chapter 3. An additional action that combats local inflammation is inhibition of phospholipase A_2. This reduces the release of arachidonic acid from tissue phospholipids and consequently reduces the formation of leukotrienes, thromboxanes, prostaglandins, and prostacyclin (see Chapter 33).

OTHER EFFECTS

Large doses of glucocorticoids inhibit growth, decrease growth hormone secretion (see Chapter 24), induce PNMT, and decrease thyroid-stimulating hormone (TSH) secretion. During fetal life, glucocorticoids accelerate the maturation of surfactant in the lungs (see Chapter 35).

REGULATION OF GLUCOCORTICOID SECRETION

ROLE OF ACTH

Both basal secretion of glucocorticoids and the increased secretion provoked by stress are dependent upon ACTH from the anterior pituitary. Angiotensin II also stimulates the adrenal cortex, but its effect is mainly on aldosterone secretion. Large doses of a number of other naturally occurring substances, including vasopressin, serotonin, and vasoactive intestinal polypeptide (VIP), are capable of stimulating the adrenal directly, but there is no evidence that these agents play any role in the physiologic regulation of glucocorticoid secretion.

CHEMISTRY & METABOLISM OF ACTH

ACTH is a single-chain polypeptide containing 39 amino acids. Its origin from proopiomelanocortin (POMC) in the pituitary is discussed in Chapter 24. The first 23 amino acids in the chain generally constitute the active "core" of the molecule. Amino acids 24–39 constitute a "tail" that stabilizes the molecule and varies slightly in composition from species to species (Figure 22–14). The ACTHs that have been isolated are generally active in all species but antigenic in heterologous species.

ACTH is inactivated in blood in vitro more slowly than in vivo; its half-life in the circulation in humans is about 10 min. A large part of an injected dose of ACTH is found in the kidneys, but neither nephrectomy nor evisceration appreciably enhances its in vivo activity, and the site of its inactivation is not known.

EFFECT OF ACTH ON THE ADRENAL

After hypophysectomy, glucocorticoid synthesis and output decline within 1 h to very low levels, although some hormone is still secreted. Within a short time after an injection of ACTH (in dogs, less than 2 min), glucocorticoid output is increased (Figure 22–15). With low doses of ACTH, the relationship between the log of the dose and the increase in glucocorticoid secretion is linear. However, the maximal rate at which glucocorticoids can be secreted is rapidly reached, and in dogs, doses larger than

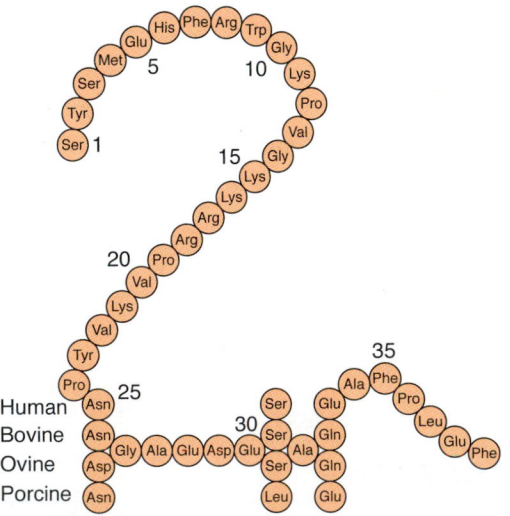

FIGURE 22–14 **Structure of ACTH.** In the species shown, the amino acid composition varies only at positions 25, 31, and 33. (Reproduced with permission from Li CH: Adrenocorticotropin 45: Revised amino acid sequences for sheep and bovine hormones. Biochem Biophys Res Commun 1972;49:835.)

FIGURE 22–15 **Changes in glucocorticoid output from the adrenal in hypophysectomized dogs following the intravenous (IV) administration of various doses of ACTH.**

10 mU only prolong the period of maximal secretion. A similar "ceiling on output" exists in humans. The effects of ACTH on adrenal morphology and the mechanism by which it increases steroid secretion have been discussed above.

ADRENAL RESPONSIVENESS

ACTH not only produces prompt increases in glucocorticoid secretion but also increases the sensitivity of the adrenal to subsequent doses of ACTH. Conversely, single doses of ACTH do not increase glucocorticoid secretion in chronically hypophysectomized animals and patients with hypopituitarism, and repeated injections or prolonged infusions of ACTH are necessary to restore normal adrenal responses to ACTH. Decreased responsiveness is also produced by doses of glucocorticoids that inhibit ACTH secretion. The decreased adrenal responsiveness to ACTH is detectable within 24 h after hypophysectomy and increases progressively with time (Figure 22–16). It is marked when the adrenal is atrophic but develops before visible changes occur in adrenal size or morphology.

CIRCADIAN RHYTHM

ACTH is secreted in irregular bursts throughout the day and plasma cortisol tends to rise and fall in response to these bursts (Figure 22–17). In humans, the bursts are most frequent in the early morning, and about 75% of the daily production of cortisol occurs between 4:00 AM and 10:00 AM. The bursts are least frequent in the evening. This **diurnal (circadian) rhythm** in ACTH secretion is present in patients with adrenal insufficiency receiving constant doses of glucocorticoids. It is not due to the stress of getting up in the morning, traumatic as that may be, because the increased ACTH secretion occurs before waking up. If the "day" is lengthened experimentally to more than 24 h, that is, if the individual is isolated and the day's activities are spread over more than 24 h, the adrenal cycle also lengthens, but the increase in ACTH secretion still occurs during the period of sleep. The biologic clock responsible

FIGURE 22–16 **Loss of ACTH responsiveness when ACTH secretion is decreased in humans.** The 1- to 24-amino-acid sequence of ACTH was infused intravenously (IV) in a dose of 250 µg over 8 hours. N, normal subjects; DX, dexamethasone 0.75 mg every 8 h for 3 days; CST, long-term corticosteroid therapy; HI, anterior pituitary insufficiency. (Reproduced with permission from Kolanowski J, et al: Adrenocortical response upon repeated stimulation with corticotropin in patients lacking endogenous corticotropin secretion. Acta Endocrinol [Kbh] 1977;85:595.)

FIGURE 22–17 **Fluctuations in plasma ACTH and glucocorticoids throughout the day in a normal girl (age 16).** The ACTH was measured by immunoassay and the glucocorticoids as 11-oxysteroids (11-OHCS). Note the greater ACTH and glucocorticoid rises in the morning, before awakening. (Reproduced, with permission, from Krieger DT, et al: Characterization of the normal temporal pattern of plasma corticosteroid levels. J Clin Endocrinol Metab 1971;32:266.)

for the diurnal ACTH rhythm is located in the suprachiasmatic nuclei of the hypothalamus (see Chapter 15).

THE RESPONSE TO STRESS

The morning plasma ACTH concentration in a healthy resting human is about 25 pg/mL (5.5 pmol/L). ACTH and cortisol values in various abnormal conditions are summarized in Figure 22–18. During severe stress, the amount of ACTH secreted exceeds the amount necessary to produce maximal glucocorticoid output. However, prolonged exposure to ACTH

in conditions such as the ectopic ACTH syndrome increases the adrenal maximum.

Increases in ACTH secretion to meet emergency situations are mediated almost exclusively through the hypothalamus via release of CRH. This polypeptide is produced by neurons in the paraventricular nuclei. It is secreted in the median eminence and transported in the portal-hypophysial vessels to the anterior pituitary, where it stimulates ACTH secretion (see Chapter 18). If the median eminence is destroyed, increased secretion in response to many different stresses is blocked. Afferent nerve pathways from many parts of the brain converge on the paraventricular nuclei. Fibers from the amygdaloid nuclei mediate responses to emotional stresses, and fear,

FIGURE 22–18 **Plasma concentrations of ACTH and cortisol in various clinical states.** (Reproduced with permission from *Textbook of Endocrinology,* 5th ed. Williams RH [editor]. Saunders, 1974.)

anxiety, and apprehension cause marked increases in ACTH secretion. Input from the suprachiasmatic nuclei provides the drive for the diurnal rhythm. Impulses ascending to the hypothalamus via the nociceptive pathways and the reticular formation trigger increased ACTH secretion in response to injury (Figure 22–18). The baroreceptors exert an inhibitory input via the nucleus of the tractus solitarius.

GLUCOCORTICOID FEEDBACK

Free glucocorticoids inhibit ACTH secretion, and the degree of pituitary inhibition is proportional to the circulating glucocorticoid level. The inhibitory effect is exerted at both the pituitary and the hypothalamic levels. The inhibition is due primarily to an action on DNA, and maximal inhibition takes several hours to develop, although more rapid "fast feedback" also occurs. The ACTH-inhibiting activity of the various steroids parallels their glucocorticoid potency. A drop in resting corticoid levels stimulates ACTH secretion, and in chronic adrenal insufficiency the rate of ACTH synthesis and secretion is markedly increased.

Thus, the rate of ACTH secretion is determined by two opposing forces: the sum of the neural and possibly other stimuli converging through the hypothalamus to increase ACTH secretion, and the magnitude of the braking action of glucocorticoids on ACTH secretion, which is proportional to their level in the circulating blood (Figure 22–19).

The dangers involved when prolonged treatment with anti-inflammatory doses of glucocorticoids is stopped deserve emphasis. Not only is the adrenal atrophic and unresponsive after such treatment, but even if its responsiveness is restored by injecting ACTH, the pituitary may be unable to secrete normal amounts of ACTH for as long as a month. The cause of the deficiency is presumably diminished ACTH synthesis. Thereafter, ACTH secretion slowly increases to supranormal levels. These in turn stimulate the adrenal, and glucocorticoid output rises, with feedback inhibition gradually reducing the elevated ACTH levels to normal (Figure 22–20). The complications of sudden cessation of steroid therapy can usually be avoided by slowly decreasing the steroid dose over a long period of time.

EFFECTS OF MINERALOCORTICOIDS

ACTIONS

Aldosterone and other steroids with mineralocorticoid activity increase the reabsorption of Na^+ from the urine, sweat, saliva, and the contents of the colon. Thus, mineralocorticoids cause retention of Na^+ in the ECF. This expands ECF volume. In the kidneys, they act primarily on the **principal cells (P cells)** of the collecting ducts (see Chapter 38). Under the influence of aldosterone, increased amounts of Na^+ are in effect

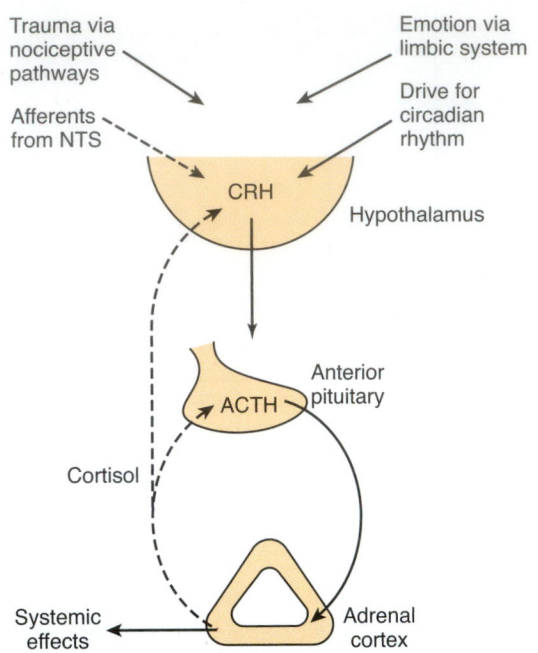

FIGURE 22–19 **Feedback control of the secretion of cortisol and other glucocorticoids via the hypothalamic-pituitary-adrenal axis.** The dashed arrows indicate inhibitory effects and the solid arrows indicate stimulating effects. NTS, nucleus tractus solitarius.

FIGURE 22–20 **Pattern of plasma ACTH and cortisol values in patients recovering from prior long-term daily treatment with large doses of glucocorticoids.** (Courtesy of R Ney.)

exchanged for K^+ and H^+ in the renal tubules, producing a K^+ diuresis (Figure 22–21) and an increase in urine acidity.

MECHANISM OF ACTION

Like many other steroids, aldosterone binds to a cytoplasmic receptor, and the receptor-hormone complex moves to the

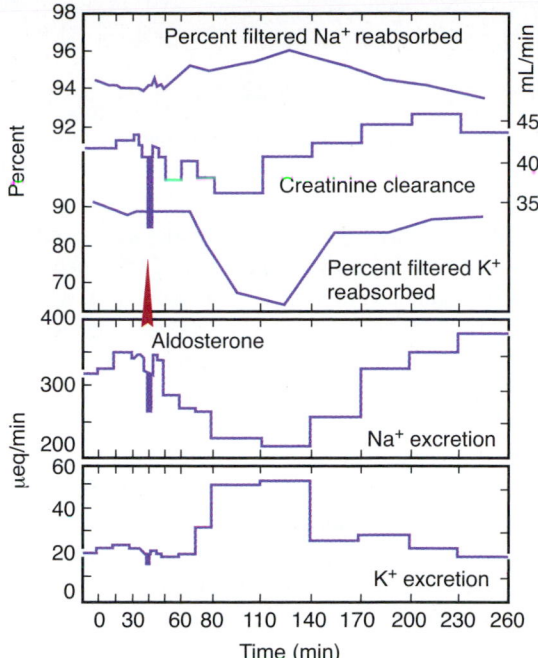

FIGURE 22–21   **Effect of aldosterone (5 μg as a single dose injected into the aorta) on electrolyte excretion in an adrenalectomized dog.** The scale for creatinine clearance is on the right.

nucleus where it alters the transcription of mRNAs. This in turn increases the production of proteins that alter cell function. The aldosterone-stimulated proteins have two effects—a rapid effect, to increase the activity of epithelial sodium channels (ENaCs) by increasing the insertion of these channels into the cell membrane from a cytoplasmic pool; and a slower effect to increase the synthesis of ENaCs. Among the genes activated by aldosterone is the gene for **serum- and glucocorticoid-regulated kinase (sgk),** a serine-threonine protein kinase. The gene for sgk is an early response gene, and sgk increases ENaC activity. Aldosterone also increases the mRNAs for the three subunits that make up ENaCs. The fact that sgk is activated by glucocorticoids as well as aldosterone is not a problem because glucocorticoids are inactivated at mineralocorticoid receptor sites. However, aldosterone activates the genes for other proteins in addition to sgk and ENaCs and inhibits others. Therefore, the exact mechanism by which aldosterone-induced proteins increase Na$^+$ reabsorption is still unsettled.

Evidence is accumulating that aldosterone also binds to the cell membrane and by a rapid, nongenomic action increases the activity of membrane Na$^+$–K$^+$ exchangers. This produces an increase in intracellular Na$^+$, and the second messenger involved is probably IP$_3$. In any case, the principal effect of aldosterone on Na$^+$ transport takes 10 to 30 min to develop and peaks even later (Figure 22–21), indicating that it depends on the synthesis of new proteins by genomic mechanism.

CLINICAL BOX 22–3

Apparent Mineralocorticoid Excess

If 11β-hydroxysteroid dehydrogenase type 2 is inhibited or absent, cortisol has marked mineralocorticoid effects. The resulting syndrome is called **apparent mineralocorticoid excess (AME).** Patients with this condition have the clinical picture of hyperaldosteronism because cortisol is acting on their mineralocorticoid receptors, and their plasma aldosterone level as well as their plasma renin activity is low. The condition can be due to congenital absence of the enzyme or to prolonged ingestion of licorice. Outside of the United States, licorice contains glycyrrhetinic acid, which inhibits 11β-hydroxysteroid dehydrogenase type 2. Individuals who eat large amounts of licorice have an increase in MR-activated sodium absorption via the epithelial sodium channel ENaC in the renal collecting duct, and blood pressure can rise.

RELATION OF MINERALOCORTICOID TO GLUCOCORTICOID RECEPTORS

It is intriguing that in vitro, the mineralocorticoid receptor has an appreciably higher affinity for glucocorticoids than the glucocorticoid receptor does, and glucocorticoids are present in large amounts in vivo. This raises the question of why glucocorticoids do not bind to the mineralocorticoid receptors in the kidneys and other locations and produce mineralocorticoid effects. At least in part, the answer is that the kidneys and other mineralocorticoid-sensitive tissues also contain the enzyme **11β-hydroxysteroid dehydrogenase type 2.** This enzyme leaves aldosterone untouched, but it converts cortisol to cortisone (Figure 22–11) and corticosterone to its 11-oxy derivative. These 11-oxy derivatives do not bind to the receptor (Clinical Box 22–3).

OTHER STEROIDS THAT AFFECT Na$^+$ EXCRETION

Aldosterone is the principal mineralocorticoid secreted by the adrenal, although corticosterone is secreted in sufficient amounts to exert a minor mineralocorticoid effect (Tables 22–1 and 22–2). Deoxycorticosterone, which is secreted in appreciable amounts only in abnormal situations, has about 3% of the activity of aldosterone. Large amounts of progesterone and some other steroids cause natriuresis, but there is little evidence that they play any normal role in the control of Na$^+$ excretion.

TABLE 22–5 Typical plasma electrolyte levels in normal humans and patients with adrenocortical diseases.

State	Plasma Electolytes (mEq/L)			
	Na$^+$	K$^+$	Cl$^-$	HCO$_3^-$
Normal	142	4.5	105	25
Adrenal insufficiency	120	6.7	85	25
Primary hyperaldosteronism	145	2.4	96	41

EFFECT OF ADRENALECTOMY

In adrenal insufficiency, Na$^+$ is lost in the urine; K$^+$ is retained, and the plasma K$^+$ rises. When adrenal insufficiency develops rapidly, the amount of Na$^+$ lost from the ECF exceeds the amount excreted in the urine, indicating that Na$^+$ also must be entering cells. When the posterior pituitary is intact, salt loss exceeds water loss, and the plasma Na$^+$ falls (Table 22–5). However, the plasma volume also is reduced, resulting in hypotension, circulatory insufficiency, and, eventually, fatal shock. These changes can be prevented to a degree by increasing the dietary NaCl intake. Rats survive indefinitely on extra salt alone, but in dogs and most humans, the amount of supplementary salt needed is so large that it is almost impossible to prevent eventual collapse and death unless mineralocorticoid treatment is also instituted (see Clinical Box 22–4).

REGULATION OF ALDOSTERONE SECRETION

STIMULI

The principal conditions that increase aldosterone secretion are summarized in Table 22–6. Some of them also increase glucocorticoid secretion; others selectively affect the output of aldosterone. The primary regulatory factors involved are ACTH from the pituitary, renin from the kidney via angiotensin II, and a direct stimulatory effect of a rise in plasma K$^+$ concentration on the adrenal cortex.

EFFECT OF ACTH

When first administered, ACTH stimulates the output of aldosterone as well as that of glucocorticoids and sex hormones. Although the amount of ACTH required to increase aldosterone output is somewhat greater than the amount that stimulates maximal glucocorticoid secretion (Figure 22–23), it is well

---

CLINICAL BOX 22–4

Secondary Effects of Excess Mineralocorticoids

A prominent feature of prolonged mineralocorticoid excess (Table 22–5) is K$^+$ depletion due to prolonged K$^+$ diuresis. H$^+$ is also lost in the urine. Na$^+$ is retained initially, but the plasma Na$^+$ is elevated only slightly if at all, because water is retained with the osmotically active sodium ions. Consequently, ECF volume is expanded and the blood pressure rises. When the ECF expansion passes a certain point, Na$^+$ excretion is usually increased in spite of the continued action of mineralocorticoids on the renal tubules. This **escape phenomenon** (Figure 22–22) is probably due to increased secretion of ANP (see Chapter 39). Because of increased excretion of Na$^+$ when the ECF volume is expanded, mineralocorticoids do not produce edema in normal individuals and patients with hyperaldosteronism. However, escape may not occur in certain disease states, and in these situations, continued expansion of ECF volume leads to edema (see Chapters 38 and 39).

---

within the range of endogenous ACTH secretion. The effect is transient, and even if ACTH secretion remains elevated, aldosterone output declines in 1 or 2 days. On the other hand, the output of the mineralocorticoid deoxycorticosterone remains elevated. The decline in aldosterone output is partly due to decreased renin secretion secondary to hypervolemia, but it is possible that some other factor also decreases the conversion of corticosterone to aldosterone. After hypophysectomy, the basal rate of aldosterone secretion is normal. The increase normally produced by surgical and other stresses is absent, but the increase produced by dietary salt restriction is unaffected for some time. Later on, atrophy of the zona glomerulosa complicates the picture in long-standing hypopituitarism, and this may lead to salt loss and hypoaldosteronism.

Normally, glucocorticoid treatment does not suppress aldosterone secretion. However, an interesting recently described syndrome is **glucocorticoid-remediable aldosteronism (GRA).** This is an autosomal dominant disorder in which the increase in aldosterone secretion produced by ACTH is no longer transient. The hypersecretion of aldosterone and the accompanying hypertension are remedied when ACTH secretion is suppressed by administering glucocorticoids. The genes encoding aldosterone synthase and 11β-hydroxylase are 95% identical and are close together on chromosome 8. In individuals with GRA, there is unequal crossing over so that the 5 regulatory region of the 11β-hydroxylase gene is fused to the coding region of the aldosterone synthase. The product of this hybrid gene is an ACTH-sensitive aldosterone synthase.

EFFECTS OF ANGIOTENSIN II & RENIN

The octapeptide angiotensin II is formed in the body from angiotensin I, which is liberated by the action of renin on

FIGURE 22–22 "Escape" from the sodium-retaining effect of desoxycorticosterone acetate (DOCA) in an adrenalectomized patient. ECF, extracellular fluid volume; TBV, total blood volume; RCV, red cell volume; PV, plasma volume. (Courtesy of EG Biglieri.)

circulating angiotensinogen (see Chapter 39). Injections of angiotensin II stimulate adrenocortical secretion and, in small doses, affect primarily the secretion of aldosterone (Figure 22–24). The sites of action of angiotensin II are both early and late in the steroid biosynthetic pathway. The early action is on the conversion of cholesterol to pregnenolone, and the late action is on the conversion of corticosterone to aldosterone (Figure 22–8). Angiotensin II does not increase the secretion of deoxycorticosterone, which is controlled by ACTH.

Renin is secreted from the juxtaglomerular cells that surround the renal afferent arterioles as they enter the glomeruli (see Chapter 39). Aldosterone secretion is regulated via the

TABLE 22–6   Conditions that increase aldosterone secretion.

Glucocorticoid secretion also increased
Surgery
Anxiety
Physical trauma
Hemorrhage
Glucocorticoid secretion unaffected
High potassium intake
Low sodium intake
Constriction of inferior vena cava in thorax
Standing
Secondary hyperaldosteronism (in some cases of congestive heart failure, cirrhosis, and nephrosis)

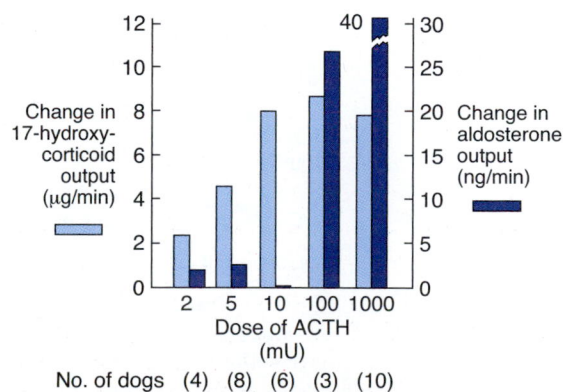

FIGURE 22–23   Changes in adrenal venous output of steroids produced by ACTH in nephrectomized hypophysectomized dogs.

FIGURE 22–24 **Changes in adrenal venous output of steroids produced by angiotensin II in nephrectomized hypophysectomized dogs.**

renin–angiotensin system in a feedback fashion (Figure 22–25). A drop in ECF volume or intra-arterial vascular volume leads to a reflex increase in renal nerve discharge and decreases renal arterial pressure. Both changes increase renin secretion, and the angiotensin II formed by the action of the renin increases the rate of secretion of aldosterone. The aldosterone causes Na⁺ and, secondarily, water retention, expanding ECF volume and shutting off the stimulus that initiated increased renin secretion.

Hemorrhage stimulates ACTH and renin secretion. Like hemorrhage, standing and constriction of the thoracic inferior vena cava decrease intrarenal arterial pressure. Dietary sodium restriction also increases aldosterone secretion via the renin–angiotensin system (Figure 22–26). Such restriction reduces ECF volume, but aldosterone and renin secretion are increased before any consistent decrease in blood pressure takes place. Consequently, the initial increase in renin secretion produced by dietary sodium restriction is probably due to a reflex increase in the activity of the renal nerves. The increase in cir-

culating angiotensin II produced by salt depletion upregulates the angiotensin II receptors in the adrenal cortex and hence increases the response to angiotensin II, whereas it down-regulates the angiotensin II receptors in the blood vessels.

ELECTROLYTES & OTHER FACTORS

An acute decline in plasma Na⁺ of about 20 mEq/L stimulates aldosterone secretion, but changes of this magnitude are rare. However, the plasma K⁺ level need increase only 1 mEq/L to stimulate aldosterone secretion, and transient increases of this magnitude may occur after a meal, particularly if it is rich in K⁺. Like angiotensin II, K⁺ stimulates the conversion of cholesterol to pregnenolone and the conversion of deoxycorticosterone to aldosterone. It appears to act by depolarizing the cell, which opens voltage-gated Ca^{2+} channels, increasing intracellular Ca^{2+}. The sensitivity of the zona glomerulosa to angiotensin II and consequently to a low-sodium diet is decreased by a low-potassium diet.

In normal individuals, plasma aldosterone concentrations increase during the portion of the day that the individual is carrying on activities in the upright position. This increase is due to a decrease in the rate of removal of aldosterone from the circulation by the liver and an increase in aldosterone secretion due to a postural increase in renin secretion. Individuals who are confined to bed show a circadian rhythm of aldosterone and renin secretion, with the highest values in the early morning before awakening.

Atrial natriuretic peptide (ANP) inhibits renin secretion and decreases the responsiveness of the zona glomerulosa to angiotensin II (see Chapter 39).

The mechanisms by which ACTH, angiotensin II, and K⁺ stimulate aldosterone secretion are summarized in Table 22–7.

FIGURE 22–25 **Feedback mechanism regulating aldosterone secretion.** The dashed arrow indicates inhibiition.

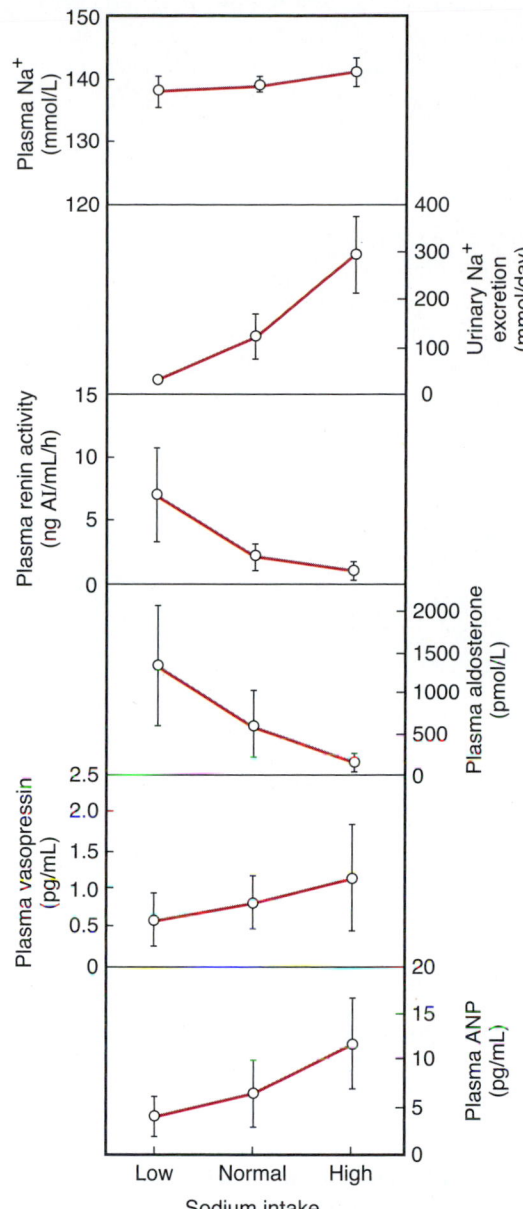

FIGURE 22–26 **Effect of low-, normal-, and high-sodium diets on sodium metabolism and plasma renin activity, aldosterone, vasopressin, and ANP in normal humans.** (Data from Sagnella GA, et al: Plasma atrial natriuretic peptide: Its relationship to changes in sodium in-take, plasma renin activity, and aldosterone in man. Clin Sci 1987;72:25.)

ROLE OF MINERALOCORTICOIDS IN THE REGULATION OF SALT BALANCE

Variation in aldosterone secretion is only one of many factors affecting Na^+ excretion. Other major factors include the glomerular filtration rate, ANP, the presence or absence of osmotic diuresis, and changes in tubular reabsorption of Na^+ independent of aldosterone. It takes some time for aldoster-

TABLE 22–7 **Second messengers involved in the regulation of aldosterone secretion.**

Secretagogue	Intracellular Mediator
ACTH	Cyclic AMP, protein kinase A
Angiotensin II	Diacylglycerol, protein kinase C
K^+	Ca^{2+} via voltage-gated Ca^{2+} channels

one to act. When one rises from the supine to the standing position, aldosterone secretion increases and Na^+ is retained from the urine. However, the decrease in Na^+ excretion develops too rapidly to be explained solely by increased aldosterone secretion. The primary function of the aldosterone-secreting mechanism is the defense of intravascular volume, but it is only one of the homeostatic mechanisms involved.

SUMMARY OF THE EFFECTS OF ADRENOCORTICAL HYPER- & HYPOFUNCTION IN HUMANS

Recapitulating the manifestations of excess and deficiency of the adrenocortical hormones in humans is a convenient way to summarize the multiple and complex actions of these steroids. A characteristic clinical syndrome is associated with excess secretion of each of the types of hormones.

Excess androgen secretion causes masculinization (**adrenogenital syndrome**) and precocious pseudopuberty or female pseudohermaphroditism.

Excess glucocorticoid secretion produces a moon-faced, plethoric appearance, with trunk obesity, purple abdominal striae, hypertension, osteoporosis, protein depletion, mental abnormalities, and, frequently, diabetes mellitus (**Cushing syndrome**). The causes of Cushing syndrome have been discussed previously.

Excess mineralocorticoid secretion leads to K^+ depletion and Na^+ retention, usually without edema but with weakness, hypertension, tetany, polyuria, and hypokalemic alkalosis (**hyperaldosteronism**). This condition may be due to primary adrenal disease (**primary hyperaldosteronism; Conn syndrome**) such as an adenoma of the zona glomerulosa, unilateral or bilateral adrenal hyperplasia, adrenal carcinoma, or GRA. In patients with primary hyperaldosteronism, renin secretion is depressed. **Secondary hyperaldosteronism** with high plasma renin activity is caused by cirrhosis, heart failure, and nephrosis. Increased renin secretion is also found in individuals with the salt-losing form of the adrenogenital syndrome (see above), because their ECF volume is low. In patients with elevated renin secretion due to renal artery constriction, aldosterone secretion is increased; in those in whom renin secretion is not elevated, aldosterone secretion is normal. The relationship of aldosterone to hypertension is discussed in Chapter 33.

FIGURE 22–27 **Pigmentation in Addison disease. A)** Tan and vitiligo. **B)** Pigmentation of scars from lesions that occurred after the development of the disease. **C)** Pigmentation of skin creases. **D)** Darkening of areolas. **E)** Pigmentation of pressure points. **F)** Pigmentation of the gums. (Reproduced with permission from Forsham PH, Di Raimondo V: *Traumatic Medicine and Surgery for the Attorney.* Butterworth, 1960.)

Primary adrenal insufficiency due to disease processes that destroy the adrenal cortex is called **Addison disease.** The condition used to be a relatively common complication of tuberculosis, and now it is usually due to autoimmune inflammation of the adrenal. Patients lose weight, are tired, and become chronically hypotensive. They have small hearts, probably because the hypotension decreases the work of the heart. Eventually they develop severe hypotension and shock (**addisonian crisis**). This is due not only to mineralocorticoid deficiency but to glucocorticoid deficiency as well. Fasting causes fatal hypoglycemia, and any stress causes collapse. Water is retained, and there is always the danger of water intoxication. Circulating ACTH levels are elevated. The diffuse tanning of the skin and the spotty pigmentation characteristic of chronic glucocorticoid deficiency (Figure 22–27) are due, at least in part, to the melanocyte-stimulating hormone (MSH) activity of the ACTH in the blood. Minor menstrual abnormalities occur in women, but the deficiency of adrenal sex hormones usually has little effect in the presence of normal testes or ovaries.

Secondary adrenal insufficiency is caused by pituitary diseases that decrease ACTH secretion, and **tertiary adrenal insufficiency** is caused by hypothalamic disorders disrupting CRH secretion. Both are usually milder than primary adrenal insufficiency because electrolyte metabolism is affected to a lesser degree. In addition, there is no pigmentation because in both of these conditions, plasma ACTH is low, not high.

Cases of isolated aldosterone deficiency have also been reported in patients with renal disease and a low circulating renin level (**hyporeninemic hypoaldosteronism**). In addition, **pseudohypoaldosteronism** is produced when there is resistance to the action of aldosterone. Patients with these syndromes have marked hyperkalemia, salt wasting, and hypotension, and they may develop metabolic acidosis.

CHAPTER SUMMARY

- The adrenal gland consists of the adrenal medulla which secretes dopamine and the catecholamines epinephrine and

- norepinephrine, and the adrenal cortex which secretes steroid hormones.
- Norepinephrine and epinephrine act on two classes of receptors, α- and β-adrenergic receptors, and exert metabolic effects that include glycogenolysis in liver and skeletal muscle, mobilization of FFA, increased plasma lactate, and stimulation of the metabolic rate.
- The hormones of the adrenal cortex are derivatives of cholesterol and include the mineralocorticoid aldosterone, the glucocorticoids cortisol and corticosterone, and the androgens dehydroepiandrosterone (DHEA) and androstenedione.
- Androgens are the hormones that exert masculinizing effects, and they promote protein anabolism and growth. The adrenal androgen androstenedione is converted to testosterone and to estrogens (aromatized) in fat and other peripheral tissues. This is an important source of estrogens in men and postmenopausal women.
- The mineralocorticoid aldosterone has effects on Na^+ and K^+ excretion and glucocorticoids affect glucose and protein metabolism.
- Glucocorticoid secretion is dependent upon ACTH from the anterior pituitary and is increased by stress. Angiotensin II increases the secretion of aldosterone.

MULTIPLE-CHOICE QUESTIONS

For all questions, select the single best answer unless otherwise directed.

1. Which of the following is produced only by *large amounts* of glucocorticoids?
   A) normal responsiveness of fat depots to norepinephrine
   B) maintenance of normal vascular reactivity
   C) increased excretion of a water load
   D) inhibition of the inflammatory response
   E) inhibition of ACTH secretion

2. Which of the following are *incorrectly* paired?
   A) gluconeogenesis : cortisol
   B) free fatty acid mobilization : dehydroepiandrosterone
   C) muscle glycogenolysis : epinephrine
   D) kaliuresis : aldosterone
   E) hepatic glycogenesis : insulin

3. Which of the following hormones has the shortest plasma half-life?
   A) corticosterone
   B) renin
   C) dehydroepiandrosterone
   D) aldosterone
   E) norepinephrine

4. Mole for mole, which of the following has the greatest effect on Na^+ excretion?
   A) progesterone
   B) cortisol
   C) vasopressin
   D) aldosterone
   E) dehydroepiandrosterone

5. Mole for mole, which of the following has the greatest effect on plasma osmolality?
   A) progesterone
   B) cortisol
   C) vasopressin
   D) aldosterone
   E) dehydroepiandrosterone

6. The secretion of which of the following would be *least* affected by a decrease in extracellular fluid volume?
   A) CRH
   B) arginine vasopressin
   C) dehydroepiandrosterone
   D) estrogens
   E) aldosterone

7. A young man presents with a blood pressure of 175/110 mm Hg. He is found to have a high circulating aldosterone but a low circulating cortisol. Glucocorticoid treatment lowers his circulating aldosterone and lowers his blood pressure to 140/85 mm Hg. He probably has an abnormal
   A) 17α-hydroxylase.
   B) 21β-hydroxylase.
   C) 3β-hydroxysteroid dehydrogenase.
   D) aldosterone synthase.
   E) cholesterol desmolase.

8. A 32-year-old woman presents with a blood pressure of 155/96 mm Hg. In response to questioning, she admits that she loves licorice and eats some at least three times a week. She probably has a low level of
   A) type 2 11β-hydroxysteroid dehydrogenase activity.
   B) ACTH.
   C) 11β-hydroxylase activity.
   D) glucuronyl transferase.
   E) norepinephrine.

9. In its action in cells, aldosterone
   A) increases transport of ENaCs from the cytoplasm to the cell membrane.
   B) does not act on the cell membrane.
   C) binds to a receptor in the nucleus.
   D) may activate a heat shock protein.
   E) also binds to glucocorticoid receptors.

CHAPTER RESOURCES

Goldstein JL, Brown MS: The cholesterol quartet. Science 2001;292:1510.

Goodman HM (editor): *Handbook of Physiology,* Section 7: *The Endocrine System.* Oxford University Press, 2000.

Larsen PR et al (editors): *Williams Textbook of Endocrinology,* 9th ed. Saunders, 2003.

Stocco DM: A review of the characteristics of the protein required for the acute regulation of steroid hormone biosynthesis: The case for the steroidogenic acute regulatory (StAR) protein. Proc Soc Exp Biol Med 1998;217:123.

White PC: Disorders of aldosterone biosynthesis and action. N Engl J Med 1994;331:250.

Hormonal Control of Calcium & Phosphate Metabolism & the Physiology of Bone

INTRODUCTION

Calcium is an essential intracellular-signaling molecule and also plays a variety of extracellular functions, thus the control of bodily calcium concentrations is vitally important. The components of the system that maintain calcium homeostasis include cell types that sense changes in extracellular calcium and release calcium-regulating hormones, and the targets of these hormones, including the kidneys, bones, and intestine, that respond with changes in calcium mobilization, excretion, or uptake. Three hormones are primarily concerned with the regulation of calcium metabolism. **1,25-Dihydroxycholecalciferol** is a steroid hormone formed from vitamin D by successive hydroxylations in the liver and kidneys. Its primary action is to increase calcium absorption from the intestine. **Parathyroid hormone (PTH)** is secreted by the parathyroid glands. Its main action is to mobilize calcium from bone and increase urinary phosphate excretion. **Calcitonin,** a calcium-lowering hormone that in mammals is secreted primarily by cells in the thyroid gland, inhibits bone resorption. Although the role of calcitonin seems to be relatively minor, all three hormones probably operate in concert to maintain the constancy of the Ca^{2+} level in the body fluids. Phosphate homeostasis is likewise critical to normal body function, particularly given its inclusion in adenosine triphosphate (ATP), its role as a biological buffer, and its role as a modifier of proteins, thereby altering their functions. Many of the systems that regulate calcium homeostasis also contribute to that of phosphate, albeit sometimes in a reciprocal fashion, and thus will also be discussed in this chapter.

CALCIUM & PHOSPHORUS METABOLISM

CALCIUM

The body of a young adult human contains about 1100 g (27.5 mol) of calcium. Ninety-nine percent of the calcium is in the skeleton. Plasma calcium, normally at a concentration of around 10 mg/dL (5 mEq/L, 2.5 mmol/L), is partly bound to protein and partly diffusible (Table 23–1). The distribution of calcium inside cells is discussed in Chapter 2.

It is the free, ionized calcium in the body fluids that is a vital second messenger (see Chapter 2) and is necessary for blood coagulation, muscle contraction, and nerve function. A decrease in extracellular Ca^{2+} exerts a net excitatory effect on nerve and muscle cells in vivo (see Chapters 4 and 5). The result is **hypocalcemic tetany,** which is characterized by extensive spasms of skeletal muscle, involving especially the muscles of the extremities and the larynx. Laryngospasm can become so severe that the airway is obstructed and fatal asphyxia is produced. Ca^{2+} also plays an important role in blood clotting (see Chapter 32), but in vivo, fatal tetany would occur before compromising the clotting reaction.

Because the extent of Ca^{2+} binding by plasma proteins is proportional to the plasma protein level, it is important to know the plasma protein level when evaluating the total plasma calcium. Other electrolytes and pH also affect the free Ca^{2+} level. Thus, for example, symptoms of tetany appear at higher total calcium levels if the patient hyperventilates, thereby increasing plasma pH. Plasma proteins are more ionized when the pH is high, providing more protein anion to bind with Ca^{2+}.

The calcium in bone is of two types: a readily exchangeable reservoir and a much larger pool of stable calcium that is only slowly exchangeable. Two independent but interacting homeostatic systems affect the calcium in bone. One is the system that regulates plasma Ca^{2+}, providing for the movement of about 500 mmol of Ca^{2+} per day into and out of the readily exchangeable pool in the bone (Figure 23–1). The other system involves bone remodeling by the constant interplay of bone resorption and deposition (see following text). However, the Ca^{2+} interchange between plasma and this stable pool of bone calcium is only about 7.5 mmol/d.

Ca^{2+} is transported across the brush border of intestinal epithelial cells via channels known as transient receptor potential vanilloid type 6 (TRPV6) and binds to an intracellular protein known as calbindin-D_{9k}. Calbindin sequesters the absorbed calcium so that it does not disturb epithelial signaling processes that involve calcium. The absorbed Ca^{2+} is thereby delivered to the basolateral membrane of the epithelial cell, from where it can be transported into the bloodstream by either a sodium/calcium exchanger (NCX1) or a calcium-dependent ATPase. Nevertheless, it should be noted that

TABLE 23–1 Distribution (mmol/L) of calcium in normal human plasma.

Total diffusible		**1.34**
Ionized (Ca^{2+})	1.18	
Complexed to HCO_3^-, citrate, etc	0.16	
Total nondiffusible (protein-bound)		**1.16**
Bound to albumin	0.92	
Bound to globulin	0.24	
Total plasma calcium		**2.50**

FIGURE 23–1 **Calcium metabolism in an adult human.** A typical daily intake of 25 mmol Ca^{2+} (1000 mg) moves through many body compartments.

recent studies indicate that some intestinal calcium uptake persists even in the absence of TRPV6 and calbindin-D_{9k}, suggesting that additional pathways are likely also involved in this critical process. The overall transport process is regulated by 1,25-dihydroxycholecalciferol (see below). As Ca^{2+} uptake rises, moreover, 1,25-dihydroxycholecalciferol levels fall in response to increased plasma Ca^{2+}.

Plasma Ca^{2+} is filtered in the kidneys, but 98–99% of the filtered Ca^{2+} is reabsorbed. About 60% of the reabsorption occurs in the proximal tubules and the remainder in the ascending limb of the loop of Henle and the distal tubule. Distal tubular reabsorption depends on the TRPV5 channel, which is related to TRPV6 discussed previously, and whose expression is regulated by parathyroid hormone.

PHOSPHORUS

Phosphate is found in ATP, cyclic adenosine monophosphate (cAMP), 2,3-diphosphoglycerate, many proteins, and other vital compounds in the body. Phosphorylation and dephosphorylation of proteins are involved in the regulation of cell function (see Chapter 2). Therefore, it is not surprising that, like calcium, phosphate metabolism is closely regulated. Total body phosphorus is 500 to 800 g (16.1–25.8 mol), 85–90% of which is in the skeleton. Total plasma phosphorus is about 12 mg/dL, with two-thirds of this total in organic compounds and the remaining inorganic phosphorus (P_i) mostly in PO_4^{3-}, HPO_4^{2-}, and $H_2PO_4^-$. The amount of phosphorus normally entering bone is about 3 mg (97 μmol)/kg/d, with an equal amount leaving via reabsorption.

P_i in the plasma is filtered in the glomeruli, and 85–90% of the filtered P_i is reabsorbed. Active transport in the proximal tubule accounts for most of the reabsorption and involves two related sodium-dependent P_i cotransporters, NaPi-IIa and NaPi-IIc. NaPi-IIa is powerfully inhibited by parathyroid hormone, which causes its internalization and degradation and thus a reduction in renal P_i reabsorption (see below).

P_i is absorbed in the duodenum and small intestine. Uptake occurs by a transporter related to those in the kidney, NaPi-IIb, that takes advantage of the low intracellular sodium concentration established by the Na, K ATPase on the basolateral membrane of intestinal epithelial cells to load P_i against its concentration gradient. However, the pathway by which P_i exits into the bloodstream is not known. Many stimuli that increase Ca^{2+} absorption, including 1,25-dihydroxycholecalciferol, also increase P_i absorption via increased NaPi-IIb expression.

VITAMIN D & THE HYDROXYCHOLECALCIFEROLS

CHEMISTRY

The active transport of Ca^{2+} and PO_4^{3-} from the intestine is increased by a metabolite of **vitamin D.** The term "vitamin D"

is used to refer to a group of closely related sterols produced by the action of ultraviolet light on certain provitamins (Figure 23–2). Vitamin D_3, which is also called cholecalciferol, is produced in the skin of mammals from 7-dehydrocholesterol by the action of sunlight. The reaction involves the rapid formation of previtamin D_3, which is then converted more slowly to vitamin D_3. Vitamin D_3 and its hydroxylated derivatives are transported in the plasma bound to a globulin vitamin D-binding protein (DBP). Vitamin D_3 is also ingested in the diet.

Vitamin D_3 is metabolized by enzymes that are members of the cytochrome P450 (CYP) superfamily (see Chapters 1 and 29). In the liver, vitamin D_3 is converted to **25-hydroxycholecalciferol** (calcidiol, 25-OHD$_3$). The 25-hydroxycholecalciferol is converted in the cells of the proximal tubules of the kidneys to the more active metabolite **1,25-dihydroxycholecalciferol,** which is also called calcitriol or 1,25-$(OH)_2D_3$. 1,25-Dihydroxycholecalciferol is also made in the placenta, in keratinocytes in the skin, and in macrophages. The normal plasma level of 25-hydroxycholecalciferol is about 30 ng/mL, and that of 1,25-dihydroxycholecalciferol is about 0.03 ng/mL (approximately 100 pmol/L). The less active metabolite 24,25-dihydroxycholecalciferol is also formed in the kidneys (Figure 23–2).

MECHANISM OF ACTION

1,25 dihydroxycholecalciferol stimulates the expression of a number of gene products involved in calcium transport and handling via its receptor, which acts as a transcriptional regulator in its ligand-bound form. One group is the family of **calbindin-D** proteins. These are members of the troponin C superfamily of Ca^{2+}-binding proteins that also includes calmodulin (see Chapter 2). Calbindin-Ds are found in human intestine, brain, and kidneys. In the intestinal epithelium and many other tissues, two calbindins are induced: calbindin-D_{9K} and calbindin-D_{28K}, with molecular weights of 9,000 and 28,000, respectively. 1,25-dihydroxycholecalciferol also increases the number of Ca^{2+}–ATPase and TRPV6 molecules in the intestinal cells, thus, the overall capacity for absorption of dietary calcium is enhanced.

In addition to increasing Ca^{2+} absorption from the intestine, 1,25-dihydroxycholecalciferol facilitates Ca^{2+} reabsorption in the kidneys via increased TRPV5 expression in the proximal tubules, increases the synthetic activity of osteoblasts, and is necessary for normal calcification of matrix (see Clinical Box 23–1). The stimulation of osteoblasts brings about a secondary increase in the activity of osteoclasts (see below).

REGULATION OF SYNTHESIS

The formation of 25-hydroxycholecalciferol does not appear to be stringently regulated. However, the formation of 1,25-dihydroxycholecalciferol in the kidneys, which is catalyzed by the renal 1α-hydroxylase, is regulated in a feedback fashion by plasma Ca^{2+} and PO_4^{3+} (Figure 23–3). When the plasma Ca^{2+} level is high, little 1,25-dihydroxycholecalciferol is produced,

FIGURE 23–2   Formation and hydroxylation of vitamin D₃. 25-hydroxylation takes place in the liver, and the other hydroxylations occur primarily in the kidneys. The formulas of 7-dehydrocholesterol, vitamin D₃, and 1,25-dihydroxycholecalciferol are also shown below.

and the kidneys produce the relatively inactive metabolite 24,25-dihydroxycholecalciferol instead. This effect of Ca^{2+} on production of 1,25-dihydroxycholecalciferol is the mechanism that brings about adaptation of Ca^{2+} absorption from the intestine (see previous text). Conversely, expression of 1α-hydroxylase is stimulated by PTH, and when the plasma Ca^{2+} level is low, PTH secretion is increased. The production of

FIGURE 23–3   Effects of PTH and 1,25-dihydroxycholecalciferol on whole body calcium homeostasis. Note that these hormones are also involved in the regulation of circulating phosphate levels. (Reproduced with permission from Widmaier EP, Raff H, Strang KT: *Vander's Human Physiology*,10th ed., McGraw-Hill, 2006.)

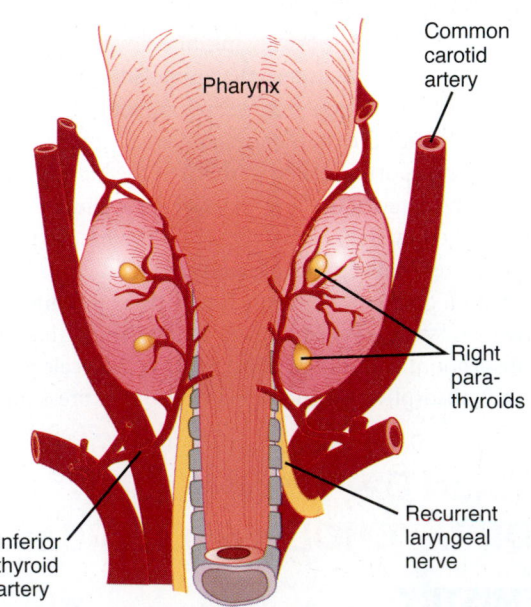

FIGURE 23–4   The human parathyroid glands, viewed from behind.

Rickets & Osteomalacia

Vitamin D deficiency causes defective calcification of bone matrix and the disease called **rickets** in children and **osteomalacia** in adults. Even though 1,25-dihydroxycholecalciferol is necessary for normal mineralization of bone matrix, the main defect in this condition is failure to deliver adequate amounts of Ca^{2+} and PO_4^{3-} to the sites of mineralization. The full-blown condition in children is characterized by weakness and bowing of weight-bearing bones, dental defects, and hypocalcemia. In adults, the condition is less obvious. It used to be most commonly due to inadequate exposure to the sun in smoggy cities, but now it is more commonly due to inadequate intake of the provitamins on which the sun acts in the skin. These cases respond to administration of vitamin D. The condition can also be caused by inactivating mutations of the gene for renal 1α-hydroxylase, in which case there is no response to vitamin D but a normal response to 1,25-dihydroxycholecalciferol **(type I vitamin D-resistant rickets).** In rare instances, it can be due to inactivating mutations of the gene for the 1,25-dihydroxycholecalciferol receptor **(type II vitamin D-resistant rickets),** in which case there is a deficient response to both vitamin D and 1,25-dihydroxycholecalciferol.

1,25-dihydroxycholecalciferol is also increased by low and inhibited by high plasma PO_4^{3-} levels, by a direct inhibitory effect of PO_4^{3-} on the 1α-hydroxylase. Additional control of 1,25-dihydroxycholecalciferol formation is exerted by a direct negative feedback effect of the metabolite on 1α-hydroxylase, a positive feedback action on the formation of 24,25-dihydroxycholecalciferol, and a direct action on the parathyroid gland to inhibit PTH expression.

An "anti-aging" protein called α-Klotho (named after Klotho, a daughter of Zeus in Greek mythology who spins the thread of life) has also recently been discovered to play important roles in calcium and phosphate homeostasis, in part by reciprocal effects on 1,25-dihydroxycholecalciferol levels. Mice deficient in α-Klotho displayed accelerated aging, decreased bone mineral density, calcifications, and hypercalcemia and hyperphosphatemia. α-Klotho plays an important role in stabilizing the membrane localization of proteins important in calcium and phosphate (re)absorption, such as TRPV5 and Na, K ATPase. Likewise, it enhances the activity of another factor, fibroblast growth factor 23 (FGF23), at its receptor. FGF23 thereby decreases renal NaPi-IIa and NaPi-IIc expression and inhibits the production of 1α-hydroxylase, reducing levels of 1,25-dihydroxycholecalciferol (Clinical Box 23–1).

THE PARATHYROID GLANDS

ANATOMY

Humans usually have four parathyroid glands: two embedded in the superior poles of the thyroid and two in its inferior poles (Figure 23–4). Each parathyroid gland is a richly vascularized disk, about $3 \times 6 \times 2$ mm, containing two distinct types of cells (Figure 23–5). The abundant **chief cells,** which contain a prominent Golgi apparatus plus endoplasmic reticulum and secretory granules, synthesize and secrete **parathyroid hormone (PTH).** The less abundant and larger **oxyphil cells** contain oxyphil granules and large numbers of mitochondria in their cytoplasm. In humans, few are seen before puberty, and thereafter they increase in number with age. Their function is unknown. Consequences of loss of parathyroid gland are discussed in Clinical Box 23–2.

SYNTHESIS & METABOLISM OF PTH

Human PTH is a linear polypeptide with a molecular weight of 9500 that contains 84 amino acid residues (Figure 23–6). It is synthesized as part of a larger molecule containing 115 amino acid residues **(preproPTH).** On entry of preproPTH into the endoplasmic reticulum, a leader sequence is removed from the amino terminal to form the 90-amino-acid polypeptide **proPTH.** Six additional amino acid residues are removed from the amino terminal of proPTH in the Golgi apparatus, and the 84-amino-acid polypeptide PTH is packaged in secretory granules and released as the main secretory product of the chief cells.

The normal plasma level of intact PTH is 10 to 55 pg/mL. The half-life of PTH is approximately 10 min, and the secreted polypeptide is rapidly cleaved by the Kupffer cells in the liver into fragments that are probably biologically inactive. PTH and these fragments are then cleared by the kidneys. Modern immunoassays for PTH are designed only to measure mature PTH (1–84) and not these fragments to obtain an accurate measure of "active" PTH.

FIGURE 23–5 **Section of human parathyroid.** (Reduced 50% from × 960.) Small cells are chief cells; large stippled cells (especially prominent in the lower left of picture) are oxyphil cells. (Reproduced with permission from Fawcett DW: *Bloom and Fawcett, A Textbook of Histology,* 11th ed. Saunders, 1986.)

CLINICAL BOX 23–2

Effects of Parathyroidectomy

Occasionally, inadvertent parathyroidectomy occurs in humans during thyroid surgery. This can have serious consequences as PTH is essential for life. After parathyroidectomy, there is a steady decline in the plasma Ca^{2+} level. Signs of neuromuscular hyperexcitability appear, followed by full-blown hypocalcemic tetany (see above). Plasma phosphate levels usually rise as the plasma calcium level falls. Symptoms usually develop 2 to 3 d postoperatively but may not appear for several weeks or more. Injections of PTH can be given to correct the chemical abnormalities, and the symptoms then disappear. Injections of Ca^{2+} salts can also give temporary relief. The signs of tetany in humans include **Chvostek's sign,** a quick contraction of the ipsilateral facial muscles elicited by tapping over the facial nerve at the angle of the jaw, and **Trousseau's sign,** a spasm of the muscles of the upper extremity that causes flexion of the wrist and thumb with extension of the fingers. In individuals with mild tetany in whom spasm is not yet evident, Trousseau sign can sometimes be produced by occluding the circulation for a few minutes with a blood pressure cuff.

CLINICAL BOX 23–3

Diseases of Parathyroid Excess

Hyperparathyroidism due to injections of parathyroid extract in animals or hypersecretion of a functioning parathyroid tumor in humans is characterized by hypercalcemia and hypophosphatemia. Humans with PTH-secreting adenomas are usually asymptomatic, with the condition detected when plasma Ca^{2+} is measured in conjunction with a routine physical examination. However, there may be minor changes in personality, and calcium-containing kidney stones occasionally form. In conditions such as chronic renal disease and rickets, in which the plasma Ca^{2+} level is chronically low, stimulation of the parathyroid glands causes compensatory parathyroid hypertrophy and secondary hyperparathyroidism. The plasma Ca^{2+} level is low in chronic renal disease primarily because the diseased kidneys lose the ability to form 1,25-dihydroxycholecalciferol. Finally, mutations in the calcium receptor, CaR, gene cause predictable long-term changes in plasma Ca^{2+}. Individuals heterozygous for inactivating mutations have familial benign hypocalciuric hypercalcemia, a condition in which there is a chronic moderate elevation in plasma Ca^{2+} because the feedback inhibition of PTH secretion by Ca^{2+} is reduced. Plasma PTH levels are normal or even elevated. However, children who are homozygous for inactivating mutations develop neonatal severe primary hyperparathyroidism. Conversely, individuals with gain-of-function mutations of the CaR gene develop familial hypercalciuric hypocalcemia due to increased sensitivity of the parathyroid glands to plasma Ca^{2+}.

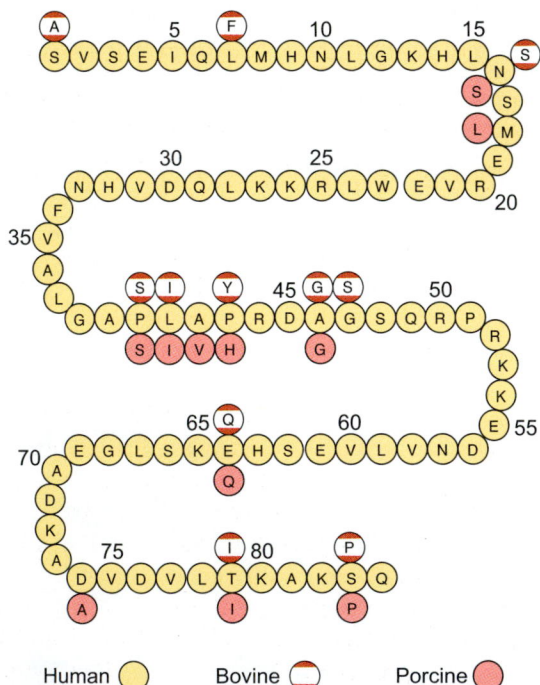

Human ◯   Bovine ◑   Porcine ●

FIGURE 23–6   Parathyroid hormone. The symbols above and below the human structure show where amino acid residues are different in bovine and porcine PTH. (Reproduced with permission from Keutmann HT, et al: Complete amino acid sequence of human parathyroid hormone. Biochemistry 1978;17:5723. Copyright © 1978 by the American Chemical Society.)

ACTIONS

PTH acts directly on bone to increase bone resorption and mobilize Ca^{2+}. In addition to increasing the plasma Ca^{2+}, PTH increases phosphate excretion in the urine and thereby depresses plasma phosphate levels. This **phosphaturic action** is due to a decrease in reabsorption of phosphate via effects on NaPi-IIa in the proximal tubules, as discussed previously. PTH also increases reabsorption of Ca^{2+} in the distal tubules, although Ca^{2+} excretion in the urine is often increased in hyperparathyroidism because the increase in the load of filtered calcium overwhelms the effect on reabsorption (Clinical Box 23-3). PTH also increases the formation of 1,25-dihydroxycholecalciferol, and this increases Ca^{2+} absorption from the intestine. On a longer time scale, PTH stimulates both osteoblasts and osteoclasts.

MECHANISM OF ACTION

It now appears that there are at least three different PTH receptors. One also binds parathyroid hormone-related protein (PTHrP; see below) and is known as the hPTH/PTHrP receptor.

FIGURE 23–7 **Signal transduction pathways activated by PTH or PTHrP binding to the hPTH/hPTHrP receptor.** Intracellular cAMP is increased via G_s and adenylyl cyclase (AC). Diacylglycerol and IP$_3$ (1,4,5-InsP$_3$) are increased via G_q and phospholipase C (PLC). (Modified and reproduced with permission from McPhee SJ, Lingappa VR, Ganong WF [editors]: *Pathophysiology of Disease*, 4th ed. McGraw-Hill, 2003.)

FIGURE 23–8 **Relation between plasma Ca^{2+} concentration and PTH response in humans.** The set point is the plasma Ca^{2+} at which half the maximal response occurred (ie, 1.2 mmol/L). (Modified and reproduced with permission from Brown E: Extracellular Ca2+ sensing, regulation of parathyroid cell functions, and role of Ca2+ and other ions as extracellular (first) messengers. Physiol Rev 1991;71:371.)

A second receptor, PTH2 (hPTH2-R), does not bind PTHrP and is found in the brain, placenta, and pancreas. In addition, there is evidence for a third receptor, CPTH, which reacts with the carboxyl terminal rather than the amino terminal of PTH. The first two receptors are coupled to G_s, and via this heterotrimeric G protein they activate adenylyl cyclase, increasing intracellular cAMP. The hPTH/PTHrP receptor also activates PLC via G_q, increasing intracellular Ca^{2+} and activating protein kinase C (Figure 23–7). However, the way these second messengers affect Ca^{2+} in bone is unsettled.

In the disease called **pseudohypoparathyroidism,** the signs and symptoms of hypoparathyroidism develop but the circulating level of PTH is normal or elevated. Because the tissues fail to respond to the hormone, this is a receptor disease. There are two forms. In the more common form, a congenital 50% reduction of the activity of G_s occurs and PTH fails to produce a normal increase in cAMP concentration. In a different, less common form, the cAMP response is normal but the phosphaturic action of the hormone is defective.

REGULATION OF SECRETION

Circulating ionized calcium acts directly on the parathyroid glands in a negative feedback fashion to regulate the secretion of PTH (Figure 23–8). The key to this regulation is a cell membrane Ca^{2+} receptor, CaR. Activation of this G-protein coupled receptor leads to phosphoinositide turnover in many tissues. In the parathyroid, its activation inhibits PTH secretion. In this way, when the plasma Ca^{2+} level is high, PTH secretion is inhibited and the Ca^{2+} is deposited in the bones. When it is low, secretion is increased and Ca^{2+} is mobilized from the bones.

1,25-dihydroxycholecalciferol acts directly on the parathyroid glands to decrease preproPTH mRNA. Increased plasma phosphate stimulates PTH secretion by lowering plasma levels of free Ca^{2+} and inhibiting the formation of 1,25-dihydroxycholecalciferol. Magnesium is required to maintain normal parathyroid secretory responses. Impaired PTH release along with diminished target organ responses to PTH account for the hypocalcemia that occasionally occurs in magnesium deficiency (Clinical Box 23–2 and Clinical Box 23–3).

PTHrP

Another protein with PTH activity, **parathyroid hormone-related protein (PTHrP),** is produced by many different tissues in the body. It has 140 amino acid residues, compared with 84 in PTH, and is encoded by a gene on human chromosome 12, whereas PTH is encoded by a gene on chromosome 11. PTHrP and PTH have marked homology at their amino terminal ends and they both bind to the hPTH/ PTHrP receptor, yet their physiologic effects are very different. How is this possible when they bind to the same receptor? For one thing, PTHrP is primarily a paracrine factor, acting close to where it is produced. It may be that circulating PTH cannot reach at least some of these sites. Second, subtle conformational differences may be produced by binding of PTH versus PTHrP to their receptor, despite their structural similarities. Another possibility is action of one or the other hormone on other, more selective receptors.

PTHrP has a marked effect on the growth and development of cartilage in utero. Mice in which both alleles of the PTHrP gene are knocked out have severe skeletal deformities and die soon after birth. In normal animals, on the other hand, PTHrP-stimulated cartilage cells proliferate and their terminal differentiation is inhibited. PTHrP is also expressed in the brain, where evidence indicates that it inhibits excitotoxic damage to developing neurons. In addition, there is evidence that it is involved in Ca^{2+} transport in the placenta. PTHrP is also found in keratinocytes in the skin, in smooth muscle, and in the teeth, where it is present in the enamel epithelium that caps each tooth. In the absence of PTHrP, teeth cannot erupt.

HYPERCALCEMIA OF MALIGNANCY

Hypercalcemia is a common metabolic complication of cancer. About 20% of hypercalcemic patients have bone metastases that produce the hypercalcemia by eroding bone (**local osteolytic hypercalcemia**). Evidence suggests that this erosion is produced by prostaglandins such as prostaglandin E_2 from the tumor. The hypercalcemia in the remaining 80% of the patients is due to elevated circulating levels of PTHrP (**humoral hypercalcemia of malignancy**). The tumors responsible for the hypersecretion include cancers of the breast, kidney, ovary, and skin.

CALCITONIN

ORIGIN

In dogs, perfusion of the thyroparathyroid region with solutions containing high concentrations of Ca^{2+} leads to a fall in peripheral plasma calcium, and after damage to this region, Ca^{2+} infusions cause a greater increase in plasma Ca^{2+} than they do in control animals. These and other observations led to the discovery that a Ca^{2+}-lowering as well as a Ca^{2+}-elevating hormone was secreted by structures in the neck. The Ca^{2+}-lowering hormone has been named **calcitonin.** In mammals, calcitonin is produced by the **parafollicular cells** of the thyroid gland, which are also known as the clear or C cells.

STRUCTURE

Human calcitonin has a molecular weight of 3500 and contains 32 amino acid residues (Figure 23–9). Much of the mRNA transcribed from the calcitonin gene is processed to a different mRNA in the nervous system, so that **calcitonin gene-related peptide (CGRP)** is formed rather than calcitonin (see Chapter 4).

SECRETION & METABOLISM

Secretion of calcitonin is increased when the thyroid gland is exposed to plasma calcium level of approximately 9.5 mg/dL.

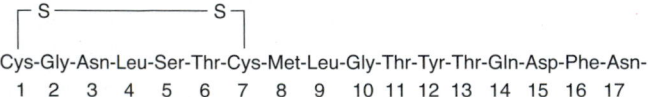

Cys-Gly-Asn-Leu-Ser-Thr-Cys-Met-Leu-Gly-Thr-Tyr-Thr-Gln-Asp-Phe-Asn-
 1   2   3   4   5   6   7   8   9 10 11 12 13 14 15 16 17

Lys-Phe-His-Thr-Phe-Pro-Gln-Thr-Ala-Ile-Gly-Val-Gly-Ala-Pro-NH$_2$
 18 19 20 21 22 23 24 25 26 27 28 29 30 31 32

FIGURE 23–9   Human calcitonin. The sequence is shown using the three letter abbreviations for constituent amino acids.

Above this level, plasma calcitonin is directly proportionate to plasma calcium. β-adrenergic agonists, dopamine, and estrogens also stimulate calcitonin secretion. Gastrin, cholecystokinin (CCK), glucagon, and secretin have all been reported to stimulate calcitonin secretion, with gastrin being the most potent stimulus (see Chapter 26). Thus, the plasma calcitonin level is elevated in Zollinger–Ellison syndrome and in pernicious anemia (see Chapter 26). However, the dose of gastrin needed to stimulate calcitonin secretion is supraphysiological and not seen after eating in normal individuals, so dietary calcium in the intestine probably does not induce secretion of a calcium-lowering hormone prior to the calcium being absorbed. In any event, the actions of calcitonin are short-lived because it has a half-life of less than 10 min in humans.

ACTIONS

Receptors for calcitonin are found in bones and the kidneys. Calcitonin lowers circulating calcium and phosphate levels. It exerts its calcium-lowering effect by inhibiting bone resorption. This action is direct, and calcitonin inhibits the activity of osteoclasts in vitro. It also increases Ca^{2+} excretion in the urine.

The exact physiologic role of calcitonin is uncertain. The calcitonin content of the human thyroid is low, and after thyroidectomy, bone density and plasma Ca^{2+} level are normal as long as the parathyroid glands are intact. In addition, there are only transient abnormalities of Ca^{2+} metabolism when a Ca^{2+} load is injected after thyroidectomy. This may be explained in part by secretion of calcitonin from tissues other than the thyroid. However, there is general agreement that the hormone has little long-term effect on the plasma Ca^{2+} level in adult animals and humans. Further, unlike PTH and 1,25-dihydroxycholecalciferol, calcitonin does not appear to be involved in phosphate homeostasis. Moreover, patients with medullary carcinoma of the thyroid have a very high circulating calcitonin level but no symptoms directly attributable to the hormone, and their bones are essentially normal. No syndrome due to calcitonin deficiency has been described. More hormone is secreted in young individuals, and it may play a role in skeletal development. In addition, it may protect the bones of the mother from excess calcium loss during pregnancy. Bone formation in the infant and lactation are major drains on Ca^{2+} stores, and plasma concentrations of 1,25-dihydroxycholecalciferol are elevated in pregnancy. They would cause bone loss in the mother if bone resorption were not simultaneously inhibited by an increase in the plasma calcitonin level.

SUMMARY

The actions of the three principal hormones that regulate the plasma concentration of Ca^{2+} can now be summarized. PTH increases plasma Ca^{2+} by mobilizing this ion from bone. It increases Ca^{2+} reabsorption in the kidney, but this may be offset by the increase in filtered Ca^{2+}. It also increases the formation of 1,25-dihydroxycholecalciferol. 1,25-Dihydroxycholecalciferol increases Ca^{2+} absorption from the intestine and increases Ca^{2+} reabsorption in the kidneys. Calcitonin inhibits bone resorption and increases the amount of Ca^{2+} in the urine.

EFFECTS OF OTHER HORMONES & HUMORAL AGENTS ON CALCIUM METABOLISM

Calcium metabolism is affected by various hormones in addition to 1,25-dihydroxycholecalciferol, PTH, and calcitonin. **Glucocorticoids** lower plasma Ca^{2+} levels by inhibiting osteoclast formation and activity, but over long periods they cause osteoporosis by decreasing bone formation and increasing bone resorption. They decrease bone formation by inhibiting protein synthesis in osteoblasts. They also decrease the absorption of Ca^{2+} and PO_4^{3-} from the intestine and increase the renal excretion of these ions. The decrease in plasma Ca^{2+} concentration also increases the secretion of PTH, and bone resorption is facilitated. **Growth hormone** increases calcium excretion in the urine, but it also increases intestinal absorption of Ca^{2+}, and this effect may be greater than the effect on excretion, with a resultant positive calcium balance. Insulin-like growth factor I (IGF-I) generated by the action of growth hormone stimulates protein synthesis in bone. As noted previously, **thyroid hormones** may cause hypercalcemia, hypercalciuria, and, in some instances, osteoporosis. **Estrogens** prevent osteoporosis by inhibiting the stimulatory effects of certain cytokines on osteoclasts. **Insulin** increases bone formation, and there is significant bone loss in untreated diabetes.

BONE PHYSIOLOGY

Bone is a special form of connective tissue with a collagen framework impregnated with Ca^{2+} and PO_4^{3-} salts, particularly **hydroxyapatites,** which have the general formula $Ca_{10}(PO_4)_6(OH)_2$. Bone is also involved in overall Ca^{2+} and PO_4^{3-} homeostasis. It protects vital organs, and the rigidity it provides permits locomotion and the support of loads against gravity. Old bone is constantly being resorbed and new bone formed, permitting remodeling that allows it to respond to the stresses and strains that are put upon it. It is a living tissue that is well vascularized and has a total blood flow of 200 to 400 mL/min in adult humans.

STRUCTURE

Bone in children and adults is of two types: **compact** or **cortical bone,** which makes up the outer layer of most bones (Figure 23–10) and accounts for 80% of the bone in the body; and **trabecular** or **spongy bones** inside the cortical bone, which makes up the remaining 20% of bone in the body. In compact bone, the surface-to-volume ratio is low, and bone cells lie in lacunae. They receive nutrients by way of canaliculi that ramify through-

out the compact bone (Figure 23–10). Trabecular bone is made up of spicules or plates, with a high surface to volume ratio and many cells sitting on the surface of the plates. Nutrients diffuse from bone extracellular fluid (ECF) into the trabeculae, but in compact bone, nutrients are provided via **haversian canals** (Figure 23–10), which contain blood vessels. Around each Haversian canal, collagen is arranged in concentric layers, forming cylinders called **osteons** or **haversian systems.**

The protein in bone matrix is over 90% type I collagen, which is also the major structural protein in tendons and skin. This collagen, which weight for weight is as strong as steel, is made up of a triple helix of three polypeptides bound tightly together. Two of these are identical α_1 polypeptides encoded by one gene, and one is an α_2 polypeptide encoded by a different gene. Collagens make up a family of structurally related proteins that maintain the integrity of many different organs. Fifteen different types of collagens encoded by more than 20 different genes have so far been identified.

BONE GROWTH

During fetal development, most bones are modeled in cartilage and then transformed into bone by ossification (**enchondral bone formation**). The exceptions are the clavicles, the mandibles, and certain bones of the skull in which mesenchymal cells form bone directly (**intramembranous bone formation**).

During growth, specialized areas at the ends of each long bone (**epiphyses**) are separated from the shaft of the bone by a plate of actively proliferating cartilage, the **epiphysial plate** (Figure 23–11). The bone increases in length as this plate lays down new bone on the end of the shaft. The width of the epiphysial plate is proportionate to the rate of growth. The width is affected by a number of hormones, but most markedly by the pituitary growth hormone and IGF-I (see Chapter 24).

Linear bone growth can occur as long as the epiphyses are separated from the shaft of the bone, but such growth ceases after the epiphyses unite with the shaft (**epiphysial closure**). The cartilage cells stop proliferating, become hypertrophic, and secrete vascular endothelial growth factor (VEGF), leading to vascularization and ossification. The epiphyses of the various bones close in an orderly temporal sequence, the last epiphyses closing after puberty. The normal age at which each of the epiphyses closes is known, and the "bone age" of a young individual can be determined by x-raying the skeleton and noting which epiphyses are open and which are closed.

BONE FORMATION & RESORPTION

The cells responsible for bone formation are **osteoblasts** and the cells responsible for bone resorption are **osteoclasts.**

Osteoblasts are modified fibroblasts. Their early development from the mesenchyme is the same as that of fibroblasts, with extensive growth factor regulation. Later, ossification-specific transcription factors, such as Cbfa1/Runx2, contribute to their differentiation. The importance of this transcription

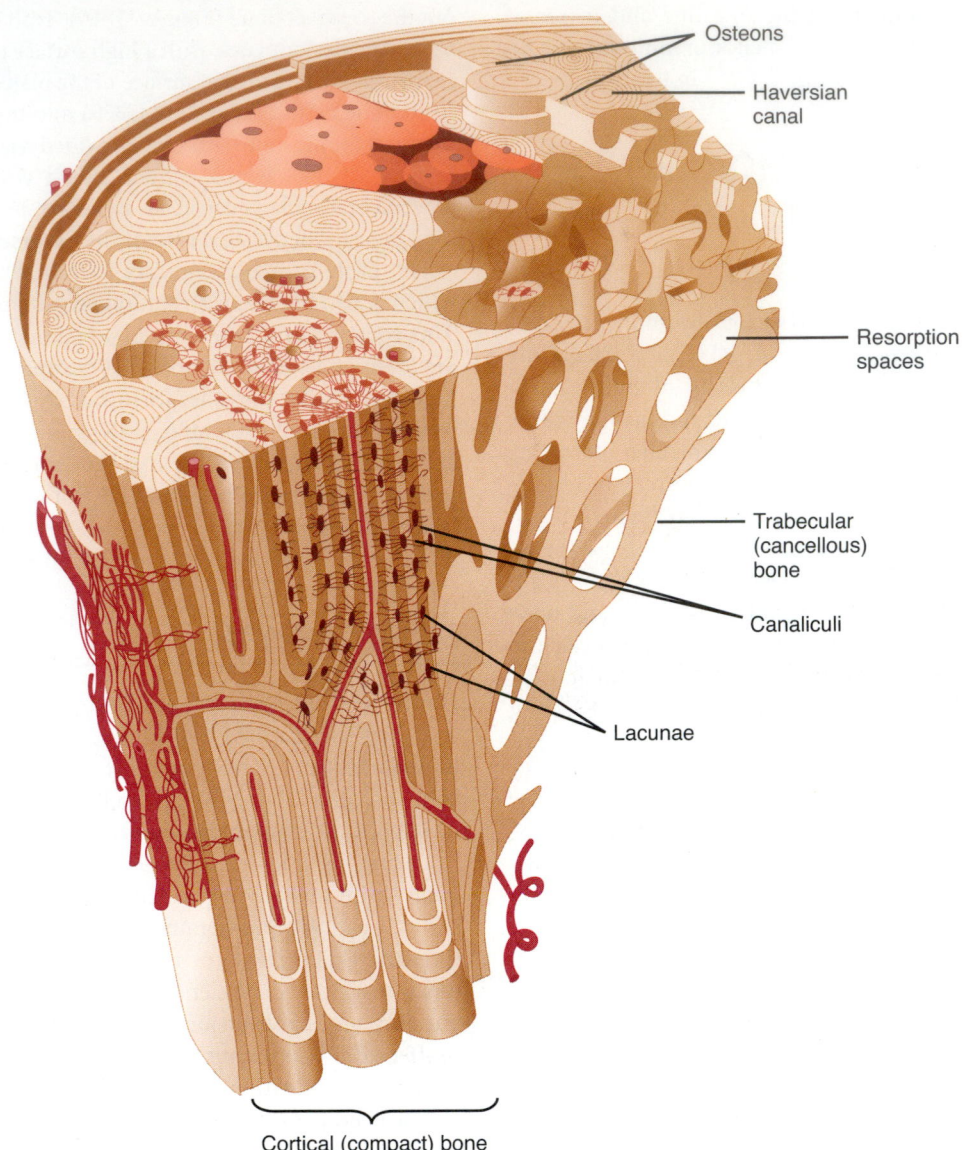

FIGURE 23–10 **Structure of compact and trabecular bone.** The compact bone is shown in horizontal section **(top)** and vertical section **(bottom).** (Reproduced with permission from Williams PL et al (editors): *Gray's Anatomy,* 37th ed. Churchill Livingstone, 1989.)

factor in bone development is underscored in knockout mice deficient for the Cbfa1/Runx gene. These mice develop to term with their skeletons made exclusively of cartilage; no ossification occurs. Normal osteoblasts are able to lay down type 1 collagen and form new bone.

Osteoclasts, on the other hand, are members of the monocyte family. Stromal cells in the bone marrow, osteoblasts, and T lymphocytes all express receptor activator for nuclear factor kappa beta ligand (RANKL) on their surface. When these cells come in contact with appropriate monocytes expressing RANK (ie, the RANKL receptor) two distinct signaling pathways are initiated: (1) there is a RANKL/RANK interaction between the cell pairs, (2) mononuclear phagocyte colony stimulating factor (M-CSF) is secreted by the nonmonocytic cells and it binds to its corresponding receptor on the monocytes (c-fins). The combination of these two signaling events

leads to differentiation of the monocytes into osteoclasts. The precursor cells also secrete **osteoprotegerin (OPG),** which controls for differentiation of the monocytes by competing with RANK for binding of RANKL.

Osteoclasts erode and absorb previously formed bone. They become attached to bone via integrins in a membrane extension called the **sealing zone.** This creates an isolated area between the bone and a portion of the osteoclast. Proton pumps (ie, H^+-dependent ATPases) then move from endosomes into the cell membrane apposed to the isolated area, and they acidify the area to approximately pH 4.0. Similar proton pumps are found in the endosomes and lysosomes of all eukaryotic cells, but in only a few instances do they move into the cell membrane. Note in this regard that the sealed-off space formed by the osteoclast resembles a large lysosome. The acidic pH dissolves hydroxyapatite, and acid

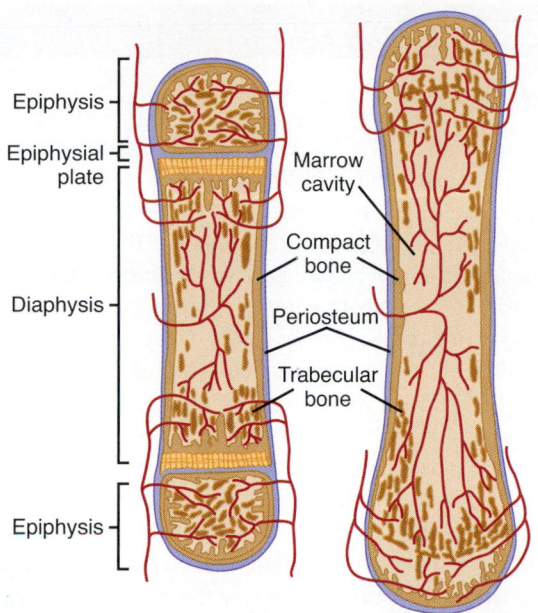

FIGURE 23–11 **Structure of a typical long bone before (left) and after (right) epiphysial closure.** Note the rearrangement of cells and growth of the bone as the epiphysial plate closes (see text for details).

proteases secreted by the cell break down collagen, forming a shallow depression in the bone (Figure 23–12). The products of digestion are then endocytosed and move across the osteoclast by transcytosis (see Chapter 2), with release into the interstitial fluid. The collagen breakdown products have pyridinoline structures, and pyridinolines can be measured in the urine as an index of the rate of bone resorption.

Throughout life, bone is being constantly resorbed and new bone is being formed. The calcium in bone turns over at a rate of 100% per year in infants and 18% per year in adults. Bone

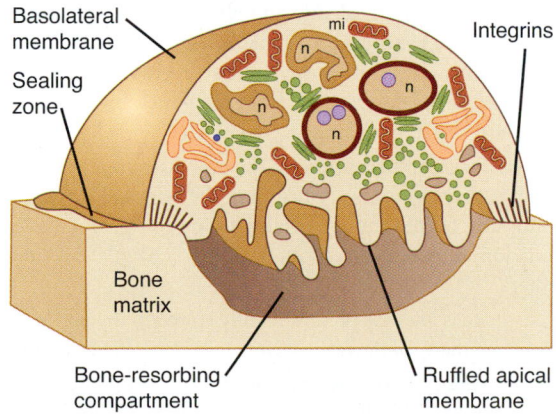

FIGURE 23–12 **Osteoclast resorbing bone.** The edges of the cell are tightly sealed to bone, permitting secretion of acid from the ruffled apical membrane and consequent erosion of the bone underneath the cell. Note the multiple nuclei (n) and mitochondria (mi). (Courtesy of R Baron.)

remodeling is mainly a local process carried out in small areas by populations of cells called bone-remodeling units. First, osteoclasts resorb bone, and then osteoblasts lay down new bone in the same general area. This cycle takes about 100 days. Modeling drifts also occur in which the shapes of bones change as bone is resorbed in one location and added in another. Osteoclasts tunnel into cortical bone followed by osteoblasts, whereas trabecular bone remodeling occurs on the surface of the trabeculae. About 5% of the bone mass is being remodeled by about 2 million bone-remodeling units in the human skeleton at any one time. The renewal rate for bone is about 4% per year for compact bone and 20% per year for trabecular bone. The remodeling is related in part to the stresses and strains imposed on the skeleton by gravity.

At the cell–cell level, there is some regulation of osteoclast formation by osteoblasts via the RANKL–RANK and the M-CSF–OPG mechanism; however, specific feedback mechanisms of osteoclasts on osteoblasts are not well defined. In a broader sense, the bone remodeling process is primarily under endocrine control. Parathyroid hormone accelerates bone resorption, and estrogens slow bone resorption by inhibiting the production of bone-eroding cytokines. An interesting new observation is that intracerebroventricular but not intravenous leptin decreases bone formation. This finding is consistent with the observations that obesity protects against bone loss and that most obese humans are resistant to the effects of leptin on appetite. Thus, there may be neuroendocrine regulation of bone mass via leptin.

BONE DISEASE

The diseases produced by selective abnormalities of the cells and processes discussed above illustrate the interplay of factors that maintain normal bone function.

In **osteopetrosis,** a rare and often severe disease, the osteoclasts are defective and are unable to resorb bone in their usual fashion so the osteoblasts operate unopposed. The result is a steady increase in bone density, neurologic defects due to narrowing and distortion of foramina through which nerves normally pass, and hematologic abnormalities due to crowding out of the marrow cavities. Mice lacking the protein encoded by the immediate-early gene *c-fos* develop osteopetrosis, and osteopetrosis also occurs in mice lacking the PU.1 transcription factor. This suggests that all these factors are involved in normal osteoclast development and function.

On the other hand, **osteoporosis** is caused by a relative excess of osteoclastic function. Loss of bone matrix in this condition (Figure 23–13) is marked, and the incidence of fractures is increased. Fractures are particularly common in the distal forearm (Colles fracture), vertebral body, and hip. All of these areas have a high content of trabecular bone, and because trabecular bone is more active metabolically, it is lost more rapidly. Fractures of the vertebrae with compression cause kyphosis, with the production of a typical "widow's hump" that is common in elderly women with osteoporosis. Fractures of

FIGURE 23–13   **Normal trabecular bone (left) compared with trabecular bone from a patient with osteoporosis (right).** The loss of mass in osteoporosis leaves bones more susceptible to breakage.

the hip in elderly individuals are associated with a mortality rate of 12–20%, and half of those who survive require prolonged expensive care.

Osteoporosis has multiple causes, but by far the most common form is **involutional osteoporosis.** All normal humans gain bone early in life, during growth. After a plateau, they begin to lose bone as they grow older (Figure 23–14). When this loss is accelerated or exaggerated, it leads to osteoporosis (see Clinical Box 23–4). Increased intake of calcium, particularly from natural sources such as milk, and moderate exercise may help prevent or slow the progress of osteoporosis, although their effects are not great. Bisphosphonates such as etidronate, which inhibit osteoclastic activity, increase the mineral content of bone when administered in a cyclic fashion and decrease the rate of new vertebral fractures. Fluoride

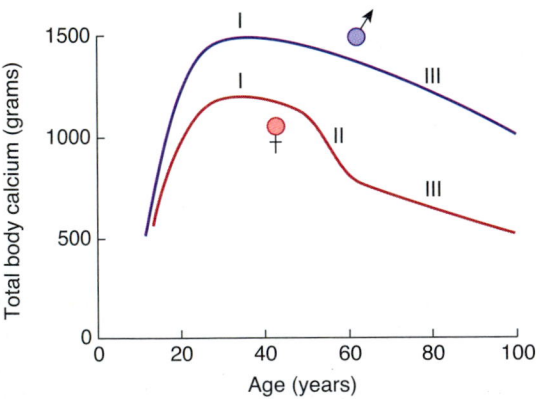

FIGURE 23–14   Total body calcium, an index of bone mass, at various ages in men and women. Note the rapid increase to young adult levels (phase I) followed by the steady loss of bone with advancing age in both sexes (phase III) and the superimposed rapid loss in women after menopause (phase II). (Reproduced by permission of Oxford University Press from Riggs BL, Melton LJ III: Involutional osteoporosis. In Evans TG, Williams TF (editors): *Oxford Textbook of Geriatric Medicine.* Oxford University Press, 1992.)

CLINICAL BOX 23–4

Osteoporosis

Adult women have less bone mass than adult men, and after menopause they initially lose it more rapidly than men of comparable age do. Consequently, they are more prone to development of serious osteoporosis. The cause of the bone loss after menopause is primarily estrogen deficiency, and estrogen treatment arrests the progress of the disease. Estrogens inhibit secretion of cytokines such as interleukin-1 (IL-1), IL-6, and tumor necrosis factor (TNF-α), and these cytokines foster the development of osteoclasts. Estrogen also stimulates production of transforming growth factor (TGF-β), and this cytokine increases apoptosis of osteoclasts. However, it now appears that even small doses of estrogens may increase the incidence of uterine and breast cancer, and in carefully controlled studies, estrogens do not protect against cardiovascular disease. Therefore, the decision to treat a postmenopausal woman with estrogens depends on a careful weighing of the risk–benefit ratio. Bone loss can also occur in both men and women as a result of inactivity. In patients who are immobilized for any reason, and during space flight, bone resorption exceeds bone formation and disuse osteoporosis develops. The plasma calcium level is not markedly elevated, but plasma concentrations of parathyroid hormone and 1,25-dihydroxycholecalciferol fall and large amounts of calcium are lost in the urine.

stimulates osteoblasts, making bone more dense, but it has proved to be of little value in the treatment of the disease.

CHAPTER SUMMARY

- Circulating levels of calcium and phosphate ions are controlled by cells that sense the levels of these electrolytes in the blood and release hormones, and effects of these hormones are evident in mobilization of the minerals from the bones, intestinal absorption, and/or renal wasting.

- The majority of the calcium in the body is stored in the bones but it is the free, ionized calcium in the cells and extracellular fluids that fulfills physiological roles in cell signaling, nerve function, muscle contraction, and blood coagulation, among others.

- Phosphate is likewise predominantly stored in the bones and regulated by many of the same factors that influence calcium levels.

- The two major hormones regulating calcium and phosphate homeostasis are 1,25-dihydroxycholecalciferol (a derivative of vitamin D) and parathyroid hormone; calcitonin is also capable of regulating levels of these ions, but its full physiologic contribution is unclear.

- 1,25-dihydroxycholecalciferol acts to elevate plasma calcium and phosphate by predominantly transcriptional mechanisms, whereas parathyroid hormone elevates calcium but decreases phosphate by increasing the latter's renal excretion. Calcitonin lowers both calcium and phosphate levels.

- Deficiencies of 1,25-dihydroxycholecalciferol or mutations in its receptor, lead to decreases in circulating calcium, defective calcification of the bones, and bone weakness. Disease states also result from either deficiencies or overproduction of parathyroid hormone, with reciprocal effects on calcium and phosphate.

- Bone is a highly structured mass with outer cortical and inner trabecular layers. The larger cortical layer has a high surface to volume layer with haversian canals that provide nutrients and gaps (lacunae) inhabited by bone cells that are connected by a canaliculi network. The smaller trabecular layer has a much higher surface to volume layer that relies on diffusion for nutrients supply.

- Regulated bone growth through puberty occurs through epiphysial plates. These plates are located near the end of the bone shaft and fuse with the shaft of the bone to cease linear bone growth.

- Bone is constantly remodeled by osteoclasts, which erode and absorb bone, and osteoblasts, which lay down new bone.

MULTIPLE-CHOICE QUESTIONS

For all questions, select the single best answer unless otherwise directed.

1. A patient with parathyroid deficiency 10 days after inadvertent damage to the parathyroid glands during thyroid surgery would probably have
   A) low plasma phosphate and Ca^{2+} levels and tetany.
   B) low plasma phosphate and Ca^{2+} levels and tetanus.
   C) a low plasma Ca^{2+} level, increased muscular excitability, and a characteristic spasm of the muscles of the upper extremity (Trousseau sign).
   D) high plasma phosphate and Ca^{2+} levels and bone demineralization.
   E) increased muscular excitability, a high plasma Ca^{2+} level, and bone demineralization.

2. A high plasma Ca^{2+} level causes
   A) bone demineralization.
   B) increased formation of 1,25-dihydroxycholecalciferol.
   C) decreased secretion of calcitonin.
   D) decreased blood coagulability.
   E) increased formation of 24,25-dihydroxycholecalciferol.

3. Which of the following is *not* involved in regulating plasma Ca^{2+} levels?
   A) kidneys
   B) skin
   C) liver
   D) lungs
   E) intestine

4. 1,25-dihydroxycholecalciferol affects intestinal Ca^{2+} absorption through a mechanism that
   A) includes alterations in the activity of genes.
   B) activates adenylyl cyclase.
   C) decreases cell turnover.
   D) changes gastric acid secretion.
   E) is comparable to the action of polypeptide hormones.

5. Which of the following would you expect to find in a patient whose diet has been low in calcium for 2 mo?
   A) increased formation of 24,25-dihydroxycholecalciferol
   B) decreased amounts of calcium-binding protein in intestinal epithelial cells
   C) increased parathyroid hormone secretion
   D) a high plasma calcitonin concentration
   E) increased plasma phosphates

6. In osteopetrosis, which of the following is defective?
   A) phosphate deposition in trabecular bone
   B) structure of parathyroid hormone related protein (PTHrP)
   C) osteoblasts
   D) osteoclasts
   E) bone collagen

7. At epiphysial closure
   A) cortical bone and trabecular bone merge.
   B) osteoclasts and osteoblasts undergo differentiation.
   C) there is an extended amount of proliferating cartilage that contributes to bone elongation.
   D) lacunae meet the trabecular bone.
   E) ephyses unite with the shaft to end normal linear bone growth.

CHAPTER RESOURCES

Brown EM: The calcium-sensing receptor: Physiology, pathophysiology and CaR-based therapeutics. Subcell Biochem 2007;45:139.

Murer H, Hernanado N, Forster L, Biber J: Molecular mechanisms in proximal tubular and small intestinal phosphate reabsorption. Mol Membr Biol 2001;18:3.

Nijenhuis T, Hoenderop JGJ, Bindels RJM: TRPV5 and TRPV6 in Ca^{2+} (re)absorption: Regulating Ca^{2+} entry at the gate. Pflugers Arch Eur J Physiol 2005;451:181.

Renkema KY, Alexander RT, Bindels FJ, Hoenderop JF: Calcium and phosphate homeostasis: Concerted interplay of new regulators. Ann Med 2008;40:82.

24

The Pituitary Gland

After studying this chapter, you should be able to:

- Describe the structure of the pituitary gland and how it relates to its function.
- Define the cell types present in the anterior pituitary and understand how their numbers are controlled in response to physiologic demands.
- Understand the function of hormones derived from proopiomelanocortin in humans, and how they are involved in regulating pigmentation in humans, other mammals, and lower vertebrates.
- Define the effects of the growth hormone in growth and metabolic function, and how insulin-like growth factor I (IGF-I) may mediate some of its actions in the periphery.
- List the stimuli that regulate growth hormone secretion and define their underlying mechanisms.
- Understand the basis of conditions where pituitary function and growth hormone secretion and function are abnormal, and how they can be treated.

INTRODUCTION

The pituitary gland, or hypophysis, lies in a pocket of the sphenoid bone at the base of the brain. It is a coordinating center for control of many downstream endocrine glands, some of which are discussed in other chapters. In many ways, it can be considered to consist of at least two (and in some species, three) separate endocrine organs that contain a plethora of hormonally active substances. The anterior pituitary secretes **thyroid-stimulating hormone (TSH, thyrotropin), adrenocorticotropic hormone (ACTH), luteinizing hormone (LH), follicle-stimulating hormone (FSH), prolactin,** and **growth hormone** (see Figure 18–9), and receives almost all of its blood supply from the portal hypophysial vessels that pass initially through the median eminence, a structure immediately below the hypothalamus. This vascular arrangement positions the cells of the anterior pituitary to respond efficiently to regulatory factors released from the hypothalamus. Of the listed hormones, prolactin acts on the breast. The remaining five are, at least in part, **tropic hormones;** that is, they stimulate secretion of hormonally active substances by other endocrine glands or, in the case of growth hormone, the liver and other tissues (see below). The hormones tropic for a particular endocrine gland are discussed in the chapter on that gland: TSH in Chapter 20; ACTH in Chapter 22; and the gonadotropins FSH and LH in Chapter 25, along with prolactin.

The posterior pituitary in mammals consists predominantly of nerves that have their cell bodies in the hypothalamus, and stores **oxytocin** and **vasopressin** in the termini of these neurons, to be released into the bloodstream. The secretion of these hormones, as well as a discussion of the overall role of the hypothalamus and median eminence in regulating both the anterior and posterior pituitary, were covered in Chapter 18. Finally, in some species there is a well-developed intermediate lobe of the pituitary, whereas in humans it is rudimentary. Nevertheless, the intermediate lobe, as well as the anterior pituitary, contain hormonally active derivatives of the proopiomelanocortin molecule that regulate skin pigmentation, among other functions (see below). To avoid redundancy, this chapter will focus particularly on growth hormone and its role in growth and facilitating the activity of other hormones, along with a number of general considerations about the pituitary. The melanocyte-stimulating hormones (MSHs) of the intermediate lobe of the pituitary, α-MSH and β-MSH, will also be touched upon.

MORPHOLOGY

GROSS ANATOMY

The anatomy of the pituitary gland is summarized in Figure 24–1 and discussed in detail in Chapter 18. The posterior pituitary is made up largely of the endings of axons from the supraoptic and paraventricular nuclei of the hypothalamus and arises initially as an extension of this structure. The anterior pituitary, on the other hand, contains endocrine cells that store its characteristic hormones and arises embryologically as an invagination of the pharynx (**Rathke's pouch**). In species where it is well developed, the intermediate lobe is formed in the embryo from the dorsal half of Rathke's pouch, but is closely adherent to the posterior lobe in the adult. It is separated from the anterior lobe by the remains of the cavity in Rathke's pouch, the **residual cleft.**

HISTOLOGY

In the posterior lobe, the endings of the supraoptic and paraventricular axons can be observed in close relation to blood vessels. **Pituicytes,** stellate cells that are modified astrocytes, are also present.

As noted above, the intermediate lobe is rudimentary in humans and a few other mammalian species. In these species, most of its cells are incorporated in the anterior lobe. Along the residual cleft are small thyroid-like follicles, some containing a little colloid. The function of the colloid, if any, is unknown.

The anterior pituitary is made up of interlacing cell cords and an extensive network of sinusoidal capillaries. The endothelium of the capillaries is fenestrated, like that in other endocrine organs. The cells contain granules of stored hormone that are extruded from the cells by exocytosis. Their constituents then enter the capillaries to be conveyed to target tissues.

CELL TYPES IN THE ANTERIOR PITUITARY

Five types of secretory cells have been identified in the anterior pituitary by immunocytochemistry and electron microscopy. Traditionally, they were also characterized by their affinity for either acidic or basic histological stains. The cell types are the somatotropes, which secrete growth hormone; the lactotropes (also called mammotropes), which secrete prolactin; the corticotropes, which secrete ACTH; the thyrotropes, which secrete TSH; and the gonadotropes, which secrete FSH and LH. The characteristics of these cells are summarized in Table 24–1. Some cells may contain two or more hormones. It is also notable that the three pituitary glycoprotein hormones, FSH, LH, and TSH, while being made up of two subunits, all share a common α subunit that is the product of a single gene and has the same amino acid composition in each hormone, although their carbohydrate residues vary. The α subunit must be combined with a β subunit characteristic of each hormone for maximal physiologic activity. The β subunits, which are produced by separate genes and differ in structure, confer hormonal specificity. The α subunits are remarkably interchangeable and hybrid molecules can be created. In addition, the placental glycoprotein gonadotropin human chorionic gonadotropin (hCG) has α and β subunits (see Chapter 25).

The anterior pituitary also contains folliculostellate cells that send processes between the granulated secretory cells. These cells produce paracrine factors that regulate the growth and function of the secretory cells discussed above. Indeed, the anterior pituitary can adjust the relative proportion of secretory cell types to meet varying requirements for different hormones at different life stages. This plasticity has recently been ascribed to the presence of a small number of pluripotent stem cells that persist in the adult gland.

PROOPIOMELANOCORTIN & DERIVATIVES

BIOSYNTHESIS

Intermediate-lobe cells and corticotropes of the anterior lobe both synthesize a large precursor protein that is cleaved to form a family of hormones. After removal of the signal peptide, this prohormone is known as **proopiomelanocortin (POMC).** This molecule is also synthesized in the hypothalamus, the lungs, the gastrointestinal tract, and the placenta. The structure of POMC, as well as its derivatives, is shown in Figure 24–2. In corticotropes, it is hydrolyzed to ACTH and a polypeptide of unknown function called β-lipotropin (LPH),

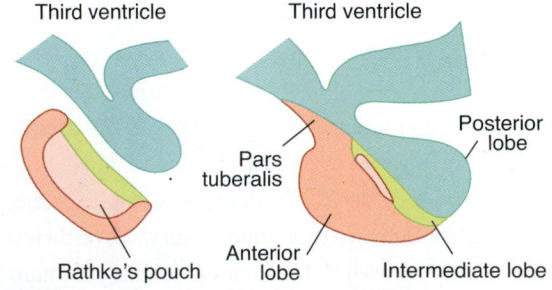

FIGURE 24–1 Diagrammatic outline of the formation of the pituitary (left) and the various parts of the organ in the adult (right).

TABLE 24–1   Hormone-secreting cells of the human anterior pituitary gland.

Cell Type	Hormones Secreted	% of Total Secretory Cells	Stain Affinity	Diameter of Secretory Granules (nm)
Somatotrope	Growth hormone	50	Acidophilic	300–400
Lactotrope	Prolactin	10–30	Acidophilic	200
Corticotrope	ACTH	10	Basophilic	400–550
Thyrotrope	TSH	5	Basophilic	120–200
Gonadotrope	FSH, LH	20	Basophilic	250–400

plus a small amount of β-endorphin, and these substances are secreted. In the intermediate lobe cells, POMC is hydrolyzed to corticotropin-like intermediate-lobe peptide (CLIP), γ-LPH, and appreciable quantities of β-endorphin. The functions, if any, of CLIP and γ-LPH are unknown, whereas β-endorphin is an opioid peptide (see Chapter 7) that has the five amino acid residues of met-enkephalin at its amino terminal end. The **melanotropins** α- and β-MSH are also formed. However, the intermediate lobe in humans is rudimentary, and it appears that neither α-MSH nor β-MSH is secreted in adults. In some species, however, the melanotropins have important physiological functions, as discussed below.

CONTROL OF SKIN COLORATION & PIGMENT ABNORMALITIES

Fish, reptiles, and amphibia change the color of their skin for thermoregulation, camouflage, and behavioral displays. They do this in part by moving black or brown granules into or out of the periphery of pigment cells called **melanophores.** The granules are made up of **melanins,** which are synthesized from dopamine (see Chapter 7) and dopaquinone. The movement of these granules is controlled by a variety of hormones and neurotransmitters, including α- and β-MSH, melanin-concentrating hormone, melatonin, and catecholamines.

Mammals have no melanophores containing pigment granules that disperse and aggregate, but they do have **melanocytes,** which have multiple processes containing melanin granules. Melanocytes express **melanotropin-1** receptors. Treatment with MSHs accelerates melanin synthesis, causing readily detectable darkening of the skin in humans in 24 h. As noted

FIGURE 24–2   Schematic representation of the preproopiomelanocortin molecule formed in pituitary cells, neurons, and other tissues. The numbers in parentheses identify the amino acid sequences in each of the polypeptide fragments. For convenience, the amino acid sequences are numbered from the amino terminal of ACTH and read toward the carboxyl terminal portion of the parent molecule, whereas the amino acid sequences in the other portion of the molecule read to the left to –131, the amino terminal of the parent molecule. The locations of Lys–Arg and other pairs of basic amino acids residues are also indicated; these are the sites of proteolytic cleavage in the formation of the smaller fragments of the parent molecule. AL, anterior lobe; IL, intermediate lobe.

above, α- and β-MSH do not circulate in adult humans, and their function is unknown. However, ACTH binds to melanotropin-1 receptors. Indeed, the pigmentary changes in several human endocrine diseases are due to changes in circulating ACTH. For example, abnormal pallor is a hallmark of hypopituitarism. Hyperpigmentation occurs in patients with adrenal insufficiency due to primary adrenal disease. Indeed, the presence of hyperpigmentation in association with adrenal insufficiency rules out the possibility that the insufficiency is secondary to pituitary or hypothalamic disease because in these conditions, plasma ACTH is not increased (see Chapter 22). Other disorders of pigmentation result from peripheral mechanisms. Thus, **albinos** have a congenital inability to synthesize melanin. This can result from a variety of different genetic defects in the pathways for melanin synthesis. **Piebaldism** is characterized by patches of skin that lack melanin as a result of congenital defects in the migration of pigment cell precursors from the neural crest during embryonic development. Not only the condition but also the precise pattern of the loss is passed from one generation to the next. **Vitiligo** involves a similar patchy loss of melanin, but the loss develops progressively after birth secondary to an autoimmune process that targets melanocytes.

GROWTH HORMONE

BIOSYNTHESIS & CHEMISTRY

The long arm of human chromosome 17 contains the growth hormone-hCS cluster that contains five genes: one, *hGH-N*, codes for the most abundant ("normal") form of growth hormone; a second, *hGH-V*, codes for the variant form of growth hormone (see below); two code for human chorionic somatomammotropin (hCS) (see Chapter 25); and the fifth is probably an hCS pseudogene.

The structure of hGH-N is shown in Figure 24–3, where it is also compared with that of hCS. Growth hormone that is secreted into the circulation by the pituitary gland consists of a complex mixture of hGH-N, peptides derived from this molecule with varying degrees of post-translational modifications, such as glycosylation, and a splice variant of hGH-N that lacks amino acids 32–46. The physiologic significance of this complex array of hormones has yet to be fully understood, particularly since their structural similarities make it difficult to assay for each species separately. Nevertheless, there is emerging evidence that while the various peptides share a broad range of functions, they may occasionally exert actions in opposition to one another. hGH-V and hCS, on the other hand, are primarily products of the placenta, and as a

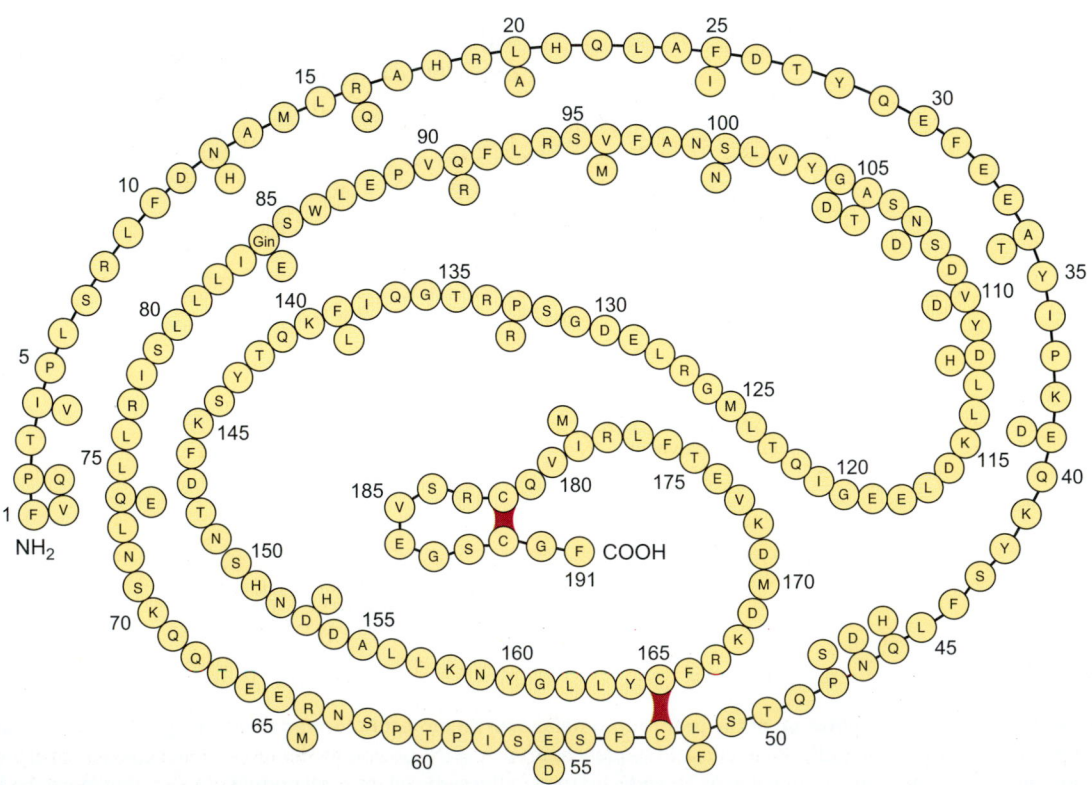

FIGURE 24–3   **Structure of the principal human growth hormone (continuous chain).** The red bars indicate disulfide bridges. The 29 residues alongside the chain identify residues that differ in human chorionic somatomammotropin (hCS; see Chapter 23). All the other residues in hCS are the same, and hCS also has 191 amino acid residues.

consequence are only found in appreciable quantities in the circulation during pregnancy (see Chapter 25).

SPECIES SPECIFICITY

The structure of growth hormone varies considerably from one species to another. Porcine and simian growth hormones have only a transient effect in the guinea pig. In monkeys and humans, bovine and porcine growth hormones do not even have a transient effect on growth, although monkey and human growth hormones are fully active in both monkeys and humans. These facts are relevant to public health discussions surrounding the presence of bovine growth hormones (used to increase milk production) in dairy products, as well as the popularity of growth hormone supplements, marketed via the Internet, with body builders. Controversially, recombinant human growth hormone has also been given to children who are short in stature, but otherwise healthy (ie, without growth hormone deficiency), with apparently limited results.

PLASMA LEVELS, BINDING, & METABOLISM

A portion of circulating growth hormone is bound to a plasma protein that is a large fragment of the extracellular domain of the growth hormone receptor (see below). It appears to be produced by cleavage of receptors in humans, and its concentration is an index of the number of growth hormone receptors in the tissues. Approximately 50% of the circulating pool of growth hormone activity is in the bound form, providing a reservoir of the hormone to compensate for the wide fluctuations that occur in secretion (see below).

The basal plasma growth hormone level measured by radioimmunoassay in adult humans is normally less than 3 ng/mL. This represents both the protein-bound and free forms. Growth hormone is metabolized rapidly, probably at least in part in the liver. The half-life of circulating growth hormone in humans is 6–20 min, and the daily growth hormone output has been calculated to be 0.2–1.0 mg/d in adults.

GROWTH HORMONE RECEPTORS

The growth hormone receptor is a 620-amino-acid protein with a large extracellular portion, a transmembrane domain, and a large cytoplasmic portion. It is a member of the cytokine receptor superfamily, which is discussed in Chapter 3. Growth hormone has two domains that can bind to its receptor, and when it binds to one receptor, the second binding site attracts another, producing a homodimer (Figure 24–4). Dimerization is essential for receptor activation.

FIGURE 24–4 **Some of the principal signaling pathways activated by the dimerized growth hormone receptor (GHR).** Solid arrows indicate established pathways; dashed arrows indicate probable pathways. The details of the PLC pathway and the pathway from Grb2 to MAP K are shown in Chapter 2. GLE-1 and GLE-2, interferon γ-activated response elements; IRS, insulin receptor substrate; p90^{RSK}, an S6 kinase; PLA$_2$, phospholipase A$_2$; SIE, Sis-induced element; SRE, serum response element; SRF, serum response factor; TCF, ternary complex factor. The variants of IGR-II are also shown: a 21-amino-acid extension of the carboxyl terminal, a tetrapeptide substitution at Ser-29, and a tripeptide substitution of Ser-33.

Growth hormone has widespread effects in the body (see below), so even though it is not yet possible precisely to correlate intracellular and whole body effects, it is not surprising that, like insulin, growth hormone activates many different intracellular signaling cascades (Figure 24–4). Of particular note is its activation of the JAK2–STAT pathway. JAK2 is a member of the Janus family of cytoplasmic tyrosine kinases. STATs (for signal transducers and activators of transcription) are a family of inactive cytoplasmic transcription factors that, upon phosphorylation by JAK kinases, migrate to the nucleus and activate various genes. JAK–STAT pathways are known also to mediate the effects of prolactin and various other growth factors.

EFFECTS ON GROWTH

In young animals in which the epiphyses have not yet fused to the long bones (see Chapter 23), growth is inhibited by hypophysectomy and stimulated by growth hormone. Chondrogenesis is accelerated, and as the cartilaginous epiphysial plates widen, they lay down more bone matrix at the ends of long bones. In this way, stature is increased. Prolonged treatment of animals with growth hormone leads to gigantism.

When the epiphyses are closed, linear growth is no longer possible. In this case, an overabundance of growth hormone produces the pattern of bone and soft tissue deformities known in humans as **acromegaly.** The sizes of most of the viscera are increased. The protein content of the body is increased, and the fat content is decreased (see Clinical Box 24–1).

EFFECTS ON PROTEIN & ELECTROLYTE METABOLISM

Growth hormone is a protein anabolic hormone and produces a positive nitrogen and phosphorus balance, a rise in plasma phosphorus, and a fall in blood urea nitrogen and amino acid levels. In adults with growth hormone deficiency, recombinant human growth hormone produces an increase in lean body mass and a decrease in body fat, along with an increase in metabolic rate and a fall in plasma cholesterol. Gastrointestinal absorption of Ca^{2+} is increased. Na^+ and K^+ excretion is reduced by an action independent of the adrenal glands, probably because these electrolytes are diverted from the kidneys to the growing tissues. On the other hand, excretion of the amino acid 4-hydroxyproline is increased during this growth, reflective of the ability of growth hormone to stimulate the synthesis of soluble collagen.

EFFECTS ON CARBOHYDRATE & FAT METABOLISM

The actions of growth hormone on carbohydrate metabolism are discussed in Chapter 21. At least some forms of growth hormone are diabetogenic because they increase hepatic glu-

CLINICAL BOX 24–1

Gigantism & Acromegaly

Tumors of the somatotropes of the anterior pituitary (pituitary adenoma) secrete large amounts of growth hormone, leading in children to **gigantism** and in adults to **acromegaly.** If the tumor arises before puberty, the individual may grow to an extraordinary height. After linear growth is no longer possible, on the other hand, the characteristic features of acromegaly arise, including greatly enlarged hands and feet, vertebral changes attributable to osteoarthritis, soft tissue swelling, hirsutism, and protrusion of the brow and jaw. Abnormal growth of internal organs may eventually impair their function such that the condition, which has an insidious onset, can prove fatal if left untreated. Hypersecretion of growth hormone is accompanied by hypersecretion of prolactin in 20–40% of patients with acromegaly. About 25% of patients have abnormal glucose tolerance tests, and 4% develop lactation in the absence of pregnancy. Acromegaly can be caused by extra-pituitary as well as intrapituitary growth hormone-secreting tumors and by hypothalamic tumors that secrete GHRH, but the latter are rare. Treatment involves surgical removal of the tumor where possible, the use of long-acting analogues of somatostatin, or both.

cose output and exert an anti-insulin effect in muscle. Growth hormone is also ketogenic and increases circulating free fatty acid (FFA) levels. The increase in plasma FFA, which takes several hours to develop, provides a ready source of energy for the tissues during hypoglycemia, fasting, and stressful stimuli. Growth hormone does not stimulate beta cells of the pancreas directly, but it increases the ability of the pancreas to respond to insulinogenic stimuli such as arginine and glucose. This is an additional way growth hormone promotes growth, since insulin has a protein anabolic effect (see Chapter 21).

SOMATOMEDINS

The effects of growth hormone on growth, cartilage, and protein metabolism depend on an interaction between growth hormone and **somatomedins,** which are polypeptide growth factors secreted by the liver and other tissues. The first of these factors isolated was called sulfation factor because it stimulated the incorporation of sulfate into cartilage. However, it also stimulated collagen formation, and its name was changed to somatomedin. It then became clear that there are a variety of different somatomedins and that they are members of an increasingly large family of **growth factors** that affect many different tissues and organs.

The principal (and in humans probably the only) circulating somatomedins are **insulin-like growth factor I (IGF-I, somatomedin C)** and **insulin-like growth factor II (IGF-II).** These factors are closely related to insulin, except that their C

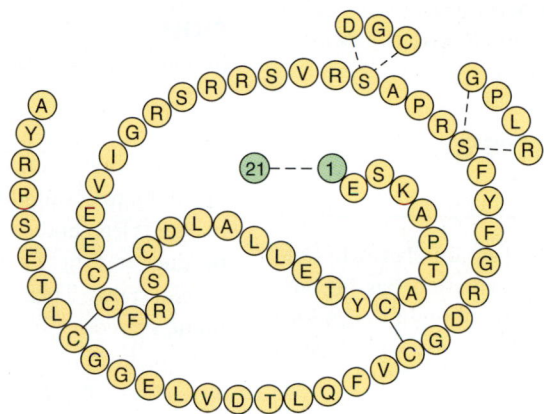

FIGURE 24-5 **Structure of human IGF-I, IGF-II, and insulin (ins) (top).** The lower panel shows the structure of human IGF-II with its disulfide bonds, as well as three variant structures shown: a 21-aa extension of the c-terminus, a tetrapeptide substitution at Ser-29, and a tripeptide substitution of Ser-33.

chains are not separated (Figure 24–5) and they have an extension of the A chain called the D domain. The hormone relaxin (see Chapter 25) is also a member of this family. Humans have two related relaxin isoforms, and both resemble IGF-II. In humans a variant form of IGF-I lacking three amino terminal amino acid residues has been found in the brain, and there are several variant forms of human IGF-II (Figure 24–5). The mRNAs for IGF-I and IGF-II are found in the liver, in cartilage, and in many other tissues, indicating that they are synthesized in these tissues.

The properties of IGF-I, IGF-II, and insulin are compared in Table 24–2. Both are tightly bound to proteins in the plasma, and, at least for IGF-I, this prolongs their half-life in the circulation. Six different IGF-binding proteins, with different patterns of distribution in various tissues, have been identified. All are present in plasma, with IGF-binding protein-3 (IGFBP-3) accounting for 95% of the binding in the circulation. The contribution of the IGFs to the insulin-like activity in blood is discussed in Chapter 21. The IGF-I receptor is very similar to the insulin receptor and probably uses similar or identical intracellular signaling pathways. The IGF-II receptor has a

TABLE 24–2 **Comparison of insulin and the insulin-like growth factors.**

	Insulin	IGF-I	IGF-II
Other names	...	Somatomedin C	Multiplication-stimulating activity (MSA)
Number of amino acids	51	70	67
Source	Pancreatic B cells	Liver and other tissues	Diverse tissues
Level regulated by	Glucose	Growth hormone after birth, nutritional status	Unknown
Plasma levels	0.3–2 ng/mL	10–700 ng/mL; peaks at puberty	300–800 ng/mL
Plasma-binding proteins	No	Yes	Yes
Major physiologic role	Control of metabolism	Skeletal and cartilage growth	Growth during fetal development

distinct structure (see Figure 21–5) and is involved in the intracellular targeting of acid hydrolases and other proteins to intracellular organelles. Secretion of IGF-I is independent of growth hormone before birth but is stimulated by growth hormone after birth, and it has pronounced growth-stimulating activity. Its concentration in plasma rises during childhood and peaks at the time of puberty, then declines to low levels in old age. IGF-II is largely independent of growth hormone and plays a role in the growth of the fetus before birth. In human fetuses in which it is overexpressed, growth of organs, especially the tongue, other muscles, kidneys, heart, and liver, is disproportionate. In adults, the gene for IGF-II is expressed only in the choroid plexus and meninges.

DIRECT & INDIRECT ACTIONS OF GROWTH HORMONE

Our understanding of the mechanism of action of growth hormone has evolved recently as new information has become available. Growth hormone was originally thought to produce growth by a direct action on tissues, then later was believed to act solely through its ability to induce somatomedins. However, if growth hormone is injected into one proximal tibial epiphysis, a unilateral increase in cartilage width is produced, and cartilage, like other tissues, makes IGF-I. A current hypothesis to explain these results holds that growth hormone acts on cartilage to convert stem cells into cells that respond to IGF-I and then locally produced and circulating IGF-I makes the cartilage grow. However, the independent role of circulating IGF-I remains important, since infusion of IGF-I to hypophysectomized rats restores bone and body growth. Overall, it seems that growth hormone and somatomedins can act both in cooperation and independently to stimulate pathways that lead to growth. The situation is almost certainly complicated further by the existence of multiple forms of growth hormone in the circulation that can, in some situations, have opposing actions.

Figure 24–6 is a summary of current views of the other actions of growth hormone and IGF-I. However, growth hormone probably combines with circulating and locally produced IGF-I in various proportions to produce at least some of these effects. Indeed, while the mainstay of therapy for acromegaly remains somatostatin analogues that inhibit the secretion of growth hormone, a growth hormone receptor antagonist has recently become available and has been found to reduce plasma IGF-I and produce clinical improvement in cases of acromegaly that fail to respond to other treatments.

HYPOTHALAMIC & PERIPHERAL CONTROL OF GROWTH HORMONE SECRETION

The secretion of growth hormone is not stable over time. Adolescents have the highest circulating levels of growth hormone, followed by children and finally adults. Levels decline in old age,

FIGURE 24–6 **Actions believed to be mediated by growth hormone (GH) and IGF-I.** (Courtesy of R Clark and N Gesundheit.)

and there has been considerable interest in injecting growth hormone to counterbalance the effects of aging. The hormone increases lean body mass and decreases body fat, but it does not produce statistically significant increases in muscle strength or mental status. There are also diurnal variations in growth hormone secretion superimposed on these developmental stages. Growth hormone is found at relatively low levels during the day, unless specific triggers for its release are present (see below). During sleep, on the other hand, large pulsatile bursts of growth hormone secretion occur. Therefore, it is not surprising that the secretion of growth hormone is under hypothalamic control. The hypothalamus controls growth hormone production by secreting growth hormone-releasing hormone (GHRH) as well as somatostatin, which inhibits growth hormone release (see Chapter 18). Thus, the balance between the effects of these hypothalamic factors on the pituitary will determine the level of growth hormone release. The stimuli of growth hormone secretion discussed as follows can therefore act by increasing hypothalamic secretion of GHRH, decreasing secretion of somatostatin, or both. A third regulator of growth hormone secretion is **ghrelin.** The main site of ghrelin synthesis and secretion is the stomach, but it is also produced in the hypothalamus and has marked growth hormone-stimulating activity. In addition, it appears to be involved in the regulation of food intake.

Growth hormone secretion is under feedback control, like the secretion of other anterior pituitary hormones. It acts on the hypothalamus to antagonize GHRH release. Growth hormone also increases circulating IGF-I, and IGF-I in turn exerts a direct inhibitory action on growth hormone secretion from the pituitary. It also stimulates somatostatin secretion (Figure 24–7).

STIMULI AFFECTING GROWTH HORMONE SECRETION

The basal plasma growth hormone concentration ranges from 0–3 ng/mL in normal adults. However, secretory rates cannot be estimated from single values because of their irregular nature. Thus, average values over 24 h (see below) and peak values may be more meaningful, albeit difficult to assess in the clinical setting. The stimuli that increase growth hormone

FIGURE 24–7 **Feedback control of growth hormone secretion.** Solid arrows represent positive effects and dashed arrows represent inhibition.

TABLE 24–3 **Stimuli that affect growth hormone secretion in humans.**

Stimuli that increase secretion
Hypoglycemia
2-Deoxyglucose
Exercise
Fasting
Increase in circulating levels of certain amino acids
Protein meal
Infusion of arginine and some other amino acids
Glucagon
Stressful stimuli
Pyrogen
Lysine vasopressin
Various psychologic stresses
Going to sleep
L-Dopa and α-adrenergic agonists that penetrate the brain
Apomorphine and other dopamine receptor agonists
Estrogens and androgens
Stimuli that decrease secretion
REM sleep
Glucose
Cortisol
FFA
Medroxyprogesterone
Growth hormone and IGF-I

secretion are summarized in Table 24–3. Most of them fall into three general categories: (1) conditions such as hypoglycemia and fasting in which there is an actual or threatened decrease in the substrate for energy production in cells, (2) conditions in which the amounts of certain amino acids are increased in the plasma, and (3) stressful stimuli. The response to glucagon has been used as a test of growth hormone reserve. Growth hormone secretion is also increased in subjects deprived of rapid eye movement (REM) sleep (see Chapter 15) and inhibited during normal REM sleep.

Glucose infusions lower plasma growth hormone levels and inhibit the response to exercise. The increase produced by 2-deoxyglucose is presumably due to intracellular glucose deficiency, since this compound blocks the catabolism of glucose 6-phosphate. Sex hormones induce growth hormone secretion, increase growth hormone responses to provocative stimuli such as arginine and insulin, and also serve as permissive factors for the action of growth hormone in the periphery. This likely contributes to the relatively high levels of circulating growth hormone and associated growth spurt in puberty. Growth hormone secretion is also induced by thyroid hormones. Growth hormone secretion is inhibited, on the other hand, by cortisol, FFA, and medroxyprogesterone.

Growth hormone secretion is increased by L-dopa, which increases the release of dopamine and norepinephrine in the brain, and by the dopamine receptor agonist apomorphine.

PHYSIOLOGY OF GROWTH

Growth hormone, while being essentially unimportant for fetal development, is the most important hormone for postnatal growth. However, growth overall is a complex phenomenon that is affected not only by growth hormone and somatomedins, but also, as would be predicted by the previous discussion,

by thyroid hormones, androgens, estrogens, glucocorticoids, and insulin. It is also affected, of course, by genetic factors, and it depends on adequate nutrition. It is normally accompanied by an orderly sequence of maturational changes, and it involves accretion of protein and increase in length and size, not just an increase in weight (which could reflect the formation of fat or retention of salt and water rather than growth per se).

ROLE OF NUTRITION

The food supply is the most important extrinsic factor affecting growth. The diet must be adequate not only in protein content but also in essential vitamins and minerals (see Chapter 27) and in calories, so that ingested protein is not burned for energy. However, the age at which a dietary deficiency occurs appears to be an important consideration. For example, once the pubertal growth spurt has commenced, considerable linear growth continues even if caloric intake is reduced. Injury and disease likewise stunt growth because they increase protein catabolism.

GROWTH PERIODS

Patterns of growth vary somewhat from species to species. Rats continue to grow, although at a declining rate, throughout life. In humans, two periods of rapid growth occur (Figure 24–8): the first in infancy and the second in late puberty just before growth stops. The first period of accelerated growth is partly a continuation of the fetal growth period. The second growth spurt, at the time of puberty, is due to growth hormone, androgens, and estrogens, and the subsequent cessation of growth is due in large part to closure of the epiphyses in the long bones by estrogens (see Chapter 25). After this time, further increases in height are not possible. Because girls mature earlier than boys, this growth spurt appears earlier in girls. Of course, in both sexes the rate of growth of individual tissues varies (Figure 24–9).

It is interesting that at least during infancy, growth is not a continuous process but is episodic or saltatory. Increases in length of human infants of 0.5 to 2.5 cm in a few days are separated by periods of 2 to 63 d during which no measurable growth can be detected. The cause of the episodic growth is unknown.

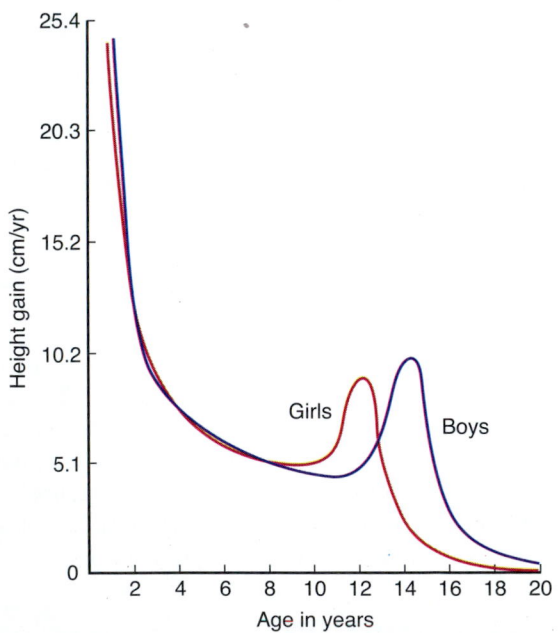

FIGURE 24–8   Rate of growth in boys and girls from birth to age 20.

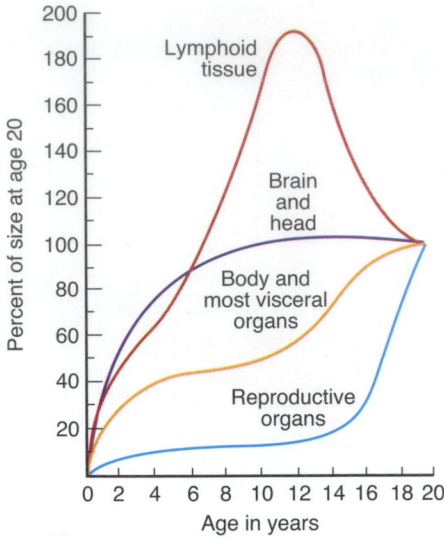

FIGURE 24–9   **Growth of different tissues at various ages as a percentage of size at age 20.** The curves are composites that include data for both boys and girls.

HORMONAL EFFECTS

The contributions of hormones to growth after birth are shown diagrammatically in Figure 24–10. Plasma growth hormone is elevated in newborns. Subsequently, average resting levels fall but the spikes of growth hormone secretion are larger, especially during puberty, so the mean plasma level over 24 h is increased; it is 2 to 4 ng/mL in normal adults, but 5 to 8 ng/mL in children. One of the factors stimulating IGF-I secretion is growth hormone, and plasma IGF-I levels rise during childhood, reaching a peak at 13 to 17 years of age. In contrast, IGF-II levels are constant throughout postnatal growth.

The growth spurt that occurs at the time of puberty (Figure 24–8) is due in part to the protein anabolic effect of androgens, and the secretion of adrenal androgens increases at this time in both sexes; however, it is also due to an interaction among sex steroids, growth hormone, and IGF-I. Treatment

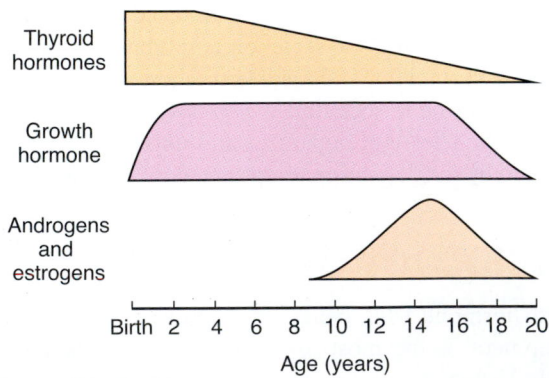

FIGURE 24–10   **Relative importance of hormones in human growth at various ages.**   (Courtesy of DA Fisher.)

with estrogens and androgens increases the secretion of growth hormone in response to various stimuli and increases plasma IGF-I secondary to this increase in circulating growth hormone. This, in turn, causes growth.

Although androgens and estrogens initially stimulate growth, estrogens ultimately terminate growth by causing the epiphyses to fuse to the long bones (epiphysial closure). Once the epiphyses have closed, linear growth ceases (see Chapter 23). This is why patients with sexual precocity are apt to be dwarfed. On the other hand, men who were castrated before puberty tend to be tall because their estrogen production is decreased and their epiphyses remain open, allowing some growth to continue past the normal age of puberty.

When growth hormone is administered to hypophysecto-mized animals, the animals do not grow as rapidly as they do when treated with growth hormone plus thyroid hormones. Thyroid hormones alone have no effect on growth in this situation. Their action is therefore permissive to that of growth hormone, possibly via potentiation of the actions of somatomedins. Thyroid hormones also appear to be necessary for a completely normal rate of growth hormone secretion; basal growth hormone levels are normal in hypothyroidism, but the response to hypoglycemia is frequently blunted. Thyroid hormones have widespread effects on the ossification of cartilage, the growth of teeth, the contours of the face, and the proportions of the body. Hypothyroid dwarfs (also known as **cretins**) therefore have infantile features (Figure 24–11). Patients who are dwarfed because of panhypopituitarism have features consistent with their chronologic age until puberty, but since they do not mature sexually, they have juvenile features in adulthood (Clinical Box 24–2).

The effect of insulin on growth is discussed in Chapter 21. Diabetic animals fail to grow, and insulin causes growth in hypophysectomized animals. However, the growth is appreciable only when large amounts of carbohydrate and protein are supplied with the insulin.

Adrenocortical hormones other than androgens exert a permissive action on growth in the sense that adrenalecto-mized animals fail to grow unless their blood pressures and circulations are maintained by replacement therapy. On the other hand, glucocorticoids are potent inhibitors of growth because of their direct action on cells, and treatment of children with pharmacologic doses of steroids slows or stops growth for as long as the treatment is continued.

CATCH-UP GROWTH

Following illness or starvation in children, a period of **catch-up growth** (Figure 24–12) takes place during which the growth rate is greater than normal. The accelerated growth

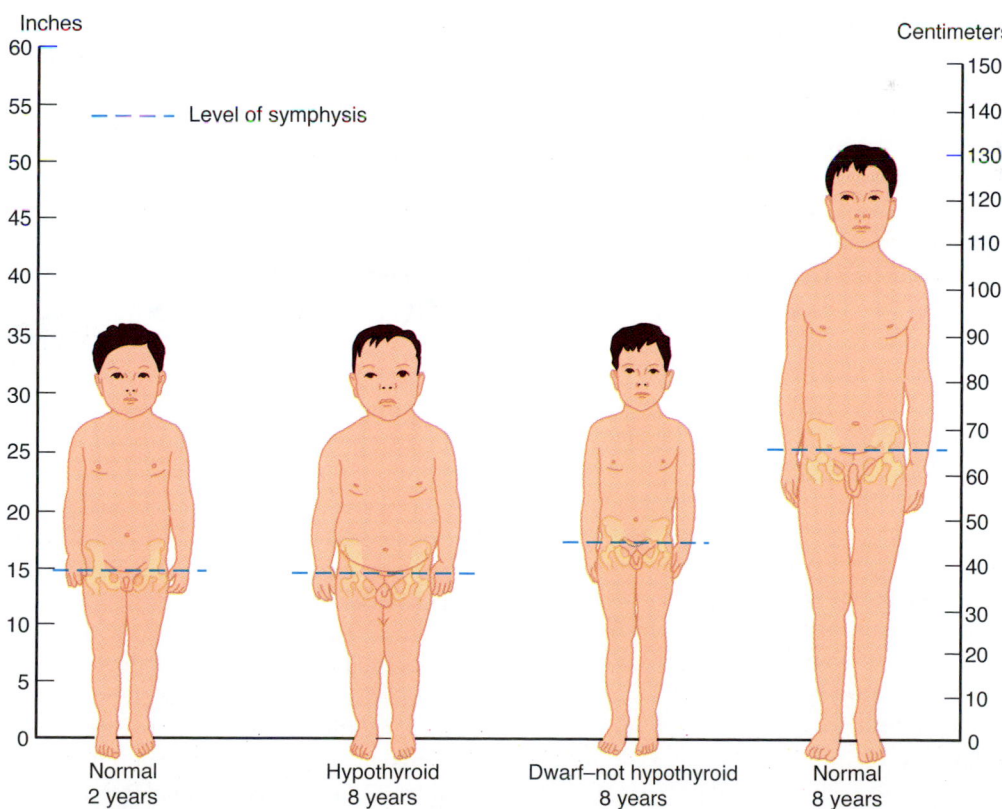

FIGURE 24–11 **Normal and abnormal growth.** Hypothyroid dwarfs (cretins) retain their infantile proportions, whereas dwarfs of the constitutional type and, to a lesser extent, of the hypopituitary type have proportions characteristic of their chronologic age. (Reproduced, with permission, from Wilkins L: *The Diagnosis and Treatment of Endocrine Disorders in Childhood and Adolescence,* 3rd ed. Thomas, 1966.)

Dwarfism

The accompanying discussion of growth control should suggest several possible etiologies of short stature. It can be due to GHRH deficiency, growth hormone deficiency, or deficient secretion of IGF-I. Isolated growth hormone deficiency is often due to GHRH deficiency, and in these instances, the growth hormone response to GHRH is normal. However, some patients with isolated growth hormone deficiency have abnormalities of their growth hormone secreting cells. In another group of dwarfed children, the plasma growth hormone concentration is normal or elevated but their growth hormone receptors are unresponsive as a result of loss-of-function mutations. The resulting condition is known as **growth hormone insensitivity** or **Laron dwarfism.** Plasma IGF-I is markedly reduced, along with IGFBP 3, which is also growth hormone-dependent. African pygmies have normal plasma growth hormone levels and a modest reduction in the plasma level of growth hormone-binding protein. However, their plasma IGF-I concentration fails to increase at the time of puberty and they experience less growth than non-pygmy controls throughout the prepubertal period.

Short stature may also be caused by mechanisms independent of specific defects in the growth hormone axis. It is characteristic of childhood hypothyroidism (cretinism) and occurs in patients with precocious puberty. It is also part of the syndrome of **gonadal dysgenesis** seen in patients who have an XO chromosomal pattern instead of an XX or XY pattern (see Chapter 25). Various bone and metabolic diseases also cause stunted growth, and in many cases there is no known cause ("constitutional delayed growth"). Chronic abuse and neglect can also cause dwarfism in children, independent of malnutrition. This condition is known as **psychosocial dwarfism** or the **Kaspar Hauser syndrome,** named for the patient with the first reported case. Finally, **achondroplasia,** the most common form of dwarfism in humans, is characterized by short limbs with a normal trunk. It is an autosomal dominant condition caused by a mutation in the gene that codes for **fibroblast growth factor receptor 3 (FGFR3).** This member of the fibroblast growth receptor family is normally expressed in cartilage and the brain.

The treatment of dwarfism is dictated by its underlying cause. If treatment is commenced promptly in childhood, almost normal stature can often be attained. The availability of recombinant forms of growth hormone and IGF-I has greatly improved treatment in cases where these hormones are deficient.

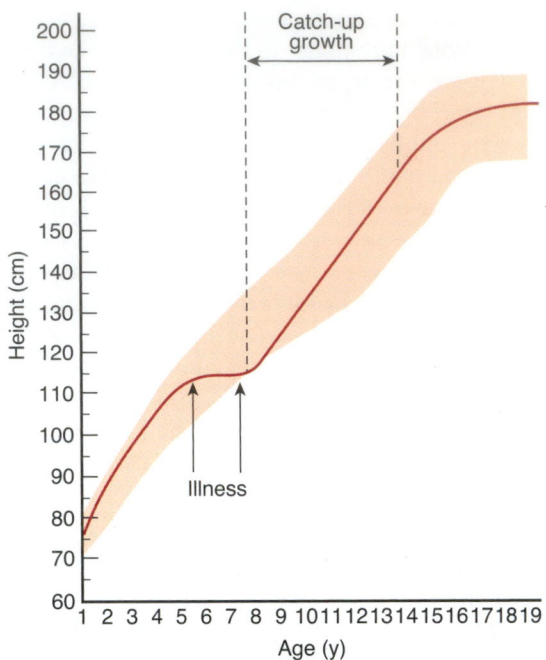

FIGURE 24–12   Growth curve for a normal boy who had an illness beginning at age 5 and ending at age 7. Catch-up growth eventually returned his height to his previous normal growth curve. (Modified from Boersma B, Wit JM: Catch-up growth. Endocr Rev 1997;18:646.)

usually continues until the previous growth curve is reached, then slows to normal. The mechanisms that bring about and control catch-up growth are unknown.

EFFECTS OF PITUITARY INSUFFICIENCY

CHANGES IN OTHER ENDOCRINE GLANDS

The widespread changes that develop when the pituitary is removed surgically or destroyed by disease in humans or animals are predictable in terms of the known hormonal functions of the gland. In hypopituitarism, the adrenal cortex atrophies, and the secretion of adrenal glucocorticoids and sex hormones falls to low levels. Stress induced increases in aldosterone secretion are absent, but basal aldosterone secretion and increases induced by salt depletion are normal, at least for some time. Since no mineralocorticoid deficiency is present, salt loss and hypovolemic shock do not develop, but the inability to increase glucocorticoid secretion makes patients with pituitary insufficiency sensitive to stress. The development of salt loss in long-standing hypopituitarism is discussed in Chapter 22. Growth is inhibited (see above). Thyroid function is depressed to low levels, and cold is tolerated poorly. The gonads atrophy, sexual cycles stop, and some of the secondary sex characteristics disappear.

INSULIN SENSITIVITY

Hypophysectomized animals have a tendency to become hypoglycemic, especially when fasted. Hypophysectomy ameliorates diabetes mellitus (see Chapter 21) and markedly increases the hypoglycemic effect of insulin. This is due in part to the deficiency of adrenocortical hormones, but hypophysectomized animals are more sensitive to insulin than adrenalectomized animals because they also lack the anti-insulin effect of growth hormone.

WATER METABOLISM

Although selective destruction of the supraoptic–posterior pituitary causes diabetes insipidus (see Chapter 18), removal of both the anterior and posterior pituitary usually causes no more than a transient polyuria. In the past, there was speculation that the anterior pituitary secreted a "diuretic hormone," but the amelioration of the diabetes insipidus is actually explained by a decrease in the osmotic load presented for excretion. Osmotically active particles hold water in the renal tubules (see Chapter 38). Because of the ACTH deficiency, the rate of protein catabolism is decreased in hypophysectomized animals. Because of the TSH deficiency, the metabolic rate is low. Consequently, fewer osmotically active products of catabolism are filtered and urine volume declines, even in the absence of vasopressin. Growth hormone deficiency contributes to the depression of the glomerular filtration rate in hypophysectomized animals, and growth hormone increases the glomerular filtration rate and renal plasma flow in humans. Finally, because of the glucocorticoid deficiency, there is the same defective excretion of a water load that is seen in adrenalectomized animals. The "diuretic" activity of the anterior pituitary can thus be explained in terms of the actions of ACTH, TSH, and growth hormone.

OTHER DEFECTS

When growth hormone deficiency develops in adulthood, it is usually accompanied by deficiencies in other anterior pituitary hormones. The deficiency of ACTH and other pituitary hormones with MSH activity may be responsible for the pallor of the skin in patients with hypopituitarism. There may be some loss of protein in adults, but wasting is not a feature of hypopituitarism in humans, and most patients with pituitary insufficiency are well nourished.

CAUSES OF PITUITARY INSUFFICIENCY IN HUMANS

Tumors of the anterior pituitary cause pituitary insufficiency. Suprasellar cysts, remnants of Rathke's pouch that enlarge and compress the pituitary, are another cause of hypopituitarism. In women who have an episode of shock due to postpartum hemorrhage, the pituitary may become infarcted, with the subsequent development of postpartum necrosis (**Sheehan syndrome**). The blood supply to the anterior lobe is vulnerable because it descends on the pituitary stalk through the rigid diaphragma sellae, and during pregnancy the pituitary is enlarged. Pituitary infarction is usually extremely rare in men.

CHAPTER SUMMARY

- The pituitary gland plays a critical role in regulating the function of downstream glands, and also exerts independent endocrine actions on a wide variety of peripheral organs and tissues. It consists of two functional sections in humans: the anterior pituitary, which secretes mainly tropic hormones; and the posterior pituitary, which contains nerve endings that release oxytocin and vasopressin. The intermediate lobe is prominent in lower vertebrates but not in humans or other mammals.

- Corticotropes of the anterior lobe synthesize proopiomelanocortin, the precursor of ACTH, endorphins, and melanocortins. The latter have a critical role in the control of skin coloration, whereas ACTH is a primary regulator of skin pigmentation in mammals.

- Growth hormone is synthesized by somatotropes and is highly species-specific. It is secreted in an episodic fashion in response to hypothalamic factors, and secretion is subject to feedback inhibition. A portion of the circulating pool is protein-bound.

- Growth hormone activates growth and influences protein, carbohydrate, and fat metabolism to react to stressful conditions. Many, but not all, of the peripheral actions of growth hormone can be attributed to its ability to stimulate production of IGF-I.

- Growth reflects a complex interplay of growth hormone, IGF-I, and many other hormones as well as extrinsic influences and genetic factors. The consequences of over- or underproduction of such influences depends on whether this occurs before or after puberty. Deficiencies in components of the growth hormone pathway in childhood lead to dwarfism; overproduction results in gigantism, acromegaly, or both.

MULTIPLE-CHOICE QUESTIONS

For all questions, select the single best answer unless otherwise directed.

1. Which of the following hormones exerts the least effect on growth?
   A) growth hormone
   B) testosterone
   C) T$_4$
   D) insulin
   E) vasopressin

2. Which of the following pituitary hormones is an opioid peptide?
   A) α-melanocyte-stimulating hormone (α-MSH)
   B) β-MSH
   C) ACTH
   D) growth hormone
   E) β-endorphin

3. Which of the following is *not* characteristic of hypopituitarism?

    A) cachexia
    B) infertility
    C) pallor
    D) low basal metabolic rate
    E) intolerance to stress

4. A scientist finds that infusion of growth hormone into the median eminence of the hypothalamus in experimental animals inhibits the secretion of growth hormone and concludes that this proves that growth hormone feeds back to inhibit GHRH secretion. Do you accept this conclusion?

    A) No, because growth hormone does not cross the blood–brain barrier.
    B) No, because the infused growth hormone could be stimulating dopamine secretion.
    C) No, because substances placed in the median eminence could be transported to the anterior pituitary.
    D) Yes, because systemically administered growth hormone inhibits growth hormone secretion.
    E) Yes, because growth hormone binds GHRH, inactivating it.

5. The growth hormone receptor

    A) activates G_s.
    B) requires dimerization to exert its effects.
    C) must be internalized to exert its effects.
    D) resembles the IGF-I receptor.
    E) resembles the ACTH receptor.

CHAPTER RESOURCES

Ayuk J, Sheppard MC: Growth hormone and its disorders. Postgrad Med J 2006;82:24.

Boissy RE, Nordlund JJ: Molecular basis of congenital hypopigmentary disorders in humans: A review. Pigment Cell Res 1997;10:12.

Buzi F, Mella P, Pilotta A, Prandi E, Lanfranchi F, Carapella T: Growth hormone receptor polymorphisms. Endocr Dev 2007;11:28.

Fauquier T, Rizzoti K, Dattani M, Lovell-Badge R, Robinson ICAF: SOX2-expressing progenitor cells generate all of the major cell types in the adult mouse pituitary gland. Proc Natl Acad Sci USA 2008;105:2907.

Hindmarsh PC, Dattani MT: Use of growth hormone in children. Nat Clin Pract Endocrinol Metab 2006;2:260.

The Gonads: Development & Function of the Reproductive System

OBJECTIVES

After studying this chapter, you should be able to:

- Name the key hormones secreted by Leydig cells and Sertoli cells of the testes and by graafian follicles and corpora lutea of the ovaries.
- Outline the role of chromosomes, hormones, and related factors in sex determination and development.
- Summarize the hormonal changes that occur at puberty in males and females.
- Outline the hormonal changes and their physiologic effects during perimenopause and menopause.
- List the physiologic stimuli and the drugs that affect prolactin secretion.
- Outline the steps involved in spermatogenesis and the mechanisms that produce erection and ejaculation.
- Know the general structure of testosterone, and describe its biosynthesis, transport, metabolism, and actions.
- Describe the processes involved in regulation of testosterone secretion.
- Describe the physiologic changes that occur in the female reproductive organs during the menstrual cycle.
- Know the general structures of 17β-estradiol and progesterone, and describe their biosynthesis, transport, metabolism, and actions.
- Describe the roles of the pituitary and the hypothalamus in the regulation of ovarian function, and the role of feedback loops in this process.
- Describe the hormonal changes that accompany pregnancy and parturition.
- Outline the processes involved in lactation.

INTRODUCTION

Modern genetics and experimental embryology make it clear that, in most species of mammals, the multiple differences between the male and the female depend primarily on a single chromosome (the Y chromosome) and a single pair of endocrine structures, the testes in the male and the ovaries in the female. The differentiation of the primitive gonads into testes or ovaries in utero is genetically determined in humans, but the formation of male genitalia depends on the presence of a functional, secreting testis; in the absence of testicular tissue, development is female. Evidence indicates that male sexual behavior and, in some species, the male pattern of gonadotropin secretion are due to the action of male hormones on the brain in early development. After birth, the gonads remain quiescent until adolescence, when they are activated by gonadotropins from the anterior pituitary. Hormones secreted by the gonads at this time cause the appearance of features typical of the adult male or female and the onset of the sexual cycle in the female. In human females, ovarian function regresses after a number of years and sexual cycles cease (the menopause). In males, gonadal function slowly declines with advancing age, but the ability to produce viable gametes persists.

In both sexes, the gonads have a dual function: the production of germ cells (**gametogenesis**) and the secretion of **sex hormones.** The **androgens** are the steroid sex hormones that are masculinizing in their action; the **estrogens** are those that are feminizing. Both types of hormones are normally secreted in both sexes. The testes secrete large amounts of androgens, principally **testosterone,** but they also secrete small amounts of estrogens. The ovaries secrete large amounts of estrogens and small amounts of androgens. Androgens are secreted from the adrenal cortex in both sexes, and some of the androgens are converted to estrogens in fat and other extragonadal and extra-adrenal tissues. The ovaries also secrete **progesterone,** a steroid that has special functions in preparing the uterus for pregnancy.

Particularly during pregnancy, the ovaries secrete the polypeptide hormone **relaxin,** which loosens the ligaments of the pubic symphysis and softens the cervix, facilitating delivery of the fetus. In both sexes, the gonads secrete other polypeptides, including **inhibin B,** a polypeptide that inhibits follicle-stimulating hormone (FSH) secretion.

The secretory and gametogenic functions of the gonads are both dependent on the secretion of the anterior pituitary gonadotropins, FSH, and luteinizing hormone (LH). The sex hormones and inhibin B feed back to inhibit gonadotropin secretion. In males, gonadotropin secretion is noncyclic; but in postpubertal females an orderly, sequential secretion of gonadotropins is necessary for the occurrence of menstruation, pregnancy, and lactation.

SEX DIFFERENTIATION & DEVELOPMENT

CHROMOSOMAL SEX

The Sex Chromosomes

Sex is determined genetically by two chromosomes, called the **sex chromosomes,** to distinguish them from the **somatic chromosomes (autosomes).** In humans and many other mammals, the sex chromosomes are called X and Y chromosomes. The Y chromosome is necessary and sufficient for the production of testes, and the testis-determining gene product is called SRY (for sex-determining region of the Y chromosome). SRY is a DNA-binding regulatory protein. It bends the DNA and acts as a transcription factor that initiates transcription of a cascade of genes necessary for testicular differentiation, including the gene for **müllerian inhibiting substance (MIS;** see below). The gene for SRY is located near the tip of the short arm of the human Y chromosome. Male cells with the diploid number of chromosomes contain an X and a Y chromosome (XY pattern), whereas female cells contain two X chromosomes (XX pattern). As a consequence of meiosis during gametogenesis, each normal ovum contains a single X chromosome, but half of the normal sperm contain an X chromosome and half contain a Y chromosome (Figure 25–1). When a sperm containing a Y chromosome fertilizes an ovum, an XY pattern results and the zygote develops into a **genetic male.** When fertilization occurs with an X-containing sperm, an XX pattern and a **genetic female** result. Cell division and the chemical nature of chromosomes are discussed in Chapter 1.

Human Chromosomes

Human chromosomes can be studied in detail. Human cells are grown in tissue culture; treated with the drug colchicine, which arrests mitosis at the metaphase; exposed to a hypotonic solution that makes the chromosomes swell and disperse; and then "squashed" onto slides. Staining techniques make it possible to identify the individual chromosomes and study them in detail (Figure 25–2). There are 46 chromosomes: in males, 22 pairs of autosomes plus an X chromosome and a Y chromosome; in females, 22 pairs of autosomes plus two X chromosomes. The individual chromosomes are usually arranged in an arbitrary pattern (**karyotype**). The individual autosome pairs are identified by the numbers 1–22 on the basis of their morphologic characteristics.

Sex Chromatin

Soon after cell division has started during embryonic development, one of the two X chromosomes of the somatic cells in normal females becomes functionally inactive. In abnormal individuals with more than two X chromosomes, only one remains active. The process that is normally responsible for inactivation is initiated in an X-inactivation center in the chromosome, probably via the transactivating factor CTCF (for CCCTC-binding factor), which is also induced in gene imprinting. However, the details of the inactivation process are still incompletely understood. The choice of which X chromosome remains active is random, so normally one X chromosome remains active in approximately half of the cells and the other X chromosome is active in the other half. The selection persists through subsequent divisions of these cells, and consequently some of the somatic cells in adult females contain an active X chromosome of paternal origin and some contain an active X chromosome of maternal origin.

In normal cells, the inactive X chromosome condenses and can be seen in various types of cells, usually near the nuclear membrane, as the **Barr body,** also called sex chromatin (Figure 25–3). Thus, there is a Barr body for each X chromosome in excess of one in the cell. The inactive X chromosome is also visible as a small "drumstick" of chromatin projecting from the nuclei of 1–15% of the polymorphonuclear leukocytes in females but not in males (Figure 25–3).

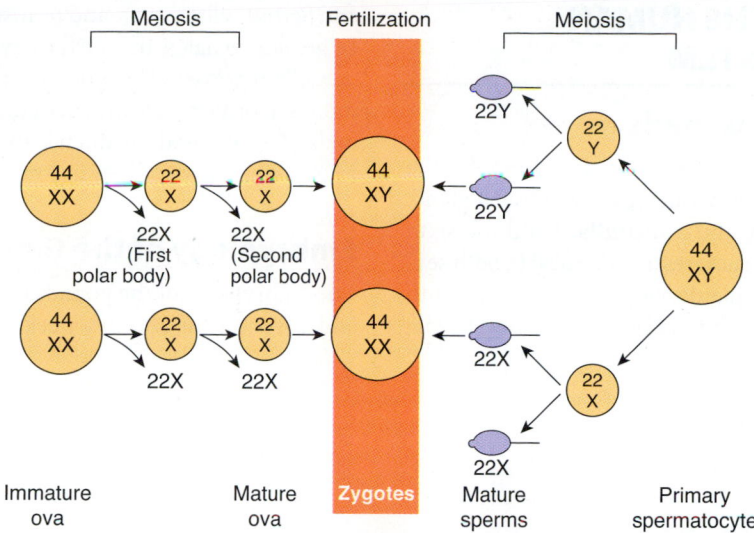

FIGURE 25–1 **Basis of genetic sex determination.** In the two-stage meiotic division in the female, only one cell survives as the mature ovum. In the male, the meiotic division results in the formation of four sperms, two containing the X and two the Y chromosome. Fertilization thus produces a male zygote with 22 pairs of autosomes plus an X and a Y or a female zygote with 22 pairs of autosomes and two X chromosomes. Note that for clarity, this figure and Figures 25–6 and 25–7 differ from the current international nomenclature for karyotypes, which lists the total number of chromosomes followed by the sex chromosome pattern. Thus, XO is 45, X; XY is 46, XY; XXY is 47, XXY, and so on.

FIGURE 25–2 **Karyotype of chromosomes from a normal male.** The chromosomes have been stained with Giemsa's stain, which produces a characteristic banding pattern. (Reproduced with permission, from Lingappa VJ, Farey K: *Physiological Medicine.* McGraw-Hill, 2000.)

FIGURE 25–3 **Left: Barr body (arrows) in the epidermal spinous cell layer. Right: Nuclear appendage ("drumstick") identified by arrow in white blood cells.** (Reproduced with permission from Grumbach MM, Barr ML: Cytologic tests of chromosomal sex in relation to sex anomalies in man. Recent Prog Horm Res 1958;14:255.)

EMBRYOLOGY OF THE HUMAN REPRODUCTIVE SYSTEM

Development of the Gonads

On each side of the embryo, a primitive gonad arises from the genital ridge, a condensation of tissue near the adrenal gland. The gonad develops a **cortex** and a **medulla.** Until the sixth week of development, these structures are identical in both sexes. In genetic males, the medulla develops during the seventh and eighth weeks into a testis, and the cortex regresses. Leydig and Sertoli cells appear, and testosterone and MIS are secreted. In genetic females, the cortex develops into an ovary and the medulla regresses. The embryonic ovary does not secrete hormones. Hormonal treatment of the mother has no effect on gonadal (as opposed to ductal and genital) differentiation in humans, although it does in some experimental animals.

Embryology of the Genitalia

The embryology of the gonads is summarized in Figures 25–4 and 25–5. In the seventh week of gestation, the embryo has

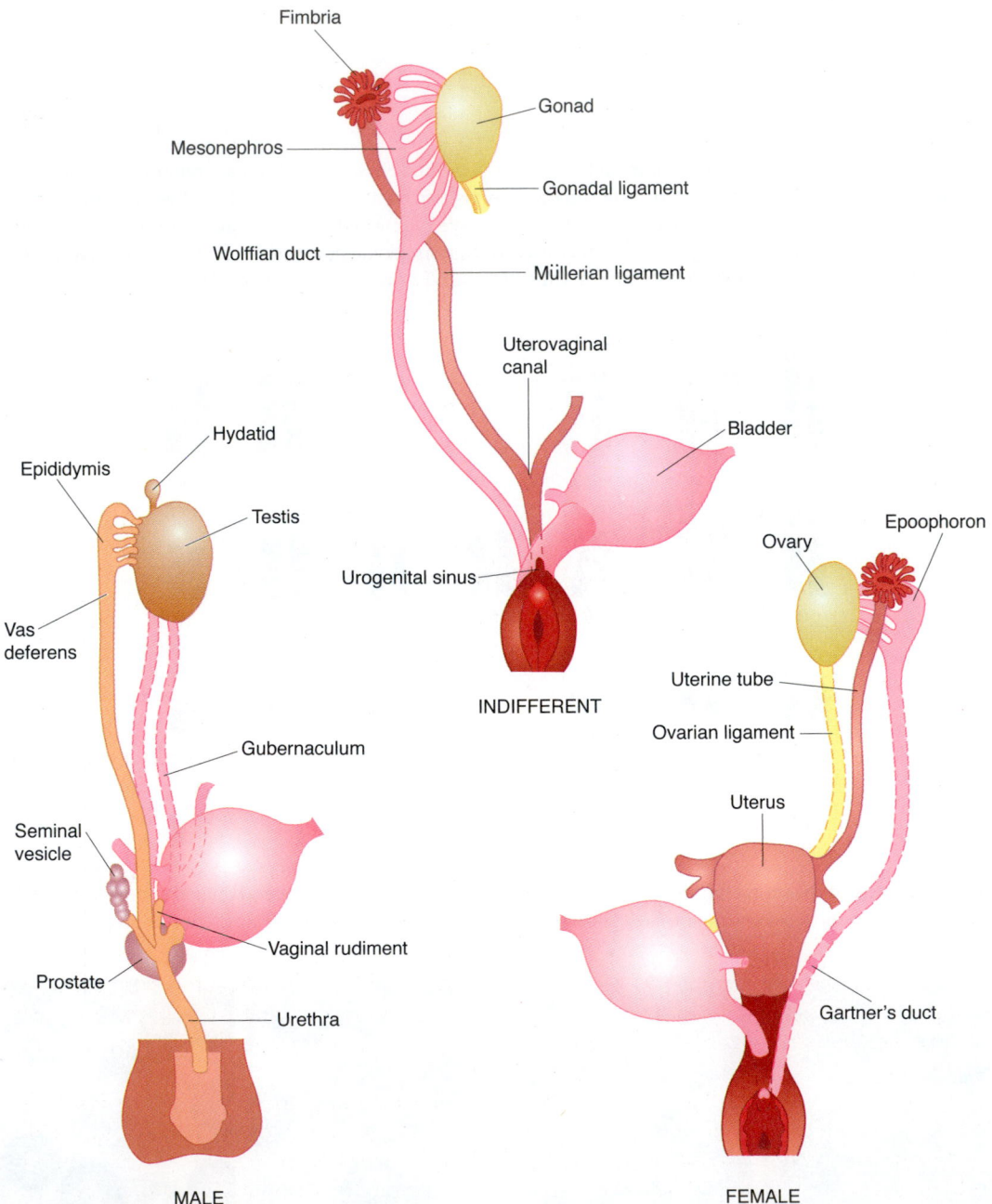

FIGURE 25–4 **Embryonic differentiation of male and female internal genitalia (genital ducts) from wolffian (male) and müllerian (female) primordia.** (After Corning HK, Wilkins L. Redrawn and reproduced with permission from *Williams Textbook of Endocrinology,* 7th ed. Wilson JD, Foster DW [editors]. Saunders, 1985.)

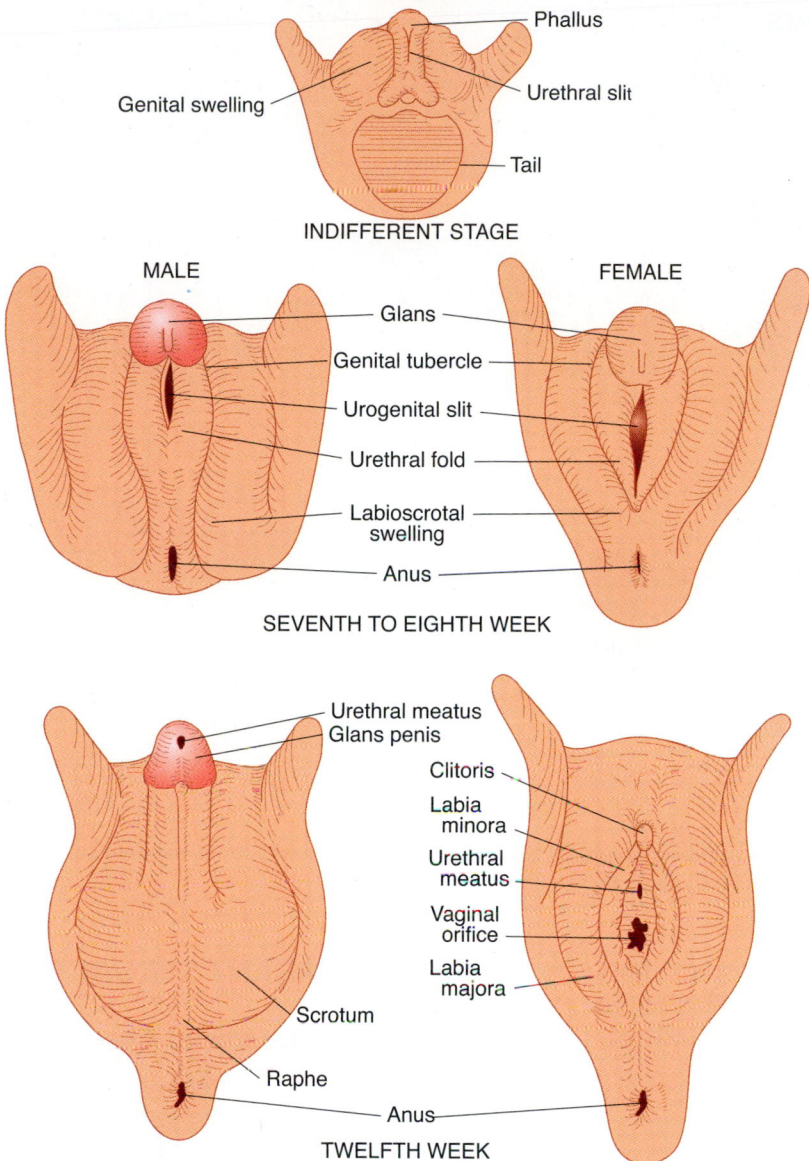

FIGURE 25–5 Differentiation of male and female external genitalia from indifferent primordial structures in the embryo.

both male and female primordial genital ducts (Figure 25–4). In a normal female fetus, the müllerian duct system then develops into uterine tubes (oviducts) and a uterus. In the normal male fetus, the wolffian duct system on each side develops into the epididymis and vas deferens. The external genitalia are similarly bipotential until the eighth week (Figure 25–5). Thereafter, the urogenital slit disappears and male genitalia form, or, alternatively, it remains open and female genitalia form.

When the embryo has functional testes, male internal and external genitalia develop. The Leydig cells of the fetal testis secrete testosterone, and the Sertoli cells secrete **müllerian inhibiting substance** (**MIS;** also called müllerian regression factor, or MRF). MIS is a 536-amino-acid homodimer that is a member of the transforming growth factor β (TGF-β) superfamily of growth factors, which includes inhibins and activins.

In their effects on the internal as opposed to the external genitalia, MIS and testosterone act unilaterally. MIS causes regression of the müllerian ducts by apoptosis on the side on which it is secreted, and testosterone fosters the development of the vas deferens and related structures from the wolffian ducts. The testosterone metabolite dihydrotestosterone induces the formation of male external genitalia and male secondary sex characteristics (Figure 25–6).

MIS continues to be secreted by the Sertoli cells, and it reaches mean values of 48 ng/mL in plasma in 1- to 2-year-old boys. Thereafter, it declines to low levels by the time of puberty and persists at low but detectable levels throughout life. In girls, MIS is produced by granulosa cells in small follicles in the ovaries, but plasma levels are very low or undetectable until puberty. Thereafter, plasma MIS is about the same

FIGURE 25–6 **Diagrammatic summary of normal sex determination, differentiation, and development in humans.** MIS, müllerian inhibiting substance; T, testosterone; DHT, dihydrotestosterone.

as in adult men, that is, about 2 ng/mL. The functions of MIS after early embryonic life are unsettled, but it is probably involved in germ cell maturation in both sexes and in control of testicular descent in boys.

Development of the Brain

At least in some species, the development of the brain as well as the external genitalia is affected by androgens early in life. In rats, a brief exposure to androgens during the first few days of life causes the male pattern of sexual behavior and the male pattern of hypothalamic control of gonadotropin secretion to develop after puberty. In the absence of androgens, female patterns develop (see Chapter 18). In monkeys, similar effects on sexual behavior are produced by exposure to androgens in utero, but the pattern of gonadotropin secretion remains cyclical. Early exposure of female human fetuses to androgens also appears to cause subtle but significant masculinizing effects on behavior. However, women with adrenogenital syndrome due to congenital adrenocortical enzyme deficiency (see Chapter 22) develop normal menstrual cycles when treated with cortisol. Thus, the human, like the monkey, appears to retain the cyclical pattern of gonadotropin secretion despite exposure to androgens in utero.

ABERRANT SEXUAL DIFFERENTIATION

Chromosomal Abnormalities

From the preceding discussion, it might be expected that abnormalities of sexual development could be caused by genetic or hormonal abnormalities as well as by other nonspecific teratogenic influences, and this is indeed the case. The major classes of abnormalities are listed in Table 25–1.

Nondisjunction of sex chromosomes during the first division in meiosis results in distinct defects (see Clinical Box 25–1). Meiosis is a two-stage process, and although nondisjunction usually occurs during the first meiotic division, it can occur in the second, producing more complex chromosomal abnormalities. In addition, nondisjunction or simple loss of a sex chromosome can occur during the early mitotic divisions after fertilization. The result of faulty mitoses in the early zygote is the production of **mosaicism,** in which two or more populations of cells have different chromosome complements. **True hermaphroditism,** the condition in which the individual has both ovaries and testes, is probably due to XX/XY mosaicism and related mosaic patterns, although other genetic aberrations are possible.

Chromosomal abnormalities also include transposition of parts of chromosomes to other chromosomes. Rarely, genetic males are found to have the XX karyotype because the short

TABLE 25–1 Classification of the major disorders of sex differentiation in humans.*

Chromosomal disorders
Gonadal dysgenesis (XO and variants)
"Superfemales" (XXX)
Seminiferous tubule dysgenesis (XXY and variants)
True hermaphroditism
Developmental disorders
Female pseudohermaphroditism
Congenital virilizing adrenal hyperplasia of fetus
Maternal androgen excess
Virilizing ovarian tumor
Iatrogenic: Treatment with androgens or certain synthetic progestational drugs
Male pseudohermaphroditism
Androgen resistance
Defective testicular development
Congenital 17α-hydroxylase deficiency
Congenital adrenal hyperplasia due to blockade of pregnenolone formation
Various nonhormonal anomalies

*Many of these syndromes can have great variation in degree and, consequently, in manifestations.

arm of their father's Y chromosome was transposed to their father's X chromosome during meiosis and they received that X chromosome along with their mother's. Similarly, deletion of the small portion of the Y chromosome containing SRY produces females with the XY karyotype.

Hormonal Abnormalities

Development of the male external genitalia occurs normally in genetic males in response to androgen secreted by the embryonic testes, but male genital development may also occur in genetic females exposed to androgens from some other source during the 8th to the 13th weeks of gestation. The syndrome that results is **female pseudohermaphroditism.** A pseudohermaphrodite is an individual with the genetic constitution and gonads of one sex and the genitalia of the other. After the 13th week, the genitalia are fully formed, but exposure to androgens can cause hypertrophy of the clitoris. Female pseudohermaphroditism may be due to congenital virilizing adrenal hyperplasia (see Chapter 22), or it may be caused by androgens administered to the mother. Conversely, one cause of the development of female external genitalia in genetic males (**male**

CLINICAL BOX 25–1

Chromosomal Abnormalities

An established defect in gametogenesis is **nondisjunction,** a phenomenon in which a pair of chromosomes fail to separate, so that both go to one of the daughter cells during meiosis. Four of the abnormal zygotes that can form as a result of nondisjunction of one of the X chromosomes during oogenesis are shown in Figure 25–7. In individuals with the XO chromosomal pattern, the gonads are rudimentary or absent, so that female external genitalia develop, stature is short, other congenital abnormalities are often present, and no sexual maturation occurs at puberty. This syndrome is called **gonadal dysgenesis** or, alternatively, **ovarian agenesis** or **Turner syndrome.** Individuals with the XXY pattern, the most common sex chromosome disorder, have the genitalia of a normal male. Testosterone secretion at puberty is often great enough for the development of male characteristics, however, the seminiferous tubules are abnormal, and the incidence of mental retardation is higher than normal. This syndrome is known as **seminiferous tubule dysgenesis** or **Klinefelter syndrome.** The XXX ("superfemale") pattern is second in frequency only to the XXY pattern and may be even more common in the general population, since it does not seem to be associated with any characteristic abnormalities. The YO combination is probably lethal.

Nondisjunction of chromosome 21 produces **trisomy 21,** the chromosomal abnormality associated with **Down syndrome** (mongolism). The additional chromosome 21 is normal, so Down syndrome is a pure case of gene excess causing abnormalities.

Many other chromosomal abnormalities occur as well as numerous diseases due to defects in single genes. These conditions are generally diagnosed in utero by analysis of fetal cells in a sample of amniotic fluid collected by inserting a needle through the abdominal wall (**amniocentesis**) or, earlier in pregnancy, by examining fetal cells obtained by a needle biopsy of chorionic villi (**chorionic villus sampling**).

pseudohermaphroditism) is defective testicular development. Because the testes also secrete MIS, genetic males with defective testes have female internal genitalia.

Another cause of male pseudohermaphroditism is **androgen resistance,** in which, as a result of various congenital abnormalities, male hormones cannot exert their full effects on the tissues. One form of androgen resistance is a 5α-**reductase deficiency,** in which the enzyme responsible for the formation of dihydrotestosterone, the active form of testosterone, is decreased. The consequences of this deficiency are discussed in the section on the male reproductive system. Other forms of androgen resistance are due to various mutations in the androgen receptor gene, and the resulting defects in receptor function range from minor to severe. Mild defects cause

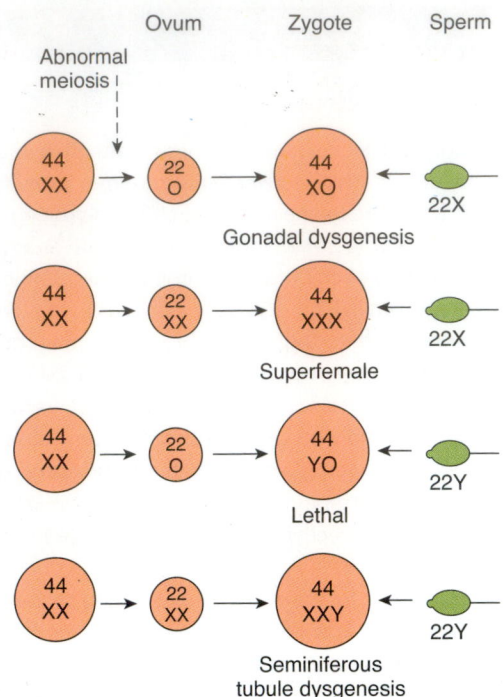

FIGURE 25–7   Summary of four possible defects produced by maternal nondisjunction of the sex chromosomes at the time of meiosis. The YO combination is believed to be lethal, and the fetus dies in utero.

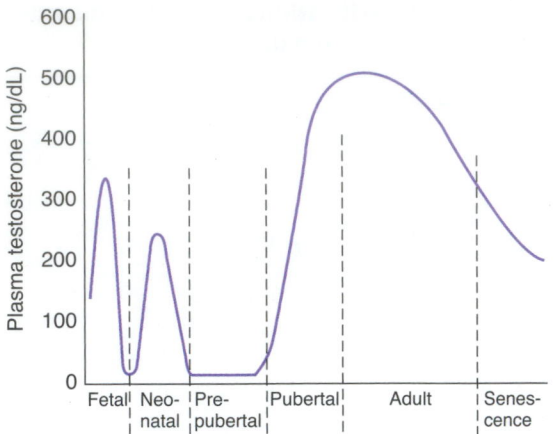

FIGURE 25–8   Plasma testosterone levels at various ages in human males.

infertility with or without gynecomastia. When the loss of receptor function is complete, the **testicular feminizing syndrome,** now known as **complete androgen resistance syndrome,** results. In this condition, MIS is present and testosterone is secreted at normal or even elevated rates. The external genitalia are female, but the vagina ends blindly because there are no female internal genitalia. Individuals with this syndrome develop enlarged breasts at puberty and usually are considered to be normal women until they are diagnosed when they seek medical advice because of lack of menstruation.

It is worth noting that genetic males with congenital blockage of the formation of pregnenolone are pseudohermaphrodites because testicular as well as adrenal androgens are normally formed from pregnenolone. Male pseudohermaphroditism also occurs when there is a congenital deficiency of 17α-hydroxylase (see Chapter 22).

PUBERTY

As noted above, a burst of testosterone secretion occurs in male fetuses before birth (Figure 25–8). In the neonatal period there is another burst, with unknown function, but thereafter the Leydig cells become quiescent. There follows in all mammals a period in which the gonads of both sexes are quiescent until they are activated by gonadotropins from the pituitary to bring about the final maturation of the reproductive system. This period of final maturation is known as **adolescence.** It is often also called **puberty,** although puberty, strictly defined, is

the period when the endocrine and gametogenic functions of the gonads have first developed to the point where reproduction is possible. In girls, the first event is **thelarche,** the development of breasts, followed by **pubarche,** the development of axillary and pubic hair, and then by **menarche,** the first menstrual period. Initial menstrual periods are generally anovulatory, and regular ovulation appears about a year later. In contrast to the situation in adulthood, removal of the gonads during the period from soon after birth to puberty causes only a small increase in gonadotropin secretion, so gonadotropin secretion is not being held in check by the gonadal hormones. In children between the ages of 7 and 10, a slow increase in estrogen and androgen secretion precedes the more rapid rise in the early teens (Figure 25–9).

The age at the time of puberty is variable. In Europe and the United States, it has been declining at the rate of 1 to 3 mo per decade for more than 175 y. In the United States in recent years, puberty generally occurs between the ages of 8 and 13 in girls and 9 and 14 in boys.

Another event that occurs in humans at the time of puberty is an increase in the secretion of adrenal androgens (see Figure 22–12). The onset of this increase is called **adrenarche.** It occurs at age 8 to 10 y in girls and age 10 to 12 y in boys. Dehydroepiandrosterone (DHEA) values peak at about age 25 in females and slightly later than that in males. They then decline slowly to low values in old age. The rise appears to be due to an increase in the activity of 17 α-hydroxylase.

Control of the Onset of Puberty

The gonads of children can be stimulated by gonadotropins; their pituitaries contain gonadotropins and their hypothalami contain gonadotropin-releasing hormone (GnRH) (see Chapter 18). However, their gonadotropins are not secreted. In immature monkeys, normal menstrual cycles can be brought on by pulsatile injection of GnRH, and they persist as long as the pulsatile injection is continued. Thus, it seems clear that pulsatile secretion of GnRH brings on puberty. During the period

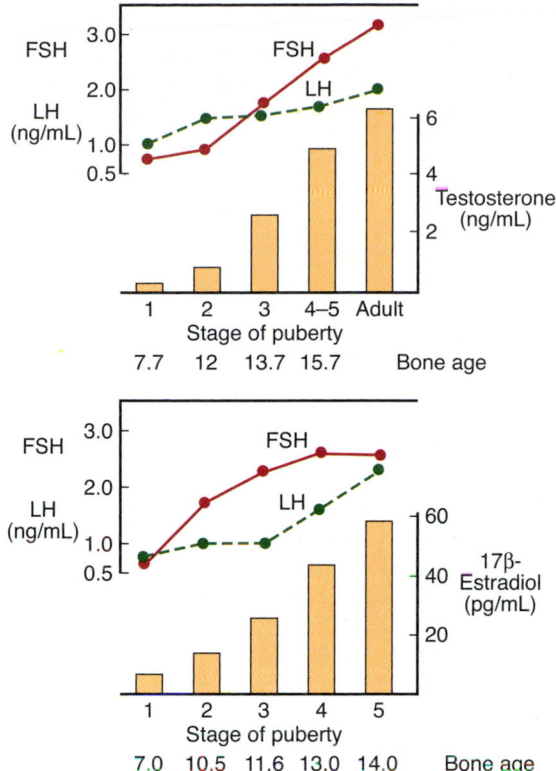

FIGURE 25–9 **Changes in plasma hormone concentrations during puberty in boys (top) and girls (bottom).** Stage 1 of puberty is preadolescence in both sexes. In boys, stage 2 is characterized by beginning enlargement of the testes, stage 3 by penile enlargement, stage 4 by growth of the glans penis, and stage 5 by adult genitalia. In girls, stage 2 is characterized by breast buds, stage 3 by elevation and enlargement of the breasts, stage 4 by projection of the areolas, and stage 5 by adult breasts. (Modified and reproduced with permission from Berenberg SR [editor]: *Puberty: Biologic and Psychosocial Components.* HE Stenfoert Kroese BV, 1975.)

from birth to puberty, a neural mechanism is operating to prevent the normal pulsatile release of GnRH. The nature of the mechanism inhibiting the GnRH pulse generator is unknown. However, one or more genes produce products that stimulate secretion of GnRH, and inhibition of these genes before puberty is an interesting possibility (see Clinical Box 25–2).

PRECOCIOUS & DELAYED PUBERTY

Sexual Precocity

The major causes of precocious sexual development in humans are listed in Table 25–2. Early development of secondary sexual characteristics without gametogenesis is caused by abnormal exposure of immature males to androgen or females to estrogen. This syndrome should be called **precocious pseudopuberty** to distinguish it from **true precocious puberty** due to an early but otherwise normal pubertal pattern of gonadotropin secretion from the pituitary.

CLINICAL BOX 25–2

Leptin

It has been argued for some time that a critical body weight must normally be reached for puberty to occur. Thus, for example, young women who engage in strenuous athletics lose weight and stop menstruating, as do girls with anorexia nervosa. If these girls start to eat and gain weight, they menstruate again, that is, they "go back through puberty." It now appears that leptin, the satiety-producing hormone secreted by fat cells, may be the link between body weight and puberty. Obese ob/ob mice that cannot make leptin are infertile, and their fertility is restored by injections of leptin. Leptin treatment also induces precocious puberty in immature female mice. However, the way that leptin fits into the overall control of puberty remains to be determined.

Constitutional precocious puberty; that is, precocious puberty in which no cause can be determined, is more common in girls than in boys. In both sexes, tumors or infections involving the hypothalamus cause precocious puberty. Indeed, in one large series of cases, precocious puberty was the most common endocrine symptom of hypothalamic disease. In experimental

TABLE 25–2 **Classification of the causes of precocious sexual development in humans.**

True precocious puberty
Constitutional
Cerebral: Disorders involving posterior hypothalamus
Tumors
Infections
Developmental abnormalities
Gonadotropin-independent precocity
Precocious pseudopuberty (no spermatogenesis or ovarian development)
Adrenal
Congenital virilizing adrenal hyperplasia
Androgen-secreting tumors (in males)
Estrogen-secreting tumors (in females)
Gonadal
Leydig cell tumors of testis
Granulosa cell tumors of ovary
Miscellaneous

animals, precocious puberty can be produced by hypothalamic lesions. Apparently the lesions interrupt a pathway that normally holds pulsatile GnRH secretion in check. Pineal tumors are sometimes associated with precocious puberty, but evidence indicates that these tumors are associated with precocity only when there is secondary damage to the hypothalamus.

Precocious gametogenesis and steroidogenesis can occur without the pubertal pattern of gonadotropin secretion (gonadotropin-independent precocity). At least in some cases of this condition, the sensitivity of LH receptors to gonadotropins is increased because of an activating mutation in the G protein that couples the receptors to adenylyl cyclase.

Delayed or Absent Puberty

The normal variation in the age at which adolescent changes occur is so wide that puberty cannot be considered to be pathologically delayed until the menarche has failed to occur by the age of 17 or testicular development by the age of 20. Failure of maturation due to panhypopituitarism is associated with dwarfing and evidence of other endocrine abnormalities. Patients with the XO chromosomal pattern and gonadal dysgenesis are also dwarfed. In some individuals, puberty is delayed even though the gonads are present and other endocrine functions are normal. In males, this clinical picture is called **eunuchoidism.** In females, it is called **primary amenorrhea.**

MENOPAUSE

The human ovaries become unresponsive to gonadotropins with advancing age, and their function declines, so that sexual cycles disappear **(menopause).** This unresponsiveness is associated with and probably caused by a decline in the number of primordial follicles, which becomes precipitous at the time of menopause (Figure 25–10). The ovaries no longer secrete progesterone and 17β-estradiol in appreciable quantities, and estrogen is formed only in small amounts by aromatization of androstenedione in peripheral tissues (see Chapter 22). The uterus and the vagina gradually become atrophic. As the negative feedback effect of estrogens and progesterone is reduced, secretion of FSH is increased, and plasma FSH increases to high levels, LH levels are moderately high. Old female mice and rats have long periods of diestrus and increased levels of gonadotropin secretion. In women, a period called perimenopause precedes menopause, and can last up to 10 y. During perimenopause the menses become irregular and the level of inhibins decrease, usually between the ages of 45 and 55. The average age at onset of the menopause has been increasing since the end of the 19th century and is currently 52 y.

The loss of ovarian function causes many symptoms such as sensations of warmth spreading from the trunk to the face (hot flushes; also called hot flashes) and night sweats. In addition, the onset of menopause increases the risk of many diseases such as osteoporosis, ischemic heart disease, and renal disease.

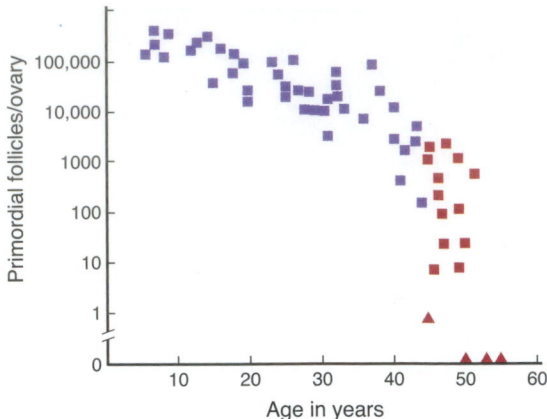

FIGURE 25–10   **Number of primordial follicles per ovary in women at various ages.** Blue squares, premenopausal women (regular menses); red squares, perimenopausal women (irregular menses for at least 1 y); red triangles, postmenopausal women (no menses for at least 1 y). Note that the vertical scale is a log scale and that the values are from one rather than two ovaries.   (Redrawn by PM Wise and reproduced with permission from Richardson SJ, Senikas V, Nelson JF: Follicular depletion during the menopausal transition: Evidence for accelerated loss and ultimate exhaustion. J Clin Endocrinol Metab 1987;65:1231.)

Hot flushes are said to occur in 75% of menopausal women and may continue intermittently for as long as 40 y. They also occur when early menopause is produced by bilateral ovariectomy, and they are prevented by estrogen treatment. In addition, they occur after castration in men. Their cause is unknown. However, they coincide with surges of LH secretion. LH is secreted in episodic bursts at intervals of 30 to 60 min or more **(circhoral secretion),** and in the absence of gonadal hormones these bursts are large. Each hot flush begins with the start of a burst. However, LH itself is not responsible for the symptoms, because they can continue after removal of the pituitary. Instead, it appears that some estrogen-sensitive event in the hypothalamus initiates both the release of LH and the episode of flushing.

Although the function of the testes tends to decline slowly with advancing age, the evidence is unclear whether there is a "male menopause" **(andropause)** similar to that occurring in women.

PITUITARY GONADOTROPINS & PROLACTIN

CHEMISTRY

FSH and LH are each made up of an α and a β subunit whose nature is discussed in Chapter 24. They are glycoproteins that contain the hexoses mannose and galactose, the hexosamines N-acetylgalactosamine and N-acetylglycosamine, and the methylpentose fucose. They also contain sialic acid. The carbohydrate in the gonadotropin molecules increases their potency by markedly slowing their metabolism. The half-life of human FSH is about 170 min; the half-life of LH is about 60 min. Loss-of-function mutations in the FSH receptor cause hypogonadism. Gain-of-function mutations cause a spontaneous form of **ovarian hyperstimulation syndrome,** a condition in which many follicles are stimulated and cytokines are released from the ovary, causing increased vascular permeability and shock.

Human pituitary prolactin contains 199 amino acid residues and three disulfide bridges and has considerable structural similarity to human growth hormone and human chorionic somatomammotropin (hCS). The half-life of prolactin, like that of growth hormone, is about 20 min. Structurally similar prolactins are secreted by the endometrium and by the placenta.

RECEPTORS

The receptors for FSH and LH are G-protein coupled receptors coupled to adenylyl cyclase through a stimulatory G protein (G_s; see Chapter 2). In addition, each has an extended, glycosylated extracellular domain.

The human prolactin receptor resembles the growth hormone receptor and is one of the superfamily of receptors that includes the growth hormone receptor and receptors for many cytokines and hematopoietic growth factors (see Chapters 2 and 3). It dimerizes and activates the Janus kinase/signal transducers and activators of transcription (JAK–STAT) pathway and other intracellular enzyme cascades.

ACTIONS

The testes and ovaries become atrophic when the pituitary is removed or destroyed. The actions of prolactin and the gonadotropins FSH and LH, as well as those of the gonadotropin secreted by the placenta, are described in detail in succeeding sections of this chapter. In brief, FSH helps maintain the spermatogenic epithelium by stimulating Sertoli cells in the male and is responsible for the early growth of ovarian follicles in the female. LH is tropic to the Leydig cells and, in females, is responsible for the final maturation of the ovarian follicles and estrogen secretion from them. It is also responsible for ovulation, the initial formation of the corpus luteum, and secretion of progesterone.

Prolactin causes milk secretion from the breast after estrogen and progesterone priming. Its effect on the breast involves increased action of mRNA and increased production of casein and lactalbumin. However, the action of the hormone is not exerted on the cell nucleus and is prevented by inhibitors of microtubules. Prolactin also inhibits the effects of gonadotropins, possibly by an action at the level of the ovary. Its role in preventing ovulation in lactating women is discussed below. The function of prolactin in normal males is unsettled, but excess prolactin secreted by tumors causes impotence.

REGULATION OF PROLACTIN SECRETION

The normal plasma prolactin concentration is approximately 5 ng/mL in men and 8 ng/mL in women. Secretion is tonically inhibited by the hypothalamus, and section of the pituitary stalk leads to an increase in circulating prolactin. Thus, the effect of the hypothalamic prolactin-inhibiting hormone (PIH) dopamine is normally greater than the effects of the various hypothalamic peptides with prolactin-releasing activity. In humans, prolactin secretion is increased by exercise, surgical and psychologic stresses, and stimulation of the nipple (Table 25–3). The plasma prolactin level rises during sleep, the rise starting after the onset of sleep and persisting throughout the sleep period. Secretion is increased during pregnancy, reaching a peak at the time of parturition. After delivery, the plasma concentration falls to nonpregnant levels in about 8 days. Suckling produces a prompt increase in secretion, but the magnitude of this rise gradually declines after a woman has been nursing for more than 3 months. With prolonged lactation, milk secretion occurs with prolactin levels that are in the normal range.

TABLE 25–3 Factors affecting the secretion of human prolactin and growth hormone.

Factor	Prolactin[a]	Growth Hormone[a]
Sleep	I+	I+
Nursing	I++	N
Breast stimulation in nonlactating women	I	N
Stress	I+	I+
Hypoglycemia	I	I+
Strenuous exercise	I	I
Sexual intercourse in women	I	N
Pregnancy	I++	N
Estrogens	I	I
Hypothyroidism	I	N
TRH	I+	N
Phenothiazines, butyrophenones	I+	N
Opioids	I	I
Glucose	N	D
Somatostatin	N	D+
L-Dopa	D+	I+
Apomorphine	D+	I+
Bromocriptine and related ergot derivatives	D+	I

[a]I, moderate increase; I+, marked increase; I++, very marked increase; N, no change; D, moderate decrease; D+, marked decrease; TRH, thyrotropin-releasing hormone.

L-Dopa decreases prolactin secretion by increasing the formation of dopamine; bromocriptine and other dopamine agonists inhibit secretion because they stimulate dopamine receptors. Chlorpromazine and related drugs that block dopamine receptors increase prolactin secretion. Thyrotropin-releasing hormone (TRH) stimulates the secretion of prolactin in addition to thyroid-stimulating hormone (TSH), and additional polypeptides with prolactin-releasing activity are present in hypothalamic tissue. Estrogens produce a slowly developing increase in prolactin secretion as a result of a direct action on the lactotropes.

It has now been established that prolactin facilitates the secretion of dopamine in the median eminence. Thus, prolactin acts in the hypothalamus in a negative feedback fashion to inhibit its own secretion (see Clinical Box 25–3).

THE MALE REPRODUCTIVE SYSTEM

STRUCTURE

The testes are made up of loops of convoluted **seminiferous tubules,** in the walls of which the spermatozoa are formed from the primitive germ cells (**spermatogenesis**). Both ends of each loop drain into a network of ducts in the head of the **epididymis.** From there, spermatozoa pass through the tail of the epididymis into the **vas deferens.** They enter through the **ejaculatory ducts** into the urethra in the body of the **prostate** at the time of ejaculation (Figure 25–11). Between the tubules in the testes are nests of cells containing lipid granules, the **interstitial cells of Leydig** (Figures 25–12 and 25–13), which secrete testosterone into the bloodstream. The spermatic arteries to the testes are tortuous, and blood in them runs parallel but in the opposite direction to blood in the pampiniform plexus of spermatic veins. This anatomic arrangement may permit countercurrent exchange of heat and testosterone. The principles of countercurrent exchange are considered in detail in relation to the kidney in Chapter 38.

GAMETOGENESIS & EJACULATION

Blood–Testis Barrier

The walls of the seminiferous tubules are lined by primitive germ cells and **Sertoli cells,** large, complex glycogen-containing cells that stretch from the basal lamina of the tubule to the lumen (Figure 25–13). Germ cells must stay in contact with Sertoli cells to survive, and this contact is maintained by

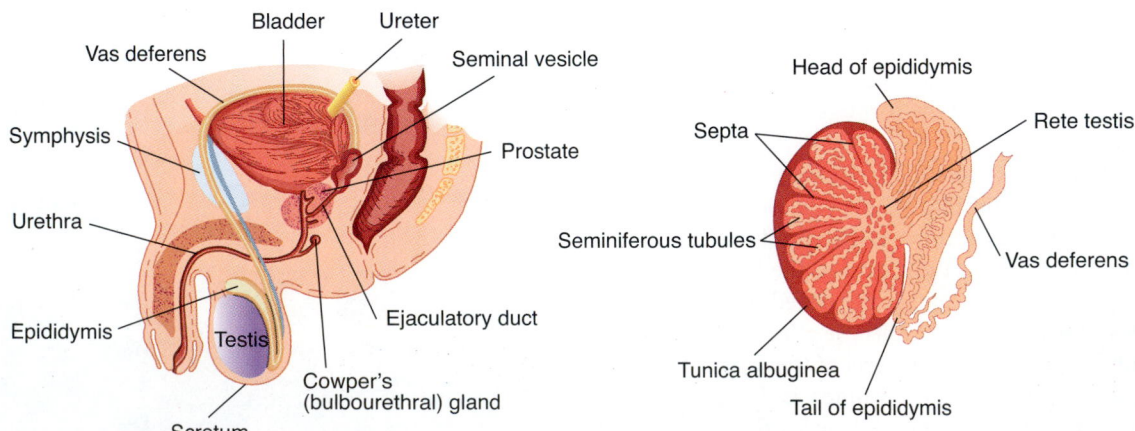

FIGURE 25–11 **Anatomical features of the male reproductive system. Left:** Male reproductive system. **Right:** Duct system of the testis.

FIGURE 25–12 **Section of human testis.**

cytoplasmic bridges. Tight junctions between adjacent Sertoli cells near the basal lamina form a **blood–testis barrier** that prevents many large molecules from passing from the interstitial tissue and the part of the tubule near the basal lamina (basal compartment) to the region near the tubular lumen (adluminal compartment) and the lumen. However, steroids penetrate this barrier with ease, and evidence suggests that some proteins pass from the Sertoli cells to the Leydig cells and vice versa in a paracrine fashion. In addition, maturing germ cells must pass through the barrier as they move to the lumen. This appears to occur without disruption of the barrier by progressive breakdown of the tight junctions above the germ cells, with concomitant formation of new tight junctions below them.

The fluid in the lumen of the seminiferous tubules is quite different from plasma; it contains very little protein and glucose but is rich in androgens, estrogens, K^+, inositol, and glutamic and aspartic acids. Maintenance of its composition presumably depends on the blood–testis barrier. The barrier also protects the germ cells from bloodborne noxious agents, prevents antigenic products of germ cell division and maturation from entering the circulation and generating an autoimmune response, and may help establish an osmotic gradient that facilitates movement of fluid into the tubular lumen.

Spermatogenesis

Spermatogonia, the primitive germ cells next to the basal lamina of the seminiferous tubules, mature into **primary spermatocytes** (Figure 25–13). This process begins during adolescence.

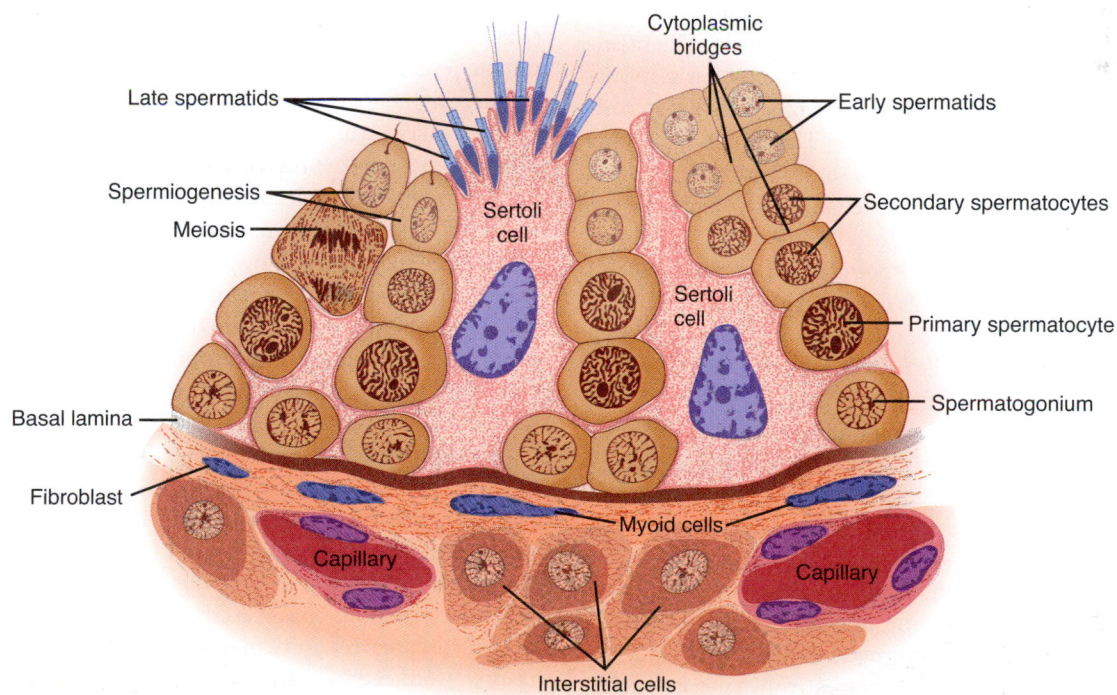

FIGURE 25–13 **Seminiferous epithelium.** Note that maturing germ cells remain connected by cytoplasmic bridges through the early spermatid stage and that these cells are closely invested by Sertoli cell cytoplasm as they move from the basal lamina to the lumen. (Reproduced with permission from Junqueira LC, Carneiro J: *Basic Histology: Text & Atlas,* 10th ed. McGraw-Hill, 2003.)

FIGURE 25–14 **Human spermatozoon, profile view.** Note the acrosome, an organelle that covers half the sperm head inside the plasma membrane of the sperm. (Reproduced with permission from Junqueira LC, Carneiro J: *Basic Histology: Text & Atlas*, 10th ed. McGraw-Hill, 2003.)

The primary spermatocytes undergo meiotic division, reducing the number of chromosomes. In this two-stage process, they divide into **secondary spermatocytes** and then into **spermatids,** which contain the haploid number of 23 chromosomes. The spermatids mature into **spermatozoa (sperms).** As a single spermatogonium divides and matures, its descendants remain tied together by cytoplasmic bridges until the late spermatid stage. This apparently ensures synchrony of the differentiation of each clone of germ cells. The estimated number of spermatids formed from a single spermatogonium is 512. In humans, it takes an average of 74 d to form a mature sperm from a primitive germ cell by this orderly process of spermatogenesis.

Each sperm is an intricate motile cell, rich in DNA, with a head that is made up mostly of chromosomal material (Figure 25–14). Covering the head like a cap is the **acrosome,** a lysosome-like organelle rich in enzymes involved in sperm penetration of the ovum and other events involved in fertilization. The motile tail of the sperm is wrapped in its proximal portion by a sheath holding numerous mitochondria. The membranes of late spermatids and spermatozoa contain a special small form of angiotensin-converting enzyme called **germinal angiotensin-converting enzyme**. The function of this enzyme in the sperms is unknown, although male mice in which the function of the angiotensin-converting enzyme gene has been disrupted have reduced fertility.

The spermatids mature into spermatozoa in deep folds of the cytoplasm of the Sertoli cells (Figure 25–13). Mature spermatozoa are released from the Sertoli cells and become free in the lumens of the tubules. The Sertoli cells secrete **androgen-binding protein (ABP), inhibin,** and **MIS.** They do not synthesize androgens, but they contain **aromatase (CYP19),** the enzyme responsible for conversion of androgens to estrogens, and they can produce estrogens. ABP probably functions to maintain a high, stable supply of androgen in the tubular fluid. Inhibin inhibits FSH secretion.

FSH and androgens maintain the gametogenic function of the testis. After hypophysectomy, injection of LH produces a high local concentration of androgen in the testes, and this maintains spermatogenesis. The stages from spermatogonia to spermatids appear to be androgen-independent. However, the maturation from spermatids to spermatozoa depends on androgen acting on the Sertoli cells in which the developing spermatozoa are embedded. FSH acts on the Sertoli cells to facilitate the last stages of spermatid maturation. In addition, it promotes the production of ABP.

An interesting observation is that the estrogen content of the fluid in the rete testis (Figure 25–11) is high, and the walls of the rete contain numerous ERα estrogen receptors. In this region, fluid is reabsorbed and the spermatozoa are concentrated. If this does not occur, the sperm entering the epididymis are diluted in a large volume of fluid, and infertility results.

Further Development of Spermatozoa

Spermatozoa leaving the testes are not fully mobile. They continue their maturation and acquire motility during their passage through the epididymis. Motility is obviously important in vivo, but fertilization occurs in vitro if an immotile spermatozoon from the head of the epididymis is microinjected directly into an ovum. The ability to move forward (**progressive motility),** which is acquired in the epididymis, involves activation of a unique protein called **CatSper,** which is localized to the principal piece of the sperm tail. This protein appears to be a Ca^{2+} ion channel that permits cAMP-generalized Ca^{2+} influx. In addition, spermatozoa express olfactory receptors, and ovaries produce odorant-like molecules. Recent evidence indicates that these molecules and their receptors interact, fostering movement of the spermatozoa toward the ovary (chemotaxis).

Ejaculation of the spermatozoon involves contractions of the vas deferens mediated in part by P2X receptors, ligand-gated cation channels that respond to ATP (see Chapter 7), and fertility is reduced in mice in which these receptors are knocked out.

Once ejaculated into the female, the spermatozoa move up the uterus to the isthmus of the uterine tubes, where they slow down and undergo **capacitation.** This further maturation process involves two components: increasing the motility of the spermatozoa and facilitating their preparation for the acrosome reaction. However, the role of capacitation appears to be facilitatory rather than obligatory, because fertilization is readily produced in vitro. From the isthmuses the capacitated spermatozoa move rapidly to the tubal ampullas, where fertilization takes place.

Effect of Temperature

Spermatogenesis requires a temperature considerably lower than that of the interior of the body. The testes are normally maintained at a temperature of about 32 °C. They are kept cool by air circulating around the scrotum and probably by heat exchange in a countercurrent fashion between the spermatic arteries and veins. When the testes are retained in the abdomen or when, in experimental animals, they are held close to the body by tight cloth binders, degeneration of the tubular walls and sterility result. Hot baths (43–45 °C for 30 min/d) and insulated athletic supporters reduce the sperm count in humans, in some cases by 90%. However, the reductions produced in this manner are not consistent enough to make the procedures reliable forms of male contraception. In addition, evidence suggests a seasonal effect in men, with sperm counts being greater in the winter regardless of the temperature to which the scrotum is exposed.

Semen

The fluid that is ejaculated at the time of orgasm, the **semen,** contains sperms and the secretions of the seminal vesicles, prostate, Cowper's glands, and, probably, the urethral glands (Table 25–4). An average volume per ejaculate is 2.5 to 3.5 mL after several days of abstinence. The volume of semen and the sperm count decrease rapidly with repeated ejaculation. Even though it takes only one sperm to fertilize the ovum, each milliliter of semen normally contains about 100 million sperms. Fifty percent of men with counts of 20 to 40 million/mL and essentially all of those with counts under 20 million/mL are sterile. The presence of many morphologically abnormal or immotile spermatozoa also correlates with infertility. The **prostaglandins** in semen, which actually come from the seminal vesicles, are in high concentration, but the function of these fatty acid derivatives in semen is unknown.

Human sperms move at a speed of about 3 mm/min through the female genital tract. Sperms reach the uterine tubes 30 to 60 min after copulation. In some species, contractions of the female organs facilitate the transport of the sperms to the uterine tubes, but it is unknown if such contractions are important in humans.

Erection

Erection is initiated by dilation of the arterioles of the penis. As the erectile tissue of the penis fills with blood, the veins are compressed, blocking outflow and adding to the turgor of the organ. The integrating centers in the lumbar segments of the spinal cord are activated by impulses in afferents from the genitalia and descending tracts that mediate erection in response to erotic psychologic stimuli. The efferent parasympathetic fibers are in the pelvic splanchnic nerves (**nervi erigentes**). The fibers presumably release acetylcholine and the vasodilator vasoactive intestinal polypeptide (VIP) as cotransmitters (see Chapter 7).

TABLE 25–4 Composition of human semen.

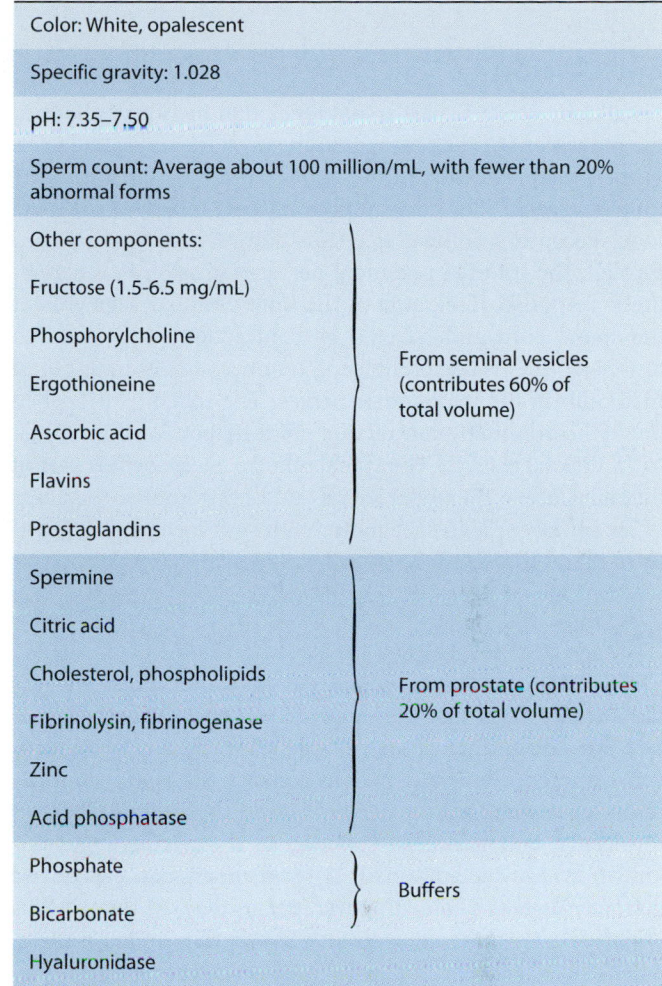

Color: White, opalescent	
Specific gravity: 1.028	
pH: 7.35–7.50	
Sperm count: Average about 100 million/mL, with fewer than 20% abnormal forms	
Other components:	
Fructose (1.5–6.5 mg/mL)	
Phosphorylcholine	
Ergothioneine	From seminal vesicles (contributes 60% of total volume)
Ascorbic acid	
Flavins	
Prostaglandins	
Spermine	
Citric acid	
Cholesterol, phospholipids	From prostate (contributes 20% of total volume)
Fibrinolysin, fibrinogenase	
Zinc	
Acid phosphatase	
Phosphate	Buffers
Bicarbonate	
Hyaluronidase	

Nonadrenergic noncholinergic fibers are also present in the nervi erigentes, and these contain large amounts of **NO synthase,** the enzyme that catalyzes the formation of nitric oxide (NO; see Chapter 33). NO activates guanylyl cyclase, resulting in increased production of cyclic GMP (cGMP), and cGMP is a potent vasodilator. Injection of inhibitors of NO synthase prevents the erection normally produced by stimulation of the pelvic nerve in experimental animals. Thus, it seems clear that NO plays a prominent role in the production of erection. The drugs sildenafil, tadalafil, and vardenafil all inhibit the breakdown of cGMP by phosphodiesterases and have gained worldwide fame for the treatment of impotence. The multiple phosphodiesterases (PDEs) in the body have been divided into seven isoenzyme families, and these drugs are all most active against PDE V, the type of phosphodiesterase found in the corpora cavernosa. It is worth noting, however, that these drugs can also produces significant inhibition of PDE VI (and others, if taken at high doses). Phosphodiesterase VI is found in the retina, and one of the side effects of these drugs is a transient loss of the ability to discriminate between blue and green (see Chapter 12).

Normally, erection is terminated by sympathetic vasoconstrictor impulses to the penile arterioles.

Ejaculation

Ejaculation is a two-part spinal reflex that involves **emission**, the movement of the semen into the urethra; and **ejaculation** proper, the propulsion of the semen out of the urethra at the time of orgasm. The afferent pathways are mostly fibers from touch receptors in the glans penis that reach the spinal cord through the internal pudendal nerves. Emission is a sympathetic response, integrated in the upper lumbar segments of the spinal cord and effected by contraction of the smooth muscle of the vasa deferentia and seminal vesicles in response to stimuli in the hypogastric nerves. The semen is propelled out of the urethra by contraction of the bulbocavernosus muscle, a skeletal muscle. The spinal reflex centers for this part of the reflex are in the upper sacral and lowest lumbar segments of the spinal cord, and the motor pathways traverse the first to third sacral roots and the internal pudendal nerves.

PSA

The prostate produces and secretes into the semen and the bloodstream a 30 kDa serine protease generally called **prostate-specific antigen (PSA).** The gene for PSA has two androgen response elements. PSA hydrolyzes the sperm motility inhibitor semenogelin in semen, and it has several substrates in plasma, but its precise function in the circulation is unknown. An elevated plasma PSA occurs in prostate cancer and is widely used as a screening test for this disease, though PSA is also elevated in benign prostatic hyperplasia and prostatitis.

Vasectomy

Bilateral ligation of the vas deferens (vasectomy) has proved to be a relatively safe and convenient contraceptive procedure. However, it has proven difficult to restore the patency of the vas in those wishing to restore fertility, and the current success rate for such operations, as measured by the subsequent production of pregnancy, is about 50%. Half of the men who have been vasectomized develop antibodies against spermatozoa, and in monkeys, the presence of such antibodies is associated with a higher incidence of infertility after restoration of the patency of the vas. However, the anti-sperm antibodies do not appear to have any other adverse effects.

ENDOCRINE FUNCTION OF THE TESTES

Chemistry & Biosynthesis of Testosterone

Testosterone, the principal hormone of the testes, is a C_{19} steroid with an –OH group in the 17 position (Figure 25–15). It is synthesized from cholesterol in the Leydig cells and is also formed from androstenedione secreted by the adrenal cortex. The biosynthetic pathways in all endocrine organs that form steroid hormones are similar, the organs differing only in the enzyme systems they contain. In the Leydig cells, the 11- and 21-hydroxylases found in the adrenal cortex (see Figure 22–7) are absent, but 17α-hydroxylase is present. Pregnenolone is therefore hydroxylated in the 17 position and then subjected to side chain cleavage to form dehydroepiandrosterone. Androstenedione is also formed via progesterone and 17-hydroxyprogesterone, but this pathway is less prominent in humans. Dehydroepiandrosterone and androstenedione are then converted to testosterone.

The secretion of testosterone is under the control of LH, and the mechanism by which LH stimulates the Leydig cells involves

FIGURE 25–15 Biosynthesis of testosterone. The formulas of the precursor steroids are shown in Figure 22–7. Although the main secretory product of the Leydig cells is testosterone, some of the precursors also enter the circulation.

increased formation of cAMP via the G protein-coupled LH receptor and G_s. Cyclic AMP increases the formation of cholesterol from cholesteryl esters and the conversion of cholesterol to pregnenolone via the activation of protein kinase A.

Secretion

The testosterone secretion rate is 4 to 9 mg/d (13.9–31.33 µmol/d) in normal adult males. Small amounts of testosterone are also secreted in females, with the major source being the ovary, but possibly from the adrenal as well.

Transport & Metabolism

Ninety-eight percent of the testosterone in plasma is bound to protein: 65% is bound to a β-globulin called **gonadal steroid-binding globulin (GBG)** or **sex steroid-binding globulin,** and 33% to albumin (Table 25–5). GBG also binds estradiol. The plasma testosterone level (free and bound) is 300 to 1000 ng/dL (10.4–34.7 nmol/L) in adult men (Figure 25–8), compared with 30 to 70 ng/dL (1.04–2.43 nmol/L) in adult women. It declines somewhat with age in males.

A small amount of circulating testosterone is converted to estradiol, but most of the testosterone is converted to 17-ketosteroids, principally androsterone and its isomer etiocholanolone (Figure 25–16), and excreted in the urine. About two thirds of the urinary 17-ketosteroids are of adrenal origin, and one third are of testicular origin. Although most of the 17-ketosteroids are weak androgens (they have 20% or less the potency of testosterone), it is worth emphasizing that not all 17-ketosteroids are androgens and not all androgens are 17-ketosteroids. Etiocholanolone, for example, has no androgenic activity, and testosterone itself is not a 17-ketosteroid.

Actions

In addition to their actions during development, testosterone and other androgens exert an inhibitory feedback effect on pituitary LH secretion; develop and maintain the male secon-

TABLE 25–5 **Distribution of gonadal steroids and cortisol in plasma.**

Steroid	% Free	% Bound to		
		CBG	GBG	Albumin
Testosterone	2	0	65	33
Androstenedione	7	0	8	85
Estradiol	2	0	38	60
Progesterone	2	18	0	80
Cortisol	4	90	0	6

CBG, corticosteroid-binding globulin; GBG, gonadal steroid-binding globulin.
(Courtesy of S Munroe.)

Androsterone

Etiocholanolone

FIGURE 25–16 **Two 17-ketosteroid metabolites of testosterone.**

dary sex characteristics; exert an important protein-anabolic, growth-promoting effect; and, along with FSH, maintain spermatogenesis.

Secondary Sex Characteristics

The widespread changes in hair distribution, body configuration, and genital size that develop in boys at puberty—the male **secondary sex characteristics**—are summarized in Table 25–6. The prostate and seminal vesicles enlarge, and the seminal vesicles begin to secrete fructose. This sugar appears to function as the main nutritional supply for the spermatozoa. The psychic effects of testosterone are difficult to define in humans, but in experimental animals, androgens provoke boisterous and aggressive play. The effects of androgens and estrogens on sexual behavior are considered in detail in

TABLE 25–6 **Changes at puberty in boys (male secondary sex characteristics).**

External genitalia: Penis increases in length and width. Scrotum becomes pigmented and rugose.

Internal genitalia: Seminal vesicles enlarge and secrete and begin to form fructose. Prostate and bulbourethral glands enlarge and secrete.

Voice: Larynx enlarges, vocal cords increase in length and thickness, and voice becomes deeper.

Hair growth: Beard appears. Hairline on scalp recedes anterolaterally. Pubic hair grows with male (triangle with apex up) pattern. Hair appears in axillas, on chest, and around anus; general body hair increases.

Mental: More aggressive, active attitude. Interest in opposite sex develops.

Body conformation: Shoulders broaden, muscles enlarge.

Skin: Sebaceous gland secretion thickens and increases (predisposing to acne).

FIGURE 25–17   **Hairline in children and adults.** The hairline of the woman is like that of the child, whereas that of the man is indented in the lateral frontal region.

Chapter 15. Although body hair is increased by androgens, scalp hair is decreased (Figure 25–17). Hereditary baldness often fails to develop unless dihydrotestosterone is present.

Anabolic Effects

Androgens increase the synthesis and decrease the breakdown of protein, leading to an increase in the rate of growth. It used to be argued that they cause the epiphyses to fuse to the long bones, thus eventually stopping growth, but it now appears that epiphysial closure is due to estrogens (see Chapter 23). Secondary to their anabolic effects, androgens cause moderate Na^+, K^+, H_2O, Ca^{2+}, SO_4^-, and PO_4^- retention; and they also increase the size of the kidneys. Doses of exogenous testosterone that exert significant anabolic effects are also masculinizing and increase libido, which limits the usefulness of the hormone as an anabolic agent in patients with wasting diseases. Attempts to develop synthetic steroids in which the anabolic action is divorced from the androgenic action have not been successful.

Mechanism of Action

Like other steroids, testosterone binds to an intracellular receptor, and the receptor–steroid complex then binds to DNA in the nucleus, facilitating transcription of various genes. In addition, testosterone is converted to **dihydrotestosterone (DHT)** by 5α-reductase in some target cells (Figures 25–15 and 25–18), and DHT binds to the same intracellular receptor as testosterone. DHT also circulates, with a plasma level that is about 10% of the testosterone level. Testosterone–receptor complexes are less stable than DHT–receptor complexes in target cells, and they conform less well to the DNA-binding state. Thus, DHT formation is a way of amplifying the action of testosterone in target tissues. Humans have two 5α-reductases, encoded by different genes. Type 1 5α-reductase is present in skin throughout the body and is the dominant enzyme in the scalp. Type 2 5α-reductase is present in genital skin, the prostate, and other genital tissues.

Testosterone–receptor complexes are responsible for the maturation of wolffian duct structures and consequently for the formation of male internal genitalia during development, but DHT–receptor complexes are needed to form male external genitalia (Figure 25–18). DHT–receptor complexes are

FIGURE 25–18   **Schematic diagram of the actions of testosterone (solid arrows) and dihydrotestosterone (dashed arrows).** Note that they both bind to the same receptor, but DHT binds more effectively. (Reproduced with permission from Wilson JD, Griffin JE, Russell W: Steroid 5α-reductase 2 deficiency. Endocr Rev 1993;14:577. Copyright © 1993 by The Endocrine Society.)

Congenital 5α-Reductase Deficiency

Congenital 5α-reductase deficiency, in which the gene for type 2 5α-reductase is mutated, is common in certain parts of the Dominican Republic. It produces an interesting form of male pseudohermaphroditism. Individuals with this syndrome are born with male internal genitalia including testes, but they have female external genitalia and are usually raised as girls. However, when they reach puberty, LH secretion and circulating testosterone levels are increased. Consequently, they develop male body contours and male libido. At this point, they usually change their gender identities and "become boys." The clitoris enlarges ("penis-at-12 syndrome") to the point that some of the individuals can have intercourse with women. This enlargement probably occurs because with the high LH, enough testosterone is produced to overcome the need for DHT amplification in the genitalia.

5α-Reductase-inhibiting drugs are now being used clinically to treat benign prostatic hyperplasia, and **finasteride,** the most extensively used drug, has its greatest effect on type 2 5α-reductase.

also primarily responsible for enlargement of the prostate and probably of the penis at the time of puberty, as well as for the facial hair, the acne, and the temporal recession of the hairline. On the other hand, the increase in muscle mass and the development of male sex drive and libido depend primarily on testosterone rather than DHT (see Clinical Box 25–4).

Testicular Production of Estrogens

Over 80% of the estradiol and 95% of the estrone in the plasma of adult men is formed by extragonadal and extraadrenal aromatization of circulating testosterone and androstenedione. The remainder comes from the testes. Some of the estradiol in testicular venous blood comes from the Leydig cells, but some is also produced by aromatization of androgens in Sertoli cells. In men, the plasma estradiol level is 20 to 50 pg/mL (73–184 pmol/L) and the total production rate is approximately 50 μg/d (184 nmol/d). In contrast to the situation in women, estrogen production moderately increases with advancing age in men.

CONTROL OF TESTICULAR FUNCTION

FSH is tropic for Sertoli cells, and FSH and androgens maintain the gametogenic function of the testes. FSH also stimulates the secretion of ABP and inhibin. Inhibin feeds back to inhibit FSH secretion. LH is tropic for Leydig cells and stimulates the secretion of testosterone, which in turn feeds back to inhibit LH secretion. Hypothalamic lesions in animals and hypothalamic disease in humans lead to atrophy of the testes and loss of their function.

Inhibins

Testosterone reduces plasma LH but, except in large doses, it has no effect on plasma FSH. Plasma FSH is elevated in patients who have atrophy of the seminiferous tubules but normal levels of testosterone and LH secretion. These observations led to the search for **inhibin,** a factor of testicular origin that inhibits FSH secretion. There are two inhibins in extracts of testes in men and in antral fluid from ovarian follicles in women. They are formed from three polypeptide subunits: a glycosylated α subunit with a molecular weight of 18,000; and two nonglycosylated β subunits, β_A and β_B, each with a molecular weight of 14,000. The subunits are formed from precursor proteins (Figure 25–19). The α subunit combines with β_A to form a heterodimer and with β_B to form another heterodimer, with the subunits linked by disulfide bonds. Both $\alpha\beta_A$ (inhibin A) and $\alpha\beta_B$ (inhibin B) inhibit FSH secretion by a direct action on the pituitary, though it now appears that it is inhibin B that is the FSH-regulating inhibin in adult men and women. Inhibins are produced by Sertoli cells in males and granulosa cells in females.

The heterodimer $\beta_A\beta_B$ and the homodimers $\beta_A\beta_A$ and $\beta_B\beta_B$ are also formed. They stimulate rather than inhibit FSH secretion and consequently are called **activins.** Their function in reproduction is unsettled. However, the inhibins and activins are members of the TGFβ superfamily of dimeric growth factors that also includes MIS. **Activin receptors** have been identified and belong to the serine/threonine kinase receptor family. Inhibins and activins are found not only in the gonads but also in the brain and many other tissues. In the bone marrow, activins are involved in the development of white blood cells. In embryonic life, activins are involved in the formation of mesoderm. All mice with a targeted deletion of the α-inhibin subunit gene initially exhibit normal growth but then develop gonadal stromal tumors, so the gene is a tumor suppressor gene.

In plasma, α_2-macroglobulin binds activins and inhibins. In tissues, activins bind to a family of four glycoproteins called **follistatins.** Binding of the activins inactivates their

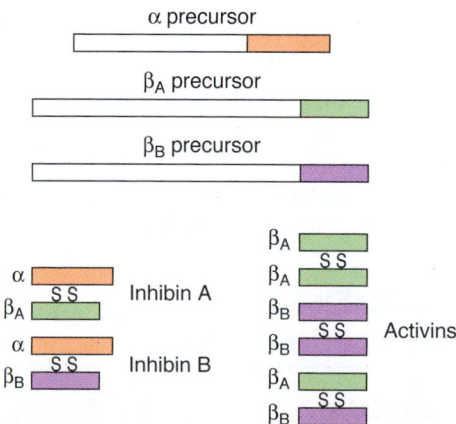

FIGURE 25–19 **Inhibin precursor proteins and the various inhibins and activins that are formed from the carboxyl terminal regions of these precursors.** SS, disulfide bonds.

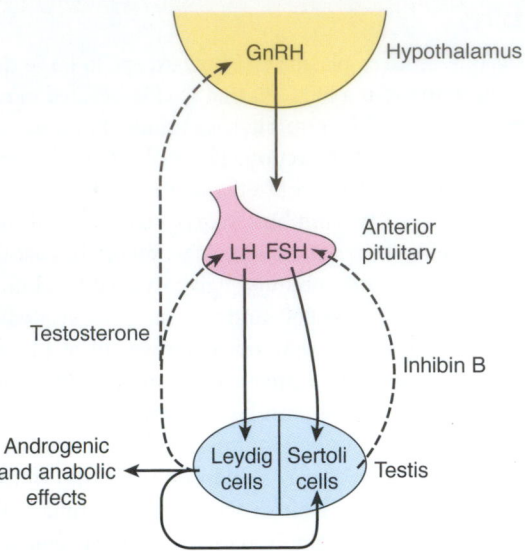

FIGURE 25–20 **Postulated interrelationships between the hypothalamus, anterior pituitary, and testes.** Solid arrows indicate excitatory effects; dashed arrows indicate inhibitory effects.

biologic activity, but the relation of follistatins to inhibin and their physiologic function remain unsettled.

Steroid Feedback

A current "working hypothesis" of the way the functions of the testes are regulated by steroids is shown in Figure 25–20. Castration is followed by a rise in the pituitary content and secretion of FSH and LH, and hypothalamic lesions prevent this rise. Testosterone inhibits LH secretion by acting directly on the anterior pituitary and by inhibiting the secretion of GnRH from the hypothalamus. Inhibin acts directly on the anterior pituitary to inhibit FSH secretion.

In response to LH, some of the testosterone secreted from the Leydig cells bathes the seminiferous epithelium and provides the high local concentration of androgen to the Sertoli cells that is necessary for normal spermatogenesis. Systemically administered testosterone does not raise the androgen level in the testes to as great a degree, and it inhibits LH secretion. Consequently, the net effect of systemically administered testosterone is generally a decrease in sperm count. Testosterone therapy has been suggested as a means of male contraception. However, the dose of testosterone needed to suppress spermatogenesis causes sodium and water retention. The possible use of inhibins as male contraceptives is now being explored.

ABNORMALITIES OF TESTICULAR FUNCTION

Cryptorchidism

The testes develop in the abdominal cavity and normally migrate to the scrotum during fetal development. **Testicular descent** to the inguinal region depends on MIS, and descent from the inguinal region to the scrotum depends on other factors. Descent is incomplete on one or, less commonly, both sides in 10% of newborn males, with the testes remaining in the abdominal cavity or inguinal canal. Gonadotropic hormone treatment speeds descent in some cases, or the defect can be corrected surgically. Spontaneous descent of the testes is the rule, and the proportion of boys with undescended testes (**cryptorchidism**) falls to 2% at age 1 y and 0.3% after puberty. However, early treatment is now recommended despite these figures because the incidence of malignant tumors is higher in undescended than in scrotal testes and because after puberty the higher temperature in the abdomen eventually causes irreversible damage to the spermatogenic epithelium.

Male Hypogonadism

The clinical picture of male hypogonadism depends on whether testicular deficiency develops before or after puberty. In adults, if it is due to testicular disease, circulating gonadotropin levels are elevated (**hypergonadotropic hypogonadism**); if it is secondary to disorders of the pituitary or the hypothalamus (eg, Kallmann syndrome), circulating gonadotropin levels are depressed (**hypogonadotropic hypogonadism**). If the endocrine function of the testes is lost in adulthood, the secondary sex characteristics regress slowly because it takes very little androgen to maintain them once they are established. The growth of the larynx during adolescence is permanent, and the voice remains deep. Men castrated in adulthood suffer some loss of libido, although the ability to copulate persists for some time. They occasionally have hot flushes and are generally more irritable, passive, and depressed than men with intact testes. When the Leydig cell deficiency dates from childhood, the clinical picture is that of **eunuchoidism.** Eunuchoid individuals over the age of 20 are characteristically tall, although not as tall as hyperpituitary giants, because their epiphyses remain open and some growth continues past the normal age of puberty. They have narrow shoulders and small muscles, a body configuration resembling that of the adult female. The genitalia are small and the voice high-pitched. Pubic hair and axillary hair are present because of adrenocortical androgen secretion. However, the hair is sparse, and the pubic hair has the female "triangle with the base up" distribution rather than the "triangle with the base down" pattern (male escutcheon) seen in normal males.

Androgen-Secreting Tumors

"Hyperfunction" of the testes in the absence of tumor formation is not a recognized entity. Androgen-secreting Leydig cell tumors are rare and cause detectable endocrine symptoms only in prepubertal boys, who develop precocious pseudopuberty (Table 25–2).

THE FEMALE REPRODUCTIVE SYSTEM

THE MENSTRUAL CYCLE

The reproductive system of women (Figure 25–21), unlike that of men, shows regular cyclic changes that teleologically may be regarded as periodic preparations for fertilization and pregnancy. In humans and other primates, the cycle is a **menstrual** cycle, and its most conspicuous feature is the periodic vaginal bleeding that occurs with the shedding of the uterine mucosa (**menstruation**). The length of the cycle is notoriously variable in women, but an average figure is 28 days from the start of one menstrual period to the start of the next. By common usage, the days of the cycle are identified by number, starting with the first day of menstruation.

Ovarian Cycle

From the time of birth, there are many **primordial follicles** under the ovarian capsule. Each contains an immature ovum (Figure 25–21). At the start of each cycle, several of these follicles enlarge, and a cavity forms around the ovum (**antrum formation**). This cavity is filled with follicular fluid. In humans, usually one of the follicles in one ovary starts to grow

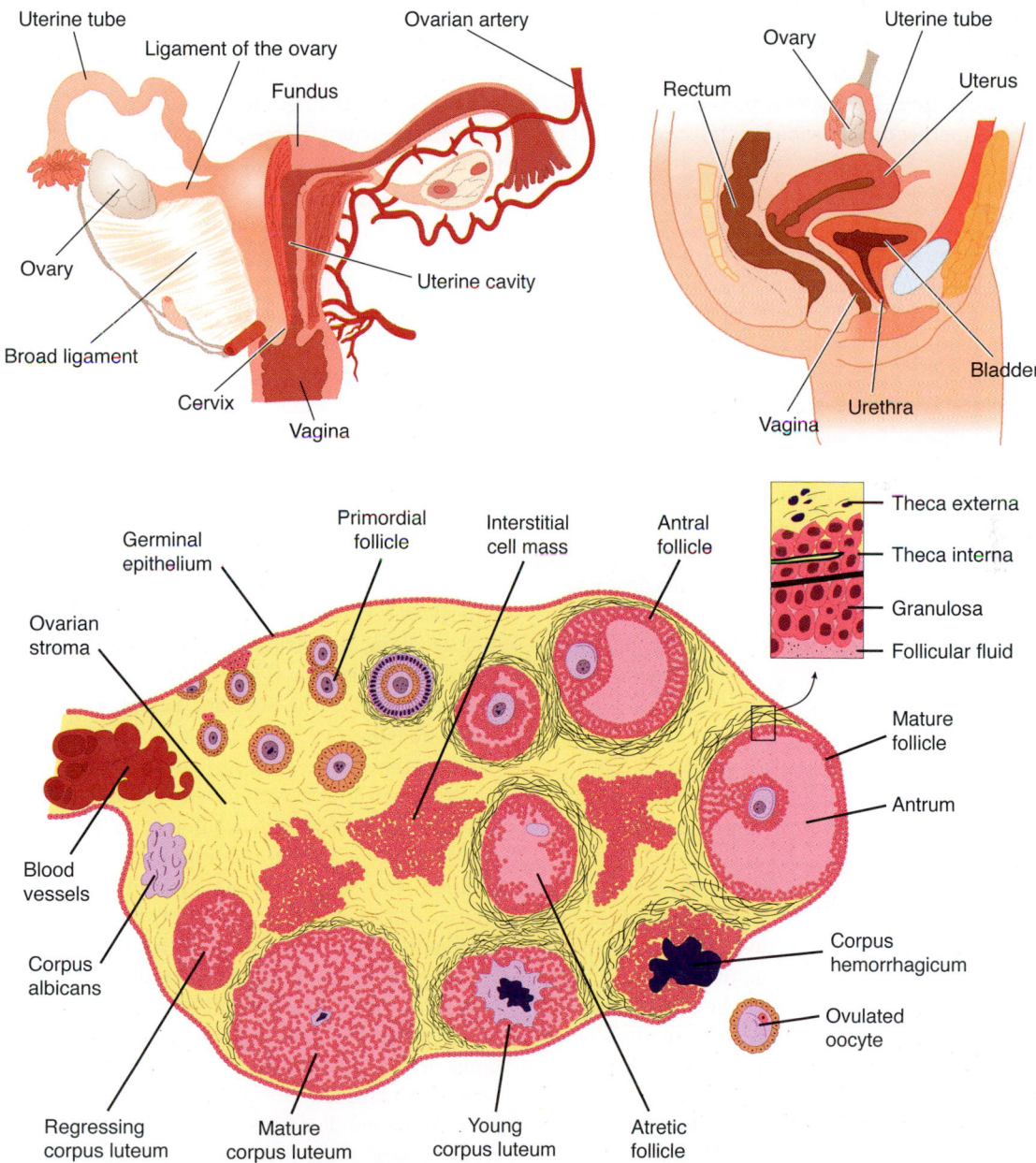

FIGURE 25–21 **Functional anatomy of the female reproductive tract.** The female reproductive organs include the ovaries, the uterus and the fallopian tubes, and the breast/mammary glands. The sequential development of a follicle, the formation of a corpus luteum and follicular atresia are shown.

rapidly on about the sixth day and becomes the **dominant follicle,** while the others regress, forming **atretic follicles.** The atretic process involves apoptosis. It is uncertain how one follicle is selected to be the dominant follicle in this **follicular phase** of the menstrual cycle, but it seems to be related to the ability of the follicle to secrete the estrogen inside it that is needed for final maturation. When women are given highly purified human pituitary gonadotropin preparations by injection, many follicles develop simultaneously.

The structure of a maturing ovarian (**graafian**) follicle is shown in Figure 25–21. The primary source of circulating estrogen is the granulosa cells of the ovaries; however, the cells of the **theca interna** of the follicle are necessary for the production of estrogen as they secrete androgens that are aromatized to estrogen by the granulosa cells.

At about the 14th day of the cycle, the distended follicle ruptures, and the ovum is extruded into the abdominal cavity. This is the process of **ovulation.** The ovum is picked up by the fimbriated ends of the uterine tubes (oviducts). It is transported to the uterus and, unless fertilization occurs, out through the vagina.

The follicle that ruptures at the time of ovulation promptly fills with blood, forming what is sometimes called a **corpus hemorrhagicum.** Minor bleeding from the follicle into the abdominal cavity may cause peritoneal irritation and fleeting lower abdominal pain ("mittelschmerz"). The granulosa and theca cells of the follicle lining promptly begin to proliferate, and the clotted blood is rapidly replaced with yellowish, lipid-rich **luteal cells,** forming the **corpus luteum.** This initiates the **luteal phase** of the menstrual cycle, during which the luteal cells secrete estrogen and progesterone. Growth of the corpus luteum depends on its developing an adequate blood supply, and there is evidence that vascular endothelial growth factor (VEGF) (see Chapter 32) is essential for this process.

If pregnancy occurs, the corpus luteum persists and usually there are no more periods until after delivery. If pregnancy does not occur, the corpus luteum begins to degenerate about 4 d before the next menses (24th day of the cycle) and is eventually replaced by scar tissue, forming a **corpus albicans.**

The ovarian cycle in other mammals is similar, except that in many species more than one follicle ovulates and multiple births are the rule. Corpora lutea form in some submammalian species but not in others.

In humans, no new ova are formed after birth. During fetal development, the ovaries contain over 7 million primordial follicles. However, many undergo atresia (involution) before birth and others are lost after birth. At the time of birth, there are 2 million ova, but 50% of these are atretic. The million that are normal undergo the first part of the first meiotic division at about this time and enter a stage of arrest in prophase in which those that survive persist until adulthood. Atresia continues during development, and the number of ova in both of the ovaries at the time of puberty is less than 300,000 (Figure 25–10). Only one of these ova per cycle (or about 500 in the course of a normal reproductive life) normally reaches maturity; the remainder degenerate. Just before ovulation, the first meiotic division is completed. One of the daughter cells, the **secondary oocyte,** receives most of the cytoplasm, while the other, the **first polar body,** fragments and disappears. The secondary oocyte immediately begins the second meiotic division, but this division stops at metaphase and is completed only when a sperm penetrates the oocyte. At that time, the **second polar body** is cast off and the fertilized ovum proceeds to form a new individual. The arrest in metaphase is due, at least in some species, to formation in the ovum of the protein $pp39^{mos}$, which is encoded by the **c-mos** protooncogene. When fertilization occurs, the $pp39^{mos}$ is destroyed within 30 min by **calpain,** a calcium-dependent cysteine protease.

Uterine Cycle

At the end of menstruation, all but the deep layers of the endometrium have sloughed. A new endometrium then regrows under the influence of estrogens from the developing follicle. The endometrium increases rapidly in thickness from the 5th to the 14th days of the menstrual cycle. As the thickness increases, the uterine glands are drawn out so that they lengthen (Figure 25–22), but they do not become convoluted or secrete to any degree. These endometrial changes are called proliferative, and this part of the menstrual cycle is sometimes called the **proliferative phase.** It is also called the preovulatory or follicular phase of the cycle. After ovulation, the endometrium becomes more highly vascularized and slightly edematous under the influence of estrogen and progesterone from the corpus luteum. The glands become coiled and tortuous and they begin to secrete a clear fluid. Consequently, this phase of the cycle is called the **secretory** or **luteal phase.** Late in the luteal phase, the endometrium, like the anterior pituitary, produces prolactin, but the function of this endometrial prolactin is unknown.

The endometrium is supplied by two types of arteries. The superficial two thirds of the endometrium that is shed during menstruation, the **stratum functionale,** is supplied by long, coiled **spiral arteries** (Figure 25–23), whereas the deep layer that is not shed, the **stratum basale,** is supplied by short, straight **basilar arteries.**

When the corpus luteum regresses, hormonal support for the endometrium is withdrawn. The endometrium becomes thinner, which adds to the coiling of the spiral arteries. Foci of necrosis appear in the endometrium, and these coalesce. In addition, spasm and degeneration of the walls of the spiral arteries take place, leading to spotty hemorrhages that become confluent and produce the menstrual flow.

The vasospasm is probably produced by locally released prostaglandins. Large quantities of prostaglandins are present in the secretory endometrium and in menstrual blood, and infusions of prostagladin $F_{2\alpha}$ ($PGF_{2\alpha}$) produce endometrial necrosis and bleeding.

From the point of view of endometrial function, the proliferative phase of the menstrual cycle represents restoration of the epithelium from the preceding menstruation, and the secretory phase represents preparation of the uterus for implantation of the fertilized ovum. The length of the secretory

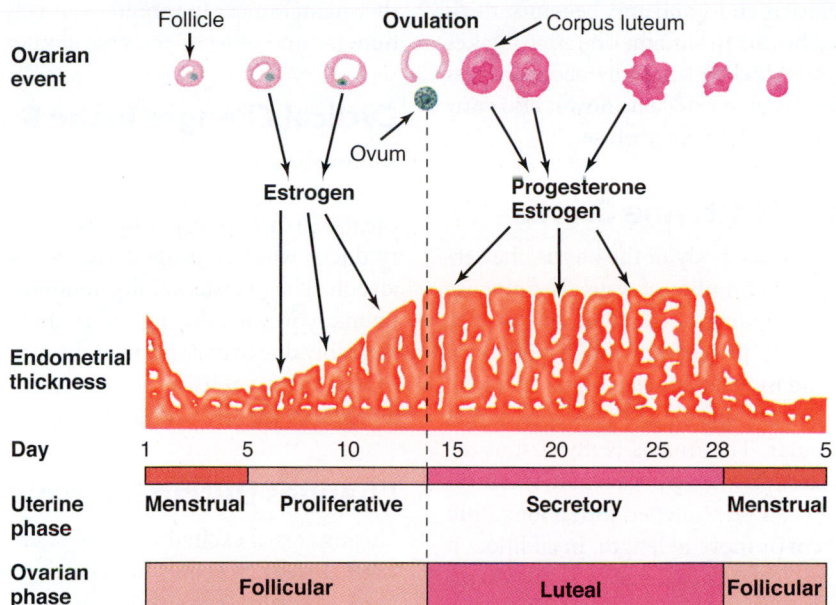

FIGURE 25–22 **Relationship between ovarian and uterine changes during the menstrual cycle.** (Reproduced with permission from Windmaier EP, Raff H, Strang KT: *Vander's Human Physiology: The Mechanisms of Body Function*, 11th ed. McGraw-Hill, 2008.)

FIGURE 25–23 **Spiral artery of endometrium.** Drawing of a spiral artery **(left)** and two uterine glands **(right)** from the endometrium of a rhesus monkey; early secretory phase. (Reproduced with permission from Daron GH: The arterial pattern of the tunica mucosa of the uterus in the *Macacus rhesus.* Am J Anat 1936;58:349.)

phase is remarkably constant at about 14 d, and the variations seen in the length of the menstrual cycle are due for the most part to variations in the length of the proliferative phase. When fertilization fails to occur during the secretory phase, the endometrium is shed and a new cycle starts.

Normal Menstruation

Menstrual blood is predominantly arterial, with only 25% of the blood being of venous origin. It contains tissue debris, prostaglandins, and relatively large amounts of fibrinolysin from endometrial tissue. The fibrinolysin lyses clots, so that menstrual blood does not normally contain clots unless the flow is excessive.

The usual duration of the menstrual flow is 3 to 5 d, but flows as short as 1 d and as long as 8 d can occur in normal women. The amount of blood lost may range normally from slight spotting to 80 mL; the average amount lost is 30 mL. Loss of more than 80 mL is abnormal. Obviously, the amount of flow can be affected by various factors, including the thickness of the endometrium, medication, and diseases that affect the clotting mechanism.

Anovulatory Cycles

In some instances, ovulation fails to occur during the menstrual cycle. Such anovulatory cycles are common for the first 12 to 18 mo after menarche and again before the onset of the menopause. When ovulation does not occur, no corpus luteum is formed and the effects of progesterone on the endometrium are absent. Estrogens continue to cause growth, however, and the proliferative endometrium becomes thick enough to break down and begins to slough. The time it takes for bleeding to occur is variable, but it usually occurs in less than 28 d from the last menstrual period. The flow is also variable and ranges from scanty to relatively profuse.

Cyclical Changes in the Uterine Cervix

Although it is continuous with the body of the uterus, the cervix of the uterus is different in a number of ways. The mucosa of the uterine cervix does not undergo cyclical desquamation, but there are regular changes in the cervical mucus. Estrogen makes the mucus thinner and more alkaline, changes that promote the survival and transport of sperms. Progesterone makes it thick, tenacious, and cellular. The mucus is thinnest at the time of ovulation, and its elasticity, or **spinnbarkeit,** increases so that by midcycle, a drop can be stretched into a long, thin thread that may be 8 to 12 cm or more in length. In addition, it dries in an arborizing, fern-like pattern (Figure 25–24) when a thin layer is spread on a slide. After ovulation and during pregnancy, it becomes thick and fails to form the fern pattern.

Vaginal Cycle

Under the influence of estrogens, the vaginal epithelium becomes cornified, and cornified epithelial cells can be identified in the vaginal smear. Under the influence of progesterone, a thick mucus is secreted, and the epithelium proliferates and becomes infiltrated with leukocytes. The cyclical changes in

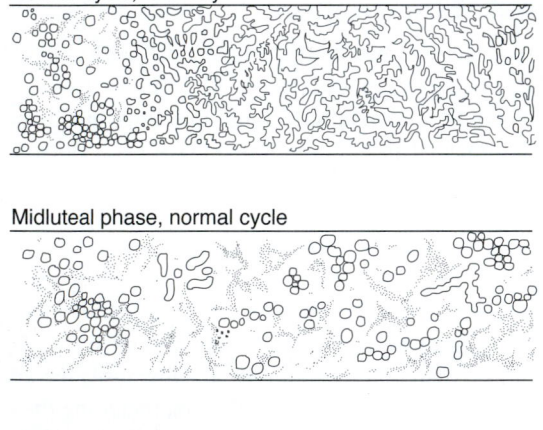

Normal cycle, 14th day

Midluteal phase, normal cycle

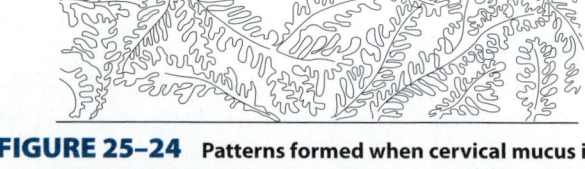

Anovulatory cycle with estrogen present

FIGURE 25–24   **Patterns formed when cervical mucus is smeared on a slide, permitted to dry, and examined under the microscope.** Progesterone makes the mucus thick and cellular. In the smear from a patient who failed to ovulate **(bottom),** no progesterone is present to inhibit the estrogen-induced fern pattern.

the vaginal smear in rats are relatively marked. The changes in humans and other species are similar but not so clear-cut.

Cyclical Changes in the Breasts

Although lactation normally does not occur until the end of pregnancy, cyclical changes take place in the breasts during the menstrual cycle. Estrogens cause proliferation of mammary ducts, whereas progesterone causes growth of lobules and alveoli. The breast swelling, tenderness, and pain experienced by many women during the 10 d preceding menstruation are probably due to distention of the ducts, hyperemia, and edema of the interstitial tissue of the breast. All these changes regress, along with the symptoms, during menstruation.

Changes During Intercourse

During sexual excitement in women, fluid is secreted onto the vaginal walls, probably because of release of VIP from vaginal nerves. A lubricating mucus is also secreted by the vestibular glands. The upper part of the vagina is sensitive to stretch, while tactile stimulation from the labia minora and clitoris adds to the sexual excitement. These stimuli are reinforced by tactile stimuli from the breasts and, as in men, by visual, auditory, and olfactory stimuli, which may build to the crescendo known as orgasm. During orgasm, autonomically mediated rhythmic contractions occur in the vaginal walls. Impulses also travel via the pudendal nerves and produce rhythmic contraction of the bulbocavernosus and ischiocavernosus muscles. The vaginal contractions may aid sperm transport but are not essential for it, since fertilization of the ovum is not dependent on orgasm.

Indicators of Ovulation

Knowing when during the menstrual cycle ovulation occurs is important in increasing fertility or, conversely, in family planning. A convenient and reasonably reliable indicator of the time of ovulation is a change—usually a rise—in the basal body temperature (Figure 25–25). The rise starts 1 to 2 d after ovulation. Women interested in obtaining an accurate temperature chart should use a digital thermometer and take their temperatures (oral or rectal) in the morning before getting out of bed. The cause of the temperature change at the time of ovulation is probably the increase in progesterone secretion, since progesterone is thermogenic.

A surge in LH secretion triggers ovulation, and ovulation normally occurs about 9 h after the peak of the LH surge at midcycle (Figure 25–25). The ovum lives for approximately 72 h after it is extruded from the follicle, but it is fertilizable for a much shorter time than this. In a study of the relation of isolated intercourse to pregnancy, 36% of women had a detected pregnancy following intercourse on the day of ovulation, but with intercourse on days after ovulation, the percentage was zero. Isolated intercourse on the first and second day before ovulation also led to pregnancy in about 36% of

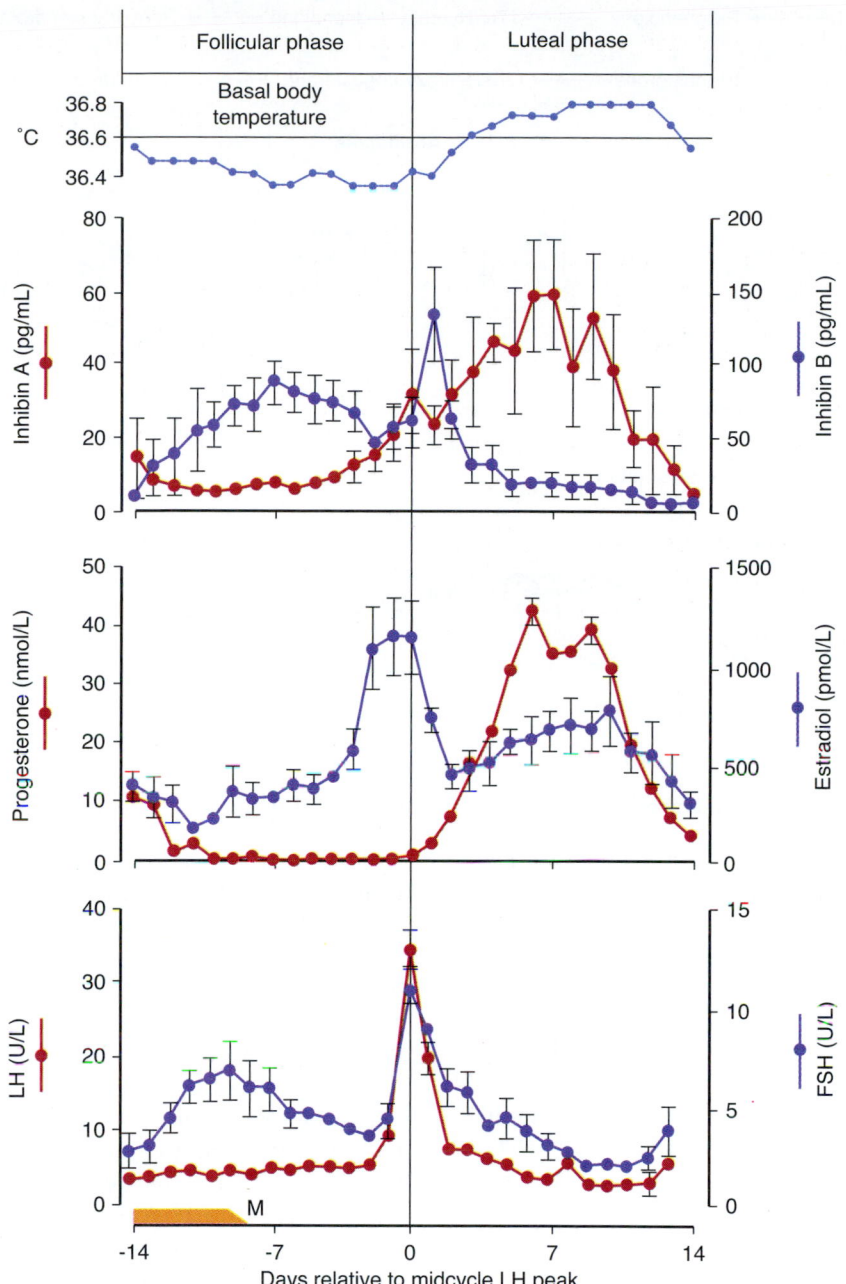

FIGURE 25–25 **Basal body temperature and plasma hormone concentrations (mean ± standard error) during the normal human menstrual cycle.** Values are aligned with respect to the day of the midcycle LH peak. FSH, follicle-stimulating hormone; LH, luteinizing hormone; M, menses.

women. A few pregnancies resulted from isolated intercourse on day 3, 4, or 5 before ovulation, although the percentage was much lower, for example, 8% on day 5 before ovulation. Thus, some sperms can survive in the female genital tract and fertilize the ovum for up to 120 h before ovulation, but the most fertile period is clearly the 48 h before ovulation. However, for those interested in the "rhythm method" of contraception, it should be noted that there are rare but documented cases in the literature of pregnancy resulting from isolated coitus on every day of the cycle.

The Estrous Cycle

Mammals other than primates do not menstruate, and their sexual cycle is called an **estrous cycle.** It is named for the conspicuous period of "heat" (**estrus**) at the time of ovulation, normally the only time during which the sexual interest of the female is aroused. In spontaneously ovulating species with estrous cycles, such as the rat, no episodic vaginal bleeding occurs but the underlying endocrine events are essentially the same as those in the menstrual cycle. In other species, ovulation is induced by copulation (reflex ovulation).

FIGURE 25–26 **Biosynthesis and metabolism of estrogens.** The formulas of the precursor steroids are shown in Figure 22–7.

OVARIAN HORMONES

Chemistry, Biosynthesis, & Metabolism of Estrogens

The naturally occurring estrogens are **17β-estradiol, estrone,** and **estriol** (Figure 25–26). They are C18 steroids which do not have an angular methyl group attached to the 10 position or a Δ^4-3-keto configuration in the A ring. They are secreted primarily by the granulosa cells of the ovarian follicles, the corpus luteum, and the placenta. Their biosynthesis depends on the enzyme **aromatase** (CYP19), which converts testosterone to estradiol and androstenedione to estrone (Figure 25–26). The latter reaction also occurs in fat, liver, muscle, and the brain.

Theca interna cells have many LH receptors, and LH acts via cAMP to increase conversion of cholesterol to androstenedione. The theca interna cells supply androstenedione to the granulosa cells. The granulosa cells make estradiol when provided with androgens (Figure 25–27), and it appears that the estradiol they form in primates is secreted into the follicular fluid. Granulosa cells have many FSH receptors, and FSH facilitates their secretion of estradiol by acting via cAMP to increase their aromatase activity. Mature granulosa cells also acquire LH receptors, and LH also stimulates estradiol production.

Two percent of the circulating estradiol is free, and the remainder is bound to protein: 60% to albumin and 38% to the same gonadal steroid-binding globulin (GBG) that binds testosterone (Table 25–5).

In the liver, estradiol, estrone, and estriol are converted to glucuronide and sulfate conjugates. All these compounds, along with other metabolites, are excreted in the urine. Appreciable amounts are secreted in the bile and reabsorbed into the bloodstream (enterohepatic circulation).

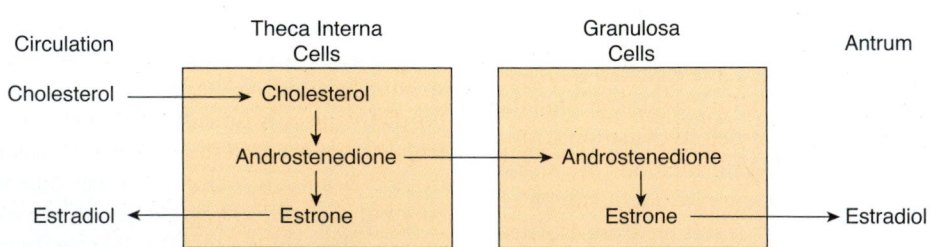

FIGURE 25–27 **Interactions between theca and granulosa cells in estradiol synthesis and secretion.**

TABLE 25–7 Twenty-four-hour production rates of sex steroids in women at different stages of the menstrual cycle.

Sex Steroids	Early Follicular	Preovulatory	Midluteal
Progesterone (mg)	1.0	4.0	25.0
17-hydroxyprogesterone (mg)	0.5	4.0	4.0
Dehydroepiandrosterone (mg)	7.0	7.0	7.0
Androstenedione (mg)	2.6	4.7	3.4
Testosterone (μg)	144.0	171.0	126.0
Estrone (μg)	50.0	350.0	250.0
Estradiol (μg)	36.0	380.0	250.0

Modified and reproduced, with permission, from Yen SSC, Jaffe RB, Barbieri RL: *Reproductive Endocrinology,* 4th ed. Saunders, 1999.

Secretion

The concentration of estradiol in the plasma during the menstrual cycle is shown in Figure 25–25. Almost all of this estrogen comes from the ovary, and two peaks of secretion occur: one just before ovulation and one during the midluteal phase. The estradiol secretion rate is 36 μg/d (133 nmol/d) in the early follicular phase, 380 μg/d just before ovulation, and 250 μg/d during the midluteal phase (Table 25–7). After menopause, estrogen secretion declines to low levels.

As noted previously, the estradiol production rate in men is about 50 μg/d (184 nmol/d).

Effects on the Female Genitalia

Estrogens facilitate the growth of the ovarian follicles and increase the motility of the uterine tubes. Their role in the cyclic changes in the endometrium, cervix, and vagina has been discussed previously. They increase uterine blood flow and have important effects on the smooth muscle of the uterus. In immature and castrated females, the uterus is small and the myometrium atrophic and inactive. Estrogens increase the amount of uterine muscle and its content of contractile proteins. Under the influence of estrogens, the muscle becomes more active and excitable, and action potentials in the individual fibers become more frequent. The "estrogen-dominated" uterus is also more sensitive to oxytocin.

Chronic treatment with estrogens causes the endometrium to hypertrophy. When estrogen therapy is discontinued, sloughing takes place with **withdrawal bleeding.** Some "breakthrough" bleeding may occur during treatment when estrogens are given for long periods.

Effects on Endocrine Organs

Estrogens decrease FSH secretion. Under some circumstances, they inhibit LH secretion (negative feedback); in other circumstances, they increase LH secretion (positive feedback). Women are sometimes given large doses of estrogens for 4 to 6 d to prevent conception after coitus during the fertile period (postcoital or "morning-after" contraception). However, in this instance, pregnancy is probably prevented by interference with implantation of the ovum rather than changes in gonadotropin secretion.

Estrogens cause increased secretion of angiotensinogen and thyroid-binding globulin. They exert an important protein anabolic effect in chickens and cattle, possibly by stimulating the secretion of androgens from the adrenal, and estrogen treatment has been used commercially to increase the weight of domestic animals. They cause epiphysial closure in humans (see Chapter 23).

Effects on the Central Nervous System

The estrogens are responsible for estrous behavior in animals, and they increase libido in humans. They apparently exert this action by a direct effect on certain neurons in the hypothalamus (Figure 25–28). Estrogens also increase the proliferation of dendrites on neurons and the number of synaptic knobs in rats.

Effects on the Breasts

Estrogens produce duct growth in the breasts and are largely responsible for breast enlargement at puberty in girls; they have been called the growth hormones of the breast. They are responsible for the pigmentation of the areolas, although pigmentation usually becomes more intense during the first pregnancy than it does at puberty. The role of the estrogens in the overall control of breast growth and lactation is discussed below.

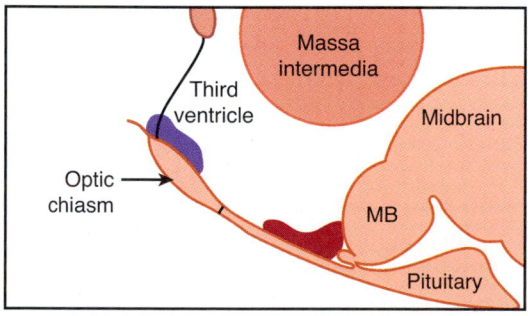

FIGURE 25–28 Loci where implantations of estrogen in the hypothalamus affect ovarian weight and sexual behavior in rats, projected on a sagittal section of the hypothalamus. The implants that stimulate sex behavior are located in the suprachiasmatic area above the optic chiasm (blue area), whereas ovarian atrophy is produced by implants in the arcuate nucleus and surrounding ventral hypothalamus (red). MB, mamillary body.

Female Secondary Sex Characteristics

The body changes that develop in girls at puberty—in addition to enlargement of breasts, uterus, and vagina—are due in part to estrogens, which are the "feminizing hormones," and in part simply to the absence of testicular androgens. Women have narrow shoulders and broad hips, thighs that converge, and arms that diverge (wide **carrying angle**). This body configuration, plus the female distribution of fat in the breasts and buttocks, is seen also in castrate males. In women, the larynx retains its prepubertal proportions and the voice stays high-pitched. Women have less body hair and more scalp hair, and the pubic hair generally has a characteristic flat-topped pattern (female escutcheon). However, growth of pubic and axillary hair in both sexes is due primarily to androgens rather than estrogens.

Other Actions

Normal women retain salt and water and gain weight just before menstruation. Estrogens cause some degree of salt and water retention. However, aldosterone secretion is slightly elevated in the luteal phase, and this also contributes to the premenstrual fluid retention.

Estrogens are said to make sebaceous gland secretions more fluid and thus to counter the effect of testosterone and inhibit formation of **comedones** ("black-heads") and acne. The liver palms, spider angiomas, and slight breast enlargement seen in advanced liver disease are due to increased circulating estrogens. The increase appears to be due to decreased hepatic metabolism of androstenedione, making more of this androgen available for conversion to estrogens.

Estrogens have a significant plasma cholesterol-lowering action, and they rapidly produce vasodilation by increasing the local production of NO. Their action on bone is discussed in Chapter 23.

Mechanism of Action

There are two principal types of nuclear estrogen receptors: estrogen receptor α (ERα) encoded by a gene on chromosome 6; and estrogen receptor β (ERβ), encoded by a gene on chromosome 14. Both are members of the nuclear receptor superfamily (see Chapter 2). After binding estrogen, they form homodimers and bind to DNA, altering its transcription. Some tissues contain one type or the other, but overlap also occurs, with some tissues containing both ERα and ERβ. ERα is found primarily in the uterus, kidneys, liver, and heart, whereas ERβ is found primarily in the ovaries, prostate, lungs, gastrointestinal tract, hemopoietic system, and central nervous system (CNS). They also form heterodimers with ERα binding to ERβ. Male and female mice in which the gene for ERα has been knocked out are sterile, develop osteoporosis, and continue to grow because their epiphyses do not close. ERβ female knockouts are infertile, but ERβ male knockouts are fertile even though they have hyperplastic prostates and loss of fat. Both receptors exist in isoforms and, like thyroid receptors, can bind to various activating and stimulating factors.

In some situations, ERβ can inhibit ERα transcription. Thus, their actions are complex, multiple, and varied.

Most of the effects of estrogens are genomic, that is, due to actions on the nucleus, but some are so rapid that it is difficult to believe they are mediated via production of mRNAs. These include effects on neuronal discharge in the brain and, possibly, feedback effects on gonadotropin secretion. Evidence is accumulating that these effects are mediated by cell membrane receptors that appear to be structurally related to the nuclear receptors and produce their effects by intracellular mitogen-activated protein kinase pathways. Similar rapid effects of progesterone, testosterone, glucocorticoids, aldosterone, and 1,25-dihydroxycholecalciferol may also be produced by membrane receptors.

Synthetic and Environmental Estrogens

The ethinyl derivative of estradiol is a potent estrogen and, unlike the naturally occurring estrogens, is relatively active when given by mouth because it is resistant to hepatic metabolism. The activity of the naturally occurring hormones is low when they are administered by mouth because the portal venous drainage of the intestine carries them to the liver, where they are inactivated before they can reach the general circulation. Some nonsteroidal substances and a few compounds found in plants have estrogenic activity. The plant estrogens are rarely a problem in human nutrition, but they may cause undesirable effects in farm animals. **Dioxins,** which are found in the environment and are produced by a variety of industrial processes, can activate estrogen response elements on genes. However, they have been reported to have antiestrogenic as well as estrogenic effects, and their role, if any, in the production of human disease remains a matter of disagreement and debate.

Because natural estrogens have undesirable as well as desirable effects (for example, they preserve bone in osteoporosis but can cause uterine and breast cancer), there has been an active search for "tailor-made" estrogens that have selective effects in humans. Two compounds, **tamoxifen** and **raloxifene,** show promise in this regard. Neither combats the symptoms of menopause, but both have the bone-preserving effects of estradiol. In addition, tamoxifen does not stimulate the breast, and raloxifene does not stimulate the breast or uterus. The way the effects of these selective estrogen receptor modulators (**SERMs**) are brought about is related to the complexity of the estrogen receptors and hence to differences in the way receptor–ligand complexes they form bind to DNA.

Chemistry, Biosynthesis, & Metabolism of Progesterone

Progesterone is a C_{21} steroid (Figure 25–29) secreted by the corpus luteum, the placenta, and (in small amounts) the follicle. It is an important intermediate in steroid biosynthesis in all tissues that secrete steroid hormones, and small amounts apparently enter the circulation from the testes and adrenal

FIGURE 25–29 **Biosynthesis of progesterone and major pathway for its metabolism.** Other metabolites are also formed.

cortex. About 2% of the circulating progesterone is free (Table 25–5), whereas 80% is bound to albumin and 18% is bound to corticosteroid-binding globulin. Progesterone has a short half-life and is converted in the liver to pregnanediol, which is conjugated to glucuronic acid and excreted in the urine.

Secretion

In men, the plasma progesterone level is approximately 0.3 ng/mL (1 nmol/L). In women, the level is approximately 0.9 ng/mL (3 nmol/L) during the follicular phase of the menstrual cycle (Figure 25–25). The difference is due to secretion of small amounts of progesterone by cells in the ovarian follicles; theca cells provide pregnenolone to the granulosa cells, which convert it to progesterone. Late in the follicular phase, progesterone secretion begins to increase. During the luteal phase, the corpus luteum produces large quantities of progesterone (Table 25–7) and plasma progesterone is markedly increased to a peak value of approximately 18 ng/mL (60 nmol/L).

The stimulating effect of LH on progesterone secretion by the corpus luteum is due to activation of adenylyl cyclase and involves a subsequent step that is dependent on protein synthesis.

Actions

The principal target organs of progesterone are the uterus, the breasts, and the brain. Progesterone is responsible for the progestational changes in the endometrium and the cyclic changes in the cervix and vagina described above. It has an antiestrogenic effect on the myometrial cells, decreasing their excitability, their sensitivity to oxytocin, and their spontaneous electrical activity while increasing their membrane potential. It also decreases the number of estrogen receptors in the endometrium and increases the rate of conversion of 17β-estradiol to less active estrogens.

In the breast, progesterone stimulates the development of lobules and alveoli. It induces differentiation of estrogen-prepared ductal tissue and supports the secretory function of the breast during lactation.

The feedback effects of progesterone are complex and are exerted at both the hypothalamic and pituitary levels. Large doses of progesterone inhibit LH secretion and potentiate the inhibitory effect of estrogens, preventing ovulation.

Progesterone is thermogenic and is probably responsible for the rise in basal body temperature at the time of ovulation. It stimulates respiration, and the alveolar PCO_2 (see Chapter 35) in women during the luteal phase of the menstrual cycle is lower than that in men. In pregnancy, the PCO_2 falls as progesterone secretion rises. However, the physiologic significance of this respiratory response is unknown.

Large doses of progesterone produce natriuresis, probably by blocking the action of aldosterone on the kidney. The hormone does not have a significant anabolic effect.

Mechanism of Action

The effects of progesterone, like those of other steroids, are brought about by an action on DNA to initiate synthesis of new mRNA. The progesterone receptor is bound to a heat shock protein in the absence of the steroid, and progesterone binding releases the heat shock protein, exposing the DNA-binding domain of the receptor. The synthetic steroid **mifepristone (RU 486)** binds to the receptor but does not release the heat shock protein, and it blocks the binding of progesterone. Because the maintenance of early pregnancy depends on the stimulatory effect of progesterone on endometrial growth and its inhibition of uterine contractility, mifepristone combined with a prostaglandin can be used to produce elective abortions.

There are two isoforms of the progesterone receptor—PR_A and PR_B—that are produced by differential processing from a single gene. PR_A is a truncated form, but it is likely that both isoforms mediate unique subsets of progesterone action.

Substances that mimic the action of progesterone are sometimes called **progestational agents, gestagens,** or **progestins.** They are used along with synthetic estrogens as oral contraceptive agents.

Relaxin

Relaxin is a polypeptide hormone that is produced in the corpus luteum, uterus, placenta, and mammary glands in women and in the prostate gland in men. During pregnancy, it relaxes the pubic symphysis and other pelvic joints and softens and dilates the uterine cervix. Thus, it facilitates delivery. It also inhibits uterine contractions and may play a role in the development of the mammary glands. In nonpregnant women, relaxin is found in the corpus luteum and the endometrium during the secretory but not the proliferative phase of the menstrual cycle. Its function in nonpregnant women is unknown. In men, it is found in semen, where it may help maintain sperm motility and aid in sperm penetration of the ovum.

In most species there is only one relaxin gene, but in humans there are two genes on chromosome 9 that code for two structurally different polypeptides that both have relaxin activity. However, only one of these genes is active in the ovary and the prostate. The structure of the polypeptide produced in these two tissues is shown in Figure 25–30.

CONTROL OF OVARIAN FUNCTION

FSH from the pituitary is responsible for the early maturation of the ovarian follicles, and FSH and LH together are responsible for their final maturation. A burst of LH secretion (Figure 25–25) is responsible for ovulation and the initial formation of the corpus luteum. A smaller midcycle burst of FSH secretion also occurs, the significance of which is uncertain. LH stimulates the secretion of estrogen and progesterone from the corpus luteum.

Hypothalamic Components

The hypothalamus occupies a key position in the control of gonadotropin secretion. Hypothalamic control is exerted by GnRH secreted into the portal hypophysial vessels. GnRH stimulates the secretion of FSH as well as LH.

GnRH is normally secreted in episodic bursts, and these bursts produce the circhoral peaks of LH secretion. They are essential for normal secretion of gonadotropins. If GnRH is administered by constant infusion, the GnRH receptors in the anterior pituitary down-regulate and LH secretion declines to zero. However, if GnRH is administered episodically at a rate of one pulse per hour, LH secretion is stimulated. This is true even when endogenous GnRH secretion has been prevented by a lesion of the ventral hypothalamus.

It is now clear not only that episodic secretion of GnRH is a general phenomenon but also that fluctuations in the frequency and amplitude of the GnRH bursts are important in generating the other hormonal changes that are responsible for the menstrual cycle. Frequency is increased by estrogens and decreased by progesterone and testosterone. The frequency increases late in the follicular phase of the cycle, culminating in the LH surge. During the secretory phase, the frequency decreases as a result of the action of progesterone (Figure 25–31), but when estrogen and progesterone secretion decrease at the end of the cycle, the frequency once again increases.

At the time of the midcycle LH surge, the sensitivity of the gonadotropes to GnRH is greatly increased because of their

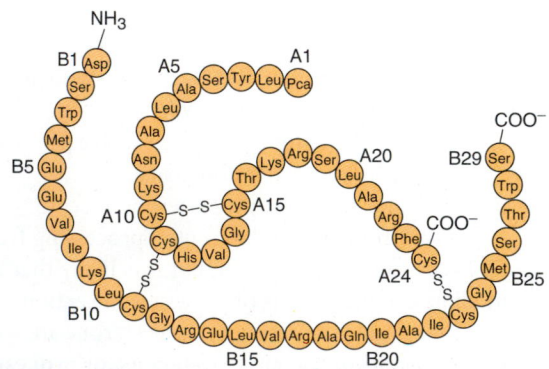

FIGURE 25–30 **Structure of human luteal and seminal relaxin.** Pca, pyroglutamic acid. (Modified and reproduced with permission from Winslow JW et al: Human seminal relaxin is a product of the same gene as human luteal relaxin. Endocrinology 1992;130:2660. Copyright © 1992 by The Endocrine Society.)

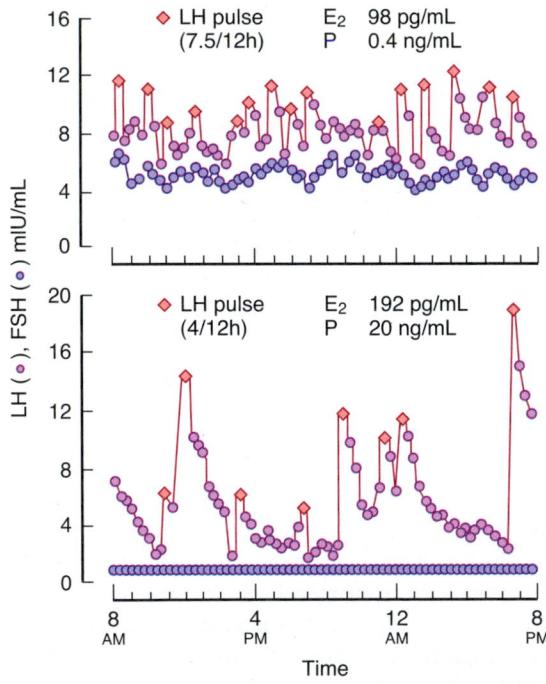

FIGURE 25–31 **Episodic secretion of LH (s) and FSH (d) during the follicular stage (top) and the luteal stage (bottom) of the menstrual cycle.** The numbers above each graph indicate the numbers of LH pulses per 12 hours and the plasma estradiol (E$_2$) and progesterone (P) concentrations at these two times of the cycle. (Reproduced with permission from Marshall JC, Kelch RO: Gonadotropin-releasing hormone: Role of pulsatile secretion in the regulation of reproduction. N Engl J Med 1986;315:1459.)

exposure to GnRH pulses of the frequency that exist at this time. This self-priming effect of GnRH is important in producing a maximum LH response.

The nature and the exact location of the GnRH pulse generator in the hypothalamus are still unsettled. However, it is known in a general way that norepinephrine and possibly epinephrine in the hypothalamus increase GnRH pulse frequencies. Conversely, opioid peptides such as the enkephalins and β-endorphin reduce the frequency of GnRH pulses.

The down-regulation of pituitary receptors and the consequent decrease in LH secretion produced by constantly elevated levels of GnRH has led to the use of long-acting GnRH analogs to inhibit LH secretion in precocious puberty and in cancer of the prostate.

Feedback Effects

Changes in plasma LH, FSH, sex steroids, and inhibin during the menstrual cycle are shown in Figure 25–25, and their feedback relations are diagrammed in Figure 25–32. During the early part of the follicular phase, inhibin B is low and FSH is modestly elevated, fostering follicular growth. LH secretion is

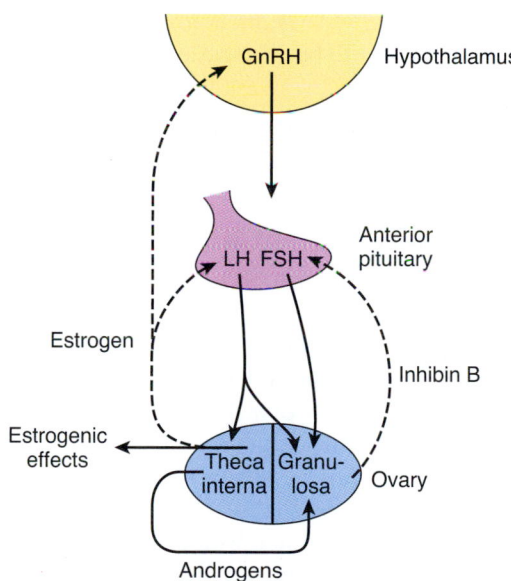

FIGURE 25–32 Feedback regulation of ovarian function.
The cells of the theca interna provide androgens to the granulosa cells, and theca cells also produce the circulating estrogens that inhibit the secretion of GnRH, LH, and FSH. Inhibin from the granulosa cells inhibits FSH secretion. LH regulates the thecal cells, whereas the granulosa cells are regulated by both LH and FSH. The dashed arrows indicate inhibitory effects and the solid arrows stimulatory effects.

held in check by the negative feedback effect of the rising plasma estrogen level. At 36 to 48 h before ovulation, the estrogen feedback effect becomes positive, and this initiates the burst of LH secretion (LH surge) that produces ovulation. Ovulation occurs about 9 h after the LH peak. FSH secretion also peaks, despite a small rise in inhibin, probably because of the strong stimulation of gonadotropes by GnRH. During the luteal phase, the secretion of LH and FSH is low because of the elevated levels of estrogen, progesterone, and inhibin.

It should be emphasized that a moderate, constant level of circulating estrogen exerts a negative feedback effect on LH secretion, whereas during the cycle, an elevated estrogen level exerts a positive feedback effect and stimulates LH secretion. It has been demonstrated that in monkeys estrogens must also be elevated for a minimum time to produce positive feedback. When circulating estrogen was increased about 300% for 24 h, only negative feedback was seen; but when it was increased about 300% for 36 h or more, a brief decline in secretion was followed by a burst of LH secretion that resembled the midcycle surge. When circulating levels of progesterone were high, the positive feedback effect of estrogen was inhibited. There is evidence that in primates, both the negative and the positive feedback effects of estrogen are exerted in the mediobasal hypothalamus, but exactly how negative feedback is switched to positive feedback and then back to negative feedback in the luteal phase remains unknown.

Control of the Cycle

In an important sense, regression of the corpus luteum (**luteolysis**) starting 3 to 4 d before menses is the key to the menstrual cycle. $PGF_{2\alpha}$ appears to be a physiologic luteolysin, but this prostaglandin is only active when endothelial cells producing ET-1 (see Chapter 33) are present. Therefore, it appears that at least in some species luteolysis is produced by the combined action of $PGF_{2\alpha}$ and ET-1. In some domestic animals, oxytocin secreted by the corpus luteum appears to exert a local luteolytic effect, possibly by causing the release of prostaglandins. Once luteolysis begins, the estrogen and progesterone levels fall and the secretion of FSH and LH increases. A new crop of follicles develops, and then a single dominant follicle matures as a result of the action of FSH and LH. Near midcycle, estrogen secretion from the follicle rises. This rise augments the responsiveness of the pituitary to GnRH and triggers a burst of LH secretion. The resulting ovulation is followed by formation of a corpus luteum. Estrogen secretion drops, but progesterone and estrogen levels then rise together, along with inhibin B. The elevated levels inhibit FSH and LH secretion for a while, but luteolysis again occurs and a new cycle starts.

Reflex Ovulation

Female cats, rabbits, mink, and some other animals have long periods of estrus, during which they ovulate only after copulation. Such **reflex ovulation** is brought about by afferent impulses from the genitalia and the eyes, ears, and nose that converge on the ventral hypothalamus and provoke an ovulation-inducing release of LH from the pituitary. In species such as rats, monkeys, and humans, ovulation is a spontaneous periodic phenomenon, but neural mechanisms are also involved. Ovulation can be delayed 24 h in rats by administering pentobarbital or various other neurally active drugs 12 h before the expected time of follicle rupture.

Contraception

Methods commonly used to prevent conception are listed in Table 25–8, along with their failure rates. Once conception has occurred, abortion can be produced by progesterone antagonists such as mifepristone.

Implantation of foreign bodies in the uterus causes changes in the duration of the sexual cycle in a number of mammalian species. In humans, such foreign bodies do not alter the menstrual cycle, but they act as effective contraceptive devices. Intrauterine implantation of pieces of metal or plastic (**intrauterine devices, IUDs**) has been used in programs aimed at controlling population growth. Although the mechanism of action of IUDs is still unsettled, they seem in general to prevent sperms from fertilizing ova. Those containing copper appear to exert a spermatocidal effect. IUDs that slowly release progesterone or synthetic progestins have the additional effect of thickening cervical mucus so that entry of sperms into the

uterus is impeded. IUDs can cause intrauterine infections, but these usually occur in the first month after insertion and in women exposed to sexually transmitted diseases.

Women undergoing long-term treatment with relatively large doses of estrogen do not ovulate, probably because they have depressed FSH levels and multiple irregular bursts of LH secretion rather than a single midcycle peak. Women treated with similar doses of estrogen plus a progestational agent do not ovulate because the secretion of both gonadotropins is suppressed. In addition, the progestin makes the cervical mucus thick and unfavorable to sperm migration, and it may also interfere with implantation. For contraception, an orally active estrogen such as ethinyl estradiol is often combined with a synthetic progestin such as norethindrone. The pills are administered for 21 d, then withdrawn for 5 to 7 d to permit menstrual flow, and started again. Like ethinyl estradiol, norethindrone has an ethinyl group on position 17 of the steroid nucleus, so it is resistant to hepatic metabolism and consequently is effective by mouth. In addition to being a progestin, it is partly metabolized to ethinyl estradiol, and for this reason it also has estrogenic activity. Small as well as large doses of estrogen are effective (Table 25–8).

Implants made up primarily of progestins such as levonorgestrel are now seeing increased use in some parts of the world. These are inserted under the skin and can prevent pregnancy for up to 5 y. They often produce amenorrhea, but otherwise they appear to be effective and well tolerated.

TABLE 25–8 Relative effectiveness of frequently used contraceptive methods.

Method	Failures per 100 Woman-Years
Vasectomy	0.02
Tubal ligation and similar procedures	0.13
Oral contraceptives	
> 50 mg estrogen and progestin	0.32
< 50 mg estrogen and progestin	0.27
Progestin only	1.2
IUD	
Copper 7	1.5
Loop D	1.3
Diaphragm	1.9
Condom	3.6
Withdrawal	6.7
Spermicide	11.9
Rhythm	15.5

Data from Vessey M, Lawless M, Yeates D: Efficacy of different contraceptive methods. Lancet 1982;1:841. Reproduced with permission.

ABNORMALITIES OF OVARIAN FUNCTION

Menstrual Abnormalities

Some women who are infertile have **anovulatory cycles;** they fail to ovulate but have menstrual periods at fairly regular intervals. As noted above, anovulatory cycles are the rule for the first 1 to 2 y after menarche and again before the menopause. **Amenorrhea** is the absence of menstrual periods. If menstrual bleeding has never occurred, the condition is called **primary amenorrhea.** Some women with primary amenorrhea have small breasts and other signs of failure to mature sexually. Cessation of cycles in a woman with previously normal periods is called **secondary amenorrhea.** The most common cause of secondary amenorrhea is pregnancy, and the old clinical maxim that "secondary amenorrhea should be considered to be due to pregnancy until proved otherwise" has considerable merit. Other causes of amenorrhea include emotional stimuli and changes in the environment, hypothalamic diseases, pituitary disorders, primary ovarian disorders, and various systemic diseases. Evidence suggests that in some women with hypothalamic amenorrhea, the frequency of GnRH pulses is slowed as a result of excess opioid activity in the hypothalamus. In encouraging preliminary studies, the frequency of GnRH pulses has been increased by administration of the orally active opioid blocker naltrexone.

The terms **hypomenorrhea** and **menorrhagia** refer to scanty and abnormally profuse flow, respectively, during regular periods. **Metrorrhagia** is bleeding from the uterus between periods, and **oligomenorrhea** is reduced frequency of periods. **Dysmenorrhea** is painful menstruation. The severe menstrual cramps that are common in young women quite often disappear after the first pregnancy. Most of the symptoms of dysmenorrhea are due to accumulation of prostaglandins in the uterus, and symptomatic relief has been obtained by treatment with inhibitors of prostaglandin synthesis.

Some women develop symptoms such as irritability, bloating, edema, decreased ability to concentrate, depression, headache, and constipation during the last 7 to 10 d of their menstrual cycles. These symptoms of the **premenstrual syndrome (PMS)** have been attributed to salt and water retention. However, it seems unlikely that this or any of the other hormonal alterations that occur in the late luteal phase are responsible because the time course and severity of the symptoms are not modified if the luteal phase is terminated early and menstruation produced by administration of mifepristone. The antidepressant fluoxetine (Prozac), which is a serotonin reuptake inhibitor, and the benzodiazepine alprazolam (Xanax) produce symptomatic relief, and so do GnRH-releasing agonists in doses that suppress the pituitary–ovarian axis. How these diverse clinical observations fit together to produce a picture of the pathophysiology of PMS is still unknown (see Clinical Box 25–5).

PREGNANCY

Fertilization & Implantation

In humans, **fertilization** of the ovum by the sperm usually occurs in the ampulla of the uterine tube. Fertilization involves (1) chemoattraction of the sperm to the ovum by substances produced by the ovum; (2) adherence to the **zona pellucida,** the membranous structure surrounding the ovum; (3) penetration of the zona pellucida and the acrosome reaction; and (4) adherence of the sperm head to the cell membrane of the ovum, with breakdown of the area of fusion and release of the sperm nucleus into the cytoplasm of the ovum (Figure 25–33). Millions of sperms are deposited in the vagina during intercourse. Eventually, 50 to 100 sperms reach the ovum, and many of them contact the zona pellucida. Sperms bind to a sperm receptor in the zona, and this is followed by the **acrosomal reaction,** that is, the breakdown of the acrosome, the lysosome-like organelle on the head of the sperm (Figure 25–14). Various enzymes are released, including the trypsin-like protease **acrosin.** Acrosin facilitates but is not required for the penetration of the sperm through the zona pellucida. When one sperm reaches the membrane of the ovum, fusion to the ovum membrane is mediated by **fertilin,** a protein on the surface of the sperm head that resembles the viral fusion proteins that permit some viruses to attack cells. The fusion provides the signal that initiates development. In addition, the fusion sets off a reduction in the membrane potential of the ovum that prevents polyspermy, the fertilization of the ovum by more than one sperm. This transient potential change is followed by a structural change in the zona pellucida that provides protection against polyspermy on a more long-term basis.

The developing embryo, now called a **blastocyst,** moves down the tube into the uterus. This journey takes about 3 d, during which the blastocyst reaches the 8- or 16-cell stage. Once in contact with the endometrium, the blastocyst becomes surrounded by an outer layer of **syncytiotrophoblast,** a multinucleate mass with no discernible cell boundaries, and an inner layer of **cytotrophoblast** made up of individual cells. The syncytiotrophoblast erodes the endometrium, and the blastocyst

FIGURE 25–33   Sequential events in fertilization in mammals.
Sperm are attracted to the ovum, bind to the zona pellucida, release acrosomal enzymes, penetrate the zona pellucida, and fuse with the membrane of the ovum, releasing the sperm nucleus into its cytoplasm. Current evidence indicates that the side, rather than the tip, of the sperm head fuses with the egg cell membrane. (Modified from Vacquier VD: Evolution of gamete recognition proteins. Science 1999;281:1995.)

burrows into it (**implantation**). The implantation site is usually on the dorsal wall of the uterus. A placenta then develops, and the trophoblast remains associated with it.

Failure to Reject the "Fetal Graft"

It should be noted that the fetus and the mother are two genetically distinct individuals, and the fetus is in effect a transplant of foreign tissue in the mother. However, the transplant is tolerated, and the rejection reaction that is characteristically produced when other foreign tissues are transplanted (see Chapter 3) fails to occur. The way the "fetal graft" is protected is unknown. However, one explanation may be that the placental trophoblast, which separates maternal and fetal tissues, does not express the polymorphic class I and class II MHC genes and instead expresses **HLA-G,** a nonpolymorphic gene. Therefore, antibodies against the fetal proteins do not develop. In addition, there is a Fas ligand on the surface of the placenta, and this bonds to T cells, causing them to undergo apoptosis (see Chapter 3).

Infertility

The vexing clinical problem of infertility often requires extensive investigation before a cause is found. In 30% of cases the problem is in the man; in 45%, the problem is in the woman; in 20%, both partners have a problem; and in 5% no cause can be found. **In vitro fertilization,** that is, removing mature ova, fertilizing them with sperm, and implanting one or more of them in the uterus at the four-cell stage is of some value in these cases. It has a 5–10% chance of producing a live birth.

Endocrine Changes

In all mammals, the corpus luteum in the ovary at the time of fertilization fails to regress and instead enlarges in response to stimulation by gonadotropic hormones secreted by the placenta. The placental gonadotropin in humans is called **human chorionic gonadotropin (hCG).** The enlarged **corpus luteum of pregnancy** secretes estrogens, progesterone, and relaxin. The relaxin helps maintain pregnancy by inhibiting myometrial contractions. In most species, removal of the ovaries at any time during pregnancy precipitates abortion. In humans, however, the placenta produces sufficient estrogen and progesterone from maternal and fetal precursors to take over the function of the corpus luteum after the sixth week of pregnancy. Ovariectomy before the sixth week leads to abortion, but ovariectomy thereafter has no effect on the pregnancy. The function of the corpus luteum begins to decline after 8 wk of pregnancy, but it persists throughout pregnancy. hCG secretion decreases after an initial marked rise, but estrogen and progesterone secretion increase until just before parturition (Table 25–9).

Human Chorionic Gonadotropin

hCG is a glycoprotein that contains galactose and hexosamine. It is produced by the syncytiotrophoblast. Like the pituitary

TABLE 25–9 Hormone levels in human maternal blood during normal pregnancy.

Hormone	Approximate Peak Value	Time of Peak Secretion
hCG	5 mg/mL	First trimester
Relaxin	1 ng/mL	First trimester
hCS	15 mg/mL	Term
Estradiol	16 ng/mL	Term
Estriol	14 ng/mL	Term
Progesterone	190 ng/mL	Term
Prolactin	200 ng/mL	Term

glycoprotein hormones, it is made up of α and β subunits. hCG-α is identical to the α subunit of LH, FSH, and TSH. The molecular weight of hCG-α is 18,000, and that of hCG-β is 28,000. hCG is primarily luteinizing and luteotropic and has little FSH activity. It can be measured by radioimmunoassay and detected in the blood as early as 6 d after conception. Its presence in the urine in early pregnancy is the basis of the various laboratory tests for pregnancy, and it can sometimes be detected in the urine as early as 14 d after conception. It appears to act on the same receptor as LH. hCG is not absolutely specific for pregnancy. Small amounts are secreted by a variety of gastrointestinal and other tumors in both sexes, and hCG has been measured in individuals with suspected tumors as a "tumor marker." It also appears that the fetal liver and kidney normally produce small amounts of hCG.

Human Chorionic Somatomammotropin

The syncytiotrophoblast also secretes large amounts of a protein hormone that is lactogenic and has a small amount of growth-stimulating activity. This hormone has been called **chorionic growth hormone-prolactin (CGP)** and **human placental lactogen (hPL),** but it is now generally called **human chorionic somatomammotropin (hCS).** The structure of hCS is very similar to that of human growth hormone (see Figure 24–3), and it appears that these two hormones and prolactin evolved from a common progenitor hormone. Large quantities of hCS are found in maternal blood, but very little reaches the fetus. Secretion of growth hormone from the maternal pituitary is not increased during pregnancy and may actually be decreased by hCS. However, hCS has most of the actions of growth hormone and apparently functions as a "maternal growth hormone of pregnancy" to bring about the nitrogen, potassium, and calcium retention, lipolysis, and decreased glucose utilization seen in this state. These latter two actions divert glucose to the fetus. The amount of hCS secreted is proportionate to the size of the placenta, which normally weighs about one-sixth as much as the fetus, and low hCS levels are a sign of placental insufficiency.

Other Placental Hormones

In addition to hCG, hCS, progesterone, and estrogens, the placenta secretes other hormones. Human placental fragments probably produce proopiomelanocortin (POMC). In culture, they release corticotropin-releasing hormone (CRH), β-endorphin, α-melanocyte-stimulating hormone (MSH), and dynorphin A, all of which appear to be identical to their hypothalamic counterparts. They also secrete GnRH and inhibin, and since GnRH stimulates and inhibin inhibits hCG secretion, locally produced GnRH and inhibin may act in a paracrine fashion to regulate hCG secretion. The trophoblast cells and amnion cells also secrete leptin, and moderate amounts of this satiety hormone enter the maternal circulation. Some also enters the amniotic fluid. Its function in pregnancy is unknown. The placenta also secretes prolactin in a number of forms.

Finally, the placenta secretes the α subunits of hCG, and the plasma concentration of free α subunits rises throughout pregnancy. These α subunits acquire a carbohydrate composition that makes them unable to combine with β subunits, and their prominence suggests that they have a function of their own. It is interesting in this regard that the secretion of the prolactin produced by the endometrium also appears to increase throughout pregnancy, and it may be that the circulating α subunits stimulate endometrial prolactin secretion.

The cytotrophoblast of the human chorion contains prorenin (see Chapter 39). A large amount of prorenin is also present in amniotic fluid, but its function in this location is unknown.

Fetoplacental Unit

The fetus and the placenta interact in the formation of steroid hormones. The placenta synthesizes pregnenolone and progesterone from cholesterol. Some of the progesterone enters the fetal circulation and provides the substrate for the formation of cortisol and corticosterone in the fetal adrenal glands (Figure 25–34). Some of the pregnenolone enters the fetus and, along with pregnenolone synthesized in the fetal liver, is the substrate for the formation of dehydroepiandrosterone sulfate (DHEAS) and 16-hydroxydehydroepiandrosterone sulfate (16-OHDHEAS) in the fetal adrenal. Some 16-hydroxylation also occurs in the fetal liver. DHEAS and 16-OHDHEAS are transported back to the placenta, where DHEAS forms estradiol and 16-OHDHEAS forms estriol. The principal estrogen formed is estriol, and since fetal 16-OHDHEAS is the principal substrate for the estrogens, the urinary estriol excretion of the mother can be monitored as an index of the state of the fetus.

Parturition

The duration of pregnancy in humans averages 270 d from fertilization (284 d from the first day of the menstrual period preceding conception). Irregular uterine contractions increase in frequency in the last month of pregnancy.

The difference between the body of the uterus and the cervix becomes evident at the time of delivery. The cervix, which

FIGURE 25–34 Interactions between the placenta and the fetal adrenal cortex in the production of steroids.

is firm in the nonpregnant state and throughout pregnancy until near the time of delivery, softens and dilates, while the body of the uterus contracts and expels the fetus.

There is still considerable uncertainty about the mechanisms responsible for the onset of labor. One factor is the increase in circulating estrogens produced by increased circulating DHEAS. This makes the uterus more excitable, increases the number of gap junctions between myometrial cells, and causes production of more prostaglandins, which in turn cause uterine contractions. In humans, CRH secretion by the fetal hypothalamus increases and is supplemented by increased placental production of CRH. This increases circulating adrenocorticotropic hormone (ACTH) in the fetus, and the resulting increase in cortisol hastens the maturation of the respiratory system. Thus, in a sense, the fetus picks the time to be born by increasing CRH secretion.

The number of oxytocin receptors in the myometrium and the decidua (the endometrium of pregnancy) increases more than 100-fold during pregnancy and reaches a peak during early labor. Estrogens increase the number of oxytocin receptors, and uterine distention late in pregnancy may also increase their formation. In early labor, the oxytocin concentration in maternal plasma is not elevated from the prelabor value of about 25 pg/mL. It is possible that the marked increase in oxytocin receptors causes the uterus to respond to normal plasma oxytocin concentrations. However, at least in rats, the amount of oxytocin mRNA in the uterus increases, reaching a peak at term; this suggests that locally produced oxytocin also participates in the process.

Premature onset of labor is a problem because premature infants have a high mortality rate and often require intensive, expensive care. Intramuscular 17α-hydroxyprogesterone causes a significant decrease in the incidence of premature labor. The mechanism by which it exerts its effect is uncertain, but it may be that the steroid provides a stable level of circulating progesterone. Progesterone relaxes uterine smooth muscle, inhibits the action of oxytocin on the muscle, and reduces

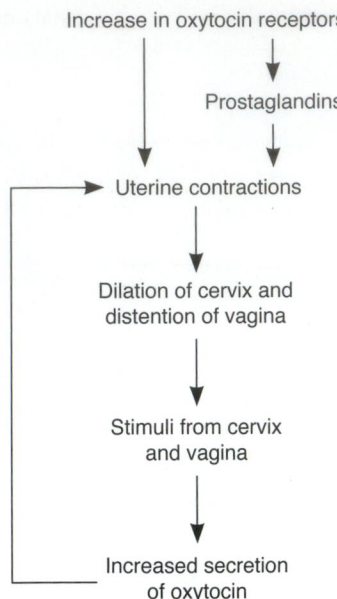

FIGURE 25–35   **Role of oxytocin in parturition.**

the formation of gap junctions between the muscle fibers. All these actions would be expected to inhibit the onset of labor.

Once labor is started, the uterine contractions dilate the cervix, and this dilation in turn sets up signals in afferent nerves that increase oxytocin secretion (Figure 25–35). The plasma oxytocin level rises and more oxytocin becomes available to act on the uterus. Thus, a positive feedback loop is established that aids delivery and terminates on expulsion of the products of conception. Oxytocin increases uterine contractions in two ways: (1) It acts directly on uterine smooth muscle cells to make them contract, and (2) it stimulates the formation of prostaglandins in the decidua. The prostaglandins enhance the oxytocin-induced contractions.

During labor, spinal reflexes and voluntary contractions of the abdominal muscles ("bearing down") also aid in delivery. However, delivery can occur without bearing down and without a reflex increase in secretion of oxytocin from the posterior pituitary gland, since paraplegic women can go into labor and deliver.

LACTATION

Development of the Breasts

Many hormones are necessary for full mammary development. In general, estrogens are primarily responsible for proliferation of the mammary ducts and progesterone for the development of the lobules. In rats, some prolactin is also needed for development of the glands at puberty, but it is not known if prolactin is necessary in humans. During pregnancy, prolactin levels increase steadily until term, and levels of estro-

gens and progesterone are elevated as well, producing full lobuloalveolar development.

Secretion & Ejection of Milk

The composition of human and cows' milk is shown in Table 25–10. In estrogen- and progesterone-primed rodents, injections of prolactin cause the formation of milk droplets and their secretion into the ducts. Oxytocin causes contraction of the myoepithelial cells lining the duct walls, with consequent ejection of the milk through the nipple.

TABLE 25–10   **Composition of colostrum and milk.***

Component	Human Colostrum	Human Milk	Cows' Milk
Water, g	...	88	88
Lactose, g	5.3	6.8	5.0
Protein, g	2.7	1.2	3.3
Casein: lactalbumin ratio	...	1:2	3:1
Fat, g	2.9	3.8	3.7
Linoleic acid	...	8.3% of fat	1.6% of fat
Sodium, mg	92	15	58
Potassium, mg	55	55	138
Chloride, mg	117	43	103
Calcium, mg	31	33	125
Magnesium, mg	4	4	12
Phosphorus, mg	14	15	100
Iron, mg	0.09[2]	0.15[a]	0.10[a]
Vitamin A, μg	89	53	34
Vitamin D, μg	...	0.03[a]	0.06[a]
Thiamine, μg	15	16	42
Riboflavin, μg	30	43	157
Nicotinic acid, μg	75	172	85
Ascorbic acid, mg	4.4[a]	4.3[a]	1.6[a]

*Weights per deciliter.

[a]Poor source.

Reproduced with permission from Findlay ALR: Lactation. Res Reprod (Nov) 1974;6(6).

Initiation of Lactation after Delivery

The breasts enlarge during pregnancy in response to high circulating levels of estrogens, progesterone, prolactin, and possibly hCG. Some milk is secreted into the ducts as early as the fifth month, but the amounts are small compared with the surge of milk secretion that follows delivery. In most animals, milk is secreted within an hour after delivery, but in women it takes 1 to 3 d for the milk to "come in."

After expulsion of the placenta at parturition, the levels of circulating estrogens and progesterone abruptly decline. The drop in circulating estrogen initiates lactation. Prolactin and estrogen synergize in producing breast growth, but estrogen antagonizes the milk-producing effect of prolactin on the breast. Indeed, in women who do not wish to nurse their babies, estrogens may be administered to stop lactation.

Suckling not only evokes reflex oxytocin release and milk ejection, it also maintains and augments the secretion of milk because of the stimulation of prolactin secretion produced by suckling.

Effect of Lactation on Menstrual Cycles

Women who do not nurse their infants usually have their first menstrual period 6 wk after delivery. However, women who nurse regularly have amenorrhea for 25 to 30 wk. Nursing stimulates prolactin secretion, and evidence suggests that prolactin inhibits GnRH secretion, inhibits the action of GnRH on the pituitary, and antagonizes the action of gonadotropins on the ovaries. Ovulation is inhibited, and the ovaries are inactive, so estrogen and progesterone output falls to low levels. Consequently, only 5–10% of women become pregnant again during the suckling period, and nursing has long been known to be an important, if only partly effective, method of birth control. Furthermore, almost 50% of the cycles in the first 6 mo after resumption of menses are anovulatory (see Clinical Box 25–6).

Gynecomastia

Breast development in the male is called **gynecomastia.** It may be unilateral but is more commonly bilateral. It is common, occurring in about 75% of newborns because of transplacental passage of maternal estrogens. It also occurs in mild, transient form in 70% of normal boys at the time of puberty and in many men over the age of 50. It occurs in androgen resistance. It is a complication of estrogen therapy and is seen in patients with estrogen-secreting tumors. It is found in a wide variety of seemingly unrelated conditions, including eunuchoidism, hyperthyroidism, and cirrhosis of the liver. Digitalis can produce it, apparently because cardiac glycosides are weakly estrogenic. It can also be caused by many other drugs. It has been seen in malnourished prisoners of war, but only after they were liberated and eating an adequate diet. A feature common to many and perhaps all cases of gynecomastia is an increase in the plasma estrogen:androgen ratio due to either increased circulating estrogens or decreased circulating androgens.

CLINICAL BOX 25–6

Chiari–Frommel Syndrome

An interesting, although rare, condition is persistence of lactation (**galactorrhea**) and amenorrhea in women who do not nurse after delivery. This condition, called the **Chiari–Frommel syndrome,** may be associated with some genital atrophy and is due to persistent prolactin secretion without the secretion of the FSH and LH necessary to produce maturation of new follicles and ovulation. A similar pattern of galactorrhea and amenorrhea with high circulating prolactin levels is seen in nonpregnant women with chromophobe pituitary tumors and in women in whom the pituitary stalk has been sectioned during treatment of cancer.

HORMONES & CANCER

About 35% of carcinomas of the breast in women of childbearing age are **estrogen-dependent;** their continued growth depends on the presence of estrogens in the circulation. The tumors are not cured by decreasing estrogen secretion, but symptoms are dramatically relieved, and the tumor regresses for months or years before recurring. Women with estrogen-dependent tumors often have a remission when their ovaries are removed. Inhibition of the action of estrogens with **tamoxifen** also produces remissions, and inhibition of estrogen formation with drugs that inhibit **aromatase** (Figure 25–26) is even more effective.

Some carcinomas of the prostate are **androgen-dependent** and regress temporarily after the removal of the testes or treatment with GnRH agonists in doses that are sufficient to produce down-regulation of the GnRH receptors on gonadotropes and decrease LH secretion.

CHAPTER SUMMARY

- Differences between males and females depend primarily on a single chromosome (the Y chromosome) and a single pair of endocrine structures (the gonads); testes in the male and ovaries in the female.

- The gonads have a dual function: the production of germ cells (gametogenesis) and the secretion of sex hormones. The testes secrete large amounts of androgens, principally testosterone, but they also secrete small amounts of estrogens. The ovaries secrete large amounts of estrogens and small amounts of androgens.

- Spermatogonia develop into mature spermatozoa that start in the seminiferous tubules in a process called spermatogenesis. This is a multistep process that includes maturation of spermatogonia into primary spermatocytes, which undergo meiotic division, resulting in haploid secondary spermatocytes and several further divisions result in spermatids. Each cell division from a spermatogonium to a spermatid is incomplete with cells remaining connected via cytoplasmic bridges. Spermatids eventually mature into motile spermatozoa to

■ complete spermatogenesis; this last part of maturation is called spermiogenesis.

■ Testosterone is the principal hormone of the testis. It is synthesized from cholesterol in Leydig cells. The secretion of testosterone from Leydig cells is under control of luteinizing hormone at a rate of 4 to 9 mg/day in adult males. Most testosterone is bound to albumin or to gonadal steroid-binding globulin in the plasma. Testosterone plays an important role in the development and maintenance of male secondary sex characteristics, as well as other defined functions.

■ Ovaries also secrete progesterone, a steroid that has special functions in preparing the uterus for pregnancy. During pregnancy the ovaries secrete relaxin, which facilitates the delivery of the fetus. In both sexes, the gonads secrete other polypeptides, including inhibin B, a polypeptide that inhibits FSH secretion.

■ In women, a period called perimenopause precedes menopause, and can last up to ten years; during this time the menstrual cycles become irregular and the level of inhibins decrease.

■ Once in menopause, the ovaries no longer secrete progesterone and 17β-estradiol and estrogen is formed only in small amounts by aromatization of androstenedione in peripheral tissues.

■ The naturally occurring estrogens are **17β-estradiol, estrone,** and **estriol.** They are secreted primarily by the granulosa cells of the ovarian follicles, the corpus luteum, and the placenta. Their biosynthesis depends on the enzyme **aromatase** (CYP19), which converts testosterone to estradiol and androstenedione to estrone. The latter reaction also occurs in fat, liver, muscle, and the brain.

MULTIPLE-CHOICE QUESTIONS

For all questions, select the single best answer unless otherwise directed.

1. If a young woman has high plasma levels of T_3, cortisol, and renin activity but her blood pressure is only slightly elevated and she has no symptoms or signs of thyrotoxicosis or Cushing syndrome, the most likely explanation is that
   A) she has been treated with TSH and ACTH.
   B) she has been treated with T_3 and cortisol.
   C) she is in the third trimester of pregnancy.
   D) she has an adrenocortical tumor.
   E) she has been subjected to chronic stress.

2. Full development and function of the seminiferous tubules require
   A) somatostatin.
   B) LH.
   C) oxytocin.
   D) FSH.
   E) androgens and FSH.

3. In humans, fertilization usually occurs in the
   A) vagina.
   B) cervix.
   C) uterine cavity.
   D) uterine tubes.
   E) abdominal cavity.

4. In human males, testosterone is produced mainly by the
   A) Leydig cells.
   B) Sertoli cells.
   C) seminiferous tubules.
   D) epididymis.
   E) vas deferens.

5. Home-use kits for determining a woman's fertile period depend on the detection of one hormone in the urine. This hormone is
   A) FSH.
   B) progesterone.
   C) estradiol.
   D) hCG.
   E) LH.

6. Which of the following is *not* a steroid?
   A) 17α-hydroxyprogesterone
   B) estrone
   C) relaxin
   D) pregnenolone
   E) etiocholanolone

7. Which of the following probably triggers the onset of labor?
   A) ACTH in the fetus
   B) ACTH in the mother
   C) prostaglandins
   D) oxytocin
   E) placental renin

CHAPTER RESOURCES

Bole-Feysot C et al: Prolactin (PRL) and its receptor: Actions, signal transduction pathways, and phenotypes observed in PRL receptor knockout mice. Endocrinol Rev 1998;19:225.

Mather JP, Moore A, Li R-H: Activins, inhibins, and follistatins: Further thoughts on a growing family of regulators. Proc Soc Exper Biol Med 1997;215:209.

Matthews J, Gustafson J-A: Estrogen signaling: A subtle balance between ER_α and ER_β. Mol Interv 2003;3:281.

McLaughlin DT, Donahoe PR: Sex determination and differentiation. N Engl J Med 2004;350:367.

Naz RK (editor): *Endocrine Disruptors.* CRC Press, 1998.

Norwitz ER, Robinson JN, Challis JRG: The control of labor. N Engl J Med 1999;341:660.

Primakoff P, Nyles DG: Penetration, adhesion, and fusion in mammalian sperm–egg interaction. Science 2002;296:2183.

Simpson ER, et al: Aromatose—A brief overview. Annu Rev Physiol 2002;64:93.

Yen SSC, Jaffe RB, Barbieri RL: *Reproductive Endocrinology: Physiology, Pathophysiology, and Clinical Management,* 4th ed. Saunders, 1999.

Overview of Gastrointestinal Function & Regulation

OBJECTIVES

After studying this chapter, you should be able to:

■ Understand the functional significance of the gastrointestinal system, and in particular, its roles in nutrient assimilation, excretion, and immunity.

■ Describe the structure of the gastrointestinal tract, the glands that drain into it, and its subdivision into functional segments.

■ List the major gastrointestinal secretions, their components, and the stimuli that regulate their production.

■ Describe water balance in the gastrointestinal tract and explain how the level of luminal fluidity is adjusted to allow for digestion and absorption.

■ Identify the major hormones, other peptides, and key neurotransmitters of the gastrointestinal system.

■ Describe the special features of the enteric nervous system and the splanchnic circulation.

INTRODUCTION

The gastrointestinal tract is a continuous tube that stretches from the mouth to the anus. Its primary function is to serve as a portal whereby nutrients and water can be absorbed into the body. In fulfilling this function, the meal is mixed with a variety of secretions that arise from both the gastrointestinal tract itself and organs that drain into it, such as the pancreas, gallbladder, and salivary glands. Likewise, the intestine displays a variety of motility patterns that serve to mix the meal with digestive secretions and move it along the length of the gastrointestinal tract. Ultimately, residues of the meal that cannot be absorbed, along with cellular debris and lipid-soluble metabolic end products that are excreted in the bile rather than

the urine, are expelled from the body. All of these functions are tightly regulated in concert with the ingestion of meals. Thus, the gastrointestinal system has evolved a large number of regulatory mechanisms that act both locally and to coordinate the function of the gut, and the organs that drain into it, over long distances.

The lumen of the gastrointestinal tract is functionally contiguous with the outside of the body. The intestine also has a very substantial surface area, which is important for its absorptive function. Finally, the gut is an unusual organ in that it becomes colonized, almost from birth, with a large number of commensal bacteria (particularly in the colon, or large intestine). This relationship is mutually beneficial, because the bacteria perform several metabolic functions that cannot be accomplished with mammalian enzymes, and likely also provide some degree of protection against subsequent infection with pathogenic microorganisms that might cause disease. Nevertheless, the constant presence of bacterial and other stimuli, as well as the large surface area that must be defended against potentially harmful substances, doubtlessly accounts for the fact that the intestine has a very well-developed local immune system that comprises both innate and adaptive immune effectors (see Chapter 3). Indeed, there are more lymphocytes in the wall of the intestine than there are circulating in the blood.

STRUCTURAL CONSIDERATIONS

The parts of the gastrointestinal tract that are encountered by the meal or its residues include, in order, the mouth, esophagus, stomach, duodenum, jejunum, ileum, cecum, colon, rectum, and anus. Throughout the length of the intestine, glandular structures deliver secretions into the lumen, particularly in the stomach and mouth. Also important in the process of digestion are secretions from the pancreas and the biliary system of the liver. The intestinal tract is also functionally divided into segments that restrict the flow of intestinal contents to optimize digestion and absorption. These sphincters include the upper and lower esophageal sphincters, the pylorus that retards emptying of the stomach, the ileocecal valve that retains colonic contents (including large numbers of bacteria) in the large intestine, and the inner and outer anal sphincters. After toilet training, the latter permit delaying the elimination of wastes until a time when it is socially convenient.

The intestine is composed of functional layers (Figure 26–1). Immediately adjacent to nutrients in the lumen is a single layer of columnar epithelial cells. This represents the barrier that nutrients must traverse to enter the body. Below the epithelium is a layer of loose connective tissue known as the lamina propria, which in turn is surrounded by concentric layers of smooth muscle, oriented circumferentially and then longitudinally to the axis of the gut (the circular and longitudinal muscle layers, respectively). The intestine is also amply supplied with blood vessels, nerve endings, and lymphatics, which are all important in its function.

The epithelium of the intestine is also further specialized in a way that maximizes the surface area available for nutrient absorption. Throughout the small intestine, it is folded up into fingerlike projections called villi (Figure 26–2). Between the villi are infoldings known as crypts. Stem cells that give rise to both crypt and villus epithelial cells reside toward the base of the crypts and are responsible for completely renewing the epithelium every few days or so. Indeed, the gastrointestinal epithelium is one of the most rapidly dividing tissues in the body. Daughter cells undergo several rounds of cell division in the crypts then migrate out onto the villi, where they are eventually shed and lost in the stool. The villus epithelial cells are also notable for the extensive microvilli that characterize their apical membranes. These microvilli are endowed with a dense glycocalyx (the brush border) that probably

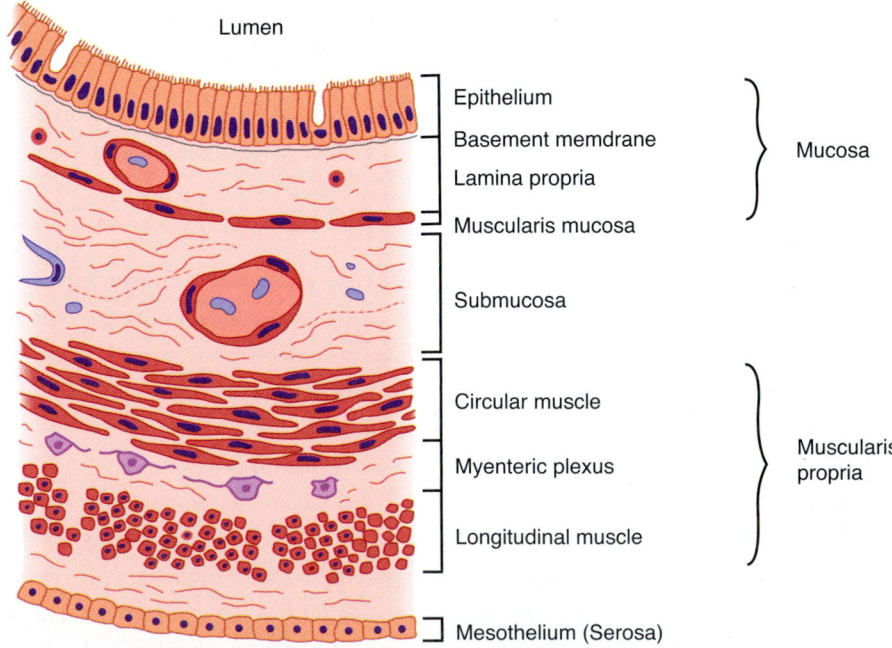

FIGURE 26–1   **Organization of the wall of the intestine into functional layers.** (Adapted from Barrett KE: *Gastrointestinal Physiology*. McGraw-Hill, 2006.)

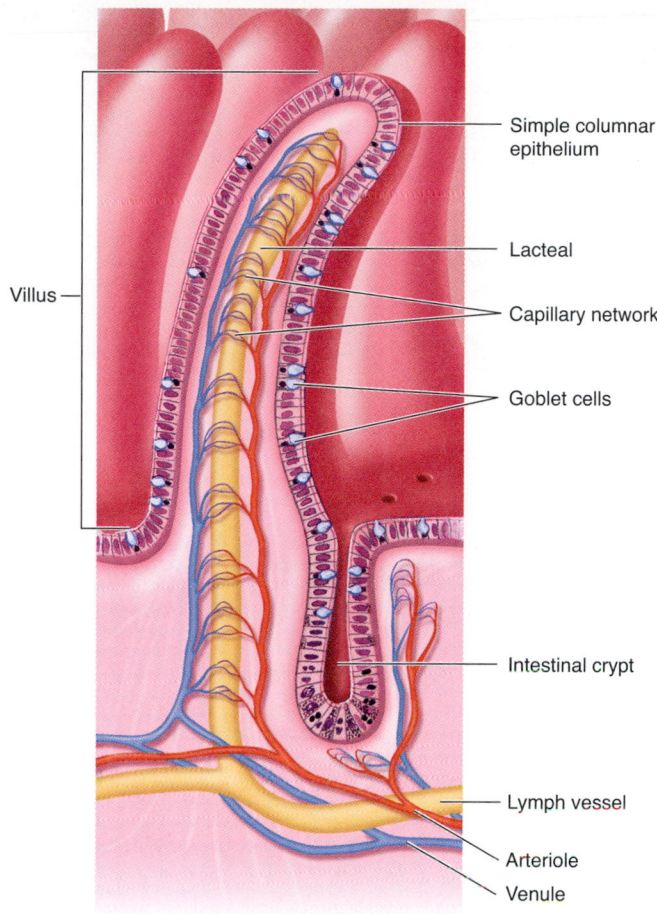

FIGURE 26–2 **The structure of intestinal villi and crypts.**
(Reproduced with permission, from Fox SI: *Human Physiology*, 10th ed. McGraw-Hill, 2008.)

Labels: Simple columnar epithelium · Lacteal · Capillary network · Goblet cells · Intestinal crypt · Lymph vessel · Arteriole · Venule · Villus

protects the cells to some extent from the effects of digestive enzymes. Some digestive enzymes are also actually part of the brush border, being membrane-bound proteins. These so-called "brush border hydrolases" perform the final steps of digestion for specific nutrients.

GASTROINTESTINAL SECRETIONS

SALIVARY SECRETION

The first secretion encountered when food is ingested is saliva. Saliva is produced by three pairs of salivary glands that drain into the oral cavity. It has a number of organic constituents that serve to initiate digestion (particularly of starch, mediated by amylase) and which also protect the oral cavity from bacteria (such as immunoglobulin A and lysozyme). Saliva also serves to lubricate the food bolus (aided by mucins). Saliva is also hypotonic compared with plasma and alkaline; the latter feature is important to neutralize any gastric secretions that reflux into the esophagus.

The salivary glands consist of blind end pieces (acini) that produce the primary secretion containing the organic constituents dissolved in a fluid that is essentially identical in its composition to plasma. The salivary glands are actually extremely active when maximally stimulated, secreting their own weight in saliva every minute. To accomplish this, they are richly endowed with surrounding blood vessels that dilate when salivary secretion is initiated. The composition of the saliva is then modified as it flows from the acini out into ducts that eventually coalesce and deliver the saliva into the mouth. Na^+ and Cl^- are extracted and K^+ and bicarbonate are added. Because the ducts are relatively impermeable to water, the loss of NaCl renders the saliva hypotonic, particularly at low secretion rates. As the rate of secretion increases, there is less time for NaCl to be extracted and the tonicity of the saliva rises, but it always stays somewhat hypotonic with respect to plasma. Overall, the three pairs of salivary glands that drain into the mouth supply 1000 to 1500 mL of saliva per day.

Salivary secretion is almost entirely controlled by neural influences, with the parasympathetic branch of the autonomic nervous system playing the most prominent role (Figure 26–3). Sympathetic input slightly modifies the composition of saliva (particularly by increasing proteinaceous content), but has little influence on volume. Secretion is triggered by reflexes that are stimulated by the physical act of chewing, but is actually initiated even before the meal is taken into the mouth as a result of central triggers that are prompted by thinking about, seeing, or smelling food. Indeed, salivary secretion can readily be conditioned, as in the classical experiments of Pavlov where dogs were conditioned to salivate in response to a ringing bell by associating this stimulus with a meal. Salivary secretion is also prompted by nausea, but inhibited by fear or during sleep.

Saliva performs a number of important functions: it facilitates swallowing, keeps the mouth moist, serves as a solvent for the molecules that stimulate the taste buds, aids speech by facilitating movements of the lips and tongue, and keeps the mouth and teeth clean. The saliva also has some antibacterial action, and patients with deficient salivation (**xerostomia**) have a higher than normal incidence of dental caries. The buffers in saliva help maintain the oral pH at about 7.0. They also help neutralize gastric acid and relieve heartburn when gastric juice is regurgitated into the esophagus.

GASTRIC SECRETION

Food is stored in the stomach; mixed with acid, mucus, and pepsin; and released at a controlled, steady rate into the duodenum (see Clinical Box 26–1).

ANATOMIC CONSIDERATIONS

The gross anatomy of the stomach is shown in Figure 26–4. The gastric mucosa contains many deep glands. In the cardia and the pyloric region, the glands secrete mucus. In the body of the stomach, including the fundus, the glands also contain **parietal (oxyntic) cells,** which secrete hydrochloric acid and intrinsic factor, and **chief (zymogen, peptic) cells,** which secrete pepsinogens (Figure 26–5) These secretions mix with

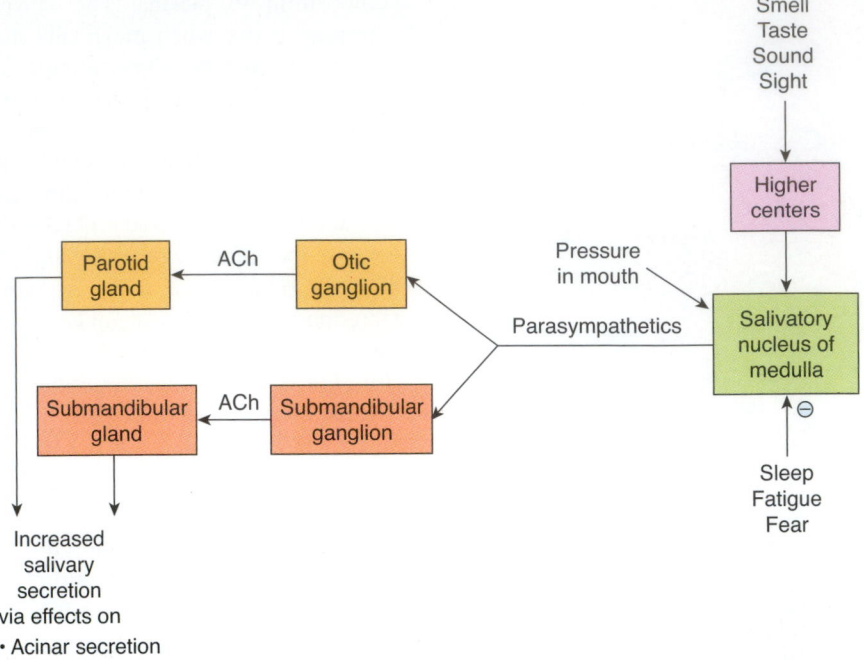

FIGURE 26–3   **Regulation of salivary secretion by the parasympathetic nervous system.** ACh, acetylcholine. (Adapted from Barrett KE: *Gastrointestinal Physiology*. McGraw-Hill, 2006.)

mucus secreted by the cells in the necks of the glands. Several of the glands open on a common chamber (**gastric pit**) that opens in turn on the surface of the mucosa. Mucus is also secreted along with HCO_3^- by mucus cells on the surface of the epithelium between glands.

The stomach has a very rich blood and lymphatic supply. Its parasympathetic nerve supply comes from the vagi and its sympathetic supply from the celiac plexus.

ORIGIN & REGULATION OF GASTRIC SECRETION

The stomach also adds a significant volume of digestive juices to the meal. Like salivary secretion, the stomach actually readies itself to receive the meal before it is actually taken in, during the so-called cephalic phase that can be influenced by food preferences. Subsequently, there is a gastric phase of secretion that is quantitatively the most significant, and finally an intestinal phase once the meal has left the stomach. Each phase is closely regulated by both local and distant triggers.

The gastric secretions (Table 26–1) arise from glands in the wall of the stomach that drain into its lumen, and also from the surface cells that secrete primarily mucus and bicarbonate to protect the stomach from digesting itself, as well as substances known as trefoil peptides that stabilize the mucus-bicarbonate layer. The glandular secretions of the stomach differ in different regions of the organ. The most characteristic secretions derive from the glands in the fundus or body of the stomach. These contain two distinctive cell types from

which the gastric secretions arise: the parietal cells, which secrete hydrochloric acid and intrinsic factor; and the chief cells, which produce pepsinogens and gastric lipase (Figure 26–5). The acid secreted by parietal cells serves to sterilize the meal and also to begin the hydrolysis of dietary macromolecules. Intrinsic factor is important for the later absorption of vitamin B_{12}, or cobalamin (Figure 26–6). Pepsinogen is the precursor of pepsin, which initiates protein digestion. Lipase similarly begins the digestion of dietary fats.

There are three primary stimuli of gastric secretion, each with a specific role to play in matching the rate of secretion to functional requirements (Figure 26–7). Gastrin is a hormone that is released by G cells in the antrum of the stomach both in response to a specific neurotransmitter released from enteric nerve endings, known as gastrin releasing peptide (GRP, or bombesin), and also in response to the presence of oligopeptides in the gastric lumen. Gastrin is then carried through the bloodstream to the fundic glands, where it binds to receptors not only on parietal (and likely, chief cells) to activate secretion, but also on so-called enterochromaffin-like cells (ECL cells) that are located in the gland, and release histamine. Histamine is also a trigger of parietal cell secretion, via binding to H_2 histamine receptors. Finally, parietal and chief cells can also be stimulated by acetylcholine, released from enteric nerve endings in the fundus.

During the cephalic phase of gastric secretion, secretion is predominantly activated by vagal input that originates from the brain region known as the dorsal vagal complex, which coordinates input from higher centers. Vagal outflow to the stomach then releases GRP and acetylcholine, thereby initiating secretory

CLINICAL BOX 26–1

Peptic Ulcer Disease

Gastric and duodenal ulceration in humans is related primarily to a breakdown of the barrier that normally prevents irritation and autodigestion of the mucosa by the gastric secretions. Infection with the bacterium *Helicobacter pylori* disrupts this barrier, as do aspirin and other nonsteroidal anti-inflammatory drugs (NSAIDs), which inhibit the production of prostaglandins and consequently decrease mucus and HCO_3^- secretion. The NSAIDs are widely used to combat pain and treat arthritis. An additional cause of ulceration is prolonged excess secretion of acid. An example of this is the ulcers that occur in the **Zollinger–Ellison syndrome.** This syndrome is seen in patients with gastrinomas. These tumors can occur in the stomach and duodenum, but most of them are found in the pancreas. The gastrin causes prolonged hypersecretion of acid, and severe ulcers are produced. Gastric and duodenal ulcers can be given a chance to heal by inhibition of acid secretion with drugs such as cimetidine that block the H_2 histamine receptors on parietal cells or omeprazole and related drugs that inhibit H^+–K^+ ATPase. *H. pylori* can be eradicated with antibiotics, and NSAID-induced ulcers can be treated by stopping the NSAID or, when this is not advisable, by treatment with the prostaglandin agonist misoprostol. Gastrinomas can sometimes be removed surgically.

function. However, before the meal enters the stomach, there are few additional triggers and thus the amount of secretion is limited. Once the meal is swallowed, on the other hand, meal constituents trigger substantial release of gastrin and the physical

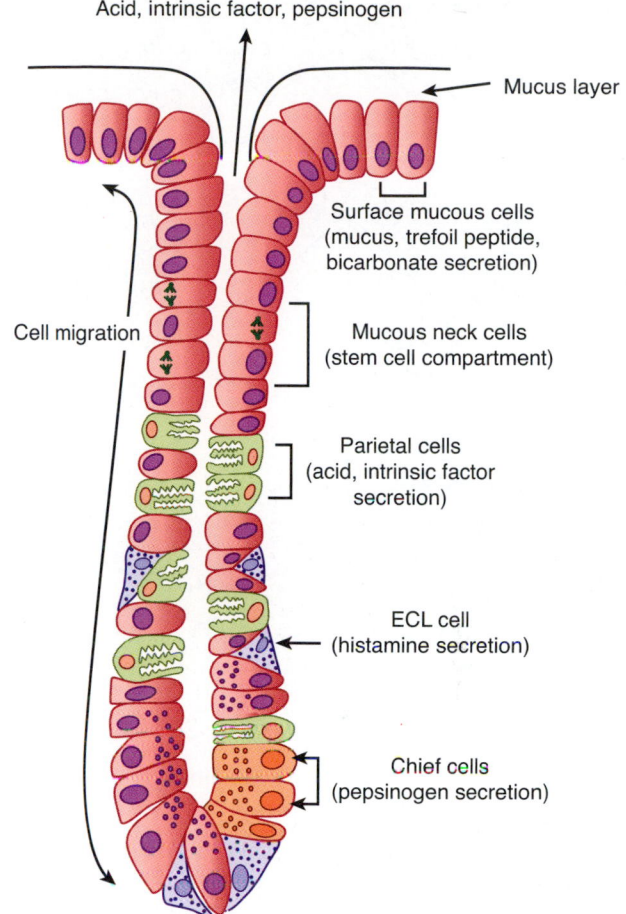

FIGURE 26–5 **Structure of a gastric gland from the fundus and body of the stomach.** These acid- and pepsinogen-producing glands are referred to as "oxyntic" glands in some sources. (Adapted from Barrett KE: *Gastrointestinal Physiology.* McGraw-Hill, 2006.)

presence of the meal also distends the stomach and activates stretch receptors, which provoke a "vago-vagal" as well as local reflexes that further amplify secretion. The presence of the meal also buffers gastric acidity that would otherwise serve as a feedback inhibitory signal to shut off secretion secondary to the release of somatostatin, which inhibits both G and ECL cells as well as secretion by parietal cells themselves (Figure 26–7). This probably represents a key mechanism whereby gastric secretion

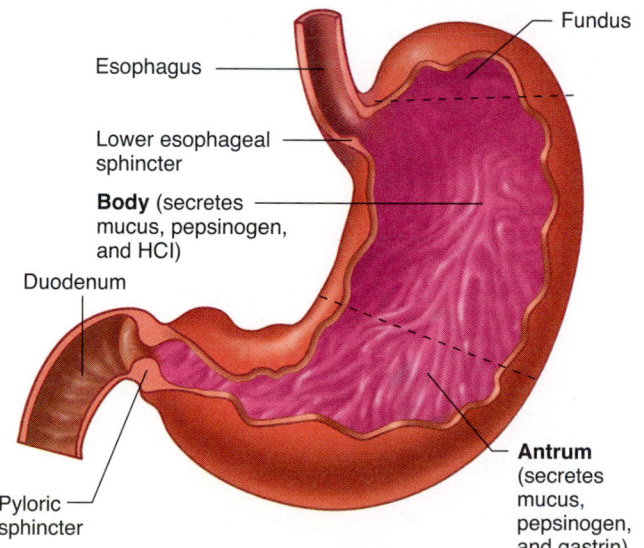

FIGURE 26–4 **Anatomy of the stomach.** The principal secretions of the body and antrum are listed in parentheses. (Reproduced with permission from Widmaier EP, Raff H, Strang KT: *Vander's Human Physiology: The Mechanisms of Body Function,* 11th ed. McGraw-Hill, 2008.)

TABLE 26–1 Contents of normal gastric juice (fasting state).

Cations: Na^+, K^+, Mg^{2+}, H^+ (pH approximately 1.0)
Anions: Cl^-, HPO_4^{2-}, SO_4^{2-}
Pepsins
Lipase
Mucus
Intrinsic factor

FIGURE 26–6 **Cyanocobalamin (vitamin B$_{12}$).**

is terminated after the meal moves from the stomach into the small intestine.

Gastric parietal cells are highly specialized for their unusual task of secreting concentrated acid (Figure 26–8). The cells are packed with mitochondria that supply energy to drive the apical H,K-ATPase, or proton pump, that moves H$^+$ ions out of the parietal cell against a concentration gradient of more than a

million-fold. At rest, the proton pumps are sequestered within the parietal cell in a series of membrane compartments known as tubulovesicles. When the parietal cell begins to secrete, on the other hand, these vesicles fuse with invaginations of the apical membrane known as canaliculi, thereby substantially amplifying the apical membrane area and positioning the proton pumps to begin acid secretion (Figure 26–9). The apical membrane also contains potassium channels, which supply the K$^+$ ions to be exchanged for H$^+$, and Cl$^-$ channels that supply the counterion for HCl secretion (Figure 26–10). The secretion of protons is also accompanied by the release of equivalent numbers of bicarbonate ions into the bloodstream, which as we will see, are later used to neutralize gastric acidity once its function is complete (Figure 26–10).

The three agonists of the parietal cell—gastrin, histamine, and acetylcholine—each bind to distinct receptors on the basolateral membrane (Figure 26–9). Gastrin and acetylcholine promote secretion by elevating cytosolic free calcium concentrations, whereas histamine increases intracellular cyclic adenosine 3',5'-monophosphate (cAMP). The net effect of these second messengers are the transport and morphological changes described above. However, it is important to be aware that the two distinct pathways for activation are synergistic, with a greater than additive effect on secretion rates when histamine plus gastrin or acetylcholine, or all three, are present simultaneously. The physiologic significance of this synergism is that high rates of secretion can be stimulated with relatively small changes in availability of each of the stimuli. Synergism is also therapeutically significant because secretion can be markedly inhibited by blocking the action of only one of the triggers (most commonly that of histamine, via H$_2$ histamine antagonists that are widely used therapies for adverse effects of excessive gastric secretion, such as reflux).

FIGURE 26–7 **Regulation of gastric acid and pepsin secretion by soluble mediators and neural input.** Gastrin is released from G cells in the antrum and travels through the circulation to influence the activity of ECL cells and parietal cells. The specific agonists of the chief cell are not well understood. Gastrin release is negatively regulated by luminal acidity via the release of somatostatin from antral D cells. (Adapted from Barrett KE: *Gastrointestinal Physiology*. McGraw-Hill, 2006.)

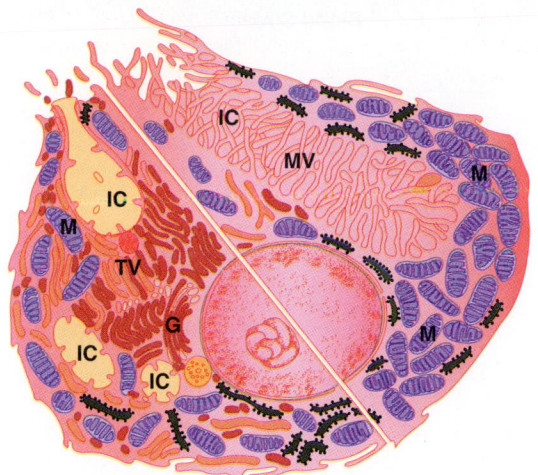

FIGURE 26–8 **Composite diagram of a parietal cell, showing the resting state (lower left) and the active state (upper right).** The resting cell has intracellular canaliculi (IC), which open on the apical membrane of the cell, and many tubulovesicular structures (TV) in the cytoplasm. When the cell is activated, the TVs fuse with the cell membrane and microvilli (MV) project into the canaliculi, so the area of cell membrane in contact with gastric lumen is greatly increased. M, mitochondrion; G, Golgi apparatus. (Adapted from Junqueira LC, Carneiro J: *Basic Histology: Text & Atlas,* 10th ed. McGraw-Hill, 2003.)

Gastric secretion adds about 2.5 L per day to the intestinal contents. However, despite their substantial volume and fine control, gastric secretions are dispensable for the full digestion and absorption of a meal, with the exception of cobalamin absorption. This illustrates an important facet of gastrointestinal physiology, that digestive and absorptive capacity are markedly in excess of normal requirements. On the other hand, if gastric secretion is chronically reduced, individuals may display increased susceptibility to infections acquired via the oral route.

PANCREATIC SECRETION

The pancreatic juice contains enzymes that are of major importance in digestion (see Table 26–2). Its secretion is controlled in part by a reflex mechanism and in part by the gastrointestinal hormones secretin and cholecystokinin (CCK).

ANATOMIC CONSIDERATIONS

The portion of the pancreas that secretes pancreatic juice is a compound alveolar gland resembling the salivary glands. Granules containing the digestive enzymes (**zymogen granules**) are formed in the cell and discharged by exocytosis (see Chapter 2) from the apexes of the cells into the lumens of the pancreatic ducts (Figure 26–11). The small duct radicles coalesce into a single duct (pancreatic duct of Wirsung), which usually joins the common bile duct to form the ampulla of Vater (Figure 26–12). The ampulla opens through the duodenal papilla, and its orifice is encircled by the sphincter of Oddi. Some individuals have an accessory pancreatic duct (duct of Santorini) that enters the duodenum more proximally.

COMPOSITION OF PANCREATIC JUICE

The pancreatic juice is alkaline (Table 26–3) and has a high HCO_3^- content (approximately 113 mEq/L vs. 24 mEq/L in plasma). About 1500 mL of pancreatic juice is secreted per day. Bile and intestinal juices are also neutral or alkaline, and these three secretions neutralize the gastric acid, raising the pH of the duodenal contents to 6.0 to 7.0. By the time the chyme reaches the jejunum, its pH is nearly neutral, but the intestinal contents are rarely alkaline.

FIGURE 26–9 **Parietal cell receptors and schematic representation of the morphological changes depicted in Figure 26–7.** Amplification of the apical surface area is accompanied by an increased density of H^+, K^+–ATPase molecules at this site. Note that acetylcholine (ACh) and gastrin signal via calcium, whereas histamine signals via cAMP. (Adapted from Barrett KE: *Gastrointestinal Physiology.* McGraw-Hill, 2006.)

FIGURE 26–10   Ion transport proteins of parietal cells. Protons are generated in the cytoplasm via the action of carbonic anhydrase II (C.A. II). Bicarbonate ions are exported from the basolateral pole of the cell either by vesicular fusion or via a chloride/bicarbonate exchanger. (Adapted from Barrett KE: *Gastrointestinal Physiology*. McGraw-Hill, 2006.)

The potential danger of the release into the pancreas of a small amount of trypsin is apparent; the resulting chain reaction would produce active enzymes that could digest the pancreas. It is therefore not surprising that the pancreas normally contains a trypsin inhibitor.

Another enzyme activated by trypsin is phospholipase A_2. This enzyme splits a fatty acid off phosphatidylcholine (PC), forming lyso-PC. Lyso-PC damages cell membranes. It has been hypothesized that in **acute pancreatitis,** a severe and sometimes fatal disease, phospholipase A_2 is activated in the pancreatic ducts, with the formation of lyso-PC from the PC that is a normal constituent of bile. This causes disruption of pancreatic tissue and necrosis of surrounding fat.

Small amounts of pancreatic digestive enzymes normally leak into the circulation, but in acute pancreatitis, the circulating levels of the digestive enzymes rise markedly. Measurement of the plasma amylase or lipase concentration is therefore of value in diagnosing the disease.

REGULATION OF THE SECRETION OF PANCREATIC JUICE

Secretion of pancreatic juice is primarily under hormonal control. Secretin acts on the pancreatic ducts to cause copious secretion of a very alkaline pancreatic juice that is rich in HCO_3^- and poor in enzymes. The effect on duct cells is due to an increase in intracellular cAMP. Secretin also stimulates bile secretion. CCK acts on the acinar cells to cause the release of zymogen granules and production of pancreatic juice rich in enzymes but low in volume. Its effect is mediated by phospholipase C (see Chapter 2).

The response to intravenous secretin is shown in Figure 26–13. Note that as the volume of pancreatic secretion increases, its Cl^- concentration falls and its HCO_3^- concentration increases. Although HCO_3^- is secreted in the small ducts, it is reabsorbed in the large ducts in exchange for Cl^- (Figure 26–14). The magnitude of the exchange is inversely proportionate to the rate of flow.

Like CCK, acetylcholine acts on acinar cells via phospholipase C to cause discharge of zymogen granules, and stimulation of the vagi causes secretion of a small amount of pancreatic juice rich in enzymes. There is evidence for vagally mediated conditioned reflex secretion of pancreatic juice in response to the sight or smell of food.

BILIARY SECRETION

An additional secretion important for gastrointestinal function, bile, arises from the liver. The bile acids contained therein are important in the digestion and absorption of fats. In addition, bile serves as a critical excretory fluid by which the body disposes of lipid soluble end products of metabolism as well as lipid soluble xenobiotics. Bile is also the only route by which the body can dispose of cholesterol—either in its native form, or following conversion to bile acids. In this chapter and the next, we will be concerned with the role of bile as a digestive fluid. In Chapter 29, a more general consideration of the transport and metabolic functions of the liver will be presented.

TABLE 26–2 Principal digestive enzymes.*

Source	Enzyme	Activator	Substrate	Catalytic Function or Products
Salivary glands	Salivary α-amylase	Cl⁻	Starch	Hydrolyzes 1:4α linkages, producing α-limit dextrins, maltotriose, and maltose
Lingual glands	Lingual lipase		Triglycerides	Fatty acids plus 1,2-diacylglycerols
Stomach	Pepsins (pepsinogens)	HCl	Proteins and polypeptides	Cleave peptide bonds adjacent to aromatic amino acids
	Gastric lipase		Triglycerides	Fatty acids and glycerol
Exocrine pancreas	Trypsin (trypsinogen)	Enteropeptidase	Proteins and polypeptides	Cleave peptide bonds on carboxyl side of basic amino acids (arginine or lysine)
	Chymotrypsins (chymotrypsinogens)	Trypsin	Proteins and polypeptides	Cleave peptide bonds on carboxyl side of aromatic amino acids
	Elastase (proelastase)	Trypsin	Elastin, some other proteins	Cleaves bonds on carboxyl side of aliphatic amino acids
	Carboxypeptidase A (procarboxypeptidase A)	Trypsin	Proteins and polypeptides	Cleave carboxyl terminal amino acids that have aromatic or branched aliphatic side chains
	Carboxypeptidase B (procarboxypeptidase B)	Trypsin	Proteins and polypeptides	Cleave carboxyl terminal amino acids that have basic side chains
	Colipase (procolipase)	Trypsin	Fat droplets	Facilitates exposure of active site of pancreaticlipase
	Pancreatic lipase	...	Triglycerides	Monoglycerides and fatty acids
	Bile salt-acid lipase		Cholesteryl esters	Cholesterol
	Cholesteryl ester hydrolase	...	Cholesteryl esters	Cholesterol
	Pancreatic α-amylase	Cl⁻	Starch	Same as salivary α-amylase
	Ribonuclease	...	RNA	Nucleotides
	Deoxyribonuclease	...	DNA	Nucleotides
	Phospholipase A_2 (pro-phospholipase A^2)	Trypsin	Phospholipids	Fatty acids, lysophospholipids
Intestinal mucosa	Enteropeptidase	...	Trypsinogen	Trypsin
	Aminopeptidases	...	Polypeptides	Cleave amino terminal amino acid from peptide
	Carboxypeptidases	...	Polypeptides	Cleave carboxyl terminal amino acid from peptide
	Endopeptidases	...	Polypeptides	Cleave between residues in midportion of peptide
	Dipeptidases	...	Dipeptides	Two amino acids
	Maltase	...	Maltose, maltotriose, α-dextrins	Glucose
	Lactase	...	Lactose	Galactose and glucose
	Sucrase[a]	...	Sucrose; also maltotriose and maltose	Fructose and glucose
	α-Dextrinase[a]	...	α-Dextrins, maltose maltotriose	Glucose
	Trehalase	...	Trehalose	Glucose
	Nuclease and related enzymes	...	Nucleic acids	Pentoses and purine and pyrimidine bases
Cytoplasm of mucosal cells	Various peptidases	...	Di-, tri-, and tetrapeptides	Amino acids

*Corresponding proenzymes, where relevant, are shown in parentheses

[a]Sucrase and a-dextrinase are separate subunits of a single protein.

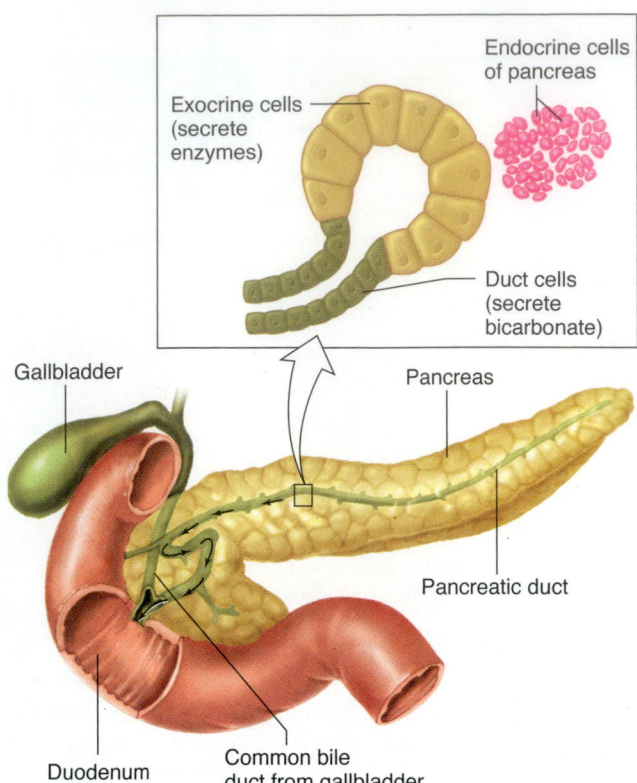

FIGURE 26–11   **Structure of the pancreas.** (Reproduced with permission from Widmaier EP, Raff H, Strang KT: *Vander's Human Physiology: The Mechanisms of Body Function*, 11th ed. McGraw-Hill, 2008.)

BILE

Bile is made up of the bile acids, bile pigments, and other substances dissolved in an alkaline electrolyte solution that resembles pancreatic juice (Table 26–4). About 500 mL is secreted per day. Some of the components of the bile are reab-

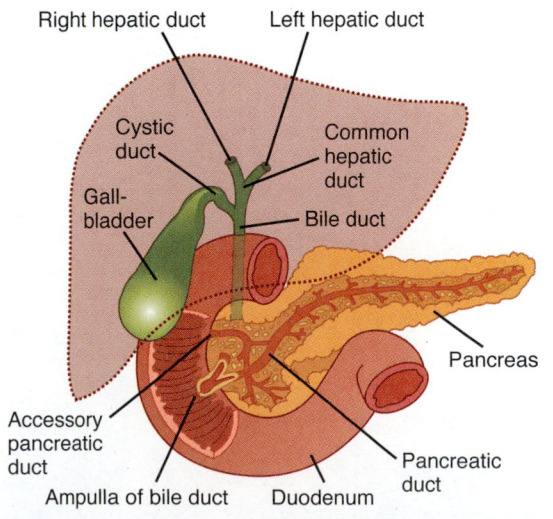

FIGURE 26–12   **Connections of the ducts of the gallbladder, liver, and pancreas.** (Adapted from Bell GH, Emslie-Smith D, Paterson CR: *Textbook of Physiology and Biochemistry*, 9th ed. Churchill Livingstone, 1976.)

TABLE 26–3   Composition of normal human pancreatic juice.

Cations: Na^+, K^+, Ca^{2+}, Mg^{2+} (pH approximately 8.0)
Anions: HCO_3^-, Cl^-, SO_4^{2-}, HPO_4^{2-}
Digestive enzymes (see Table 26–1; 95% of protein in juice)
Other proteins

sorbed in the intestine and then excreted again by the liver (**enterohepatic circulation**).

The glucuronides of the **bile pigments,** bilirubin and biliverdin, are responsible for the golden yellow color of bile. The formation of these breakdown products of hemoglobin is discussed in detail in Chapter 29, and their excretion is discussed below.

The **bile acids** secreted into the bile are conjugated to glycine or taurine, a derivative of cysteine. The bile acids are synthesized from cholesterol. The four major bile acids found in humans are listed in Figure 26–15. In common with vitamin D, cholesterol, a variety of steroid hormones, and the digitalis glycosides, the bile acids contain the steroid nucleus (see Chapter 22). The two principal (primary) bile acids formed in the liver are cholic acid and chenodeoxycholic acid. In the colon, bacteria convert cholic acid to deoxycholic acid and chenodeoxycholic acid to lithocholic acid. In addition, small quantities of ursodeoxycholic acid are formed from chenodeoxycholic acid. Ursodeoxycholic acid is a tautomer of chenodeoxycholic acid at the 7-position. Because they are formed

FIGURE 26–13   **Effect of a single dose of secretin on the composition and volume of the pancreatic juice in humans.**

FIGURE 26–14 **Ion transport pathways present in pancreatic duct cells.** CA, carbonic anhydrase; NHE-1, sodium/hydrogen exchanger-1; NBC, sodium-bicarbonate cotransporter. (Adapted from Barrett KE: *Gastrointestinal Physiology*. McGraw-Hill, 2006.)

TABLE 26–4 **Composition of human hepatic duct bile.**

Water	97.0%
Bile salts	0.7%
Bile pigments	0.2%
Cholesterol	0.06%
Inorganic salts	0.7%
Fatty acids	0.15%
Phosphatidylcholine	0.2%
Fat	0.1%
Alkaline phosphatase	...

by bacterial action, deoxycholic, lithocholic, and ursodeoxycholic acids are called secondary bile acids.

The bile salts have a number of important actions: they reduce surface tension and, in conjunction with phospholipids and monoglycerides, are responsible for the emulsification of fat preparatory to its digestion and absorption in the small intestine (see Chapter 27). They are **amphipathic,** that is, they have both hydrophilic and hydrophobic domains; one surface of the molecule is hydrophilic because the polar peptide bond and the carboxyl and hydroxyl groups are on that surface, whereas the other surface is hydrophobic. Therefore, the bile salts tend to form cylindrical disks called **micelles.** A top view of micelles is shown in Figure 26–16 and a side view of one in

Cholic acid

	Group at position			Percent in human bile
	3	**7**	**12**	
Cholic acid	OH	OH	OH	50
Chenodeoxycholic acid	OH	OH	H	30
Deoxycholic acid	OH	H	OH	15
Lithocholic acid	OH	H	H	5

FIGURE 26–15 **Human bile acids.** The numbers in the formula for cholic acid refer to the positions in the steroid ring.

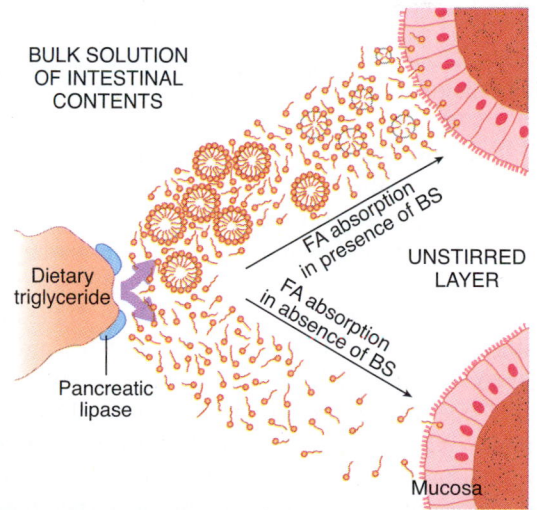

FIGURE 26–16 **Lipid digestion and passage to intestinal mucosa.** Fatty acids (FA) are liberated by the action of pancreatic lipase on dietary triglycerides and, in the presence of bile salts (BS), form micelles (the circular structures), which diffuse through the unstirred layer to the mucosal surface. (Adapted from Thomson ABR: Intestinal absorption of lipids: Influence of the unstirred water layer and bile acid micelle. In: *Disturbances in Lipid and Lipoprotein Metabolism*. Dietschy JM, Gotto AM Jr, Ontko JA [editors]: American Physiological Society, 1978.)

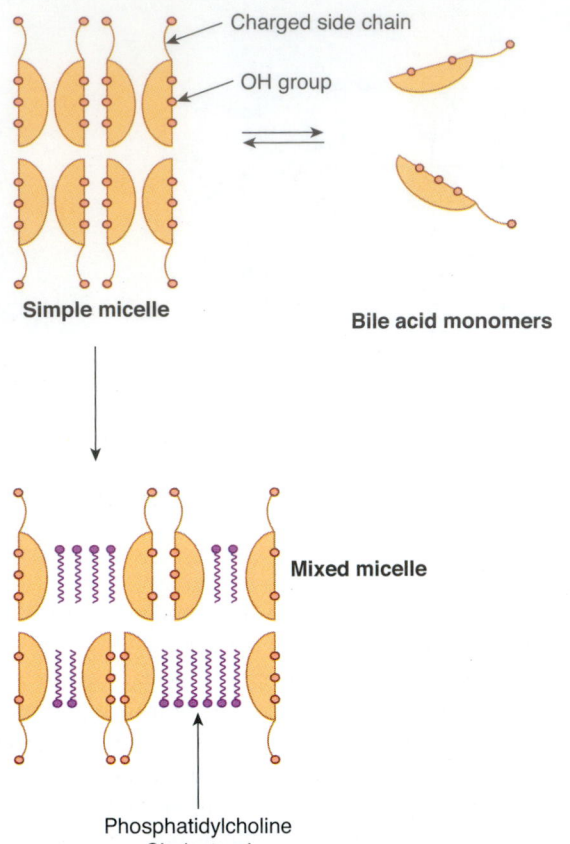

FIGURE 26–17   Physical forms adopted by bile acids in solution. Micelles are shown in cross-section, and are actually thought to be cylindrical in shape. Mixed micelles of bile acids present in hepatic bile also incorporate cholesterol and phosphatidylcholine. (Adapted from Barrett KE: *Gastrointestinal Physiology.* McGraw-Hill, 2006.)

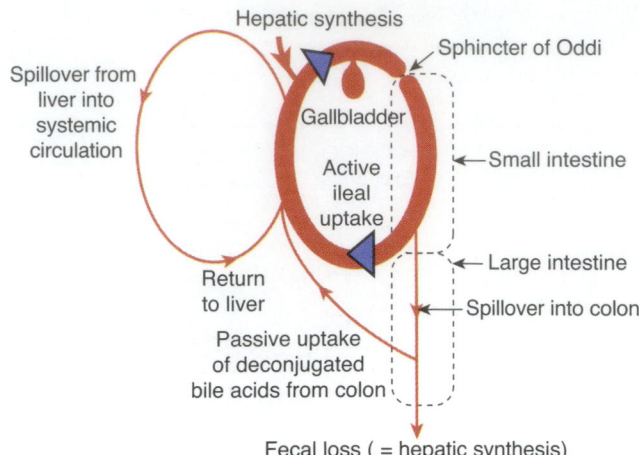

Fecal loss (= hepatic synthesis)

FIGURE 26–18   Quantitative aspects of the circulation of bile acids. The majority of the bile acid pool circulates between the small intestine and liver. A minority of the bile acid pool is in the systemic circulation (due to incomplete hepatocyte uptake from the portal blood) or spills over into the colon and is lost to the stool. Fecal loss must be equivalent to hepatic synthesis of bile acids at steady state. (Adapted from Barrett KE: *Gastrointestinal Physiology.* McGraw-Hill, 2006.)

Figure 26–17. Their hydrophilic portions face out and their hydrophobic portions face in. Above a certain concentration, called the **critical micelle concentration,** all bile salts added to a solution form micelles. Lipids collect in the micelles, with cholesterol in the hydrophobic center and amphipathic phospholipids and monoglycerides lined up with their hydrophilic heads on the outside and their hydrophobic tails in the center. The micelles play an important role in keeping lipids in solution and transporting them to the brush border of the intestinal epithelial cells, where they are absorbed (see Chapter 27).

Ninety to 95% of the bile salts are absorbed from the small intestine. Once they are deconjugated, they can be absorbed by nonionic diffusion, but most are absorbed in their conjugated forms from the terminal ileum (Figure 26–18) by an extremely efficient Na^+–bile salt cotransport system powered by basolateral Na^+–K^+ ATPase. The remaining 5–10% of the bile salts enter the colon and are converted to the salts of deoxycholic acid and lithocholic acid. Lithocholate is relatively insoluble and is mostly excreted in the stools; only 1% is absorbed. However, deoxycholate is absorbed.

The absorbed bile salts are transported back to the liver in the portal vein and reexcreted in the bile (enterohepatic circu-

lation) (Figure 26–18). Those lost in the stool are replaced by synthesis in the liver; the normal rate of bile salt synthesis is 0.2 to 0.4 g/d. The total bile salt pool of approximately 3.5 g recycles repeatedly via the enterohepatic circulation; it has been calculated that the entire pool recycles twice per meal and six to eight times per day. When bile is excluded from the intestine, up to 50% of ingested fat appears in the feces. A severe malabsorption of fat-soluble vitamins also results. When bile salt reabsorption is prevented by resection of the terminal ileum or by disease in this portion of the small intestine, the amount of fat in the stools is also increased because when the enterohepatic circulation is interrupted, the liver cannot increase the rate of bile salt production to a sufficient degree to compensate for the loss.

INTESTINAL FLUID & ELECTROLYTE TRANSPORT

The intestine itself also supplies a fluid environment in which the processes of digestion and absorption can occur. Then, when the meal has been assimilated, fluid used during digestion and absorption is reclaimed by transport back across the epithelium to avoid dehydration. Water moves passively into and out of the gastrointestinal lumen, driven by electrochemical gradients established by the active transport of ions and other solutes. In the period after a meal, much of the fluid reuptake is driven by the coupled transport of nutrients, such as glucose, with sodium ions. In the period between meals, absorptive mechanisms center exclusively around electrolytes. In both cases, secretory fluxes of fluid are largely driven by the active transport of chloride ions into the lumen, although absorption still predominates overall.

TABLE 26–5 Daily water turnover (mL) in the gastrointestinal tract.

Ingested		2000
Endogenous secretions		7000
Salivary glands	1500	
Stomach	2500	
Bile	500	
Pancreas	1500	
Intestine	+1000	
	7000	
Total input		9000
Reabsorbed		8800
Jejunum	5500	
Ileum	2000	
Colon	+1300	
	8800	
Balance in stool		200

Data from Moore EW: *Physiology of Intestinal Water and Electrolyte Absorption.* American Gastroenterological Society, 1976.

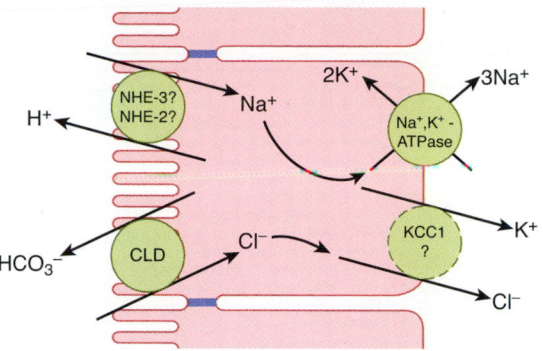

FIGURE 26–19 **Electroneutral NaCl absorption in the small intestine and colon.** NaCl enters across the apical membrane via the coupled activity of a sodium/hydrogen exchanger and a chloride/bicarbonate exchanger.

FIGURE 26–20 **Electrogenic sodium absorption in the colon.** Sodium enters the epithelial cell via epithelial sodium channels (ENaC).

Overall water balance in the gastrointestinal tract is summarized in Table 26–5. The intestines are presented each day with about 2000 mL of ingested fluid plus 7000 mL of secretions from the mucosa of the gastrointestinal tract and associated glands. Ninety-eight percent of this fluid is reabsorbed, with a daily fluid loss of only 200 mL in the stools.

In the small intestine, secondary active transport of Na^+ is important in bringing about absorption of glucose, some amino acids, and other substances such as bile acids (see above). Conversely, the presence of glucose in the intestinal lumen facilitates the reabsorption of Na^+. In the period between meals, when nutrients are not present, sodium and chloride are absorbed together from the lumen by the coupled activity of a sodium/hydrogen exchanger (NHE) and chloride/bicarbonate exchanger in the apical membrane, in a so-called electroneutral mechanism (Figure 26–19). Water then follows to maintain an osmotic balance. In the colon, moreover, an additional electrogenic mechanism for sodium absorption is expressed, particularly in the distal colon. In this mechanism, sodium enters across the apical membrane via an ENaC (epithelial sodium) channel that is identical to that expressed in the distal tubule of the kidney (Figure 26–20). This underpins the ability of the colon to desiccate the stool and ensure that only a small portion of the fluid load used daily in the digestion and absorption of meals is lost from the body. Following a low-salt diet, increased expression of ENaC in response to aldosterone increases the ability to reclaim sodium from the stool.

Despite the predominance of absorptive mechanisms, secretion also takes place continuously throughout the small intestine and colon to adjust the local fluidity of the intestinal contents as needed for mixing, diffusion, and movement of the meal and its residues along the length of the gastrointestinal tract. Cl^- normally enters enterocytes from the interstitial fluid via Na^+–K^+–$2Cl^-$ cotransporters in their basolateral membranes (Figure 26–21), and the Cl^- is then secreted into the intestinal lumen via channels that are regulated by various protein kinases. The cystic fibrosis transmembrane conductance regulator (CFTR) channel that is defective in the disease of cystic fibrosis is quantitatively most important, and is activated by protein kinase A and hence by cAMP (see Clinical Box 26–2).

Water moves into or out of the intestine until the osmotic pressure of the intestinal contents equals that of the plasma. The osmolality of the duodenal contents may be hypertonic or hypotonic, depending on the meal ingested, but by the time the meal enters the jejunum, its osmolality is close to that of

FIGURE 26–21   **Chloride secretion in the small intestine and colon.** Chloride uptake occurs via the sodium/potassium/2 chloride cotransporter, NKCC1. Chloride exit is via the cystic fibrosis transmembrane conductance regulator (CFTR) as well as perhaps via other chloride channels, not shown.

plasma. This osmolality is maintained throughout the rest of the small intestine; the osmotically active particles produced by digestion are removed by absorption, and water moves passively out of the gut along the osmotic gradient thus generated. In the colon, Na^+ is pumped out and water moves passively with it, again along the osmotic gradient. **Saline cathartics** such as magnesium sulfate are poorly absorbed salts that retain their osmotic equivalent of water in the intestine, thus increasing intestinal volume and consequently exerting a laxative effect.

Some K^+ is secreted into the intestinal lumen, especially as a component of mucus. K^+ channels are present in the luminal as well as the basolateral membrane of the enterocytes of the colon, so K^+ is secreted into the colon. In addition, K^+ moves passively down its electrochemical gradient. The accumulation of K^+ in the colon is partially offset by $H^+–K^+$ ATPase in the luminal membrane of cells in the distal colon, with resulting active transport of K^+ into the cells. Nevertheless, loss of ileal or colonic fluids in chronic diarrhea can lead to severe hypokalemia. When the dietary intake of K^+ is high for a prolonged period, aldosterone secretion is increased and more K^+ enters the colon. This is due in part to the appearance of more $Na^+–K^+$ ATPase pumps in the basolateral membranes of the cells, with a consequent increase in intracellular K^+ and K^+ diffusion across the luminal membranes of the cells.

GASTROINTESTINAL REGULATION

The various functions of the gastrointestinal tract, including secretion, digestion, and absorption (Chapter 27) and motility (Chapter 28) must be regulated in an integrated way to ensure efficient assimilation of nutrients after a meal. There are three main modalities for gastrointestinal regulation that operate in a complementary fashion to ensure that function is appropriate. First, **endocrine** regulation is mediated by the release of

hormones by triggers associated with the meal. These hormones travel through the bloodstream to change the activity of a distant segment of the gastrointestinal tract, an organ draining into it (eg, the pancreas), or both. Second, some similar mediators are not sufficiently stable to persist in the bloodstream, but instead alter the function of cells in the local area where they are released, in a **paracrine** fashion. Finally, the intestinal system is endowed with extensive neural connections. These include connections to the central nervous system (**extrinsic innervation**), but also the activity of a largely autonomous **enteric nervous system** that comprises both sensory and secreto-motor neurons. The enteric nervous system integrates central input to the gut, but can also regulate gut function independently in response to changes in the luminal environment. In some cases, the same substance can mediate regulation by endocrine, paracrine, and neurocrine pathways (eg, cholecystokinin, see below).

HORMONES/PARACRINES

Biologically active polypeptides that are secreted by nerve cells and gland cells in the mucosa act in a paracrine fashion, but they also enter the circulation. Measurement of their concentrations in blood after a meal has shed light on the roles these **gastrointestinal hormones** play in the regulation of gastrointestinal secretion and motility.

When large doses of the hormones are given, their actions overlap. However, their physiologic effects appear to be relatively discrete. On the basis of structural similarity (Table 26–6) and, to a degree, similarity of function, the key hormones fall into one of two families: the gastrin family, the primary members of which are gastrin and CCK; and the secretin family, the primary members of which are secretin, glucagon, glicentin (GLI), vasoactive intestinal peptide (VIP; actually a neurotransmitter, or neurocrine), and gastric inhibitory polypeptide (also known as glucose-dependent insulinotropic peptide, or GIP). There are also other hormones that do not fall readily into these families.

ENTEROENDOCRINE CELLS

More than 15 types of hormone-secreting **enteroendocrine cells** have been identified in the mucosa of the stomach, small intestine, and colon. Many of these secrete only one hormone and are identified by letters (G cells, S cells, etc). Others manufacture serotonin or histamine and are called **enterochromaffin** or **enterochromaffin-like (ECL) cells,** respectively.

GASTRIN

Gastrin is produced by cells called G cells in the antral portion of the gastric mucosa (Figure 26–22). G cells are flask-shaped, with a broad base containing many gastrin granules and a narrow apex that reaches the mucosal surface. Microvilli project from the apical end into the lumen. Receptors mediating gastrin responses to changes in gastric contents are present on the microvilli. Other cells in the gastrointestinal tract that secrete hormones have a similar morphology.

Gastrin is typical of a number of polypeptide hormones in that it shows both **macroheterogeneity** and **microheterogeneity.** Macroheterogeneity refers to the occurrence in tissues and body fluids of peptide chains of various lengths; microheterogeneity refers to differences in molecular structure due to derivatization of single amino acid residues. Preprogastrin is processed into fragments of various sizes. Three main fragments contain 34, 17, and 14 amino acid residues. All have the same carboxyl terminal configuration (Table 26–6). These forms are also known as G 34, G 17, and G 14 gastrins, respectively. Another form is the carboxyl terminal tetrapeptide, and there is also a large form that is extended at the amino terminal and contains more than 45 amino acid residues. One form of derivatization is sulfation of the tyrosine that is the sixth amino acid residue from the carboxyl terminal. Approximately equal amounts of nonsulfated and sulfated forms are present in blood and tissues, and they are equally active. Another derivatization is amidation of the carboxyl terminal phenylalanine.

What is the physiologic significance of this marked heterogeneity? Some differences in activity exist between the various components, and the proportions of the components also differ in the various tissues in which gastrin is found. This suggests that different forms are tailored for different actions. However, all that can be concluded at present is that G 17 is the principal form with respect to gastric acid secretion. The carboxyl terminal tetrapeptide has all the activities of gastrin but only 10% of the strength of G 17.

G 14 and G 17 have half-lives of 2 to 3 min in the circulation, whereas G 34 has a half-life of 15 min. Gastrins are inactivated primarily in the kidney and small intestine.

In large doses, gastrin has a variety of actions, but its principal physiologic actions are stimulation of gastric acid and pepsin secretion and stimulation of the growth of the mucosa of the stomach and small and large intestines (**trophic action**). Gastrin secretion is affected by the contents of the stomach, the rate of discharge of the vagus nerves, and bloodborne factors (Table 26–7). Atropine does not inhibit the gastrin response to a test meal in humans, because the transmitter secreted by the postganglionic vagal fibers that innervate the G cells is gastrin-releasing polypeptide (GRP; see below) rather than acetylcholine. Gastrin secretion is also increased by the presence of the products of protein digestion in the stomach, particularly amino acids, which act directly on the G cells. Phenylalanine and tryptophan are particularly effective.

Acid in the antrum inhibits gastrin secretion, partly by a direct action on G cells and partly by release of somatostatin, a relatively potent inhibitor of gastrin secretion. The effect of acid is the basis of a negative feedback loop regulating gastrin secretion. Increased secretion of the hormone increases acid secretion, but the acid then feeds back to inhibit further gastrin secretion. In conditions such as pernicious anemia in which the acid-secreting cells of the stomach are damaged, gastrin secretion is chronically elevated.

CHOLECYSTOKININ

Cholecystokinin (CCK) is secreted by cells in the mucosa of the upper small intestine. It has a plethora of actions in the gastrointestinal system, but the most important appear to be the stimulation of pancreatic enzyme secretion, the contraction of the gallbladder (the action for which it was named), and relaxation of the sphincter of Oddi, which allows both bile and pancreatic juice to flow into the intestinal lumen.

Like gastrin, CCK shows both macroheterogeneity and microheterogeneity. Prepro-CCK is processed into many fragments. A large CCK contains 58 amino acid residues (CCK 58). In addition, there are CCK peptides that contain 39 amino acid residues (CCK 39) and 33 amino acid residues (CCK 33), several forms that contain 12 (CCK 12) or slightly more amino acid residues, and a form that contains 8 amino acid residues (CCK 8). All of these forms have the same 5 amino acids at the carboxyl terminal as gastrin (Table 26–6).

TABLE 26–6. Structures of some of the hormonally active polypeptides secreted by cells in the human gastrointestinal tract.[a]

Gastrin Family		GIP Secretin Family				Other Polypeptides			
CCK 39	Gastrin 34	Glucagon		Secretin	VIP	Motilin	Substance P	GRP	Guanylin
Tyr		Tyr	His	His	His	Phe	Arg	Val	Pro
Ile		Ala	Ser	Ser	Ser	Val	Pro	Pro	Asn
Gln		Glu	Gln	Asp	Asp	Pro	Lys	Leu	Thr
Gln		Gly	Gly	Gly	Ala	Ile	Pro	Pro	Cys
Ala		Thr	Thr	Thr	Val	Phe	Gln	Ala	Glu
Arg	(pyro)Glu	Phe	Phe	Phe	Phe	Thr	Gln	Gly	Ile
Lys	Leu	Ile	Thr	Thr	Thr	Tyr	Phe	Gly	Cys
Ala	Gly	Ser	Ser	Ser	Asp	Gly	Phe	Gly	Ala
Pro	Pro	Asp	Asp	Glu	Asn	Glu	Gly	Thr	Tyr
Ser	Gln	Tyr	Tyr	Leu	Tyr	Leu	Leu	Val	Ala
Gly	Gly	Ser	Ser	Ser	Thr	Gln	Met-NH2	Leu	Ala
Arg	Pro	Ile	Lys	Arg	Arg	Arg		Thr	Cys
Met	Pro	Ala	Tyr	Leu	Leu	Met		Lys	Thr
Ser	His	Met	Leu	Arg	Arg	Gln		Met	Gly
Ile	Leu	Asp	Asp	Glu	Lys	Glu		Tyr	Cys
Val	Val	Lys	Ser	Gly	Gln	Lys		Pro	
Lys	Ala	Ile	Arg	Ala	Met	Glu		Arg	
Asn	Asp	His	Arg	Arg	Ala	Arg		Gly	
Leu	Pro	Gln	Ala	Leu	Val	Asn		Asn	
Gln	Ser	Gln	Gln	Gln	Lys	Lys		His	
Asn	Lys	Asp	Asp	Arg	Lys	Gly		Trp	
Leu	Lys	Phe	Phe	Leu	Tyr	Gln		Ala	
Asp	Gln	Val	Val	Leu	Leu			Val	
Pro	Gly	Asn	Gln	Gln	Asn			Gly	
Ser	Pro	Trp	Trp	Gly	Ser			His	
His	Trp	Leu	Leu	Leu	Ile			Leu	
Arg	Leu	Leu	Met	Val-NH2	Leu			Met-NH2	
Ile	Glu	Ala	Asn		Asn-NH2				
Ser	Glu	Glu	Thr						
Asp	Glu	Lys							
Arg	Glu	Gly							
Asp	Glu	Lys							
Tys	Ala	Lys							
Met	Tys	Asn							
Gly	Gly	Asp							
Trp	Trp	Trp							
Met	Met	Lys							
Asp	Asp	His							
Phe-NH2	Phe-NH2	Asn							
		Ile							
		Thr							
		Gln							

[a]Homologous amino acid residues are enclosed by the lines that generally cross from one polypeptide to another. Arrows indicate points of cleavage to form smaller variants. Tys, tyrosine sulfate. All gastrins occur in unsulfated (gastrin I) and sulfated (gastrin II) forms. Glicentin, an additional member of the secretin family, is a C-terminally extended relative of glucagon.

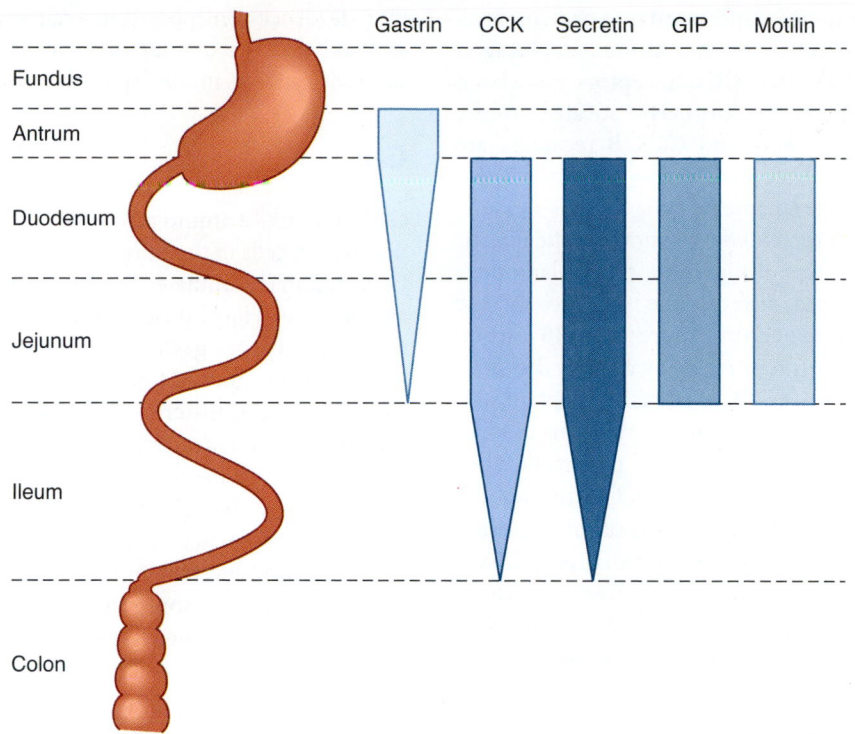

FIGURE 26–22 **Sites of production of the five gastrointestinal hormones along the length of the gastrointestinal tract.** The width of the bars reflects the relative abundance at each location.

TABLE 26–7 Stimuli that affect gastrin secretion.

Stimuli that increase gastrin secretion
Luminal
Peptides and amino acids
Distention
Neural
Increased vagal discharge via GRP
Bloodborne
Calcium
Epinephrine
Stimuli that inhibit gastrin secretion
Luminal
Acid
Somatostatin
Bloodborne
Secretin, GIP, VIP, glucagon, calcitonin

The carboxyl terminal tetrapeptide (CCK 4) also exists in tissues. The carboxyl terminal is amidated, and the tyrosine that is the seventh amino acid residue from the carboxyl terminal is sulfated. Unlike gastrin, the nonsulfated form of CCK has not been found in tissues. However, derivatization of other amino acid residues in CCK can occur. The half-life of circulating CCK is about 5 minutes, but little is known about its metabolism.

In addition to its secretion by I cells in the upper intestine, CCK is found in nerves in the distal ileum and colon. It is also found in neurons in the brain, especially the cerebral cortex, and in nerves in many parts of the body (see Chapter 7). In the brain, it may be involved in the regulation of food intake, and it appears to be related to the production of anxiety and analgesia. The CCK secreted in the duodenum and jejunum is probably mostly CCK 8 and CCK 12, although CCK 58 is also present in the intestine and circulating blood in some species. The enteric and pancreatic nerves contain primarily CCK 4. CCK 58 and CCK 8 are found in the brain.

In addition to its primary actions, CCK augments the action of secretin in producing secretion of an alkaline pancreatic juice. It also inhibits gastric emptying, exerts a trophic effect on the pancreas, increases the synthesis of enterokinase, and may enhance the motility of the small intestine and colon. There is some evidence that, along with secretin, it augments the contraction of the pyloric sphincter, thus preventing the reflux of duodenal contents into the stomach. Gastrin and CCK stimulate glucagon secretion, and since the secretion of both gastrointestinal hormones is increased by a protein meal, either or both may be the "gut factor" that stimulates glucagon secretion (see Chapter 21). Two CCK receptors have been identified. CCK-A receptors are primarily located in the periphery, whereas both CCK-A and CCK-B receptors are found in the brain. Both activate PLC, causing increased production of IP_3 and DAG (see Chapter 2).

The secretion of CCK is increased by contact of the intestinal mucosa with the products of digestion, particularly peptides and amino acids, and also by the presence in the duodenum of fatty acids containing more than 10 carbon atoms. There are also two protein releasing factors that activate CCK secretion, known as CCK-releasing peptide and monitor peptide, which derive from the intestinal mucosa and pancreas, respectively. Because the bile and pancreatic juice that enter the duodenum in response to CCK further the digestion of protein and fat, and the products of this digestion stimulate further CCK secretion, a sort of positive feedback operates in the control of the secretion of this hormone. However, the positive feedback is terminated when the products of digestion move on to the lower portions of the gastrointestinal tract, and also because CCK-releasing peptide and monitor peptide are degraded by proteolytic enzymes once these are no longer occupied in digesting dietary proteins.

SECRETIN

Secretin occupies a unique position in the history of physiology. In 1902, Bayliss and Starling first demonstrated that the excitatory effect of duodenal stimulation on pancreatic secretion was due to a bloodborne factor. Their research led to the identification of the first hormone, secretin. They also suggested that many chemical agents might be secreted by cells in the body and pass in the circulation to affect organs some distance away. Starling introduced the term **hormone** to categorize such "chemical messengers." Modern endocrinology is the proof of the correctness of this hypothesis.

Secretin is secreted by S cells that are located deep in the glands of the mucosa of the upper portion of the small intestine. The structure of secretin (Table 26–6) is different from that of CCK and gastrin, but very similar to that of glucagon, GLI, VIP, and GIP. Only one form of secretin has been isolated, and the fragments of the molecule that have been tested to date are inactive. Its half-life is about 5 minutes, but little is known about its metabolism.

Secretin increases the secretion of bicarbonate by the duct cells of the pancreas and biliary tract. It thus causes the secretion of a watery, alkaline pancreatic juice. Its action on pancreatic duct cells is mediated via cAMP. It also augments the action of CCK in producing pancreatic secretion of digestive enzymes. It decreases gastric acid secretion and may cause contraction of the pyloric sphincter.

The secretion of secretin is increased by the products of protein digestion and by acid bathing the mucosa of the upper small intestine. The release of secretin by acid is another example of feedback control: Secretin causes alkaline pancreatic juice to flood into the duodenum, neutralizing the acid from the stomach and thus inhibiting further secretion of the hormone.

GIP

GIP contains 42 amino acid residues (Table 26–6) and is produced by K cells in the mucosa of the duodenum and jejunum. Its secretion is stimulated by glucose and fat in the duodenum, and because in large doses it inhibits gastric secretion and motility, it was named gastric inhibitory peptide. However, it now appears that it does not have significant gastric inhibiting activity when administered in smaller amounts comparable to those seen after a meal. In the meantime, it was found that GIP stimulates insulin secretion. Gastrin, CCK, secretin, and glucagon also have this effect, but GIP is the only one of these that stimulates insulin secretion when administered in doses that produce blood levels comparable to those produced by oral glucose. For this reason, it is often called **glucose-dependent insulinotropic polypeptide.** The glucagon derivative GLP-1 (7–36) (see Chapter 21) also stimulates insulin secretion and is said to be more potent in this regard than GIP. Therefore, it may also be a physiologic B cell-stimulating hormone of the gastrointestinal tract.

The integrated action of gastrin, CCK, secretin, and GIP in facilitating digestion and utilization of absorbed nutrients is summarized in Figure 26–23.

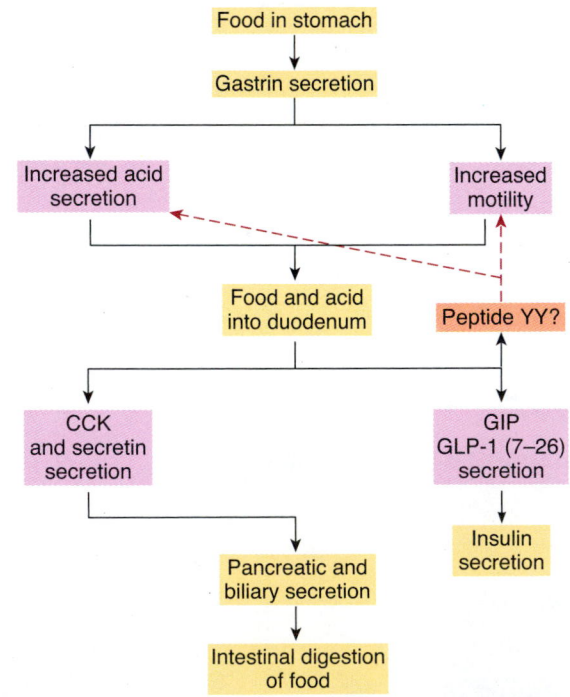

FIGURE 26–23 **Integrated action of gastrointestinal hormones in regulating digestion and utilization of absorbed nutrients.** The dashed arrows indicate inhibition. The exact identity of the hormonal factor or factors from the intestine that inhibit(s) gastric acid secretion and motility is unsettled, but it may be peptide YY.

VIP

VIP contains 28 amino acid residues (Table 26–6). It is found in nerves in the gastrointestinal tract and thus is not itself a hormone, despite its similarities to secretin. Prepro-VIP contains both VIP and a closely related polypeptide (**PHM-27** in humans, PHI-27 in other species). VIP is also found in blood, in which it has a half-life of about 2 minutes. In the intestine, it markedly stimulates intestinal secretion of electrolytes and hence of water. Its other actions include relaxation of intestinal smooth muscle, including sphincters; dilation of peripheral blood vessels; and inhibition of gastric acid secretion. It is also found in the brain and many autonomic nerves (see Chapter 7), where it often occurs in the same neurons as acetylcholine. It potentiates the action of acetylcholine in salivary glands. However, VIP and acetylcholine do not coexist in neurons that innervate other parts of the gastrointestinal tract. VIP-secreting tumors (VIPomas) have been described in patients with severe diarrhea.

MOTILIN

Motilin is a polypeptide containing 22 amino acid residues that is secreted by enterochromaffin cells and Mo cells in the stomach, small intestine, and colon. It acts on G protein-coupled receptors on enteric neurons in the duodenum and colon and on injection produces contraction of smooth muscle in the stomach and intestines. Its circulating level increases at intervals of approximately 100 min in the interdigestive state, and it is a major regulator of the migrating motor complexes (MMCs) (Figure 26–24) that control gastrointestinal motility between meals. Conversely, when a meal is ingested, secretion of motilin is suppressed until digestion and absorption are complete. The antibiotic erythromycin binds to motilin receptors, and derivatives of this compound may be of value in treating patients in whom gastrointestinal motility is decreased.

SOMATOSTATIN

Somatostatin, the growth-hormone-inhibiting hormone originally isolated from the hypothalamus, is secreted as a paracrine by D cells in the pancreatic islets (see Chapter 21) and by similar D cells in the gastrointestinal mucosa. It exists in tissues in two forms, somatostatin 14 and somatostatin 28, and both are secreted. Somatostatin inhibits the secretion of gastrin, VIP, GIP, secretin, and motilin. Its secretion is stimulated by acid in the lumen, and it probably acts in a paracrine fashion to mediate the inhibition of gastrin secretion produced by acid. It also inhibits

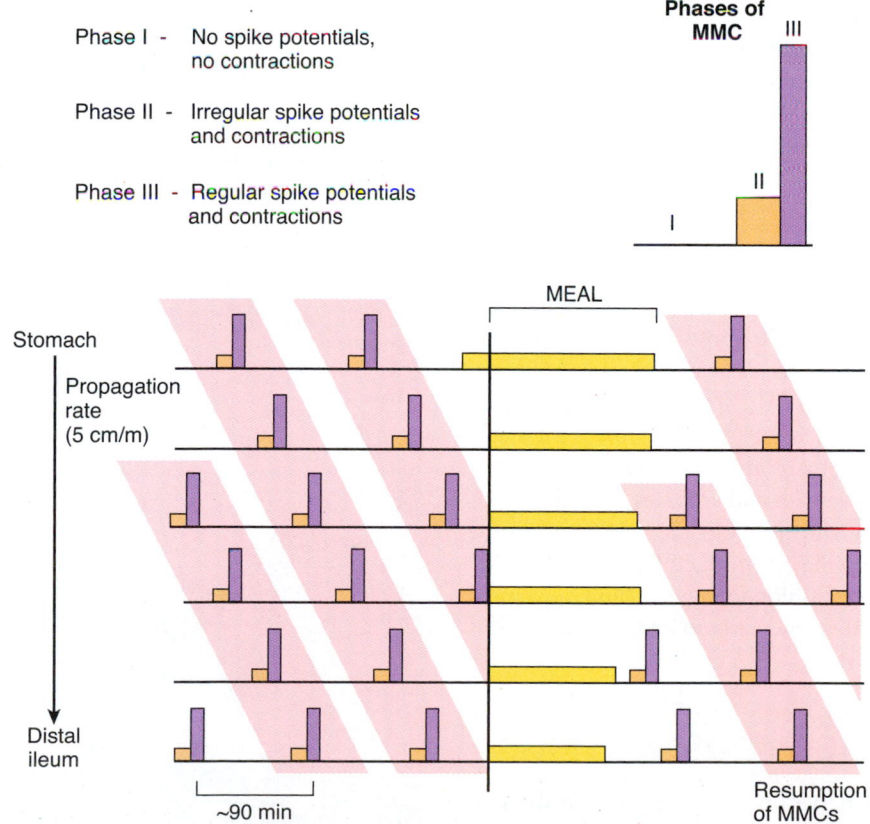

FIGURE 26–24 **Migrating motor complexes (MMCs).** Note that the complexes move down the gastrointestinal tract at a regular rate during fasting, that they are completely inhibited by a meal, and that they resume 90–120 minutes after the meal. (Reproduced with permission from Chang EB, Sitrin MD, Black DD: *Gastrointestinal, Hepatobiliary, and Nutritional Physiology.* Lippincott-Raven, 1996.)

pancreatic exocrine secretion; gastric acid secretion and motility; gallbladder contraction; and the absorption of glucose, amino acids, and triglycerides.

OTHER GASTROINTESTINAL PEPTIDES

PEPTIDE YY

The structure of peptide YY is discussed in Chapter 21. It also inhibits gastric acid secretion and motility and is a good candidate to be the gastric inhibitory peptide (Figure 26–23). Its release from the jejunum is stimulated by fat.

OTHERS

Ghrelin is secreted primarily by the stomach and appears to play an important role in the central control of food intake. It also stimulates growth hormone secretion by acting directly on receptors in the pituitary (see Chapter 24).

Substance P (Table 26–6) is found in endocrine and nerve cells in the gastrointestinal tract and may enter the circulation. It increases the motility of the small intestine. The neurotransmitter **GRP** contains 27 amino acid residues, and the 10 amino acid residues at its carboxyl terminal are almost identical to those of amphibian **bombesin.** It is present in the vagal nerve endings that terminate on G cells and is the neurotransmitter producing vagally mediated increases in gastrin secretion. **Glucagon** from the gastrointestinal tract may be responsible (at least in part) for the hyperglycemia seen after pancreatectomy.

Guanylin is a gastrointestinal polypeptide that binds to guanylyl cyclase. It is made up of 15 amino acid residues (Table 26–6) and is secreted by cells of the intestinal mucosa. Stimulation of guanylyl cyclase increases the concentration of intracellular cyclic 3′,5′-guanosine monophosphate (cGMP), and this in turn causes increased secretion of Cl^- into the intestinal lumen. Guanylin appears to act predominantly in a paracrine fashion, and it is produced in cells from the pylorus to the rectum. In an interesting example of molecular mimicry, the heat-stable enterotoxin of certain diarrhea-producing strains of *E. coli* has a structure very similar to guanylin and activates guanylin receptors in the intestine. Guanylin receptors are also found in the kidneys, the liver, and the female reproductive tract, and guanylin may act in an endocrine fashion to regulate fluid movement in these tissues as well, and particularly to integrate the actions of the intestine and kidneys.

THE ENTERIC NERVOUS SYSTEM

Two major networks of nerve fibers are intrinsic to the gastrointestinal tract: the **myenteric plexus** (Auerbach's plexus), between the outer longitudinal and middle circular muscle layers, and the **submucous plexus** (Meissner's plexus), between the middle circular layer and the mucosa (Figure 26–1). Collectively, these neurons constitute the **enteric nervous system.** The system contains about 100 million sensory neurons, interneurons, and motor neurons in humans—as many as are found in the whole spinal cord—and the system is probably best viewed as a displaced part of the central nervous system (CNS) that is concerned with the regulation of gastrointestinal function. It is sometimes referred to as the "little brain" for this reason. It is connected to the CNS by parasympathetic and sympathetic fibers but can function autonomously without these connections (see below). The myenteric plexus innervates the longitudinal and circular smooth muscle layers and is concerned primarily with motor control, whereas the submucous plexus innervates the glandular epithelium, intestinal endocrine cells, and submucosal blood vessels and is primarily involved in the control of intestinal secretion. The neurotransmitters in the system include acetylcholine, the amines norepinephrine and serotonin, the amino acid γ-aminobutyrate (GABA), the purine adenosine triphosphate (ATP), the gases NO and CO, and many different peptides and polypeptides (Table 26–8). Some of these peptides also act in a paracrine fashion, and some enter the bloodstream, becoming hormones. Not surprisingly, most of them are also found in the brain.

EXTRINSIC INNERVATION

The intestine receives a dual extrinsic innervation from the autonomic nervous system, with parasympathetic cholinergic activity generally increasing the activity of intestinal smooth

TABLE 26–8 Principal peptides found in the enteric nervous system.

CGRP
CCK
Endothelin-2
Enkephalins
Galanin
GRP
Neuropeptide Y
Neurotensin
Peptide YY
PACAP
Somatostatin
Substance P
TRH
VIP

muscle and sympathetic noradrenergic activity generally decreasing it while causing sphincters to contract. The preganglionic parasympathetic fibers consist of about 2000 vagal efferents and other efferents in the sacral nerves. They generally end on cholinergic nerve cells of the myenteric and submucous plexuses. The sympathetic fibers are postganglionic, but many of them end on postganglionic cholinergic neurons, where the norepinephrine they secrete inhibits acetylcholine secretion by activating α_2 presynaptic receptors. Other sympathetic fibers appear to end directly on intestinal smooth muscle cells. The electrical properties of intestinal smooth muscle are discussed in Chapter 5. Still other fibers innervate blood vessels, where they produce vasoconstriction. It appears that the intestinal blood vessels have a dual innervation: They have an extrinsic noradrenergic innervation and an intrinsic innervation by fibers of the enteric nervous system. VIP and NO are among the mediators in the intrinsic innervation, which seems, among other things, to be responsible for the hyperemia that accompanies digestion of food. It is unsettled whether the blood vessels have an additional cholinergic innervation.

GASTROINTESTINAL (SPLANCHNIC) CIRCULATION

A final general point that should be made about the gastrointestinal tract relates to its unusual circulatory features. The blood flow to the stomach, intestines, pancreas, and liver is arranged in a series of parallel circuits, with all the blood from the intestines and pancreas draining via the portal vein to the liver (Figure 26–25). The blood from the intestines, pancreas, and spleen drains via the hepatic portal vein to the liver and from the liver via the hepatic veins to the inferior vena cava. The viscera and the liver receive about 30% of the cardiac output via the celiac, superior mesenteric, and inferior mesenteric arteries. The liver receives about 1300 mL/min from the portal vein and 500 mL/min from the hepatic artery during fasting, and the portal supply increases still further after meals.

CHAPTER SUMMARY

- The gastrointestinal system evolved as a portal to permit controlled nutrient uptake in multicellular organisms. It is functionally continuous with the outside environment and is defended by a well-developed mucosal immune system. Nevertheless, the gut usually lives in harmony with an extensive commensal microflora, particularly in the colon.

- Digestive secretions serve to chemically alter the components of meals (particularly macromolecules) such that their constituents can be absorbed across the epithelium. Meal components are acted on sequentially by saliva, gastric juice, pancreatic juice, and bile, which contain enzymes, ions, water, and other specialized components.

- The intestine and the organs that drain into it secrete about 8 L of fluid per day, which are added to water consumed in food and

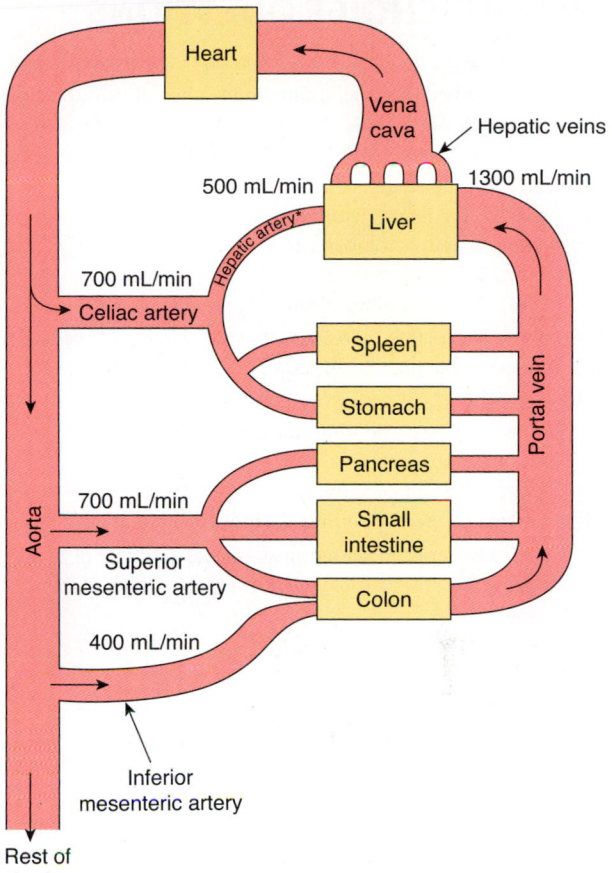

*Branches of the hepatic artery also supply the stomach, pancreas and small intestine

FIGURE 26–25 **Schematic of the splanchnic circulation under fasting conditions.** Note that even during fasting, the liver receives the majority of its blood supply via the portal vein.

beverages. Most of this fluid is reabsorbed, leaving only approximately 200 mL to be lost to the stool. Fluid secretion and absorption are both dependent on the active epithelial transport of ions, nutrients, or both.

- Gastrointestinal functions are regulated in an integrated fashion by endocrine, paracrine, and neurocrine mechanisms. Hormones and paracrine factors are released from enteroendocrine cells in response to signals coincident with the intake of meals.

- The enteric nervous system conveys information from the central nervous system to the gastrointestinal tract, but also often can activate programmed responses of secretion and motility in an autonomous fashion.

- The intestine has an unusual circulation, in that the majority of its venous outflow does not return directly to the heart, but rather is directed initially to the liver via the portal vein.

MULTIPLE-CHOICE QUESTIONS

For all questions, select the single best answer unless otherwise directed.

1. Water is absorbed in the jejunum, ileum, and colon and excreted in the feces. Arrange these in order of the amount of water absorbed or excreted from greatest to smallest.

   A) colon, jejunum, ileum, feces
   B) feces, colon, ileum, jejunum
   C) jejunum, ileum, colon, feces
   D) colon, ileum, jejunum, feces
   E) feces, jejunum, ileum, colon

2. Drugs and toxins that increase the cAMP content of the intestinal mucosa cause diarrhea because they

   A) increase Na^+–K^+ cotransport in the small intestine.
   B) increase K^+ secretion into the colon.
   C) inhibit K^+ absorption in the crypts of Lieberkühn.
   D) increase Na^+ absorption in the small intestine.
   E) increase Cl^- secretion into the intestinal lumen.

3. A patient with a tumor secreting abnormal amounts of gastrin (gastrinoma) would be most likely to exhibit which of the following?

   A) decreased chief cell exocytosis
   B) duodenal ulceration
   C) increased gastric pH in the period between meals
   D) a reduced incidence of gastroesophageal reflux disease
   E) protein malabsorption

4. Which of the following has the highest pH?

   A) gastric juice
   B) hepatic bile
   C) pancreatic juice
   D) saliva
   E) secretions of the intestinal glands

5. Which of the following would *not* be produced by total pancreatectomy?

   A) vitamin E deficiency
   B) hyperglycemia
   C) metabolic acidosis
   D) weight gain
   E) decreased absorption of amino acids

CHAPTER RESOURCES

Baron TH, Morgan DE: Current concepts: Acute necrotizing pancreatitis. N Engl J Med 1999;340:1412.

Barrett KE: *Gastrointestinal Physiology.* McGraw-Hill, 2006.

Bengmark S: Econutrition and health maintenance—A new concept to prevent GI inflammation, ulceration, and sepsis. Clin Nutr 1996;15:1.

Chong L, Marx J (editors): Lipids in the limelight. Science 2001;294:1861.

Go VLW, et al: *The Pancreas: Biology, Pathobiology and Disease,* 2nd ed. Raven Press, 1993.

Hersey SJ, Sachs G: Gastric acid secretion. Physiol Rev 1995;75:155.

Hofmann AF: Bile acids: The good, the bad, and the ugly. News Physiol Sci 1999;14:24.

Hunt RH, Tytgat GN (editors): *Helicobacter pylori: Basic Mechanisms to Clinical Cure.* Kluwer Academic, 2000.

Itoh Z: Motilin and clinical application. Peptides 1997;18:593.

Johnston DE, Kaplan MM: Pathogenesis and treatment of gallstones. N Engl J Med 1993;328:412.

Kunzelmann K, Mall M: Electrolyte transport in the mammalian colon: Mechanisms and implications for disease. Physiol Rev 2002;82:245.

Lamberts SWJ, et al: Octreotide. N Engl J Med 1996;334:246.

Lewis JH (editor): *A Pharmacological Approach to Gastrointestinal Disorders.* Williams & Wilkins, 1994.

Meier PJ, Stieger B: Molecular mechanisms of bile formation. News Physiol Sci 2000;15:89.

Montecucco C, Rappuoli R: Living dangerously: How *Helicobacter pylori* survives in the human stomach. Nat Rev Mol Cell Biol 2001;2:457.

Nakazato M: Guanylin family: New intestinal peptides regulating electrolyte and water homeostasis. J Gastroenterol 2001;36:219.

Rabon EC, Reuben MA: The mechanism and structure of the gastric H^+, K^+–ATPase. Annu Rev Physiol 1990;52:321.

Sachs G, Zeng N, Prinz C: Pathophysiology of isolated gastric endocrine cells. Annu Rev Physiol 1997;59:234.

Sellin JH: SCFAs: The enigma of weak electrolyte transport in the colon. News Physiol Sci 1999;14:58.

Specian RD, Oliver MG: Functional biology of intestinal goblet cells. Am J Med 1991;260:C183.

Topping DL, Clifton PM: Short-chain fatty acids and human colonic function: Select resistant starch and nonstarch polysaccharides. Physiol Rev 2001;81:1031.

Trauner M, Meier PJ, Boyer JL: Molecular mechanisms of cholestasis. N Engl J Med 1998;339:1217.

Walsh JH (editor): *Gastrin.* Raven Press, 1993.

Williams JA, Blevins GT Jr: Cholecystokinin and regulation of pancreatic acinar cell function. Physiol Rev 1993;73:701.

Wolfe MM, Lichtenstein DR, Singh G: Gastrointestinal toxicity of nonsteroidal anti-inflammatory drugs. N Engl J Med 1999;340:1888.

Wright EM: The intestinal Na^+/glucose cotransporter. Annu Rev Physiol 1993;55:575.

Young JA, van Lennep EW: *The Morphology of Salivary Glands.* Academic Press, 1978.

Zoetendal EG et al: Molecular ecological analysis of the gastrointestinal microbiota: A review. J Nutr 2004;134:465.

Digestion, Absorption, & Nutritional Principles

27

OBJECTIVES

After studying this chapter, you should be able to:

- Understand how nutrients are delivered to the body and the chemical processes needed to convert them to a form suitable for absorption.
- List the major dietary carbohydrates and define the luminal and brush border processes that produce absorbable monosaccharides as well as the transport mechanisms that provide for the uptake of these hydrophilic molecules.
- Understand the process of protein assimilation, and the ways in which it is comparable to, or converges from, that used for carbohydrates.
- Define the stepwise processes of lipid digestion and absorption, the role of bile acids in solubilizing the products of lipolysis, and the consequences of fat malabsorption.
- Identify the source and functions of short-chain fatty acids in the colon.
- Delineate the mechanisms of uptake for vitamins and minerals.
- Understand basic principles of energy metabolism and nutrition.

INTRODUCTION

The gastrointestinal system is the portal through which nutritive substances, vitamins, minerals, and fluids enter the body. Proteins, fats, and complex carbohydrates are broken down into absorbable units (**digested**), principally in the small intestine. The products of digestion and the vitamins, minerals, and water cross the mucosa and enter the lymph or the blood (**absorption**). The digestive and absorptive processes are the subject of this chapter.

Digestion of the major foodstuffs is an orderly process involving the action of a large number of **digestive enzymes** (Table 27–1). Enzymes from the salivary glands attack carbohydrates (and fats in some species); enzymes from the stomach attack proteins and fats; and enzymes from the exocrine portion of the pancreas attack carbohydrates, proteins, lipids, DNA, and RNA. Other enzymes that complete the digestive process are found in the luminal membranes and the cytoplasm of the cells that line the small intestine. The action of the enzymes is aided by the hydrochloric acid secreted by the stomach and the bile secreted by the liver.

Most substances pass from the intestinal lumen into the enterocytes and then out of the enterocytes to the interstitial fluid. The processes responsible for movement across the luminal cell membrane are often quite different from those responsible for movement across the basal and lateral cell membranes to the interstitial fluid.

TABLE 27–1 Normal transport of substances by the intestine and location of maximum absorption or secretion.[a]

| | Small Intestine | | | |
Absorption of:	Upper[b]	Mid	Lower	Colon
Sugars (glucose, galactose, etc)	++	+++	++	0
Amino acids	++	++	++	0
Water-soluble and fat-soluble vitamins except vitamin B_{12}	+++	++	0	0
Betaine, dimethylglycine, sarcosine	+	++	++	?
Antibodies in newborns	+	++	+++	?
Pyrimidines (thymine and uracil)	+	+	?	?
Long-chain fatty acid absorption and conversion to triglyceride	+++	++	+	0
Bile acids	+	+	+++	
Vitamin B_{12}	0	+	+++	0
Na^+	+++	++	+++	+++
K^+	+	+	+	Sec
Ca^{2+}	+++	++	+	?
Fe^{2+}	+++	+	+	?
Cl^-	+++	++	+	+
SO_4^{2-}	++	+	0	?

[a]Amount of absorption is graded + to +++. Sec, secreted when luminal K^+ is low.

[b]Upper small intestine refers primarily to jejunum, although the duodenum is similar in most cases studied (with the notable exception that the duodenum secretes HCO_3^- and shows little net absorption or secretion of NaCl).

DIGESTION & ABSORPTION: CARBOHYDRATES

DIGESTION

The principal dietary carbohydrates are polysaccharides, disaccharides, and monosaccharides. Starches (glucose polymers) and their derivatives are the only polysaccharides that are digested to any degree in the human gastrointestinal tract. Amylopectin, which constitutes 80–90% of dietary starch, is a branched molecule, whereas amylose is a straight chain with only 1:4α linkages (Figure 27–1). The disaccharides **lactose** (milk sugar) and **sucrose** (table sugar) are also ingested, along with the monosaccharides fructose and glucose.

In the mouth, starch is attacked by salivary α-amylase. However, the optimal pH for this enzyme is 6.7, and its action is inhibited by the acidic gastric juice when food enters the stomach. In the small intestine, both the salivary and the pancreatic α-amylase also act on the ingested polysaccharides. Both the salivary and the pancreatic α-amylases hydrolyze 1:4α linkages but spare 1:6α linkages and terminal 1:4α linkages. Consequently, the end products of α-amylase digestion are oligosaccharides: the disaccharide **maltose;** the trisaccha-

ride **maltotriose;** and **α-limit dextrins,** polymers of glucose containing an average of about eight glucose molecules with 1:6α linkages (Figure 27–1).

The oligosaccharidases responsible for the further digestion of the starch derivatives are located in the brush border of small intestinal epithelial cells (Figure 27–1). Some of these enzymes have more than one substrate. **Isomaltase** is mainly responsible for hydrolysis of 1:6α linkages. Along with **maltase and sucrase,** it also breaks down maltotriose and maltose. Sucrase and isomaltase are initially synthesized as a single glycoprotein chain which is inserted into the brush border membrane. It is then hydrolyzed by pancreatic proteases into sucrase and isomaltase subunits.

Sucrase hydrolyzes sucrose into a molecule of glucose and a molecule of fructose. In addition, two disaccharidases are present in the brush border: **lactase,** which hydrolyzes lactose to glucose and galactose, and **trehalase,** which hydrolyzes trehalose, a 1:1α-linked dimer of glucose, into two glucose molecules.

Deficiency of one or more of the brush border oligosaccharidases may cause diarrhea, bloating, and flatulence after ingestion of sugar (Clinical Box 27–1). The diarrhea is due to the increased number of osmotically active oligosaccharide molecules that remain in the intestinal lumen, causing the volume of the intestinal contents to increase. In the colon,

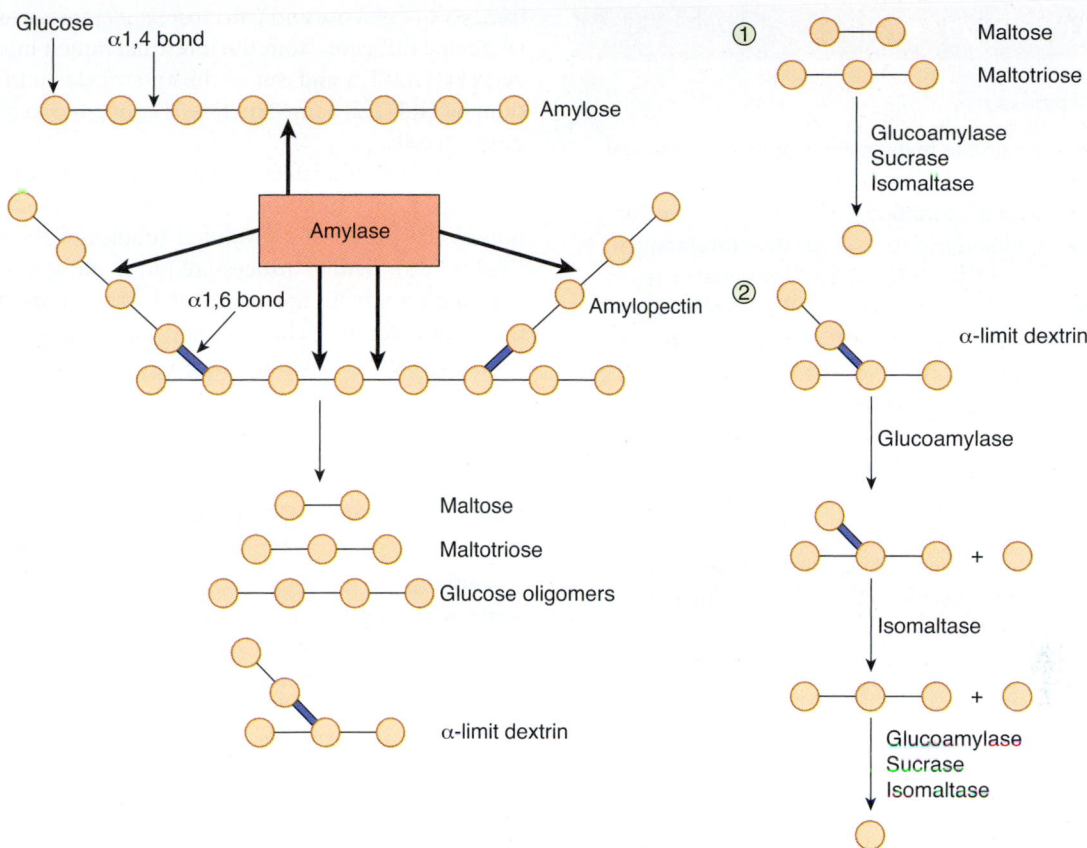

FIGURE 27–1 **Left:** Structure of amylose and amylopectin, which are polymers of glucose (indicated by circles). These molecules are partially digested by the enzyme amylase, yielding the products shown at the bottom of the figure. **Right:** Brush border hydrolases responsible for the sequential digestion of the products of luminal starch digestion (1, linear oligomers; 2, alpha-limit dextrins).

bacteria break down some of the oligosaccharides, further increasing the number of osmotically active particles. The bloating and flatulence are due to the production of gas (CO_2 and H_2) from disaccharide residues in the lower small intestine and colon.

ABSORPTION

Hexoses are rapidly absorbed across the wall of the small intestine (Table 27–1). Essentially all the hexoses are removed before the remains of a meal reach the terminal part of the ileum. The sugar molecules pass from the mucosal cells to the blood in the capillaries draining into the portal vein.

The transport of most hexoses is dependent on Na^+ in the intestinal lumen; a high concentration of Na^+ on the mucosal surface of the cells facilitates and a low concentration inhibits sugar influx into the epithelial cells. This is because glucose and Na^+ share the same **cotransporter,** or **symport,** the **sodium-dependent glucose transporter** (SGLT, Na^+ glucose cotransporter) (Figure 27–2). The members of this family of transporters, SGLT 1 and SGLT 2, resemble the glucose transporters responsible for facilitated diffusion (see Chapter 21) in that they cross the cell membrane 12 times and have their –COOH and –NH$_2$ terminals on the cytoplasmic side of the

membrane. However, there is no homology to the glucose transporter (GLUT) series of transporters. SGLT-1 is responsible for uptake of dietary glucose from the gut. The related transporter, SGLT 2, is responsible for glucose transport out of the renal tubules (see Chapter 38).

Because the intracellular Na^+ concentration is low in intestinal cells as it is in other cells, Na^+ moves into the cell along its concentration gradient. Glucose moves with the Na^+ and is released in the cell (Figure 27–2). The Na^+ is transported into the lateral intercellular spaces, and the glucose is transported by GLUT 2 into the interstitium and thence to the capillaries. Thus, glucose transport is an example of secondary active transport (see Chapter 2); the energy for glucose transport is provided indirectly, by the active transport of Na^+ out of the cell. This maintains the concentration gradient across the luminal border of the cell, so that more Na^+ and consequently more glucose enter. When the Na^+/glucose cotransporter is congenitally defective, the resulting **glucose/galactose malabsorption** causes severe diarrhea that is often fatal if glucose and galactose are not promptly removed from the diet. The use of glucose and its polymers to retain Na^+ in diarrheal disease was discussed in Chapter 26.

SGLT-1 also transports galactose, but fructose utilizes a different mechanism. Its absorption is independent of Na^+ or the

Lactose Intolerance

In most mammals and in many races of humans, intestinal lactase activity is high at birth, then declines to low levels during childhood and adulthood. The low lactase levels are associated with intolerance to milk (**lactose intolerance**). Most Europeans and their American descendants retain sufficient intestinal lactase activity in adulthood; the incidence of lactase deficiency in northern and western Europeans is only about 15%. However, the incidence in blacks, American Indians, Asians, and Mediterranean populations is 70–100%. When such individuals ingest dairy products, they are unable to digest lactose sufficiently, and so symptoms such as bloating, pain, gas, and diarrhea are produced by the unabsorbed osmoles that are subsequently digested by colonic bacteria. Milk intolerance can be ameliorated by administration of commercial lactase preparations, but this is expensive. Yogurt is better tolerated than milk in intolerant individuals because it contains its own bacterial lactase.

transport of glucose and galactose; it is transported instead by facilitated diffusion from the intestinal lumen into the enterocytes by GLUT 5 and out of the enterocytes into the interstitium by GLUT 2. Some fructose is converted to glucose in the mucosal cells.

Insulin has little effect on intestinal transport of sugars. In this respect, intestinal absorption resembles glucose reabsorption in the proximal convoluted tubules of the kidneys (see Chapter 38); neither process requires phosphorylation, and both are essentially normal in diabetes but are depressed by the drug phlorizin. The maximal rate of glucose absorption from the intestine is about 120 g/h.

PROTEINS & NUCLEIC ACIDS

PROTEIN DIGESTION

Protein digestion begins in the stomach, where pepsins cleave some of the peptide linkages. Like many of the other enzymes concerned with protein digestion, pepsins are secreted in the form of inactive precursors (**proenzymes**) and activated in the

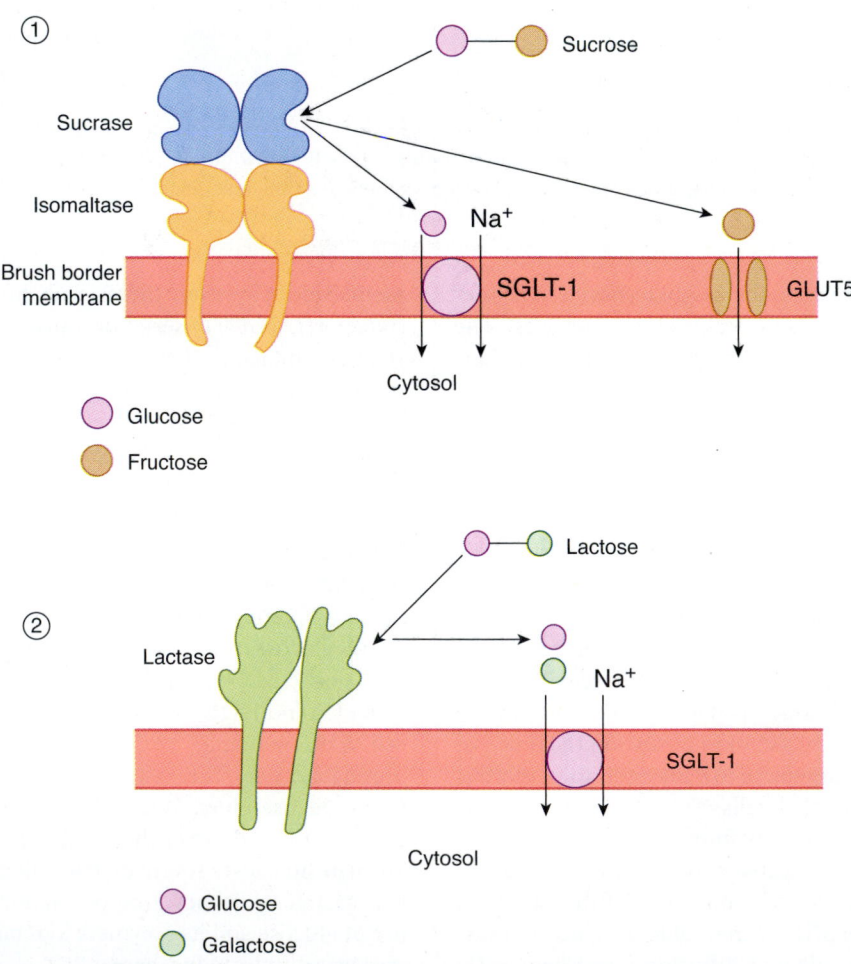

FIGURE 27–2 **Brush border digestion and assimilation of the disaccharides sucrose (panel 1) and lactose (panel 2).** SGLT-1, sodium-glucose cotransporter-1.

gastrointestinal tract. The pepsin precursors are called pepsinogens and are activated by gastric acid. Human gastric mucosa contains a number of related pepsinogens, which can be divided into two immunohistochemically distinct groups, pepsinogen I and pepsinogen II. Pepsinogen I is found only in acid-secreting regions, whereas pepsinogen II is also found in the pyloric region. Maximal acid secretion correlates with pepsinogen I levels.

Pepsins hydrolyze the bonds between aromatic amino acids such as phenylalanine or tyrosine and a second amino acid, so the products of peptic digestion are polypeptides of very diverse sizes. Because pepsins have a pH optimum of 1.6 to 3.2, their action is terminated when the gastric contents are mixed with the alkaline pancreatic juice in the duodenum and jejunum. The pH of the intestinal contents in the duodenal bulb is 2.0 to 4.0, but in the rest of the duodenum it is about 6.5.

In the small intestine, the polypeptides formed by digestion in the stomach are further digested by the powerful proteolytic enzymes of the pancreas and intestinal mucosa. Trypsin, the chymotrypsins, and elastase act at interior peptide bonds in the peptide molecules and are called **endopeptidases.** The formation of the active endopeptidases from their inactive precursors occurs only when they have reached their site of action, secondary to the action of the brush border hydrolase, **enterokinase** (Figure 27–3). The powerful protein-splitting enzymes of the pancreatic juice are secreted as inactive proenzymes. Trypsinogen is converted to the active enzyme trypsin by **enterokinase** when the pancreatic juice enters the duodenum. Enterokinase

contains 41% polysaccharide, and this high polysaccharide content apparently prevents it from being digested itself before it can exert its effect. Trypsin converts chymotrypsinogens into chymotrypsins and other proenzymes into active enzymes (Figure 27–3). Trypsin can also activate trypsinogen; therefore, once some trypsin is formed, there is an auto-catalytic chain reaction. Enterokinase deficiency occurs as a congenital abnormality and leads to protein malnutrition.

The carboxypeptidases of the pancreas are **exopeptidases** that hydrolyze the amino acids at the carboxyl ends of the polypeptides (Figure 27–4). Some free amino acids are liberated in the intestinal lumen, but others are liberated at the cell surface by the aminopeptidases, carboxypeptidases, endopeptidases, and dipeptidases in the brush border of the mucosal cells. Some di- and tripeptides are actively transported into the intestinal cells and hydrolyzed by intracellular peptidases, with the amino acids entering the bloodstream. Thus, the final digestion to amino acids occurs in three locations: the intestinal lumen, the brush border, and the cytoplasm of the mucosal cells.

ABSORPTION

At least seven different transport systems transport amino acids into enterocytes. Five of these require Na^+ and cotransport amino acids and Na^+ in a fashion similar to the cotransport of Na^+ and glucose (Figure 27–3). Two of these five also require Cl^-. In two systems, transport is independent of Na^+.

FIGURE 27–3 Mechanism to avoid activation of pancreatic proteases until they are in the duodenal lumen.

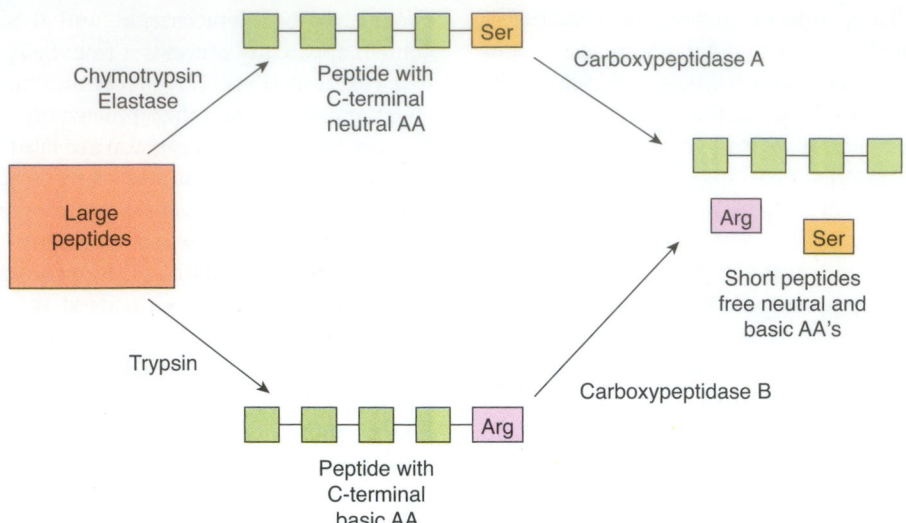

FIGURE 27–4 **Luminal digestion of peptides by pancreatic endopeptidases and exopeptidases.** Individual amino acids are shown as squares.

The di- and tripeptides are transported into enterocytes by a system known as PepT1 (or peptide transporter 1) that requires H^+ instead of Na^+ (Figure 27–5). There is very little absorption of larger peptides. In the enterocytes, amino acids released from the peptides by intracellular hydrolysis plus the amino acids absorbed from the intestinal lumen and brush border are transported out of the enterocytes along their basolateral borders by at least five transport systems. From there, they enter the hepatic portal blood.

Absorption of amino acids is rapid in the duodenum and jejunum but slow in the ileum. Approximately 50% of the digested protein comes from ingested food, 25% from proteins in digestive juices, and 25% from desquamated mucosal cells. Only 2–5% of the protein in the small intestine escapes digestion and absorption. Some of this is eventually digested by bac-

terial action in the colon. Almost all of the protein in the stools is not of dietary origin but comes from bacteria and cellular debris. Evidence suggests that the peptidase activities of the brush border and the mucosal cell cytoplasm are increased by resection of part of the ileum and that they are independently altered in starvation. Thus, these enzymes appear to be subject to homeostatic regulation. In humans, a congenital defect in the mechanism that transports neutral amino acids in the intestine and renal tubules causes **Hartnup disease.** A congenital defect in the transport of basic amino acids causes **cystinuria.** However, most patients do not experience nutritional deficiencies of these amino acids because peptide transport compensates.

In infants, moderate amounts of undigested proteins are also absorbed. The protein antibodies in maternal colostrum are largely secretory immunoglobulins (IgAs), the production of which is increased in the breast in late pregnancy. They cross the mammary epithelium by transcytosis and enter the circulation of the infant from the intestine, providing passive immunity against infections. Absorption is by endocytosis and subsequent exocytosis.

Protein absorption declines with age, but adults still absorb small quantities. Foreign proteins that enter the circulation provoke the formation of antibodies, and the antigen–antibody reaction occurring on subsequent entry of more of the same protein may cause allergic symptoms. Thus, absorption of proteins from the intestine may explain the occurrence of allergic symptoms after eating certain foods. The incidence of food allergy in children is said to be as high as 8%. Certain foods are more allergenic than others. Crustaceans, mollusks, and fish are common offenders, and allergic responses to legumes, cows' milk, and egg white are also relatively frequent.

Absorption of protein antigens, particularly bacterial and viral proteins, takes place in large **microfold cells** or **M cells,** specialized intestinal epithelial cells that overlie aggregates of lymphoid tissue (Peyer's patches). These cells pass the antigens

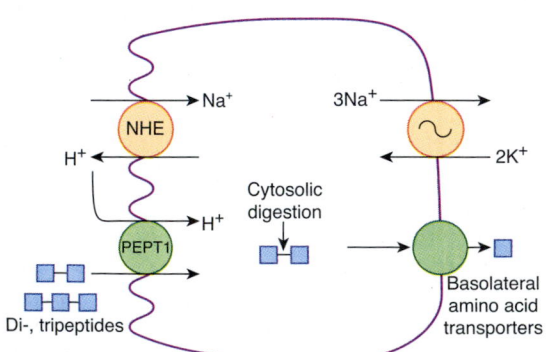

FIGURE 27–5 **Disposition of short peptides in intestinal epithelial cells.** Peptides are absorbed together with a proton supplied by an apical sodium/hydrogen exchanger (NHE) by the peptide transporter 1 (PepT1). Absorbed peptides are digested by cytosolic proteases, and any amino acids that are surplus to the needs of the epithelial cell are transported into the bloodstream by a series of basolateral transport proteins.

to the lymphoid cells, and lymphocytes are activated. The activated lymphoblasts enter the circulation, but they later return to the intestinal mucosa and other epithelia, where they secrete IgA in response to subsequent exposures to the same antigen. This **secretory immunity** is an important defense mechanism (see Chapter 3).

NUCLEIC ACIDS

Nucleic acids are split into nucleotides in the intestine by the pancreatic nucleases, and the nucleotides are split into the nucleosides and phosphoric acid by enzymes that appear to be located on the luminal surfaces of the mucosal cells. The nucleosides are then split into their constituent sugars and purine and pyrimidine bases. The bases are absorbed by active transport.

LIPIDS

FAT DIGESTION

A lingual lipase is secreted by Ebner's glands on the dorsal surface of the tongue in some species, and the stomach also secretes a lipase (Table 27–1). They are of little quantitative significance for lipid digestion other than in the setting of pancreatic insufficiency, however.

Most fat digestion therefore begins in the duodenum, pancreatic lipase being one of the most important enzymes involved. This enzyme hydrolyzes the 1- and 3-bonds of the triglycerides (triacylglycerols) with relative ease but acts on the 2-bonds at a very low rate, so the principal products of its action are free fatty acids and 2-monoglycerides (2-monoacylglycerols). It acts on fats that have been emulsified (see below). Its activity is facilitated when an amphipathic helix that covers the active site like a lid is bent back. **Colipase,** a protein with a molecular weight of about 11,000, is also secreted in the pancreatic juice, and when this molecule binds to the –COOH-terminal domain of the pancreatic lipase, opening of the lid is facilitated. Colipase is secreted in an inactive proform (Table 27–1) and is activated in the intestinal lumen by trypsin.

Another pancreatic lipase that is activated by bile salts has been characterized. This 100,000-kDa **cholesterol esterase** represents about 4% of the total protein in pancreatic juice. In adults, pancreatic lipase is 10–60 times more active, but unlike pancreatic lipase, this bile salt-activated lipase catalyzes the hydrolysis of cholesterol esters, esters of fat-soluble vitamins, and phospholipids, as well as triglycerides. A very similar enzyme is found in human milk.

Fats are relatively insoluble, which limits their ability to cross the unstirred layer and reach the surface of the mucosal cells. However, they are finely emulsified in the small intestine by the detergent action of bile salts, lecithin, and monoglycerides. When the concentration of bile salts in the intestine is high, as it is after contraction of the gallbladder, lipids and bile salts interact spontaneously to form **micelles** (Figure 26–16). These cylindrical aggregates, which are discussed in more detail in Chapter 29, take up lipids, and although their lipid concentration varies, they generally contain fatty acids, monoglycerides, and cholesterol in their hydrophobic centers. Micellar formation further solubilizes the lipids and provides a mechanism for their transport to the enterocytes. Thus, the micelles move down their concentration gradient through the unstirred layer to the brush border of the mucosal cells. The lipids diffuse out of the micelles, and a saturated aqueous solution of the lipids is maintained in contact with the brush border of the mucosal cells (Figure 26–16).

STEATORRHEA

Pancreatectomized animals and patients with diseases that destroy the exocrine portion of the pancreas have fatty, bulky, clay-colored stools (**steatorrhea**) because of the impaired digestion and absorption of fat. The steatorrhea is due mostly to lipase deficiency. However, acid inhibits the lipase, and the lack of alkaline secretion from the pancreas also contributes by lowering the pH of the intestine contents. In some cases, hypersecretion of gastric acid can cause steatorrhea. Another cause of steatorrhea is defective reabsorption of bile salts in the distal ileum (see Chapter 29).

FAT ABSORPTION

Traditionally, lipids were thought to enter the enterocytes by passive diffusion, but some evidence now suggests that carriers are involved. Inside the cells, the lipids are rapidly esterified, maintaining a favorable concentration gradient from the lumen into the cells (Figure 27–6). There are also carriers that export certain lipids back into the lumen, thereby limiting their oral availability. This is the case for plant sterols as well as cholesterol.

The fate of the fatty acids in enterocytes depends on their size. Fatty acids containing less than 10 to 12 carbon atoms are water-soluble enough that they pass through the enterocyte unmodified and are actively transported into the portal blood. They circulate as free (unesterified) fatty acids. The fatty acids containing more than 10 to 12 carbon atoms are too insoluble for this. They are reesterified to triglycerides in the enterocytes. In addition, some of the absorbed cholesterol is esterified. The triglycerides and cholesterol esters are then coated with a layer of protein, cholesterol, and phospholipid to form chylomicrons. These leave the cell and enter the lymphatics, because they are too large to pass through the junctions between capillary endothelial cells (Figure 27–6).

In mucosal cells, most of the triglyceride is formed by the acylation of the absorbed 2-monoglycerides, primarily in the smooth endoplasmic reticulum. However, some of the triglyceride is formed from glycerophosphate, which in turn is a product of glucose catabolism. Glycerophosphate is also

FIGURE 27–6 **Intracellular handling of the products of lipid digestion.** Absorbed fatty acids (FA) and monoglycerides (MG) are re-esterified to form triglyceride (TG) in the smooth endoplasmic reticulum. Apoproteins synthesized in the rough endoplasmic reticulum are coated around lipid cores, and the resulting chylomicrons are secreted from the basolateral pole of epithelial cells by exocytosis.

converted into glycerophospholipids that participate in chylomicron formation. The acylation of glycerophosphate and the formation of lipoproteins occur in the rough endoplasmic reticulum. Carbohydrate moieties are added to the proteins in the Golgi apparatus, and the finished chylomicrons are extruded by exocytosis from the basal or lateral aspects of the cell.

Absorption of long-chain fatty acids is greatest in the upper parts of the small intestine, but appreciable amounts are also absorbed in the ileum. On a moderate fat intake, 95% or more of the ingested fat is absorbed. The processes involved in fat absorption are not fully mature at birth, and infants fail to absorb 10–15% of ingested fat. Thus, they are more susceptible to the ill effects of disease processes that reduce fat absorption.

SHORT-CHAIN FATTY ACIDS IN THE COLON

Increasing attention is being focused on short-chain fatty acids (SCFAs) that are produced in the colon and absorbed from it. SCFAs are two- to five-carbon weak acids that have an average normal concentration of about 80 mmol/L in the lumen. About 60% of this total is acetate, 25% propionate, and 15% butyrate. They are formed by the action of colonic bacteria on complex carbohydrates, resistant starches, and other components of the dietary fiber, that is, the material that escapes digestion in the upper gastrointestinal tract and enters the colon.

Absorbed SCFAs are metabolized and make a significant contribution to the total caloric intake. In addition, they exert a trophic effect on the colonic epithelial cells, combat inflammation, and are absorbed in part by exchange for H^+, helping to maintain acid–base equilibrium. SCFAs are absorbed by specific transporters present in colonic epithelial cells. SCFAs

also promote the absorption of Na^+, although the exact mechanism for coupled Na^+–SCFA absorption is unsettled.

ABSORPTION OF VITAMINS & MINERALS

VITAMINS

Absorption of the fat-soluble vitamins A, D, E, and K is deficient if fat absorption is depressed because of lack of pancreatic enzymes or if bile is excluded from the intestine by obstruction of the bile duct. Most vitamins are absorbed in the upper small intestine, but vitamin B_{12} is absorbed in the ileum. This vitamin binds to intrinsic factor, a protein secreted by the stomach, and the complex is absorbed across the ileal mucosa (see Chapter 26).

Vitamin B_{12} absorption and folate absorption are Na^+-independent, but all seven of the remaining water-soluble vitamins—thiamin, riboflavin, niacin, pyridoxine, pantothenate, biotin, and ascorbic acid—are absorbed by carriers that are Na^+ cotransporters.

CALCIUM

A total of 30–80% of ingested calcium is absorbed. The absorptive process and its relation to 1,25-dihydroxycholecalciferol are discussed in Chapter 23. Through this vitamin D derivative, Ca^{2+} absorption is adjusted to body needs; absorption is increased in the presence of Ca^{2+} deficiency and decreased in the presence of Ca^{2+} excess. Ca^{2+} absorption is also facilitated by protein. It is inhibited by phosphates and oxalates because these anions form insoluble salts with Ca^{2+} in the intestine. Magnesium absorption is also facilitated by protein.

IRON

In adults, the amount of iron lost from the body is relatively small. The losses are generally unregulated, and total body stores of iron are regulated by changes in the rate at which it is absorbed from the intestine. Men lose about 0.6 mg/d, largely in the stools. Women have a variable, larger loss averaging about twice this value because of the additional iron lost during menstruation. The average daily iron intake in the United States and Europe is about 20 mg, but the amount absorbed is equal only to the losses. Thus, the amount of iron absorbed is normally about 3–6% of the amount ingested. Various dietary factors affect the availability of iron for absorption; for example, the phytic acid found in cereals reacts with iron to form insoluble compounds in the intestine, as do phosphates and oxalates.

Most of the iron in the diet is in the ferric (Fe^{3+}) form, whereas it is the ferrous (Fe^{2+}) form that is absorbed. Fe^{3+} reductase activity is associated with the iron transporter in the

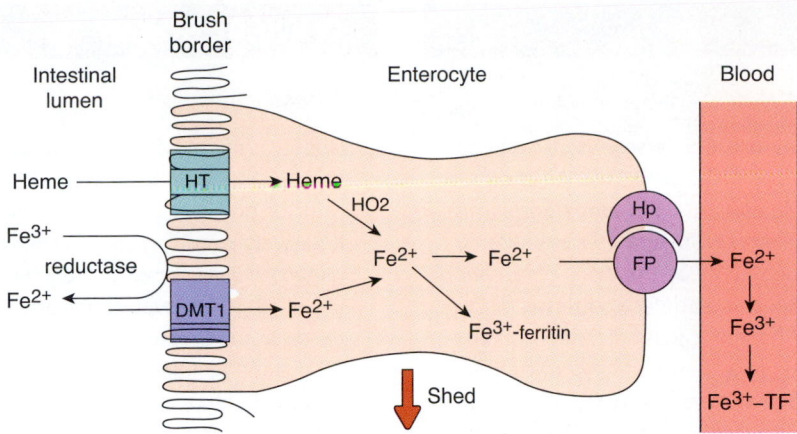

FIGURE 27–7   **Absorption of iron.** Fe^{3+} is converted to Fe^{2+} by ferric reductase, and Fe^{2+} is transported into the enterocyte by the apical membrane iron transporter DMT1. Heme is transported into the enterocyte by a separate heme transporter (HT), and HO2 releases Fe^{2+} from the heme. Some of the intracellular Fe^{2+} is converted to Fe^{3+} and bound to ferritin. The rest binds to the basolateral Fe^{2+} transporter ferroportin (FP) and is transported to the interstitial fluid. The transport is aided by hephaestin (Hp). In plasma, Fe^{2+} is converted to Fe^{3+} and bound to the iron transport protein transferrin (TF).

brush borders of the enterocytes (Figure 27–7). Gastric secretions dissolve the iron and permit it to form soluble complexes with ascorbic acid and other substances that aid its reduction to the Fe^{2+} form. The importance of this function in humans is indicated by the fact that iron deficiency anemia is a troublesome and relatively frequent complication of partial gastrectomy.

Almost all iron absorption occurs in the duodenum. Transport of Fe^{2+} into the enterocytes occurs via divalent metal transporter 1 (**DMT1**) (Figure 27–7). Some is stored in ferritin, and the remainder is transported out of the enterocytes by a basolateral transporter named **ferroportin 1.** A protein called **hephaestin (Hp)** is associated with ferroportin 1. It is not a transporter itself, but it facilitates basolateral transport. In the plasma, Fe^{2+} is converted to Fe^{3+} and bound to the iron transport protein **transferrin.** This protein has two iron-binding sites. Normally, transferrin is about 35% saturated with iron, and the normal plasma iron level is about 130 μg/dL (23 μmol/L) in men and 110 μg/dL (19 μmol/L) in women.

Heme (see Chapter 32) binds to an apical transport protein in enterocytes and is carried into the cytoplasm. In the cytoplasm, HO2, a subtype of heme oxygenase, removes Fe^{2+} from the porphyrin and adds it to the intracellular Fe^{2+} pool.

Seventy percent of the iron in the body is in hemoglobin, 3% in myoglobin, and the rest in ferritin, which is present not only in enterocytes, but also in many other cells. Apoferritin is a globular protein made up of 24 subunits. Ferritin is readily visible under the electron microscope and has been used as a tracer in studies of phagocytosis and related phenomena. Ferritin molecules in lysosomal membranes may aggregate in deposits that contain as much as 50% iron. These deposits are called **hemosiderin.**

Intestinal absorption of iron is regulated by three factors: recent dietary intake of iron, the state of the iron stores in the body, and the state of erythropoiesis in the bone marrow. The normal operation of the factors that maintain iron balance is essential for health (Clinical Box 27–2).

NUTRITIONAL PRINCIPLES & ENERGY METABOLISM

The animal organism oxidizes carbohydrates, proteins, and fats, producing principally CO_2, H_2O, and the energy necessary for life processes (Clinical Box 27–3). CO_2, H_2O, and energy are also produced when food is burned outside the body. However, in the body, oxidation is not a one-step, semiexplosive reaction but a complex, slow, stepwise process called **catabolism,** which liberates energy in small, usable amounts. Energy can be stored in the body in the form of special energy-rich phosphate compounds and in the form of proteins, fats, and complex carbohydrates synthesized from simpler molecules. Formation of these substances by processes that take up rather than liberate energy is called **anabolism.** This chapter consolidates consideration of endocrine function by providing a brief summary of the production and utilization of energy and the metabolism of carbohydrates, proteins, and fats.

METABOLIC RATE

The amount of energy liberated by the catabolism of food in the body is the same as the amount liberated when food is burned outside the body. The energy liberated by catabolic processes in the body is used for maintaining body functions, digesting and metabolizing food, thermoregulation, and physical activity. It appears as external work, heat, and energy storage:

Energy output = External work + Energy storage + Heat

The amount of energy liberated per unit of time is the **metabolic rate.** Isotonic muscle contractions perform work at a peak efficiency approximating 50%:

$$\text{Efficiency} = \frac{\text{Work done}}{\text{Total energy expended}}$$

Essentially all of the energy of isometric contractions appears as heat, because little or no external work (force multiplied by the distance that the force moves a mass) is done (see Chapter 5). Energy is stored by forming energy-rich compounds. The amount of energy storage varies, but in fasting individuals it is zero or negative. Therefore, in an adult individual who has not eaten recently and who is not moving (or growing, reproducing, or lactating), all of the energy output appears as heat.

CALORIES

The standard unit of heat energy is the **calorie (cal),** defined as the amount of heat energy necessary to raise the temperature of 1 g of water 1 degree, from 15 °C to 16 °C. This unit is also called the gram calorie, small calorie, or standard calorie. The unit commonly used in physiology and medicine is the **Calorie (kilocalorie; kcal),** which equals 1000 cal.

CALORIMETRY

The energy released by combustion of foodstuffs outside the body can be measured directly (**direct calorimetry**) by oxidizing the compounds in an apparatus such as a **bomb calorimeter,** a metal vessel surrounded by water inside an insulated container. The food is ignited by an electric spark. The change in the temperature of the water is a measure of the calories produced. Similar measurements of the energy released by combustion of compounds in living animals and humans are much more complex, but calorimeters have been constructed that can physically accommodate human beings. The heat produced by their bodies is measured by the change in temperature of the water in the walls of the calorimeter.

The caloric values of the common foodstuffs, as measured in a bomb calorimeter, are found to be 4.1 kcal/g of carbohydrate, 9.3 kcal/g of fat, and 5.3 kcal/g of protein. In the body, similar values are obtained for carbohydrate and fat, but the oxidation of protein is incomplete, the end products of protein catabolism being urea and related nitrogenous compounds in addition to CO_2 and H_2O (see below). Therefore, the caloric value of protein in the body is only 4.1 kcal/g.

INDIRECT CALORIMETRY

Energy production can also be calculated by measuring the products of the energy-producing biologic oxidations; that is, CO_2, H_2O, and the end products of protein catabolism produced, but this is difficult. However, O_2 is not stored, and except when an O_2 debt is being incurred, the amount of O_2 consumption per unit of time is proportionate to the energy liberated by metabolism. Consequently, measurement of O_2 consumption (**indirect calorimetry**) is used to determine the metabolic rate.

RESPIRATORY QUOTIENT (RQ)

The **respiratory quotient (RQ)** is the ratio in the steady state of the volume of CO_2 produced to the volume of O_2 consumed per unit of time. It should be distinguished from the **respiratory exchange ratio (R),** which is the ratio of CO_2 to O_2 at any given time whether or not equilibrium has been reached. R is affected by factors other than metabolism. RQ and R can be calculated for reactions outside the body, for individual organs and tissues, and for the whole body. The RQ of carbohydrate is 1.00, and that of fat is about 0.70. This is because H and O are present in carbohydrate in the same proportions as in water, whereas in the various fats, extra O_2 is necessary for the formation of H_2O.

Carbohydrate:

$$C_6H_{12}O_6 + 6O_2 \rightarrow 6CO_2 + 6H_2O$$
$$\text{(glucose)}$$
$$RQ = 6/6 = 1.00$$

Fat:

$$2C_{51}H_{98}O_6 + 145O_2 \rightarrow 102CO_2 + 98H_2O$$
$$\text{(tripalmitin)}$$
$$RQ = 102/145 = 0.703$$

Determining the RQ of protein in the body is a complex process, but an average value of 0.82 has been calculated. The approximate amounts of carbohydrate, protein, and fat being oxidized in the body at any given time can be calculated from the RQ and the urinary nitrogen excretion. RQ and R for the whole body differ in various conditions. For example, during hyperventilation, R rises because CO_2 is being blown off. During strenuous exercise, R may reach 2.00 because CO_2 is being blown off and lactic acid from anaerobic glycolysis is being converted to CO_2 (see below). After exercise, R may fall for a while to 0.50 or less. In metabolic acidosis, R rises because respiratory compensation for the acidosis causes the amount of CO_2 expired to rise (see Chapter 39). In severe acidosis, R may be greater than 1.00. In metabolic alkalosis, R falls.

The O_2 consumption and CO_2 production of an organ can be calculated at equilibrium by multiplying its blood flow per unit of time by the arteriovenous differences for O_2 and CO_2 across the organ, and the RQ can then be calculated. Data on the RQ of individual organs are of considerable interest in drawing inferences about the metabolic processes occurring in them. For example, the RQ of the brain is regularly 0.97–0.99, indicating that its principal but not its only fuel is carbohydrate. During secretion of gastric juice, the stomach has a negative R because it takes up more CO_2 from the arterial blood than it puts into the venous blood (see Chapter 26).

MEASURING THE METABOLIC RATE

In determining the metabolic rate, O_2 consumption is usually measured with some form of oxygen-filled spirometer and a CO_2-absorbing system. Such a device is illustrated in Figure 27–8. The spirometer bell is connected to a pen that writes on a rotating drum as the bell moves up and down. The slope of a line joining the ends of each of the spirometer excursions is proportional to the O_2 consumption. The amount of O_2 (in milliliters) consumed per unit of time is corrected to standard temperature and pressure (see Chapter 35) and then converted to energy production by multiplying by 4.82 kcal/L of O_2 consumed.

FACTORS AFFECTING THE METABOLIC RATE

The metabolic rate is affected by many factors (Table 27–2). The most important is muscular exertion. O_2 consumption is elevated not only during exertion but also for as long afterward as is necessary to repay the O_2 debt (see Chapter 5). Recently ingested foods also increase the metabolic rate because of their **specific dynamic action (SDA).** The SDA of a food is the obligatory energy expenditure that occurs during its assimilation into the body. It takes 30 kcal to assimilate the amount of protein sufficient to raise the metabolic rate 100 kcal; 6 kcal to assimilate a similar amount of carbohydrate; and 5 kcal to assimilate a similar amount of fat. The cause of the SDA, which may last up to 6 h, is uncertain.

Another factor that stimulates metabolism is the environmental temperature. The curve relating the metabolic rate to the environmental temperature is U-shaped. When the environmental temperature is lower than body temperature, heat-producing mechanisms such as shivering are activated and the metabolic rate rises. When the temperature is high enough to raise the body temperature, metabolic processes generally accelerate, and the metabolic rate rises about 14% for each degree Celsius of elevation.

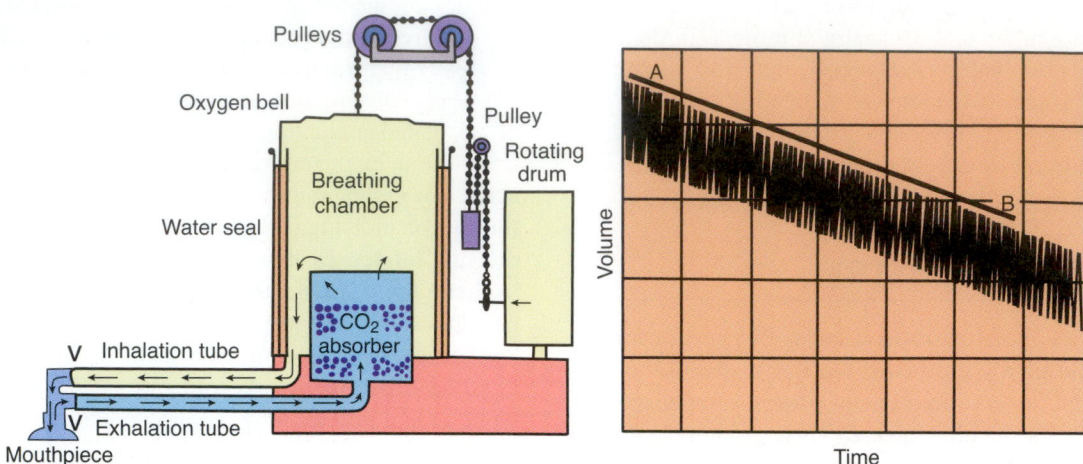

FIGURE 27–8   **Diagram of a modified Benedict apparatus, a recording spirometer used for measuring human O$_2$ consumption, and the record obtained with it.** The slope of the line AB is proportionate to the O$_2$ consumption. V: one-way check valve.

The metabolic rate determined at rest in a room at a comfortable temperature in the thermoneutral zone 12 to 14 h after the last meal is called the **basal metabolic rate (BMR).** This value falls about 10% during sleep and up to 40% during prolonged starvation. The rate during normal daytime activities is, of course, higher than the BMR because of muscular activity and food intake. The **maximum metabolic rate** reached during exercise is often said to be 10 times the BMR, but trained athletes can increase their metabolic rate as much as 20-fold.

TABLE 27–2   Factors affecting the metabolic rate.

Muscular exertion during or just before measurement
Recent ingestion of food
High or low environmental temperature
Height, weight, and surface area
Sex
Age
Growth
Reproduction
Lactation
Emotional state
Body temperature
Circulating levels of thyroid hormones
Circulating epinephrine and norepinephrine levels

The BMR of a man of average size is about 2000 kcal/d. Large animals have higher absolute BMRs, but the ratio of BMR to body weight in small animals is much greater. One variable that correlates well with the metabolic rate in different species is the body surface area. This would be expected, since heat exchange occurs at the body surface. The actual relation to body weight (W) would be

$$BMR = 3.52W^{0.67}$$

However, repeated measurements by numerous investigators have come up with a higher exponent, averaging 0.75:

$$BMR = 3.52W^{0.75}$$

Thus, the slope of the line relating metabolic rate to body weight is steeper than it would be if the relation were due solely to body area (Figure 27–9). The cause of the greater slope has been much debated but remains unsettled.

For clinical use, the BMR is usually expressed as a percentage increase or decrease above or below a set of generally used standard normal values. Thus, a value of +65 means that the individual's BMR is 65% above the standard for that age and sex.

The decrease in metabolic rate is part of the explanation of why, when an individual is trying to lose weight, weight loss is initially rapid and then slows down.

ENERGY BALANCE

The first law of thermodynamics, the principle that states that energy is neither created nor destroyed when it is converted from one form to another, applies to living organisms as well as inanimate systems. One may therefore speak of an **energy**

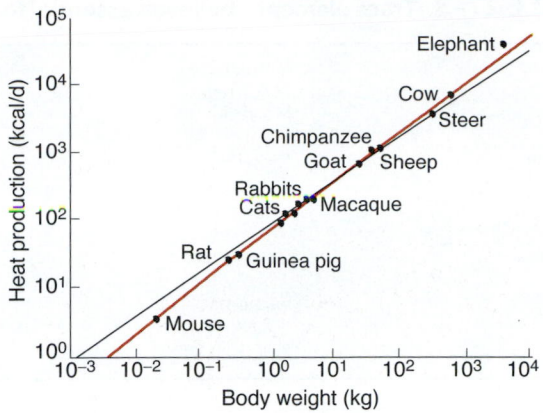

FIGURE 27–9 **Correlation between metabolic rate and body weight, plotted on logarithmic scales.** The slope of the colored line is 0.75. The black line represents the way surface area increases with weight for geometrically similar shapes and has a slope of 0.67. (Modified from Kleiber M and reproduced with permission from McMahon TA: Size and shape in biology. Science 1973;179:1201. Copyright © 1973 by the American Association for the Advancement of Science.)

balance between caloric intake and energy output. If the caloric content of the food ingested is less than the energy output, that is, if the balance is negative, endogenous stores are utilized. Glycogen, body protein, and fat are catabolized, and the individual loses weight. If the caloric value of the food intake exceeds energy loss due to heat and work and the food is properly digested and absorbed, that is, if the balance is positive, energy is stored, and the individual gains weight.

To balance basal output so that the energy-consuming tasks essential for life can be performed, the average adult must take in about 2000 kcal/d. Caloric requirements above the basal level depend on the individual's activity. The average sedentary student (or professor) needs another 500 kcal, whereas a lumberjack needs up to 3000 additional kcal per day.

NUTRITION

The aim of the science of nutrition is the determination of the kinds and amounts of foods that promote health and well-being. This includes not only the problems of undernutrition but those of overnutrition, taste, and availability (Clinical Box 27–4). However, certain substances are essential constituents of any human diet. Many of these compounds have been mentioned in previous sections of this chapter, and a brief summary of the essential and desirable dietary components is presented below.

ESSENTIAL DIETARY COMPONENTS

An optimal diet includes, in addition to sufficient water (see Chapter 38), adequate calories, protein, fat, minerals, and vitamins.

CALORIC INTAKE & DISTRIBUTION

As noted above, the caloric value of the dietary intake must be approximately equal to the energy expended if body weight is to be maintained. In addition to the 2000 kcal/d necessary to meet basal needs, 500 to 2500 kcal/d (or more) are required to meet the energy demands of daily activities.

The distribution of the calories among carbohydrate, protein, and fat is determined partly by physiologic factors and partly by taste and economic considerations. A daily protein intake of 1 g/kg body weight to supply the eight nutritionally essential amino acids and other amino acids is desirable. The source of the protein is also important. **Grade I proteins,** the animal proteins of meat, fish, dairy products, and eggs, contain amino acids in approximately the proportions required for protein synthesis and other uses. Some of the plant proteins are also grade I, but most are **grade II** because they supply different proportions of amino acid and some lack one or more of the essential amino acids. Protein needs can be met with a mixture of grade II proteins, but the intake must be large because of the amino acid wastage.

Fat is the most compact form of food, since it supplies 9.3 kcal/g. However, often it is also the most expensive. Indeed, internationally there is a reasonably good positive correlation between fat intake and standard of living. In the past, Western diets have contained large amounts (100 g/d or more). The evidence indicating that a high unsaturated/saturated fat ratio in the diet is of value in the prevention of atherosclerosis and the current interest in preventing obesity may change this. In Central and South American Indian communities where corn (carbohydrate) is the dietary staple, adults live without ill effects for years on a very low fat intake. Therefore, provided that the needs for essential fatty acids are met, a low-fat intake does not seem to be harmful, and a diet low in saturated fats is desirable.

Carbohydrate is the cheapest source of calories and provides 50% or more of the calories in most diets. In the average middle-class American diet, approximately 50% of the calories come from carbohydrate, 15% from protein, and 35% from fat. When calculating dietary needs, it is usual to meet the protein requirement first and then split the remaining calories between fat and carbohydrate, depending on taste, income, and other factors. For example, a 65-kg man who is moderately active needs about 2800 kcal/d. He should eat at least 65 g of protein daily, supplying 267 (65×4.1) kcal. Some of this should be grade I protein. A reasonable figure for fat intake is 50 to 60 g. The rest of the caloric requirement can be met by supplying carbohydrate.

MINERAL REQUIREMENTS

A number of minerals must be ingested daily for the maintenance of health. Besides those for which recommended daily dietary allowances have been set, a variety of different trace elements should be included. Trace elements are defined as elements found in tissues in minute amounts. Those believed to be essential for life, at least in experimental animals, are listed in Table 27–3. In humans, iron deficiency causes anemia. Cobalt is part of the vitamin B_{12} molecule, and vitamin B_{12} deficiency leads to megaloblastic anemia (see Chapter 32). Iodine deficiency causes thyroid disorders (see Chapter 20). Zinc deficiency causes skin ulcers, depressed immune responses, and hypogonadal dwarfism. Copper deficiency causes anemia and changes in ossification. Chromium deficiency causes insulin resistance. Fluorine deficiency increases the incidence of dental caries.

Conversely, some minerals can be toxic when present in the body in excess. For example, severe iron overload causes hemo-

TABLE 27–3 Trace elements believed essential for life.

Arsenic	Manganese
Chromium	Molybdenum
Cobalt	Nickel
Copper	Selenium
Fluorine	Silicon
Iodine	Vanadium
Iron	Zinc

chromatosis, copper excess causes brain damage (Wilson disease), and aluminum poisoning in patients with renal failure who are receiving dialysis treatment causes a rapidly progressive dementia that resembles Alzheimer disease (see Chapter 19).

Sodium and potassium are also essential minerals, but listing them is academic, because it is very difficult to prepare a sodium-free or potassium-free diet. A low-salt diet is well tolerated for prolonged periods because of the compensatory mechanisms that conserve Na^+.

VITAMINS

Vitamins were discovered when it was observed that diets adequate in calories, essential amino acids, fats, and minerals failed to maintain health. The term **vitamin** has now come to refer to any organic dietary constituent necessary for life, health, and growth that does not function by supplying energy.

Because there are minor differences in metabolism between mammalian species, some substances are vitamins in one species and not in another. The sources and functions of the major vitamins in humans are listed in Table 27–4. Most vitamins have important functions in intermediary metabolism or the special metabolism of the various organ systems. Those that are water-soluble (vitamin B complex, vitamin C) are easily absorbed, but the fat-soluble vitamins (vitamins A, D, E, and K) are poorly absorbed in the absence of bile or pancreatic lipase. Some dietary fat intake is necessary for their absorption, and in obstructive jaundice or disease of the exocrine pancreas, deficiencies of the fat-soluble vitamins can develop even if their intake is adequate. Vitamin A and vitamin D are bound to transfer proteins in the circulation. The α-tocopherol form of vitamin E is normally bound to chylomicrons. In the liver, it is transferred to very low density lipoprotein (VLDL) and distributed to tissues by an α-tocopherol transfer protein. When this protein is abnormal due to mutation of its gene in humans, there is cellular deficiency of vitamin E and the development of a condition resembling Friedreich ataxia. Two Na^+-dependent L-ascorbic acid transporters have recently been isolated. One is found in the kidneys, intestines, and liver, and the other in the brain and eyes.

TABLE 27–4   Vitamins essential or probably essential to human nutrition.[a]

Vitamin	Action	Deficiency Symptoms	Sources	Chemistry
A (A_1, A_2)	Constituents of visual pigments (see Chapter 12: Vision); necessary for fetal development and for cell development throughout life	Night blindness, dry skin	Yellow vegetables and fruit	Vitamin A_1 alcohol (retinol).
B complex				
Thiamin (vitamin B_1)	Cofactor in decarboxylations	Beriberi, neuritis	Liver, unrefined cereal grains	
Riboflavin (vitamin B_2)	Constituent of flavoproteins	Glossitis, cheilosis	Liver, milk	
Niacin	Constituent of NAD^+ and $NADP^+$	Pellagra	Yeast, lean meat, liver	Can be synthesized in body from tryptophan.
Pyridoxine (vitamin B_6)	Forms prosthetic group of certain decarboxylases and transaminases. Converted in body into pyridoxal phosphate and pyridoxamine phosphate	Convulsions, hyperirritability	Yeast, wheat, corn, liver	
Pantothenic acid	Constituent of CoA	Dermatitis, enteritis, alopecia, adrenal insufficiency	Eggs, liver, yeast	
Biotin	Catalyzes CO_2 "fixation" (in fatty acid synthesis, etc)	Dermatitis, enteritis	Egg yolk, liver, tomatoes	
Folates (folic acid) and related compounds	Coenzymes for "1-carbon" transfer; involved in methylating reactions	Sprue, anemia. Neural tube defects in children born to folate-deficient women	Leafy green vegetables	Folic acid
Cyanocobalamin (vitamin B_{12})	Coenzyme in amino acid metabolism. Stimulates erythropoiesis	Pernicious anemia (see Chapter 26: Overview of Gastrointestinal Function & Regulation)	Liver, meat, eggs, milk	Complex of four substituted pyrrole rings around a cobalt atom (see Chapter 26: Overview of Gastrointestinal Function & Regulation)
C	Maintains prosthetic metal ions in their reduced form; scavenges free radicals	Scurvy	Citrus fruits, leafy green vegetables	Ascorbic acid (synthesized in most mammals except guinea pigs and primates, including humans).

(continued)

TABLE 27–4   **Vitamins essential or probably essential to human nutrition.^{a} (Continued)**

Vitamin	Action	Deficiency Symptoms	Sources	Chemistry
D group	Increase intestinal absorption of calcium and phosphate (see Chapter 21: Hormonal Control of Calcium & Phosphate Metabolism & the Physiology of Bone)	Rickets	Fish liver	Family of sterols (see Chapter 21: Hormonal Control of Calcium & Phosphate Metabolism & the Physiology of Bone)
E group	Antioxidants; cofactors in electron transport in cytochrome chain?	Ataxia and other symptoms and signs of spinocerebellar dysfunction	Milk, eggs, meat, leafy vegetables	α-Tocopherol; β- and γ-tocopherol also active.
K group	Catalyze γ carboxylation of glutamic acid residues on various proteins concerned with blood clotting	Hemorrhagic phenomena	Leafy green vegetables	Vitamin K₃; a large number of similar compounds have biological activity.

^{a}Choline is synthesized in the body in small amounts, but it has recently been added to the list of essential nutrients.

The diseases caused by deficiency of each of the vitamins are listed in Table 27–4. It is worth remembering, however, particularly in view of the advertising campaigns for vitamin pills and supplements, that very large doses of the fat-soluble vitamins are definitely toxic. **Hypervitaminosis A** is characterized by anorexia, headache, hepatosplenomegaly, irritability, scaly dermatitis, patchy loss of hair, bone pain, and hyperostosis. Acute vitamin A intoxication was first described by Arctic explorers, who developed headache, diarrhea, and dizziness after eating polar bear liver. The liver of this animal is particularly rich in vitamin A. **Hypervitaminosis D** is associated with weight loss, calcification of many soft tissues, and eventual renal failure. **Hypervitaminosis K** is characterized by gastrointestinal disturbances and anemia. Large doses of water-soluble vitamins have been thought to be less likely to cause problems because they can be rapidly cleared from the body. However, it has been demonstrated that ingestion of megadoses of pyridoxine (vitamin B_6) can produce peripheral neuropathy.

CHAPTER SUMMARY

■ A typical mixed meal consists of carbohydrates, proteins, and lipids (the latter largely in the form of triglycerides). Each must be digested to allow its uptake into the body. Specific transporters carry the products of digestion into the body.

■ In the process of carbohydrate assimilation, the epithelium can only transport monomers, whereas for proteins, short peptides can be absorbed in addition to amino acids.

■ The protein assimilation machinery, which rests heavily on the proteases in pancreatic juice, is arranged such that these enzymes are not activated until they reach their substrates in the small intestinal lumen. This is accomplished by the restricted localization of an activating enzyme, enterokinase.

■ Lipids face special challenges to assimilation given their hydrophobicity. Bile acids solubilize the products of lipolysis in micelles and accelerate their ability to diffuse to the epithelial surface. The assimilation of triglycerides is enhanced by this mechanism, whereas that of cholesterol and fat-soluble vitamins absolutely requires it.

■ The catabolism of nutrients provides energy to the body in a controlled fashion, via stepwise oxidations and other reactions.

■ A balanced diet is important for health, and certain substances obtained from the diet are essential to life. The caloric value of dietary intake must be approximately equal to energy expenditure for homeostasis.

MULTIPLE-CHOICE QUESTIONS

For all questions, select the single best answer unless otherwise directed.

1. Maximum absorption of short-chain fatty acids produced by bacteria occurs in the
   A) stomach.
   B) duodenum.
   C) jejunum.
   D) ileum.
   E) colon.

2. Calcium absorption is increased by
   A) hypercalcemia.
   B) oxalates in the diet.
   C) iron overload.
   D) 1,25-dihydroxycholecalciferol.
   E) increased Na$^+$ absorption.

3. A decrease in which of the following would be expected in a child exhibiting a congenital absence of enterokinase?

   A) incidence of pancreatitis
   B) glucose absorption
   C) bile acid reabsorption
   D) gastric pH
   E) protein assimilation

4. In Hartnup disease (a defect in the transport of neutral amino acids), patients do not become deficient in these amino acids due to the activity of

   A) PepT1.
   B) brush border peptidases.
   C) Na^+,K^+ ATPase.
   D) cystic fibrosis transmembrane conductance regulator (CFTR).
   E) trypsin.

5. A newborn baby is brought to the pediatrician suffering from severe diarrhea that worsens with meals. The symptoms diminish when nutrients are delivered intravenously. The child most likely has a mutation in which of the following intestinal transporters?

   A) Na^+,K^+ ATPase
   B) NHE3
   C) SGLT1
   D) H^+,K^+ ATPase
   E) NKCC1

CHAPTER RESOURCES

Andrews NC: Disorders of iron metabolism. N Engl J Med 1999;341:1986.

Chong L, Marx J (editors): Lipids in the limelight. Science 2001;294:1861.

Farrell RJ, Kelly CP: Celiac sprue. N Engl J Med 2002;346:180.

Hofmann AF: Bile acids: The good, the bad, and the ugly. News Physiol Sci 1999;14:24.

Levitt MD, Bond JH: Volume, composition and source of intestinal gas. Gastroenterology 1970;59:921.

Mann NS, Mann SK: Enterokinase. Proc Soc Exp Biol Med 1994;206:114.

Meier PJ, Stieger B: Molecular mechanisms of bile formation. News Physiol Sci 2000;15:89.

Topping DL, Clifton PM: Short-chain fatty acids and human colonic function: Select resistant starch and nonstarch polysaccharides. Physiol Rev 2001;81:1031.

Wright EM: The intestinal Na^+/glucose cotransporter. Annu Rev Physiol 1993;55:575.

Gastrointestinal Motility

INTRODUCTION

The digestive and absorptive functions of the gastrointestinal system outlined in the previous chapter depend on a variety of mechanisms that soften the food, propel it through the length of the gastrointestinal tract (Table 28–1), and mix it with hepatic bile stored in the gallbladder and digestive enzymes secreted by the salivary glands and pancreas. Some of these mechanisms depend on intrinsic properties of the intestinal smooth muscle. Others involve the operation of reflexes involving the neurons intrinsic to the gut, reflexes involving the central nervous system (CNS), paracrine effects of chemi-cal messengers, and gastrointestinal hormones.

GENERAL PATTERNS OF MOTILITY

PERISTALSIS

Peristalsis is a reflex response that is initiated when the gut wall is stretched by the contents of the lumen, and it occurs in all parts of the gastrointestinal tract from the esophagus to the rec-tum. The stretch initiates a circular contraction behind the stimulus and an area of relaxation in front of it (Figure 28–1). The wave of contraction then moves in an oral-to-caudal direc-tion, propelling the contents of the lumen forward at rates that vary from 2 to 25 cm/s. Peristaltic activity can be increased or decreased by the autonomic input to the gut, but its occurrence is independent of the extrinsic innervation. Indeed, progres-sion of the contents is not blocked by removal and resuture of a segment of intestine in its original position and is blocked only if the segment is reversed before it is sewn back into place. Peristalsis is an excellent example of the integrated activity of the enteric nervous system. It appears that local stretch releases serotonin, which activates sensory neurons that activate the myenteric plexus. Cholinergic neurons passing in a retrograde direction in this plexus activate neurons that release substance P and acetylcholine, causing smooth muscle contraction. At the same time, cholinergic neurons passing in an anterograde

TABLE 28–1 Mean lengths of various segments of the gastrointestinal tract as measured by intubation in living humans.

Segment	Length (cm)
Pharynx, esophagus, and stomach	65
Duodenum	25
Jejunum and ileum	260
Colon	110

Data from Hirsch JE, Ahrens EH Jr, Blankenhorn DH: Measurement of human intestinal length in vivo and some causes of variation. Gastroenterology 1956;31:274.

direction activate neurons that secrete NO, vasoactive intestinal polypeptide (VIP), and adenosine triphosphate (ATP), producing the relaxation ahead of the stimulus.

SEGMENTATION & MIXING

When the meal is present, the enteric nervous system promotes a motility pattern that is related to peristalsis, but is designed to retard the movement of the intestinal contents along the length of the intestinal tract to provide time for digestion and absorption (Figure 28–1). This motility pattern is known as segmentation, and it provides for ample mixing of the intestinal contents (known as chyme) with the digestive juices. A segment of bowel contracts at both ends, and then a second contraction occurs in the center of the segment to force the chyme both backward and forward. Unlike peristalsis, therefore, retrograde movement of the chyme occurs routinely in

FIGURE 28–1   Patterns of gastrointestinal motility and propulsion. An isolated contraction moves contents orally and aborally. Segmentation mixes contents over a short stretch of intestine, as indicated by the time sequence from left to right. In the diagram on the left, the vertical arrows indicate the sites of subsequent contraction. Peristalsis involves both contraction and relaxation, and moves contents aborally.

the setting of segmentation. This mixing pattern persists for as long as nutrients remain in the lumen to be absorbed. It presumably reflects programmed activity of the bowel dictated by the enteric nervous system, and can occur independent of central input, although the latter can modulate it.

BASIC ELECTRICAL ACTIVITY & REGULATION OF MOTILITY

Except in the esophagus and the proximal portion of the stomach, the smooth muscle of the gastrointestinal tract has spontaneous rhythmic fluctuations in membrane potential between about –65 and –45 mV. This **basic electrical rhythm (BER)** is initiated by the **interstitial cells of Cajal,** stellate mesenchymal pacemaker cells with smooth muscle-like features that send long multiply branched processes into the intestinal smooth muscle. In the stomach and the small intestine, these cells are located in the outer circular muscle layer near the myenteric plexus; in the colon, they are at the submucosal border of the circular muscle layer. In the stomach and small intestine, there is a descending gradient in pacemaker frequency, and as in the heart, the pacemaker with the highest frequency usually dominates.

The BER itself rarely causes muscle contraction, but **spike potentials** superimposed on the most depolarizing portions of the BER waves do increase muscle tension (Figure 28–2). The depolarizing portion of each spike is due to Ca^{2+} influx, and the repolarizing portion is due to K^+ efflux. Many polypeptides and neurotransmitters affect the BER. For example, acetylcholine increases the number of spikes and the tension of the smooth muscle, whereas epinephrine decreases the number of spikes and the tension. The rate of the BER is about 4/min in the stomach. It is about 12/min in the duodenum and falls to about 8/min in the distal ileum. In the colon, the BER rate rises from about 2/min at the cecum to about 6/min at the sigmoid. The function of the BER is to coordinate peristaltic and other motor activity; contractions occur only during the depolarizing part of the waves. After vagotomy or transection of the stomach wall, for example, peristalsis in the stomach becomes irregular and chaotic.

MIGRATING MOTOR COMPLEX

During fasting between periods of digestion, the pattern of electrical and motor activity in gastrointestinal smooth muscle becomes modified so that cycles of motor activity migrate from the stomach to the distal ileum. Each cycle, or **migrating motor complex (MMC),** starts with a quiescent period (phase I), continues with a period of irregular electrical and mechanical activity (phase II), and ends with a burst of regular activity (phase III). The MMCs are initiated by motilin, migrate aborally at a rate of about 5 cm/min, and occur at intervals of approximately 90 min. Gastric secretion, bile flow, and pancreatic secretion increase during each MMC. They likely serve to clear the stomach and small intestine of luminal contents in preparation for the next meal. They are immediately stopped by ingestion of food

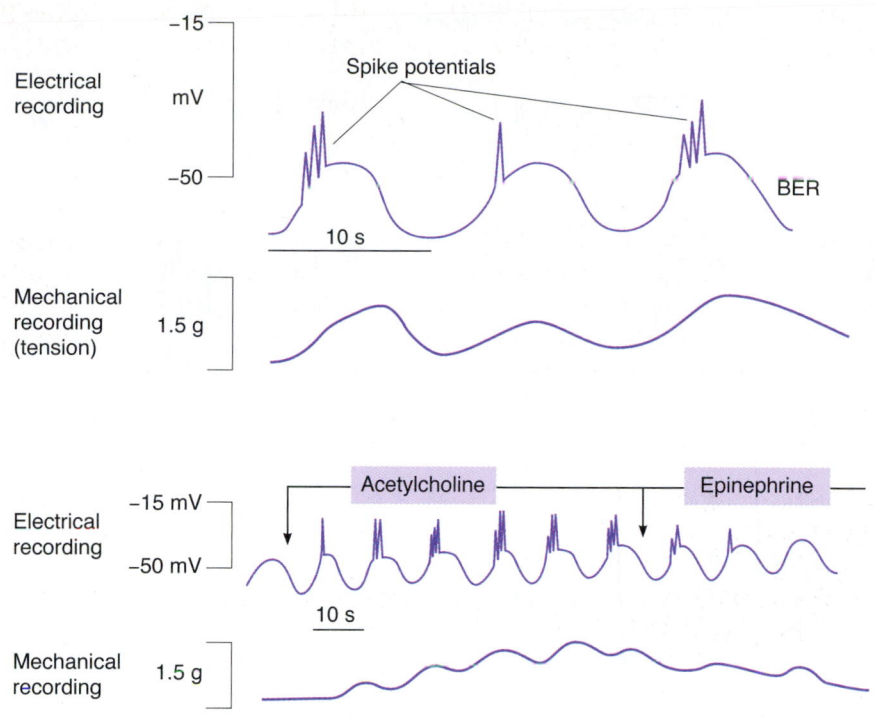

FIGURE 28–2 **Basic electrical rhythm (BER) of gastrointestinal smooth muscle. Top:** Morphology, and relation to muscle contraction. **Bottom:** Stimulatory effect of acetylcholine and inhibitory effect of epinephrine. (Modified and reproduced with permission from Chang EB, Sitrin MD, Black DD: *Gastrointestinal, Hepatobiliary, and Nutritional Physiology.* Lippincott-Raven, 1996.)

(which suppresses motilin release via mechanisms that have not yet been elucidated), with a return to peristalsis and the other forms of BER and spike potentials.

SEGMENT-SPECIFIC PATTERNS OF MOTILITY

MOUTH & ESOPHAGUS

In the mouth, food is mixed with saliva and propelled into the esophagus. Peristaltic waves in the esophagus move the food into the stomach.

MASTICATION

Chewing (**mastication**) breaks up large food particles and mixes the food with the secretions of the salivary glands. This wetting and homogenizing action aids swallowing and subsequent digestion. Large food particles can be digested, but they cause strong and often painful contractions of the esophageal musculature. Particles that are small tend to disperse in the absence of saliva and also make swallowing difficult because they do not form a bolus. The number of chews that is optimal depends on the food, but usually ranges from 20 to 25.

Edentulous patients are generally restricted to a soft diet and have considerable difficulty eating dry food.

SWALLOWING

Swallowing (deglutition) is a reflex response that is triggered by afferent impulses in the trigeminal, glossopharyngeal, and vagus nerves (Figure 28–3). These impulses are integrated in the nucleus of the tractus solitarius and the nucleus ambiguus. The efferent fibers pass to the pharyngeal musculature and the tongue via the trigeminal, facial, and hypoglossal nerves. Swallowing is initiated by the voluntary action of collecting the oral contents on the tongue and propelling them backward into the pharynx. This starts a wave of involuntary contraction in the pharyngeal muscles that pushes the material into the esophagus. Inhibition of respiration and glottic closure are part of the reflex response. A peristaltic ring contraction of the esophageal muscle forms behind the material, which is then swept down the esophagus at a speed of approximately 4 cm/s. When humans are in an upright position, liquids and semisolid foods generally fall by gravity to the lower esophagus ahead of the peristaltic wave.

LOWER ESOPHAGEAL SPHINCTER

Unlike the rest of the esophagus, the musculature of the gastroesophageal junction (**lower esophageal sphincter; LES**) is tonically active but relaxes on swallowing. The tonic activity of the LES between meals prevents reflux of gastric contents into the esophagus. The LES is made up of three components (Figure 28–4). The esophageal smooth muscle is more prominent

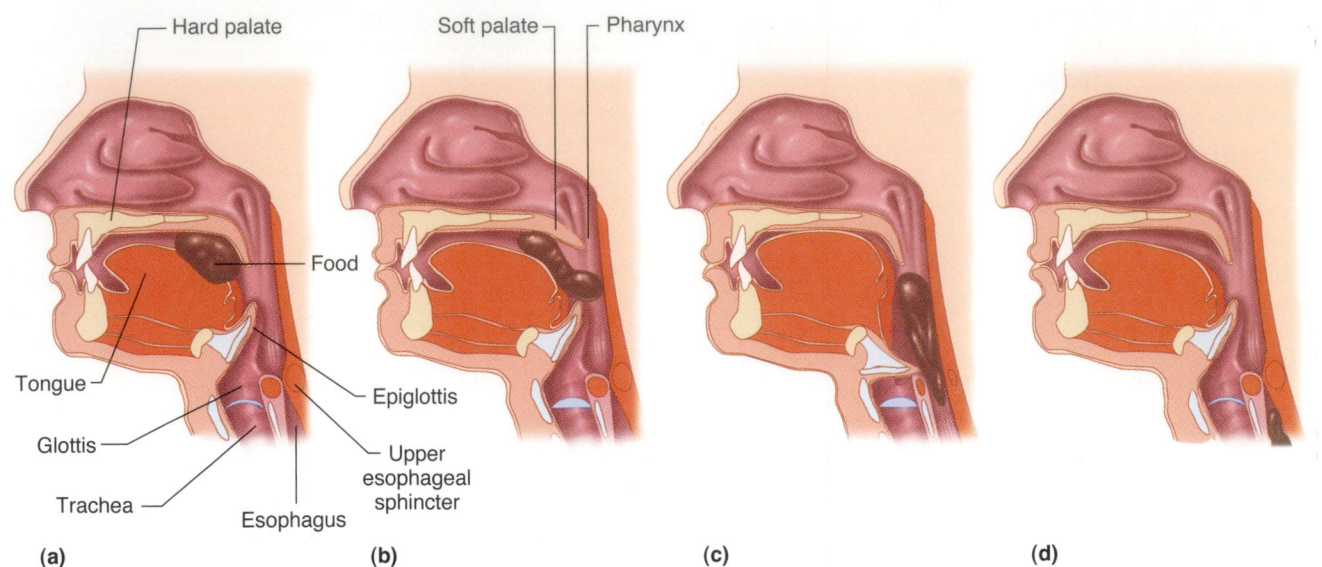

FIGURE 28–3   **Movement of food through the pharynx and upper esophagus during swallowing. (a)** The tongue pushes the food bolus to the back of the mouth. **(b)** The soft palate elevates to prevent food from entering the nasal passages. **(c)** The epiglottis covers the glottis to prevent food from entering the trachea and the upper esophageal sphincter relaxes. **(d)** Food descends into the esophagus.

at the junction with the stomach (intrinsic sphincter). Fibers of the crural portion of the diaphragm, a skeletal muscle, surround the esophagus at this point (extrinsic sphincter) and exert a pinchcock-like action on the esophagus. In addition, the oblique or sling fibers of the stomach wall create a flap valve that helps close off the esophagogastric junction and prevent regurgitation when intragastric pressure rises.

The tone of the LES is under neural control. Release of acetylcholine from vagal endings causes the intrinsic sphincter to

contract, and release of NO and VIP from interneurons innervated by other vagal fibers causes it to relax. Contraction of the crural portion of the diaphragm, which is innervated by the phrenic nerves, is coordinated with respiration and contractions of chest and abdominal muscles. Thus, the intrinsic and extrinsic sphincters operate together to permit orderly flow of food into the stomach and to prevent reflux of gastric contents into the esophagus (Clinical Box 28–1).

FIGURE 28–4   **Esophagogastric junction.** Note that the lower esophageal sphincter (intrinsic sphincter) is supplemented by the crural portion of the diaphragm (extrinsic sphincter), and that the two are anchored to each other by the phrenoesophageal ligament.   (Reproduced with permission, from Mittal RK, Balaban DH: The esophagogastric junction. N Engl J Med 1997;336:924. Copyright © 1997 by the Massachusetts Medical Society. All rights reserved.)

AEROPHAGIA & INTESTINAL GAS

Some air is unavoidably swallowed in the process of eating and drinking (**aerophagia**). Some of the swallowed air is regurgitated (belching), and some of the gases it contains are absorbed, but much of it passes on to the colon. Here, some of the oxygen is absorbed, and hydrogen, hydrogen sulfide, carbon dioxide, and methane formed by the colonic bacteria from carbohydrates and other substances are added to it. It is then expelled as **flatus.** The smell is largely due to sulfides. The volume of gas normally found in the human gastrointestinal tract is about 200 mL, and the daily production is 500 to 1500 mL. In some individuals, gas in the intestines causes cramps, **borborygmi** (rumbling noises), and abdominal discomfort.

STOMACH

Food is stored in the stomach; mixed with acid, mucus, and pepsin; and released at a controlled, steady rate into the duodenum.

GASTRIC MOTILITY & EMPTYING

When food enters the stomach, the fundus and upper portion of the body relax and accommodate the food with little if any increase in pressure (**receptive relaxation**). Peristalsis then begins in the lower portion of the body, mixing and grinding the food and permitting small, semiliquid portions of it to pass through the pylorus and enter the duodenum.

Receptive relaxation is vagally mediated and triggered by movement of the pharynx and esophagus. Peristaltic waves controlled by the gastric BER begin soon thereafter and sweep toward the pylorus. The contraction of the distal stomach caused by each wave is sometimes called **antral systole** and can last up to 10 s. Waves occur three to four times per minute.

In the regulation of gastric emptying, the antrum, pylorus, and upper duodenum apparently function as a unit. Contraction of the antrum is followed by sequential contraction of the pyloric region and the duodenum. In the antrum, partial contraction ahead of the advancing gastric contents prevents solid masses from entering the duodenum, and they are mixed and crushed instead. The more liquid gastric contents are squirted a bit at a time into the small intestine. Normally, regurgitation from the duodenum does not occur, because the contraction of the pyloric segment tends to persist slightly longer than that of the duodenum. The prevention of regurgitation may also be due to the stimulating action of cholecystokinin (CCK) and secretin on the pyloric sphincter.

REGULATION OF GASTRIC MOTILITY & EMPTYING

The rate at which the stomach empties into the duodenum depends on the type of food ingested. Food rich in carbohydrate leaves the stomach in a few hours. Protein-rich food leaves more slowly, and emptying is slowest after a meal containing fat (Figure 28–5). The rate of emptying also depends on the osmotic pressure of the material entering the duodenum. Hyperosmolality of the duodenal contents is sensed by "duodenal osmoreceptors" that initiate a decrease in gastric emptying which is probably neural in origin.

Fats, carbohydrates, and acid in the duodenum inhibit gastric acid and pepsin secretion and gastric motility via neural and hormonal mechanisms. The hormone involved is probably peptide YY. CCK has also been implicated as an inhibitor of gastric emptying (Clinical Box 28–2).

VOMITING

Vomiting is an example of central regulation of gut motility functions. Vomiting starts with salivation and the sensation of nausea. Reverse peristalsis empties material from the upper part of the small intestine into the stomach. The glottis closes, preventing aspiration of vomitus into the trachea. The breath is held in mid inspiration. The muscles of the abdominal wall contract, and because the chest is held in a fixed position, the

(300-mL liquid meals)

FIGURE 28–5   **Effect of protein and fat on the rate of emptying of the human stomach.** Subjects were fed 300-mL liquid meals. (Reproduced with permission from Brooks FP: Integrative lecture. Response of the GI tract to a meal. *Undergraduate Teaching Project.* American Gastroenterological Association, 1974.)

contraction increases intra-abdominal pressure. The lower esophageal sphincter and the esophagus relax, and the gastric contents are ejected. The "vomiting center" in the reticular formation of the medulla (Figure 28–6) consists of various scattered groups of neurons in this region that control the different components of the vomiting act.

CLINICAL BOX 28–2

Consequences of Gastric Bypass Surgery

Patients who are morbidly obese often undergo a surgical procedure in which the stomach is stapled so that most of it is bypassed, and thus the reservoir function of the stomach is lost. As a result, such patients must eat frequent small meals. If larger meals are taken, because of rapid absorption of glucose from the intestine and the resultant hyperglycemia and abrupt rise in insulin secretion, gastrectomized patients sometimes develop hypoglycemic symptoms about 2 h after meals. Weakness, dizziness, and sweating after meals, due in part to hypoglycemia, are part of the picture of the **"dumping syndrome,"** a distressing syndrome that develops in patients in whom portions of the stomach have been removed or the jejunum has been anastomosed to the stomach. Another cause of the symptoms is rapid entry of hypertonic meals into the intestine; this provokes the movement of so much water into the gut that significant hypovolemia and hypotension are produced.

Irritation of the mucosa of the upper gastrointestinal tract is one trigger for vomiting. Impulses are relayed from the mucosa to the medulla over visceral afferent pathways in the sympathetic nerves and vagi. Other causes of vomiting can

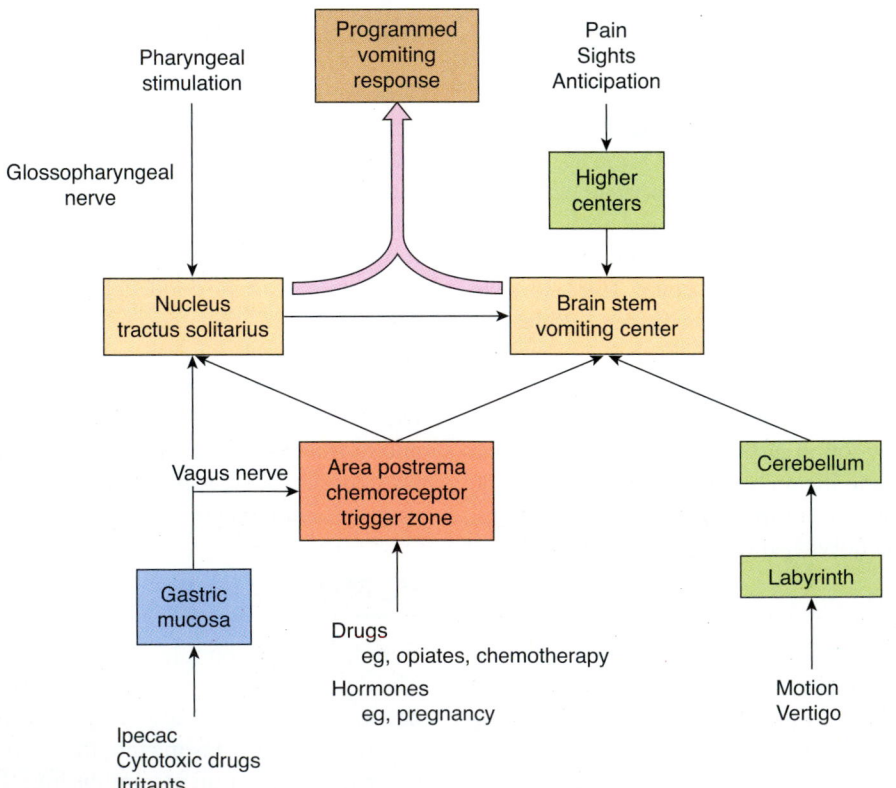

FIGURE 28–6   Neural pathways leading to the initiation of vomiting in response to various stimuli.

arise centrally. For example, afferents from the vestibular nuclei mediate the nausea and vomiting of motion sickness. Other afferents presumably reach the vomiting control areas from the diencephalon and limbic system, because emetic responses to emotionally charged stimuli also occur. Thus, we speak of "nauseating smells" and "sickening sights."

Chemoreceptor cells in the medulla can also initiate vomiting when they are stimulated by certain circulating chemical agents. The **chemoreceptor trigger zone** in which these cells are located (Figure 28–6) is in the **area postrema**, a V-shaped band of tissue on the lateral walls of the fourth ventricle near the obex. This structure is one of the circumventricular organs (see Chapter 34) and is not protected by the blood–brain barrier. Lesions of the area postrema have little effect on the vomiting response to gastrointestinal irritation or motion sickness, but abolish the vomiting that follows injection of apomorphine and a number of other emetic drugs. Such lesions also decrease vomiting in uremia and radiation sickness, both of which may be associated with endogenous production of circulating emetic substances.

Serotonin (5-HT) released from enterochromaffin cells in the small intestine appears to initiate impulses via 5-HT_3 receptors that trigger vomiting. In addition, there are dopamine D2 receptors and 5-HT_3 receptors in the area postrema and adjacent nucleus of the solitary tract. 5-HT_3 antagonists such as ondansetron and D_2 antagonists such as chlorpromazine and haloperidol are effective antiemetic agents. Corticosteroids, cannabinoids, and benzodiazepines, alone or in combination with 5-HT_3 and D_2 antagonists, are also useful in treatment of the vomiting produced by chemotherapy. The mechanisms of action of corticosteroids and cannabinoids are unknown, whereas the benzodiazepines probably reduce the anxiety associated with chemotherapy.

SMALL INTESTINE

In the small intestine, the intestinal contents are mixed with the secretions of the mucosal cells and with pancreatic juice and bile.

INTESTINAL MOTILITY

The MMCs that pass along the intestine at regular intervals in the fasting state and their replacement by peristaltic and other contractions controlled by the BER are described above. In the small intestine, there are an average of 12 BER cycles/min in the proximal jejunum, declining to 8/min in the distal ileum. There are three types of smooth muscle contractions: peristaltic waves, segmentation contractions, and tonic contractions. **Peristalsis** is described above. It propels the intestinal contents **(chyme)** toward the large intestines. **Segmentation contractions** (Figure 28–1), also described above, move the chyme to and fro and increase its exposure to the mucosal surface. These contractions are initiated by focal increases in Ca^{2+} influx with waves of increased Ca^{2+} concentration spreading

from each focus. **Tonic contractions** are relatively prolonged contractions that in effect isolate one segment of the intestine from another. Note that these last two types of contractions slow transit in the small intestine to the point that the transit time is actually longer in the fed than in the fasted state. This permits longer contact of the chyme with the enterocytes and fosters absorption (Clinical Box 28–3).

COLON

The colon serves as a reservoir for the residues of meals that cannot be digested or absorbed (Figure 28–7). Motility in this segment is likewise slowed to allow the colon to absorb water, Na^+, and other minerals. By removal of about 90% of the fluid, it converts the 1000 to 2000 mL of isotonic chyme that enters it each day from the ileum to about 200 to 250 mL of semisolid feces.

MOTILITY OF THE COLON

The ileum is linked to the colon by a structure known as the ileocecal valve, which restricts reflux of colonic contents, and particularly the large numbers of commensal bacteria, into the relatively sterile ileum. The portion of the ileum containing the ileocecal valve projects slightly into the cecum, so that increases in colonic pressure squeeze it shut, whereas increases in ileal pressure open it. It is normally closed. Each time a peristaltic wave reaches it, it opens briefly, permitting some of the ileal chyme to squirt into the cecum. When food leaves the stomach, the cecum relaxes and the passage of chyme through the ileocecal valve increases **(gastroileal reflex)**. This is presumably a vagal reflex.

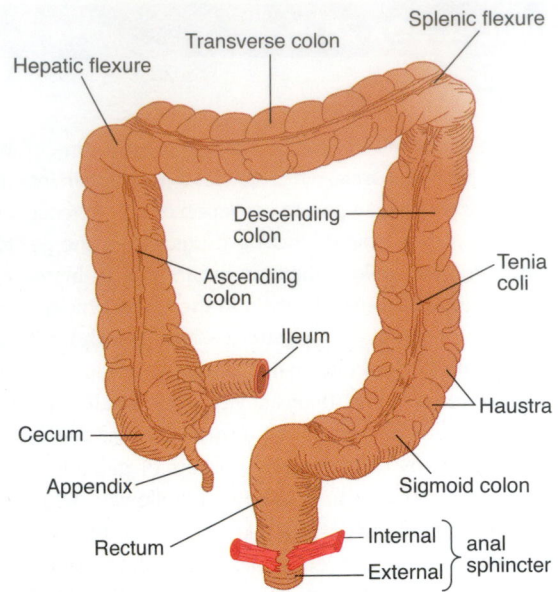

FIGURE 28–7   **The human colon.**

The movements of the colon include segmentation contractions and peristaltic waves like those occurring in the small intestine. Segmentation contractions mix the contents of the colon and, by exposing more of the contents to the mucosa, facilitate absorption. Peristaltic waves propel the contents toward the rectum, although weak antiperistalsis is sometimes seen. A third type of contraction that occurs only in the colon is the **mass action contraction,** in which there is simultaneous contraction of the smooth muscle over large confluent areas. These contractions move material from one portion of the colon to another (Clinical Box 28–4). They also move material into the rectum, and rectal distention initiates the defecation reflex (see below).

The movements of the colon are coordinated by the BER of the colon. The frequency of this wave, unlike the wave in the small intestine, increases along the colon, from about 2/min at the ileocecal valve to 6/min at the sigmoid.

TRANSIT TIME IN THE SMALL INTESTINE & COLON

The first part of a test meal reaches the cecum in about 4 h, and all the undigested portions have entered the colon in 8 or 9 h. On average, the first remnants of the meal traverse the first third of the colon in 6 h, the second third in 9 h, and reach the terminal part of the colon (the sigmoid colon) in 12 h. From the sigmoid colon to the anus, transport is much slower (Clinical Box 28–5). When small colored beads are fed with a meal, an average of 70% of them are recovered in the stool in 72 h, but total recovery requires more than a week. Transit time, pressure fluctuations, and changes in pH in the gastrointestinal tract can be observed by monitoring the progress of a small pill that contains sensors and a miniature radio transmitter.

Hirschsprung Disease

Some children present with a genetically determined condition of abnormal colonic motility known as Hirschsprung disease or **aganglionic megacolon,** which is characterized by abdominal distention, anorexia, and lassitude. The disease is typically diagnosed in infancy, and affects as many as 1 in 5000 live births. It is due to a congenital absence of the ganglion cells in both the myenteric and submucous plexuses of a segment of the distal colon, as a result of failure of the normal cranial-to-caudal migration of neural crest cells during development. The action of endothelins on the endothelin B receptor (see Chapter 7) are necessary for normal migration of certain neural crest cells, and knockout mice lacking endothelin B receptors developed megacolon. In addition, one cause of congenital aganglionic megacolon in humans appears to be a mutation in the endothelin B receptor gene. The absence of peristalsis in patients with this disorder causes feces to pass the aganglionic region with difficulty, and children with the disease may defecate as infrequently as once every 3 wk. The symptoms can be relieved completely if the aganglionic portion of the colon is resected and the portion of the colon above it anastomosed to the rectum.

DEFECATION

Distention of the rectum with feces initiates reflex contractions of its musculature and the desire to defecate. In humans, the sympathetic nerve supply to the internal (involuntary) anal sphincter is excitatory, whereas the parasympathetic supply is inhibitory. This sphincter relaxes when the rectum is distended. The nerve supply to the external anal sphincter, a skeletal muscle, comes from the pudendal nerve. The sphincter is maintained in a state of tonic contraction, and moderate distention of the rectum increases the force of its contraction (Figure 28–8). The urge to defecate first occurs when rectal pressure increases to about 18 mm Hg. When this pressure reaches 55 mm Hg, the external as well as the internal sphincter relaxes and there is reflex expulsion of the contents of the rectum. This is why reflex evacuation of the rectum can occur even in the setting of spinal injury.

Before the pressure that relaxes the external anal sphincter is reached, voluntary defecation can be initiated by straining. Normally, the angle between the anus and the rectum is approximately 90 degrees (Figure 28–9), and this plus contraction of the puborectalis muscle inhibit defecation. With straining, the abdominal muscles contract, the pelvic floor is lowered 1 to 3 cm, and the puborectalis muscle relaxes. The anorectal angle is reduced to 15 degrees or less. This is combined with relaxation of the external anal sphincter and

CLINICAL BOX 28–5

Constipation

Constipation refers to a pathological decrease in bowel movements. It was previously considered to reflect changes in motility, but the recent success of a drug designed to enhance chloride secretion for the treatment of chronic constipation suggests alterations in the balance between secretion and absorption in the colon could also contribute to symptom generation. Patients with persistent constipation, and particularly those with a recent change in bowel habits, should be examined carefully to rule out underlying organic disease. However, many normal humans defecate only once every 2–3 d, even though others defecate once a day and some as often as three times a day. Furthermore, the only symptoms caused by constipation are slight anorexia and mild abdominal discomfort and distention. These symptoms are not due to absorption of "toxic substances," because they are promptly relieved by evacuating the rectum and can be reproduced by distending the rectum with inert material. In western societies, the amount of misinformation and undue apprehension about constipation probably exceeds that about any other health topic. Symptoms other than those described above that are attributed by the lay public to constipation are due to anxiety or other causes.

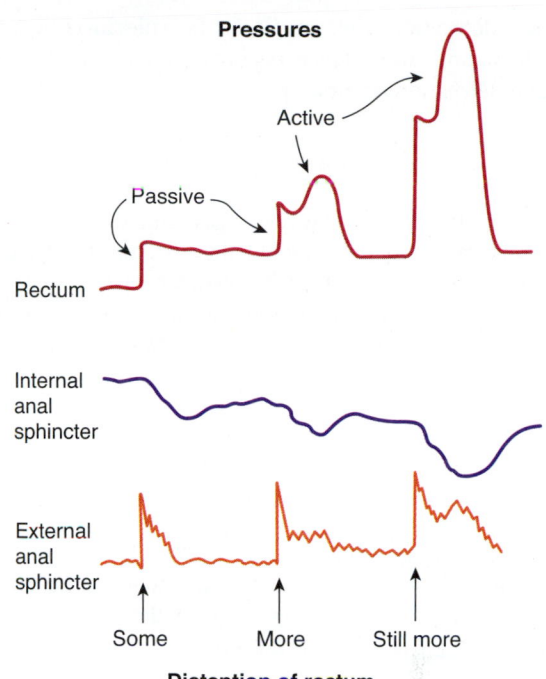

FIGURE 28–8 Responses to distention of the rectum by pressures less than 55 mm Hg. Distention produces passive tension due to stretching of the wall of the rectum, and additional active tension when the smooth muscle in the wall contracts. (Reproduced with permission from Davenport HW: *A Digest of Digestion,* 2nd ed. Year Book, 1978.)

defecation occurs. Defecation is therefore a spinal reflex that can be voluntarily inhibited by keeping the external sphincter contracted or facilitated by relaxing the sphincter and contracting the abdominal muscles.

Distention of the stomach by food initiates contractions of the rectum and, frequently, a desire to defecate. The response is called the **gastrocolic reflex,** and may be amplified by an action of gastrin on the colon. Because of the

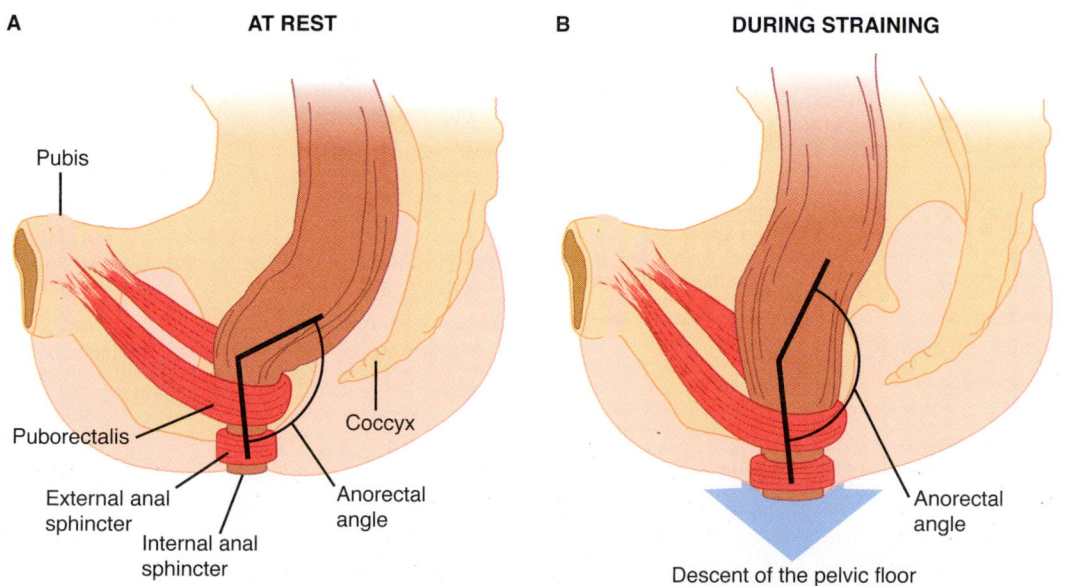

FIGURE 28–9   **Sagittal view of the anorectal area at rest (above) and during straining (below).** Note the reduction of the anorectal angle and lowering of the pelvic floor during straining. (Modified and reproduced with permission from Lembo A, Camilleri, M: Chronic constipation. N Engl J Med 2003;349:1360.)

response, defecation after meals is the rule in children. In adults, habit and cultural factors play a large role in determining when defecation occurs.

CHAPTER SUMMARY

■ The regulatory factors that govern gastrointestinal secretion also regulate its motility to soften the food, mix it with secretions, and propel it along the length of the tract.

■ Two major patterns of motility are peristalsis and segmentation, which serve to propel or retard/mix the luminal contents, respectively. Peristalsis involves coordinated contractions and relaxations above and below the food bolus.

■ The membrane potential of the majority of gastrointestinal smooth muscle undergoes rhythmic fluctuations that sweep along the length of the gut. The rhythm varies in different gut segments and is established by pacemaker cells known as interstitial cells of Cajal. This basic electrical rhythm provides for sites of muscle contraction when stimuli superimpose spike potentials on the depolarizing portion of the BER waves.

■ In the period between meals, the intestine is relatively quiescent, but every 90 min or so it is swept through by a large peristaltic wave triggered by the hormone motilin. This migrating motor complex presumably serves a "housekeeping" function.

■ Swallowing is triggered centrally and is coordinated with a peristaltic wave along the length of the esophagus that drives the food bolus to the stomach, even against gravity. Relaxation of the lower esophageal sphincter is timed to just precede the arrival of the bolus, thereby limiting reflux of the gastric contents. Nevertheless, gastroesophageal reflux disease is one of the most common gastrointestinal complaints

■ The stomach accommodates the meal by a process of receptive relaxation. This permits an increase in volume without a significant increase in pressure. The stomach then serves to mix the meal and to control its delivery to downstream segments.

■ Luminal contents move slowly through the colon, which enhances water recovery. Distension of the rectum causes reflex contraction of the internal anal sphincter and the desire to defecate. After toilet training, defecation can be delayed till a convenient time via voluntary contraction of the external anal sphincter.

MULTIPLE-CHOICE QUESTIONS

For all questions, select the single best answer unless otherwise directed.

1. In infants, defecation often follows a meal. The cause of colonic contractions in this situation is
    A) histamine.
    B) increased circulating levels of CCK.
    C) the gastrocolic reflex.
    D) increased circulating levels of somatostatin.
    E) the enterogastric reflex.

2. The symptoms of the dumping syndrome (discomfort after meals in patients with intestinal short circuits such as anastomosis of the jejunum to the stomach) are caused in part by
    A) increased blood pressure.
    B) increased secretion of glucagon.
    C) increased secretion of CCK.
    D) hypoglycemia.
    E) hyperglycemia.

3. Gastric pressures seldom rise above the levels that breach the lower esophageal sphincter, even when the stomach is filled with a meal, due to which of the following processes?
    A) peristalsis
    B) gastroileal reflex
    C) segmentation
    D) stimulation of the vomiting center
    E) receptive relaxation

4. The migrating motor complex is triggered by which of the following?
    A) motilin
    B) NO
    C) CCK
    D) somatostatin
    E) secretin

5. A patient with achalasia would be expected to exhibit a decrease in which of the following?
    A) esophageal peristalsis
    B) expression of neuronal NO synthase at the esophageal/gastric junction
    C) acetylcholine receptors
    D) substance P release
    E) contraction of the crural diaphragm

CHAPTER RESOURCES

Barrett KE: *Gastrointestinal Physiology.* McGraw-Hill, 2006.

Cohen S, Parkman HP: Heartburn—A serious symptom. N Engl J Med 1999;340:878.

Itoh Z: Motilin and clinical application. Peptides 1997;18:593.

Lembo A, Camilleri M: Chronic constipation. N Engl J Med 2003;349:1360.

Levitt MD, Bond JH: Volume, composition and source of intestinal gas. Gastroenterology 1970;59:921.

Mayer EA, Sun XP, Willenbucher RF: Contraction coupling in colonic smooth muscle. Annu Rev Physiol 1992;54:395.

Mittal RK, Balaban DH: The esophagogastric junction. N Engl J Med 1997;336:924.

Sanders KM, Warm SM: Nitric oxide as a mediator of noncholinergic neurotransmission. Am J Physiol 1992;262:G379.

Transport & Metabolic Functions of the Liver

INTRODUCTION

The liver is the largest gland in the body. It is essential for life because it conducts a vast array of biochemical and metabolic functions, including ridding the body of substances that would otherwise be injurious if allowed to accumulate, and excreting drug metabolites. It is also the first port of call for most nutrients absorbed across the gut wall, supplies most of the plasma proteins, and synthesizes the bile that optimizes the absorption of fats as well as serving as an excretory fluid. The liver and associated biliary system have therefore evolved an array of structural and physiologic features that underpin this broad range of critical functions.

THE LIVER

FUNCTIONAL ANATOMY

An important function of the liver is to serve as a filter between the blood coming from the gastrointestinal tract and the blood in the rest of the body. Blood from the intestines and other viscera reach the liver via the portal vein. This blood percolates in sinusoids between plates of hepatic cells and eventually drains into the hepatic veins, which enter the inferior vena cava. During its passage through the hepatic plates, it is extensively modified chemically. Bile is formed on the other side at each plate. The bile passes to the intestine via the hepatic duct (Figure 29–1).

In each hepatic lobule, the plates of hepatic cells are usually only one cell thick. Large gaps occur between the endothelial cells, and plasma is in intimate contact with the cells (Figure 29–2). Hepatic artery blood also enters the sinusoids. The central veins coalesce to form the hepatic veins, which drain into the inferior vena cava. The average transit time for blood

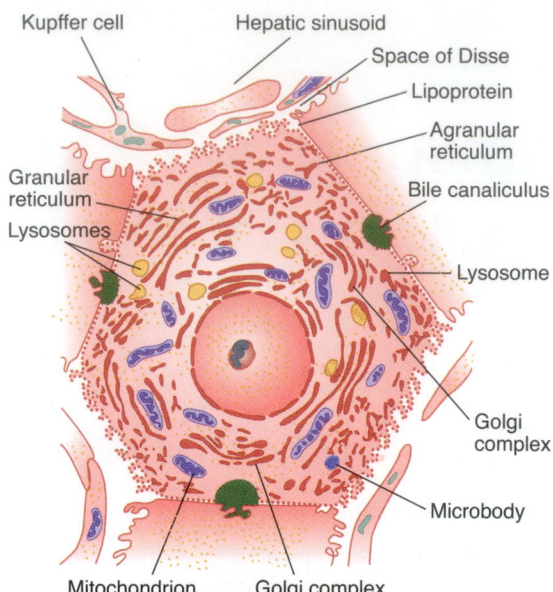

FIGURE 29–2 **Hepatocyte.** Note the relation of the cell to bile canaliculi and sinusoids. Note also the wide openings between the endothelial cells next to the hepatocyte. (Reproduced with permission from Fawcett DW: *Bloom and Fawcett, A Textbook of Histology,* 11th ed. Saunders, 1986.)

FIGURE 29–1 **Top:** Organization of the liver. CV, central vein. PS, portal space containing branches of bile duct (green), portal vein (blue), and hepatic artery (red). **Bottom:** Arrangement of plates of liver cells, sinusoids, and bile ducts in a liver lobule, showing centripetal flow of blood in sinusoids to central vein and centrifugal flow of bile in bile canaliculi to bile ducts. (Reproduced with permission from Fawcett DW: *Bloom and Fawcett, A Textbook of Histology,* 11th ed. Saunders, 1986.)

across the liver lobule from the portal venule to the central hepatic vein is about 8.4 s. Additional details of the features of the hepatic micro- and macrocirculation, which are critical to organ function, are provided below. Numerous macrophages **(Kupffer cells)** are anchored to the endothelium of the sinusoids and project into the lumen. The functions of these phagocytic cells are discussed in Chapter 3.

Each liver cell is also apposed to several bile canaliculi (Figure 29–2). The canaliculi drain into intralobular bile ducts, and these coalesce via interlobular bile ducts to form the right and left hepatic ducts. These ducts join outside the liver to

form the common hepatic duct. The cystic duct drains the gallbladder. The hepatic duct unites with the cystic duct to form the common bile duct (Figure 29–1). The common bile duct enters the duodenum at the duodenal papilla. Its orifice is surrounded by the sphincter of Oddi, and it usually unites with the main pancreatic duct just before entering the duodenum. The sphincter is usually closed, but when the gastric contents enter the duodenum, cholecystokinin (CCK) is released and the gastrointestinal hormone relaxes the sphincter and makes the gallbladder contract.

The walls of the extrahepatic biliary ducts and the gallbladder contain fibrous tissue and smooth muscle. They are lined by a layer of columnar cells with scattered mucous glands. In the gallbladder, the surface is extensively folded; this increases its surface area and gives the interior of the gallbladder a honeycombed appearance. The cystic duct is also folded to form the so-called spiral valves. This arrangement is believed to increase the turbulence of bile as it flows out of the gallbladder, thereby reducing the risk that it will precipitate and form gallstones.

HEPATIC CIRCULATION

Large gaps occur between endothelial cells in the walls of hepatic sinusoids, and the sinusoids are highly permeable. The way the intrahepatic branches of the hepatic artery and portal vein converge on the sinusoids and drain into the central lobular veins of the liver is shown in Figure 29–1. The functional unit of the liver is the acinus. Each acinus is at the end of a vascular stalk containing terminal branches of portal veins, hepatic arteries, and bile ducts. Blood flows from the center of

FIGURE 29–3 **Concept of the acinus as the functional unit of the liver.** In each acinus, blood in the portal venule and hepatic arteriole enters the center of the acinus and flows outward to the hepatic venule. (Reproduced with permission from Lautt WW, Greenway CV: Conceptual review of the hepatic vascular bed. Hepatology 1987;7:952. Copyright © 1987 by The American Association for the Study of Liver Diseases.)

this functional unit to the terminal branches of the hepatic veins at the periphery (Figure 29–3). This is why the central portion of the acinus, sometimes called zone 1, is well oxygenated, the intermediate zone (zone 2) is moderately well oxygenated, and the peripheral zone (zone 3) is least well oxygenated and most susceptible to anoxic injury. The hepatic veins drain into the inferior vena cava. The acini have been likened to grapes or berries, each on a vascular stem. The human liver contains about 100,000 acini.

Portal venous pressure is normally about 10 mm Hg in humans, and hepatic venous pressure is approximately 5 mm Hg. The mean pressure in the hepatic artery branches that converge on the sinusoids is about 90 mm Hg, but the pressure in the sinusoids is lower than the portal venous pressure, so a marked pressure drop occurs along the hepatic arterioles. This pressure drop is adjusted so that there is an inverse relationship between hepatic arterial and portal venous blood flow. This inverse relationship may be maintained in part by the rate at which adenosine is removed from the region around the arterioles. According to this hypothesis, adenosine is produced by metabolism at a constant rate. When portal flow is reduced, it is washed away more slowly, and the local accumulation of adenosine dilates the terminal arterioles. In the period between meals, moreover, many of the sinusoids are collapsed. Following a meal, on the other hand, when portal flow to the liver from the intestine increases considerably, these "reserve" sinusoids are recruited. This arrangement means that portal pressures do not increase linearly with portal flow until all sinusoids have been recruited. This may be important to prevent fluid loss from the highly permeable liver under normal conditions. Indeed, if hepatic pressures are increased in disease states (such as the hardening of the liver

that is seen in cirrhosis), many liters of fluid can accumulate in the peritoneal cavity as ascites.

The intrahepatic portal vein radicles have smooth muscle in their walls that is innervated by noradrenergic vasoconstrictor nerve fibers reaching the liver via the third to eleventh thoracic ventral roots and the splanchnic nerves. The vasoconstrictor innervation of the hepatic artery comes from the hepatic sympathetic plexus. No known vasodilator fibers reach the liver. When systemic venous pressure rises, the portal vein radicles are dilated passively and the amount of blood in the liver increases. In congestive heart failure, this hepatic venous congestion may be extreme. Conversely, when diffuse noradrenergic discharge occurs in response to a drop in systemic blood pressure, the intrahepatic portal radicles constrict, portal pressure rises, and blood flow through the liver is brisk, bypassing most of the organ. Most of the blood in the liver enters the systemic circulation. Constriction of the hepatic arterioles diverts blood from the liver, and constriction of the mesenteric arterioles reduces portal inflow. In severe shock, hepatic blood flow may be reduced to such a degree that patchy necrosis of the liver takes place.

FUNCTIONS OF THE LIVER

The liver has many complex functions that are summarized in Table 29–1. Several will be touched upon briefly here.

METABOLISM & DETOXIFICATION

It is beyond the scope of this volume to touch upon all of the metabolic functions of the liver. Instead, we will describe here those aspects most closely aligned to gastrointestinal physiology. First, the liver plays key roles in carbohydrate metabolism, including glycogen storage, conversion of galactose and fructose to glucose, and gluconeogenesis, as well as many of the reactions covered in Chapter 1. The substrates for these reactions derive from the products of carbohydrate digestion and absorption that are transported from the intestine to the liver in the portal blood. The liver also plays a major role in maintaining the stability of blood glucose levels in the postprandial period, removing excess glucose from the blood and returning it as needed—the so-called **glucose buffer function** of the liver. In liver failure, hypoglycemia is commonly seen. Similarly, the liver contributes to fat metabolism. It supports a high rate of fatty acid oxidation for energy supply to the liver itself and other organs. Amino acids and two carbon fragments derived from carbohydrates are also converted in the liver to fats for storage. The liver also synthesizes most of the lipoproteins required by the body and preserves cholesterol homeostasis by synthesizing this molecule and also converting excess cholesterol to bile acids.

The liver also detoxifies the blood of substances originating from the gut or elsewhere in the body (Clinical Box 29–1). Part of this function is physical in nature—bacteria and other

TABLE 29–1 Principal functions of the liver.

Formation and secretion of bile
Nutrient and vitamin metabolism
Glucose and other sugars
Amino acids
Lipids
Fatty acids
Cholesterol
Lipoproteins
Fat-soluble vitamins
Water-soluble vitamins
Inactivation of various substances
Toxins
Steroids
Other hormones
Synthesis of plasma proteins
Acute-phase proteins
Albumin
Clotting factors
Steroid-binding and other hormone-binding proteins
Immunity
Kupffer cells

particulates are trapped in and broken down by the strategically-located Kupffer cells. The remaining reactions are biochemical, and mediated in their first stages by the large number of cytochrome P450 enzymes expressed in hepatocytes. These convert xenobiotics and other toxins to inactive, less lipophilic metabolites. Detoxification reactions are divided into phase I (oxidation, hydroxylation, and other reactions mediated by cytochrome P450s) and phase II (esterification). Ultimately, metabolites are secreted into the bile for elimination via the gastrointestinal tract. In this regard, in addition to disposing of drugs, the liver is responsible for metabolism of essentially all steroid hormones. Liver disease can therefore result in the apparent overactivity of the relevant hormone systems.

SYNTHESIS OF PLASMA PROTEINS

The principal proteins synthesized by the liver are listed in Table 29–1. Albumin is quantitatively the most significant, and accounts for the majority of plasma oncotic pressure. Many of the products are **acute-phase proteins,** proteins synthesized

Hepatic Encephalopathy

The clinical importance of hepatic ammonia metabolism is seen in liver failure, when increased levels of circulating ammonia cause the condition of hepatic encephalopathy. Initially, patients may seem merely confused, but if untreated, the condition can progress to coma and irreversible changes in cognition. The disease results not only from the loss of functional hepatocytes, but also shunting of portal blood around the hardened liver, meaning that less ammonia is removed from the blood by the remaining hepatic mass. Additional substances that are normally detoxified by the liver likely also contribute to the mental status changes. The condition can be minimized by reducing the load of ammonia coming to the liver from the colon (eg, by feeding the nonabsorbable carbohydrate, lactulose, which is converted into short-chain fatty acids in the colonic lumen and thereby traps luminal ammonia in its ionized form). However, in severe disease, the only truly effective treatment is to perform a liver transport, although the paucity of available organs means that there is great interest in artificial liver assist devices that could clean the blood.

and secreted into the plasma on exposure to stressful stimuli (see Chapter 3). Others are proteins that transport steroids and other hormones in the plasma, and still others are clotting factors. Following blood loss, the liver replaces the plasma proteins in days to weeks. The only major class of plasma proteins not synthesized by the liver are the immunoglobulins.

BILE

Bile is made up of the bile acids, bile pigments, and other substances dissolved in an alkaline electrolyte solution that resembles pancreatic juice (Table 29–2). About 500 mL is secreted per day. Some of the components of the bile are reabsorbed in the intestine and then excreted again by the liver (**enterohepatic circulation**). In addition to its role in digestion and absorption of fats (Chapter 27), bile (and subsequently the feces) is the major excretory route for lipid-soluble waste products.

TABLE 29–2 Comparison of human hepatic duct bile and gallbladder bile.

	Hepatic Duct Bile	Gallbladder Bile
Percentage of solids	2–4	10–12
Bile acids (mmol/L)	10–20	50–200
pH	7.8–8.6	7.0–7.4

The glucuronides of the **bile pigments**, bilirubin and biliverdin, are responsible for the golden yellow color of bile. The formation of these breakdown products of hemoglobin is discussed in detail in Chapter 32, and their excretion is discussed in the following text.

BILIRUBIN METABOLISM & EXCRETION

Most of the bilirubin in the body is formed in the tissues by the break down of hemoglobin (see Chapter 32 and Figure 29–4). The bilirubin is bound to albumin in the circulation. Some of it is tightly bound, but most of it can dissociate in the liver, and free bilirubin enters liver cells via a member of the organic anion transporting polypeptide (OATP) family, and then becomes bound to cytoplasmic proteins (Figure 29–5). It is next conjugated to glucuronic acid in a reaction catalyzed by the enzyme **glucuronyl transferase** (UDP-glucuronosyltransferase). This enzyme is located primarily in the smooth endoplasmic reticulum. Each bilirubin molecule reacts with two uridine diphosphoglucuronic acid (UDPG) molecules to form bilirubin diglucuronide. This glucuronide, which is more water-soluble than the free bilirubin, is then transported against a concentra-

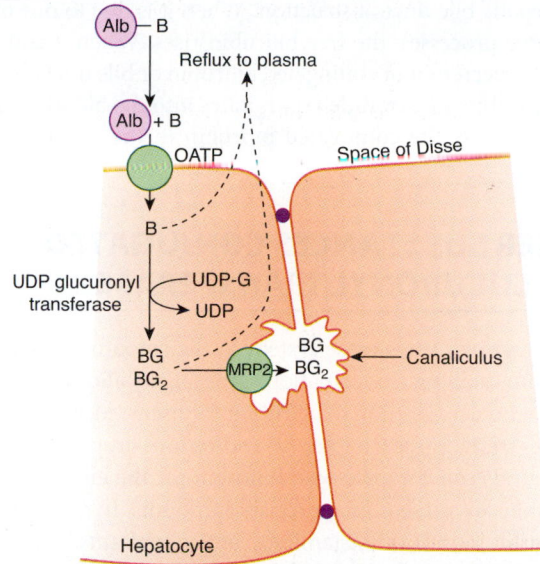

FIGURE 29–5 **Handling of bilirubin by hepatocytes.** Albumin (Alb)-bound bilirubin (B) enters the space of Disse adjacent to the basolateral membrane of hepatocytes, and bilirubin is selectively transported into the hepatocyte. Here, it is conjugated with glucuronic acid (G). The conjugates are secreted into bile via the multidrug resistance protein 2 (MRP-2). Some unconjugated and conjugated bilirubin also refluxes into the plasma. OATP, organic anion transporting polypeptide.

tion gradient most likely by an active transporter known as multidrug resistance protein-2 (MRP-2) into the bile canaliculi. A small amount of the bilirubin glucuronide escapes into the blood, where it is bound less tightly to albumin than is free bilirubin, and is excreted in the urine. Thus, the total plasma bilirubin normally includes free bilirubin plus a small amount of conjugated bilirubin. Most of the bilirubin glucuronide passes via the bile ducts to the intestine.

The intestinal mucosa is relatively impermeable to conjugated bilirubin but is permeable to unconjugated bilirubin and to urobilinogens, a series of colorless derivatives of bilirubin formed by the action of bacteria in the intestine. Consequently, some of the bile pigments and urobilinogens are reabsorbed in the portal circulation. Some of the reabsorbed substances are again excreted by the liver (enterohepatic circulation), but small amounts of urobilinogens enter the general circulation and are excreted in the urine.

JAUNDICE

When free or conjugated bilirubin accumulates in the blood, the skin, scleras, and mucous membranes turn yellow. This yellowness is known as **jaundice** (icterus) and is usually detectable when the total plasma bilirubin is greater than 2 mg/dL (34 μmol/L). Hyperbilirubinemia may be due to (1) excess production of bilirubin (hemolytic anemia, etc; see Chapter 32), (2) decreased uptake of bilirubin into hepatic cells, (3) disturbed intracellular protein binding or conjugation, (4) disturbed secretion of conjugated bilirubin into the bile canaliculi, or (5) intrahepatic or

FIGURE 29–4 **Conversion of heme to bilirubin is a two-step reaction catalyzed by heme oxygenase and biliverdin reductase.** M, methyl; P, propionate; V, vinyl.

extrahepatic bile duct obstruction. When it is due to one of the first three processes, the free bilirubin rises. When it is due to disturbed secretion of conjugated bilirubin or bile duct obstruction, bilirubin glucuronide regurgitates into the blood, and it is predominantly the conjugated bilirubin in the plasma that is elevated.

OTHER SUBSTANCES CONJUGATED BY GLUCURONYL TRANSFERASE

The glucuronyl transferase system in the smooth endoplasmic reticulum catalyzes the formation of the glucuronides of a variety of substances in addition to bilirubin. As discussed above, the list includes steroids (see Chapter 22) and various drugs. These other compounds can compete with bilirubin for the enzyme system when they are present in appreciable amounts. In addition, several barbiturates, antihistamines, anticonvulsants, and other compounds cause marked proliferation of the smooth endoplasmic reticulum in the hepatic cells, with a concurrent increase in hepatic glucuronyl transferase activity. Phenobarbital has been used successfully for the treatment of a congenital disease in which there is a relative deficiency of glucuronyl transferase (type 2 UDP-glucuronosyltransferase deficiency).

OTHER SUBSTANCES EXCRETED IN THE BILE

Cholesterol and alkaline phosphatase are excreted in the bile. In patients with jaundice due to intra- or extrahepatic obstruction of the bile duct, the blood levels of these two substances usually rise. A much smaller rise is generally seen when the jaundice is due to nonobstructive hepatocellular disease. Adrenocortical and other steroid hormones and a number of drugs are excreted in the bile and subsequently reabsorbed (enterohepatic circulation).

AMMONIA METABOLISM & EXCRETION

The liver is critical for ammonia handling in the body. Ammonia levels must be carefully controlled because it is toxic to the central nervous system (CNS), and freely permeable across the blood–brain barrier. The liver is the only organ in which the complete urea cycle (also known as the Krebs-Henseleit cycle) is expressed (Figure 29–6). This converts circulating ammonia to urea, which can then be excreted in the urine (Figure 29–7).

Ammonia in the circulation comes primarily from the colon and kidneys with lesser amounts deriving from the breakdown of red blood cells and from metabolism in the muscles. As it passes through the liver, the vast majority of

Net reaction

$$2NH_3 + CO_2 = Urea + H_2O$$

① Carbamoyl synthetase   ② Arginosuccinate synthetase   ③ Arginine succinate lyase   ④ Arginase

FIGURE 29–6 The urea cycle, which converts ammonia to urea, takes place in the mitochondria and cytosol of hepatocytes.

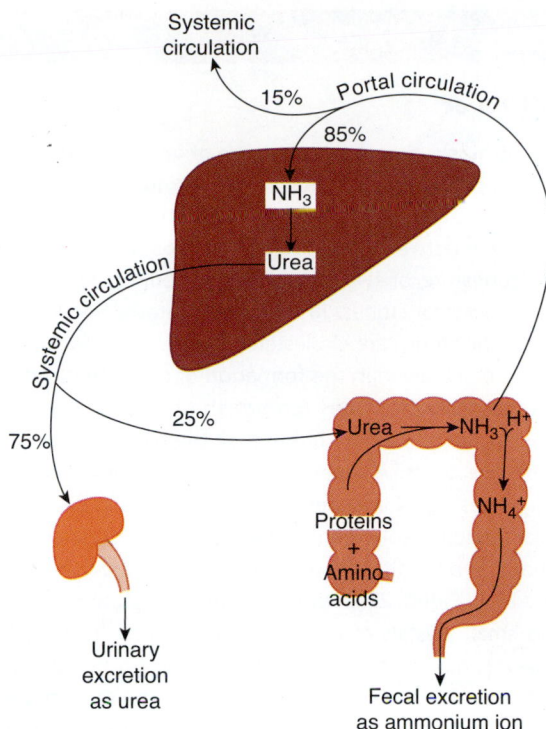

FIGURE 29–7 **Whole body ammonia homeostasis in health.** The majority of ammonia produced by the body is excreted by the kidneys in the form of urea.

ammonia in the circulation is cleared into the hepatocytes. There, it is converted in the mitochondria to carbamoyl phosphate, which in turn reacts with ornithine to generate citrulline. A series of subsequent cytoplasmic reactions eventually produce arginine, and this can be dehydrated to urea and ornithine. The latter returns to the mitochondria to begin another cycle, and urea, as a small molecule, diffuses readily back out into the sinusoidal blood. It is then filtered in the kidneys and lost from the body in the urine.

THE BILIARY SYSTEM

BILE FORMATION

Bile contains substances that are actively secreted into it across the canalicular membrane, such as bile acids, phosphatidylcholine, conjugated bilirubin, cholesterol, and xenobiotics. Each of these enters the bile by means of a specific canalicular transporter. It is the active secretion of bile acids, however, that is believed to be the primary driving force for the initial formation of canalicular bile. Because they are osmotically active, the canalicular bile is transiently hypertonic. However, the tight junctions that join adjacent hepatocytes are relatively permeable and thus a number of additional substances passively enter the bile from the plasma by diffusion. These substances include water, glucose, calcium, glutathione, amino acids, and urea.

Phosphatidylcholine that enters the bile forms mixed micelles with the bile acids and cholesterol. The ratio of bile acids:phosphatidylcholine:cholesterol in canalicular bile is approximately 10:3:1. Deviations from this ratio may cause cholesterol to precipitate, leading to one type of gallstones (Figure 29–8).

The bile is then transferred to progressively larger bile ductules and ducts, where it undergoes modification of its composition. The bile ductules are lined by cholangiocytes, specialized columnar epithelial cells. Their tight junctions are less permeable than those of the hepatocytes, although they remain freely permeable to water and thus bile remains isotonic. The ductules scavenge plasma constituents, such as glucose and amino acids, and return them to the circulation by active transport. Glutathione is also hydrolyzed to its constituent amino acids by an enzyme, gamma glutamyltranspeptidase (GGT), expressed on the apical membrane of the cholangiocytes. Removal of glucose and amino acids is likely important to prevent bacterial overgrowth of the bile, particularly during gallbladder storage (see below). The ductules also secrete bicarbonate in response to secretin in the postprandial period, as well as IgA and mucus for protection.

FUNCTIONS OF THE GALLBLADDER

In normal individuals, bile flows into the gallbladder when the sphincter of Oddi is closed (ie, the period in between meals). In the gallbladder, the bile is concentrated by absorption of water. The degree of this concentration is shown by the increase in the concentration of solids (Table 29–2); liver bile is 97% water, whereas the average water content of gallbladder bile is 89%. However, because the bile acids are a micellar

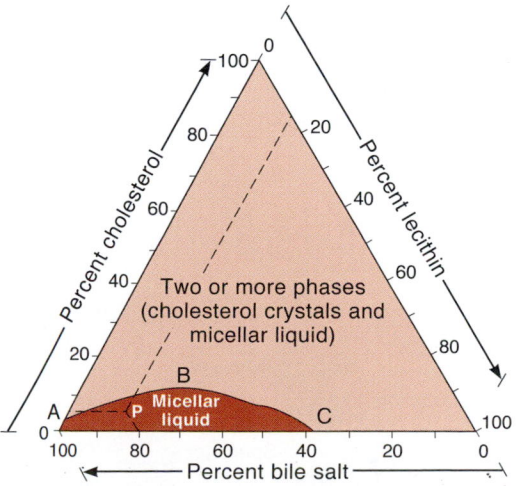

FIGURE 29–8 **Cholesterol solubility in bile as a function of the proportions of lecithin, bile salts, and cholesterol.** In bile that has a composition described by any point below line ABC (eg, point P), cholesterol is solely in micellar solution; points above line ABC describe bile in which there are cholesterol crystals as well. (Reproduced with permission from Small DM: Gallstones. N Engl J Med 1968;279:588.)

solution, the micelles simply become larger, and since osmolarity is a colligative property, bile remains isotonic. However, bile becomes slightly acidic as sodium ions are exchanged for protons (although the overall concentration of sodium ions rises with a concomitant loss of chloride and bicarbonate as the bile is concentrated).

When the bile duct and cystic duct are clamped, the intrabiliary pressure rises to about 320 mm of bile in 30 min, and bile secretion stops. However, when the bile duct is clamped and the cystic duct is left open, water is reabsorbed in the gallbladder, and the intrabiliary pressure rises only to about 100 mm of bile in several hours.

REGULATION OF BILIARY SECRETION

When food enters the mouth, the resistance of the sphincter of Oddi decreases under both neural and hormonal influences (Figure 29–9). Fatty acids and amino acids in the duodenum release CCK, which causes gallbladder contraction.

The production of bile is increased by stimulation of the vagus nerves and by the hormone secretin, which increases the water and HCO_3^- content of bile. Substances that increase the secretion of bile are known as **choleretics.** Bile acids themselves are among the most important physiologic choleretics.

EFFECTS OF CHOLECYSTECTOMY

The periodic discharge of bile from the gallbladder aids digestion but is not essential for it. Cholecystectomized patients maintain good health and nutrition with a constant slow discharge of bile into the duodenum, although eventually the bile duct becomes somewhat dilated, and more bile tends to enter the duodenum after meals than at other times. Cholecystectomized patients can even tolerate fried foods, although they generally must avoid foods that are particularly high in fat content.

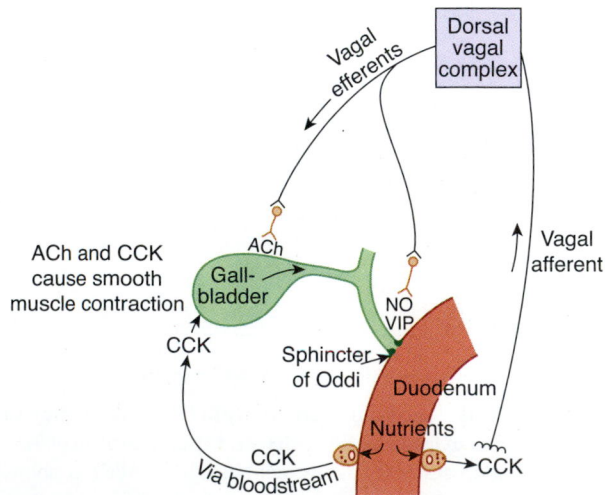

FIGURE 29–9 **Neurohumoral control of gallbladder contraction and biliary secretion.**

VISUALIZING THE GALLBLADDER

Exploration of the right upper quadrant with an ultrasonic beam (**ultrasonography**) and computed tomography (CT) have become the most widely used methods for visualizing the gallbladder and detecting gallstones. A third method of diagnosing gallbladder disease is **nuclear cholescintigraphy.** When administered intravenously, technetium-99m-labeled derivatives of iminodiacetic acid are excreted in the bile and provide excellent gamma camera images of the gallbladder and bile ducts. The response of the gallbladder to CCK can then be observed following intravenous administration of the hormone. The biliary tree can also be visualized by injecting contrast media from an endoscope channel maneuvered into the sphincter of Oddi, in a procedure known as endoscopic retrograde cholangiopancreatography (ERCP). It is even possible to insert small instruments with which to remove gallstone fragments that may be obstructing the flow of bile, the flow of pancreatic juice, or both (Clinical Box 29–2).

CHAPTER SUMMARY

- The liver conducts a huge number of metabolic reactions and serves to detoxify and dispose of many exogenous substances, as well as metabolites endogenous to the body that would be harmful if allowed to accumulate.

- The structure of the liver is such that it can filter large volumes of blood and remove even hydrophobic substances that are protein-bound. This function is provided for by a fenestrated endothelium. The liver also receives essentially all venous blood from the intestine prior to its delivery to the remainder of the body.

- The liver serves to buffer blood glucose, synthesize the majority of plasma proteins, contribute to lipid metabolism, and preserve cholesterol homeostasis.

- Bilirubin is an end product of heme metabolism that is glucuronidated by the hepatocyte to permit its excretion in bile. Bilirubin and its metabolites impart color to the bile and stools.

- The liver removes ammonia from the blood and converts it to urea for excretion by the kidneys. An accumulation of ammonia as well as other toxins causes hepatic encephalopathy in the setting of liver failure.

- Bile contains substances actively secreted across the canalicular membrane by hepatocytes, and notably bile acids, phosphatidylcholine, and cholesterol. The composition of bile is modified as it passes through the bile ducts and is stored in the gallbladder. Gallbladder contraction is regulated to coordinate bile availability with the timing of meals.

MULTIPLE-CHOICE QUESTIONS

For all questions, select the single best answer unless otherwise directed.

1. Removal of the entire colon would be expected to cause
   A) death.
   B) megaloblastic anemia.
   C) severe malnutrition.
   D) a decrease in the blood level of ammonia in patients with cirrhosis of the liver.
   E) decreased urinary urobilinogen.

2. After complete hepatectomy, a rise would be expected in the blood level of
   A) glucose.
   B) fibrinogen.
   C) 25-hydroxycholecalciferol.
   D) conjugated bilirubin.
   E) estrogens.

3. Which of the following cell types protects against sepsis secondary to translocation of intestinal bacteria?
   A) hepatic stellate cell
   B) cholangiocyte
   C) Kupffer cell
   D) hepatocyte
   E) gallbladder epithelial cell

4. P450s (CYPs) are found in many parts of the body. In which of the following do they *not* play an important role?
   A) bile acid formation
   B) carcinogenesis
   C) steroid hormone formation
   D) detoxification of drugs
   E) glycogen synthesis

CHAPTER RESOURCES

Ankoma-Sey V: Hepatic regeneration—Revising the myth of Prometheus. News Physiol Sci 1999;14:149.

Arias JM, et al (editors): *The Liver: Biology and Pathology,* 3rd ed. Raven Press, 1994.

Chong L, Marx J (editors): Lipids in the limelight. Science 2001;294:1861.

Hofmann AF: Bile acids: The good, the bad, and the ugly. News Physiol Sci 1999;14:24.

Lee WM: Drug-induced hepatoxicity. N Engl J Med 2003;349:474.

Meier PJ, Stieger B: Molecular mechanisms of bile formation. News Physiol Sci 2000;15:89.

Michalopoulos GK, DeFrances MC: Liver regeneration. Science 1997;276:60.

Trauner M, Meier PJ, Boyer JL: Molecular mechanisms of cholestasis. N Engl J Med 1998;339:1217.

Origin of the Heartbeat & the Electrical Activity of the Heart

OBJECTIVES

After studying this chapter, you should be able to:

- Describe the structure and function of the conduction system of the heart and compare the action potentials in each part.

- Describe the way the electrocardiogram (ECG) is recorded, the waves of the ECG, and the relationship of the ECG to the electrical axis of the heart.

- Name the common cardiac arrhythmias and describe the processes that produce them.

- List the principal early and late ECG manifestations of myocardial infarction and explain the early changes in terms of the underlying ionic events that produce them.

- Describe the ECG changes and the changes in cardiac function produced by alterations in the ionic composition of the body fluids.

INTRODUCTION

The parts of the heart normally beat in orderly sequence: Contraction of the atria (**atrial systole**) is followed by contraction of the ventricles (**ventricular systole**), and during **diastole** all four chambers are relaxed. The heartbeat originates in a specialized **cardiac conduction system** and spreads via this system to all parts of the myocardium. The structures that make up the conduction system (Figure 30–1) are the **sinoatrial node (SA node)**, the **internodal atrial pathways**, the **atrioventricular node (AV node)**, the **bundle of His** and its branches, and the **Purkinje system.** The various parts of the conduction system and, under abnormal conditions, parts of the myocardium, are capable of spontaneous discharge. However, the SA node normally discharges most rapidly, with depolarization spreading from it to the other regions before they discharge spontaneously. The SA node is therefore the normal **cardiac pacemaker,** with its rate of discharge determining the rate at which the heart beats. Impulses generated in the SA node pass through the atrial pathways to the AV node, through this node to the bundle of His, and through the branches of the bundle of His via the Purkinje system to the ventricular muscle.

FIGURE 30–1 **Conducting system of the heart. Left:** Anatomical depiction of the human heart with additional focus on areas of the conduction system. **Right:** Typical transmembrane action potentials for the SA and AV nodes, other parts of the conduction system, and the atrial and ventricular muscles are shown along with the correlation to the extracellularly recorded electrical activity, that is, the electrocardiogram (ECG). The action potentials and ECG are plotted on the same time axis but with different zero points on the vertical scale. LAF, left anterior fascicle.

ORIGIN & SPREAD OF CARDIAC EXCITATION

ANATOMIC CONSIDERATIONS

In the human heart, the SA node is located at the junction of the superior vena cava with the right atrium. The AV node is located in the right posterior portion of the interatrial septum (Figure 30–1). There are three bundles of atrial fibers that contain Purkinje-type fibers and connect the SA node to the AV node: the anterior internodal tract of Bachman, the middle internodal tract of Wenckebach, and the posterior internodal tract of Thorel. Conduction also occurs through atrial myocytes, but it is more rapid in these bundles. The AV node is continuous with the bundle of His, which gives off a left bundle branch at the top of the interventricular septum and continues as the right bundle branch. The left bundle branch divides into an anterior fascicle and a posterior fascicle. The branches and fascicles run subendocardially down either side of the septum and come into contact with the Purkinje system, whose fibers spread to all parts of the ventricular myocardium.

The histology of cardiac muscle is described in Chapter 5. The conduction system is composed, for the most part, of modified cardiac muscle that has fewer striations and indistinct boundaries. The SA node and, to a lesser extent, the AV node also contain small round cells with few organelles, which are connected by gap junctions. These are probably the actual pacemaker cells, and therefore they are called **P cells.** The atrial muscle fibers are separated from those of the ven-

tricles by a fibrous tissue ring, and normally the only conducting tissue between the atria and ventricles is the bundle of His.

The SA node develops from structures on the right side of the embryo and the AV node from structures on the left. This is why in the adult the right vagus is distributed mainly to the SA node and the left vagus mainly to the AV node. Similarly, the sympathetic innervation on the right side is distributed primarily to the SA node and the sympathetic innervation on the left side primarily to the AV node. On each side, most sympathetic fibers come from the stellate ganglion. Noradrenergic fibers are epicardial, whereas the vagal fibers are endocardial. However, connections exist for reciprocal inhibitory effects of the sympathetic and parasympathetic innervation of the heart on each other. Thus, acetylcholine acts presynaptically to reduce norepinephrine release from the sympathetic nerves, and conversely, neuropeptide Y released from noradrenergic endings may inhibit the release of acetylcholine.

PROPERTIES OF CARDIAC MUSCLE

The electrical responses of cardiac muscle and nodal tissue and the ionic fluxes that underlie them are discussed in detail in Chapter 5 and are briefly reviewed here for comparison with the pacemaker cells below. Myocardial fibers have a resting membrane potential of approximately –90 mV (Figure 30–2A). The individual fibers are separated by membranes, but depolarization spreads radially through them as if they were a syncytium because of the presence of gap junctions. The transmembrane action potential of single cardiac muscle cells

FIGURE 30–2 **Comparison of action potentials in ventricular muscle and diagram of the membrane potential of pacemaker tissue.**
A) Phases of action potential in ventricular myocyte (0–4, see text for details) are superimposed with principal changes in current that contribute to changes in membrane potential. **B)** The principal current responsible for each part of the potential of pacemaker tissue is shown under or beside the component. L, long-lasting; T, transient. Other ion channels contribute to the electrical response. Note that the resting membrane potential of pacemaker tissue is somewhat lower than that of atrial and ventricular muscle.

is characterized by rapid depolarization (phase 0), an initial rapid repolarization (phase 1), a plateau (phase 2), and a slow repolarization process (phase 3) that allows return to the resting membrane potential (phase 4). The initial depolarization is due to Na^+ influx through rapidly opening Na^+ channels (the Na^+ current, I_{Na}). The inactivation of Na^+ channels contributes to the rapid repolarization phase. Ca^{2+} influx through more slowly opening Ca^{2+} channels (the Ca^{2+} current, I_{Ca}) produces the plateau phase, and repolarization is due to net K^+ efflux through multiple types of K^+ channels. Recorded extracellularly, the summed electrical activity of all the cardiac muscle fibers is the electrocardiogram (ECG). The timing of the discharge of the individual units relative to the ECG is shown in Figure 30–1.

PACEMAKER POTENTIALS

Rhythmically discharging cells have a membrane potential that, after each impulse, declines to the firing level. Thus, this **prepotential** or **pacemaker potential** (Figure 30–2B) triggers the next impulse. At the peak of each impulse, I_K begins and brings about repolarization. I_K then declines, and a channel that can pass both Na^+ and K^+ is activated. Because this channel is activated following hyperpolarization, it is referred to as an "h" channel; however, because of its unusual (funny) activation this has also been dubbed an "f" channel. As I_h increases, the membrane begins to depolarize, forming the first part of the prepotential. Ca^{2+} channels then open. These are of two types in the heart, the **T** (for transient) **channels** and the **L** (for long-lasting) **channels.** The calcium current (I_{Ca}) due to opening of T channels completes the prepotential, and I_{Ca} due to opening of L channels produces the impulse. Other ion channels are also involved, and there is evidence that local Ca^{2+} release from the sarcoplasmic reticulum (**Ca^{2+} sparks**) occurs during the prepotential.

The action potentials in the SA and AV nodes are largely due to Ca^{2+}, with no contribution by Na^+ influx. Consequently, there is no sharp, rapid depolarizing spike before the plateau, as there is in other parts of the conduction system and the atrial and ventricular fibers. In addition, prepotentials are normally prominent only in the SA and AV nodes. However, "latent pacemakers" are present in other portions of the conduction system that can take over when the SA and AV nodes are depressed or conduction from them is blocked. Atrial and ventricular muscle fibers do not have prepotentials, and they discharge spontaneously only when injured or abnormal.

When the cholinergic vagal fibers to nodal tissue are stimulated, the membrane becomes hyperpolarized and the slope of the prepotentials is decreased (Figure 30–3) because the acetylcholine released at the nerve endings increases the K^+ conductance of nodal tissue. This action is mediated by M_2 muscarinic receptors, which, via the βγ subunit of a G protein, open a special set of K^+ channels. The resulting I_{KAch} slows the depolarizing effect of I_h. In addition, activation of the M_2

FIGURE 30–3 **Effect of sympathetic (noradrenergic) and vagal (cholinergic) and sympathetic (noradrenergic) stimulation on the membrane potential of the SA node.** Note the reduced slope of the prepotential after vagal stimulation and the increased spontaneous discharge after sympathetic stimulation.

receptors decreases cyclic adenosine 3',5'-monophosphate (cAMP) in the cells, and this slows the opening of the Ca^{2+} channels. The result is a decrease in firing rate. Strong vagal stimulation may abolish spontaneous discharge for some time.

Conversely, stimulation of the sympathetic cardiac nerves speeds the depolarizing effect of I_h, and the rate of spontaneous discharge increases (Figure 30–3). Norepinephrine secreted by the sympathetic endings binds to β_1 receptors, and the resulting increase in intracellular cAMP facilitates the opening of L channels, increasing I_{Ca} and the rapidity of the depolarization phase of the impulse.

The rate of discharge of the SA node and other nodal tissue is influenced by temperature and by drugs. The discharge frequency is increased when the temperature rises, and this may contribute to the tachycardia associated with fever. Digitalis depresses nodal tissue and exerts an effect like that of vagal stimulation, particularly on the AV node.

SPREAD OF CARDIAC EXCITATION

Depolarization initiated in the SA node spreads radially through the atria, then converges on the AV node. Atrial depolarization is complete in about 0.1 s. Because conduction in the AV node is slow (Table 30–1), a delay of about 0.1 s (**AV nodal delay**) occurs before excitation spreads to the ventricles. It is interesting to note here that when there is a lack of contribution of I_{Na} in the depolarization (phase 0) of the action potential, a marked loss of conduction is observed. This delay is shortened by stimulation of the sympathetic nerves to the heart and lengthened by stimulation of the vagi. From the top of the septum, the wave of depolarization spreads in the rapidly conducting Purkinje fibers to all parts of the ventricles in 0.08–0.1 s. In humans, depolarization of the ventricular muscle starts at the left side of the interventricular septum and moves first to the right across the mid portion of the septum. The wave of depolarization then spreads down the septum to the apex of the heart. It returns along the ventricular walls to the AV groove, proceeding from the endocardial to the epicardial surface (Figure 30–4). The last parts of the heart to be depolarized are the posterobasal portion of the left ventricle, the pulmonary conus, and the uppermost portion of the septum.

TABLE 30–1 Conduction speeds in cardiac tissue.

Tissue	Conduction Rate (m/s)
SA node	0.05
Atrial pathways	1
AV node	0.05
Bundle of His	1
Purkinje system	4
Ventricular muscle	1

THE ELECTROCARDIOGRAM

Because the body fluids are good conductors (ie, because the body is a **volume conductor**), fluctuations in potential that represent the algebraic sum of the action potentials of myocardial fibers can be recorded extracellularly. The record of these potential fluctuations during the cardiac cycle is the **electrocardiogram (ECG)**.

The ECG may be recorded by using an **active or exploring electrode** connected to an indifferent electrode at zero potential (**unipolar recording**) or by using two active electrodes (**bipolar recording**). In a volume conductor, the sum of the potentials at the points of an equilateral triangle with a current source in the center is zero at all times. A triangle with the heart at its center (**Einthoven's triangle**) can be approximated by placing electrodes on both arms and on the left leg. These are the three **standard limb leads** used in electrocardiography. If these electrodes are connected to a common terminal, an indifferent electrode that stays near zero potential is obtained. Depolarization moving toward an active electrode in a volume conductor produces a positive deflection, whereas depolarization moving in the opposite direction produces a negative deflection.

The names of the various waves and segments of the ECG in humans are shown in Figure 30–5. By convention, an upward deflection is written when the active electrode becomes positive relative to the indifferent electrode, and a downward deflection is written when the active electrode becomes negative. The P wave is produced by atrial depolarization, the QRS complex by ventricular depolarization, and the T wave by ventricular repolarization. The U wave is an inconstant finding, believed to be due to slow repolarization of the papillary muscles. The intervals between the various waves of the ECG and the events in the heart that occur during these intervals are shown in Table 30–2.

BIPOLAR LEADS

Bipolar leads were used before unipolar leads were developed. The **standard limb leads**—leads I, II, and III—each record the differences in potential between two limbs. Because current flows only in the body fluids, the records obtained are those that would be obtained if the electrodes were at the points of attachment of the limbs, no matter where on the limbs the electrodes are placed. In lead I, the electrodes are connected so that an upward deflection is inscribed when the left arm becomes positive relative to the right (left arm positive). In lead II, the electrodes are on the right arm and left leg, with the leg positive; and in lead III, the electrodes are on the left arm and left leg, with the leg positive.

UNIPOLAR (V) LEADS

An additional nine unipolar leads, that is, leads that record the potential difference between an exploring electrode and an

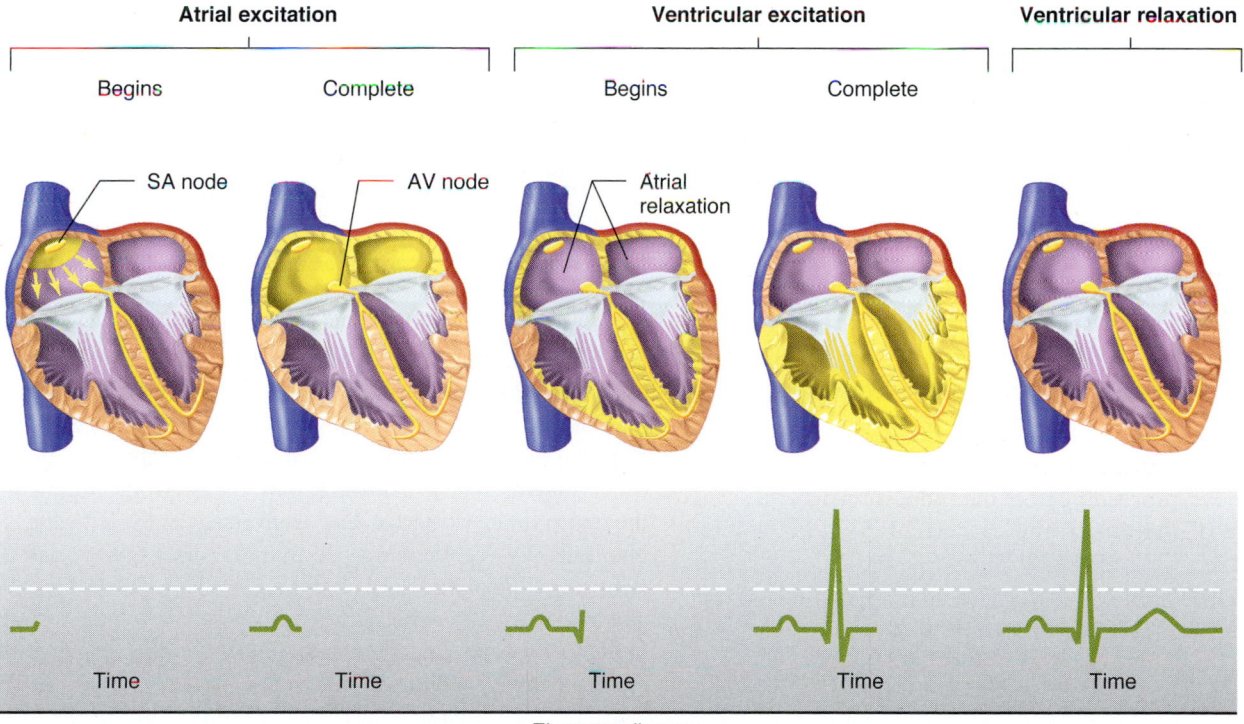

FIGURE 30–4 **Normal spread of electrical activity in the heart. A)** Conducting system of the heart. **B)** Sequence of cardiac excitation.
Top: Anatomical position of electrical activity. **Bottom:** corresponding electrocardiogram. The yellow color denotes areas that are depolarized.
(Reproduced with permission from Goldman MJ: *Principles of Clinical Electrocardiography,* 12th ed. Originally published by Appleton & Lange. Copyright © 1986 by McGraw-Hill.)

FIGURE 30–5   Waves of the ECG.

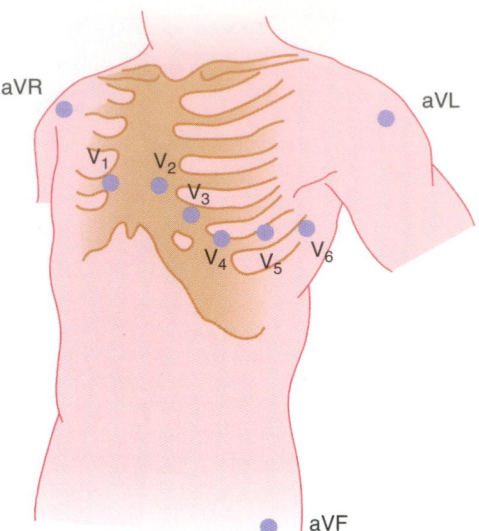

FIGURE 30–6   Unipolar electrocardiographic leads.

indifferent electrode, are commonly used in clinical electro-cardiography. There are six unipolar chest leads (precordial leads) designated V_1–V_6 (Figure 30–6) and three unipolar limb leads: VR (right arm), VL (left arm), and VF (left foot). **Augmented limb leads,** designated by the letter *a* (aVR, aVL, aVF), are generally used. The augmented limb leads are recordings between one limb and the other two limbs. This increases the size of the potentials by 50% without any change in configuration from the nonaugmented record.

Unipolar leads can also be placed at the tips of catheters and inserted into the esophagus or heart.

NORMAL ECG

The ECG of a normal individual is shown in Figure 30–7. The sequence in which the parts of the heart are depolarized (Figure

TABLE 30–2   ECG intervals.

Intervals	Normal Durations		Events in the Heart during Interval
	Average	**Range**	
PR interval[a]	0.18[b]	0.12–0.20	Atrial depolarization and conduction through AV node
QRS duration	0.08	to 0.10	Ventricular depolarization and atrial repolarization
QT interval	0.40	to 0.43	Ventricular depolarization plus ventricular repolarization
ST interval (QT minus QRS)	0.32	...	Ventricular repolarization (during T wave)

[a]Measured from the beginning of the P wave to the beginning of the QRS complex.

[b]Shortens as heart rate increases from average of 0.18 s at a rate of 70 beats/min to 0.14 s at a rate of 130 beats/min.

30–4) and the position of the heart relative to the electrodes are the important considerations in interpreting the configurations of the waves in each lead. The atria are located posteriorly in the chest. The ventricles form the base and anterior surface of the heart, and the right ventricle is anterolateral to the left. Thus, aVR "looks at" the cavities of the ventricles. Atrial depolarization, ventricular depolarization, and ventricular repolarization move away from the exploring electrode, and the P wave, QRS complex, and T wave are therefore all negative (downward) deflections; aVL and aVF look at the ventricles, and the deflections are therefore predominantly positive or biphasic. There is no Q wave in V_1 and V_2, and the initial portion of the QRS complex is a small upward deflection because ventricular depolarization first moves across the midportion of the septum from left to right toward the exploring electrode. The wave of excitation then moves down the septum and into the left ventricle away from the electrode, producing a large S wave. Finally, it moves back along the ventricular wall toward the electrode, producing the return to the isoelectric line. Conversely, in the left ventricular leads (V_4–V_6) there may be an initial small Q wave (left to right septal depolarization), and there is a large R wave (septal and left ventricular depolarization) followed in V_4 and V_5 by a moderate S wave (late depolarization of the ventricular walls moving back toward the AV junction).

There is considerable variation in the position of the normal heart, and the position affects the configuration of the electrocardiographic complexes in the various leads.

BIPOLAR LIMB LEADS & THE CARDIAC VECTOR

Because the standard limb leads are records of the potential differences between two points, the deflection in each lead at any instant indicates the magnitude and direction in the axis of the lead of the electromotive force generated in the heart (**cardiac**

FIGURE 30–7   Normal ECG. (Reproduced with permission from Goldman MJ: *Principles of Clinical Electrocardiography,* 12th ed. Originally published by Appleton & Lange. Copyright © 1986 by McGraw-Hill.)

vector or **axis**). The vector at any given moment in the two dimensions of the frontal plane can be calculated from any two standard limb leads (Figure 30–8) if it is assumed that the three electrode locations form the points of an equilateral triangle (Einthoven's triangle) and that the heart lies in the center of the triangle. These assumptions are not completely warranted, but calculated vectors are useful approximations. An approximate **mean QRS vector** ("electrical axis of the heart") is often plotted by using the average QRS deflection in each lead, as shown in Figure 30–8. This is a **mean** vector as opposed to an **instantaneous** vector, and the average QRS deflections should be measured by integrating the QRS complexes. However, they can be approximated by measuring the net differences between the positive and negative peaks of the QRS. The normal direction

of the mean QRS vector is generally said to be –30 to +110 degrees on the coordinate system shown in Figure 30–8. **Left** or **right axis deviation** is said to be present if the calculated axis falls to the left of –30 degrees or to the right of +110 degrees, respectively. Right axis deviation suggests right ventricular hypertrophy, and left axis deviation may be due to left ventricular hypertrophy, but there are better and more reliable electrocardiographic criteria for ventricular hypertrophy.

VECTORCARDIOGRAPHY

If the tops of the arrows representing all the instantaneous cardiac vectors in the frontal plane during the cardiac cycle are connected, from first to last, the line connecting them forms a

FIGURE 30–8   **Cardiac vector. Left:** Einthoven's triangle. Perpendiculars dropped from the midpoints of the sides of the equilateral triangle intersect at the center of electrical activity. RA, right arm; LA, left arm; LL, left leg. **Center:** Calculation of mean QRS vector. In each lead, distances equal to the height of the R wave minus the height of the largest negative deflection in the QRS complex are measured off from the midpoint of the side of the triangle representing that lead. An arrow drawn from the center of electrical activity to the point of intersection of perpendiculars extended from the distances measured off on the sides represents the magnitude and direction of the mean QRS vector. **Right:** Reference axes for determining the direction of the vector.

series of three loops: one for the P wave, one for the QRS complex, and one for the T wave. This can be done electronically and the loops, called **vectorcardiograms,** are projected on the face of an oscilloscope.

His BUNDLE ELECTROGRAM

In patients with heart block, the electrical events in the AV node, bundle of His, and Purkinje system are frequently studied with a catheter containing an electrode at its tip that is passed through a vein to the right side of the heart and manipulated into a position close to the tricuspid valve. Three or more standard electrocardiographic leads are recorded simultaneously. The record of the electrical activity obtained with the catheter (Figure 30–9) is the **His bundle electrogram (HBE).** It normally shows an A deflection when the AV node is activated, an H spike during transmission through the His

bundle, and a V deflection during ventricular depolarization. With the HBE and the standard electrocardiographic leads, it is possible to accurately time three intervals: (1) the PA interval, the time from the first appearance of atrial depolarization to the A wave in the HBE, which represents conduction time from the SA node to the AV node; (2) the AH interval, from the A wave to the start of the H spike, which represents the AV nodal conduction time; and (3) the HV interval, the time from the start of the H spike to the start of the QRS deflection in the ECG, which represents conduction in the bundle of His and the bundle branches. The approximate normal values for these intervals in adults are PA, 27 ms; AH, 92 ms; and HV, 43 ms. These values illustrate the relative slowness of conduction in the AV node (Table 30–1).

MONITORING

The ECG is often recorded continuously in hospital coronary care units, with alarms arranged to sound at the onset of life-threatening arrhythmias. Using a small portable tape recorder **(Holter monitor),** it is also possible to record the ECG in ambulatory individuals as they go about their normal activities. The recording is later played back at high speed and analyzed. Long-term continuous records can be obtained. Recordings obtained with monitors have proved valuable in the diagnosis of arrhythmias and in planning the treatment of patients recovering from myocardial infarctions.

FIGURE 30–9   **Normal His bundle electrogram (HBE) with simultaneously recorded ECG.**

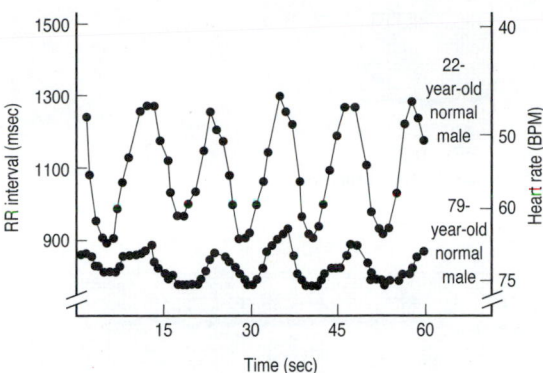

FIGURE 30–10 **Sinus arrhythmia in a young man and an old man.** Each subject breathed five times per minute. With each inspiration the RR interval (the interval between R waves) declined, indicating an increase in heart rate. Note the marked reduction in the magnitude of the arrhythmia in the older man. These records were obtained after β-adrenergic blockade, but would have been generally similar in its absence. (Reproduced with permission from Pfeifer MA et al: Differential changes of autonomic nervous system function with age in man. Am J Med 1983;75:249.)

CLINICAL APPLICATIONS: CARDIAC ARRHYTHMIAS

NORMAL CARDIAC RATE

In the normal human heart, each beat originates in the SA node (**normal sinus rhythm, NSR**). The heart beats about 70 times a minute at rest. The rate is slowed (**bradycardia**) during sleep and accelerated (**tachycardia**) by emotion, exercise, fever, and many other stimuli. In healthy young individuals breathing at a normal rate, the heart rate varies with the phases of respiration: It accelerates during inspiration and decelerates during expiration, especially if the depth of breathing is increased. This **sinus arrhythmia** (Figure 30–10) is a normal phenomenon and is due primarily to fluctuations in parasympathetic output to the heart. During inspiration, impulses in the vagi from the stretch receptors in the lungs inhibit the cardio-inhibitory area in the medulla oblongata. The tonic vagal discharge that keeps the heart rate slow decreases, and the heart rate rises.

Disease processes affecting the sinus node lead to marked bradycardia accompanied by dizziness and syncope (**sick sinus syndrome**).

ABNORMAL PACEMAKERS

The AV node and other portions of the conduction system can, in abnormal situations, become the cardiac pacemaker. In addition, diseased atrial and ventricular muscle fibers can have their membrane potentials reduced and discharge repetitively.

As noted above, the discharge rate of the SA node is more rapid than that of the other parts of the conduction system, and this is why the SA node normally controls the heart rate. When conduction from the atria to the ventricles is completely interrupted, **complete (third-degree) heart block** results, and the ventricles beat at a low rate (**idioventricular rhythm**) independently of the atria (Figure 30–11). The block may be due to disease in the AV node (**AV nodal block**) or in the conducting system below the node (**infranodal block**). In patients with AV nodal block, the remaining nodal tissue becomes the pacemaker and the rate of the idioventricular rhythm is approximately 45 beats/min. In patients with infranodal block due to disease in the bundle of His, the ventricular pacemaker is located more peripherally in the conduction system and the ventricular rate is lower; it averages 35 beats/min, but in individual cases it can be as low as 15 beats/min. In such individuals, there may also be periods of asystole lasting a minute or more. The resultant cerebral ischemia causes dizziness and fainting (**Stokes–Adams syndrome**). Causes of third-degree heart block include septal myocardial infarction and damage to the bundle of His during surgical correction of congenital interventricular septal defects.

When conduction between the atria and ventricles is slowed but not completely interrupted, **incomplete heart block** is present. In the form called **first-degree heart block,** all the atrial impulses reach the ventricles but the PR interval is abnormally long. In the form called **second-degree heart block,** not all atrial impulses are conducted to the ventricles. For example, a ventricular beat may follow every second or every third atrial beat (2:1 block, 3:1 block, etc). In another form of incomplete heart block, there are repeated sequences of beats in which the PR interval lengthens progressively until a ventricular beat is dropped (**Wenckebach phenomenon**). The PR interval of the cardiac cycle that follows each dropped beat is usually normal or only slightly prolonged (Figure 30–11).

Sometimes one branch of the bundle of His is interrupted, causing **right** or **left bundle branch block.** In bundle branch block, excitation passes normally down the bundle on the intact side and then sweeps back through the muscle to activate the ventricle on the blocked side. The ventricular rate is therefore normal, but the QRS complexes are prolonged and deformed (Figure 30–11). Block can also occur in the anterior or posterior fascicle of the left bundle branch, producing the condition called **hemiblock** or **fascicular block.** Left anterior hemiblock produces abnormal left axis deviation in the ECG, whereas left posterior hemiblock produces abnormal right axis deviation. It is not uncommon to find combinations of fascicular and branch blocks (**bifascicular** or **trifascicular block**). The His bundle electrogram permits detailed analysis of the site of block when there is a defect in the conduction system.

IMPLANTED PACEMAKERS

When there is marked bradycardia in patients with sick sinus syndrome or third-degree heart block, an electronic pacemaker is frequently implanted. These devices, which have become sophisticated and reliable, are useful in patients with sinus node dysfunction, AV block, and bifascicular or trifascicular block. They are useful also in patients with severe neurogenic

PR = 0.16 s
Normal complex

PR = 0.38 s
First-degree heart block

Second-degree heart block
(2:1 heart block)

Second-degree heart block
(Wenckebach phenomenon)

Complete heart block. Atrial rate, 107; ventricular rate, 43

Two V leads in left
bundle branch block

FIGURE 30–11   Heart block.

syncope in whom carotid sinus stimulation produces pauses of more than 3 s between heartbeats.

ECTOPIC FOCI OF EXCITATION

Normally, myocardial cells do not discharge spontaneously, and the possibility of spontaneous discharge of the His bundle and Purkinje system is low because the normal pacemaker discharge of the SA node is more rapid than their rate of spontaneous discharge. However, in abnormal conditions, the His–Purkinje fibers or the myocardial fibers may discharge spontaneously. In these conditions, **increased automaticity** of the heart is said to be present. If an irritable **ectopic focus** discharges once, the result is a beat that occurs before the expected next normal beat and transiently interrupts the cardiac rhythm (atrial, nodal, or ventricular **extrasystole** or **premature beat**). If the focus discharges repetitively at a rate higher than that of the SA node, it produces rapid, regular tachycardia (atrial, ventricular, or nodal **paroxysmal tachycardia** or **atrial flutter**).

REENTRY

A more common cause of paroxysmal arrhythmias is a defect in conduction that permits a wave of excitation to propagate continuously within a closed circuit (**circus movement**). For example, if a transient block is present on one side of a portion of the conducting system, the impulse can go down the other side. If

the block then wears off, the impulse may conduct in a retrograde direction in the previously blocked side back to the origin and then descend again, establishing a circus movement. An example of this in a ring of tissue is shown in Figure 30–12. If the reentry is in the AV node, the reentrant activity depolarizes the atrium, and the resulting atrial beat is called an echo beat. In addition, the reentrant activity in the node propagates back down to the ventricle, producing paroxysmal nodal tachycardia. Circus movements can also become established in the atrial or ventricular muscle fibers. In individuals with an abnormal extra bundle of conducting tissue connecting the atria to the ventricles (bundle of Kent), the circus activity can pass in one direction through the AV node and in the other direction through the bundle, thus involving both the atria and the ventricles.

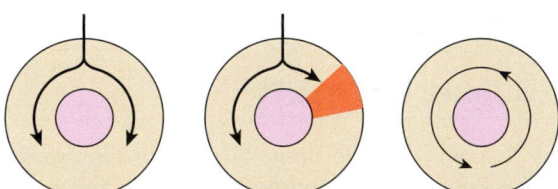

FIGURE 30–12   Depolarization of a ring of cardiac tissue.
Normally, the impulse spreads in both directions in the ring **(left)** and the tissue immediately behind each branch of the impulse is refractory. When a transient block occurs on one side **(center),** the impulse on the other side goes around the ring, and if the transient block has now worn off **(right),** the impulse passes this area and continues to circle indefinitely (circus movement).

ATRIAL ARRHYTHMIAS

Excitation spreading from an independently discharging focus in the atria stimulates the AV node prematurely and is conducted to the ventricles. The P waves of atrial extrasystoles are abnormal, but the QRST configurations are usually normal (Figure 30–13). The excitation may depolarize the SA node, which must repolarize and then depolarize to the firing level before it can initiate the next normal beat. Consequently, a pause occurs between the extrasystole and the next normal beat that is usually equal in length to the interval between the normal beats preceding the extrasystole, and the rhythm is "reset" (see below).

Atrial tachycardia occurs when an atrial focus discharges regularly or there is reentrant activity producing atrial rates up to 220/min. Sometimes, especially in digitalized patients, some degree of atrioventricular block is associated with the tachycardia (**paroxysmal atrial tachycardia with block**).

In atrial flutter, the atrial rate is 200 to 350/min (Figure 30–13). In the most common form of this arrhythmia, there is large counterclockwise circus movement in the right atrium. This produces a characteristic sawtooth pattern of flutter waves due to atrial contractions. It is almost always associated with 2:1 or greater AV block, because in adults the AV node cannot conduct more than about 230 impulses per minute.

In **atrial fibrillation,** the atria beat very rapidly (300–500/min) in a completely irregular and disorganized fashion. Because the AV node discharges at irregular intervals, the ventricles beat at a completely irregular rate, usually 80 to 160/min (Figure 30–13). The condition can be paroxysmal or chronic, and in some cases there appears to be a genetic predisposition. The cause of atrial fibrillation is still a matter of debate, but in most cases it appears to be due to multiple concurrently circulating reentrant excitation waves in both atria. However, some cases of paroxysmal atrial fibrillation seem to be produced by discharge of one or more ectopic foci. Many of these foci appear to be located in the pulmonary veins as much as 4 cm from the heart. Atrial muscle fibers extend along the pulmonary veins and are the origin of these discharges.

CONSEQUENCES OF ATRIAL ARRHYTHMIAS

Occasional atrial extrasystoles occur from time to time in most normal humans and have no pathologic significance. In paroxysmal atrial tachycardia and flutter, the ventricular rate may be so high that diastole is too short for adequate filling of the ventricles with blood between contractions. Consequently, cardiac output is reduced and symptoms of heart failure appear. Heart failure may also complicate atrial fibrillation when the ventricular rate is high. Acetylcholine liberated at vagal endings depresses conduction in the atrial musculature and AV node. This is why stimulating reflex vagal discharge by pressing on the eyeball (**oculocardiac reflex**) or massaging the carotid sinus often converts tachycardia and sometimes converts atrial flutter to normal sinus rhythm. Alternatively, vagal stimulation increases the degree of AV block, abruptly lowering the ventricular rate. Digitalis also depresses AV conduction and is used to lower a rapid ventricular rate in atrial fibrillation.

VENTRICULAR ARRHYTHMIAS

Premature beats that originate in an ectopic ventricular focus usually have bizarrely shaped prolonged QRS complexes (Figure 30–14) because of the slow spread of the impulse from the focus through the ventricular muscle to the rest of the ventricle. They are usually incapable of exciting the bundle of His, and retrograde conduction to the atria therefore does not occur. In the meantime, the next succeeding normal SA nodal impulse depolarizes the atria. The P wave is usually buried in the QRS of the extrasystole. If the normal impulse reaches the ventricles, they are still in the refractory period following depolarization from the ectopic focus.

Atrial extrasystole

Atrial tachycardia

Atrial flutter

Atrial fibrillation

FIGURE 30–13 **Atrial arrhythmias.** The illustration shows an atrial premature beat with its P wave superimposed on the T wave of the preceding beat (*arrow*); atrial tachycardia; atrial flutter with 4:1 AV block; and atrial fibrillation with a totally irregular ventricular rate. (Tracings reproduced with permission from Goldschlager N, Goldman MJ: *Principles of Clinical Electrocardiography,* 13th ed. Originally published by Appleton & Lange. Copyright © 1989 by McGraw-Hill.)

FIGURE 30–14 **Top:** Ventricular premature beats (VPB). The lines under the tracing illustrate the compensatory pause and show that the duration of the premature beat plus the preceding normal beat is equal to the duration of two normal beats. **Bottom:** Ventricular tachycardia.

premature beats do not interrupt the regular discharge of the SA node, whereas atrial premature beats often interrupt and "reset" the normal rhythm.

Atrial and ventricular premature beats are not strong enough to produce a pulse at the wrist if they occur early in diastole, when the ventricles have not had time to fill with blood and the ventricular musculature is still in its relatively refractory period. They may not even open the aortic and pulmonary valves, in which case there is, in addition, no second heart sound.

Paroxysmal ventricular tachycardia (Figure 30–14) is in effect a series of rapid, regular ventricular depolarizations usually due to a circus movement involving the ventricles. **Torsade de pointes** is a form of ventricular tachycardia in which the QRS morphology varies (Figure 30–15). Tachycardias originating above the ventricles (supraventricular tachycardias such as paroxysmal nodal tachycardia) can be distinguished from paroxysmal ventricular tachycardia by use of the HBE; in supraventricular tachycardias, a His bundle H deflection is present, whereas in ventricular tachycardias, there is none. Ventricular premature beats are common and, in the absence of ischemic heart disease, usually benign. Ventricular tachycardia is more serious because cardiac output is decreased, and ventricular fibrillation is an occasional complication of ventricular tachycardia.

In **ventricular fibrillation** (Figure 30–15), the ventricular muscle fibers contract in a totally irregular and ineffective way because of the very rapid discharge of multiple ventricular

However, the second succeeding impulse from the SA node produces a normal beat. Thus, ventricular premature beats are followed by a **compensatory pause** that is often longer than the pause after an atrial extrasystole. Furthermore, ventricular

FIGURE 30–15 Record obtained from an implanted cardioverter–defibrillator in a 12-year-old boy with congenital long QT syndrome who collapsed while answering a question in school. **Top:** Normal sinus rhythm with long QT interval. **Middle:** Torsade de pointes. **Bottom:** Ventricular fibrillation with discharge of defibrillator, as programmed 7.5 s after the start of ventricular tachycardia, converting the heart to normal sinus rhythm. The boy recovered consciousness in 2 min and had no neurologic sequelae. (Reproduced with permission from Moss AJ, Daubert JP: Images in clinical medicine. Internal ventricular fibrillation. N Engl J Med 2000;342:398.)

ectopic foci or a circus movement. The fibrillating ventricles, like the fibrillating atria, look like a quivering "bag of worms." Ventricular fibrillation can be produced by an electric shock or an extrasystole during a critical interval, the **vulnerable period.** The vulnerable period coincides in time with the midportion of the T wave; that is, it occurs at a time when some of the ventricular myocardium is depolarized, some is incompletely repolarized, and some is completely repolarized. These are excellent conditions in which to establish reentry and a circus movement. The fibrillating ventricles cannot pump blood effectively, and circulation of the blood stops. Therefore, in the absence of emergency treatment, ventricular fibrillation that lasts more than a few minutes is fatal. The most frequent cause of sudden death in patients with myocardial infarcts is ventricular fibrillation.

LONG QT SYNDROME

An indication of vulnerability of the heart during repolarization is the fact that in patients in whom the QT interval is prolonged, cardiac repolarization is irregular and the incidence of ventricular arrhythmias and sudden death increases. The syndrome can be caused by a number of different drugs, by electrolyte abnormalities, and by myocardial ischemia. It can also be congenital. Mutations of eight different genes have been reported to cause the syndrome. Six cause reduced function of various K^+ channels by alterations in their structure; one inhibits a K^+ channel by reducing the amount of the ankyrin isoform that links it to the cytoskeleton; and one increases the function of the cardiac Na^+ channel.

ACCELERATED AV CONDUCTION

An interesting condition seen in some otherwise normal individuals who are prone to attacks of paroxysmal atrial arrhythmias is **accelerated AV conduction (Wolff–Parkinson–White syndrome).** Normally, the only conducting pathway between the atria and the ventricles is the AV node. Individuals with Wolff–Parkinson–White syndrome have an additional aberrant muscular or nodal tissue connection (**bundle of Kent**) between the atria and ventricles. This conducts more rapidly than the slowly conducting AV node, and one ventricle is excited early. The manifestations of its activation merge with the normal QRS pattern, producing a short PR interval and a prolonged QRS deflection slurred on the upstroke (Figure 30–16), with a normal interval between the start of the P wave and the end of the QRS complex ("PJ interval"). The paroxysmal atrial tachycardias seen in this syndrome often follow an atrial premature beat. This beat conducts normally down the AV node but spreads to the ventricular end of the aberrant bundle, and the impulse is transmitted retrograde to the atrium. A circus movement is thus established. Less commonly, an atrial premature beat finds the AV node refractory but reaches the ventricles via the bundle of Kent, setting up a circus movement in which the impulse passes from the ventricles to the atria via the AV node.

FIGURE 30–16 **Accelerated AV conduction. Top:** Normal sinus beat. **Middle:** Short PR interval; wide, slurred QRS complex; normal PJ interval (Wolff–Parkinson–White syndrome). **Bottom:** Short PR interval, normal QRS complex (Lown–Ganong–Levine syndrome). (Reproduced with permission from Goldschlager N, Goldman MJ: *Principles of Clinical Electrocardiography,* 13th ed. Originally published by Appleton & Lange. Copyright © 1989 by McGraw-Hill.)

In some instances, the Wolff–Parkinson–White syndrome is familial. In two such families, there is a mutation in a gene that codes for an AMP-activated protein kinase. Presumably, this kinase is normally involved in suppressing abnormal atrioventricular pathways during fetal development.

Attacks of paroxysmal supraventricular tachycardia, usually nodal tachycardia, are seen in individuals with short PR intervals and normal QRS complexes (**Lown–Ganong–Levine syndrome**). In this condition, depolarization presumably passes from the atria to the ventricles via an aberrant bundle that bypasses the AV node but enters the intraventricular conducting system distal to the node.

ANTIARRHYTHMIC DRUGS

Many different drugs have been developed that are used in the treatment of arrhythmias because they slow conduction in the conduction system and the myocardium. This depresses ectopic activity and reduces the discrepancy between normal and reentrant paths so that reentry does not occur. However, it has now become clear that in some patients any of these drugs can be **proarrhythmic** rather than antiarrhythmic—that is, they can also cause various arrhythmias. Therefore, they are increasingly being replaced by radiofrequency catheter ablation for the treatment of arrhythmias.

RADIOFREQUENCY CATHETER ABLATION OF REENTRANT PATHWAYS

Catheters with electrodes at the tip can now be inserted into the chambers of the heart and its environs and used to map the exact location of an ectopic focus or accessory bundle that is responsible for the production of reentry and supraventricular tachycardia. The pathway can then be ablated by passing radiofrequency current with the catheter tip placed close to the bundle or focus. In skilled hands, this form of treatment can be very effective and is associated with few complications. It is particularly useful in conditions that cause supraventricular tachycardias, including Wolff–Parkinson–White syndrome and atrial flutter. It has also been used with success to ablate foci in the pulmonary veins causing paroxysmal atrial fibrillation.

ELECTROCARDIOGRAPHIC FINDINGS IN OTHER CARDIAC & SYSTEMIC DISEASES

MYOCARDIAL INFARCTION

When the blood supply to part of the myocardium is interrupted, profound changes take place in the myocardium that lead to irreversible changes and death of muscle cells. The ECG is very useful for diagnosing ischemia and locating areas of infarction. The underlying electrical events and the resulting electrocardiographic changes are complex, and only a brief review can be presented here.

The three major abnormalities that cause electrocardiographic changes in acute myocardial infarction are summarized in Table 30–3. The first change—abnormally rapid repolarization after discharge of the infarcted muscle fibers as a result of accelerated opening of K^+ channels—develops seconds after occlusion of a coronary artery in experimental animals. It lasts only a few minutes, but before it is over the resting membrane potential of the infarcted fibers declines because of the loss of intracellular K^+. Starting about 30 min later, the infarcted fibers also begin to depolarize more slowly than the surrounding normal fibers.

All three of these changes cause current flow that produces elevation of the ST segment in electrocardiographic leads recorded with electrodes over the infarcted area (Figure 30–17). Because of the rapid repolarization in the infarct, the membrane potential of the area is greater than it is in the normal area during the latter part of repolarization, making the normal region negative relative to the infarct. Extracellularly, current therefore flows out of the infarct into the normal area (since, by convention, current flow is from positive to negative). This current flows toward electrodes over the injured area, causing increased positivity between the S and T waves of the ECG. Similarly, the delayed depolarization of the infarcted cells causes the infarcted area to be positive relative to the healthy tissue (Table 30–3) during the early part of repolarization, and the result is also ST segment elevation. The remaining change—the decline in resting membrane potential during diastole—causes a current flow into the infarct during ventricular diastole. The result of this current flow is a depression of the TQ segment of the ECG. However, the electronic arrangement in electrocardiographic recorders is such that a TQ segment depression is recorded as an ST segment elevation. Thus, the hallmark of acute myocardial infarction is elevation of the ST segments in the leads overlying the area of infarction (Figure 30–17). Leads on the opposite side of the heart show ST segment depression.

After some days or weeks, the ST segment abnormalities subside. The dead muscle and scar tissue become electrically silent. The infarcted area is therefore negative relative to the normal myocardium during systole, and it fails to contribute its share of positivity to the electrocardiographic complexes. The manifestations of this negativity are multiple and subtle. Common changes include the appearance of a Q wave in some of the leads in which it was not previously present and an increase in the size of the normal Q wave in some of the other leads, although so-called non-Q-wave infarcts are also seen. These infarcts tend to be less severe, but there is a high incidence of subsequent reinfarction. Another finding in infarction of the anterior left ventricle is "failure of progression of the R wave"; that is, the R wave fails to become successively larger in the precordial leads as the electrode is moved from right to left over the left ventricle. If the septum is infarcted, the conduction system may be damaged, causing bundle branch block or other forms of heart block.

Myocardial infarctions are often complicated by serious ventricular arrhythmias, with the threat of ventricular fibrillation and death. In experimental animals, and presumably in humans, ventricular arrhythmias occur during three periods. During the first 30 min of an infarction, arrhythmias due to reentry are common. There follows a period relatively free from arrhythmias, but, starting 12 h after infarction, arrhythmias occur as a result of increased automaticity. Arrhythmias occurring 3 d to several weeks after infarction are once again usually due to reentry. It is worth noting in this regard that infarcts that damage the epicardial portions of the myocardium interrupt sympathetic nerve fibers, producing denervation super-sensitivity to catecholamines in the area beyond the infarct. Alternatively, endocardial lesions can selectively interrupt vagal fibers, leaving the actions of sympathetic fibers unopposed.

TABLE 30–3 Summary of the three major abnormalities of membrane polarization associated with acute myocardial infarction.

Defect in Infarcted Cells	Current Flow	Resultant ECG Change in Leads Over Infarct
Rapid repolarization	Out of infarct	ST segment elevation
Decreased resting membrane potential	Into infarct	TQ segment depression (manifested as ST segment elevation)
Delayed depolarization	Out of infarct	ST segment elevation

FIGURE 30–17 **Diagrammatic illustration of serial electrocardiographic patterns in anterior infarction. A)** Normal tracing. **B)** Very early pattern (hours after infarction): ST segment elevation in I, aVL, and V_{3-6}; reciprocal ST depression in II, III, and aVF. **C)** Later pattern (many hours to a few days): Q waves have appeared in I, aVL, and V_{5-6}. QS complexes are present in V_{3-4}. This indicates that the major transmural infarction is underlying the area recorded by V_{3-4}; ST segment changes persist but are of lesser degree, and the T waves are beginning to invert in the leads in which the ST segments are elevated. **D)** Late established pattern (many days to weeks): The Q waves and QS complexes persist, the ST segments are isoelectric, and the T waves are symmetric and deeply inverted in leads that had ST elevation and tall in leads that had ST depression. This pattern may persist for the remainder of the patient's life. **E)** Very late pattern: This may occur many months to years after the infarction. The abnormal Q waves and QS complexes persist. The T waves have gradually returned to normal. (Reproduced with permission from Goldschlager N, Goldman MJ: *Principles of Clinical Electrocardiography*, 13th ed. Originally published by Appleton & Lange. Copyright © 1989 by McGraw-Hill.)

EFFECTS OF CHANGES IN THE IONIC COMPOSITION OF THE BLOOD

Changes in ECF Na^+ and K^+ concentration would be expected to affect the potentials of the myocardial fibers, because the electrical activity of the heart depends upon the distribution of these ions across the muscle cell membranes. Clinically, a fall in the plasma level of Na^+ may be associated with low-voltage electrocardiographic complexes, but changes in the plasma K^+ level produce severe cardiac abnormalities. Hyperkalemia is a very dangerous and potentially lethal condition because of its effects on the heart. As the plasma K^+ level rises, the first change in the ECG is the

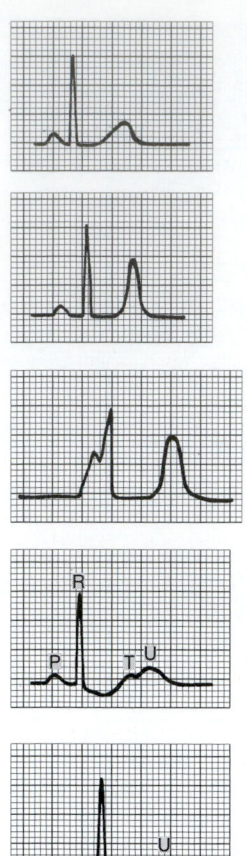

Normal tracing (plasma K⁺ 4–5.5 meq/L). PR interval = 0.16 s; QRS interval = 0.06 s; QT interval = 0.4 s (normal for an assumed heart rate of 60).

Hyperkalemia (plasma K⁺ ±7.0 meq/L). The PR and QRS intervals are within normal limits. Very tall, slender peaked T waves are now present.

Hyperkalemia (plasma K⁺ ±8.5 meq/L). There is no evidence of atrial activity; the QRS complex is broad and slurred and the QRS interval has widened to 0.2 s. The T waves remain tall and slender. Further elevation of the plasma K⁺ level may result in ventricular tachycardia and ventricular fibrillation.

Hypokalemia (plasma K⁺ ±3.5 meq/L). PR interval = 0.2 s; QRS interval = 0.06 s; ST segment depression. A prominent U wave is now present immediately following the T. The actual QT interval remains 0.4 s. If the U wave is erroneously considered a part of the T, a falsely prolonged QT interval of 0.6 s will be measured.

Hypokalemia (plasma K⁺ ±2.5 meq/L). The PR interval is lengthened to 0.32 s; the ST segment is depressed; the T wave is inverted; a prominent U wave is seen. The true QT interval remains normal.

FIGURE 30–18 **Correlation of plasma K⁺ level and the ECG, assuming that the plasma Ca²⁺ level is normal.** The diagrammed complexes are left ventricular epicardial leads. (Reproduced with permission from Goldman MJ: *Principles of Clinical Electrocardiography,* 12th ed. Originally published by Appleton & Lange. Copyright © 1986 by McGraw-Hill.)

appearance of tall peaked T waves, a manifestation of altered repolarization (Figure 30–18). At higher K⁺ levels, paralysis of the atria and prolongation of the QRS complexes occur. Ventricular arrhythmias may develop. The resting membrane potential of the muscle fibers decreases as the extracellular K⁺ concentration increases. The fibers eventually become unexcitable, and the heart stops in diastole. Conversely, a decrease in the plasma K⁺ level causes prolongation of the PR interval, prominent U waves, and, occasionally, late T wave inversion in the precordial leads. If the T and U waves merge, the apparent QT interval is often prolonged; if the T and U waves are separated, the true QT interval is seen to be of normal duration. Hypokalemia is a serious condition, but it is not as rapidly fatal as hyperkalemia.

Increases in extracellular Ca²⁺ concentration enhance myocardial contractility. When large amounts of Ca²⁺ are infused into experimental animals, the heart relaxes less during diastole and eventually stops in systole (**calcium rigor**). However, in clinical conditions associated with hypercalcemia, the plasma calcium level is rarely if ever high enough to affect the heart. Hypocalcemia causes prolongation of the ST segment and consequently of the QT interval, a change that is also produced by phenothiazines and tricyclic antidepressant drugs and by various diseases of the central nervous system.

CHAPTER SUMMARY

- Contractions in the heart are controlled via a well-regulated electrical signaling cascade that originates in pacemaker cells in the sinoatrial (SA) node and is passed via internodal atrial pathways to the atrioventrical (AV) node, the bundle of His, the Purkinje system, and to all parts of the ventricle.

- Most cardiac cells have an action potential that includes a rapid depolarization, an initial rapid repolarization, a plateau, and a slow repolarization process to return to resting potential. These changes are defined by sequential activation and inactivation of Na⁺, Ca²⁺, and K⁺ channels.

- Pacemaker cells have a slightly different sequence of events. After repolarization to the resting potential, there is a slow depolarization that occurs due to a channel that can pass both Na⁺ and K⁺. As this "funny" current continues to depolarize the cell, Ca²⁺ channels are activated to rapidly depolarize the cell. The hyperpolarization phase is again dominated by K⁺ current.

- Spread of the electrical signal from cell to cell is via gap junctions. The rate of spread is dependent on anatomical features, but also can be altered (to a certain extent) via neural input.

- The electrocardiogram (ECG) is an algebraic sum of the electrical activity in the heart. The normal ECG includes well-defined waves and segments, including the P wave (atrial depolarization), the QRS complex (ventricular depolarization), and the T wave (ventricular hyperpolarization). Various arrhythmias can be detected in irregular ECG recordings.

■ Because of the contribution of ionic movement to cardiac muscle contraction, heart tissue is sensitive to ionic composition of the blood. Most serious are increases in $[K^+]$ that can produce severe cardiac abnormalities, including paralysis of the atria and ventricular arrhythmias.

MULTIPLE-CHOICE QUESTIONS

For all questions, select the single best answer unless otherwise directed.

1. Which part of the ECG (eg, Figure 30–5) corresponds to ventricular repolarization?
   A) the P wave
   B) the QRS duration
   C) the T wave
   D) the U wave
   E) the PR interval

2. Which of the following normally has a slowly depolarizing "prepotential"?
   A) sinoatrial node
   B) atrial muscle cells
   C) bundle of His
   D) Purkinje fibers
   E) ventricular muscle cells

3. In second-degree heart block
   A) the ventricular rate is lower than the atrial rate.
   B) the ventricular ECG complexes are distorted.
   C) there is a high incidence of ventricular tachycardia.
   D) stroke volume is decreased.
   E) cardiac output is increased.

4. Currents caused by opening of which of the following channels contribute to the repolarization phase of the action potential of ventricular muscle fibers?
   A) Na^+ channels
   B) Cl^- channels
   C) Ca^{2+} channels
   D) K^+ channels
   E) HCO_3^- channels

5. In complete heart block
   A) fainting may occur because the atria are unable to pump blood into the ventricles.
   B) ventricular fibrillation is common.
   C) the atrial rate is lower than the ventricular rate.
   D) fainting may occur because of prolonged periods during which the ventricles fail to contract.

CHAPTER RESOURCES

Hile B: *Ionic Channels of Excitable Membranes*, 3rd ed. Sinauer Associates, Inc., 2001.

Jackson WF: Ion channels and vascular tone. Hypertension 2000;35:173.

Jessup M, Brozena S: Heart failure. N Engl J Med 2003;348:2007.

Morady F: Radiofrequency ablation as treatment for cardiac arrhythmias. N Engl J Med 1999;340:534.

Nabel EG: Genomic medicine: cardiovascular disease. N Engl J Med 2003;349:60.

Roder DM: Drug-induced prolongation of the Q-T interval. N Engl J Med 2004;350:1013.

Rowell LB: *Human Cardiovascular Control*. Oxford University Press, 1993.

Wagner GS: *Marriott's Practical Electrocardiography*, 10th ed. Lippincott Williams and Wilkins, 2000.

The Heart as a Pump

INTRODUCTION

Of course, the electrical activity of the heart discussed in the previous chapter is designed to subserve the heart's primary physiological role—to pump blood through the lungs, where gas exchange can occur, and thence to the remainder of the body (Clinical Box 31–1). This is accomplished when the orderly depolarization process described in the previous chapter triggers a wave of contraction that spreads through the myocardium. In single muscle fibers, contraction starts just after depolarization and lasts until about 50 ms after repolarization is completed (see Figure 5–15). Atrial systole starts after the P wave of the electrocardiogram (ECG); ven-tricular systole starts near the end of the R wave and ends just after the T wave. In this chapter, we will consider how these changes in contraction produce sequential changes in pressures and flows in the heart chambers and blood vessels, and thereby propel blood appropriately as needed by whole body demands for oxygen and nutrients. As an aside, it should be noted that the term **systolic pressure** in the vascular system refers to the peak pressure reached during systole, not the mean pressure; similarly, the **diastolic pressure** refers to the lowest pressure during diastole.

MECHANICAL EVENTS OF THE CARDIAC CYCLE

EVENTS IN LATE DIASTOLE

Late in diastole, the mitral (bicuspid) and tricuspid valves between the atria and ventricles (atrioventricular [AV] valves) are open and the aortic and pulmonary valves are closed. Blood flows into the heart throughout diastole, filling the atria and ventricles. The rate of filling declines as the ventricles become distended, and, especially when the heart rate is low, the cusps of the AV valves drift toward the closed position (Figure 31–1). The pressure in the ventricles remains low. About 70% of the ventricular filling occurs passively during diastole.

CLINICAL BOX 31–1

Heart Failure

Heart failure occurs when the heart is unable to put out an amount of blood that is adequate for the needs of the tissues. It can be acute and associated with sudden death, or chronic. The failure may involve primarily the right ventricle (cor pulmonale), but much more commonly it involves the larger, thicker left ventricle or both ventricles. Heart failure may also be systolic or diastolic. In **systolic failure,** stroke volume is reduced because ventricular contraction is weak. This causes an increase in the end-systolic ventricular volume, so that the **ejection fraction** falls from 65% to as low as 20%. The initial response to failure is activation of the genes that cause cardiac myocytes to hypertrophy, and thickening of the ventricular wall **(cardiac remodeling).** The incomplete filling of the arterial system leads to increased discharge of the sympathetic nervous system and increased secretion of renin and aldosterone, so Na^+ and water are retained. These responses are initially compensatory, but eventually the failure worsens and the ventricles dilate.

In **diastolic failure,** the ejection fraction is initially maintained, but the elasticity of the myocardium is reduced so filling during diastole is reduced. This leads to inadequate stroke volume and the same cardiac remodeling and Na^+ and water retention that occur in systolic failure. It should be noted that the inadequate cardiac output in failure may be relative rather than absolute. When a large arterior venous fistula is present, in thyrotoxicosis and in thiamine deficiency, cardiac output may be elevated in absolute terms but still be inadequate to meet the needs of the tissues **(high-output failure).**

Treatment of congestive heart failure is aimed at improving cardiac contractility, treating the symptoms, and decreasing the load on the heart. Currently, the most effective treatment in general use is inhibition of the production of angiotensin II with angiotensin-converting enzyme (ACE) inhibitors. Blockade of the effects of angiotensin II on AT_1 receptors with nonpeptide antagonists is also of value. Blocking the production of angiotensin II or its effects also reduces the circulating aldosterone level and decreases blood pressure, reducing the afterload against which the heart pumps. The effects of aldosterone can be further reduced by administering aldosterone receptor blockers. Reducing venous tone with nitrates or hydralazine increases venous capacity so that the amount of blood returned to the heart is reduced, lowering the preload. Diuretics reduce the fluid overload. Drugs that block β-adrenergic receptors have been shown to decrease mortality and morbidity. Digitalis derivatives such as digoxin have classically been used to treat congestive heart failure because of their ability to increase intracellular Ca^{2+} and hence exert a positive inotropic effect, but they are now used in a secondary role to treat systolic dysfunction and slow the ventricular rate in patients with atrial fibrillation.

ATRIAL SYSTOLE

Contraction of the atria propels some additional blood into the ventricles. Contraction of the atrial muscle narrows the orifices of the superior and inferior vena cava and pulmonary veins, and the inertia of the blood moving toward the heart tends to keep blood in it. However, despite these inhibitory influences, there is some regurgitation of blood into the veins.

VENTRICULAR SYSTOLE

At the start of ventricular systole, the AV valves close. Ventricular muscle initially shortens relatively little, but intraventricular pressure rises sharply as the myocardium presses on the blood in the ventricle (Figure 31–2). This period of **isovolumetric (isovolumic, isometric) ventricular contraction** lasts about 0.05 s, until the pressures in the left and right ventricles exceed the pressures in the aorta (80 mm Hg; 10.6 kPa) and pulmonary artery (10 mm Hg) and the aortic and pulmonary valves open. During isovolumetric contraction, the AV valves bulge into the atria, causing a small but sharp rise in atrial pressure (Figure 31–3).

When the aortic and pulmonary valves open, the phase of **ventricular ejection** begins. Ejection is rapid at first, slowing down as systole progresses. The intraventricular pressure rises to a maximum and then declines somewhat before ventricular systole ends. Peak pressures in the left and right ventricles are about 120 and 25 mm Hg, respectively. Late in systole, pressure in the aorta actually exceeds that in the left ventricle, but for a short period momentum keeps the blood moving forward. The AV valves are pulled down by the contractions of the ventricular muscle, and atrial pressure drops. The amount of blood ejected by each ventricle per stroke at rest is 70 to 90 mL. The **end-diastolic ventricular volume** is about 130 mL. Thus, about 50 mL of blood remains in each ventricle at the end of systole **(end-systolic ventricular volume),** and the **ejection fraction,** the percent of the end-diastolic ventricular volume that is ejected with each stroke, is about 65%. The ejection fraction is a valuable index of ventricular function. It can be measured by injecting radionuclide-labeled red blood cells and imaging the cardiac blood pool at the end of diastole and the end of systole (equilibrium radionuclide angiocardiography), or by computed tomography.

A Systole

Isovolumetric ventricular contraction

Atria relaxed

Ventricles contract

Ventricular ejection
Blood flows out of ventricle

Atria relaxed

Ventricles contract

AV valves:	Closed	Closed
Aortic and pulmonary valves:	Closed	Open

B Diastole

Isovolumetric ventricular relaxation

Atria relaxed

Ventricles relaxed

Ventricular filling
Blood flows into ventricles

Atrial contraction

Atria relaxed

Ventricles relaxed

Atria contract

Ventricles relaxed

AV valves:	Closed	Open	Open
Aortic and pulmonary valves:	Closed	Closed	Closed

FIGURE 31–1 **Divisions of the cardiac cycle: A) systole and B) diastole.** The phases of the cycle are identical in both halves of the heart. The direction in which the pressure difference favors flow is denoted by an arrow; note, however, that flow will not actually occur if valve prevents it.

EARLY DIASTOLE

Once the ventricular muscle is fully contracted, the already falling ventricular pressures drop more rapidly. This is the period of **protodiastole,** which lasts about 0.04 s. It ends when the momentum of the ejected blood is overcome and the aortic and pulmonary valves close, setting up transient vibrations in the blood and blood vessel walls. After the valves are closed, pressure continues to drop rapidly during the period of **isovolumetric ventricular relaxation.** Isovolumetric relaxation

ends when the ventricular pressure falls below the atrial pressure and the AV valves open, permitting the ventricles to fill. Filling is rapid at first, then slows as the next cardiac contraction approaches. Atrial pressure continues to rise after the end of ventricular systole until the AV valves open, then drops and slowly rises again until the next atrial systole.

FIGURE 31–2 **Pressure–volume loop of the left ventricle. During diastole, the ventricle fills and pressure increases from d to a.** Pressure then rises sharply from a to b during isovolumetric contraction and from b to c during ventricular ejection. At c, the aortic valves close and pressure falls during isovolumetric relaxation from c back to d. (Reproduced with permission from McPhee SJ, Lingappa VR, Ganong WF [editors]: *Pathophysiology of Disease,* 4th ed. McGraw-Hill, 2003.)

PERICARDIUM

The myocardium is covered by a fibrous layer known as the epicardium. This, in turn, is surrounded by the pericardium, which separates the heart from the rest of the thoracic viscera. The space between the epicardium and pericardium (the **pericardial sac**) normally contains 5 to 30 mL of clear fluid, which lubricates the heart and permits it to contract with minimal friction.

TIMING

Although events on the two sides of the heart are similar, they are somewhat asynchronous. Right atrial systole precedes left atrial systole, and contraction of the right ventricle starts after that of the left (see Chapter 30). However, since pulmonary arterial pressure is lower than aortic pressure, right ventricular ejection begins before that of the left. During expiration, the pulmonary and aortic valves close at the same time; but during inspiration, the aortic valve closes slightly before the pulmonary. The slower closure of the pulmonary valve is due to lower impedance of the pulmonary vascular tree. When measured over a period of minutes, the outputs of the two ventricles are, of course, equal, but transient differences in output during the respiratory cycle occur in normal individuals.

LENGTH OF SYSTOLE & DIASTOLE

Cardiac muscle has the unique property of contracting and repolarizing faster when the heart rate is high (see Chapter 5),

and the duration of systole decreases from 0.27 s at a heart rate of 65 to 0.16 s at a rate of 200 beats/min (Table 31–1). The shortening is due mainly to a decrease in the duration of systolic ejection. However, the duration of systole is much more fixed than that of diastole, and when the heart rate is increased, diastole is shortened to a much greater degree. For example, at a heart rate of 65, the duration of diastole is 0.62 s, whereas at a heart rate of 200, it is only 0.14 s. This fact has important physiologic and clinical implications. It is during diastole that the heart muscle rests, and coronary blood flow to the subendocardial portions of the left ventricle occurs only during diastole (see Chapter 34). Furthermore, most of the ventricular filling occurs in diastole. At heart rates up to about 180, filling is adequate as long as there is ample venous return, and cardiac output per minute is increased by an increase in rate. However, at very high heart rates, filling may be compromised to such a degree that cardiac output per minute falls.

Because it has a prolonged action potential, cardiac muscle cannot contract in response to a second stimulus until near the end of the initial contraction (see Figure 5–15). Therefore, cardiac muscle cannot be tetanized like skeletal muscle. The highest rate at which the ventricles can contract is theoretically about 400/min, but in adults the AV node will not conduct more than about 230 impulses/min because of its long refractory period. A ventricular rate of more than 230 is seen only in paroxysmal ventricular tachycardia (see Chapter 30).

Exact measurement of the duration of isovolumetric ventricular contraction is difficult in clinical situations, but it is relatively easy to measure the duration of **total electromechanical systole (QS$_2$), the preejection period (PEP),** and the **left ventricular ejection time (LVET)** by recording the ECG, phonocardiogram, and carotid pulse simultaneously. QS$_2$ is the period from the onset of the QRS complex to the closure of the aortic valves, as determined by the onset of the second heart sound. LVET is the period from the beginning of the carotid pressure rise to the dicrotic notch (see below). PEP is the difference between QS$_2$ and LVET and represents the time for the electrical as well as the mechanical events that precede systolic ejection. The ratio PEP/LVET is normally about 0.35, and it increases without a change in QS$_2$ when left ventricular performance is compromised in a variety of cardiac diseases.

ARTERIAL PULSE

The blood forced into the aorta during systole not only moves the blood in the vessels forward but also sets up a pressure wave that travels along the arteries. The pressure wave expands the arterial walls as it travels, and the expansion is palpable as the **pulse.** The rate at which the wave travels, which is independent of and much higher than the velocity of blood flow, is about 4 m/s in the aorta, 8 m/s in the large arteries, and 16 m/s in the small arteries of young adults. Consequently, the pulse is felt in the radial artery at the wrist about 0.1 s after the peak of systolic ejection into the aorta (Figure 31–3). With advancing age, the arteries become more rigid, and the pulse wave moves faster.

FIGURE 31-3 **Events of the cardiac cycle at a heart rate of 75 beats/min.** The phases of the cardiac cycle identified by the numbers at the bottom are as follows: 1, atrial systole; 2, isovolumetric ventricular contraction; 3, ventricular ejection; 4, isovolumetric ventricular relaxation; 5, ventricular filling. Note that late in systole, aortic pressure actually exceeds left ventricular pressure. However, the momentum of the blood keeps it flowing out of the ventricle for a short period. The pressure relationships in the right ventricle and pulmonary artery are similar. Atr. syst., atrial systole; Ventric. syst., ventricular systole.

TABLE 31–1 Variation in length of action potential and associated phenomena with cardiac rate.[a]

	Heart Rate 75/min	Heart Rate 200/min	Skeletal Muscle
Duration, each cardiac cycle	0.80	0.30	…
Duration of systole	0.27	0.16	…
Duration of action potential	0.25	0.15	0.007
Duration of absolute refractory period	0.20	0.13	0.004
Duration of relative refractory period	0.05	0.02	0.003
Duration of diastole	0.53	0.14	…

[a]All values are in seconds.

Courtesy of AC Barger and GS Richardson.

The strength of the pulse is determined by the pulse pressure and bears little relation to the mean pressure. The pulse is weak ("thready") in shock. It is strong when stroke volume is large; for example, during exercise or after the administration of histamine. When the pulse pressure is high, the pulse waves may be large enough to be felt or even heard by the individual (palpitation, "pounding heart"). When the aortic valve is incompetent (aortic insufficiency), the pulse is particularly strong, and the force of systolic ejection may be sufficient to make the head nod with each heartbeat. The pulse in aortic insufficiency is called a **collapsing, Corrigan,** or **water-hammer pulse.**

The **dicrotic notch,** a small oscillation on the falling phase of the pulse wave caused by vibrations set up when the aortic valve snaps shut (Figure 31–3), is visible if the pressure wave is recorded but is not palpable at the wrist. The pulmonary artery pressure curve also has a dicrotic notch produced by the closure of the pulmonary valves.

ATRIAL PRESSURE CHANGES & THE JUGULAR PULSE

Atrial pressure rises during atrial systole and continues to rise during isovolumetric ventricular contraction when the AV valves bulge into the atria. When the AV valves are pulled down by the contracting ventricular muscle, pressure falls rapidly and then rises as blood flows into the atria until the AV valves open early in diastole. The return of the AV valves to their relaxed position also contributes to this pressure rise by reducing atrial capacity. The atrial pressure changes are transmitted to the great veins, producing three characteristic waves in the record of jugular pressure (Figure 31–3). The **a wave** is due to atrial systole. As noted above, some blood regurgitates into the great veins when the atria contract. In addition, venous inflow stops, and the resultant rise in venous pressure contributes to the a wave. The **c wave** is the transmitted manifestation of the rise in atrial pres-

sure produced by the bulging of the tricuspid valve into the atria during isovolumetric ventricular contraction. The **v wave** mirrors the rise in atrial pressure before the tricuspid valve opens during diastole. The jugular pulse waves are superimposed on the respiratory fluctuations in venous pressure. Venous pressure falls during inspiration as a result of the increased negative intrathoracic pressure and rises again during expiration.

HEART SOUNDS

Two sounds are normally heard through a stethoscope during each cardiac cycle. The first is a low, slightly prolonged "lub" (**first sound**), caused by vibrations set up by the sudden closure of the AV valves at the start of ventricular systole (Figure 31–3). The second is a shorter, high-pitched "dup" (**second sound**), caused by vibrations associated with closure of the aortic and pulmonary valves just after the end of ventricular systole. A soft, low-pitched **third sound** is heard about one third of the way through diastole in many normal young individuals. It coincides with the period of rapid ventricular filling and is probably due to vibrations set up by the inrush of blood. A **fourth sound** can sometimes be heard immediately before the first sound when atrial pressure is high or the ventricle is stiff in conditions such as ventricular hypertrophy. It is due to ventricular filling and is rarely heard in normal adults.

The first sound has a duration of about 0.15 s and a frequency of 25 to 45 Hz. It is soft when the heart rate is low, because the ventricles are well filled with blood and the leaflets of the AV valves float together before systole. The second sound lasts about 0.12 s, with a frequency of 50 Hz. It is loud and sharp when the diastolic pressure in the aorta or pulmonary artery is elevated, causing the respective valves to shut briskly at the end of systole. The interval between aortic and pulmonary valve closure during inspiration is frequently long enough for the second sound to be reduplicated (physiologic splitting of the second sound). Splitting also occurs in various diseases. The third sound, when present, has a duration of 0.1 s.

MURMURS

Murmurs, or **bruits,** are abnormal sounds heard in various parts of the vascular system. The two terms are used interchangeably, though "murmur" is more commonly used to denote noise heard over the heart than over blood vessels. As discussed in detail in Chapter 32, blood flow is laminar, nonturbulent, and silent up to a critical velocity; above this velocity and beyond an obstruction, blood flow is turbulent and creates sounds. Blood flow speeds up when an artery or a heart valve is narrowed.

Examples of vascular sounds outside the heart are the bruit heard over a large, highly vascular goiter, the bruit heard over a carotid artery when its lumen is narrowed and distorted by atherosclerosis, and the murmurs heard over an aneurysmal dilation of one of the large arteries, an arteriovenous (A-V) fistula, or a patent ductus arteriosus.

The major—but certainly not the only—cause of cardiac murmurs is disease of the heart valves. When the orifice of a valve is narrowed (**stenosis**), blood flow through it is accelerated and turbulent. When a valve is incompetent, blood flows through it backward (**regurgitation** or **insufficiency**), again through a narrow orifice that accelerates flow. The timing (systolic or diastolic) of a murmur due to any particular valve (Table 31–2) can be predicted from a knowledge of the mechanical events of the cardiac cycle. Murmurs due to disease of a particular valve can generally be heard best when the stethoscope is directly over the valve. There are also other aspects of the duration, character, accentuation, and transmission of the sound that help to locate its origin in one valve or another. One of the loudest murmurs is that produced when blood flows backward in diastole through a hole in a cusp of the aortic valve. Most murmurs can be heard only with the aid of the stethoscope, but this high-pitched musical diastolic murmur is sometimes audible to the unaided ear several feet from the patient.

In patients with congenital interventricular septal defects, flow from the left to the right ventricle causes a systolic murmur. Soft murmurs may also be heard in patients with interatrial septal defects, although they are not a constant finding.

Soft systolic murmurs are also common in individuals, especially children, who have no cardiac disease. Systolic murmurs are also heard in anemic patients as a result of the low viscosity of the blood and associated rapid flow (see Chapter 32).

ECHOCARDIOGRAPHY

Wall movement and other aspects of cardiac function can be evaluated by the noninvasive technique of **echocardiography.** Pulses of ultrasonic waves are emitted from a transducer that also functions as a receiver to detect waves reflected back from various parts of the heart. Reflections occur wherever acoustic impedance changes, and a recording of the echoes displayed against time on an oscilloscope provides a record of the movements of the ventricular wall, septum, and valves during the cardiac cycle. When combined with Doppler techniques, echocardiography can be used to measure velocity and volume of flow through valves. It has considerable clinical usefulness, particularly in evaluating and planning therapy in patients with valvular lesions.

CARDIAC OUTPUT
METHODS OF MEASUREMENT

In experimental animals, cardiac output can be measured with an electromagnetic flow meter placed on the ascending aorta. Two methods of measuring output that are applicable to humans, in addition to Doppler combined with echocardiography, are the **direct Fick method** and the **indicator dilution method.**

The **Fick principle** states that the amount of a substance taken up by an organ (or by the whole body) per unit of time is equal to the arterial level of the substance minus the venous level (**A-V difference**) times the blood flow. This principle

TABLE 31–2 Heart murmurs.

Valve	Abnormality	Timing of Murmur
Aortic or pulmonary	Stenosis	Systolic
	Insufficiency	Diastolic
Mitral or tricuspid	Stenosis	Diastolic
	Insufficiency	Systolic

can be applied, of course, only in situations in which the arterial blood is the sole source of the substance taken up. The principle can be used to determine cardiac output by measuring the amount of O_2 consumed by the body in a given period and dividing this value by the A-V difference across the lungs. Because systemic arterial blood has the same O_2 content in all parts of the body, the arterial O_2 content can be measured in a sample obtained from any convenient artery. A sample of venous blood in the pulmonary artery is obtained by means of a cardiac catheter. It has now become commonplace to insert a long catheter through a forearm vein and to guide its tip into the heart with the aid of a fluoroscope. The procedure is generally benign. Catheters can be inserted through the right atrium and ventricle into the small branches of the pulmonary artery. An example of the calculation of cardiac output using a typical set of values is as follows:

$$\text{Output of left ventricle} = \frac{O_2 \text{ consumption (mL/min)}}{[A_{O_2}] - [V_{O_2}]}$$

$$= \frac{250 \text{ mL/min}}{\begin{array}{c}190 \text{ mL/L arterial blood} - \\ 140 \text{ mL/L venous blood in} \\ \text{pulmonary artery}\end{array}}$$

$$= \frac{250 \text{ mL/min}}{50 \text{ mL/L}}$$

$$= 5 \text{ L/min}$$

In the indicator dilution technique, a known amount of a substance such as a dye or, more commonly, a radioactive isotope is injected into an arm vein and the concentration of the indicator in serial samples of arterial blood is determined. The output of the heart is equal to the amount of indicator injected divided by its average concentration in arterial blood after a single circulation through the heart (Figure 31–4). The indicator must, of course, be a substance that stays in the bloodstream during the test and has no harmful or hemodynamic effects. In practice, the log of the indicator concentration in the serial arterial samples is plotted against time as the concentration rises, falls, and then rises again as the indicator recirculates. The initial decline in concentration, linear on a semilog plot, is extrapolated to the abscissa, giving the time for first passage of the indicator through the circulation. The cardiac output for that period is calculated (Figure 31–4) and then converted to output per minute.

$$F = \frac{E}{\int_{o}^{\alpha} Cdt}$$

F = flow
E = amount of indicator injected
C = instantaneous concentration of indicator in arterial blood

In the **rest** example above,

$$\begin{array}{c} \text{Flow in 39 s} \\ \text{(time of first passage)} \end{array} = \frac{\begin{array}{c} \text{5 mg injection} \end{array}}{\begin{array}{c} \text{1.6 mg/L} \\ \text{(avg concentration)} \end{array}}$$

Flow = 3.1 L in 39 s
Flow (cardiac output)/min = $3.1 \times \frac{60}{39} = 4.7$ L

For the **exercise** example,

$$\text{Flow in 9 s} = \frac{\text{5 mg}}{\text{1.51 mg/L}} = 3.3 \text{ L}$$

$$\text{Flow/min} = 3.3 \times \frac{60}{9} = 22.0 \text{ L}$$

FIGURE 31–4 **Determination of cardiac output by indicator (dye) dilution.**

A popular indicator dilution technique is **thermodilution,** in which the indicator used is cold saline. The saline is injected into the right atrium through one channel of a double-lumen catheter, and the temperature change in the blood is recorded in the pulmonary artery, using a thermistor in the other, longer side of the catheter. The temperature change is inversely proportionate to the amount of blood flowing through the pulmonary artery; that is, to the extent that the cold saline is diluted by blood. This technique has two important advantages: (1) the saline is completely innocuous; and (2) the cold is dissipated in the tissues so recirculation is not a problem, and it is easy to make repeated determinations.

CARDIAC OUTPUT IN VARIOUS CONDITIONS

The amount of blood pumped out of the heart per beat, the **stroke volume,** is about 70 mL from each ventricle in a resting man of average size in the supine position. The output of the heart per unit of time is the **cardiac output.** In a resting, supine man, it averages about 5.0 L/min (70 mL × 72 beats/min). There is a correlation between resting cardiac output and body surface area. The output per minute per square meter of body surface (the **cardiac index**) averages 3.2 L. The effects of various conditions on cardiac output are summarized in Table 31–3.

FACTORS CONTROLLING CARDIAC OUTPUT

Predictably, changes in cardiac output that are called for by physiologic conditions can be produced by changes in cardiac rate or stroke volume or both (Figure 31–5). The cardiac rate is controlled primarily by the autonomic nerves, with sympathetic stimulation increasing the rate and parasympathetic stimulation decreasing it (see Chapter 30). Stroke volume is also determined in part by neural input, with sympathetic stimuli making the myocardial muscle fibers contract with greater strength at any given length and parasympathetic stimuli having the opposite effect. When the strength of contraction increases without an increase in fiber length, more of the blood that normally remains in the ventricles is expelled; that is, the ejection fraction increases. The cardiac accelerator action of the catecholamines liberated by sympathetic stimulation is referred to as their **chronotropic action,** whereas their effect on the strength of cardiac contraction is called their **inotropic action.**

TABLE 31–3 **Effect of various conditions on cardiac output.**

	Condition or Factor[a]
No change	Sleep
	Moderate changes in environmental temperature
Increase	Anxiety and excitement (50–100%)
	Eating (30%)
	Exercise (up to 700%)
	High environmental temperature
	Pregnancy
	Epinephrine
Decrease	Sitting or standing from lying position (20–30%)
	Rapid arrhythmias
	Heart disease

[a]Approximate percent changes are shown in parentheses.

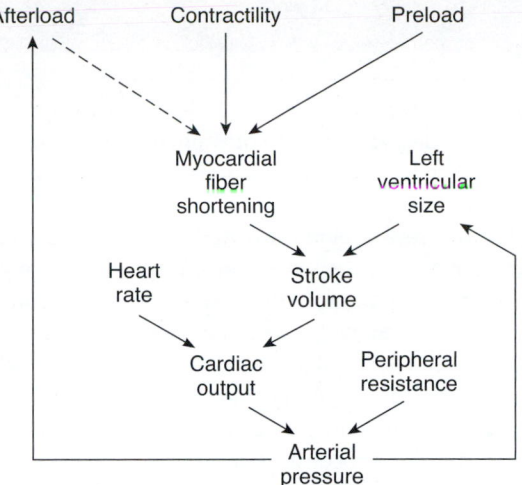

FIGURE 31–5 **Interactions between the components that regulate cardiac output and arterial pressure.** Solid arrows indicate increases, and the dashed arrow indicates a decrease.

The force of contraction of cardiac muscle depends on its preloading and its afterloading. These factors are illustrated in Figure 31–6, in which a muscle strip is stretched by a load (the **preload**) that rests on a platform. The initial phase of the contraction is isometric; the elastic component in series with the contractile element is stretched, and tension increases until it is sufficient to lift the load. The tension at which the load is lifted is the **afterload.** The muscle then contracts isotonically without developing further tension. In vivo, the preload is the degree to which the myocardium is stretched before it contracts and the afterload is the resistance against which blood is expelled.

RELATION OF TENSION TO LENGTH IN CARDIAC MUSCLE

The length–tension relationship in cardiac muscle (see Figure 5–17) is similar to that in skeletal muscle (see Figure 5–11); as

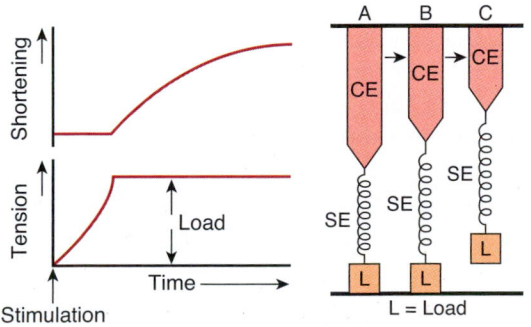

FIGURE 31–6 **Model for contraction of afterloaded muscles. A:** Rest. **B:** Partial contraction of the contractile element of the muscle (CE), with stretching of the series elastic element (SE) but no shortening. **C:** Complete contraction, with shortening. (Reproduced with permission from Sonnenblick EH in: *The Myocardial Cell: Structure, Function and Modification.* Briller SA, Conn HL [editors]. University Pennsylvania Press, 1966.)

the muscle is stretched, the developed tension increases to a maximum and then declines as stretch becomes more extreme. Starling pointed this out when he stated that the "energy of contraction is proportional to the initial length of the cardiac muscle fiber" (**Starling's law of the heart** or the **Frank–Starling law**). For the heart, the length of the muscle fibers (ie, the extent of the preload) is proportional to the end-diastolic volume. The relation between ventricular stroke volume and end-diastolic volume is called the Frank–Starling curve.

Regulation of cardiac output as a result of changes in cardiac muscle fiber length is sometimes called **heterometric regulation,** whereas regulation due to changes in contractility independent of length is sometimes called **homometric regulation.**

FACTORS AFFECTING END-DIASTOLIC VOLUME

Alterations in systolic and diastolic function have different effects on the heart. When systolic contractions are reduced, there is a primary reduction in stroke volume. Diastolic function also affects stroke volume, but in a different way.

An increase in intrapericardial pressure limits the extent to which the ventricle can fill (eg, as a result of infection or pressure from a tumor), as does a decrease in ventricular compliance; that is, an increase in ventricular stiffness produced by myocardial infarction, infiltrative disease, and other abnormalities. Atrial contractions aid ventricular filling. Factors affecting the amount of blood returning to the heart likewise influence the degree of cardiac filling during diastole. An increase in total blood volume increases venous return (Clinical Box 31–2). Constriction of the veins reduces the size of the venous reservoirs, decreasing venous pooling and thus increasing venous return. An increase in the normal negative intrathoracic pressure increases the pressure gradient along which blood flows to the heart, whereas a decrease impedes venous return. Standing decreases venous return, and muscular activity increases it as a result of the pumping action of skeletal muscle.

The effects of systolic and diastolic dysfunction on the pressure–volume loop of the left ventricle are summarized in Figure 31–7.

MYOCARDIAL CONTRACTILITY

The contractility of the myocardium exerts a major influence on stroke volume. When the sympathetic nerves to the heart are stimulated, the whole length–tension curve shifts upward and to the left (Figure 31–8). The positive inotropic effect of norepinephrine liberated at the nerve endings is augmented by circulating norepinephrine, and epinephrine has a similar effect. Conversely, there is a negative inotropic effect of vagal stimulation on both atrial and (to a lesser extent) ventricular muscle.

Changes in cardiac rate and rhythm also affect myocardial contractility (known as the force–frequency relation, Figure 31–8). Ventricular extrasystoles condition the myocardium in such a way that the next succeeding contraction is stronger

CLINICAL BOX 31–2

Shock

Circulatory shock comprises a collection of different entities that share certain common features; however, the feature that is common to all the entities is inadequate tissue perfusion with a relatively or absolutely inadequate cardiac output. The cardiac output may be inadequate because the amount of fluid in the vascular system is inadequate to fill it **(hypovolemic shock).** Alternatively, it may be inadequate in the relative sense because the size of the vascular system is increased by vasodilation even though the blood volume is normal **(distributive, vasogenic,** or **low-resistance shock).** Shock may also be caused by inadequate pumping action of the heart as a result of myocardial abnormalities **(cardiogenic shock),** and by inadequate cardiac output as a result of obstruction of blood flow in the lungs or heart **(obstructive shock).**

Hypovolemic shock is also called "cold shock." It is characterized by hypotension; a rapid, thready pulse; cold, pale, clammy skin; intense thirst; rapid respiration; and restlessness or, alternatively, torpor. None of these findings, however, are invariably present. Hypovolemic shock is commonly subdivided into categories on the basis of cause. Of these, it is useful to consider the effects of hemorrhage in some detail because of the multiple compensatory reactions that come into play to defend extracellular fluid (ECF) volume. Thus, the decline in blood volume produced by bleeding decreases venous return, and cardiac output falls. The heart rate is increased, and with severe hemorrhage, a fall in blood pressure always occurs. With moderate hemorrhage (5–15 mL/kg body weight), pulse pressure is reduced but mean arterial pressure may be normal. The blood pressure changes vary from individual to individual, even when exactly the same amount of blood is lost. The skin is cool and pale and may have a grayish tinge because of stasis in the capillaries and a small amount of cyanosis. Inadequate perfusion of the tissues leads to increased anaerobic glycolysis, with the production of large amounts of lactic acid. In severe cases, the blood lactate level rises from the normal value of about 1 mmol/L to 9 mmol/L or more. The resulting **lactic acidosis** depresses the myocardium, decreases peripheral vascular responsiveness to catecholamines, and may be severe enough to cause coma. When blood volume is reduced and venous return is decreased, moreover, stimulation of arterial baroreceptors is reduced, increasing sympathetic output. Even if there is no drop in mean arterial pressure, the decrease in pulse pressure decreases the rate of discharge in the arterial baroreceptors, and reflex tachycardia and vasoconstriction result.

With more severe blood loss, tachycardia is replaced by bradycardia; this occurs while shock is still reversible. With even greater hemorrhage, the heart rate rises again. The bradycardia is presumably due to unmasking a vagally mediated depressor reflex, and the response may have evolved as a mechanism for stopping further blood loss. Vasoconstriction is generalized, sparing only the vessels of the brain and heart. A widespread reflex venoconstriction also helps maintain the filling pressure of the heart. In the kidneys, both afferent and efferent arterioles are constricted, but the efferent vessels are constricted to a greater degree. The glomerular filtration rate is depressed, but renal plasma flow is decreased to a greater extent, so that the filtration fraction increases. Na^+ retention is marked, and the nitrogenous products of metabolism are retained in the blood **(azotemia** or **uremia).** If the hypotension is prolonged, renal tubular damage may be severe **(acute renal failure).** After a moderate hemorrhage, the circulating plasma volume is restored in 12 to 72 h. Preformed albumin also enters rapidly from extravascular stores, but most of the tissue fluids that are mobilized are protein-free. After the initial influx of preformed albumin, the rest of the plasma protein losses are replaced, presumably by hepatic synthesis, over a period of 3 to 4 d. Erythropoietin appears in the circulation, and the reticulocyte count increases, reaching a peak in 10 d. The red cell mass is restored to normal in 4 to 8 wk.

The treatment of shock is aimed at correcting the cause and helping the physiologic compensatory mechanisms to restore an adequate level of tissue perfusion. If the primary cause of the shock is blood loss, the treatment should include early and rapid transfusion of adequate amounts of compatible whole blood. In shock due to burns and other conditions in which there is hemoconcentration, plasma is the treatment of choice to restore the fundamental defect, the loss of plasma. Concentrated human serum albumin and other hypertonic solutions expand the blood volume by drawing fluid out of the interstitial spaces. They are valuable in emergency treatment but have the disadvantage of further dehydrating the tissues of an already dehydrated patient.

---

than the preceding normal contraction. This **postextrasystolic potentiation** is independent of ventricular filling, since it occurs in isolated cardiac muscle and is due to increased availability of intracellular Ca^{2+}. A sustained increment in contractility can be produced therapeutically by delivering paired electrical stimuli to the heart in such a way that the second stimulus is delivered shortly after the refractory period of the first. It has also been shown that myocardial contractility increases as the heart rate increases, although this effect is relatively small.

Catecholamines exert their inotropic effect via an action on cardiac β_1-adrenergic receptors and Gs, with resultant activation of adenylyl cyclase and increased intracellular cyclic adenosine

FIGURE 31–7 **Effect of systolic and diastolic dysfunction on the pressure–volume loop of the left ventricle. Left:** Systolic dysfunction shifts the isovolumic pressure–volume curve (see Figure 31–2) to the right, decreasing the stroke volume from b–c to b'–c'. **Right:** Diastolic dysfunction increases end-diastolic volume and shifts the diastolic pressure–volume relationship upward and to the left. This reduces the stroke volume from b–c to b'–c'. (Reproduced with permission from McPhee SJ, Lingappa VR, Ganong WF [editors]: *Pathophysiology of Disease*, 4th ed. McGraw-Hill, 2003.)

3′,5′-monophosphate (cAMP). Xanthines such as caffeine and theophylline that inhibit the breakdown of cAMP are predictably positively inotropic. The positively inotropic effect of digitalis and related drugs (Figure 31–8), on the other hand, is due to their inhibitory effect on the Na^+–K^+ ATPase in the myocardium (see Chapter 5). Hypercapnia, hypoxia, acidosis, and drugs such as quinidine, procainamide, and barbiturates depress myocardial contractility. The contractility of the myocardium is also reduced

in heart failure (intrinsic depression). The causes of this depression are not fully understood, but may reflect down-regulation of β-adrenergic receptors and associated signaling pathways and impaired calcium liberation from the sarcoplasmic reticulum. In acute heart failure, such as that associated with sepsis, this response could be considered an appropriate adaptation to a situation where energy supply to the heart is limited, thereby reducing energy expenditure and avoiding cell death.

FIGURE 31–8 **Effect of changes in myocardial contractility on the Frank–Starling curve.** The curve shifts downward and to the right as contractility is decreased. The major factors influencing contractility are summarized on the right. The dashed lines indicate portions of the ventricular function curves where maximum contractility has been exceeded; that is, they identify points on the "descending limb" of the Frank–Starling curve. EDV, end-diastolic volume. (Reproduced with permission from Braunwald E, Ross J, Sonnenblick EH: Mechanisms of contraction of the normal and failing heart. N Engl J Med 1967;277:794. Courtesy of Little, Brown.)

Circulatory Changes during Exercise

The blood flow of resting skeletal muscle is low (2–4 mL/100 g/min). When a muscle contracts, it compresses the vessels in it if it develops more than 10% of its maximal tension; when it develops more than 70% of its maximal tension, blood flow is completely stopped. Between contractions, however, flow is so greatly increased that blood flow per unit of time in a rhythmically contracting muscle is increased as much as 30-fold. Local mechanisms maintaining a high blood flow in exercising muscle include a fall in tissue PO_2, a rise in tissue PCO_2, and accumulation of K^+ and other vasodilator metabolites. The temperature rises in active muscle, and this further dilates the vessels. Dilation of the arterioles and precapillary sphincters causes a 10- to 100-fold increase in the number of open capillaries. The average distance between the blood and the active cells—and the distance O_2 and metabolic products must diffuse—is thus greatly decreased. The dilation increases the cross-sectional area of the vascular bed, and the velocity of flow therefore decreases.

The systemic cardiovascular response to exercise that provides for the additional blood flow to contracting muscle depends on whether the muscle contractions are primarily isometric or primarily isotonic with the performance of external work. With the start of an isometric muscle contraction, the heart rate rises, probably as a result of psychic stimuli acting on the medulla oblongata. The increase is largely due to decreased vagal tone, although increased discharge of the cardiac sympathetic nerves plays some role. Within a few seconds of the onset of an isometric muscle contraction, systolic and diastolic blood pressures rise sharply. Stroke volume changes relatively little, and blood flow to the steadily contracting muscles is reduced as a result of compression of their blood vessels. The response to exercise involving isotonic muscle contraction is similar in that there is a prompt increase in heart rate, but different in that a marked increase in stroke volume occurs. In addition, there is a net fall in total peripheral resistance due to vasodilation in exercising muscles. Consequently, systolic blood pressure rises only moderately, whereas diastolic pressure usually remains unchanged or falls.

The difference in response to isometric and isotonic exercise is explained in part by the fact that the active muscles are tonically contracted during isometric exercise and consequently contribute to increased total peripheral resistance. Cardiac output is increased during isotonic exercise to values that may exceed 35 L/min, the amount being proportionate to the increase in O_2 consumption. The maximal heart rate achieved during exercise decreases with age. In children, it rises to 200 or more beats/min; in adults it rarely exceeds 195 beats/min, and in elderly individuals the rise is even smaller. Both at rest and at any given level of exercise, trained athletes have a larger stroke volume and lower heart rate than untrained individuals and they tend to have larger hearts. Training increases the maximal oxygen consumption (VO_{2max}) that can be produced by exercise in an individual. VO_{2max} averages about 38 mL/kg/min in active healthy men and about 29 mL/kg/min in active healthy women. It is lower in sedentary individuals. VO_{2max} is the product of maximal cardiac output and maximal O_2 extraction by the tissues, and both increase with training.

A great increase in venous return also takes place with exercise, although the increase in venous return is not the primary cause of the increase in cardiac output. Venous return is increased by the activity of the muscle and thoracic pumps; by mobilization of blood from the viscera; by increased pressure transmitted through the dilated arterioles to the veins; and by noradrenergically mediated venoconstriction, which decreases the volume of blood in the veins. Blood mobilized from the splanchnic area and other reservoirs may increase the amount of blood in the arterial portion of the circulation by as much as 30% during strenuous exercise. After exercise, the blood pressure may transiently drop to subnormal levels, presumably because accumulated metabolites keep the muscle vessels dilated for a short period. However, the blood pressure soon returns to the pre-exercise level. The heart rate returns to normal more slowly.

INTEGRATED CONTROL OF CARDIAC OUTPUT

The mechanisms listed above operate in an integrated way to maintain cardiac output. For example, during muscular exercise, there is increased sympathetic discharge, so that myocardial contractility is increased and the heart rate rises. The increase in heart rate is particularly prominent in normal individuals, and there is only a modest increase in stroke volume (see Table 31–4 and Clinical Box 31–3). However, patients with transplanted hearts are able to increase their cardiac output during exercise in the absence of cardiac innervation through the operation of the Frank–Starling mechanism (Figure 31–9). Circulating catecholamines also contribute. If venous return increases and there is no change in sympathetic tone, venous pressure rises, diastolic inflow is greater, ventricular end-diastolic pressure increases, and the heart muscle contracts more forcefully. During muscular exercise, venous return is increased by the pumping action of the muscles and the increase in respiration (see Chapter 33). In addition, because of vasodilation in the contracting muscles, peripheral resistance and, consequently, afterload are decreased. The end

TABLE 31–4 Changes in cardiac function with exercise. Note that stroke volume levels off, then falls somewhat (as a result of the shortening of diastole) when the heart rate rises to high values.

Work (kg-m/min)	O$_2$ Usage (mL/min)	Pulse Rate (per min)	Cardiac Output (L/min)	Stroke Volume (mL)	A-V O$_2$ Difference (mL/dL)
Rest	267	64	6.4	100	4.3
288	910	104	13.1	126	7.0
540	1430	122	15.2	125	9.4
900	2143	161	17.8	110	12.3
1260	3007	173	20.9	120	14.5

Reproduced with permission from Asmussen E, Nielsen M: The cardiac output in rest and work determined by the acetylene and the dye injection methods. Acta Physiol Scand 1952;27:217.

result in both normal and transplanted hearts is thus a prompt and marked increase in cardiac output.

One of the differences between untrained individuals and trained athletes is that the athletes have lower heart rates, greater end-systolic ventricular volumes, and greater stroke volumes at rest. Therefore, they can potentially achieve a given increase in cardiac output by further increases in stroke volume without increasing their heart rate to as great a degree as an untrained individual.

OXYGEN CONSUMPTION BY THE HEART

The basal O$_2$ consumption by the myocardium is about 2 mL/ 100 g/min. This value is considerably higher than that of resting skeletal muscle. O$_2$ consumption by the beating heart is about 9 mL/100 g/min at rest. Increases occur during exercise and in a number of different states. Cardiac venous O$_2$ tension is low, and little additional O$_2$ can be extracted from the blood

FIGURE 31–9 Cardiac responses to moderate supine exercise in normal humans and patients with transplanted and hence denervated hearts. (Reproduced with permission from Kent KM, Cooper T: The denervated heart. N Engl J Med 1974;291:1017.)

in the coronaries, so increases in O$_2$ consumption require increases in coronary blood flow. The regulation of coronary flow is discussed in Chapter 34.

O$_2$ consumption by the heart is determined primarily by the intramyocardial tension, the contractile state of the myocardium, and the heart rate. Ventricular work per beat correlates with O$_2$ consumption. The work is the product of stroke volume and mean arterial pressure in the pulmonary artery or the aorta (for the right and left ventricle, respectively). Because aortic pressure is 7 times greater than pulmonary artery pressure, the stroke work of the left ventricle is approximately 7 times the stroke work of the right. In theory, a 25% increase in stroke volume without a change in arterial pressure should produce the same increase in O$_2$ consumption as a 25% increase in arterial pressure without a change in stroke volume. However, for reasons that are incompletely understood, pressure work produces a greater increase in O$_2$ consumption than volume work. In other words, an increase in afterload causes a greater increase in cardiac O$_2$ consumption than does an increase in preload. This is why angina pectoris due to deficient delivery of O$_2$ to the myocardium is more common in aortic stenosis than in aortic insufficiency. In aortic stenosis, intraventricular pressure must be increased to force blood through the stenotic valve, whereas in aortic insufficiency, regurgitation of blood produces an increase in stroke volume with little change in aortic impedance.

It is worth noting that the increase in O$_2$ consumption produced by increased stroke volume when the myocardial fibers are stretched is an example of the operation of the law of Laplace. This law, which is discussed in detail in Chapter 32, states that the tension developed in the wall of a hollow viscus is proportionate to the radius of the viscus, and the radius of a dilated heart is increased. O$_2$ consumption per unit time increases when the heart rate is increased by sympathetic stimulation because of the increased number of beats and the increased velocity and strength of each contraction. However, this is somewhat offset by the decrease in end-systolic volume and hence in the radius of the heart.

CHAPTER SUMMARY

■ Blood flows into the atria and then the ventricles of the heart during diastole and atrial systole, and is ejected during systole when the ventricles contract and pressure exceeds the pressures in the pulmonary artery and aorta.

■ Careful timing of the opening and closing of the atrioventricular (AV), pulmonary, and aortic valves allows blood to move in an appropriate direction through the heart with minimal regurgitation.

■ The proportion of blood leaving the ventricles in each cardiac cycle is called the ejection fraction and is a sensitive indicator of cardiac health.

■ The arterial pulse represents a pressure wave set up when blood is forced into the aorta; it travels much faster than the blood itself.

■ Heart sounds reflect the normal vibrations set up by abrupt valve closures; heart murmurs can arise from abnormal flow often (although not exclusively) caused by diseased valves.

■ Changes in cardiac output reflect variations in heart rate, stroke volume, or both; these are controlled, in turn, by neural and hormonal input to cardiac myocytes.

■ Cardiac output is strikingly increased during exercise.

■ In heart failure, the ejection fraction of the heart is reduced due to impaired contractility in systole or reduced filling during diastole; this results in inadequate blood supplies to meet the body's needs. Initially, this is manifested only during exercise, but eventually the heart will not be able to supply sufficient blood flow even at rest.

MULTIPLE-CHOICE QUESTIONS

For all questions, select the single best answer unless otherwise directed.

1. The second heart sound is caused by
   A) closure of the aortic and pulmonary valves.
   B) vibrations in the ventricular wall during systole.
   C) ventricular filling.
   D) closure of the mitral and tricuspid valves.
   E) retrograde flow in the vena cava.

2. The fourth heart sound is caused by
   A) closure of the aortic and pulmonary valves.
   B) vibrations in the ventricular wall during systole.
   C) ventricular filling.
   D) closure of the mitral and tricuspid valves.
   E) retrograde flow in the vena cava.

3. The dicrotic notch on the aortic pressure curve is caused by
   A) closure of the mitral valve.
   B) closure of the tricuspid valve.
   C) closure of the aortic valve.
   D) closure of the pulmonary valve.
   E) rapid filling of the left ventricle.

4. During exercise, a man consumes 1.8 L of oxygen per minute. His arterial O_2 content is 190 mL/L, and the O_2 content of his mixed venous blood is 134 mL/L. His cardiac output is approximately
   A) 3.2 L/min.
   B) 16 L/min.
   C) 32 L/min.
   D) 54 L/min.
   E) 160 mL/min.

5. The work performed by the left ventricle is substantially greater than that performed by the right ventricle, because in the left ventricle
   A) the contraction is slower.
   B) the wall is thicker.
   C) the stroke volume is greater.
   D) the preload is greater.
   E) the afterload is greater.

6. Starling's law of the heart
   A) does not operate in the failing heart.
   B) does not operate during exercise.
   C) explains the increase in heart rate produced by exercise.
   D) explains the increase in cardiac output that occurs when venous return is increased.
   E) explains the increase in cardiac output when the sympathetic nerves supplying the heart are stimulated.

CHAPTER RESOURCES

Leach JK, Priola DV, Grimes LA, Skipper BJ: Shortening deactivation of cardiac muscle: Physiological mechanisms and clinical implications. J Investig Med 1999;47:369.

Overgaard CB, Dzavik V: Inotropes and vasopressors: Review of physiology and clinical use in cardiovascular disease. Circulation 2008;118:1047.

Rudiger A., Singer M: Mechanisms of sepsis-induced cardiac dysfunction. Crit Care Med 2007;35:1599.

Blood as a Circulatory Fluid & the Dynamics of Blood & Lymph Flow

O B J E C T I V E S

After studying this chapter, you should be able to:

- Describe the components of blood and lymph, their origins, and the role of hemoglobin in transporting oxygen in red blood cells.

- Understand the molecular basis of blood groups and the reasons for transfusion reactions.

- Delineate the process of hemostasis that restricts blood loss when vessels are damaged, and the adverse consequences of intravascular thrombosis.

- Identify the types of blood and lymphatic vessels that make up the circulatory system and the regulation and function of their primary constituent cell types.

- Describe how physical principles dictate the flow of blood and lymph around the body.

- Understand the basis of methods used to measure blood flow and blood pressure in various vascular segments.

- Understand the basis of disease states where components of the blood and vasculature are abnormal, dysregulated, or both.

INTRODUCTION

The **circulatory system** supplies O_2 and substances absorbed from the gastrointestinal tract to the tissues, returns CO_2 to the lungs and other products of metabolism to the kidneys, functions in the regulation of body temperature, and distributes hormones and other agents that regulate cell function. The blood, the carrier of these substances, is pumped through a closed system of blood vessels by the heart. From the left ventricle, blood is pumped through the arteries and arterioles to the capillaries, where it equilibrates with the interstitial fluid. The capillaries drain through venules into the veins and back to the right atrium. Some tissue fluids enter another system of closed vessels, the lymphatics, which drain lymph via the thoracic duct and the right lymphatic duct into the venous system. The circulation is controlled by multiple regulatory systems that function in general to maintain adequate capillary blood flow when possible in all organs, but particularly in the heart and brain.

Blood flows through the circulation primarily because of the forward motion imparted to it by the pumping of the heart, although in the case of the systemic circulation, diastolic recoil of the walls of the arteries, compression of the veins by skeletal muscles during exercise, and the negative pressure in the thorax during inspiration also move the blood forward. The resistance to flow depends to a minor degree on the viscosity of the blood but mostly on the diameter of the vessels, principally the arterioles. The blood flow to each tissue is regulated by local chemical and general neural and humoral mechanisms that dilate or constrict the vessels of the tissue. All the blood flows through the lungs, but the systemic circulation is made up of numerous different circuits in parallel (Figure 32–1). The arrangement permits wide variations in regional blood flow without changing total systemic flow.

This chapter is concerned with blood and lymph and with the multiple functions of the cells they contain. It will also address general principles that apply to all parts of the circulation and with pressure and flow in the systemic circulation. The homeostatic mechanisms operating to adjust flow are the subject of Chapter 33. The special characteristics of pulmonary and renal circulation are discussed in Chapters 35 and 38. Likewise, the role of blood as the carrier of many immune effector cells will not be discussed here, but rather will be covered in Chapter 33.

BLOOD AS A CIRCULATORY FLUID

Blood consists of a protein-rich fluid known as plasma, in which are suspended cellular elements: white blood cells, red blood cells, and platelets. The normal total circulating blood volume is about 8% of the body weight (5600 mL in a 70-kg man). About 55% of this volume is plasma.

BONE MARROW

In the adult, red blood cells, many white blood cells, and platelets are formed in the bone marrow. In the fetus, blood cells are also formed in the liver and spleen, and in adults such **extramedullary hematopoiesis** may occur in diseases in which the bone marrow becomes destroyed or fibrosed. In children, blood cells are actively produced in the marrow cavities of all the bones. By age 20, the marrow in the cavities of the long bones, except for the upper humerus and femur, has become inactive (Figure 32–2). Active cellular marrow is called **red marrow;** inactive marrow that is infiltrated with fat is called **yellow marrow.**

The bone marrow is actually one of the largest organs in the body, approaching the size and weight of the liver. It is also one of the most active. Normally, 75% of the cells in the marrow belong to the white blood cell-producing myeloid series and only 25% are maturing red cells, even though there are

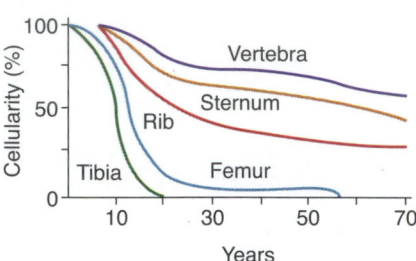

FIGURE 32–2   **Changes in red bone marrow cellularity with age.** 100% equals the degree of cellularity at birth. (Reproduced with permission from Whitby LEH, Britton CJC: *Disorders of the Blood,* 10th ed. Churchill Livingstone, 1969.)

over 500 times as many red cells in the circulation as there are white cells. This difference in the marrow reflects the fact that the average life span of white cells is short, whereas that of red cells is long.

Hematopoietic stem cells (HSCs) are bone marrow cells that are capable of producing all types of blood cells. They differentiate into one or another type of committed stem cells **(progenitor cells).** These in turn form the various differentiated types of blood cells. There are separate pools of progenitor cells for megakaryocytes, lymphocytes, erythrocytes, eosinophils, and basophils; neutrophils and monocytes arise from a common precursor. The bone marrow stem cells are also the source of osteoclasts (see Chapter 23), Kupffer cells (see Chapter 29), mast cells, dendritic cells, and Langerhans cells. The HSCs are few in number but are capable of completely replacing the bone marrow when injected into a host whose own bone marrow has been completely destroyed.

The HSCs are derived from uncommitted, totipotent stem cells that can be stimulated to form any cell in the body. Adults have a few of these, but they are more readily obtained from the blastocysts of embryos. There is not surprisingly immense interest in stem cell research due to its potential to regenerate diseased tissues, but ethical issues are involved, and debate on these issues will undoubtedly continue.

WHITE BLOOD CELLS

Normally, human blood contains 4000 to 11,000 white blood cells per microliter (Table 32–1). Of these, the **granulocytes (polymorphonuclear leukocytes, PMNs)** are the most numerous. Young granulocytes have horseshoe-shaped nuclei that become multilobed as the cells grow older (Figure 32–3).

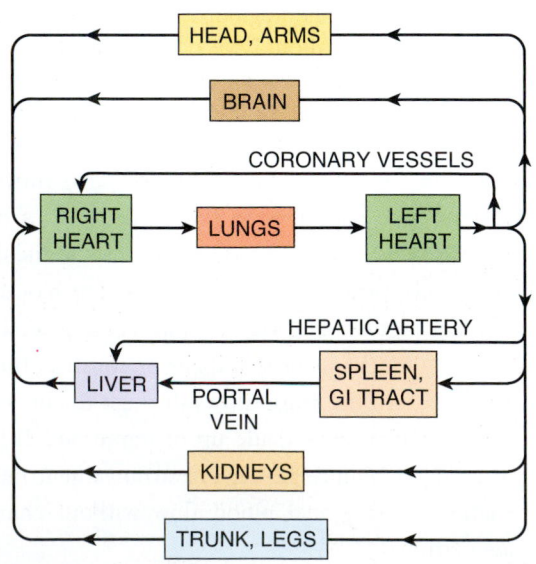

FIGURE 32–1   **Diagram of the circulation in the adult.**

TABLE 32–1 Normal values for the cellular elements in human blood.

Cell	Cells/μL (average)	Approximate Normal Range	Percentage of Total White Cells
Total white blood cells	9000	4000–11,000	...
Granulocytes			
Neutrophils	5400	3000–6000	50–70
Eosinophils	275	150–300	1–4
Basophils	35	0–100	0.4
Lymphocytes	2750	1500–4000	20–40
Monocytes	540	300–600	2–8
Erythrocytes			
Females	4.8×10^6	...	...
Males	5.4×10^6	...	...
Platelets	300,000	200,000–500,000	...

Most of them contain neutrophilic granules (**neutrophils**), but a few contain granules that stain with acidic dyes (**eosinophils**), and some have basophilic granules (**basophils**). The other two cell types found normally in peripheral blood are **lymphocytes**, which have large round nuclei and scanty cytoplasm, and **monocytes**, which have abundant agranular cytoplasm and kidney-shaped nuclei (Figure 32–3). Acting together, these cells provide the body with powerful defenses against tumors and viral, bacterial, and parasitic infections that was discussed in Chapter 3.

PLATELETS

Platelets are small, granulated bodies that aggregate at sites of vascular injury. They lack nuclei and are 2–4 μm in diameter (Figure 32–3). There are about 300,000/μL of circulating blood, and they normally have a half-life of about 4 d. The **megakaryocytes**, giant cells in the bone marrow, form platelets by pinching off bits of cytoplasm and extruding them into the circulation. Between 60% and 75% of the platelets that have been extruded from the bone marrow are in the circulating blood, and the remainder are mostly in the spleen. Splenectomy causes an increase in the platelet count (**thrombocytosis**).

RED BLOOD CELLS

The red blood cells (**erythrocytes**) carry hemoglobin in the circulation. They are biconcave disks (Figure 32–4) that are manufactured in the bone marrow. In mammals, they lose their nuclei before entering the circulation. In humans, they survive in the circulation for an average of 120 d. The average normal red blood cell count is 5.4 million/μL in men and 4.8 million/μL in women. Each human red blood cell is about 7.5 μm in diameter and 2 μm thick, and each contains approximately 29 pg of hemoglobin (Table 32–2). There are thus about 3×10^{13} red blood cells and about 900 g of hemoglobin in the circulating blood of an adult man (Figure 32–5).

The feedback control of erythropoiesis by erythropoietin is discussed in Chapter 39, and the role of IL-1, IL-3, IL-6 (interleukin), and GM-CSF (granulocyte-macrophage colony-stimulating factor) in development of the relevant erythroid stem cells is shown in Figure 32–3.

ROLE OF THE SPLEEN

The spleen is an important blood filter that removes aged or abnormal red cells. It also contains many platelets and plays a significant role in the immune system. Abnormal red cells are removed if they are not as flexible as normal red cells and consequently are unable to squeeze through the slits between the endothelial cells that line the splenic sinuses (see Clinical Box 32–1).

HEMOGLOBIN

The red, oxygen-carrying pigment in the red blood cells of vertebrates is **hemoglobin,** a protein with a molecular weight of 64,450. Hemoglobin is a globular molecule made up of four subunits (Figure 32–6). Each subunit contains a **heme** moiety conjugated to a polypeptide. Heme is an iron-containing porphyrin derivative (Figure 32–7). The polypeptides are referred to collectively as the **globin** portion of the hemoglobin molecule. There are two pairs of polypeptides in each hemoglobin molecule. In normal adult human hemoglobin (**hemoglobin A**), the two polypeptides are called α chains, each of which contains 141 amino acid residues, and β chains, each of which contains 146 amino acid residues. Thus, hemoglobin A is designated $\alpha_2\beta_2$. Not all the hemoglobin in the blood of normal adults is hemoglobin A. About 2.5% of the hemoglobin is hemoglobin A_2, in which β chains are replaced by δ chains ($\alpha_2\delta_2$). The δ chains also contain 146 amino acid residues, but 10 individual residues differ from those in the β chains.

There are small amounts of hemoglobin A derivatives closely associated with hemoglobin A that represent glycated hemoglobins. One of these, hemoglobin A_{1c} (HbA_{1c}), has a glucose attached to the terminal valine in each β chain and is of special interest because it increases in the blood of patients with poorly controlled diabetes mellitus (see Chapter 21).

REACTIONS OF HEMOGLOBIN

Hemoglobin binds O_2 to form **oxyhemoglobin,** O_2 attaching to the Fe^{2+} in the heme. The affinity of hemoglobin for O_2 is affected by pH, temperature, and the concentration in the red cells of 2,3-bisphosphoglycerate (2,3-BPG). 2,3-BPG and H^+

FIGURE 32–3   Development of various formed elements of the blood from bone marrow cells. Cells below the horizontal line are found in normal peripheral blood. The principal sites of action of erythropoietin (erythro) and the various colony-stimulating factors (CSF) that stimulate the differentiation of the components are indicated. G, granulocyte; M, macrophage; IL, interleukin; thrombo, thrombopoietin; SCF, stem cell factor.

compete with O_2 for binding to deoxygenated hemoglobin, decreasing the affinity of hemoglobin for O_2 by shifting the positions of the four peptide chains (quaternary structure). The details of the oxygenation and deoxygenation of hemoglobin and the physiologic role of these reactions in O_2 transport are discussed in Chapter 36.

When blood is exposed to various drugs and other oxidizing agents in vitro or in vivo, the ferrous iron (Fe^{2+}) that is nor-

mally in the molecule is converted to ferric iron (Fe^{3+}), forming **methemoglobin.** Methemoglobin is dark-colored, and when it is present in large quantities in the circulation, it causes a dusky discoloration of the skin resembling cyanosis (see Chapter 36). Some oxidation of hemoglobin to methemoglobin occurs normally, but an enzyme system in the red cells, the dihydronicotinamide adenine dinucleotide (NADH)-methemoglobin reductase system, converts methemoglobin

FIGURE 32–4 Human red blood cells and fibrin fibrils. Blood was placed on a polyvinyl chloride surface, fixed, and photographed with a scanning electron microscope. Reduced from ×2590. (Courtesy of NF Rodman.)

back to hemoglobin. Congenital absence of this system is one cause of hereditary methemoglobinemia.

Carbon monoxide reacts with hemoglobin to form **carbon monoxyhemoglobin (carboxyhemoglobin)**. The affinity of hemoglobin for O_2 is much lower than its affinity for carbon monoxide, which consequently displaces O_2 on hemoglobin, reducing the oxygen-carrying capacity of blood (see Chapter 36).

TABLE 32–2 Characteristics of human red cells.[a]

		Male	Female
Hematocrit (Hct) (%)		47	42
Red blood cells (RBC) ($10^6/\mu L$)		5.4	4.8
Hemoglobin (Hb) (g/dL)		16	14
Mean corpuscular volume (MCV) (fL)	$= \dfrac{Hct \times 10}{RBC\ (10^6/\mu L)}$	87	87
Mean corpuscular hemoglobin (MCH) (pg)	$= \dfrac{Hb \times 10}{RBC\ (10^6/\mu L)}$	29	29
Mean corpuscular hemoglobin concentration (MCHC) (g/dL)	$= \dfrac{Hb \times 100}{Hct}$	34	34
Mean cell diameter (MCD) (μm)	= Mean diameter of 500 cells in smear	7.5	7.5

[a]Cells with MCVs > 95 fL are called macrocytes; cells with MCVs < 80 fL are called microcytes; cells with MCHs < 25 g/dL are called hypochromic.

HEMOGLOBIN IN THE FETUS

The blood of the human fetus normally contains **fetal hemoglobin (hemoglobin F)**. Its structure is similar to that of hemoglobin A except that the β chains are replaced by γ chains; that is, hemoglobin F is $\alpha_2\gamma_2$. The γ chains also contain 146 amino acid residues but have 37 that differ from those in the β chain. Fetal hemoglobin is normally replaced by adult hemoglobin soon after birth (Figure 32–8). In certain individuals, it fails to disappear and persists throughout life. In the body, its O_2 content at a given PO_2 is greater than that of adult hemoglobin because it binds 2,3-BPG less avidly. Hemoglobin F is critical to facilitate movement of O_2 from the maternal to the fetal circulation, particularly at later stages of gestation where oxygen demand increases (see Chapter 34). In young embryos there are, in addition, ζ and ε chains, forming Gower 1

FIGURE 32–5 **Red cell formation and destruction.** RBC, red blood cells.

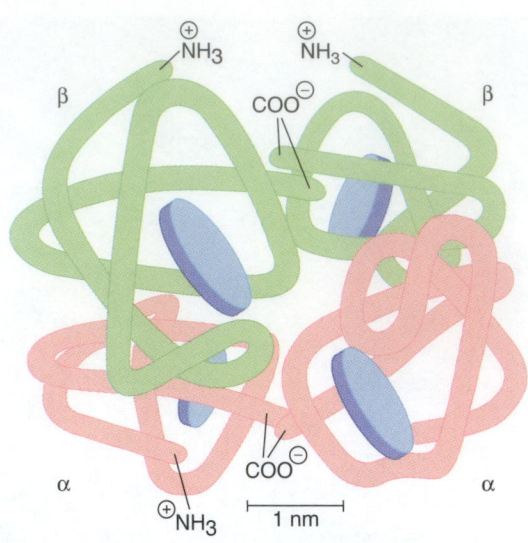

FIGURE 32–6 **Diagrammatic representation of a molecule of hemoglobin A, showing the four subunits.** There are two α and two β polypeptide chains, each containing a heme moiety. These moieties are represented by the disks. (Reproduced with permission from Harper HA et al: *Physiologische Chemie.* Springer, 1975.)

hemoglobin ($\zeta_2\varepsilon_2$) and Gower 2 hemoglobin ($\alpha_2\varepsilon_2$). There are two copies of the α globin gene on human chromosome 16. In addition, there are five globin genes in tandem on chromosome 11 that encode β, γ, and δ globin chains and the two chains normally found only during fetal life. Switching from one form of hemoglobin to another during development seems to be regulated largely by oxygen availability, with relative hypoxia favoring the production of hemoglobin F both via direct effects on globin gene expression, as well as up-regulated production of erythropoietin.

SYNTHESIS OF HEMOGLOBIN

The average normal hemoglobin content of blood is 16 g/dL in men and 14 g/dL in women, all of it in red cells. In the body of a 70-kg man, there are about 900 g of hemoglobin, and 0.3 g of hemoglobin is destroyed and 0.3 g synthesized every hour

(Figure 32–5). The heme portion of the hemoglobin molecule is synthesized from glycine and succinyl-CoA (see Clinical Box 32–2).

CATABOLISM OF HEMOGLOBIN

When old red blood cells are destroyed by tissue macrophages, the globin portion of the hemoglobin molecule is split off, and the heme is converted to **biliverdin.** The enzyme involved is a subtype of heme oxygenase (see Figure 29–4), and CO is formed in the process. CO may be an intercellular messenger, like NO (see Chapters 2 and 3).

FIGURE 32–7 **Reaction of heme with O₂.** The abbreviations M, V, and P stand for the groups shown on the molecule on the left.

Abnormalities of Hemoglobin Production

There are two major types of inherited disorders of hemoglobin in humans: the **hemoglobinopathies,** in which abnormal globin polypeptide chains are produced, and the **thalassemias** and related disorders, in which the chains are normal in structure but produced in decreased amounts or absent because of defects in the regulatory portion of the globin genes. Mutant genes that cause the production of abnormal hemoglobins are widespread, and over 1000 abnormal hemoglobins have been described in humans. In one of the most common examples, hemoglobin S, the α chains are normal but the β chains have a single substitution of a valine residue for one glutamic acid, leading to **sickle cell anemia** (Table 32–3). When an abnormal gene inherited from one parent dictates formation of an abnormal hemoglobin (ie, when the individual is heterozygous), half the circulating hemoglobin is abnormal and half is normal. When identical abnormal genes are inherited from both parents, the individual is homozygous and all the hemoglobin is abnormal. It is theoretically possible to inherit two different abnormal hemoglobins, one from the father and one from the mother. Studies of the inheritance and geographic distribution of abnormal hemoglobins have made it possible in some cases to decide where the mutant gene originated and approximately how long ago the mutation occurred. In general, harmful mutations tend to die out, but mutant genes that confer traits with survival value persist and spread in the population. Many of the abnormal hemoglobins are harmless; however, some have abnormal O_2 equilibriums, while others cause anemia. For example, hemoglobin S polymerizes at low O_2 tensions, and this causes the red cells to become sickle-shaped, hemolyze, and form aggregates that block blood vessels. The sickle cell gene is an example of a gene that has persisted and spread in the population due to its beneficial effect when present in heterozygous form. It originated in Africa, and confers resistance to one type of malaria. In some parts of Africa, 40% of the population is heterozygous for hemoglobin S. There is a corresponding prevalence of 10% among African Americans in the United States. Hemoglobin F decreases the polymerization of deoxygenated hemoglobin S, and hydroxyurea stimulates production of hemoglobin F in children and adults. It has proved to be a very valuable agent for the treatment of sickle cell disease. In patients with severe sickle cell disease, bone marrow transplantation has also been shown to have some benefit.

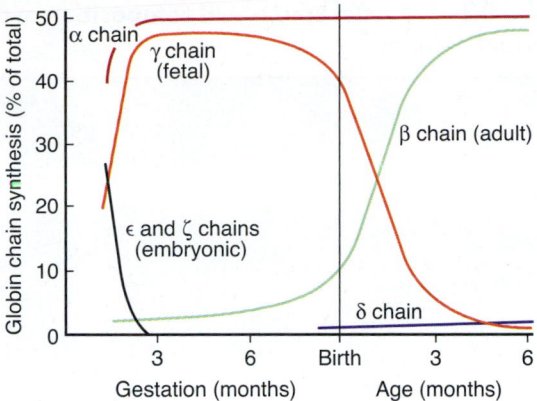

FIGURE 32–8   Development of human hemoglobin chains.

therapy (exposure to light) is of value in treating infants with jaundice due to hemolysis. Iron is essential for hemoglobin synthesis; if blood is lost from the body and the iron deficiency is not corrected, **iron deficiency anemia** results. The metabolism of iron is discussed in Chapter 27.

BLOOD TYPES

The membranes of human red cells contain a variety of **blood group antigens,** which are also called **agglutinogens.** The most important and best known of these are the A and B antigens, but there are many more.

THE ABO SYSTEM

The A and B antigens are inherited as mendelian dominants, and individuals are divided into four major **blood types** on this basis. Type A individuals have the A antigen, type B have the B, type AB have both, and type O have neither. The A and B antigens are complex oligosaccharides that differ in their terminal sugar. An *H* gene codes for a fucose transferase that adds a terminal fucose, forming the H antigen that is usually present in individuals of all blood types (Figure 32–10). Individuals who are type A also express a second transferase that catalyzes placement of a terminal *N*-acetylgalactosamine on the H antigen, whereas individuals who are type B express a transferase that places a terminal galactose. Individuals who are type AB have both transferases. Individuals who are type O have neither, so the H antigen persists.

Antibodies against red cell agglutinogens are called **agglutinins.** Antigens very similar to A and B are common in intestinal bacteria and possibly in foods to which newborn individuals are exposed. Therefore, infants rapidly develop antibodies against the antigens not present in their own cells. Thus, type A individuals develop anti-B antibodies, type B individuals develop anti-A antibodies, type O individuals develop both, and type AB individuals develop neither (Table 32–4). When the plasma of a type A individual is mixed with type B red cells, the anti-B antibodies cause the type B red

In humans, most of the biliverdin is converted to **bilirubin** (Figure 32–9) and excreted in the bile (see Chapter 29). The iron from the heme is reused for hemoglobin synthesis.

Exposure of the skin to white light converts bilirubin to lumirubin, which has a shorter half-life than bilirubin. **Photo-**

TABLE 32–3 Partial amino acid composition of normal human β chain, and some hemoglobins with abnormal β chains.[a]

Hemoglobin	Positions on Polypeptide Chain of Hemoglobin						
	1 2 3	6 7	26	63	67	121	146
A (normal)	Val-His-Leu	Glu-Glu	Glu	His	Val	Glu	His
S (sickle cell)		Val					
C		Lys					
G$_{San Jose}$		Gly					
E			Lys				
M$_{Saskatoon}$				Tyr			
M$_{Milwaukee}$					Glu		
O$_{Arabia}$						Lys	

[a]Other hemoglobins have abnormal α chains. Abnormal hemoglobins that are very similar electrophoretically but differ slightly in composition are indicated by the same letter and a subscript indicating the geographic location where they were first discovered; hence, M$_{Saskatoon}$ and M$_{Milwaukee}$.

FIGURE 32–9 **Bilirubin.** The abbreviations M, V, and P stand for the groups shown on the molecule on the left in Figure 32–7.

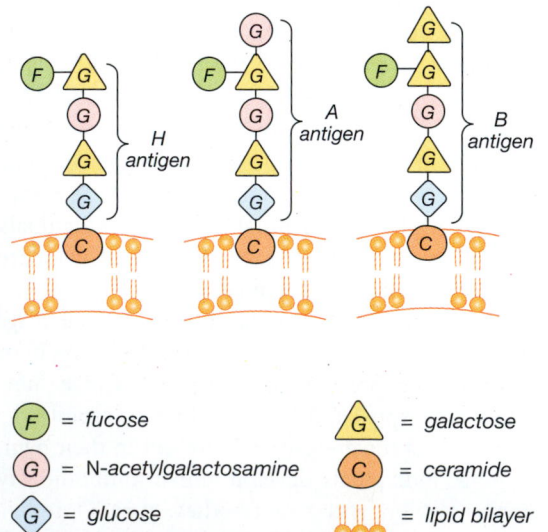

F	= fucose	G (galactose)	= galactose
G	= N-acetylgalactosamine	C	= ceramide
G	= glucose		= lipid bilayer

FIGURE 32–10 Antigens of the ABO system on the surface of red blood cells.

cells to clump (agglutinate), as shown in Figure 32–11. The other agglutination reactions produced by mismatched plasma and red cells are summarized in Table 32–4. **Blood typing** is performed by mixing an individual's red blood cells with antisera containing the various agglutinins on a slide and seeing whether agglutination occurs.

TRANSFUSION REACTIONS

Dangerous **hemolytic transfusion reactions** occur when blood is transfused into an individual with an incompatible blood type; that is, an individual who has agglutinins against the red cells in the transfusion. The plasma in the transfusion is usually so diluted in the recipient that it rarely causes agglutination even when the titer of agglutinins against the recipient's cells is high. However, when the recipient's plasma has agglutinins against the donor's red cells, the cells agglutinate and hemolyze. Free hemoglobin is liberated into the plasma. The severity of the resulting transfusion reaction may vary from an asymptomatic minor rise in the plasma bilirubin level to severe jaundice and renal tubular damage leading to anuria and death.

Incompatibilities in the ABO blood group system are summarized in Table 32–4. Persons with type AB blood are "universal recipients" because they have no circulating agglutinins and can be given blood of any type without developing a transfusion reaction due to ABO incompatibility. Type O individuals are "universal donors" because they lack A and B antigens, and type O blood can be given to anyone without producing a transfusion reaction due to ABO incompatibility. This does not mean, however, that blood should ever be transfused without being cross-matched except in the most extreme emergencies, since the possibility of reactions or sensitization due to incompatibilities in systems other than ABO systems always

FIGURE 32–11   Red cell agglutination in incompatible plasma.

exists. In cross-matching, donor red cells are mixed with recipient plasma on a slide and checked for agglutination. It is advisable to check the action of the donor's plasma on the recipient cells in addition, even though, as noted above, this is rarely a source of trouble.

A procedure that has recently become popular is to withdraw the patient's own blood in advance of elective surgery and then infuse this blood back (**autologous transfusion**) if a transfusion is needed during the surgery. With iron treatment, 1000 to 1500 mL can be withdrawn over a 3-wk period. The popularity of banking one's own blood is due primarily to fear of transmission of infectious diseases by heterologous transfu-

sions, but of course another advantage is elimination of the risk of transfusion reactions.

INHERITANCE OF A & B ANTIGENS

The A and B antigens are inherited as mendelian allelomorphs, A and B being dominants. For example, an individual with type B blood may have inherited a B antigen from each parent or a B antigen from one parent and an O from the other; thus, an individual whose **phenotype** is B may have the **genotype** BB (**homozygous**) or BO (**heterozygous**).

When the blood types of the parents are known, the possible genotypes of their children can be stated. When both parents are type B, they could have children with genotype BB (B antigen from both parents), BO (B antigen from one parent, O from the other heterozygous parent), or OO (O antigen from both parents, both being heterozygous). When the blood types of a mother and her child are known, typing can prove that a man cannot be the father, although it cannot prove that he is the father. The predictive value is increased if the blood typing of the parties concerned includes identification of antigens other than the ABO agglutinogens. With the use of DNA fingerprinting (see Chapter 1), the exclusion rate for paternity rises to close to 100%.

TABLE 32–4   Summary of ABO system.

Blood Type	Agglutinins in Plasma	Frequency in United States (%)	Plasma Agglutinates Red Cells of Type:
O	Anti-A, anti-B	45	A, B, AB
A	Anti-B	41	B, AB
B	Anti-A	10	A, AB
AB	None	4	None

OTHER AGGLUTINOGENS

In addition to the ABO system of antigens in human red cells, there are systems such as the Rh, MNSs, Lutheran, Kell, Kidd, and many others. There are over 500 billion possible known blood group phenotypes, and because undiscovered antigens undoubtedly exist, it has been calculated that the number of phenotypes is actually in the trillions.

The number of blood groups in animals is as large as it is in humans. An interesting question is why this degree of polymorphism developed and has persisted through evolution. Certain diseases are more common in individuals with one blood type or another, but the differences are not great. One, the Duffy antigen, is a chemokine receptor. Many of the others seem to be cell recognition molecules, but the significance of a recognition code of this complexity is unknown.

THE RH GROUP

Aside from the antigens of the ABO system, those of the Rh system are of the greatest clinical importance. The Rh factor, named for the rhesus monkey because it was first studied using the blood of this animal, is a system composed primarily of the C, D, and E antigens, although it actually contains many more. Unlike the ABO antigens, the system has not been detected in tissues other than red cells. D is by far the most antigenic component, and the term Rh-positive as it is generally used means that the individual has agglutinogen D. The D protein is not glycosylated, and its function is unknown. The Rh-negative individual has no D antigen and forms the anti-D agglutinin when injected with D-positive cells. The Rh typing serum used in routine blood typing is anti-D serum. Eighty-five percent of Caucasians are D-positive and 15% are D-negative; over 99% of Asians are D-positive. Unlike the antibodies of the ABO system, anti-D antibodies do not develop without exposure of a D-negative individual to D-positive red cells by transfusion or entrance of fetal blood into the maternal circulation. However, D-negative individuals who have received a transfusion of D-positive blood (even years previously) can have appreciable anti-D titers and thus may develop transfusion reactions when transfused again with D-positive blood.

HEMOLYTIC DISEASE OF THE NEWBORN

Another complication due to Rh incompatibility arises when an Rh-negative mother carries an Rh-positive fetus. Small amounts of fetal blood leak into the maternal circulation at the time of delivery, and some mothers develop significant titers of anti-Rh agglutinins during the postpartum period. During the next pregnancy, the mother's agglutinins cross the placenta to the fetus. In addition, there are some cases of fetal–maternal hemorrhage during pregnancy, and sensitization can occur during pregnancy. In any case, when anti-Rh agglutinins cross the placenta to an Rh-positive fetus, they can cause hemolysis and various forms of **hemolytic disease of the newborn (erythroblastosis fetalis).** If hemolysis in the fetus is severe, the infant may die in utero or may develop anemia, severe jaundice, and edema **(hydrops fetalis). Kernicterus,** a neurologic syndrome in which unconjugated bilirubin is deposited in the basal ganglia, may also develop, especially if birth is complicated by a period of hypoxia. Bilirubin rarely penetrates the brain in adults, but it does in infants with erythroblastosis, possibly in part because the blood–brain barrier is more permeable in infancy. However, the main reasons that the concentration of unconjugated bilirubin is very high in this condition are that production is increased and the bilirubin-conjugating system is not yet mature.

About 50% of Rh-negative individuals are sensitized (develop an anti-Rh titer) by transfusion of Rh-positive blood. Because sensitization of Rh-negative mothers by carrying an Rh-positive fetus generally occurs at birth, the first child is usually normal. However, hemolytic disease occurs in about 17% of the Rh-positive fetuses born to Rh-negative mothers who have previously been pregnant one or more times with Rh-positive fetuses. Fortunately, it is usually possible to prevent sensitization from occurring the first time by administering a single dose of anti-Rh antibodies in the form of Rh immune globulin during the postpartum period. Such passive immunization does not harm the mother and has been demonstrated to prevent active antibody formation by the mother. In obstetric clinics, the institution of such treatment on a routine basis to unsensitized Rh-negative women who have delivered an Rh-positive baby has reduced the overall incidence of hemolytic disease by more than 90%. In addition, fetal Rh typing with material obtained by amniocentesis or chorionic villus sampling is now possible, and treatment with a small dose of Rh immune serum will prevent sensitization during pregnancy.

PLASMA

The fluid portion of the blood, the **plasma,** is a remarkable solution containing an immense number of ions, inorganic molecules, and organic molecules that are in transit to various parts of the body or aid in the transport of other substances. Normal plasma volume is about 5% of body weight, or roughly 3500 mL in a 70-kg man. Plasma clots on standing, remaining fluid only if an anticoagulant is added. If whole blood is allowed to clot and the clot is removed, the remaining fluid is called **serum.** Serum has essentially the same composition as plasma, except that its fibrinogen and clotting factors II, V, and VIII (Table 32–5) have been removed and it has a higher serotonin content because of the breakdown of platelets during clotting.

PLASMA PROTEINS

The plasma proteins consist of **albumin, globulin,** and **fibrinogen** fractions. Most capillary walls are relatively impermeable to the proteins in plasma, and the proteins therefore exert

TABLE 32–5 System for naming blood-clotting factors.

Factor[a]	Names
I	Fibrinogen
II	Prothrombin
III	Thromboplastin
IV	Calcium
V	Proaccelerin, labile factor, accelerator globulin
VII	Proconvertin, SPCA, stable factor
VIII	Antihemophilic factor (AHF), antihemophilic factor A, antihemophilic globulin (AHG)
IX	Plasma thromboplastic component (PTC), Christmas factor, antihemophilic factor B
X	Stuart–Prower factor
XI	Plasma thromboplastin antecedent (PTA), antihemophilic factor C
XII	Hageman factor, glass factor
XIII	Fibrin-stabilizing factor, Laki–Lorand factor
HMW-K	High-molecular-weight kininogen, Fitzgerald factor
Pre-K_a	Prekallikrein, Fletcher factor
Ka	Kallikrein
PL	Platelet phospholipid

[a]Factor VI is not a separate entity and has been dropped.

an osmotic force of about 25 mm Hg across the capillary wall (**oncotic pressure;** see Chapter 1) that pulls water into the blood. The plasma proteins are also responsible for 15% of the buffering capacity of the blood (see Chapter 39) because of the weak ionization of their substituent COOH and NH_2 groups. At the normal plasma pH of 7.40, the proteins are mostly in the anionic form (see Chapter 1). Plasma proteins may have specific functions (eg, antibodies and the proteins concerned with blood clotting), whereas others function as carriers for various hormones, other solutes, and drugs.

ORIGIN OF PLASMA PROTEINS

Circulating antibodies are manufactured by lymphocytes. Most of the other plasma proteins are synthesized in the liver. These proteins and their principal functions are listed in Table 32–6.

Data on the turnover of albumin show that its synthesis plays an important role in the maintenance of normal levels. In normal adult humans, the plasma albumin level is 3.5 to 5.0 g/dL, and the total exchangeable albumin pool is 4.0 to 5.0 g/kg body weight; 38–45% of this albumin is intravascular, and much of the rest of it is in the skin. Between 6% and 10% of the exchangeable pool is degraded per day, and the degraded albumin is replaced by hepatic synthesis of 200 to 400 mg/kg/d. The albumin is probably transported to the extravascular areas by vesicular transport across the walls of the capillaries (see Chapter 2). Albumin synthesis is carefully regulated. It is decreased during fasting and increased in conditions such as nephrosis in which there is excessive albumin loss.

HYPOPROTEINEMIA

Plasma protein levels are maintained during starvation until body protein stores are markedly depleted. However, in prolonged starvation and in malabsorption syndromes due to intestinal diseases, plasma protein levels are low (**hypoproteinemia**). They are also low in liver disease, because hepatic protein synthesis is depressed, and in nephrosis, because large amounts of albumin are lost in the urine. Because of the decrease in the plasma oncotic pressure, edema tends to develop. Rarely, there is congenital absence of one or another plasma protein. An example of congenital protein deficiency is the congenital form of **afibrinogenemia,** characterized by defective blood clotting.

HEMOSTASIS

Hemostasis is the process of forming clots in the walls of damaged blood vessels and preventing blood loss while maintaining blood in a fluid state within the vascular system. A collection of complex interrelated systemic mechanisms operates to maintain a balance between coagulation and anticoagulation.

RESPONSE TO INJURY

When a small blood vessel is transected or damaged, the injury initiates a series of events (Figure 32–12) that lead to the formation of a clot. This seals off the damaged region and prevents further blood loss. The initial event is constriction of the vessel and formation of a temporary **hemostatic plug** of platelets that is triggered when platelets bind to collagen and aggregate. This is followed by conversion of the plug into the definitive clot. The constriction of an injured arteriole or small artery may be so marked that its lumen is obliterated, at least temporarily. The vasoconstriction is due to serotonin and other vasoconstrictors liberated from platelets that adhere to the walls of the damaged vessels.

THE CLOTTING MECHANISM

The loose aggregation of platelets in the temporary plug is bound together and converted into the definitive clot by **fibrin.** Fibrin formation involves a cascade of enzymatic reactions and a series of numbered clotting factors (Table 32–5). The fundamental reaction is conversion of the soluble plasma protein fibrinogen to insoluble fibrin (Figure 32–13). The process

TABLE 32–6 Some of the proteins synthesized by the liver: Physiologic functions and properties.

Name	Principal Function	Binding Characteristics	Serum or Plasma Concentration
Albumin	Binding and carrier protein; osmotic regulator	Hormones, amino acids, steroids, vitamins, fatty acids	4500–5000 mg/dL
Orosomucoid	Uncertain; may have a role in inflammation		Trace; rises in inflammation
α_1-Antiprotease	Trypsin and general protease inhibitor	Proteases in serum and tissue secretions	1.3–1.4 mg/dL
α-Fetoprotein	Osmotic regulation; binding and carrier protein[a]	Hormones, amino acids	Found normally in fetal blood
α_2-Macroglobulin	Inhibitor of serum endoproteases	Proteases	150–420 mg/dL
Antithrombin-III	Protease inhibitor of intrinsic coagulation system	1:1 binding to proteases	17–30 mg/dL
Ceruloplasmin	Transport of copper	Six atoms copper/mol	15–60 mg/dL
C-reactive protein	Uncertain; has role in tissue inflammation	Complement C1q	< 1 mg/dL; rises in inflammation
Fibrinogen	Precursor to fibrin in hemostasis		200–450 mg/dL
Haptoglobin	Binding, transport of cell-free hemoglobin	Hemoglobin 1:1 binding	40–180 mg/dL
Hemopexin	Binds to porphyrins, particularly heme for heme recycling	1:1 with heme	50–100 mg/dL
Transferrin	Transport of iron	Two atoms iron/mol	3.0–6.5 mg/dL
Apolipoprotein B	Assembly of lipoprotein particles	Lipid carrier	
Angiotensinogen	Precursor to pressor peptide angiotensin II		
Proteins, coagulation factors II, VII, IX, X	Blood clotting		20 mg/dL
Antithrombin C, protein C	Inhibition of blood clotting		
Insulinlike growth factor I	Mediator of anabolic effects of growth hormone	IGF-I receptor	
Steroid hormone-binding globulin	Carrier protein for steroids in bloodstream	Steroid hormones	3.3 mg/dL
Thyroxine-binding globulin	Carrier protein for thyroid hormone in bloodstream	Thyroid hormones	1.5 mg/dL
Transthyretin (thyroid-binding prealbumin)	Carrier protein for thyroid hormone in bloodstream	Thyroid hormones	25 mg/dL

[a]The function of alpha-fetoprotein is uncertain, but because of its structural homology to albumin it is often assigned these functions.

involves the release of two pairs of polypeptides from each fibrinogen molecule. The remaining portion, **fibrin monomer,** then polymerizes with other monomer molecules to form **fibrin.** The fibrin is initially a loose mesh of interlacing strands. It is converted by the formation of covalent cross-linkages to a dense, tight aggregate (stabilization). This latter reaction is catalyzed by activated factor XIII and requires Ca^{2+}.

The conversion of fibrinogen to fibrin is catalyzed by thrombin. Thrombin is a serine protease that is formed from its circulating precursor, prothrombin, by the action of activated factor X. It has additional actions, including activation

of platelets, endothelial cells, and leukocytes via so-called proteinase activated receptors, which are G protein-coupled.

Factor X can be activated by either of two systems, known as intrinsic and extrinsic (Figure 32–13). The initial reaction in the **intrinsic system** is conversion of inactive factor XII to active factor XII (XIIa). This activation, which is catalyzed by high-molecular-weight kininogen and kallikrein (see Chapter 33), can be brought about in vitro by exposing the blood to glass, or in vivo by collagen fibers underlying the endothelium. Active factor XII then activates factor XI, and active factor XI activates factor IX. Activated factor IX forms a complex with

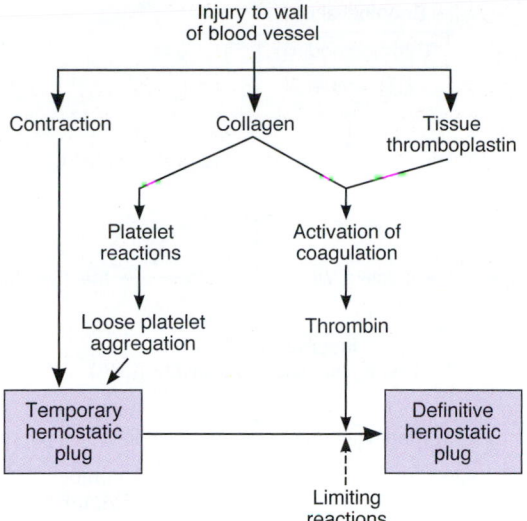

FIGURE 32–12 **Summary of reactions involved in hemostasis.** The dashed arrow indicates inhibition. (Modified from Deykin D: Thrombogenesis, N Engl J Med 1967;267:622.)

FIGURE 32–13 **The clotting mechanism.** a, active form of clotting factor. TPL, tissue thromboplastin; TFI, tissue factor pathway inhibitor. For other abbreviations, see Table 32–5.

active factor VIII, which is activated when it is separated from von Willebrand factor. The complex of IXa and VIIIa activate factor X. Phospholipids from aggregated platelets (PL) and Ca^{2+} are necessary for full activation of factor X. The **extrinsic system** is triggered by the release of tissue thromboplastin, a protein–phospholipid mixture that activates factor VII. Tissue thromboplastin and factor VII activate factors IX and X. In the presence of PL, Ca^{2+}, and factor V, activated factor X catalyzes the conversion of prothrombin to thrombin. The extrinsic pathway is inhibited by a **tissue factor pathway inhibitor** that forms a quaternary structure with tissue thromboplastin (TPL), factor VIIa, and factor Xa.

ANTICLOTTING MECHANISMS

The tendency of blood to clot is balanced in vivo by reactions that prevent clotting inside the blood vessels, break down any clots that do form, or both. These reactions include the interaction between the platelet-aggregating effect of thromboxane A$_2$ and the antiaggregating effect of prostacyclin, which causes clots to form at the site when a blood vessel is injured but keeps the vessel lumen free of clot (see Chapter 33 and Clinical Box 32–3).

Antithrombin III is a circulating protease inhibitor that binds to serine proteases in the coagulation system, blocking their activity as clotting factors. This binding is facilitated by **heparin,** a naturally occurring anticoagulant that is a mixture of sulfated polysaccharides with molecular weights averaging 15,000–18,000. The clotting factors that are inhibited are the active forms of factors IX, X, XI, and XII.

The endothelium of the blood vessels also plays an active role in preventing the extension of clots. All endothelial cells except those in the cerebral microcirculation produce **thrombomodulin,** a thrombin-binding protein, on their surfaces. In circulating blood, thrombin is a procoagulant that activates factors V and VIII, but when it binds to thrombomodulin, it becomes an anticoagulant in that the thrombomodulin–thrombin complex activates protein C (Figure 32–14). Activated protein C (APC), along with its cofactor protein S, inactivates factors V and VIII and inactivates an inhibitor of tissue plasminogen activator, increasing the formation of plasmin.

Plasmin (fibrinolysin) is the active component of the **plasminogen (fibrinolytic) system** (Figure 32–14). This enzyme lyses fibrin and fibrinogen, with the production of fibrinogen degradation products (FDP) that inhibit thrombin. Plasmin is formed from its inactive precursor, plasminogen, by the action of thrombin and **tissue-type plasminogen activator (t-PA).** It is also activated by **urokinase-type plasminogen activator (u-PA).** If the t-PA gene or the u-PA gene is knocked out in mice, some fibrin deposition occurs and clot lysis is slowed. However, when both are knocked out, spontaneous fibrin deposition is extensive.

Human plasminogen consists of a 560-amino-acid heavy chain and a 241-amino-acid light chain. The heavy chain, with glutamate at its amino terminal, is folded into five loop structures, each held together by three disulfide bonds (Figure 32–15). These loops are called kringles because of their

CLINICAL BOX 32–3

Abnormalities of Hemostasis

In addition to clotting abnormalities due to platelet disorders, hemorrhagic diseases can be produced by selective deficiencies of most of the clotting factors (Table 32–7). Hemophilia A, which is caused by factor VIII deficiency, is relatively common. The disease has been treated with factor VIII-rich preparations made from plasma, or, more recently, factor VIII produced by recombinant DNA techniques. von Willebrand factor deficiency likewise causes a bleeding disorder (von Willebrand disease) by reducing platelet adhesion and by lowering plasma factor VIII. The condition can be congenital or acquired. The large von Willebrand molecule is subject to cleavage and resulting inactivation by the plasma metalloprotease ADAM 13 in vascular areas where fluid shear stress is elevated. Finally, when absorption of vitamin K is depressed along with absorption of other fat-soluble vitamins (see Chapter 27), the resulting clotting factor deficiencies may cause the development of a significant bleeding tendency.

Formation of clots inside blood vessels is called **thrombosis** to distinguish it from the normal extravascular clotting of blood. Thromboses are a major medical problem. They are particularly prone to occur where blood flow is sluggish because the slow flow permits activated clotting factors to accumulate instead of being washed away. They also occur in vessels when the intima is damaged by atherosclerotic plaques, and over areas of damage to the endocardium. They frequently occlude the arterial supply to the organs in which they form, and bits of thrombus **(emboli)** sometimes break off and travel in the bloodstream to distant sites, damaging other organs. An example is obstruction of the pulmonary artery or its branches by thrombi from the leg veins **(pulmonary embolism).** Congenital absence of protein C leads to uncontrolled intravascular coagulation and, in general, death in infancy. If this condition is diagnosed and treatment is instituted, the coagulation defect disappears. Resistance to activated protein C is another cause of thrombosis, and this condition is common. It is due to a point mutation in the gene for factor V, which prevents activated protein C from inactivating the factor. Mutations in protein S and antithrombin III may less commonly increase the incidence of thrombosis.

Disseminated intravascular coagulation is another serious complication of septicemia, extensive tissue injury, and other diseases in which fibrin is deposited in the vascular system and many small- and medium-sized vessels are thrombosed. The increased consumption of platelets and coagulation factors causes bleeding to occur at the same time. The cause of the condition appears to be increased generation of thrombin due to increased TPL activity without adequate tissue factor inhibitory pathway activity.

FIGURE 32–14   **The fibrinolytic system and its regulation by protein C.**

resemblance to a Danish pastry of the same name. The kringles are lysine-binding sites by which the molecule attaches to fibrin and other clot proteins, and they are also found in prothrombin. Plasminogen is converted to active plasmin when t-PA hydrolyzes the bond between Arg 560 and Val 561.

Plasminogen receptors are located on the surfaces of many different types of cells and are plentiful on endothelial cells. When plasminogen binds to its receptors, it becomes activated, so intact blood vessel walls are provided with a mechanism that discourages clot formation.

Human t-PA is now produced by recombinant DNA techniques for clinical use in myocardial infarction and stroke. Streptokinase, a bacterial enzyme, is also fibrinolytic and is also used in the treatment of early myocardial infarction (see Chapter 34).

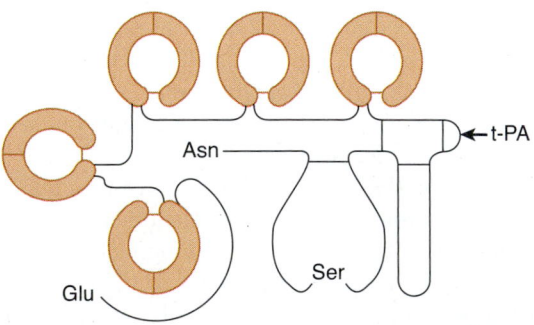

FIGURE 32–15   **Structure of human plasminogen.** Note the Glu at the amino terminal, the Asn at the carboxyl terminal, and five uniquely shaped loop structures (kringles). Hydrolysis by t-PA at the arrow separates the carboxyl terminal light chain from the amino terminal heavy chain but leaves the disulfide bonds intact. This activates the molecule. (Modified and reproduced with permission from Bachman F, in: *Thrombosis and Hemostasis.* Verstraete M et al [editors]. Leuven University Press, 1987.)

TABLE 32–7 Examples of diseases due to deficiency of clotting factors.

Deficiency of Factor:	Clinical Syndrome	Cause
I	Afibrinogenemia	Depletion during pregnancy with premature separation of placenta; also congenital (rare)
II	Hypoprothrombinemia (hemorrhagic tendency in liver disease)	Decreased hepatic synthesis, usually secondary to vitamin K deficiency
V	Parahemophilia	Congenital
VII	Hypoconvertinemia	Congenital
VIII	Hemophilia A (classic hemophilia)	Congenital defect due to various abnormalities of the gene on X chromosome that codes for factor VIII; disease is therefore inherited as sex-linked characteristic
IX	Hemophilia B (Christmas disease)	Congenital
X	Stuart–Power factor deficiency	Congenital
XI	PTA deficiency	Congenital
XII	Hageman trait	Congenital

ANTICOAGULANTS

As noted above, heparin is a naturally occurring anticoagulant that facilitates the action of antithrombin III. Low-molecular-weight fragments with an average molecular weight of 5000 have been produced from unfractionated heparin, and these low-molecular-weight heparins are seeing increased clinical use because they have a longer half-life and produce a more predictable anticoagulant response than unfractionated heparin. The highly basic protein protamine forms an irreversible complex with heparin and is used clinically to neutralize heparin.

In vivo, a plasma Ca^{2+} level low enough to interfere with blood clotting is incompatible with life, but clotting can be prevented in vitro if Ca^{2+} is removed from the blood by the addition of substances such as oxalates, which form insoluble salts with Ca^{2+}, or **chelating agents,** which bind Ca^{2+}. Coumarin derivatives such as **dicumarol** and **warfarin** are also effective anticoagulants. They inhibit the action of vitamin K, which is a necessary cofactor for the enzyme that catalyzes the conversion of glutamic acid residues to γ-carboxyglutamic acid residues. Six of the proteins involved in clotting require conversion of a number of glutamic acid residues to γ-carboxyglutamic acid residues before being released into the circulation, and hence all six are vitamin K-dependent. These proteins are factors II (prothrombin), VII, IX, and X, protein C, and protein S (see above).

LYMPH

Lymph is tissue fluid that enters the lymphatic vessels. It drains into the venous blood via the thoracic and right lymphatic ducts. It contains clotting factors and clots on standing in vitro. In most locations, it also contains proteins that traverse capillary walls and return to the blood via the lymph. Its protein content is generally lower than that of plasma, which contains about 7 g/dL, but lymph protein content varies with the region from which the lymph drains (Table 32–8). Water-insoluble fats are absorbed from the intestine into the lymphatics, and the lymph in the thoracic duct after a meal is milky because of its high fat content (see Chapter 27). Lymphocytes enter the circulation principally through the lymphatics, and there are appreciable numbers of lymphocytes in thoracic duct lymph.

STRUCTURAL FEATURES OF THE CIRCULATION

Here, we will first describe the two major cell types that make up the blood vessels and then how they are arranged into the various vessel types that subserve the needs of the circulation.

ENDOTHELIUM

Located between the circulating blood and the media and adventitia of the blood vessels, the endothelial cells constitute a large and important organ. They respond to flow changes, stretch, a variety of circulating substances, and inflammatory mediators. They secrete growth regulators and vasoactive substances (see below and Chapter 33).

TABLE 32–8 Probable approximate protein content of lymph in humans.

Source of Lymph	Protein Content (g/dL)
Choroid plexus	0
Ciliary body	0
Skeletal muscle	2
Skin	2
Lung	4
Gastrointestinal tract	4.1
Heart	4.4
Liver	6.2

Data largely from JN Diana.

VASCULAR SMOOTH MUSCLE

The smooth muscle in blood vessel walls has been one of the most-studied forms of visceral smooth muscle because of its importance in the regulation of blood pressure and hypertension. The membranes of the muscle cells contain various types of K^+, Ca^{2+}, and Cl^- channels. Contraction is produced primarily by the myosin light chain mechanism described in Chapter 5. However, vascular smooth muscle also undergoes prolonged contractions that determine vascular tone. These may be due in part to the latch-bridge mechanism (see Chapter 5), but other factors also play a role. Some of the molecular mechanisms that appear to be involved in contraction and relaxation are shown in Figure 32–16.

Vascular smooth muscle cells provide an interesting example of the way high and low cytosolic Ca^{2+} can have different and even opposite effects (see Chapter 2). In these cells, influx of Ca^{2+} via voltage-gated Ca^{2+} channels produces a diffuse increase in cytosolic Ca^{2+} that initiates contraction. However, the Ca^{2+} influx also initiates Ca^{2+} release from the sarcoplasmic reticulum via ryanodine receptors (see Chapter 5), and the high local Ca^{2+} concentration produced by these Ca^{2+} sparks increases the activity of **Ca^{2+}-activated K^+ channels** in the cell membrane. These are also known as big K or **BK channels** because K^+ flows through them at a high rate.

The increased K^+ efflux increases the membrane potential, shutting off voltage-gated Ca^{2+} channels and producing relaxation. The site of action of the Ca^{2+} sparks is the β_1-subunit of the BK channel, and mice in which this subunit is knocked out develop increased vascular tone and blood pressure. Obviously, therefore, the sensitivity of the β_1 subunit to Ca^{2+} sparks plays an important role in the control of vascular tone.

ARTERIES & ARTERIOLES

The characteristics of the various types of blood vessels are listed in Table 32–9. The walls of all arteries are made up of an outer layer of connective tissue, the adventitia; a middle layer of smooth muscle, the media; and an inner layer, the intima, made up of the endothelium and underlying connective tissue (Figure 32–17). The walls of the aorta and other arteries of large diameter contain a relatively large amount of elastic tissue, primarily located in the inner and external elastic laminas. They are stretched during systole and recoil on the blood during diastole. The walls of the arterioles contain less elastic tissue but much more smooth muscle. The muscle is innervated by noradrenergic nerve fibers, which function as constrictors, and in some instances by cholinergic fibers, which dilate the

FIGURE 32–16   **Some of the established and postulated mechanisms involved in the contraction and relaxation of vascular smooth muscle.** A, agonist; AA, arachidonic acid; BK, Ca^+-activated K^+ channel; G, heterotrimeric G protein; MLC, myosin light chain; MLCK, myosin light chain kinase; PLD, phospholipase D; R, receptor; SF, sarcoplasmic reticulum; VGCC, voltage-gated Ca^{2+} channel; RR, ryanodine receptors. For other abbreviations, see Chapter 2 and Appendix. (Modified from Khahl R: Mechanisms of vascular smooth muscle contraction. Council for High Blood Pressure Newsletter, Spring 2001.)

TABLE 32–9 Characteristics of various types of blood vessels in humans.

Vessel	Lumen Diameter	Wall Thickness	All Vessels of Each Type	
			Approximate Total Cross Sectional Area (cm²)	Percentage of Blood Volume Contained[a]
Aorta	2.5 cm	2 mm	4.5	2
Artery	0.4 cm	1 mm	20	8
Arteriole	30 μm	20 μm	400	1
Capillary	5 μm	1 μm	4500	5
Venule	20 μm	2 μm	4000 }	
Vein	0.5 cm	0.5 mm	40 }	54
Vena cava	3 cm	1.5 mm	18 }	

[a]In systemic vessels; there is an additional 12% in the heart and 18% in the pulmonary circulation.

vessels. The arterioles are the major site of the resistance to blood flow, and small changes in their caliber cause large changes in the total peripheral resistance.

CAPILLARIES

The arterioles divide into smaller muscle-walled vessels, sometimes called **metarterioles,** and these in turn feed into capillaries (Figure 32–18). The openings of the capillaries are surrounded on the upstream side by minute smooth muscle **precapillary sphincters.** It is unsettled whether the metarterioles are innervated, and it appears that the precapillary sphincters are not. However, they can of course respond to local or circulating vasoconstrictor substances. The capillaries are about 5 μm in diameter at the arterial end and 9 μm in diameter at the venous end. When the sphincters are dilated, the

FIGURE 32–17   Structure of normal muscle artery. (Reproduced with permission from Ross R, Glomset JA: The pathogenesis of atherosclerosis. N Engl J Med 1976;295:369.)

Labels in figure: Endothelium; Internal elastic lamina; External elastic lamina; Intima; Media; Adventitia

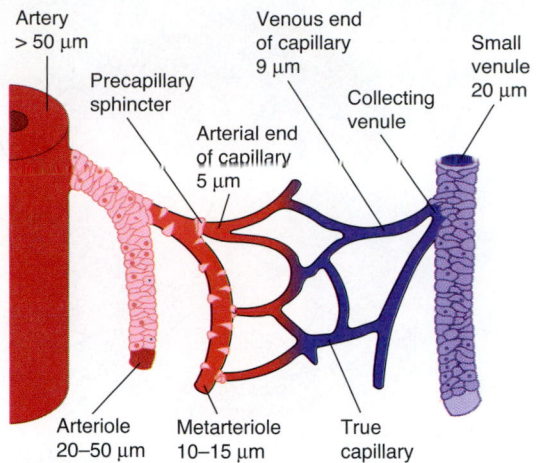

FIGURE 32–18   The microcirculation. Arterioles give rise to metarterioles, which give rise to capillaries. The capillaries drain via short collecting venules to the venules. The walls of the arteries, arterioles, and small venules contain relatively large amounts of smooth muscle. There are scattered smooth muscle cells in the walls of the metarterioles, and the openings of the capillaries are guarded by muscular precapillary sphincters. The diameters of the various vessels are also shown.   (Courtesy of JN Diana.)

Labels in figure: Artery >50 μm; Precapillary sphincter; Venous end of capillary 9 μm; Collecting venule; Small venule 20 μm; Arterial end of capillary 5 μm; Arteriole 20–50 μm; Metarteriole 10–15 μm; True capillary

diameter of the capillaries is just sufficient to permit red blood cells to squeeze through in "single file." As they pass through the capillaries, the red cells become thimble- or parachute-shaped, with the flow pushing the center ahead of the edges. This configuration appears to be due simply to the pressure in the center of the vessel whether or not the edges of the red blood cell are in contact with the capillary walls.

The total area of all the capillary walls in the body exceeds 6300 m² in the adult. The walls, which are about 1 μm thick, are made up of a single layer of endothelial cells. The structure of the walls varies from organ to organ. In many beds, including those in skeletal, cardiac, and smooth muscle, the junctions between the endothelial cells (Figure 32–19) permit the passage of molecules up to 10 nm in diameter. It also appears that plasma and its dissolved proteins are taken up by endocytosis, transported across the endothelial cells, and discharged by exocytosis (**vesicular transport;** see Chapter 2). However, this process can account for only a small portion of the transport across the endothelium. In the brain, the capillaries resemble the capillaries in muscle, but the junctions between endothelial cells are tighter, and transport across them is largely limited to small molecules. In most endocrine glands, the intestinal villi, and parts of the kidneys, the cytoplasm of the endothelial cells is attenuated to form gaps called **fenestrations.** These fenestrations are 20 to 100 nm in diameter and may permit the passage of larger molecules, although they appear to be closed by a thin membrane. An exception to this, however, is found in the liver, where the sinusoidal capillaries are extremely porous, the endothelium is discontinuous, and gaps occur between endothelial cells that are not closed by membranes (see Figure 29–2). Some of the gaps are 600 nm in diameter, and others may be as large

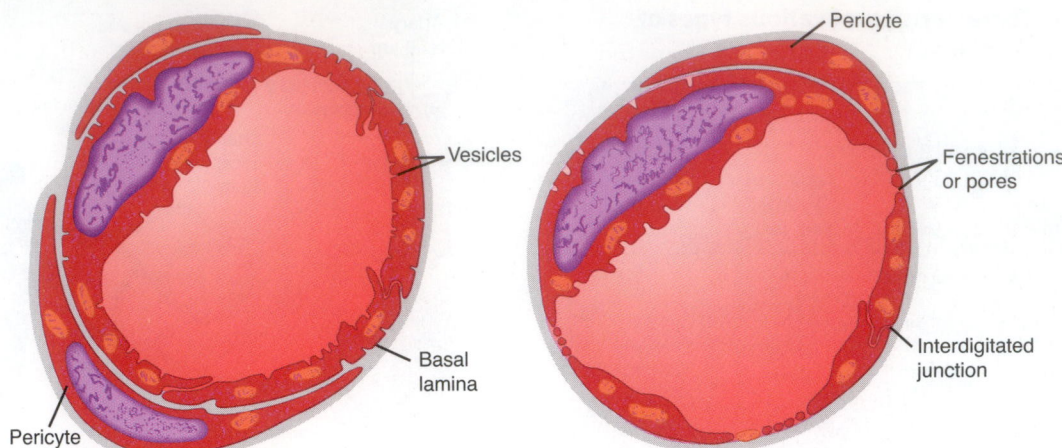

FIGURE 32–19 **Cross-sections of capillaries. Left:** Type of capillary found in muscle. **Right:** Fenestrated type of capillary. (Reproduced with permission from Fawcett DW: *Bloom and Fawcett, Textbook of Histology,* 11th ed. Saunders, 1986.)

as 3000 nm. They therefore permit the passage of large molecules, including plasma proteins, which is important for hepatic function (see Chapter 29). The permeabilities of capillaries in various parts of the body, expressed in terms of their hydraulic conductivity, are summarized in Table 32–10.

Capillaries and postcapillary venules have **pericytes** around their endothelial cells (Figure 32–19). These cells have long processes that wrap around the vessels. They are contractile and release a wide variety of vasoactive agents. They also synthesize and release constituents of the basement membrane and extracellular matrix. One of their physiologic functions appears to be regulation of flow through the junctions between endothelial cells, particularly in the presence of inflammation. They are closely related to the mesangial cells in the renal glomeruli (see Chapter 38).

TABLE 32–10 Hydraulic conductivity of capillaries in various parts of the body.

Organ	Conductivity[a]	Type of Endothelium
Brain (excluding circumventricular organs)	3	
Skin	100	Continuous
Skeletal muscle	250	
Lung	340	
Heart	860	
Gastrointestinal tract (intestinal mucosa)	13,000	
		Fenestrated
Glomerulus in kidney	15,000	

[a]Units of conductivity are 10^-13 cm^3 s^{-1} $dyne^{-1}$.

Data courtesy of JN Diana.

LYMPHATICS

The lymphatics serve to collect plasma and its constituents that have exuded from the capillaries into the interstitial space. They drain from the body tissues via a system of vessels that coalesce and eventually enter the right and left subclavian veins at their junctions with the respective internal jugular veins. The lymph vessels contain valves and regularly traverse lymph nodes along their course. The ultrastructure of the small lymph vessels differs from that of the capillaries in several details: No fenestrations are visible in the lymphatic endothelium; very little if any basal lamina is present under the endothelium; and the junctions between endothelial cells are open, with no tight intercellular connections.

ARTERIOVENOUS ANASTOMOSES

In the fingers, palms, and ear lobes, short channels connect arterioles to venules, bypassing the capillaries. These **arteriovenous (A-V) anastomoses,** or **shunts,** have thick, muscular walls and are abundantly innervated, presumably by vasoconstrictor nerve fibers.

VENULES & VEINS

The walls of the venules are only slightly thicker than those of the capillaries. The walls of the veins are also thin and easily distended. They contain relatively little smooth muscle, but considerable venoconstriction is produced by activity in the noradrenergic nerves to the veins and by circulating vasoconstrictors such as endothelins. Variations in venous tone are important in circulatory adjustments.

The intima of the limb veins is folded at intervals to form **venous valves** that prevent retrograde flow. The way these valves function was first demonstrated by William Harvey in the 17th century. No valves are present in the very small veins, the great veins, or the veins from the brain and viscera.

ANGIOGENESIS

When tissues grow, blood vessels must proliferate if the tissue is to maintain a normal blood supply. Therefore, angiogenesis, the formation of new blood vessels, is important during fetal life and growth to adulthood. It is also important in adulthood for processes such as wound healing, formation of the corpus luteum after ovulation, and formation of new endometrium after menstruation. Abnormally, it is important in tumor growth; if tumors do not develop a blood supply, they do not grow.

During embryonic development, a network of leaky capillaries is formed in tissues from angioblasts: this process is sometimes called **vasculogenesis.** Vessels then branch off from nearby vessels, hook up with the capillaries, and provide them with smooth muscle, which brings about their maturation. Angiogenesis in adults is presumably similar, but consists of new vessel formation by branching from pre-existing vessels rather than from angioblasts.

Many factors are involved in angiogenesis. A key compound is the protein growth factor **vascular endothelial growth factor (VEGF).** This factor exists in multiple isoforms, and there are three VEGF receptors that are tyrosine kinases, which also cooperate with nonkinase co-receptors known as neuropilins in some cell types. VEGF appears to be primarily responsible for vasculogenesis, whereas the budding of vessels that connect to the immature capillary network is regulated by other as yet unidentified factors. Some of the VEGF isoforms and receptors may play a more prominent role in the formation of lymphatic vessels (**lymphangiogenesis**) than that of blood vessels.

The actions of VEGF and related factors have received considerable attention in recent years because of the requirement for angiogenesis in the development of tumors. VEGF antagonists and other angiogenesis inhibitors have now entered clinical practice as adjunctive therapies for many malignancies and are being tested as first line therapies as well.

BIOPHYSICAL CONSIDERATIONS FOR CIRCULATORY PHYSIOLOGY

FLOW, PRESSURE, & RESISTANCE

Blood always flows, of course, from areas of high pressure to areas of low pressure, except in certain situations when momentum transiently sustains flow (see Figure 31–3). The relationship between mean flow, mean pressure, and resistance in the blood vessels is analogous in a general way to the relationship between the current, electromotive force, and resistance in an electrical circuit expressed in Ohm's law:

$$\text{Current (I)} = \frac{\text{Electromotive force (E)}}{\text{Resistance (R)}}$$

$$\text{Flow (F)} = \frac{\text{Pressure (P)}}{\text{Resistance (R)}}$$

Flow in any portion of the vascular system is equal to the **effective perfusion pressure** in that portion divided by the **resistance.** The effective perfusion pressure is the mean intraluminal pressure at the arterial end minus the mean pressure at the venous end. The units of resistance (pressure divided by flow) are dynes/cm^5. To avoid dealing with such complex units, resistance in the cardiovascular system is sometimes expressed in **R units,** which are obtained by dividing pressure in mm Hg by flow in mL/s (see also Table 34–1). Thus, for example, when the mean aortic pressure is 90 mm Hg and the left ventricular output is 90 mL/s, the total peripheral resistance is

$$\frac{90 \text{ mm Hg}}{90 \text{ mL/s}} = 1 \text{ R unit}$$

METHODS FOR MEASURING BLOOD FLOW

Blood flow can be measured by cannulating a blood vessel, but this has obvious limitations. Various noninvasive devices have therefore been developed to measure flow. Most commonly, blood velocity can be measured with **Doppler flow meters.** Ultrasonic waves are sent into a vessel diagonally, and the waves reflected from the red and white blood cells are picked up by a downstream sensor. The frequency of the reflected waves is higher by an amount that is proportionate to the rate of flow toward the sensor because of the Doppler effect.

Indirect methods for measuring the blood flow of various organs in humans include adaptations of the Fick and indicator dilution techniques described in Chapter 31. One example is the use of the Kety N_2O method for measuring cerebral blood flow (see Chapter 34). Another is determination of the renal blood flow by measuring the clearance of *para*-aminohippuric acid (see Chapter 38). A considerable amount of data on blood flow in the extremities has been obtained by **plethysmography** (Figure 32–20). The forearm, for example, is sealed in a watertight chamber (**plethysmograph**). Changes in the volume of the forearm, reflecting changes in the amount of blood and interstitial fluid it contains, displace the water, and this displacement is measured with a volume recorder. When the venous drainage of the forearm is occluded, the rate of increase in the volume of the forearm is a function of the arterial blood flow (**venous occlusion plethysmography**).

APPLICABILITY OF PHYSICAL PRINCIPLES TO FLOW IN BLOOD VESSELS

Physical principles and equations that describe the behavior of perfect fluids in rigid tubes have often been used indiscriminately to explain the behavior of blood in blood vessels. Blood vessels are not rigid tubes, and the blood is not a perfect fluid but a two-phase system of liquid and cells. Therefore, the behavior of the circulation deviates, sometimes markedly, from that predicted by these principles. However, the physical

FIGURE 32–20   Plethysmography.

principles are of value when used as an aid to understanding what goes on in the body.

LAMINAR FLOW

The flow of blood in straight blood vessels, like the flow of liquids in narrow rigid tubes, is normally **laminar.** Within the blood vessels, an infinitely thin layer of blood in contact with the wall of the vessel does not move. The next layer within the vessel has a low velocity, the next a higher velocity, and so forth, velocity being greatest in the center of the stream (Figure 32–21). Laminar flow occurs at velocities up to a certain **critical velocity.** At or above this velocity, flow is turbulent. Laminar flow is silent, but turbulent flow creates sounds.

The probability of turbulence is also related to the diameter of the vessel and the viscosity of the blood. This probability can be expressed by the ratio of inertial to viscous forces as follows:

$$Re = \frac{\rho D \dot{V}}{\eta}$$

where Re is the Reynolds number, named for the man who described the relationship; ρ is the density of the fluid; D is the diameter of the tube under consideration; V is the velocity of the flow; and η is the viscosity of the fluid. The higher the value of Re, the greater the probability of turbulence. When D is in cm, V is in cm/s^{-1}, and η is in poise; flow is usually not turbulent if Re is less than 2000. When Re is more than 3000, turbulence is almost always present. Laminar flow can be disturbed at the branching points of arteries, and the resulting turbulence may increase the likelihood that atherosclerotic plaques will be deposited. Constriction of an artery likewise increases the velocity of blood flow through the constriction, producing turbulence and sound beyond the constriction (Figure 32–22). Examples are bruits heard over arteries constricted by atherosclerotic plaques and the sounds of Korotkoff heard when measuring blood pressure (see below).

In humans, the critical velocity is sometimes exceeded in the ascending aorta at the peak of systolic ejection, but it is usually exceeded only when an artery is constricted. Turbulence occurs more frequently in anemia because the viscosity of the blood is lower. This may be the explanation of the systolic murmurs that are common in anemia.

SHEAR STRESS & GENE ACTIVATION

Flowing blood creates a force on the endothelium that is parallel to the long axis of the vessel. This **shear stress** (γ) is proportionate to viscosity (η) times the shear rate (dy/dr), which

FIGURE 32–21   **Diagram of the velocities of concentric laminas of a viscous fluid flowing in a tube, illustrating the parabolic distribution of velocities.**

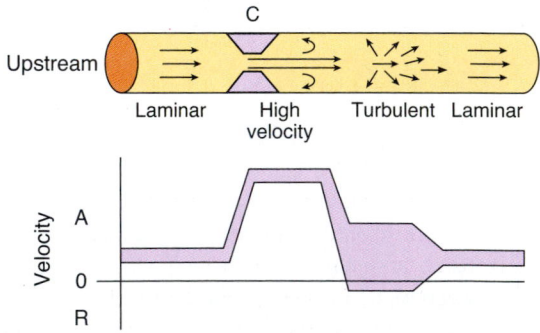

FIGURE 32–22   **Top:** Effect of constriction (C) on the profile of velocities in a blood vessel. The arrows indicate direction of velocity components, and their length is proportionate to their magnitude. **Bottom:** Range of velocities at each point along the vessel. In the area of turbulence, there are many different anterograde (A) and some retrograde (R) velocities.   (Modified and reproduced with permission from Richards KE: Doppler echocardiography in diagnosis and quantification of vascular disease. Mod Concepts Cardiovasc Dis 1987;56:43. By permission of the American Heart Association.)

TABLE 32–11 Genes in human, bovine, and rabbit endothelial cells that are affected by shear stress, and transcription factors involved.[a]

Gene	Transcription Factors
Endothelin 1	AP-1
VCAM-1	AP-1, NF-κB
ACE	SSRE, AP-1, Egr-1
Tissue factor	SP1, Egr-1
TM	AP-1
PDGF-α	SSRE, Egr-1
PDGF-β	SSRE
ICAM-1	SSRE, AP-1, NF-κB
TGF-β	SSRE, AP-1, NF-κB
Egr-1	SREs
c-fos	SSRE
c-jun	SSRE, AP-1
NOS 3	SSRE, AP-1, NF-κB
MCP-1	SSRE, AP-1, NF-κB

[a]Acronyms are expanded in the Appendix.

Modified from Braddock M et al: Fluid shear stress modulation of gene expression in endothelial cells. News Physiol Sci 1998;13:241.

is the rate at which the axial velocity increases from the vessel wall toward the lumen.

$$\gamma = \eta \, (dy/dr)$$

Change in shear stress and other physical variables, such as cyclic strain and stretch, produce marked changes in the expression of genes by endothelial cells. The genes that are activated include those that produce growth factors, integrins, and related molecules (Table 32–11).

AVERAGE VELOCITY

When considering flow in a system of tubes, it is important to distinguish between velocity, which is displacement per unit time (eg, cm/s), and flow, which is volume per unit time (eg, cm³/s). Velocity (V) is proportional to flow (Q) divided by the area of the conduit (A):

$$\dot{V} = \frac{Q}{A}$$

Therefore, $Q = A \times V$, and if flow stays constant, velocity increases in direct proportion to any decrease in A (Figure 32–22).

The average velocity of fluid movement at any point in a system of tubes in parallel is inversely proportional to the *total*

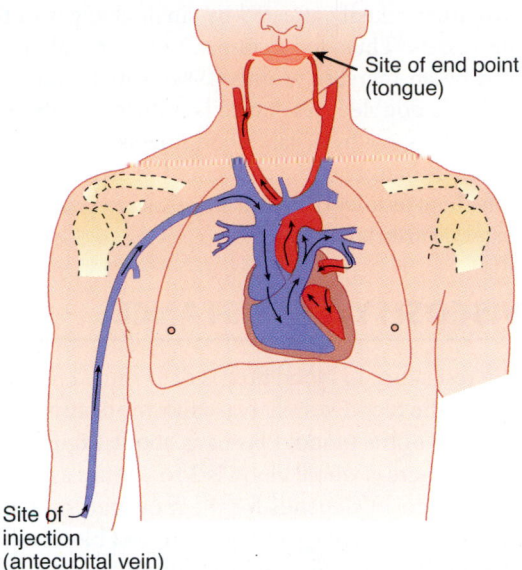

FIGURE 32–23 Pathway traversed by the injected material when the arm-to-tongue circulation time is measured.

cross-sectional area at that point. Therefore, the average velocity of the blood is high in the aorta, declines steadily in the smaller vessels, and is lowest in the capillaries, which have 1000 times the *total* cross-sectional area of the aorta (Table 32–9). The average velocity of blood flow increases again as the blood enters the veins and is relatively high in the vena cava, although not so high as in the aorta. Clinically, the velocity of the circulation can be measured by injecting a bile salt preparation into an arm vein and timing the first appearance of the bitter taste it produces (Figure 32–23). The average normal arm-to-tongue **circulation time** is 15 s.

POISEUILLE–HAGEN FORMULA

The relationship between the flow in a long narrow tube, the viscosity of the fluid, and the radius of the tube is expressed mathematically in the **Poiseuille–Hagen formula:**

$$F = (P_A - P_B) \times \left(\frac{\pi}{8}\right) \times \left(\frac{1}{\eta}\right) \times \left(\frac{r^4}{L}\right)$$

where

F = flow

$P_A - P_B$ = pressure difference between two ends of the tube

η = viscosity

r = radius of tube

L = length of tube

Because flow is equal to pressure difference divided by resistance (R),

$$R = \frac{8\eta L}{\pi r^4}$$

Because flow varies directly and resistance inversely with the fourth power of the radius, blood flow and resistance in vivo are markedly affected by small changes in the caliber of the vessels. Thus, for example, flow through a vessel is doubled by an increase of only 19% in its radius; and when the radius is doubled, resistance is reduced to 6% of its previous value. This is why organ blood flow is so effectively regulated by small changes in the caliber of the arterioles and why variations in arteriolar diameter have such a pronounced effect on systemic arterial pressure.

VISCOSITY & RESISTANCE

The resistance to blood flow is determined not only by the radius of the blood vessels (**vascular hindrance**) but also by the viscosity of the blood. Plasma is about 1.8 times as viscous as water, whereas whole blood is 3 to 4 times as viscous as water. Thus, viscosity depends for the most part on the **hematocrit,** that is, the percentage of the volume of blood occupied by red blood cells. The effect of viscosity in vivo deviates from that predicted by the Poiseuille–Hagen formula. In large vessels, increases in hematocrit cause appreciable increases in viscosity. However, in vessels smaller than 100 μm in diameter—that is, in arterioles, capillaries, and venules—the viscosity change per unit change in hematocrit is much less than it is in large-bore vessels. This is due to a difference in the nature of flow through the small vessels. Therefore, the net change in viscosity per unit change in hematocrit is considerably smaller in the body than it is in vitro (Figure 32–24). This is why hematocrit changes have relatively little effect on the peripheral resistance except when the changes are large. In severe polycythemia, the increase in resistance does increase the work of the heart. Con-

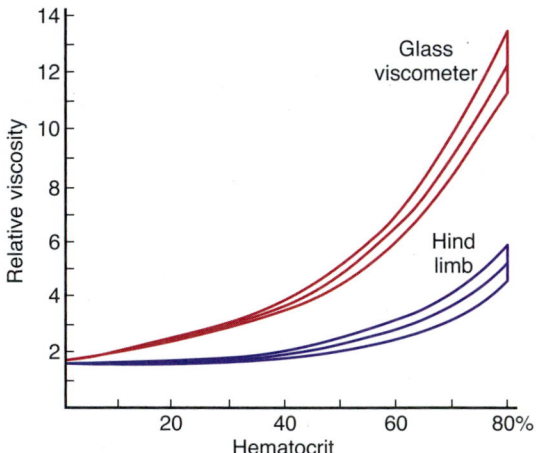

FIGURE 32–24 **Effect of changes in hematocrit on the relative viscosity of blood measured in a glass viscometer and in the hind leg of a dog.** In each case, the middle line represents the mean and the upper and lower lines the standard deviation. (Reproduced with permission from Whittaker SRF, Winton FR: The apparent viscosity of blood flowing in the isolated hind limb of the dog, and its variation with corpuscular concentration. J Physiol [Lond] 1933;78:338.)

FIGURE 32–25 Relation of pressure to flow in thin-walled blood vessels.

versely, in marked anemia, peripheral resistance is decreased, in part because of the decline in viscosity. Of course, the decrease in hemoglobin decreases the O_2-carrying ability of the blood, but the improved blood flow due to the decrease in viscosity partially compensates for this.

Viscosity is also affected by the composition of the plasma and the resistance of the cells to deformation. Clinically significant increases in viscosity are seen in diseases in which plasma proteins such as the immunoglobulins are markedly elevated as well as when red blood cells are abnormally rigid (hereditary spherocytosis).

CRITICAL CLOSING PRESSURE

In rigid tubes, the relationship between pressure and flow of homogeneous fluids is linear, but in thin-walled blood vessels in vivo it is not. When the pressure in a small blood vessel is reduced, a point is reached at which no blood flows, even though the pressure is not zero (Figure 32–25). This is because the vessels are surrounded by tissues that exert a small but definite pressure on them, and when the intraluminal pressure falls below the tissue pressure, they collapse. In inactive tissues, for example, the pressure in many capillaries is low because the precapillary sphincters and metarterioles are constricted, and many of these capillaries are collapsed. The pressure at which flow ceases is called the **critical closing pressure.**

LAW OF LAPLACE

The relationship between distending pressure and tension is shown diagrammatically in Figure 32–26. It is perhaps surprising that structures as thin-walled and delicate as the capillaries are not more prone to rupture. The principal reason for their relative invulnerability is their small diameter. The protective effect of small size in this case is an example of the operation of the **law of Laplace,** an important physical principle with several other applications in physiology. This law states that tension in the wall of a cylinder (T) is equal to the product of the transmural pressure (P) and the radius (r) divided by the wall thickness (w):

$$T = Pr/w$$

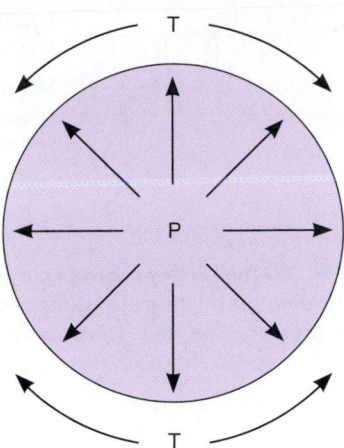

FIGURE 32–26 Relationship between distending pressure (P) and wall tension (T) in a hollow viscus.

The **transmural pressure** is the pressure inside the cylinder minus the pressure outside the cylinder, but because tissue pressure in the body is low, it can generally be ignored and P equated to the pressure inside the viscus. In a thin-walled viscus, w is very small and it too can be ignored, but it becomes a significant factor in vessels such as arteries. Therefore, in a thin-walled viscus, P = T divided by the two principal radii of curvature of the viscus:

$$P = T\left(\frac{1}{r_1} + \frac{1}{r_2}\right)$$

In a sphere, $r_1 = r_2$, so

$$P = \frac{2T}{r}$$

In a cylinder such as a blood vessel, one radius is infinite, so

$$P = \frac{T}{r}$$

Consequently, the smaller the radius of a blood vessel, the lower the tension in the wall necessary to balance the distending pressure. In the human aorta, for example, the tension at normal pressures is about 170,000 dynes/cm, and in the vena cava it is about 21,000 dynes/cm; but in the capillaries, it is approximately 16 dynes/cm.

The law of Laplace also makes clear a disadvantage faced by dilated hearts. When the radius of a cardiac chamber is increased, a greater tension must be developed in the myocardium to produce any given pressure; consequently, a dilated heart must do more work than a nondilated heart. In the lungs, the radii of curvature of the alveoli become smaller during expiration, and these structures would tend to collapse because of the pull of surface tension if the tension were not reduced by the surface-tension-lowering agent, surfactant (see Chapter 35). Another example of the operation of this law is seen in the urinary bladder (see Chapter 38).

RESISTANCE & CAPACITANCE VESSELS

In vivo, the veins are an important blood reservoir. Normally, they are partially collapsed and oval in cross-section. A large amount of blood can be added to the venous system before the veins become distended to the point where further increments in volume produce a large rise in venous pressure. The veins are therefore called **capacitance vessels.** The small arteries and arterioles are referred to as **resistance vessels** because they are the principal site of the peripheral resistance (see below).

At rest, at least 50% of the circulating blood volume is in the systemic veins, 12% is in the heart cavities, and 18% is in the low-pressure pulmonary circulation. Only 2% is in the aorta, 8% in the arteries, 1% in the arterioles, and 5% in the capillaries (Table 32–9). When extra blood is administered by transfusion, less than 1% of it is distributed in the arterial system (the **"high-pressure system"**), and all the rest is found in the systemic veins, pulmonary circulation, and heart chambers other than the left ventricle (the **"low-pressure system"**).

ARTERIAL & ARTERIOLAR CIRCULATION

The pressure and velocities of the blood in the various parts of the systemic circulation are summarized in Figure 32–27. The general relationships in the pulmonary circulation are similar, but the pressure in the pulmonary artery is 25/10 mm Hg or less.

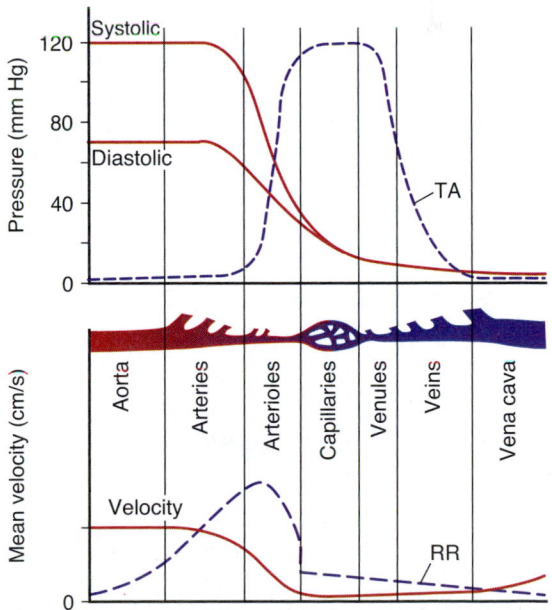

FIGURE 32–27 Diagram of the changes in pressure and velocity as blood flows through the systemic circulation. TA, total cross-sectional area of the vessels, which increases from 4.5 cm^2 in the aorta to 4500 cm^2 in the capillaries (Table 32–9). RR, relative resistance, which is highest in the arterioles.

VELOCITY & FLOW OF BLOOD

Although the mean velocity of the blood in the proximal portion of the aorta is 40 cm/s, the flow is phasic, and velocity ranges from 120 cm/s during systole to a negative value at the time of the transient backflow before the aortic valve closes in diastole. In the distal portions of the aorta and in the large arteries, velocity is also much greater in systole than it is in diastole. However, the vessels are elastic, and forward flow is continuous because of the recoil during diastole of the vessel walls that have been stretched during systole (Figure 32–28). Pulsatile flow appears to maintain optimal function of the tissues, apparently via distinct effects on gene transcription. If an organ is perfused with a pump that delivers a nonpulsatile flow, inflammatory markers are produced, there is a gradual rise in vascular resistance, and ultimately tissue perfusion fails.

ARTERIAL PRESSURE

The pressure in the aorta and in the brachial and other large arteries in a young adult human rises to a peak value (**systolic pressure**) of about 120 mm Hg during each heart cycle and falls to a minimum (**diastolic pressure**) of about 70 mm Hg. The arterial pressure is conventionally written as systolic pressure over diastolic pressure, for example, 120/70 mm Hg. One millimeter of mercury equals 0.133 kPa, so in SI units (see Appendix) this value is 16.0/9.3 kPa. The **pulse pressure,** the difference between the systolic and diastolic pressures, is normally about 50

FIGURE 32–28   **Changes in blood flow during the cardiac cycle in the dog.** Diastole is followed by systole starting at 0.1 and again at 0.5 s. Flow patterns in humans are similar. Ao, aorta; PA, pulmonary artery; PV, pulmonary vein; IVC, inferior vena cava; RA, renal artery.   (Reproduced with permission from Milnor WR: Pulsatile blood flow. N Engl J Med 1972;287:27.)

FIGURE 32–29   **Brachial artery pressure curve of a normal young human, showing the relation of systolic and diastolic pressure to mean pressure.** The shaded area above the mean pressure line is equal to the shaded area below it.

mm Hg. The **mean pressure** is the average pressure throughout the cardiac cycle. Because systole is shorter than diastole, the mean pressure is slightly less than the value halfway between systolic and diastolic pressure. It can actually be determined only by integrating the area of the pressure curve (Figure 32–29); however, as an approximation, mean pressure equals the diastolic pressure plus one-third of the pulse pressure.

The pressure falls very slightly in the large- and medium-sized arteries because their resistance to flow is small, but it falls rapidly in the small arteries and arterioles, which are the main sites of the peripheral resistance against which the heart pumps. The mean pressure at the end of the arterioles is 30 to 38 mm Hg. Pulse pressure also declines rapidly to about 5 mm Hg at the ends of the arterioles (Figure 32–26). The magnitude of the pressure drop along the arterioles varies considerably depending on whether they are constricted or dilated.

EFFECT OF GRAVITY

The pressures in Figure 32–28 are those in blood vessels at heart level. The pressure in any vessel below heart level is increased and that in any vessel above heart level is decreased by the effect of gravity. The magnitude of the gravitational effect is 0.77 mm Hg/cm of vertical distance above or below the heart at the density of normal blood. Thus, in an adult human in the upright position, when the mean arterial pressure at heart level is 100 mm Hg, the mean pressure in a large artery in the head (50 cm above the heart) is 62 mm Hg (100 – [0.77 × 50]) and the pressure in a large artery in the foot (105 cm below the heart) is 180 mm Hg (100 + [0.77 × 105]). The effect of gravity on venous pressure is similar (Figure 32–30).

METHODS OF MEASURING BLOOD PRESSURE

If a cannula is inserted into an artery, the arterial pressure can be measured directly with a mercury manometer or a suitably calibrated strain gauge. When an artery is tied off beyond the point at which the cannula is inserted, an **end pressure** is recorded, flow in the artery is interrupted, and all the kinetic energy of flow is converted into pressure energy. If, alternatively,

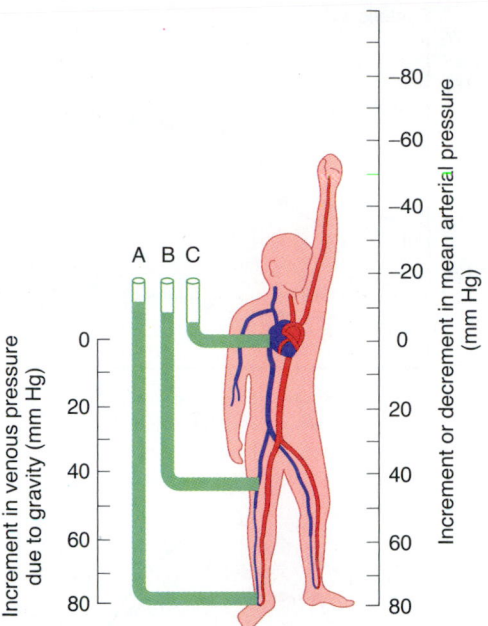

FIGURE 32–30   Effects of gravity on arterial and venous pressure. The scale on the right indicates the increment (or decrement) in mean pressure in a large artery at each level. The mean pressure in all large arteries is approximately 100 mm Hg when they are at the level of the left ventricle. The scale on the left indicates the increment in venous pressure at each level due to gravity. The manometers on the left of the figure indicate the height to which a column of blood in a tube would rise if connected to an ankle vein (A), the femoral vein (B), or the right atrium (C), with the subject in the standing position. The approximate pressures in these locations in the recumbent position; that is, when the ankle, thigh, and right atrium are at the same level, are A, 10 mm Hg; B, 7.5 mm Hg; and C, 4.6 mm Hg.

a T tube is inserted into a vessel and the pressure is measured in the side arm of the tube, the recorded **side pressure,** under conditions where pressure drop due to resistance is negligible, is lower than the end pressure by the kinetic energy of flow. This is because in a tube or a blood vessel the total energy—the sum of the kinetic energy of flow and the potential energy—is constant **(Bernoulli's principle).**

It is worth noting that the pressure drop in any segment of the arterial system is due both to resistance and to conversion of potential into kinetic energy. The pressure drop due to energy lost in overcoming resistance is irreversible, since the energy is dissipated as heat; but the pressure drop due to conversion of potential to kinetic energy as a vessel narrows is reversed when the vessel widens out again (Figure 32–31).

Bernoulli's principle also has a significant application in pathophysiology. According to the principle, the greater the velocity of flow in a vessel, the lower the lateral pressure distending its walls. When a vessel is narrowed, the velocity of flow in the narrowed portion increases and the distending pressure decreases. Therefore, when a vessel is narrowed by a pathologic process such as an atherosclerotic plaque, the lat-

eral pressure at the constriction is decreased and the narrowing tends to maintain itself.

AUSCULTATORY METHOD

The arterial blood pressure in humans is routinely measured by the **auscultatory method.** An inflatable cuff (**Riva–Rocci cuff**) attached to a mercury manometer (**sphygmomanometer**) is wrapped around the arm and a stethoscope is placed over the brachial artery at the elbow. The cuff is rapidly inflated until the pressure is well above the expected systolic pressure in the brachial artery. The artery is occluded by the cuff, and no sound is heard with the stethoscope. The pressure in the cuff is then lowered slowly. At the point at which systolic pressure in the artery just exceeds the cuff pressure, a spurt of blood passes through with each heartbeat and, synchronously with each beat, a tapping sound is heard below the cuff. The cuff pressure at which the sounds are first heard is the systolic pressure. As the cuff pressure is lowered further, the sounds become louder, then dull and muffled. These are the **sounds of Korotkoff.** Finally, in most individuals, they disappear. When direct and indirect blood pressure measurements are made simultaneously, the diastolic pressure in resting adults correlates best with the pressure at which the sound disappears. However, in adults after exercise and in children, the diastolic pressure correlates best with the pressure at which the sounds become muffled. This is also true in diseases such as hyperthyroidism and aortic insufficiency.

The sounds of Korotkoff are produced by turbulent flow in the brachial artery. When the artery is narrowed by the cuff, the velocity of flow through the constriction exceeds the **critical velocity** and turbulent flow results (Figure 32–22). At cuff pressures just below the systolic pressure, flow through the artery occurs only at the peak of systole, and the intermittent turbulence produces a tapping sound. As long as the pressure in the cuff is above the diastolic pressure in the artery, flow is

FIGURE 32–31   Bernoulli's principle. When fluid flows through the narrow portion of the tube, the kinetic energy of flow is increased as the velocity increases, and the potential energy is reduced. Consequently, the measured pressure (P) is lower than it would have been at that point if the tube had not been narrowed. The dashed line indicates what the pressure drop due to frictional forces would have been if the tube had been of uniform diameter.

interrupted at least during part of diastole, and the intermittent sounds have a staccato quality. When the cuff pressure is near the arterial diastolic pressure, the vessel is still constricted, but the turbulent flow is continuous. Continuous sounds have a muffled rather than a staccato quality.

NORMAL ARTERIAL BLOOD PRESSURE

The blood pressure in the brachial artery in young adults in the sitting position at rest is approximately 120/70 mm Hg. Because the arterial pressure is the product of the cardiac output and the peripheral resistance, it is affected by conditions that affect either or both of these factors. Emotion increases the cardiac output and peripheral resistance, and about 20% of hypertensive patients have blood pressures that are higher in the doctor's office than at home, going about their regular daily activities ("white coat hypertension"). Blood pressure normally falls up to 20 mm Hg during sleep. This fall is reduced or absent in hypertension.

There is general agreement that blood pressure rises with advancing age, but the magnitude of this rise is uncertain because hypertension is a common disease and its incidence increases with advancing age (see Clinical Box 32–4). Individuals who have systolic blood pressures < 120 mm Hg at age 50 to 60 and never develop clinical hypertension still have systolic pressures that rise throughout life (Figure 32–32). This rise may be the closest approximation to the rise in normal individuals. Individuals with mild hypertension that is untreated show a significantly more rapid rise in systolic pressure. In both groups, diastolic pressure also rises, but then starts to fall in middle age as the stiffness of arteries increases. Consequently, pulse pressure rises with advancing age.

It is interesting that systolic and diastolic blood pressures are lower in young women than in young men until age 55 to 65, after which they become comparable. Because there is a positive correlation between blood pressure and the incidence of heart attacks and strokes (see below), the lower blood pressure before menopause in women may be one reason that, on average, they live longer than men.

CAPILLARY CIRCULATION

At any one time, only 5% of the circulating blood is in the capillaries, but this 5% is in a sense the most important part of the blood volume because it is the only pool from which O_2 and nutrients can enter the interstitial fluid and into which CO_2 and waste products can enter the bloodstream. Exchange across the capillary walls is essential to the survival of the tissues.

METHODS OF STUDY

It is difficult to obtain accurate measurements of capillary pressures and flows. Capillary pressure has been estimated by determining the amount of external pressure necessary to

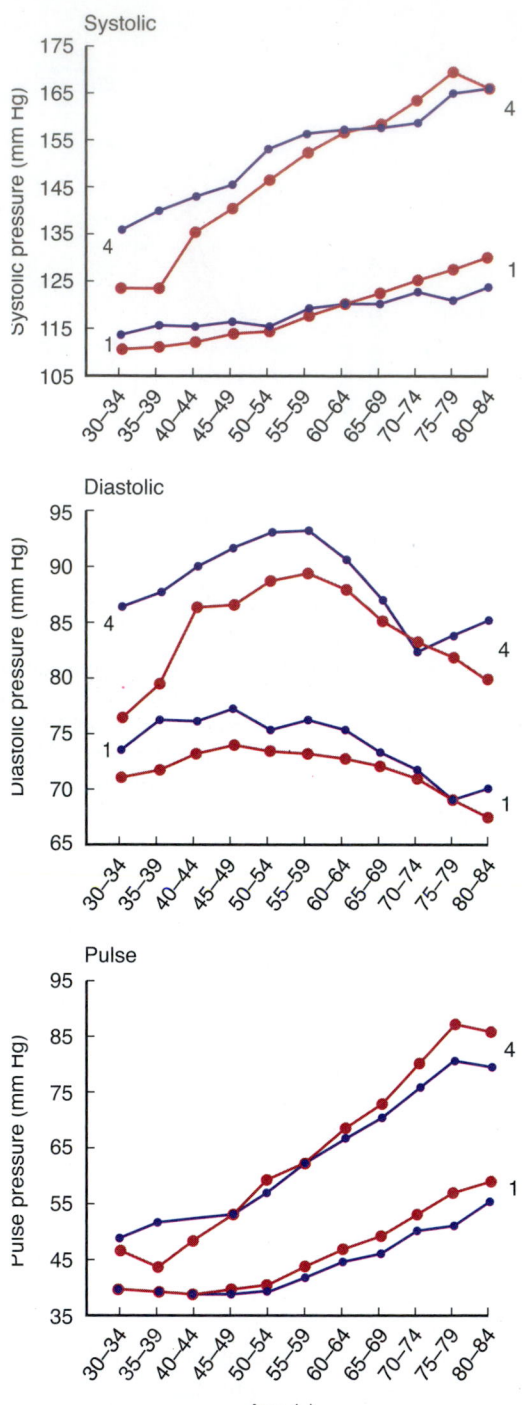

FIGURE 32–32   Effects of age and sex on arterial pressure components in humans. Data are from a large group of individuals who were studied every 2 y throughout their adult lives. Group 1: Individuals who had systolic blood pressures < 120 mm Hg at age 50 to 60. Group 4: Individuals who had systolic blood pressure ≥ 160 mm Hg at age 50 to 60, that is, individuals with mild, untreated hypertension. The red line shows the values for women, and the blue line shows the values for men.   (Modified and reproduced with permission from Franklin SS et al: Hemodynamic patterns of age-related changes in blood pressure: The Framingham Heart Study. Circulation 1997;96:308.)

Hypertension

Hypertension is a sustained elevation of the systemic arterial pressure. It is most commonly due to increased peripheral resistance and is a very common abnormality in humans. It can be produced by many diseases (Table 32–12) and causes a number of serious disorders. When the resistance against which the left ventricle must pump (afterload) is elevated for a long period, the cardiac muscle hypertrophies. The initial response is activation of immediate-early genes in the ventricular muscle, followed by activation of a series of genes involved in growth during fetal life. Left ventricular hypertrophy is associated with a poor prognosis. The total O_2 consumption of the heart, already increased by the work of expelling blood against a raised pressure (see Chapter 31), is increased further because there is more muscle. Therefore, any decrease in coronary blood flow has more serious consequences in hypertensive patients than it does in normal individuals, and degrees of coronary vessel narrowing that do not produce symptoms when the size of the heart is normal may produce myocardial infarction when the heart is enlarged.

The incidence of atherosclerosis increases in hypertension, and myocardial infarcts are common even when the heart is not enlarged. Eventually, the ability to compensate for the high peripheral resistance is exceeded, and the heart fails. Hypertensive individuals are also predisposed to thromboses of cerebral vessels and cerebral hemorrhage. An additional complication is renal failure. However, the incidence of heart failure, strokes, and renal failure can be markedly reduced by active treatment of hypertension, even when the hypertension is relatively mild. In about 88% of patients with elevated blood pressure, the cause of the hypertension is unknown, and they are said to have **essential hypertension.** At present, essential hypertension is treatable but not curable. Effective lowering of the blood pressure can be produced by drugs that block α-adrenergic receptors, either in the periphery or in the central nervous system; drugs that block β-adrenergic receptors; drugs that inhibit the activity of angiotensin-converting enzyme; and calcium channel blockers that relax vascular smooth muscle. Essential hypertension is probably polygenic in origin, and environmental factors are also involved.

In other, less common forms of hypertension, the cause is known. A review of these is helpful because it emphasizes ways disordered physiology can lead to disease. Pathology that compromises the renal blood supply leads to renal hypertension, as does narrowing (coarctation) of the thoracic aorta, which both increases renin secretion and increases peripheral resistance. Pheochromocytomas, adrenal medullary tumors that secrete norepinephrine and epinephrine, can cause sporadic or sustained hypertension (see Chapter 22). Estrogens increase angiotensinogen secretion, and contraceptive pills containing large amounts of estrogen cause hypertension (pill hypertension) on this basis (see Chapter 25). Increased secretion of aldosterone or other mineralocorticoids causes renal Na^+ retention, which leads to hypertension. A primary increase in plasma mineralocorticoids inhibits renin secretion. For unknown reasons, plasma renin is also low in 10–15% of patients with essential hypertension and normal circulating mineralocortical levels (low renin hypertension). Mutations in a number of single genes are also known to cause hypertension. These cases of monogenic hypertension are rare but informative. One of these is glucocorticoid-remediable aldosteronism (GRA), in which a hybrid gene encodes an adrenocorticotropic hormone (ACTH)-sensitive aldosterone synthase, with resulting hyperaldosteronism (see Chapter 22). 11-β hydroxylase deficiency also causes hypertension by increasing the secretion of deoxycorticosterone (see Chapter 22). Normal blood pressure is restored when ACTH secretion is inhibited by administering a glucocorticoid. Mutations that decrease 11-β hydroxysteroid dehydrogenase cause loss of specificity of the mineralocorticoid receptors (see Chapter 22) with stimulation of them by cortisol and, in pregnancy, by the elevated circulating levels of progesterone. Finally, mutations of the genes for ENaCs that reduce degradation of the β or γ subunits increase ENaC activity and lead to excess renal Na^+ retention and hypertension (Liddle syndrome; see Chapter 38).

---

occlude the capillaries or the amount of pressure necessary to make saline start to flow through a micropipette inserted so that its tip faces the arteriolar end of the capillary.

CAPILLARY PRESSURE & FLOW

Capillary pressures vary considerably, but typical values in human nail bed capillaries are 32 mm Hg at the arteriolar end and 15 mm Hg at the venous end. The pulse pressure is approximately 5 mm Hg at the arteriolar end and zero at the venous end. The capillaries are short, but blood moves slowly (about 0.07 cm/s) because the total cross-sectional area of the capillary bed is large. Transit time from the arteriolar to the venular end of an average-sized capillary is 1 to 2 s.

TABLE 32–12 Estimated frequency of various forms of hypertension in the general hypertensive population.

	Percentage of Population
Essential hypertension	88
Renal hypertension	
Renovascular	2
Parenchymal	3
Endocrine hypertension	
Primary aldosteronism	5
Cushing syndrome	0.1
Pheochromocytoma	0.1
Other adrenal forms	0.2
Estrogen treatment ("pill hypertension")	1
Miscellaneous (Liddle syndrome, coarctation of the aorta, etc)	0.6

Reproduced with permission from McPhee SJ, Lingappa V, Ganong WF: *Pathophysiology of Disease,* 4th ed. McGraw-Hill, 2003.

EQUILIBRATION WITH INTERSTITIAL FLUID

As noted above, the capillary wall is a thin membrane made up of endothelial cells. Substances pass through the junctions between endothelial cells and through fenestrations when they are present. Some also pass through the cells by vesicular transport.

The factors other than vesicular transport that are responsible for transport across the capillary wall are diffusion and filtration (see Chapter 1). Diffusion is quantitatively much more important. O_2 and glucose are in higher concentration in the bloodstream than in the interstitial fluid and diffuse into the interstitial fluid, whereas CO_2 diffuses in the opposite direction.

The rate of filtration at any point along a capillary depends on a balance of forces sometimes called the **Starling forces,** after the physiologist who first described their operation in detail. One of these forces is the **hydrostatic pressure gradient** (the hydrostatic pressure in the capillary minus the hydrostatic pressure of the interstitial fluid) at that point. The interstitial fluid pressure varies from one organ to another, and there is considerable evidence that it is subatmospheric (about –2 mm Hg) in subcutaneous tissue. It is, however, positive in the liver and kidneys and as high as 6 mm Hg in the brain. The other force is the **osmotic pressure gradient** across the capillary wall (colloid osmotic pressure of plasma minus colloid osmotic pressure of interstitial fluid). This component is directed inward.

Thus:

$$\text{Fluid movement} = k[(P_c - P_i) - (\pi_c - \pi_i)]$$

where

k = capillary filtration coefficient

P_c = capillary hydrostatic pressure

P_i = interstitial hydrostatic pressure

π_c = capillary colloid osmotic pressure

π_i = interstitial colloid osmotic pressure

π_i is usually negligible, so the osmotic pressure gradient $(\pi_c - \pi_i)$ usually equals the oncotic pressure. The capillary filtration coefficient takes into account, and is proportional to, the permeability of the capillary wall and the area available for filtration. The magnitude of the Starling forces along a typical muscle capillary is shown in Figure 32–33. Fluid moves into the interstitial space at the arteriolar end of the capillary and into the capillary at the venular end. In other capillaries, the balance of Starling forces may be different. For example, fluid moves out of almost the entire length of the capillaries in the renal glomeruli. On the other hand, fluid moves into the capillaries through almost their entire length in the intestines. About 24 L of fluid is filtered through the capillaries per day. This is about 0.3% of the cardiac output. About 85% of the filtered fluid is reabsorbed into the capillaries, and the remainder returns to the circulation via the lymphatics.

It is worth noting that small molecules often equilibrate with the tissues near the arteriolar end of each capillary. In this situation, total diffusion can be increased by increasing blood flow; that is, exchange is **flow-limited** (Figure 32–34). Conversely, transfer of substances that do not reach equilibrium with the tissues during their passage through the capillaries is said to be **diffusion-limited.**

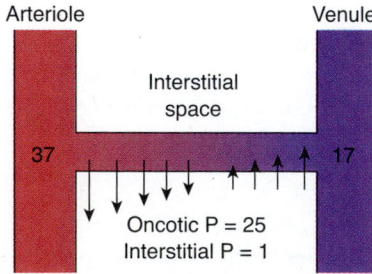

FIGURE 32–33 Schematic representation of pressure gradients across the wall of a muscle capillary. The numbers at the arteriolar and venular ends of the capillary are the hydrostatic pressures in mm Hg at these locations. The arrows indicate the approximate magnitude and direction of fluid movement. In this example, the pressure differential at the arteriolar end of the capillary is 11 mm Hg ([37 – 1] – 25) outward; at the opposite end, it is 9 mm Hg (25 – [17 – 1]) inward.

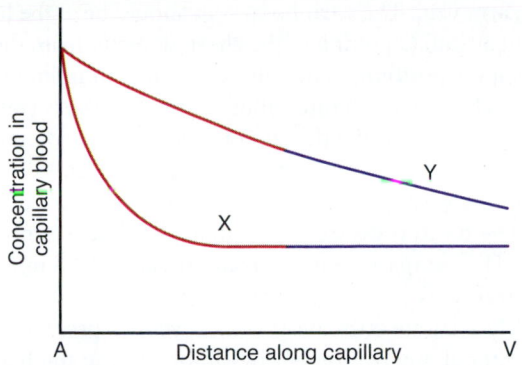

FIGURE 32–34 Flow-limited and diffusion-limited exchange across capillary walls. A and V indicate the arteriolar and venular ends of the capillary. Substance X equilibrates with the tissues (movement into the tissues equals movement out) well before the blood leaves the capillary, whereas substance Y does not equilibrate. If other factors stay constant, the amount of X entering the tissues can be increased only by increasing blood flow; that is, it is flow-limited. The movement of Y is diffusion-limited.

ACTIVE & INACTIVE CAPILLARIES

In resting tissues, most of the capillaries are collapsed. In active tissues, the metarterioles and the precapillary sphincters dilate. The intracapillary pressure rises, overcoming the critical closing pressure of the vessels, and blood flows through all of the capillaries. Relaxation of the smooth muscle of the metarterioles and precapillary sphincters is due to the action of vasodilator metabolites formed in active tissue (see Chapter 33).

After noxious stimulation, substance P released by the axon reflex (see Chapter 34) increases capillary permeability. Bradykinin and histamine also increase capillary permeability. When capillaries are stimulated mechanically, they empty (white reaction; see Chapter 34), probably due to contraction of the precapillary sphincters.

VENOUS CIRCULATION

Blood flows through the blood vessels, including the veins, primarily because of the pumping action of the heart. However, venous flow is aided by the heartbeat, the increase in the negative intrathoracic pressure during each inspiration, and contractions of skeletal muscles that compress the veins (**muscle pump**).

VENOUS PRESSURE & FLOW

The pressure in the venules is 12 to 18 mm Hg. It falls steadily in the larger veins to about 5.5 mm Hg in the great veins outside the thorax. The pressure in the great veins at their entrance into the right atrium (**central venous pressure**) averages 4.6 mm Hg, but fluctuates with respiration and heart action.

Peripheral venous pressure, like arterial pressure, is affected by gravity. It is increased by 0.77 mm Hg for each centimeter below the right atrium and decreased by a like amount for each centimeter above the right atrium the pressure is measured (Figure 32–30). Thus, on a proportional basis, gravity has a greater effect on venous than on arterial pressures.

When blood flows from the venules to the large veins, its average velocity increases as the total cross-sectional area of the vessels decreases. In the great veins, the velocity of blood is about one fourth that in the aorta, averaging about 10 cm/s.

THORACIC PUMP

During inspiration, the intrapleural pressure falls from –2.5 to –6 mm Hg. This negative pressure is transmitted to the great veins and, to a lesser extent, the atria, so that central venous pressure fluctuates from about 6 mm Hg during expiration to approximately 2 mm Hg during quiet inspiration. The drop in venous pressure during inspiration aids venous return. When the diaphragm descends during inspiration, intra-abdominal pressure rises, and this also squeezes blood toward the heart because backflow into the leg veins is prevented by the venous valves.

EFFECTS OF HEARTBEAT

The variations in atrial pressure are transmitted to the great veins, producing the **a, c,** and **v waves** of the venous pressure-pulse curve (see Chapter 31). Atrial pressure drops sharply during the ejection phase of ventricular systole because the atrioventricular valves are pulled downward, increasing the capacity of the atria. This action sucks blood into the atria from the great veins. The sucking of the blood into the atria during systole contributes appreciably to the venous return, especially at rapid heart rates.

Close to the heart, venous flow becomes pulsatile. When the heart rate is slow, two periods of peak flow are detectable, one during ventricular systole, due to pulling down of the atrioventricular valves, and one in early diastole, during the period of rapid ventricular filling (Figure 32–28).

MUSCLE PUMP

In the limbs, the veins are surrounded by skeletal muscles, and contraction of these muscles during activity compresses the veins. Pulsations of nearby arteries may also compress veins. Because the venous valves prevent reverse flow, the blood moves toward the heart. During quiet standing, when the full effect of gravity is manifest, venous pressure at the ankle is 85–90 mm Hg (Figure 32–30). Pooling of blood in the leg veins reduces venous return, with the result that cardiac output is reduced, sometimes to the point where fainting occurs. Rhythmic contractions of the leg muscles while the person is standing serve to lower the venous pressure in the legs to less than 30 mm Hg by propelling blood toward the heart. This heartward movement of the blood is decreased in patients with **varicose veins** because their valves are incompetent. These patients may develop stasis and ankle edema. However, even when the valves are incompetent, muscle contractions continue to produce a basic heartward movement of the blood because the resistance of the larger veins in the direction of the heart is less than the resistance of the small vessels away from the heart.

VENOUS PRESSURE IN THE HEAD

In the upright position, the venous pressure in the parts of the body above the heart is decreased by the force of gravity. The neck veins collapse above the point where the venous pressure is close to zero. However, the dural sinuses have rigid walls and cannot collapse. The pressure in them in the standing or sitting position is therefore subatmospheric. The magnitude of the negative pressure is proportional to the vertical distance above the top of the collapsed neck veins, and in the superior sagittal sinus may be as much as –10 mm Hg. This fact must be kept in mind by neurosurgeons. Neurosurgical procedures are sometimes performed with the patient seated. If one of the sinuses is opened during such a procedure it sucks air, causing **air embolism.**

AIR EMBOLISM

Because air, unlike fluid, is compressible, its presence in the circulation has serious consequences. The forward movement of the blood depends on the fact that blood is incompressible. Large amounts of air fill the heart and effectively stop the circulation, causing sudden death because most of the air is compressed by the contracting ventricles rather than propelled into the arteries. Small amounts of air are swept through the heart with the blood, but the bubbles lodge in the small blood vessels. The surface capillarity of the bubbles markedly increases the resistance to blood flow, and flow is reduced or abolished. Blockage of small vessels in the brain leads to serious and even fatal neurologic abnormalities. Treatment with hyperbaric oxygen (see Chapter 37) is of value because the pressure reduces the size of the gas emboli. In experimental animals, the amount of air that produces fatal air embolism varies considerably, depending in part on the rate at which it enters the veins. Sometimes as much as 100 mL can be injected without ill effects, whereas at other times as little as 5 mL is lethal.

MEASURING VENOUS PRESSURE

Central venous pressure can be measured directly by inserting a catheter into the thoracic great veins. **Peripheral venous pressure** correlates well with central venous pressure in most conditions. To measure peripheral venous pressure, a needle attached to a manometer containing sterile saline is inserted into an arm vein. The peripheral vein should be at the level of the right atrium (a point half the chest diameter from the back in the supine position). The values obtained in millimeters of saline can be converted into millimeters of mercury (mm Hg) by dividing by 13.6 (the density of mercury). The amount by which peripheral venous pressure exceeds central venous pressure increases with the distance from the heart along the veins. The mean pressure in the antecubital vein is normally 7.1 mm Hg, compared with a mean pressure of 4.6 mm Hg in the central veins.

A fairly accurate estimate of central venous pressure can be made without any equipment by simply noting the height to which the external jugular veins are distended when the subject lies with the head slightly above the heart. The vertical distance between the right atrium and the place the vein collapses (the place where the pressure in it is zero) is the venous pressure in mm of blood.

Central venous pressure is decreased during negative pressure breathing and shock. It is increased by positive pressure breathing, straining, expansion of the blood volume, and heart failure. In advanced congestive heart failure or obstruction of the superior vena cava, the pressure in the antecubital vein may reach values of 20 mm Hg or more.

LYMPHATIC CIRCULATION & INTERSTITIAL FLUID VOLUME

LYMPHATIC CIRCULATION

Fluid efflux normally exceeds influx across the capillary walls, but the extra fluid enters the lymphatics and drains through them back into the blood. This keeps the interstitial fluid pressure from rising and promotes the turnover of tissue fluid. The normal 24-h lymph flow is 2 to 4 L.

Lymphatic vessels can be divided into two types: initial lymphatics and collecting lymphatics (Figure 32–35). The former lack valves and smooth muscle in their walls, and they are found in regions such as the intestine or skeletal muscle. Tissue fluid appears to enter them through loose junctions between the endothelial cells that form their walls. The fluid in them apparently is massaged by muscle contractions of the organs and contraction of arterioles and venules, with which they are often associated. They drain into the collecting lymphatics, which have valves and smooth muscle in their walls and contract in a peristaltic fashion, propelling the lymph along the vessels. Flow in the collecting lymphatics is further aided by movements of skeletal muscle, the negative intrathoracic pressure during inspiration, and the suction effect of high-velocity flow of blood in the veins in which the lymphatics terminate. However, the contractions are the principal factor propelling the lymph.

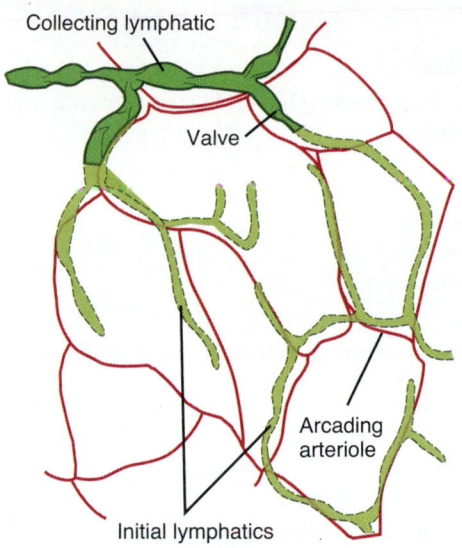

FIGURE 32–35 **Initial lymphatics draining into collecting lymphatics in the mesentery.** Note the close association with arcading arterioles, indicated by the single red lines. (Reproduced with permission from Schmid Schönbein GW, Zeifach BW: Fluid pump mechanisms in initial lymphatics. News Physiol Sci 1994;9:67.)

TABLE 32–13 Causes of increased interstitial fluid volume and edema.

Increased filtration pressure
Venular constriction
Increased venous pressure (heart failure, incompetent valves, venous obstruction, increased total ECF volume, effect of gravity, etc)
Decreased osmotic pressure gradient across capillary
Decreased plasma protein level
Accumulation of osmotically active substances in interstitial space
Increased capillary permeability
Substance P
Histamine and related substances
Kinins, etc
Inadequate lymph flow

OTHER FUNCTIONS OF THE LYMPHATIC SYSTEM

Appreciable quantities of protein enter the interstitial fluid in the liver and intestine, and smaller quantities enter from the blood in other tissues. The macromolecules enter the lymphatics, presumably at the junctions between the endothelial cells, and the proteins are returned to the bloodstream via the lymphatics. The amount of protein returned in this fashion in 1 d is equal to 25–50% of the total circulating plasma protein. The transport of absorbed long-chain fatty acids and cholesterol from the intestine via the lymphatics has been discussed in Chapter 27.

INTERSTITIAL FLUID VOLUME

The amount of fluid in the interstitial spaces depends on the capillary pressure, the interstitial fluid pressure, the oncotic pressure, the capillary filtration coefficient, the number of active capillaries, the lymph flow, and the total extracellular fluid (ECF) volume. The ratio of precapillary to postcapillary venular resistance is also important. Precapillary constriction lowers filtration pressure, whereas postcapillary constriction raises it. Changes in any of these variables lead to changes in the volume of interstitial fluid. Factors promoting an increase in this volume are summarized in Table 32–13. **Edema** is the accumulation of interstitial fluid in abnormally large amounts.

In active tissues, capillary pressure rises, often to the point where it exceeds the oncotic pressure throughout the length of the capillary. In addition, osmotically active metabolites may temporarily accumulate in the interstitial fluid because they cannot be washed away as rapidly as they are formed. To the extent that they accumulate, they exert an osmotic effect that decreases the magnitude of the osmotic gradient due to the oncotic pressure. The amount of fluid leaving the capillaries is therefore markedly increased and the amount entering them reduced. Lymph flow is increased, decreasing the degree to which the fluid would otherwise accumulate, but exercising muscle, for example, still increases in volume by as much as 25%.

Interstitial fluid tends to accumulate in dependent parts because of the effect of gravity. In the upright position, the capillaries in the legs are protected from the high arterial pressure by the arterioles, but the high venous pressure is transmitted to them through the venules. Skeletal muscle contractions keep the venous pressure low by pumping blood toward the heart (see above) when the individual moves about; however, if one stands still for long periods, fluid accumulates and edema eventually develops. The ankles also swell during long trips when travelers sit for prolonged periods with their feet in a dependent position. Venous obstruction may contribute to the edema in these situations.

Whenever there is abnormal retention of salt in the body, water is also retained. The salt and water are distributed throughout the ECF, and since the interstitial fluid volume is therefore increased, there is a predisposition to edema. Salt and water retention is a factor in the edema seen in heart failure, nephrosis, and cirrhosis, but there are also variations in the mechanisms that govern fluid movement across the capillary walls in these diseases. In congestive heart failure, for example, venous pressure is usually elevated, with a consequent elevation in capillary pressure. In cirrhosis of the liver, oncotic pressure is low because hepatic synthesis of plasma proteins is depressed; and in nephrosis, oncotic pressure is low because large amounts of protein are lost in the urine.

Another cause of edema is inadequate lymphatic drainage. Edema caused by lymphatic obstruction is called **lymphedema,** and the edema fluid has a high protein content. If it persists, it causes a chronic inflammatory condition that leads to fibrosis of the interstitial tissue. One cause of lymphedema is radical mastectomy, during which removal of the axillary lymph nodes leads to reduced lymph drainage. In filariasis, parasitic worms migrate into the lymphatics and obstruct them. Fluid accumulation plus tissue reaction lead in time to massive swelling, usually of the legs or scrotum (**elephantiasis**).

CHAPTER SUMMARY

- Blood consists of a suspension of red blood cells (erythrocytes), white blood cells, and platelets in a protein-rich fluid known as plasma.
- Blood cells arise in the bone marrow and are subject to regular renewal; the majority of plasma proteins are synthesized by the liver.
- Hemoglobin, stored in red blood cells, transports oxygen to peripheral tissues. Fetal hemoglobin is specialized to facilitate diffusion of oxygen from mother to fetus during development. Mutated forms of hemoglobin lead to red cell abnormalities and anemia.
- Complex oligosaccharide structures, specific to groups of individuals, form the basis of the ABO blood group system. AB blood group oligosaccharides, as well as other blood group molecules, can trigger the production of antibodies in naïve individuals following inappropriate transfusions, with potentially serious consequences due to erythrocyte agglutination.
- Blood flows from the heart to arteries and arterioles, thence to capillaries, and eventually to venules and veins and back to the heart. Each segment of the vasculature has specific contractile properties and regulatory mechanisms that subserve physiologic function. Physical principles of pressure, wall tension, and vessel caliber govern the flow of blood through each segment of the circulation.
- Transfer of oxygen and nutrients from the blood to tissues, as well as collection of metabolic wastes, occurs exclusively in the capillary beds.
- Fluid also leaves the circulation across the walls of capillaries. Some is reabsorbed; the remainder enters the lymphatic system, which eventually drains into the subclavian veins to return fluid to the bloodstream.
- Hypertension is an increase in mean blood pressure that is usually chronic and is common in humans. Hypertension can result in serious health consequences if left untreated. The majority of hypertension is of unknown cause, but several gene mutations underlie rare forms of the disease and are informative about mechanisms that control the dynamics of the circulatory system and its integration with other organs.

MULTIPLE-CHOICE QUESTIONS

For all questions, select the single best answer unless otherwise directed.

1. Which of the following has the highest *total* cross-sectional area in the body?
   A) arteries
   B) arterioles
   C) capillaries
   D) venules
   E) veins

2. Lymph flow from the foot is
   A) increased when an individual rises from the supine to the standing position.
   B) increased by massaging the foot.
   C) increased when capillary permeability is decreased.
   D) decreased when the valves of the leg veins are incompetent.
   E) decreased by exercise.

3. The pressure in a capillary in skeletal muscle is 35 mm Hg at the arteriolar end and 14 mm Hg at the venular end. The interstitial pressure is 0 mm Hg. The colloid osmotic pressure is 25 mm Hg in the capillary and 1 mm Hg in the interstitium. The net force producing fluid movement across the capillary wall at its arteriolar end is
   A) 3 mm Hg out of the capillary.
   B) 3 mm Hg into the capillary.
   C) 10 mm Hg out of the capillary.
   D) 11 mm Hg out of the capillary.
   E) 11 mm Hg into the capillary.

4. The velocity of blood flow
   A) is higher in the capillaries than the arterioles.
   B) is higher in the veins than in the venules.
   C) is higher in the veins than the arteries.
   D) falls to zero in the descending aorta during diastole.
   E) is reduced in a constricted area of a blood vessel.

5. When the radius of the resistance vessels is increased, which of the following is increased?
   A) systolic blood pressure
   B) diastolic blood pressure
   C) viscosity of the blood
   D) hematocrit
   E) capillary blood flow

6. When the viscosity of the blood is increased, which of the following is increased?
   A) mean blood pressure
   B) radius of the resistance vessels
   C) radius of the capacitance vessels
   D) central venous pressure
   E) capillary blood flow

7. A pharmacologist discovers a drug that stimulates the production of VEGF receptors. He is excited because his drug might be of value in the treatment of
   A) coronary artery disease.
   B) cancer.
   C) emphysema.
   D) diabetes insipidus.
   E) dysmenorrhea.

8. Why is the dilator response to injected acetylcholine changed to a constrictor response when the endothelium is damaged?

 A) More Na^+ is generated.

 B) More bradykinin is generated.

 C) The damage lowers the pH of the remaining layers of the artery.

 D) The damage augments the production of endothelin by the endothelium.

 E) The damage interferes with the production of NO by the endothelium.

CHAPTER RESOURCES

de Montalembert M: Management of sickle cell disease. Brit Med J 2008;337:626.

Miller JL: Signaled expression of fetal hemoglobin during development. Transfusion 2005;45:1229.

Perrotta S, Gallagher PG, Mohandas N: Hereditary spherocytosis. Lancet 2008;372:1411.

Semenza GL: Vasculogenesis, angiogenesis, and arteriogenesis: Mechanisms of blood vessel formation and remodeling. J Cell Biochem 2007;102:840.

Cardiovascular Regulatory Mechanisms

INTRODUCTION

In humans and other mammals, multiple cardiovascular regulatory mechanisms have evolved. These mechanisms increase the blood supply to active tissues and increase or decrease heat loss from the body by redistributing the blood. In the face of challenges such as hemorrhage, they maintain the blood flow to the heart and brain. When the challenge faced is severe, flow to these vital organs is maintained at the expense of the circulation to the rest of the body.

Circulatory adjustments are effected by altering the output of the pump (the heart), changing the diameter of the resistance vessels (primarily the arterioles), or altering the amount of blood pooled in the capacitance vessels (the veins). Regulation of cardiac output is discussed in Chapter 31. The caliber of the arterioles is adjusted in part by autoregulation (Table 33–1). It is also increased in active tissues by locally produced vasodilator metabolites, is affected by substances secreted by the endothelium, and is regulated systemically by circulating vasoactive substances and the nerves that innervate the arterioles. The caliber of the capacitance vessels is also affected by circulating vasoactive substances and by vasomotor nerves. The systemic regulatory mechanisms synergize with the local mechanisms and adjust vascular responses throughout the body.

The terms **vasoconstriction** and **vasodilation** are generally used to refer to constriction and dilation of the resistance vessels. Changes in the caliber of the veins are referred to specifically as **venoconstriction** or **venodilation.**

NEURAL CONTROL OF THE CARDIOVASCULAR SYSTEM

NEURAL REGULATORY MECHANISMS

Although the arterioles and the other resistance vessels are most densely innervated, all blood vessels except capillaries and venules contain smooth muscle and receive motor nerve fibers from the sympathetic division of the autonomic nervous system. The fibers to the resistance vessels regulate tissue blood flow and arterial pressure. The fibers to the venous capacitance vessels vary the volume of blood "stored" in the veins. The innervation of most veins is sparse, but the splanchnic veins are well innervated. Venoconstriction is produced by stimuli that also activate the vasoconstrictor nerves to the arterioles. The resultant decrease in venous capacity increases venous return, shifting blood to the arterial side of the circulation.

INNERVATION OF THE BLOOD VESSELS

Sympathetic noradrenergic fibers end on blood vessels in all parts of the body to mediate vasoconstriction. In addition to their vasoconstrictor innervation, resistance vessels in skeletal muscles are innervated by vasodilator fibers, which, although they travel with the sympathetic nerves, are cholinergic (**sympathetic cholinergic vasodilator system**). There is no tonic activity in the vasodilator fibers, but the vasoconstrictor fibers to most vascular beds have some tonic activity. When the sympathetic nerves are cut (**sympathectomy**), the blood vessels dilate. In most tissues, vasodilation is produced by decreasing the rate of tonic discharge in the vasoconstrictor nerves, although in skeletal muscles it can also be produced by activating the sympathetic cholinergic vasodilator system (Table 33–1).

CARDIAC INNERVATION

Impulses in the sympathetic nerves to the heart increase the cardiac rate (**chronotropic effect**), rate of transmission in the cardiac conductive tissue (**dromotropic effect**), and the force of contraction (**inotropic effect**). They also inhibit the effects of vagal parasympathetic stimulation, probably by release of neuropeptide Y, which is a cotransmitter in the sympathetic endings. Impulses in vagal fibers decrease heart rate. A moderate amount of tonic discharge takes place in the cardiac sympathetic nerves at rest, but there is a good deal of tonic vagal discharge (**vagal tone**) in humans and other large animals. After the administration of parasympatholytic drugs such as atropine, the heart rate in humans increases from 70, its normal resting value, to 150 to 180 beats/min because the sympathetic tone is unopposed. In humans in whom both noradrenergic and cholinergic systems are blocked, the heart rate is approximately 100 beats/min.

CARDIOVASCULAR CONTROL

The cardiovascular system is under neural influences coming from several parts of the brain (see Figure 17–6), which in turn receive feedback from sensory receptors in the vasculature (eg, baroreceptors). A simplified model of the feedback control circuit is shown in Figure 33–1. An increase in neural output from the brain stem to sympathetic nerves leads to a decrease in blood vessel diameter (arteriolar constriction) and increases in stroke volume and heart rate, which contribute to a rise in blood pressure. This in turn causes an increase in baroreceptor activity, which signals the brain stem to reduce the neural output to sympathetic nerves.

Venoconstriction and a decrease in the stores of blood in the venous reservoirs usually accompany increases in arteriolar constriction, although changes in the capacitance vessels

TABLE 33–1 Summary of factors affecting the caliber of the arterioles.

Constriction	Dilation
Local factors	
Decreased local temperature	Increased CO_2 and decreased O_2
Autoregulation	Increased K^+, adenosine, lactate, etc
	Decreased local pH
	Increased local temperature
Endothelial products	
Endothelin-1	NO
Locally released platelet serotonin	Kinins
Thromboxane A_2	Prostacyclin
Circulating hormones	
Epinephrine (except in skeletal muscle and liver)	Epinephrine in skeletal muscle and liver
Norepinephrine	CGRPα
AVP	Substance P
Angiotensin II	Histamine
Circulating Na^+-K^+ ATPase inhibitor	ANP
Neuropeptide Y	VIP
Neural factors	
Increased discharge of sympathetic nerves	Decreased discharge of sympathetic nerves
	Activation of sympathetic cholinergic vasodilator nerves to skeletal muscle

do not always parallel changes in the resistance vessels. In the presence of an increase in sympathetic nerve activity to the heart and vasculature, there is usually an associated decrease in the activity of vagal fibers to the heart. Conversely, a decrease in sympathetic activity causes vasodilation, a fall in blood pressure, and an increase in the storage of blood in the venous reservoirs. There is usually a concomitant decrease in heart rate, but this is mostly due to stimulation of the vagal innervation of the heart.

MEDULLARY CONTROL OF THE CARDIOVASCULAR SYSTEM

One of the major sources of excitatory input to sympathetic nerves controlling the vasculature is neurons located near the pial surface of the medulla in the rostral ventrolateral medulla

FIGURE 33–1 **Feedback control of blood pressure.** Brain stem excitatory input to sympathetic nerves to the heart and vasculature increases heart rate and stroke volume and reduces vessel diameter. Together these increase blood pressure, which activates the baroreceptor reflex to reduce the activity in the brain stem.

(RVLM; Figure 33–2). This region is sometimes called a vasomotor area. The axons of RVLM neurons course dorsally and medially and then descend in the lateral column of the spinal cord to the thoracolumbar intermediolateral gray column (IML). They contain phenylethanolamine-*N*-methyltransferase (PNMT; see Chapter 7), but it appears that the excitatory transmitter they secrete is glutamate rather than epinephrine. Neurovascular compression of the RVLM has been linked to some cases of essential hypertension in humans (see Clinical Box 33–1).

The activity of RVLM neurons is determined by many factors (see Table 33–2). They include not only the very important fibers from arterial and venous baroreceptors, but also fibers from other parts of the nervous system and from the carotid and aortic chemoreceptors. In addition, some stimuli act directly on the vasomotor area.

There are descending tracts to the vasomotor area from the cerebral cortex (particularly the limbic cortex) that relay in the hypothalamus. These fibers are responsible for the blood

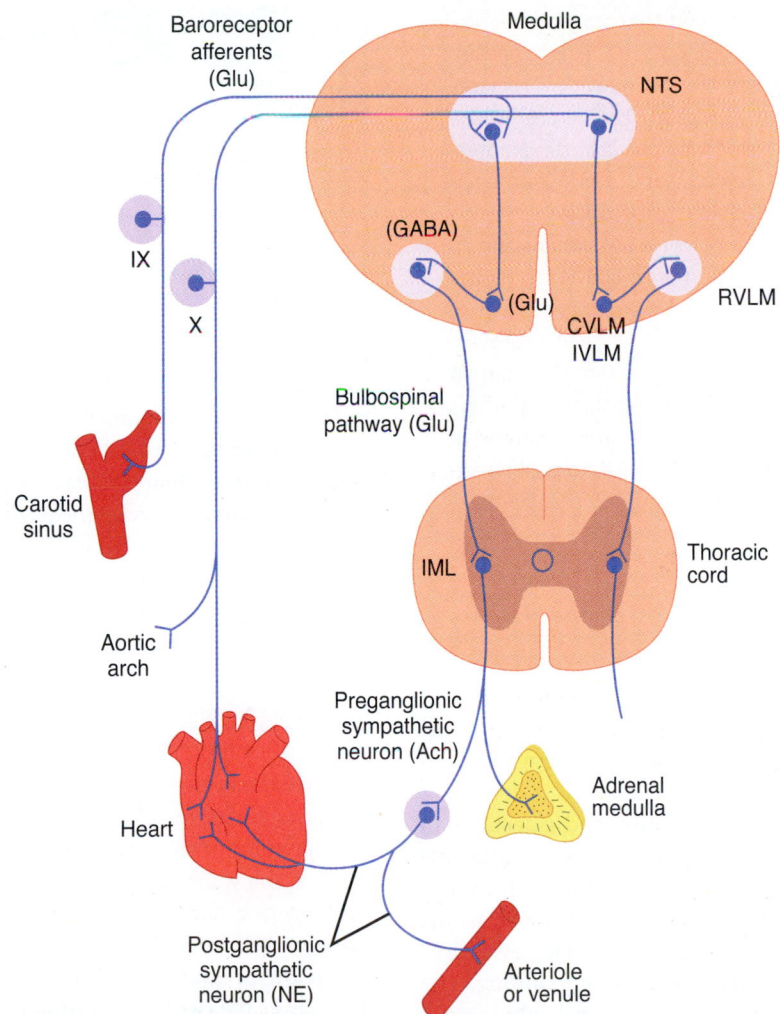

FIGURE 33–2 **Basic pathways involved in the medullary control of blood pressure.** The vagal efferent pathways that slow the heart are not shown. The putative neurotransmitters in the pathways are indicated in parentheses. Glu, glutamate; GABA, γ-aminobutyric acid; Ach, acetylcholine; NE, norepinephrine; IML, intermediolateral gray column; NTS, nucleus of the tractus solitarius; CVLM, IVLM, RVLM, caudal, intermediate, and rostral ventrolateral medulla; IX and X, glossopharyngeal and vagus nerves.

CLINICAL BOX 33–1

Essential Hypertension & Neurovascular Compression of the RVLM

In about 88% of patients with elevated blood pressure, the cause of the hypertension is unknown, and they are said to have **essential hypertension.** There are data available to support the view that **neurovascular compression** of the RVLM is associated with essential hypertension in some subjects. In the 1970s, Dr. Peter Jannetta, a neurosurgeon in Pittsburgh, PA, developed a technique for "microvascular decompression" of the medulla to treat trigeminal neuralgia and hemifacial spasm, which he attributed to pulsatile compression of the vertebral and posterior inferior cerebellar arteries impinging on the fifth and seventh cranial nerves. Moving the arteries away from the nerves led to reversal of the neurologic symptoms in many cases. Some of these patients were also hypertensive, and they showed reductions in blood pressure postoperatively. Later, a few human studies claimed that surgical decompression of the RVLM could sometimes relieve hypertension. There are several reports of patients with a schwannoma or meningioma lying close to the RVLM whose hypertension has been reversed by surgical decompression. Magnetic resonance angiography (MRA) has been used to compare the incidence of neurovascular compression in hypertensive and normotensive individuals and to correlate indices of sympathetic nerve activity with the presence or absence of compression. Some of these studies showed a higher incidence of coexistence of neurovascular compression with essential hypertension than in other forms of hypertension or normotension, but others showed the presences of a compression in normotensive subjects. On the other hand, there was a strong positive relationship between the presence of neurovascular compression and increased sympathetic activity.

TABLE 33–2 Factors affecting the activity of the RVLM.

Direct stimulation
CO_2
Hypoxia
Excitatory inputs
Cortex via hypothalamus
Mesencephalic periaqueductal gray
Brain stem reticular formation
Pain pathways
Somatic afferents (somatosympathetic reflex)
Carotid and aortic chemoreceptors
Inhibitory inputs
Cortex via hypothalamus
Caudal ventrolateral medulla
Caudal medullary raphé nuclei
Lung inflation afferents
Carotid, aortic, and cardiopulmonary baroreceptors

Unlike the vasculature, the heart is controlled by both sympathetic and parasympathetic (vagal) nerves. The medulla is also a major site of origin of excitatory input to cardiac vagal motor neurons in the nucleus ambiguus (Figure 33–3). Table 33–3 is a summary of conditions that affect the heart rate. In general, stimuli that increase the heart rate also increase blood pressure, whereas those that decrease the heart rate lower blood pressure. However, there are exceptions, such as the production of hypotension and tachycardia by stimulation of atrial stretch receptors and the production of hypertension and bradycardia by increased intracranial pressure.

BARORECEPTORS

The **baroreceptors** are stretch receptors in the walls of the heart and blood vessels. The **carotid sinus** and **aortic arch** receptors monitor the arterial circulation. Receptors are also located in the walls of the right and left atria at the entrance of the superior and inferior venae cavae and the pulmonary veins, as well as in the pulmonary circulation. These receptors in the low-pressure part of the circulation are referred to collectively as the **cardiopulmonary receptors.**

The carotid sinus is a small dilation of the internal carotid artery just above the bifurcation of the common carotid into external and internal carotid branches (Figure 33–4). Baroreceptors are located in this dilation. They are also found in the

pressure rise and tachycardia produced by emotions such as sexual excitement and anger. The connections between the hypothalamus and the vasomotor area are reciprocal, with afferents from the brain stem closing the loop.

Inflation of the lungs causes vasodilation and a decrease in blood pressure. This response is mediated via vagal afferents from the lungs that inhibit vasomotor discharge. Pain usually causes a rise in blood pressure via afferent impulses in the reticular formation converging in the RVLM. However, prolonged severe pain may cause vasodilation and fainting. The activity in afferents from exercising muscles probably exerts a similar pressor effect via pathway to the RVLM. The pressor response to stimulation of somatic afferent nerves is called the **somatosympathetic reflex.**

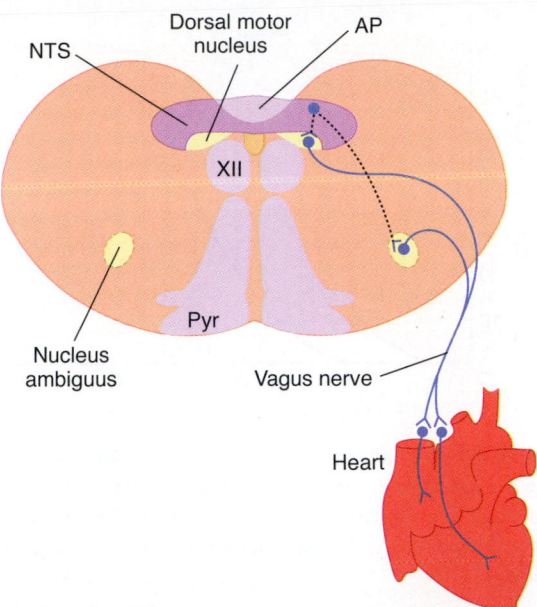

FIGURE 33–3 Basic pathways involved in the medullary control of heart rate by the vagus nerves. NTS neurons (dashed lines) project to and inhibit cardiac preganglionic parasympathetic neurons primarily in the nucleus ambiguus. Some are also located in the dorsal motor nucleus of the vagus; however, this nucleus primarily contains vagal motor neurons that project to the gastrointestinal tract. AP, area postrema; Pyr, pyramid; XII, hypoglossal nucleus.

TABLE 33–3 Factors affecting heart rate.

Heart rate accelerated by:
Decreased activity of arterial baroreceptors
Increased activity of atrial stretch receptors
Inspiration
Excitement
Anger
Most painful stimuli
Hypoxia
Exercise
Thyroid hormones
Fever
Heart rate slowed by:
Increased activity of arterial baroreceptors
Expiration
Fear
Grief
Stimulation of pain fibers in trigeminal nerve
Increased intracranial pressure

wall of the arch of the aorta. The receptors are located in the adventitia of the vessels. The afferent nerve fibers from the carotid sinus form a distinct branch of the glossopharyngeal nerve, the **carotid sinus nerve.** The fibers from the aortic arch form a branch of the vagus nerve, the **aortic depressor nerve.**

The baroreceptors are stimulated by distention of the structures in which they are located, and so they discharge at an increased rate when the pressure in these structures rises. Their afferent fibers pass via the glossopharyngeal and vagus nerves to the medulla. Most of them end in the nucleus of the tractus solitarius (NTS), and the excitatory transmitter they secrete is glutamate (Figure 33–2). Excitatory (glutamate) projections extend from the NTS to the caudal ventrolateral medulla (CVLM), where they stimulate γ-aminobutyrate (GABA)-secreting inhibitory neurons that project to the RVLM. Excitatory projections also extend from the NTS to the vagal motor neurons in the nucleus ambiguus and dorsal motor nucleus (Figure 33–3). Thus, increased baroreceptor discharge *inhibits* the tonic discharge of sympathetic nerves and *excites* the vagal innervation of the heart. These neural changes produce vasodilation, venodilation, a drop in blood pressure, bradycardia, and a decrease in cardiac output.

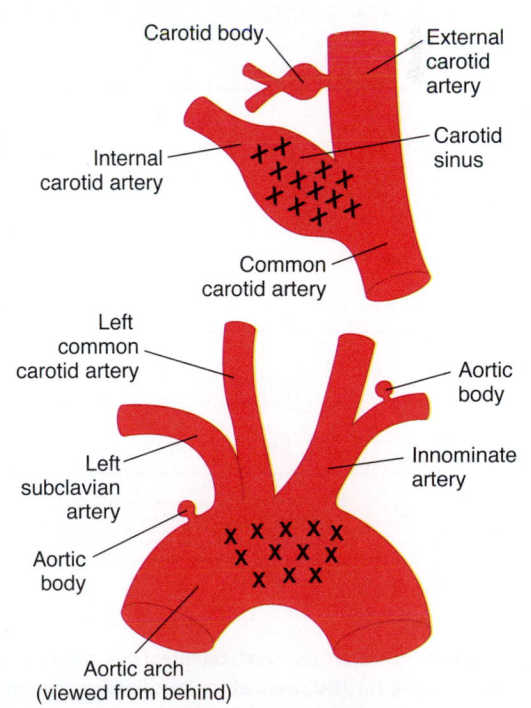

FIGURE 33–4 Baroreceptor areas in the carotid sinus and aortic arch. X, sites where receptors are located. The carotid and aortic bodies, which contain chemoreceptors, are also shown.

BARORECEPTOR NERVE ACTIVITY

Baroreceptors are more sensitive to pulsatile pressure than to constant pressure. A decline in pulse pressure without any change in mean pressure decreases the rate of baroreceptor discharge and provokes a rise in systemic blood pressure and tachycardia. At normal blood pressure levels (about 100 mm Hg mean pressure), a burst of action potentials appears in a single baroreceptor fiber during systole, but there are few action potentials in early diastole (Figure 33–5). At lower mean pressures, this phasic change in firing is even more dramatic with activity only occurring during systole. At these lower pressures, the overall firing rate is considerably reduced. The threshold for eliciting activity in the carotid sinus nerve is about 50 mm Hg; maximal activity occurs at about 200 mm Hg.

When one carotid sinus is isolated and perfused and the other baroreceptors are denervated, there is no discharge in the afferent fibers from the perfused sinus and no drop in the animal's arterial pressure or heart rate when the perfusion pressure is below 30 mm Hg (Figure 33–6). At carotid sinus perfusion pressures of 70–110 mm Hg, there is a near linear relationship between perfusion pressure and the fall in systemic blood pressure and heart rate. At perfusion pressures above 150 mm Hg there is no further increase in response, presumably because the rate of baroreceptor discharge and the degree of inhibition of sympathetic nerve activity are maximal.

From the foregoing discussion, it is apparent that the baroreceptors on the arterial side of the circulation, their afferent connections to the medullary cardiovascular areas, and the efferent pathways from these areas constitute a reflex feedback mechanism that operates to stabilize blood pressure and heart

FIGURE 33–6   **Fall in systemic blood pressure produced by raising the pressure in the isolated carotid sinus to various values. Solid line:** Response in a normal monkey. **Dashed line:** Response in a hypertensive monkey, demonstrating baroreceptor resetting (arrow).

rate. Any drop in systemic arterial pressure decreases the inhibitory discharge in the buffer nerves, and there is a compensatory rise in blood pressure and cardiac output. Any rise in pressure produces dilation of the arterioles and decreases cardiac output until the blood pressure returns to its previous normal level.

BARORECEPTOR RESETTING

In chronic hypertension, the baroreceptor reflex mechanism is "reset" to maintain an elevated rather than a normal blood pressure. In perfusion studies on hypertensive experimental animals, raising the pressure in the isolated carotid sinus lowers the elevated systemic pressure, and decreasing the perfusion pressure raises the elevated pressure (Figure 33–6). Little is known about how and why this occurs, but resetting occurs rapidly in experimental animals. It is also rapidly reversible, both in experimental animals and in clinical situations.

ROLE OF BARORECEPTORS IN SHORT-TERM CONTROL OF BLOOD PRESSURE

The changes in pulse rate and blood pressure that occur in humans on standing up or lying down are due for the most part to baroreceptor reflexes. The function of the receptors can be tested by monitoring changes in heart rate as a function of increasing arterial pressure during infusion of the α-adrenergic agonist phenylephrine. A normal response is shown in Figure 33–7; from a systolic pressure of about 120 to 150 mm Hg, there is a linear relation between pressure and lowering of the heart rate (greater RR interval). Baroreceptors are very important in short-term control of arterial pressure. Activation of the reflex allows for rapid adjustments in blood pressure in

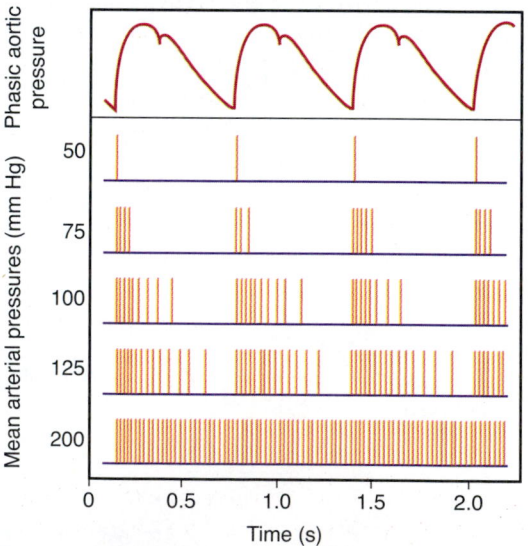

FIGURE 33–5   **Discharges (vertical lines) in a single afferent nerve fiber from the carotid sinus at various levels of mean arterial pressures, plotted against changes in aortic pressure with time.** Baroreceptors are very sensitive to changes in pulse pressure as shown by the record of phasic aortic pressure. (Reproduced with permission from Berne RM, Levy MN: *Cardiovascular Physiology*, 3rd ed. Mosby, 1977.)

FIGURE 33–7   Baroreflex-mediated lowering of the heart rate during infusion of phenylephrine in a human subject. Note that the values for the RR interval of the electrocardiogram, which are plotted on the vertical axis, are inversely proportionate to the heart rate. (Reproduced with permission from Kotrly K et al: Effects of fentanyl-diazepam-nitrous oxide anaesthesia on arterial baroreflex control of heart rate in man. Br J Anaesth 1986;58:406.)

response to abrupt changes in blood volume, cardiac output, or peripheral resistance during exercise.

Blood pressure initially rises dramatically after bilateral section of baroreceptor nerves or bilateral lesions of the NTS. However, after a period of time, mean blood pressure returns to near control levels, but there are huge fluctuations in pressure during the course of a day. Removal of the baroreceptor reflex prevents an individual from responding to stimuli that cause abrupt changes in blood volume, cardiac output, or peripheral resistance, including exercise and postural changes. A long-term change in blood pressure resulting from loss of baroreceptor reflex control is called **neurogenic hypertension.**

ATRIAL STRETCH RECEPTORS

The stretch receptors in the atria are of two types: those that discharge primarily during atrial systole (type A), and those that discharge primarily late in diastole, at the time of peak atrial filling (type B). The discharge of type B baroreceptors is increased when venous return is increased and decreased by positive-pressure breathing, indicating that these baroreceptors respond primarily to distention of the atrial walls. The reflex circulatory adjustments initiated by increased discharge from most if not all of these receptors include vasodilation and a fall in blood pressure. However, the heart rate is increased rather than decreased.

CARDIOPULMONARY RECEPTORS

Receptors in the endocardial surfaces of the ventricles are activated during ventricular distention. The response is a vagal bradycardia and hypotension, comparable to a baroreceptor reflex. Left ventricular stretch receptors may play a role in the maintenance of vagal tone that keeps the heart rate low at rest.

CLINICAL BOX 33-2

Cardiopulmonary Chemosensitive Receptors

For nearly 150 years, it has been known that activation of chemosensitive vagal C fibers in the cardiopulmonary region (eg, juxtacapillary region of alveoli, ventricles, atria, great veins, and pulmonary artery) causes profound bradycardia, hypotension, and a brief period of apnea followed by rapid shallow breathing. This response pattern is called the **Bezold–Jarisch reflex** and was named after the individuals who first reported these findings. This reflex can be elicited by a variety of substances including capsaicin, serotonin, phenylbiguanide, and veratridine in cats, rabbits, and rodents. Although originally viewed as a pharmacologic curiosity, there is a growing body of evidence supporting the view that the Bezold–Jarisch reflex is activated during certain pathophysiologic conditions. For example, this reflex may be activated during myocardial ischemia and reperfusion as a result of increased production of oxygen radicals and by agents used as radio-contrast for coronary angiography. This can contribute to the hypotension that is frequently a stubborn complication of this disease. Activation of cardiopulmonary chemosensitive receptors may also be part of a defense mechanism protecting individuals from toxic chemical hazards. Activation of cardiopulmonary reflexes may help reduce the amount of inspired pollutants that get absorbed into the blood, protecting vital organs from potential toxicity of these pollutants, and facilitating the elimination of the pollutants. Finally, the syndrome of cardiac slowing with hypotension **(vasovagal syncope)** has also been attributed to activation of the Bezold–Jarisch reflex. Vasovagal syncope can occur after prolonged upright posture that results in pooling of blood in the lower extremities and diminished intracardiac blood volume (also called **postural syncope**). This phenomenon is exaggerated if combined with dehydration. The resultant arterial hypotension is sensed in the carotid sinus baroreceptors, and afferent fibers from these receptors trigger autonomic signals that increase cardiac rate and contractility. However, pressure receptors in the wall of the left ventricle respond by sending signals that trigger paradoxical bradycardia and decreased contractility, resulting in sudden marked hypotension. The individual also feels lightheaded and may experience a brief episode of loss of consciousness.

Various chemicals are known to elicit reflexes due to activation of cardiopulmonary chemoreceptors and may play a role in various cardiovascular disorders (see Clinical Box 33–2).

VALSALVA MANEUVER

The function of the receptors can also be tested by monitoring the changes in pulse and blood pressure that occur in response

FIGURE 33–8 **Diagram of the response to straining (the Valsalva maneuver) in a normal man, recorded with a needle in the brachial artery.** Blood pressure rises at the onset of straining because increased intrathoracic pressure is added to the pressure of the blood in the aorta. It then falls because the high intrathoracic pressure compresses veins, decreasing venous return and cardiac output. (Courtesy of M McIlroy.)

to brief periods of straining (forced expiration against a closed glottis: the **Valsalva maneuver**). Valsalva maneuvers occur regularly during coughing, defecation, and heavy lifting. The blood pressure rises at the onset of straining (Figure 33–8) because the increase in intrathoracic pressure is added to the pressure of the blood in the aorta. It then falls because the high intrathoracic pressure compresses the veins, decreasing venous return and cardiac output. The decreases in arterial pressure and pulse pressure inhibit the baroreceptors, causing tachycardia and a rise in peripheral resistance. When the glottis is opened and the intrathoracic pressure returns to normal, cardiac output is restored but the peripheral vessels are constricted. The blood pressure therefore rises above normal, and this stimulates the baroreceptors, causing bradycardia and a drop in pressure to normal levels.

In sympathectomized patients, heart rate changes still occur because the baroreceptors and the vagi are intact. However, in patients with autonomic insufficiency, a syndrome in which autonomic function is widely disrupted, the heart rate changes are absent. For reasons that are still obscure, patients with primary hyperaldosteronism also fail to show the heart rate changes and the blood pressure rise when the intrathoracic pressure returns to normal. Their response to the Valsalva maneuver returns to normal after removal of the aldosterone-secreting tumor.

PERIPHERAL CHEMORECEPTOR REFLEX

Peripheral arterial chemoreceptors in the **carotid and aortic bodies** (Figure 33–2) have very high rates of blood flow. These receptors are primarily activated by a reduction in partial pressure of oxygen (PaO_2), but they also respond to an increase in the partial pressure of carbon dioxide ($PaCO_2$) and pH. Chemoreceptors exert their main effects on respiration; however, their activation also leads to vasoconstriction. Heart rate changes are variable and depend on various factors, including

changes in respiration. A direct effect of chemoreceptor activation is to increase vagal nerve activity. However, hypoxia also produces hyperpnea and increased catecholamine secretion from the adrenal medulla, both of which produce tachycardia and an increase in cardiac output. Hemorrhage that produces hypotension leads to chemoreceptor stimulation due to decreased blood flow to the chemoreceptors and consequent stagnant anoxia of these organs. Chemoreceptor discharge may also contribute to the production of **Mayer waves.** These should not be confused with **Traube–Hering waves,** which are fluctuations in blood pressure synchronized with respiration. The Mayer waves are slow, regular oscillations in arterial pressure that occur at the rate of about one per 20–40 s during hypotension. Under these conditions, hypoxia stimulates the chemoreceptors. The stimulation raises the blood pressure, which improves the blood flow in the receptor organs and eliminates the stimulus to the chemoreceptors, so that the pressure falls and a new cycle is initiated.

DIRECT EFFECTS ON THE RVLM

When intracranial pressure is increased, the blood supply to RVLM neurons is compromised, and the local hypoxia and hypercapnia increase their discharge. The resultant rise in systemic arterial pressure (**Cushing reflex**) tends to restore the blood flow to the medulla and over a considerable range, the blood pressure rise is proportional to the increase in intracranial pressure. The rise in blood pressure causes a reflex decrease in heart rate via the arterial baroreceptors. This is why bradycardia rather than tachycardia is characteristically seen in patients with increased intracranial pressure.

A rise in arterial PCO_2 stimulates the RVLM, but the direct peripheral effect of hypercapnia is vasodilation. Therefore, the peripheral and central actions tend to cancel each other out. Moderate hyperventilation, which significantly lowers the CO_2 tension of the blood, causes cutaneous and cerebral vasoconstriction in humans, but there is little change in blood pressure. Exposure to high concentrations of CO_2 is associated with marked cutaneous and cerebral vasodilation, but vasoconstriction occurs elsewhere and usually there is a slow rise in blood pressure.

LOCAL REGULATION

AUTOREGULATION

The capacity of tissues to regulate their own blood flow is referred to as **autoregulation.** Most vascular beds have an intrinsic capacity to compensate for moderate changes in perfusion pressure by changes in vascular resistance, so that blood flow remains relatively constant. This capacity is well developed in the kidneys (see Chapter 38), but it has also been observed in the mesentery, skeletal muscle, brain, liver, and myocardium. It is probably due in part to the intrinsic contractile response of smooth muscle to stretch **(myogenic theory of autoregulation).** As the pressure rises, the blood vessels are distended and the vascular smooth muscle fibers that surround the vessels contract. If it is postulated that the muscle responds to the tension in the vessel wall, this theory could explain the greater degree of contraction at higher pressures; the wall tension is proportional to the distending pressure times the radius of the vessel (law of Laplace; see Chapter 32), and the maintenance of a given wall tension as the pressure rises would require a decrease in radius. Vasodilator substances tend to accumulate in active tissues, and these "metabolites" also contribute to autoregulation **(metabolic theory of autoregulation).** When blood flow decreases, they accumulate and the vessels dilate; when blood flow increases, they tend to be washed away.

VASODILATOR METABOLITES

The metabolic changes that produce vasodilation include, in most tissues, decreases in O_2 tension and pH. These changes cause relaxation of the arterioles and precapillary sphincters. A local fall in O_2 tension, in particular, can initiate a program of vasodilatory gene expression secondary to production of hypoxia-inducible factor-1α (HIF-1α), a transcription factor with multiple targets. Increases in CO_2 tension and osmolality also dilate the vessels. The direct dilator action of CO_2 is most pronounced in the skin and brain. The neurally mediated vasoconstrictor effects of systemic as opposed to local hypoxia and hypercapnia have been discussed above. A rise in temperature exerts a direct vasodilator effect, and the temperature rise in active tissues (due to the heat of metabolism) may contribute to the vasodilation. K^+ is another substance that accumulates locally, and has demonstrated dilator activity secondary to the hyperpolarization of vascular smooth muscle cells. Lactate may also contribute to the dilation. In injured tissues, histamine released from damaged cells increases capillary permeability. Thus, it is probably responsible for some of the swelling in areas of inflammation. Adenosine may play a

vasodilator role in cardiac muscle but not in skeletal muscle. It also inhibits the release of norepinephrine.

LOCALIZED VASOCONSTRICTION

Injured arteries and arterioles constrict strongly. The constriction appears to be due in part to the local liberation of serotonin from platelets that stick to the vessel wall in the injured area. Injured veins also constrict.

A drop in tissue temperature causes vasoconstriction, and this local response to cold plays a part in temperature regulation (see Chapter 18).

SUBSTANCES SECRETED BY THE ENDOTHELIUM

ENDOTHELIAL CELLS

As noted in Chapter 32, the endothelial cells constitute a large and important tissue. They secrete many growth factors and vasoactive substances. The vasoactive substances include prostaglandins and thromboxanes, nitric oxide, and endothelins.

PROSTACYCLIN & THROMBOXANE A$_2$

Prostacyclin is produced by endothelial cells and thromboxane A_2 by platelets from their common precursor arachidonic acid via the cyclooxygenase pathway. Thromboxane A_2 promotes platelet aggregation and vasoconstriction, whereas prostacyclin inhibits platelet aggregation and promotes vasodilation. The balance between platelet thromboxane A_2 and prostacyclin fosters localized platelet aggregation and consequent clot formation (see Chapter 32) while preventing excessive extension of the clot and maintaining blood flow around it.

The thromboxane A_2–prostacyclin balance can be shifted toward prostacyclin by administration of low doses of aspirin. Aspirin produces irreversible inhibition of cyclooxygenase by acetylating a serine residue in its active site. Obviously, this reduces production of both thromboxane A_2 and prostacyclin. However, endothelial cells produce new cyclooxygenase in a matter of hours, whereas platelets cannot manufacture the enzyme, and the level rises only as new platelets enter the circulation. This is a slow process because platelets have a half-life of about 4 days. Therefore, administration of small amounts of aspirin for prolonged periods reduces clot formation and has been shown to be of value in preventing myocardial infarctions, unstable angina, transient ischemic attacks, and stroke.

NITRIC OXIDE

A chance observation two decades ago led to the discovery that the endothelium plays a key role in vasodilation. Many different stimuli act on the endothelial cells to produce **endo-thelium-derived relaxing factor (EDRF),** a substance that is now known to be **nitric oxide (NO).** NO is synthesized from arginine (Figure 33–9) in a reaction catalyzed by nitric oxide synthase (NO synthase, NOS). Three isoforms of NOS have been identified: NOS 1, found in the nervous system; NOS 2, found in macrophages and other immune cells; and NOS 3, found in endothelial cells. NOS 1 and NOS 3 are activated by agents that increase intracellular Ca^{2+} concentrations, including the vasodilators acetylcholine and bradykinin. The NOS in immune cells is not activated by Ca^{2+} but is induced by cytokines. The NO that is formed in the endothelium diffuses to smooth muscle cells, where it activates soluble guanylyl cyclase, producing cyclic 3,5-guanosine monophosphate (cGMP; see Figure 33–9), which in turn mediates the relaxation of vascular smooth muscle. NO is inactivated by hemoglobin.

Adenosine, atrial natriuretic peptide (ANP), and histamine via H_2 receptors produce relaxation of vascular smooth muscle that is independent of the endothelium. However, acetylcholine, histamine via H_1 receptors, bradykinin, vasoactive intestinal peptide (VIP), substance P, and some other polypeptides act via the endothelium, and various vasoconstrictors that act directly on vascular smooth muscle would produce much greater constriction if their effects were not limited by their ability simultaneously to cause release of NO. When flow to a tissue is suddenly increased by arteriolar dilation, the large arteries to the tissue also dilate. This flow-induced dilation is due to local release of NO. Products of platelet aggregation also cause release of NO, and the resulting vasodilation helps keep blood vessels with an intact endothelium patent. This is in contrast to injured blood vessels, where the endothelium is damaged at the site of injury and platelets therefore aggregate and produce vasoconstriction (see Chapter 32).

Further evidence for a physiologic role of NO is the observation that mice lacking NOS 3 are hypertensive. This suggests that tonic release of NO is necessary to maintain normal blood pressure.

NO is also involved in vascular remodeling and angiogenesis, and NO may be involved in the pathogenesis of atherosclerosis. It is interesting in this regard that some patients with heart transplants develop an accelerated form of atherosclerosis in the vessels of the transplant, and there is reason to believe that this is triggered by endothelial damage. Nitroglycerin and other nitrovasodilators that are of great value in the treatment of angina act by stimulating guanylyl cyclase in the same manner as NO.

Penile erection is also produced by release of NO, with consequent vasodilation and engorgement of the corpora cavernosa (see Chapter 25). This accounts for the efficacy of drugs such as Viagra, which slow the breakdown of cGMP.

OTHER FUNCTIONS OF NO

NO is present in the brain and, acting via cGMP, it is important in brain function (see Chapter 7). It is necessary for the antimicrobial and cytotoxic activity of various inflammatory cells, although the net effect of NO in inflammation and tissue injury depends on the amount and kinetics of release, which in turn may depend on the specific NOS isoform involved. In the gastrointestinal tract, it is important in the relaxation of smooth muscle. Other functions of NO are mentioned in other parts of this book.

CARBON MONOXIDE

The production of carbon monoxide (CO) from heme is shown in Figure 29–4. HO2, the enzyme that catalyzes the reaction, is present in cardiovascular tissues, and there is growing evidence that CO as well as NO produces local dilation in blood vessels. Interestingly, hydrogen sulfide is likewise emerging as a third gaseotransmitter that regulates vascular tone, although the relative roles of NO, CO, and H_2S have yet to be established.

ENDOTHELINS

Endothelial cells also produce **endothelin-1,** one of the most potent vasoconstrictor agents yet isolated. Endothelin-1 (ET-1), endothelin-2 (ET-2), and endothelin-3 (ET-3) are the members of a family of three similar 21-amino-acid polypeptides (Figure 33–10). Each is encoded by a different gene. The unique structure of the endothelins resembles that of the sarafotoxins, polypeptides found in the venom of a snake, the Israeli burrowing asp.

FIGURE 33–9 **Synthesis of NO from arginine in endothelial cells and its action via stimulation of soluble guanylyl cyclase and generation of cGMP to produce relaxation in vascular smooth muscle cells.** The endothelial form of nitric oxide synthase (NOS) is activated by increased intracellular Ca^{2+} concentration, and an increase is produced by acetylcholine (Ach), bradykinin, or shear stress acting on the cell membrane. Thiol, tetrahydrobiopterin, FAD, and FMN are requisite cofactors.

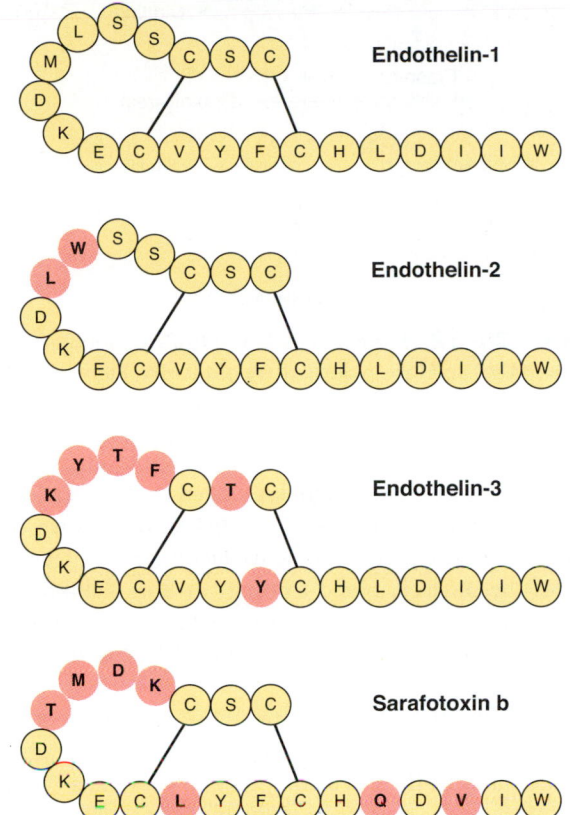

FIGURE 33–10 **Structure of human endothelins and one of the snake venom sarafotoxins.** The amino acid residues that differ from endothelin-1 are indicated in pink.

ENDOTHELIN-1

In endothelial cells, the product of the endothelin-1 gene is processed to a 39-amino-acid prohormone, **big endothelin-1,** which has about 1% of the activity of endothelin-1. The prohormone is cleaved at a tryptophan-valine (Trp-Val) bond to form endothelin-1 by **endothelin-converting enzyme.** Small amounts of big endothelin-1 and endothelin-1 are secreted into the blood, but for the most part, they are secreted locally and act in a paracrine fashion.

Two different endothelin receptors have been cloned, both of which are coupled via G proteins to phospholipase C (see Chapter 2). The ET_A receptor, which is specific for endothelin-1, is found in many tissues and mediates the vasoconstriction produced by endothelin-1. The ET_B receptor responds to all three endothelins, and is coupled to G_i. It may mediate vasodilation, and it appears to mediate the developmental effects of the endothelins (see below).

REGULATION OF SECRETION

Endothelin-1 is not stored in secretory granules, and most regulatory factors alter the transcription of its gene, with changes in secretion occurring promptly thereafter. Factors activating and inhibiting the gene are summarized in Table 33–4.

TABLE 33–4 Regulation of endothelin-1 secretion via transcription of its gene.

Stimulators
Angiotensin II
Catecholamines
Growth factors
Hypoxia
Insulin
Oxidized LDL
HDL
Shear stress
Thrombin
Inhibitors
NO
ANP
PGE_2
Prostacyclin

CARDIOVASCULAR FUNCTIONS

As noted above, endothelin-1 appears to be primarily a paracrine regulator of vascular tone. However, endothelin-1 is not increased in hypertension, and in mice in which one allele of the endothelin-1 gene is knocked out, blood pressure is actually elevated rather than reduced. The concentration of circulating endothelin-1 is, however, elevated in congestive heart failure and after myocardial infarction, so it may play a role in the pathophysiology of these diseases.

OTHER FUNCTIONS OF ENDOTHELINS

Endothelin-1 is found in the brain and kidneys as well as the endothelial cells. Endothelin-2 is produced primarily in the kidneys and intestine. Endothelin-3 is present in the blood and is found in high concentrations in the brain. It is also found in the kidneys and gastrointestinal tract. In the brain, endothelins are abundant and, in early life, are produced by both astrocytes and neurons. They are found in the dorsal root ganglia, ventral horn cells, the cortex, the hypothalamus, and cerebellar Purkinje cells. They also play a role in regulating transport across the blood–brain barrier. There are endothelin receptors on mesangial cells (see Chapter 38), and the polypeptide participates in tubuloglomerular feedback.

Mice that have both alleles of the endothelin-1 gene deleted have severe craniofacial abnormalities and die of respiratory failure at birth. They also have megacolon (Hirschsprung disease), apparently because the cells that normally form the myenteric plexus fail to migrate to the distal colon. In addition, endothelins play a role in closing the ductus arteriosus at birth.

SYSTEMIC REGULATION BY HORMONES

Many circulating hormones affect the vascular system. The vasodilator hormones include kinins, VIP, and ANP. Circulating vasoconstrictor hormones include vasopressin, norepinephrine, epinephrine, and angiotensin II.

KININS

Two related vasodilator peptides called **kinins** are found in the body. One is the nonapeptide **bradykinin,** and the other is the decapeptide **lysylbradykinin,** also known as **kallidin** (Figure 33–11). Lysylbradykinin can be converted to bradykinin by aminopeptidase. Both peptides are metabolized to inactive fragments by **kininase I,** a carboxypeptidase that removes the carboxyl terminal arginine (Arg). In addition, the dipeptidyl-carboxypeptidase **kininase II** inactivates bradykinin and lysylbradykinin by removing phenylalanine-arginine (Phe-Arg) from the carboxyl terminal. Kininase II is the same enzyme as **angiotensin-converting enzyme** (see Chapter 39), which removes histidine-leucine (His-Leu) from the carboxyl terminal end of angiotensin I.

Bradykinin and lysylbradykinin are formed from two precursor proteins: **high-molecular-weight kininogen** and **low-molecular-weight kininogen** (Figure 33–12). They are formed by alternative splicing of a single gene located on chromosome 3. Proteases called **kallikreins** release the peptides from their precursors. They are produced in humans by a family of three genes located on chromosome 19. There are two types of kallikreins: **plasma kallikrein,** which circulates in an inactive form, and **tissue kallikrein,** which appears to be located primarily on the apical membranes of cells concerned with transcellular electrolyte transport. Tissue kallikrein is found in many tissues, including sweat and salivary glands, the pancreas, the prostate, the intestine, and the kidneys. Tissue kallikrein acts on high-molecular-weight kininogen to form bradykinin and low-molecular-weight kininogen

FIGURE 33–11   Kinins. Lysylbradykinin **(top)** can be converted to bradykinin **(bottom)** by aminopeptidase. The peptides are inactivated by kininase I (KI) or kininase II (KII) at the sites indicated by the short arrows.

FIGURE 33–12   Formation of kinins from high-molecular-weight (HMW) and low-molecular-weight (LMW) kininogens.

to form lysylbradykinin. When activated, plasma kallikrein acts on high-molecular-weight kininogen to form bradykinin.

Inactive plasma kallikrein (**prekallikrein**) is converted to the active form, kallikrein, by active factor XII, the factor that initiates the intrinsic blood clotting cascade. Kallikrein also activates factor XII in a positive feedback loop, and high-molecular-weight kininogen has a factor XII-activating action (see Figure 32–13).

The actions of both kinins resemble those of histamine. They are primarily tissue hormones, although small amounts are also found in the circulating blood. They cause contraction of visceral smooth muscle, but they relax vascular smooth muscle via NO, lowering blood pressure. They also increase capillary permeability, attract leukocytes, and cause pain upon injection under the skin. They are formed during active secretion in sweat glands, salivary glands, and the exocrine portion of the pancreas, and they are probably responsible for the increase in blood flow when these tissues are actively secreting their products.

Two bradykinin receptors, B_1 and B_2, have been identified. Their amino acid residues are 36% identical, and both are coupled to G proteins. The B_1 receptor may mediate the pain-producing effects of the kinins, but little is known about its distribution and function. The B_2 receptor has strong homology to the H_2 receptor and is found in many different tissues.

NATRIURETIC HORMONES

There is a family of natriuretic peptides involved in vascular regulation, including atrial natriuretic peptide (ANP) secreted by the heart, brain natriuretic peptide (BNP), and C-type natriuretic peptide (CNP). They are released in response to hypervolemia. ANP and BNP circulate, whereas CNP acts predominantly in a paracrine fashion. In general, these peptides antagonize the action of various vasoconstrictor agents and lower blood pressure. ANP and BNP also serve to coordinate the control of vascular tone with fluid and electrolyte homeostasis via actions on the kidney.

CIRCULATING VASOCONSTRICTORS

Vasopressin is a potent vasoconstrictor, but when it is injected in normal individuals, there is a compensating decrease in cardiac output, so that there is little change in blood pressure. Its role in blood pressure regulation is discussed in Chapter 18.

Norepinephrine has a generalized vasoconstrictor action, whereas epinephrine dilates the vessels in skeletal muscle and the liver. The relative unimportance of circulating norepinephrine, as opposed to norepinephrine released from vasomotor nerves, is pointed out in Chapter 22, where the cardiovascular actions of catecholamines are discussed in detail.

Angiotensin II has a generalized vasoconstrictor action. It is formed by the action of angiotensin converting enzyme (ACE) on angiotensin I, which itself is liberated by the action of renin from the kidney on circulating angiotensinogen (see Chapter 39). Renin secretion, in turn, is increased when the blood pressure falls or extracellular fluid (ECF) volume is reduced, and angiotensin II therefore helps to maintain blood pressure. Angiotensin II also increases water intake and stimulates aldosterone secretion, and increased formation of angiotensin II is part of a homeostatic mechanism that operates to maintain ECF volume (see Chapter 22). In addition, there are renin–angiotensin systems in many different organs, and there may be one in the walls of blood vessels. Angiotensin II produced in blood vessel walls could be important in some forms of clinical hypertension. The role of angiotensin II in cardiovascular regulation is also amply demonstrated in the widespread use of so-called ACE inhibitors as antihypertensive medications.

Urotensin-II, a polypeptide first isolated from the spinal cord of fish, is present in human cardiac and vascular tissue. It is one of the most potent mammalian vasoconstrictors known, but its pathophysiogic and physiologic roles are currently the subject of intense interest.

CHAPTER SUMMARY

- RVLM neurons project to the thoracolumbar IML and release glutamate on preganglionic sympathetic neurons that innervate the heart and vasculature.
- The NTS is the major excitatory input to cardiac vagal motor neurons in the nucleus ambiguus.
- Carotid sinus and aortic depressor baroreceptors are innervated by branches of the 9th and 10th cranial nerves, respectively (glossopharyngeal and aortic depressor nerves). These receptors are most sensitive to changes in pulse pressure but also respond to changes in mean arterial pressure.
- Baroreceptor nerves terminate in the NTS and release glutamate. NTS neurons project to the CVLM and nucleus ambiguus and release glutamate. CVLM neurons project to RVLM and release GABA. This leads to a reduction in sympathetic activity and an increase in vagal activity (ie, the baroreceptor reflex).
- Activation of peripheral chemoreceptors in the carotid and aortic bodies by a reduction in PaO_2 or an increase in $PaCO_2$ leads to an increase in vasoconstriction. Heart rate changes are variable and depend on a number of factors including changes in respiration.
- In addition to various neural inputs, RVLM neurons are directly activated by hypoxia and hypercapnia.
- Most vascular beds have an intrinsic capacity to respond to changes in blood pressure within a certain range by altering vascular resistance to maintain stable blood flow. This property is known as autoregulation.
- Local factors such as oxygen tension, pH, temperature, and metabolic products contribute to vascular regulation; many produce vasodilation to restore blood flow.
- The endothelium is an important source of vasoactive mediators that act to either contract or relax vascular smooth muscle.
- Three gaseous mediators—NO, CO, and H_2S—are important regulators of vasodilation.
- Endothelins and angiotensin II induce vasoconstriction and may be involved in the pathogenesis of some forms of hypertension.

MULTIPLE-CHOICE QUESTIONS

For all questions, select the single best answer unless otherwise directed.

1. When a pheochromocytoma (tumor of the adrenal medulla) suddenly discharges a large amount of epinephrine into the circulation, the patient's heart rate would be expected to
    A) increase, because the increase in blood pressure stimulates the carotid and aortic baroreceptors.
    B) increase, because epinephrine has a direct chronotropic effect on the heart.
    C) increase, because of increased tonic parasympathetic discharge to the heart.
    D) decrease, because the increase in blood pressure stimulates the carotid and aortic chemoreceptors.
    E) decrease, because of increased tonic parasympathetic discharge to the heart.

2. Activation of the baroreceptor reflex
    A) is primarily involved in short-term regulation of systemic blood pressure.
    B) leads to an increase in heart rate because of inhibition of the vagal cardiac motor neurons.
    C) inhibits neurons in the CVLM.
    D) excites neurons in the RVLM.
    E) all of the above

3. Sympathetic nerve activity would be expected to increase
    A) if glutamate receptors were blocked in the NTS.
    B) if GABA receptors were blocked in the RVLM.
    C) if there was a compression of the RVLM.
    D) during hypoxia.
    E) for all of the above.

CHAPTER RESOURCES

Ahluwalia A, MacAllister RJ, Hobbs AJ: Vascular actions of natriuretic peptides. Cyclic GMP-dependent and -independent mechanisms. Basic Res Cardiol 2004;99:83.

Benarroch EE: *Central Autonomic Network. Functional Organization and Clinical Correlations.* Futura Publishing, 1997.

Chapleau MW, Abboud F (editors): Neuro-cardiovascular regulation: From molecules to man. Ann NY Acad Sci 2001;940.

de Burgh Daly M: *Peripheral Arterial Chemoreceptors and Respiratory-Cardiovascular Integration.* Clarendon Press, 1997.

Haddy FJ, Vanhouttee PM, Feletou M: Role of potassium in regulating blood flow and blood pressure. Am J Physiol Regul Integr Comp Physiol 2006;290:R546.

Loewy AD, Spyer KM (editors): *Central Regulation of Autonomic Function.* Oxford University Press, 1990.

Marshall JM: Peripheral chemoreceptors and cardiovascular regulation. Physiol Rev 1994;74:543.

Paffett ML, Walker BR: Vascular adaptations to hypoxia: Molecular and cellular mechanisms regulating vascular tone. Essays Biochem 2007;43:105.

Squire LR, Bloom FE, Spitzer NC, du Lac S, Ghosh A, Berg D (editors): *Fundamental Neuroscience,* 3rd ed. Academic Press, 2008.

Trouth CO, Millis RM, Kiwull-Schöne HF, Schläfke ME: *Ventral Brainstem Mechanisms and Control of Respiration and Blood Pressure.* Marcel Dekker, 1995.

Circulation Through Special Regions

INTRODUCTION

The distribution of the cardiac output to various parts of the body at rest in a normal man is shown in Table 34–1. The general principles described in preceding chapters apply to the circulation of all these regions, but the vascular supplies of many organs have additional special features that are important to their physiology. The portal circulation of the anterior pituitary is discussed in Chapter 24, the pulmonary circulation in Chapter 35, the renal circulation in Chapter 38, and the circulation of the splanchnic area, particularly the intestines and liver, in Chapters 26 and 29. This chapter is concerned with the special circulations of the brain, the heart, and the skin, as well as the placenta and fetus.

CEREBRAL CIRCULATION: ANATOMIC CONSIDERATIONS

VESSELS

The principal arterial inflow to the brain in humans is via four arteries: two internal carotids and two vertebrals. In humans, the carotid arteries are quantitatively the most significant. The vertebral arteries unite to form the basilar artery, and the basilar artery and the carotids form the **circle of Willis** below the hypothalamus. The circle of Willis is the origin of the six large vessels supplying the cerebral cortex. Substances injected into one carotid artery are distributed almost exclusively to the cerebral hemisphere on that side. Normally no crossing over occurs, probably because the pressure is equal on both sides. Even when it is not, the anastomotic channels in the circle do not permit a very large flow. Occlusion of one carotid artery, particularly in older patients, often causes serious symptoms of cerebral ischemia. There are precapillary anastomoses between the cerebral vessels, but flow through these channels is generally insufficient to maintain the circulation and prevent infarction when a cerebral artery is occluded.

Venous drainage from the brain by way of the deep veins and dural sinuses empties principally into the internal jugular veins in humans, although a small amount of venous blood

TABLE 34–1 Resting blood flow and O_2 consumption of various organs in a 63-kg adult man with a mean arterial blood pressure of 90 mm Hg and an O_2 consumption of 250 mL/min.

| Region | Mass (kg) | Blood Flow | | Arteriovenous Oxygen Difference (mL/L) | Oxygen Consumption | | Resistance (R units)[a] | | Percentage of Total | |
		mL/min	mL/100 g/min		mL/min	mL/100 g/min	Absolute	per kg	Cardiac Output	Oxygen Consumption
Liver	2.6	1500	57.7	34	51	2.0	3.6	9.4	27.8	20.4
Kidneys	0.3	1260	420.0	14	18	6.0	4.3	1.3	23.3	7.2
Brain	1.4	750	54.0	62	46	3.3	7.2	10.1	13.9	18.4
Skin	3.6	462	12.8	25	12	0.3	11.7	42.1	8.6	4.8
Skeletal muscle	31.0	840	2.7	60	50	0.2	6.4	198.4	15.6	20.0
Heart muscle	0.3	250	84.0	114	29	9.7	21.4	6.4	4.7	11.6
Rest of body	23.8	336	1.4	129	44	0.2	16.1	383.2	6.2	17.6
Whole body	63.0	5400	8.6	46	250	0.4	1.0	63.0	100.0	100.0

[a]R units are pressure (mm Hg) divided by blood flow (mL/s).

Reproduced with permission from Bard P (editor): *Medical Physiology,* 11th ed. Mosby, 1961.

drains through the ophthalmic and pterygoid venous plexuses, through emissary veins to the scalp, and down the system of paravertebral veins in the spinal canal.

The cerebral vessels have a number of unique anatomic features. In the choroid plexuses, there are gaps between the endothelial cells of the capillary wall, but the choroid epithelial cells that separate them from the cerebrospinal fluid (CSF) are connected to one another by tight junctions. The capillaries in the brain substance resemble nonfenestrated capillaries in muscle (see Chapter 32), but there are tight junctions between the endothelial cells that limit the passage of substances through the junctions. In addition, there are relatively few vesicles in the endothelial cytoplasm, and presumably little vesicular transport. However, multiple transport systems are present in the capillary cells. The brain capillaries are surrounded by the endfeet of astrocytes (Figure 34–1). These endfeet are closely applied to the basal lamina of the capillaries, but they do not cover the entire capillary wall, and gaps of about 20 nm occur between endfeet (Figure 34–2). However, the endfeet induce the tight junctions in the capillaries (see Chapter 32). The protoplasm of astrocytes is also found around synapses, where it appears to isolate the synapses in the brain from one another.

INNERVATION

Three systems of nerves innervate the cerebral blood vessels. Postganglionic sympathetic neurons have their cell bodies in the superior cervical ganglia, and their endings contain norepinephrine. Many also contain neuropeptide Y. Cholinergic neurons that probably originate in the sphenopalatine ganglia

also innervate the cerebral vessels, and the postganglionic cholinergic neurons on the blood vessels contain acetylcholine. Many also contain vasoactive intestinal peptide (VIP) and peptide histidyl methionine (PHM-27) (see Chapter 7). These nerves end primarily on large arteries. Sensory nerves are found on more distal arteries. They have their cell bodies in the trigeminal ganglia and contain substance P, neurokinin A, and calcitonin gene-related peptide (CGRP). Substance P, CGRP, VIP, and PHM-27 cause vasodilation, whereas

FIGURE 34–1   **Relation of fibrous astrocyte (3) to a capillary (2) and neuron (4) in the brain.** The endfeet of the astrocyte processes form a discontinuous membrane around the capillary (1). Astrocyte processes also envelop the neuron. (Adapted from Krstic RV: *Die Gewebe des Menschen und der Säugetiere.* Springer, 1978.)

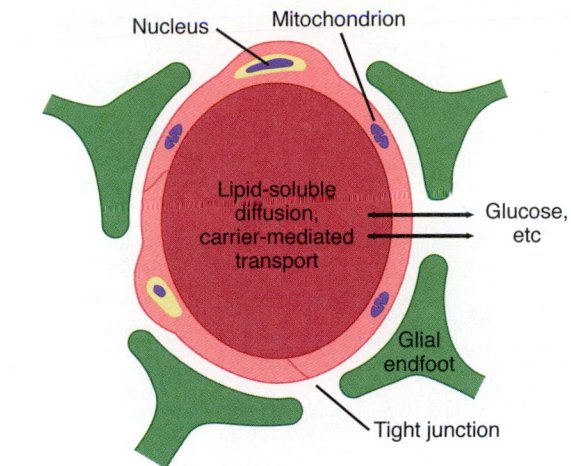

FIGURE 34-2 Transport across cerebral capillaries.

neuropeptide Y is a vasoconstrictor. Touching or pulling on the cerebral vessels causes pain.

CEREBROSPINAL FLUID

FORMATION & ABSORPTION

CSF fills the ventricles and subarachnoid space. In humans, the volume of CSF is about 150 mL and the rate of CSF production is about 550 mL/d. Thus the CSF turns over about 3.7 times a day. In experiments on animals, it has been estimated that 50–70% of the CSF is formed in the choroid plexuses and the remainder is formed around blood vessels and along ventricular walls. Presumably, the situation in humans is similar. The CSF in the ventricles flows through the foramens of Magendie and Luschka to the subarachnoid space and is absorbed through the **arachnoid villi** into veins, primarily the cerebral venous sinuses. The villi consist of projections of the fused arachnoid membrane and endothelium of the sinuses into the venous sinuses. Similar, smaller villi project into veins around spinal nerve routes. These projections may contribute to the outflow of CSF into venous blood by a process known as **bulk flow**, which is unidirectional. However, recent studies suggest that, at least in animals, a more important route for CSF reabsorption into the bloodstream in health is via the cribriform plate above the nose and thence into the cervical lymphatics. However, reabsorption via one-way valves (of uncertain structural basis) in the arachnoid villi may assume a greater role if CSF pressure is elevated. Likewise, when CSF builds up abnormally, aquaporin water channels may be expressed in the choroid plexus and brain microvessels as a compensatory adaptation.

CSF is formed continuously by the choroid plexus in two stages. First, plasma is passively filtered across the choroidal capillary endothelium. Next, secretion of water and ions across the choroidal epithelium provides for active control of CSF composition and quantity. Bicarbonate, chloride, and potassium ions enter the CSF via channels in the epithelial cell

TABLE 34–2 Concentration of various substances in human CSF and plasma.

Substance		CSF	Plasma	Ratio CSF/Plasma
Na⁺	(meq/kg H₂O)	147.0	150.0	0.98
K⁺	(meq/kg H₂O)	2.9	4.6	0.62
Mg²⁺	(meq/kg H₂O)	2.2	1.6	1.39
Ca²⁺	(meq/kg H₂O)	2.3	4.7	0.49
Cl⁻	(meq/kg H₂O)	113.0	99.0	1.14
HCO₃⁻	(meq/L)	25.1	24.8	1.01
Pco₂	(mm Hg)	50.2	39.5	1.28
pH		7.33	7.40	...
Osmolality	(mosm/kg H₂O)	289.0	289.0	1.00
Protein	(mg/dL)	20.0	6000.0	0.003
Glucose	(mg/dL)	64.0	100.0	0.64
Inorganic P	(mg/dL)	3.4	4.7	0.73
Urea	(mg/dL)	12.0	15.0	0.80
Creatinine	(mg/dL)	1.5	1.2	1.25
Uric acid	(mg/dL)	1.5	5.0	0.30
Cholesterol	(mg/dL)	0.2	175.0	0.001

apical membranes. Aquaporins provide for water movement to balance osmotic gradients. The composition of CSF (Table 34–2) is essentially the same as that of brain extracellular fluid (ECF), which in living humans makes up 15% of the brain volume. In adults, free communication appears to take place between the brain interstitial fluid and CSF, although the diffusion distances from some parts of the brain to the CSF are appreciable. Consequently, equilibration may take some time to occur, and local areas of the brain may have extracellular microenvironments that are transiently different from CSF.

Lumbar CSF pressure is normally 70 to 180 mm H₂O. Up to pressures well above this range, the rate of CSF formation is independent of intraventricular pressure. However, absorption is proportional to the pressure (Figure 34–3). At a pressure of 112 mm H₂O, which is the average normal CSF pressure, filtration and absorption are equal. Below a pressure of approximately 68 mm H₂O, absorption stops. Large amounts of fluid accumulate when the capacity for CSF reabsorption is decreased (**external hydrocephalus, communicating hydrocephalus**). Fluid also accumulates proximal to the block and distends the ventricles when the foramens of Luschka and Magendie are blocked or there is obstruction within the ventricular system (**internal hydrocephalus, noncommunicating hydrocephalus**).

FIGURE 34–3 **CSF formation and absorption in humans at various CSF pressures.** Note that at 112 mm CSF, formation and absorption are equal, and at 68 mm CSF, absorption is zero. (Modified and reproduced with permission from Cutler RWP, et al: Formation and absorption of cerebrospinal fluid in man. Brain 1968;91:707.)

PROTECTIVE FUNCTION

The most critical role for CSF (and the meninges) is to protect the brain. The dura is attached firmly to bone. Normally, there is no "subdural space," with the arachnoid being held to the dura by the surface tension of the thin layer of fluid between the two membranes. As shown in Figure 34–4, the brain itself is supported within the arachnoid by the blood vessels and nerve roots and by the multiple fine fibrous **arachnoid trabeculae.** The brain weighs about 1400 g in air, but in its "water bath" of CSF it has a net weight of only 50 g. The buoyancy of the brain in the CSF permits its relatively flimsy attachments to suspend it very effectively. When the head receives a blow, the arachnoid slides on the dura and the brain moves, but its motion is gently checked by the CSF cushion and by the arachnoid trabeculae.

The pain produced by spinal fluid deficiency illustrates the importance of CSF in supporting the brain. Removal of CSF during lumbar puncture can cause a severe headache after the fluid is removed, because the brain hangs on the vessels and nerve roots, and traction on them stimulates pain fibers. The pain can be relieved by intrathecal injection of sterile isotonic saline.

HEAD INJURIES

Without the protection of the spinal fluid and the meninges, the brain would probably be unable to withstand even the minor traumas of everyday living; but with the protection afforded, it takes a fairly severe blow to produce cerebral damage. The brain is damaged most commonly when the skull is fractured and bone is driven into neural tissue (depressed skull fracture), when the brain moves far enough to tear the delicate bridging veins from the cortex to the bone, or when the brain is accelerated by a blow on the head and is driven against the skull or the tentorium at a point opposite where the blow was struck (**contrecoup injury**).

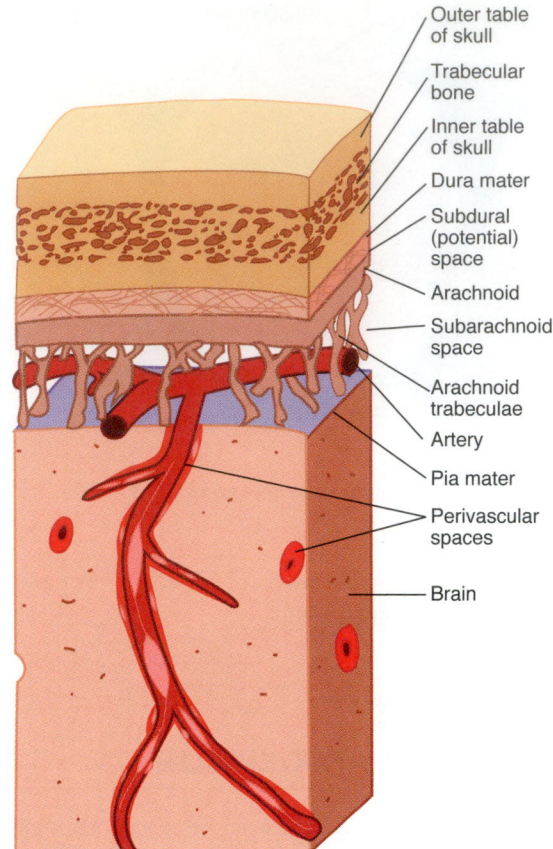

FIGURE 34–4 **Investing membranes of the brain, showing their relation to the skull and to brain tissue.** (Reproduced with permission from Wheater PR et al: *Functional Histology.* Churchill Livingstone, 1979.)

THE BLOOD–BRAIN BARRIER

The tight junctions between capillary endothelial cells in the brain and between the epithelial cells in the choroid plexus effectively prevent proteins from entering the brain in adults and slow the penetration of some smaller molecules as well. An example is the slow penetration of urea (Figure 34–5). This uniquely limited exchange of substances into the brain is referred to as the **blood–brain barrier,** a term most commonly used to encompass this barrier overall and more specifically the barrier in the choroid epithelium between blood and CSF.

Passive diffusion across the tight cerebral capillaries is very limited, and little vesicular transport takes place. However, there are numerous carrier-mediated and active transport systems in the cerebral capillaries. These move substances out of as well as into the brain, though movement out of the brain is generally more free than movement into it.

PENETRATION OF SUBSTANCES INTO THE BRAIN

Water, CO_2, and O_2 penetrate the brain with ease, as do the lipid-soluble free forms of steroid hormones, whereas their

FIGURE 34–5 **Penetration of urea into muscle, brain, spinal cord, and CSF.** Urea was administered by constant infusion.

FIGURE 34–6 **Localization of the various GLUT transporters in the brain.** (Adapted from Maher F, Vannucci SJ, Simpson IA: Glucose transporter proteins in brain. FASEB J 1994;8:1003.)

protein-bound forms and, in general, all proteins and polypeptides do not. The rapid passive penetration of CO_2 contrasts with the regulated transcellular penetration of H^+ and HCO_3^- and has physiologic significance in the regulation of respiration (see Chapter 37).

Glucose is the major ultimate source of energy for nerve cells. Its diffusion across the blood–brain barrier would be very slow, but the rate of transport into the CSF is markedly enhanced by the presence of specific transporters, including the glucose transporter 1 (GLUT 1). The brain contains two forms of GLUT 1: GLUT 1 55K and GLUT 1 45K. Both are encoded by the same gene, but they differ in the extent to which they are glycosylated. GLUT 1 55K is present in high concentration in brain capillaries (Figure 34–6). Infants with congenital GLUT 1 deficiency develop low CSF glucose concentrations in the presence of normal plasma glucose, and they have seizures and delayed development. In addition, transporters for thyroid hormones; several organic acids; choline; nucleic acid precursors; and neutral, basic, and acidic amino acids are present at the blood–brain barrier.

A variety of drugs and peptides actually cross the cerebral capillaries but are promptly transported back into the blood by a multidrug nonspecific transporter in the apical membranes of the endothelial cells. This **P-glycoprotein** is a member of the family of adenosine triphosphate (ATP) binding cassettes that transport various proteins and lipids across cell membranes (see Chapter 2). In the absence of this transporter in mice, much larger proportions of systemically administered doses of various chemotherapeutic drugs, analgesics, and opioid peptides are found in the brain than in controls. If pharmacologic agents that inhibit this transporter can be developed, they could be of value in the treatment of brain tumors and other central nervous system (CNS) diseases in which it is difficult to introduce adequate amounts of therapeutic agents into the brain.

CIRCUMVENTRICULAR ORGANS

When dyes that bind to proteins in the plasma are injected, they stain many tissues but spare most of the brain. However, four small areas in or near the brain stem do take up the stain. These areas are (1) the **posterior pituitary** (neurohypophysis) and the adjacent ventral part of the **median eminence** of the hypothalamus, (2) the **area postrema,** (3) the **organum vasculosum of the lamina terminalis (OVLT,** supraoptic crest), and (4) the **subfornical organ (SFO).**

These areas are referred to collectively as the **circumventricular organs** (Figure 34–7). All have fenestrated capillaries, and because of their permeability they are said to be "outside the blood–brain barrier." Some of them function as **neurohemal organs**; that is, areas in which polypeptides secreted by neurons enter the circulation. Others contain receptors for many different peptides and other substances, and function as chemoreceptor zones in which substances in the circulating blood can act to trigger changes in brain function without penetrating the blood–brain barrier. For example, the area postrema is a chemoreceptor trigger zone that initiates vomiting in response to chemical changes in the plasma (see Chapter 28). It is also concerned with cardiovascular control, and in many species circulating angiotensin II acts on the area postrema to produce a neurally mediated increase in blood pressure. Angiotensin II also acts on the SFO and possibly on the OVLT to increase water intake. In addition, it appears that the OVLT is the site of the osmoreceptor controlling vasopressin secretion (see Chapter 39), and evidence suggests that circulating interleukin-1 (IL-1) produces fever by acting here too.

The subcommissural organ (Figure 34–7) is closely associated with the pineal gland and histologically resembles the circumventricular organs. However, it does not have fenestrated capillaries, is not highly permeable, and has no established

FIGURE 34–7   Circumventricular organs. The neurohypophysis (NH), organum vasculosum of the lamina terminalis (OVLT, organum vasculosum of the lamina terminalis), subfornical organ (SFO), and area postrema (AP) are shown projected on a sagittal section of the human brain. SCO, subcommissural organ; PI, pineal.

function. Conversely, the pineal and the anterior pituitary do have fenestrated capillaries and are outside the blood–brain barrier, but both are endocrine glands and are not part of the brain.

FUNCTION OF THE BLOOD–BRAIN BARRIER

The blood–brain barrier strives to maintain the constancy of the environment of the neurons in the central nervous system (see Clinical Box 34–1). Even minor variations in the concentrations of K^+, Ca^{2+}, Mg^{2+}, H^+, and other ions can have far-reaching consequences. The constancy of the composition of the ECF in all parts of the body is maintained by multiple homeostatic mechanisms (see Chapters 1 and 39), but because of the sensitivity of the cortical neurons to ionic change, it is not surprising that an additional defense has evolved to protect them. Other functions of the blood–brain barrier include protection of the brain from endogenous and exogenous toxins in the blood and prevention of the escape of neurotransmitters into the general circulation.

DEVELOPMENT OF THE BLOOD–BRAIN BARRIER

In experimental animals, many small molecules penetrate the brain more readily during the fetal and neonatal period than they do in the adult. On this basis, it is often stated that the blood–brain barrier is immature at birth. Humans are more

CLINICAL BOX 34–1

Clinical Implications of the Blood–Brain Barrier

Physicians must know the degree to which drugs penetrate the brain in order to treat diseases of the nervous system intelligently. For example, it is clinically relevant that the amines dopamine and serotonin penetrate brain tissue to a very limited degree but their corresponding acid precursors, L-dopa and 5-hydroxytryptophan, respectively, enter with relative ease (see Chapters 7 and 16). Another important clinical consideration is the fact that the blood–brain barrier tends to break down in areas of infection or injury. Tumors develop new blood vessels, and the capillaries that are formed lack contact with normal astrocytes. Therefore, there are no tight junctions, and the vessels may even be fenestrated. The lack of a barrier helps in identifying the location of tumors; substances such as radioactive iodine-labeled albumin penetrate normal brain tissue very slowly, but they enter tumor tissue, making the tumor stand out as an island of radioactivity in the surrounding normal brain. The blood–brain barrier can also be temporarily disrupted by sudden marked increases in blood pressure or by intravenous injection of hypertonic fluids.

mature at birth than rats and various other experimental animals, and detailed data on passive permeability of the human blood–brain barrier are not available. However, in severely jaundiced infants with high plasma levels of free bilirubin and an immature hepatic bilirubin-conjugating system, free bilirubin enters the brain and, in the presence of asphyxia, damages the basal ganglia (**kernicterus**). The counterpart of this situation in later life is the Crigler–Najjar syndrome in which there is a congenital deficiency of glucuronyl transferase. These individuals can have very high free bilirubin levels in the blood and develop encephalopathy. In other conditions, free bilirubin levels are generally not high enough to produce brain damage.

CEREBRAL BLOOD FLOW & ITS REGULATION

KETY METHOD

According to the **Fick principle** (see Chapter 31), the blood flow of any organ can be measured by determining the amount of a given substance (Q_x) removed from the bloodstream by the organ per unit of time and dividing that value by the difference between the concentration of the substance in arterial blood and the concentration in the venous blood from the organ ($[A_x] - [V_x]$). Thus:

$$\text{Cerebral blood flow (CBF)} = \frac{Q_x}{[A_x] - [V_x]}$$

This can be applied clinically using inhaled nitrous oxide (N₂O) (**Kety method**). The average cerebral blood flow in young adults is 54 mL/100 g/min. The average adult brain weighs about 1400 g, so the flow for the whole brain is about 756 mL/min. Note that the Kety method provides an average value for perfused areas of brain because it gives no information about regional differences in blood flow. It also can only measure flow to perfused parts of the brain. If the blood flow to a portion of the brain is occluded, the measured flow does not change because the nonperfused area does not take up any N₂O.

In spite of the marked local fluctuations in brain blood flow with neural activity, the cerebral circulation is regulated in such a way that total blood flow remains relatively constant. The factors involved in regulating the flow are summarized in Figure 34–8.

ROLE OF INTRACRANIAL PRESSURE

In adults, the brain, spinal cord, and spinal fluid are encased, along with the cerebral vessels, in a rigid bony enclosure. The cranial cavity normally contains a brain weighing approximately 1400 g, 75 mL of blood, and 75 mL of spinal fluid. Because brain tissue and spinal fluid are essentially incompressible, the volume of blood, spinal fluid, and brain in the cranium at any time must be relatively constant (**Monro–Kellie doctrine**). More importantly, the cerebral vessels are compressed whenever the intracranial pressure rises. Any change in venous pressure promptly causes a similar change in intracranial pressure. Thus, a rise in venous pressure decreases cerebral blood flow both by decreasing the effective perfusion pressure and by compressing the cerebral vessels. This relationship helps to compensate for changes in arterial blood pressure at the level of the head. For example, if the body is accelerated upward (positive g), blood moves toward the feet and arterial pressure at the level of the head decreases. However, venous pressure also falls and intracranial pressure falls, so that the pressure on the vessels de-

creases and blood flow is much less severely compromised than it would otherwise be. Conversely, during acceleration downward, force acting toward the head (negative g) increases arterial pressure at head level, but intracranial pressure also rises, so that the vessels are supported and do not rupture. The cerebral vessels are protected during the straining associated with defecation or delivery in the same way.

AUTOREGULATION

As seen in other vascular beds, autoregulation is prominent in the brain (Figure 34–9). This process, by which the flow to many tissues is maintained at relatively constant levels despite variations in perfusion pressure, is discussed in Chapter 32. In the brain, autoregulation maintains a normal cerebral blood flow at arterial pressures of 65 to 140 mm Hg.

ROLE OF VASOMOTOR & SENSORY NERVES

The innervation of large cerebral blood vessels by postganglionic sympathetic and parasympathetic nerves and the additional distal innervation by sensory nerves have been described above. The nerves may also modulate tone indirectly, via the release of paracine substances from astrocytes. The precise role of these nerves, however, remains a matter of debate. It has been argued that noradrenergic discharge occurs when the blood pressure is markedly elevated. This reduces the resultant passive increase in blood flow and helps protect the blood–brain barrier from the disruption that could otherwise occur (see above). Thus, vasomotor discharges affect autoregulation. With sympathetic stimulation, the constant-flow, or plateau, part of the pressure-flow curve is extended to the right (Figure 34–9); that is, greater increases in pressure can occur without an increase in flow. On the other hand, the vasodilator hydralazine and the angiotensin-converting enzyme (ACE) inhibitor captopril reduce the length of the plateau. Finally, neurovascular coupling may adjust local perfusion in response to changes in brain activity (see below).

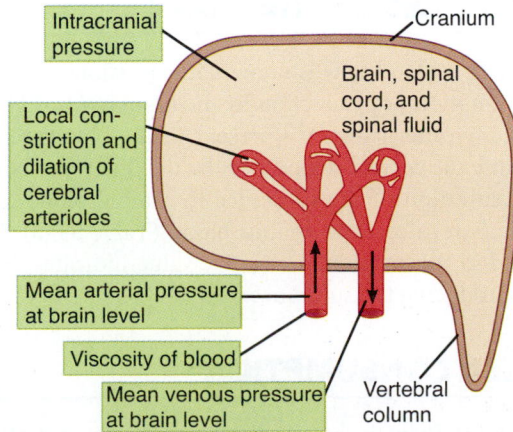

FIGURE 34–8 **Diagrammatic summary of the factors affecting overall cerebral blood flow.**

FIGURE 34–9 **Autoregulation of cerebral blood flow (CBF) during steady-state conditions.** The blue line shows the alteration produced by sympathetic stimulation during autoregulation.

FIGURE 34–10   **Activity in the human brain at five different horizontal levels while a subject generates a verb that is appropriate for each noun presented by an examiner.** This mental task activates the frontal cortex (slices 1–4), anterior cingulate gyrus (slice 1), and posterior temporal lobe (slice 3) on the left side and the cerebellum (slices 4 and 5) on the right side. Light purple, moderate activation; dark purple, marked activation.   (Based on PET scans in Posner MI, Raichle ME: *Images of Mind.* Scientific American Library, 1994.)

BLOOD FLOW IN VARIOUS PARTS OF THE BRAIN

A major advance in recent decades has been the development of techniques for monitoring regional blood flow in living, conscious humans. Among the most valuable methods are **positron emission tomography (PET)** and related techniques in which a short-lived radioisotope is used to label a compound and the compound is injected. The arrival and clearance of the tracer are monitored by scintillation detectors placed over the head. Because blood flow is tightly coupled to brain metabolism, local uptake of 2-deoxyglucose is also a good index of blood flow (see below and Chapter 1). If the 2-deoxyglucose is labeled with a short-half-life positron emitter such as ^{18}F, ^{11}O, or ^{15}O, its concentration in any part of the brain can be monitored.

Another valuable technique involves magnetic resonance imaging (MRI). MRI is based on detecting resonant signals from different tissues in a magnetic field. **Functional magnetic resonance imaging (fMRI)** measures the amount of blood in a tissue area. When neurons become active, their increased discharge alters the local field potential. A still unsettled mechanism triggers an increase in local blood flow and oxygen. The increase in oxygenated blood is detected by fMRI. PET scanning can be used to measure not only blood flow but the concentration of molecules, such as dopamine, in various regions of the living brain. On the other hand, fMRI does not involve the use of radioactivity. Consequently, it can be used at frequent intervals to measure changes in regional blood flow in a single individual.

In resting humans, the average blood flow in gray matter is 69 mL/100 g/min compared with 28 mL/100 g/min in white matter. A striking feature of cerebral function is the marked variation in local blood flow with changes in brain activity. An example is shown in Figure 34–10. In subjects who are awake but at rest, blood flow is greatest in the premotor and frontal regions. This is the part of the brain that is believed to be concerned with decoding and analyzing afferent input and with intellectual activity. During voluntary clenching of the right hand, flow is increased in the hand area of the left motor cortex and the corresponding sensory areas in the postcentral gyrus.

Especially when the movements being performed are sequential, the flow is also increased in the supplementary motor area. When subjects talk, there is a bilateral increase in blood flow in the face, tongue, and mouth-sensory and motor areas and the upper premotor cortex in the categorical (usually the left) hemisphere. When the speech is stereotyped, Broca's and Wernicke's areas do not show increased flow, but when the speech is creative—that is, when it involves ideas—flow increases in both these areas. Reading produces widespread increases in blood flow. Problem solving, reasoning, and motor ideation without movement produce increases in selected areas of the premotor and frontal cortex. In anticipation of a cognitive task, many of the brain areas that will be activated during the task are activated beforehand, as if the brain produces an internal model of the expected task. In right-handed individuals, blood flow to the left hemisphere is greater when a verbal task is being performed and blood flow to the right hemisphere is greater when a spatial task is being performed (see Clinical Box 34–2).

BRAIN METABOLISM & OXYGEN REQUIREMENTS

UPTAKE & RELEASE OF SUBSTANCES BY THE BRAIN

If the cerebral blood flow is known, it is possible to calculate the consumption or production by the brain of O_2, CO_2, glucose, or any other substance present in the bloodstream by multiplying the cerebral blood flow by the difference between the concentration of the substance in arterial blood and its concentration in cerebral venous blood (Table 34–3). When calculated in this fashion, a negative value indicates that the brain is producing the substance.

OXYGEN CONSUMPTION

O_2 consumption by the human brain (**cerebral metabolic rate for O_2,** $CMRO_2$) averages approximately 20% of the total body resting O_2 consumption (Table 34–1). The brain is

CLINICAL BOX 34–2

Changes in Cerebral Blood Flow in Disease

Several disease states are now known to be associated with localized or general changes in cerebral blood flow, as revealed by PET scanning and fMRI techniques. For example, epileptic foci are hyperemic during seizures, whereas flow is reduced in other parts of the brain. Between seizures, flow is sometimes reduced in the foci that generate the seizures. Parietooccipital flow is decreased in patients with symptoms of agnosia (see Chapter 14). In Alzheimer disease, the earliest change is decreased metabolism and blood flow in the superior parietal cortex, with later spread to the temporal and finally the frontal cortex. The pre- and postcentral gyri, basal ganglia, thalamus, brain stem, and cerebellum are relatively spared. In Huntington disease, blood flow is reduced bilaterally in the caudate nucleus, and this alteration in flow occurs early in the disease. In manic depressives (but interestingly, not in patients with unipolar depression), there is a general decrease in cortical blood flow when the patients are depressed. In schizophrenia, some evidence suggests decreased blood flow in the frontal lobes, temporal lobes, and basal ganglia. Finally, during the aura in patients with migraine, a bilateral decrease in blood flow starts in the occipital cortex and spreads anteriorly to the temporal and parietal lobes.

extremely sensitive to hypoxia, and occlusion of its blood supply produces unconsciousness in a period as short as 10 s. The vegetative structures in the brain stem are more resistant to hypoxia than the cerebral cortex, and patients may recover from accidents such as cardiac arrest and other conditions causing fairly prolonged hypoxia with normal vegetative functions but severe, permanent intellectual deficiencies. The basal

TABLE 34–3 Utilization and production of substances by the adult human brain in vivo.

Substance	Uptake (+) or Output (−) per 100 g of Brain/min	Total/min
Substances utilized		
Oxygen	+3.5 mL	+49 mL
Glucose	+5.5 mg	+77 mg
Glutamate	+0.4 mg	+5.6 mg
Substances produced		
Carbon dioxide	−3.5 mL	−49 mL
Glutamine	−0.6 mL	−8.4 mg

Substances not used or produced in the fed state: lactate, pyruvate, total ketones, and α-ketoglutarate.

ganglia use O_2 at a very high rate, and symptoms of Parkinson disease as well as intellectual deficits can be produced by chronic hypoxia. The thalamus and the inferior colliculus are also very susceptible to hypoxic damage (see Clinical Box 34–3).

ENERGY SOURCES

Glucose is the major ultimate source of energy for the brain; under normal conditions, 90% of the energy needed to maintain ion gradients across cell membranes and transmit electrical impulses comes from this source. Glucose enters the brain via GLUT 1 in cerebral capillaries (see above). Other transporters then distribute it to neurons and glial cells.

Glucose is taken up from the blood in large amounts, and the RQ (respiratory quotient; see Chapter 21) of cerebral tissue is 0.95–0.99 in normal individuals. Importantly, insulin is not required for most cerebral cells to utilize glucose. In general, glucose utilization at rest parallels blood flow and O_2 consumption. This does not mean that the total source of energy is always glucose. During prolonged starvation, appreciable utilization of other substances occurs. Indeed, evidence indicates that as much as 30% of the glucose taken up under normal conditions is converted to amino acids, lipids, and proteins, and that substances other than glucose are metabolized for energy during convulsions. Some utilization of amino acids from the circulation may also take place even though the amino acid arteriovenous difference across the brain is normally minute.

The consequences of hypoglycemia in terms of neural function are discussed in Chapter 21.

GLUTAMATE & AMMONIA REMOVAL

The brain's uptake of glutamate is approximately balanced by its output of glutamine. Glutamate entering the brain takes up ammonia and leaves as glutamine. The glutamate–glutamine conversion in the brain—the opposite of the reaction in the kidney that produces some of the ammonia entering the tubules—serves as a detoxifying mechanism to keep the brain free of ammonia. Ammonia is very toxic to nerve cells, and ammonia intoxication is believed to be a major cause of the bizarre neurologic symptoms in hepatic coma (see Chapter 29).

CORONARY CIRCULATION

ANATOMIC CONSIDERATIONS

The two coronary arteries that supply the myocardium arise from the sinuses behind two of the cusps of the aortic valve at the root of the aorta (Figure 34–11). Eddy currents keep the valves away from the orifices of the arteries, and they are patent throughout the cardiac cycle. Most of the venous blood returns to the heart through the coronary sinus and anterior

Stroke

When the blood supply to a part of the brain is interrupted, ischemia damages or kills the cells in the area, producing the signs and symptoms of a stroke. There are two general types of strokes: hemorrhagic and ischemic. Hemorrhagic stroke occurs when a cerebral artery or arteriole ruptures, sometimes but not always at the site of a small aneurysm. Ischemic stroke occurs when flow in a vessel is compromised by atherosclerotic plaques on which thrombi form. Thrombi may also be produced elsewhere (eg, in the atria in patients with atrial fibrillation) and pass to the brain as emboli where they then lodge and interrupt flow. In the past, little could be done to modify the course of a stroke and its consequences. However, it has now become clear that in the penumbra, the area surrounding the most severe brain damage, ischemia reduces glutamate uptake by astrocytes, and the increase in local glutamate causes excitotoxic damage and death to neurons (see Chapter 7). In experimental animals, and perhaps in humans, drugs that prevent this excitotoxic damage significantly reduce the effects of strokes. In addition, clot-lysing drugs such as tissue-type plasminogen activator (t-PA) (see Chapter 32) are of benefit in ischemic strokes. Both antiexcitotoxic treatment and t-PA must be given early in the course of a stroke to be of maximum benefit, and this is why stroke has become a condition in which rapid diagnosis and treatment have become important. In addition, of course, it is important to determine if a stroke is thrombotic or hemorrhagic, since clot lysis is contraindicated in the latter.

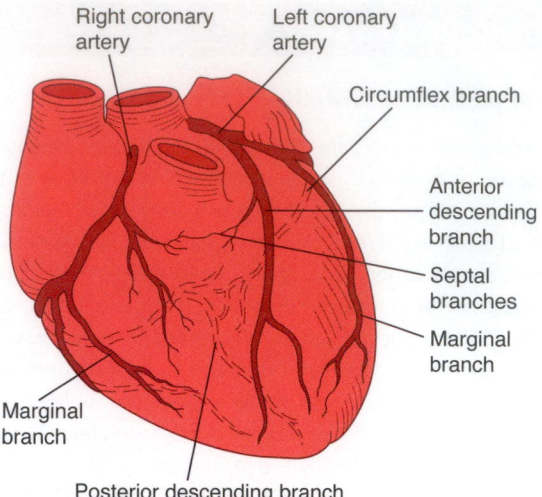

FIGURE 34–11 **Coronary arteries and their principal branches in humans.** (Reproduced with permission from Ross G: The cardiovascular system. In: *Essentials of Human Physiology.* Ross G [editor]. Copyright © 1978 by Year Book Medical Publishers.)

cardiac veins (Figure 34–12), which drain into the right atrium. In addition, there are other vessels that empty directly into the heart chambers. These include **arteriosinusoidal vessels,** sinusoidal capillary-like vessels that connect arterioles to the chambers; **thebesian veins** that connect capillaries to the chambers; and a few **arterioluminal vessels** that are small arteries draining directly into the chambers. A few anastomoses occur between the coronary arterioles and extracardiac arterioles, especially around the mouths of the great veins. Anastomoses between coronary arterioles in humans only pass particles less than 40 μm in diameter, but evidence indicates that these channels enlarge and increase in number in patients with coronary artery disease.

PRESSURE GRADIENTS & FLOW IN THE CORONARY VESSELS

The heart is a muscle that, like skeletal muscle, compresses its blood vessels when it contracts. The pressure inside the left ventricle is slightly higher than in the aorta during systole

(Table 34–4). Consequently, flow occurs in the arteries supplying the subendocardial portion of the left ventricle only during diastole, although the force is sufficiently dissipated in the more superficial portions of the left ventricular myocardium to permit some flow in this region throughout the cardiac cycle. Because diastole is shorter when the heart rate is high, left ventricular coronary flow is reduced during tachycardia. On the other hand, the pressure differential between the aorta and the right ventricle, and the differential between the aorta and the atria, are somewhat greater during systole than during diastole. Consequently, coronary flow in those parts of the heart is not appreciably reduced during systole. Flow in the right and left coronary arteries is shown in Figure 34–13. Because no blood flow occurs during systole in the subendocardial portion of the left ventricle, this region is prone to ischemic damage and is the most common site of myocardial infarction. Blood flow to the left ventricle is decreased in

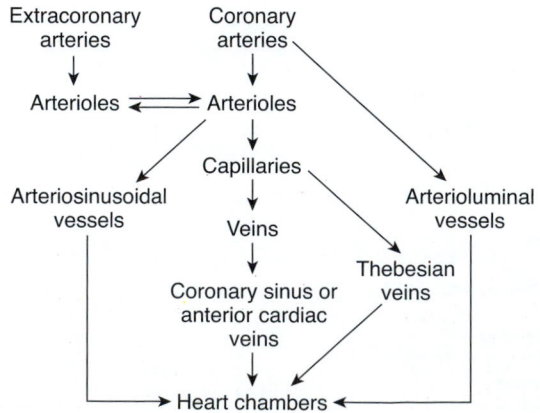

FIGURE 34–12 **Diagram of the coronary circulation.**

TABLE 34–4 Pressure in aorta and left and right ventricles (vent) in systole and diastole.

	Pressure (mm Hg) in			Pressure Differential (mm Hg) between Aorta and	
	Aorta	Left Vent	Right Vent	Left Vent	Right Vent
Systole	120	121	25	–1	95
Diastole	80	0	0	80	80

patients with stenotic aortic valves because the pressure in the left ventricle must be much higher than that in the aorta to eject the blood. Consequently, the coronary vessels are severely compressed during systole. Patients with this disease are particularly prone to develop symptoms of myocardial ischemia, in part because of this compression and in part because the myocardium requires more O_2 to expel blood through the stenotic aortic valve. Coronary flow is also decreased when the aortic diastolic pressure is low. The rise in venous pressure in conditions such as congestive heart failure reduces coronary flow because it decreases effective coronary perfusion pressure (see Clinical Box 34–4).

Coronary blood flow has been measured by inserting a catheter into the coronary sinus and applying the Kety method to the heart on the assumption that the N_2O content of coronary venous blood is typical of the entire myocardial effluent. Coronary flow at rest in humans is about 250 mL/min (5% of the cardiac output). A number of techniques utilizing **radionuclides,** radioactive tracers that can be detected with radiation detectors over the chest, have been used to study regional blood flow in the heart and to detect areas of ischemia and infarct as well as to evaluate ventricular function. Radionuclides such as thallium-201 ($^{201}T1$) are pumped into cardiac muscle cells by Na, K ATPase and equilibrate with the intracellular K^+ pool. For the first 10–15 min after intravenous injection, $^{201}T1$ distribution is directly proportional to myocardial blood flow, and areas of ischemia can be detected by their low uptake. The uptake of this isotope is often determined soon after exercise and again several hours later to bring out areas in which exertion leads to compromised flow. Conversely, radiopharmaceuticals such as technetium-99m stannous pyrophosphate (^{99m}Tc-PYP) are selectively taken up by infarcted tissue by an incompletely understood mechanism and make infarcts stand out as "hot spots" on scintigrams of the chest. Coronary angiography can be combined with measurement of ^{133}Xe washout (see above) to provide detailed analysis of coronary blood flow. Radiopaque contrast medium is first injected into the coronary arteries, and x-rays are used to outline their distribution. The angiographic camera is then replaced with a scintillation camera, and ^{133}Xe washout is measured.

VARIATIONS IN CORONARY FLOW

At rest, the heart extracts 70–80% of the O_2 from each unit of blood delivered to it (Table 34–1). O_2 consumption can be increased significantly only by increasing blood flow. Therefore, it is not surprising that blood flow increases when the metabolism of the myocardium is increased. The caliber of the coronary vessels, and consequently the rate of coronary blood flow, is influenced not only by pressure changes in the aorta but also by chemical and neural factors. The coronary circulation also shows considerable autoregulation.

CHEMICAL FACTORS

The close relationship between coronary blood flow and myocardial O_2 consumption indicates that one or more of the products of metabolism cause coronary vasodilation. Factors suspected of playing this role include O_2 lack and increased local concentrations of CO_2, H^+, K^+, lactate, prostaglandins, adenine nucleotides, and adenosine. Likely several or all of these vasodilator metabolites act in an integrated fashion, redundant fashion, or both. Asphyxia, hypoxia, and intracoronary injections of cyanide all increase coronary blood flow 200–300% in denervated as well as intact hearts, and the feature common to these three stimuli is hypoxia of the myocardial fibers. A similar increase in flow is produced in the area supplied by a coronary artery if the artery is occluded and then released. This **reactive hyperemia** is similar to that seen in the

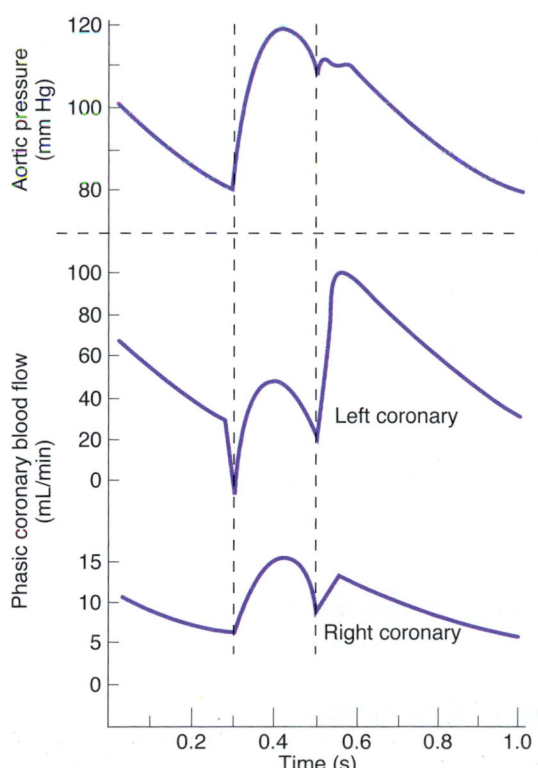

FIGURE 34–13 Blood flow in the left and right coronary arteries during various phases of the cardiac cycle. Systole occurs between the two vertical dashed lines. (Reproduced with permission from Berne RM, Levy MN: *Physiology,* 2nd ed. Mosby, 1988.)

Coronary Artery Disease

When flow through a coronary artery is reduced to the point that the myocardium it supplies becomes hypoxic, **angina pectoris** develops (see Chapter 31). If the myocardial ischemia is severe and prolonged, irreversible changes occur in the muscle, and the result is **myocardial infarction.** Many individuals have angina only on exertion, and blood flow is normal at rest. Others have more severe restriction of blood flow and have anginal pain at rest as well. Partially occluded coronary arteries can be constricted further by vasospasm, producing myocardial infarction. However, it is now clear that the most common cause of myocardial infarction is rupture of an **atherosclerotic plaque,** or hemorrhage into it, which triggers the formation of a coronary-occluding clot at the site of the plaque. The electrocardiographic changes in myocardial infarction are discussed in Chapter 30. When myocardial cells actually die, they leak enzymes into the circulation, and measuring the rises in serum enzymes and isoenzymes produced by infarcted myocardial cells also plays an important role in the diagnosis of myocardial infarction. The enzymes most commonly measured today are the MB isomer of creatine kinase (CK-MB), troponin T, and troponin I. Myocardial infarction is a very common cause of death in developed countries because of the widespread occurrence of atherosclerosis. In addition, there is a relation between atherosclerosis and circulating levels of **lipoprotein(a) (Lp[a]).** Lp(a) has an outer coat count of apo(a). It interferes with fibrinolysis by down-regulating plasmin generation (see Chapter 32). There is also a strong positive correlation between atherosclerosis and circulating levels of homocysteine. This substance damages endothelial cells. It is converted to nontoxic methionine in the presence of folate and vitamin B_{12}, and clinical trials are under way to determine whether supplements of folate and B_{12} lower the incidence of coronary disease. It now appears that atherosclerosis has an important inflammatory component as well. The lesions of the disease contain inflammatory cells, and there is a positive correlation between increased levels of C-reactive protein and other **inflammatory markers** in the circulation and subsequent myocardial infarction. Treatment of myocardial infarction aims to restore flow to the affected area as rapidly as possible while minimizing reperfusion injury. Needless to say, it should be started as promptly as possible to avoid irreversible changes in heart function.

skin (see below). Evidence suggests that in the heart it is due to release of adenosine.

NEURAL FACTORS

The coronary arterioles contain α-adrenergic receptors, which mediate vasoconstriction, and β-adrenergic receptors, which mediate vasodilation. Activity in the noradrenergic nerves to the heart and injections of norepinephrine cause coronary vasodilation. However, norepinephrine increases the heart rate and the force of cardiac contraction, and the vasodilation is due to production of vasodilator metabolites in the myocardium secondary to the increase in its activity. When the inotropic and chronotropic effects of noradrenergic discharge are blocked by a β-adrenergic blocking drug, stimulation of the noradrenergic nerves or injection of norepinephrine in unanesthetized animals elicits coronary vasoconstriction. Thus, the direct effect of noradrenergic stimulation is constriction rather than dilation of the coronary vessels. On the other hand, stimulation of vagal fibers to the heart dilates the coronaries.

When the systemic blood pressure falls, the overall effect of the reflex increase in noradrenergic discharge is increased coronary blood flow secondary to the metabolic changes in the myocardium at a time when the cutaneous, renal, and splanchnic vessels are constricted. In this way the circulation of the heart, like that of the brain, is preserved when flow to other organs is compromised.

CUTANEOUS CIRCULATION

The amount of heat lost from the body is regulated to a large extent by varying the amount of blood flowing through the skin. The fingers, toes, palms, and earlobes contain well-innervated anastomotic connections between arterioles and venules (arteriovenous anastomoses; see Chapter 32). Blood flow in response to thermoregulatory stimuli can vary from 1 to as much as 150 mL/100 g of skin/min, and it has been postulated that these variations are possible because blood can be shunted through the anastomoses. The subdermal capillary and venous plexus is a blood reservoir of some importance, and the skin is one of the few places where the reactions of blood vessels can be observed visually.

WHITE REACTION

When a pointed object is drawn lightly over the skin, the stroke lines become pale (**white reaction**). The mechanical stimulus apparently initiates contraction of the precapillary sphincters, and blood drains out of the capillaries and small veins. The response appears in about 15 s.

TRIPLE RESPONSE

When the skin is stroked more firmly with a pointed instrument, instead of the white reaction there is reddening at the site that appears in about 10 s (**red reaction**). This is followed in a few minutes by local swelling and diffuse, mottled reddening around the injury. The initial redness is due to capillary dilation, a direct response of the capillaries to pressure. The swelling (**wheal**) is local edema due to increased permeability of the capillaries and postcapillary venules, with consequent extravasation of fluid. The redness spreading out from the injury (**flare**) is due to arteriolar dilation. This three-part response—the red reaction, wheal, and flare—is called the **triple response** and is part of the normal reaction to injury (see Chapter 3). It persists after total sympathectomy. On the other hand, the flare is absent in locally anesthetized skin and in denervated skin after the sensory nerves have degenerated, but it is present immediately after nerve block or section above the site of the injury. This, plus other evidence, indicates that it is due to an **axon reflex,** a response in which impulses initiated in sensory nerves by the injury are relayed antidromically down other branches of the sensory nerve fibers (Figure 34–14). This is the one situation in the body in which there is substantial evidence for a physiologic effect due to antidromic conduction. The transmitter released at the central termination of the sensory C fiber neurons is substance P (see Chapter 7), and substance P and CGRP are present in all parts of the neurons. Both dilate arterioles and, in addition, substance P causes extravasation of fluid. Effective nonpeptide antagonists to substance P have now been developed, and they reduce the extravasation. Thus, it appears that these peptides produce the wheal.

REACTIVE HYPEREMIA

A response of the blood vessels that occurs in many organs but is visible in the skin is **reactive hyperemia,** an increase in the amount of blood in a region when its circulation is reestablished after a period of occlusion. When the blood supply to a limb is occluded, the cutaneous arterioles below the occlusion dilate. When the circulation is reestablished, blood flowing into the dilated vessels makes the skin become fiery red. O_2 in the atmosphere can diffuse a short distance through the skin, and reactive hyperemia is prevented if the circulation of the limb is occluded in an atmosphere of 100% O_2. Therefore, the arteriolar dilation is apparently due to a local effect of hypoxia.

GENERALIZED RESPONSES

Noradrenergic nerve stimulation and circulating epinephrine and norepinephrine constrict cutaneous blood vessels. No known vasodilator nerve fibers extend to the cutaneous vessels, and thus vasodilation is brought about by a decrease in constrictor tone as well as the local production of vasodilator metabolites. Skin color and temperature also depend on the state of the capillaries and venules. A cold blue or gray skin is one in which the arterioles are constricted and the capillaries dilated; a warm red skin is one in which both are dilated.

Because painful stimuli cause diffuse noradrenergic discharge, a painful injury causes generalized cutaneous vasoconstriction in addition to the local triple response. When the body temperature rises during exercise, the cutaneous blood vessels dilate in spite of continuing noradrenergic discharge in other parts of the body. Dilation of cutaneous vessels in response to a rise in hypothalamic temperature overcomes other reflex activity. Cold causes cutaneous vasoconstriction; however, with severe cold, superficial vasodilation may supervene. This vasodilation is the cause of the ruddy complexion seen on a cold day.

Shock is more profound in patients with elevated temperatures because of cutaneous vasodilation, and patients in shock should not be warmed to the point that their body temperature rises. This is sometimes a problem because well-meaning laymen have read in first-aid books that "injured patients should be kept warm," and they pile blankets on accident victims who are in shock.

PLACENTAL & FETAL CIRCULATION

UTERINE CIRCULATION

The blood flow of the uterus parallels the metabolic activity of the myometrium and endometrium and undergoes cyclic fluctuations that correlate with the menstrual cycle in nonpregnant women. The function of the spiral and basilar arteries of the endometrium in menstruation is discussed in Chapter 25. During pregnancy, blood flow increases rapidly as the uterus increases in size (Figure 34–15). Vasodilator metabolites are undoubtedly produced in the uterus, as they are in other active tissues. In early pregnancy, the arteriovenous O_2 difference across the uterus is small, and it has been suggested that estrogens act on the blood vessels to increase uterine blood flow in excess of tissue O_2 needs. However, even though uterine blood flow increases 20-fold during pregnancy, the

Spinal cord   **Sensory neuron**

Endings in skin

→ Orthodromic conduction
- - -> Antidromic conduction
······> Direction taken by impulses

Endings near arteriole

FIGURE 34–14   Axon reflex.

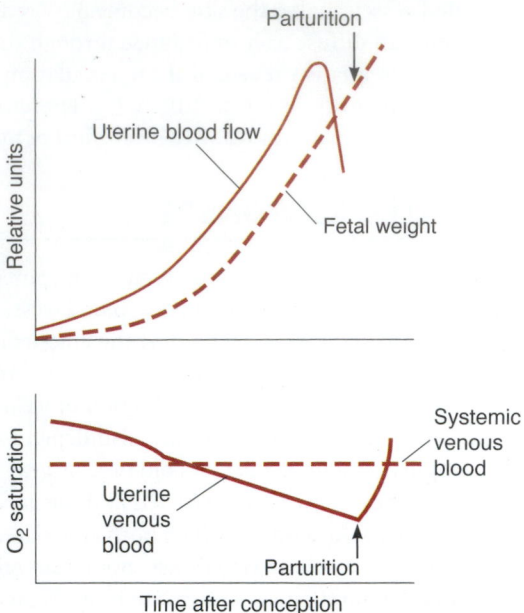

FIGURE 34–15 **Changes in uterine blood flow and the amount of O$_2$ in uterine venous blood during pregnancy.** (After Barcroft H. Modified and redrawn with permission from Keele CA, Neil E: *Samson Wright's Applied Physiology,* 12th ed. Oxford University Press, 1971.)

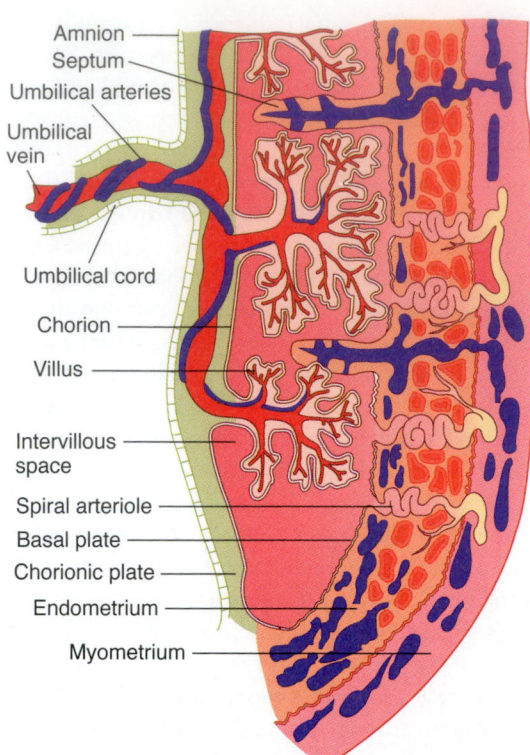

FIGURE 34–16 **Diagram of a section through the human placenta, showing the way the fetal villi project into the maternal sinuses.** (Reproduced with permission from Benson RC: *Handbook of Obstetrics and Gynecology,* 8th ed. Originally published by Appleton & Lange. Copyright © 1983 McGraw-Hill.)

size of the conceptus increases much more, changing from a single cell to a fetus plus a placenta that weighs 4 to 5 kg at term in humans. Consequently, more O$_2$ is extracted from the uterine blood during the latter part of pregnancy, and the O$_2$ saturation of uterine blood falls. Corticotrophin-releasing hormone appears to play an important role in up-regulating uterine blood flow, as well as in the eventual timing of birth.

PLACENTA

The placenta is the "fetal lung" (Figures 34–16 and 34–17). Its maternal portion is in effect a large blood sinus. Into this "lake" project the villi of the fetal portion containing the small branches of the fetal umbilical arteries and vein (Figure 34–16). O$_2$ is taken up by the fetal blood and CO$_2$ is discharged into the maternal circulation across the walls of the villi in a fashion analogous to O$_2$ and CO$_2$ exchange in the lungs (see Chapter 36). However, the cellular layers covering the villi are thicker and less permeable than the alveolar membranes in the lungs, and exchange is much less efficient. The placenta is also the route by which all nutritive materials enter the fetus and by which fetal wastes are discharged to the maternal blood.

FETAL CIRCULATION

The arrangement of the circulation in the fetus is shown diagrammatically in Figure 34–17. Fifty-five percent of the fetal cardiac output goes through the placenta. The blood in the umbilical vein in humans is believed to be about 80% saturated with O$_2$, compared with 98% saturation in the arterial circulation of the adult. The **ductus venosus** (Figure 34–18) diverts

some of this blood directly to the inferior vena cava, and the remainder mixes with the portal blood of the fetus. The portal and systemic venous blood of the fetus is only 26% saturated, and the saturation of the mixed blood in the inferior vena cava is approximately 67%. Most of the blood entering the heart through the inferior vena cava is diverted directly to the left atrium via the patent foramen ovale. Most of the blood from the superior vena cava enters the right ventricle and is expelled into the pulmonary artery. The resistance of the collapsed lungs is high, and the pressure in the pulmonary artery is several mm Hg higher than it is in the aorta, so that most of the blood in the pulmonary artery passes through the **ductus arteriosus** to the aorta. In this fashion, the relatively unsaturated blood from the right ventricle is diverted to the trunk and lower body of the fetus, while the head of the fetus receives the better-oxygenated blood from the left ventricle. From the aorta, some of the blood is pumped into the umbilical arteries and back to the placenta. The O$_2$ saturation of the blood in the lower aorta and umbilical arteries of the fetus is approximately 60%.

FETAL RESPIRATION

The tissues of fetal and newborn mammals have a remarkable but poorly understood resistance to hypoxia. However, the O$_2$ saturation of the maternal blood in the placenta is so low that the fetus might suffer hypoxic damage if fetal red cells did not

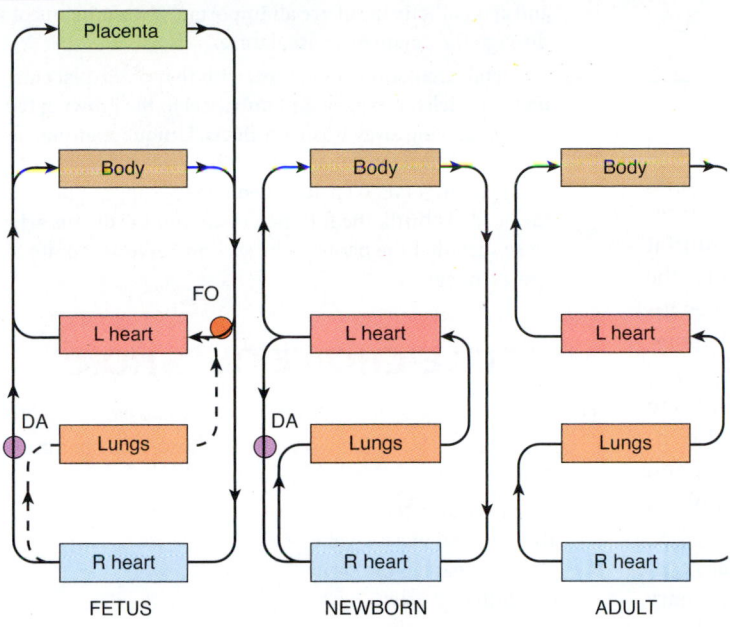

FIGURE 34–17 Diagram of the circulation in the fetus, the newborn infant, and the adult. DA, ductus arteriosus; FO, foramen ovale. (Redrawn and reproduced with permission from Born GVR et al: Changes in the heart and lungs at birth. Cold Spring Harbor Symp Quant Biol 1954;19:102.)

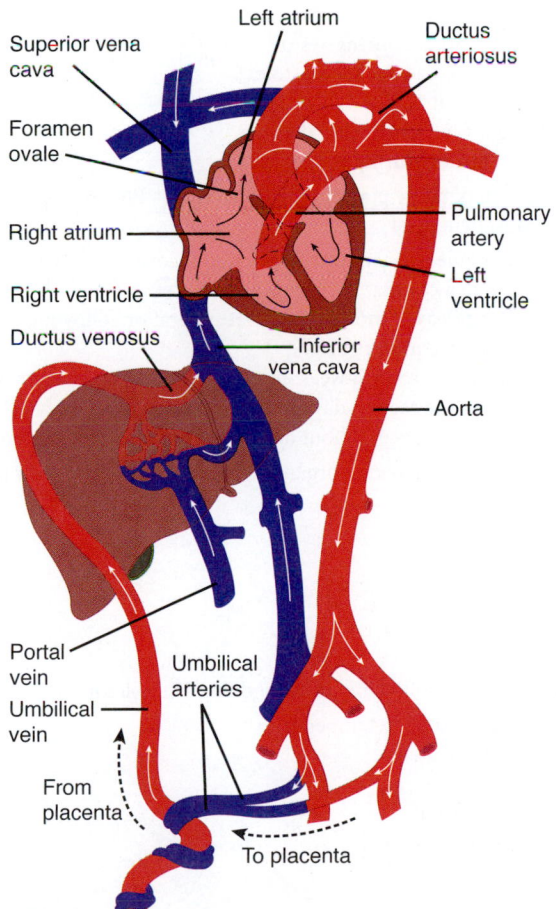

FIGURE 34–18 Circulation in the fetus. Most of the oxygenated blood reaching the heart via the umbilical vein and inferior vena cava is diverted through the foramen ovale and pumped out the aorta to the head, while the deoxygenated blood returned via the superior vena cava is mostly pumped through the pulmonary artery and ductus arteriosus to the feet and the umbilical arteries.

have a greater O_2 affinity than adult red cells (Figure 34–19). The fetal red cells contain fetal hemoglobin (hemoglobin F), whereas the adult cells contain adult hemoglobin (hemoglobin A). The cause of the difference in O_2 affinity between the two is that hemoglobin F binds 2, 3-DPG less effectively than hemoglobin A does. The decrease in O_2 affinity due to the binding of 2, 3-DPG is discussed in Chapter 32).

Some hemoglobin A is present in blood during fetal life (see Chapter 32). After birth, production of hemoglobin F normally ceases, and by the age of 4 mo 90% of the circulating hemoglobin is hemoglobin A.

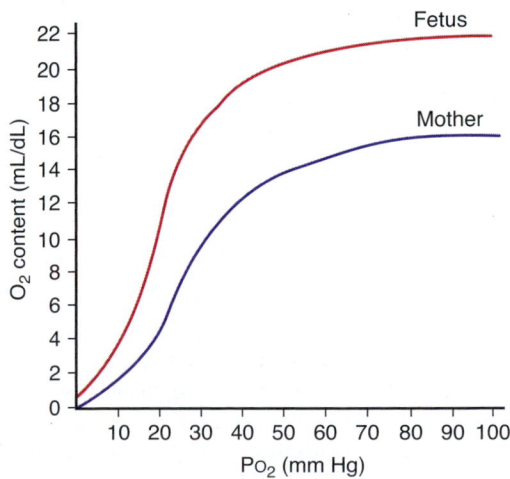

FIGURE 34–19 Dissociation curves of hemoglobin in human maternal and fetal blood.

CHANGES IN FETAL CIRCULATION & RESPIRATION AT BIRTH

Because of the patent ductus arteriosus and foramen ovale (Figure 34–18), the left heart and right heart pump in parallel in the fetus rather than in series as they do in the adult. At birth, the placental circulation is cut off and the peripheral resistance suddenly rises. The pressure in the aorta rises until it exceeds that in the pulmonary artery. Meanwhile, because the placental circulation has been cut off, the infant becomes increasingly asphyxial. Finally, the infant gasps several times, and the lungs expand. The markedly negative intrapleural pressure (–30 to –50 mm Hg) during the gasps contributes to the expansion of the lungs, but other factors are likely also involved. The sucking action of the first breath plus constriction of the umbilical veins squeezes as much as 100 mL of blood from the placenta (the "placental transfusion").

Once the lungs are expanded, the pulmonary vascular resistance falls to less than 20% of the value in utero, and pulmonary blood flow increases markedly. Blood returning from the lungs raises the pressure in the left atrium, closing the foramen ovale by pushing the valve that guards it against the interatrial septum. The ductus arteriosus constricts within a few hours after birth, producing functional closure, and permanent anatomic closure follows in the next 24–48 h due to extensive intimal thickening. The mechanism producing the initial constriction is not completely understood, but the increase in arterial O_2 tension plays an important role. Relatively high concentrations of vasodilators are present in the ductus in utero—especially prostaglandin F_{2a}—and synthesis of these prostaglandins is blocked by inhibition of cyclooxygenase at birth. In many premature infants the ductus fails to close spontaneously, but closure can be produced by infusion of drugs that inhibit cyclooxygenase. NO may also be involved in maintaining ductal patency in this setting.

CHAPTER SUMMARY

- Cerebrospinal fluid is produced predominantly in the choroid plexus of the brain, in part via active transport mechanisms in the choroid epithelial cells. Fluid is reabsorbed into the bloodstream to maintain appropriate pressure in the setting of continuous production.

- The permeation of circulating substances into the brain is tightly controlled. Water, CO_2, and O_2 permeate freely. Other substances (such as glucose) require specific transport mechanisms, whereas entry of macromolecules is negligible. The effectiveness of the blood–brain barrier in preventing entry of xenobiotics is bolstered by active efflux mediated by P-glycoprotein.

- The coronary circulation supplies oxygen to the contracting myocardium. Metabolic products and neural input induce vasodilation as needed for oxygen demand. Blockage of coronary arteries may lead to irreversible injury to heart tissue.

- Control of cutaneous blood flow is a key facet of temperature regulation, and is underpinned by varying levels of shunting through arteriovenous anastomoses. Hypoxia, axon reflexes, and sympathetic input are all important determinants of flow through the cutaneous vasculature.

- The fetal circulation cooperates with that of the placenta and uterus to deliver oxygen and nutrients to the growing fetus, as well as carrying away waste products. Unique anatomic features of the fetal circulation as well as biochemical properties of fetal hemoglobin serve to ensure adequate O_2 supply, particularly to the head. At birth, the foramen ovale and the ductus arteriosus close such that the neonatal lungs now serve as the site for oxygen exchange.

MULTIPLE-CHOICE QUESTIONS

For all questions, select the single best answer unless otherwise directed.

1. Blood in which of the following vessels normally has the lowest P_{O_2}?
   A) maternal artery
   B) maternal uterine vein
   C) maternal femoral vein
   D) umbilical artery
   E) umbilical vein

2. The pressure differential between the heart and the aorta is least in the
   A) left ventricle during systole.
   B) left ventricle during diastole.
   C) right ventricle during systole.
   D) right ventricle during diastole.
   E) left atrium during systole.

3. Injection of tissue plasminogen activator (t-PA) would probably be most beneficial
   A) after at least 1 y of uncomplicated recovery following occlusion of a coronary artery.
   B) after at least 2 mo of rest and recuperation following occlusion of a coronary artery.
   C) during the second week after occlusion of a coronary artery.
   D) during the second day after occlusion of a coronary artery.
   E) during the second hour after occlusion of a coronary artery.

4. Which of the following organs has the greatest blood flow per 100 g of tissue?
   A) brain
   B) heart muscle
   C) skin
   D) liver
   E) kidneys

5. Which of the following does *not* dilate arterioles in the skin?
   A) increased body temperature
   B) epinephrine
   C) bradykinin
   D) substance P
   E) vasopressin

6. A baby boy is brought to the hospital because of convulsions. In the course of a workup, his body temperature and plasma glucose are found to be normal, but his cerebrospinal fluid glucose is 12 mg/dL (normal, 65 mg/dL). A possible explanation of his condition is
   A) constitutive activation of GLUT 3 in neurons.
   B) SGLT 1 deficiency in astrocytes.
   C) GLUT 5 deficiency in cerebral capillaries.
   D) GLUT 1 55K deficiency in cerebral capillaries.
   E) GLUT 1 45K deficiency in microglia.

CHAPTER RESOURCES

Begley DJ, Bradbury MW, Kreater J (editors): *The Blood–Brain Barrier and Drug Delivery to the CNS.* Marcel Dekker, 2000.

Birmingham K (editor): The heart. Nature 2002;415:197.

Duncker DJ, Bache RJ: Regulation of coronary blood flow during exercise. Physiol Rev 2008;88:1009.

Hamel E: Perivascular nerves and the regulation of cerebrovascular tone. J Appl Physiol 2006;100:1059.

Johanson CE, et al: Multiplicity of cerebrospinal fluid functions: New challenges in health and disease. Cerebrospinal Fluid Res 2008;5:10.

Ward JPT: Oxygen sensing in context. Biochim Biophys Acta 2008;1777:1.

C H A P T E R

35

Pulmonary Function

OBJECTIVES

After studying this chapter, you should be able to:

- Define partial pressure and calculate the partial pressure of each of the important gases in the atmosphere at sea level.
- List the passages through which air passes from the exterior to the alveoli, and describe the cells that line each of them.
- List the major muscles involved in respiration, and state the role of each.
- Define the basic measures of lung volume and give approximate values for each in a normal adult.
- Define compliance, and give examples of diseases in which it is abnormal.
- Describe the chemical composition and function of surfactant.
- List the factors that determine alveolar ventilation.
- Define diffusion capacity, and compare the diffusion of O_2 with that of CO_2 in the lungs.
- Compare the pulmonary and systemic circulations, listing the main differences between them.
- Describe basic lung defense and metabolic functions.

INTRODUCTION

Respiration, as the term is generally used, includes two processes: **external respiration,** the absorption of O_2 and removal of CO_2 from the body as a whole; and **internal respiration,** the utilization of O_2 and production of CO_2 by cells and the gaseous exchanges between the cells and their fluid medium. Aspects of external respiratory physiology are presented throughout this section. In this chapter, the processes responsible for the uptake of O_2 and excretion of CO_2 in the lungs are explored. The next chapter is concerned with the transport of O_2 and CO_2 to and from the tissues. The final chapter in this section examines some key factors that regulate respiration. Throughout each chapter, clinical implications of specific physiology will be presented.

PROPERTIES OF GASES

The pressure of a gas is proportional to its temperature and the number of moles per volume:

$$P = \frac{nRT}{V} \quad \text{(from equation of state of ideal gas)}$$

where

P = Pressure

n = Number of moles

R = Gas constant

T = Absolute temperature

V = Volume

PARTIAL PRESSURES

Unlike liquids, gases expand to fill the volume available to them, and the volume occupied by a given number of gas molecules at a given temperature and pressure is (ideally) the same regardless of the composition of the gas. Therefore, the pressure exerted by any one gas in a mixture of gases (its **partial pressure**) is equal to the total pressure times the fraction of the total amount of gas it represents.

The composition of dry air is 20.98% O_2, 0.04% CO_2, 78.06% N_2, and 0.92% other inert constituents such as argon and helium. The barometric pressure (PB) at sea level is 760 mm Hg (1 atmosphere). The partial pressure (indicated by the symbol P) of O_2 in dry air is therefore 0.21 × 760, or 160 mm Hg at sea level. The P_{N_2} and the other inert gases is 0.79 × 760, or 600 mm Hg; and the P_{CO_2} is 0.0004 × 760, or 0.3 mm Hg. The water vapor in the air in most climates reduces these percentages, and therefore the partial pressures, to a slight degree. Air equilibrated with water is saturated with water vapor, and inspired air is saturated by the time it reaches the lungs. The P_{H_2O} at body temperature (37 °C) is 47 mm Hg. Therefore, the partial pressures at sea level of the other gases in the air reaching the lungs are P_{O_2}, 149 mm Hg; P_{CO_2}, 0.3 mm Hg; and P_{N_2} (including the other inert gases), 564 mm Hg.

Gas diffuses from areas of high pressure to areas of low pressure, with the rate of diffusion depending on the concentration gradient and the nature of the barrier between the two areas. When a mixture of gases is in contact with and permitted to equilibrate with a liquid, each gas in the mixture dissolves in the liquid to an extent determined by its partial pressure and its solubility in the fluid. The partial pressure of a gas in a liquid is the pressure that, in the gaseous phase in equilibrium with the liquid, would produce the concentration of gas molecules found in the liquid.

METHODS OF QUANTITATING RESPIRATORY PHENOMENA

Modern spirometers permit direct measurement of gas intake and output. Since gas volumes vary with temperature and

TABLE 35–1 Standard conditions to which measurements involving gas volumes are corrected.

STPD	0 °C, 760 mm Hg, dry (standard temperature and pressure, dry)
BTPS	Body temperature and pressure, saturated with water vapor
ATPD	Ambient temperature and pressure, dry
ATPS	Ambient temperature and pressure, saturated with water vapor

pressure and since the amount of water vapor in them varies, these devices have the ability to correct respiratory measurements involving volume to a stated set of standard conditions. The four most commonly used standards and their abbreviations are shown in Table 35–1. It should be noted that correct measurements are highly dependent on the ability for the practitioner to properly encourage the patient to fully utilize the device. Modern techniques for gas analysis make possible rapid, reliable measurements of the composition of gas mixtures and the gas content of body fluids. For example, O_2 and CO_2 electrodes, small probes sensitive to O_2 or CO_2, can be inserted into the airway or into blood vessels or tissues and the P_{O_2} and P_{CO_2} recorded continuously. Chronic assessment of oxygenation is carried out noninvasively with a **pulse oximeter,** which is usually attached to the ear.

ANATOMY OF THE LUNGS

THE RESPIRATORY SYSTEM

The respiratory system is made up of a gas-exchanging organ (the lungs) and a "pump" that ventilates the lungs. The pump consists of the chest wall; the respiratory muscles, which increase and decrease the size of the thoracic cavity; the areas in the brain that control the muscles; and the tracts and nerves that connect the brain to the muscles. At rest, a normal human breathes 12 to 15 times a minute. About 500 mL of air per breath, or 6 to 8 L/min, is inspired and expired. This air mixes with the gas in the alveoli, and, by simple diffusion, O_2 enters the blood in the pulmonary capillaries while CO_2 enters the alveoli. In this manner, 250 mL of O_2 enters the body per minute and 200 mL of CO_2 is excreted.

Traces of other gases, such as methane from the intestines, are also found in expired air. Alcohol and acetone are expired when present in appreciable quantities in the body. Indeed, over 250 different volatile substances have been identified in human breath.

AIR PASSAGES

After passing through the nasal passages and pharynx, where it is warmed and takes up water vapor, the inspired air passes

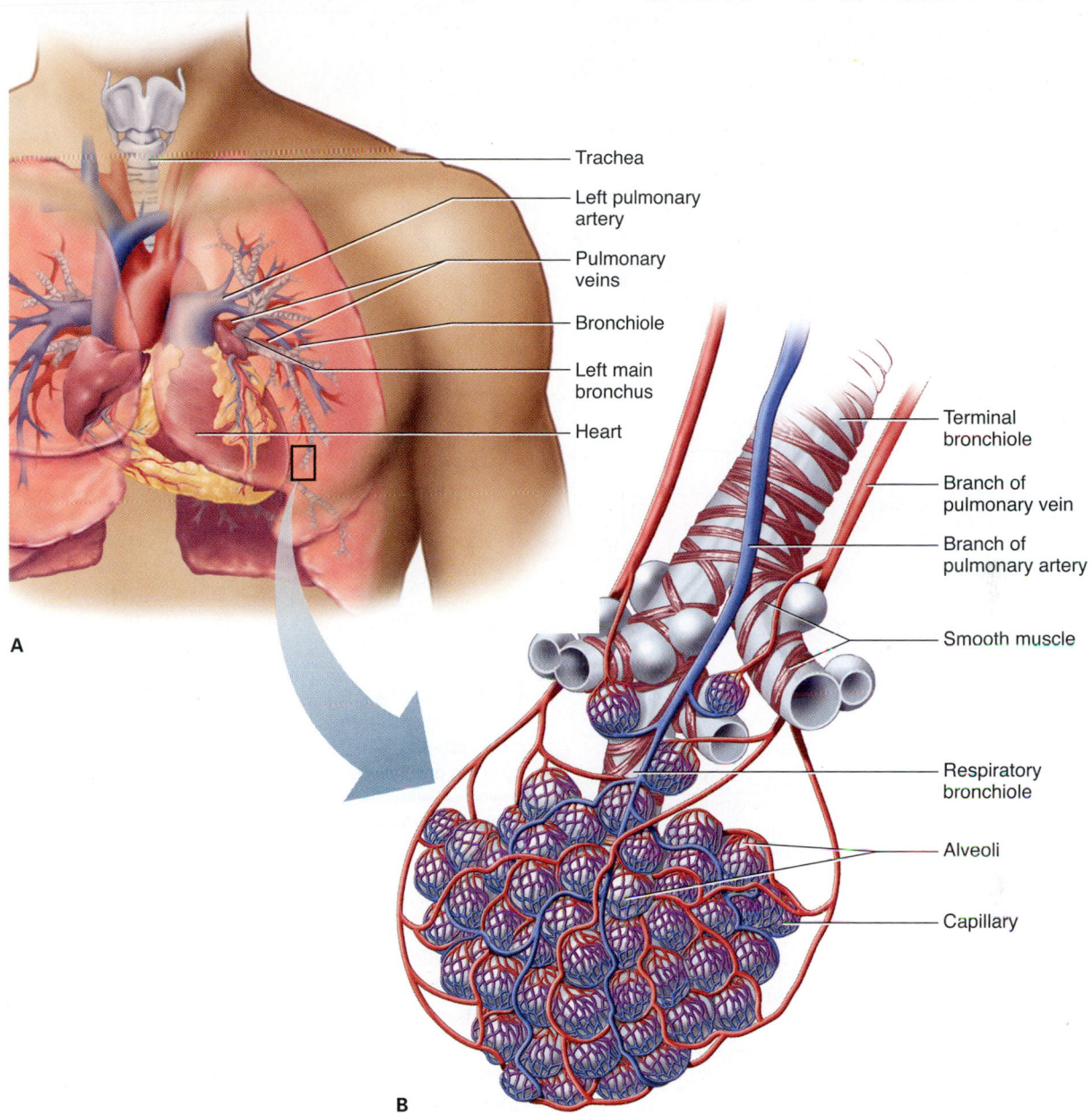

A

B

FIGURE 35–1 **Structure of the respiratory system. A)** The respiratory system is diagrammed with a transparent lung to emphasize the flow of air into and out of the system. **B)** Enlargement of boxed area from (A) shows transition from conducting airway to the respiratory airway, with emphasis on the anatomy of the alveoli. Red and blue represent oxygenated and deoxygenated blood, respectively. (*Continued*)

down the trachea and through the bronchioles, respiratory bronchioles, and alveolar ducts to the alveoli, where gas exchange occurs (Figure 35–1). Between the trachea and the alveolar sacs, the airways divide 23 times. The first 16 generations of passages form the conducting zone of the airways that transports gas from and to the exterior. They are made up of bronchi, bronchioles, and terminal bronchioles. The remaining seven generations form the transitional and respiratory zones where gas exchange occurs; they are made up of respiratory bronchioles, alveolar ducts, and alveoli. These multiple divisions greatly increase the total cross-sectional area of the air-

ways, from 2.5 cm^2 in the trachea to 11,800 cm^2 in the alveoli (Figure 35–2). Consequently, the velocity of air flow in the small airways declines to very low values.

The alveoli are surrounded by pulmonary capillaries (Figure 35–1). In most areas, air and blood are separated only by the alveolar epithelium and the capillary endothelium, so they are about 0.5 μm apart (Figure 35–3). Humans have 300 million alveoli, and the total area of the alveolar walls in contact with capillaries in both lungs is about 70 m^2.

The alveoli are lined by two types of epithelial cells. **Type I cells** are flat cells with large cytoplasmic extensions and are

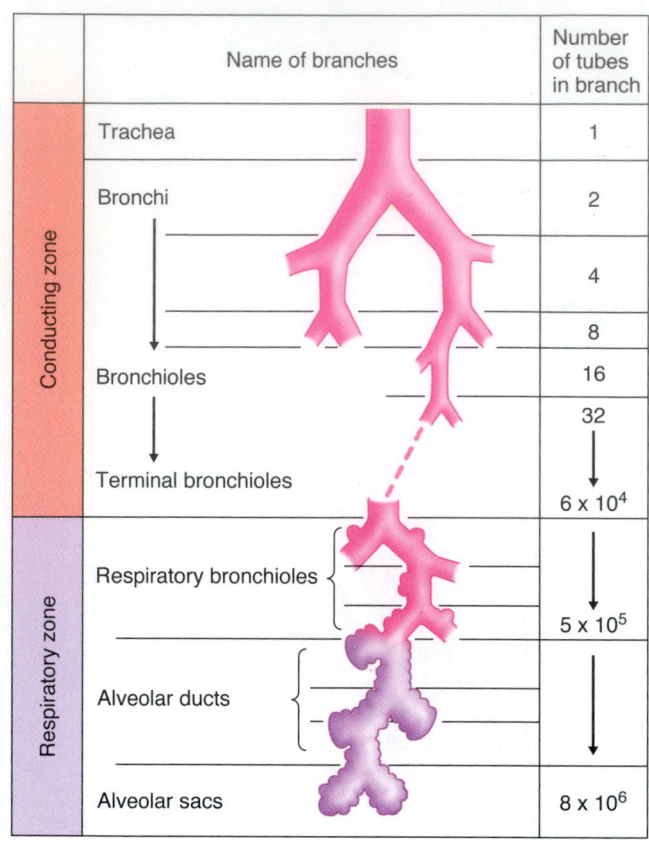

C

FIGURE 35–1 (*Continued*) **C)** The branching patterns of the airway during the transition form conducting to respiratory airway are drawn (not all divisions are drawn, and drawings are not to scale).

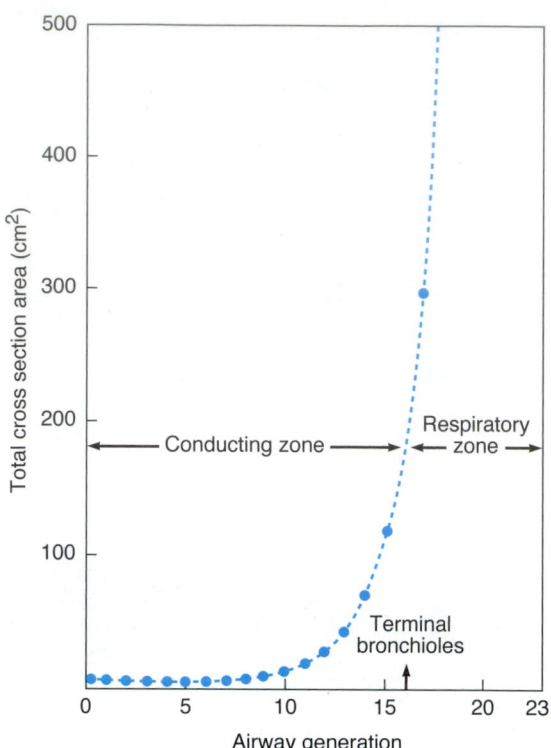

FIGURE 35–2 **Total airway cross-sectional area as a function of airway generation.** Note the extremely rapid increase in total cross-sectional area in the respiratory zone. As a result, forward velocity of gas during inspiration falls to a very low level in this zone. (Reproduced with permission from West JB: *Respiratory Physiology: The Essentials,* 4th ed. Williams & Wilkins, 1991.)

the primary lining cells of the alveoli, covering approximately 95% of the alveolar epithelial surface area. **Type II cells (granular pneumocytes)** are thicker and contain numerous lamellar inclusion bodies. A primary function of these cells is to secrete surfactant; however, they are also important in alveolar repair as well as other cellular physiology. Although these cells make up approximately 5% of the surface area, they represent approximately 60% of the epithelial cells in the alveoli. The alveoli also contain other specialized cells, including pulmonary alveolar macrophages (PAMs, or AMs), lymphocytes, plasma cells, neuroendocrine cells, and mast cells. The mast cells contain heparin, various lipids, histamine, and various proteases that participate in allergic reactions (see Chapter 3).

THE BRONCHI & THEIR INNERVATION

The trachea and bronchi have cartilage in their walls but relatively little smooth muscle. They are lined by a ciliated epithelium that contains mucous and serous glands. Cilia are present as far as the respiratory bronchioles, but glands are absent from the epithelium of the bronchioles and terminal bronchioles, and their walls do not contain cartilage. However, their walls contain more smooth muscle, of which the largest

amount relative to the thickness of the wall is present in the terminal bronchioles.

The walls of the bronchi and bronchioles are innervated by the autonomic nervous system. Muscarinic receptors are abundant, and cholinergic discharge causes bronchoconstriction. The bronchial epithelium and smooth muscle contain β_2-adrenergic receptors. Many of these are not innervated. Some may be located on cholinergic endings, where they inhibit acetylcholine release. The β_2 receptors mediate bronchodilation. They increase bronchial secretion, while α_1 adrenergic receptors inhibit secretion. There is, in addition, a **noncholinergic, nonadrenergic innervation** of the bronchioles that produces bronchodilation, and evidence suggests that vasoactive intestinal polypeptide (VIP) is the mediator responsible for the dilation.

ANATOMY OF BLOOD FLOW IN THE LUNG

Both the **pulmonary circulation** and the **bronchial circulation** contribute to blood flow in the lung. In the pulmonary circulation, almost all the blood in the body passes via the pulmonary artery to the pulmonary capillary bed, where it is oxygenated and returned to the left atrium via the pulmonary

FIGURE 35–3 **Portion of an interalveolar septum in the adult human lung. A)** A cross-section of the respiratory zone shows the relationship between capillaries and the airway epithelium. Only 4 of the 18 alveoli are labeled. **B)** Enlargement of the boxed area from (A) displaying intimate relationship between capillaries, the interstitium, and the alveolar epithelium. **C)** Electron micrograph displaying area depicted in (B). The pulmonary capillary (cap) in the septum contains plasma with red blood cells apposed to the thin epithelial cells that line the alveoli. Note the closely apposed endothelial wall and pulmonary epithelium, separated at places by connective tissue fibers (cf); en, nucleus of endothelial cell; epl, nucleus of type I alveolar epithelial cell; a, alveolar space; ma, alveolar macrophage. (Reproduced with permission from (A, B) Widmaier EP, Raff H, Strang KT: *Vander's Human Physiology: The Mechanisms of Body Function*, 11th ed. McGraw-Hill, 2008; and (C) Burri PA: Development and growth of the human lung. In: *Handbook of Physiology*, Section 3, *The Respiratory System*. Fishman AP, Fisher AB [editors]. American Physiological Society, 1985.)

veins (Figure 35–4). The separate and much smaller bronchial circulation includes the bronchial arteries that come from systemic arteries. They form capillaries, which drain into bronchial veins or anastomose with pulmonary capillaries or veins (Figure 35–5). The bronchial veins drain into the azygos vein. The bronchial circulation nourishes the trachea down to the terminal bronchioles and also supplies the pleura and hilar lymph nodes. It should be noted that lymphatic channels are more abundant in the lungs than in any other organ.

MECHANICS OF RESPIRATION

INSPIRATION & EXPIRATION

The lungs and the chest wall are elastic structures. Normally, no more than a thin layer of fluid is present between the lungs and the chest wall (intrapleural space). The lungs slide easily on the chest wall, but resist being pulled away from it in the same way that two moist pieces of glass slide on each other but resist separation. The pressure in the "space" between the lungs and chest wall (intrapleural pressure) is subatmospheric (Figure 35–6). The lungs are stretched when they expand at birth, and at the end of quiet expiration their tendency to recoil from the chest wall is just balanced by the tendency of the

FIGURE 35–4 Pulmonary and systemic circulations. Representative areas of blood flow are labeled with corresponding blood pressure (mm Hg). (Modified from Comroe JH Jr.: *Physiology of Respiration,* 2nd ed. Year Book, 1974.)

FIGURE 35–6 Pressure in the alveoli and the plural space relative to atmospheric pressure during inspiration and expiration. The dashed line indicates what the intrapleural pressure would be in the absence of airway and tissue resistance; the actual curve (solid line) is skewed to the left by the resistance. Volume of breath during inspiration/expiration is graphed for comparison.

chest wall to recoil in the opposite direction. If the chest wall is opened, the lungs collapse; and if the lungs lose their elasticity, the chest expands and becomes barrel-shaped.

Inspiration is an active process. The contraction of the inspiratory muscles increases intrathoracic volume. The intrapleural pressure at the base of the lungs, which is normally about –2.5 mm Hg (relative to atmospheric) at the start of inspiration, decreases to about –6 mm Hg. The lungs

are pulled into a more expanded position. The pressure in the airway becomes slightly negative, and air flows into the lungs. At the end of inspiration, the lung recoil begins to pull the chest back to the expiratory position, where the recoil pressures of the lungs and chest wall balance. The pressure in the airway becomes slightly positive, and air flows out of the lungs. Expiration during quiet breathing is passive in the sense that no muscles that decrease intrathoracic volume contract. However, some contraction of the inspiratory muscles occurs in the early part of expiration. This contraction exerts a braking action on the recoil forces and slows expiration.

Strong inspiratory efforts reduce intrapleural pressure to values as low as –30 mm Hg, producing correspondingly greater degrees of lung inflation. When ventilation is increased, the extent of lung deflation is also increased by active contraction of expiratory muscles that decrease intrathoracic volume.

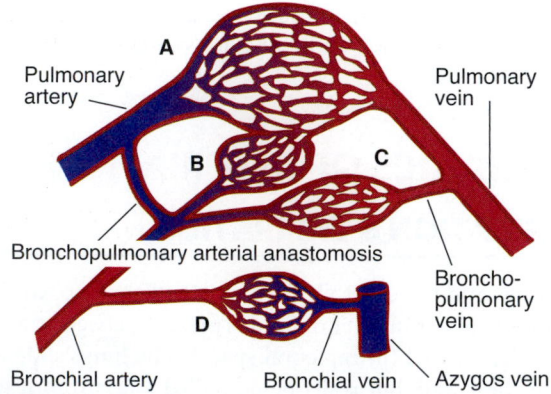

FIGURE 35–5 Relationship between the bronchial and pulmonary circulations. The pulmonary artery supplies pulmonary capillary network **A.** The bronchial artery supplies capillary networks **B, C,** and **D.** Blue-colored areas represent blood of low O_2 content. (Reproduced with permission from Murray JF: *The Normal Lung.* Saunders, 1986.)

LUNG VOLUMES

The amount of air that moves into the lungs with each inspiration (or the amount that moves out with each expiration) is called the **tidal volume.** The air inspired with a maximal inspiratory effort in excess of the tidal volume is the **inspiratory reserve volume.** The volume expelled by an active expiratory effort after passive expiration is the **expiratory reserve volume,** and the air left in the lungs after a maximal expiratory effort is the **residual volume.** Normal values for these lung volumes, and names applied to combinations of them, are shown in Figure 35–7. The space in the conducting zone of the airways occupied by gas that does not exchange with blood in the pulmonary vessels is the **respiratory dead space.** The **forced vital capacity (FVC),** the largest amount of air that can be expired after a maximal inspiratory effort, is frequently measured clinically as an index of pulmonary function. It gives useful information about the strength of the respiratory muscles and other aspects of pulmonary function. The fraction of the vital capacity expired during the first second of a forced expiration is referred to as FEV_1 (formerly the timed vital capacity) (Figure 35–8). The FEV_1 to FVC ratio (FEV_1/FVC) is a useful tool in the diagnosis of airway disease (Clinical Box 35–1). The amount of air inspired per minute (**pulmonary ventilation, respiratory minute volume**) is normally about 6 L (500 mL/ breath × 12 breaths/min). The **maximal voluntary ventilation (MVV)** is the largest volume of gas that can be moved into and out of the lungs in 1 min by voluntary effort. The normal MVV is 125 to 170 L/min.

Respiratory Volumes and Capacities for an Average Young Adult Male		
Measurement	Typical Value	Definition
Respiratory Volumes		
① Tidal volume (TV)	500 ml	Amount of air inhaled or exhaled in one breath during relaxed, quiet breathing
② Inspiratory reserve volume (IRV)	3000 ml	Amount of air in excess of tidal inspiration that can be inhaled with maximum effort
③ Expiratory reserve volume (ERV)	1200 ml	Amount of air in excess of tidal expiration that can be exhaled with maximum effort
④ Residual volume (RV)	1200 ml	Amount of air remaining in the lungs after maximum expiration; keeps alveoli inflated between breaths and mixes with fresh air on next inspiration
Respiratory Capacities		
⑤ Vital capacity (VC)	4700 ml	Amount of air that can be exhaled with maximum effort after maximum inspiration (ERV + TV + IRV); used to assess strength of thoracic muscles as well as pulmonary function
⑥ Inspiratory capacity (IC)	3500 ml	Maximum amount of air that can be inhaled after a normal tidal expiration (TV + IRV)
⑦ Functional residual capacity (FRC)	2400 ml	Amount of air remaining in the lungs after a normal tidal expiration (RV + ERV)
⑧ Total lung capacity (TLC)	5900 ml	Maximum amount of air the lungs can contain (RV + VC)

FIGURE 35–7   Lung volumes and capacity measurements. Top left: A cartoon figure representing lung space divided into lung volumes. Dead space refers to areas where gas exchange does not occur; all other spaces are defined in the accompanying table. **Top right:** Spirometer recordings are shown with marked lung volumes and capacities. Table at bottom defines individual measurements and values from the top graphs. Note that residual volume, total lung capacity, and function residual capacity cannot be measure with a spirometer. (Right figure reproduced with permission from Widmaier EP, Raff H, Strang KT: *Vander's Human Physiology: The Mechanisms of Body Function,* 11th ed. McGraw-Hill, 2008.)

FIGURE 35–8 **Volume of gas expired by a normal adult man during a forced expiration, demonstrating the FEV$_1$ and the total vital capacity (VC).** (Reproduced, with permission, from Crapo RO: Pulmonary-function testing. N Engl J Med 1994;331:25. Copyright © 1994, Massachusetts Medical Society.)

RESPIRATORY MUSCLES

Movement of the **diaphragm** accounts for 75% of the change in intrathoracic volume during quiet inspiration. Attached around the bottom of the thoracic cage, this muscle arches over the liver and moves downward like a piston when it contracts. The distance it moves ranges from 1.5 cm to as much as 7 cm with deep inspiration (Figure 35–9).

The diaphragm has three parts: the costal portion, made up of muscle fibers that are attached to the ribs around the bottom of the thoracic cage; the crural portion, made up of fibers that are attached to the ligaments along the vertebrae; and the central tendon, into which the costal and the crural fibers insert. The central tendon is also the inferior part of the pericardium. The crural fibers pass on either side of the esophagus and can compress it when they contract. The costal and crural portions are innervated by different parts of the phrenic nerve and can contract separately. For example, during vomiting and eructation, intra-abdominal pressure is increased by contraction of the costal fibers but the crural fibers remain relaxed, allowing material to pass from the stomach into the esophagus.

The other important **inspiratory muscles** are the **external intercostal muscles,** which run obliquely downward and forward from rib to rib. The ribs pivot as if hinged at the back, so that when the external intercostals contract they elevate the lower ribs. This pushes the sternum outward and increases the anteroposterior diameter of the chest. The transverse diameter also increases, but to a lesser degree. Either the diaphragm or the external intercostal muscles alone can maintain adequate ventilation at rest. Transection of the spinal cord above the third cervical segment is fatal without artificial respiration, but transection below the fifth cervical segment is not, because it leaves the phrenic nerves that innervate the diaphragm intact; the phrenic nerves arise from cervical segments 3–5. Conversely, in patients with bilateral phrenic nerve palsy but intact innervation of their intercostal muscles, respiration is somewhat labored but adequate to maintain life.

The scalene and sternocleidomastoid muscles in the neck are accessory inspiratory muscles that help to elevate the thoracic cage during deep labored respiration.

A decrease in intrathoracic volume and forced expiration result when the **expiratory muscles** contract. The internal intercostals have this action because they pass obliquely downward and posteriorly from rib to rib and therefore pull the rib cage downward when they contract. Contractions of the muscles of the anterior abdominal wall also aid expiration by pulling the rib cage downward and inward and by increasing the intra-abdominal pressure, which pushes the diaphragm upward.

GLOTTIS

The abductor muscles in the larynx contract early in inspiration, pulling the vocal cords apart and opening the glottis. During swallowing or gagging, a reflex contraction of the adductor muscles closes the glottis and prevents aspiration of food, fluid, or vomitus into the lungs. In unconscious or anesthetized patients, glottic closure may be incomplete and vomitus may enter the trachea, causing an inflammatory reaction in the lung (**aspiration pneumonia**).

The laryngeal muscles are supplied by the vagus nerves. When the abductors are paralyzed, there is inspiratory stridor. When the adductors are paralyzed, food and fluid enter the trachea, causing aspiration pneumonia and edema. Bilateral cervical vagotomy in animals causes the slow development of fatal pulmonary congestion and edema. The edema is due at least in part to aspiration, although some edema develops even if a tracheostomy is performed before the vagotomy.

BRONCHIAL TONE

In general, the smooth muscle in the bronchial walls aids respiration. The bronchi dilate during inspiration and constrict during expiration. Dilation is produced by sympathetic discharge and constriction by parasympathetic discharge. Stimulation of sensory receptors in the airways by irritants and chemicals such as sulfur dioxide produces reflex bronchoconstriction that is mediated via cholinergic pathways. Cool air also causes bronchoconstriction, and so does exercise, possibly because the increased respiration associated with it cools the airways. In addition, the bronchial muscles protect the bronchi during coughing. There is a circadian rhythm in bronchial tone, with maximal constriction at about 6:00 AM and maximal dilation at about 6:00 PM. Many chemical substances including VIP, substance P, adenosine, and many cytokines and inflammatory modulators can affect bronchial tone, although their full roles in the physiologic regulation of bronchial tone is still unsettled.

CLINICAL BOX 35–1

Airway Diseases That Alter Airflow

Obstructive Disease: Asthma

Asthma is characterized by episodic or chronic wheezing, cough, and a feeling of tightness in the chest as a result of bronchoconstriction. Although the disease is not fully understood, three airway abnormalities are present: **airway obstruction** that is at least partially reversible, airway inflammation, and airway hyperresponsiveness to a variety of stimuli. A link to allergy has long been recognized, and plasma IgE levels are often elevated. Proteins released from eosinophils in the inflammatory reaction may damage the airway epithelium and contribute to the hyperresponsiveness. Leukotrienes are released from eosinophils and mast cells, and can enhance bronchoconstriction. Numerous other amines, neuropeptides, chemokines, and interleukins have effects on bronchial smooth muscle or produce inflammation, and they may be involved in asthma.

Because β_2-adrenergic receptors mediate bronchodilation, β_2-adrenergic agonists have long been the mainstay of treatment for mild to moderate asthma attacks. Inhaled and systemic steroids are used even in mild to moderate cases to reduce inflammation; they are very effective, but their side effects can be a problem. Agents that block synthesis of leukotrienes or their $CysLT_1$ receptor have also proved useful in some cases.

Restrictive Disease: Emphysema

Emphysema is a degenerative and potentially fatal pulmonary disease that is characterized by a loss of lung elasticity and replacement of alveoli with large air sacs. This loss of elasticity prevents full expansion of the lung, or **airway restriction,** during breathing. The most common cause of emphysema is heavy cigarette smoking. The smoke causes an increase in the number of pulmonary alveolar macrophages, and these macrophages release a chemical substance that attracts leukocytes to the lungs. The leukocytes in turn release proteases including elastase, which attacks the elastic tissue in the lungs. At the same time, α_1-antitrypsin, a plasma protein that normally inactivates elastase and other proteases, is itself inhibited. The α_1-antitrypsin is inactivated by oxygen radicals, and these are released by the leukocytes. The final result is a protease–antiprotease imbalance with increased destruction of lung tissue. Similar protease–antiprotease imbalance can occur through congenital deficiency α_1-antitrypsin.

Airflow Measurements of Obstructive & Restrictive Disease

In a healthy normal adult male, FVC is approximately 5.0 L, FEV_1 is approximately 4.0 L, and thus, the calculated FEV_1/FVC is 80%. As would be expected, patients with obstructive or restrictive diseases display reduced FVC, on the order of 3.0 L, and this measurement alone does not differentiate between the two. However, measurement of FEV_1 can significantly vary between the two diseases. In obstructive disorders, patients tend to show a slow, steady slope to the FVC, resulting in a small FEV_1, on the order of 1.3 L. However, in the restrictive disorder patients, air flow tends to be fast at first, and then due to the loss of elasticity, quickly levels out to approach FVC. The resultant FEV_1 is much greater, on the order of 2.8 L, even though FVC is equivalent. A quick calculation of FEV_1/FVC for obstructive (42%) versus restrictive (90%) patients defines the hallmark measurements in evaluating these two diseases. Obstructive disorders result in a marked decrease in both FVC and FEV_1/FVC, whereas restrictive disorders result in a loss of FVC without loss in FEV_1/FVC.

COMPLIANCE OF THE LUNGS & CHEST WALL

The interaction between the recoil of the lungs and recoil of the chest can be demonstrated in living subjects through a spirometer that has a valve just beyond the mouthpiece. The mouthpiece contains a pressure-measuring device. After the subject inhales a given amount, the valve is shut, closing off the airway. The respiratory muscles are then relaxed while the pressure in the airway is recorded. The procedure is repeated after inhaling or actively exhaling various volumes. The curve of airway pressure obtained in this way, plotted against volume, is the **relaxation pressure curve** of the total respiratory system (Figure 35–10). The pressure is zero at a lung volume that corresponds to the volume of gas in the lungs at the end of quiet expiration (**functional residual capacity,** or **FRC;** also known as relaxation volume). It is positive at greater volumes and negative at smaller volumes. The change in lung volume per unit change in airway pressure ($\Delta V/\Delta P$) is the **compliance** (stretchability) of the lungs and chest wall. It is normally measured in the pressure range where the relaxation pressure curve is steepest, and the normal value is approximately 0.2 L/cm H_2O. However, compliance depends on lung volume; an individual with only one lung has approximately half the ΔV for a given ΔP. Compliance is also slightly greater when measured during deflation than when measured during inflation. Consequently, it is more informative to examine the whole pressure–volume curve. The curve is shifted downward and to the right (compliance is decreased) by pulmonary congestion and interstitial pulmonary fibrosis (Figure 35–11). Pulmonary fibrosis is a progressive restrictive airway disease of unknown cause in which there is stiffening and scarring of

FIGURE 35–9   X-ray of chest in full expiration (left) and full inspiration (right). The dashed white line on the right is an outline of the lungs in full expiration. Note the difference in intrathoracic volume. (Reproduced with permission from Comroe JH Jr.: *Physiology of Respiration*, 2nd ed., Year Book, 1974.)

the lung. The curve is shifted upward and to the left (compliance is increased) in emphysema. It should be noted that compliance is a static measure of lung and chest recoil. The **resistance** of the lung and chest is the pressure difference required for a unit of air flow; this measurement, which is dynamic rather than static, also takes into account the resistance to air flow in the airways.

ALVEOLAR SURFACE TENSION

An important factor affecting the compliance of the lungs is the surface tension of the film of fluid that lines the alveoli. The magnitude of this component at various lung volumes can be measured by removing the lungs from the body of an experimental

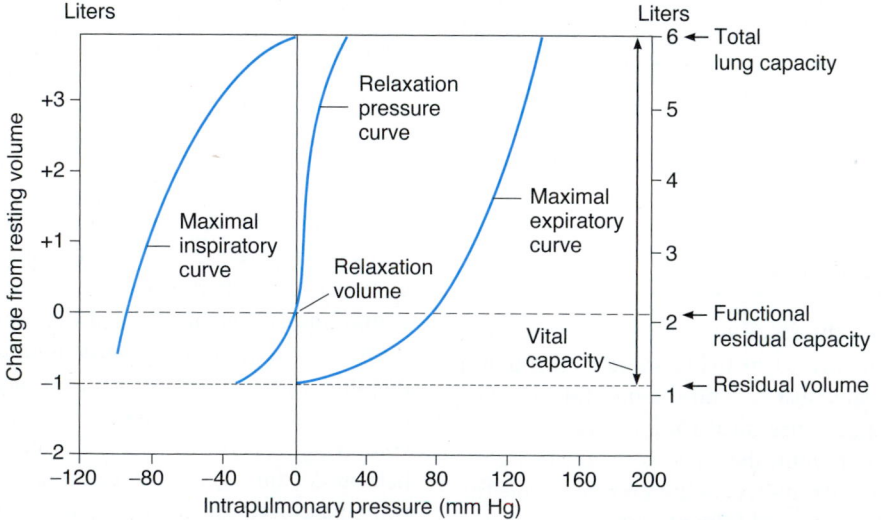

FIGURE 35–10   Intrapulmonary pressure and volume relationship, the relaxation pressure curve. The middle curve is the relaxation pressure curve of the total respiratory system; that is, the static pressure curve of values obtained when the lungs are inflated or deflated by various amounts and the intrapulmonary pressure (elastic recoil pressure) is measured with the airway closed. The relaxation volume is the point where the recoil of the chest and the recoil of the lungs balance. The slope of the curve is the compliance of the lungs and chest wall. The maximal inspiratory and expiratory curves are the airway pressures that can be developed during maximal inspiratory and expiratory efforts.

FIGURE 35–11 **Static expiratory pressure–volume curves of lungs in normal subjects and subjects with severe emphysema and pulmonary fibrosis.** (Modified and reproduced with permission from Pride NB, Macklem PT: Lung mechanics in disease. In: *Handbook of Physiology.* Section 3, *The Respiratory System.* Vol III, part 2. Fishman AP [editor]. American Physiological Society, 1986.)

animal and distending them alternately with saline and with air while measuring the intrapulmonary pressure. Because saline reduces the surface tension to nearly zero, the pressure–volume curve obtained with saline measures only the tissue elasticity (Figure 35–12), whereas the curve obtained with air measures both tissue elasticity and surface tension. The difference between the two curves, the elasticity due to surface tension, is much smaller at small than at large lung volumes. The surface

FIGURE 35–12 **Pressure–volume relations in the lungs of a cat after removal from the body. Saline:** lungs inflated and deflated with saline to reduce surface tension, resulting in a measurement of tissue elasticity. **Air:** lungs inflated (Inf) and deflated (Def) with air results in a measure of both tissue elasticity and surface tension. (Reproduced with permission from Morgan TE: Pulmonary surfactant. N Engl J Med 1971;284:1185.)

TABLE 35–2 Approximate composition of surfactant.

Component	Percentage Composition
Dipalmitoylphosphatidylcholine	62
Phosphatidylglycerol	5
Other phospholipids	10
Neutral lipids	13
Proteins	8
Carbohydrate	2

tension is also much lower than the expected surface tension at a water–air interface of the same dimensions.

SURFACTANT

The low surface tension when the alveoli are small is due to the presence in the fluid lining the alveoli of **surfactant,** a lipid surface-tension-lowering agent. Surfactant is a mixture of dipalmitoylphosphatidylcholine (DPPC), other lipids, and proteins (Table 35–2). If the surface tension is not kept low when the alveoli become smaller during expiration, they collapse in accordance with the law of Laplace. In spherical structures like the alveoli, the distending pressure equals two times the tension divided by the radius ($P = 2T/r$); if T is not reduced as r is reduced, the tension overcomes the distending pressure. Surfactant also helps to prevent pulmonary edema. It has been calculated that if it were not present, the unopposed surface tension in the alveoli would produce a 20 mm Hg force favoring transudation of fluid from the blood into the alveoli.

Surfactant is produced by type II alveolar epithelial cells (Figure 35–13). Typical **lamellar bodies,** membrane-bound organelles containing whorls of phospholipid, are formed in these cells and secreted into the alveolar lumen by exocytosis. Tubes of lipid called **tubular myelin** form from the extruded bodies, and the tubular myelin in turn forms the phospholipid film. Following secretion, the phospholipids of surfactant line up in the alveoli with their hydrophobic fatty acid tails facing the alveolar lumen. Surface tension is inversely proportional to their concentration per unit area. The surfactant molecules move further apart as the alveoli enlarge during inspiration, and surface tension increases, whereas it decreases when they move closer together during expiration. Some of the protein–lipid complexes in surfactant are taken up by endocytosis in type II alveolar cells and recycled.

Formation of the phospholipid film is greatly facilitated by the proteins in surfactant. This material contains four unique proteins: surfactant protein (SP)-A, SP-B, SP-C, and SP-D. SP-A is a large glycoprotein and has a collagen-like domain within its structure. It has multiple functions, including regulation of the feedback uptake of surfactant by the type II alveolar

FIGURE 35–13 **Formation and metabolism of surfactant.**
Lamellar bodies (LB) are formed in type II alveolar epithelial cells and secreted by exocytosis into the fluid lining the alveoli. The released lamellar body material is converted to tubular myelin (TM), and the TM is the source of the phospholipid surface film (SF). Surfactant is taken up by endocytosis into alveolar macrophages and type II epithelial cells. N, nucleus; RER, rough endoplasmic reticulum; CB, composite body. (Reproduced with permission from Wright JR: Metabolism and turnover of lung surfactant. Am Rev Respir Dis 1987;136:426.)

epithelial cells that secrete it. SP-B and SP-C are smaller proteins, which facilitate formation of the monomolecular film of phospholipid. A mutation of the gene for SP-C has been reported to be associated with familial interstitial lung disease. Like SP-A, SP-D is a glycoprotein. Its full function is uncertain. However, SP-A and SP-D are members of the collectin family of proteins that are involved in innate immunity in the conducting airway as well as in the alveoli. For other roles of surfactant, see Clinical Box 35–2.

WORK OF BREATHING

Work is performed by the respiratory muscles in stretching the elastic tissues of the chest wall and lungs (elastic work; approximately 65% of the total work), moving inelastic tissues (viscous resistance; 7% of total), and moving air through the respiratory passages (airway resistance; 28% of total). Because pressure times volume ($g/cm^2 \times cm^3 = g \times cm$) has the same dimensions as work (force × distance), the work of breathing can be calculated from the relaxation pressure curve (Figures 35–10 and 35–14). The total elastic work required for inspiration is represented by the area ABCA in Figure 35–14. Note that the relaxation pressure curve of the total respiratory system differs from that of the lungs alone. The actual elastic work required to increase the volume of the lungs alone is area ABDEA. The amount of elastic work required to inflate the whole respiratory system is less than the amount required to inflate the lungs alone because part of the work comes from elastic energy stored in the thorax. The elastic energy lost from the thorax (area AFGBA) is equal to that gained by the lungs (area AEDCA).

FIGURE 35–14 **Relaxation pressure curves in the lung.** The relaxation pressure curves of the total respiratory system (P_{TR}), the lungs (P_L), and the chest (P_W) are plotted together with standard volumes for functional residual capacity and tidal volume. The transmural pressure is intrapulmonary pressure minus intrapleural pressure in the case of the lungs, intrapleural pressure minus outside (barometric) pressure in the case of the chest wall, and intrapulmonary pressure minus barometric pressure in the case of the total respiratory system. From these curves, the total and actual elastic work associated with breathing can be derived (see text). (Modified from Mines AH: *Respiratory Physiology*, 3rd ed. Raven Press, 1993.)

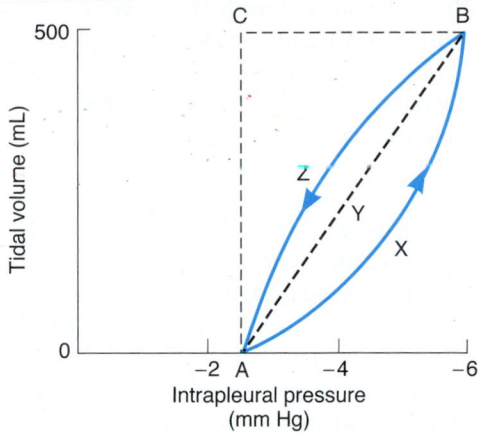

FIGURE 35–15 **Pressure volume relationships in breathing.** Diagrammatic representation of pressure and volume changes during quiet inspiration (line AXB) and expiration (line BZA). Line AYB is the compliance line.

FIGURE 35–16 **Intrapleural pressures in the upright position and their effect on ventilation.** Note that because intrapulmonary pressure is atmospheric, the more negative intrapleural pressure at the apex holds the lung in a more expanded position at the start of inspiration. Further increases in volume per unit increase in intrapleural pressure are smaller than at the base because the expanded lung is stiffer. (Reproduced with permission from West JB: *Ventilation/Blood Flow and Gas Exchange*, 3rd ed. Blackwell, 1977.)

The frictional resistance to air movement is relatively small during quiet breathing, but it does cause the intrapleural pressure changes to lead the lung volume changes during inspiration and expiration (Figure 35–6), producing a **hysteresis loop** rather than a straight line when pressure is plotted against volume (Figure 35–15). In this diagram, area AXBYA represents the work done to overcome airway resistance and lung viscosity. If the air flow becomes turbulent during rapid respiration, the energy required to move the air is greater than when the flow is laminar.

Estimates of the total work of quiet breathing range from 0.3 up to 0.8 kg-m/min. The value rises markedly during exercise, but the energy cost of breathing in normal individuals represents less than 3% of the total energy expenditure during exercise. The work of breathing is greatly increased in diseases such as emphysema, asthma, and congestive heart failure with dyspnea and orthopnea. The respiratory muscles have length–tension relations like those of other skeletal and cardiac muscles, and when they are severely stretched, they contract with less strength. They can also become fatigued and fail (pump failure), leading to inadequate ventilation.

DIFFERENCES IN VENTILATION & BLOOD FLOW IN DIFFERENT PARTS OF THE LUNG

In the upright position, ventilation per unit lung volume is greater at the base of the lung than at the apex. The reason for this is that at the start of inspiration, intrapleural pressure is less negative at the base than at the apex (Figure 35–16), and since the intrapulmonary intrapleural pressure difference is less than at the apex, the lung is less expanded. Conversely, at the apex, the lung is more expanded; that is, the percentage of maximum lung volume is greater. Because of the stiffness of the lung, the increase in lung volume per unit increase in pressure is smaller when the lung is initially more expanded, and

ventilation is consequently greater at the base. Blood flow is also greater at the base than the apex. The relative change in blood flow from the apex to the base is greater than the relative change in ventilation, so the ventilation/perfusion ratio is low at the base and high at the apex.

The ventilation and perfusion differences from the apex to the base of the lung have usually been attributed to gravity; they tend to disappear in the supine position, and the weight of the lung would be expected to make the intrapleural pressure lower at the base in the upright position. However, the inequalities of ventilation and blood flow in humans were found to persist to a remarkable degree in the weightlessness of space. Therefore, other factors also play a role in producing the inequalities.

DEAD SPACE & UNEVEN VENTILATION

Because gaseous exchange in the respiratory system occurs only in the terminal portions of the airways, the gas that occupies the rest of the respiratory system is not available for gas exchange with pulmonary capillary blood. Normally, the volume (in mL) of this **anatomic dead space** is approximately equal to the body weight in pounds. As an example, in a man who weighs 150 lb (68 kg), only the first 350 mL of the 500 mL inspired with each breath at rest mixes with the air in the alveoli. Conversely, with each expiration, the first 150 mL expired is gas that occupied the dead space, and only the last 350 mL is gas from the alveoli. Consequently, the **alveolar ventilation;**

TABLE 35–3 Effect of variations in respiratory rate and depth on alveolar ventilation.

Respiratory rate	30/min	10/min
Tidal volume	200 mL	600 mL
Minute volume	6 L	6 L
Alveolar ventilation	$(200 - 150) \times 30 =$ 1500 mL	$(600 - 150) \times 10 =$ 4500 mL

that is, the amount of air reaching the alveoli per minute, is less than the respiratory minute volume. Note in addition that because of the dead space, rapid shallow breathing produces much less alveolar ventilation than slow deep breathing at the same respiratory minute volume (Table 35–3).

It is important to distinguish between the **anatomic dead space** (respiratory system volume exclusive of alveoli) and the **total (physiologic) dead space** (volume of gas not equilibrating with blood; ie, wasted ventilation). In healthy individuals, the two dead spaces are identical and can be estimated by body weight. However, in disease states, no exchange may take place between the gas in some of the alveoli and the blood, and some of the alveoli may be overventilated. The volume of gas in non-perfused alveoli and any volume of air in the alveoli in excess of that necessary to arterialize the blood in the alveolar capillaries is part of the dead space (nonequilibrating) gas volume. The anatomic dead space can be measured by analysis of the single-breath N_2 curves (Figure 35–17). From mid-inspiration, the subject takes as deep a breath as possible of pure O_2, then exhales steadily while the N_2 content of the expired gas is continuously measured. The initial gas exhaled (phase I) is the gas that filled the dead space and that consequently contains no N_2. This is followed by a mixture of dead space and alveolar gas (phase II) and then by alveolar gas (phase III). The volume of the dead space is the volume of the gas expired from peak inspiration to the midportion of phase II.

Phase III of the single-breath N_2 curve terminates at the **closing volume (CV)** and is followed by phase IV, during which the N_2 content of the expired gas is increased. The CV is the lung volume above residual volume at which airways in the lower, dependent parts of the lungs begin to close off because of the lesser transmural pressure in these areas. The gas in the upper portions of the lungs is richer in N_2 than the gas in the lower, dependent portions because the alveoli in the upper portions are more distended at the start of the inspiration of O_2 and, consequently, the N_2 in them is less diluted with O_2. It is also worth noting that in most normal individuals, phase III has a slight positive slope even before phase IV is reached. This indicates that even during phase III there is a gradual increase in the proportion of the expired gas coming from the relatively N_2-rich upper portions of the lungs.

The total dead space can be calculated from the PCO_2 of expired air, the PCO_2 of arterial blood, and the tidal volume. The tidal volume (V_T) times the PCO_2 of the expired gas ($PECO_2$) equals the arterial PCO_2 ($PaCO_2$) times the difference between the tidal volume and the dead space (V_D) plus the PCO_2 of inspired air ($PICO_2$) times V_D (**Bohr's equation**):

$$PECO_2 \times V_T = PaCO_2 \times (V_T - V_D) + PICO_2 \times V_D$$

The term $PICO_2 \times V_D$ is so small that it can be ignored and the equation solved for V_D. If, for example,

$PECO_2 = 28$ mm Hg

$PaCO_2 = 40$ mm Hg

$V_T = 500$ mL

then,

$V_d = 150$ mL

The equation can also be used to measure the anatomic dead space if one replaces $PaCO_2$ with alveolar PCO_2 ($PACO_2$), which is the PCO_2 of the last 10 mL of expired gas. PCO_2 is an average of gas from different alveoli in proportion to their ventilation regardless of whether they are perfused. This is in contrast to $PaCO_2$, which is gas equilibrated only with perfused alveoli, and consequently, in individuals with unperfused alveoli, is greater than PCO_2.

GAS EXCHANGE IN THE LUNGS

SAMPLING ALVEOLAR AIR

Theoretically, all but the first 150 mL expired from a healthy 150-lb man (ie, the dead space) with each expiration is the gas that was in the alveoli (**alveolar air**), but some mixing always occurs at the interface between the dead-space gas and the alveolar air (Figure 35–17). A later portion of expired air is therefore the portion taken for analysis. Using modern apparatus with a suitable automatic valve, it is possible to collect the last 10 mL expired during quiet breathing. The composition of alveolar gas is compared with that of inspired and expired air in Figure 35–18.

FIGURE 35–17 **Single-breath N_2 curve.** From mid-inspiration, the subject takes a deep breath of pure O_2 then exhales steadily. The changes in the N_2 concentration of expired gas during expiration are shown, with the various phases of the curve indicated by roman numerals. Notably, region I is representative of the dead space (DS); from I–II is a mixture of DS and alveolar gas; the transition form III–IV is the closing volume (CV), and the end of IV is the residual volume (RV).

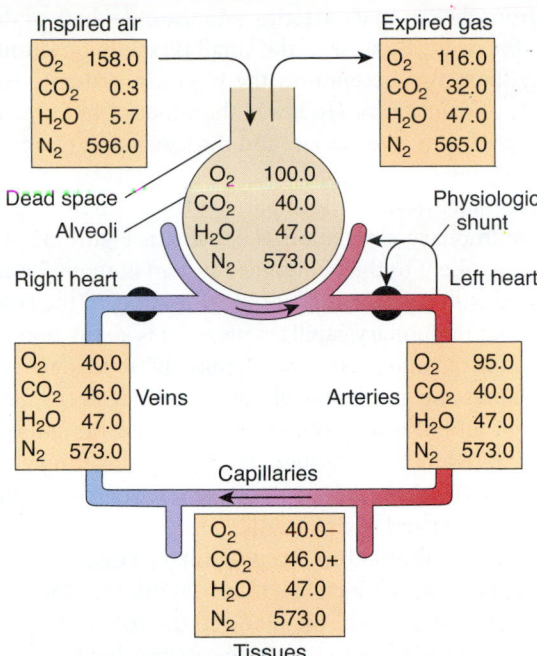

FIGURE 35–18 **Partial pressures of gases (mm Hg) in various parts of the respiratory system and in the circulatory system.**

PAO_2 can also be calculated from the **alveolar gas equation:**

$$PAO_2 = PIO_2 - PACO_2 \left(FIO_2 + \frac{1 - FIO_2}{R} \right)$$

where FIO_2 is the fraction of O_2 molecules in the dry gas, PIO_2 is the inspired PO_2, and R is the respiratory exchange ratio; that is, the flow of CO_2 molecules across the alveolar membrane per minute divided by the flow of O_2 molecules across the membrane per minute.

COMPOSITION OF ALVEOLAR AIR

Oxygen continuously diffuses out of the gas in the alveoli into the bloodstream, and CO_2 continuously diffuses into the alveoli from the blood. In the steady state, inspired air mixes with the alveolar gas, replacing the O_2 that has entered the blood and diluting the CO_2 that has entered the alveoli. Part of this mixture is expired. The O_2 content of the alveolar gas then falls and its CO_2 content rises until the next inspiration. Because the volume of gas in the alveoli is about 2 L at the end of expiration (functional residual capacity), each 350 mL increment of inspired and expired air has relatively little effect on PO_2 and PCO_2. Indeed, the composition of alveolar gas remains remarkably constant, not only at rest but also under a variety of other conditions.

DIFFUSION ACROSS THE ALVEOLOCAPILLARY MEMBRANE

Gases diffuse from the alveoli to the blood in the pulmonary capillaries or vice versa across the thin alveolocapillary membrane made up of the pulmonary epithelium, the capillary endothelium, and their fused basement membranes (Figure 35–3). Whether or not substances passing from the alveoli to the capillary blood reach equilibrium in the 0.75 s that blood takes to traverse the pulmonary capillaries at rest depends on their reaction with substances in the blood. Thus, for example, the anesthetic gas nitrous oxide (N_2O) does not react and reaches equilibrium in about 0.1 s (Figure 35–19). In this situation, the amount of N_2O taken up is not limited by diffusion but by the amount of blood flowing through the pulmonary capillaries; that is, it is **flow-limited.** On the other hand, carbon monoxide (CO) is taken up by hemoglobin in the red blood cells at such a high rate that the partial pressure of CO in the capillaries stays very low and equilibrium is not reached in the 0.75 s the blood is in the pulmonary capillaries. Therefore, the transfer of CO is not limited by perfusion at rest and instead is **diffusion-limited.** O_2 is intermediate between N_2O and CO; it is taken up by hemoglobin, but much less avidly than CO, and it reaches equilibrium with capillary blood in about 0.3 s. Thus, its uptake is **perfusion-limited.**

The **diffusing capacity** of the lung for a given gas is directly proportionate to the surface area of the alveolocapillary membrane and inversely proportionate to its thickness. The diffusing capacity for CO ($DLCO$) is measured as an index of diffusing capacity because its uptake is diffusion-limited. $DLCO$ is proportionate to the amount of CO entering the blood (VCO) divided by the partial pressure of CO in the alveoli minus the partial pressure of CO in the blood entering the pulmonary capillaries. Except in habitual cigarette smokers,

FIGURE 35–19 **Uptake of various substances during the 0.75 s they are in transit through a pulmonary capillary.** N_2O is not bound in blood, so its partial pressure in blood rises rapidly to its partial pressure in the alveoli. Conversely, CO is avidly taken up by red blood cells, so its partial pressure reaches only a fraction of its partial pressure in the alveoli. O_2 is intermediate between the two.

this latter term is close to zero, so it can be ignored and the equation becomes:

$$DLCO = \frac{\dot{V}CO}{PACO}.$$

The normal value of DLCO at rest is about 25 mL/min/mm Hg. It increases up to threefold during exercise because of capillary dilation and an increase in the number of active capillaries.

The PO_2 of alveolar air is normally 100 mm Hg (Figure 35–18), and the PO_2 of the blood entering the pulmonary capillaries is 40 mm Hg. The diffusing capacity for O_2, like that for CO at rest, is about 25 mL/min/mm Hg, and the PO_2 of blood is raised to 97 mm Hg, a value just under the alveolar PO_2. This falls to 95 mm Hg in the aorta because of the physiologic shunt. DLO_2 increases to 65 mL/min/mm Hg or more during exercise and is reduced in diseases such as sarcoidosis and beryllium poisoning (berylliosis) that cause fibrosis of the alveolar walls.

The PCO_2 of venous blood is 46 mm Hg, whereas that of alveolar air is 40 mm Hg, and CO_2 diffuses from the blood into the alveoli along this gradient. The PCO_2 of blood leaving the lungs is 40 mm Hg. CO_2 passes through all biological membranes with ease, and the diffusing capacity of the lung for CO_2 is much greater than the capacity for O_2. It is for this reason that CO_2 retention is rarely a problem in patients with alveolar fibrosis even when the reduction in diffusing capacity for O_2 is severe.

PULMONARY CIRCULATION

PULMONARY BLOOD VESSELS

The pulmonary vascular bed resembles the systemic one, except that the walls of the pulmonary artery and its large branches are about 30% as thick as the wall of the aorta, and the small arterial vessels, unlike the systemic arterioles, are endothelial tubes with relatively little muscle in their walls. The walls of the postcapillary vessels also contain some smooth muscle. The pulmonary capillaries are large, and there are multiple anastomoses, so that each alveolus sits in a capillary basket.

PRESSURE, VOLUME, & FLOW

With two quantitatively minor exceptions, the blood put out by the left ventricle returns to the right atrium and is ejected by the right ventricle, making the pulmonary vasculature unique in that it accommodates a blood flow that is almost equal to that of all the other organs in the body. One of the exceptions is part of the bronchial blood flow. As shown in Figure 35–5, there are anastomoses between the bronchial capillaries and the pulmonary capillaries and veins, and although some of the bronchial blood enters the bronchial veins, some enters the pulmonary capillaries and veins, bypassing the right ventricle. The other exception is blood that

flows from the coronary arteries into the chambers of the left side of the heart. Because of the small **physiologic shunt** created by those two exceptions, the blood in systemic arteries has a PO_2 about 2 mm Hg lower than that of blood that has equilibrated with alveolar air, and the saturation of hemoglobin is 0.5% less.

The pressure in the various parts of the pulmonary portion of the pulmonary circulation is shown in Figure 35–4. The pressure gradient in the pulmonary system is about 7 mm Hg, compared with a gradient of about 90 mm Hg in the systemic circulation. Pulmonary capillary pressure is about 10 mm Hg, whereas the oncotic pressure is 25 mm Hg, so that an inward-directed pressure gradient of about 15 mm Hg keeps the alveoli free of all but a thin film of fluid. When the pulmonary capillary pressure is more than 25 mm Hg—as it may be, for example, in "backward failure" of the left ventricle—pulmonary congestion and edema result.

The volume of blood in the pulmonary vessels at any one time is about 1 L, of which less than 100 mL is in the capillaries. The mean velocity of the blood in the root of the pulmonary artery is the same as that in the aorta (about 40 cm/s). It falls off rapidly, then rises slightly again in the larger pulmonary veins. It takes a red cell about 0.75 s to traverse the pulmonary capillaries at rest and 0.3 s or less during exercise.

EFFECT OF GRAVITY

Gravity has a relatively marked effect on the pulmonary circulation. In the upright position, the upper portions of the lungs are well above the level of the heart, and the bases are at or below it. Consequently, in the upper part of the lungs, the blood flow is less, the alveoli are larger, and ventilation is less than at the base (Figure 35–20). The pressure in the capillaries at the top of the lungs is close to the atmospheric pressure in the alveoli. Pulmonary arterial pressure is normally just sufficient to maintain perfusion, but if it is reduced or if alveolar pressure is increased, some of the capillaries collapse. Under these circumstances, no gas exchange takes place in the affected alveoli and they become part of the physiologic dead space.

In the middle portions of the lungs, the pulmonary arterial and capillary pressure exceeds alveolar pressure, but the pressure in the pulmonary venules may be lower than alveolar pressure during normal expiration, so they are collapsed. Under these circumstances, blood flow is determined by the pulmonary artery–alveolar pressure difference rather than the pulmonary artery–pulmonary vein difference. Beyond the constriction, blood "falls" into the pulmonary veins, which are compliant and take whatever amount of blood the constriction lets flow into them. This has been called the **waterfall effect.** Obviously, the compression of vessels produced by alveolar pressure decreases and pulmonary blood flow increases as the arterial pressure increases toward the base of the lung.

In the lower portions of the lungs, alveolar pressure is lower than the pressure in all parts of the pulmonary circulation and blood flow is determined by the arterial–venous pressure

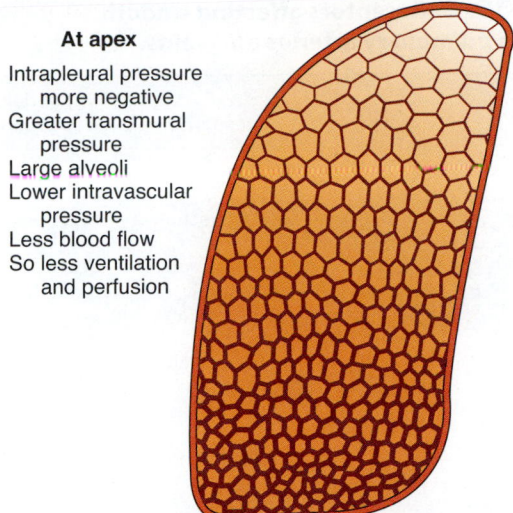

At apex

Intrapleural pressure
more negative
Greater transmural
pressure
Large alveoli
Lower intravascular
pressure
Less blood flow
So less ventilation
and perfusion

FIGURE 35–20 **Diagram of normal differences in ventilation and perfusion of the lung in the upright position.** Outlined areas are representative of changes in alveolar size (not actual size). Note the gradual change in alveolar size from top (apex) to bottom. Characteristic differences of alveoli at the apex of the lung are stated. (Modified from Levitsky, MG: *Pulmonary Physiology,* 6th ed. McGraw-Hill, 2003).

difference. Examples of diseases affecting pulmonary circulation are given in Clinical Box 35–3.

VENTILATION/PERFUSION RATIOS

The ratio of pulmonary ventilation to pulmonary blood flow for the whole lung at rest is about 0.8 (4.2 L/min ventilation divided by 5.5 L/min blood flow). However, relatively marked differences occur in this **ventilation/perfusion ratio** in various parts of the normal lung as a result of the effect of gravity, and local changes in the ventilation/perfusion ratio are common in disease. If the ventilation to an alveolus is reduced relative to its perfusion, the P_{O_2} in the alveolus falls because less O_2 is delivered to it and the P_{CO_2} rises because less CO_2 is expired. Conversely, if perfusion is reduced relative to ventilation, the P_{CO_2} falls because less CO_2 is delivered and the P_{O_2} rises because less O_2 enters the blood. These effects are summarized in Figure 35–21.

As noted above, ventilation, as well as perfusion in the upright position, declines in a linear fashion from the bases to the apices of the lungs. However, the ventilation/perfusion ratios are high in the upper portions of the lungs. When widespread, nonuniformity of ventilation and perfusion in the lungs can cause CO_2 retention and declines in systemic arterial P_{O_2}.

PULMONARY RESERVOIR

Because of their distensibility, the pulmonary veins are an important blood reservoir. When a normal individual lies down, the pulmonary blood volume increases by up to 400 mL, and

Diseases Affecting the Pulmonary Circulation

Pulmonary Hypertension

Sustained primary pulmonary hypertension can occur at any age. Like systemic arterial hypertension, it is a syndrome with multiple causes. However, the causes are different from those causing systemic hypertension. They include hypoxia, inhalation of cocaine, treatment with dexfenfluramine and related appetite-suppressing drugs that increase extracellular serotonin, and systemic lupus erythematosus. Some cases are familial and appear to be related to mutations that increase the sensitivity of pulmonary vessels to growth factors or cause deformations in the pulmonary vascular system.

All these conditions lead to increased pulmonary vascular resistance. If appropriate therapy is not initiated, the increased right ventricular afterload can lead eventually to right heart failure and death. Treatment with vasodilators such as prostacyclin and prostacyclin analogs is effective. Until recently, these had to be administered by continuous intravenous infusion, but aerosolized preparations that appear to be effective are now available.

Pulmonary Embolization

One of the normal functions of the lungs is to filter out small blood clots, and this occurs without any symptoms. When emboli block larger branches of the pulmonary artery, they provoke a rise in pulmonary arterial pressure and rapid, shallow respiration **(tachypnea)**. The rise in pulmonary arterial pressure may be due to reflex vasoconstriction via the sympathetic nerve fibers, but reflex vasoconstriction appears to be absent when large branches of the pulmonary artery are blocked. The tachypnea is a reflex response to activation of vagally innervated pulmonary receptors close to the vessel walls. These appear to be activated at the site of the embolization.

when the person stands up this blood is discharged into the general circulation. This shift is the cause of the decrease in vital capacity in the supine position and is responsible for the occurrence of orthopnea in heart failure.

REGULATION OF PULMONARY BLOOD FLOW

It is unsettled whether pulmonary veins and pulmonary arteries are regulated separately, although constriction of the veins increases pulmonary capillary pressure and constriction of pulmonary arteries increases the load on the right side of the heart.

Pulmonary blood flow is affected by both active and passive factors. There is an extensive autonomic innervation of the pulmonary vessels, and stimulation of the cervical sympathetic

FIGURE 35–21 **Effects of decreasing or increasing the ventilation/perfusion ratio ($\dot{V}_A/\dot{Q}$) on the P_{CO_2} and P_{O_2} in an alveolus.** The drawings above the curve represent an alveolus and a pulmonary capillary, and the dark red areas indicate sites of blockage. With complete obstruction of the airway to the alveolus, P_{CO_2} and P_{O_2} approximate the values in mixed venous blood ($\bar{V}$). With complete block of perfusion, P_{CO_2} and P_{O_2} approximate the values in inspired air. (Reproduced with permission from West JB: *Ventilation/Blood Flow and Gas Exchange,* 3rd ed. Blackwell, 1977.)

ganglia reduces pulmonary blood flow by as much as 30%. The vessels also respond to circulating humoral agents. Several of the receptors involved and their effect on pulmonary smooth muscle are summarized in Table 35–4. Many of the dilator responses are endothelium-dependent and presumably operate via release of nitric oxide (NO).

Passive factors such as cardiac output and gravitational forces also have significant effects on pulmonary blood flow. Local adjustments of perfusion to ventilation are determined by local effects of O_2 (or the lack of O_2). With exercise, cardiac output increases and pulmonary arterial pressure rises proportionately with little or no vasodilation. More red cells move through the lungs without any reduction in the O_2 saturation of the hemoglobin in them, and consequently, the total amount of O_2 delivered to the systemic circulation is increased. Capillaries dilate, and previously underperfused capillaries are "recruited" to carry blood. The net effect is a marked increase in pulmonary blood flow with few, if any, alterations in autonomic outflow to the pulmonary vessels.

When a bronchus or a bronchiole is obstructed, hypoxia develops in the underventilated alveoli beyond the obstruction. The O_2 deficiency apparently acts directly on vascular smooth muscle in the area to produce constriction, shunting blood away from the hypoxic area. Accumulation of CO_2 leads to a drop in pH in the area, and a decline in pH also produces vasoconstriction in the lungs, as opposed to the vasodilation it produces in other tissues. Conversely, reduction of the blood flow to a portion of the lung lowers the alveolar P_{CO_2} in that area, and this leads to constriction of the bronchi supplying it, shifting ventilation away from the poorly perfused area. Systemic hypoxia also causes the pulmonary arterioles to constrict, with a resultant increase in pulmonary arterial pressure.

TABLE 35–4 Receptors affecting smooth muscle in pulmonary arteries and veins.

Receptor	Subtype	Response	Endothelium Dependency
Autonomic			
Adrenergic	α_1	Contraction	No
	α_2	Relaxation	Yes
	β_2	Relaxation	Yes
Muscarinic	M_3	Relaxation	Yes
Purinergic	P_{2x}	Contraction	No
	P_{2y}	Relaxation	Yes
Tachykinin	NK_1	Relaxation	Yes
	NK_2	Contraction	No
VIP	?	Relaxation	?
CGRP	?	Relaxation	No
Humoral			
Adenosine	A_1	Contraction	No
	A_2	Relaxation	No
Angiotensin II	AT_1	Contraction	No
ANP	ANP_A	Relaxation	No
	ANP_B	Relaxation	No
Bradykinin	B_1?	Relaxation	Yes
	B_2	Relaxation	Yes
Endothelin	ET_A	Contraction	No
	ET_B	Relaxation	Yes
Histamine	H_1	Relaxation	Yes
	H_2	Relaxation	No
5-HT	5-HT_1	Contraction	No
	5-HT_{1C}	Relaxation	Yes
Thromboxane	TP	Contraction	No
Vasopressin	V_1	Relaxation	Yes

Modified and reproduced with permission from Barnes PJ, Lin SF: Regulation of pulmonary vascular tone. Pharmacol Rev 1995;47:88.

OTHER FUNCTIONS OF THE RESPIRATORY SYSTEM

LUNG DEFENSE MECHANISMS

The respiratory passages that lead from the exterior to the alveoli do more than serve as gas conduits. They humidify and cool or warm the inspired air so that even very hot or very cold air is at or near body temperature by the time it reaches the alveoli. Airway epithelial cells can secrete a variety of molecules that aid in lung defense. Secretory immunoglobulins (IgA), collectins (including Surfactant A and D), defensins and other peptides and proteases, reactive oxygen species, and reactive nitrogen species are all generated by airway epithelial cells. These secretions can act directly as antimicrobials to help keep the airway free of infection. Airway epithelial cells also secrete a variety of chemokines and cytokines that recruit the traditional immune cells and others to site of infections.

Various mechanisms operate to prevent foreign matter from reaching the alveoli. The hairs in the nostrils strain out many particles larger than 10 μm in diameter. Most of the remaining particles of this size settle on mucous membranes in the nose and pharynx; because of their momentum, they do not follow the airstream as it curves downward into the lungs, and they impact on or near the **tonsils** and **adenoids,** large collections of immunologically active lymphoid tissue in the back of the pharynx. Particles 2 to 10 μm in diameter generally fall on the walls of the bronchi as the air flow slows in the smaller passages. There they can initiate reflex bronchial constriction and coughing. Alternatively, they can be moved away from the lungs by the "mucociliary escalator." The epithelium of the respiratory passages from the anterior third of the nose to the beginning of the respiratory bronchioles is ciliated. The cilia are bathed in a periciliary fluid where they typically beat at rates of 10–15 Hz. On top of the periciliary layer and the beating cilia rests a mucus layer, a complex mixture of proteins and polysaccharides secreted from specialized cells, glands, or both in the conducting airway. This combination allows for the trapping of foreign particles (in the mucus) and their transport out of the airway (powered by ciliary beat). The ciliary mechanism is capable of moving particles away from the lungs at a rate of at least 16 mm/min. When ciliary motility is defective, as can occur from smoking, other environmental conditions, or genetic deficiency, mucus transport is virtually absent. This can lead to chronic sinusitis, recurrent lung infections, and bronchiectasis. Some of these symptoms are evident in cystic fibrosis (Clinical Box 35–4).

The pulmonary alveolar macrophages (PAMs) are another important component of the pulmonary defense system. Like other macrophages , these cells come originally from the bone marrow. Particles less than 2 μm in diameter can evade the mucociliary escalator and reach the alveoli. PAMs are actively phagocytic and ingest these small particles. They also help process inhaled antigens for immunologic attack, and they secrete substances that attract granulocytes to the lungs as well as substances that stimulate granulocyte and monocyte formation in the bone marrow. When the PAMs ingest large amounts of the substances in cigarette smoke or other irritants, they may also release lysosomal products into the extracellular space to cause inflammation.

METABOLIC & ENDOCRINE FUNCTIONS OF THE LUNGS

In addition to their functions in gas exchange, the lungs have a number of metabolic functions. They manufacture surfactant for local use, as noted above. They also contain a fibrinolytic system that lyses clots in the pulmonary vessels. They release a variety of substances that enter the systemic arterial blood (Table 35–5), and they remove other substances from the systemic venous blood that reach them via the pulmonary artery. Prostaglandins are removed from the circulation, but they are also synthesized in the lungs and released into the blood when lung tissue is stretched.

TABLE 35–5 Biologically active substances metabolized by the lungs.

Synthesized and used in the lungs
Surfactant
Synthesized or stored and released into the blood
Prostaglandins
Histamine
Kallikrein
Partially removed from the blood
Prostaglandins
Bradykinin
Adenine nucleotides
Serotonin
Norepinephrine
Acetylcholine
Activated in the lungs
Angiotensin I → angiotensin II

The lungs also activate one hormone; the physiologically inactive decapeptide angiotensin I is converted to the pressor, aldosterone-stimulating octapeptide angiotensin II in the pulmonary circulation. The reaction occurs in other tissues as well, but it is particularly prominent in the lungs. Large amounts of the angiotensin-converting enzyme responsible for this activation are located on the surface of the endothelial cells of the pulmonary capillaries. The converting enzyme also inactivates bradykinin. Circulation time through the pulmonary capillaries is less than 1 s, yet 70% of the angiotensin I reaching the lungs is converted to angiotensin II in a single trip through the capillaries. Four other peptidases have been identified on the surface of the pulmonary endothelial cells, but their full physiologic role is unsettled.

Removal of serotonin and norepinephrine reduces the amounts of these vasoactive substances reaching the systemic circulation. However, many other vasoactive hormones pass through the lungs without being metabolized. These include epinephrine, dopamine, oxytocin, vasopressin, and angiotensin II. In addition, various amines and polypeptides are secreted by neuroendocrine cells in the lungs.

CHAPTER SUMMARY

- The pressure exerted by any one gas in a mixture of gases is defined as its partial pressure. Partial pressures (P) of gases in air at sea level are as follows: $P_{O_2} = 149$ mm Hg; $P_{CO_2} = 0.3$ mm Hg; P_{N_2} (including other gases) = 564 mm Hg.

- Air enters the respiratory system in the upper airway, then proceeds to the conducting airway and on to the respiratory airway that ends in the alveoli. In the upper airway, air is humidified and warmed. The cross sectional area of the airway gradually increases through the conducting zone, then rapidly increases during the transition from conducting to respiratory zones

- The epithelium that line the conducting airway include ciliated cells that keep particulates from reaching the respiratory zone. The epithelium that lines the alveoli consist of two cell types: alveolar type I cells and alveolar type II cells. Type I cells are flattened epithelial cells that provide approximately 95% of the alveolar surface area and are the site of gas exchange. Type II cells are cuboidal epithelial cells that secrete surfactants that line the alveolar surface.

- There are several important measures of lung volume, including: tidal volume; inspiratory volume; expiratory reserve volume; forced vital capacity (FVC); the forced expiratory volume in one second (FEV_1); respiratory minute volume and maximal voluntary ventilation.

- Lung compliance refers to the ability of lungs to stretch. However, many normal factors affect lung compliance and it is best represented by a whole pressure-volume curve.

- Surfactant is a lipid-protein mixture that is in the fluid lining the alveolar epithelium. A primary function of surfactant is to increase surface tension in the alveoli to keep them from deflating.

- Both ventilation and perfusion are greater at the base of the lung and lower at the apex of the lung. The ventilation/perfusion ratio is lower at the base compared to the apex of the lung.

- Not all air that enters the airway is available for gas exchange. The regions where gas is not exchanged in the airway are termed "dead space." The conducting airway represents anatomical dead space. Increased dead space can occur in response to disease that affects air exchange in the respiratory zone.

- The pressure gradient in the pulmonary circulation system is much less than that in the systemic circulation. Because pulmonary capillary pressure is much lower than oncotic pressure in the plasma, fluid remains in the plasma as it traverses the lung.

- The mucociliary escalator in the conducting airway helps to keep particulates out of the respiratory zone.

- There are a variety of biologically activated substances that are metabolized in the lung. These include substances that are made and function in the lung (eg, surfactant), substances that are released or removed from the blood (eg, prostaglandins), and substances that are activated as they pass through the lung (eg, angiotensin II).

MULTIPLE-CHOICE QUESTIONS

For all questions, select the single best answer unless otherwise directed.

1. On the summit of Mt. Everest, where the barometric pressure is about 250 mm Hg, the partial pressure of O_2 is about
   A) 0.1 mm Hg.
   B) 0.5 mm Hg.
   C) 5 mm Hg.
   D) 50 mm Hg.
   E) 100 mm Hg.

2. The forced vital capacity is
   A) the amount of air that normally moves into (or out of) the lung with each respiration.
   B) the amount of air that enters the lung but does not participate in gas exchange.
   C) the largest amount of air expired after maximal expiratory effort.
   D) the largest amount of gas that can be moved into and out of the lungs in 1 min.

3. The tidal volume is
   A) the amount of air that normally moves into (or out of) the lung with each respiration.
   B) the amount of air that enters the lung but does not participate in gas exchange.
   C) the largest amount of air expired after maximal expiratory effort.
   D) the largest amount of gas that can be moved into and out of the lungs in 1 min.

4. Which of the following is responsible for the movement of O_2 from the alveoli into the blood in the pulmonary capillaries?
   A) active transport
   B) filtration
   C) secondary active transport
   D) facilitated diffusion
   E) passive diffusion

5. Which of the following causes relaxation of bronchial smooth muscle?
   A) leukotrienes
   B) vasoactive intestinal polypeptide
   C) acetylcholine
   D) cool air
   E) sulfur dioxide

6. Airway resistance
   A) is increased if the lungs are removed and inflated with saline.
   B) does not affect the work of breathing.
   C) is increased in paraplegic patients.
   D) is increased in asthma.
   E) makes up 80% of the work of breathing.

7. Surfactant lining the alveoli
   A) helps prevent alveolar collapse.
   B) is produced in alveolar type I cells and secreted into the alveolus.
   C) is increased in the lungs of heavy smokers.
   D) is a glycolipid complex.

CHAPTER RESOURCES

Barnes PJ: Chronic obstructive pulmonary disease. N Engl J Med 2000;343:269.

Budhiraja R, Tudor RM, Hassoun PM: Endothelial dysfunction in pulmonary hypertension. Circulation 2004;88:159.

Crystal RG, West JB (editors): *The Lung: Scientific Foundations*, 2nd ed. Raven Press, 1997.

Fishman AP, et al (editors): *Fishman's Pulmonary Diseases and Disorders*, 4th ed. McGraw-Hill, 2008.

Levitzky MG: *Pulmonary Physiology*, 7th ed. McGraw-Hill, 2007.

Prisk GK, Paiva M, West JB (editors): *Gravity and the Lung: Lessons from Micrography*. Marcel Dekker, 2001.

West JB: *Pulmonary Pathophysiology*, 5th ed. McGraw-Hill, 1995.

Wright JR: Immunoregulatory functions of surfactant proteins. Nat Rev Immunol 2005;5:58.

Gas Transport & pH in the Lung

OBJECTIVES

After studying this chapter, you should be able to:

- Describe the manner in which O_2 flows "downhill" from the lungs to the tissues and CO_2 flows "downhill" from the tissues to the lungs.

- Describe the reactions of O_2 with hemoglobin and the oxygen–hemoglobin dissociation curve.

- List the important factors affecting the affinity of hemoglobin for O_2 and the physiologic significance of each.

- List the reactions that increase the amount of CO_2 in the blood, and draw the CO_2 dissociation curve for arterial and venous blood.

- List the principal buffers in blood and, using the Henderson–Hasselbalch equation, describe what is unique about the bicarbonate buffer system.

- Define alkalosis and acidosis and outline respiratory and renal compensatory mechanisms in response to alkalosis and acidosis.

- Define hypoxia and describe its four principal forms.

- List and explain the effects of carbon monoxide on the body.

- Describe the effects of hypercapnia and hypocapnia, and give examples of conditions that can cause them.

INTRODUCTION

The partial pressure gradients for O_2 and CO_2, plotted in graphic form in Figure 36–1, emphasize that they are the key to gas movement and that O_2 "flows downhill" from the air through the alveoli and blood into the tissues, whereas CO_2 "flows downhill" from the tissues to the alveoli. However, the amount of both of these gases transported to and from the tissues would be grossly inadequate if it were not that about 99% of the O_2 that dissolves in the blood combines with the O_2-carrying protein hemoglobin and that about 94.5% of the CO_2 that dissolves enters into a series of reversible chemical reactions that convert it into other compounds. Thus, the presence of hemoglobin increases the O_2-carrying capacity of the blood 70-fold, and the reactions of CO_2 increase the blood CO_2 content 17-fold. In this chapter, physiologic details that underlie O_2 and CO_2 movement under various conditions are discussed.

OXYGEN TRANSPORT

OXYGEN DELIVERY TO THE TISSUES

The O_2 delivery system in the body consists of the lungs and the cardiovascular system. O_2 delivery to a particular tissue depends on the amount of O_2 entering the lungs, the adequacy of pulmonary gas exchange, the blood flow to the tissue, and the capacity of the blood to carry O_2. The blood flow depends on the degree of constriction of the vascular bed in the tissue and the cardiac output. The amount of O_2 in the blood is determined by the amount of dissolved O_2, the amount of

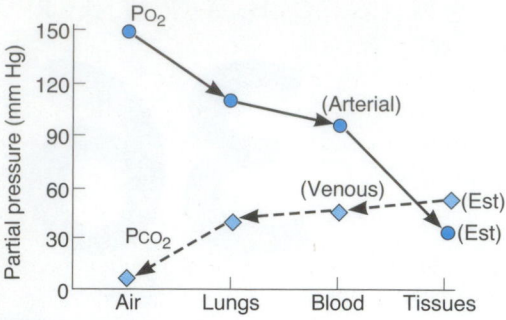

FIGURE 36–1 P_{O_2} **and** P_{CO_2} **values in air, lungs, blood, and tissues.** Note that both O_2 and CO_2 diffuse "downhill" along gradients of decreasing partial pressure. (Redrawn and reproduced with permission from Kinney JM: Transport of carbon dioxide in blood. Anesthesiology 1960;21:615.)

hemoglobin in the blood, and the affinity of the hemoglobin for O_2.

REACTION OF HEMOGLOBIN & OXYGEN

The dynamics of the reaction of hemoglobin with O_2 make it a particularly suitable O_2 carrier. Hemoglobin is a protein made up of four subunits, each of which contains a **heme** moiety attached to a polypeptide chain. In normal adults, most of the hemoglobin molecules contain two α and two β chains. Heme (see Figure 32–7) is a porphyrin ring complex that includes one atom of ferrous iron. Each of the four iron atoms in hemoglobin can reversibly bind one O_2 molecule. The iron stays in the ferrous state, so that the reaction is **oxygenation**, not oxidation. It has been customary to write the reaction of hemoglobin with O_2 as $Hb + O_2 \rightleftarrows HbO_2$. Because it contains four deoxyhemoglobin (Hb) units, the hemoglobin molecule can also be represented as Hb_4, and it actually reacts with four molecules of O_2 to form Hb_4O_8.

$$Hb_4 + O_2 \rightleftarrows Hb_4O_2$$

$$Hb_4O_2 + O_2 \rightleftarrows Hb_4O_4$$

$$Hb_4O_4 + O_2 \rightleftarrows Hb_4O_6$$

$$Hb_4O_6 + O_2 \rightleftarrows Hb_4O_8$$

The reaction is rapid, requiring less than 0.01 s. The deoxygenation (reduction) of Hb_4O_8 is also very rapid.

The quaternary structure of hemoglobin determines its affinity for O_2. In deoxyhemoglobin, the globin units are tightly bound in a **tense (T) configuration,** which reduces the affinity of the molecule for O_2. When O_2 is first bound, the bonds holding the globin units are released, producing a **relaxed (R) configuration,** which exposes more O_2 binding sites. The net result is a 500-fold increase in O_2 affinity. In tissue, these reactions are reversed, releasing O_2. The transition from one state to another has been calculated to occur about 10^8 times in the life of a red blood cell.

The **oxygen–hemoglobin dissociation curve** relates percentage saturation of the O_2 carrying power of hemoglobin to

FIGURE 36–2 **Oxygen–hemoglobin dissociation curve.** pH 7.40, temperature 38 °C. Inset table notes the percentage of saturated hemoglobin to P_{O_2} and dissolved O_2. (Redrawn and reproduced with permission from Comroe JH Jr., et al: *The Lung: Clinical Physiology and Pulmonary Function Tests,* 2nd ed. Year Book, 1962.)

the P_{O_2} (Figure 36–2). This curve has a characteristic sigmoid shape due to the T–R interconversion. Combination of the first heme in the Hb molecule with O_2 increases the affinity of the second heme for O_2, and oxygenation of the second increases the affinity of the third, and so on, so that the affinity of Hb for the fourth O_2 molecule is many times that for the first.

When blood is equilibrated with 100% O_2 (P_{O_2} = 760 mm Hg), the normal hemoglobin becomes 100% saturated. When fully saturated, each gram of normal hemoglobin contains 1.39 mL of O_2. However, blood normally contains small amounts of inactive hemoglobin derivatives, and the measured value in vivo is lower. The traditional figure is 1.34 mL of O_2. The hemoglobin concentration in normal blood is about 15 g/dL (14 g/dL in women and 16 g/dL in men). Therefore, 1 dL of blood contains 20.1 mL (1.34 mL × 15) of O_2 bound to hemoglobin when the hemoglobin is 100% saturated. The amount of dissolved O_2 is a linear function of the P_{O_2} (0.003 mL/dL blood/mm Hg P_{O_2}).

In vivo, the hemoglobin in the blood at the ends of the pulmonary capillaries is about 97.5% saturated with O_2 (P_{O_2} = 97 mm Hg). Because of a slight admixture with venous blood that bypasses the pulmonary capillaries (physiologic shunt), the hemoglobin in systemic arterial blood is only 97% saturated. The arterial blood therefore contains a total of about 19.8 mL of O_2 per dL: 0.29 mL in solution and 19.5 mL bound to hemoglobin. In venous blood at rest, the hemoglobin is 75% saturated and the total O_2 content is about 15.2 mL/dL: 0.12 mL in solution and 15.1 mL bound to hemoglobin. Thus, at rest the tissues remove about 4.6 mL of O_2 from each deciliter of blood passing through them (Table 36–1); 0.17 mL of this total represents O_2 that was in solution in the blood, and the remainder represents O_2 that was liberated from hemoglobin. In this way, 250 mL of O_2 per minute is transported from the blood to the tissues at rest.

TABLE 36–1 Gas content of blood.

	mL/dL of Blood Containing 15 g of Hemoglobin			
	Arterial Blood (Po$_2$ 95 mm Hg; Pco$_2$ 40 mm Hg; Hb 97% Saturated)		Venous Blood (Po$_2$ 40 mm Hg; Pco$_2$ 46 mm Hg; Hb 75% Saturated)	
Gas	Dissolved	Combined	Dissolved	Combined
O$_2$	0.29	19.5	0.12	15.1
CO$_2$	2.62	46.4	2.98	49.7
N$_2$	0.98	0	0.98	0

FACTORS AFFECTING THE AFFINITY OF HEMOGLOBIN FOR OXYGEN

Three important conditions affect the oxygen–hemoglobin dissociation curve: the **pH,** the **temperature,** and the concentration of **2,3-biphosphoglycerate (BPG; 2,3-BPG).** A rise in temperature or a fall in pH shifts the curve to the right (Figure 36–3). When the curve is shifted in this direction, a higher PO$_2$ is required for hemoglobin to bind a given amount of O$_2$. Conversely, a fall in temperature or a rise in pH shifts the curve to the left, and a lower PO$_2$ is required to bind a given amount of O$_2$. A convenient index for comparison of such shifts is the P$_{50}$, the PO$_2$ at which hemoglobin is half saturated with O$_2$. The higher the P$_{50}$, the lower the affinity of hemoglobin for O$_2$.

The decrease in O$_2$ affinity of hemoglobin when the pH of blood falls is called the **Bohr effect** and is closely related to the fact that deoxygenated hemoglobin (deoxyhemoglobin) binds H$^+$ more actively than does oxygenated hemoglobin (oxyhemoglobin). The pH of blood falls as its CO$_2$ content increases, so that when the PCO$_2$ rises, the curve shifts to the right and the P$_{50}$ rises. Most of the unsaturation of hemoglobin that occurs in the tissues is secondary to the decline in the PO$_2$, but an extra 1–2% unsaturation is due to the rise in PCO$_2$ and consequent shift of the dissociation curve to the right.

2,3-BPG is very plentiful in red cells. It is formed from 3-phosphoglyceraldehyde, which is a product of glycolysis via the Embden–Meyerhof pathway (Figure 36–4). It is a highly charged anion that binds to the β chains of deoxyhemoglobin. One mole of deoxyhemoglobin binds 1 mol of 2,3-BPG. In effect,

$$HbO_2 + 2,3\text{-BPG} \rightleftarrows Hb - 2,3\text{-BPG} + O_2$$

In this equilibrium, an increase in the concentration of 2,3-BPG shifts the reaction to the right, causing more O$_2$ to be liberated.

Because acidosis inhibits red cell glycolysis, the 2,3-BPG concentration falls when the pH is low. Conversely, thyroid hormones, growth hormones, and androgens can all increase the concentration of 2,3-BPG and the P$_{50}$.

Exercise has been reported to produce an increase in 2,3-BPG within 60 min, although the rise may not occur in trained athletes. The P$_{50}$ is also increased during exercise, because the temperature rises in active tissues and CO$_2$ and metabolites accumulate, lowering the pH. In addition, much more O$_2$ is removed from each unit of blood flowing through active tissues because the tissues' PO$_2$ declines. Finally, at low PO$_2$ values, the oxygen–hemoglobin dissociation curve is steep, and large amounts of O$_2$ are liberated per unit drop in PO$_2$. Some clinical features of hemoglobin are discussed in Clinical Box 36–1.

MYOGLOBIN

Myoglobin is an iron-containing pigment found in skeletal muscle. It resembles hemoglobin but binds 1 rather than 4 mol of O$_2$ per mole. Its dissociation curve is a rectangular hyperbola rather than a sigmoid curve. Because its curve is to the left of the hemoglobin curve (Figure 36–5), it takes up O$_2$ from hemoglobin in the blood. It releases O$_2$ only at low PO$_2$ values, but the PO$_2$ in exercising muscle is close to zero. The myoglobin content is greatest in muscles specialized for sustained contraction. The muscle blood supply is compressed during such contractions, and myoglobin may provide O$_2$ when blood flow is cut off.

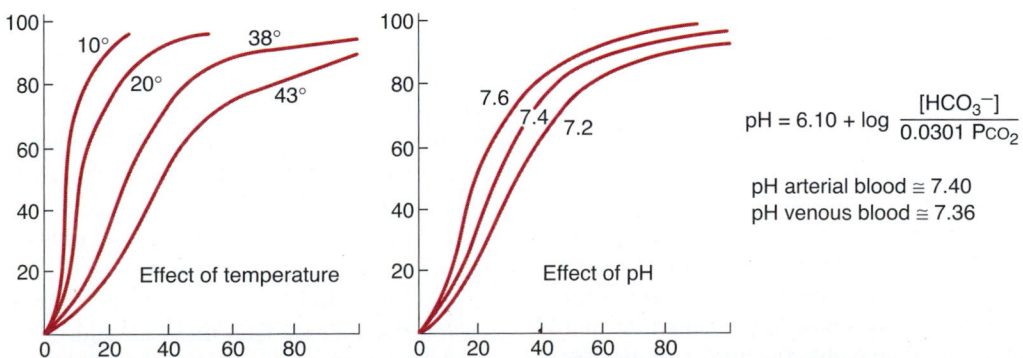

FIGURE 36–3 Effects of temperature and pH on the oxygen–hemoglobin dissociation curve. Both changes in temperature **(left)** and pH **(right)** can alter the affinity of hemoglobin for O$_2$. Plasma pH can be estimated using the modified Henderson–Hasselbalch equation, as shown. (Redrawn and reproduced with permission from Comroe JH Jr., et al: *The Lung: Clinical Physiology and Pulmonary Function Tests,* 2nd ed. Year Book, 1962.)

FIGURE 36–4 **Formation and catabolism of 2,3-BPG.** Note that 2,3 BPG can be associated with the Embden–Meyerhoff pathway (see Chapter 1).

CARBON DIOXIDE TRANSPORT

FATE OF CARBON DIOXIDE IN BLOOD

The solubility of CO_2 in blood is about 20 times that of O_2; therefore, considerably more CO_2 than O_2 is present in simple solution at equal partial pressures. The CO_2 that diffuses into red blood cells is rapidly hydrated to H_2CO_3 because of the presence of carbonic anhydrase. The H_2CO_3 dissociates to H^+ and HCO_3^-, and the H^+ is buffered, primarily by hemoglobin, while the HCO_3^- enters the plasma. Some of the CO_2 in the red cells reacts with the amino groups of hemoglobin and other proteins (R), forming **carbamino compounds:**

$$CO_2 + R-N \overset{H}{\underset{H}{}} \rightleftarrows R-N \overset{H}{\underset{COOH}{}}$$

Because deoxyhemoglobin binds more H^+ than oxyhemoglobin does and forms carbamino compounds more readily, binding of O_2 to hemoglobin reduces its affinity for CO_2 **(Haldane effect).** Consequently, venous blood carries more CO_2 than arterial blood, CO_2 uptake is facilitated in the tissues, and CO_2 release is facilitated in the lungs. About 11% of the CO_2 added to the blood in the systemic capillaries is carried to the lungs as carbamino-CO_2.

CHLORIDE SHIFT

Because the rise in the HCO_3^- content of red cells is much greater than that in plasma as the blood passes through the capillaries, about 70% of the HCO_3^- formed in the red cells enters the plasma. The excess HCO_3^- leaves the red cells in exchange for Cl^- (Figure 36–6). This process is mediated by **anion exchanger 1** (**AE1;** formerly called Band 3), a major membrane protein in the red blood cell. Because of this **chloride shift,** the Cl^- content of the red cells in venous blood is significantly greater than that in arterial blood. The chloride shift occurs rapidly and is essentially complete within 1 s.

Note that for each CO_2 molecule added to a red cell, there is an increase of one osmotically active particle in the cell—either an HCO_3^- or a Cl^- in the red cell (Figure 36–6). Consequently, the red cells take up water and increase in size. For

FIGURE 36–5 **Dissociation curve of hemoglobin and myoglobin.** The myoglobin binding curve (B) lacks the sigmoidal shape of the hemoglobin binding curve (A) because of the single O_2 binding site in each molecule. Myoglobin also has greater affinity for O_2 than hemoglobin (curve shifted left) and thus can store O_2 in muscle.

this reason, plus the fact that a small amount of fluid in the arterial blood returns via the lymphatics rather than the veins, the hematocrit of venous blood is normally 3% greater than that of the arterial blood. In the lungs, the Cl^- moves out of the cells and they shrink.

SUMMARY OF CARBON DIOXIDE TRANSPORT

For convenience, the various fates of CO_2 in the plasma and red cells are summarized in Table 36–2. The extent to which they increase the capacity of the blood to carry CO_2 is indicated by the difference between the lines indicating the dissolved CO_2 and the total CO_2 in the dissociation curves for CO_2 shown in Figure 36–7.

Of the approximately 49 mL of CO_2 in each deciliter of arterial blood (Table 36–1), 2.6 mL is dissolved, 2.6 mL is in carbamino compounds, and 43.8 mL is in HCO_3^-. In the tissues, 3.7 mL of CO_2 per deciliter of blood is added; 0.4 mL stays in solution, 0.8 mL forms carbamino compounds, and

FIGURE 36–6 **Fate of CO_2 in the red blood cell.** Upon entering the red blood cell, CO_2 is rapidly hydrated to H_2CO_3 by carbonic anhydrase. H_2CO_3 is in equilibrium with H^+ and its conjugate base, HCO_3^-. H^+ can interact with deoxyhemoglobin, whereas HCO_3^- can be transported outside of the cell via AE1 (Band 3). In effect, for each CO_2 molecule that enters the red cell, there is an additional HCO_3^- or Cl^- in the cell.

TABLE 36–2 Fate of CO_2 in blood.

In plasma
1. Dissolved
2. Formation of carbamino compounds with plasma protein
3. Hydration, H^+ buffered, HCO_3^- in plasma
In red blood cells
1. Dissolved
2. Formation of carbamino-Hb
3. Hydration, H^+ buffered, 70% of HCO_3^- enters the plasma
4. Cl^- shifts into cells; mOsm in cells increases

2.5 mL forms HCO_3^-. The pH of the blood drops from 7.40 to 7.36. In the lungs, the processes are reversed, and the 3.7 mL of CO_2 is discharged into the alveoli. In this fashion, 200 mL of CO_2 per minute at rest and much larger amounts during exercise are transported from the tissues to the lungs and excreted. It is worth noting that this amount of CO_2 is equivalent in 24 hours to over 12,500 mEq of H^+.

ACID–BASE BALANCE & GAS TRANSPORT

The major source of acids in the blood under normal conditions is through cellular metabolism. The CO_2 formed by metabolism in the tissues is in large part hydrated to H_2CO_3, and the total H^+ load from this source is over 12,500 mEq/d. However, most of the CO_2 is excreted in the lungs, and the small quantities of the remaining H^+ are excreted by the kidneys.

FIGURE 36–7 **CO_2 dissociation curves.** The arterial point (a) and the venous point (v) indicate the total CO_2 content found in arterial blood and venous blood of normal resting humans. Note the low amount of CO_2 that is dissolved (orange trace) compared to that which can be carried by other means (Table 36–2). (Modified and reproduced with permission from Schmidt RF, Thews G [editors]: *Human Physiology.* Springer, 1983.)

Fruits are the main dietary source of alkali. They contain Na^+ and K^+ salts of weak organic acids, and the anions of these salts are metabolized to CO_2, leaving $NaHCO_3$ and $KHCO_3$ in the body. Such ingestion contributes little to changes in pH and a more common cause of alkalosis is loss of acid from the body as a result of vomiting of gastric juice rich in HCl. This is, of course, equivalent to adding alkali to the body.

BUFFERING IN THE BLOOD

Acid and base shifts in the blood are largely controlled by three main buffers in blood: (1) proteins, (2) hemoglobin, and (3) the carbonic acid–bicarbonate system. Plasma **proteins** are effective buffers because both their free carboxyl and their free amino groups dissociate:

$$RCOOH \rightleftarrows RCOO^- + H^+$$

$$pH = pK'_{RCOOH} + \log \frac{[RCOO^-]}{[RCOOH]}$$

$$RNH_3^+ \rightleftarrows RNH_2 + H^+$$

$$pH = pK'_{RNH_3} + \log \frac{[RNH_2]}{[RNH_3^+]}$$

The second buffer system is provided by the dissociation of the imidazole groups of the histidine residues in **hemoglobin:**

In the pH 7.0–7.7 range, the free carboxyl and amino groups of hemoglobin contribute relatively little to its buffering capacity. However, the hemoglobin molecule contains 38 histidine residues, and on this basis—plus the fact that hemoglobin is present in large amounts—the hemoglobin in blood has six times the buffering capacity of the plasma proteins. In addition, the action of hemoglobin is unique because the imidazole groups of deoxyhemoglobin (Hb) dissociate less than those of oxyhemoglobin (HbO_2), making Hb a weaker acid and therefore a better buffer than HbO_2. Titration curves for Hb and HbO_2 are shown in Figure 36–8.

The third and major buffer system in blood is the **carbonic acid–bicarbonate system:**

$$H_2CO_3 \rightleftarrows H^+ + HCO_3^-$$

The Henderson–Hasselbalch equation for this system is

$$pH = pK + \log \frac{[HCO_3^-]}{[H_2CO_3]}$$

FIGURE 36–8 **Titration curves for hemoglobin.** Individual titration curves for deoxygenated hemoglobin (Hb) and oxygenated hemoglobin (HbO_2) are shown. The arrow from a to c indicates the number of millimoles of H that can be added without pH shift. The arrow from a to b indicates the pH shift on deoxygenation.

The pK for this system in an ideal solution is low (about 3), and the amount of H_2CO_3 is small and hard to measure accurately. However, in the body, H_2CO_3 is in equilibrium with CO_2:

$$H_2CO_3 \rightleftarrows CO_2 + H_2O$$

If the pK is changed to pK' (apparent ionization constant; distinguished from the true pK due to less than ideal conditions for the solution) and $[CO_2]$ is substituted for $[H_2CO_3]$, the pK' is 6.1:

$$pH = 6.10 + \log \frac{[HCO_3^-]}{[CO2]}$$

The clinically relevant form of this equation is:

$$pH = 6.10 + \log \frac{[HCO_3^-]}{0.0301\ P_{CO_2}}$$

since the amount of dissolved CO_2 is proportional to the partial pressure of CO_2 and the solubility coefficient of CO_2 in mmol/L/mm Hg is 0.0301. $[HCO_3^-]$ cannot be measured directly, but pH and P_{CO_2} can be measured with suitable accuracy with pH and P_{CO_2} glass electrodes, and $[HCO_3^-]$ can then be calculated.

The pK' of this system is still low relative to the pH of the blood, but the system is one of the most effective buffer systems in the body because the amount of dissolved CO_2 is controlled by respiration. Additional control of the plasma concentration of HCO_3^- is provided by the kidneys. When H^+ is added to the blood, HCO_3^- declines as more H_2CO_3 is formed. If the extra H_2CO_3 were not converted to CO_2 and H_2O and the CO_2 excreted in the lungs, the H_2CO_3 concentration would rise. When enough H^+ has been added to halve the plasma HCO_3^-, the pH would have dropped from 7.4 to 6.0. However, not only is all the extra H_2CO_3 that is formed removed, but also the H^+ rise stimulates respiration and therefore produces a drop in

FIGURE 36–9 **Buffering by the H_2CO_3–HCO_3^- system in blood.** The bars are drawn as if buffering occurred in separate steps over time (left to right) in order to show the effect of the initial reaction, the reduction of H_2CO_3 to its previous value, and its further reduction by the increase in ventilation. In this case, [H_2CO_3] is actually the concentration of dissolved CO_2, so that the mEq/L values for it are arbitrary.

TABLE 36–3 Plasma pH, HCO_3^-, and PCO_2 values in various typical disturbances of acid–base balance.[a]

Condition	Arterial Plasma			Cause
	pH	HCO_3^- (mEq/L)	PCO_2 (mm Hg)	
Normal	7.40	24.1	40	
Metabolic acidosis	7.28	18.1	40	NH_4 Cl ingestion
	6.96	5.0	23	Diabetic acidosis
Metabolic alkalosis	7.50	30.1	40	$NaHCO_3^-$ ingestion
	7.56	49.8	58	Prolonged vomiting
Respiratory acidosis	7.34	25.0	48	Breathing 7% CO_2
	7.34	33.5	64	Emphysema
Respiratory alkalosis	7.53	22.0	27	Voluntary hyper-ventilation
	7.48	18.7	26	Three-week residence at 4000-m altitude

[a]In the diabetic acidosis and prolonged vomiting examples, respiratory compensation for primary metabolic acidosis and alkalosis has occurred, and the PCO_2 has shifted from 40 mm Hg. In the emphysema and high-altitude examples, renal compensation for primary respiratory acidosis and alkalosis has occurred and has made the deviations from normal of the plasma HCO_3^- larger than they would otherwise be.

PCO_2, so that some additional H_2CO_3 is removed. The pH thus falls only to 7.2 or 7.3 (Figure 36–9).

There are two additional factors that make the carbonic-acid-bicarbonate system such a good biological buffer. First, the reaction $CO_2 + H_2O \rightleftarrows H_2CO_3$ proceeds slowly in either direction unless the enzyme **carbonic anhydrase** is present. There is no carbonic anhydrase in plasma, but there is an abundant supply in red blood cells. Second, the presence of hemoglobin in the blood increases the buffering of the system by binding free H^+ produced by the hydration of CO_2 and allowing for movement of the HCO_3^- into the plasma.

ACIDOSIS & ALKALOSIS

The pH of the arterial plasma is normally 7.40 and that of venous plasma slightly lower. A decrease in pH below the norm (**acidosis**) is technically present whenever the arterial pH is below 7.40 and an increase in pH (**alkalosis**) is technically present whenever pH is above 7.40. In practice, variations of up to 0.05 pH unit occur without untoward effects.

Acid–base disorders are split into four categories: respiratory acidosis, respiratory alkalosis, metabolic acidosis, and metabolic alkalosis. In addition, these disorders can occur in combination. Some examples of acid–base disturbances are shown in Table 36–3.

RESPIRATORY ACIDOSIS

Any short-term rise in arterial PCO_2 (ie, above 40 mm Hg) due to decreased ventilation results in **respiratory acidosis.** The CO_2 that is retained is in equilibrium with H_2CO_3, which in turn is in equilibrium with HCO_3^-, so that the plasma HCO_3^- rises and a new equilibrium is reached at a lower pH. This can

be indicated graphically on a plot of plasma HCO_3^- concentration versus pH (Figure 36–10). The pH change observed at any increase in PCO_2 during respiratory acidosis is dependent on the buffering capacity of the blood. The initial changes shown in Figure 36–10 are those that occur independently of any compensatory mechanism; that is, they are those of **uncompensated respiratory acidosis.**

RESPIRATORY ALKALOSIS

Any short-term decrease in ventilation that lowers PCO_2 below what is needed for proper CO_2 exchange (ie, below 35 mm Hg) results in **respiratory alkalosis.** The decreased CO_2 shifts the equilibrium of the carbonic acid–bicarbonate system to effectively lower the [H^+] and increase the pH. As in respiratory acidosis, initial pH changes corresponding to respiratory alkalosis (Figure 36–10) are those that occur independently of any compensatory mechanism and are thus **uncompensated respiratory alkalosis.**

METABOLIC ACIDOSIS & ALKALOSIS

Blood pH changes can also arise by nonrespiratory mechanism. **Metabolic acidosis** (or nonrespiratory acidosis) occurs when strong acids are added to blood. If, for example, a large amount of acid is ingested (eg, aspirin overdose), acids in the blood are

FIGURE 36–10   Acid–base nomogram. Changes in the P_{CO_2} (curved lines), plasma HCO_3^-, and pH (or [H^+]) of arterial blood in respiratory and metabolic acidosis are shown. Note the shifts in HCO_3^- and pH as acute respiratory acidosis and alkalosis are compensated, producing their chronic counterparts. (Reproduced with permission from Cogan MG, Rector FC Jr.: Acid–base disorders. In: *The Kidney*, 4th ed. Brenner BM, Rector FC Jr. [editors]. Saunders, 1991.)

FIGURE 36–11   Acid–base paths during metabolic acidosis. Changes in true plasma pH, HCO_3^-, and P_{CO_2} at rest, during metabolic acidosis and alkalosis, and following respiratory compensation are plotted. Metabolic acidosis or alkalosis causes changes in pH along the P_{CO_2} isobar line. Respiratory compensation moves pH towards normal by altering P_{CO_2}. (This is called a Davenport diagram and is based on Davenport HW: *The ABC of Acid–Base Chemistry*, 6th ed. University of Chicago Press, 1974.)

quickly increased, lowering the available Hb^-, $Prot^-$, and HCO_3^- buffers. The H_2CO_3 that is formed is converted to H_2O and CO_2, and the CO_2 is rapidly excreted via the lungs. This is the situation in **uncompensated metabolic acidosis** (Figure 36–10). Note that in contrast to respiratory acidosis, P_{CO_2} is unchanged and the shift toward metabolic acidosis occurs along the isobar line (Figure 36–11). When the free [H^+] level falls as a result of addition of alkali, or more commonly, the removal of large amounts of acid (eg, following vomiting), **metabolic alkalosis** results. In uncompensated metabolic alkalosis the pH rises along the isobar line (Figures 36–10 and 36–11).

RESPIRATORY & RENAL COMPENSATION

Uncompensated acidosis and alkalosis as described above are seldom seen because of compensation systems. The two main compensatory systems are **respiratory compensation** and **renal compensation.**

The respiratory system compensates for metabolic acidosis or alkalosis by altering ventilation, and consequently, the P_{CO_2}, which can directly change blood pH. Respiratory mechanisms tend to be fast. In response to metabolic acidosis, ventilation is increased, resulting in a decrease of P_{CO_2} (eg, from 40 mm Hg to 20 mm Hg) and a subsequent increase in pH toward normal (Figure 36–11). In response to metabolic alkalosis, ventilation is decreased, P_{CO_2} is increased, and a subsequent decrease in pH occurs. Because respiratory compensation is a quick response, the graphical representation in Figure 36–11 overstates the two-step adjustment in blood pH. In actuality, as

soon as metabolic acidosis begins, respiratory compensation is invoked and pH is kept from the large shifts depicted.

For complete compensation from respiratory or metabolic acidosis/alkalosis, renal compensatory mechanisms are invoked. The kidney responds to acidosis by actively secreting fixed acids while retaining filtered HCO_3^-. In contrast, the kidney responds to alkalosis by decreasing H^+ secretion and by decreasing the retention of filtered HCO_3^-.

Renal tubule cells in the kidney have active carbonic anhydrase and thus can produce H^+ and HCO_3^- from CO_2. In response to acidosis, these cells secrete H^+ into the tubular fluid in exchange for Na^+ while the HCO_3^- is actively reabsorbed into the peritubular capillary; for each H^+ secreted, one Na^+ and one HCO_3^- are added to the blood. The result of this renal compensation for respiratory acidosis is shown graphically in the shift from acute to chronic respiratory acidosis in Figure 36–10.

Conversely, in response to alkalosis, the kidney decreases H^+ secretion and depresses HCO_3^- reabsorption. The kidney tends to reabsorb HCO_3^- until the level in plasma exceeds 26–28 mEq/L (normal is 24 mEq/L). Above this threshold, HCO_3^- appears in the urine. The result of this renal compensation for respiratory alkalosis is shown graphically in the shift from acute to chronic respiratory alkalosis in Figure 36–10. Clinical evaluations of acid–base status are discussed in Clinical Box 36–2.

HYPOXIA

Hypoxia is O_2 deficiency at the tissue level. It is a more correct term than **anoxia,** with there rarely being no O_2 at all left in the tissues.

CLINICAL BOX 36–2

Clinical Evaluation of Acid–Base Status

In evaluating disturbances of acid–base balance, it is important to know the pH and HCO_3^- content of arterial plasma. Reliable pH determinations can be made with a pH meter and a glass pH electrode. Using pH and a direct measurement of the P_{CO_2} with a CO_2 electrode, HCO_3^- concentration can be calculated. The P_{CO_2} is 7 to 8 mm Hg higher and the pH 0.03 to 0.04 unit lower in venous than arterial plasma because venous blood contains the CO_2 being carried from the tissues to the lungs. Therefore, the calculated HCO_3^- concentration is about 2 mmol/L higher. However, if this is kept in mind, free-flowing venous blood can be substituted for arterial blood in most clinical situations.

A measurement that is of some value in the differential diagnosis of metabolic acidosis is the **anion gap.** This gap, which is something of a misnomer, refers to the difference between the concentration of cations other than Na^+ and the concentration of anions other than Cl^- and HCO_3^- in the plasma. It consists for the most part of proteins in the anionic form, HPO_4^{2-}, SO_4^{2-}, and organic acids, and a normal value is about 12 mEq/L. It is increased when the plasma concentration of K^+, Ca^{2+}, or Mg^+ is decreased; when the concentration of or the charge on plasma proteins is increased; or when organic anions such as lactate or foreign anions accumulate in blood. It is decreased when cations are increased or when plasma albumin is decreased. The anion gap is increased in metabolic acidosis due to ketoacidosis, lactic acidosis, and other forms of acidosis in which organic anions are increased.

CLINICAL BOX 36–3

Effects of Hypoxia on Cells and Selected Tissues

Effects on Cells

Hypoxia causes the production of transcription factors **(hypoxia-inducible factors; HIFs).** These are made up of α and β subunits. In normally oxygenated tissues, the α subunits are rapidly ubiquitinated and destroyed. However, in hypoxic cells, the α subunits dimerize with β subunits, and the dimers activate genes that produce angiogenic factors and erythropoietin.

Effects on the Brain

In hypoxic hypoxia and the other generalized forms of hypoxia, the brain is affected first. A sudden drop in the inspired P_{O_2} to less than 20 mm Hg, which occurs, for example, when cabin pressure is suddenly lost in a plane flying above 16,000 m, causes loss of consciousness in 10 to 20 s and death in 4 to 5 min. Less severe hypoxia causes a variety of mental aberrations not unlike those produced by alcohol: impaired judgment, drowsiness, dulled pain sensibility, excitement, disorientation, loss of time sense, and headache. Other symptoms include anorexia, nausea, vomiting, tachycardia, and, when the hypoxia is severe, hypertension. The rate of ventilation is increased in proportion to the severity of the hypoxia of the carotid chemoreceptor cells.

Respiratory Stimulation

Dyspnea is by definition difficult or labored breathing in which the subject is conscious of shortness of breath; **hyperpnea** is the general term for an increase in the rate or depth of breathing regardless of the patient's subjective sensations. **Tachypnea** is rapid, shallow breathing. In general, a normal individual is not conscious of respiration until ventilation is doubled, and breathing is not uncomfortable until ventilation is tripled or quadrupled. Whether or not a given level of ventilation is uncomfortable also appears to depend on a variety of other factors. Hypercapnia and, to a lesser extent, hypoxia cause dyspnea. An additional factor is the effort involved in moving the air in and out of the lungs (the work of breathing).

Traditionally, hypoxia has been divided into four types. Numerous other classifications have been used, but the four-type system still has considerable utility if the definitions of the terms are kept clearly in mind. The four categories are (1) **hypoxic hypoxia,** in which the P_{O_2} of the arterial blood is reduced; (2) **anemic hypoxia,** in which the arterial P_{O_2} is normal but the amount of hemoglobin available to carry O_2 is reduced; (3) **stagnant** or **ischemic hypoxia,** in which the blood flow to a tissue is so low that adequate O_2 is not delivered to it despite a normal P_{O_2} and hemoglobin concentration; and (4) **histotoxic hypoxia,** in which the amount of O_2 delivered to a tissue is adequate but, because of the action of a toxic agent, the tissue cells cannot make use of the O_2 supplied to them. Some specific effects of hypoxia on cells and tissues are discussed in Clinical Box 36–3.

HYPOXIC HYPOXIA

By definition, hypoxic hypoxia is a condition of reduced arterial P_{O_2}. Hypoxic hypoxia is a problem in normal individuals at high altitudes and is a complication of pneumonia and a variety of other diseases of the respiratory system.

EFFECTS OF DECREASED BAROMETRIC PRESSURE

The composition of air stays the same, but the total barometric pressure falls with increasing altitude (Figure 36–12). Therefore, the P_{O_2} also falls. At 3000 m (approximately 10,000 ft) above sea level, the alveolar P_{O_2} is about 60 mm Hg and there is enough hypoxic stimulation of the chemoreceptors

FIGURE 36–12   **Composition of alveolar air in individuals breathing air (0–6100 m) and 100% O$_2$ (6100–13,700 m).** The minimal alveolar PO$_2$ that an unacclimatized subject can tolerate without loss of consciousness is about 35–40 mm Hg. Note that with increasing altitude, the alveolar PCO$_2$ drops because of the hyperventilation due to hypoxic stimulation of the carotid and aortic chemoreceptors. The fall in barometric pressure with increasing altitude is not linear, because air is compressible.

to definitely increase ventilation. As one ascends higher, the alveolar PO$_2$ falls less rapidly and the alveolar PCO$_2$ declines somewhat because of the hyperventilation. The resulting fall in arterial PCO$_2$ produces respiratory alkalosis.

HYPOXIC SYMPTOMS BREATHING AIR

A number of compensatory mechanisms operate over a period of time to increase altitude tolerance **(acclimatization),** but in unacclimatized subjects, mental symptoms such as irritability appear at about 3700 m. At 5500 m, the hypoxic symptoms are severe; and at altitudes above 6100 m (20,000 ft), consciousness is usually lost.

HYPOXIC SYMPTOMS BREATHING OXYGEN

The total atmospheric pressure becomes the limiting factor in altitude tolerance when breathing 100% O$_2$.

The partial pressure of water vapor in the alveolar air is constant at 47 mm Hg, and that of CO$_2$ is normally 40 mm Hg, so that the lowest barometric pressure at which a normal alveolar PO$_2$ of 100 mm Hg is possible is 187 mm Hg, the pressure at about 10,400 m (34,000 ft). At greater altitudes, the increased ventilation due to the decline in alveolar PO$_2$ lowers the alveolar PCO$_2$ somewhat, but the maximum alveolar PO$_2$

that can be attained when breathing 100% O$_2$ at the ambient barometric pressure of 100 mm Hg at 13,700 m is about 40 mm Hg. At about 14,000 m, consciousness is lost in spite of the administration of 100% O$_2$. At 19,200 m, the barometric pressure is 47 mm Hg, and at or below this pressure the body fluids boil at body temperature. The point is largely academic, however, because any individual exposed to such a low pressure would be dead of hypoxia before the bubbles of steam could cause death.

Of course, an artificial atmosphere can be created around an individual; in a pressurized suit or cabin supplied with O$_2$ and a system to remove CO$_2$, it is possible to ascend to any altitude and to live in the vacuum of interplanetary space. Some delayed effects of high altitude are discussed in Clinical Box 36–4.

ACCLIMATIZATION

Acclimatization to altitude is due to the operation of a variety of compensatory mechanisms. The respiratory alkalosis produced by the hyperventilation shifts the oxygen–hemoglobin dissociation curve to the left, but a concomitant increase in red blood cell 2,3-BPG tends to decrease the O$_2$ affinity of hemoglobin. The net effect is a small increase in P$_{50}$. The decrease in O$_2$ affinity makes more O$_2$ available to the tissues. However, the value of the increase in P$_{50}$ is limited because when the

Delayed Effects of High Altitude

When they first arrive at a high altitude, many individuals develop transient "mountain sickness." This syndrome develops 8 to 24 h after arrival at altitude and lasts 4 to 8 d. It is characterized by headache, irritability, insomnia, breathlessness, and nausea and vomiting. Its cause is unsettled, but it appears to be associated with cerebral edema. The low PO_2 at high altitude causes arteriolar dilation, and if cerebral autoregulation does not compensate, there is an increase in capillary pressure that favors increased transudation of fluid into brain tissue. Individuals who do not develop mountain sickness have a diuresis at high altitude, and urine volume is decreased in individuals who develop the condition.

High-altitude illness includes not only mountain sickness but also two more serious syndromes that complicate it: **high-altitude cerebral edema** and **high-altitude pulmonary edema.** In high-altitude cerebral edema, the capillary leakage in mountain sickness progresses to frank brain swelling, with ataxia, disorientation, and in some cases coma and death due to herniation of the brain through the tentorium. High-altitude pulmonary edema is a patchy edema of the lungs that is related to the marked pulmonary hypertension that develops at high altitude. It has been argued that it occurs because not all pulmonary arteries have enough smooth muscle to constrict in response to hypoxia, and in the capillaries supplied by those arteries, the general rise in pulmonary arterial pressure causes a capillary pressure increase that disrupts their walls (stress failure).

All forms of high-altitude illness are benefited by descent to lower altitude and by treatment with the diuretic acetazolamide. This drug inhibits carbonic anhydrase, producing increased HCO_3^- excretion in the urine, stimulating respiration, increasing $PaCO_2$, and reducing the formation of CSF. When cerebral edema is marked, large doses of glucocorticoids are often administered as well. Their mechanism of action is unsettled. In high-altitude pulmonary edema, prompt treatment with O_2 is essential—and, if available, use of a hyperbaric chamber. Portable hyperbaric chambers are now available in a number of mountain areas. Nifedipine, a Ca^{2+} channel blocker that lowers pulmonary artery pressure, is also useful.

FIGURE 36–13 **Effect of acclimatization on the ventilatory response at various altitudes.** $\dot{V}_E \dot{V}_{O_2}$ is the ventilatory equivalent, the ratio of expired minute volume ($\dot{V}_E$) to the O_2 consumption ($\dot{V}_{O_2}$). (Reproduced with permission from Lenfant C, Sullivan K: Adaptation to high altitude. N Engl J Med 1971;284:1298.)

tory response begins to decline slowly, but it takes years of residence at higher altitudes for it to decline to the initial level. Associated with this decline is a gradual desensitization to the stimulatory effects of hypoxia.

Erythropoietin secretion increases promptly on ascent to high altitude and then falls somewhat over the following 4 d as the ventilatory response increases and the arterial PO_2 rises. The increase in circulating red blood cells triggered by the erythropoietin begins in 2 to 3 d and is sustained as long as the individual remains at high altitude.

Compensatory changes also occur in the tissues. The mitochondria, which are the site of oxidative reactions, increase in number, and myoglobin increases, which facilitates the movement of O_2 into the tissues. The tissue content of cytochrome oxidase also increases.

The effectiveness of the acclimatization process is indicated by the fact that permanent human habitations exist in the Andes and Himalayas at elevations above 5500 m (18,000 ft). The natives who live in these villages are barrel-chested and markedly polycythemic. They have low alveolar PO_2 values, but in most other ways they are remarkably normal.

DISEASES CAUSING HYPOXIC HYPOXIA

Hypoxic hypoxia is the most common form of hypoxia seen clinically. The diseases that cause it can be roughly divided into those in which the gas exchange apparatus fails, those such as congenital heart disease in which large amounts of blood are shunted from the venous to the arterial side of the circulation, and those in which the respiratory pump fails. Lung failure occurs when conditions such as pulmonary fibrosis produce alveolar–capillary block, or there is ventilation–perfusion imbalance. Pump failure can be due to fatigue of the respiratory muscles in conditions in which the work of breathing is increased or to a

arterial PO_2 is markedly reduced, the decreased O_2 affinity also interferes with O_2 uptake by hemoglobin in the lungs.

The initial ventilatory response to increased altitude is relatively small, because the alkalosis tends to counteract the stimulating effect of hypoxia. However, ventilation steadily increases over the next 4 d (Figure 36–13) because the active transport of H^+ into cerebrospinal fluid (CSF), or possibly a developing lactic acidosis in the brain, causes a fall in CSF pH that increases the response to hypoxia. After 4 d, the ventilavariety of mechanical defects such as pneumothorax or bronchial obstruction that limit ventilation. It can also be caused by abnormalities of the neural mechanisms that control ventilation, such as depression of the respiratory neurons in the medulla by morphine and other drugs. Some specific causes of hypoxic hypoxia are discussed in the following text.

VENTILATION–PERFUSION IMBALANCE

Patchy ventilation–perfusion imbalance is by far the most common cause of hypoxic hypoxia in clinical situations. In disease processes that prevent ventilation of some of the alveoli, the ventilation–blood flow ratios in different parts of the lung determine the extent to which systemic arterial PO_2 declines. If nonventilated alveoli are perfused, the nonventilated but perfused portion of the lung is in effect a right-to-left shunt, dumping unoxygenated blood into the left side of the heart. Lesser degrees of ventilation–perfusion imbalance are more common. In the example illustrated in Figure 36–14, the underventilated alveoli (B) have a low alveolar PO_2, whereas the overventilated alveoli (A) have a high alveolar PO_2. However, the unsaturation of the hemoglobin of the blood coming from B is not completely compensated by the greater saturation of the blood coming from A, because hemoglobin is normally nearly saturated in the lungs and the higher alveolar PO_2 adds only a little more O_2 to the hemoglobin than it normally carries. Consequently, the arterial blood is unsaturated. On the other hand, the CO_2 content of the arterial blood is generally normal in such situations, since extra loss of CO_2 in overventilated regions can balance diminished loss in underventilated areas.

VENOUS-TO-ARTERIAL SHUNTS

When a cardiovascular abnormality such as an interatrial septal defect permits large amounts of unoxygenated venous blood to bypass the pulmonary capillaries and dilute the oxygenated blood in the systemic arteries ("right-to-left shunt"), chronic hypoxic hypoxia and cyanosis (**cyanotic congenital heart disease**) result. Administration of 100% O_2 raises the O_2 content of alveolar air and improves the hypoxia due to hypoventilation, impaired diffusion, or ventilation–perfusion imbalance (short of perfusion of totally unventilated segments) by increasing the amount of O_2 in the blood leaving the lungs. However, in patients with venous-to-arterial shunts and normal lungs, any beneficial effect of 100% O_2 is slight and is due solely to an increase in the amount of dissolved O_2 in the blood.

	A	B	A + B		A	B	A + B
Alveolar ventilation (L/min)	2.0	2.0	4.0	Alveolar ventilation (L/min)	3.2	0.8	4.0
Pulmonary blood flow (L/min)	2.5	2.5	5.0	Pulmonary blood flow (L/min)	2.5	2.5	5.0
Ventilation/blood flow ratio	0.8	0.8	0.8	Ventilation/blood flow ratio	1.3	0.3	0.8
Mixed venous O_2 saturation (%)	75.0	75.0	75.0	Mixed venous O_2 saturation (%)	75.0	75.0	75.0
Arterial O_2 saturation (%)	97.4	97.4	97.4	Arterial O_2 saturation (%)	98.2	91.7	95.0
Mixed venous O_2 tension (mm Hg)	40.0	40.0	40.0	Mixed venous O_2 tension (mm Hg)	40.0	40.0	40.0
Alveolar O_2 tension (mm Hg)	104.0	104.0	104.0	Alveolar O_2 tension (mm Hg)	116.0	66.0	106.0
Arterial O_2 tension (mm Hg)	104.0	104.0	104.0	Arterial O_2 tension (mm Hg)	116.0	66.0	84.0

FIGURE 36–14 **Comparison of ventilation/blood flow relationships in health and disease. Left:** "Ideal" ventilation/blood flow relationship. **Right:** Nonuniform ventilation and uniform blood flow, uncompensated. $\dot{V}_A$, alveolar ventilation; MV, respiratory minute volume. (Reproduced with permission from Comroe JH Jr., et al: *The Lung: Clinical Physiology and Pulmonary Function Tests,* 2nd ed. Year Book, 1962.)

OTHER FORMS OF HYPOXIA

ANEMIC HYPOXIA

Hypoxia due to anemia is not severe at rest unless the hemoglobin deficiency is marked, because red blood cell 2,3-DPG increases. However, anemic patients may have considerable difficulty during exercise because of limited ability to increase O_2 delivery to the active tissues (Figure 36–15).

CARBON MONOXIDE POISONING

Small amounts of carbon monoxide (CO) are formed in the body, and this gas may function as a chemical messenger in the brain and elsewhere. In larger amounts, it is poisonous. Outside the body, it is formed by incomplete combustion of carbon. It was used by the Greeks and Romans to execute criminals, and today it causes more deaths than any other gas. CO poisoning has become less common in the United States, since natural gas, which does not contain CO, replaced artificial gases such as coal gas, which contains large amounts. However, the exhaust of gasoline engines is 6% or more CO.

CO is toxic because it reacts with hemoglobin to form **carbon monoxyhemoglobin (carboxyhemoglobin, COHb)**, and COHb cannot take up O_2 (Figure 36–15). Carbon monoxide poisoning is often listed as a form of anemic hypoxia because the amount of hemoglobin that can carry O_2 is reduced, but the total hemoglobin content of the blood is unaffected by CO. The affinity of hemoglobin for CO is 210 times its affinity for O_2, and COHb liberates CO very slowly. An additional difficulty is that when COHb is present the dissociation curve of the remaining HbO_2 shifts to the left, decreasing the amount of O_2 released. This is why an anemic individual who has 50% of the normal amount of HbO_2 may be able to perform moderate work, whereas an individual whose HbO_2 is reduced to the same level because of the formation of COHb is seriously incapacitated.

Because of the affinity of CO for hemoglobin, progressive COHb formation occurs when the alveolar P_{CO} is greater than 0.4 mm Hg. However, the amount of COHb formed depends on the duration of exposure to CO as well as the concentration of CO in the inspired air and the alveolar ventilation.

CO is also toxic to the cytochromes in the tissues, but the amount of CO required to poison the cytochromes is 1000 times the lethal dose; tissue toxicity thus plays no role in clinical CO poisoning.

The symptoms of CO poisoning are those of any type of hypoxia, especially headache and nausea, but there is little stimulation of respiration, since in the arterial blood, P_{O_2} remains normal and the carotid and aortic chemoreceptors are not stimulated. The cherry-red color of COHb is visible in the skin, nail beds, and mucous membranes. Death results when about 70–80% of the circulating hemoglobin is converted to COHb. The symptoms produced by chronic exposure to sublethal concentrations of CO are those of progressive brain damage, including mental changes and, sometimes, a parkinsonism-like state.

Treatment of CO poisoning consists of immediate termination of the exposure and adequate ventilation, by artificial respiration if necessary. Ventilation with O_2 is preferable to ventilation with fresh air, since O_2 hastens the dissociation of COHb. Hyperbaric oxygenation (see below) is useful in this condition.

HYPOPERFUSION HYPOXIA

Hypoperfusion hypoxia, or stagnant hypoxia, is due to slow circulation and is a problem in organs such as the kidneys and heart during shock. The liver and possibly the brain are damaged by hypoperfusion hypoxia in congestive heart failure. The blood flow to the lung is normally very large, and it takes prolonged hypotension to produce significant damage. However, acute respiratory distress syndrome (ARDS) can develop when there is prolonged circulatory collapse.

HISTOTOXIC HYPOXIA

Hypoxia due to inhibition of tissue oxidative processes is most commonly the result of cyanide poisoning. Cyanide inhibits cytochrome oxidase and possibly other enzymes. Methylene blue or nitrites are used to treat cyanide poisoning. They act by forming **methemoglobin,** which then reacts with cyanide to form **cyanmethemoglobin,** a nontoxic compound. The extent of treatment with these compounds is, of course, limited by the amount of methemoglobin that can be safely formed. Hyperbaric oxygenation may also be useful.

OXYGEN TREATMENT OF HYPOXIA

Administration of oxygen-rich gas mixtures is of very limited value in hypoperfusion, anemic, and histotoxic hypoxia

FIGURE 36–15 **Effects of anemia and CO on hemoglobin binding of O_2.** Normal oxyhemoglobin (14g/dL hemoglobin) dissociation curve compared with anemia (7 g/dL hemoglobin) and with oxyhemoglobin dissociation curves in CO poisoning (50% carboxyhemoglobin). Note that the CO-poisoning curve is shifted to the left of the anemia curve. (Reproduced with permission from Leff AR, Schumacker PT: *Respiratory Physiology: Basics and Applications.* Saunders, 1993.)

CLINICAL BOX 36–5

Administration of Oxygen & Its Potential Toxicity

It is interesting that while O_2 is necessary for life in aerobic organisms, it is also toxic. Indeed, 100% O_2 has been demonstrated to exert toxic effects not only in animals but also in bacteria, fungi, cultured animal cells, and plants. The toxicity seems to be due to the production of reactive oxygen species including superoxide anion (O_2^-) and H_2O_2. When 80–100% O_2 is administered to humans for periods of 8 h or more, the respiratory passages become irritated, causing substernal distress, nasal congestion, sore throat, and coughing.

Some infants treated with O_2 for respiratory distress syndrome develop a chronic condition characterized by lung cysts and densities **(bronchopulmonary dysplasia).** This syndrome may be a manifestation of O_2 toxicity. Another complication in these infants is **retinopathy of prematurity (retrolental fibroplasia),** the formation of opaque vascular tissue in the eyes, which can lead to serious visual defects. The retinal receptors mature from the center to the periphery of the retina, and they use considerable O_2. This causes the retina to become vascularized in an orderly fashion. Oxygen treatment before maturation is complete provides the needed O_2 to the photoreceptors, and consequently the normal vascular pattern fails to develop. Evidence indicates that this condition can be prevented or ameliorated by treatment with vitamin E, which exerts an antioxidant effect, and, in animals, by growth hormone inhibitors.

Administration of 100% O_2 at increased pressure accelerates the onset of O_2 toxicity, with the production not only of tracheobronchial irritation but also of muscle twitching, ringing in the ears, dizziness, convulsions, and coma. The speed with which these symptoms develop is proportional to the pressure at which the O_2 is administered; for example, at 4 atmospheres, symptoms develop in half the subjects in 30 min, whereas at 6 atmospheres, convulsions develop in a few minutes.

On the other hand, exposure to 100% O_2 at 2 to 3 atmospheres can increase dissolved O_2 in arterial blood to the point that arterial O_2 tension is greater than 2000 mm Hg and tissue O_2 tension is 400 mm Hg. If exposure is limited to 5 h or less at these pressures, O_2 toxicity is not a problem. Therefore, **hyperbaric O_2** therapy in closed tanks is used to treat diseases in which improved oxygenation of tissues cannot be achieved in other ways. It is of demonstrated value in carbon monoxide poisoning, radiation-induced tissue injury, gas gangrene, very severe blood loss anemia, diabetic leg ulcers and other wounds that are slow to heal, and rescue of skin flaps and grafts in which the circulation is marginal. It is also the primary treatment for decompression sickness and air embolism.

In hypercapnic patients in severe pulmonary failure, the CO_2 level may be so high that it depresses rather than stimulates respiration. Some of these patients keep breathing only because the carotid and aortic chemoreceptors drive the respiratory center. If the hypoxic drive is withdrawn by administering O_2, breathing may stop. During the resultant apnea, the arterial P_{O_2} drops but breathing may not start again, as P_{CO_2} further depresses the respiratory center. Therefore, O_2 therapy in this situation must be started with care.

---

because all that can be accomplished in this way is an increase in the amount of dissolved O_2 in the arterial blood. This is also true in hypoxic hypoxia when it is due to shunting of unoxygenated venous blood past the lungs. In other forms of hypoxic hypoxia, O_2 is of great benefit. Treatment regimens that deliver less than 100% O_2 are of value both acutely and chronically, and administration of O_2 24 h/d for 2 y in this fashion has been shown to significantly decrease the mortality of chronic obstructive pulmonary disease. O_2 toxicity and therapy are discussed in Clinical Box 36–5.

HYPERCAPNIA & HYPOCAPNIA

HYPERCAPNIA

Retention of CO_2 in the body **(hypercapnia)** initially stimulates respiration. Retention of larger amounts produces symptoms due to depression of the central nervous system: confusion, diminished sensory acuity, and, eventually, coma with respiratory depression and death. In patients with these symptoms, the P_{CO_2} is markedly elevated, severe respiratory acidosis is present, and the plasma HCO_3^- may exceed 40 mEq/L. Large amounts of HCO_3^- are excreted, but more HCO_3^- is reabsorbed, raising the plasma HCO_3^- and partially compensating for the acidosis.

CO_2 is so much more soluble than O_2 that hypercapnia is rarely a problem in patients with pulmonary fibrosis. However, it does occur in ventilation–perfusion inequality and when for any reason alveolar ventilation is inadequate in the various forms of pump failure. It is exacerbated when CO_2 production is increased. For example, in febrile patients there is a 13% increase in CO_2 production for each 1°C rise in temperature, and a high carbohydrate intake increases CO_2 production because of the increase in the respiratory quotient. Normally, alveolar ventilation increases and the extra CO_2 is expired, but it accumulates when ventilation is compromised.

HYPOCAPNIA

Hypocapnia is the result of hyperventilation. During voluntary hyperventilation, the arterial P_{CO_2} falls from 40 to as low as 15 mm Hg while the alveolar P_{O_2} rises to 120 to 140 mm Hg.

The more chronic effects of hypocapnia are seen in neurotic patients who chronically hyperventilate. Cerebral blood flow may be reduced 30% or more because of the direct constrictor effect of hypocapnia on the cerebral vessels. The cerebral ischemia causes light-headedness, dizziness, and paresthesias. Hypocapnia also increases cardiac output. It has a direct constrictor effect on many peripheral vessels, but it depresses the vasomotor center, so that the blood pressure is usually unchanged or only slightly elevated.

Other consequences of hypocapnia are due to the associated respiratory alkalosis, the blood pH being increased to 7.5 or 7.6. The plasma HCO_3^- level is low, but HCO_3^- reabsorption is decreased because of the inhibition of renal acid secretion by the low P_{CO_2}. The plasma total calcium level does not change, but the plasma Ca^{2+} level falls and hypocapnic individuals develop carpopedal spasm, a positive Chvostek sign, and other signs of tetany.

CHAPTER SUMMARY

- Partial pressure differences between air and blood for O_2 and CO_2 dictate a net flow of O_2 into the blood and CO_2 out of the blood in the pulmonary system. However, this flow is greatly enhanced by the ability for hemoglobin to bind O_2 and chemical reactions that increase CO_2 in the blood (eg, carbonic anhydrase).

- The amount of O_2 in the blood is determined by the amount dissolved (minor) and the amount bound (major) to hemoglobin. Each hemoglobin molecule contains four subunits that each can bind O_2. Binding of the first O_2 to hemoglobin increases the affinity for the second O_2, and this pattern is continued until four O_2 are bound. Hemoglobin O_2 binding is also affected by pH, temperature, and the concentration of 2,3-biphospholycerate (2,3-BPG).

- CO_2 in blood is rapidly converted into H_2CO_3 due to the activity of carbonic anhydrase. CO_2 also readily forms carbamino compounds with blood proteins (including hemoglobin). The rapid net loss of CO_2 allows more CO_2 to dissolve in blood.

- The pH of plasma is 7.4. A decrease in plasma pH is termed acidosis and an increase of plasma pH is termed alkalosis. Acid and base shifts in the blood are controlled by proteins, including hemoglobin, and principally by the carbonic acid-bicarbonate buffering system. The carbonic acid-bicarbonate buffering system is effective because dissolved CO_2 can be controlled by respiration.

- A short-term change in arterial P_{CO_2} due to decreased ventilation results in respiratory acidosis. A short-term change in arterial P_{CO_2} due to increased ventilation results in respiratory alkalosis. Metabolic acidosis occurs when strong acids are added to the blood, and metabolic alkalosis occurs when strong bases are added to (or strong acids are removed from) the blood.

- Respiratory compensation to acidosis or alkalosis involves quick changes in ventilation. Such changes effectively change the P_{CO_2} in the blood plasma. Renal compensation mechanisms are much slower and involve H^+ secretion or HCO_3^- reabsorption.

- Hypoxia is a deficiency of O_2 at the tissue level. Hypoxia has powerful consequences at the cellular, tissue, and organ level: It can alter cellular transcription factors and thus protein expression; it can quickly alter brain function and produce symptoms similar to alcohol (eg, dizziness, impaired mental function, drowsiness, headache); and it can affect ventilation. Long-term hypoxia results in cell and tissue death.

MULTIPLE-CHOICE QUESTIONS

For all questions, select the single best answer unless otherwise directed.

1. Most of the CO_2 transported in the blood is
   A) dissolved in plasma.
   B) in carbamino compounds formed from plasma proteins.
   C) in carbamino compounds formed from hemoglobin.
   D) bound to Cl^-.
   E) in HCO_3^-.

2. Which of the following has the greatest effect on the ability of blood to transport oxygen?
   A) capacity of the blood to dissolve oxygen
   B) amount of hemoglobin in the blood
   C) pH of plasma
   D) CO_2 content of red blood cells
   E) temperature of the blood

3. Which of the following is *not* true of the system?
   $$CO_2 + H_2O \rightleftharpoons^1 H_2CO_3 \rightleftharpoons^2 H^+ + HCO_3^-$$
   A) Reaction 1 is catalyzed by carbonic anhydrase.
   B) Because of reaction 2, the pH of blood declines during breath holding.
   C) Reaction 1 occurs in the kidneys.
   D) Reaction 1 occurs primarily in plasma.
   E) The reactions move to the left when there is excess H^+ in the tissues.

4. Uncompensated respiratory acidosis differs from uncompensated metabolic acidosis in that
   A) plasma pH change is always greater in uncompensated respiratory acidosis compared to uncompensated metabolic acidosis.
   B) there are no compensation mechanisms for respiratory acidosis, whereas there is respiratory compensation for metabolic acidosis.
   C) uncompensated respiratory acidosis involves changes in plasma $[HCO_3^-]$, whereas plasma $[HCO_3^-]$ is unchanged in uncompensated metabolic acidosis.
   D) uncompensated respiratory acidosis is associated with a change in P_{CO_2}, whereas uncompensated metabolic acidosis occurs along the isobar line for P_{CO_2}.

5. O_2 delivery to the tissues would be reduced to the greatest extent in
   A) a normal subject breathing 100% O_2 on top of Mt. Everest.
   B) a normal subject running a marathon at sea level.
   C) a patient with carbon monoxide poisoning.
   D) a patient who has ingested cyanide.
   E) a patient with moderately severe metabolic acidosis.

CHAPTER RESOURCES

Crystal RG, West JB (editors): *The Lung: Scientific Foundations*, 2nd ed. Raven Press, 1997.

Fishman AP, et al (editors): *Fishman's Pulmonary Diseases and Disorders,* 4th ed. McGraw-Hill, 2008.

Hackett PH, Roach RC: High-altitude illness. N Engl J Med 2001;345:107.

Laffey JG, Kavanagh BP: Hypocapnia. N Engl J Med 2002;347:43.

Levitzky, MG: *Pulmonary Physiology,* 7th ed. McGraw-Hill, 2007.

Prisk GK, Paiva M, West JB (editors): *Gravity and the Lung: Lessons from Micrography.* Marcel Dekker, 2001.

Voelkel NF: High-altitude pulmonary edema. N Engl J Med 2002;346:1607.

West JB: *Pulmonary Pathophysiology,* 5th ed. McGraw-Hill, 1995.

Regulation of Respiration

INTRODUCTION

Spontaneous respiration is produced by rhythmic discharge of motor neurons that innervate the respiratory muscles. This discharge is totally dependent on nerve impulses from the brain; breathing stops if the spinal cord is transected above the origin of the phrenic nerves. The rhythmic discharges from the brain that produce spontaneous respiration are regulated by alterations in arterial P_{O_2}, P_{CO_2}, and H^+ concentration, and this chemical control of breathing is supplemented by a number of non-chemical influences. The physiological bases for these phenomena are discussed in this chapter.

NEURAL CONTROL OF BREATHING

CONTROL SYSTEMS

Two separate neural mechanisms regulate respiration. One is responsible for voluntary control and the other for automatic control. The voluntary system is located in the cerebral cortex and sends impulses to the respiratory motor neurons via the corticospinal tracts. The automatic system is driven by a group of pacemaker cells in the medulla. Impulses from these cells activate motor neurons in the cervical and thoracic spinal cord that innervate inspiratory muscles. Those in the cervical cord activate the diaphragm via the phrenic nerves, and those in the thoracic spinal cord activate the external intercostal muscles. However, the impulses also reach the innervation of the internal intercostal muscles and other expiratory muscles.

The motor neurons to the expiratory muscles are inhibited when those supplying the inspiratory muscles are active, and vice versa. Although spinal reflexes contribute to this **reciprocal innervation,** it is due primarily to activity in descending pathways. Impulses in these descending pathways excite agonists and inhibit antagonists. The one exception to

FIGURE 37–1 **Pacemaker cells in the pre-Bötzinger complex (pre-BÖTC). Top:** Anatomical diagram of the pre-BÖTC from a neonatal rat. **Bottom:** Sample rhythmic discharge tracing of neurons in the pre-BÖTC complex from a brain slice of a neonatal rat. IO, inferior olive; LRN, lateral reticular nucleus; NA, nucleus ambiguus; XII, nucleus of 12th cranial nerve; 5SP, spinal nucleus of trigeminal nerve. (Modified from Feldman JC, Gray PA: Sighs and gasps in a dish. Nat Neurosci 2000;3:531.)

the reciprocal inhibition is a small amount of activity in phrenic axons for a short period after inspiration. The function of this post-inspiratory output appears to be to brake the lung's elastic recoil and make respiration smooth.

MEDULLARY SYSTEMS

The main components of the **respiratory control pattern generator** responsible for automatic respiration are located in the medulla. Rhythmic respiration is initiated by a small group of synaptically coupled pacemaker cells in the **pre-Bötzinger complex** (pre-BÖTC) on either side of the medulla between the nucleus ambiguus and the lateral reticular nucleus (Figure 37–1). These neurons discharge rhythmically, and they produce rhythmic discharges in phrenic motor neurons that are abolished by sections between the pre-Bötzinger complex and these motor neurons. They also contact the hypoglossal nuclei, and the tongue is involved in the regulation of airway resistance.

Neurons in the pre-Bötzinger complex discharge rhythmically in brain slice preparations in vitro, and if the slices become hypoxic, discharge changes to one associated with gasping. Addition of cadmium to the slices causes occasional sigh-like discharge patterns. There are NK1 receptors and μ-opioid receptors on these neurons, and, in vivo, substance P stimulates and opioids inhibit respiration. Depression of respiration is a side effect that limits the use of opioids in the treatment of pain. However, it is now known that $5HT_4$ receptors are present in the pre-Bötzinger complex and treatment with $5HT_4$ agonists blocks the inhibitory effect of opiates on respiration in experimental animals, without inhibiting their analgesic effect.

In addition, dorsal and ventral groups of respiratory neurons are present in the medulla (Figure 37–2). However, lesions of these neurons do not abolish respiratory activity, and they apparently project to the pre-Bötzinger pacemaker neurons.

FIGURE 37–2 **Respiratory neurons in the brain stem.** Dorsal view of brain stem; cerebellum removed. The effects of various lesions and brain stem transections are shown; the spirometer tracings at the right indicate the depth and rate of breathing. If a lesion is introduced at D, breathing ceases. The effects of higher transections, with and without vagus nerves transection, are shown (see text for details). DRG, dorsal group of respiratory neurons; VRG, ventral group of respiratory neurons; NPBL, nucleus parabrachialis (pneumotaxic center); 4th vent, fourth ventricle; IC, inferior colliculus; CP, middle cerebellar peduncle. The roman numerals identify cranial nerves. (Modified from Mitchell RA, Berger A: State of the art: Review of neural regulation of respiration. Am Rev Respir Dis 1975;111:206.)

PONTINE & VAGAL INFLUENCES

Although the rhythmic discharge of medullary neurons concerned with respiration is spontaneous, it is modified by neurons in the pons and afferents in the vagus from receptors in the airways and lungs. An area known as the **pneumotaxic center** in the medial parabrachial and Kölliker–Fuse nuclei of the dorsolateral pons contains neurons active during inspiration and neurons active during expiration. When this area is damaged, respiration becomes slower and tidal volume greater, and when the vagi are also cut in anesthetized animals, there are prolonged inspiratory spasms that resemble breath holding (**apneusis**; section B in Figure 37–2). The normal function of the pneumotaxic center is unknown, but it may play a role in switching between inspiration and expiration.

Stretching of the lungs during inspiration initiates impulses in afferent pulmonary vagal fibers. These impulses inhibit inspiratory discharge. This is why the depth of inspiration is increased after vagotomy (Figure 37–2) and apneusis develops if the vagi are cut after damage to the pneumotaxic center. Vagal feedback activity does not alter the rate of rise of the neural activity in respiratory motor neurons (Figure 37–3).

When the activity of the inspiratory neurons is increased in intact animals, the rate and the depth of breathing are increased. The depth of respiration is increased because the lungs are stretched to a greater degree before the amount of vagal and pneumotaxic center inhibitory activity is sufficient to overcome the more intense inspiratory neuron discharge. The respiratory rate is increased because the after-discharge in the vagal and possibly the pneumotaxic afferents to the medulla is rapidly overcome.

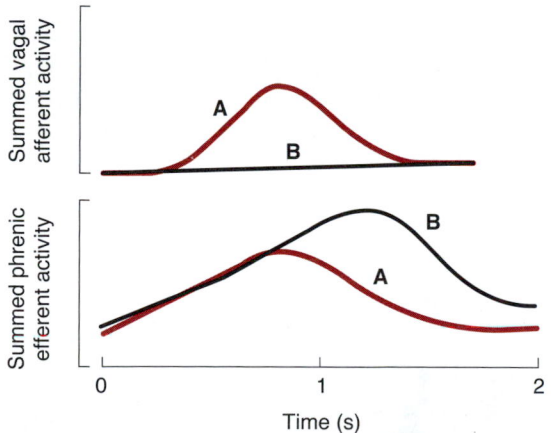

FIGURE 37–3 **Afferent vagal fibers inhibit inspiratory discharge.** Superimposed records of two breaths: **(A)** with and **(B)** without feedback vagal afferent activity from stretch receptors in the lungs. Note that the rate of rise in phrenic nerve activity to the diaphragm is unaffected but the discharge is prolonged in the absence of vagal input.

REGULATION OF RESPIRATORY ACTIVITY

A rise in the P_{CO_2} or H^+ concentration of arterial blood or a drop in its P_{O_2} increases the level of respiratory neuron activity in the medulla, and changes in the opposite direction have a slight inhibitory effect. The effects of variations in blood chemistry on ventilation are mediated via respiratory **chemoreceptors**—the carotid and aortic bodies and collections of cells in the medulla and elsewhere that are sensitive to changes in the chemistry of the blood. They initiate impulses that stimulate the respiratory center. Superimposed on this basic **chemical control of respiration,** other afferents provide non-chemical controls that affect breathing in particular situations (Table 37–1).

CHEMICAL CONTROL OF BREATHING

The chemical regulatory mechanisms adjust ventilation in such a way that the alveolar P_{CO_2} is normally held constant, the effects of excess H^+ in the blood are combated, and the P_{O_2} is raised when it falls to a potentially dangerous level. The respiratory minute volume is proportional to the metabolic rate, but the link between metabolism and ventilation is CO_2, not O_2. The receptors in the carotid and aortic bodies are stimulated by a rise in the P_{CO_2} or H^+ concentration of arterial blood or a decline in its P_{O_2}. After denervation of the carotid chemoreceptors, the response to a drop in P_{O_2} is abolished; the predominant effect of hypoxia after denervation of the carotid bodies is a direct depression of the respiratory center. The response to changes in arterial blood H^+ concentration in the pH 7.3–7.5 range is also abolished, although larger changes exert some effect. The response to changes in arterial P_{CO_2}, on the other hand, is affected only slightly; it is reduced no more than 30–35%.

TABLE 37–1 **Stimuli affecting the respiratory center.**

Chemical control
CO_2 (via CSF and brain interstitial fluid H^+ concentration)
$\left.\begin{array}{l} O_2 \\ H^+ \end{array}\right\}$ (via carotid and aortic bodies)
Non-chemical control
Vagal afferents from receptors in the airways and lungs
Afferents from the pons, hypothalamus, and limbic system
Afferents from proprioceptors
Afferents from baroreceptors: arterial, atrial, ventricular, pulmonary

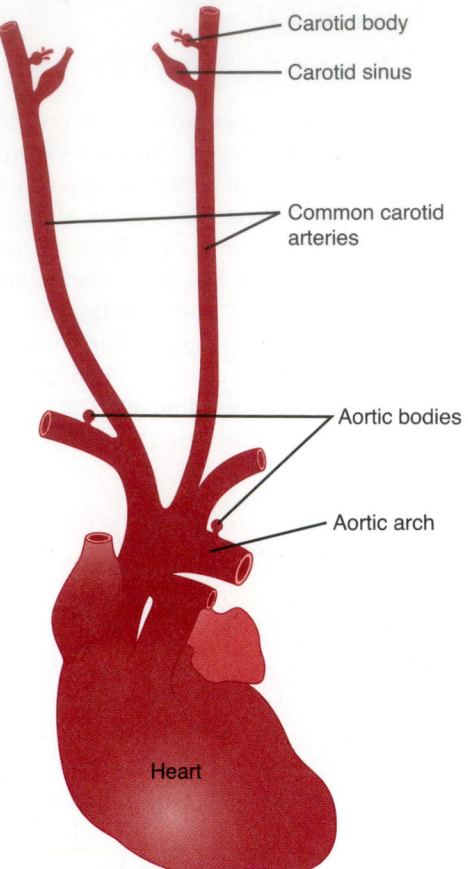

FIGURE 37–4   Location of carotid and aortic bodies. Carotid bodies are positioned near a major arterial baroreceptor, the carotid sinus. Two aortic bodies are shown near the aortic arch.

CAROTID & AORTIC BODIES

There is a carotid body near the carotid bifurcation on each side, and there are usually two or more aortic bodies near the arch of the aorta (Figure 37–4). Each carotid and aortic body (**glomus**) contains islands of two types of cells, type I and type II cells, surrounded by fenestrated sinusoidal capillaries. The type I or **glomus cells** are closely associated with cuplike endings of the afferent nerves (Figure 37–5). The glomus cells resemble adrenal chromaffin cells and have dense-core granules containing catecholamines that are released upon exposure to hypoxia and cyanide. The cells are excited by hypoxia, and the principal transmitter appears to be dopamine, which excites the nerve endings by way of D_2 receptors. The type II cells are glia-like, and each surrounds four to six type I cells. Their function is probably sustentacular.

Outside the capsule of each body, the nerve fibers acquire a myelin sheath; however, they are only 2 to 5 μm in diameter and conduct at the relatively low rate of 7 to 12 m/s. Afferents from the carotid bodies ascend to the medulla via the carotid sinus and glossopharyngeal nerves, and fibers from the aortic bodies ascend in the vagi. Studies in which one carotid body has been isolated and perfused while recordings are being taken from its

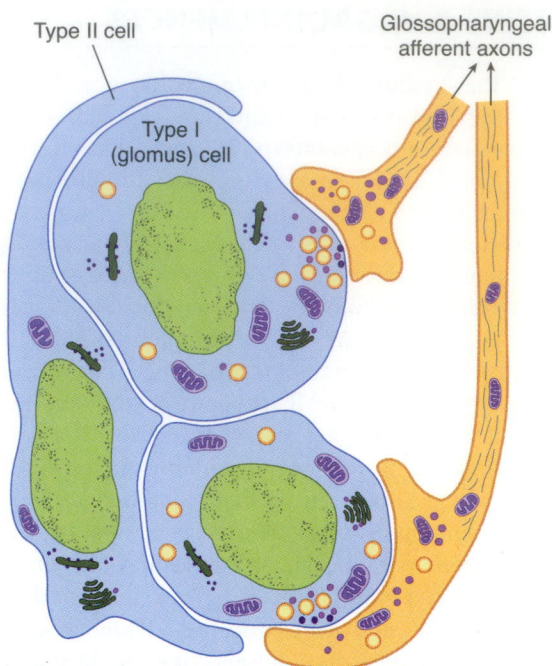

FIGURE 37–5   Organization of the carotid body. Type I (glomus) cells contain catecholamines. When exposed to hypoxia, they release their catecholamines, which stimulate the cuplike endings of the carotid sinus nerve fibers in the glossopharyngeal nerve. The glia-like type II cells surround the type I cells and probably have a sustentacular function.

afferent nerve fibers show that there is a graded increase in impulse traffic in these afferent fibers as the PO_2 of the perfusing blood is lowered (Figure 37–6) or the PCO_2 is raised.

Type I glomus cells have O_2-sensitive K^+ channels, whose conductance is reduced in proportion to the degree of hypoxia to which they are exposed. This reduces the K^+ efflux, depolarizing the cell and causing Ca^{2+} influx, primarily via L-type Ca^{2+} channels. The Ca^{2+} influx triggers action potentials and transmitter release, with consequent excitation

FIGURE 37–6   Effect of PCO_2 on afferent nerve firing. The rate of discharge of a single afferent fiber from the carotid body is plotted at several PO_2 (circles) and fitted to a line. A sharp increase in firing rate is observed as PO_2 falls below normal resting levels (ie, near 100 mm Hg). (Courtesy of S Sampson.)

of the afferent nerve endings. The smooth muscle of pulmonary arteries contains similar O_2-sensitive K^+ channels, which mediate the vasoconstriction caused by hypoxia. This is in contrast to systemic arteries, which contain adenosine triphosphate (ATP) dependent K^+ channels that permit more K^+ efflux with hypoxia and consequently cause vasodilation instead of vasoconstriction.

The blood flow in each 2-mg carotid body is about 0.04 mL/min, or 2000 mL/100 g of tissue/min compared with a blood flow 54 mL or 420 mL per 100 g/min in the brain and kidneys, respectively. Because the blood flow per unit of tissue is so enormous, the O_2 needs of the cells can be met largely by dissolved O_2 alone. Therefore, the receptors are not stimulated in conditions such as anemia or carbon monoxide poisoning, in which the amount of dissolved O_2 in the blood reaching the receptors is generally normal, even though the combined O_2 in the blood is markedly decreased. The receptors are stimulated when the arterial P_{O_2} is low or when, because of vascular stasis, the amount of O_2 delivered to the receptors per unit time is decreased. Powerful stimulation is also produced by cyanide, which prevents O_2 utilization at the tissue level. In sufficient doses, nicotine and lobeline activate the chemoreceptors. It has also been reported that infusion of K^+ increases the discharge rate in chemoreceptor afferents, and because the plasma K^+ level is increased during exercise, the increase may contribute to exercise-induced hyperpnea.

Because of their anatomic location, the aortic bodies have not been studied in as great detail as the carotid bodies. Their responses are probably similar but of lesser magnitude. In humans in whom both carotid bodies have been removed but the aortic bodies left intact, the responses are essentially the same as those following denervation of both carotid and aortic bodies in animals: little change in ventilation at rest, but the ventilatory response to hypoxia is lost and the ventilatory response to CO_2 is reduced by 30%.

Neuroepithelial bodies composed of innervated clusters of amine-containing cells are found in the airways. These cells have an outward K^+ current that is reduced by hypoxia, and this would be expected to produce depolarization. However, the function of these hypoxia-sensitive cells is uncertain because, as noted above, removal of the carotid bodies alone abolishes the respiratory response to hypoxia.

CHEMORECEPTORS IN THE BRAIN STEM

The chemoreceptors that mediate the hyperventilation produced by increases in arterial P_{CO_2} after the carotid and aortic bodies are denervated are located in the medulla oblongata and consequently are called **medullary chemoreceptors.** They are separate from the dorsal and ventral respiratory neurons and are located on the ventral surface of the medulla (Figure 37–7). Recent evidence indicates that additional chemoreceptors are located in the vicinity of the solitary tract nuclei, the locus ceruleus, and the hypothalamus.

FIGURE 37–7   Rostral (R) and caudal (C) chemosensitive areas on the ventral surface of the medulla.

The chemoreceptors monitor the H^+ concentration of cerebrospinal fluid (CSF), including the brain interstitial fluid. CO_2 readily penetrates membranes, including the blood–brain barrier, whereas H^+ and HCO_3^- penetrate slowly. The CO_2 that enters the brain and CSF is promptly hydrated. The H_2CO_3 dissociates, so that the local H^+ concentration rises. The H^+ concentration in brain interstitial fluid parallels the arterial P_{CO_2}. Experimentally produced changes in the P_{CO_2} of CSF have minor, variable effects on respiration as long as the H^+ concentration is held constant, but any increase in spinal fluid H^+ concentration stimulates respiration. The magnitude of the stimulation is proportional to the rise in H^+ concentration. Thus, the effects of CO_2 on respiration are mainly due to its movement into the CSF and brain interstitial fluid, where it increases the H^+ concentration and stimulates receptors sensitive to H^+.

VENTILATORY RESPONSES TO CHANGES IN ACID–BASE BALANCE

In metabolic acidosis due, for example, to the accumulation of acid ketone bodies in the circulation in diabetes mellitus, there is pronounced respiratory stimulation (Kussmaul breathing). The hyperventilation decreases alveolar P_{CO_2} ("blows off CO_2") and thus produces a compensatory fall in blood H^+ concentration. Conversely, in metabolic alkalosis due, for example, to protracted vomiting with loss of HCl from the body, ventilation is depressed and the arterial P_{CO_2} rises, raising the H^+ concentration toward normal. If there is an increase in ventilation that is not secondary to a rise in arterial H^+ concentration, the drop in P_{CO_2} lowers the H^+ concentration below normal **(respiratory alkalosis);** conversely, hypoventilation that is not secondary to a fall in plasma H^+ concentration causes **respiratory acidosis.**

VENTILATORY RESPONSES TO CO$_2$

The arterial PCO$_2$ is normally maintained at 40 mm Hg. When arterial PCO$_2$ rises as a result of increased tissue metabolism, ventilation is stimulated and the rate of pulmonary excretion of CO$_2$ increases until the arterial PCO$_2$ falls to normal, shutting off the stimulus. The operation of this feedback mechanism keeps CO$_2$ excretion and production in balance.

When a gas mixture containing CO$_2$ is inhaled, the alveolar PCO$_2$ rises, elevating the arterial PCO$_2$ and stimulating ventilation as soon as the blood that contains more CO$_2$ reaches the medulla. CO$_2$ elimination is increased, and the alveolar PCO$_2$ drops toward normal. This is why relatively large increments in the PCO$_2$ of inspired air (eg, 15 mm Hg) produce relatively slight increments in alveolar PCO$_2$ (eg, 3 mm Hg). However, the PCO$_2$ does not drop to normal, and a new equilibrium is reached at which the alveolar PCO$_2$ is slightly elevated and the hyperventilation persists as long as CO$_2$ is inhaled. The essentially linear relationship between respiratory minute volume and the alveolar PCO$_2$ is shown in Figure 37–8.

Of course, this linearity has an upper limit. When the PCO$_2$ of the inspired gas is close to the alveolar PCO$_2$, elimination of CO$_2$ becomes difficult. When the CO$_2$ content of the inspired gas is more than 7%, the alveolar and arterial PCO$_2$ begin to rise

abruptly in spite of hyperventilation. The resultant accumulation of CO$_2$ in the body (**hypercapnia**) depresses the central nervous system, including the respiratory center, and produces headache, confusion, and eventually coma (**CO$_2$ narcosis**).

VENTILATORY RESPONSE TO OXYGEN LACK

When the O$_2$ content of the inspired air is decreased, respiratory minute volume is increased. The stimulation is slight when the PO$_2$ of the inspired air is more than 60 mm Hg, and marked stimulation of respiration occurs only at lower PO$_2$ values (Figure 37–9). However, any decline in arterial PO$_2$ below 100 mm Hg produces increased discharge in the nerves from the carotid and aortic chemoreceptors. There are two reasons why this increase in impulse traffic does not increase ventilation to any extent in normal individuals until the PO$_2$ is less than 60 mm Hg. Because Hb is a weaker acid than HbO$_2$, there is a slight decrease in the H$^+$ concentration of arterial blood when the arterial PO$_2$ falls and hemoglobin becomes less saturated with O$_2$. The fall in H$^+$ concentration tends to inhibit respiration. In addition, any increase in ventilation that does

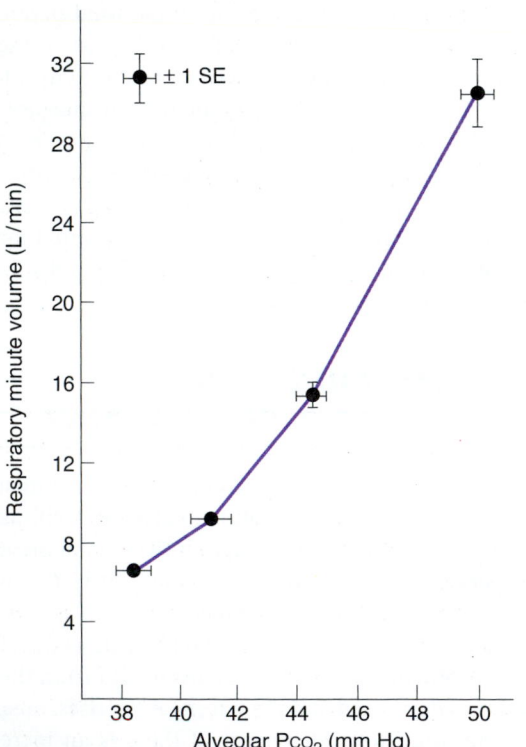

FIGURE 37–8 **Responses of normal subjects to inhaling O$_2$ and approximately 2, 4, and 6% CO$_2$.** The relatively linear increase in respiratory minute volume in response to increased CO$_2$ is due to an increase in both the depth and rate of respiration. (Reproduced with permission from Lambertsen CJ in: *Medical Physiology*, 13th ed. Mountcastle VB [editor]. Mosby, 1974.)

FIGURE 37–9 **Top:** Average respiratory minute volume during the first half hour of exposure to gases containing various amounts of O$_2$. Marked changes in ventilation occur at PO$_2$ values lower than 60 mm Hg. The horizontal line in each case indicates the mean; the vertical bar indicates one standard deviation. **Bottom:** Alveolar PO$_2$ and PCO$_2$ values when breathing air at various barometric pressures. The two graphs are aligned so that the PO$_2$ of the inspired gas mixtures in the upper graph correspond to the PO$_2$ at the various barometric pressures in the lower graph. (Courtesy of RH Kellogg.)

FIGURE 37–10 **Ventilation at various alveolar PO_2 values when PCO_2 is held constant at 49, 44, or 37 mm Hg.** Note the dramatic effect on the ventilatory response to PO_2 when PCO is increased above normal. (Data from Loeschke HH and Gertz KH.)

FIGURE 37–11 **Fan of lines showing CO_2 response curves at various fixed values of alveolar PO_2.** Decreased PAO_2 results in a more sensitive response to $PACO_2$.

occur lowers the alveolar PCO_2, and this also tends to inhibit respiration. Therefore, the stimulatory effects of hypoxia on ventilation are not clearly manifest until they become strong enough to override the counterbalancing inhibitory effects of a decline in arterial H^+ concentration and PCO_2.

The effects on ventilation of decreasing the alveolar PO_2 while holding the alveolar PCO_2 constant are shown in Figure 37–10. When the alveolar PCO_2 is stabilized at a level 2 to 3 mm Hg above normal, there is an inverse relationship between ventilation and the alveolar PO_2 even in the 90 to 110 mm Hg range; but when the alveolar PCO_2 is fixed at lower than normal values, there is no stimulation of ventilation by hypoxia until the alveolar PO_2 falls below 60 mm Hg.

EFFECTS OF HYPOXIA ON THE CO_2 RESPONSE CURVE

When the converse experiment is performed—that is, when the alveolar PO_2 is held constant while the response to varying amounts of inspired CO_2 is tested—a linear response is obtained (Figure 37–11). When the CO_2 response is tested at different fixed PO_2 values, the slope of the response curve changes, with the slope increased when alveolar PO_2 is decreased. In other words, hypoxia makes the individual more sensitive to increases in arterial PCO_2. However, the alveolar PCO_2 level at which the curves in Figure 37–11 intersect is unaffected. In the normal individual, this threshold value is just below the normal alveolar PCO_2, indicating that normally there is a very slight but definite "CO_2 drive" of the respiratory area.

EFFECT OF H^+ ON THE CO_2 RESPONSE

The stimulatory effects of H^+ and CO_2 on respiration appear to be additive and not, like those of CO_2 and O_2, complexly interrelated. In metabolic acidosis, the CO_2 response curves are similar to those in Figure 37–11, except that they are shifted to the left. In other words, the same amount of respiratory stimulation is produced by lower arterial PCO_2 levels. It has been calculated that the CO_2 response curve shifts 0.8 mm Hg to the left for each nanomole rise in arterial H^+. About 40% of the ventilatory response to CO_2 is removed if the increase in arterial H^+ produced by CO_2 is prevented. As noted above, the remaining 60% is probably due to the effect of CO_2 on spinal fluid or brain interstitial fluid H^+ concentration.

BREATH HOLDING

Respiration can be voluntarily inhibited for some time, but eventually the voluntary control is overridden. The point at which breathing can no longer be voluntarily inhibited is called the **breaking point**. Breaking is due to the rise in arterial PCO_2 and the fall in PO_2. Individuals can hold their breath longer after removal of the carotid bodies. Breathing 100% oxygen before breath holding raises alveolar PO_2 initially, so that the breaking point is delayed. The same is true of hyperventilating room air, because CO_2 is blown off and arterial PCO_2 is lower at the start. Reflex or mechanical factors appear to influence the breaking point, since subjects who hold their breath as long as possible and then breathe a gas mixture low in O_2 and high in CO_2 can hold their breath for an additional 20 s or more. Psychological factors also play a role, and subjects can hold their breath longer when they are told their performance is very good than when they are not.

NON-CHEMICAL INFLUENCES ON RESPIRATION

RESPONSES MEDIATED BY RECEPTORS IN THE AIRWAYS & LUNGS

Receptors in the airways and lungs are innervated by myelinated and unmyelinated vagal fibers. The unmyelinated fibers are C fibers. The receptors innervated by myelinated fibers are commonly divided into **slowly adapting receptors** and **rapidly adapting receptors** on the basis of whether sustained stimulation leads to prolonged or transient discharge in their afferent nerve fibers (Table 37–2). The other group of receptors presumably consists of the endings of C fibers, and they are divided into pulmonary and bronchial subgroups on the basis of their location.

The shortening of inspiration produced by vagal afferent activity (Figure 37–3) is mediated by slowly adapting receptors, as are the **Hering–Breuer reflexes.** The Hering–Breuer inflation reflex is an increase in the duration of expiration produced by steady lung inflation, and the Hering–Breuer deflation reflex is a decrease in the duration of expiration produced by marked deflation of the lung. Because the rapidly adapting receptors are stimulated by chemicals such as histamine, they have been called **irritant receptors.** Activation of rapidly adapting receptors in the trachea causes coughing, bronchoconstriction, and mucus secretion, and activation of rapidly adapting receptors in the lung may produce hyperpnea.

Because the C fiber endings are close to pulmonary vessels, they have been called J (juxtacapillary) receptors. They are stimulated by hyperinflation of the lung, but they respond as well to intravenous or intracardiac administration of chemicals such as capsaicin. The reflex response that is produced is apnea followed by rapid breathing, bradycardia, and hypotension (**pulmonary chemoreflex**). A similar response is produced by receptors in the heart (**Bezold–Jarisch reflex** or the **coronary chemoreflex**). The physiologic role of this reflex is uncertain, but it probably occurs in pathologic states such as pulmonary congestion or embolization, in which it is produced by endogenously released substances.

COUGHING & SNEEZING

Coughing begins with a deep inspiration followed by forced expiration against a closed glottis. This increases the intrapleural pressure to 100 mm Hg or more. The glottis is then suddenly opened, producing an explosive outflow of air at velocities up to 965 km (600 mi) per hour. Sneezing is a similar expiratory effort with a continuously open glottis. These reflexes help expel irritants and keep airways clear. Other aspects of innervation are considered in a special case (Clinical Box 37–1).

AFFERENTS FROM PROPRIOCEPTORS

Carefully controlled experiments have shown that active and passive movements of joints stimulate respiration, presumably

TABLE 37–2 Airway and lung receptors.

Vagal Innervation	Type	Location in Interstitium	Stimulus	Response
Myelinated	Slowly adapting	Among airway smooth muscle cells (?)	Lung inflation	Inspiratory time shortening
				Hering–Breuer inflation and deflation reflexes
				Bronchodilation
				Tachycardia
				Hyperpnea
	Rapidly adapting	Among airway epithelial cells	Lung hyperinflation	Cough
			Exogenous and endogenous substances (eg, histamine, prostaglandins)	Bronchoconstriction
				Mucus secretion
Unmyelinated C fibers	Pulmonary C fibers	Close to blood vessels	Lung hyperinflation	Apnea followed by rapid breathing
	Bronchial C fibers		Exogenous and endogenous substances (eg, capsaicin, bradykinin, serotonin)	Bronchoconstriction
				Bradycardia
				Hypotension
				Mucus secretion

Modified and reproduced with permission from Berger AJ, Hornbein TF: Control of respiration. In: *Textbook of Physiology,* 21st ed. Vol. 2. Patton HD, et al (editors). Saunders, 1989.

Lung Innervation & Patients with Heart–Lung Transplants

Transplantation of the heart and lungs is now an established treatment for severe pulmonary disease and other conditions. In individuals with transplants, the recipient's right atrium is sutured to the donor heart, and the donor heart does not reinnervate, so the resting heart rate is elevated. The donor trachea is sutured to the recipient's just above the carina, and afferent fibers from the lungs do not regrow. Consequently, healthy patients with heart–lung transplants provide an opportunity to evaluate the role of lung innervation in normal physiology. Their cough responses to stimulation of the trachea are normal because the trachea remains innervated, but their cough responses to stimulation of the smaller airways are absent. Their bronchi tend to be dilated to a greater degree than normal. In addition, they have the normal number of yawns and sighs, indicating that these do not depend on innervation of the lungs. Finally, they lack Hering–Breuer reflexes, but their pattern of breathing at rest is normal, indicating that these reflexes do not play an important role in the regulation of resting respiration in humans.

Afferents from "Higher Centers"

Pain and emotional stimuli affect respiration, so there must also be afferents from the limbic system and hypothalamus to the respiratory neurons in the brain stem. In addition, even though breathing is not usually a conscious event, both inspiration and expiration are under voluntary control. The pathways for voluntary control pass from the neocortex to the motor neurons innervating the respiratory muscles, bypassing the medullary neurons.

Because voluntary and automatic control of respiration are separate, automatic control is sometimes disrupted without loss of voluntary control. The clinical condition that results has been called **Ondine's curse.** In German legend, Ondine was a water nymph who had an unfaithful mortal lover. The king of the water nymphs punished the lover by casting a curse on him that took away all his automatic functions. In this state, he could stay alive only by staying awake and remembering to breathe. He eventually fell asleep from sheer exhaustion, and his respiration stopped. Patients with this intriguing condition generally have bulbar poliomyelitis or disease processes that compress the medulla.

because impulses in afferent pathways from proprioceptors in muscles, tendons, and joints stimulate the inspiratory neurons. This effect probably helps increase ventilation during exercise. Other afferents are considered in Clinical Box 37–2.

RESPIRATORY COMPONENTS OF VISCERAL REFLEXES

Inhibition of respiration and closure of the glottis during vomiting, swallowing, and sneezing not only prevent the aspiration of food or vomitus into the trachea but, in the case of vomiting, fix the chest so that contraction of the abdominal muscles increases the intra-abdominal pressure. Similar glottic closure and inhibition of respiration occur during voluntary and involuntary straining.

Hiccup is a spasmodic contraction of the diaphragm and other inspiratory muscles that produces an inspiration during which the glottis suddenly closes. The glottic closure is responsible for the characteristic sensation and sound. Hiccups occur in the fetus in utero as well as throughout extrauterine life. Their function is unknown. Most attacks of hiccups are usually of short duration, and they often respond to breath holding or other measures that increase arterial P_{CO_2}. Intractable hiccups, which can be debilitating, sometimes respond to dopamine antagonists and perhaps to some centrally acting analgesic compounds.

Yawning is a peculiar "infectious" respiratory act whose physiologic basis and significance are uncertain. Like hiccup-

ing, it occurs in utero, and it occurs in fish and tortoises as well as mammals. The view that it is needed to increase O_2 intake has been discredited. Underventilated alveoli have a tendency to collapse, and it has been suggested that the deep inspiration and stretching them open prevents the development of atelectasis. However, in actual experiments, no atelectasis-preventing effect of yawning could be demonstrated. Yawning increases venous return to the heart, which may benefit the circulation. It has been suggested that yawning is a nonverbal signal used for communication between monkeys in a group, and one could argue that on a different level, the same thing is true in humans.

RESPIRATORY EFFECTS OF BARORECEPTOR STIMULATION

Afferent fibers from the baroreceptors in the carotid sinuses, aortic arch, atria, and ventricles relay to the respiratory neurons, as well as the vasomotor and cardioinhibitory neurons in the medulla. Impulses in them inhibit respiration, but the inhibitory effect is slight and of little physiologic importance. The hyperventilation in shock is due to chemoreceptor stimulation caused by acidosis and hypoxia secondary to local stagnation of blood flow, and is not baroreceptor-mediated. The activity of inspiratory neurons affects blood pressure and heart rate, and activity in the vasomotor and cardiac areas in the medulla may have minor effects on respiration.

EFFECTS OF SLEEP

Respiration is less rigorously controlled during sleep than in the waking state, and brief periods of apnea occur in normal sleeping adults. Changes in the ventilatory response to hypoxia vary. If the P_{CO_2} falls during the waking state, various stimuli from proprioceptors and the environment maintain respiration, but during sleep, these stimuli are decreased and a decrease in P_{CO_2} can cause apnea. During rapid eye movement (REM) sleep, breathing is irregular and the CO_2 response is highly variable.

RESPIRATORY ABNORMALITIES

ASPHYXIA

In asphyxia produced by occlusion of the airway, acute hypercapnia and hypoxia develop together. Stimulation of respiration is pronounced, with violent respiratory efforts. Blood pressure and heart rate rise sharply, catecholamine secretion is increased, and blood pH drops. Eventually the respiratory efforts cease, the blood pressure falls, and the heart slows. Asphyxiated animals can still be revived at this point by artificial respiration, although they are prone to ventricular fibrillation, probably because of the combination of hypoxic myocardial damage and high circulating catecholamine levels. If artificial respiration is not started, cardiac arrest occurs in 4 to 5 min.

DROWNING

Drowning is asphyxia caused by immersion, usually in water. In about 10% of drownings, the first gasp of water after the losing struggle not to breathe triggers laryngospasm, and death results from asphyxia without any water in the lungs. In the remaining cases, the glottic muscles eventually relax and fluid enters the lungs. Fresh water is rapidly absorbed, diluting the plasma and causing intravascular hemolysis. Ocean water is markedly hypertonic and draws fluid from the vascular system into the lungs, decreasing plasma volume. The immediate goal in the treatment of drowning is, of course, resuscitation, but long-term treatment must also take into account the circulatory effects of the water in the lungs.

PERIODIC BREATHING

The acute effects of voluntary hyperventilation demonstrate the interaction of the chemical mechanisms regulating respiration. When a normal individual hyperventilates for 2 to 3 min, then stops and permits respiration to continue without exerting any voluntary control over it, a period of apnea occurs. This is followed by a few shallow breaths and then by another period of apnea, followed again by a few breaths (**periodic breathing**). The cycles may last for some time before normal breathing is resumed (Figure 37–12). The apnea

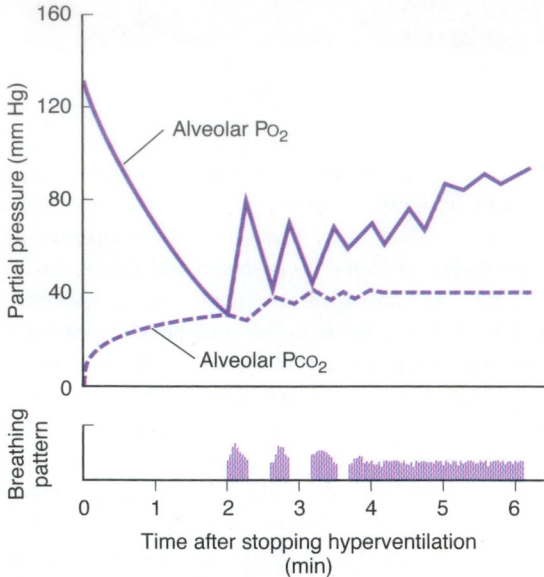

FIGURE 37–12 **Changes in breathing and composition of alveolar air after forced hyperventilation for 2 min.** Bars in bottom indicate breathing, whereas blank spaces are indicative of apnea.

apparently is due to a lack of CO_2 because it does not occur following hyperventilation with gas mixtures containing 5% CO_2. During the apnea, the alveolar P_{O_2} falls and the P_{CO_2} rises. Breathing resumes because of hypoxic stimulation of the carotid and aortic chemoreceptors before the CO_2 level has returned to normal. A few breaths eliminate the hypoxic stimulus, and breathing stops until the alveolar P_{O_2} falls again. Gradually, however, the P_{CO_2} returns to normal, and normal breathing resumes. Changes in breathing patterns can be symptomatic of disease (Clinical Box 37–3).

EFFECTS OF EXERCISE

Exercise provides a physiological example to explore many of the control systems discussed above. Of course, many cardiovascular and respiratory mechanisms must operate in an integrated fashion if the O_2 needs of the active tissue are to be met and the extra CO_2 and heat removed from the body during exercise. Circulatory changes increase muscle blood flow while maintaining adequate circulation in the rest of the body. In addition, there is an increase in the extraction of O_2 from the blood in exercising muscles and an increase in ventilation. This provides extra O_2, eliminates some of the heat, and excretes extra CO_2. A focus on regulation of ventilation and tissue O_2 is presented below, as many other aspects of regulation have been presented in previous chapters.

CHANGES IN VENTILATION

During exercise, the amount of O_2 entering the blood in the lungs is increased because the amount of O_2 added to each

Periodic Breathing in Disease

Cheyne–Stokes Respiration

Periodic breathing occurs in various disease states and is often called **Cheyne–Stokes respiration.** It is seen most commonly in patients with congestive heart failure and uremia, but it occurs also in patients with brain disease and during sleep in some normal individuals. Some of the patients with Cheyne–Stokes respiration have increased sensitivity to CO_2. The increased response is apparently due to disruption of neural pathways that normally inhibit respiration. In these individuals, CO_2 causes relative hyperventilation, lowering the arterial P_{CO_2}. During the resultant apnea, the arterial P_{CO_2} again rises to normal, but the respiratory mechanism again overresponds to CO_2. Breathing ceases, and the cycle repeats.

Another cause of periodic breathing in patients with cardiac disease is prolongation of the lung-to-brain circulation time, so that it takes longer for changes in arterial gas tensions to affect the respiratory area in the medulla. When individuals with a slower circulation hyperventilate, they lower the P_{CO_2} of the blood in their lungs, but it takes longer than normal for the blood with a low P_{CO_2} to reach the brain. During this time, the P_{CO_2} in the pulmonary capillary blood continues to be lowered, and when this blood reaches the brain, the low P_{CO_2} inhibits the respiratory area, producing apnea. In other words, the respiratory control system oscillates because the negative feedback loop from lungs to brain is abnormally long.

Sleep Apnea

Episodes of apnea during sleep can be central in origin; that is, due to failure of discharge in the nerves producing respiration, or they can be due to airway obstruction (**obstructive sleep apnea**). This can occur at any age and is produced when the pharyngeal muscles relax during sleep. In some cases, failure of the genioglossus muscles to contract during inspiration contributes to the blockage; these muscles pull the tongue forward, and when they do not contract the tongue falls back and obstructs the airway. After several increasingly strong respiratory efforts, the patient wakes up, takes a few normal breaths, and falls back to sleep. Not surprisingly, the apneic episodes are most common during REM sleep, when the muscles are most hypotonic. The symptoms are loud snoring, morning headaches, fatigue, and daytime sleepiness. When severe and prolonged, the condition apparently causes hypertension and its complications. In addition, the incidence of motor vehicle accidents in sleep apnea patients is 7 times greater than it is in the general driving population.

FIGURE 37–13 **Relation between work load, blood lactate level, and O_2 uptake.** I–VI, increasing work loads produced by increasing the speed and grade of a treadmill on which the subjects worked. (Reproduced with permission from Mitchell JH, Blomqvist G: Maximal oxygen uptake. N Engl J Med 1971;284:1018.)

unit of blood and the pulmonary blood flow per minute are increased. The P_{O_2} of blood flowing into the pulmonary capillaries falls from 40 to 25 mm Hg or less, so that the alveolar–capillary P_{O_2} gradient is increased and more O_2 enters the blood. Blood flow per minute is increased from 5.5 L/min to as much as 20 to 35 L/min. The total amount of O_2 entering the blood therefore increases from 250 mL/min at rest to values as high as 4000 mL/min. The amount of CO_2 removed from each unit of blood is increased, and CO_2 excretion increases from 200 mL/min to as much as 8000 mL/min. The increase in O_2 uptake is proportional to work load, up to a maximum. Above this maximum, O_2 consumption levels off and the blood lactate level continues to rise (Figure 37–13). The lactate comes from muscles in which aerobic resynthesis of energy stores cannot keep pace with their utilization, and an **oxygen debt** is being incurred.

Ventilation increases abruptly with the onset of exercise, which is followed after a brief pause by a further, more gradual increase (Figure 37–14). With moderate exercise, the increase is due mostly to an increase in the depth of respiration; this is accompanied by an increase in the respiratory rate when the exercise is more strenuous. Ventilation abruptly decreases when exercise ceases, which is followed after a brief pause by a

FIGURE 37–14 **Diagrammatic representation of changes in ventilation during exercise.** See text for details.

more gradual decline to pre-exercise values. The abrupt increase at the start of exercise is presumably due to psychic stimuli and afferent impulses from proprioceptors in muscles, tendons, and joints. The more gradual increase is presumably humoral, even though arterial pH, PCO_2, and PO_2 remain constant during moderate exercise. The increase in ventilation is proportional to the increase in O_2 consumption, but the mechanisms responsible for the stimulation of respiration are still the subject of much debate. The increase in body temperature may play a role. Exercise increases the plasma K^+ level, and this increase may stimulate the peripheral chemoreceptors. In addition, it may be that the sensitivity of the neurons controlling the response to CO_2 is increased or that the respiratory fluctuations in arterial PCO_2 increase so that, even though the mean arterial PCO_2 does not rise, it is CO_2 that is responsible for the increase in ventilation. O_2 also seems to play some role, despite the lack of a decrease in arterial PO_2, since during the performance of a given amount of work, the increase in ventilation while breathing 100% O_2 is 10–20% less than the increase while breathing air. Thus, it currently appears that a number of different factors combine to produce the increase in ventilation seen during moderate exercise.

When exercise becomes more vigorous, buffering of the increased amounts of lactic acid that are produced liberates more CO_2, and this further increases ventilation. The response to graded exercise is shown in Figure 37–15. With increased production of acid, the increases in ventilation and CO_2 production remain proportional, so alveolar and arterial CO_2 change relatively little **(isocapnic buffering)**. Because of the hyperventilation, alveolar PO_2 increases. With further accumulation of lactic acid, the increase in ventilation outstrips CO_2 production and alveolar PCO_2 falls, as does arterial PCO_2. The decline in arterial PCO_2 provides respiratory compensation for the metabolic acidosis produced by the additional lactic acid. The additional increase in ventilation produced by the acidosis is dependent on the carotid bodies and does not occur if they are removed.

The respiratory rate after exercise does not reach basal levels until the O_2 debt is repaid. This may take as long as 90 min. The stimulus to ventilation after exercise is not the arterial PCO_2, which is normal or low, or the arterial PO_2, which is normal or high, but the elevated arterial H^+ concentration due to the lactic acidemia. The magnitude of the O_2 debt is the amount by which O_2 consumption exceeds basal consumption from the end of exertion until the O_2 consumption has returned to pre-exercise basal levels. During repayment of the O_2 debt, the O_2 concentration in muscle myoglobin rises slightly. ATP and phosphorylcreatine are resynthesized, and lactic acid is removed. Eighty percent of the lactic acid is converted to glycogen and 20% is metabolized to CO_2 and H_2O.

Because of the extra CO_2 produced by the buffering of lactic acid during strenuous exercise, the ratio of CO_2 to O_2 (respiratory exchange ratio; R) rises, reaching 1.5 to 2.0. After exertion, while the O_2 debt is being repaid, the R falls to 0.5 or less.

FIGURE 37–15 **Physiologic responses to work rate during exercise.** Changes in alveolar PCO_2, alveolar PO, ventilation ($\dot{V}_E$), CO_2 production ($\dot{V}_{CO_2}$), O_2 consumption ($\dot{V}_{O_2}$), arterial HCO_3^-, and arterial pH with graded increases in work by an adult male on a bicycle ergometer. Resp comp, respiratory compensation. See text for details. (Reproduced with permission from Wasserman K, Whipp BJ, Casaburi R: Respiratory control during exercise. In: *Handbook of Physiology*. Section 3, *The Respiratory System*. Vol II, part 2. Fishman AP [editor]. American Physiological Society, 1986.)

CHANGES IN THE TISSUES

Maximum O_2 uptake during exercise is limited by the maximum rate at which O_2 is transported to the mitochondria in the exercising muscle. However, this limitation is not normally due to deficient O_2 uptake in the lungs, and hemoglobin in arterial blood is saturated even during the most severe exercise.

During exercise, the contracting muscles use more O_2, and the tissue PO_2 and the PO_2 in venous blood from exercising muscle fall nearly to zero. More O_2 diffuses from the blood, the blood PO_2 of the blood in the muscles drops, and more O_2 is removed from hemoglobin. Because the capillary bed of contracting muscle is dilated and many previously closed capillaries are open, the mean distance from the blood to the tissue cells is greatly decreased; this facilitates the movement of O_2 from blood to cells. The oxygen–hemoglobin dissociation curve is steep in the PO_2 range below 60 mm Hg, and a relatively large amount of O_2 is supplied for each drop of 1 mm Hg in PO_2 (see Figure 36–2). Additional O_2 is supplied because, as a result of the accumulation of CO_2 and the rise in temperature in active tissues—and perhaps because of a rise in red blood cell 2,3-biphosphoglycerate (2,3-BPG)—the dissociation curve shifts to the right. The net effect is a threefold increase in O_2 extraction from each unit of blood (see Figure 36–3). Because this increase is accompanied by a 30-fold or greater increase in blood flow, it permits the metabolic rate of muscle to rise as much as 100-fold during exercise.

EXERCISE TOLERANCE & FATIGUE

What determines the maximum amount of exercise that can be performed by an individual? Obviously, exercise tolerance has a time as well as an intensity dimension. For example, a fit young man can produce a power output on a bicycle of about 700 watts for 1 min, 300 watts for 5 min, and 200 watts for 40 min. It used to be argued that the limiting factors in exercise performance were the rate at which O_2 could be delivered to the tissues or the rate at which O_2 could enter the body in the lungs. These factors play a role, but it is clear that other factors also contribute and that exercise stops when the sensation of **fatigue** progresses to the sensation of exhaustion. Fatigue is produced in part by bombardment of the brain by neural impulses from muscles, and the decline in blood pH produced by lactic acidosis also makes one feel tired, as do the rise in body temperature, dyspnea, and, perhaps, the uncomfortable sensations produced by activation of the J receptors in the lungs.

CHAPTER SUMMARY

- Breathing is under both voluntary control (located in the cerebral cortex) and automatic control (driven by pacemaker cells in the medulla). There is a reciprocal innervation to expiratory and inspiratory muscles in that motor neurons supplying expiratory muscles are inactive when motor neurons supplying inspiratory muscles are active, and vice versa.

- The pre-Bötzinger complex on either side of the medulla contains synaptically coupled pacemaker cells that allow for rhythmic generation of breathing. The spontaneous activity of these neurons can be altered by neurons in the pneumotaxic center, although the full regulatory function of these neurons on normal breathing is not understood.

- Breathing patterns are sensitive to chemicals in the blood through activation of respiratory chemoreceptors. There are chemoreceptors in the carotid and aortic bodies and in collections of cells in the medulla. These chemoreceptors respond to changes in PO_2 and PCO_2 as well as H^+ to regulate breathing.

- Receptors in the airway are additionally innervated by slowly adapting and rapidly adapting myelinated vagal fibers. Slowly adapting receptors can be activated by lung inflation. Rapidly adapting receptors, or irritant receptors, can be activated by chemicals such as histamine and result in cough or even hyperpnea.

- Receptors in the airway are also innervated by unmyelinated vagal fibers (C fibers) that are typically found next to pulmonary vessels. They are stimulated by hyperinflation (or exogenous substances including capsaicin) and lead to the pulmonary chemoreflex. The physiologic role for this response is not fully understood.

MULTIPLE-CHOICE QUESTIONS

For all questions, select the single best answer unless otherwise directed.

1. The main respiratory control neurons
   A) send out regular bursts of impulses to expiratory muscles during quiet respiration.
   B) are unaffected by stimulation of pain receptors.
   C) are located in the pons.
   D) send out regular bursts of impulses to inspiratory muscles during quiet respiration.
   E) are unaffected by impulses from the cerebral cortex.

2. Intravenous lactic acid increases ventilation. The receptors responsible for this effect are located in the
   A) medulla oblongata.
   B) carotid bodies.
   C) lung parenchyma.
   D) aortic baroreceptors.
   E) trachea and large bronchi.

3. Spontaneous respiration ceases after
   A) transection of the brain stem above the pons.
   B) transection of the brain stem at the caudal end of the medulla.
   C) bilateral vagotomy.
   D) bilateral vagotomy combined with transection of the brain stem at the superior border of the pons.
   E) transection of the spinal cord at the level of the first thoracic segment.

4. The following physiologic events that occur in vivo are listed in random order: **(1)** decreased CSF pH; **(2)** increased arterial PCO_2; **(3)** increased CSF PCO_2; **(4)** stimulation of medullary chemoreceptors; **(5)** increased alveolar PCO_2.
   What is the usual sequence in which they occur when they affect respiration?
   A) 1, 2, 3, 4, 5
   B) 4, 1, 3, 2, 5
   C) 3, 4, 5, 1, 2
   D) 5, 2, 3, 1, 4
   E) 5, 3, 2, 4, 1

5. The following events that occur in the carotid bodies when they are exposed to hypoxia are listed in random order: **(1)** depolarization of type I glomus cells; **(2)** excitation of afferent nerve endings; **(3)** reduced conductance of hypoxia-sensitive K^+ channels in type I glomus cells; **(4)** Ca^{2+} entry into type I glomus cells; **(5)** decreased K^+ efflux.
   What is the usual sequence in which they occur on exposure to hypoxia?
   A) 1, 3, 4, 5, 2
   B) 1, 4, 2, 5, 3
   C) 3, 4, 5, 1, 2
   D) 3, 1, 4, 5, 2
   E) 3, 5, 1, 4, 2

6. Stimulation of the central (proximal) end of a cut vagus nerve would be expected to
   A) increase heart rate.
   B) stimulate inspiration.
   C) inhibit coughing.
   D) raise blood pressure.
   E) cause apnea.

7. Injection of a drug that stimulates the carotid bodies would be expected to cause

    A) a decrease in the pH of arterial blood.
    B) a decrease in the P_{CO_2} of arterial blood.
    C) an increase in the HCO_3^- concentration of arterial blood.
    D) an increase in urinary Na^+ excretion.
    E) an increase in plasma Cl^-.

8. Variations in which of the following components of blood or CSF do *not* affect respiration?

    A) arterial HCO_3^- concentration
    B) arterial H^+ concentration
    C) arterial Na^+ concentration
    D) CSF CO_2 concentration
    E) CSF H^+ concentration

CHAPTER RESOURCES

Barnes PJ: Chronic obstructive pulmonary disease. N Engl J Med 2000;343:269.

Crystal RG, West JB (editors): *The Lung: Scientific Foundations,* 2nd ed. Lippincott-Raven, 1997.

Fishman AP, et al (editors): *Fishman's Pulmonary Diseases and Disorders,* 4th ed. McGraw-Hill, 2008.

Hackett PH, Roach RC: High-altitude illness. N Engl J Med 2001;345:107.

Jones NL, Killian KJ: Exercise limitation in health and disease. N Engl J Med 2000;343:632.

Laffey JG, Kavanagh BP: Hypocapnia. N Engl J Med 2002;347:43.

Levitzky, MG: *Pulmonary Physiology,* 7th ed. McGraw Hill, 2007.

Prisk GK, Paiva M, West JB (editors): *Gravity and the Lung: Lessons from Micrography.* Marcel Dekker, 2001.

Putnam RW, Dean JB, Ballantyne D (editors): Central chemosensitivity. Respir Physiol 2001;129:1.

Rekling JC, Feldman JL: Pre-Bötzinger complex and pacemaker neurons: hypothesized site and kernel for respiratory rhythm generation. Annu Rev Physiol 1998;60:385.

Tobin MJ: Advances in mechanical ventilation. N Engl J Med 2001;344:1986.

Voelkel NF: High-altitude pulmonary edema. N Engl J Med 2002;346:1607.

Ware LB, Matthay MA: The acute respiratory distress syndrome. N Engl J Med 2000;342:1334.

West JB: *Pulmonary Pathophysiology,* 5th ed. McGraw-Hill, 1995.

C H A P T E R

38

Renal Function & Micturition

INTRODUCTION

In the kidneys, a fluid that resembles plasma is filtered through the glomerular capillaries into the renal tubules (**glomerular filtration**). As this glomerular filtrate passes down the tubules, its volume is reduced and its composition altered by the processes of **tubular reabsorption** (removal of water and solutes from the tubular fluid) and **tubular secretion** (secretion of solutes into the tubular fluid) to form the urine that enters the renal pelvis. A comparison of the composition of the plasma and an average urine specimen illustrates the magnitude of some of these changes (Table 38–1). It emphasizes the manner by which water and important electrolytes and metabolites are conserved while wastes are eliminated in the urine. Furthermore, the composition of the urine can be varied to maintain whole body fluid homeostasis (extracellular fluid [ECF]). This is achieved via many homeostatic regulatory mechanisms that function to change the amount of water and solutes in the urine. From the renal pelvis, the urine passes to the bladder and is expelled to the exterior by the process of urination, or **micturition.** The kidneys are also endocrine organs, making kinins (see Chapter 33) and 1, 25-dihydroxycholecalciferol (see Chapter 23), and making and secreting renin (see Chapter 39).

TABLE 38–1 Typical urinary and plasma concentrations of some physiologically important substances.

Substance	Concentration in		
	Urine (U)	Plasma (P)	U/P Ratio
Glucose (mg/dL)	0	100	0
Na$^+$ (mEq/L)	90	140	0.6
Urea (mg/dL)	900	15	60
Creatinine (mg/dL)	150	1	150

FUNCTIONAL ANATOMY

THE NEPHRON

Each individual renal tubule and its glomerulus is a unit **(nephron).** The size of the kidneys between species varies, as does the number of nephrons they contain. Each human kidney has approximately 1.3 million nephrons. The specific structures of the nephron are shown in diagrammatic fashion in Figure 38–1.

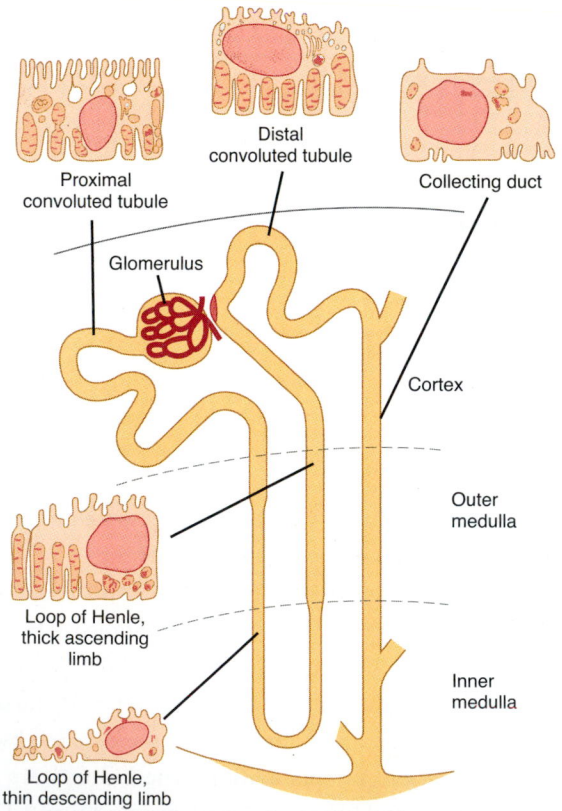

FIGURE 38–1   Diagram of a juxtamedullary nephron. The main histologic features of the cells that make up each portion of the tubule are also shown.

Labels in figure:
Proximal convoluted tubule
Distal convoluted tubule
Collecting duct
Glomerulus
Cortex
Outer medulla
Loop of Henle, thick ascending limb
Inner medulla
Loop of Henle, thin descending limb

The glomerulus, which is about 200 μm in diameter, is formed by the invagination of a tuft of capillaries into the dilated, blind end of the nephron **(Bowman's capsule).** The capillaries are supplied by an **afferent arteriole** and drained by a slightly smaller **efferent arteriole** (Figure 38–2), and it is from the glomerulus that the filtrate is formed. Two cellular layers separate the blood from the glomerular filtrate in Bowman's capsule: the capillary endothelium and the specialized epithelium of the capsule. The endothelium of the glomerular capillaries is fenestrated, with pores that are 70 to 90 nm in diameter. The endothelium of the glomerular capillaries is completely surrounded by the glomerular basement membrane along with specialized cells called podocytes. **Podocytes** have numerous pseudopodia that interdigitate (Figure 38–2) to form **filtration slits** along the capillary wall. The slits are approximately 25 nm wide, and each is closed by a thin membrane. The glomerular basement membrane, the basal lamina, does not contain visible gaps or pores. Stellate cells called **mesangial cells** are located between the basal lamina and the endothelium. They are similar to cells called **pericytes,** which are found in the walls of capillaries elsewhere in the body. Mesangial cells are especially common between two neighboring capillaries, and in these locations the basal membrane forms a sheath shared by both capillaries (Figure 38–2). The mesangial cells are contractile and play a role in the regulation of glomerular filtration. Mesangial cells secrete the extracellular matrix, take up immune complexes, and are involved in the progression of glomerular disease.

Functionally, the glomerular membrane permits the free passage of neutral substances up to 4 nm in diameter and almost totally excludes those with diameters greater than 8 nm. However, the charges on molecules as well as their diameters affect their passage into Bowman's capsule. The total area of glomerular capillary endothelium across which filtration occurs in humans is about 0.8 m^2.

The general features of the cells that make up the walls of the tubules are shown in Figure 38–1; however, there are cell subtypes in all segments, and the anatomic differences between them correlate with differences in function.

The human **proximal convoluted tubule** is about 15 mm long and 55 μm in diameter. Its wall is made up of a single layer of cells that interdigitate with one another and are united by apical tight junctions. Between the bases of the cells are extensions of the extracellular space called the **lateral intercellular spaces.** The luminal edges of the cells have a striate **brush border** due to the presence of many microvilli.

The convoluted proximal tubule straightens and the next portion of each nephron is the **loop of Henle.** The descending portion of the loop and the proximal portion of the ascending limb are made up of thin, permeable cells. On the other hand, the thick portion of the ascending limb (Figure 38–1) is made up of thick cells containing many mitochondria. The nephrons with glomeruli in the outer portions of the renal cortex have short loops of Henle **(cortical nephrons),** whereas those with glomeruli in the juxtamedullary region of the cortex **(juxtamedullary nephrons)** have long loops extending down

FIGURE 38–2   **Structural details of glomerulus. A)** Section through vascular pole, showing capillary loops. **B)** Relation of mesangial cells and podocytes to glomerular capillaries. **C)** Detail of the way podocytes form filtration slits on the basal lamina, and the relation of the lamina to the capillary endothelium. **D)** Enlargement of the rectangle in **C** to show the podocyte processes. The fuzzy material on their surfaces is glomerular polyanion.

into the medullary pyramids. In humans, only 15% of the nephrons have long loops.

The thick end of the ascending limb of the loop of Henle reaches the glomerulus of the nephron from which the tubule arose and nestles between its afferent and efferent arterioles. Specialized cells at the end form the **macula densa,** which is close to the efferent and particularly the afferent arteriole (Figure 38–2). The macula, the neighboring **lacis cells,** and the renin-secreting **juxtaglomerular cells** in the afferent arteriole form the **juxtaglomerular apparatus** (see Figure 39–9).

The **distal convoluted tubule,** which starts at the macula densa, is about 5 mm long. Its epithelium is lower than that of

the proximal tubule, and although a few microvilli are present, there is no distinct brush border. The distal tubules coalesce to form **collecting ducts** that are about 20 mm long and pass through the renal cortex and medulla to empty into the pelvis of the kidney at the apexes of the medullary pyramids. The epithelium of the collecting ducts is made up of **principal cells (P cells)** and **intercalated cells (I cells).** The P cells, which predominate, are relatively tall and have few organelles. They are involved in Na^+ reabsorption and vasopressin-stimulated water reabsorption. The I cells, which are present in smaller numbers and are also found in the distal tubules, have more microvilli, cytoplasmic vesicles, and mitochondria. They are concerned with acid secretion and HCO_3^- transport. The total length of the nephrons, including the collecting ducts, ranges from 45 to 65 mm.

Cells in the kidneys that appear to have a secretory function include not only the juxtaglomerular cells but also some of the cells in the interstitial tissue of the medulla. These cells are called **type I medullary interstitial cells.** They contain lipid droplets and probably secrete prostaglandins, predominantly PGE_2. PGE_2 is also secreted by the cells in the collecting ducts; prostacyclin (PGI_2) and other prostaglandins are secreted by the arterioles and glomeruli.

BLOOD VESSELS

The renal circulation is diagrammed in Figure 38–3. The **afferent arterioles** are short, straight branches of the interlobu-

lar arteries. Each divides into multiple capillary branches to form the tuft of vessels in the glomerulus. The capillaries coalesce to form the **efferent arteriole,** which in turn breaks up into capillaries that supply the tubules (**peritubular capillaries**) before draining into the interlobular veins. The arterial segments between glomeruli and tubules are thus technically a portal system, and the glomerular capillaries are the only capillaries in the body that drain into arterioles. However, there is relatively little smooth muscle in the efferent arterioles.

The capillaries draining the tubules of the cortical nephrons form a peritubular network, whereas the efferent arterioles from the juxtamedullary glomeruli drain not only into a peritubular network, but also into vessels that form hairpin loops (the **vasa recta**). These loops dip into the medullary pyramids alongside the loops of Henle (Figure 38–3). The descending vasa recta have a nonfenestrated endothelium that contains a

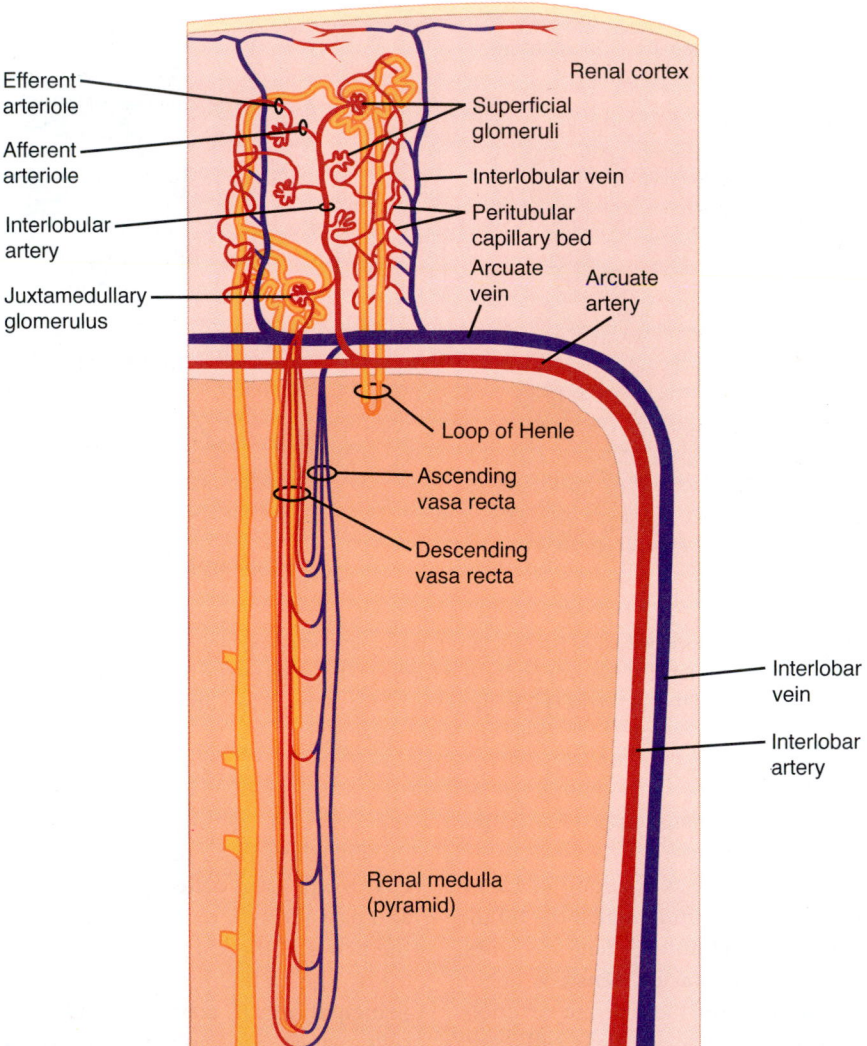

FIGURE 38–3   Renal circulation. Interlobar arteries divide into arcuate arteries, which give off interlobular arteries in the cortex. The interlobular arteries provide an afferent arteriole to each glomerulus. The efferent arteriole from each glomerulus breaks up into capillaries that supply blood to the renal tubules. Venous blood enters interlobular veins, which in turn flow via arcuate veins to the interlobar veins.   (Modified from Boron WF, Boulpaep EL: *Medical Physiology.* Saunders, 2003.)

facilitated transporter for urea, and the ascending vasa recta have a fenestrated endothelium, consistent with their function in conserving solutes.

The efferent arteriole from each glomerulus breaks up into capillaries that supply a number of different nephrons. Thus, the tubule of each nephron does not necessarily receive blood solely from the efferent arteriole of the same nephron. In humans, the total surface of the renal capillaries is approximately equal to the total surface area of the tubules, both being about 12 m^2. The volume of blood in the renal capillaries at any given time is 30 to 40 mL.

LYMPHATICS

The kidneys have an abundant lymphatic supply that drains via the thoracic duct into the venous circulation in the thorax.

CAPSULE

The renal capsule is thin but tough. If the kidney becomes edematous, the capsule limits the swelling, and the tissue pressure (**renal interstitial pressure**) rises. This decreases the glomerular filtration rate and is claimed to enhance and prolong anuria in acute renal failure.

INNERVATION OF THE RENAL VESSELS

The renal nerves travel along the renal blood vessels as they enter the kidney. They contain many postganglionic sympathetic efferent fibers and a few afferent fibers. There also appears to be a cholinergic innervation via the vagus nerve, but its function is uncertain. The sympathetic preganglionic innervation comes primarily from the lower thoracic and upper lumbar segments of the spinal cord, and the cell bodies of the postganglionic neurons are in the sympathetic ganglion chain, in the superior mesenteric ganglion, and along the renal artery. The sympathetic fibers are distributed primarily to the afferent and efferent arterioles, the proximal and distal tubules, and the juxtaglomerular cells (see Chapter 39). In addition, there is a dense noradrenergic innervation of the thick ascending limb of the loop of Henle.

Nociceptive afferents that mediate pain in kidney disease parallel the sympathetic efferents and enter the spinal cord in the thoracic and upper lumbar dorsal roots. Other renal afferents presumably mediate a **renorenal reflex** by which an increase in ureteral pressure in one kidney leads to a decrease in efferent nerve activity to the contralateral kidney, and this decrease permits an increase in its excretion of Na$^+$ and water.

RENAL CIRCULATION

BLOOD FLOW

In a resting adult, the kidneys receive 1.2 to 1.3 L of blood per minute, or just under 25% of the cardiac output. Renal blood flow can be measured with electromagnetic or other types of flow meters, or it can be determined by applying the Fick principle (see Chapter 33) to the kidney; that is, by measuring the amount of a given substance taken up per unit of time and dividing this value by the arteriovenous difference for the substance across the kidney. Because the kidney filters plasma, the **renal plasma flow** equals the amount of a substance excreted per unit of time divided by the renal arteriovenous difference as long as the amount in the red cells is unaltered during passage through the kidney. Any excreted substance can be used if its concentration in arterial and renal venous plasma can be measured and if it is not metabolized, stored, or produced by the kidney and does not itself affect blood flow.

Renal plasma flow can be measured by infusing *p*-aminohippuric acid (PAH) and determining its urine and plasma concentrations. PAH is filtered by the glomeruli and secreted by the tubular cells, so that its **extraction ratio** (arterial concentration minus renal venous concentration divided by arterial concentration) is high. For example, when PAH is infused at low doses, 90% of the PAH in arterial blood is removed in a single circulation through the kidney. It has therefore become commonplace to calculate the "renal plasma flow" by dividing the amount of PAH in the urine by the plasma PAH level, ignoring the level in renal venous blood. Peripheral venous plasma can be used because its PAH concentration is essentially identical to that in the arterial plasma reaching the kidney. The value obtained should be called the **effective renal plasma flow (ERPF)** to indicate that the level in renal venous plasma was not measured. In humans, ERPF averages about 625 mL/min.

$$ERPF = \frac{U_{PAH}\dot{V}}{P_{PAH}} = \text{Clearance of PAH } (C_{PAH})$$

Example:
Concentration of PAH in urine (U_{PAH}): 14 mg/mL
Urine flow ($\dot{V}$): 0.9 mL/min
Concentration of PAH in plasma (P_{PAH}): 0.02 mg/mL

$$ERPF = \frac{14 \times 0.9}{0.02}$$

$$= 630 \text{ mL/min}$$

It should be noted that the ERPF determined in this way is the **clearance** of PAH. The concept of clearance is discussed in detail below.

ERPF can be converted to actual renal plasma flow (RPF):

Average PAH extraction ratio: 0.9

$$\frac{ERP}{\text{Extraction ration}} = \frac{630}{0.9} = \text{Actual RPF} = 700 \text{ mL/min}$$

From the renal plasma flow, the renal blood flow can be calculated by dividing by 1 minus the hematocrit:

Hematocrit (Hct): 45%

$$\text{Renal blood flow} = RPF \times \frac{1}{1-\text{Hct}}$$

$$= 700 \times \frac{1}{0.55}$$

$$= 1273 \text{ mL/min}$$

PRESSURE IN RENAL VESSELS

The pressure in the glomerular capillaries has been measured directly in rats and has been found to be considerably lower than predicted on the basis of indirect measurements. When the mean systemic arterial pressure is 100 mm Hg, the glomerular capillary pressure is about 45 mm Hg. The pressure drop across the glomerulus is only 1 to 3 mm Hg, but a further drop occurs in the efferent arteriole so that the pressure in the peritubular capillaries is about 8 mm Hg. The pressure in the renal vein is about 4 mm Hg. Pressure gradients are similar in squirrel monkeys and presumably in humans, with a glomerular capillary pressure that is about 40% of systemic arterial pressure.

REGULATION OF THE RENAL BLOOD FLOW

Norepinephrine (noradrenaline) constricts the renal vessels, with the greatest effect of injected norepinephrine being exerted on the interlobular arteries and the afferent arterioles. Dopamine is made in the kidney and causes renal vasodilation and natriuresis. Angiotensin II exerts a constrictor effect on both the afferent and efferent arterioles. Prostaglandins increase blood flow in the renal cortex and decrease blood flow in the renal medulla. Acetylcholine also produces renal vasodilation. A high-protein diet raises glomerular capillary pressure and increases renal blood flow.

FUNCTIONS OF THE RENAL NERVES

Stimulation of the renal nerves increases renin secretion by a direct action of released norepinephrine on β_1-adrenergic receptors on the juxtaglomerular cells (see Chapter 39) and it increases Na^+ reabsorption, probably by a direct action of norepinephrine on renal tubular cells. The proximal and distal tubules and the thick ascending limb of the loop of Henle are richly innervated. When the renal nerves are stimulated to increasing extents in experimental animals, the first response is an increase in the sensitivity of the juxtaglomerular cells (Table 38–2), followed by increased renin secretion, then increased Na^+ reabsorption, and finally, at the highest threshold, renal vasoconstriction with decreased glomerular filtration and renal blood flow. It is still unsettled whether the effect on

TABLE 38–2 Renal responses to graded renal nerve stimulation.

Renal Nerve Stimulation Frequency (Hz)	RSR[a]	$U_{NA}V$	GFR	RBF[a]
0.25	No effect on basal values; augments RSR mediated by nonneural stimuli.	0	0	0
0.50	Increased without changing $U_{NA}V$, GFR, or RBF.	0	0	0
1.0	Increased with decreased without changing GFR or RBF.	↓	0	0
2.50	Increased with decreased $U_{NA}V$, GFR, and RBF.	↓	↓	↓

[a]RSR, renin secretion rate; , urinary sodium excretion; RBF, renal blood flow.

Reproduced from DiBona GF: Neural control of renal function: Cardiovascular implications. Hypertension 1989;13:539. By permission of the American Heart Association.

Na^+ reabsorption is mediated via α- or β-adrenergic receptors, and it may be mediated by both. The physiologic role of the renal nerves in Na^+ metabolism is also unsettled, in part because most renal functions appear to be normal in patients with transplanted kidneys, and it takes some time for transplanted kidneys to acquire a functional innervation.

Strong stimulation of the sympathetic noradrenergic nerves to the kidneys causes a marked decrease in renal blood flow. This effect is mediated by α_1-adrenergic receptors and to a lesser extent by postsynaptic α_2-adrenergic receptors. Some tonic discharge takes place in the renal nerves at rest in animals and humans. When systemic blood pressure falls, the vasoconstrictor response produced by decreased discharge in the baroreceptor nerves includes renal vasoconstriction. Renal blood flow is decreased during exercise and, to a lesser extent, on rising from the supine position.

AUTOREGULATION OF RENAL BLOOD FLOW

When the kidney is perfused at moderate pressures (90–220 mm Hg in the dog), the renal vascular resistance varies with the pressure so that renal blood flow is relatively constant (Figure 38–4). Autoregulation of this type occurs in other organs, and several factors contribute to it (see Chapter 33). Renal autoregulation is present in denervated and in isolated, perfused kidneys, but is prevented by the administration of drugs that paralyze vascular smooth muscle. It is probably produced in part by a direct contractile response to stretch of the smooth muscle of the afferent arteriole. NO may also be involved. At low perfusion pressures, angiotensin II also appears to play a role by constricting the efferent arterioles, thus

FIGURE 38–4 Autoregulation in the kidneys.

maintaining the glomerular filtration rate. This is believed to be the explanation of the renal failure that sometimes develops in patients with poor renal perfusion who are treated with drugs that inhibit angiotensin-converting enzyme.

REGIONAL BLOOD FLOW & OXYGEN CONSUMPTION

The main function of the renal cortex is filtration of large volumes of blood through the glomeruli, so it is not surprising that the renal cortical blood flow is relatively great and little oxygen is extracted from the blood. Cortical blood flow is about 5 mL/g of kidney tissue/min (compared with 0.5 mL/g/min in the brain), and the arteriovenous oxygen difference for the whole kidney is only 14 mL/L of blood, compared with 62 mL/L for the brain and 114 mL/L for the heart (see Table 34–1). The P_{O_2} of the cortex is about 50 mm Hg. On the other hand, maintenance of the osmotic gradient in the medulla requires a relatively low blood flow. It is not surprising, therefore, that the blood flow is about 2.5 mL/g/min in the outer medulla and 0.6 mL/g/min in the inner medulla. However, metabolic work is being done, particularly to reabsorb Na^+ in the thick ascending limb of Henle, so relatively large amounts of O_2 are extracted from the blood in the medulla. The P_{O_2} of the medulla is about 15 mm Hg. This makes the medulla vulnerable to hypoxia if flow is reduced further. NO, prostaglandins, and many cardiovascular peptides in this region function in a paracrine fashion to maintain the balance between low blood flow and metabolic needs.

GLOMERULAR FILTRATION

MEASURING GFR

The **glomerular filtration rate (GFR)** can be measured in intact experimental animals and humans by measuring the excretion and plasma level of a substance that is freely filtered through the glomeruli and neither secreted nor reabsorbed by the tubules. The amount of such a substance in the urine per unit of time must have been provided by filtering exactly the number of milliliters of plasma that contained this amount. Therefore, if the substance is designated by the letter X, the GFR is equal to the concentration of X in urine (U_X) times the **urine flow** per unit of time ($\dot{V}$) divided by the **arterial plasma level** of X (P_X), or $U_X\dot{V}/P_X$. This value is called the clearance of X (C_X). P_X is, of course, the same in all parts of the arterial circulation, and if X is not metabolized to any extent in the tissues, the level of X in peripheral venous plasma can be substituted for the arterial plasma level.

SUBSTANCES USED TO MEASURE GFR

In addition to the requirement that it be freely filtered and neither reabsorbed nor secreted in the tubules, a substance suitable for measuring the GFR should be nontoxic and not metabolized by the body. Inulin, a polymer of fructose with a molecular weight of 5200 that is found in Jerusalem artichokes (Helianthus tuberosus), meets these criteria in humans and most animals and is extensively used to measure GFR. In practice, a loading dose of inulin is administered intravenously, followed by a sustaining infusion to keep the arterial plasma level constant. After the inulin has equilibrated with body fluids, an accurately timed urine specimen is collected and a plasma sample obtained halfway through the collection. Plasma and urinary inulin concentrations are determined and the clearance calculated:

$$U_{IN} = 35 \text{ mg/mL}$$
$$\dot{V} = 0.9 \text{ mL/min}$$
$$P_{IN} = 0.25 \text{ mg/mL}$$
$$C_{IN} = \frac{U_{IN}\dot{V}}{P_{IN}} = \frac{35 \times 0.9}{0.25}$$
$$C_{IN} = 126 \text{ mL/min}$$

In dogs, cats, rabbits, and a number of other mammalian species, clearance of creatinine (C_{Cr}) can also be used to determine the precise GFR, but in primates, including humans, some creatinine is secreted by the tubules and some may be reabsorbed. In addition, plasma creatinine determinations are inaccurate at low creatinine levels because the method for determining creatinine measures small amounts of other plasma constituents. In spite of this, the clearance of endogenous creatinine is frequently measured in patients. The values agree quite well with the GFR values measured with inulin because, although the value for $U_{Cr}\dot{V}$ is high as a result of tubular secretion, the value for P_{Cr} is also high as a result of nonspecific chromogens, and the errors thus tend to cancel. Endogenous creatinine clearance is easy to measure and is a worthwhile index of renal function, but when precise measurements of GFR are needed it seems unwise to rely on a method that owes what accuracy it has to compensating errors.

NORMAL GFR

The GFR in a healthy person of average size is approximately 125 mL/min. Its magnitude correlates fairly well with surface area, but values in women are 10% lower than those in men even after correction for surface area. A rate of 125 mL/min is 7.5 L/h, or 180 L/d, whereas the normal urine volume is about 1 L/d. Thus, 99% or more of the filtrate is normally reabsorbed. At the rate of 125 mL/min, in 1 day the kidneys filter an amount of fluid equal to 4 times the total body water, 15 times the ECF volume, and 60 times the plasma volume.

CONTROL OF GFR

The factors governing filtration across the glomerular capillaries are the same as those governing filtration across all other capillaries (see Chapter 32), that is, the size of the capillary bed, the permeability of the capillaries, and the hydrostatic and osmotic pressure gradients across the capillary wall. For each nephron:

$$GFR = K_f [(P_{GC} - P_T) - (\pi_{GC} - \pi_T)]$$

K_f, the glomerular ultrafiltration coefficient, is the product of the glomerular capillary wall hydraulic conductivity (ie, its permeability) and the effective filtration surface area. P_{GC} is the mean hydrostatic pressure in the glomerular capillaries, P_T the mean hydrostatic pressure in the tubule (Bowman's space), π_{GC} the oncotic pressure of the plasma in the glomerular capillaries, and π_T the oncotic pressure of the filtrate in the tubule (Bowman's space).

PERMEABILITY

The permeability of the glomerular capillaries is about 50 times that of the capillaries in skeletal muscle. Neutral substances with effective molecular diameters of less than 4 nm are freely filtered, and the filtration of neutral substances with diameters of more than 8 nm approaches zero (Figure 38–5). Between these values, filtration is inversely proportionate to diameter. However, sialoproteins in the glomerular capillary wall are negatively charged, and studies with anionically charged and cationically charged dextrans indicate that the negative charges repel negatively charged substances in blood, with the result that filtration of anionic substances 4 nm in diameter is less than half that of neutral substances of the same size. This probably explains why albumin, with an effective molecular diameter of approximately 7 nm, normally has a glomerular concentration only 0.2% of its plasma concentration rather than the higher concentration that would be expected on the basis of diameter alone; circulating albumin is negatively charged. Filtration of cationic substances is greater than that of neutral substances.

The amount of protein in the urine is normally less than 100 mg/d, and most of this is not filtered but comes from shed tubular cells. The presence of significant amounts of albumin in the urine is called **albuminuria.** In nephritis, the negative charges in the glomerular wall are dissipated, and albumin-

FIGURE 38–5   **Effect of electric charge on the fractional clearance of dextran molecules of various sizes in rats.** The negative charges in the glomerular membrane retard the passage of negatively charged molecules (anionic dextran) and facilitate the passage of positively charged molecules (cationic dextran).   (Reproduced with permission from Brenner BM, Beeuwkes R: The renal circulations. Hosp Pract [July] 1978;13:35.)

uria can occur for this reason without an increase in the size of the "pores" in the membrane.

SIZE OF THE CAPILLARY BED

K_f can be altered by the mesangial cells, with contraction of these cells producing a decrease in K_f that is largely due to a reduction in the area available for filtration. Contraction of points where the capillary loops bifurcate probably shifts flow away from some of the loops, and elsewhere, contracted mesangial cells distort and encroach on the capillary lumen. Agents that have been shown to affect the mesangial cells are listed in Table 38–3. Angiotensin II is an important regulator of mesangial contraction, and there are angiotensin II receptors in the glomeruli. In addition, some evidence suggests that mesangial cells make renin.

TABLE 38–3 **Agents causing contraction or relaxation of mesangial cells.**

Contraction	Relaxation
Endothelins	ANP
Angiotensin II	Dopamine
Vasopressin	PGE_2
Norepinephrine	cAMP
Platelet-activating factor	
Platelet-derived growth factor	
Thromboxane A_2	
PGF_2	
Leukotrienes C_4 and D_4	
Histamine	

HYDROSTATIC & OSMOTIC PRESSURE

The pressure in the glomerular capillaries is higher than that in other capillary beds because the afferent arterioles are short, straight branches of the interlobular arteries. Furthermore, the vessels "downstream" from the glomeruli, the efferent arterioles, have a relatively high resistance. The capillary hydrostatic pressure is opposed by the hydrostatic pressure in Bowman's capsule. It is also opposed by the oncotic pressure gradient across the glomerular capillaries ($\pi_{GC} - \pi_T$). π_T is normally negligible, and the gradient is essentially equal to the oncotic pressure of the plasma proteins.

The actual pressures in one strain of rats are shown in Figure 38–6. The net filtration pressure (P_{UF}) is 15 mm Hg at the afferent end of the glomerular capillaries, but it falls to zero—that is, filtration equilibrium is reached—proximal to the efferent end of the glomerular capillaries. This is because fluid leaves the plasma and the oncotic pressure rises as blood passes through the glomerular capillaries. The calculated change in $\Delta\pi$ along an idealized glomerular capillary is also shown in Figure 38–6. It is apparent that portions of the glomerular capillaries do not normally contribute to the formation of the glomerular ultrafiltrate; that is, exchange across the glomerular capillaries is flow-limited rather than diffusion-limited. It is also apparent that a decrease in the rate of rise of the Δ curve produced by an increase in renal plasma flow would increase filtration because it would increase the distance along the capillary in which filtration was taking place.

There is considerable species variation in whether filtration equilibrium is reached, and some uncertainties are inherent in

TABLE 38–4 Factors affecting the GFR.

Changes in renal blood flow
Changes in glomerular capillary hydrostatic pressure
Changes in systemic blood pressure
Afferent or efferent arteriolar constriction
Changes in hydrostatic pressure in Bowman's capsule
Ureteral obstruction
Edema of kidney inside tight renal capsule
Changes in concentration of plasma proteins: dehydration, hypoproteinemia, etc (minor factors)
Changes in K_f
Changes in glomerular capillary permeability
Changes in effective filtration surface area

the measurement of K_f. It is uncertain whether filtration equilibrium is reached in humans.

CHANGES IN GFR

Variations in the factors discussed in the preceding paragraphs and listed in Table 38–4 have predictable effects on the GFR. Changes in renal vascular resistance as a result of autoregulation tend to stabilize filtration pressure, but when the mean systemic arterial pressure drops below the autoregulatory range (Figure 38–4), GFR drops sharply. The GFR tends to be maintained when efferent arteriolar constriction is greater than afferent constriction, but either type of constriction decreases blood flow to the tubules.

FILTRATION FRACTION

The ratio of the GFR to the RPF, the **filtration fraction,** is normally 0.16 to 0.20. The GFR varies less than the RPF. When there is a fall in systemic blood pressure, the GFR falls less than the RPF because of efferent arteriolar constriction, and consequently the filtration fraction rises.

TUBULAR FUNCTION

GENERAL CONSIDERATIONS

The amount of any substance (X) that is filtered is the product of the GFR and the plasma level of the substance ($C_{ln}P_X$). The tubular cells may add more of the substance to the filtrate (tubular secretion), may remove some or all of the substance from the filtrate (tubular reabsorption), or may do both. The amount of the substance excreted per unit of time ($U_X\dot{V}$)

	(mm Hg)	
	Afferent end	Efferent end
P_{GC}	45	45
P_T	10	10
π_{GC}	20	35
P_{UF}	15	0

$$P_{UF} = P_{GC} - P_T - \pi_{GC}$$

FIGURE 38–6 Hydrostatic pressure (P_{GC}) and osmotic pressure (π_{GC}) in a glomerular capillary in the rat. P_T, pressure in Bowman's capsule; P_{UF}, net filtration pressure. π_T is normally negligible, so $\Delta\pi = \pi_{GC}$. $\Delta P = P_{GC} - P_T$. (Reproduced with permission from Mercer PF, Maddox DA, Brenner BM: Current concepts of sodium chloride and water transport by the mammalian nephron. West J Med 1974;120:33.)

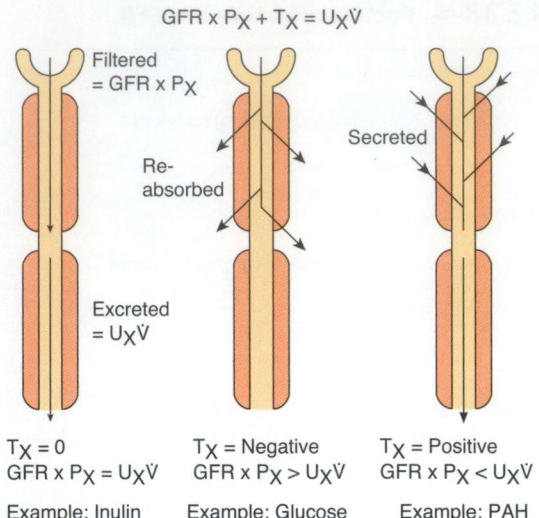

$$GFR \times P_X + T_X = U_X V$$

Filtered
$= GFR \times P_X$

Re-
absorbed

Secreted

Excreted
$= U_X V$

$T_X = 0$
$GFR \times P_X = U_X V$

Example: Inulin

$T_X = $ Negative
$GFR \times P_X > U_X V$

Example: Glucose

$T_X = $ Positive
$GFR \times P_X < U_X V$

Example: PAH

FIGURE 38–7   Tubular function. For explanation of symbols, see text.

equals the amount filtered plus the **net amount transferred** by the tubules. This latter quantity is conveniently indicated by the symbol T_X (Figure 38–7). The clearance of the substance equals the GFR if there is no net tubular secretion or reabsorption, exceeds the GFR if there is net tubular secretion, and is less than the GFR if there is net tubular reabsorption.

Much of our knowledge about glomerular filtration and tubular function has been obtained by using micropuncture techniques. Micropipettes can be inserted into the tubules of the living kidney and the composition of aspirated tubular fluid determined by the use of microchemical techniques. In addition, two pipettes can be inserted in a tubule and the tubule perfused in vivo. Alternatively, isolated perfused segments of tubules can be studied in vitro, and tubular cells can be grown and studied in culture.

MECHANISMS OF TUBULAR REABSORPTION & SECRETION

Small proteins and some peptide hormones are reabsorbed in the proximal tubules by endocytosis. Other substances are secreted or reabsorbed in the tubules by passive diffusion between cells and through cells by facilitated diffusion down chemical or electrical gradients or active transport against such gradients. Movement is by way of ion channels, exchangers, cotransporters, and pumps. Many of these have now been cloned, and their regulation is being studied.

It is important to note that the pumps and other units in the luminal membrane are different from those in the basolateral membrane. It is this different distribution that makes possible net movement of solutes across the epithelia.

Like transport systems elsewhere, renal active transport systems have a maximal rate, or **transport maximum (Tm)**, at

which they can transport a particular solute. Thus, the amount of a particular solute transported is proportional to the amount present up to the Tm for the solute, but at higher concentrations, the transport mechanism is **saturated** and there is no appreciable increment in the amount transported. However, the Tms for some systems are high, and it is difficult to saturate them.

It should also be noted that the tubular epithelium, like that of the small intestine, is a **leaky epithelium** in that the tight junctions between cells permit the passage of some water and electrolytes. The degree to which leakage by this **paracellular pathway** contributes to the net flux of fluid and solute into and out of the tubules is controversial since it is difficult to measure, but current evidence seems to suggest that it is a significant factor in the proximal tubule. One indication of this is that paracellin-1, a protein localized to tight junctions, is related to Mg^{2+} reabsorption, and a loss-of-function mutation of its gene causes severe Mg^{2+} and Ca^{2+} loss in the urine.

The effects of tubular reabsorption and secretion on substances of major physiologic interest are summarized in Table 38–5.

Na⁺ REABSORPTION

The reabsorption of Na^+ and Cl^- plays a major role in body electrolyte and water homeostasis. In addition, Na^+ transport is coupled to the movement of H^+, glucose, amino acids, organic acids, phosphate, and other electrolytes and substances across the tubule walls. The principal cotransporters and exchangers in the various parts of the nephron are listed in Table 38–6. In the proximal tubules, the thick portion of the ascending limb of the loop of Henle, the distal tubules, and the collecting ducts, Na^+ moves by cotransport or exchange from the tubular lumen into the tubular epithelial cells down its concentration and electrical gradients, and is then actively pumped from these cells into the interstitial space. Na^+ is pumped into the interstitium by Na, K ATPase in the basolateral membrane. Thus, Na^+ is actively transported out of all parts of the renal tubule except the thin portions of the loop of Henle. The operation of the ubiquitous Na^+ pump is considered in detail in Chapter 2. It extrudes three Na^+ in exchange for two K^+ that are pumped into the cell.

The tubular cells along the nephron are connected by tight junctions at their luminal edges, but there is space between the cells along the rest of their lateral borders. Much of the Na^+ is actively transported into these extensions of the interstitial space, the **lateral intercellular spaces** (Figure 38–8).

Normally about 60% of the filtered Na^+ is reabsorbed in the proximal tubule, primarily by Na–H exchange. Another 30% is absorbed via the Na–2Cl–K cotransporter in the thick ascending limb of the loop of Henle, and about 7% is absorbed by Na–Cl cotransporter in the distal convoluted tubule. The remainder of the filtered Na^+, about 3%, is absorbed via the ENaC channels in the collecting ducts, and this is the portion that is regulated by aldosterone in the production of homeostatic adjustments in Na^+ balance.

TABLE 38–5 Renal handling of various plasma constituents in a normal adult human on an average diet.

Substance	Per 24 Hours				
	Filtered	Reabsorbed	Secreted	Excreted	Percentage Reabsorbed
Na$^+$ (mEq)	26,000	25,850		150	99.4
K$^+$ (mEq)	600	560[a]	502	90	93.3
Cl$^-$ (mEq)	18,000	17,850		150	99.2
HCO$_3^-$ (mEq)	4,900	4,900		0	100
Urea (mmol)	870	460[b]		410	53
Creatinine (mmol)	12	1[c]	1[c]	12	
Uric acid (mmol)	50	49	4	5	98
Glucose (mmol)	800	800		0	100
Total solute (mOsm)	54,000	53,400	100	700	98.9
Water (mL)	180,000	179,000		1000	99.4

[a]K$^+$ is both reabsorbed and secreted.

[b]Urea moves into as well as out of some portions of the nephron.

[c]Variable secretion and probable reabsorption of creatinine in humans.

TABLE 38–6 Transport proteins involved in the movement of Na$^+$ and Cl$^-$ across the apical membranes of renal tubular cells.[a]

Site	Apical Transporter	Function
Proximal tubule	Na/glucose CT	Na$^+$ uptake, glucose uptake
	Na$^+$/P$_i$ CT	Na$^+$ uptake, Pi uptake
	Na$^+$ amino acid CT	Na$^+$ uptake, amino acid uptake
	Na/lactate CT	Na$^+$ uptake, lactate uptake
	Na/H exchanger	Na$^+$ uptake, H$^+$ extrusion
	Cl/base exchanger	Cl$^-$ uptake
Thick ascending limb	Na–K–2Cl CT	Na$^+$ uptake, Cl$^-$ uptake, K$^+$ uptake
	Na/H exchanger	Na$^+$ uptake, H$^+$ extrusion
	K$^+$ channels	K$^+$ extrusion (recycling)
Distal convoluted tubule	NaCl CT	Na$^+$ uptake, Cl$^-$ uptake
Collecting duct	Na$^+$ channel (ENaC)	Na$^+$ uptake

[a]Uptake indicates movement from tubular lumen to cell interior, extrusion is movement from cell interior to tubular lumen. CT, cotransporter; P$_i$, inorganic phosphate.

Modified with permission from Schnermann JB, Sayegh EI: *Kidney Physiology.* Lippincott-Raven, 1998.

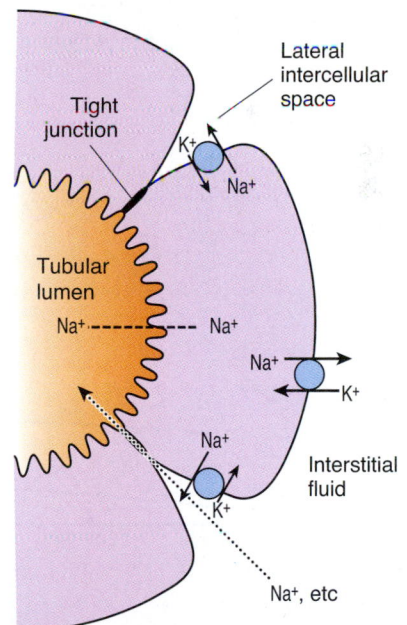

FIGURE 38–8 Mechanism of Na$^+$ reabsorption in the proximal tubule. Na$^+$ moves out of the tubular lumen by cotransport and exchange mechanism through the apical membrane of the tubule (dashed line). The Na$^+$ is then actively transported into the interstitial fluid by Na, K ATPase in the basolateral membrane (solid lines). K$^+$ enters the interstitial fluid via K$^+$ channels. A small amount of Na$^+$, other solutes, and H$_2$O re-enter the tubular lumen by passive transport through the tight junctions (dotted lines).

GLUCOSE REABSORPTION

Glucose, amino acids, and bicarbonate are reabsorbed along with Na^+ in the early portion of the proximal tubule (Figure 38–9). Farther along the tubule, Na^+ is reabsorbed with Cl^-. Glucose is typical of substances removed from the urine by secondary active transport. It is filtered at a rate of approximately 100 mg/min (80 mg/dL of plasma × 125 mL/min). Essentially all of the glucose is reabsorbed, and no more than a few milligrams appear in the urine per 24 h. The amount reabsorbed is proportional to the amount filtered and hence to the plasma glucose level (P_G) times the GFR up to the transport maximum (Tm_G). When the Tm_G is exceeded, the amount of glucose in the urine rises (Figure 38–10). The Tm_G is about 375 mg/min in men and 300 mg/min in women.

The **renal threshold** for glucose is the plasma level at which the glucose first appears in the urine in more than the normal minute amounts. One would predict that the renal threshold would be about 300 mg/dL, that is, 375 mg/min (Tm_G) divided by 125 mL/min (GFR). However, the actual renal threshold is about 200 mg/dL of arterial plasma, which corresponds to a venous level of about 180 mg/dL. Figure 38–10 shows why the actual renal threshold is less than the predicted threshold. The "ideal" curve shown in this diagram would be obtained if the Tm_G in all the tubules was identical and if all the glucose were removed from each tubule when the amount filtered was below the Tm_G. This is not the case, and in humans, for example, the actual curve is rounded and devi-

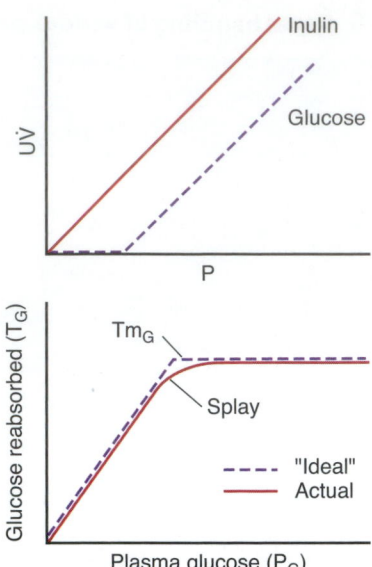

FIGURE 38–10 **Top:** Relation between the plasma level (P) and excretion (UV) of glucose and inulin. **Bottom:** Relation between the plasma glucose level (P_G) and amount of glucose reabsorbed (T_G).

ates considerably from the "ideal" curve. This deviation is called **splay.** The magnitude of the splay is inversely proportionate to the avidity with which the transport mechanism binds the substance it transports.

GLUCOSE TRANSPORT MECHANISM

Glucose reabsorption in the kidneys is similar to glucose reabsorption in the intestine (see Chapter 27). Glucose and Na^+ bind to the sodium-dependent glucose transporter (SGLT) 2 in the apical membrane, and glucose is carried into the cell as Na^+ moves down its electrical and chemical gradient. The Na^+ is then pumped out of the cell into the interstitium, and the glucose is transported by glucose transporter (GLUT) 2 into the interstitial fluid. At least in the rat, there is some transport by SGLT 1 and GLUT 1 as well.

SGLT 2 specifically binds the d isomer of glucose, and the rate of transport of d-glucose is many times greater than that of l-glucose. Glucose transport in the kidneys is inhibited, as it is in the intestine, by the plant glucoside **phlorhizin,** which competes with d-glucose for binding to the carrier.

ADDITIONAL EXAMPLES OF SECONDARY ACTIVE TRANSPORT

Like glucose reabsorption, amino acid reabsorption is most marked in the early portion of the proximal convoluted tubule. Absorption in this location resembles absorption in the intestine (see Chapter 27). The main carriers in the apical membrane cotransport Na^+, whereas the carriers in the basolateral membranes are not Na^+-dependent. Na^+ is pumped

FIGURE 38–9 **Reabsorption of various solutes in the proximal tubule.** TF/P, tubular fluid:plasma concentration ratio. (Courtesy of FC Rector Jr.)

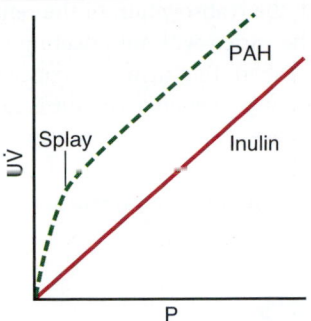

FIGURE 38–11 Relation between plasma levels (P) and excretion (UV) of PAH and inulin.

out of the cells by Na, K ATPase and the amino acids leave by passive or facilitated diffusion to the interstitial fluid.

Some Cl^- is reabsorbed with Na^+ and K^+ in the thick ascending limb of the loop of Henle. In addition, two members of the family of **Cl channels** have been identified in the kidney. Mutations in the gene for one of the renal channels is associated with Ca^{2+}-containing kidney stones and hypercalciuria **(Dent disease)**, but how tubular transport of Ca^{2+} and Cl^- are linked is still unsettled.

PAH TRANSPORT

The dynamics of PAH transport illustrate the operation of the active transport mechanisms that secrete substances into the tubular fluid (see Clinical Box 38–1). The filtered load of PAH is a linear function of the plasma level, but PAH secretion increases as P_{PAH} rises only until a maximal secretion rate (Tm_{PAH}) is reached (Figure 38–11). When P_{PAH} is low, C_{PAH} is high; but as P_{PAH} rises above Tm_{PAH}, C_{PAH} falls progressively. It eventually approaches the clearance of inulin (C_{In}) (Figure 38–12), because the amount of PAH secreted becomes a smaller and smaller fraction of the total amount excreted.

FIGURE 38–12 Clearance of inulin, glucose, and PAH at various plasma levels of each substance in humans.

Conversely, the clearance of glucose is essentially zero at P_G levels below the renal threshold; but above the threshold, C_G rises to approach C_{In} as P_G is raised.

The use of C_{PAH} to measure ERPF is discussed above.

TUBULOGLOMERULAR FEEDBACK & GLOMERULOTUBULAR BALANCE

Signals from the renal tubule in each nephron feed back to affect filtration in its glomerulus. As the rate of flow through the ascending limb of the loop of Henle and first part of the distal tubule increases, glomerular filtration in the same nephron decreases, and, conversely, a decrease in flow increases the GFR (Figure 38–13). This process, which is called

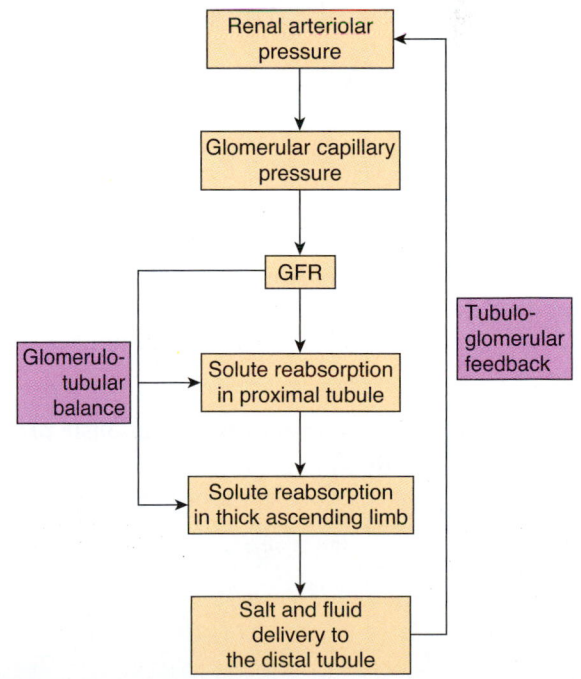

FIGURE 38–13 Mechanisms of glomerulotubular balance and tubuloglomerular feedback.

tubuloglomerular feedback, tends to maintain the constancy of the load delivered to the distal tubule.

The sensor for this response is the **macula densa.** The amount of fluid entering the distal tubule at the end of the thick ascending limb of the loop of Henle depends on the amount of Na^+ and Cl^- in it. The Na^+ and Cl^- enter the macula densa cells via the Na–K–2Cl cotransporter in their apical membranes. The increased Na^+ causes increased Na, K ATPase activity and the resultant increased ATP hydrolysis causes more adenosine to be formed. Presumably, adenosine is secreted from the basal membrane of the cells. It acts via adenosine A_1 receptors on the macula densa cells to increase their release of Ca^{2+} to the vascular smooth muscle in the afferent arterioles. This causes afferent vasoconstriction and a resultant decrease in GFR. Presumably, a similar mechanism generates a signal that decreases renin secretion by the adjacent juxtaglomerular cells in the afferent arteriole (see Chapter 39), but this remains unsettled.

Conversely, an increase in GFR causes an increase in the reabsorption of solutes, and consequently of water, primarily in the proximal tubule, so that in general the percentage of the solute reabsorbed is held constant. This process is called **glomerulotubular balance,** and it is particularly prominent for Na^+. The change in Na^+ reabsorption occurs within seconds after a change in filtration, so it seems unlikely that an extrarenal humoral factor is involved. One factor is the oncotic pressure in the peritubular capillaries. When the GFR is high, there is a relatively large increase in the oncotic pressure of the plasma leaving the glomeruli via the efferent arterioles and hence in their capillary branches. This increases the reabsorption of Na^+ from the tubule. However, other as yet unidentified intrarenal mechanisms are also involved.

WATER TRANSPORT

Normally, 180 L of fluid is filtered through the glomeruli each day, while the average daily urine volume is about 1 L. The same load of solute can be excreted per 24 h in a urine volume of 500 mL with a concentration of 1400 mOsm/kg or in a volume of 23.3 L with a concentration of 30 mOsm/kg (Table 38–7). These figures demonstrate two important facts: First, at least 87% of the filtered water is reabsorbed, even when the urine volume is

23 L; and second, the reabsorption of the remainder of the filtered water can be varied without affecting total solute excretion. Therefore, when the urine is concentrated, water is retained in excess of solute; and when it is dilute, water is lost from the body in excess of solute. Both facts have great importance in the regulation of the osmolality of the body fluids. A key regulator of water output is vasopressin acting on the collecting ducts.

AQUAPORINS

Rapid diffusion of water across cell membranes depends on the presence of water channels, integral membrane proteins called **aquaporins.** To date, 13 aquaporins have been cloned; however, only 4 aquaporins (aquaporin-1, aquaporin-2, aquaporin-3, and aquaporin-4) play a key role in the kidney. The roles played by aquaporin-1 and aquaporin-2 in renal water transport are discussed below.

PROXIMAL TUBULE

Active transport of many substances occurs from the fluid in the proximal tubule, but micropuncture studies have shown that the fluid remains essentially iso-osmotic to the end of the proximal tubule (Figure 38–9). **Aquaporin-1** is localized to both the basolateral and apical membrane of the proximal tubules and its presence allows water to move rapidly out of the tubule along the osmotic gradients set up by active transport of solutes, and isotonicity is maintained. Because the ratio of the concentration in tubular fluid to the concentration in plasma (TF/P) of the nonreabsorbable substance inulin is 2.5 to 3.3 at the end of the proximal tubule, it follows that 60–70% of the filtered solute and 60–70% of the filtered water have been removed by the time the filtrate reaches this point (Figure 38–14).

When aquaporin-1 was knocked out in mice, proximal tubular water permeability was reduced by 80%. When the mice were subjected to dehydration, their urine osmolality did not increase (<700 mOsm/kg), even though other renal aquaporins were present. In humans with mutations that eliminate aquaporin-1 activity, the defect in water metabolism is not as severe, though their response to dehydration is defective.

TABLE 38–7 Alterations in water metabolism produced by vasopressin in humans. In each case, the osmotic load excreted is 700 mOsm/d.

	GFR (mL/min)	Percentage of Filtered Water Reabsorbed	Urine Volume (L/d)	Urine Concentration (mOsm/kg H_2O)	Gain or Loss of Water in Excess of Solute (L/d)
Urine isotonic to plasma	125	98.7	2.4	290	...
Vasopressin (maximal antidiuresis)	125	99.7	0.5	1400	1.9 gain
No vasopressin ("complete" diabetes insipidus)	125	87.1	23.3	30	20.9 loss

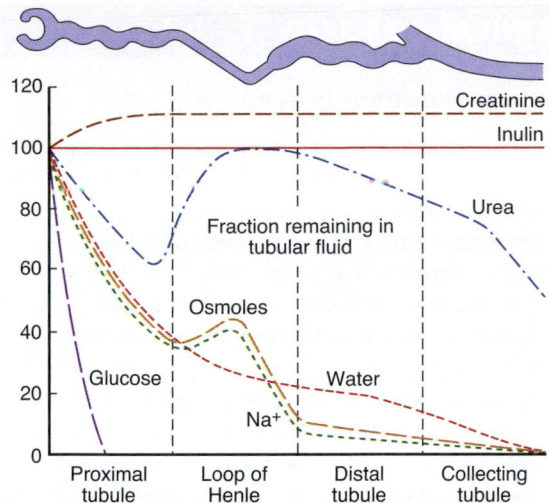

FIGURE 38–14 **Changes in the percentage of the filtered amount of substances remaining in the tubular fluid along the length of the nephron in the presence of vasopressin.** (Modified from Sullivan LP, Grantham JJ: *Physiology of the Kidney*, 2nd ed. Lea & Febiger, 1982.)

LOOP OF HENLE

As noted above, the loops of Henle of the juxtamedullary nephrons dip deeply into the medullary pyramids before draining into the distal convoluted tubules in the cortex, and all the collecting ducts descend back through the medullary pyramids to drain at the tips of the pyramids into the renal pelvis. There is a graded increase in the osmolality of the interstitium of the pyramids in humans: The osmolality at the tips of the papillae can reach about 1200 mOsm/kg of H_2O, approximately four times that of plasma. The descending limb of the loop of Henle is permeable to water, due to the presence of **aquaporin-1** in both the apical and basolateral membrane, but the ascending limb is impermeable to water (Table 38–8). Na^+, K^+, and Cl^- are cotransported out of the thick segment of the ascending limb. Therefore, the fluid in the descending limb of the loop of Henle becomes **hypertonic** as water moves out of the tubule into the hypertonic interstitium. In the ascending limb it becomes more dilute because of the movement of Na^+ and Cl^- out of the tubular lumen, and when fluid reaches the top of the ascending limb (called the diluting segment) it is now **hypotonic** to plasma. In passing through the descending loop of Henle, another 15% of the filtered water is removed, so approximately 20% of the filtered water enters the distal tubule, and the TF/P of inulin at this point is about 5.

In the thick ascending limb, a carrier cotransports one Na^+, one K^+, and $2Cl^-$ from the tubular lumen into the tubular cells. This is another example of secondary active transport; the Na^+ is actively transported from the cells into the interstitium by Na, K ATPase in the basolateral membranes of the cells, keeping the intracellular Na^+ low. The Na–K–2Cl transporter has 12 transmembrane domains with intracellular amino and carboxyl terminals. It is a member of a family of

TABLE 38–8 Permeability and transport in various segments of the nephron.[a]

	Permeability			Active Transport of Na^+
	H_2O	Urea	NaCl	
Loop of Henle				
Thin descending limb	4+	+	±	0
Thin ascending limb	0	+	4+	0
Thick ascending limb	0	±	±	4+
Distal convoluted tubule	±	±	±	3+
Collecting tubule				
Cortical portion	3+*	0	±	2+
Outer medullary portion	3+*	0	±	1+
Inner medullary portion	3+*	3+	±	1+

[a]Data are based on studies of rabbit and human kidneys. Values indicated by asterisks are in the presence of vasopressin. These values are 1+ in the absence of vasopressin.

Modified and reproduced with permission from Kokko JP: Renal concentrating and diluting mechanisms. Hosp Pract [Feb] 1979;110:14.

transporters found in many other locations, including salivary glands, the gastrointestinal tract, and the airways.

The K^+ diffuses back into the tubular lumen and back into the interstitium via ROMK and other K^+ channels. The Cl^- moves into the interstitium via ClC-Kb channels (Figure 38–15).

DISTAL TUBULE

The distal tubule, particularly its first part, is in effect an extension of the thick segment of the ascending limb. It is relatively impermeable to water, and continued removal of the solute in excess of solvent further dilutes the tubular fluid.

COLLECTING DUCTS

The collecting ducts have two portions: a cortical portion and a medullary portion. The changes in osmolality and volume in the collecting ducts depend on the amount of vasopressin acting on the ducts. This antidiuretic hormone from the posterior pituitary gland increases the permeability of the collecting ducts to water. The key to the action of vasopressin on the collecting ducts is aquaporin-2. Unlike the other aquaporins, this aquaporin is stored in vesicles in the cytoplasm of principal cells. Vasopressin causes rapid insertion of these vesicles into the apical membrane of cells. The effect is mediated via the

FIGURE 38–15 **NaCl transport in the thick ascending limb of the loop of Henle.** The Na–K–2Cl cotransporter moves these ions into the tubular cell by secondary active transport. Na^+ is transported out of the cell into the interstitium by Na, K ATPase in the basolateral membrane of the cell. Cl^- exits in basolateral ClC-Kb Cl^- channels. Barttin, a protein in the cell membrane, is essential for normal ClC-Kb function. K^+ moves from the cell to the interstitium and the tubular lumen by ROMK and other K^+ channels (see Clinical Box 38–2).

vasopressin V_2 receptor, cyclic adenosine 5-monophosphate (cAMP) and protein kinase A. Cytoskeletal elements are involved, including microtubule-based motor proteins (dynein and dynactin) as well as actin filament-binding proteins such as myosin-1.

In the presence of enough vasopressin to produce maximal antidiuresis, water moves out of the hypotonic fluid entering the cortical collecting ducts into the interstitium of the cortex, and the tubular fluid becomes isotonic. In this fashion, as much as 10% of the filtered water is removed. The isotonic fluid then enters the medullary collecting ducts with a TF/P inulin of about 20. An additional 4.7% or more of the filtrate is reabsorbed into the hypertonic interstitium of the medulla, producing a concentrated urine with a TF/P inulin of over 300. In humans, the osmolality of urine may reach 1400 mOsm/kg of H_2O, almost five times the osmolality of plasma, with a total of 99.7% of the filtered water being reabsorbed (Table 38–7). In other species, the ability to concentrate urine is even greater. Maximal urine osmolality is about 2500 mOsm/kg in dogs, about 3200 mOsm/kg in laboratory rats, and as high as 5000 mOsm/kg in certain desert rodents.

When vasopressin is absent, the collecting duct epithelium is relatively impermeable to water. The fluid therefore remains hypotonic, and large amounts flow into the renal pelvis. In humans, the urine osmolality may be as low as 30 mOsm/kg of H_2O. The impermeability of the distal portions of the nephron is not absolute; along with the salt that is pumped out of the collecting duct fluid, about 2% of the filtered water is reabsorbed in the absence of vasopressin. However, as much as 13% of the filtered water may be excreted, and urine flow may reach 15 mL/min or more.

THE COUNTERCURRENT MECHANISM

The concentrating mechanism depends upon the maintenance of a gradient of **increasing osmolality** along the medullary pyramids. This gradient is produced by the operation of the loops of Henle as **countercurrent multipliers** and maintained by the operation of the vasa recta as **countercurrent exchangers.** A countercurrent system is a system in which the inflow runs parallel to, counter to, and in close proximity to the outflow for some distance. This occurs for both the loops of Henle and the vasa recta in the renal medulla (Figure 38–3).

The operation of each loop of Henle as a countercurrent multiplier depends on the high permeability of the thin descending limb to water (via aquaporin-1), the active transport of Na^+ and Cl^- out of the thick ascending limb, and the inflow of tubular fluid from the proximal tubule, with outflow into the distal tubule. The process can be explained using hypothetical steps

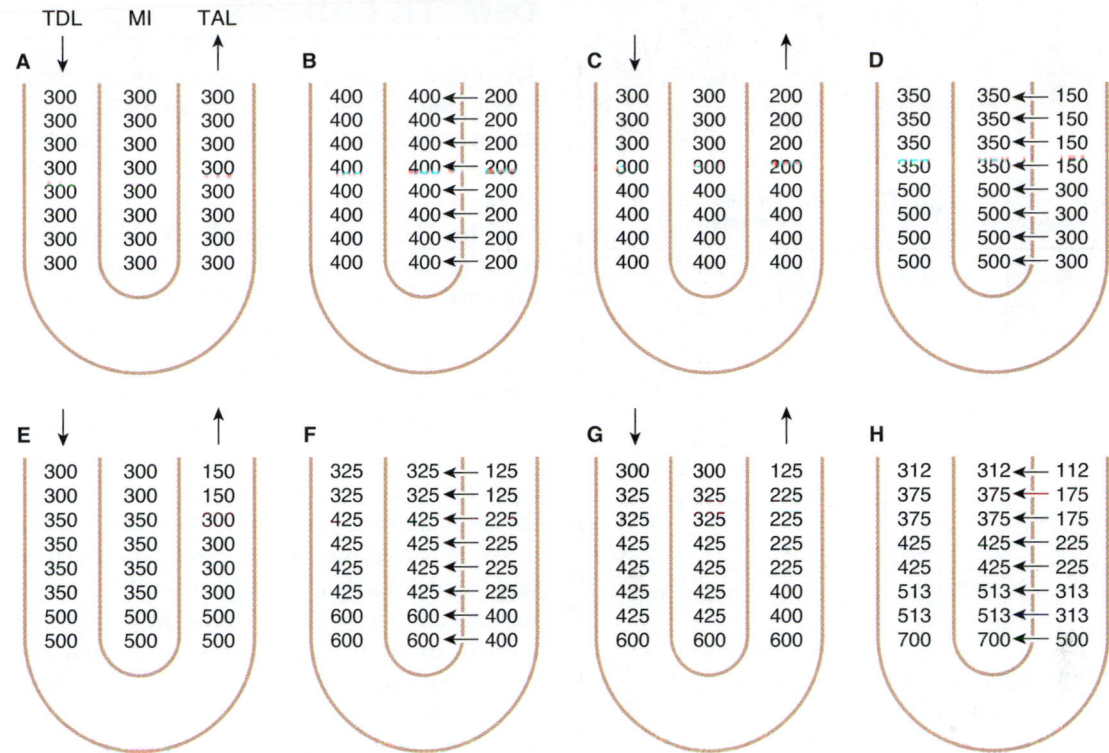

FIGURE 38–16 **Operation of the loop of Henle as a countercurrent multiplier producing a gradient of hyperosmolarity in the medullary interstitium (MI).** TDL, thin descending limb; TAL, thick ascending limb. The process of generation of the gradient is illustrated as occurring in hypothetical steps, starting at A, where osmolality in both limbs and the interstitium is 300 mOsm/kg of water. The pumps in the thick ascending limb move Na^+ and Cl^- into the interstitium, increasing its osmolality to 400 mOsm/kg, and this equilibrates with the fluid in the thin descending limb. However, isotonic fluid continues to flow into the thin descending limb and hypotonic fluid out of the thick ascending limb. Continued operation of the pumps makes the fluid leaving the thick ascending limb even more hypotonic, while hypertonicity accumulates at the apex of the loop. (Modified and reproduced with permission from Johnson LR [editor]: *Essential Medical Physiology*, Raven Press, 1992.)

leading to the normal equilibrium condition, although the steps do not occur in vivo. It is also important to remember that the equilibrium is maintained unless the osmotic gradient is washed out. These steps are summarized in Figure 38–16 for a cortical nephron with no thin ascending limb. Assume first a condition in which osmolality is 300 mOsm/kg of H_2O throughout the descending and ascending limbs and the medullary interstitium (Figure 38–16A). Assume in addition that the pumps in the thick ascending limb can pump 100 mOsm/kg of Na^+ and Cl^- from the tubular fluid to the interstitium, increasing interstitial osmolality to 400 mOsm/kg of H_2O. Water then moves out of the thin descending limb, and its contents equilibrate with the interstitium (Figure 38–16B). However, fluid containing 300 mOsm/kg of H_2O is continuously entering this limb from the proximal tubule (Figure 38–16C), so the gradient against which the Na^+ and Cl^- are pumped is reduced and more enters the interstitium (Figure 38–16D). Meanwhile, hypotonic fluid flows into the distal tubule, and isotonic and subsequently hypertonic fluid flows into the ascending thick limb. The process keeps repeating, and the final result is a gradient of osmolality from the top to the bottom of the loop.

In juxtamedullary nephrons with longer loops and thin ascending limbs, the osmotic gradient is spread over a greater

distance and the osmolality at the tip of the loop is greater. This is because the thin ascending limb is relatively impermeable to water but permeable to Na^+ and Cl^-. Therefore, Na^+ and Cl^- move down their concentration gradients into the interstitium, and there is additional passive countercurrent multiplication. The greater the length of the loop of Henle, the greater the osmolality that can be reached at the tip of the medulla.

The osmotic gradient in the medullary pyramids would not last long if the Na^+ and urea in the interstitial spaces were removed by the circulation. These solutes remain in the pyramids primarily because the vasa recta operate as countercurrent exchangers (Figure 38–17). The solutes diffuse out of the vessels conducting blood toward the cortex and into the vessels descending into the pyramid. Conversely, water diffuses out of the descending vessels and into the fenestrated ascending vessels. Therefore, the solutes tend to recirculate in the medulla and water tends to bypass it, so that hypertonicity is maintained. The water removed from the collecting ducts in the pyramids is also removed by the vasa recta and enters the general circulation. Countercurrent exchange is a passive process; it depends on movement of water and could not maintain the osmotic gradient along the pyramids if the process of countercurrent multiplication in the loops of Henle were to cease.

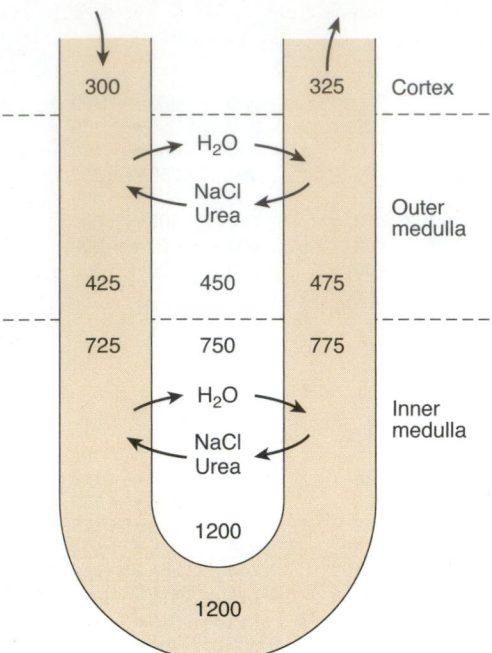

**FIGURE 38–17   Operation of the vasa recta as
countercurrent exchangers in the kidney.** NaCl and urea diffuse out
of the ascending limb of the vessel and into the descending limb,
whereas water diffuses out of the descending and into the ascending
limb of the vascular loop.

It is worth noting that there is a very large osmotic gradient
in the loop of Henle and, in the presence of vasopressin, in the
collecting ducts. It is the countercurrent system that makes
this gradient possible by spreading it along a system of tubules
1 cm or more in length, rather than across a single layer of
cells that is only a few micrometers thick. There are other
examples of the operation of countercurrent exchangers in
animals. One is the heat exchange between the arteries and
venae comitantes of the limbs. To a minor degree in humans,
but to a major degree in mammals living in cold water, heat is
transferred from the arterial blood flowing into the limbs to
the adjacent veins draining blood back into the body, making
the tips of the limbs cold while conserving body heat.

ROLE OF UREA

Urea contributes to the establishment of the osmotic gradient
in the medullary pyramids and to the ability to form a concentrated urine in the collecting ducts. Urea transport is mediated
by urea transporters, presumably by facilitated diffusion.
There are at least four isoforms of the transport protein UT-A
in the kidneys (UT-A1 to UT-A4); UT-B is found in erythrocytes. The amount of urea in the medullary interstitium and,
consequently, in the urine varies with the amount of urea filtered, and this in turn varies with the dietary intake of protein.
Therefore, a high-protein diet increases the ability of the kidneys to concentrate the urine.

OSMOTIC DIURESIS

The presence of large quantities of unreabsorbed solutes in the
renal tubules causes an increase in urine volume called
osmotic diuresis. Solutes that are not reabsorbed in the proximal tubules exert an appreciable osmotic effect as the volume
of tubular fluid decreases and their concentration rises. Therefore, they "hold water in the tubules." In addition, the concentration gradient against which Na^+ can be pumped out of the
proximal tubules is limited. Normally, the movement of water
out of the proximal tubule prevents any appreciable gradient
from developing, but Na^+ concentration in the fluid falls when
water reabsorption is decreased because of the presence in the
tubular fluid of increased amounts of unreabsorbable solutes.
The limiting concentration gradient is reached, and further
proximal reabsorption of Na^+ is prevented; more Na^+ remains
in the tubule, and water stays with it. The result is that the loop
of Henle is presented with a greatly increased volume of isotonic fluid. This fluid has a decreased Na^+ concentration, but
the total amount of Na^+ reaching the loop per unit time is increased. In the loop, reabsorption of water and Na^+ is decreased because the medullary hypertonicity is decreased. The
decrease is due primarily to decreased reabsorption of Na^+,
K^+, and Cl^- in the ascending limb of the loop because the limiting concentration gradient for Na^+ reabsorption is reached.
More fluid passes through the distal tubule, and because of the
decrease in the osmotic gradient along the medullary pyramids, less water is reabsorbed in the collecting ducts. The result is a marked increase in urine volume and excretion of Na^+
and other electrolytes.

Osmotic diuresis is produced by the administration of compounds such as mannitol and related polysaccharides that are
filtered but not reabsorbed. It is also produced by naturally
occurring substances when they are present in amounts
exceeding the capacity of the tubules to reabsorb them. For
example, in **diabetes mellitus,** if blood glucose is high, glucose in the glomerular filtrate is high, thus the filtered load
will exceed the Tm_G and glucose will remain in the tubules
causing polyuria. Osmotic diuresis can also be produced by
the infusion of large amounts of sodium chloride or urea.

It is important to recognize the difference between osmotic
diuresis and water diuresis. In water diuresis, the amount of
water reabsorbed in the proximal portions of the nephron is
normal, and the maximal urine flow that can be produced is
about 16 mL/min. In osmotic diuresis, increased urine flow
is due to decreased water reabsorption in the proximal
tubules and loops and very large urine flows can be produced.
As the load of excreted solute is increased, the concentration
of the urine approaches that of plasma (Figure 38–18) in spite
of maximal vasopressin secretion, because an increasingly
large fraction of the excreted urine is isotonic proximal tubular fluid. If osmotic diuresis is produced in an animal with
diabetes insipidus, the urine concentration rises for the same
reason.

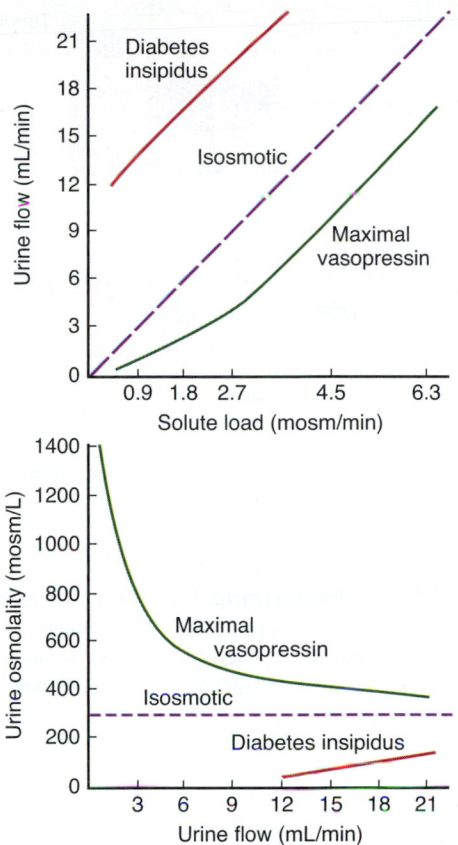

FIGURE 38–18 **Approximate relationship between urine concentration and urine flow in osmotic diuresis in humans.** The dashed line in the lower diagram indicates the concentration at which the urine is isosmotic with plasma. (Reproduced with permission from Berliner RW, Giebisch G in: *Best and Taylor's Physiological Basis of Medical Practice,* 9th ed. Brobeck JR [editor]. Williams & Wilkins, 1979.)

RELATION OF URINE CONCENTRATION TO GFR

The magnitude of the osmotic gradient along the medullary pyramids is increased when the rate of flow of fluid through the loops of Henle is decreased. A reduction in GFR such as that caused by dehydration produces a decrease in the volume of fluid presented to the countercurrent mechanism, so that the rate of flow in the loops declines and the urine becomes more concentrated. When the GFR is low, the urine can become quite concentrated in the absence of vasopressin. If one renal artery is constricted in an animal with diabetes insipidus, the urine excreted on the side of the constriction becomes hypertonic because of the reduction in GFR, whereas that excreted on the opposite side remains hypotonic.

"FREE WATER CLEARANCE"

In order to quantitate the gain or loss of water by excretion of a concentrated or dilute urine, the "free water clearance"

(C_{H_2O}) is sometimes calculated. This is the difference between the urine volume and the clearance of osmoles (C_{Osm}):

$$C_{H_2O} = \dot{V} - \frac{U_{Osm}\,\dot{V}}{P_{Osm}}$$

where $\dot{V}$ is the urine flow rate and U_{Osm} and P_{Osm} the urine and plasma osmolality, respectively. C_{Osm} is the amount of water necessary to excrete the osmotic load in a urine that is isotonic with plasma. Therefore, C_{H_2O} is negative when the urine is hypertonic and positive when the urine is hypotonic. For example, using the data in Table 38–7, the values for C_{H_2O} are –1.3 mL/min (–1.9 L/d) during maximal antidiuresis and 14.5 mL/min (20.9 L/d) in the absence of vasopressin.

REGULATION OF Na$^+$ EXCRETION

Na$^+$ is filtered in large amounts, but it is actively transported out of all portions of the tubule except the descending thin limb of Henle's loop. Normally, 96% to well over 99% of the filtered Na$^+$ is reabsorbed. Because Na$^+$ is the most abundant cation in ECF and because Na$^+$ salts account for over 90% of the osmotically active solute in the plasma and interstitial fluid, the amount of Na$^+$ in the body is a prime determinant of the ECF volume. Therefore, it is not surprising that multiple regulatory mechanisms have evolved in terrestrial animals to control the excretion of this ion. Through the operation of these regulatory mechanisms, the amount of Na$^+$ excreted is adjusted to equal the amount ingested over a wide range of dietary intakes, and the individual stays in Na$^+$ balance. Thus, urinary Na$^+$ output ranges from less than 1 mEq/d on a low-salt diet to 400 mEq/d or more when the dietary Na$^+$ intake is high. In addition, there is a natriuresis when saline is infused intravenously and a decrease in Na$^+$ excretion when ECF volume is reduced.

MECHANISMS

Variations in Na$^+$ excretion are brought about by changes in GFR (Table 38–9) and changes in tubular reabsorption, primarily in the 3% of filtered Na$^+$ that reaches the collecting ducts. The factors affecting the GFR, including tubuloglomerular feedback, have been discussed previously. Factors affecting Na$^+$ reabsorption include the circulating level of aldosterone and other adrenocortical hormones, the circulating level of ANP and other natriuretic hormones, and the rate of tubular secretion of H$^+$ and K$^+$.

EFFECTS OF ADRENOCORTICAL STEROIDS

Adrenal mineralocorticoids such as aldosterone increase tubular reabsorption of Na$^+$ in association with secretion of K$^+$ and H$^+$ and also Na$^+$ reabsorption with Cl$^-$. When these hormones are injected into adrenalectomized animals, a latent

TABLE 38–9 Changes in Na⁺ excretion that would occur as a result of changes in GFR if there were no concomitant changes in Na⁺ reabsorption.

GFR (mL/min)	Plasma Na$^+$ (μEq/mL)	Amount Filtered (μEq/min)	Amount Reabsorbed (μEq/min)	Amount Excreted (μEq/min)
125	145	18,125	18,000	125
127	145	18,415	18,000	415
124.1	145	18,000	18,000	0

FIGURE 38–19 **Renal Principal cell.** Na⁺ enters via the ENaCs in the apical membrane and is pumped into the interstitial fluid by Na, K ATPases in the basolateral membrane. Aldosterone activates the genome to produce serum- and glucocorticoid-regulated kinase (sgk) and other proteins, and the number of active ENaCs is increased.

period of 10 to 30 min occurs before their effects on Na⁺ reabsorption become manifest, because of the time required for the steroids to alter protein synthesis via their action on DNA. Mineralocorticoids may also have more rapid membrane-mediated effects, but these are not apparent in terms of Na⁺ excretion in the whole animal. The mineralocorticoids act primarily in the collecting ducts to increase the number of active epithelial sodium channels (ENaCs) in this part of the nephron. The molecular mechanisms believed to be involved are discussed in Chapter 22 and summarized in Figure 38–19.

In Liddle syndrome, mutations in the genes that code for the β subunit and less commonly the γ subunit of the ENaCs cause them to become constitutively active in the kidney. This leads to Na⁺ retention and hypertension.

OTHER HUMORAL EFFECTS

Reduction of dietary intake of salt increases aldosterone secretion (see Figure 22-26), producing marked but slowly developing decreases in Na⁺ excretion. A variety of other humoral factors affect Na⁺ reabsorption. PGE₂ causes a natriuresis, possibly by inhibiting Na, K ATPase and possibly by increasing intracellular Ca²⁺, which in turn inhibits Na⁺ transport via ENaCs. Endothelin and IL-1 cause natriuresis, probably by increasing the formation of PGE₂. ANP and related molecules increase intracellular cyclic 3',5'-guanosine monophosphate (cGMP), and this inhibits transport via the ENaCs. Inhibition of Na, K ATPase by another natriuretic hormone, which appears to be endogenously produced ouabain, also increases Na⁺ excretion. Angiotensin II increases reabsorption of Na⁺ and HCO₃⁻ by an action on the proximal tubules. There is an appreciable amount of angiotensin-converting enzyme in the kidneys, and the kidneys convert 20% of the circulating angiotensin I reaching them to angiotensin II. In addition, angiotensin I is generated in the kidneys.

Prolonged exposure to high levels of circulating mineralocorticoids does not cause edema in otherwise normal individuals because eventually the kidneys escape from the effects of the steroids. This **escape phenomenon,** which may be due to increased secretion of ANP, is discussed in Chapter 22. It appears to be reduced or absent in nephrosis, cirrhosis, and heart failure, and patients with these diseases continue to

retain Na⁺ and become edematous when exposed to high levels of mineralocorticoids.

REGULATION OF WATER EXCRETION

WATER DIURESIS

The feedback mechanism controlling vasopressin secretion and the way vasopressin secretion is stimulated by a rise and inhibited by a drop in the effective osmotic pressure of the plasma are discussed in Chapter 18. The **water diuresis** produced by drinking large amounts of hypotonic fluid begins about 15 min after ingestion of a water load and reaches its maximum in about 40 min. The act of drinking produces a small decrease in vasopressin secretion before the water is absorbed, but most of the inhibition is produced by the decrease in plasma osmolality after the water is absorbed.

WATER INTOXICATION

During excretion of an average osmotic load, the maximal urine flow that can be produced during a water diuresis is about 16 mL/min. If water is ingested at a higher rate than this for any length of time, swelling of the cells because of the uptake of water from the hypotonic ECF becomes severe and, rarely, the symptoms of **water intoxication** may develop. Swelling of the cells in the brain causes convulsions and coma and leads eventually to death. Water intoxication can also occur when water intake is not reduced after administration of exogenous vasopressin or when secretion of endogenous vasopressin occurs in response to non-osmotic stimuli such as surgical trauma.

REGULATION OF K⁺ EXCRETION

Much of the filtered K^+ is removed from the tubular fluid by active reabsorption in the proximal tubules (Table 38–5), and K^+ is then secreted into the fluid by the distal tubular cells. The rate of K^+ secretion is proportional to the rate of flow of the tubular fluid through the distal portions of the nephron, because with rapid flow there is less opportunity for the tubular K^+ concentration to rise to a value that stops further secretion. In the absence of complicating factors, the amount secreted is approximately equal to the K^+ intake, and K^+ balance is maintained. In the collecting ducts, Na^+ is generally reabsorbed and K^+ is secreted. There is no rigid one-for-one exchange, and much of the movement of K^+ is passive. However, there is electrical coupling in the sense that intracellular migration of Na^+ from the lumen tends to lower the potential difference across the tubular cell, and this favors movement of K^+ into the tubular lumen. Because Na^+ is also reabsorbed in association with H^+ secretion, there is competition for the Na^+ in the tubular fluid. K^+ excretion is decreased when the amount of Na^+ reaching the distal tubule is small, and it is also decreased when H^+ secretion is increased.

DIURETICS

Although a detailed discussion of diuretic agents is outside the scope of this book, consideration of their mechanisms of action constitutes an informative review of the factors affecting urine volume and electrolyte excretion. These mechanisms are summarized in Table 38–10. Water, alcohol, osmotic diuretics, xanthines, and acidifying salts have limited clinical usefulness, and the vasopressin antagonists are currently undergoing clinical trials. However, many of the other agents on the list are used extensively in medical practice.

The carbonic anhydrase-inhibiting drugs are only moderately effective as diuretic agents, but because they inhibit acid secretion by decreasing the supply of carbonic acid, they have far-reaching effects. Not only is Na^+ excretion increased because H^+ secretion is decreased, but also HCO_3^- reabsorption is depressed; and because H^+ and K^+ compete with each other and with Na^+, the decrease in H^+ secretion facilitates the secretion and excretion of K^+.

Furosemide and the other loop diuretics inhibit the Na–K–2Cl cotransporter in the thick ascending limb of Henle's loop. They cause a marked natriuresis and kaliuresis. Thiazides act by inhibiting Na–Cl cotransport in the distal tubule. The diuresis they cause is less marked, but both loop diuretics and thiazides cause increased delivery of Na^+ (and fluid) to the collecting ducts, facilitating K^+ excretion. Thus, over time, K^+

TABLE 38–10 Mechanism of action of various diuretics.

Agent	Mechanism of Action
Water	Inhibits vasopressin secretion.
Ethanol	Inhibits vasopressin secretion.
Antagonists of V_2 vasopressin receptors such as astolvaptan	Inhibit action of vasopressin on collecting duct.
Large quantities of osmotically active substances such as mannitol and glucose	Produce osmotic diuresis.
Xanthines such as caffeine and theophylline	Decrease tubular reabsorption of Na^+ and increase GFR.
Acidifying salts such as $CaCl_2$ and NH_4Cl	Supply acid load; H^+ is buffered, but an anion is excreted with Na^+ when the ability of the kidneys to replace Na^+ with H^+ is exceeded.
Carbonic anhydrase inhibitors such as acetazolamide (Diamox)	Decrease H^+ secretion, with resultant increase in Na^+ and K^+ excretion.
Metolazone (Zaroxolyn), thiazides such as chlorothiazide (Diuril)	Inhibit the Na–Cl cotransporter in the early portion of the distal tubule.
Loop diuretics such as furosemide (Lasix), ethacrynic acid (Edecrin), and bumetanide	Inhibit the Na–K–2Cl cotransporter in the medullary thick ascending limb of the loop of Henle
K^+-retaining natriuretics such as spironolactone (Aldactone), triamterene (Dyrenium), and amiloride (Midamor)	Inhibit Na^+–K^+ "exchange" in the collecting ducts by inhibiting the action of aldosterone (spironolactone) or by inhibiting the ENaCs (amiloride).

depletion and hypokalemia are common complications in those who use them if they do not supplement their K^+ intake. On the other hand, the so-called K^+-sparing diuretics act in the collecting duct by inhibiting the action of aldosterone or blocking ENaCs.

EFFECTS OF DISORDERED RENAL FUNCTION

A number of abnormalities are common to many different types of renal disease. The secretion of renin by the kidneys and the relation of the kidneys to hypertension are discussed in Chapter 39. A frequent finding in various forms of renal disease is the presence in the urine of protein, leukocytes, red cells, and **casts,** which are proteinaceous material precipitated in the tubules and washed into the bladder. Other important consequences of renal disease are loss of the ability to concentrate or dilute the urine, uremia, acidosis, and abnormal retention of Na^+ (see Clinical Box 38–3).

Proteinuria

In many renal diseases and in one benign condition, the permeability of the glomerular capillaries is increased, and protein is found in the urine in more than the usual trace amounts (**proteinuria**). Most of this protein is albumin, and the defect is commonly called **albuminuria.** The relation of charges on the glomerular membrane to albuminuria has been discussed above. The amount of protein in the urine may be very large, and especially in nephrosis, the urinary protein loss may exceed the rate at which the liver can synthesize plasma proteins. The resulting hypoproteinemia reduces the oncotic pressure, and the plasma volume declines, sometimes to dangerously low levels, while edema fluid accumulates in the tissues.

A benign condition that causes proteinuria is a poorly understood change in renal hemodynamics, which in some otherwise normal individuals, causes protein to appear in urine when they are in the standing position (**orthostatic albuminuria).** Urine formed when these individuals are lying down is protein-free.

LOSS OF CONCENTRATING & DILUTING ABILITY

In renal disease, the urine becomes less concentrated and urine volume is often increased, producing the symptoms of **polyuria** and **nocturia** (waking up at night to void). The ability to form a dilute urine is often retained, but in advanced renal disease, the osmolality of the urine becomes fixed at about that of plasma, indicating that the diluting and concentrating functions of the kidney have both been lost. The loss is due in part to disruption of the countercurrent mechanism, but a more important cause is a loss of functioning nephrons. When one kidney is removed surgically, the number of functioning nephrons is halved. The number of osmoles excreted is not reduced to this extent, and so the remaining nephrons must each be filtering and excreting more osmotically active substances, producing what is in effect an osmotic diuresis. In osmotic diuresis, the osmolality of the urine approaches that of plasma. The same thing happens when the number of functioning nephrons is reduced by disease. The increased filtration in the remaining nephrons eventually damages them, and thus more nephrons are lost. The damage resulting from increased filtration may be due to progressive fibrosis in the proximal tubule cells, but this is unsettled. However, the eventual result of this positive feedback is loss of so many nephrons that complete renal failure with **oliguria** or even **anuria** results.

UREMIA

When the breakdown products of protein metabolism accumulate in the blood, the syndrome known as **uremia** develops. The symptoms of uremia include lethargy, anorexia, nausea and vomiting, mental deterioration and confusion, muscle twitching, convulsions, and coma. The blood urea nitrogen (BUN) and creatinine levels are high, and the blood levels of these substances are used as an index of the severity of the uremia. It probably is not the accumulation of urea and creatinine per se but rather the accumulation of other toxic substances—possibly organic acids or phenols—that produces the symptoms of uremia.

The toxic substances that cause the symptoms of uremia can be removed by dialyzing the blood of uremic patients against a bath of suitable composition in an artificial kidney (**hemodialysis**). Patients can be kept alive and in reasonable health for many months on dialysis, even when they are completely anuric or have had both kidneys removed. However, the treatment of choice today is certainly transplantation of a kidney from a suitable donor.

Other features of chronic renal failure include anemia, which is caused primarily by failure to produce erythropoietin, and secondary hyperparathyroidism due to 1,25-dihydroxycholecalciferol deficiency (see Chapter 23).

ACIDOSIS

Acidosis is common in chronic renal disease because of failure to excrete the acid products of digestion and metabolism (see Chapter 40). In the rare syndrome of **renal tubular acidosis,** there is specific impairment of the ability to make the urine acidic, and other renal functions are usually normal. However, in most cases of chronic renal disease the urine is maximally acidified, and acidosis develops because the total amount of H^+ that can be secreted is reduced because of impaired renal tubular production of NH_4^+.

ABNORMAL Na⁺ HANDLING

Many patients with renal disease retain excessive amounts of Na^+ and become edematous. Na^+ retention in renal disease has at least three causes. In acute glomerulonephritis, a disease that affects primarily the glomeruli, the amount of Na^+ filtered is decreased markedly. In the nephrotic syndrome, an increase in aldosterone secretion contributes to the salt retention. The plasma protein level is low in this condition, and so fluid moves from the plasma into the interstitial spaces and the plasma volume falls. The decline in plasma volume triggers the increase in aldosterone secretion via the renin–angiotensin system. A third cause of Na^+ retention and edema in renal disease is **heart failure.** Renal disease predisposes to heart failure, partly because of the hypertension it frequently produces.

THE BLADDER

FILLING

The walls of the ureters contain smooth muscle arranged in spiral, longitudinal, and circular bundles, but distinct layers of muscle are not seen. Regular peristaltic contractions occurring one to five times per minute move the urine from the renal pelvis to the bladder, where it enters in spurts synchronous with each peristaltic wave. The ureters pass obliquely through the bladder wall and, although there are no ureteral sphincters as such, the oblique passage tends to keep the ureters closed except during peristaltic waves, preventing reflux of urine from the bladder.

EMPTYING

The smooth muscle of the bladder, like that of the ureters, is arranged in spiral, longitudinal, and circular bundles. Contraction of the circular muscle, which is called the **detrusor muscle,** is mainly responsible for emptying the bladder during urination (micturition). Muscle bundles pass on either side of the urethra, and these fibers are sometimes called the **internal urethral sphincter,** although they do not encircle the urethra. Farther along the urethra is a sphincter of skeletal muscle, the sphincter of the membranous urethra (**external urethral sphincter**). The bladder epithelium is made up of a superficial layer of flat cells and a deep layer of cuboidal cells. The innervation of the bladder is summarized in Figure 38–20.

The physiology of bladder emptying and the physiologic basis of its disorders are subjects about which there is much confusion. Micturition is fundamentally a spinal reflex facilitated and inhibited by higher brain centers and, like defecation, subject to voluntary facilitation and inhibition. Urine enters the bladder without producing much increase in intravesical pressure until the viscus is well filled. In addition, like other types of smooth muscle, the bladder muscle has the property of plasticity; when it is stretched, the tension initially produced is not maintained. The relation between intravesical pressure and volume can be studied by inserting a catheter and emptying the bladder, then recording the pressure while the bladder is filled with 50-mL increments of water or air (**cystometry**). A plot of intravesical pressure against the volume of fluid in the bladder is called a **cystometrogram** (Figure 38–21). The curve shows an initial slight rise in pressure when the first increments in volume are produced; a long, nearly flat segment as further increments are produced; and a sudden, sharp rise in pressure as the micturition reflex is triggered. These three components are sometimes called segments Ia, Ib, and II. The first urge to void is felt at a bladder volume of about 150 mL, and a marked sense of fullness at about 400 mL. The flatness of segment Ib is a manifestation of the law of Laplace. This law states that the pressure in a spherical viscus is equal to twice the wall tension divided by the radius. In the case of the bladder, the tension increases as the organ fills, but so does the radius. Therefore, the pressure increase is slight until the organ is relatively full.

During micturition, the perineal muscles and external urethral sphincter are relaxed, the detrusor muscle contracts, and urine passes out through the urethra. The bands of smooth muscle on either side of the urethra apparently play no role in micturition, and their main function in males is believed to be the prevention of reflux of semen into the bladder during ejaculation.

The mechanism by which voluntary urination is initiated remains unsettled. One of the initial events is relaxation of the muscles of the pelvic floor, and this may cause a sufficient

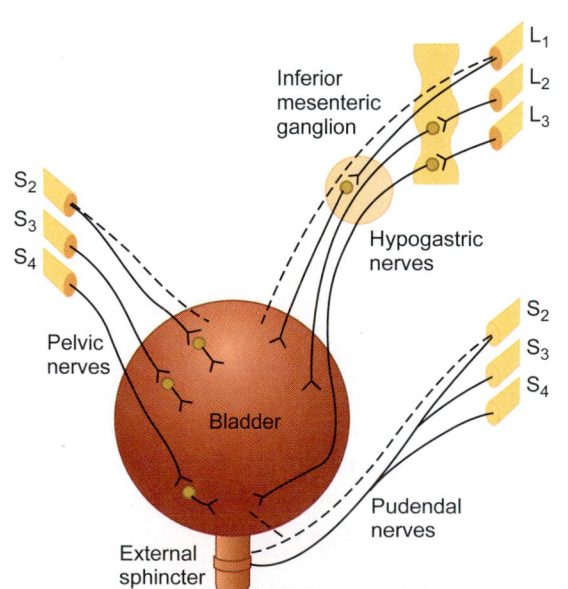

FIGURE 38–20 Innervation of the bladder. Dashed lines indicate sensory nerves. Parasympathetic innervation is shown at the left, sympathetic at the upper right, and somatic at the lower right.

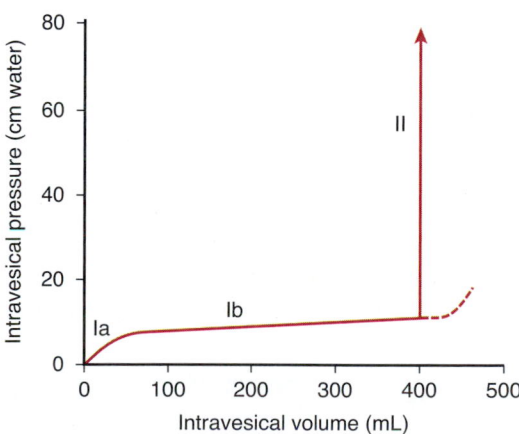

FIGURE 38–21 Cystometrogram in a normal human. The numerals identify the three components of the curve described in the text. The dashed line indicates the pressure–volume relations that would have been found had micturition not occurred and produced component II. (Modified and reproduced with permission from Tanagho EA, McAninch JW: *Smith's General Urology,* 15th ed. McGraw-Hill, 2000.)

downward tug on the detrusor muscle to initiate its contraction. The perineal muscles and external sphincter can be contracted voluntarily, preventing urine from passing down the urethra or interrupting the flow once urination has begun. It is through the learned ability to maintain the external sphincter in a contracted state that adults are able to delay urination until the opportunity to void presents itself. After urination, the female urethra empties by gravity. Urine remaining in the urethra of the male is expelled by several contractions of the bulbocavernosus muscle.

REFLEX CONTROL

The bladder smooth muscle has some inherent contractile activity; however, when its nerve supply is intact, stretch receptors in the bladder wall initiate a reflex contraction that has a lower threshold than the inherent contractile response of the muscle. Fibers in the pelvic nerves are the afferent limb of the voiding reflex, and the parasympathetic fibers to the bladder that constitute the efferent limb also travel in these nerves. The reflex is integrated in the sacral portion of the spinal cord. In the adult, the volume of urine in the bladder that normally initiates a reflex contraction is about 300 to 400 mL. The sympathetic nerves to the bladder play no part in micturition, but in males they do mediate the contraction of the bladder muscle that prevents semen from entering the bladder during ejaculation.

The stretch receptors in the bladder wall have no small motor nerve system. However, the threshold for the voiding reflex, like the stretch reflexes, is adjusted by the activity of facilitatory and inhibitory centers in the brainstem. There is a facilitatory area in the pontine region and an inhibitory area in the midbrain. After transection of the brain stem just above the pons, the threshold is lowered and less bladder filling is required to trigger it, whereas after transection at the top of the midbrain, the threshold for the reflex is essentially normal. There is another facilitatory area in the posterior hypothalamus. Humans with lesions in the superior frontal gyrus have a reduced desire to urinate and difficulty in stopping micturition once it has commenced. However, stimulation experiments in animals indicate that other cortical areas also affect the process. The bladder can be made to contract by voluntary facilitation of the spinal voiding reflex when it contains only a few milliliters of urine. Voluntary contraction of the abdominal muscles aids the expulsion of urine by increasing the intra-abdominal pressure, but voiding can be initiated without straining even when the bladder is nearly empty.

EFFECTS OF DEAFFERENTATION

When the sacral dorsal roots are cut in experimental animals or interrupted by diseases of the dorsal roots, such as **tabes dorsalis** in humans, all reflex contractions of the bladder are abolished. The bladder becomes distended, thin-walled, and hypotonic, but some contractions occur because of the intrinsic response of the smooth muscle to stretch.

CLINICAL BOX 38–4

Abnormalities of Micturition

Three major types of bladder dysfunction are due to neural lesions: (1) the type due to interruption of the afferent nerves from the bladder, (2) the type due to interruption of both afferent and efferent nerves, and (3) the type due to interruption of facilitatory and inhibitory pathways descending from the brain. In all three types the bladder contracts, but the contractions are generally not sufficient to empty the viscus completely, and residual urine is left in the bladder.

EFFECTS OF DENERVATION

When the afferent and efferent nerves are both destroyed, as they may be by tumors of the cauda equina or filum terminale, the bladder is flaccid and distended for a while. Gradually, however, the muscle of the "decentralized bladder" becomes active, with many contraction waves that expel dribbles of urine out of the urethra. The bladder becomes shrunken and the bladder wall hypertrophied. The reason for the difference between the small, hypertrophic bladder seen in this condition and the distended, hypotonic bladder seen when only the afferent nerves are interrupted is not known. The hyperactive state in the former condition suggests the development of denervation hypersensitization even though the neurons interrupted are preganglionic rather than postganglionic (see Clinical Box 38–4).

EFFECTS OF SPINAL CORD TRANSECTION

During spinal shock, the bladder is flaccid and unresponsive. It becomes overfilled, and urine dribbles through the sphincters (**overflow incontinence**). After spinal shock has passed, the voiding reflex returns, although there is, of course, no voluntary control and no inhibition or facilitation from higher centers when the spinal cord is transected. Some paraplegic patients train themselves to initiate voiding by pinching or stroking their thighs, provoking a mild mass reflex (see Chapter 16). In some instances, the voiding reflex becomes hyperactive, bladder capacity is reduced, and the wall becomes hypertrophied. This type of bladder is sometimes called the **spastic neurogenic bladder.** The reflex hyperactivity is made worse by, and may be caused by, infection in the bladder wall.

CHAPTER SUMMARY

- Plasma enters the kidneys and is filtered in the glomerulus. As the filtrate passes down the nephron and through the tubules its volume is reduced and water and solutes are removed (tubular reabsorption) and waste products are secreted (tubular secretion).

- A nephron consists of an individual renal tubule and its glomerulus. Each tubule has several segments, beginning with the proximal tubule, followed by the loop of Henle (descending and ascending limbs), the distal convoluted tubule, the connecting tubule, and the collecting duct.

- The kidneys receive just under 25% of the cardiac output and renal plasma flow can be measured by infusing p-aminohippuric acid (PAH) and determining its urine and plasma concentrations.

- Renal blood flow enters the glomerulus via the afferent arteriole and leaves via the efferent arteriole (whose diameter is smaller). Renal blood flow is regulated by norepinephrine (constriction, reduction of flow), dopamine (vasodilation, increases flow), angiotensin II (constricts), prostaglandins (dilation in the renal cortex and constriction in the renal medulla), and acetylcholine (vasodilation).

- Glomerular filtration rate can be measured by a substance that is freely filtered and neither reabsorbed nor secreted in the tubules, is nontoxic, and is not metabolized by the body. Inulin meets these criteria and is extensively used to measure GFR.

- Urine is stored in the bladder before voiding (micturition). The micturition response involves reflex pathways, but is under voluntary control.

MULTIPLE-CHOICE QUESTIONS

For all questions, select the single best answer unless otherwise directed.

1. In the presence of vasopressin, the greatest fraction of filtered water is absorbed in the
   A) proximal tubule.
   B) loop of Henle.
   C) distal tubule.
   D) cortical collecting duct.
   E) medullary collecting duct.

2. In the absence of vasopressin, the greatest fraction of filtered water is absorbed in the
   A) proximal tubule.
   B) loop of Henle.
   C) distal tubule.
   D) cortical collecting duct.
   E) medullary collecting duct.

3. If the clearance of a substance which is freely filtered is less than that of inulin,
   A) there is net reabsorption of the substance in the tubules.
   B) there is net secretion of the substance in the tubules.
   C) the substance is neither secreted nor reabsorbed in the tubules.
   D) the substance becomes bound to protein in the tubules.
   E) the substance is secreted in the proximal tubule to a greater degree than in the distal tubule.

4. Glucose reabsorption occurs in the
   A) proximal tubule.
   B) loop of Henle.
   C) distal tubule.
   D) cortical collecting duct.
   E) medullary collecting duct.

5. On which of the following does aldosterone exert its greatest effect?
   A) glomerulus
   B) proximal tubule
   C) thin portion of the loop of Henle
   D) thick portion of the loop of Henle
   E) cortical collecting duct

6. What is the clearance of a substance when its concentration in the plasma is 10 mg/dL, its concentration in the urine is 100 mg/dL, and urine flow is 2 mL/min?
   A) 2 mL/min
   B) 10 mL/min
   C) 20 mL/min
   D) 200 mL/min
   E) Clearance cannot be determined from the information given.

7. As urine flow increases during osmotic diuresis
   A) the osmolality of urine falls below that of plasma.
   B) the osmolality of urine increases because of the increased amounts of nonreabsorbable solute in the urine.
   C) the osmolality of urine approaches that of plasma because plasma leaks into the tubules.
   D) the osmolality of urine approaches that of plasma because an increasingly large fraction of the excreted urine is isotonic proximal tubular fluid.
   E) the action of vasopressin on the renal tubules is inhibited.

CHAPTER RESOURCES

Anderson K-E: Pharmacology of lower urinary tract smooth muscles and penile erectile tissue. Pharmacol Rev 1993;45:253.

Brenner BM, Rector FC Jr. (editors): *The Kidney,* 6th ed. 2 vols. Saunders, 1999.

Brown D: The ins and outs of aquaporin-2 trafficking. Am J Physiol Renal Physiol 2003;284:F893.

Brown D, Stow JL: Protein trafficking and polarity in kidney epithelium: From cell biology to physiology. Physiol Rev 1996;76:245.

DiBona GF, Kopp UC: Neural control of renal function. Physiol Rev 1997; 77:75.

Garcia NH, Ramsey CR, Knox FG: Understanding the role of paracellular transport in the proximal tubule. News Physiol Sci 1998;13:38.

Nielsen S, et al: Aquaporins in the kidney: From molecules to medicine. Physiol Rev 2002;82:205.

Spring KR: Epithelial fluid transport: A century of investigation. News Physiol Sci 1999;14:92.

Valten V: Tubuloglomerular feedback and the control of glomerular filtration rate. News Physiol Sci 2003;18:169.

Regulation of Extracellular Fluid Composition & Volume

INTRODUCTION

This chapter is a review of the major homeostatic mechanisms that operate, primarily through the kidneys and the lungs, to maintain the **tonicity,** the **volume,** and the **specific ionic composition** of the extracellular fluid (ECF). The interstitial portion of this fluid is the fluid environment of the cells, and life depends upon the constancy of this "internal sea" (see Chapter 1).

DEFENSE OF TONICITY

The defense of the tonicity of the ECF is primarily the function of the vasopressin-secreting and thirst mechanisms. The total body osmolality is directly proportional to the total body sodium plus the total body potassium divided by the total body water, so that changes in the osmolality of the body fluids occur when a disproportion exists between the amount of these electrolytes and the amount of water ingested or lost from the body. When the effective osmotic pressure of the plasma rises, vasopressin secretion is increased and the thirst mechanism is stimulated; water is retained in the body, diluting the hypertonic plasma; and water intake is increased (Figure 39–1). Conversely, when the plasma becomes hypotonic, vasopressin secretion is decreased and "solute-free water" (water in excess of solute) is excreted. In this way, the tonicity of the body fluids is maintained within a narrow normal range. In health, plasma osmolality ranges from 280 to 295 mOsm/kg of H_2O, with vasopressin secretion maximally inhibited at 285 mOsm/kg and stimulated at higher values (Figure 39–2).

VASOPRESSIN RECEPTORS

There are at least three kinds of vasopressin receptors: V_{1A}, V_{1B}, and V_2. All are G protein-coupled. The V_{1A} and V_{1B}

FIGURE 39–1 **Mechanisms for defending ECF tonicity.** The dashed arrow indicates inhibition. (Courtesy of J Fitzsimmons.)

receptors act through phosphatidylinositol hydrolysis to increase the intracellular Ca^{2+} concentration. The V_2 receptors act through G_s to increase cyclic adenosine 3',5'-monophosphate (cAMP) levels.

EFFECTS OF VASOPRESSIN

Because one of its principal physiologic effects is the retention of water by the kidney, vasopressin is often called the **antidiuretic hormone (ADH).** It increases the permeability of the collecting ducts of the kidney, so that water enters the hypertonic interstitium of the renal pyramids. The urine becomes concentrated, and its volume decreases. The overall effect is therefore retention of water in excess of solute; consequently, the effective osmotic pressure of the body fluids is decreased. In the absence of vasopressin, the urine is hypotonic to plasma, urine volume is increased, and there is a net water loss. Consequently, the osmolality of the body fluid rises.

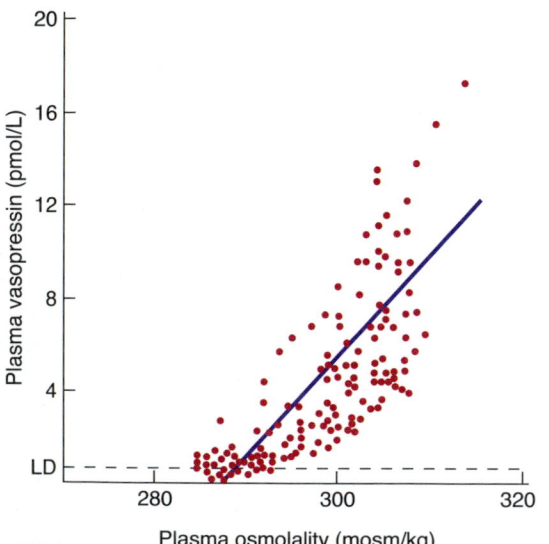

FIGURE 39–2 **Relation between plasma osmolality and plasma vasopressin in healthy adult humans during infusion of hypertonic saline.** LD, limit of detection. (Reproduced with permission from Thompson CJ et al: The osmotic thresholds for thirst and vasopressin are similar in healthy humans. Clin Sci [Colch] 1986;71:651.)

The mechanism by which vasopressin exerts its antidiuretic effect is activated by **V₂ receptors** and involves the insertion of proteins called water channels into the apical (luminal) membranes of the principal cells of the collecting ducts. Movement of water across membranes by simple diffusion is now known to be augmented by movement through water channels called **aquaporins,** and to date 13 (AQP0–AQP12) have been identified and water channels are now known to be expressed in almost all tissues in the body. The vasopressin-responsive water channel in the collecting ducts is aquaporin-2. These channels are stored in endosomes inside the cells, and vasopressin causes their rapid translocation to the luminal membranes.

V_{1A} receptors mediate the vasoconstrictor effect of vasopressin, and vasopressin is a potent stimulator of vascular smooth muscle in vitro. However, relatively large amounts of vasopressin are needed to raise blood pressure in vivo, because vasopressin also acts on the brain to cause a decrease in cardiac output. The site of this action is the **area postrema,** one of the circumventricular organs (see Chapter 34). Hemorrhage is a potent stimulus for vasopressin secretion, and the blood pressure fall after hemorrhage is more marked in animals that have been treated with synthetic peptides that block the pressor action of vasopressin. Consequently, it appears that vasopressin does play a role in blood pressure homeostasis.

V_{1A} receptors are also found in the liver and the brain. Vasopressin causes glycogenolysis in the liver, and, as noted above, it is a neurotransmitter in the brain and spinal cord.

The V_{1B} receptors (also called V_3 receptors) appear to be unique to the anterior pituitary, where they mediate increased adrenocorticotropic hormone (ACTH) secretion from the corticotropes.

METABOLISM

Circulating vasopressin is rapidly inactivated, principally in the liver and kidneys. It has a **biologic half-life** (time required for inactivation of half a given amount) of approximately 18 min in humans.

CONTROL OF VASOPRESSIN SECRETION: OSMOTIC STIMULI

Vasopressin is stored in the posterior pituitary and released into the bloodstream by impulses in the nerve fibers that contain the hormone. The factors affecting its secretion are summarized in Table 39–1. When the effective osmotic pressure of the plasma is increased above the normal 285 mOsm/kg, the rate of discharge of these neurons increases and vasopressin secretion is increased (Figure 39–2). At 285 mOsm/kg, plasma vasopressin is at or near the limits of detection by available assays, but a further decrease probably takes place when plasma osmolality is below this level. Vasopressin secretion is regulated by osmoreceptors located in the anterior hypothalamus. They are outside the blood–brain barrier and appear to be located in the circumventricular organs, primarily the organum

TABLE 39–1 Summary of stimuli affecting vasopressin secretion.

Vasopressin Secretion Increased	Vasopressin Secretion Decreased
Increased effective osmotic pressure of plasma	Decreased effective osmotic pressure of plasma
Decreased ECF volume	Increased ECF volume
Pain, emotion, "stress," exercise	Alcohol
Nausea and vomiting	
Standing	
Clofibrate, carbamazepine	
Angiotensin II	

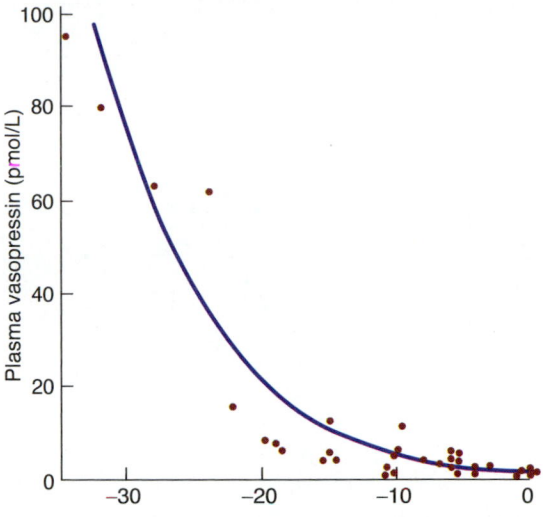

FIGURE 39–3 **Relation of mean arterial blood pressure to plasma vasopressin in healthy adult humans in whom a progressive decline in blood pressure was induced by infusion of graded doses of the ganglionic blocking drug trimethaphan.** The relation is exponential rather than linear. (Drawn from data in Baylis PH: Osmoregulation and control of vasopressin secretion in healthy humans. Am J Physiol 1987;253:R671.)

vasculosum of the lamina terminalis (OVLT) (see Chapter 34). The osmotic threshold for thirst (Figure 39–1) is the same as or slightly greater than the threshold for increased vasopressin secretion (Figure 39–2), and it is still uncertain whether the same osmoreceptors mediate both effects.

Vasopressin secretion is thus controlled by a delicate feedback mechanism that operates continuously to defend the osmolality of the plasma. Significant changes in secretion occur when osmolality is changed as little as 1%. In this way, the osmolality of the plasma in normal individuals is maintained very close to 285 mOsm/L.

VOLUME EFFECTS ON VASOPRESSIN SECRETION

ECF volume also affects vasopressin secretion. Vasopressin secretion is increased when ECF volume is low and decreased when ECF volume is high (Table 39–1). There is an inverse relationship between the rate of vasopressin secretion and the rate of discharge in afferents from stretch receptors in the low- and high-pressure portions of the vascular system. The low-pressure receptors are those in the great veins, right and left atria, and pulmonary vessels; the high-pressure receptors are those in the carotid sinuses and aortic arch (see Chapter 33). The exponential increases in plasma vasopressin produced by decreases in blood pressure are documented in Figure 39–3. However, the low-pressure receptors monitor the fullness of the vascular system, and moderate decreases in blood volume that decrease central venous pressure without lowering arterial pressure can also increase plasma vasopressin.

Thus, the low-pressure receptors are the primary mediators of volume effects on vasopressin secretion. Impulses pass from them via the vagi to the nucleus of the tractus solitarius (NTS). An inhibitory pathway projects from the NTS to the caudal ventrolateral medulla (CVLM), and there is a direct excitatory pathway from the CVLM to the hypothalamus. Angiotensin II reinforces the response to hypovolemia and

hypotension by acting on the circumventricular organs to increase vasopressin secretion (see Chapter 34).

Hypovolemia and hypotension produced by conditions such as hemorrhage release large amounts of vasopressin, and in the presence of hypovolemia, the osmotic response curve is shifted to the left (Figure 39–4). Its slope is also increased. The result is water retention and reduced plasma osmolality. This includes hyponatremia, since Na^+ is the most abundant osmotically active component of the plasma.

OTHER STIMULI AFFECTING VASOPRESSIN SECRETION

A variety of stimuli in addition to osmotic pressure changes and ECF volume aberrations increase vasopressin secretion. These include pain, nausea, surgical stress, and some emotions (Table 39–1). Nausea is associated with particularly large increases in vasopressin secretion. Alcohol decreases vasopressin secretion.

CLINICAL IMPLICATIONS

In various clinical conditions, volume and other non-osmotic stimuli bias the osmotic control of vasopressin secretion. For example, patients who have had surgery may have elevated levels of plasma vasopressin because of pain and hypovolemia, and this may cause them to develop a low plasma osmolality and dilutional hyponatremia (see Clinical Box 39–1).

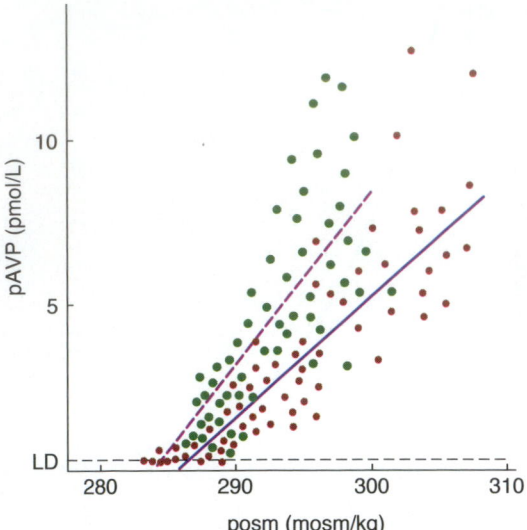

FIGURE 39–4 **Effect of hypovolemia and hypervolemia on the relation between plasma vasopressin (pAVP) and plasma osmolality (posm).** Seven blood samples were drawn at various times from 10 normal men when hypovolemia was induced by water deprivation (green circles, dashed line) and again when hypervolemia was induced by infusion of hypertonic saline (red circles, solid line). Linear regression analysis defined the relationship pAVP = 0.52 (posm – 283.5) for water deprivation and pAVP = 0.38 (posm – 285.6) for hypertonic saline. LD, limit of detection. Note the steeper curve as well as the shift of the intercept to the left during hypovolemia. (Courtesy of CJ Thompson.)

Patients with inappropriate hypersecretion of vasopressin have been successfully treated with demeclocycline, an antibiotic that reduces the renal response to vasopressin.

Diabetes insipidus is the syndrome that results when there is a vasopressin deficiency (**central diabetes insipidus**) or when the kidneys fail to respond to the hormone (**nephrogenic diabetes insipidus**).

Causes of vasopressin deficiency include disease processes in the supraoptic and paraventricular nuclei, the hypothalamohypophysial tract, or the posterior pituitary gland. It has been estimated that 30% of the clinical cases are due to neoplastic lesions of the hypothalamus, either primary or metastatic; 30% are posttraumatic; 30% are idiopathic; and the remainder are due to vascular lesions, infections, systemic diseases such as sarcoidosis that affect the hypothalamus, or mutations in the gene for prepropressophysin. The disease that develops after surgical removal of the posterior lobe of the pituitary may be temporary if only the distal ends of the supraoptic and paraventricular fibers are damaged, because the fibers recover, make new vascular connections, and begin to secrete vasopressin again.

The symptoms of diabetes insipidus are passage of large amounts of dilute urine (**polyuria**) and the drinking of large amounts of fluid (**polydipsia**), provided the thirst mechanism is intact. It is the polydipsia that keeps these patients healthy. If their sense of thirst is depressed for any reason and their

intake of dilute fluid decreases, they develop dehydration that can be fatal.

Another cause of diabetes insipidus is inability of the kidneys to respond to vasopressin (**nephrogenic diabetes insipidus**). Two forms of this disease have been described. In one form, the gene for the V_2 receptor is mutated, making the receptor unresponsive. The V_2 receptor gene is on the X chromosome, thus this condition is X-linked and inheritance is sex-linked recessive. In the other form of the condition, mutations occur in the autosomal gene for aquaporin-2 and produce nonfunctional versions of this water channel, many of which do not reach the apical membrane of the collecting duct but are trapped in intracellular locations.

The amelioration of diabetes insipidus produced by the development of concomitant anterior pituitary insufficiency is discussed in Chapter 24.

SYNTHETIC AGONISTS & ANTAGONISTS

Synthetic peptides that have selective actions and are more active than naturally occurring vasopressin and oxytocin have been produced by altering the amino acid residues. For example, 1-deamino-8-D-arginine vasopressin (desmopressin; dDAVP) has very high antidiuretic activity with little pressor activity, making it valuable in the treatment of vasopressin deficiency.

DEFENSE OF VOLUME

The volume of the ECF is determined primarily by the total amount of osmotically active solute in the ECF. The composition of the ECF is discussed in Chapter 2. Because Na^+ and Cl^- are by far the most abundant osmotically active solutes in ECF, and because changes in Cl^- are to a great extent secondary to changes in Na^+, the amount of Na^+ in the ECF is the most important determinant of ECF volume. Therefore, the mechanisms that control Na^+ balance are the major mechanisms defending ECF volume. However, there is volume control of water excretion as well; a rise in ECF volume inhibits

vasopressin secretion, and a decline in ECF volume produces an increase in the secretion of this hormone. Volume stimuli override the osmotic regulation of vasopressin secretion. Angiotensin II stimulates aldosterone and vasopressin secretion. It also causes thirst and constricts blood vessels, which help to maintain blood pressure. Thus, angiotensin II plays a key role in the body's response to hypovolemia (Figure 39–5). In addition, expansion of the ECF volume increases the secretion of atrial natriuretic peptide (ANP) and brain natriuretic peptide (BNP) by the heart, and this causes natriuresis and diuresis.

In disease states, loss of water from the body (**dehydration**) causes a moderate decrease in ECF volume, because water is

FIGURE 39–5 **Summary of the renin–angiotensin system and the stimulation of aldosterone secretion by angiotensin II.** The plasma concentration of renin is the rate-limiting step in the renin–angiotensin system; therefore, it is the major determinant of plasma angiotensin II concentration.

lost from both the intracellular and extracellular fluid compartments; but loss of Na^+ in the stools (diarrhea), urine (severe acidosis, adrenal insufficiency), or sweat (heat prostration) decreases ECF volume markedly and eventually leads to shock. The immediate compensations in shock operate principally to maintain intravascular volume, but they also affect Na^+ balance. In adrenal insufficiency, the decline in ECF volume is due not only to loss of Na^+ in the urine but also to its movement into cells. Because of the key position of Na^+ in volume homeostasis, it is not surprising that more than one mechanism has evolved to control the excretion of this ion.

The filtration and reabsorption of Na^+ in the kidneys and the effects of these processes on Na^+ excretion are discussed in Chapter 38. When ECF volume is decreased, blood pressure falls, glomerular capillary pressure declines, and the glomerular filtration rate (GFR) therefore falls, reducing the amount of Na^+ filtered. Tubular reabsorption of Na^+ is increased, in part because the secretion of aldosterone is increased. Aldosterone secretion is controlled in part by a feedback system in which the change that initiates increased secretion is a decline in mean intravascular pressure. Other changes in Na^+ excretion occur too rapidly to be due solely to changes in aldosterone secretion. For example, rising from the supine to the standing position increases aldosterone secretion. However, Na^+ excretion is decreased within a few minutes, and this rapid change in Na^+ excretion occurs in adrenalectomized subjects. It is probably due to hemodynamic changes and possibly to decreased ANP secretion.

The kidneys produce three hormones: 1,25-dihydroxycholecalciferol (see Chapter 23), renin, and erythropoietin. Natriuretic peptides, substances secreted by the heart and other tissues, increase excretion of sodium by the kidneys, and an additional natriuretic hormone inhibits Na, K ATPase.

THE RENIN–ANGIOTENSIN SYSTEM

RENIN

The rise in blood pressure produced by injection of kidney extracts is due to **renin,** an acid protease secreted by the kidneys into the bloodstream. This enzyme acts in concert with angiotensin-converting enzyme to form angiotensin II (Figure 39–6). It is a glycoprotein with a molecular weight of 37,326 in humans. The molecule is made up of two lobes, or domains, between which the active site of the enzyme is located in a deep cleft. Two aspartic acid residues, one at position 104 and one at position 292 (residue numbers from human preprorenin), are juxtaposed in the cleft and are essential for activity. Thus, renin is an aspartyl protease.

Like other hormones, renin is synthesized as a large preprohormone. Human **preprorenin** contains 406 amino acid residues. The **prorenin** that remains after removal of a leader sequence of 23 amino acid residues from the amino terminal contains 383 amino acid residues, and after removal of the pro

FIGURE 39–6   Formation and metabolism of circulating angiotensins.

sequence from the amino terminal of prorenin, active **renin** contains 340 amino acid residues. Prorenin has little if any biologic activity.

Some prorenin is converted to renin in the kidneys, and some is secreted. Prorenin is secreted by other organs, including the ovaries. After nephrectomy, the prorenin level in the circulation is usually only moderately reduced and may actually rise, but the active-renin level falls to essentially zero. Thus, very little prorenin is converted to renin in the circulation, and active renin is a product primarily, if not exclusively, of the kidneys. Prorenin is secreted constitutively, whereas active renin is formed in the secretory granules of the juxtaglomerular cells, the cells in the kidneys that produce renin (see below). Active renin has a half-life in the circulation of 80 min or less. Its only known function is to split the decapeptide **angiotensin I** from the amino terminal end of **angiotensinogen (renin substrate)** (Figure 39–7).

ANGIOTENSINOGEN

Circulating angiotensinogen is found in the α_2-globulin fraction of the plasma (Figure 39–6). It contains about 13% carbohydrate and is made up of 453 amino acid residues. It is synthesized in the liver with a 32-amino-acid signal sequence that is removed in the endoplasmic reticulum. Its circulating level is increased by glucocorticoids, thyroid hormones, estrogens, several cytokines, and angiotensin II.

ANGIOTENSIN-CONVERTING ENZYME & ANGIOTENSIN II

Angiotensin-converting enzyme (ACE) is a dipeptidyl carboxypeptidase that splits off histidyl-leucine from the physiologically inactive angiotensin I, forming the octapeptide **angiotensin II** (Figure 39–7). The same enzyme inactivates bradykinin (Figure 39–6). Increased tissue bradykinin produced when ACE is inhibited acts on B_2 receptors to produce the cough that is an annoying side effect in up to 20% of gle gene. However, the gene has two different promoters, producing two different mRNAs. In male mice in which the ACE gene has been knocked out, blood pressure is lower than normal, but in females it is normal. In addition, fertility is reduced in males but not in females.

METABOLISM OF ANGIOTENSIN II

Angiotensin II is metabolized rapidly; its half-life in the circulation in humans is 1 to 2 min. It is metabolized by various peptidases. An aminopeptidase removes the aspartic acid (Asp) residue from the amino terminal of the peptide (Figure 39–7). The resulting heptapeptide has physiologic activity and is sometimes called **angiotensin III**. Removal of a second amino terminal residue from angiotensin III produces the hexapeptide sometimes called angiotensin IV, which is also said to have some activity. Most, if not all, of the other peptide fragments that are formed are inactive. In addition, aminopeptidase can act on angiotensin I to produce (des-Asp[1]) angiotensin I, and this compound can be converted directly to angiotensin III by the action of ACE. Angiotensin-metabolizing activity is found in red blood cells and many tissues. In addition, angiotensin II appears to be removed from the circulation by some sort of trapping mechanism in the vascular beds of tissues other than the lungs.

Renin is usually measured by incubating the sample to be assayed and measuring by immunoassay the amount of angiotensin I generated. This measures the **plasma renin activity (PRA)** of the sample. Deficiency of angiotensinogen as well as renin can cause low PRA values, and to avoid this problem, exogenous angiotensinogen is often added, so that **plasma renin concentration (PRC)** rather than PRA is measured. The normal PRA in supine subjects eating a normal amount of

patients treated with ACE inhibitors (see Clinical Box 39–2). Most of the converting enzyme that forms angiotensin II in the circulation is located in endothelial cells. Much of the conversion occurs as the blood passes through the lungs, but conversion also occurs in many other parts of the body.

ACE is an ectoenzyme that exists in two forms: a **somatic** form found throughout the body and a **germinal** form found solely in postmeiotic spermatogenic cells and spermatozoa (see Chapter 25). Both ACEs have a single transmembrane domain and a short cytoplasmic tail. However, somatic ACE is a 170-kDa protein with two homologous extracellular domains, each containing an active site (Figure 39–8). Germinal ACE is a 90-kDa protein that has only one extracellular domain and active site. Both enzymes are formed from a sin-

FIGURE 39–7 **Structure of the amino terminal end of angiotensinogen and angiotensins I, II, and III in humans.** R, remainder of protein. After removal of a 24-amino-acid leader sequence, angiotensinogen contains 453 amino acid residues. The structure of angiotensin II in dogs, rats, and many other mammals is the same as that in humans. Bovine and ovine angiotensin II have valine instead of isoleucine at position 5.

Extracellular
extension

Zn²⁺   Zn²⁺

Amino terminal
catalytic site

Carboxyl terminal
catalytic site

NH₂

Trans-
membrane
domain

Intracellular
extension

COOH

FIGURE 39–8 **Diagrammatic representation of the structure of the somatic form of angiotensin-converting enzyme.** Note the short cytoplasmic tail of the molecule and the two extracellular catalytic sites, each of which binds a zinc ion (Zn^{2+}). (Reproduced with permission from Johnston CI: Tissue angiotensin-converting enzyme in cardiac and vascular hypertrophy, repair, and remodeling. Hypertension 1994;23:258. Copyright © 1994 by The American Heart Association.)

sodium is approximately 1 ng of angiotensin I generated per milliliter per hour. The plasma angiotensin II concentration in such subjects is about 25 pg/mL (approximately 25 pmol/L).

ACTIONS OF ANGIOTENSINS

Angiotensin I appears to function solely as the precursor of angiotensin II and does not have any other established action.

Angiotensin II—previously called hypertensin or angiotonin—produces arteriolar constriction and a rise in systolic and diastolic blood pressure. It is one of the most potent vasoconstrictors known, being four to eight times as active as norepinephrine on a weight basis in normal individuals. However, its pressor activity is decreased in Na^+-depleted individuals and in patients with cirrhosis and some other diseases. In these conditions, circulating angiotensin II is increased, and this down-regulates the angiotensin receptors in vascular smooth muscle. Consequently, there is less response to injected angiotensin II.

Angiotensin II also acts directly on the adrenal cortex to increase the secretion of aldosterone, and the renin–angiotensin system is a major regulator of aldosterone secretion. Additional actions of angiotensin II include facilitation of the release of norepinephrine by a direct action on postganglionic sympathetic neurons, contraction of mesangial cells with a resultant decrease in glomerular filtration rate (see Chapter 38), and a direct effect on the renal tubules to increase Na^+ reabsorption.

Angiotensin II also acts on the brain to decrease the sensitivity of the baroreflex, and this potentiates the pressor effect of angiotensin II. In addition, it acts on the brain to increase water intake and increase the secretion of vasopressin and ACTH. It does not penetrate the blood–brain barrier, but it triggers these responses by acting on the circumventricular organs, four small structures in the brain that are outside the blood–brain barrier (see Chapter 34). One of these structures, the area postrema, is primarily responsible for the pressor potentiation, whereas two of the others, the subfornical organ (SFO) and the organum vasculosum of the lamina terminalis (OVLT), are responsible for the increase in water intake (dipsogenic effect). It is not certain which of the circumventricular organs are responsible for the increases in vasopressin and ACTH secretion.

Angiotensin III [(des-Asp¹) angiotensin II] has about 40% of the pressor activity of angiotensin II, but 100% of the aldosterone-stimulating activity. It has been suggested that angiotensin III is the natural aldosterone-stimulating peptide, whereas angiotensin II is the blood-pressure-regulating peptide. However, this appears not to be the case, and instead angiotensin III is simply a breakdown product with some biologic activity. The same is probably true of angiotensin IV, though some researchers have argued that it has unique effects in the brain.

TISSUE RENIN–ANGIOTENSIN SYSTEMS

In addition to the system that generates circulating angiotensin II, many different tissues contain independent renin–angiotensin systems that generate angiotensin II, apparently for local use. Components of the renin–angiotensin system are found in the walls of blood vessels and in the uterus, the placenta, and the fetal membranes. Amniotic fluid has a high concentration of prorenin. In addition, tissue renin–angiotensin systems, or at least several components of the renin–angiotensin system, are present in the eyes, exocrine portion of the pancreas, heart, fat, adrenal cortex, testis, ovary, anterior and intermediate lobes of the pituitary, pineal, and brain. Tissue renin contributes very little to the circulating renin pool, because plasma renin activity falls to undetectable levels after the kidneys are removed. The functions of these tissue renin–angiotensin systems are unsettled, though evidence is accumulating that angiotensin II is a significant growth factor in the heart and blood vessels. ACE inhibitors or AT_1 receptor blockers are now the treatment of choice for congestive heart failure, and part of their value may be due to inhibition of the growth effects of angiotensin II.

ANGIOTENSIN II RECEPTORS

There are at least two classes of angiotensin II receptors. AT_1 receptors are serpentine receptors coupled by a G protein (G_q) to phospholipase C, and angiotensin II increases the cytosolic free Ca^{2+} level. It also activates numerous tyrosine kinases. In vascular smooth muscle, AT_1 receptors are associated with caveolae (see Chapter 2), and AII increases production of caveolin-1, one of the three isoforms of the protein that is

FIGURE 39–9 **Left:** Diagram of glomerulus, showing the juxtaglomerular apparatus. **Right:** Phase contrast photomicrograph of afferent arteriole in an unstained, freeze-dried preparation of the kidney of a mouse. Note the red blood cell in the lumen of the arteriole and the granulated juxtaglomerular cells in the wall. (Courtesy of C Peil.)

characteristic of caveolae. In rodents, two different but closely related AT$_1$ subtypes, AT$_{1A}$ and AT$_{1B}$, are coded by two separate genes. The AT$_{1A}$ subtype is found in blood vessel walls, the brain, and many other organs. It mediates most of the known effects of angiotensin II. The AT$_{1B}$ subtype is found in the anterior pituitary and the adrenal cortex. In humans, an AT$_1$ receptor gene is present on chromosome 3. There may be a second AT$_1$ type, but it is still unsettled whether distinct AT$_{1A}$ and AT$_{1B}$ subtypes occur.

There are also AT$_2$ receptors, which are coded in humans by a gene on the X chromosome. Like the AT$_1$ receptors, they have seven transmembrane domains, but their actions are different. They act via a G protein to activate various phosphatases which in turn antagonize growth effects and open K$^+$ channels. In addition, AT$_2$ receptor activation increases the production of NO and therefore increases intracellular cyclic 3,5-guanosine monophosphate (cGMP). The overall physiologic consequences of these second-messenger effects are unsettled. AT$_2$ receptors are more plentiful in fetal and neonatal life, but they persist in the brain and other organs in adults.

The AT$_1$ receptors in the arterioles and the AT$_1$ receptors in the adrenal cortex are regulated in opposite ways: an excess of angiotensin II down-regulates the vascular receptors, but it up-regulates the adrenocortical receptors, making the gland more sensitive to the aldosterone-stimulating effect of the peptide.

THE JUXTAGLOMERULAR APPARATUS

The renin in kidney extracts and the bloodstream is produced by the **juxtaglomerular cells (JG cells).** These epitheloid cells are located in the media of the afferent arterioles as they enter the glomeruli (Figure 39–9). The membrane-lined secretory granules in them have been shown to contain renin. Renin is also found in agranular **lacis cells** that are located in the junction between the afferent and efferent arterioles, but its significance in this location is unknown.

At the point where the afferent arteriole enters the glomerulus and the efferent arteriole leaves it, the tubule of the nephron touches the arterioles of the glomerulus from which it arose. At this location, which marks the start of the distal convolution, there is a modified region of tubular epithelium called the **macula densa** (Figure 39–9). The macula densa is in close proximity to the JG cells. The lacis cells, the JG cells, and the macula densa constitute the **juxtaglomerular apparatus.**

REGULATION OF RENIN SECRETION

Several different factors regulate renin secretion (Table 39–2), and the rate of renin secretion at any given time is determined by the summed activity of these factors. One factor is an intrarenal baroreceptor mechanism that causes renin secretion to decrease when arteriolar pressure at the level of the JG cells increases and to increase when arteriolar pressure at this level falls. Another renin-regulating sensor is in the macula densa. Renin secretion is inversely proportional to the amount of Na$^+$ and Cl$^-$ entering the distal renal tubules from the loop of Henle. Presumably, these electrolytes enter the macula densa cells via the Na–K–2Cl$^-$ transporters in their apical membranes, and the increase in some fashion triggers a signal that decreases renin secretion in the juxtaglomerular cells in the adjacent afferent arterioles. A possible mediator is NO, but the identity of the signal remains unsettled. Renin secretion also varies inversely with the plasma K$^+$ level, but the effect of K$^+$ appears to be mediated by the changes it produces in Na$^+$ and Cl$^-$ delivery to the macula densa.

TABLE 39–2 Factors that affect renin secretion.

Stimulatory
Increased sympathetic activity via renal nerves
Increased circulating catecholamines
Prostaglandins
Inhibitory
Increased Na$^+$ and Cl$^-$ reabsorption across macula densa
Increased afferent arteriolar pressure
Angiotensin II
Vasopressin

Angiotensin II feeds back to inhibit renin secretion by a direct action on the JG cells. Vasopressin also inhibits renin secretion in vitro and in vivo, although there is some debate about whether its in vivo effect is direct or indirect.

Finally, increased activity of the sympathetic nervous system increases renin secretion. The increase is mediated both by increased circulating catecholamines and by norepinephrine secreted by postganglionic renal sympathetic nerves. The catecholamines act mainly on β_1-adrenergic receptors on the JG cells and renin release is mediated by an increase in intracellular cAMP.

The principal conditions that increase renin secretion in humans are listed in Table 39–3. Most of the listed conditions decrease central venous pressure, which triggers an increase in sympathetic activity, and some also decrease renal arteriolar pressure (see Clinical Box 39–3). Renal artery constriction and constriction of the aorta proximal to the renal arteries produces a decrease in renal arteriolar pressure. Psychologic stimuli increase the activity of the renal nerves.

TABLE 39–3 Conditions that increase renin secretion.

Na$^+$ depletion
Diuretics
Hypotension
Hemorrhage
Upright posture
Dehydration
Cardiac failure
Cirrhosis
Constriction of renal artery or aorta
Various psychologic stimuli

Role of Renin in Clinical Hypertension

Constriction of one renal artery causes a prompt increase in renin secretion and the development of sustained hypertension (**renal** or **Goldblatt hypertension**). Removal of the ischemic kidney or the arterial constriction cures the hypertension if it has not persisted too long. In general, the hypertension produced by constricting one renal artery with the other kidney intact (one-clip, two-kidney Goldblatt hypertension) is associated with increased circulating renin. The clinical counterpart of this condition is **renal hypertension** due to atheromatous narrowing of one renal artery or other abnormalities of the renal circulation. However, plasma renin activity is usually normal in one-clip one-kidney Goldblatt hypertension. The explanation of the hypertension in this situation is unsettled. However, many patients with hypertension respond to treatment with ACE inhibitors or losartan even when their renal circulation appears to be normal and they have normal or even low plasma renin activity.

HORMONES OF THE HEART & OTHER NATRIURETIC FACTORS

STRUCTURE

The existence of various **natriuretic hormones** has been postulated for some time. Two of these are secreted by the heart. The muscle cells in the atria and, to a much lesser extent in the ventricles, contain secretory granules (Figure 39–10). The granules increase in number when NaCl intake is increased and ECF expanded, and extracts of atrial tissue cause natriuresis.

The first natriuretic hormone isolated from the heart was **atrial natriuretic peptide (ANP),** a polypeptide with a characteristic 17-amino-acid ring formed by a disulfide bond between two cysteines. The circulating form of this polypeptide has 28 amino acid residues (Figure 39–11). It is formed from a large precursor molecule that contains 151 amino acid residues, including a 24-amino-acid signal peptide. ANP was subsequently isolated from other tissues, including the brain, where it exists in two forms that are smaller than circulating ANP. A second natriuretic polypeptide was isolated from porcine brain and named **brain natriuretic peptide (BNP;** also known as **B-type natriuretic peptide).** It is also present in the brain in humans, but more is present in the human heart, including the ventricles. The circulating form of this hormone contains 32 amino acid residues. It has the same 17-member ring as ANP, though some of the amino acid residues in the ring are different (Figure 39–11). A third member of this

FIGURE 39–10 **ANP granules (g) interspersed between mitochondria (m) in rat atrial muscle cell.** G, Golgi complex; N, nucleus. The granules in human atrial cells are similar (× 17,640). (Courtesy of M Cantin.)

family has been named **C-type natriuretic peptide (CNP)** because it was the third in the sequence to be isolated. It contains 22 amino acid residues (Figure 39–11), and there is also a larger 53-amino-acid form. CNP is present in the brain, the pituitary, the kidneys, and vascular endothelial cells. However, very little is present in the heart and the circulation, and it appears to be primarily a paracrine mediator.

ACTIONS

ANP and BNP in the circulation act on the kidneys to increase Na^+ excretion, and injected CNP has a similar effect. They appear to produce this effect by dilating afferent arterioles and relaxing mesangial cells. Both of these actions increase glomerular filtration (see Chapter 38). In addition, they act on the renal tubules to inhibit Na^+ reabsorption. Other actions include an increase in capillary permeability, leading to extravasation of fluid and a decline in blood pressure. In addition, they relax vascular smooth muscle in arterioles and venules. CNP has a greater dilator effect on veins than ANP and BNP. These peptides also inhibit renin secretion and counteract the pressor effects of catecholamines and angiotensin II.

In the brain, ANP is present in neurons, and an ANP-containing neural pathway projects from the anteromedial part of the hypothalamus to the areas in the lower brain stem that are concerned with neural regulation of the cardiovascular system. In general, the effects of ANP in the brain are opposite to those of angiotensin II, and ANP-containing neural circuits appear to be involved in lowering blood pressure and promoting natriuresis. CNP and BNP in the brain probably have functions similar to those of ANP, but detailed information is not available.

NATRIURETIC PEPTIDE RECEPTORS

Three different natriuretic peptide receptors (NPR) have been isolated and characterized (Figure 39–12). The NPR-A and NPR-B receptors both span the cell membrane and have cytoplasmic domains that are guanylyl cyclases. ANP has the greatest affinity for the NPR-A receptor, and CNP has the greatest affinity for the NPR-B receptor. The third receptor, NPR-C, binds all three natriuretic peptides but has a markedly truncated cytoplasmic domain. Some evidence suggests that it acts via G proteins to activate phospholipase C and inhibit adenylyl cyclase. However, it has also been argued that this

FIGURE 39–11 **Human ANP, BNP, and CNP. Top:** Single-letter codes for amino acid residues aligned to show common sequences (colored). **Bottom:** Shape of molecules. Note that one cysteine is the carboxyl terminal amino acid residue in CNP, so there is no carboxyl terminal extension from the 17-member ring. (Modified from Imura H, Nakao K, Itoh H: The natriuretic peptide system in the brain: Implication in the central control of cardiovascular and neuroendocrine functions. Front Neuroendocrinol 1992;13:217.)

FIGURE 39–12 **Diagrammatic representation of natriuretic peptide receptors.** The NPR-A and NPR-B receptor molecules have intracellular guanylyl cyclase domains, whereas the putative clearance receptor, NPR-C, has only a small cytoplasmic domain. CM, cell membrane.

receptor does not trigger any intracellular change and is instead a **clearance receptor** that removes natriuretic peptides from the bloodstream and then releases them later, helping to maintain a steady blood level of the hormones.

SECRETION & METABOLISM

The concentration of ANP in plasma is about 5 fmol/mL in normal humans ingesting moderate amounts of NaCl. ANP secretion is increased when the ECF volume is increased by infusion of isotonic saline and when the atria are stretched. BNP secretion is increased when the ventricles are stretched. ANP secretion is also increased by immersion in water up to the neck (Figure 39–13), a procedure that counteracts the effect of gravity on the circulation and increases central venous and consequently atrial pressure. Note that immersion also decreases the secretion of renin and aldosterone. Conversely, a small but measurable decrease in plasma ANP occurs in association with a decrease in central venous pressure on rising from the supine to the standing position. Thus, it seems clear that the atria respond directly to stretch in vivo and that the rate of ANP secretion is proportional to the degree to which the atria are stretched by increases in central venous pressure. Similarly, BNP secretion is proportional to the degree to which the ventricles are stretched. Plasma levels of both hormones are elevated in congestive heart failure, and their measurement is seeing increasing use in the diagnosis of this condition.

Circulating ANP has a short half-life. It is metabolized by neutral endopeptidase (NEP), which is inhibited by thiorphan. Therefore, administration of thiorphan increases circulating ANP.

Na, K ATPase-INHIBITING FACTOR

Another natriuretic factor is present in blood. This factor produces natriuresis by inhibiting Na, K ATPase and raises rather than lowers blood pressure. Current evidence indicates that it may well be the digitalis-like steroid **ouabain** and that it comes from the adrenal glands. However, its physiologic significance is not yet known.

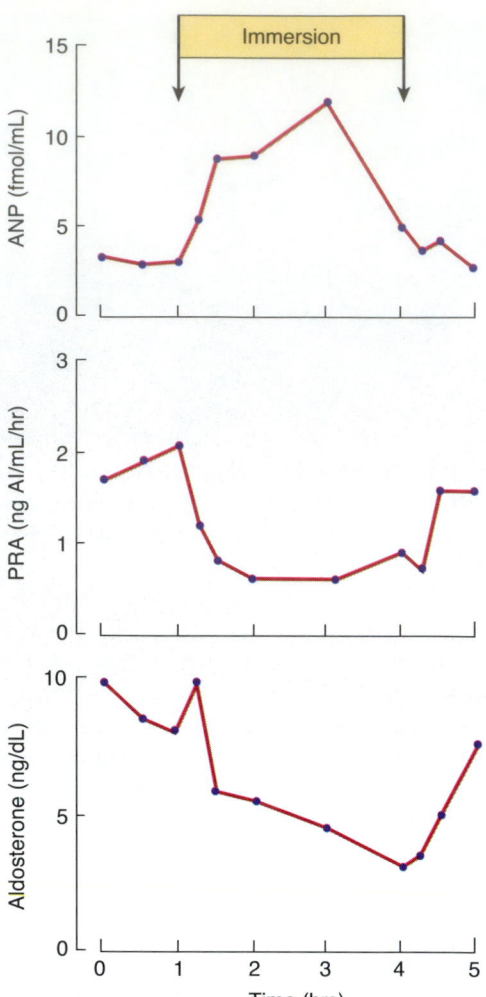

FIGURE 39–13 **Effect of immersion in water up to the neck for 3 h on plasma concentrations of ANP, PRA, and aldosterone.** (Modified and reproduced with permission from Epstein M, et al: Increases in circulating atrial natriuretic factor during immersion-induced central hypervolaemia in normal humans. Hypertension 1986;4 [Suppl 2]:593.)

DEFENSE OF SPECIFIC IONIC COMPOSITION

Special regulatory mechanisms maintain the levels of certain specific ions in the ECF as well as the levels of glucose and other nonionized substances important in metabolism (see Chapter 1). The feedback of Ca^{2+} on the parathyroids and the calcitonin-secreting cells to adjust their secretion maintains the ionized calcium level of the ECF (see Chapter 23). The Mg^{2+} concentration is subject to close regulation, but the mechanisms controlling Mg^+ metabolism are incompletely understood.

The mechanisms controlling Na^+ and K^+ content are part of those determining the volume and tonicity of ECF and have been discussed above. The levels of these ions are also dependent on the H^+ concentration, and pH is one of the major factors affecting the anion composition of ECF. This will be discussed in Chapter 40.

ERYTHROPOIETIN

STRUCTURE & FUNCTION

When an individual bleeds or becomes hypoxic, hemoglobin synthesis is enhanced, and production and release of red blood cells from the bone marrow (**erythropoiesis**) are increased (see Chapter 32). Conversely, when the red cell volume is increased above normal by transfusion, the erythropoietic activity of the bone marrow decreases. These adjustments are brought about by changes in the circulating level of **erythropoietin,** a circulating glycoprotein that contains 165 amino acid residues and four oligosaccharide chains that are necessary for its activity in vivo. Its blood level is markedly increased in anemia (Figure 39–14).

Erythropoietin increases the number of erythropoietin-sensitive committed stem cells in the bone marrow that are converted to red blood cell precursors and subsequently to mature erythrocytes (see Chapter 32). The receptor for erythropoietin is a linear protein with a single transmembrane domain that is a member of the cytokine receptor superfamily (see Chapter 3). The receptor has tyrosine kinase activity, and it activates a cascade of serine and threonine kinases, resulting in inhibited apoptosis of red cells and their increased growth and development.

The principal site of inactivation of erythropoietin is the liver, and the hormone has a half-life in the circulation of about 5 h. However, the increase in circulating red cells that it triggers takes 2 to 3 d to appear, since red cell maturation is a relatively slow process. Loss of even a small portion of the sialic acid residues in the carbohydrate moieties that are part of the erythropoietin molecule shortens its half-life to 5 min, making it biologically ineffective.

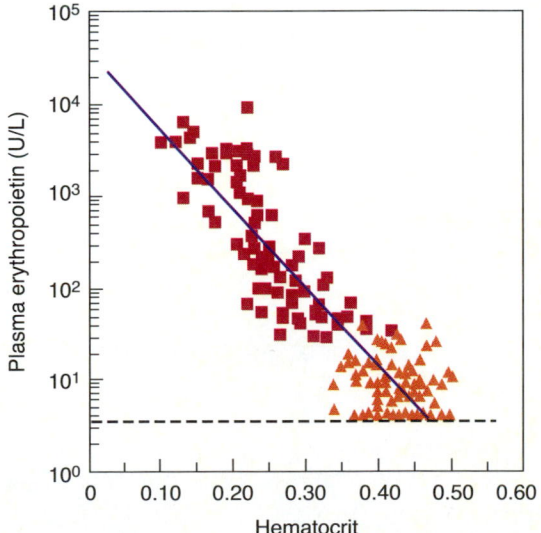

FIGURE 39–14 **Plasma erythropoietin levels in normal blood donors (triangles) and patients with various forms of anemia (squares).** (Reproduced with permission from Erslev AJ: Erythropoietin. N Engl J Med 1991;324:1339.)

SOURCES

In adults, about 85% of the erythropoietin comes from the kidneys and 15% from the liver. Both these organs contain the mRNA for erythropoietin. Erythropoietin can also be extracted from the spleen and salivary glands, but these tissues do not contain the mRNA and consequently do not appear to manufacture the hormone. When renal mass is reduced in adults by renal disease or nephrectomy, the liver cannot compensate and anemia develops.

Erythropoietin is produced by interstitial cells in the peritubular capillary bed of the kidneys and by perivenous hepatocytes in the liver. It is also produced in the brain, where it exerts a protective effect against excitotoxic damage triggered by hypoxia; and in the uterus and oviducts, where it is induced by estrogen and appears to mediate estrogen-dependent angiogenesis.

The gene for the hormone has been cloned, and recombinant erythropoietin produced in animal cells is available for clinical use as epoetin alfa. The recombinant erythropoietin is of value in the treatment of the anemia associated with renal failure; 90% of the patients with end-stage renal failure who are on dialysis are anemic as a result of erythropoietin deficiency. Erythropoietin is also used to stimulate red cell production in individuals who are banking a supply of their own blood in preparation for autologous transfusions during elective surgery (see Chapter 32).

REGULATION OF SECRETION

The usual stimulus for erythropoietin secretion is hypoxia, but secretion of the hormone can also be stimulated by cobalt salts and androgens. Recent evidence suggests that the O_2 sensor regulating erythropoietin secretion in the kidneys and the liver is a heme protein that in the deoxy form stimulates and in the oxy form inhibits transcription of the erythropoietin gene to form erythropoietin mRNA. Secretion of the hormone is facilitated by the alkalosis that develops at high altitudes. Like renin secretion, erythropoietin secretion is facilitated by catecholamines via a β-adrenergic mechanism, although the renin–angiotensin system is totally separate from the erythropoietin system.

CHAPTER SUMMARY

■ Total body osmolality is directly proportional to the total body sodium plus the total body potassium divided by the total body water. Changes in the osmolality of the body fluids occur when a disproportion exists between the amount of these electrolytes and the amount of water ingested or lost from the body.

■ Vasopressin's main physiologic effect is the retention of water by the kidney by increasing the water permeability of the renal collecting ducts. Water is absorbed from the urine, the urine becomes concentrated, and its volume decreases.

■ Vasopressin is stored in the posterior pituitary and released into the bloodstream in response to the stimulation of osmoreceptors or baroreceptors. Increases in secretion occur when osmolality is changed as little as 1%, thus keeping the osmolality of the plasma very close to 285 mOsm/L.

■ The amount of Na^+ in the ECF is the most important determinant of ECF volume, and mechanisms that control Na^+ balance are the major mechanisms defending ECF volume. The main mechanism regulating sodium balance is the renin–angiotensin system, a hormone system that regulates blood pressure.

■ The kidneys secrete the enzyme renin and renin acts in concert with angiotensin-converting enzyme to form angiotensin II. Angiotensin II acts directly on the adrenal cortex to increase the secretion of aldosterone. Aldosterone increases the retention of sodium from the urine via action on the renal collecting duct.

MULTIPLE-CHOICE QUESTIONS

For all questions, select the single best answer unless otherwise directed.

1. Dehydration increases the plasma concentration of all the following hormones *except*

   A) vasopressin.
   B) angiotensin II.
   C) aldosterone.
   D) norepinephrine.
   E) atrial natriuretic peptide.

2. In a patient who has become dehydrated, body water should be replaced by intravenous infusion of

   A) distilled water.
   B) 0.9% sodium chloride solution.
   C) 5% glucose solution.
   D) hyperoncotic albumin.
   E) 10% glucose solution.

3. Renin is secreted by

   A) cells in the macula densa.
   B) cells in the proximal tubules.
   C) cells in the distal tubules.
   D) juxtaglomerular cells.
   E) cells in the peritubular capillary bed.

4. Erythropoietin is secreted by

   A) cells in the macula densa.
   B) cells in the proximal tubules.
   C) cells in the distal tubules.
   D) juxtaglomerular cells.
   E) cells in the peritubular capillary bed.

5. When a woman who has been on a low-sodium diet for 8 d is given an intravenous injection of captopril, a drug that inhibits angiotensin-converting enzyme, one would expect

   A) her blood pressure to rise because her cardiac output would fall.
   B) her blood pressure to rise because her peripheral resistance would fall.
   C) her blood pressure to fall because her cardiac output would fall.
   D) her blood pressure to fall because her peripheral resistance would fall.
   E) her plasma renin activity to fall because her circulating angiotensin I level would rise.

6. Which of the following would *not* be expected to increase renin secretion?

   A) administration of a drug that blocks angiotensin-converting enzyme
   B) administration of a drug that blocks AT1 receptors
   C) administration of a drug that blocks β-adrenergic receptors
   D) constriction of the aorta between the celiac artery and the renal arteries
   E) administration of a drug that reduces ECF volume

7. Which of the following is *least* likely to contribute to the beneficial effects of angiotensin-converting enzyme inhibitors in the treatment of congestive heart failure?

   A) vasodilation
   B) decreased cardiac growth
   C) decreased cardiac afterload
   D) increased plasma renin activity
   E) decreased plasma aldosterone

CHAPTER RESOURCES

Adrogue HJ, Madias NE: Hypernatremia. N Engl J Med 2000;342:1493.

Adrogue HJ, Madias NE: Hyponatremia. N Engl J Med 2000;342:101.

Corvol P, Jeunemaitre X: Molecular genetics of human hypertension: Role of angiotensinogen. Endocr Rev 1997;18:662.

Morel F: Sites of hormone action in the mammalian nephron. Am J Physiol 1981;240:F159.

McKinley MS, Johnson AK: The physiologic regulation of thirst and fluid intake. News Physiol Sci 2004;19:1.

Robinson AG, Verbalis JG: Diabetes insipidus. Curr Ther Endocrinol Metab 1997;6:1.

Verkman AS: Mammalian aquaporins: Diverse physiological roles and potential clinical significance. Expert Rev Mol Med. 2008;10:13.

Zeidel ML: Hormonal regulation of inner medullary collecting duct sodium transport. Am J Physiol 1993;265:F159.

Acidification of the Urine & Bicarbonate Excretion

RENAL H+ SECRETION

The cells of the proximal and distal tubules, like the cells of the gastric glands, secrete hydrogen ions (see Chapter 26). Acidification also occurs in the collecting ducts. The reaction that is primarily responsible for H^+ secretion in the proximal tubules is Na–H exchange (Figure 40–1). This is an example of secondary active transport; extrusion of Na^+ from the cells into the interstitium by Na, K ATPase lowers intracellular Na^+, and this causes Na^+ to enter the cell from the tubular lumen, with coupled extrusion of H^+. The H^+ comes from intracellular dissociation of H_2CO_3, and the HCO_3^- that is formed diffuses into the interstitial fluid. Thus, for each H^+ ion secreted, one Na^+ ion and one HCO_3^- ion enter the interstitial fluid.

Carbonic anhydrase catalyzes the formation of H_2CO_3, and drugs that inhibit carbonic anhydrase depress both secretion of acid by the proximal tubules and the reactions which depend on it.

Some evidence suggests that H^+ is secreted in the proximal tubules by other types of pumps, but the evidence for these additional pumps is controversial, and in any case, their contribution is small relative to that of the Na–H exchange mechanism. This is in contrast to what occurs in the distal tubules

and collecting ducts, where H^+ secretion is relatively independent of Na^+ in the tubular lumen. In this part of the tubule, most H^+ is secreted by an ATP-driven proton pump. Aldosterone acts on this pump to increase distal H^+ secretion. The I cells in this part of the renal tubule secrete acid and, like the parietal cells in the stomach, contain abundant carbonic anhydrase and numerous tubulovesicular structures. There is evidence that the H^+-translocating ATPase that produces H^+ secretion is located in these vesicles as well as in the luminal cell membrane and that, in acidosis, the number of H^+ pumps is increased by insertion of these tubulovesicles into the luminal cell membrane. Some of the H^+ is also secreted by $H–K^+$ ATPase. The I cells contain **Band 3,** an anion exchange protein, in their basolateral cell membranes, and this protein may function as a Cl/HCO_3 exchanger for the transport of HCO_3^- to the interstitial fluid.

FATE OF H+ IN THE URINE

The amount of acid secreted depends upon the subsequent events in the tubular urine. The maximal H^+ gradient against which the transport mechanisms can secrete in humans corresponds to a urine pH of about 4.5; that is, an H^+ concentration in

FIGURE 40–1 **Secretion of acid by proximal tubular cells in the kidney.** H^+ is transported into the tubular lumen by an antiport in exchange for Na^+. Active transport by Na, K ATPase is indicated by arrows in the circle. Dashed arrows indicate diffusion.

the urine that is 1000 times the concentration in plasma. pH 4.5 is thus the **limiting pH.** This is normally reached in the collecting ducts. If there were no buffers that "tied up" H^+ in the urine, this pH would be reached rapidly, and H^+ secretion would stop. However, three important reactions in the tubular fluid remove free H^+, permitting more acid to be secreted (Figure 40–2). These are the reactions with HCO_3^- to form CO_2 and H_2O, with HPO_4^{2-} to form $H_2PO_4^-$, and with NH_3 to form NH_4^+.

REACTION WITH BUFFERS

The dynamics of buffering are discussed in Chapter 1 and below. The pK' of the bicarbonate system is 6.1, that of the dibasic phosphate system is 6.8, and that of the ammonia system is 9.0. The concentration of HCO_3^- in the plasma, and consequently in the glomerular filtrate, is normally about 24 mEq/L, whereas that of phosphate is only 1.5 mEq/L. Therefore, in the proximal tubule, most of the secreted H^+ reacts with HCO_3^- to form H_2CO_3 (Figure 40–2). The H_2CO_3 breaks down to form CO_2 and H_2O. In the proximal (but not in the distal) tubule, there is carbonic anhydrase in the brush border of the cells; this facilitates the formation of CO_2 and H_2O in the tubular fluid. The CO_2, which diffuses readily across all biological membranes, enters the tubular cells, where it adds to the pool of CO_2 available to form H_2CO_3. Because most of the H^+ is removed from the tubule, the pH of the fluid is changed very little. This is the mechanism by which HCO_3^- is reabsorbed; for each mole of HCO_3^- removed from the tubular fluid, 1 mol of HCO_3^- diffuses from the tubular cells into the blood, even though it is not the same mole that disappeared from the tubular fluid.

Secreted H^+ also reacts with dibasic phosphate (HPO_4^{2-}) to form monobasic phosphate ($H_2PO_4^-$). This happens to the greatest extent in the distal tubules and collecting ducts,

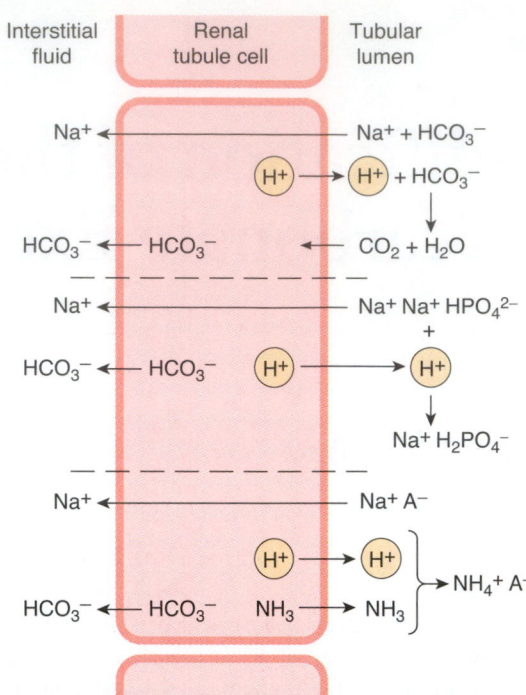

FIGURE 40–2 **Fate of H^+ secreted into a tubule in exchange for Na^+. Top:** Reabsorption of filtered bicarbonate via CO_2. **Middle:** Formation of monobasic phosphate. **Bottom:** Ammonium formation. Note that in each instance one Na^+ ion and one HCO_3^- ion enter the bloodstream for each H^+ ion secreted. A^-, anion.

because it is here that the phosphate that escapes proximal reabsorption is greatly concentrated by the reabsorption of water. The reaction with NH_3 occurs in the proximal and distal tubules. H^+ also combines to a minor degree with other buffer anions.

Each H^+ ion that reacts with the buffers contributes to the urinary **titratable acidity,** which is measured by determining the amount of alkali that must be added to the urine to return its pH to 7.4, the pH of the glomerular filtrate. However, the titratable acidity obviously measures only a fraction of the acid secreted, since it does not account for the H_2CO_3 that has been converted to H_2O and CO_2. In addition, the pK' of the ammonia system is 9.0, and the ammonia system is titrated only from the pH of the urine to pH 7.4, so it contributes very little to the titratable acidity.

AMMONIA SECRETION

Reactions in the renal tubular cells produce NH_4^+ and HCO_3^-. NH_4^+ is in equilibrium with NH_3 and H^+ in the cells. Because the pK' of this reaction is 9.0, the ratio of NH_3 to NH_4^+ at pH 7.0 is 1:100 (Figure 40–3). However, NH_3 is lipid-soluble and diffuses across the cell membranes down its concentration gradient into the interstitial fluid and tubular urine. In the urine it reacts with H^+ to form NH_4^+, and the NH_4^+ remains in the urine.

$$NH_4^+ \rightleftharpoons NH_3 + H^+$$

$$pH = pK' + \log \frac{[NH_3]}{[NH_4^+]}$$

$$Glutamine \xrightarrow{Glutaminase} Glutamate + NH_4^+$$

$$Glutamate \xrightarrow[dehydrogenase]{Glutamic} \alpha\text{-Ketoglutarate} + NH_4^+$$

FIGURE 40–3 Major reactions involved in ammonia production in the kidneys.

The principal reaction producing NH_4^+ in cells is conversion of glutamine to glutamate. This reaction is catalyzed by the enzyme **glutaminase,** which is abundant in renal tubular cells (Figure 40–3). **Glutamic dehydrogenase** catalyzes the conversion of glutamate to α-ketoglutarate, with the production of more NH_4^+. Subsequent metabolism of α-ketoglutarate utilizes $2H^+$, freeing $2HCO_3^-$.

In chronic acidosis, the amount of NH_4^+ excreted at any given urine pH also increases, because more NH_3 enters the tubular urine. The effect of this **adaptation** of NH_3 secretion, the cause of which is unsettled, is a further removal of H^+ from the tubular fluid and consequently a further enhancement of H^+ secretion.

The process by which NH_3 is secreted into the urine and then changed to NH_4^+, maintaining the concentration gradient for diffusion of NH_3, is called **nonionic diffusion** (see Chapter 2). Salicylates and a number of other drugs that are weak bases or weak acids are also secreted by nonionic diffusion. They diffuse into the tubular fluid at a rate that depends on the pH of the urine, so the amount of each drug excreted varies with the pH of the urine.

pH CHANGES ALONG THE NEPHRONS

A moderate drop in pH occurs in the proximal tubular fluid, but, as noted above, most of the secreted H^+ has little effect on luminal pH because of the formation of CO_2 and H_2O from H_2CO_3. In contrast, the distal tubule has less capacity to secrete H^+, but secretion in this segment has a greater effect on urinary pH.

FACTORS AFFECTING ACID SECRETION

Renal acid secretion is altered by changes in the intracellular P_{CO_2}, K^+ concentration, carbonic anhydrase level, and adrenocortical hormone concentration. When the P_{CO_2} is high (**respiratory acidosis**), more intracellular H_2CO_3 is available to buffer the hydroxyl ions and acid secretion is enhanced, whereas the reverse is true when the P_{CO_2} falls. K^+ depletion enhances acid secretion, apparently because the loss of K^+ causes intracellular acidosis even though the plasma pH may be ele-

FIGURE 40–4 Effect of ECF volume on HCO_3^- filtration, reabsorption, and excretion in rats. The pattern of HCO_3^- excretion is similar in humans. The plasma HCO_3^- concentration is normally about 24 mEq/L. (Reproduced with permission from Valtin H: *Renal Function,* 2nd ed. Little, Brown, 1983.)

vated. Conversely, K^+ excess in the cells inhibits acid secretion. When carbonic anhydrase is inhibited, acid secretion is inhibited because the formation of H_2CO_3 is decreased. Aldosterone and the other adrenocortical steroids that enhance tubular reabsorption of Na^+ also increase the secretion of H^+ and K^+.

BICARBONATE EXCRETION

Although the process of HCO_3^- reabsorption does not actually involve transport of this ion into the tubular cells, HCO_3^- reabsorption is proportional to the amount filtered over a relatively wide range. There is no demonstrable Tm, but HCO_3^- reabsorption is decreased by an unknown mechanism when the extracellular fluid (ECF) volume is expanded (Figure 40–4). When the plasma HCO_3^- concentration is low, all the filtered HCO_3^- is reabsorbed; but when the plasma HCO_3^- concentration is high; that is, above 26 to 28 mEq/L (the renal threshold for HCO_3^-), HCO_3^- appears in the urine and the urine becomes alkaline. Conversely, when the plasma HCO_3^- falls below about 26 mEq/L, the value at which all the secreted H^+ is being used to reabsorb HCO_3^-, more H^+ becomes available to combine with other buffer anions. Therefore, the lower the plasma HCO_3^- concentration drops, the more acidic the urine becomes and the greater its NH_4^+ content (see Clinical Box 40–1).

DEFENSE OF H⁺ CONCENTRATION

The mystique that envelopes the subject of acid–base balance makes it necessary to point out that the core of the problem is not "buffer base" or "fixed cation" or the like, but simply the

Implications of Urinary pH Changes

Depending on the rates of the interrelated processes of acid secretion, NH_4^+ production, and HCO_3^- excretion, the pH of the urine in humans varies from 4.5 to 8.0. Excretion of urine that is at a pH different from that of the body fluids has important implications for the body's electrolyte and acid–base economy. Acids are buffered in the plasma and cells, the overall reaction being $HA + NaH_3 \rightarrow NaA + H_2CO_3$. The H_2CO_3 forms CO_2 and H_2O, and the CO_2 is expired, while the NaA appears in the glomerular filtrate. To the extent that the Na^+ is replaced by H^+ in the urine, Na^+ is conserved in the body. Furthermore, for each H^+ ion excreted with phosphate or as NH_4^+, there is a net gain of one HCO_3^- ion in the blood, replenishing the supply of this important buffer anion. Conversely, when base is added to the body fluids, the OH^- ions are buffered, raising the plasma HCO_3^-. When the plasma level exceeds 28 mEq/L, the urine becomes alkaline and the extra HCO_3^- is excreted in the urine. Because the rate of maximal H^+ secretion by the tubules varies directly with the arterial P_{CO_2}, HCO_3^- reabsorption also is affected by the P_{CO_2}. This relationship has been discussed in more detail in the text.

TABLE 40–1 H^+ concentration and pH of body fluids.

		H⁺ Concentration		
		mEq/L	mol/L	pH
Gastric HCl		150	0.15	0.8
Maximal urine acidity		0.03	3×10^{-5}	4.5
Plasma	Extreme acidosis	0.0001	1×10^{-7}	7.0
	Normal	0.00004	4×10^{-8}	7.4
	Extreme alkalosis	0.00002	2×10^{-8}	7.7
Pancreatic juice		0.00001	1×10^{-8}	8.0

maintenance of the H^+ concentration of the ECF. The mechanisms regulating the composition of the ECF are particularly important as far as this specific ion is concerned, because the machinery of the cells is very sensitive to changes in H^+ concentration. Intracellular H^+ concentration, which can be measured by using microelectrodes, pH-sensitive fluorescent dyes, and phosphorus magnetic resonance, is different from extracellular pH and appears to be regulated by a variety of intracellular processes. However, it is sensitive to changes in ECF H^+ concentration.

The pH notation is a useful means of expressing H^+ concentrations in the body, because the H^+ concentrations happen to be low relative to those of other cations. Thus, the normal Na^+ concentration of arterial plasma that has been equilibrated with red blood cells is about 140 mEq/L, whereas the H^+ concentration is 0.00004 mEq/L (Table 40–1). The pH, the negative logarithm of 0.00004, is therefore 7.4. Of course, a decrease in pH of 1 unit, for example, from 7.0 to 6.0, represents a 10-fold increase in H^+ concentration. It is important to remember that the pH of blood is the pH of **true plasma**—plasma that has been in equilibrium with red cells—because the red cells contain hemoglobin, which is quantitatively one of the most important blood buffers (see Chapter 36).

H^+ BALANCE

The pH of the arterial plasma is normally 7.40 and that of venous plasma slightly lower. Technically, **acidosis** is present

whenever the arterial pH is below 7.40, and **alkalosis** is present whenever it is above 7.40, although variations of up to 0.05 pH unit occur without untoward effects. The H^+ concentrations in the ECF that are compatible with life cover an approximately fivefold range, from 0.00002 mEq/L (pH 7.70) to 0.0001 mEq/L (pH 7.00).

Amino acids are utilized in the liver for gluconeogenesis, leaving NH_4^+ and HCO_3^- as products from their amino and carboxyl groups (Figure 40–5). The NH_4^+ is incorporated into urea and the protons that are formed are buffered intracellularly by HCO_3^-, so little NH_4^+ and HCO_3^- escape into the circulation. However, metabolism of sulfur-containing amino acids produces H_2SO_4, and metabolism of phosphorylated amino acids such as phosphoserine produces H_3PO_4. These strong acids enter the circulation and present a major H^+ load to the buffers in the ECF. The H^+ load from amino acid metabolism is normally about 50 mEq/d. The CO_2 formed by

FIGURE 40–5 Role of the liver and kidneys in the handling of metabolically produced acid loads. Sites where regulation occurs are indicated by asterisks. (Modified and reproduced with permission from Knepper MA, et al: Ammonium, urea, and systemic pH regulation. Am J Physiol 1987;235:F199.)

metabolism in the tissues is in large part hydrated to H_2CO_3 (see Chapter 36), and the total H^+ load from this source is over 12,500 mEq/d. However, most of the CO_2 is excreted in the lungs, and only small quantities of the H^+ remain to be excreted by the kidneys. Common sources of extra acid loads are strenuous exercise (lactic acid), diabetic ketosis (acetoacetic acid and β-hydroxybutyric acid), and ingestion of acidifying salts such as NH_4Cl and $CaCl_2$, which in effect add HCl to the body. Failure of diseased kidneys to excrete normal amounts of acid is also a cause of acidosis. Fruits are the main dietary source of alkali. They contain Na^+ and K^+ salts of weak organic acids, and the anions of these salts are metabolized to CO_2, leaving $NaHCO_3$ and $KHCO_3$ in the body. $NaHCO_3$ and other alkalinizing salts are sometimes ingested in large amounts, but a more common cause of alkalosis is loss of acid from the body as a result of vomiting of gastric juice rich in HCl. This is, of course, equivalent to adding alkali to the body.

BUFFERING

Buffering is of key importance in maintaining H^+ homeostasis. It is defined in Chapter 1 and discussed in Chapter 36 in the context of gas transport, with an emphasis on roles for proteins, hemoglobin and the carbonic anhydrase system in the blood. Carbonic anhydrase is also found in high concentration in gastric acid-secreting cells (see Chapter 26) and in renal tubular cells (see Chapter 38). Carbonic anhydrase is a protein with a molecular weight of 30,000 that contains an atom of zinc in each molecule. It is inhibited by cyanide, azide, and sulfide. The sulfonamides also inhibit this enzyme, and sulfonamide derivatives have been used clinically as diuretics because of their inhibitory effects on carbonic anhydrase in the kidney (see Chapter 38).

Buffering in vivo is, of course, not limited to the blood. The principal buffers in the blood, interstitial fluid, and intracellular fluid are listed in Table 40–2. The principal buffers in cerebrospinal fluid (CSF) and urine are the bicarbonate and phosphate systems. In metabolic acidosis, only 15–20% of the acid load is buffered by the H_2CO_3–HCO_3^- system in the ECF, and most of the remainder is buffered in cells. In metabolic alkalosis, about 30–35% of the OH^- load is buffered in cells, whereas in respiratory acidosis and alkalosis, almost all the buffering is intracellular.

TABLE 40–2 Principal buffers in body fluids.

Blood	$H_2CO_3 \rightleftarrows H^+ + HCO_3^-$
	$HProt \rightleftarrows H^+ + Prot^-$
	$HHb \rightleftarrows H^+ + Hb^-$
Interstitial fluid	$H_2CO_3 \rightleftarrows H^+ + HCO_3^-$
Intracellular fluid	$HProt \rightleftarrows H^+ + Prot^-$
	$H_2PO_4^- \rightleftarrows H^+ + HPO_4^{2-}$

In animal cells, the principal regulators of intracellular pH are HCO_3^- transporters. Those characterized to date include the Cl^--HCO_3^- exchanger **AE1** (formerly band 3), three Na^+–HCO_3^- cotransporters, and a K^+–HCO_3^- cotransporter.

SUMMARY

When a strong acid is added to the blood, the major buffer reactions are driven to the left. The blood levels of the three "buffer anions" Hb^- (hemoglobin), $Prot^-$ (protein), and HCO_3^- consequently drop. The anions of the added acid are filtered into the renal tubules. They are accompanied ("covered") by cations, particularly Na^+, because electrochemical neutrality is maintained. By processes that have been discussed above, the tubules replace the Na^+ with H^+ and in so doing reabsorb equimolar amounts of Na^+ and HCO_3^-, thus conserving the cations, eliminating the acid, and restoring the supply of buffer anions to normal. When CO_2 is added to the blood, similar reactions occur, except that since it is H_2CO_3 that is formed, the plasma HCO_3^- rises rather than falls.

RENAL COMPENSATION TO RESPIRATORY ACIDOSIS AND ALKALOSIS

As noted in Chapter 36, a rise in arterial P_{CO_2} due to decreased ventilation causes **respiratory acidosis** and conversely, a decline in P_{CO_2} causes **respiratory alkalosis.** The initial changes shown in Figure 40–6 are those that occur independently of any compensatory mechanism; that is, they are those of **uncompensated** respiratory acidosis or alkalosis. In either situation, changes are produced in the kidneys, which then tend to **compensate** for the acidosis or alkalosis, adjusting the pH toward normal.

HCO_3^- reabsorption in the renal tubules depends not only on the filtered load of HCO_3^-, which is the product of the glomerular filtration rate (GFR) and the plasma HCO_3^- level, but also on the rate of H^+ secretion by the renal tubular cells, since HCO_3^- is reabsorbed by exchange for H^+. The rate of H^+ secretion—and hence the rate of HCO_3^- reabsorption—is proportional to the arterial P_{CO_2}, probably because the more CO_2 that is available to form H_2CO_3 in the cells, the more H^+ can be secreted. Furthermore, when the P_{CO_2} is high, the interior of most cells becomes more acidic. In respiratory acidosis, renal tubular H^+ secretion is therefore increased, removing H^+ from the body; and even though the plasma HCO_3^- is elevated, HCO_3^- reabsorption is increased, further raising the plasma HCO_3^-. This renal compensation for respiratory acidosis is shown graphically in the shift from acute to chronic respiratory acidosis in Figure 40–6. Cl^- excretion is increased, and plasma Cl^- falls as plasma HCO_3^- is increased. Conversely, in respiratory alkalosis, the low P_{CO_2} hinders renal H^+ secretion, HCO_3^- reabsorption is depressed, and HCO_3^- is excreted, further

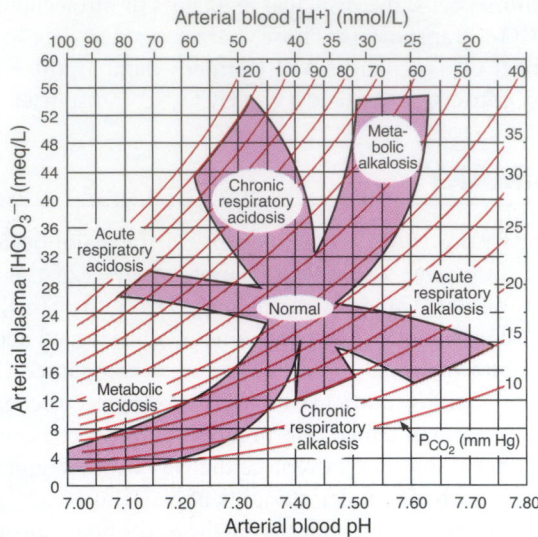

FIGURE 40–6 Acid–base nomogram showing changes in the CO_2 (curved lines), plasma HCO_3^-, and pH of arterial blood in respiratory and metabolic acidosis. Note the shifts in HCO_3^- and pH as acute respiratory acidosis and alkalosis are compensated, producing their chronic counterparts. (Reproduced with permission from Cogan MG, Rector FC Jr: Acid–base disorders. In: *The Kidney*, 4th ed. Brenner BM, Rector FC Jr [editors]. Saunders, 1991.)

reducing the already low plasma HCO_3^- and lowering the pH toward normal.

METABOLIC ACIDOSIS

When acids stronger than HHb and the other buffer acids are added to blood, **metabolic acidosis** is produced; and when the free H^+ level falls as a result of addition of alkali or removal of acid, **metabolic alkalosis** results. Following the example from Chapter 36, if H_2SO_4 is added, the H^+ is buffered and the Hb^-, $Prot^-$, and HCO_3^- levels in plasma drop. The H_2CO_3 formed is converted to H_2O and CO_2, and the CO_2 is rapidly excreted via the lungs. This is the situation in **uncompensated** metabolic acidosis. Actually, the rise in plasma H^+ stimulates respiration, so that the P_{CO_2}, instead of rising or remaining constant, is reduced. This **respiratory compensation** raises the pH even further. The **renal** compensatory mechanisms then bring about the excretion of the extra H^+ and return the buffer systems to normal.

RENAL COMPENSATION

The anions that replace HCO_3^- in the plasma in metabolic acidosis are filtered, each with a cation (principally Na^+), thus maintaining electrical neutrality. The renal tubular cells secrete H^+ into the tubular fluid in exchange for Na^+; and for each H^+ secreted, one Na^+ and one HCO_3^- are added to the blood. The limiting urinary pH of 4.5 would be reached rapidly and the total amount of H^+ secreted would be small if no buffers were present in the urine to "tie up" H^+. However, se-

creted H^+ reacts with HCO_3^- to form CO_2 and H_2O (bicarbonate reabsorption); with HPO_4^{2-} to form $H_2PO_4^-$; and with NH_3 to form NH_4^+. In this way, large amounts of H^+ can be secreted, permitting correspondingly large amounts of HCO_3^- to be returned to (in the case of bicarbonate reabsorption) or added to the depleted body stores and large numbers of the cations to be reabsorbed. It is only when the acid load is very large that cations are lost with the anions, producing diuresis and depletion of body cation stores. In chronic acidosis, glutamine synthesis in the liver is increased, using some of the NH_4^+ that usually is converted to urea (Figure 40–5), and the glutamine provides the kidneys with an additional source of NH_4^+. NH_3 secretion increases over a period of days (adaptation of NH_3 secretion), further improving the renal compensation for acidosis. In addition, the metabolism of glutamine in the kidneys produces α-ketoglutarate, and this in turn is decarboxylated, producing HCO_3^-, which enters the bloodstream and helps buffer the acid load (Figure 40–5).

The overall reaction in blood when a strong acid such as H_2SO_4 is added is:

$$2NaHCO_3 + H_2SO_4 \rightarrow Na_2SO_4 + 2H_2CO_3$$

For each mole of H^+ added, 1 mole of $NaHCO_3$ is lost. The kidney in effect reverses the reaction:

$$Na_2SO_4 + 2H_2CO_3 \rightarrow 2NaHCO_3 + 2H^+ + SO_4^{2-}$$

and the H^+ and SO_4^{2-} are excreted. Of course, H_2SO_4 is not excreted as such, the H^+ appearing in the urine as titratable acidity and NH_4^+.

In metabolic acidosis, the respiratory compensation tends to inhibit the renal response in the sense that the induced drop in P_{CO_2} hinders acid secretion, but it also decreases the filtered load of HCO_3^- and so its net inhibitory effect is not great.

METABOLIC ALKALOSIS

In metabolic alkalosis, the plasma HCO_3^- level and pH rise (Figure 40–7). The respiratory compensation is a decrease in ventilation produced by the decline in H^+ concentration, and this elevates the P_{CO_2}. This brings the pH back toward normal while elevating the plasma HCO_3^- level still further. The magnitude of this compensation is limited by the carotid and aortic chemoreceptor mechanisms, which drive the respiratory center if any appreciable fall occurs in the arterial PO_2. In metabolic alkalosis, more renal H^+ secretion is expended in reabsorbing the increased filtered load of HCO_3^-; and if the HCO_3^- level in plasma exceeds 26–28 mEq/L, HCO_3^- appears in the urine. The rise in P_{CO_2} inhibits the renal compensation by facilitating acid secretion, but its effect is relatively slight.

THE SIGGAARD–ANDERSEN CURVE NOMOGRAM

Use of the Siggaard–Andersen curve nomogram (Figure 40–7) to plot the acid–base characteristics of arterial blood is helpful

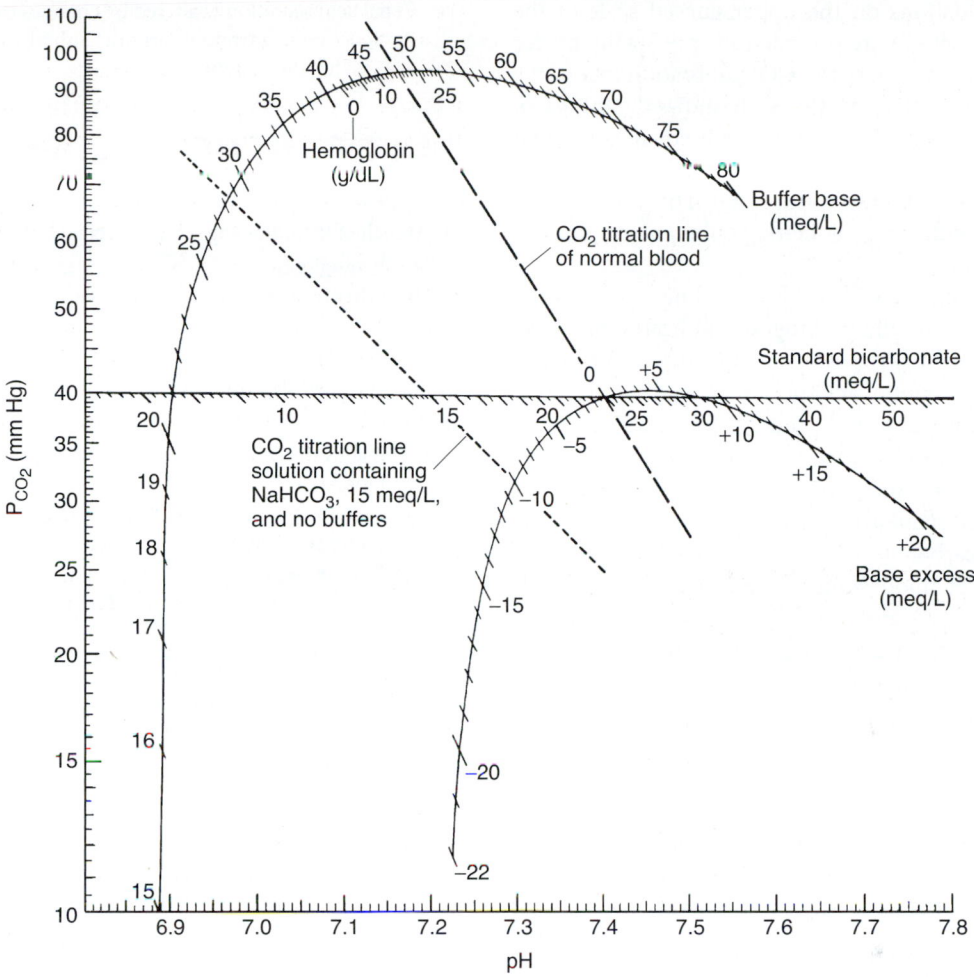

FIGURE 40–7 **Siggaard–Andersen curve nomogram.** (Courtesy of O Siggaard–Andersen and Radiometer, Copenhagen, Denmark.)

in clinical situations. This nomogram has P_{CO_2} plotted on a log scale on the vertical axis and pH on the horizontal axis. Thus, any point to the left of a vertical line through pH 7.40 indicates acidosis, and any point to the right indicates alkalosis. The position of the point above or below the horizontal line through a P_{CO_2} of 40 mm Hg defines the effective degree of hypoventilation or hyperventilation.

If a solution containing $NaHCO_3$ and no buffers were equilibrated with gas mixtures containing various amounts of CO_2, the pH and P_{CO_2} values at equilibrium would fall along the dashed line on the left in Figure 40–7 or a line parallel to it. If buffers were present, the slope of the line would be greater; and the greater the buffering capacity of the solution, the steeper the line. For normal blood containing 15 g of hemoglobin/dL, the CO_2 **titration line** passes through the 15-g/dL mark on the hemoglobin scale (on the underside of the upper curved scale) and the point where the P_{CO_2} = 40 mm Hg and pH = 7.40 lines intersect, as shown in Figure 40–7. When the hemoglobin content of the blood is low, there is significant loss of buffering capacity, and the slope of the CO_2 titration line diminishes. However, blood of course contains buffers in addition to hemoglobin, so that even the line drawn

from the zero point on the hemoglobin scale through the normal P_{CO_2}–pH intercept is steeper than the curve for a solution containing no buffers.

For clinical use, arterial blood or arterialized capillary blood is drawn anaerobically and its pH measured. The pHs of the same blood after equilibration with each of two gas mixtures containing different known amounts of CO_2 are also determined. The pH values at the known P_{CO_2} levels are plotted and connected to provide the CO_2 titration line for the blood sample. The pH of the blood sample before equilibration is plotted on this line, and the P_{CO_2} of the sample is read off the vertical scale. The **standard bicarbonate** content of the sample is indicated by the point at which the CO_2 titration line intersects the bicarbonate scale on the P_{CO_2} = 40 mm Hg line. The standard bicarbonate is not the actual bicarbonate concentration of the sample but, rather, what the bicarbonate concentration would be after elimination of any respiratory component. It is a measure of the alkali reserve of the blood, except that it is measured by determining the pH rather than the total CO_2 content of the sample after equilibration. Like the alkali reserve, it is an index of the degree of metabolic acidosis or alkalosis present.

Additional graduations on the upper curved scale of the nomogram (Figure 40–7) are provided for measuring **buffer base** content; the point where the CO_2 calibration line of the arterial blood sample intersects this scale shows the mEq/L of buffer base in the sample. The buffer base is equal to the total number of buffer anions (principally $Prot^-$, HCO_3^-, and Hb^-) that can accept hydrogen ions in the blood. The normal value in an individual with 15 g of hemoglobin per deciliter of blood is 48 mEq/L.

The point at which the CO_2 calibration line intersects the lower curved scale on the nomogram indicates the **base excess.** This value, which is positive in alkalosis and negative in acidosis, is the amount of acid or base that would restore 1 L of blood to normal acid–base composition at a PCO_2 of 40 mm Hg. It should be noted that a base deficiency cannot be completely corrected simply by calculating the difference between the normal standard bicarbonate (24 mEq/L) and the actual standard bicarbonate and administering this amount of $NaHCO_3$ per liter of blood; some of the added HCO_3^- is converted to CO_2 and H_2O, and the CO_2 is lost in the lungs. The actual amount that must be added is roughly 1.2 times the standard bicarbonate deficit, but the lower curved scale on the nomogram, which has been developed empirically by analyzing many blood samples, is more accurate.

In treating acid–base disturbances, one must, of course, consider not only the blood but also all the body fluid compartments. The other fluid compartments have markedly different concentrations of buffers. It has been determined empirically that administration of an amount of acid (in alkalosis) or base (in acidosis) equal to 50% of the body weight in kilograms times the blood base excess per liter will correct the acid–base disturbance in the whole body. At least when the abnormality is severe, however, it is unwise to attempt such a large correction in a single step; instead, about half the indicated amount should be given and the arterial blood acid–base values determined again. The amount required for final correction can then be calculated and administered. It is also worth noting that, at least in lactic acidosis, $NaHCO_3$ decreases cardiac output and lowers blood pressure, so it should be used with caution.

CHAPTER SUMMARY

■ The cells of the proximal and distal tubules secrete hydrogen ions. Acidification also occurs in the collecting ducts. The reaction that is primarily responsible for H^+ secretion in the proximal tubules is Na^+–H^+ exchange. Na is absorbed from the lumen of the tubule and H is excreted.

■ The maximal H^+ gradient against which the transport mechanisms can secrete in humans corresponds to a urine pH of about 4.5. However, three important reactions in the tubular fluid remove free H^+, permitting more acid to be secreted. These are the reactions with HCO_3^- to form CO_2 and H_2O, with HPO_4^{2-} to form $H_2PO_4^-$, and with NH_3 to form NH_4^+.

■ Carbonic anhydrase catalyzes the formation of H_2CO_3, and drugs that inhibit carbonic anhydrase depress secretion of acid by the proximal tubules.

■ Renal acid secretion is altered by changes in the intracellular PCO_2, K^+ concentration, carbonic anhydrase level, and adrenocortical hormone concentration.

MULTIPLE-CHOICE QUESTIONS

For all questions, select the single best answer unless otherwise directed.

1. Which of the following is the principal buffer in interstitial fluid?
   A) hemoglobin
   B) other proteins
   C) carbonic acid
   D) H_2PO_4
   E) compounds containing histidine

2. Increasing alveolar ventilation increases the blood pH because
   A) it activates neural mechanisms that remove acid from the blood.
   B) it makes hemoglobin a stronger acid.
   C) it increases the PO_2 of the blood.
   D) it decreases the PCO_2 in the alveoli.
   E) the increased muscle work of increased breathing generates more CO_2.

3. In uncompensated metabolic alkalosis
   A) the plasma pH, the plasma HCO_3^- concentration, and the arterial PCO_2 are all low.
   B) the plasma pH is high and the plasma HCO_3^- concentration and arterial PCO_2 are low.
   C) the plasma pH and the plasma HCO_3^- concentration are low and the arterial PCO_2 is normal.
   D) the plasma pH and the plasma HCO_3^- concentration are high and the arterial PCO_2 is normal.
   E) the plasma pH is low, the plasma HCO_3^- concentration is high, and the arterial PCO_2 is normal.

4. In a patient with a plasma pH of 7.10, the $[HCO_3^-]/[H_2CO_3]$ ratio in plasma is
   A) 20.
   B) 10.
   C) 2.
   D) 1.
   E) 0.1.

CHAPTER RESOURCES

Adrogué HJ, Madius NE: Management of life-threatening acid–base disorders. N Engl J Med 1998;338:26.

Brenner BM, Rector FC Jr. (editors): *The Kidney,* 6th ed. 2 vols. Saunders, 1999.

Davenport HW: *The ABC of Acid–Base Chemistry,* 6th ed. University of Chicago Press, 1974.

Halperin ML: *Fluid, Electrolyte, and Acid–Base Physiology,* 3rd ed. Saunders, 1998.

Lemann J Jr., Bushinsky DA, Hamm LL: Bone buffering of acid and base in humans. Am J Physiol Renal Physiol 2003;285:F811. Review.

Vize PD, Wolff AS, Bard JBL (editors): *The Kidney: From Normal Development to Congenital Disease.* Academic Press, 2003.

Answers to Multiple Choice Questions

Chapter 1
1. B   2. C   3. B   4. C   5. C   6. D   7. E   8. E.

Chapter 2
1. A   2. D   3. D   4. B   5. C   6. C   7. B.

Chapter 3
1. C   2. E   3. B   4. C

Chapter 4
1. B   2. E   3. B   4. C   5. B

Chapter 5
1. B   2. D   3. B   4. C   5. C.

Chapter 6
1. E   2. C   3. E   4. B

Chapter 7
1. C   2. D   3. C   4. B   5. A

Chapter 8
1. C   2. D   3. A   4. E   5. C   6. E   7. A

Chapter 9
1. C   2. E   3. C   4. E

Chapter 10
1. C   2. C   3. A   4. D

Chapter 11
1. D   2. B   3. E   4. D

Chapter 12
1. D   2. D   3. B   4. E   5. D   6. B   7. D   8. D

Chapter 13
1. E   2. B   3. C   4. E   5. D   6. D   7. E   8. A

Chapter 14
1. D   2. C   3. D   4. D   5. D   6. C   7. D   8. E

Chapter 15
1. C   2. D   3. C   4. D   5. A   6. C

Chapter 16
1. C   2. E   3. E   4. D   5. B   6. A

Chapter 17
1. A   2. B   3. D   4. D

Chapter 18
1. B   2. E   3. B   4. A   5. A   6. B   7. D   8. D

Chapter 19
1. A   2. E   3. C   4. D   5. B   6. D   7. D

Chapter 20
1. C   2. D   3. A   4. C   5. B   6. D   7. A   8. C   9. E

Chapter 21
1. E   2. D   3. D   4. C   5. E   6. D   7. C

Chapter 22
1. D   2. B   3. E   4. D   5. C   6. D   7. D   8. A   9. A

Chapter 23
1. C   2. E   3. D   4. A   5. C   6. D   7. E

Chapter 24
1. E   2. E   3. A   4. C   5. B

Chapter 25
1. C   2. E   3. D   4. A   5. E   6. C   7. A

Chapter 26
1. C   2. E   3. B   4. C   5. D

Chapter 27
1. E   2. D   3. E   4. A   5. C

Chapter 28
1. C   2. D   3. E   4. A   5. B

Chapter 29
1. E   2. E   3. C   4. E

Chapter 30
1. B   2. A   3. A   4. D   5. D

Chapter 31
1. A   2. C   3. C   4. C   5. E   6. D

Chapter 32
1. C   2. B   3. D   4. B   5. E   6. A   7. A   8. E

Chapter 33
1. B   2. A   3. E

Chapter 34
1. D   2. A   3. E   4. E   5. E   6. D

Chapter 35
1. D   2. C   3. A   4. E   5. B   6. D   7. A

Chapter 36
1. E   2. B   3. D   4. D   5. C

Chapter 37
1. D   2. B   3. B   4. D   5. E   6. E   7. B   8. C

Chapter 38
1. A   2. A   3. A   4. A   5. E   6. C   7. D

Chapter 39
1. E   2. C   3. D   4. E   5. D   6. C   7. D

Chapter 40
1. C   2. D   3. D   4. B

Subject Index

Sep 6 '91